The International Halley Watch Atlas of Large-Scale Phenomena

The International Halley Watch Atlas of Large-Scale Phenomena

John C. Brandt

Laboratory for Atmospheric and Space Physics
University of Colorado

Malcolm B. Niedner, Jr.

Laboratory for Astronomy and Solar Physics
NASA Goddard Space Flight Center

Jürgen Rahe

Solar System Exploration Division
National Aeronautics and Space Administration

ISBN 1-880768-00-3

University of Colorado-Boulder
Laboratory for Atmospheric and Space Physics
Box 590
Boulder, CO 80309-0590

Printed by
Johnson Printing Co.
Boulder, CO

Table of Contents

Preface

This Atlas is one of the archive products of the International Halley Watch (IHW)[1,2,3], and specifically of the IHW's Large-Scale Phenomenon (LSP) Discipline. It presents a comprehensive record of the large-scale structure of Halley's Comet in 1985 and 1986 obtained by a world-wide network of observers. The organization of the Atlas, described below, presents the evolving appearance of the comet throughout the apparition, and shows short-term changes and the effects of different observing techniques. An atlas of the 1910 return — the first to be recorded photographically — is also available.[4]

Many individuals and organizations contributed to this enterprise. First and foremost, we thank the many observers who took the images and forwarded them in good order to the prime LSP archive and Discipline Center located at the NASA Goddard Space Flight Center (GSFC), Greenbelt, Maryland, USA. At GSFC, we thank Barbara Pfarr, Joan Isensee, and Archie Warnock, in particular, for creating and maintaining this LSP archive. LSP Discipline activities at NASA/GSFC are described in detail by Niedner.[5] As with any undertaking of this complexity, errors can occur, and users of this Atlas will undoubtedly uncover some. Please inform the authors with specifics so we can include them in future errata lists.

A duplicate LSP archive is located at the Laboratory for Atmospheric and Space Physics, University of Colorado, Boulder, Colorado, USA, and the principal preparation of the Atlas was carried out there. We thank Carolyn Collins Petersen, Martin Snow, Yu Yi, Cora E. Randall, Lynn Teets, Sunne Pollart, Lynn Laubisch, and Rionda Osman-Jouchoux. Photo Craft, Inc. of Boulder provided excellent and timely black-and-white printing services; we thank especially Kathy Scrimgeour and Roy McCutchen. The printing was done by Johnson Publishing Co. of Boulder, Colorado; we appreciate the efforts of Johnson's Rusty Sandberg. We gratefully acknowledge the total support of Ray Newburn and Murray Geller at the Lead Center for the IHW, at the NASA Jet Propulsion Laboratory, Pasadena, California, USA. Finally, we thank the Solar System Exploration Division of NASA Headquarters, Washington, D.C., USA, for financial support.

Our goal was to provide a scientifically faithful and attractive record of the large-scale structure of this famous comet, obtained through a remarkable exercise of international cooperation. We hope that the readers enjoy and benefit from this Atlas as much as we have enjoyed preparing it.

John C. Brandt
Malcolm B. Niedner, Jr.
Jürgen Rahe
LSP Discipline Specialists
April 1992

Introduction

The principal goals of the Large-Scale Phenomena (LSP) Discipline of the International Halley Watch (IHW) were: (1) the establishment of a global network of observers and observatories committed to wide-field imaging of comet P/Halley in 1985-86; (2) the collection of the resulting data; and (3) the participation — within the larger IHW community — in the preparation and distribution of a complete and substantive archive. This photographic Atlas of comet P/Halley is one component of that archive. Various references in the Bibliography contain more details on the rationale and history of the LSP Discipline and of the IHW itself.

Most of the images reproduced in this Atlas were obtained when the comet was a bright, large-scale object — usually, but not always, showing some tail structure — i.e., from late-1985 through mid-1986. Although the research-oriented reader of this book is left to make an individual scientific interpretation of the photographic record, it is important to describe the Atlas organization, the IHW's Large-Scale Phenomena Network (LSPN), the specific details concerning the Atlas production, and (briefly) the large-scale structure and evolution of the comet during the apparition.

General Organization of the Atlas

Following this introduction is a short section showing the structure and evolution of the comet as recorded on color films and emulsions. Although substantially more scientific work is performed using black and white materials, color imagery reveals better that plasma and dust tails radiate in different regions of the visible spectrum (the plasma tail appears bluish, the dust tail yellowish). Next is the main body of the photographic record, which we call the Daily Record. It contains 757 black-and-white images which show the day-to-day evolution of P/Halley on a large scale. To maintain our publication schedule, the final contents of the Daily Record were decided relatively early so that work at Johnson Publishing Company could be carried out while the rest of the Atlas was being assembled. The Supplemental Daily Record, which follows the Daily Record, contains images not available or not ready for inclusion when the daily record was completed. Images out of time sequence are denoted by an asterisk in the caption.

Although most of the telescopes and cameras used to obtain the images reproduced in this book have fields of view of several degrees or more, the comet sometimes showed tail phenomena of large enough angular extent that offset imagery was required to record the 'full comet'. Several observatories utilized this 'mosaic' technique (when required), and a special Mosaic section has been devoted to these images. In a few cases, head-region images were not available to us even though offset frames were. Some of these 'tail-only' images have been included in the Supplement to the Daily Record. The Sequences section shows the evolution of the comet on several timescales, in a much more compact presentation than that of the Daily Record. The final Atlas section contains appendix material including the Bibliography, lists of observers and observatories, Halley ephemeris information for the times the photographs were taken, and information on the spectral response of the emulsions and filters used by the LSPN observers. Photometry of the digital images should be possible.[6]

The Large-Scale Phenomena Network (LSPN)

The specific goals of the LSP Discipline Specialist Team were: (1) to obtain (as far as practical) nearly continuous coverage of the comet's large-scale structure during the time that tail structure might reasonably be expected — i.e., late 1985 through mid-1986; (2) to supply approximately 1000 images to the IHW Lead Center in digital form; and (3) to publish an atlas of the 1000 best images. To assist in coordinating the LSPN observations and maintain communication with observers, the LSP Discipline Team issued six newsletters, and several progress reports in the IHW Newsletters.[7-15] Individuals and organizations around the world were extremely helpful in obtaining excellent images and submitting them to the LSP Team in good form.

The 1123 images (taken from 1024 separate photos) used in this Atlas are a subset of the total photographic data set received by the LSP Discipline Specialist Team. We have tried to select images showing various aspects of the large-scale structure and evolution of comet Halley. In selecting the images, we tried to include one for each day between mid-November 1985 and mid-June 1986, choosing the best images available. See Appendix B for graphs of the coverage provided by Atlas images for the entire apparition of the comet, including the time of flybys by the different Halley missions.

The total submission to the LSP Discipline Team consisted of more than 3,500 images, of which 3,383 have been archived on compact disc (CD-ROM) in either digital or "dataless" (FITS header only) form. The 3,383 CD-ROM images were taken on 103 different observatory/telescope/location combinations.[5]

Because Halley's Comet went to southerly declinations during the spring of 1986, and also due to the relative lack of land masses and observatories in the southern hemisphere, the goals of achieving essentially continuous coverage during the entire apparition and of supporting the Halley spacecraft encounters during March 6-14, 1986, placed a difficult requirement on the Large-Scale Phenomena Network. In order to overcome this obstacle, an Island Network was created within the LSPN. Celestron Schmidt cameras were installed at six different observing sites which were favorable for filling coverage gaps. The "island network" evolved to become a "network" with observers at the following locations: (1) Tahiti, French Polynesia; (2) Easter Island; (3) British Antarctic Survey's Faraday Station; (4) Sutherland, South Africa; (5) Cederberg, South Africa; and (6) Reunion Island.[16] The observers in this network made important contributions to the Large-Scale Phenomena Network. In particular, we note the extraordinary work done by Dr. William Liller on Easter Island. He was aided by good weather and contributed the major portion of the color section of this Atlas.

During certain time periods, observing conditions were not very favorable. These occurred during the 6-week period approximately centered on perihelion (February 9, 1986), when the comet was essentially on the other side of the sun (see discussion by Yeomans[17]), and during full moon intervals. The observing condition for full moon was improved once during a total lunar eclipse, and indeed, one image obtained during that time is included (see Fig. 666, page 491).

Although the position of the comet near perihelion made ground-based observations nearly impossible, a few images of the comet were recorded from spacecraft. First, the Pioneer Venus Orbiter (PVO) had a favorable viewing angle and was not blinded by the sun. PVO has no visual wavelength instruments onboard, but scans by the UV photometer carried out between February 2 and 6 produced a Lyman- image showing large-scale structure. In addition, the Solar Maximum Mission spacecraft in Earth orbit carried a coronagraph which permitted images of comet Halley to be obtained while the comet was extremely close to the sun.[18] Some of these observations are included in this Atlas.

A few images of Halley's Comet which are not cosmetically optimal are nevertheless included here, since the goal of the Atlas is to provide a comprehensive scientific record of Halley's Comet. It should be kept in mind that wide-field imagery of comets is subject to the kinds of mistakes that plague other forms of photography: lack of proper exposure, scratched or lightstruck emulsions, non-uniform development, etc. There is also a great deal of pressure to record as long an interval as possible each clear night (due to the comet's unpredictable variability), even if this results in some images being taken during astronomical twilight. In addition, there is the threat of poor observing conditions — clouds, city lights, moonlight, etc. An image of comet Halley which appears to the untrained reader to be less than perfect is often scientifically important, and so is included in the Atlas.

Atlas Production

As expected, the majority of the images included in the Atlas are the result of routine submissions to the LSP Discipline Specialist Team. These images are identified by their LSPN numbers. (The LSPN numbers are internal network numbers and should not be confused with any other numbers or filenames.) Three additional sources of images were used: (1) the Near Nucleus Studies Network provided images showing inner, head-region structure. They were supplied by Stephen Larson, Lunar and Planetary Laboratory, University of Arizona, and are identified by their Near Nucleus Studies Network (NNSN) numbers; (2) the Amateur Observations Network of the IHW supplied selected amateur observations. These were made available through the cooperation of Stephen Edberg, Jet Propulsion Laboratory, and are identified by Amateur Observations Network (AON) numbers; (3) several observations came from direct appeals to individual observers.[19-21] These last images have no reference numbers and are credited directly to the observers.

Except as noted below, scales and orientation for each image were determined by measuring at least two stars of known position and the coordinates of the comet's head (as a check); knowledge of the Earth-comet distance and the angle between the line of sight (LOS) and the prolonged radius vector (PRV) then yielded a linear distance scale for plasma-tail projections along the PRV. The geometrical parameters (below) for the observations are illustrated in Appendix B:

r is the sun-comet distance in AU;
Δ is the Earth-comet distance in AU; and
β is the foreshortening angle in degrees (angle LOS-PRV).

For each measured image, the caption gives a running figure number; the date in decimals of a day (to 0.00001 day) of the exposure midpoint; the length of the exposure (to 0.1 minute); the emulsion/filter combination; r and Δ (to 0.01 AU); β (to 0.1°); the LSPN, NNSN or AON number; and credit to the observer (and organization, where appropriate). Exceptions to the general approach are as follows: (1) for some photographs, scales and orientation are not useful and are not included — these images can easily be recognized; (2) images from the former Soviet Union were received in digital form. The standard procedure was followed but scales and orientation were determined from information given in the headers; (3) scales and orientations are not included for the color images, the mosaics, the sequences and selected "tail-only" images. For these, scales and orientation can be determined by a comparison with similar images in the Daily Record or the Supplemental Daily Record.

Introduction

Nearly all images were received at the primary archive at the Goddard Space Flight Center. High quality film copies were made and sent to the duplicate archive located at the University of Colorado. Almost all 1024 final images selected for the Atlas were made from contact prints at the University of Colorado or its contractors. In order to ensure a scientifically faithful record of the comet's appearance, special efforts were made to preserve the contrast of the original images; we consciously avoided attempting to enhance any images. The images are usually presented as negatives because detail is easier to see in this form; a few images had to be reversed (black for white) to achieve this.

The comet can show a vastly different appearance when observed with various emission/filter combinations. Therefore, we have included tables in Appendix B indicating the approximate response of the emulsions, as well as the different emulsion/filter combinations.

The Atlas itself was prepared with its role as an archival record in mind. The materials used have been specified in accordance with state-of-the-art knowledge of preservation techniques. Specifically, acid-free paper (Lustro Offset Enamel 70 Gloss) and other acid-free materials are used throughout. We have tried to follow the guidelines established by the Committee on Production Guidelines, Council on Library Resources, and published in Publisher's Weekly, May 29, 1981 and July 2, 1982. We used 1-inch wide margins to reduce fingerprint contact with the image area and to facilitate rebinding, if that should become necessary. Hand binding is believed to offer the ultimate in preservation potential. Accordingly, in addition to the main casebound copies, a number of Atlas copies were hand bound and deposited in research libraries throughout the world. These libraries often have controlled environments and limited access. With these special Atlas copies we believe that all reasonable effort has been made for the preservation of this record.

Description of Large-Scale Structure 1985 - 1986

Figure I shows Halley's Comet at recovery on October 16, 1982.[22] At this time, the comet was located at $r = 11.05$ AU and $\Delta = 10.94$ AU. One of the fascinations of comet work is the combination of physics and geometry that produces the development of a major comet from the faint speck shown here, to the dramatic, sky-filling display shown in Figure II. As Halley's Comet was developing, it shared the sky briefly (Figure III) with P/Giacobini-Zinner, the first comet visited by a spacecraft (the International Cometary Explorer, September 11, 1985).

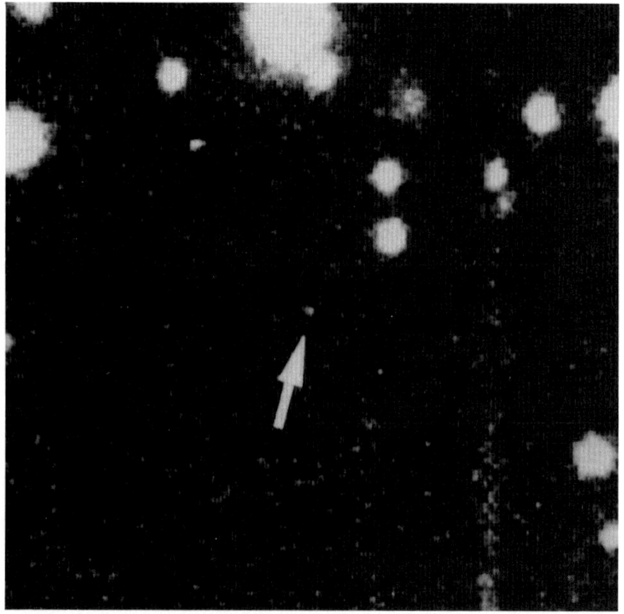

Fig. I. Recovery image of Comet Halley on Oct. 16, 1982, obtained by G.E. Danielson and D. Jewitt.

Fig. II. Comet Halley and the Milky Way on March 12, 1986. (Courtesy R.H. McNaught, Siding Spring Observatory)

This impressive record of large-scale structure together with the in situ measurements carried out in 1985 and 1986 is a powerful database with which to attack many problems of cometary plasma physics and the comet/solar-wind interaction. Possible areas of interest include the beginning and end of plasma activity, the occurrence of waves and motions in the tail, the evolution of tail rays, and disconnection events. In these areas, Halley's Comet did not disappoint us; plasma tail activity was extensive and spectacular. For the dust tail, areas of interest include the size distribution and composition of individual dust grains; the role of impulsive events in the formation of structures not rooted to the head (rotational bursting of grains, etc.); synchronic band structure; neckline structures and anti-tails.

Some of the changes in the appearance of the comet are due to its orbit and the geometrical conditions when observed from Earth. The orbit of comet Halley at this apparition as projected on the plane of the ecliptic is shown in Appendix B. The dates of closest approach to Earth are November 27, 1985 (Δ = 0.6194 AU) and April 11, 1986 (Δ = 0.4172 AU). Perihelion occurred on February 9, 1986. The geometry of the plasma and dust tails as seen on the plane of the sky is given by Yeomans[17] and Niedner.[14] Descriptions of the comet in 1985-1986 begin with Niedner.[23] Also see Brandt and Niedner[24,25] and Brandt.[26]

Initial Plasma-Tail Turn-on: Mid November 1985

Practically all of the wide-field images known to us and taken before November 13, 1985 (r = 1.75 AU) show no tail, which qualitatively is in agreement with the behavior this comet exhibited during its 1910 apparition. Of particular interest is a trio of wide-band (103a-E) images taken by L. Kohoutek at the National Observatory Station at Calar Alto, Spain, on November 10, 13, and 18, 1985. The images show no plasma tail on the 10th and 18th, whereas on November 13, a 21-arcminute-long tail is discernable; the permanently-visible plasma tail developed around December 1.

December 1985 - January 1986

With the development of a "permanent" plasma tail around December 1, and with dark skies in early December, Halley showed the kind of activity which fully justified the formation of an extensive Large-Scale Phenomena Network (103 observatories). The first clear disconnection event (DE) was photographed by a number of observatories on December 4-5, 1985. Two days later, the comet developed a system of helical waves[27] similar to those seen in comets Arend-Roland 1957 III and Kohoutek 1973 XII. There was a prolonged interval of tail disconnection activity, some of it prominent, during December 13-15. A fairly spectacular DE occurred on December 31 - January 1. Additional DEs occurred on January 8 and 10, with the latter easily being the most impressive among the pre-perihelion events.

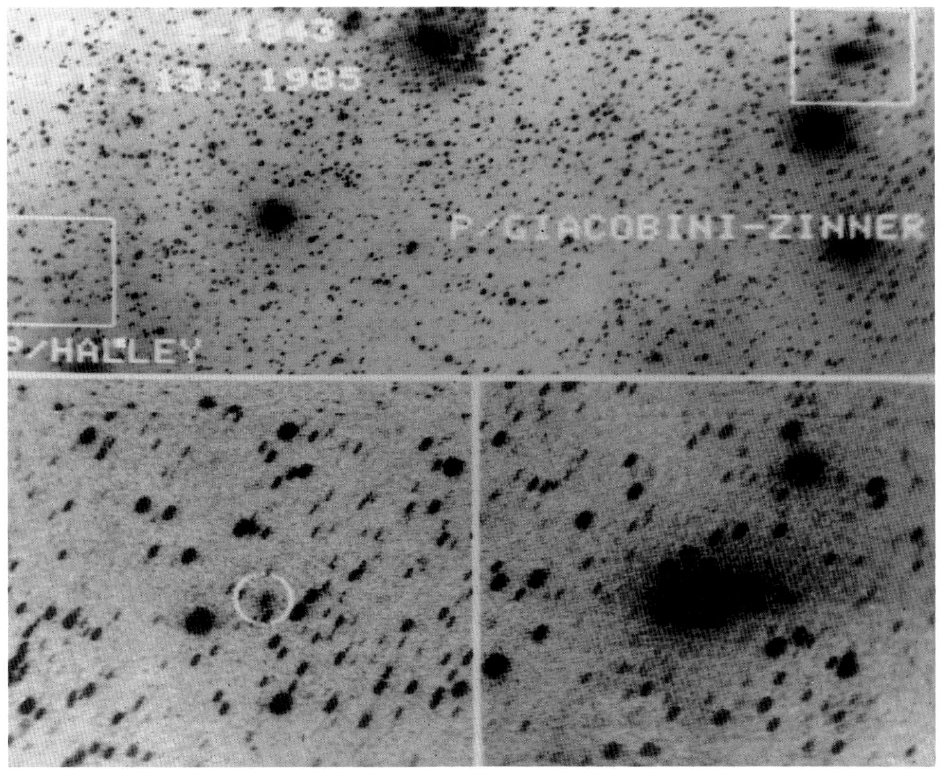

Fig. III. Comets Halley and Giacobini-Zinner 1985 on Sept. 13, 1985. (Photograph by E. Moore, Joint Observatory for Cometary Research. LSPN-43)

February - April 1986

The post-perihelion plasma tail activities, especially disconnection events, were consistently more impressive than their pre-perihelion counterparts. The first known post-perihelion event occurred on about February 22 as seen in the spectacular UK Schmidt image (Fig IV). This reproduction has been specially processed by copying the original onto high-contrast film and making the prints from the copy (compare with LSPN 2376, p. 132). Narrowband CO^+ images[28] obtained with the European Southern Observatory (ESO) wide-field CCD camera on March 4 show a disturbance in the outer tail, which may be an advanced state of a DE that occurred several days earlier.

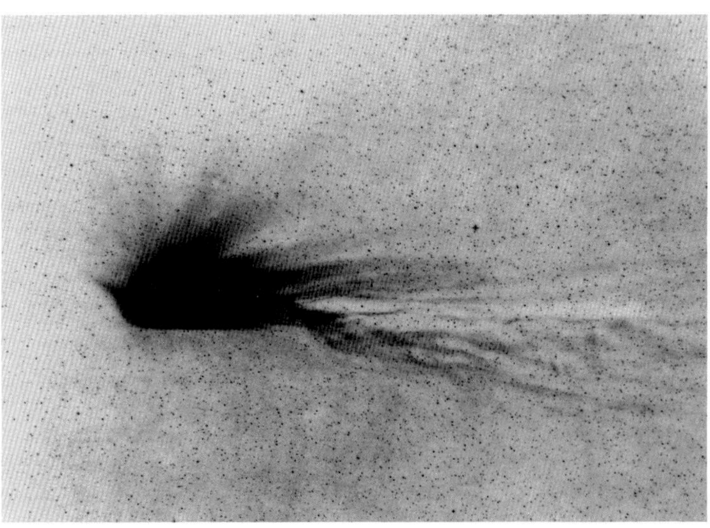

Fig. IV. Comet Halley on Feb. 22, 1986; see text. (UK Schmidt Telescope Unit. LSPN 2376)

To the best of our knowledge, the encounter of the Soviet VEGA-1 mission with Halley on March 6 occurred when the comet was not undergoing (or had not recently undergone) a DE or similar tail disturbance. The same cannot be said, however, for the VEGA-2 encounter with Halley on March 9. Plates taken on March 10 by the ESO Schmidt[28], and by F.D. Miller on the Cerro Tololo Inter-American Observatory (CTIO) Schmidt, show a truly spectacular event. E. E. Barnard Observatory photographs show this event occurring close to the cometary head on March 9.47 UT, only 4 hours after VEGA-2's encounter, and the very early stages of it can just be made out in photographs taken elsewhere on March 8.9 UT. Niedner and Schwingenschuh[29] attributed the development of the DE to the passage of the comet through a reversal in the interplanetary magnetic field detected by VEGA-1 when it was extremely close to Halley.

An additional example of image enhancing techniques is shown in the April 12 image taken by R. T. Jones (Figure V). The processing technique was to produce a low contrast but high density positive. A high contrast duplicate negative was made from this positive, and then the final print.

May - July 1986

The large-scale appearance of the comet before its perihelion passage was almost entirely due to the plasma tail. Post-perihelion, February through April, the comet displayed both plasma and dust tails. Both tails persisted into early May, but then the plasma structure faded away leaving a narrow, unchanging tail structure which consisted of dust (probably a neck-line structure)[30]. The large-scale structure in June was dust and observations were not extensive. The dust structure persisted at least to 6 July 1986 (LSPN 2453). At this time, $r = 2.80$ AU and $\Delta = 2.50$ AU.

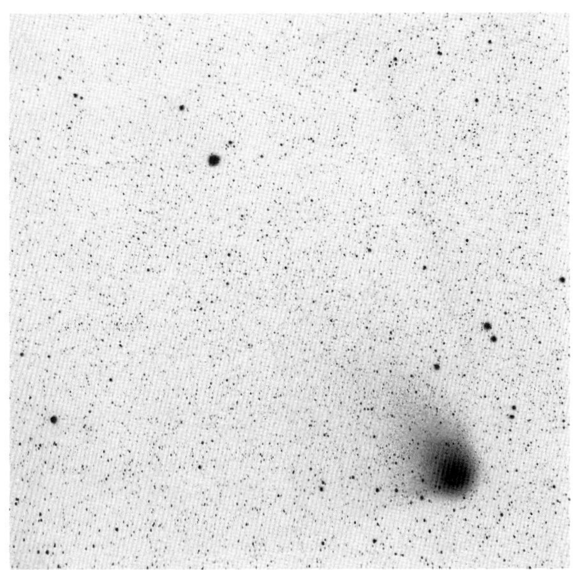
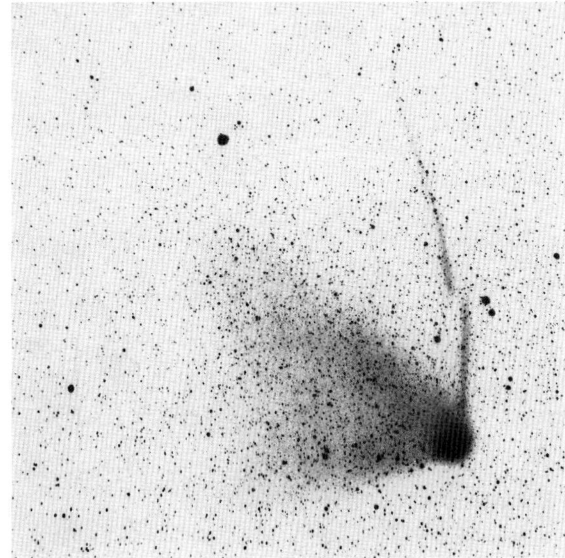

Fig. V. Comet Halley on Apr. 12, 1986 as photographed by R.T. Jones, United Kingdom. Normal processing is shown at left and enhanced processing (see text) at right.

Plasma Tail Turn-Off: Early May 1986

The time of plasma turn-off is not entirely straightforward to establish. The visible plasma structures faded away during the first week of May, probably around May 3. However, some plasma activity may have persisted after this time because the spectrum continued to show bands from ionized species until at least mid-June, 1986.

Outburst 1991

Halley's Comet displayed an enormous outburst on February 12, 1991 (Figure VI). The field of view at the comet's distance was approximately 700,000 by 700,000 km, and the central cloud (presumably dust) exceeded 300,000 km in diameter. The outburst persisted for months; the details were reported by West, Hainaut, and Smette.[31] A major outburst at a heliocentric distance of 14.4 AU is not easily explained and beautifully illustrates the challenges of understanding cometary phenomena.

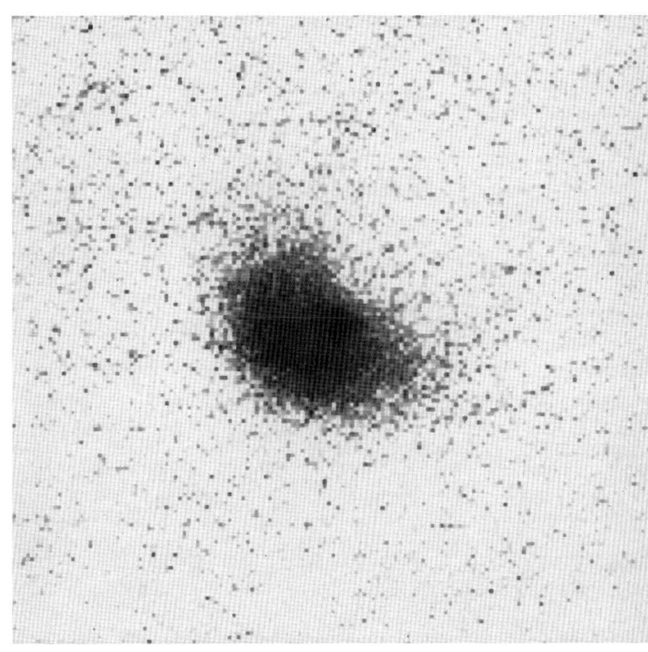

Fig. VI. Outburst of comet Halley on Feb. 12, 1991 observed by O. Hainaut and A. Smette, European Southern Observatory.

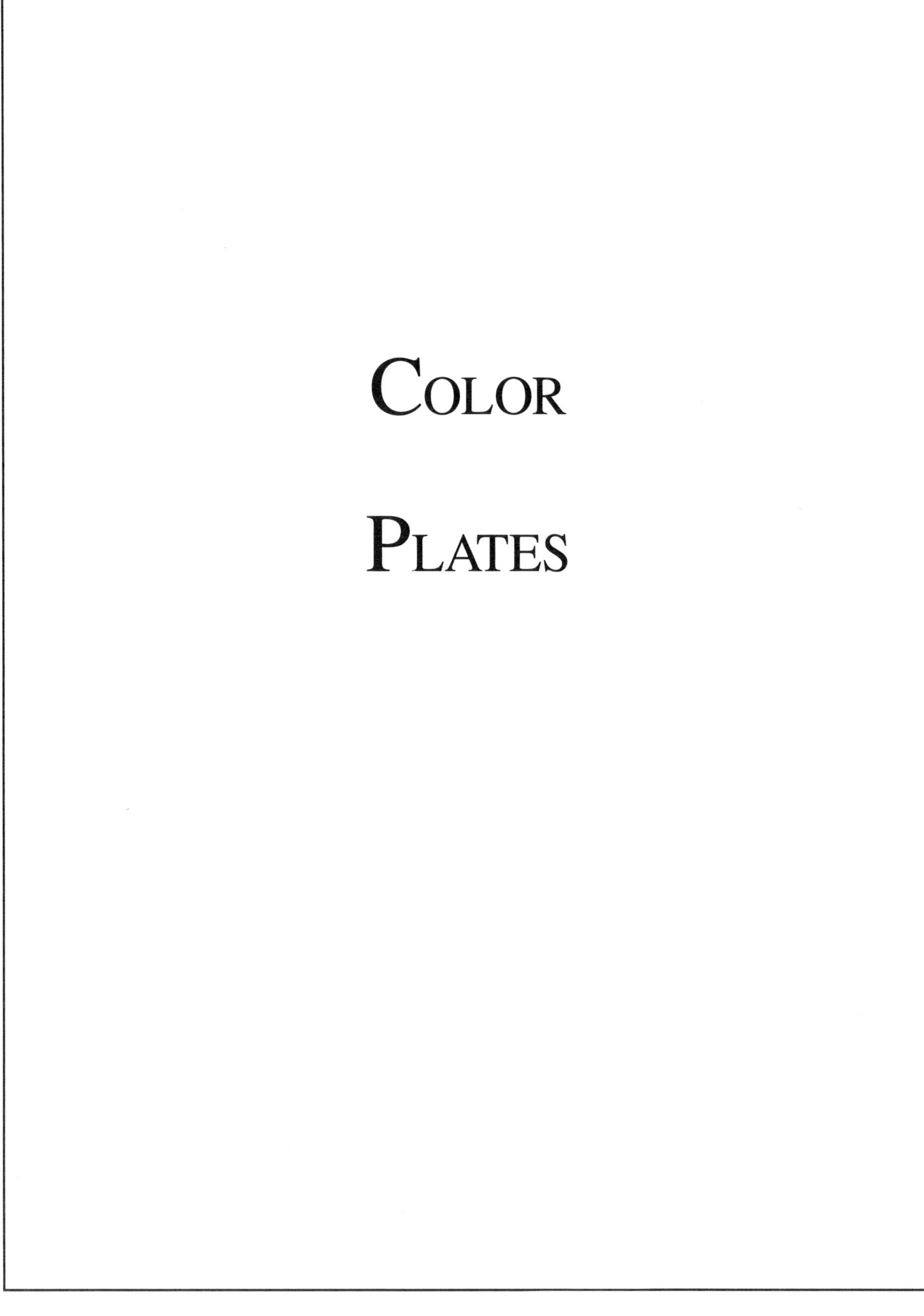

COLOR PLATES

March 7, 1986. Courtesy of Greg Polus, Deer, AR, USA.

March 20, 1986. Courtesy of Greg Polus, Deer, AR, USA.

December 9, 1985. ©ROE/AAT Board. Photograph made from UK Schmidt telescope plates by D. Malin.

March 8, 1986; LSPN-1725. Photograph by W. Liller, Easter Island, LSPN Island Network.

March 10, 1986; LSPN-1726. Photograph by W. Liller, Easter Island, LSPN Island Network.

March 11, 1986; LSPN-1727. Photograph by W. Liller, Easter Island, LSPN Island Network.

March 12, 1986; LSPN-1728. Photograph by W. Liller, Easter Island, LSPN Island Network.

March 13, 1986; LSPN-1729. Photograph by W. Liller, Easter Island, LSPN Island Network.

March 14, 1986; LSPN-1730. Photograph by W. Liller, Easter Island, LSPN Island Network.

March 15, 1986; LSPN-1731. Photograph by W. Liller, Easter Island, LSPN Island Network.

March 16, 1986; LSPN-1732. Photograph by W. Liller, Easter Island, LSPN Island Network.

March 17, 1986; LSPN-1733. Photograph by W. Liller, Easter Island, LSPN Island Network.

March 18, 1986; LSPN-1734. Photograph by W. Liller, Easter Island, LSPN Island Network.

March 19, 1986; LSPN-1735. Photograph by W. Liller, Easter Island, LSPN Island Network.

March 19, 1986; LSPN-602. Photograph from Société Astronomique de France, Reunion Island, LSPN Island Network.

March 22, 1986; LSPN-1736. Photograph by W. Liller, Easter Island, LSPN Island Network.

March 23, 1986; LSPN-1737. Photograph by W. Liller, Easter Island, LSPN Island Network.

April 4, 1986; LSPN-1740. Photograph by W. Liller, Easter Island, LSPN Island Network.

April 6, 1986; LSPN-1742. Photograph by W. Liller, Easter Island, LSPN Island Network.

April 8, 1986; LSPN-1743. Photograph by W. Liller, Easter Island, LSPN Island Network.

April 11, 1986; LSPN-1746. Photograph by W. Liller, Easter Island, LSPN Island Network.

April 12, 1986; LSPN-1747. Photograph by W. Liller, Easter Island, LSPN Island Network.

April 14, 1986; LSPN-1748. Photograph by W. Liller, Easter Island, LSPN Island Network.

April 15, 1986; LSPN-1749. Photograph by W. Liller, Easter Island, LSPN Island Network.

April 17, 1986; LSPN-1751. Photograph by W. Liller, Easter Island, LSPN Island Network.

April 18, 1986; LSPN-1752. Photograph by W. Liller, Easter Island, LSPN Island Network.

April 30, 1986; LSPN-1753. Photograph by W. Liller, Easter Island, LSPN Island Network.

April 8, 1986. Courtesy of C. E. Williams Middle School, Can-Do Project, South Carolina, USA, flown on Kuiper Airborne Observatory over New Zealand.

March 21, 1986. Courtesy of the European Southern Observatory.

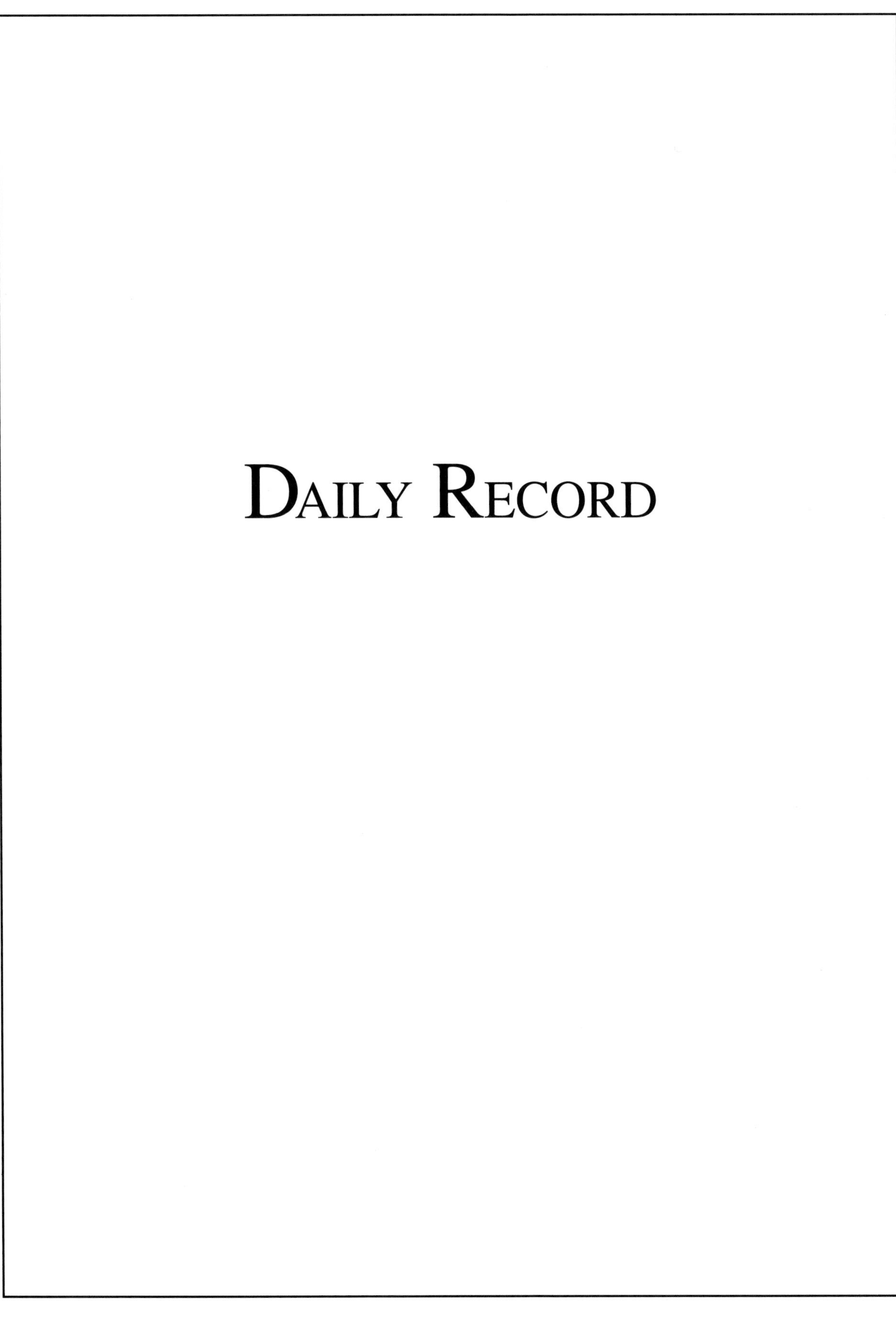

Daily Record

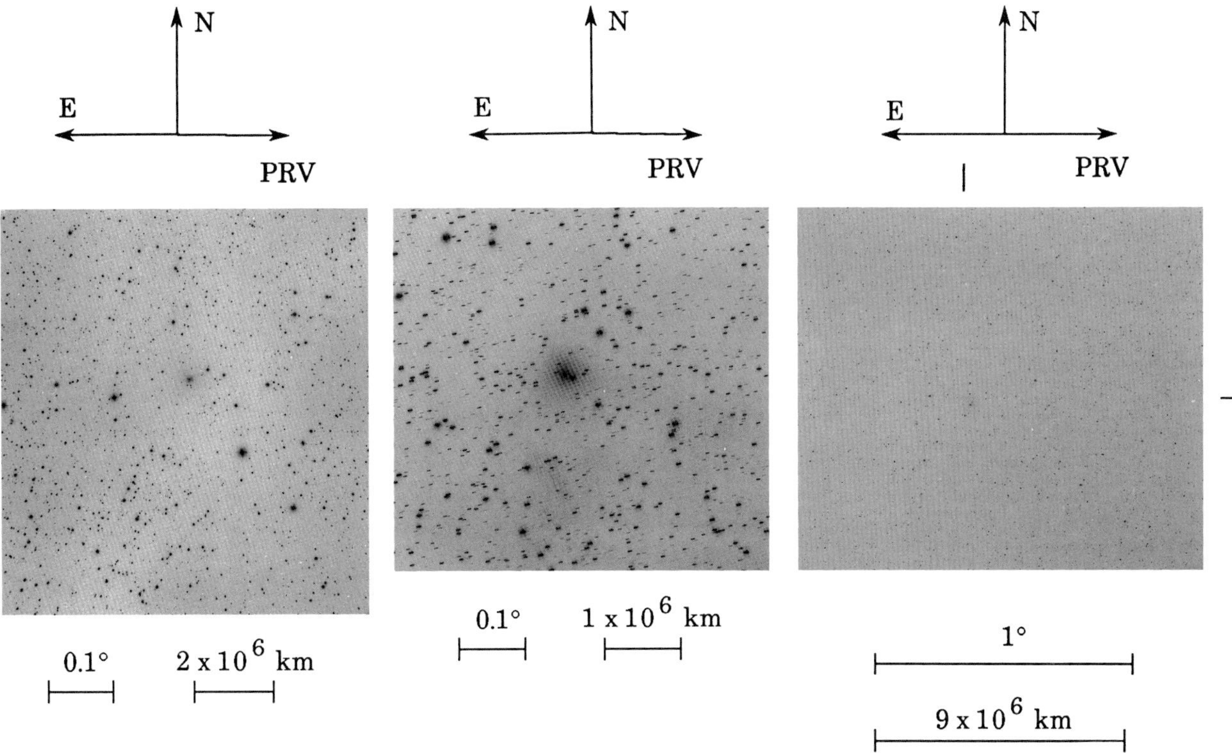

(left) Fig. 1: 1985 Sep 17.46736 UT; exposure 30.0 minutes on 103a-O emulsion with no filter; r=2.52, Δ=2.47, β=156.8°; LSPN-768.
Photograph by E. Harlan, Lick Observatory.

(center) Fig. 2: 1985 Oct 13.50000 UT; exposure 30.0 minutes on 103a-O emulsion with no filter; r=2.18, Δ=1.64, β=154.3°; LSPN-771.
Photograph by E. Harlan, Lick Observatory.

(right) Fig. 3: 1985 Oct 19.26040 UT; exposure 10.0 minutes on IIa-D emulsion with no filter; r=2.10, Δ=1.46, β=154.8°; LSPN-720.
Photograph by C. Torres/H. Wroblewski, Cerro el Roble Astronomical Observatory.

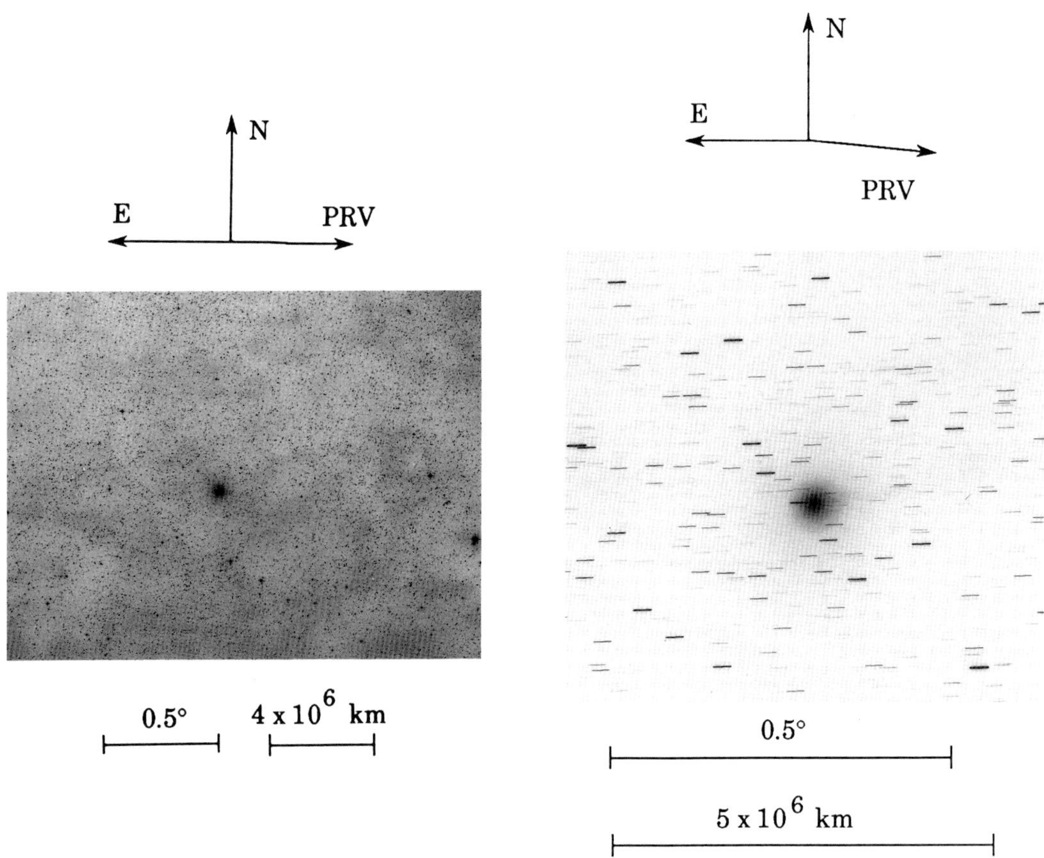

(left) Fig. 4: 1985 Oct 22.02434 UT; exposure 30.0 minutes on 103a-O emulsion with no filter; r=2.06, Δ=1.37, β=155.4°; LSPN-3431.
Photograph by I. Toth, Konkoly Observatory, Piszkesteto Station.

(right) Fig. 5: 1985 Nov 6.29514 UT; exposure 30.0 minutes on 103a-O emulsion with no filter; r=1.85, Δ=0.93, β=163.3°; LSPN-777.
Photograph by E. Harlan, Lick Observatory.

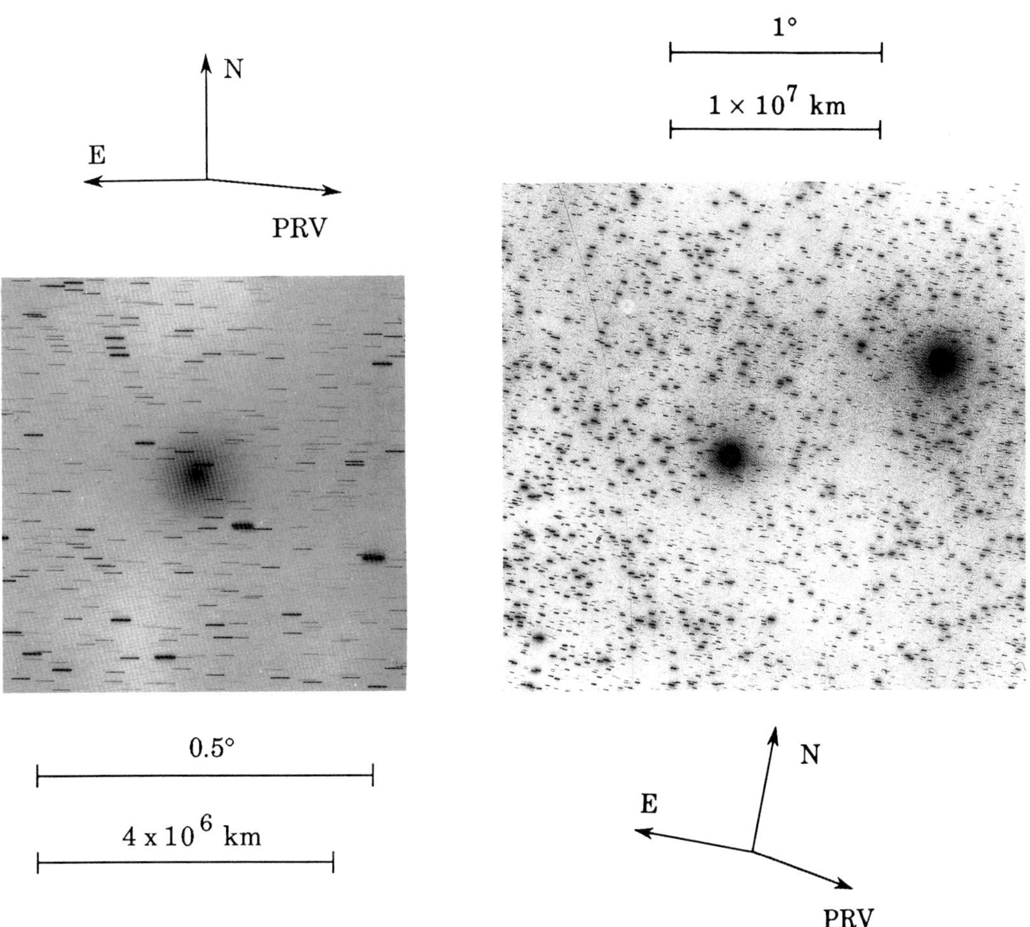

(left) Fig. 6: 1985 Nov 7.28125 UT; exposure 30.0 minutes on 103a-O emulsion with no filter; r=1.83, Δ=0.91, β=164.2°; LSPN-778.
Photograph by E. Harlan, Lick Observatory.

(right) Fig. 7: 1985 Nov 8.95625 UT; exposure 20.0 minutes on IIa-O emulsion with no filter; r=1.81, Δ=0.86, β=166.0°; LSPN-2455.
Photograph by V. Ivanova et al., Bulgarian National Observatory.

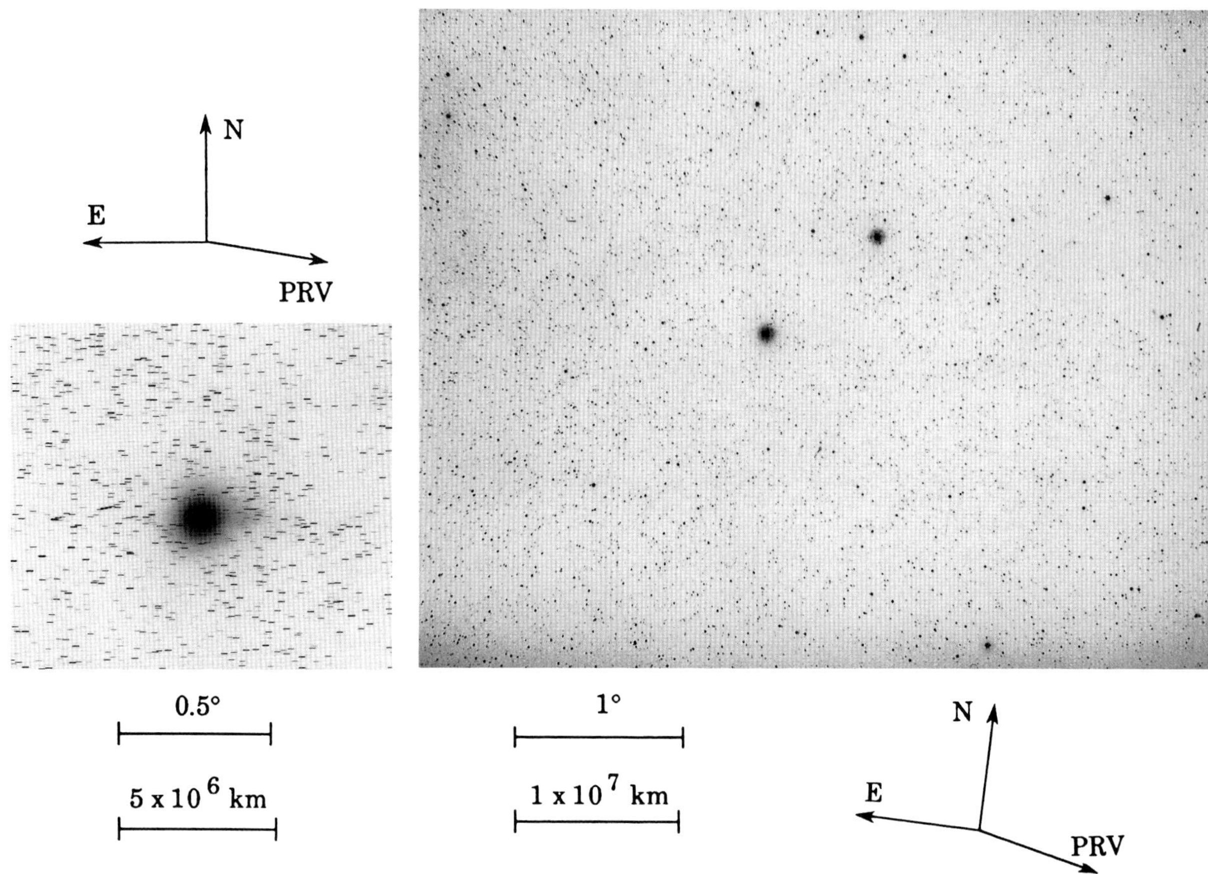

(left) Fig. 8: 1985 Nov 9.04579 UT; exposure 20.0 minutes on ORWO ZU-21 emulsion with no filter; r=1.81, Δ=0.86, β=166.1°; LSPN-1963.
Photograph by I. Platais, Riga Radio-Astrophysical Observatory.

(right) Fig. 9: 1985 Nov 9.14604 UT; exposure 3.0 minutes on Tech. Pan 2415 emulsion; r=1.80, Δ=0.86, β=166.2°; AON-850104.
Photograph by R. Dilsizian, USA.

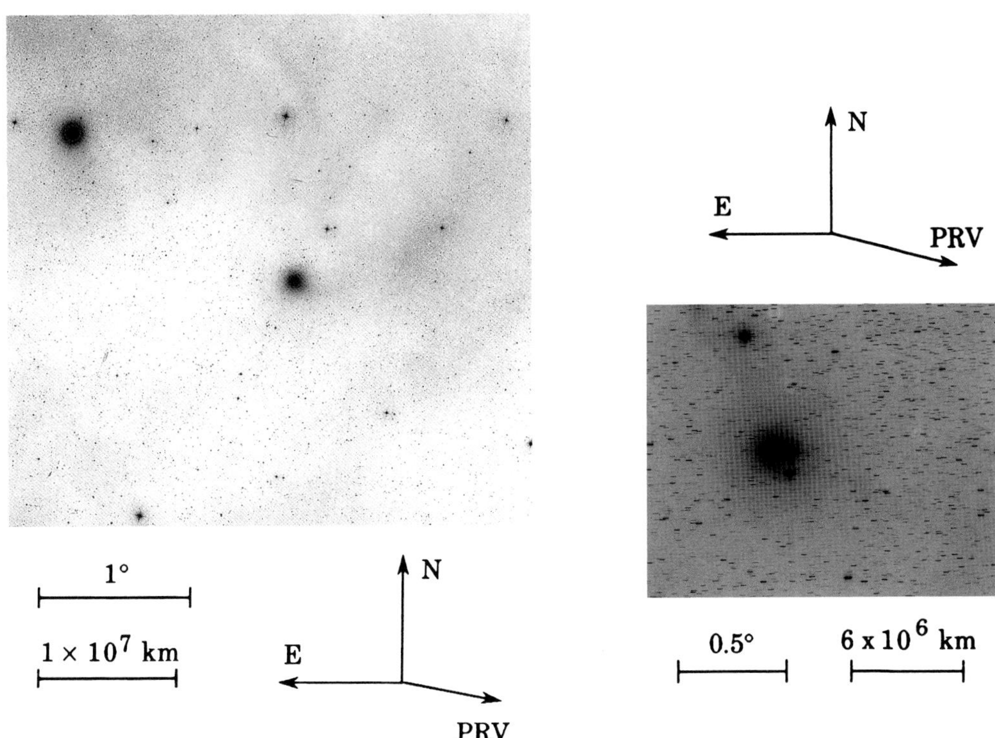

(left) Fig. 10: 1985 Nov 10.57361 UT; exposure 20.0 minutes on IIa-O emulsion with no filter; r=1.78, Δ=0.83, β=167.9°; LSPN-1323.
Photograph by T. Tsujimura, University Kyoto Observatory.

(right) Fig. 11: 1985 Nov 11.91042 UT; exposure 20.0 minutes on ORWO ZU-21 emulsion with no filter; r=1.76, Δ=0.80, β=169.7°; LSPN-2310.
Photograph by V. Shkodrov et al., Bulgarian National Observatory.

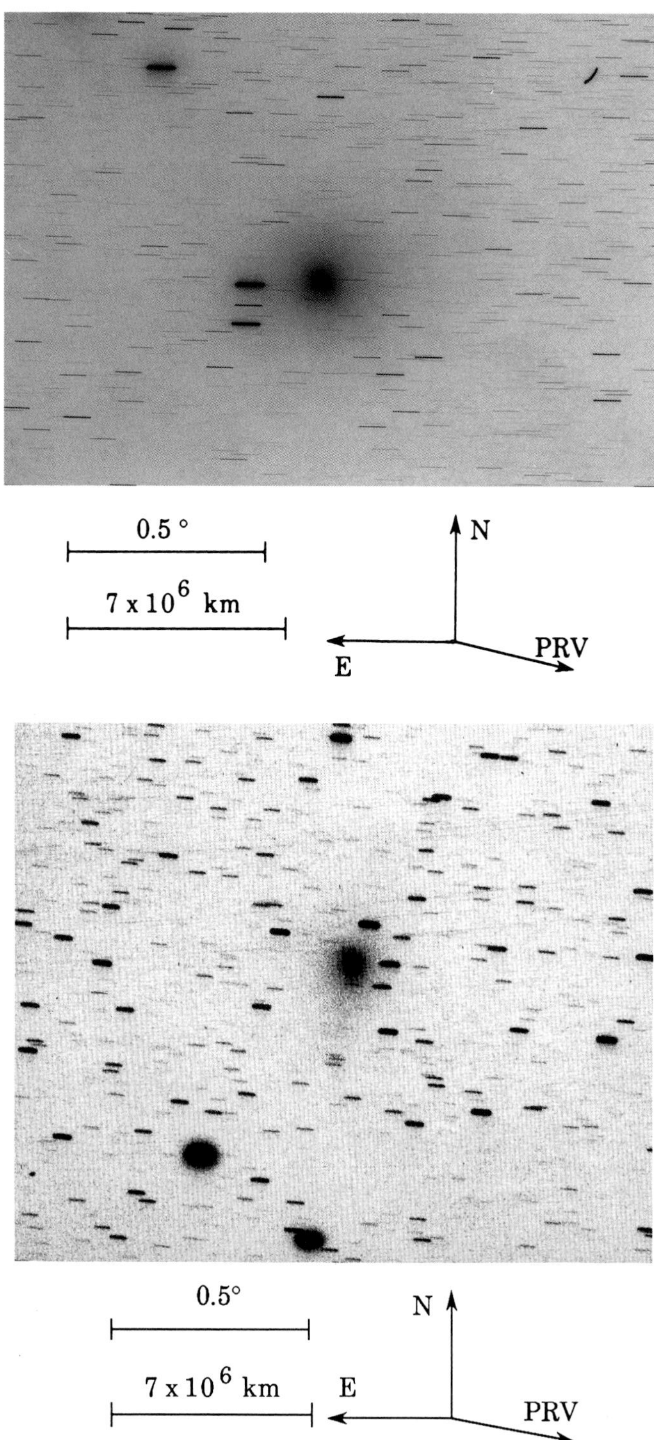

(top) Fig. 12: 1985 Nov 12.06493 UT; exposure 61.0 minutes on ORWO ZU-21 emulsion with UG2d filter; r=1.76, Δ=0.79, β=169.9°; LSPN-3458.
Photograph by H. Meusinger/K. Mau, Karl Schwarzschild Observatory.

(botttom) Fig. 13: 1985 Nov 12.91100 UT; exposure 30.0 minutes on IIa-F emulsion; r=1.75, Δ=0.78, β=171.1°; AON-850154.
Photograph by H. B. Ridley, UK.

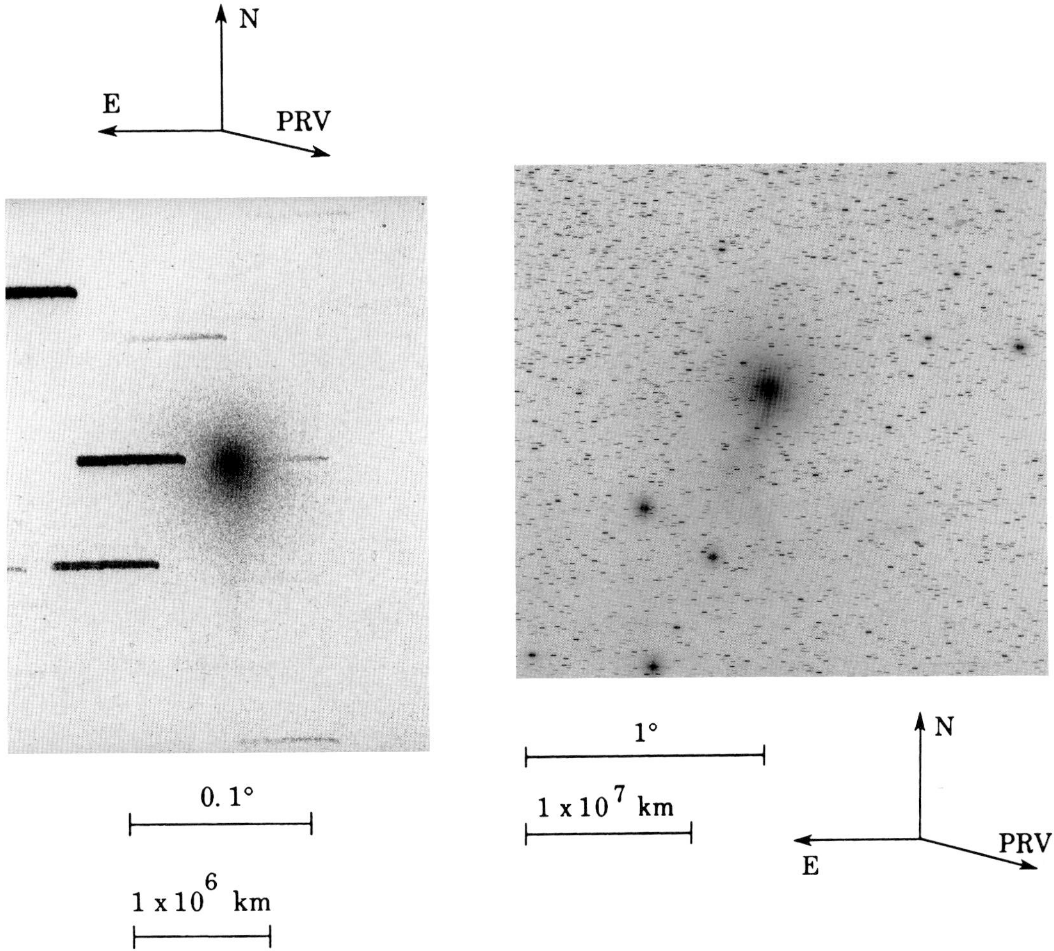

(left) Fig. 14: 1985 Nov 12.98703 UT; exposure 46.0 minutes on 3M 1000 emulsion; r=1.75, $\Delta=0.78$, $\beta=171.2°$; AON-850155.
Photograph by M. Mobberley, UK.

(right) Fig. 15: 1985 Nov 13.00069 UT; exposure 14.0 minutes on 103a-E emulsion with no filter; r=1.75, $\Delta=0.78$, $\beta=171.3°$; LSPN-3828.
Photograph by L. Kohoutek, Calar Alto Station, German National Ast. Observatory.

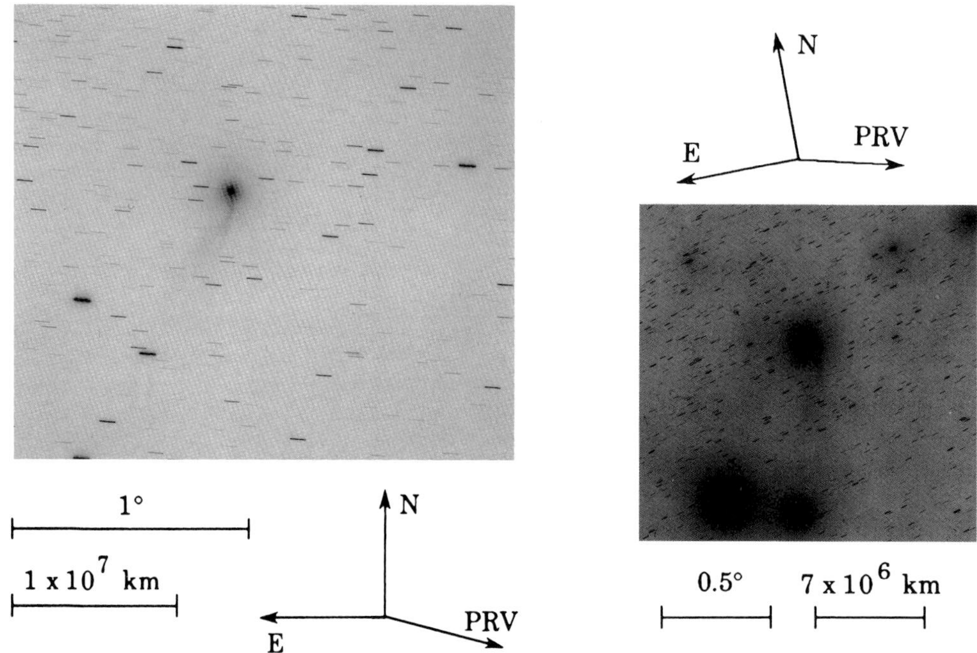

(left) Fig. 16: 1985 Nov 13.04028 UT; exposure 50.0 minutes on IIIa-J emulsion; r=1.75, Δ=0.77, β=171.3°; LSPN-3829.
Photograph by L. Kohoutek, Calar Alto Station, German National Ast. Observatory.

(right) Fig. 17: 1985 Nov 13.06806 UT; exposure 20.0 minutes on ORWO ZU-21 emulsion with no filter; r=1.75, Δ=0.77, β=171.4°; LSPN-2312.
Photograph by V. Ivanova et al., Bulgarian National Observatory.

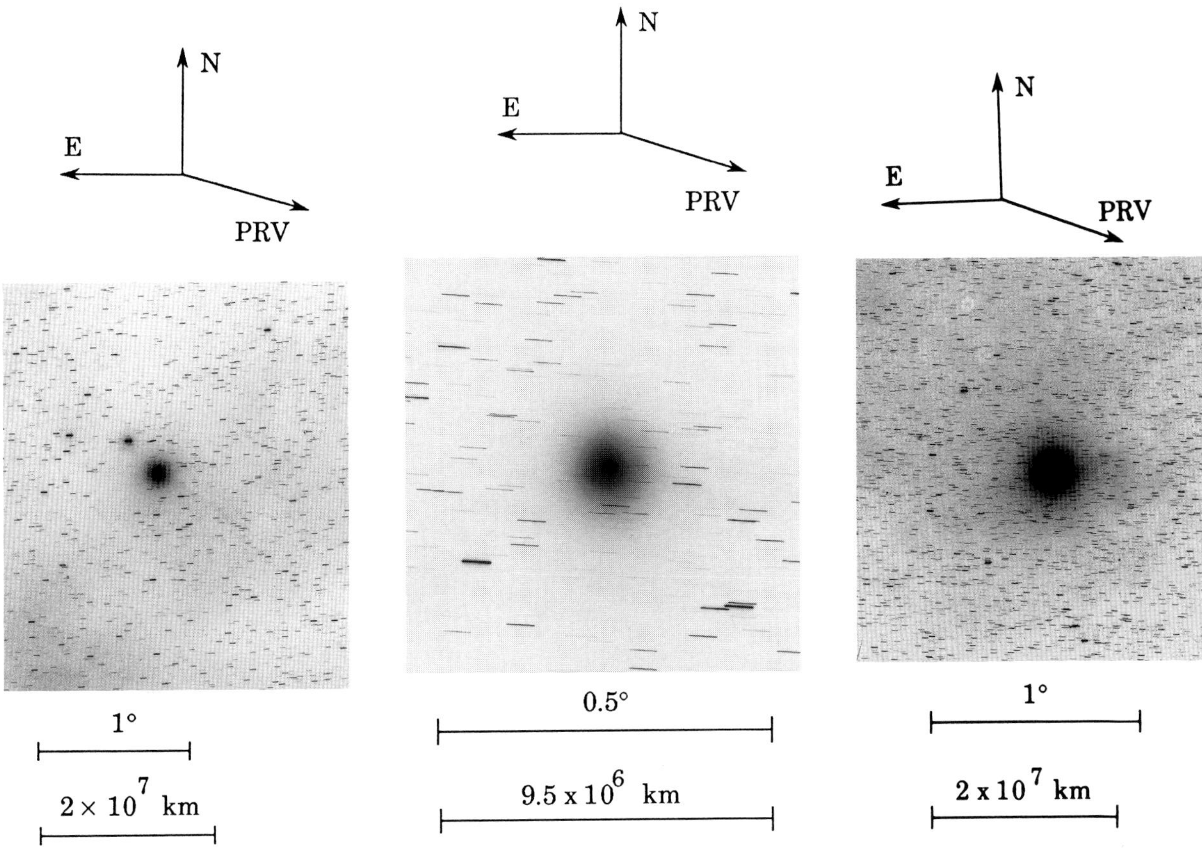

(left) Fig. 18: 1985 Nov 13.66944 UT; exposure 30.0 minutes on IIa-O emulsion with no filter; r=1.74, Δ=0.76, β=172.3°; LSPN-1328.
Photograph by T. Tsujimura, University Kyoto Observatory.

(center) Fig. 19: 1985 Nov 14.22500 UT; exposure 30.0 minutes on 103a-O emulsion with no filter; r=1.73, Δ=0.75, β=173.1°; LSPN-779.
Photograph by E. Harlan, Lick Observatory.

(right) Fig. 20: 1985 Nov 15.02708 UT; exposure 20.0 minutes on ORWO ZU-21 emulsion with no filter; r=1.72, Δ=0.74, β=174.4°; LSPN-2314.
Photograph by V. Ivanova et al., Bulgarian National Observatory.

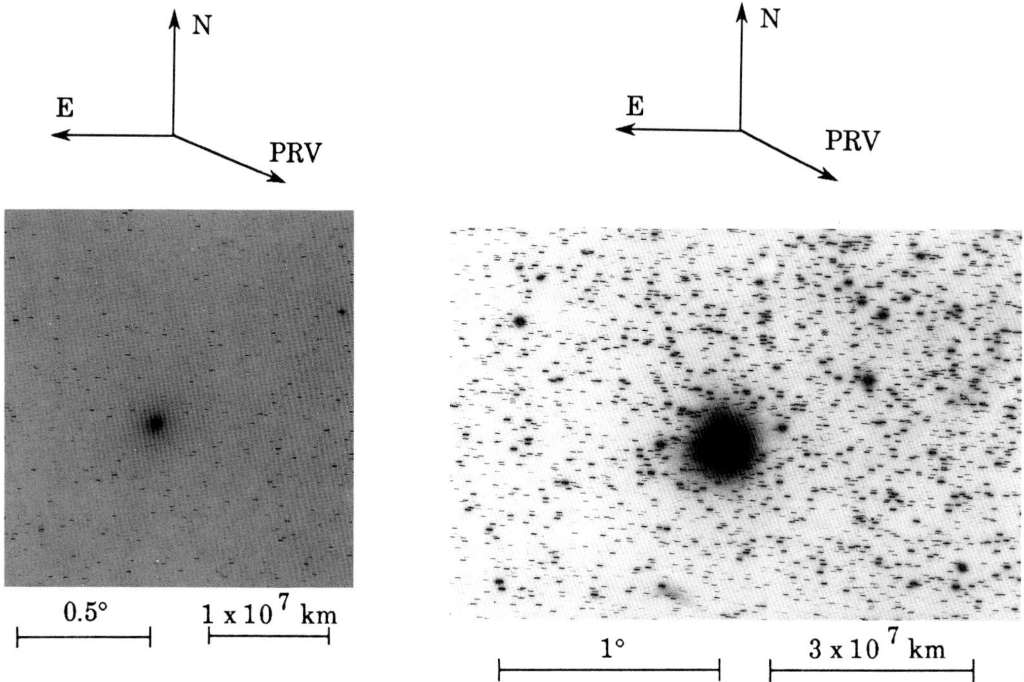

(left) Fig. 21: 1985 Nov 15.09306 UT; exposure 10.0 minutes on 103a-O emulsion with no filter; r=1.72, Δ=0.74, β=174.5°; LSPN-721.
Photograph by C. Torres/H. Wroblewski, Cerro el Roble Astronomical Observatory.

(right) Fig. 22: 1985 Nov 15.89375 UT; exposure 20.0 minutes on ORWO ZU-21 emulsion with no filter; r=1.71, Δ=0.72, β=175.8°; LSPN-2459.
Photograph by V. Ivanova et al., Bulgarian National Observatory.

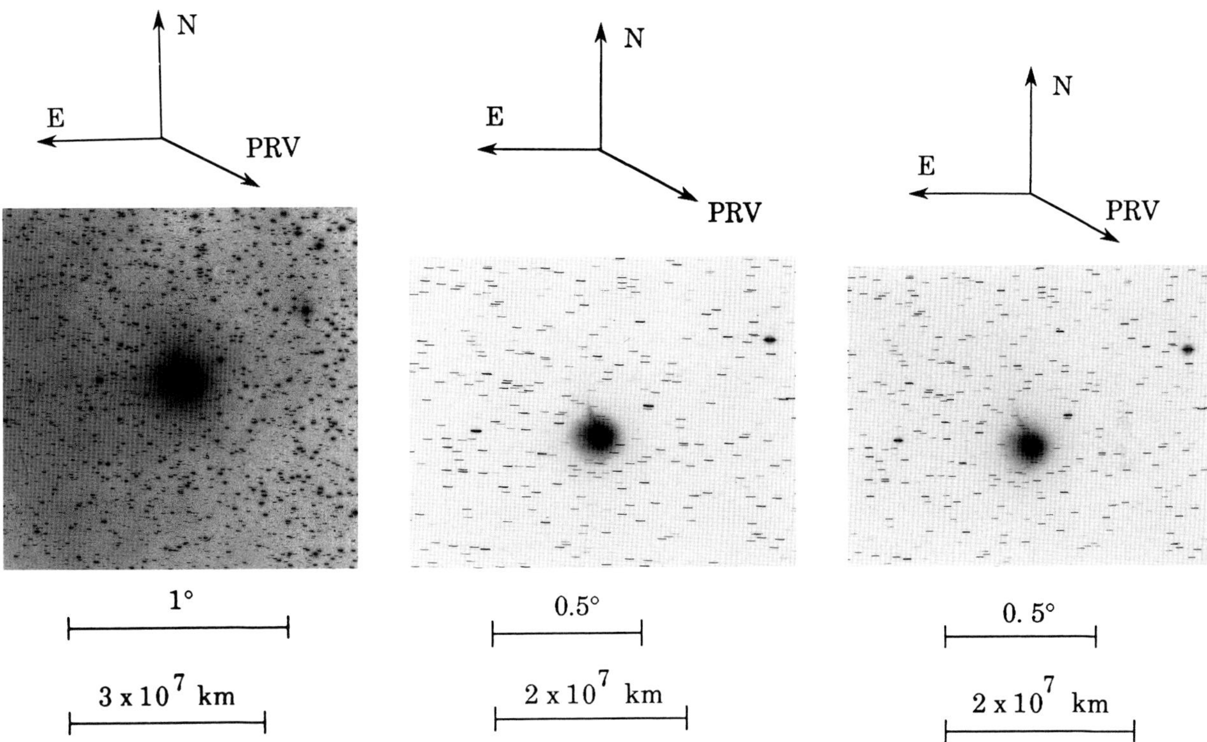

(left) Fig. 23: 1985 Nov 15.93715 UT; exposure 13.1 minutes on ORWO ZU-21 emulsion with no filter; r=1.71, Δ=0.72, β=175.9°; LSPN-2460.
Photograph by V. Ivanova et al., Bulgarian National Observatory.

(center) Fig. 24: 1985 Nov 15.94094 UT; exposure 20.0 minutes on ORWO ZU-21 emulsion with GG-13 filter; r=1.71, Δ=0.72, β=175.9°; LSPN-1965.
Photograph by I. Platais, Riga Radio-Astrophysical Observatory.

(right) Fig. 25: 1985 Nov 15.95863 UT; exposure 17.0 minutes on ORWO ZU-21 emulsion with GG-13 filter; r=1.71, Δ=0.72, β=176.0°; LSPN-1966.
Photograph by I. Platais, Riga Radio-Astrophysical Observatory.

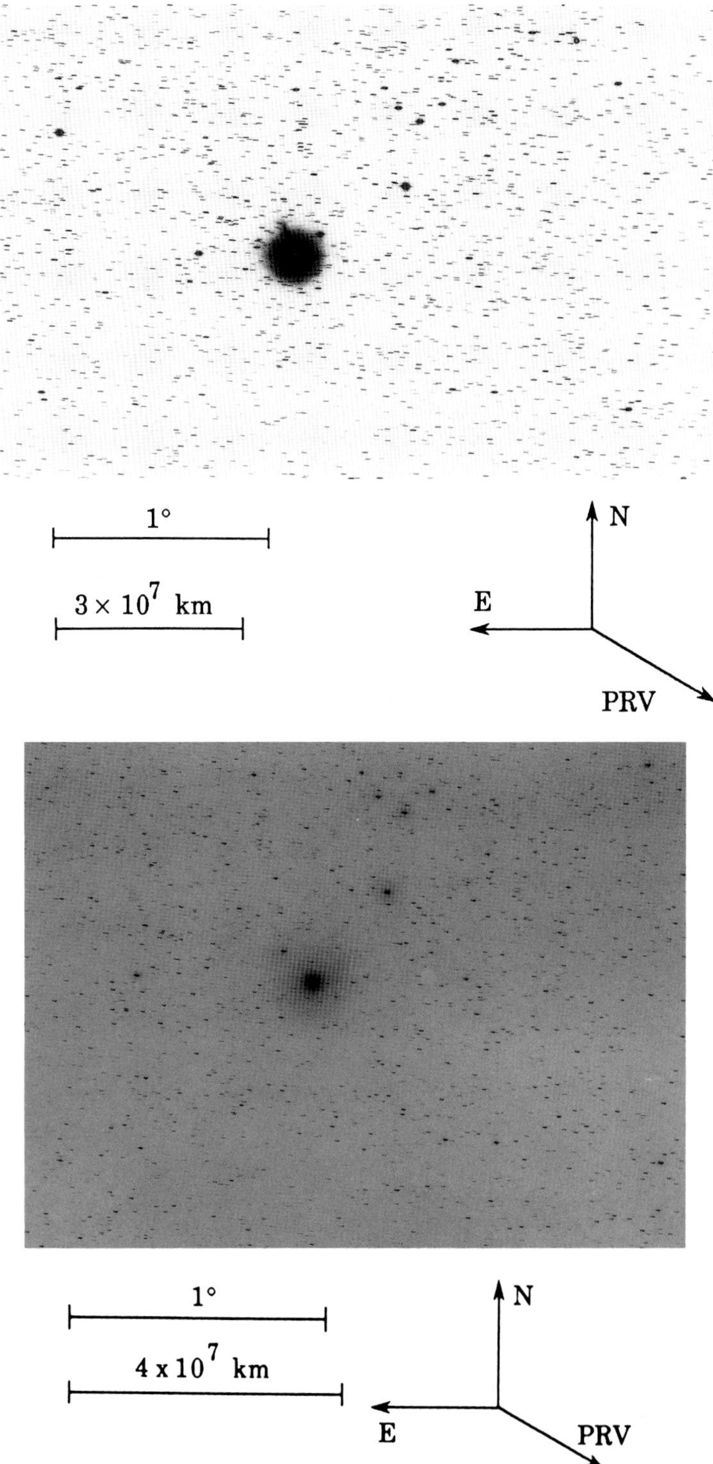

(top) Fig. 26: 1985 Nov 15.96597 UT; exposure 20.0 minutes on ORWO ZU-21 emulsion with no filter; r=1.71, Δ=0.72, β=176.0°; LSPN-2461.
Photograph by V. Ivanova et al., Bulgarian National Observatory.

(botttom) Fig. 27: 1985 Nov 16.06597 UT; exposure 10.0 minutes on 103a-O emulsion with no filter; r=1.70, Δ=0.72, β=176.1°; LSPN-722.
Photograph by C. Torres/H. Wroblewski, Cerro el Roble Astronomical Observatory.

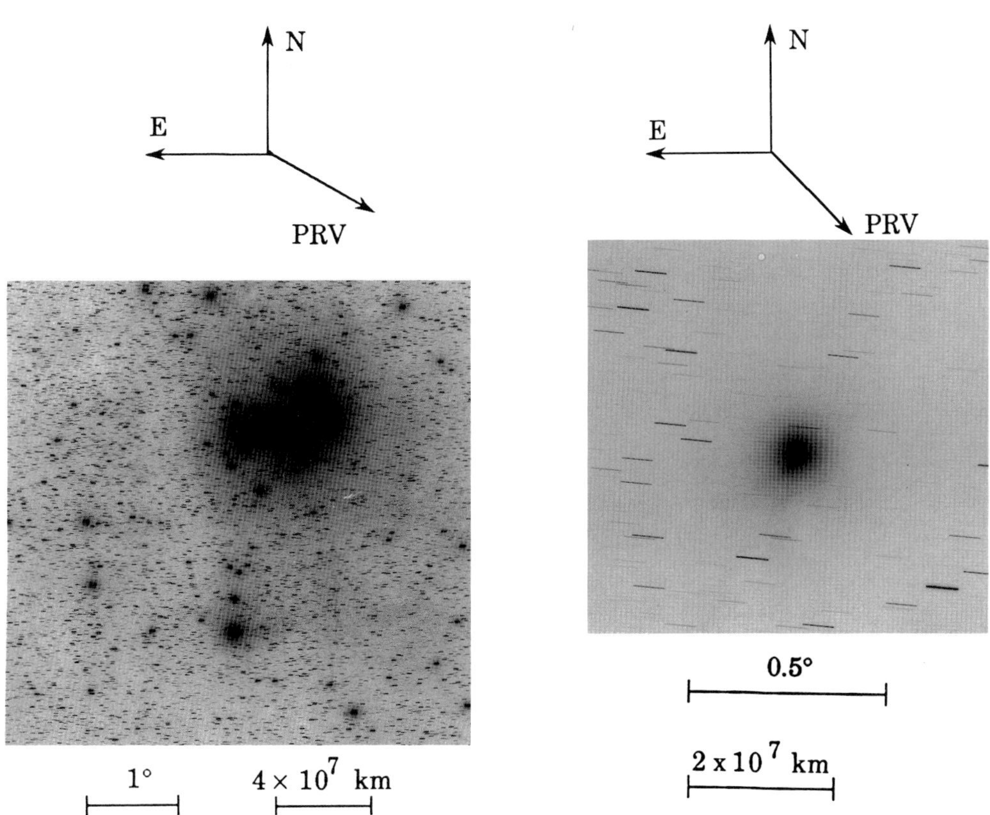

(left) Fig. 28: 1985 Nov 16.18958 UT; exposure 30.0 minutes on IIa-O emulsion with no filter; r=1.70, Δ=0.72, β=176.3°; LSPN-1686.
Photograph by E. Moore/L. Bair, Joint Observatory for Cometary Research.

(right) Fig. 29: 1985 Nov 17.04097 UT; exposure 50.0 minutes on ORWO ZU-21 emulsion with UG2d filter; r=1.69, Δ=0.70, β=177.7°; LSPN-3460.
Photograph by R. Ziener/K. Mau, Karl Schwarzschild Observatory.

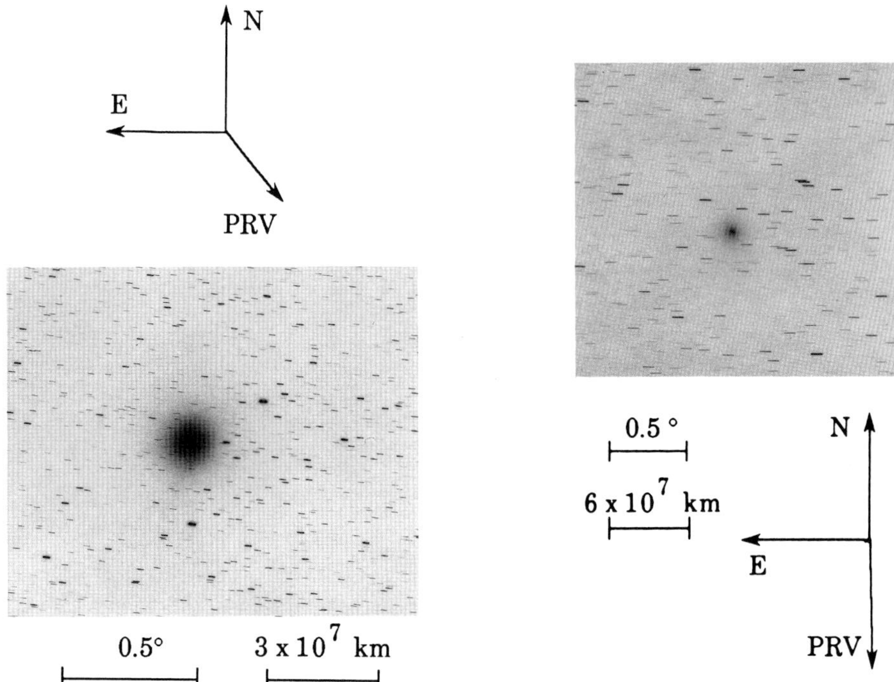

(left) Fig. 30: 1985 Nov 17.20174 UT; exposure 15.0 minutes on IIa-O emulsion with no filter; r=1.69, Δ=0.70, β=178.0°; LSPN-143.
Photograph by W. Liller, University of Michigan/CTIO.

(right) Fig. 31: 1985 Nov 17.91007 UT; exposure 45.0 minutes on IIa-O emulsion with no filter; r=1.68, Δ=0.69, β=178.6°; LSPN-2841.
Photograph by G. Malcolm, Boyden Observatory.

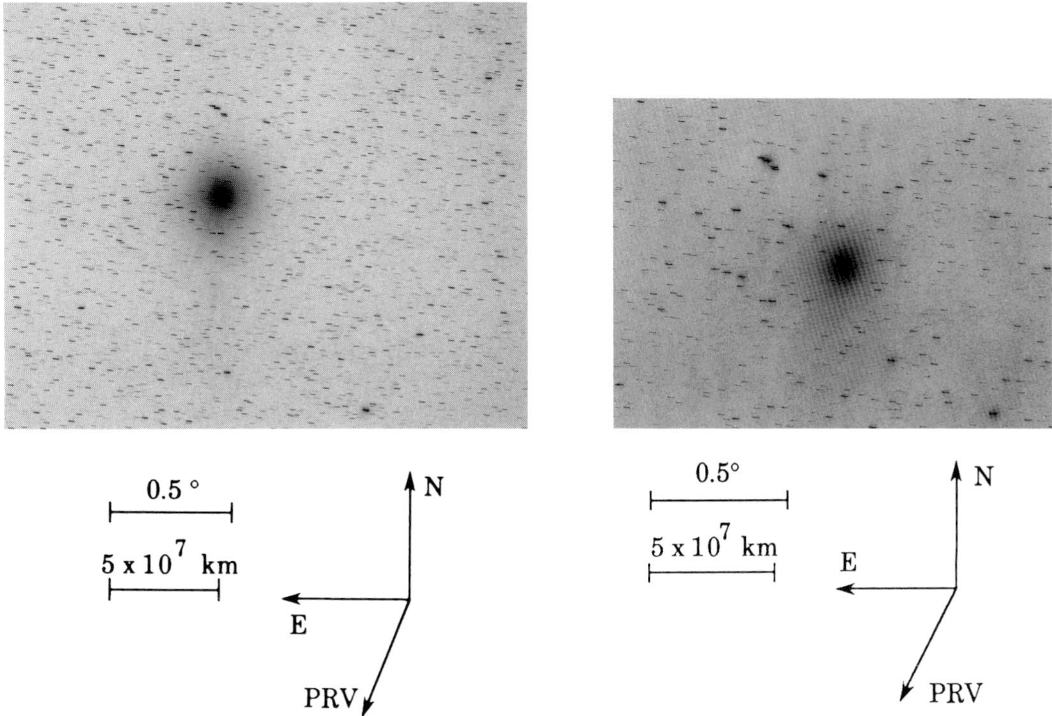

(left) Fig. 32: 1985 Nov 18.09694 UT; exposure 14.0 minutes on IIIa-F emulsion with no filter; r=1.67, Δ=0.69, β=178.6°; LSPN-3834.
Photograph by L. Kohoutek, Calar Alto Station, German National Ast. Observatory.

(right) Fig. 33: 1985 Nov 18.22951 UT; exposure 15.0 minutes on IIa-O emulsion with no filter; r=1.67, Δ=0.68, β=178.6°; LSPN-146.
Photograph by W. Liller, University of Michigan/CTIO.

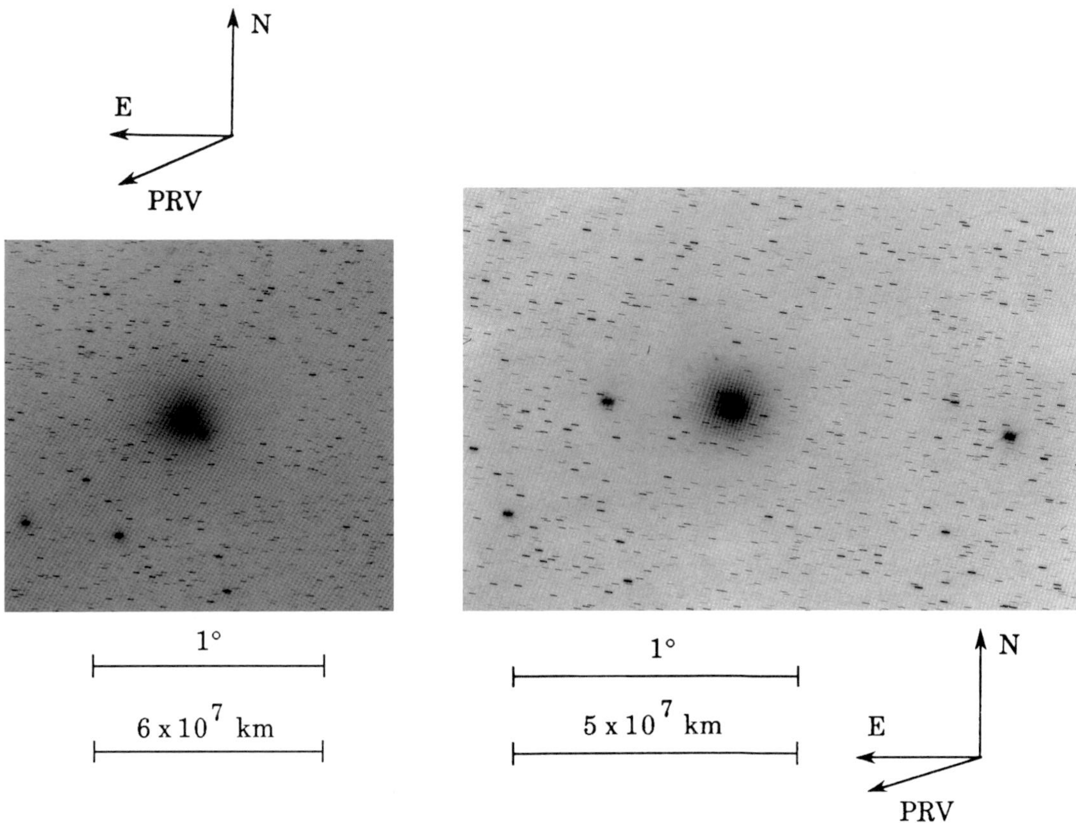

(left) Fig. 34: 1985 Nov 19.08403 UT; exposure 14.0 minutes on 103a-E emulsion with no filter; r=1.66, Δ=0.67, β=177.4°; LSPN-3838.
Photograph by L. Kohoutek, Calar Alto Station, German National Ast. Observatory.

(right) Fig. 35: 1985 Nov 19.29618 UT; exposure 15.0 minutes on IIa-O emulsion with no filter; r=1.66, Δ=0.67, β=177.0°; LSPN-149.
Photograph by W. Liller, University of Michigan/CTIO.

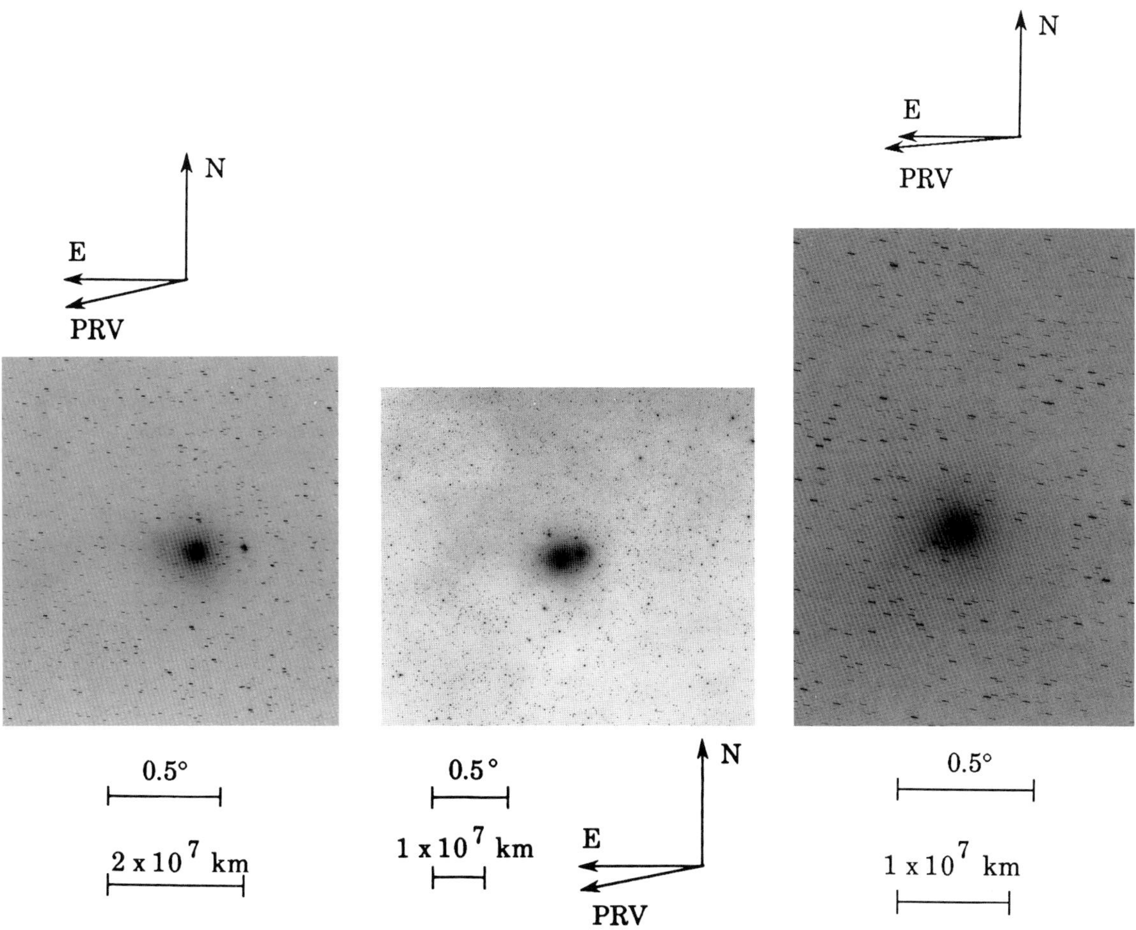

(left) Fig. 36: 1985 Nov 19.62470 UT; exposure 10.0 minutes on IIa-O emulsion with no filter; r=1.65, Δ=0.67, β=176.4°; LSPN-1600.
Photograph by H. Maehara, Kiso Observatory.

(center) Fig. 37: 1985 Nov 19.66042 UT; exposure 10.0 minutes on IIa-O emulsion with no filter; r=1.65, Δ=0.67, β=176.3°; LSPN-1331.
Photograph by T. Tsujimura, University Kyoto Observatory.

(right) Fig. 38: 1985 Nov 20.26285 UT; exposure 15.0 minutes on IIa-O emulsion with no filter; r=1.64, Δ=0.66, β=175.1°; LSPN-150.
Photograph by W. Liller, University of Michigan/CTIO.

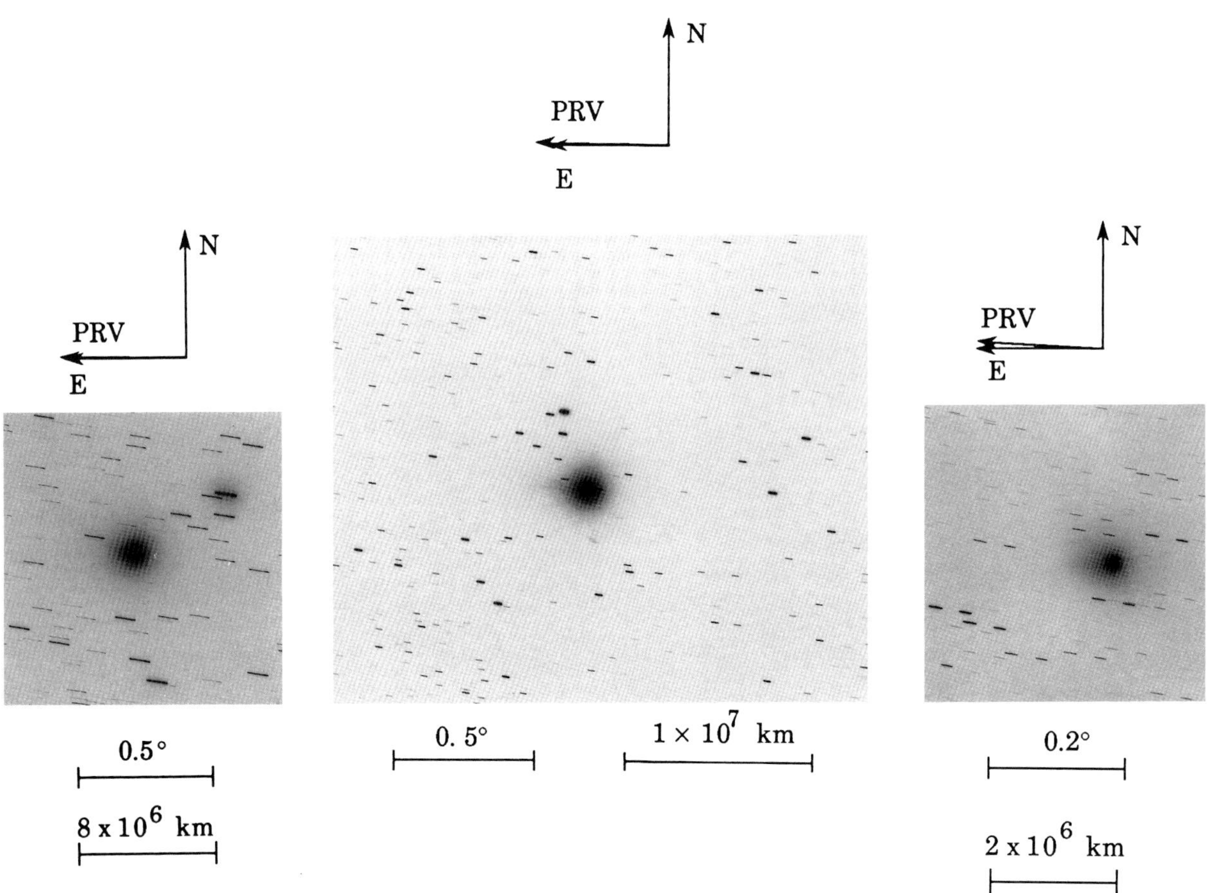

(left) Fig. 39: 1985 Nov 21.12569 UT; exposure 40.0 minutes on IIIa-J emulsion with no filter; r=1.63, Δ=0.65, β=173.3°; LSPN-3179.
Photograph by F. Dossin et al., Observatory of Haute-Provence.

(center) Fig. 40: 1985 Nov 21.28924 UT; exposure 15.0 minutes on IIa-O emulsion with no filter; r=1.63, Δ=0.65, β=172.9°; LSPN-154.
Photograph by W. Liller, University of Michigan/CTIO.

(right) Fig. 41: 1985 Nov 22.28090 UT; exposure 20.0 minutes on 103a-D emulsion with GG-495 filter; r=1.61, Δ=0.64, β=170.7°; LSPN-783.
Photograph by E. Harlan, Lick Observatory.

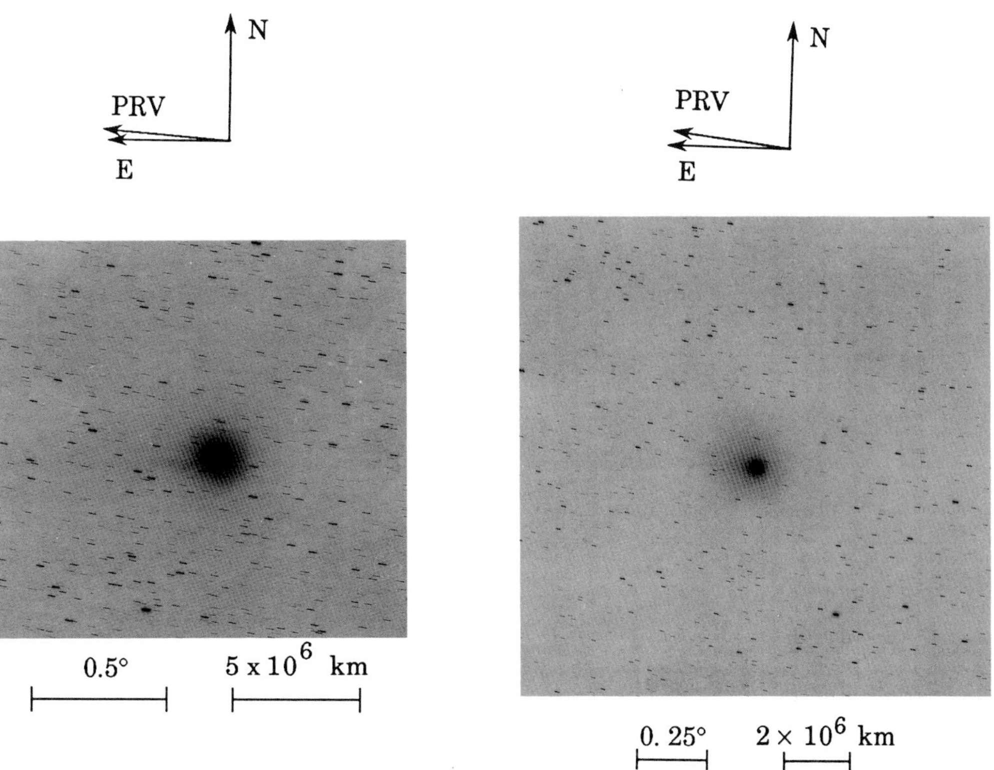

(left) Fig. 42: 1985 Nov 22.30729 UT; exposure 15.0 minutes on IIa-O emulsion with no filter; r=1.61, Δ=0.64, β=170.6°; LSPN-157.
Photograph by W. Liller, University of Michigan/CTIO.

(right) Fig. 43: 1985 Nov 23.31319 UT; exposure 10.0 minutes on IIa-O emulsion with no filter; r=1.60, Δ=0.63, β=168.3°; LSPN-159.
Photograph by W. Liller, University of Michigan/CTIO.

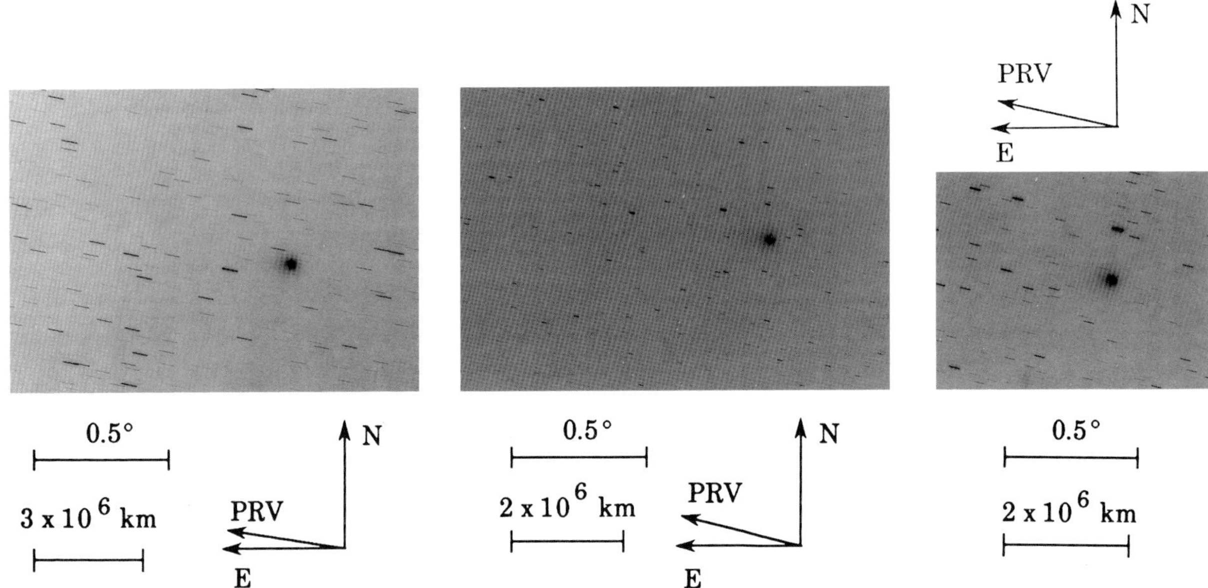

(left) Fig. 44: 1985 Nov 24.15000 UT; exposure 30.0 minutes on 153-01 emulsion with RG-630 filter; r=1.59, Δ=0.63, β=166.3°; LSPN-160.
Photograph by W. Liller, University of Michigan/CTIO.

(center) Fig. 45: 1985 Nov 27.10486 UT; exposure 10.0 minutes on 153-01 emulsion with RG-630 filter; r=1.54, Δ=0.62, β=159.3°; LSPN-163.
Photograph by W. Liller, University of Michigan/CTIO.

(right) Fig. 46: 1985 Nov 28.10347 UT; exposure 20.0 minutes on 153-01 emulsion with RG-630 filter; r=1.53, Δ=0.62, β=156.9°; LSPN-164.
Photograph by W. Liller, University of Michigan/CTIO.

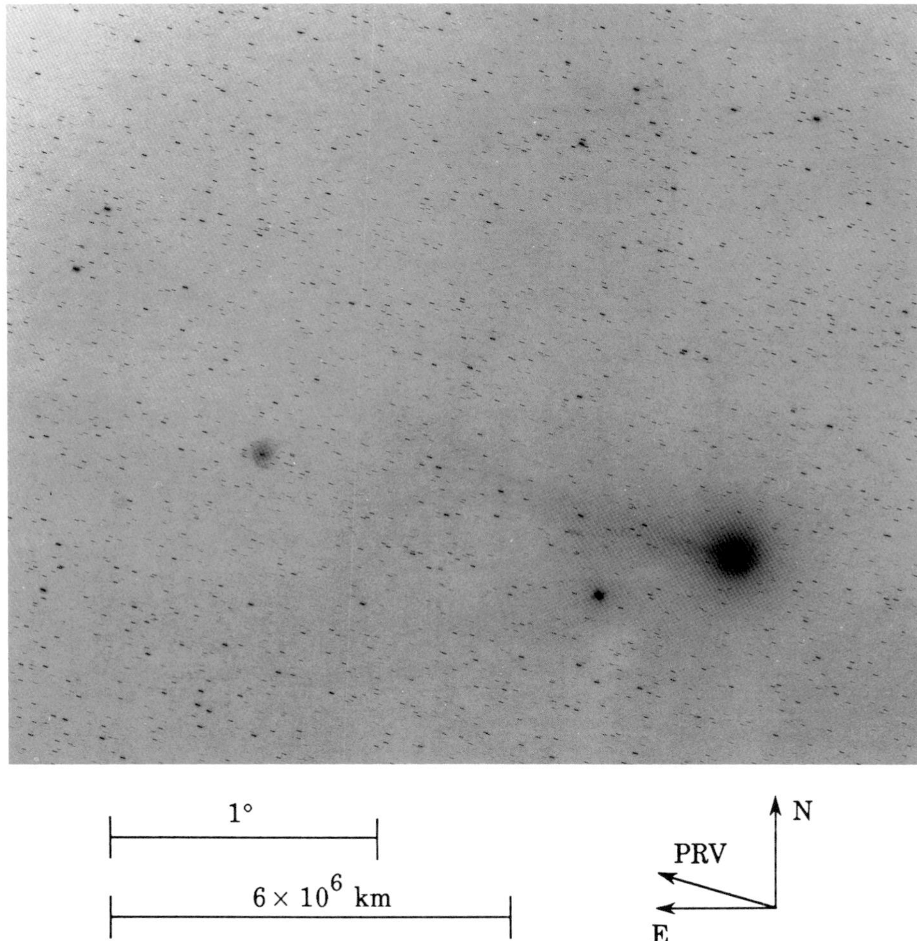

Fig. 47: 1985 Nov 29.03125 UT; exposure 10.0 minutes on IIa-O emulsion with no filter; r=1.51, Δ=0.62, β=154.7°; LSPN-165.
Photograph by W. Liller, University of Michigan/CTIO.

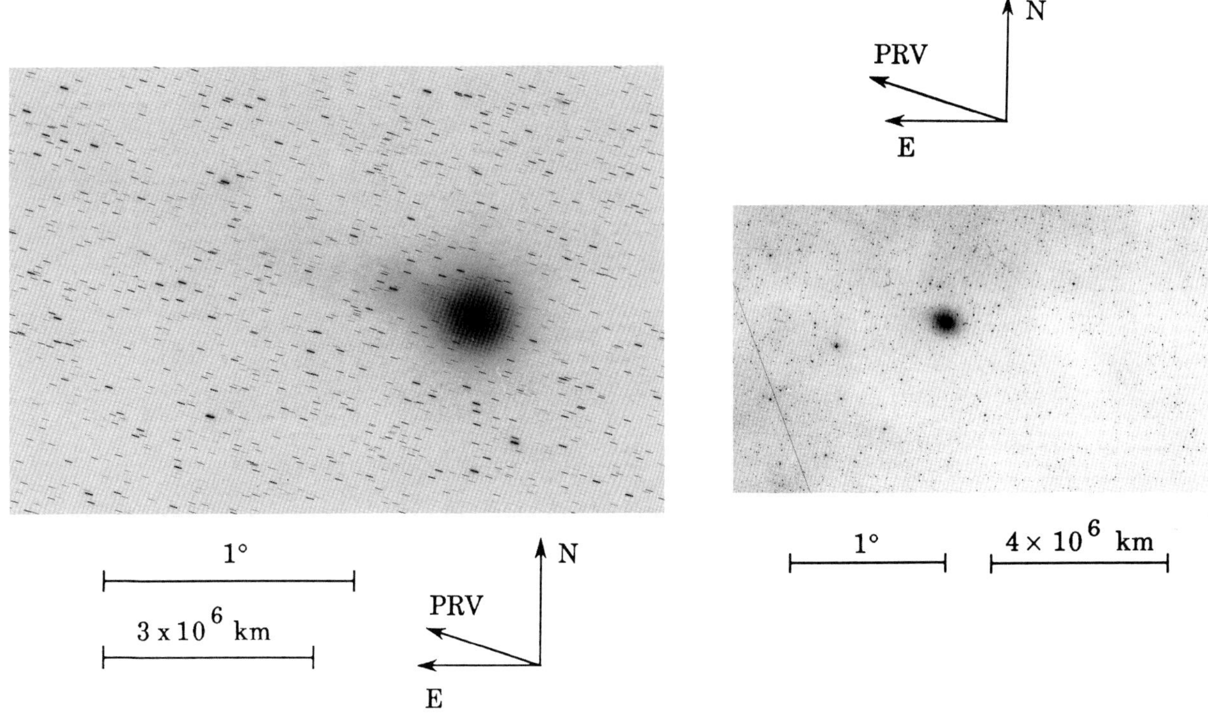

(left) Fig. 48: 1985 Nov 30.06007 UT; exposure 15.0 minutes on IIa-O emulsion with no filter; r=1.50, Δ=0.63, β=152.3°; LSPN-168.
Photograph by W. Liller, University of Michigan/CTIO.

(right) Fig. 49: 1985 Nov 30.38681 UT; exposure 10.0 minutes on 103a-F emulsion with Y-50 filter; r=1.49, Δ=0.63, β=151.6°; LSPN-1333.
Photograph by T. Tsujimura, University Kyoto Observatory.

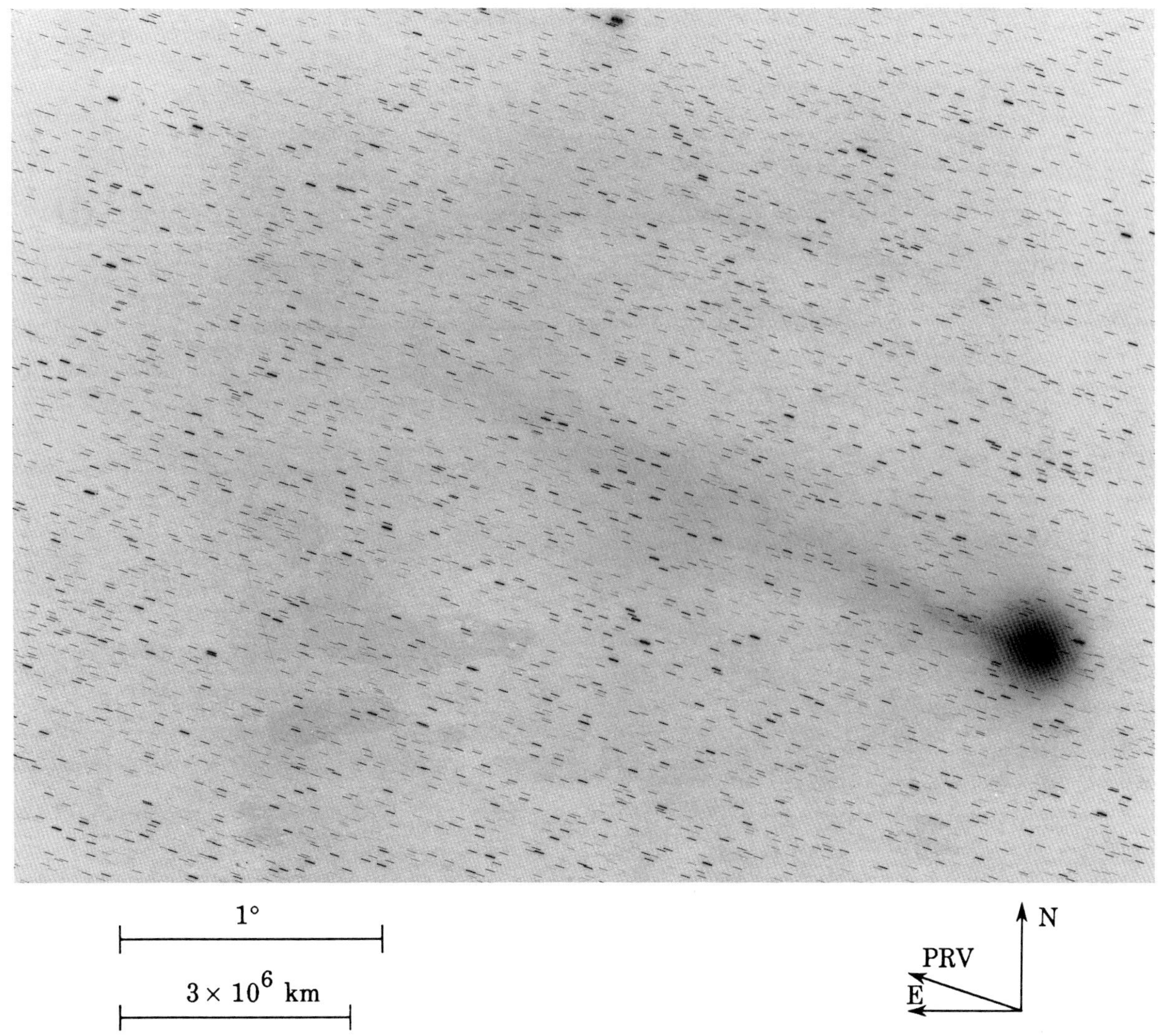

Fig. 50: 1985 Dec 1.08194 UT; exposure 20.0 minutes on IIa-O emulsion with no filter; r=1.48, Δ=0.63, β=150.1°; LSPN-170.
Photograph by W. Liller, University of Michigan/CTIO.

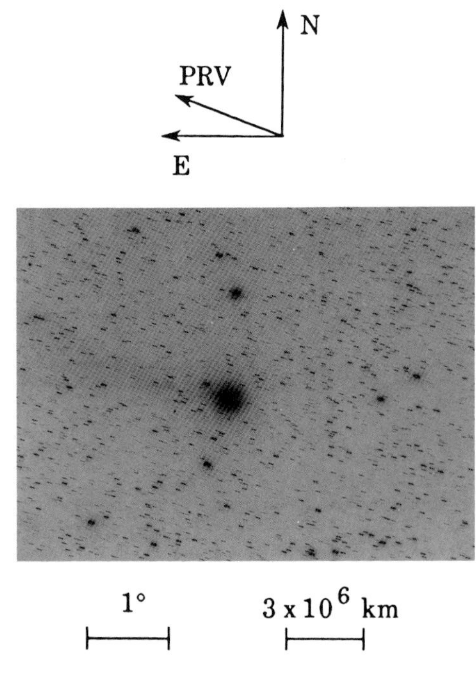

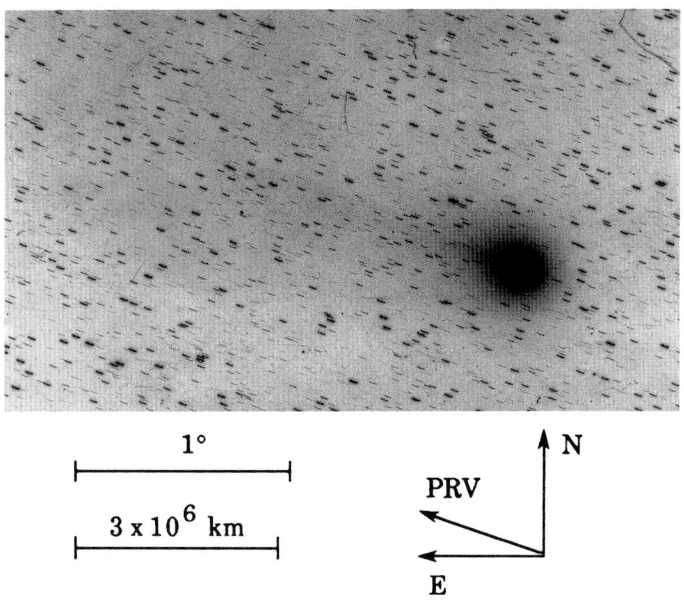

(top) Fig. 51: 1985 Dec 1.26037 UT; exposure 25.0 minutes on IIIa-J emulsion with no filter; r=1.48, Δ=0.63, β=149.7°; LSPN-1862.
Photograph by D. Cruikshank, Mauna Kea Observatory.

(bottom) Fig. 52: 1985 Dec 2.72153 UT; exposure 20.0 minutes on ORWO ZU-21 emulsion with no filter; r=1.46, Δ=0.64, β=146.6°; LSPN-2465.
Photograph by V. Ivanova et al., Bulgarian National Observatory.

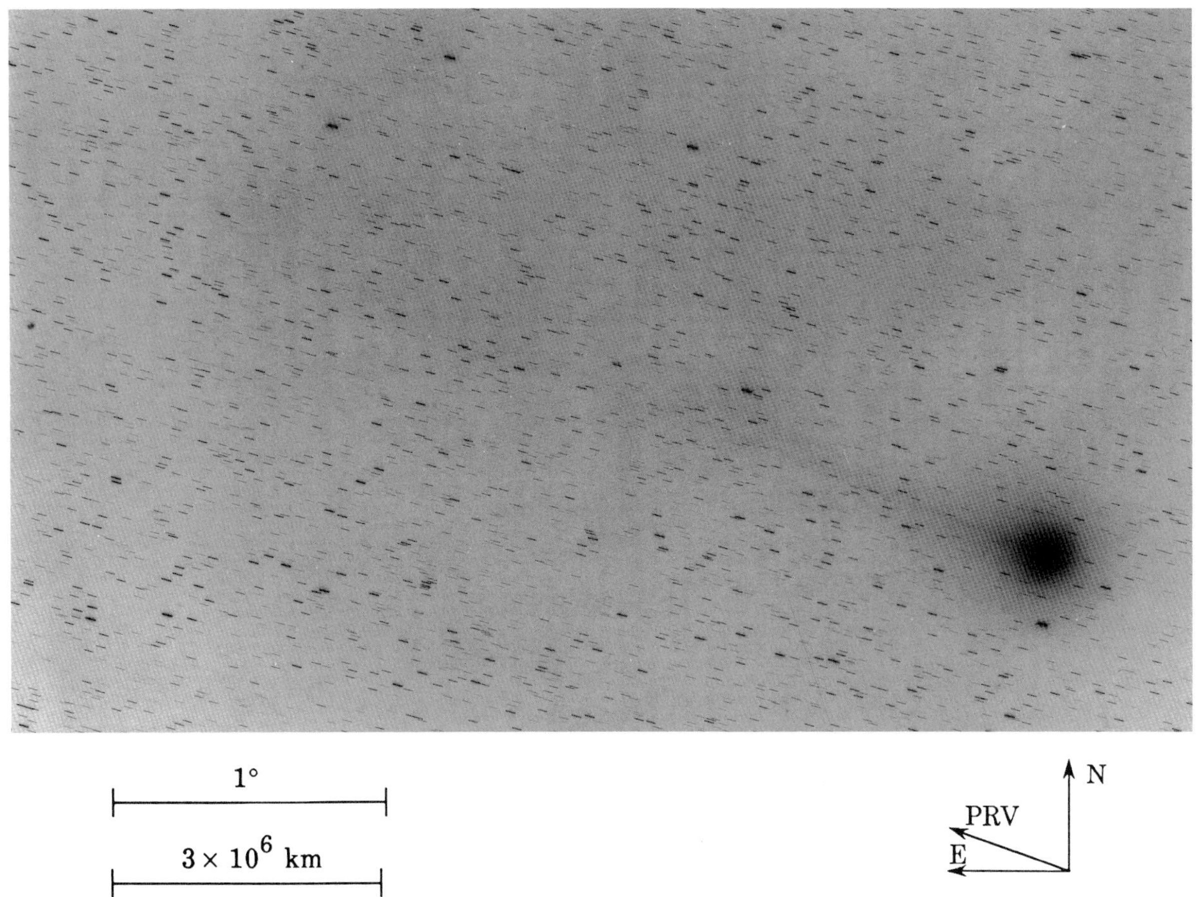

Fig. 53: 1985 Dec 3.07778 UT; exposure 20.0 minutes on IIa-O emulsion with no filter; r=1.45, Δ=0.65, β=145.8°; LSPN-175.
Photograph by W. Liller, University of Michigan/CTIO.

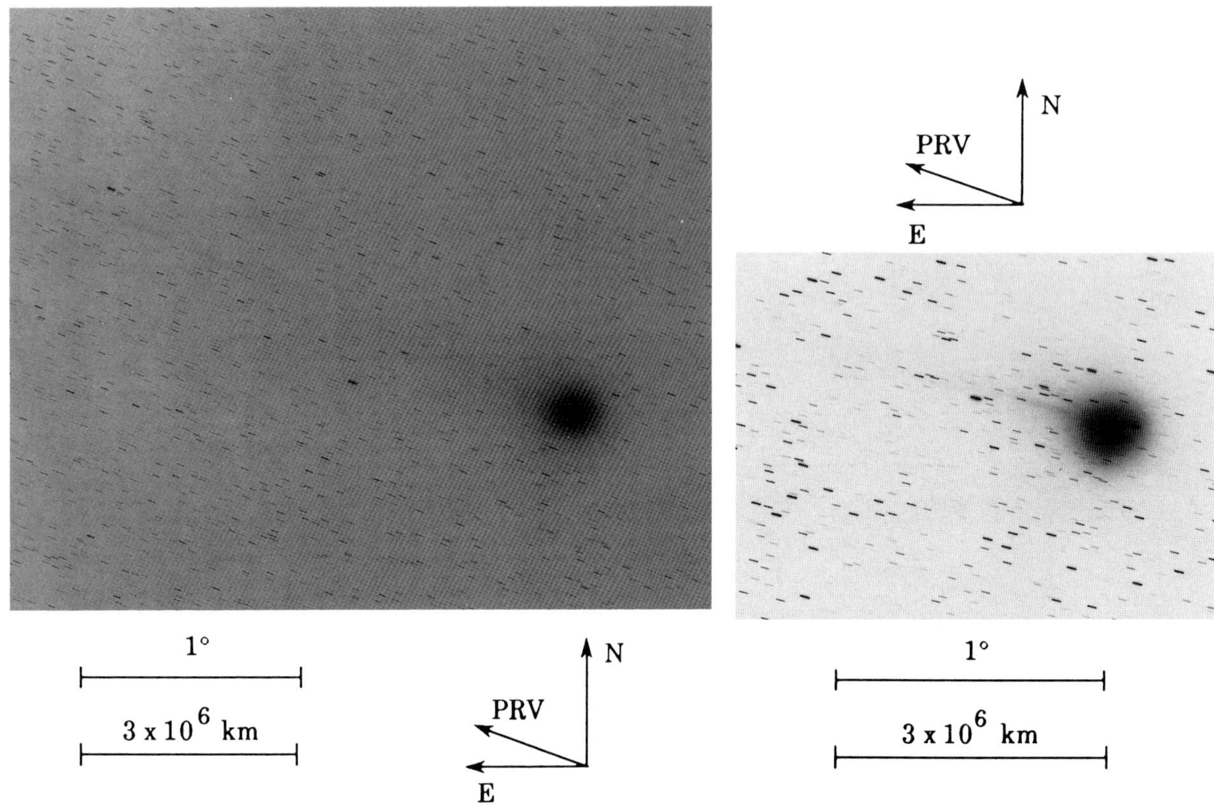

(left) Fig. 54: 1985 Dec 3.49847 UT; exposure 20.0 minutes on IIa-O emulsion with no filter; r=1.45, Δ=0.65, β=145.0°; LSPN-1606.
Photograph by H. Maehara, Kiso Observatory.

(right) Fig. 55: 1985 Dec 4.05417 UT; exposure 20.0 minutes on IIa-O emulsion with no filter; r=1.44, Δ=0.66, β=143.9°; LSPN-178.
Photograph by W. Liller, University of Michigan/CTIO.

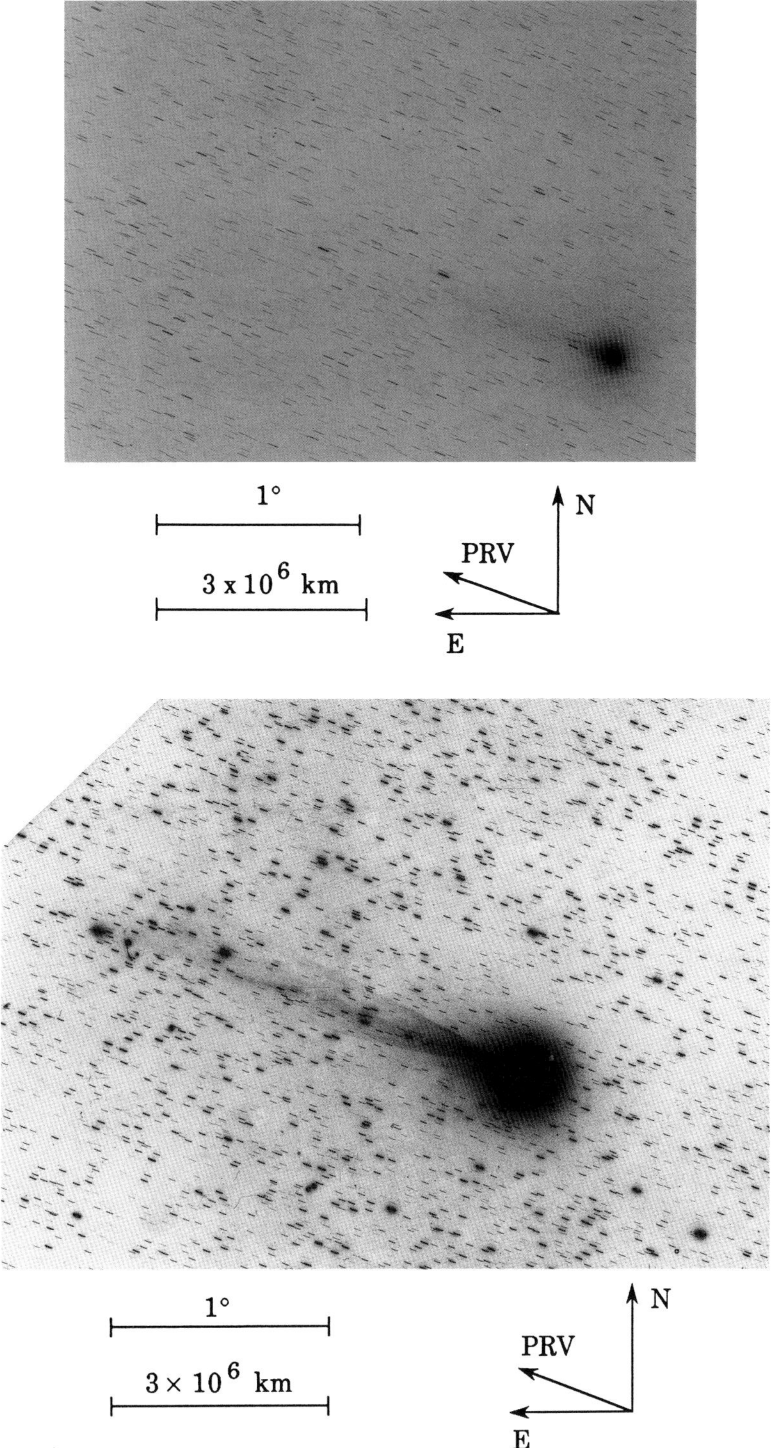

(top) Fig. 56: 1985 Dec 4.45830 UT; exposure 30.0 minutes on IIa-O emulsion with GG-395 filter; r=1.43, Δ=0.66, β=143.1°; LSPN-2371.
Photograph from Royal Observatory/UK Schmidt Telescope Unit.

(bottom) Fig. 57: 1985 Dec 4.71084 UT; exposure 20.0 minutes on ORWO ZU-21 emulsion with no filter; r=1.43, Δ=0.66, β=142.7°; LSPN-2468.
Photograph by V. Ivanova et al., Bulgarian National Observatory.

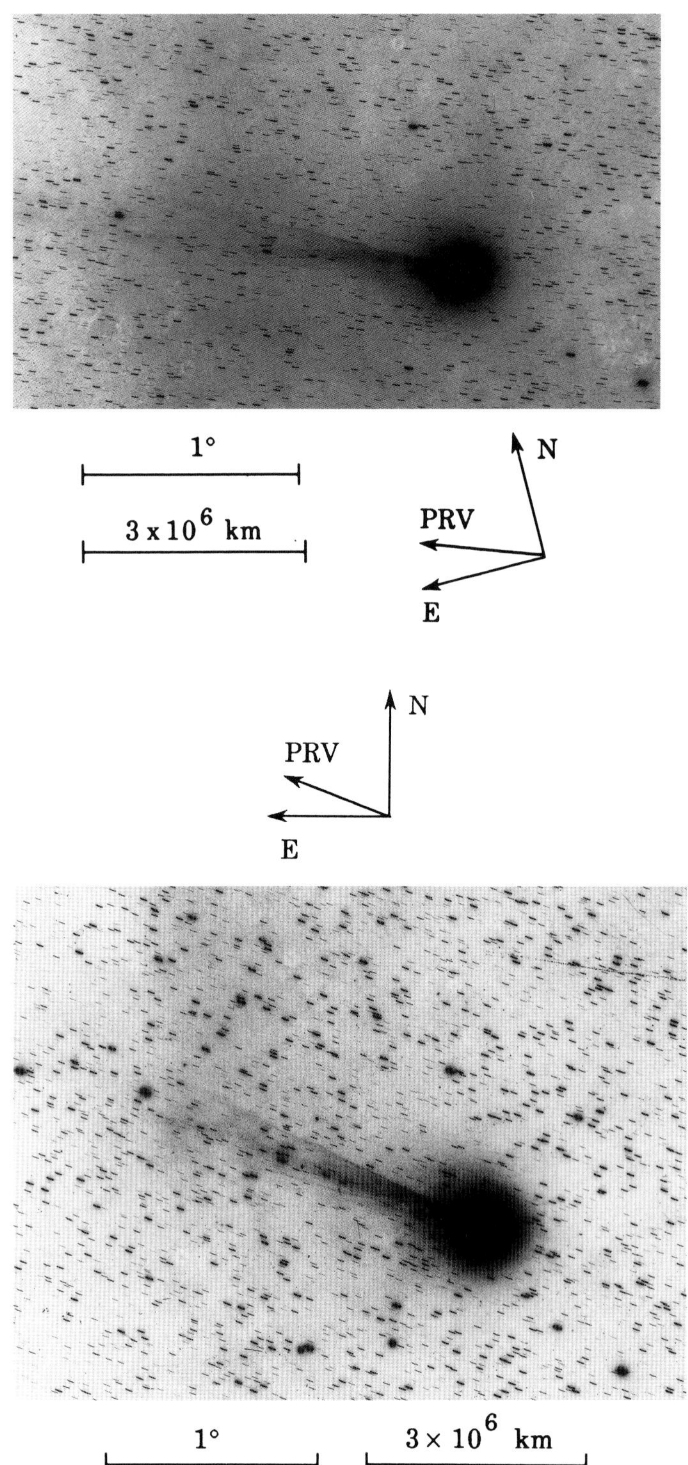

(top) Fig. 58: 1985 Dec 4.75830 UT; exposure 20.0 minutes on ORWO ZU-21 emulsion with no filter; r=1.43, Δ=0.66, β=142.6°; LSPN-2469.
Photograph by V. Ivanova et al., Bulgarian National Observatory.

(bottom) Fig. 59: 1985 Dec 4.79163 UT; exposure 20.0 minutes on ORWO ZU-21 emulsion with no filter; r=1.43, Δ=0.66, β=142.5°; LSPN-2470.
Photograph by V. Ivanova et al., Bulgarian National Observatory.

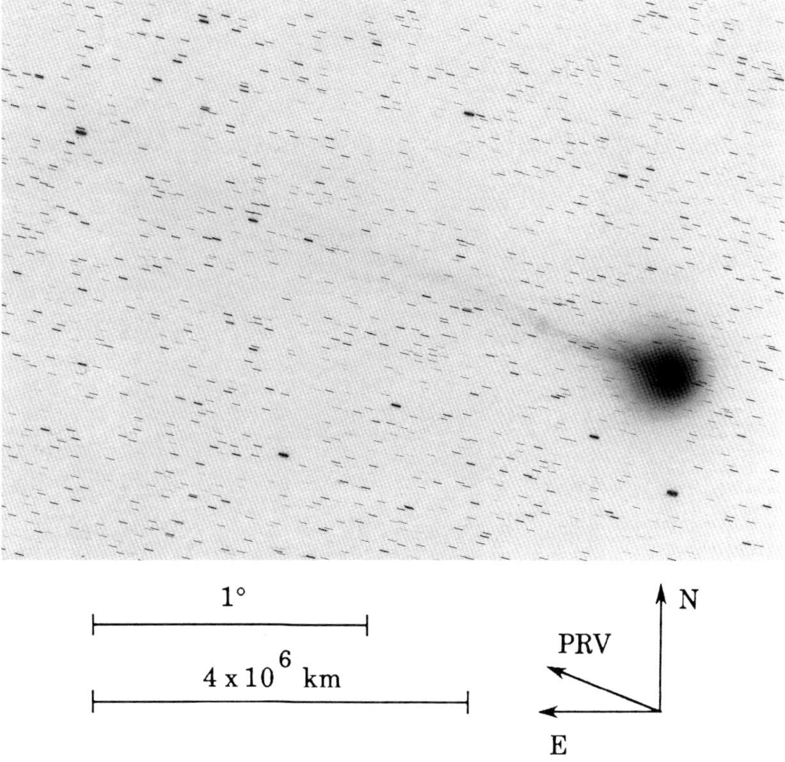

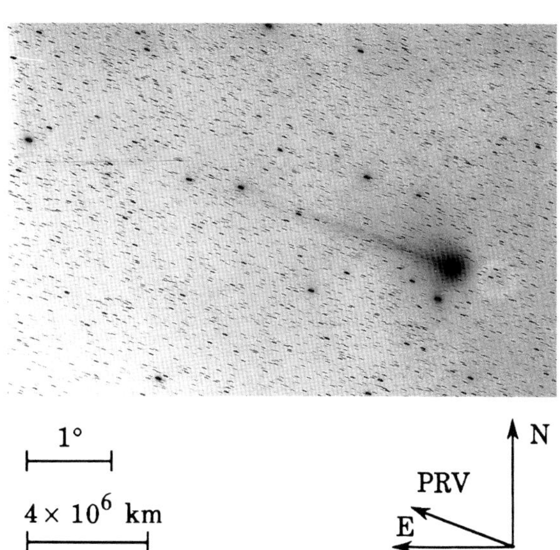

(top) Fig. 60: 1985 Dec 5.06042 UT; exposure 20.0 minutes on IIa-O emulsion with no filter; r=1.42, Δ=0.67, β=142.0°; LSPN-181.
Photograph by W. Liller, University of Michigan/CTIO.

(bottom) Fig. 61: 1985 Dec 5.15139 UT; exposure 30.0 minutes on IIa-O emulsion with no filter; r=1.42, Δ=0.67, β=141.8°; LSPN-1675.
Photograph by E. Moore/L. Bair, Joint Observatory for Cometary Research.

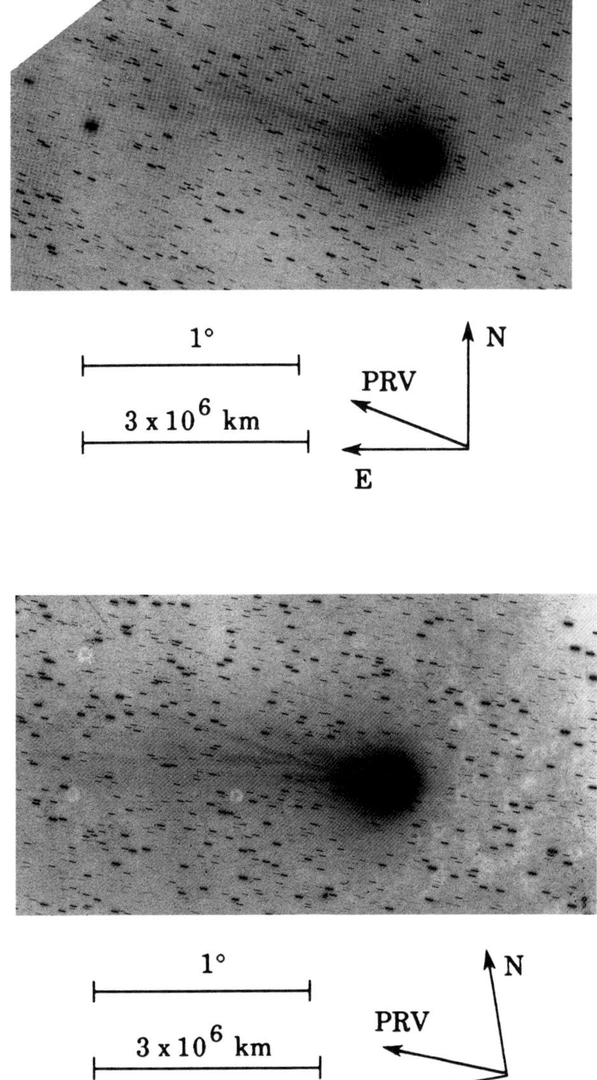

(top) Fig. 62: 1985 Dec 5.75830 UT; exposure 22.0 minutes on IIa-O emulsion with no filter; r=1.41, Δ=0.67, β=140.8°; LSPN-2472.
Photograph by V. Ivanova et al., Bulgarian National Observatory.

(bottom) Fig. 63*: 1985 Dec 5.92149 UT; exposure 20.0 minutes on ORWO ZU-21 emulsion with no filter; r=1.41, Δ=0.68, β=140.5°; LSPN-2476.
Photograph by V. Ivanova et al., Bulgarian National Observatory.

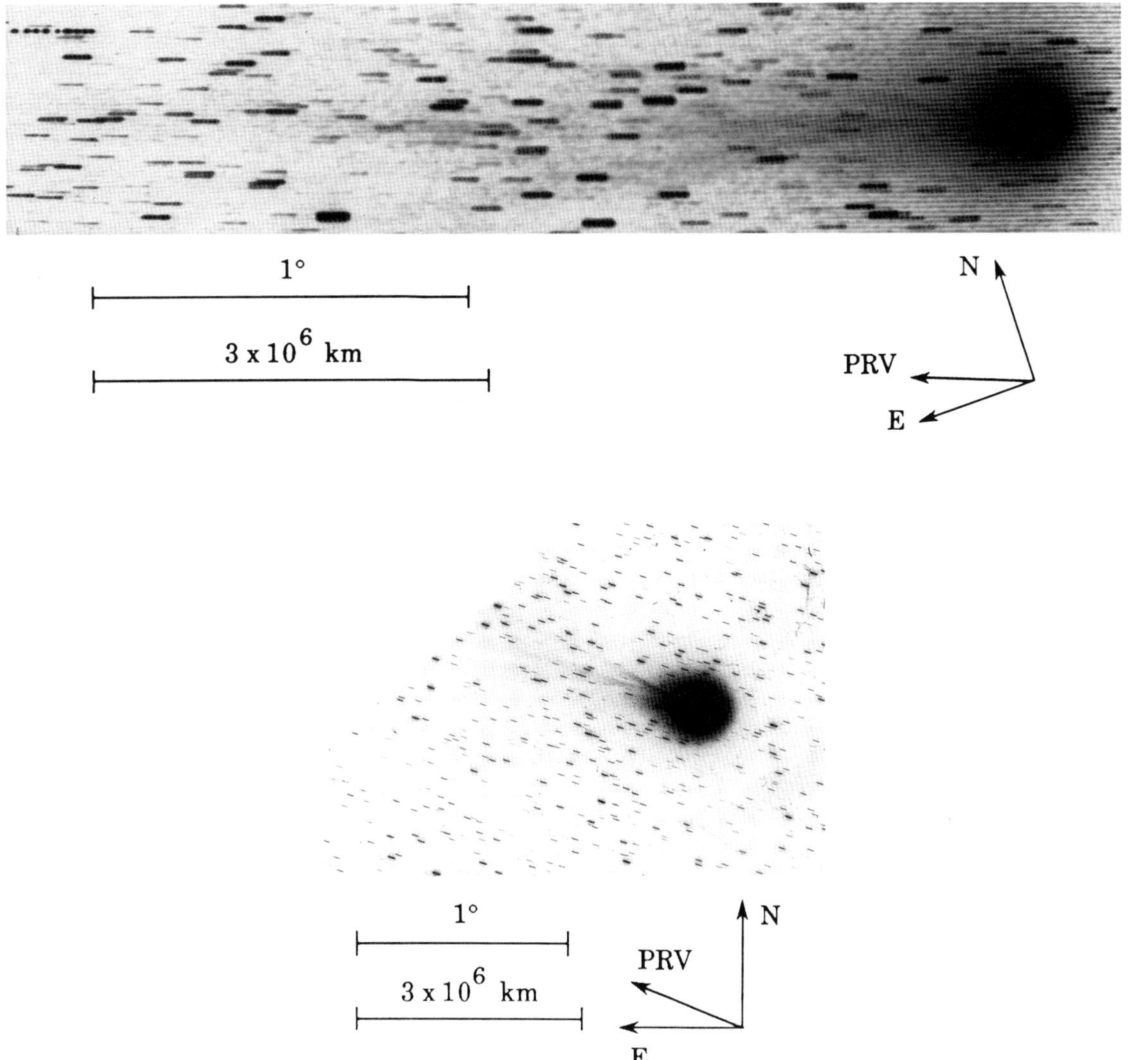

(top) Fig. 64: 1985 Dec 5.77236 UT; exposure 45.0 minutes on 103a-O emulsion with no filter; r=1.41, Δ=0.68, β=140.7°; LSPN-3435.
Photograph by N. Chernykh, Crimean Astrophysical Observatory.

(bottom) Fig. 65: 1985 Dec 5.80483 UT; exposure 20.0 minutes on ORWO ZU-21 emulsion with no filter; r=1.41, Δ=0.68, β=140.7°; LSPN-2473.
Photograph by V. Ivanova et al., Bulgarian National Observatory.

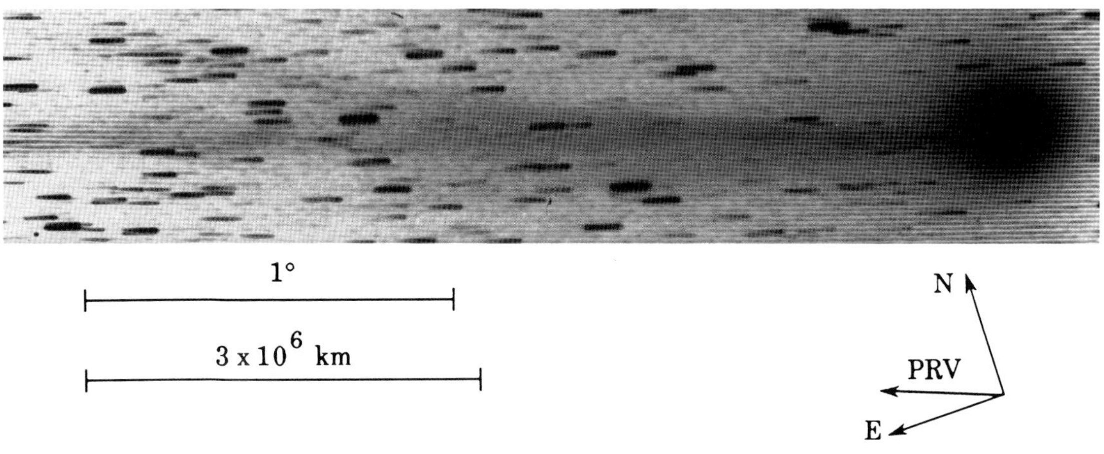

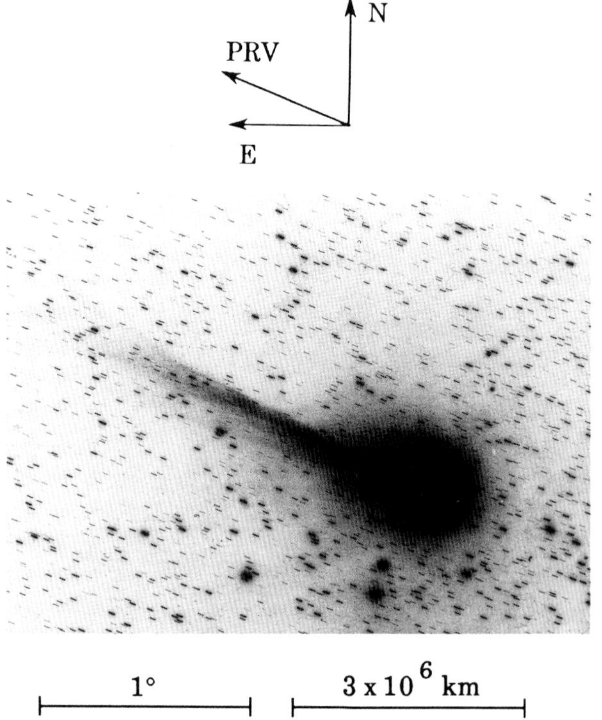

(top) Fig. 66: 1985 Dec 6.72897 UT; exposure 60.0 minutes on 103a-O emulsion with no filter; r=1.40, Δ=0.69, β=139.1°; LSPN-3436.
Photograph by N. Chernykh, Crimean Astrophysical Observatory.

(bottom) Fig. 67: 1985 Dec 6.73052 UT; exposure 20.0 minutes on ORWO ZU-21 emulsion with no filter; r=1.40, Δ=0.69, β=139.1°; LSPN-2477.
Photograph by V. Ivanova et al., Bulgarian National Observatory.

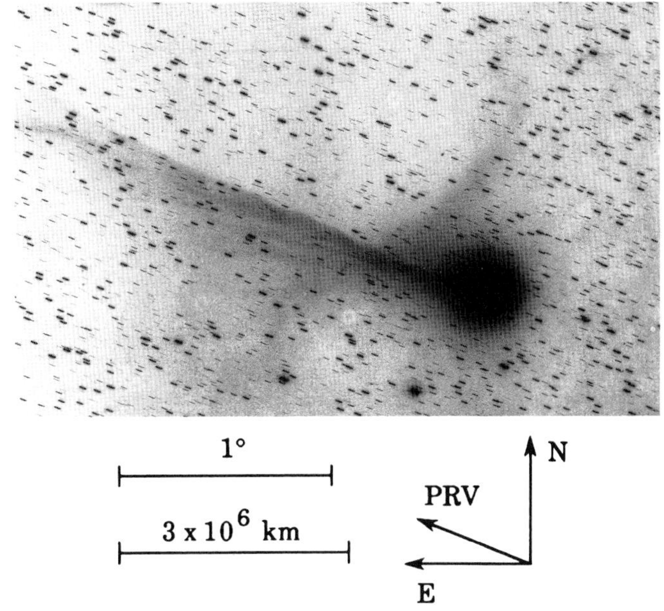

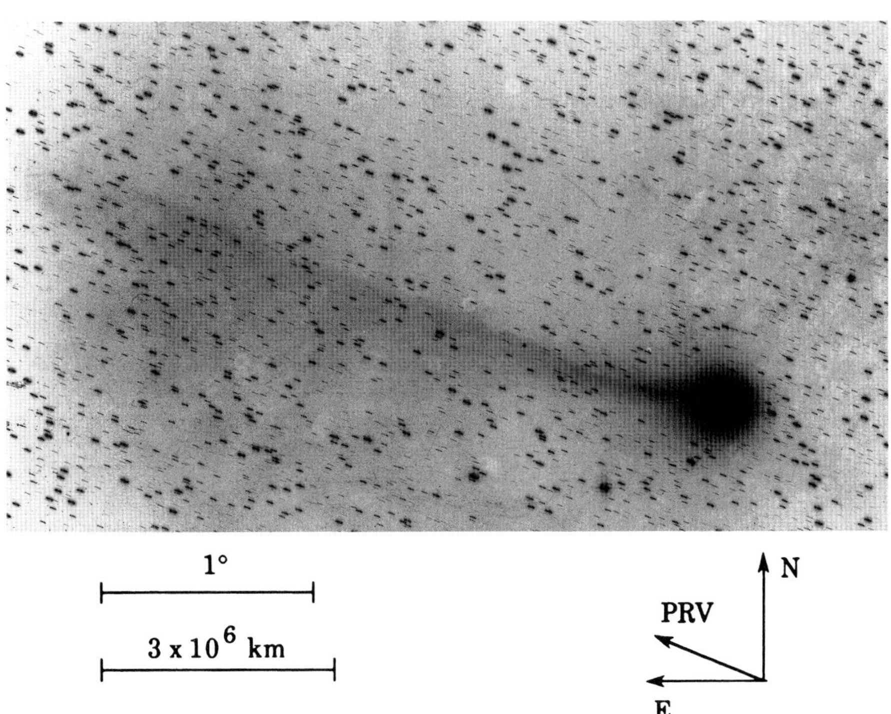

(top) Fig. 68: 1985 Dec 6.77705 UT; exposure 20.0 minutes on ORWO ZU-21 emulsion with no filter; r=1.40, Δ=0.69, β=139.0°; LSPN-2478.
Photograph by V. Ivanova et al., Bulgarian National Observatory.

(bottom) Fig. 69: 1985 Dec 6.89163 UT; exposure 20.0 minutes on ORWO ZU-21 emulsion with no filter; r=1.39, Δ=0.69, β=138.8°; LSPN-2481.
Photograph by V. Ivanova et al., Bulgarian National Observatory.

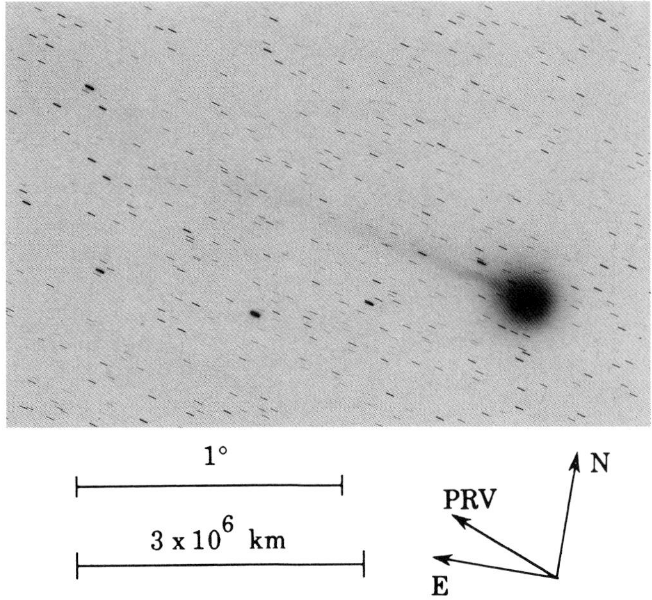

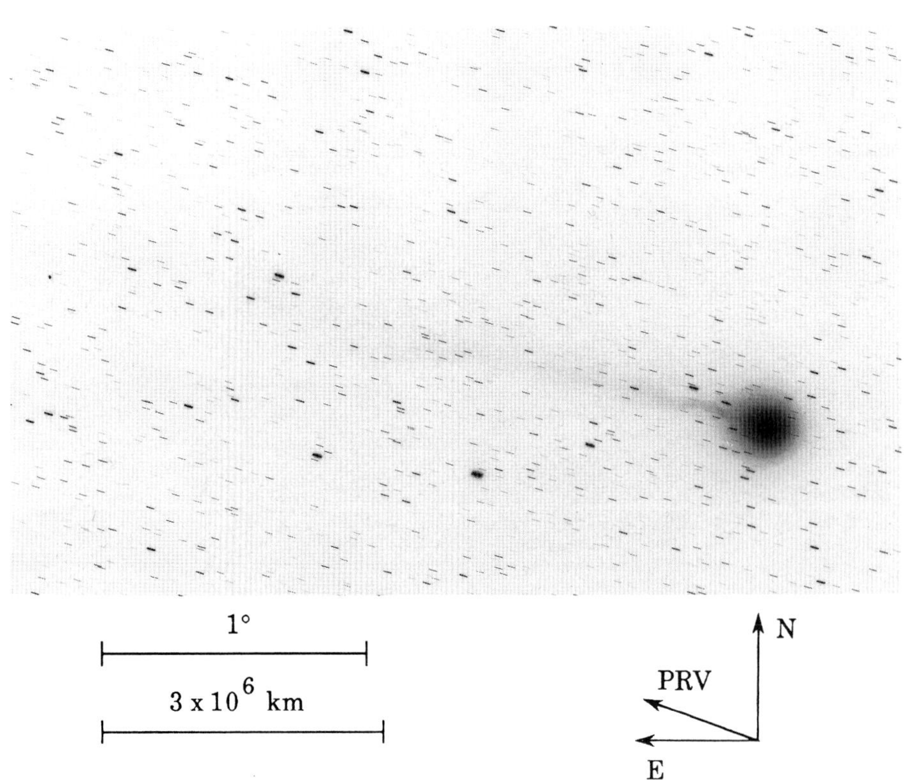

(top) Fig. 70: 1985 Dec 7.09028 UT; exposure 20.0 minutes on IIa-O emulsion with no filter; r=1.39, Δ=0.69, β=138.5°; LSPN-186.
Photograph by W. Liller, University of Michigan/CTIO.

(bottom) Fig. 71: 1985 Dec 7.13194 UT; exposure 20.0 minutes on IIa-O emulsion with no filter; r=1.39, Δ=0.69, β=138.5°; LSPN-187.
Photograph by W. Liller, University of Michigan/CTIO.

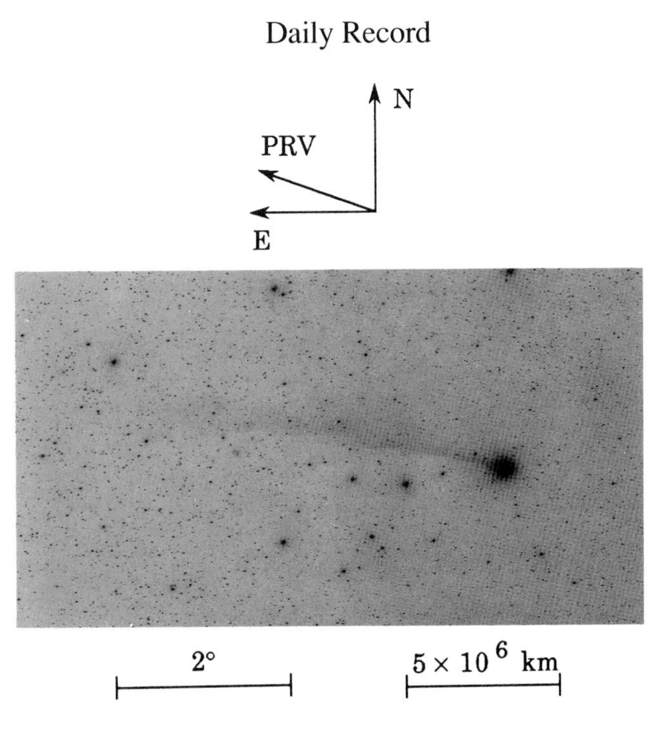

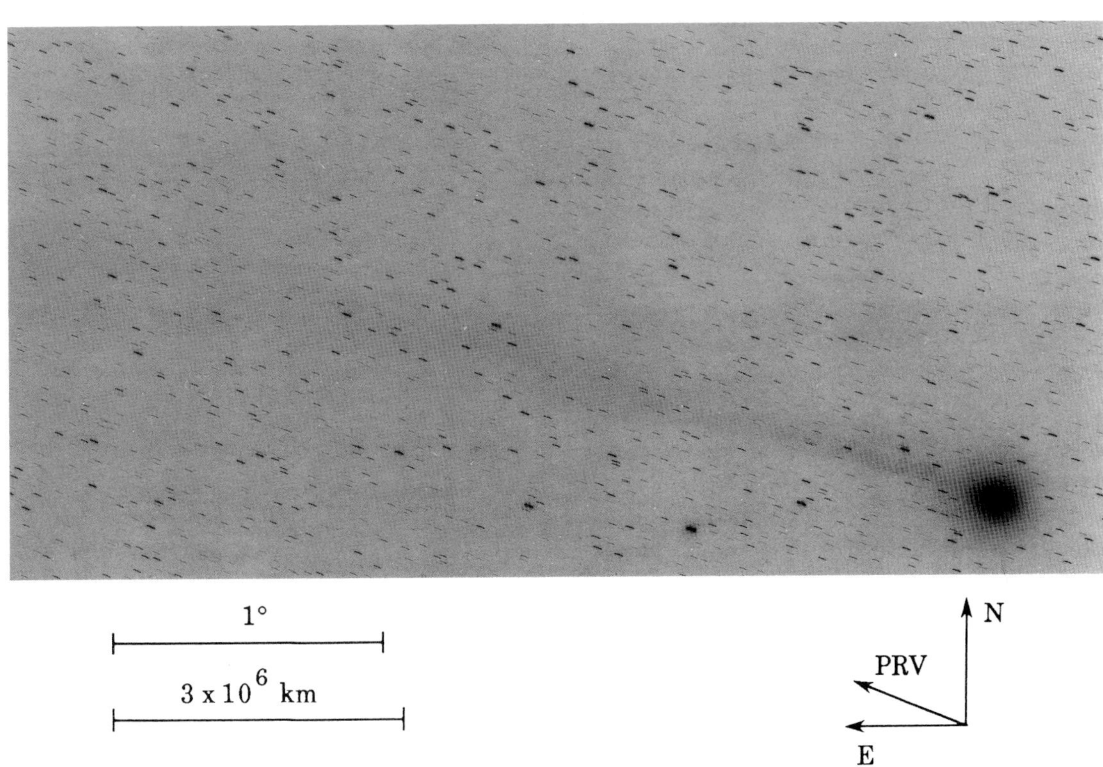

(top) Fig. 72: 1985 Dec 7.16667 UT; exposure 10.0 minutes on IIa-O emulsion with no filter; r=1.39, Δ=0.69, β=138.4°; LSPN-1673.
Photograph by E. Moore/L. Bair, Joint Observatory for Cometary Research.

(bottom) Fig. 73: 1985 Dec 7.17361 UT; exposure 20.0 minutes on IIa-O emulsion with no filter; r=1.39, Δ=0.69, β=138.4°; LSPN-188.
Photograph by W. Liller, University of Michigan/CTIO.

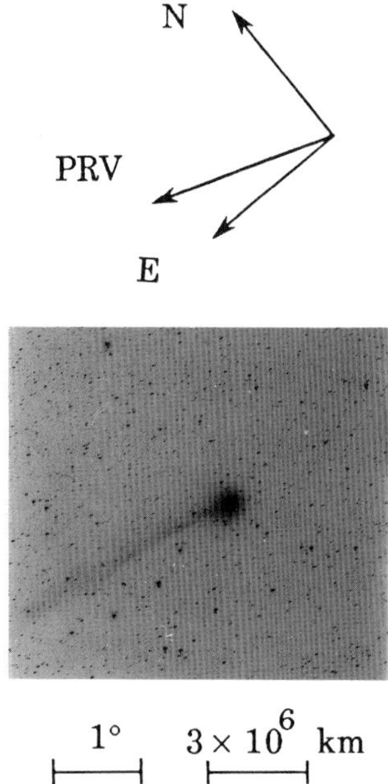

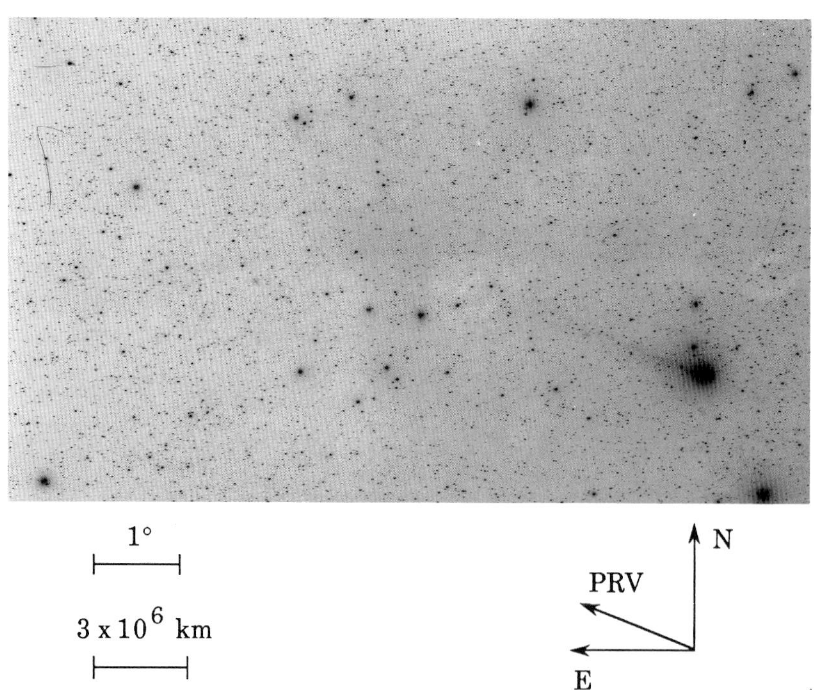

(top) Fig. 74: 1985 Dec 7.18333 UT; exposure 13.0 minutes on IIa-O emulsion with no filter; r=1.39, Δ=0.69, β=138.4°; LSPN-122.
Photograph by G. Emerson, E. E. Barnard Observatory.

(bottom) Fig. 75: 1985 Dec 8.23021 UT; exposure 15.0 minutes on IIa-O emulsion with no filter; r=1.37, Δ=0.71, β=136.8°; LSPN-1671.
Photograph by E. Moore et al., Joint Observatory for Cometary Research.

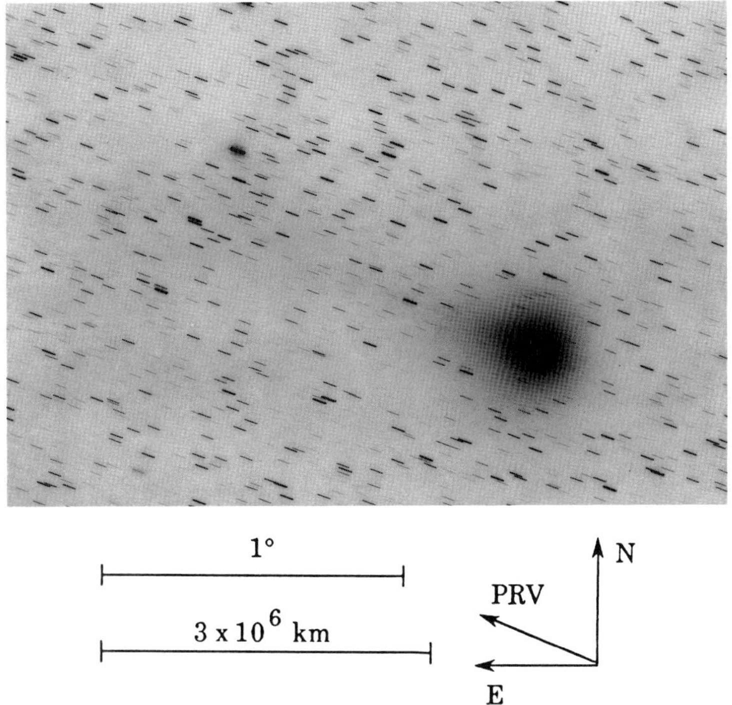

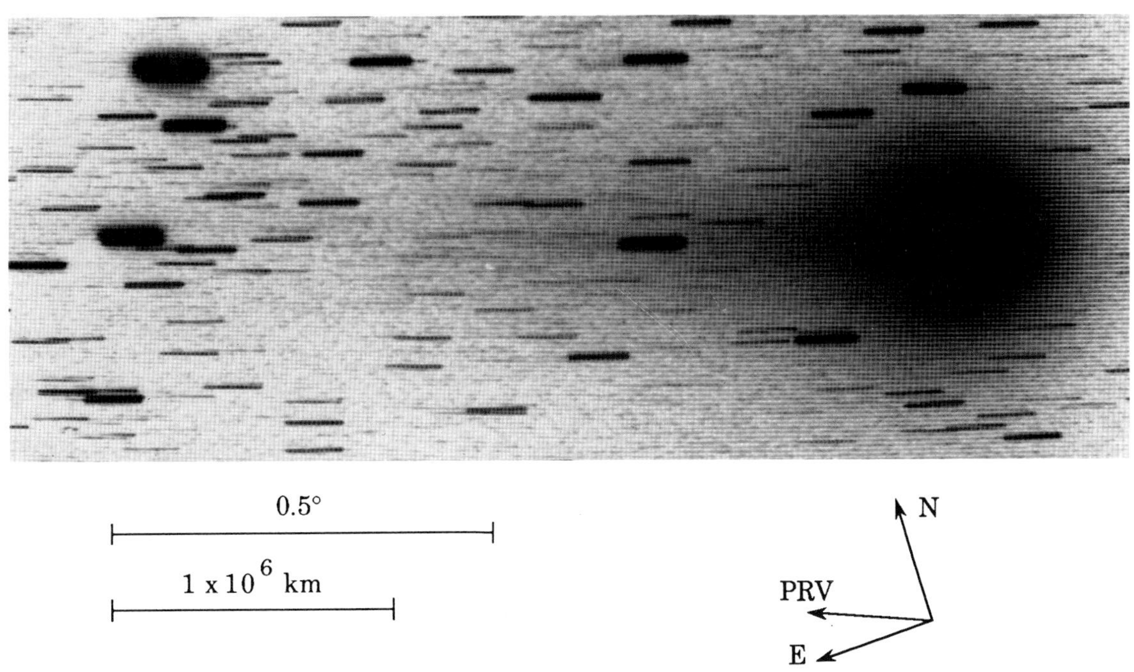

(top) Fig. 76: 1985 Dec 8.68607 UT; exposure 30.0 minutes on ORWO ZU-21 emulsion with GG-13 filter; r=1.37, Δ=0.71, β=136.1°; LSPN-1970.
Photograph by I. Eglitis, Riga Radio-Astrophysical Observatory.

(bottom) Fig. 77: 1985 Dec 8.69087 UT; exposure 60.0 minutes on 103a-O emulsion with no filter; r=1.37, Δ=0.71, β=136.1°; LSPN-3437.
Photograph by N. Chernykh, Crimean Astrophysical Observatory.

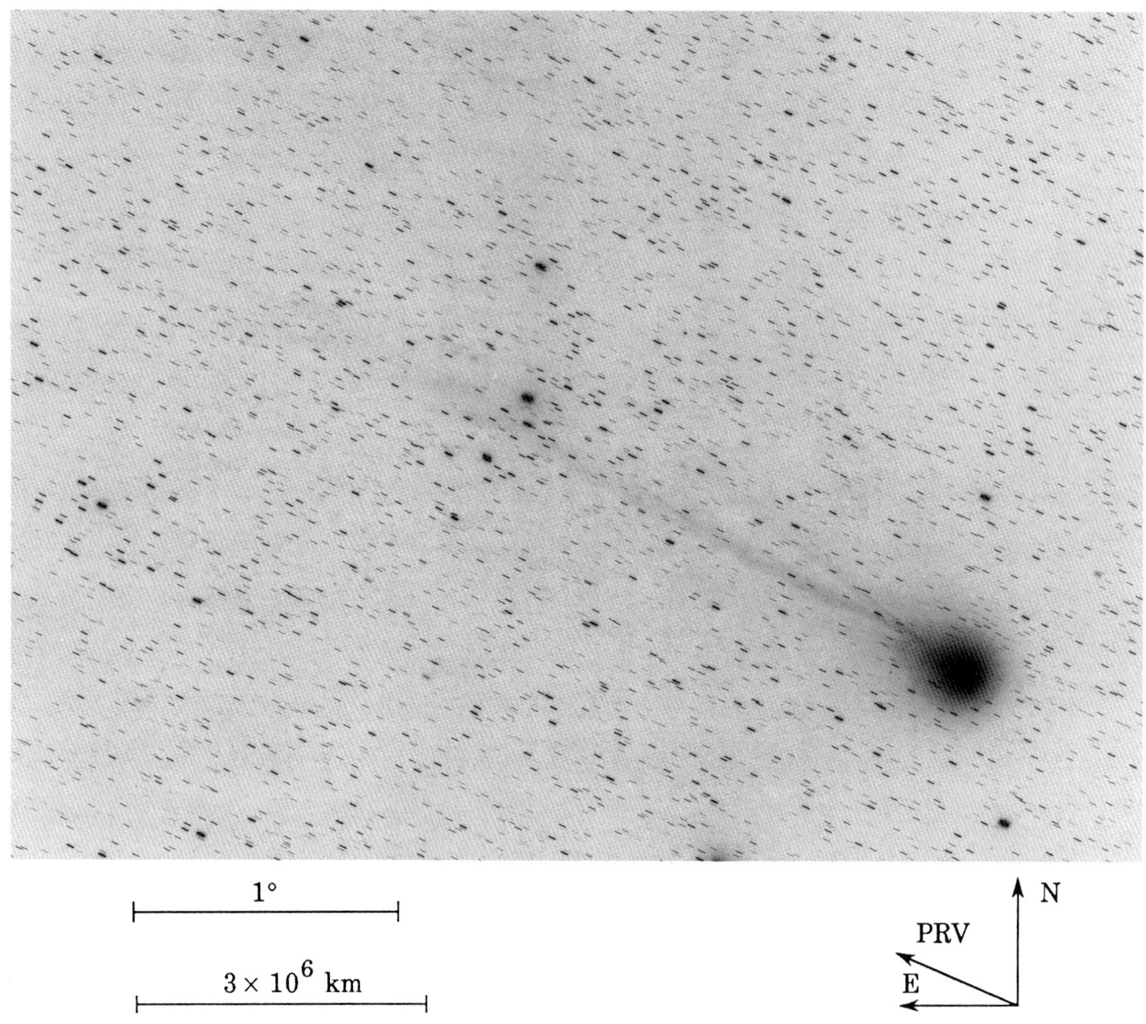

Fig. 78: 1985 Dec 9.05556 UT; exposure 20.0 minutes on IIa-O emulsion with no filter; r=1.36, Δ=0.72, β=135.6°; LSPN-191.
Photograph by W. Liller, University of Michigan/CTIO.

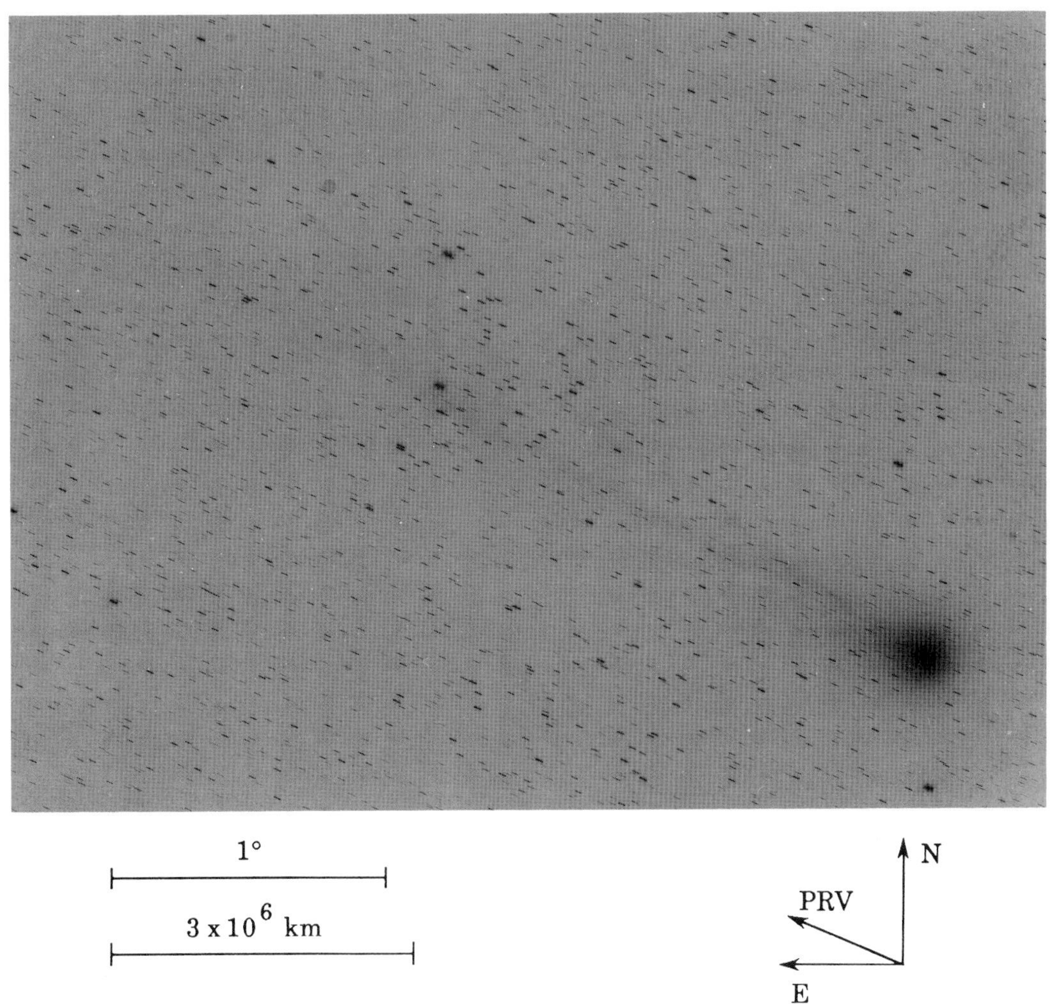

Fig. 79: 1985 Dec 9.13750 UT; exposure 20.0 minutes on IIa-O emulsion with no filter; r=1.36, Δ=0.72, β=135.5°; LSPN-193.
Photograph by W. Liller, University of Michigan/CTIO.

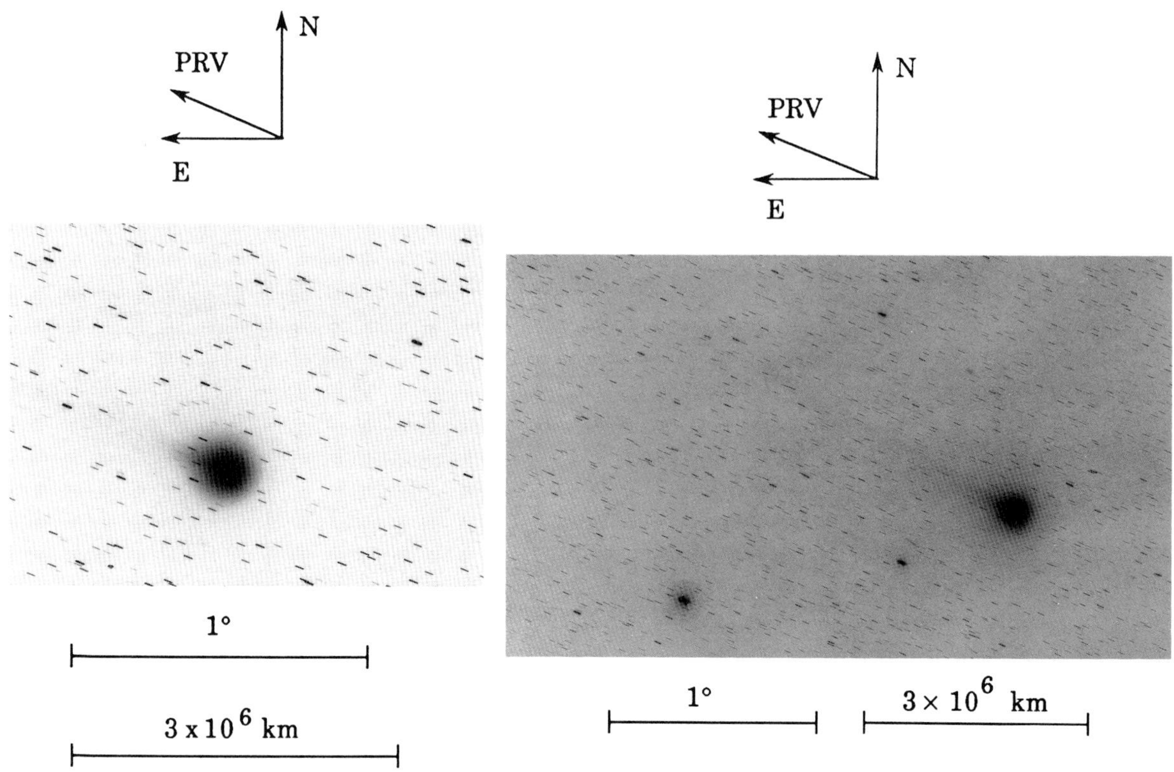

(left) Fig. 80*: 1985 Dec 8.74579 UT; exposure 20.0 minutes on ORWO ZU-21 emulsion with GG-13 filter; r=1.37, Δ=0.72, β=136.0°; LSPN-1971.
Photograph by I. Eglitis, Riga Radio-Astrophysical Observatory.

(right) Fig. 81: 1985 Dec 9.44845 UT; exposure 25.0 minutes on IIa-O emulsion with GG-385 filter; r=1.35, Δ=0.73, β=135.0°; LSPN-2372.
Photograph from Royal Observatory/UK Schmidt Telescope Unit.

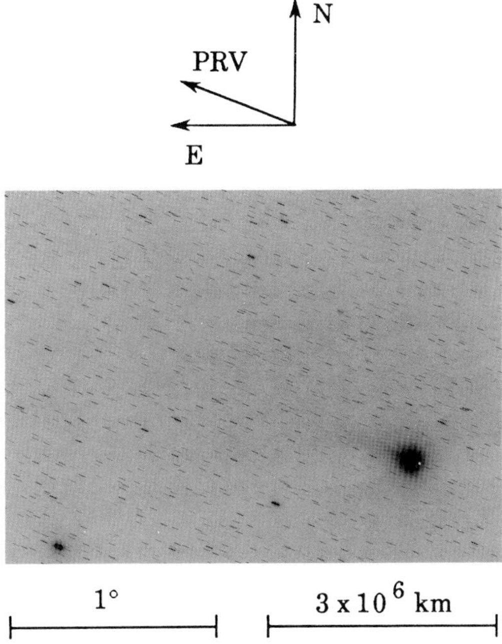

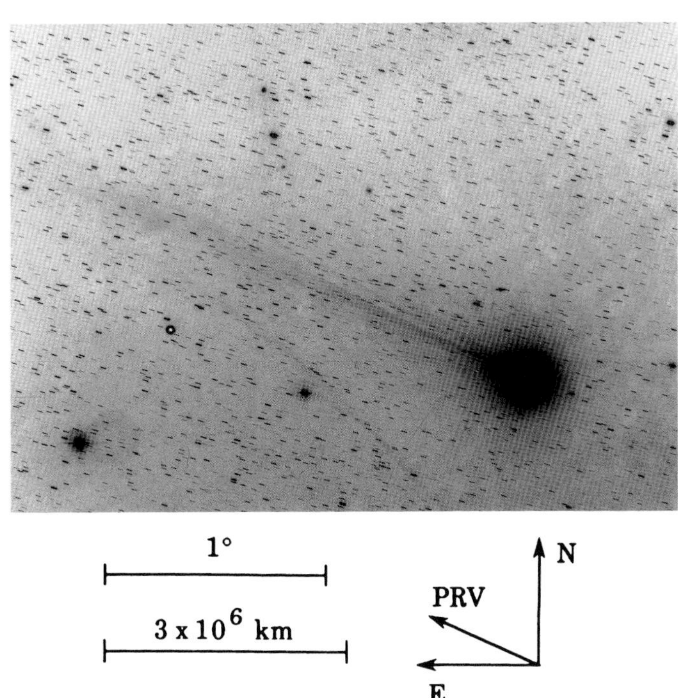

(top) Fig. 82: 1985 Dec 9.50940 UT; exposure 25.0 minutes on IIa-D emulsion with GG-495 filter; r=1.35, Δ=0.73, β=135.0°; LSPN-2374.
Photograph from Royal Observatory/UK Schmidt Telescope Unit.

(bottom) Fig. 83: 1985 Dec 9.73537 UT; exposure 20.0 minutes on ORWO ZU-21 emulsion with no filter; r=1.35, Δ=0.73, β=134.7°; LSPN-2319.
Photograph by V. Shkodrov et al., Bulgarian National Observatory.

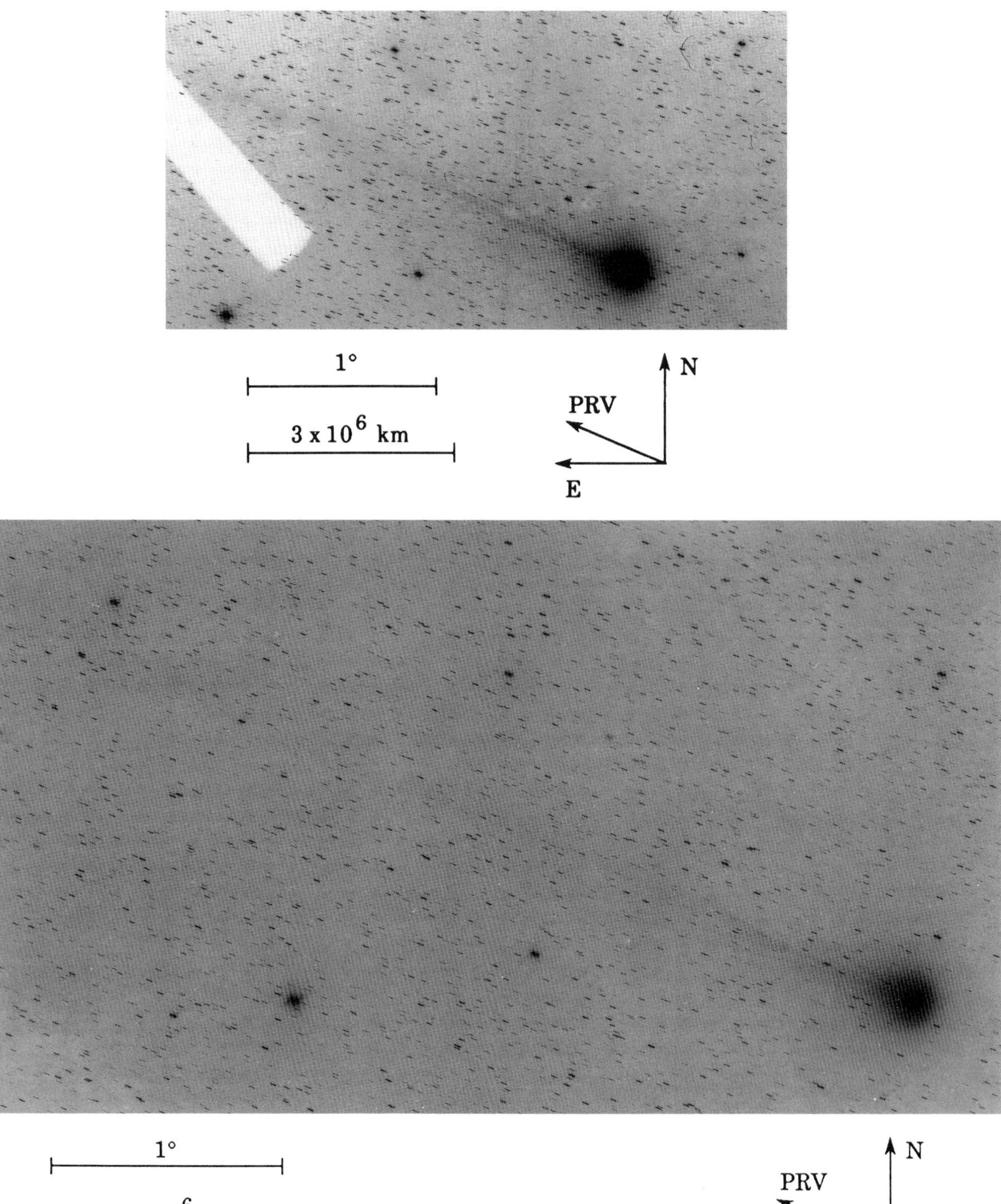

(top) Fig. 84: 1985 Dec 9.78190 UT; exposure 20.0 minutes on ORWO ZU-21 emulsion with no filter; r=1.35, Δ=0.73, β=134.6°; LSPN-2320.
Photograph by V. Shkodrov et al., Bulgarian National Observatory.

(bottom) Fig. 85: 1985 Dec 10.05625 UT; exposure 20.0 minutes on IIa-O emulsion with no filter; r=1.35, Δ=0.74, β=134.2°; LSPN-195.
Photograph by W. Liller, University of Michigan/CTIO.

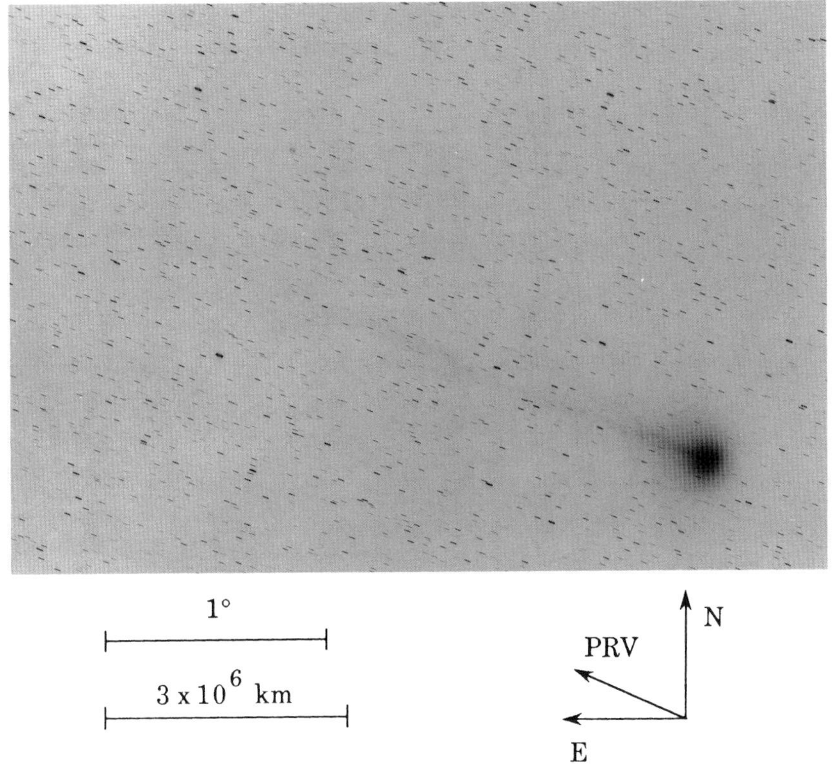

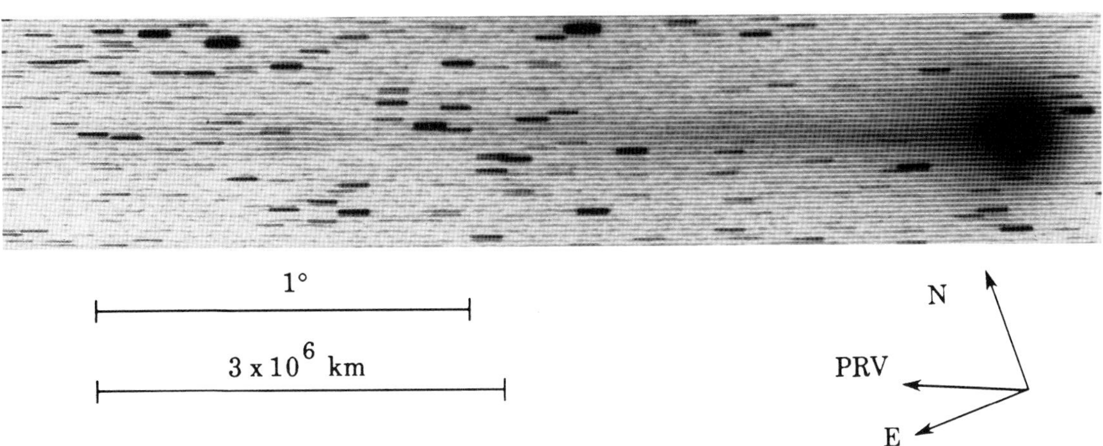

(top) Fig. 86: 1985 Dec 10.39634 UT; exposure 20.0 minutes on IIa-O emulsion with no filter; r=1.34, Δ=0.74, β=133.8°; LSPN-1610.
Photograph by H. Maehara, Kiso Observatory.

(bottom) Fig. 87: 1985 Dec 10.68730 UT; exposure 60.0 minutes on 103a-O emulsion with no filter; r=1.34, Δ=0.75, β=133.5°; LSPN-3438.
Photograph by N. Chernykh, Crimean Astrophysical Observatory.

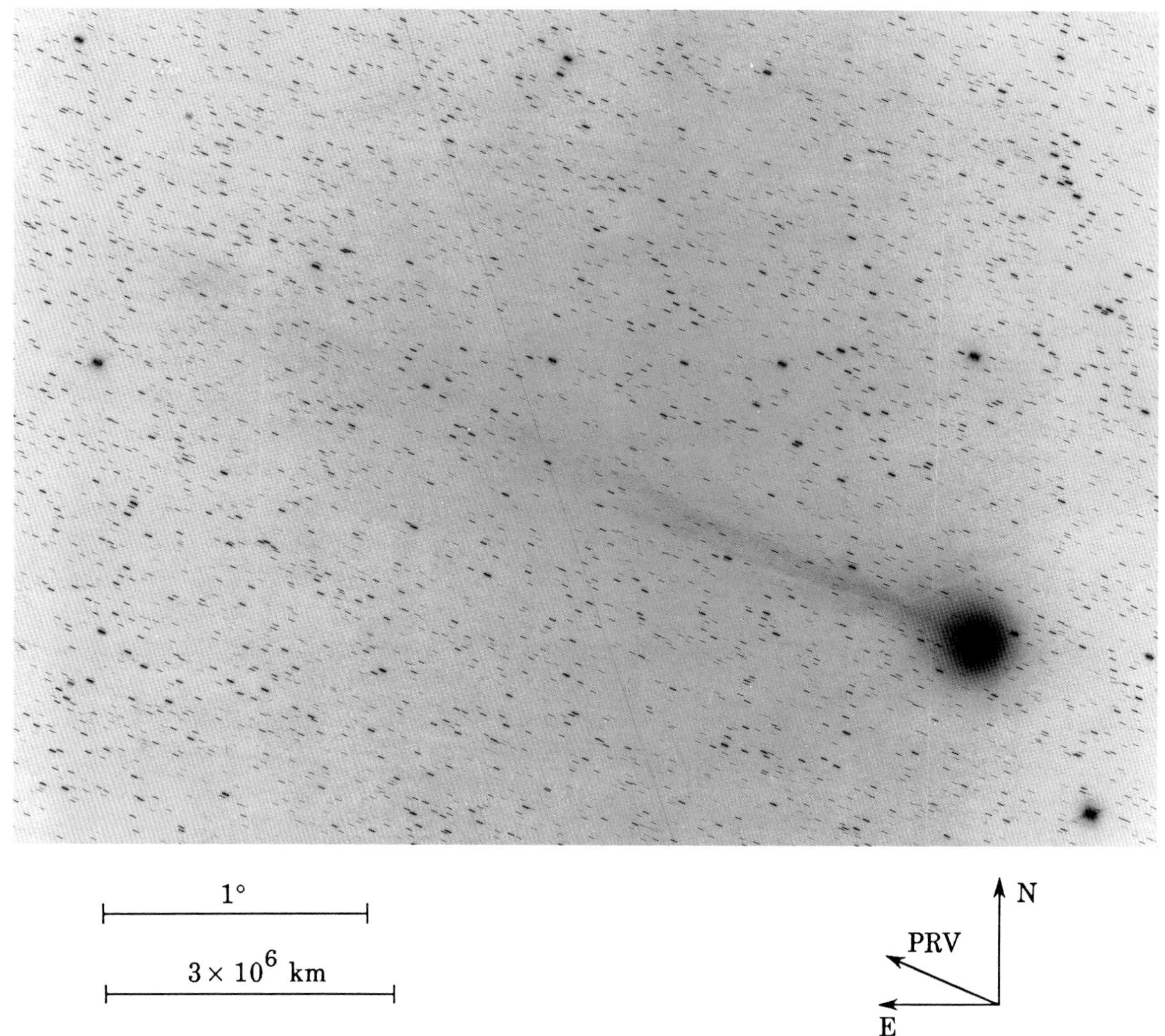

Fig. 88: 1985 Dec 11.07639 UT; exposure 20.0 minutes on IIa-O emulsion with no filter; r=1.33, Δ=0.75, β=133.0°; LSPN-201.
Photograph by W. Liller, University of Michigan/CTIO.

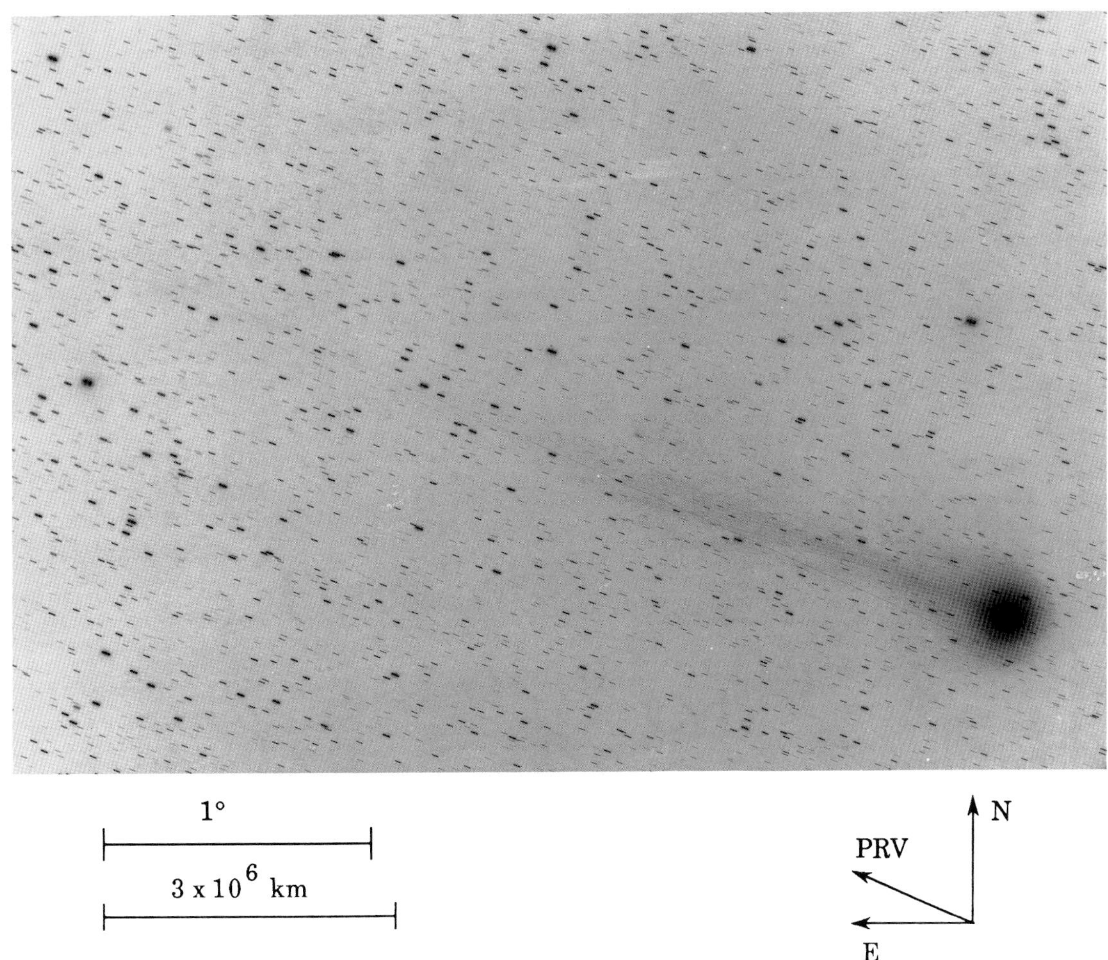

Fig. 89: 1985 Dec 11.11736 UT; exposure 20.0 minutes on IIa-O emulsion with no filter; r=1.33, Δ=0.75, β=132.9°; LSPN-203.
Photograph by W. Liller, University of Michigan/CTIO.

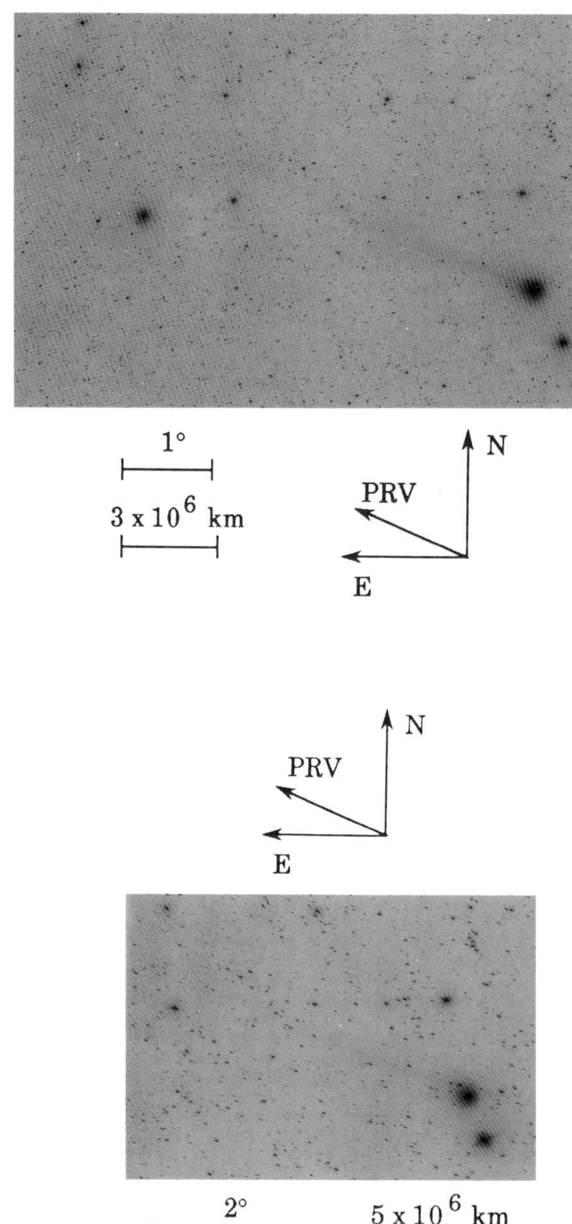

(top) Fig. 90: 1985 Dec 11.13785 UT; exposure 7.0 minutes on IIa-O emulsion with no filter; r=1.33, Δ=0.75, β=132.9°; LSPN-1663.
Photograph by E. Moore et al., Joint Observatory for Cometary Research.

(bottom) Fig. 91: 1985 Dec 11.21875 UT; exposure 30.0 minutes on IIIa-J emulsion with no filter; r=1.33, Δ=0.75, β=132.8°; LSPN-1867.
Photograph by A. Storrs, Mauna Kea Observatory.

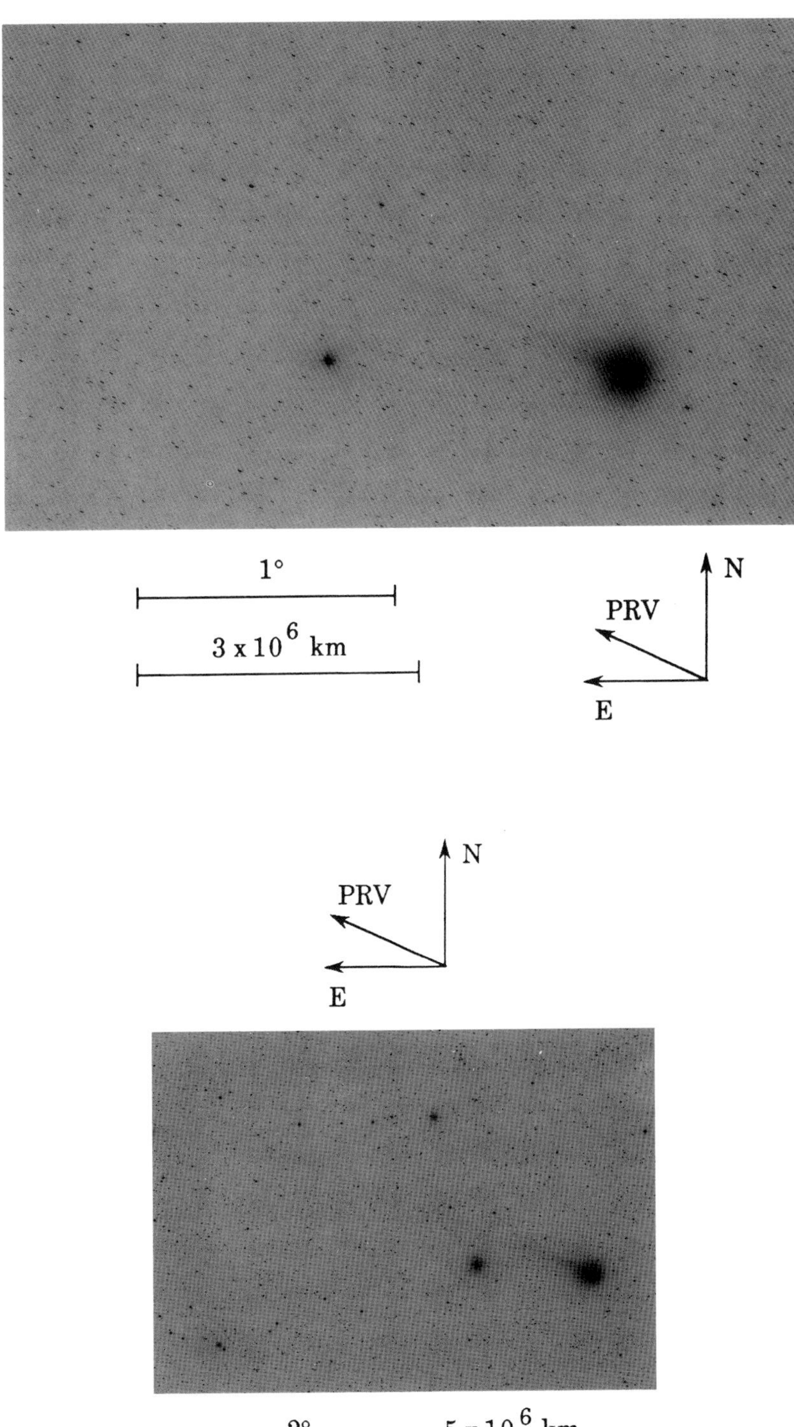

(top) Fig. 92: 1985 Dec 12.07431 UT; exposure 10.0 minutes on 103a-O emulsion with no filter; r=1.31, Δ=0.77, β=131.9°; LSPN-723.
Photograph by C. Torres/H. Wroblewski, Cerro el Roble Astronomical Observatory.

(bottom) Fig. 93*: 1985 Dec 12.17535 UT; exposure 7.0 minutes on IIa-O emulsion with no filter; r=1.31, Δ=0.77, β=131.8°; LSPN-1662.
Photograph by L. Bair/E. Marr, Joint Observatory for Cometary Research.

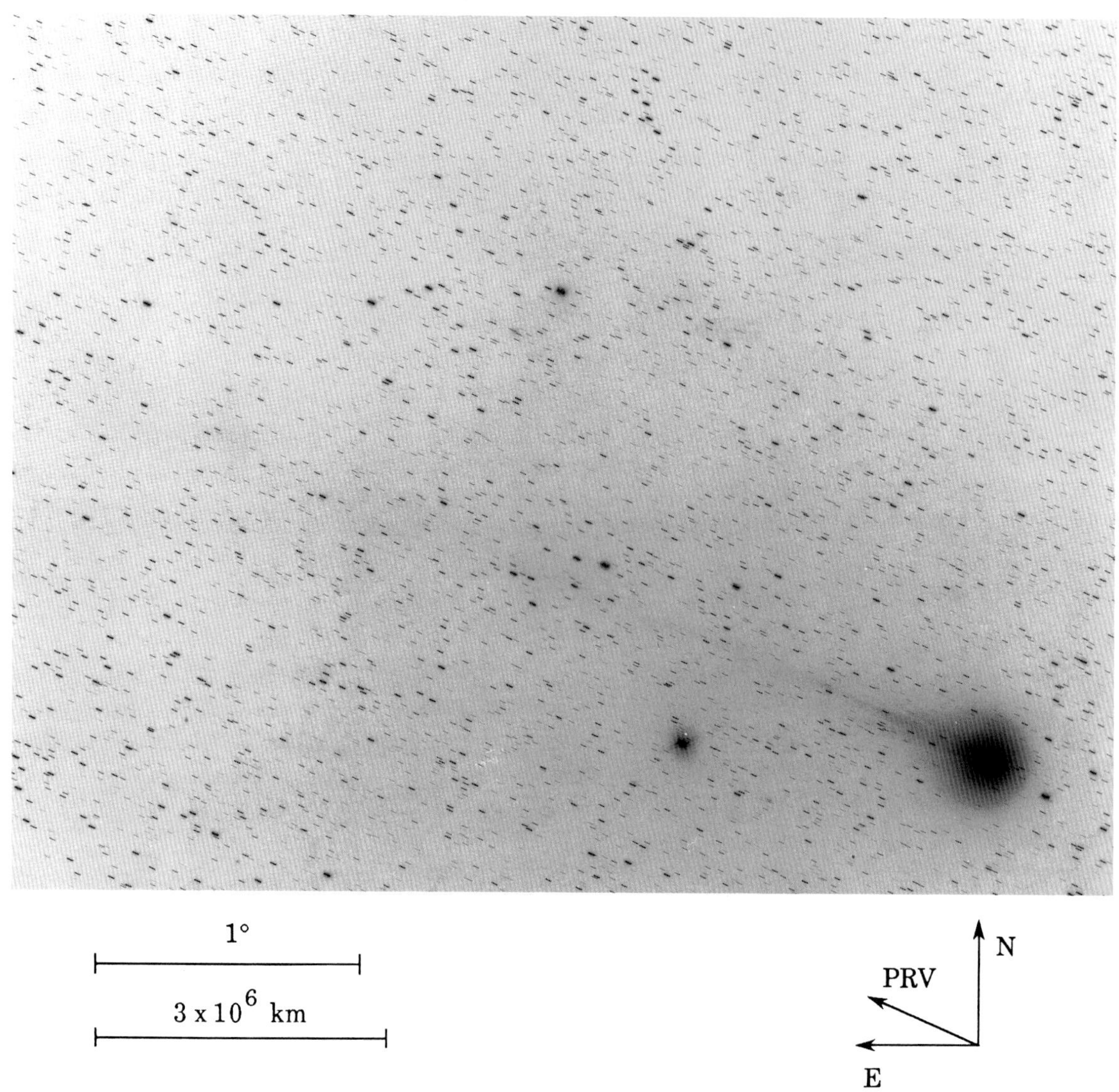

Fig. 94: 1985 Dec 12.08056 UT; exposure 20.0 minutes on IIa-O emulsion with no filter; r=1.31, Δ=0.77, β=131.9°; LSPN-206.
Photograph by W. Liller, University of Michigan/CTIO.

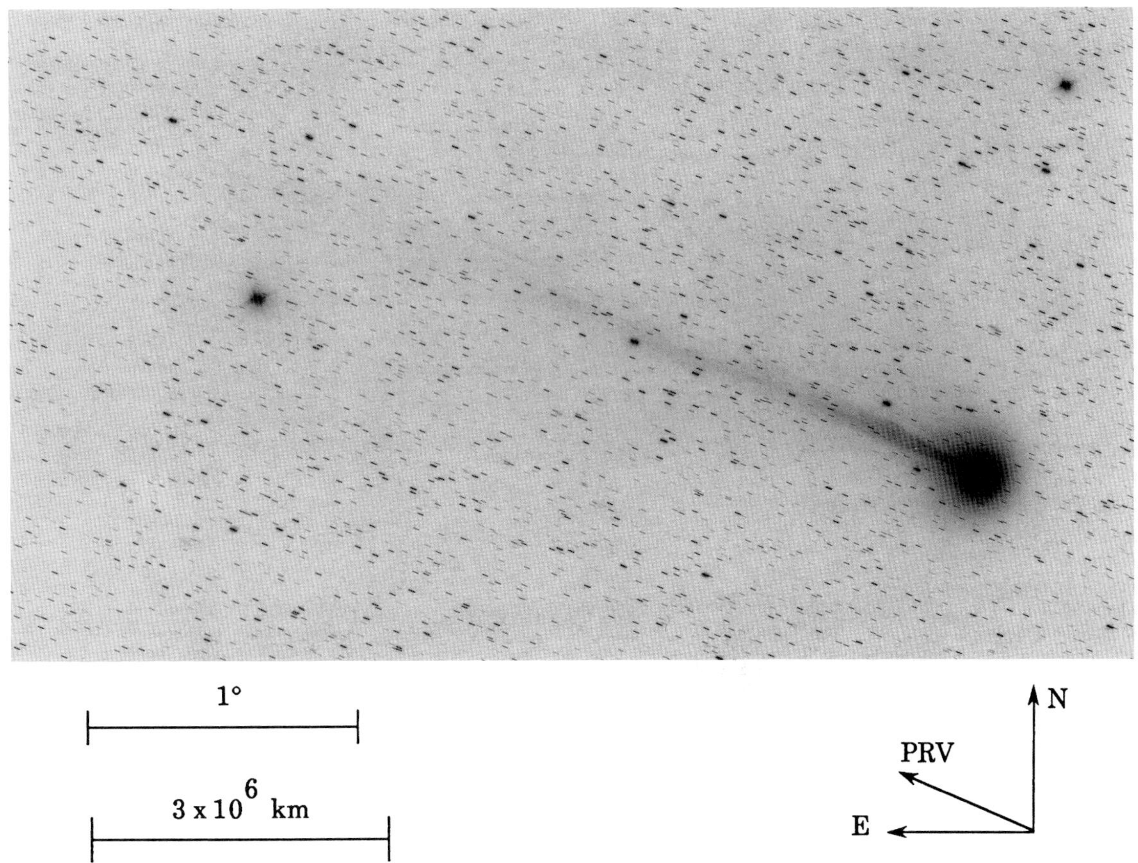

Fig. 99: 1985 Dec 13.07014 UT; exposure 20.0 minutes on IIa-O emulsion with no filter; r=1.30, Δ=0.79, β=130.9°; LSPN-211.
Photograph by W. Liller, University of Michigan/CTIO.

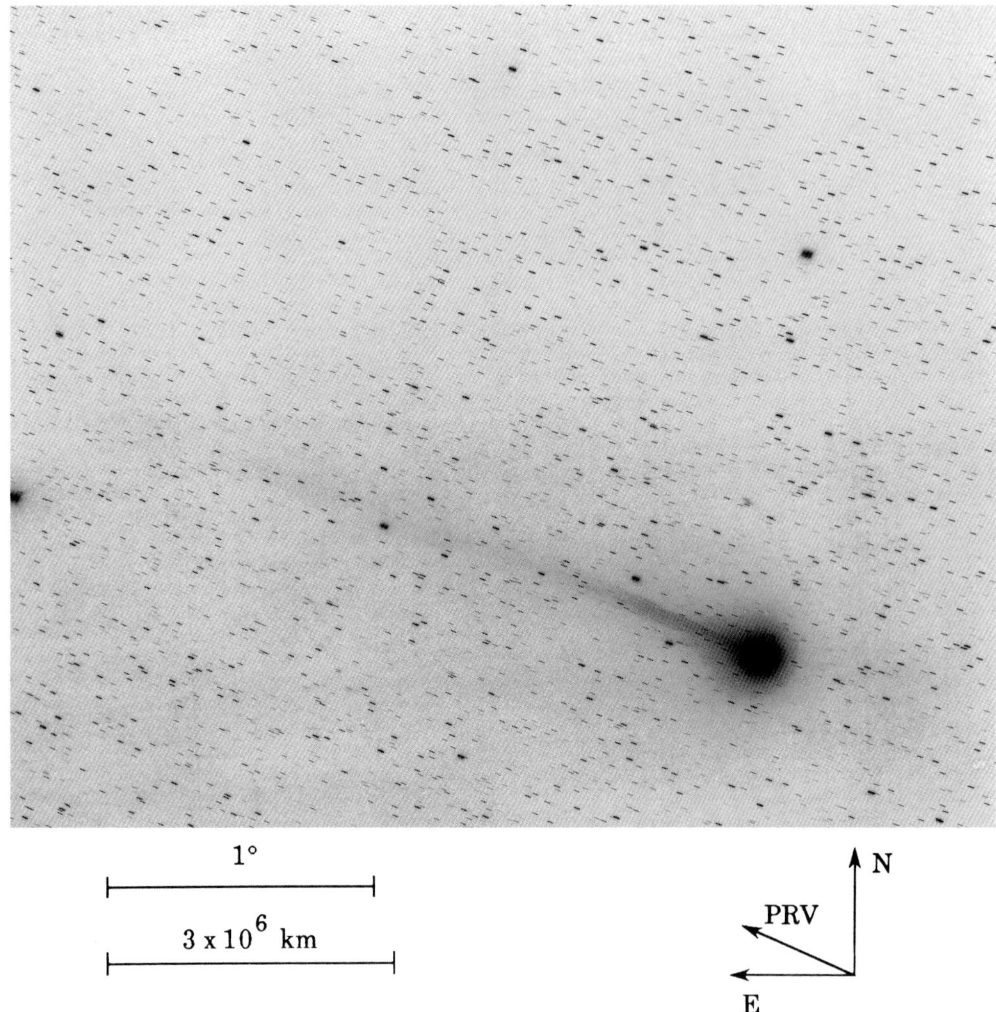

Fig. 100: 1985 Dec 13.13125 UT; exposure 20.0 minutes on IIa-O emulsion with no filter; r=1.30, Δ=0.79, β=130.8°; LSPN-214.
Photograph by W. Liller, University of Michigan/CTIO.

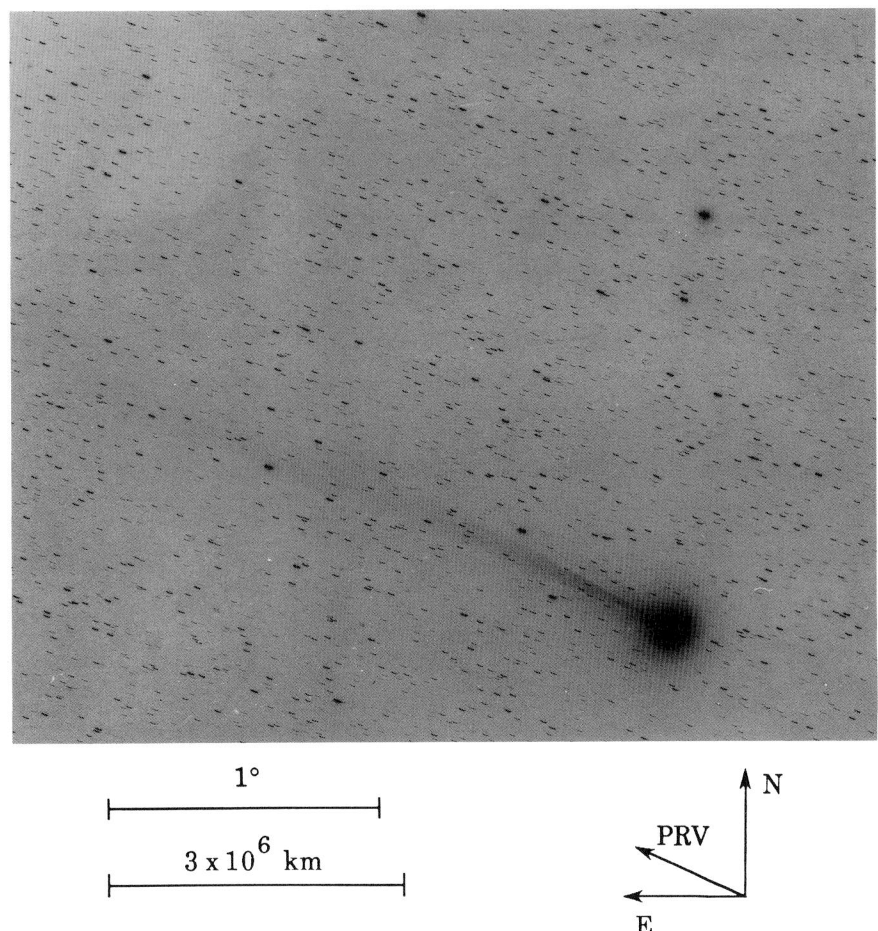

Fig. 101: 1985 Dec 13.20972 UT; exposure 20.0 minutes on IIa-O emulsion with no filter; r=1.30, Δ=0.79, β=130.7°; LSPN-2580.
Photograph by R. Hill, Warner and Swasey Observatory.

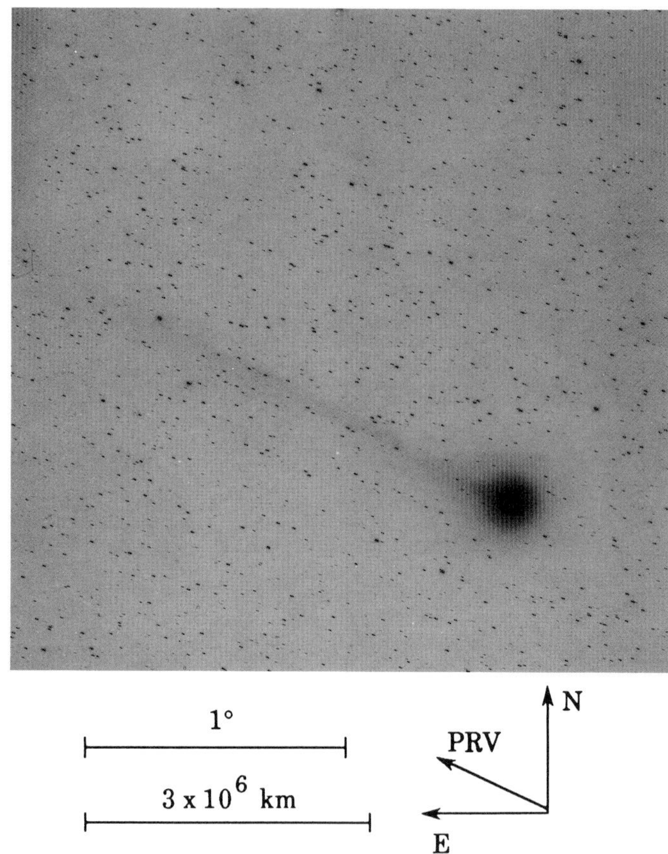

Fig. 102: 1985 Dec 13.75139 UT; exposure 10.0 minutes on IIIa-J emulsion with no filter; r=1.29, Δ=0.80, β=130.2°; LSPN-3186.
Photograph by F. Dossin et al., Observatory of Haute-Provence.

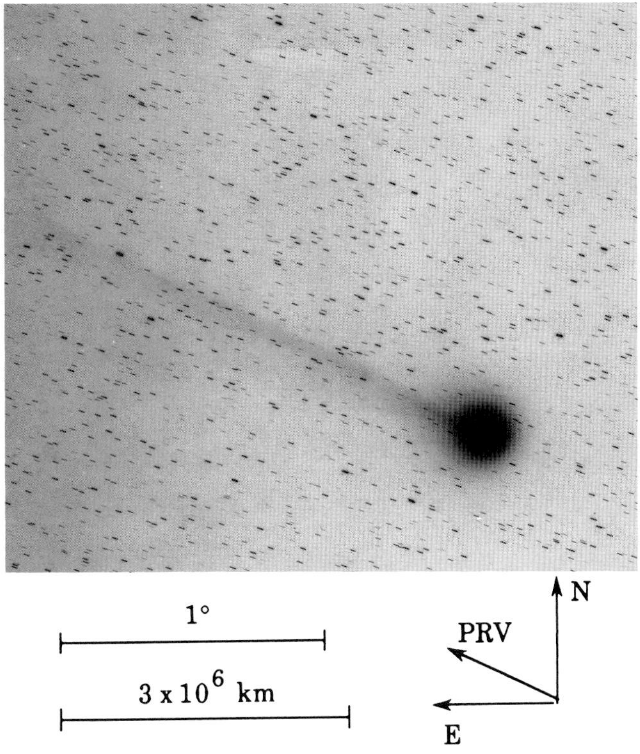

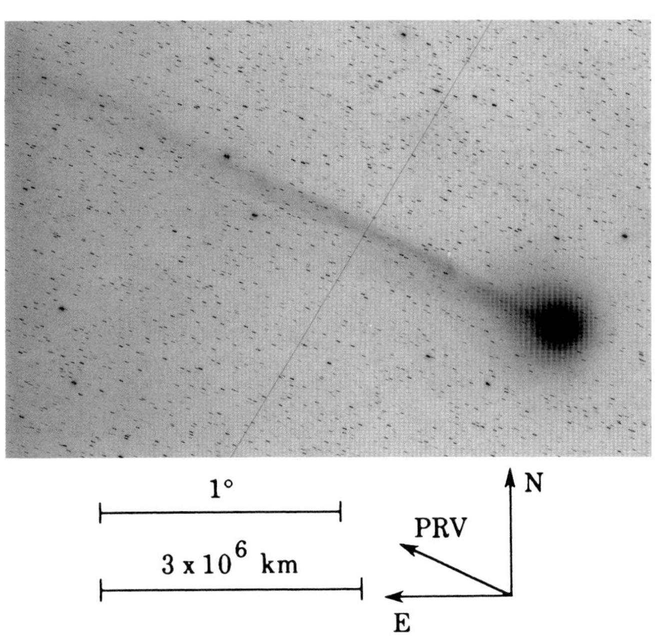

(top) Fig. 103: 1985 Dec 13.76528 UT; exposure 20.0 minutes on IIIa-J emulsion with no filter; r=1.29, Δ=0.80, β=130.2°; LSPN-3187.
Photograph by F. Dossin et al., Observatory of Haute-Provence.

(bottom) Fig. 104: 1985 Dec 13.78403 UT; exposure 14.0 minutes on 103a-E emulsion with no filter; r=1.29, Δ=0.80, β=130.2°; LSPN-3839.
Photograph by L. Kohoutek, Calar Alto Station, German National Ast. Observatory.

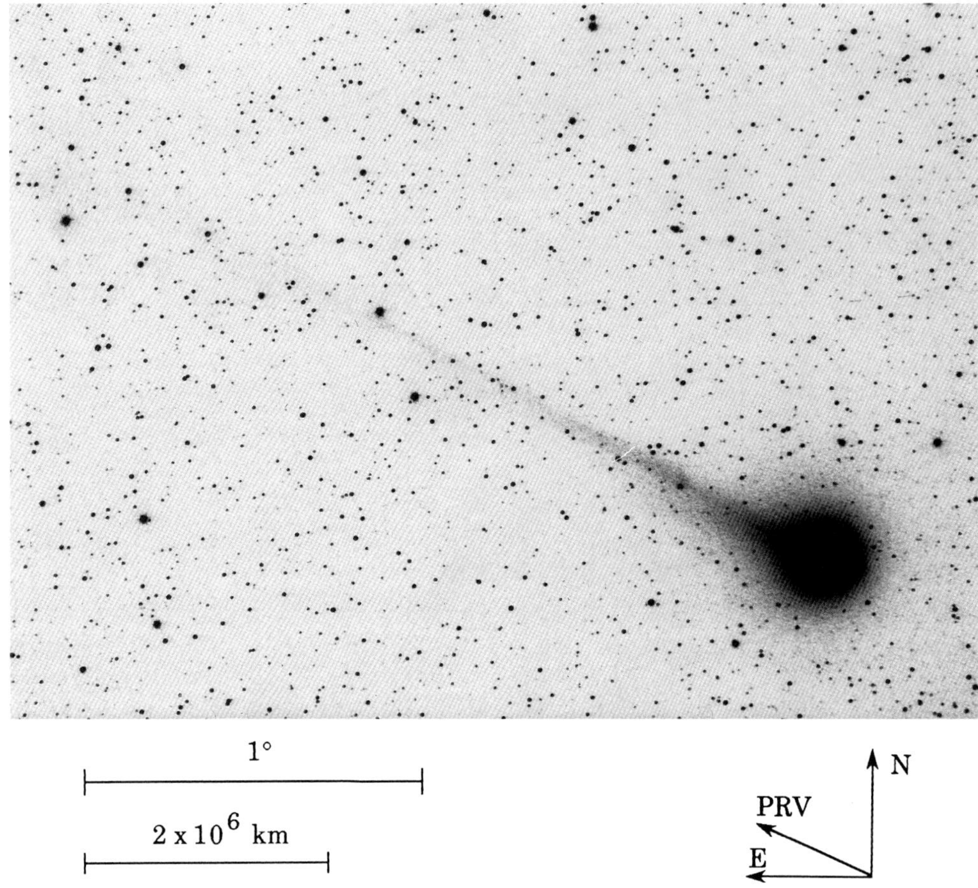

Fig. 105*: 1985 Dec 13.78351 UT; exposure 4.5 minutes on Tech. Pan 2415 emulsion with no filter; r=1.29, Δ=0.80, β=130.2°; AON-850459.
Photograph by M. Jäger, Austria.

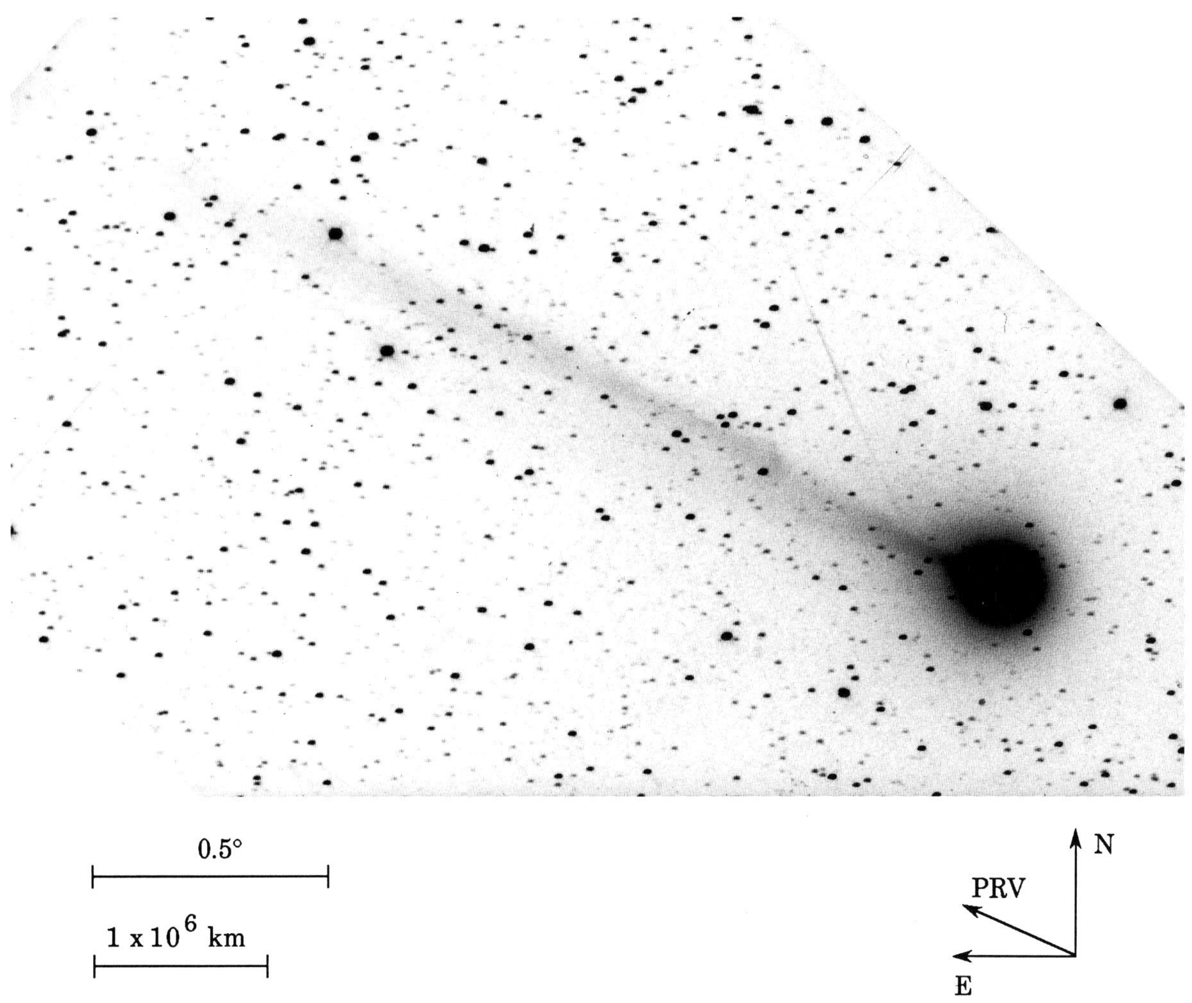

Fig. 106: 1985 Dec 13.80000 UT; exposure 10.0 minutes on Tech. Pan 2415 emulsion with no filter; r=1.29, Δ=0.80, β=130.2°; LSPN-1769.
Photograph by F. Van Wyk, Sutherland, LSPN Island Network.

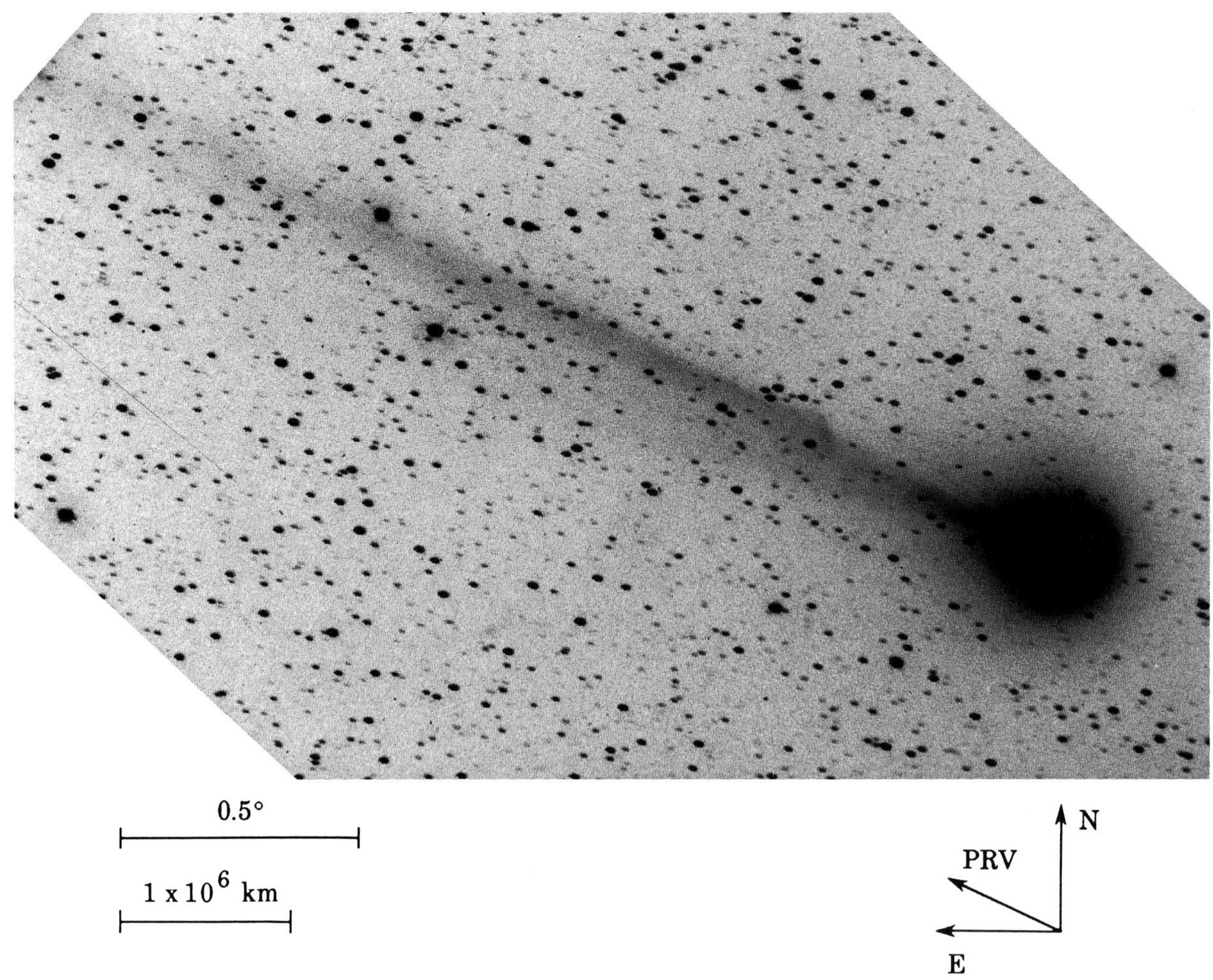

Fig. 107: 1985 Dec 13.81146 UT; exposure 15.0 minutes on Tech. Pan 2415 emulsion with no filter; r=1.29, Δ=0.80, β=130.2°; LSPN-1770.
Photograph by F. Van Wyk, Sutherland, LSPN Island Network.

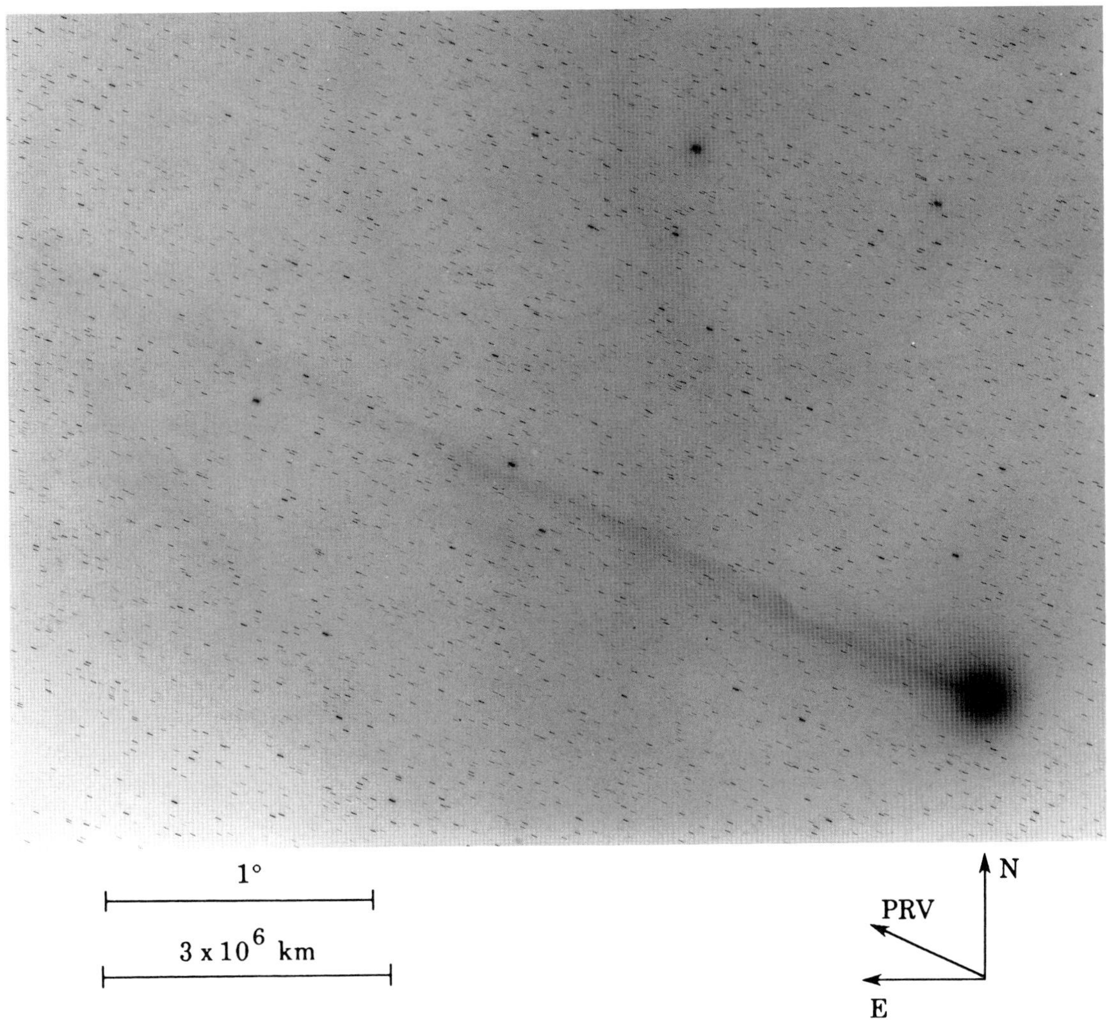

Fig. 108: 1985 Dec 14.05069 UT; exposure 20.0 minutes on IIa-O emulsion with no filter; r=1.28, Δ=0.80, β=130.0°; LSPN-215.
Photograph by W. Liller, University of Michigan/CTIO.

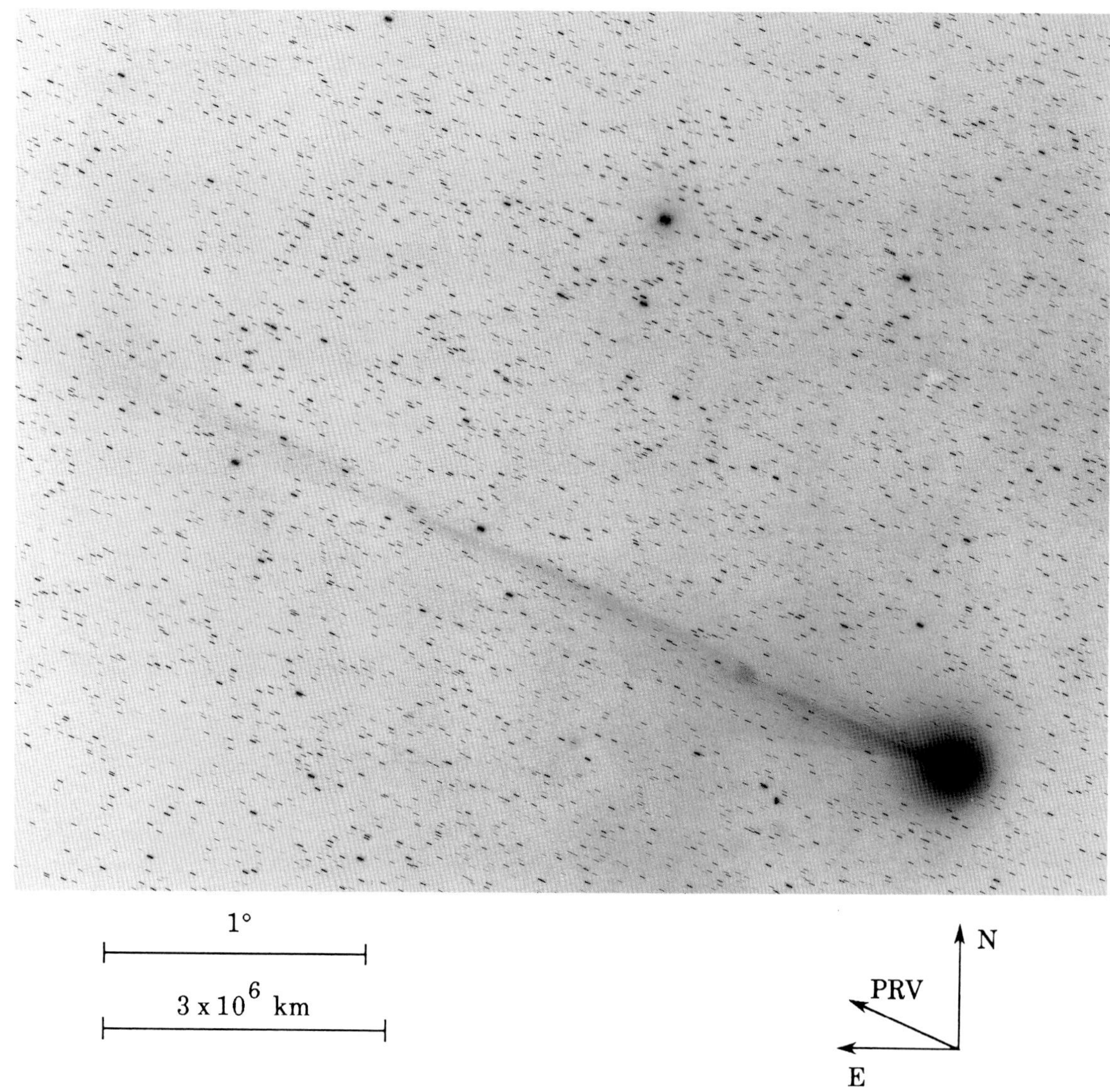

Fig. 109: 1985 Dec 14.06944 UT; exposure 20.0 minutes on IIa-O emulsion with no filter; r=1.28, Δ=0.80, β=129.9°; LSPN-216.
Photograph by W. Liller, University of Michigan/CTIO.

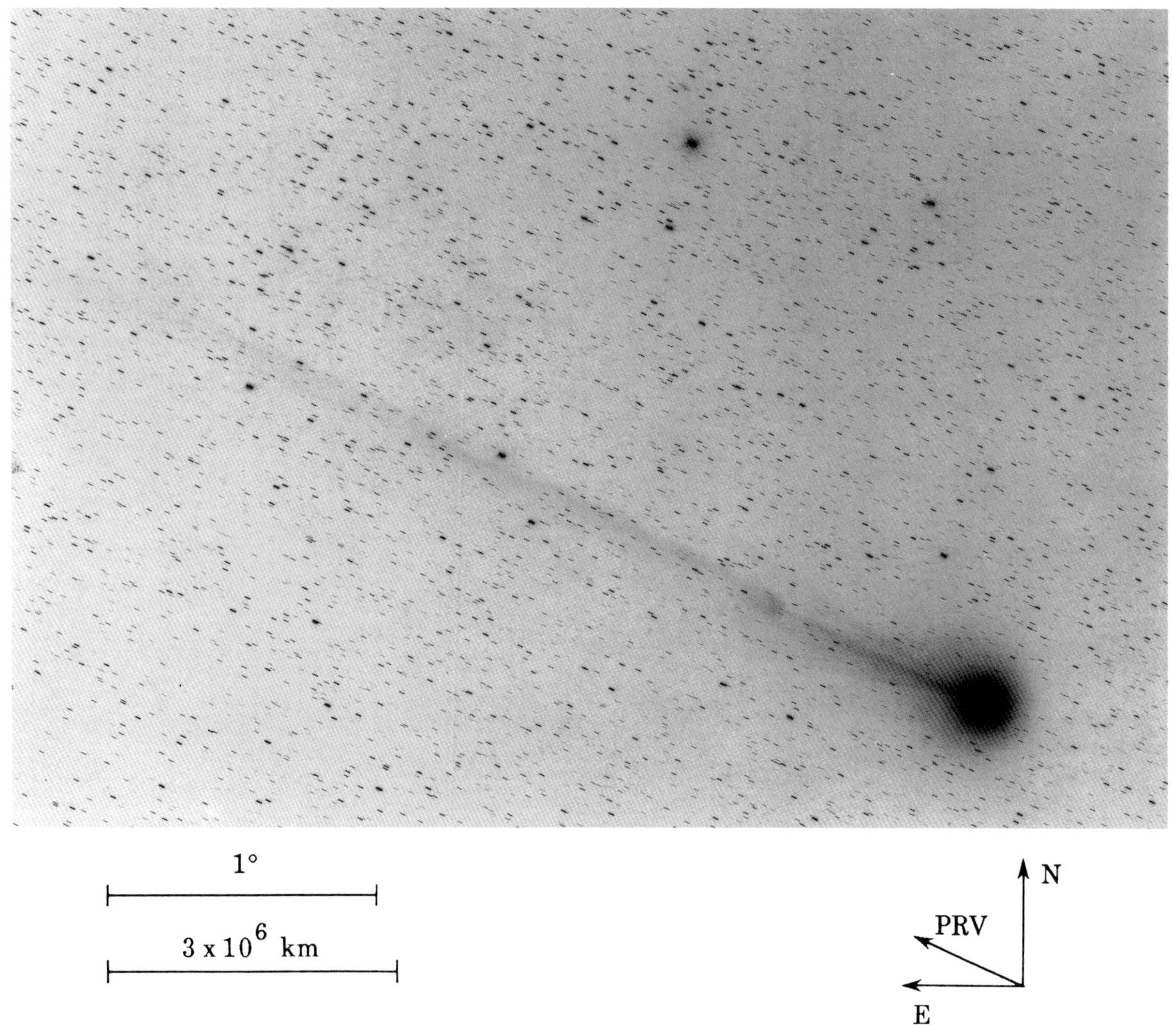

Fig. 110: 1985 Dec 14.08819 UT; exposure 20.0 minutes on IIa-O emulsion with no filter; r=1.28, Δ=0.81, β=129.9°; LSPN-217.
Photograph by W. Liller, University of Michigan/CTIO.

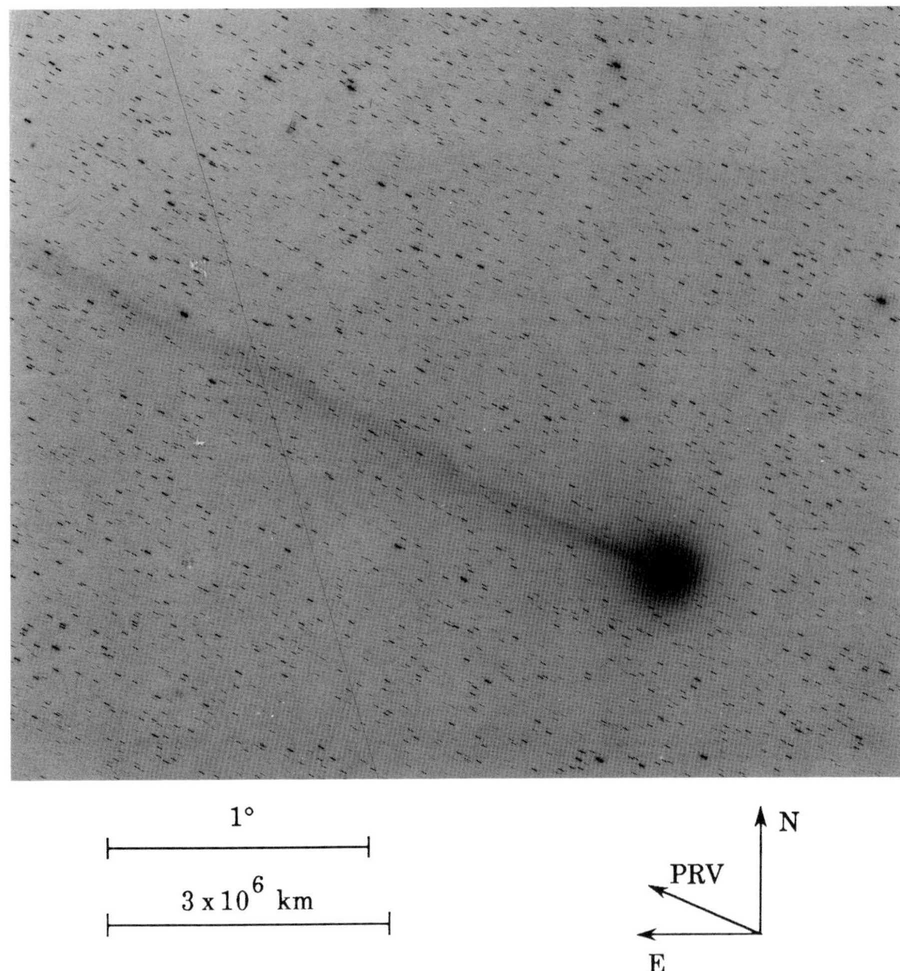

Fig. 111: 1985 Dec 14.09931 UT; exposure 20.0 minutes on IIa-O emulsion with no filter; r=1.28, Δ=0.81, β=129.9°; LSPN-2582.
Photograph by R. Hill, Warner and Swasey Observatory.

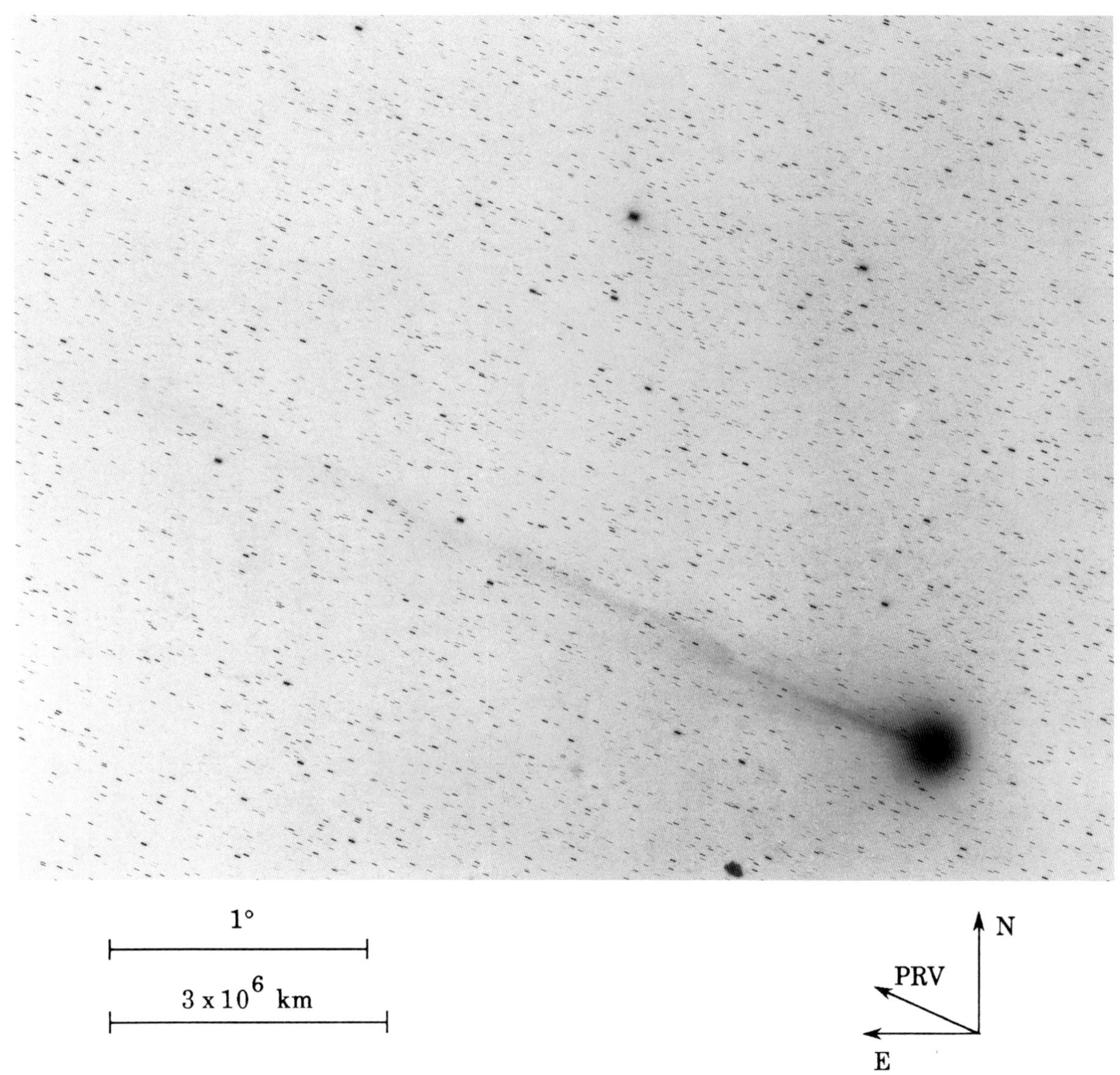

Fig. 112: 1985 Dec 14.10833 UT; exposure 20.0 minutes on IIa-O emulsion with no filter; r=1.28, Δ=0.81, β=129.9°; LSPN-218.
Photograph by W. Liller, University of Michigan/CTIO.

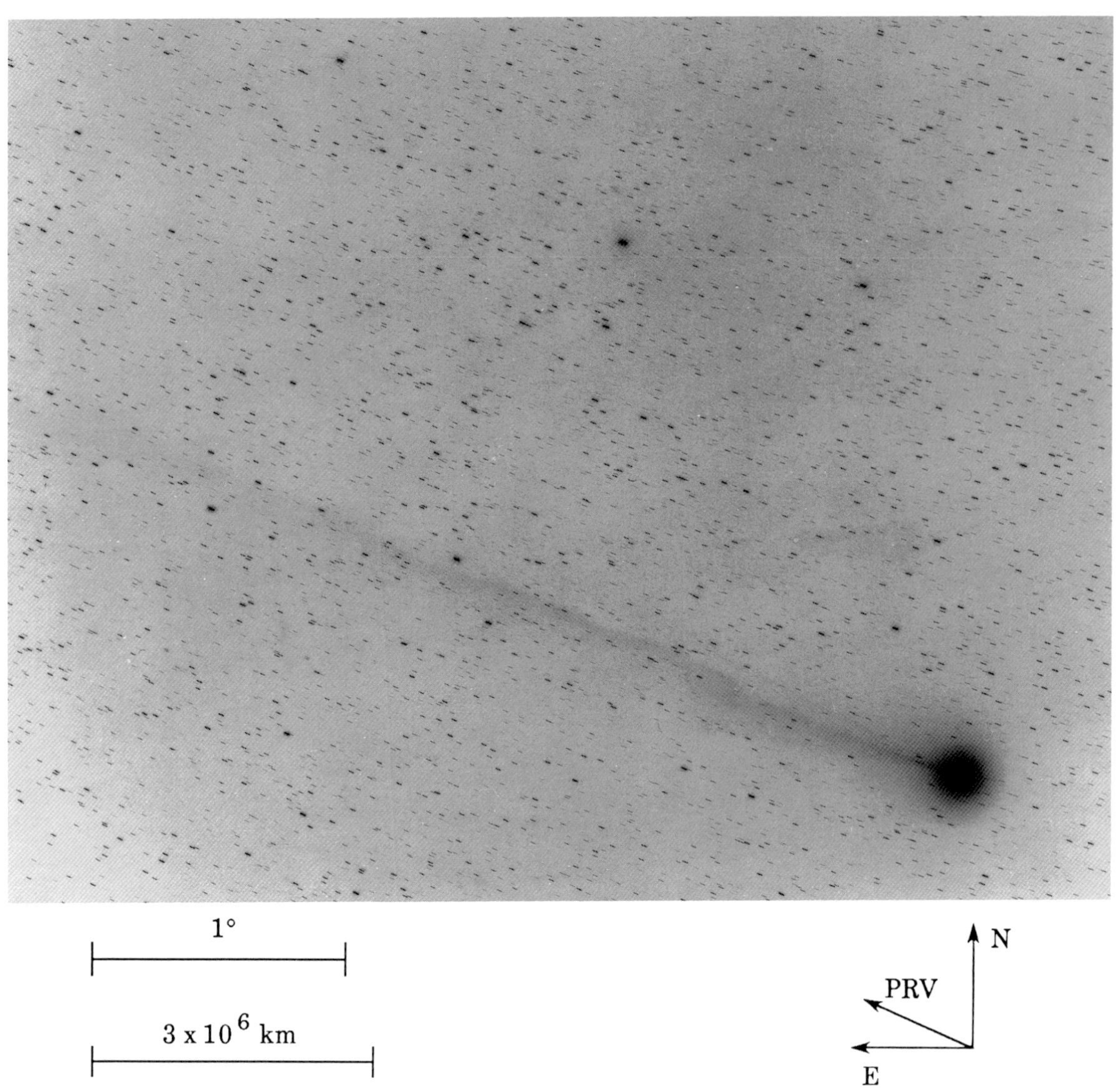

Fig. 113: 1985 Dec 14.12986 UT; exposure 20.0 minutes on IIa-O emulsion with no filter; r=1.28, Δ=0.81, β=129.9°; LSPN-219.
Photograph by W. Liller, University of Michigan/CTIO.

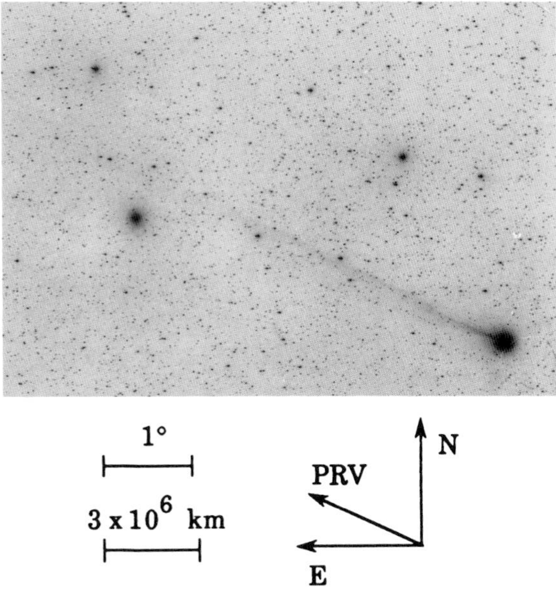

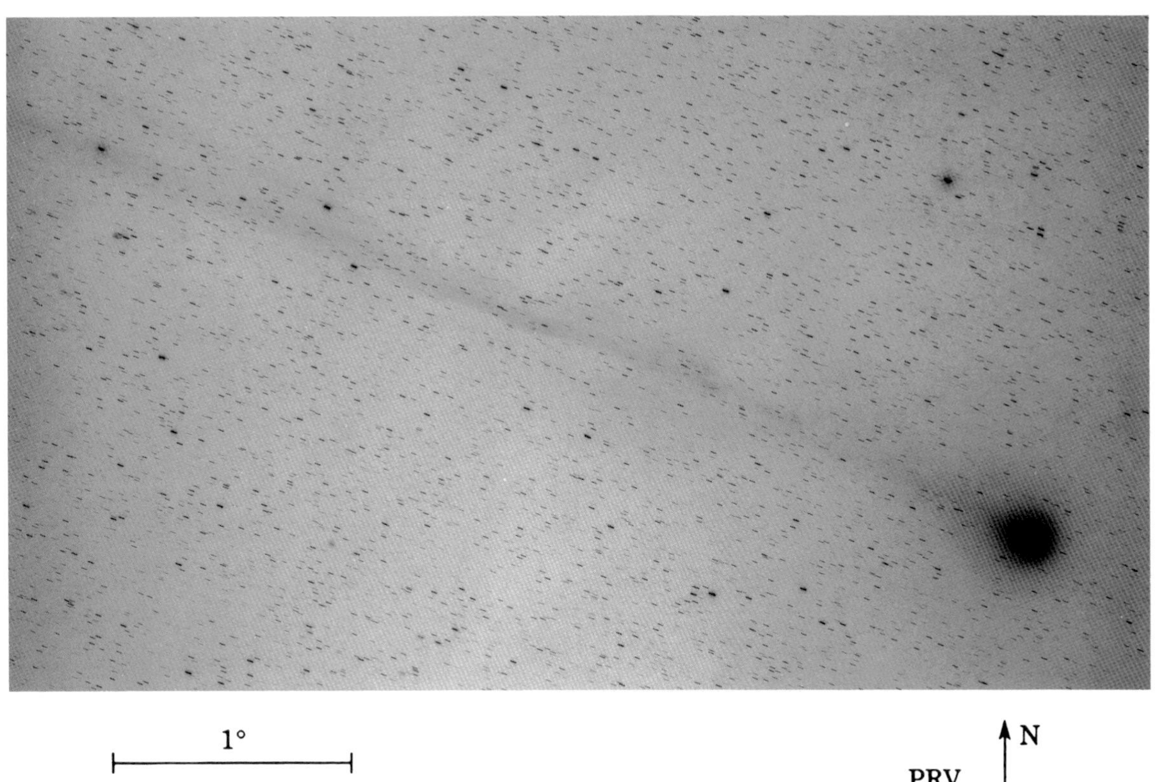

(top) Fig. 114: 1985 Dec 14.13715 UT; exposure 9.0 minutes on IIa-O emulsion with no filter; r=1.28, Δ=0.81, β=129.9°; LSPN-1723.
Photograph by E. Moore et al., Joint Observatory for Cometary Research.

(bottom) Fig. 115*: 1985 Dec 14.91285 UT; exposure 20.0 minutes on 103a-E emulsion with no filter; r=1.27, Δ=0.82, β=129.2°; LSPN-3843.
Photograph by L. Kohoutek, Calar Alto Station, German National Ast. Observatory.

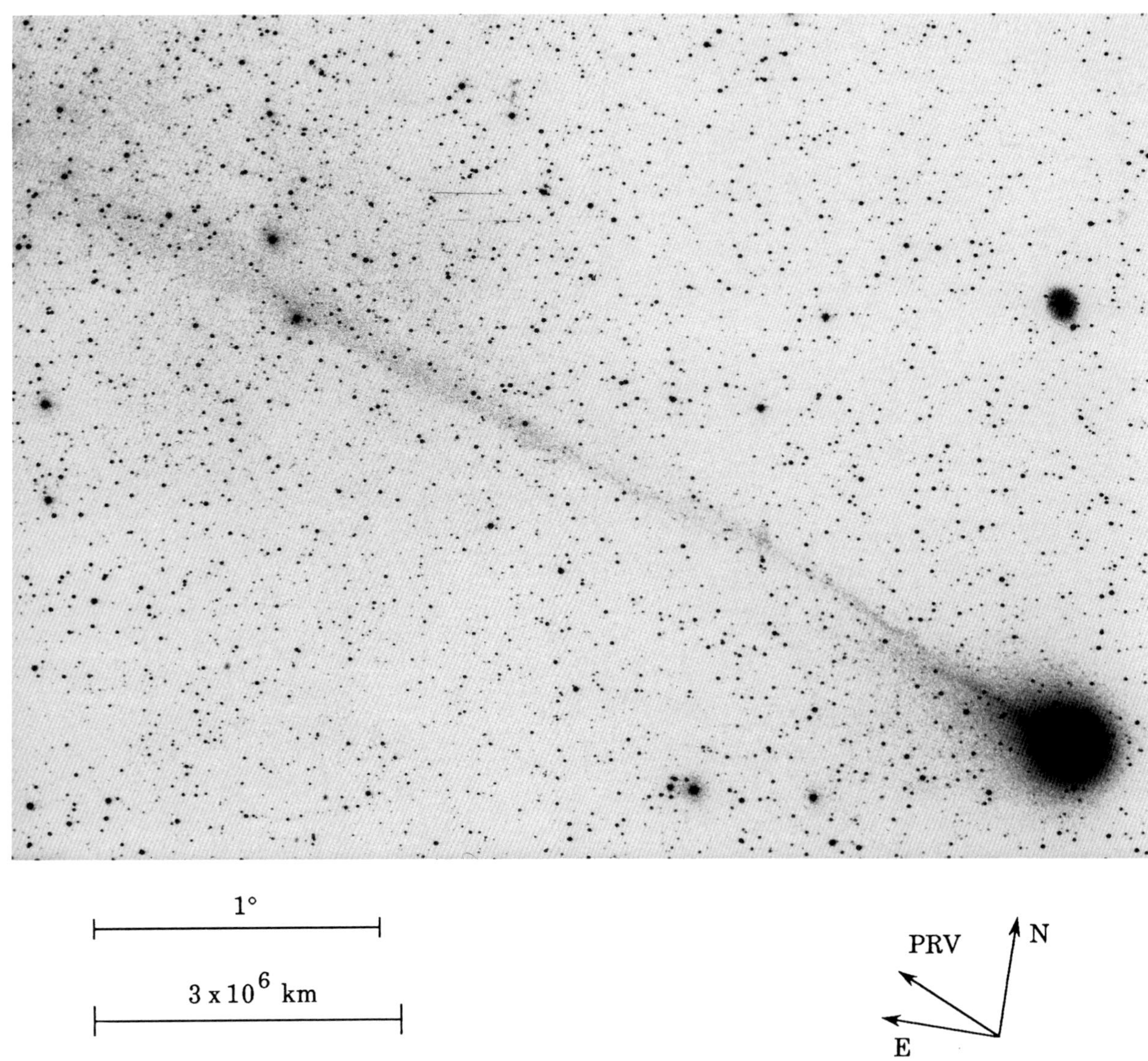

Fig. 116: 1985 Dec 14.81000 UT; exposure 5.0 minutes on Tech. Pan 2415 emulsion; r=1.27, Δ=0.82, β=129.3°; AON-850472.
Photograph by M. Jäger, Austria.

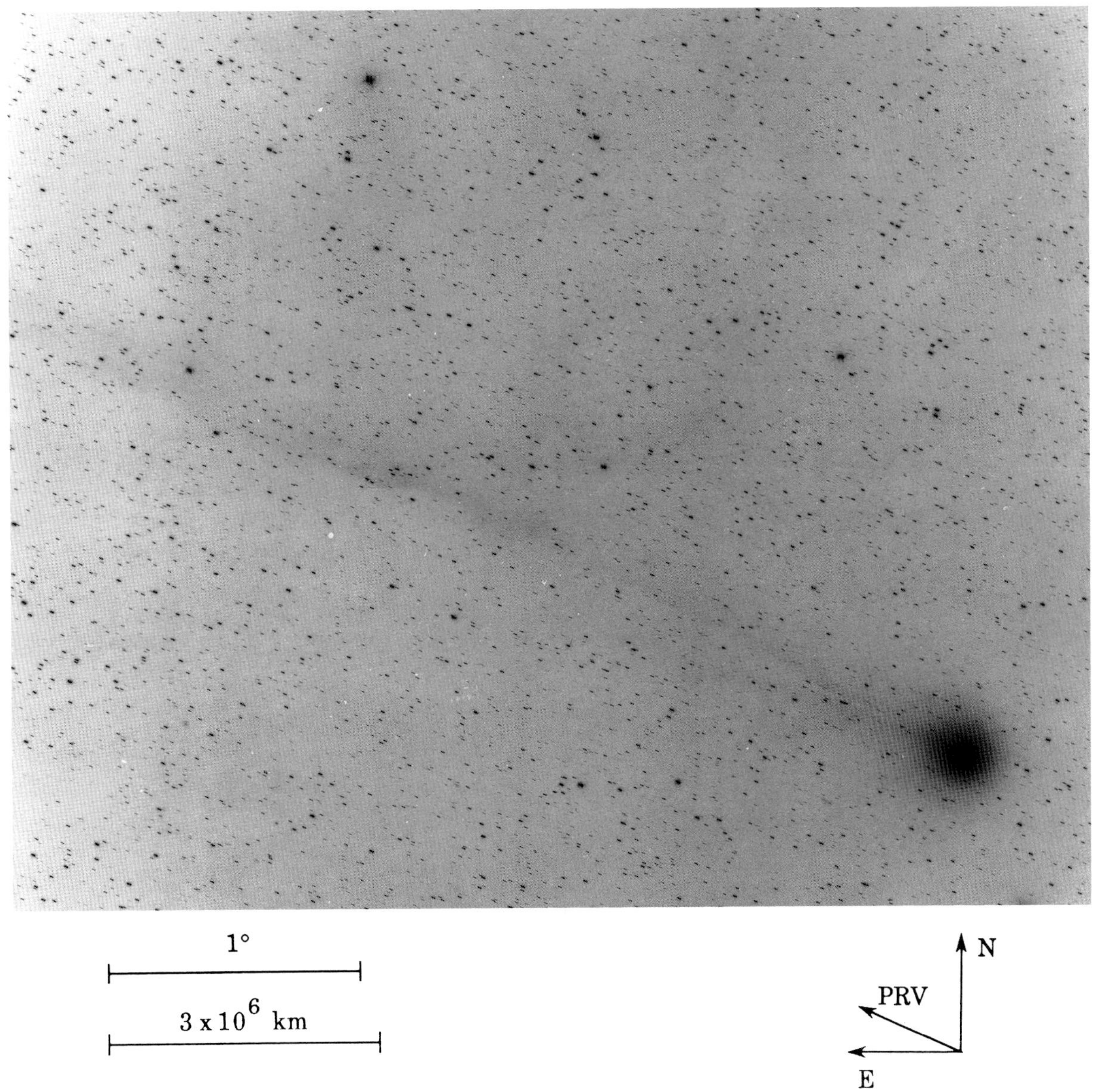

Fig. 117: 1985 Dec 15.05104 UT; exposure 15.0 minutes on IIa-O emulsion with no filter; r=1.27, Δ=0.82, β=129.1°; LSPN-220.
Photograph by W. Liller, University of Michigan/CTIO.

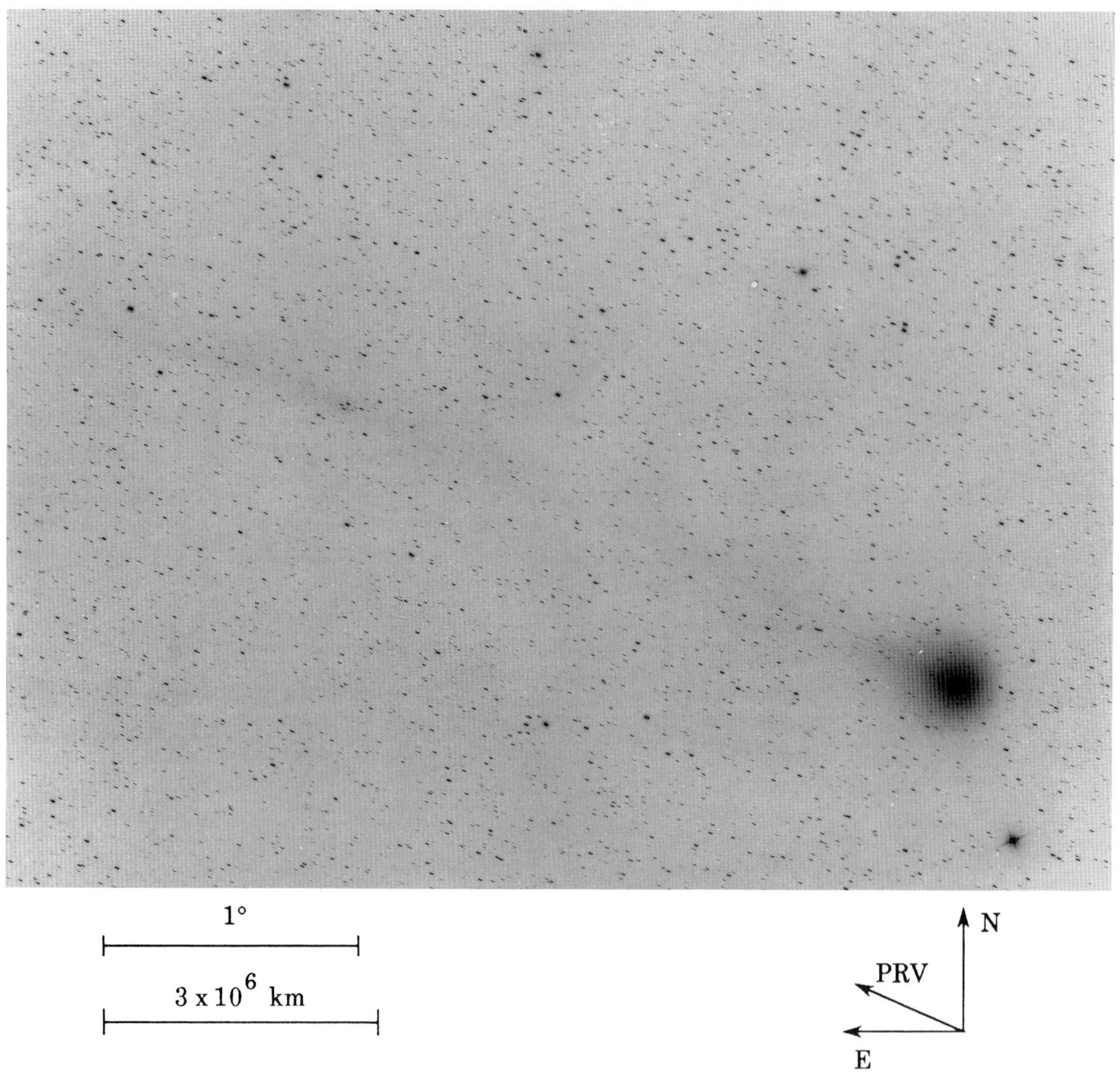

Fig. 118: 1985 Dec 15.08715 UT; exposure 15.0 minutes on IIa-O emulsion with no filter; r=1.27, Δ=0.82, β=129.1°; LSPN-222.
Photograph by W. Liller, University of Michigan/CTIO.

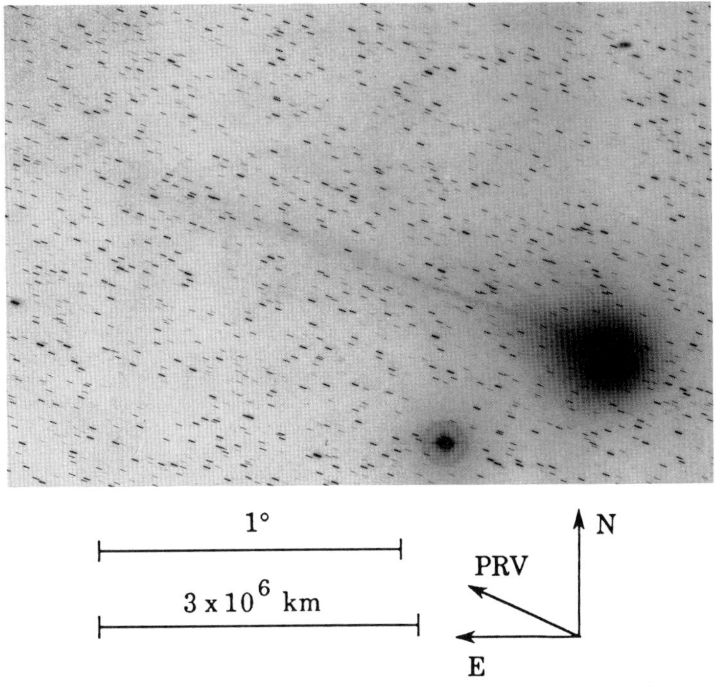

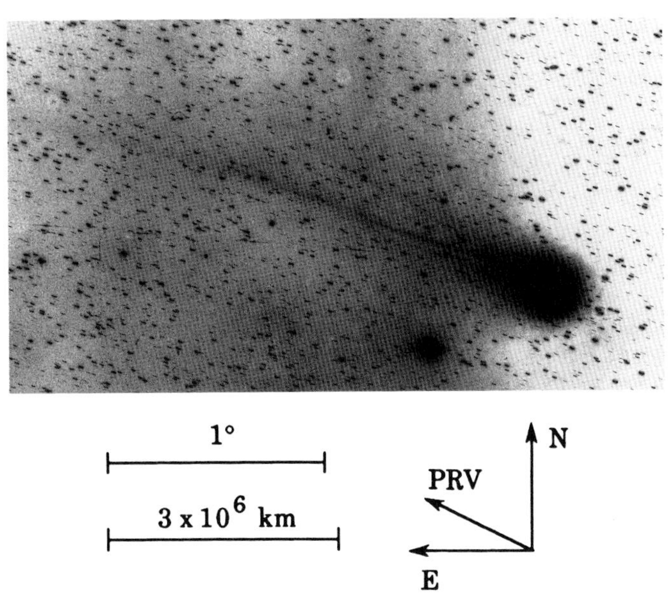

(top) Fig. 119: 1985 Dec 15.67307 UT; exposure 22.0 minutes on ORWO ZU-21 emulsion with GG-13 filter; r=1.26, Δ=0.83, β=128.7°; LSPN-1972.
Photograph by A. Alksnis, Riga Radio-Astrophysical Observatory.

(bottom) Fig. 120: 1985 Dec 15.69784 UT; exposure 20.0 minutes on ORWO ZU-21 emulsion with no filter; r=1.26, Δ=0.83, β=128.6°; LSPN-2489.
Photograph by V. Ivanova et al., Bulgarian National Observatory.

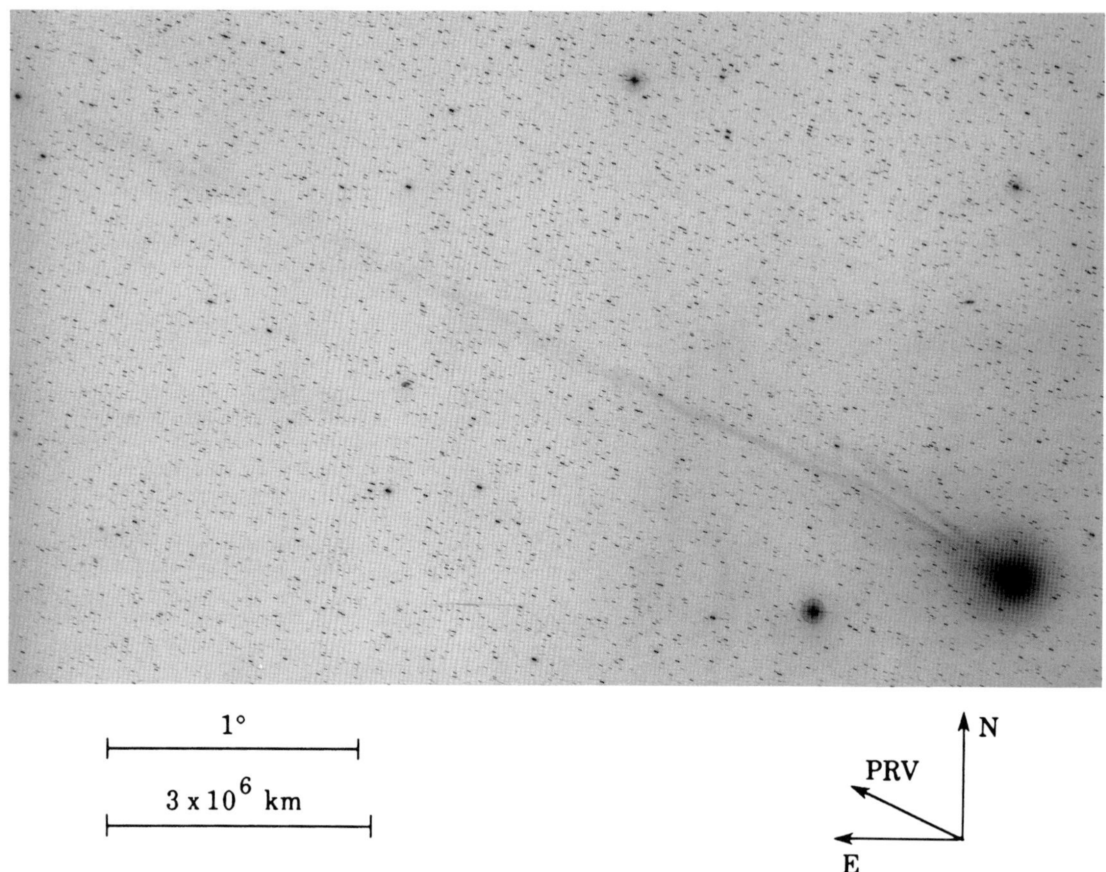

Fig. 121: 1985 Dec 15.89096 UT; exposure 14.0 minutes on 103a-E emulsion with no filter; r=1.26, Δ=0.84, β=128.5°; LSPN-3846.
Photograph by L. Kohoutek, Calar Alto Station, German National Ast. Observatory.

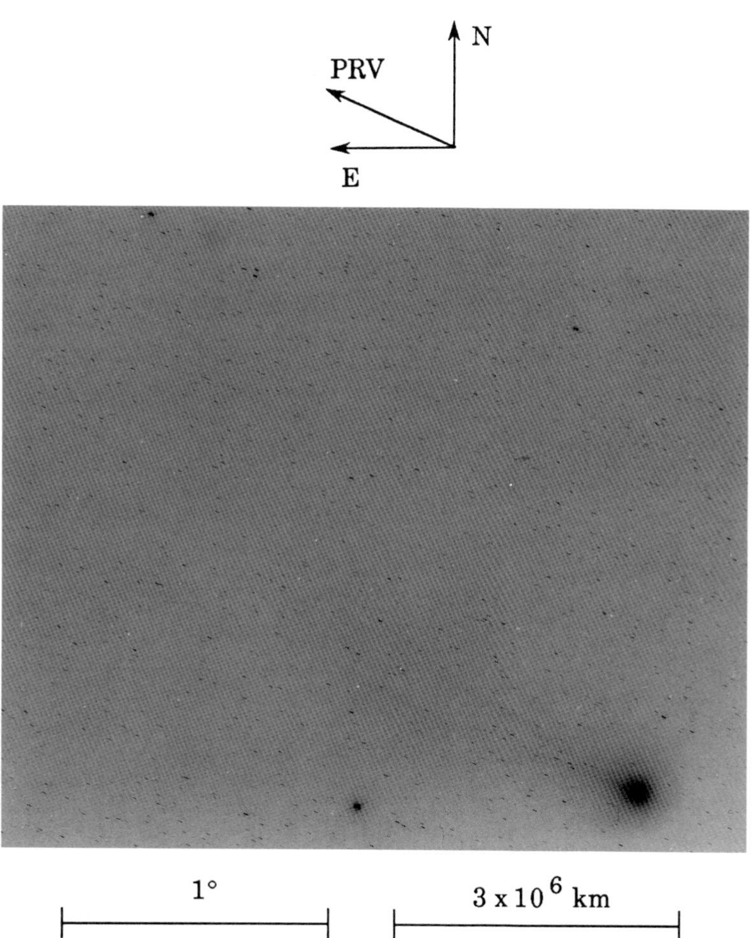

Fig. 122: 1985 Dec 16.05382 UT; exposure 15.0 minutes on IIa-O emulsion with no filter; r=1.25, Δ=0.84, β=128.4°; LSPN-225.
Photograph by W. Liller, University of Michigan/CTIO.

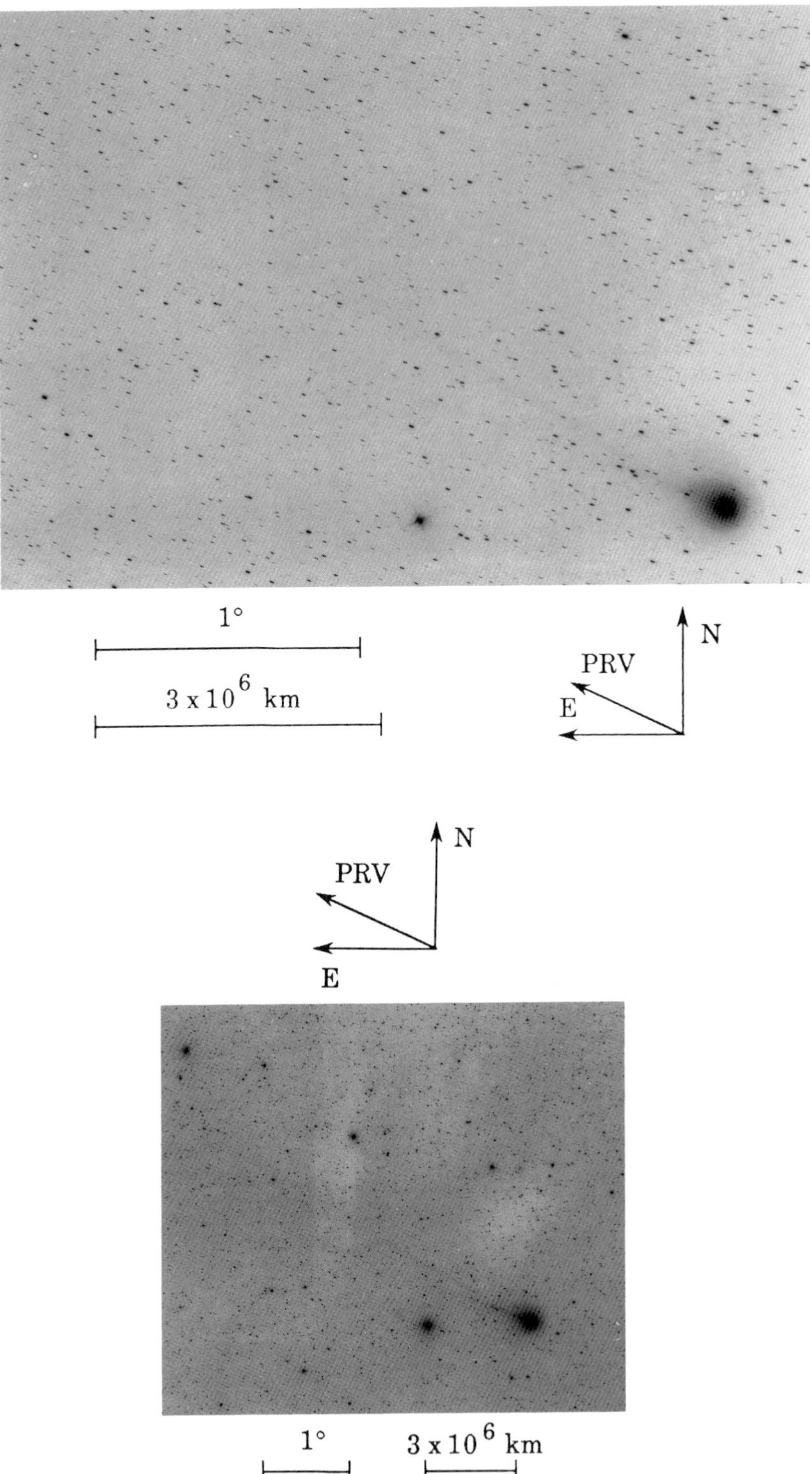

(top) Fig. 123: 1985 Dec 16.12257 UT; exposure 15.0 minutes on IIa-O emulsion with no filter; r=1.25, Δ=0.84, β=128.3°; LSPN-229.
Photograph by W. Liller, University of Michigan/CTIO.

(bottom) Fig. 124: 1985 Dec 16.16736 UT; exposure 10.0 minutes on IIa-O emulsion with no filter; r=1.25, Δ=0.84, β=128.3°; LSPN-1718.
Photograph by E. Moore et al., Joint Observatory for Cometary Research.

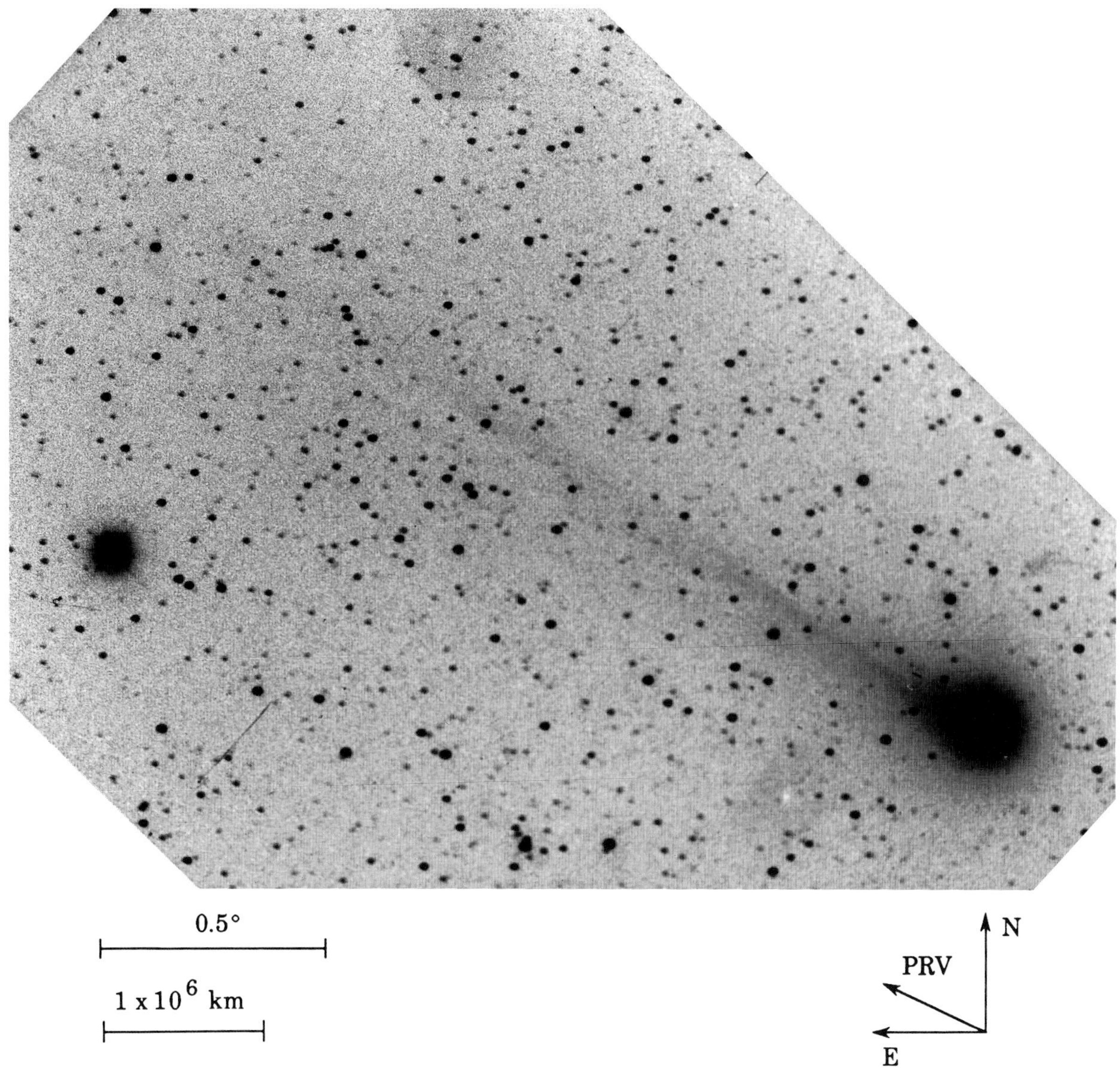

Fig. 125: 1985 Dec 16.80868 UT; exposure 5.0 minutes on Tech. Pan 2415 emulsion with no filter; r=1.24, Δ=0.86, β=127.9°; LSPN-1776.
Photograph by F. Van Wyk, Sutherland, LSPN Island Network.

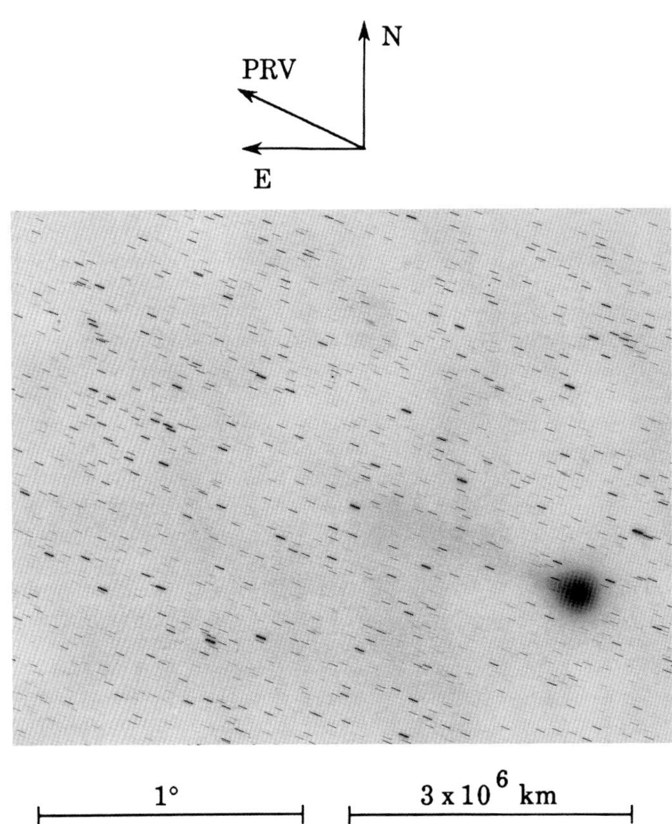

Fig. 126: 1985 Dec 17.08958 UT; exposure 40.0 minutes on IIa-O emulsion with GG-495 filter; r=1.24, Δ=0.86, β=127.7°; LSPN-231.
Photograph by W. Liller, University of Michigan/CTIO.

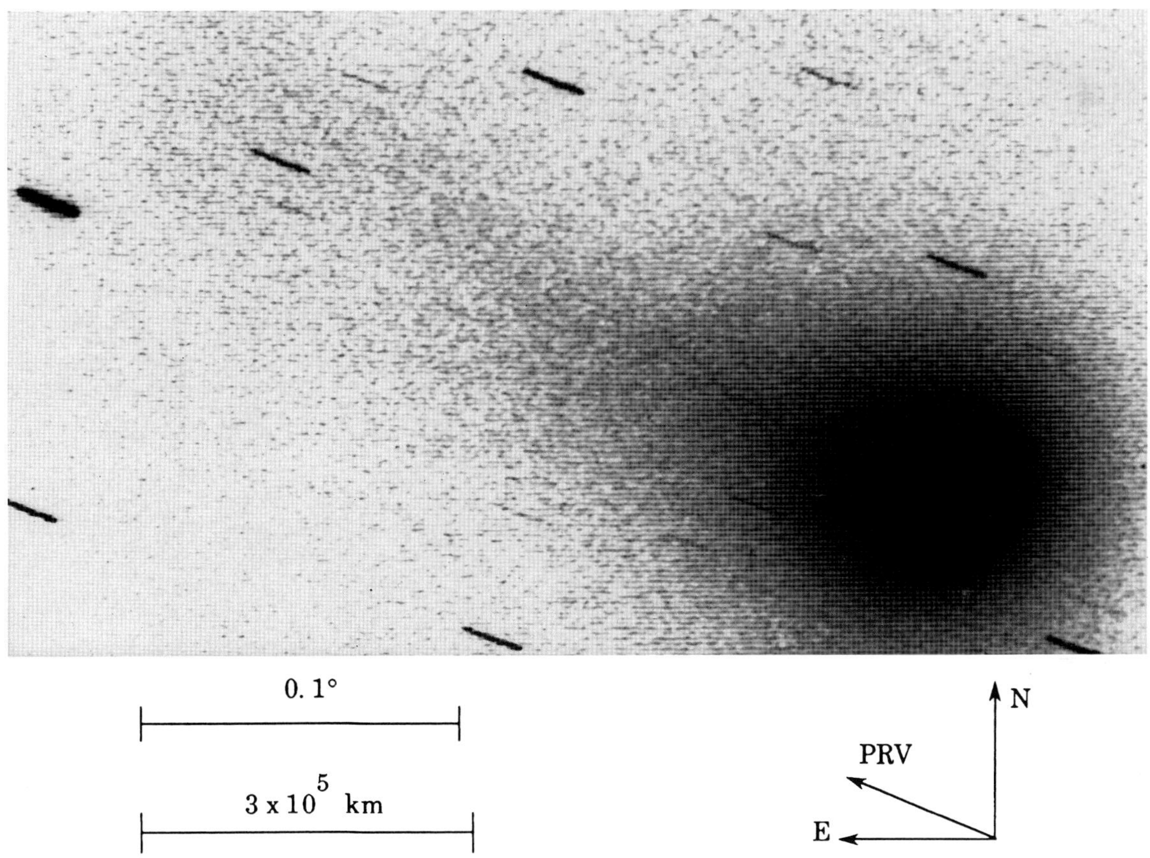

Fig. 127: 1985 Dec 17.57957 UT; exposure 3.0 minutes on 103a-O emulsion with no filter; r=1.23, Δ=0.87, β=127.4°; LSPN-3397.
Photograph by F. Rspaev, Assa Observatory.

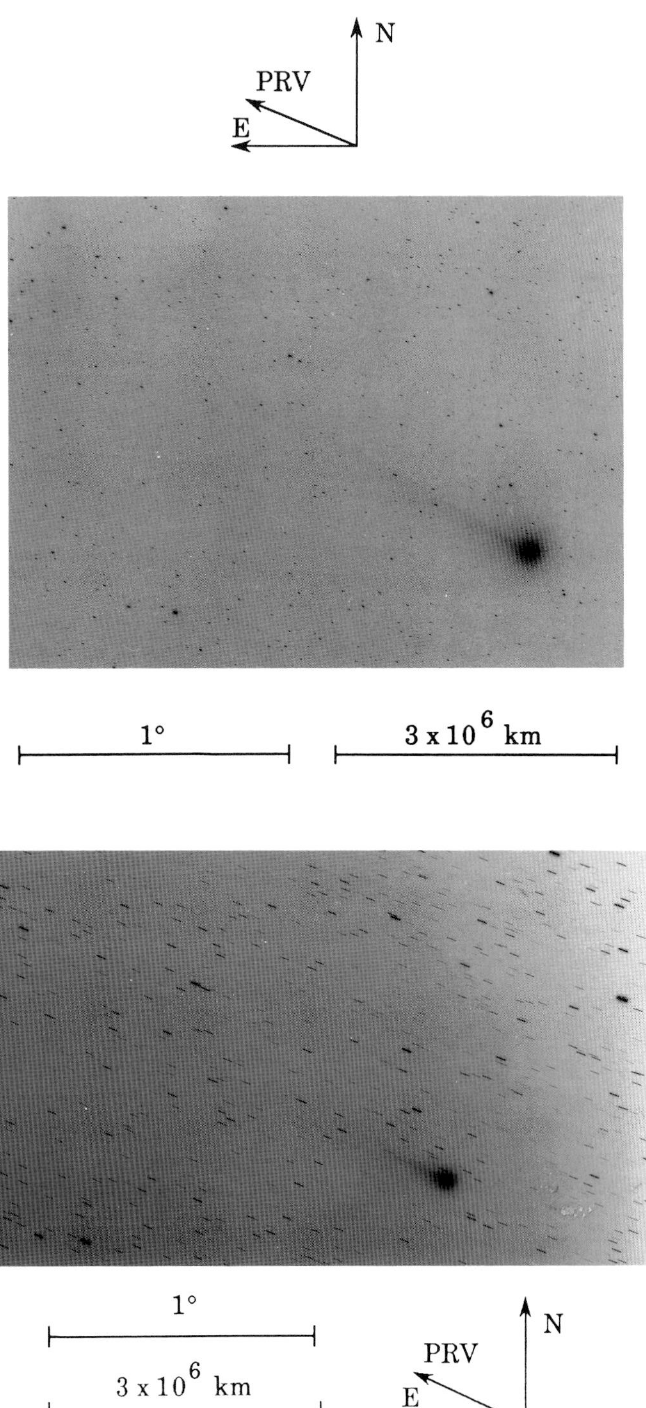

(top) Fig. 128: 1985 Dec 18.05556 UT; exposure 10.0 minutes on IIa-O emulsion with GG-385 filter; r=1.22, Δ=0.88, β=127.2°; LSPN-232.
Photograph by W. Liller, University of Michigan/CTIO.

(bottom) Fig. 129: 1985 Dec 18.08889 UT; exposure 40.0 minutes on 153-01 emulsion with RG-610 filter; r=1.22, Δ=0.88, β=127.2°; LSPN-233.
Photograph by W. Liller, University of Michigan/CTIO.

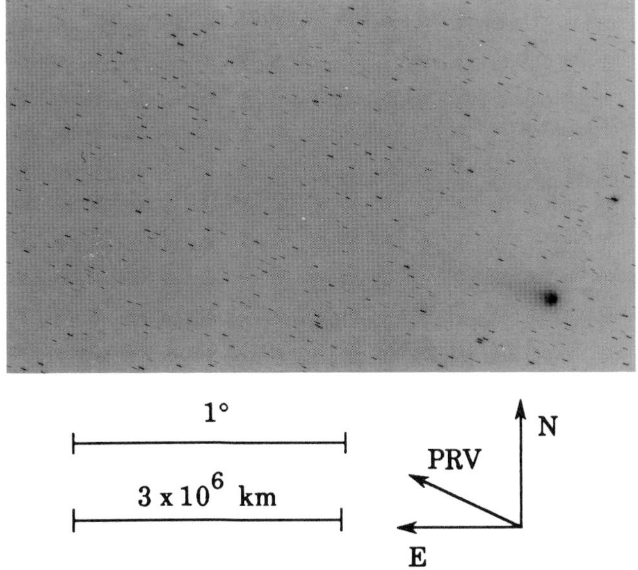

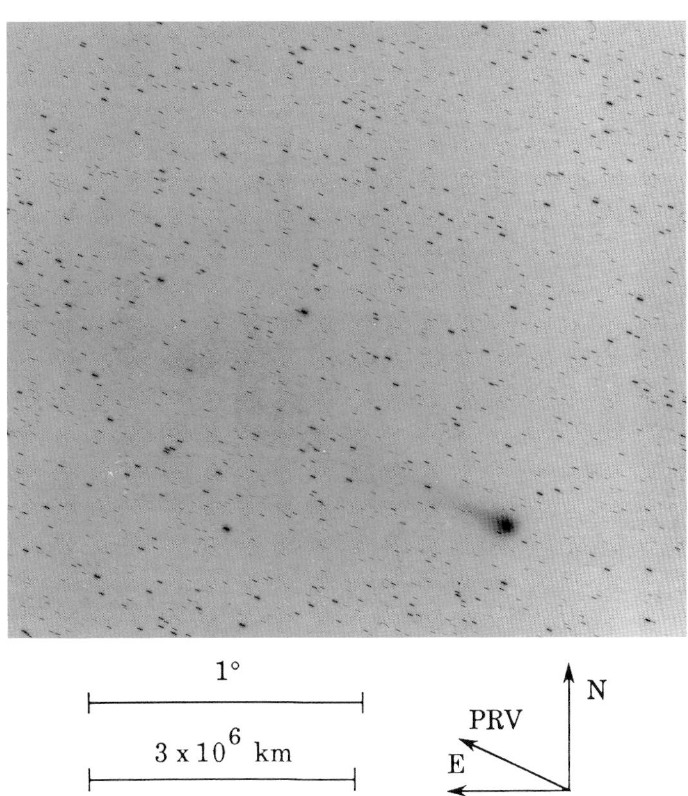

(top) Fig. 130: 1985 Dec 20.05868 UT; exposure 25.0 minutes on 153-01 emulsion with RG-610 filter; r=1.19, Δ=0.96, β=126.3°; LSPN-240.
Photograph by W. Liller, University of Michigan/CTIO.

(bottom) Fig. 131: 1985 Dec 21.07187 UT; exposure 25.0 minutes on 153-01 emulsion with RG-610 filter; r=1.18, Δ=0.92, β=126.0°; LSPN-239.
Photograph by W. Liller, University of Michigan/CTIO.

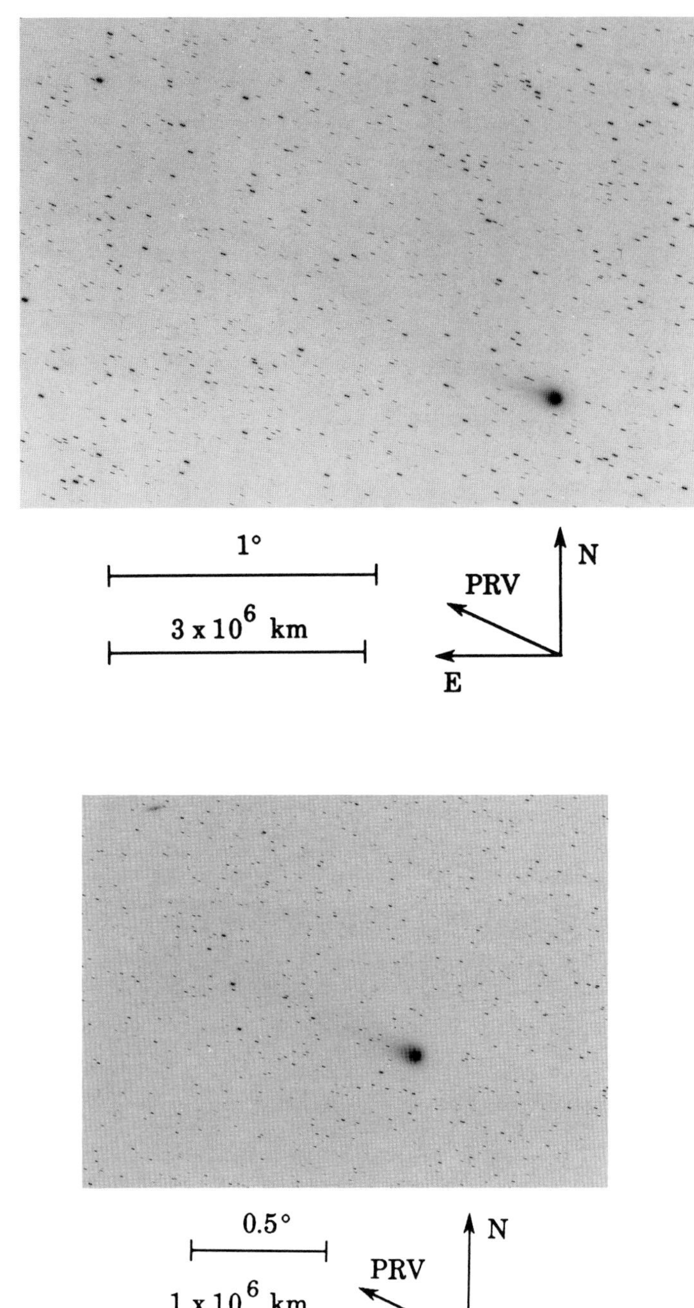

(top) Fig. 132: 1985 Dec 22.07465 UT; exposure 25.0 minutes on 153-01 emulsion with RG-610 filter; r=1.16, Δ=0.96, β=125.7°; LSPN-237.
Photograph by W. Liller, University of Michigan/CTIO.

(bottom) Fig. 133: 1985 Dec 24.05903 UT; exposure 16.0 minutes on 098-04 emulsion with RG-610 filter; r=1.13, Δ=1.00, β=125.4°; LSPN-243.
Photograph by W. Liller, University of Michigan/CTIO.

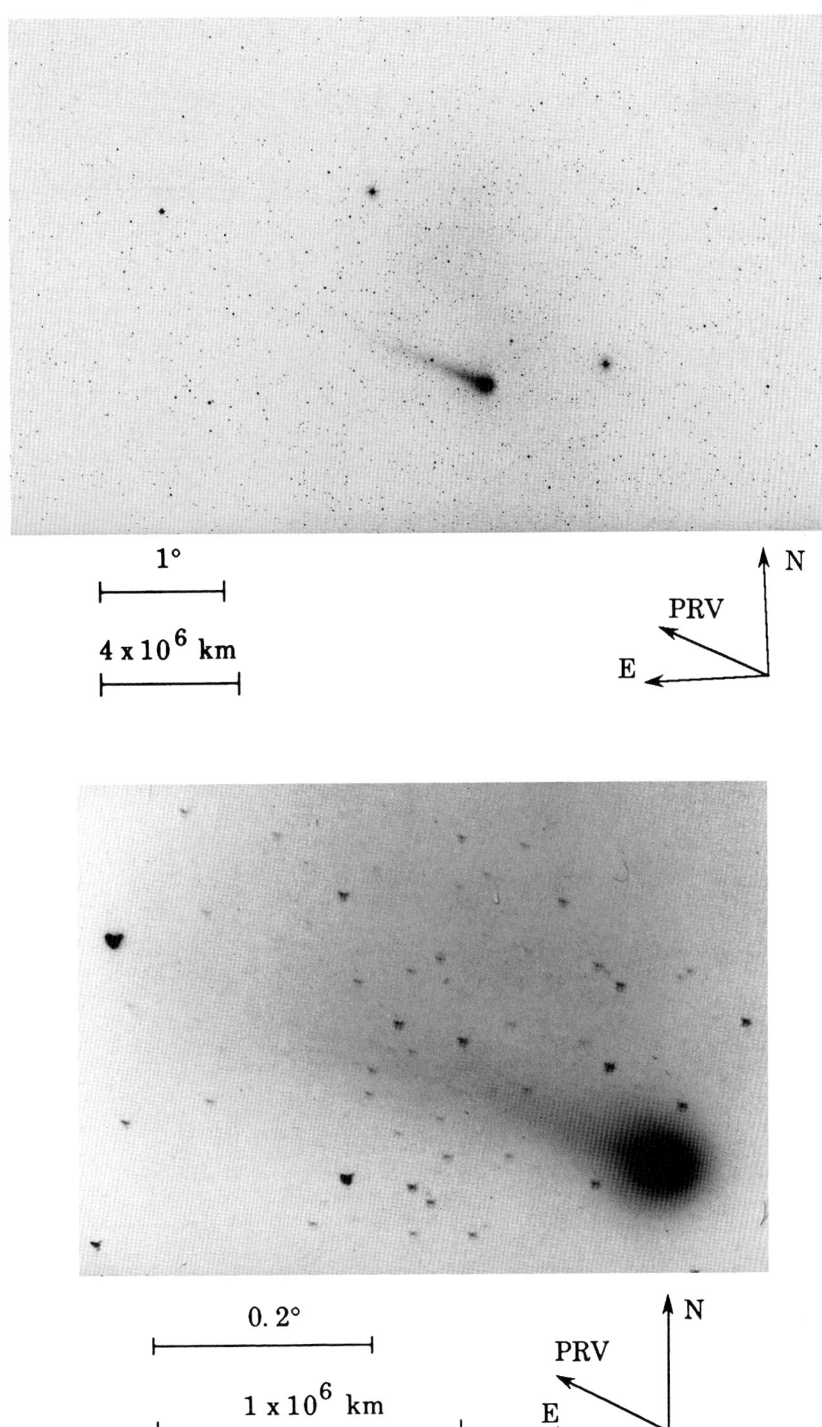

(top) Fig. 134: 1985 Dec 28.89210 UT; exposure 2.3 minutes on Tech. Pan 2415 emulsion; r=1.05, Δ=1.10, β=125.7°; AON-850541.
Photograph by R. Dilsizian, USA.

(bottom) Fig. 135: 1985 Dec 29.05021 UT; three exposures 5.0 minutes each on Tech. Pan 2415 emulsion; r=1.05, Δ=1.10, β=125.7°; AON-850545.
Photograph by R. B. Minton, USA.

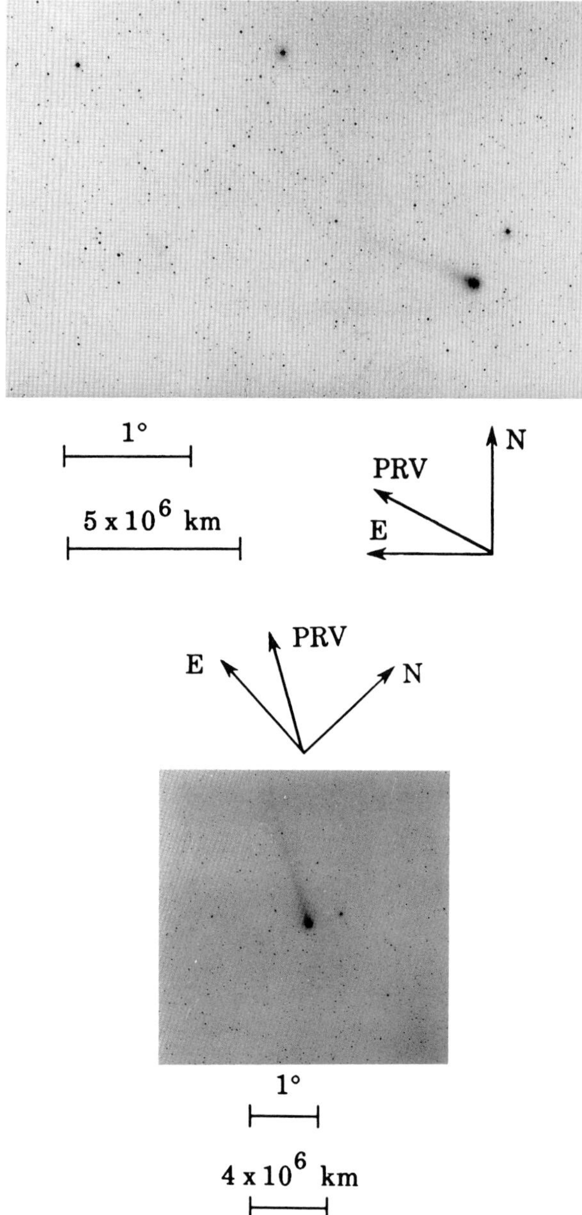

(top) Fig. 136: 1985 Dec 29.97400 UT; exposure 2.0 minutes on Tech Pan 2415 emulsion; r=1.04, Δ=1.12, β=125.9°; AON-850552.
Photograph by R. Dilsizian, USA.

(bottom) Fig. 137: 1985 Dec 30.08438 UT; exposure 3.0 minutes on IIa-O emulsion with no filter; r=1.04, Δ=1.12, β=126.0°; LSPN-126.
Photograph by G. Emerson, E. E. Barnard Observatory.

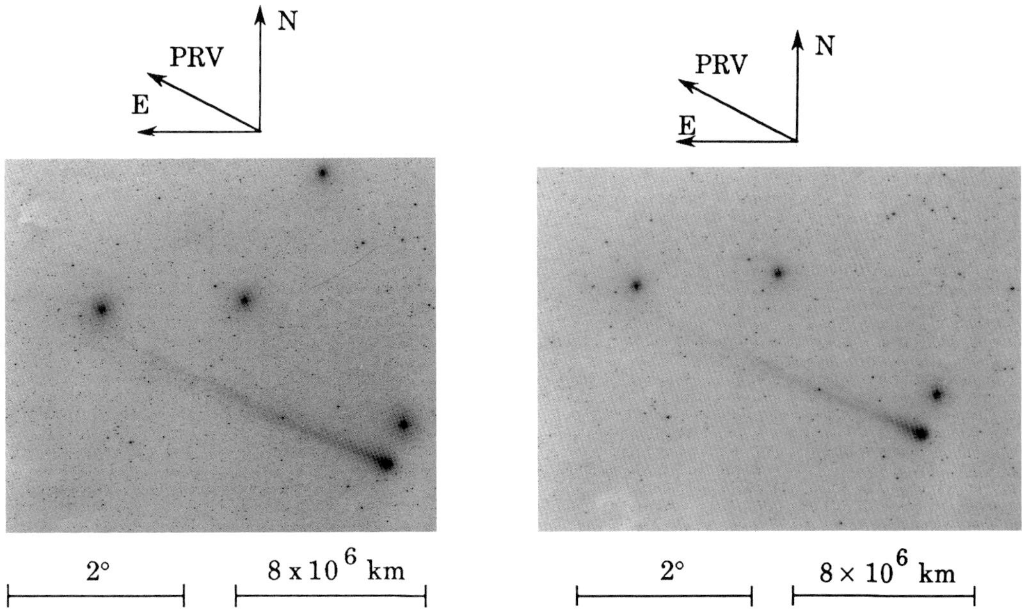

(left) Fig. 138: 1985 Dec 30.09444 UT; exposure 6.0 minutes on IIa-O emulsion with no filter; r=1.04, Δ=1.12, β=126.0°; LSPN-844.
Photograph by E. Moore/E. Marr, Joint Observatory for Cometary Research.

(right) Fig. 139: 1985 Dec 30.09931 UT; exposure 6.0 minutes on IIa-O emulsion with no filter; r=1.04, Δ=1.12, β=126.0°; LSPN-845.
Photograph by E. Moore/E. Marr, Joint Observatory for Cometary Research.

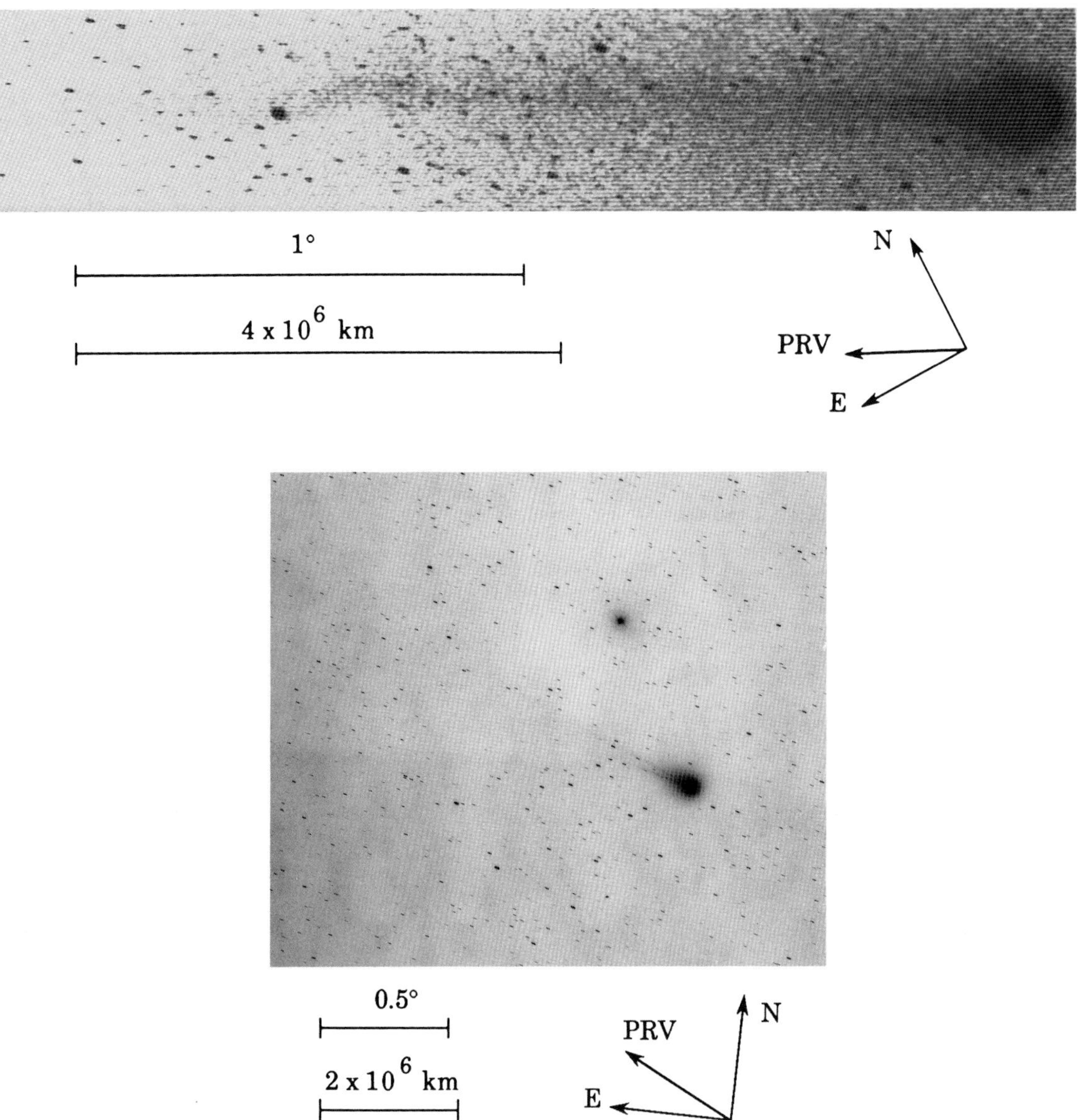

(top) Fig. 140: 1985 Dec 30.58681 UT; exposure 40.0 minutes on ORWO ZU-21 emulsion with no filter; r=1.03, Δ=1.13, β=126.1°; LSPN-3287.
Photograph by S. Gerasimenko, Gissar Observatory.

(bottom) Fig. 141: 1985 Dec 30.80208 UT; exposure 30.0 minutes on IIa-O emulsion with no filter; r=1.03, Δ=1.14, β=126.2°; LSPN-3195.
Photograph by A. Laugier, Observatory of Haute-Provence.

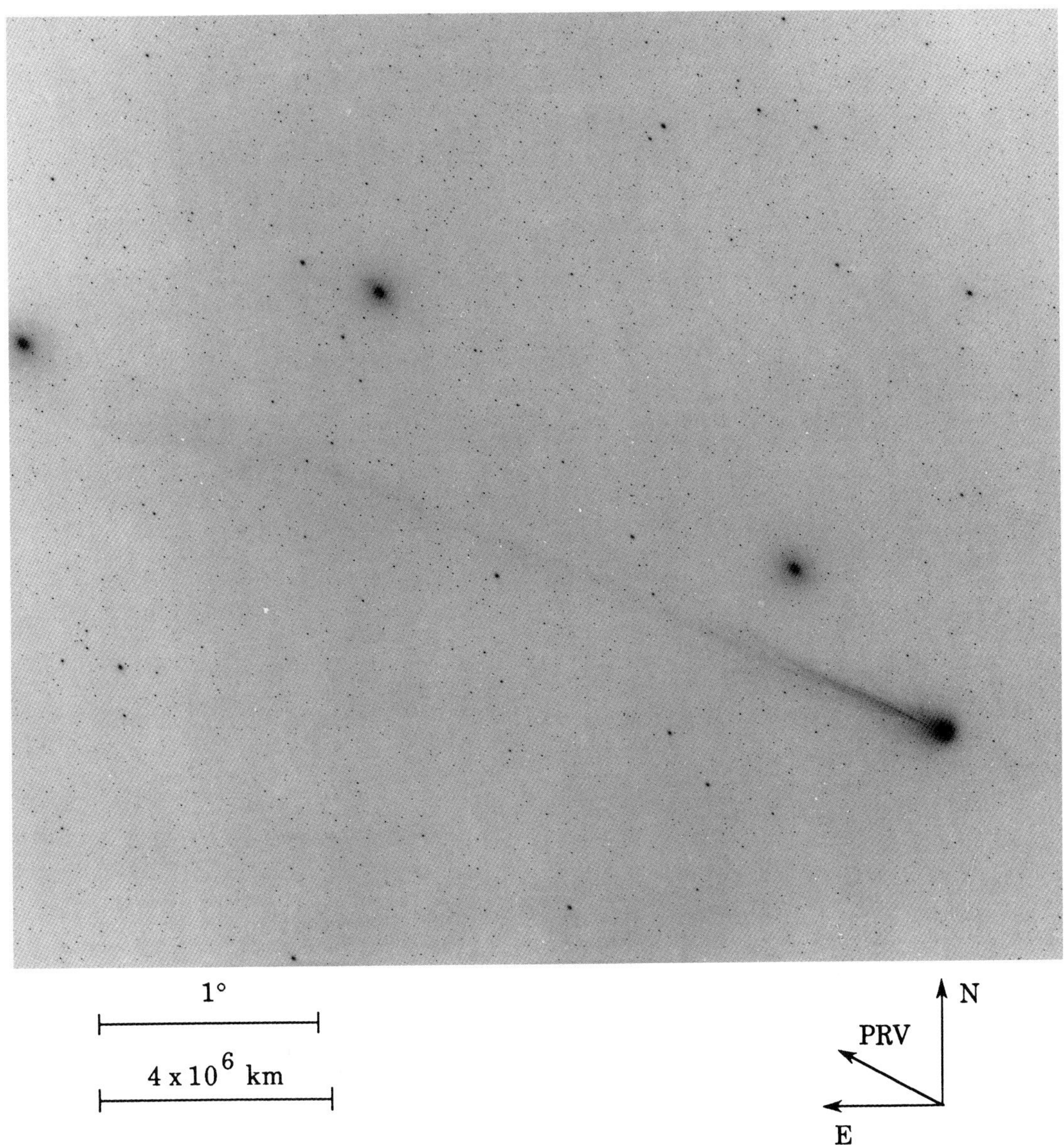

Fig. 142: 1985 Dec 31.38922 UT; exposure 10.0 minutes on IIa-O emulsion with no filter; r=1.02, Δ=1.15, β=126.4°; LSPN-1620.
Photograph by H. Maehara, Kiso Observatory.

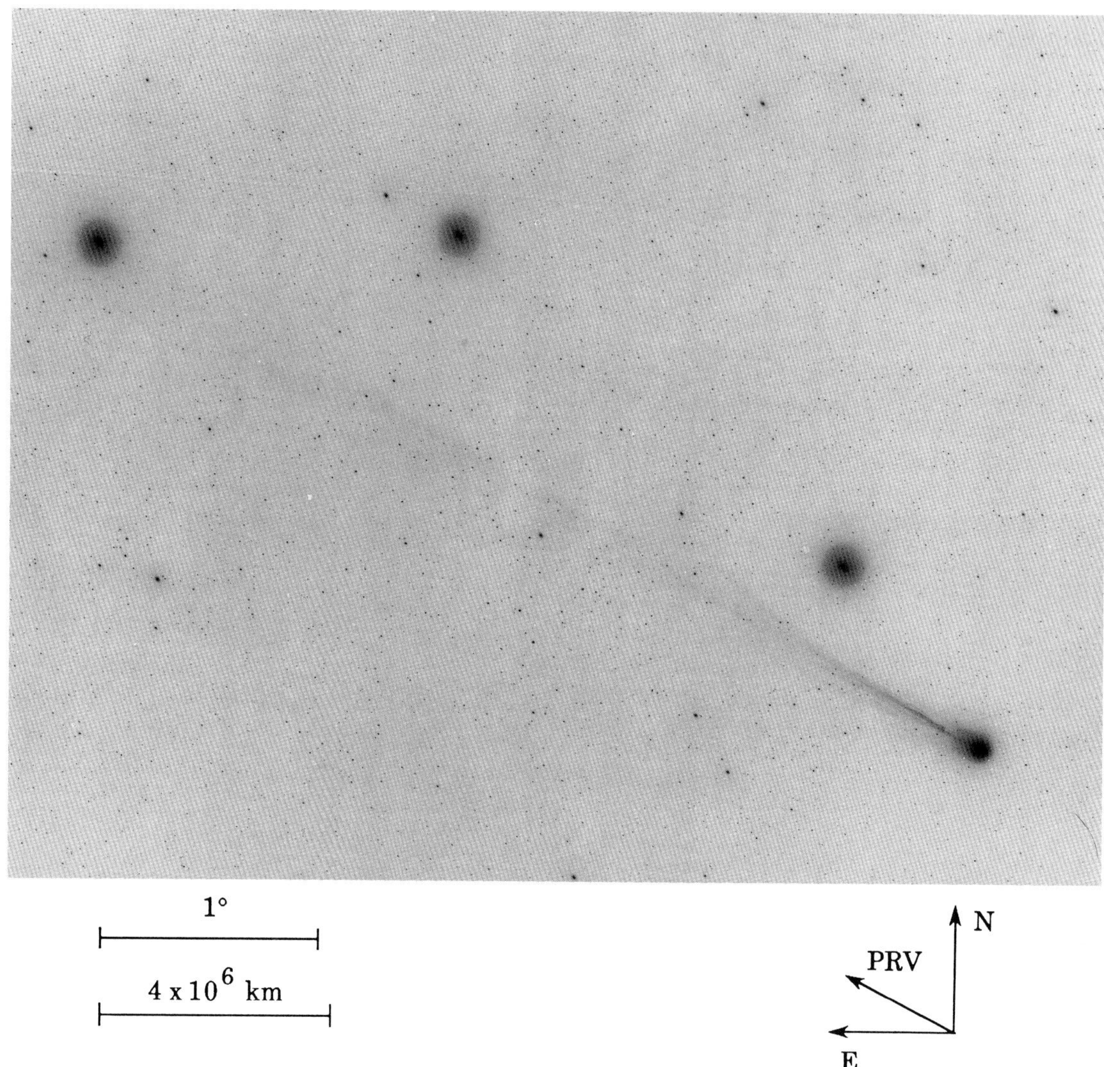

Fig. 143: 1985 Dec 31.40644 UT; exposure 10.0 minutes on IIa-O emulsion with no filter; r=1.02, Δ=1.15, β=126.4°; LSPN-1621.
Photograph by H. Maehara, Kiso Observatory.

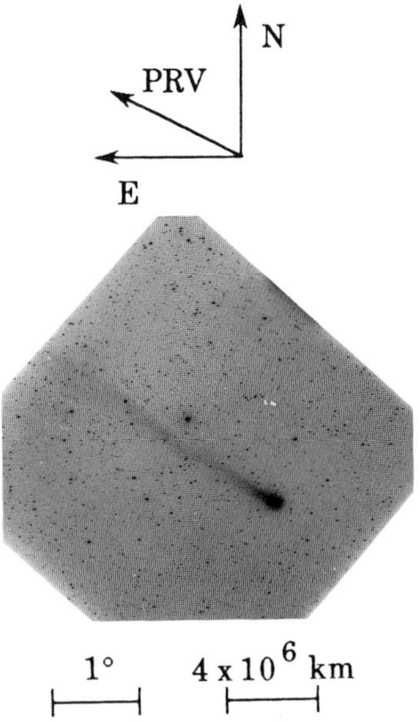

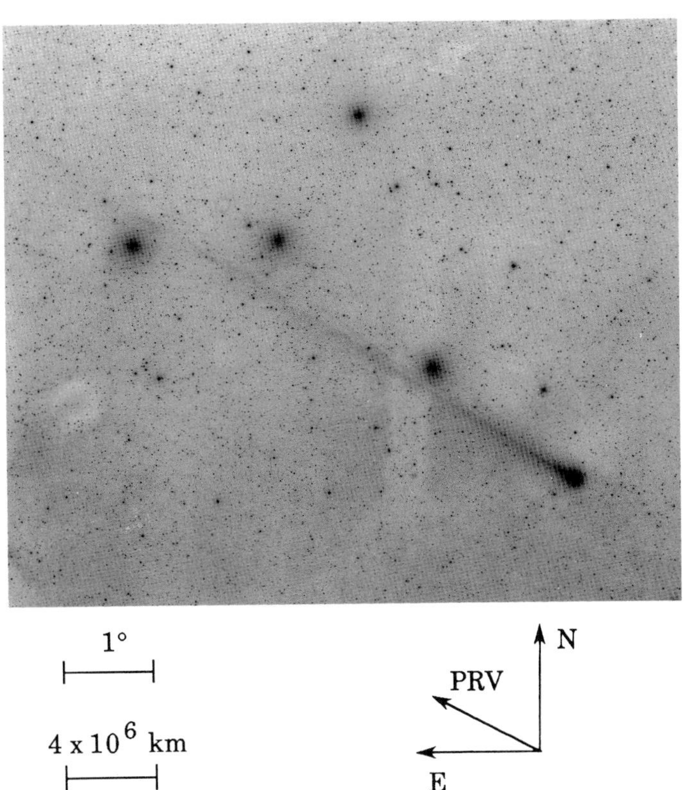

(top) Fig. 144: 1986 Jan 1.07153 UT; exposure 6.0 minutes on IIa-O emulsion with no filter; r=1.01, Δ=1.16, β=126.6°; LSPN-127.
Photograph by G. Emerson, E. E. Barnard Observatory.

(bottom) Fig. 145: 1986 Jan 2.11458 UT; exposure 10.0 minutes on IIa-O emulsion with no filter; r=0.99, Δ=1.18, β=127.0°; LSPN-850.
Photograph by E. Moore/E. Marr, Joint Observatory for Cometary Research.

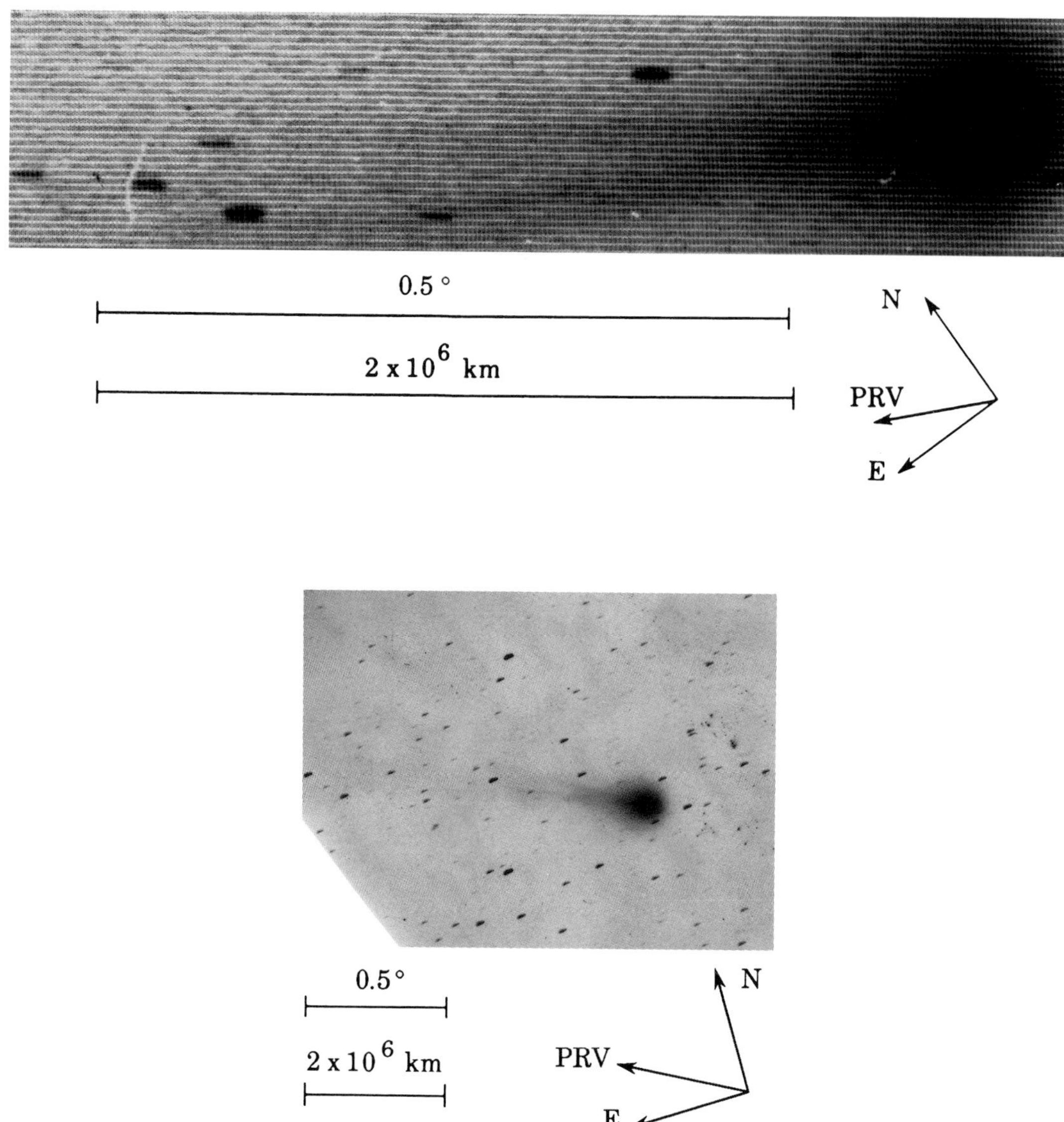

(top) Fig. 146: 1986 Jan 2.58966 UT; exposure 50.0 minutes on ORWO ZU-21 emulsion with no filter; r=0.98, Δ=1.19, β=127.2°; LSPN-3266.
Photograph by S. Gerasimenko, Gissar Observatory.

(bottom) Fig. 147: 1986 Jan 2.60694 UT; exposure 52.0 minutes on 103a-O emulsion with no filter; r=0.98, Δ=1.19, β=127.2°; LSPN-3615.
Photograph submitted by K. Sivaraman, Indian Institute for Astrophysics, Kodaikanal Station.

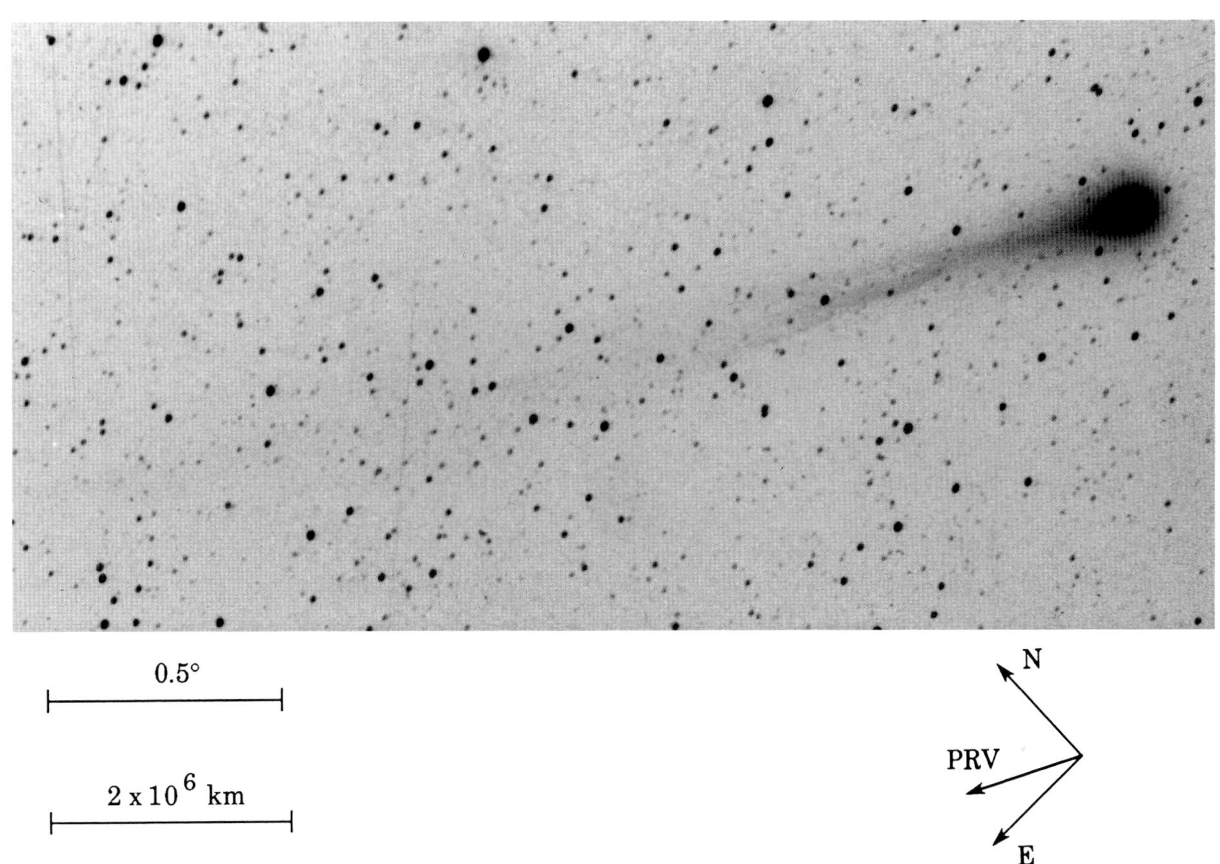

Fig. 148: 1986 Jan 2.80764 UT; exposure 10.0 minutes on Tech. Pan 2415 emulsion with no filter; r=0.98, Δ=1.19, β=127.3°; LSPN-1779.
Photograph by F. Van Wyk, Sutherland, LSPN Island Network.

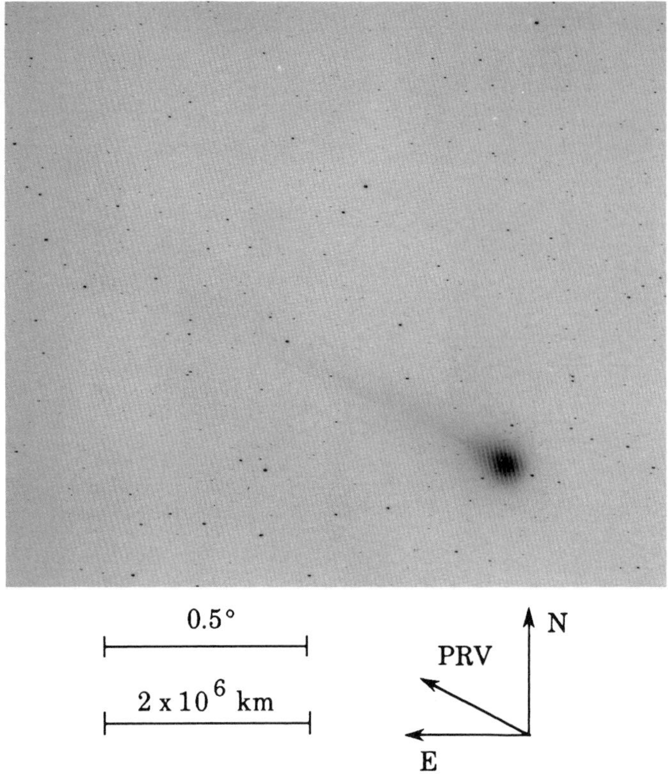

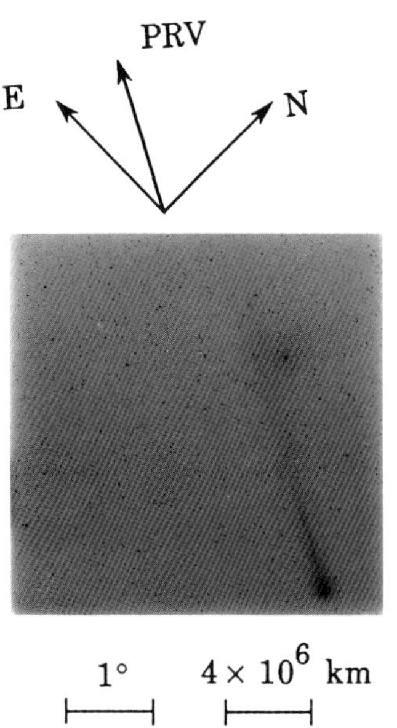

(top) Fig. 149*: 1986 Jan 2.71615 UT; exposure 7.0 minutes on ORWO ZU-21 emulsion with GG-13 filter; r=0.98, Δ=1.19, β=127.3°; LSPN-3462.
Photograph by F. Börngen/F. Ludwig, Karl Schwarzschild Observatory.

(bottom) Fig. 150: 1986 Jan 3.05694 UT; exposure 6.0 minutes on IIa-O emulsion with no filter; r=0.98, Δ=1.20, β=127.4°; LSPN-128.
Photograph by G. Emerson, E. E. Barnard Observatory.

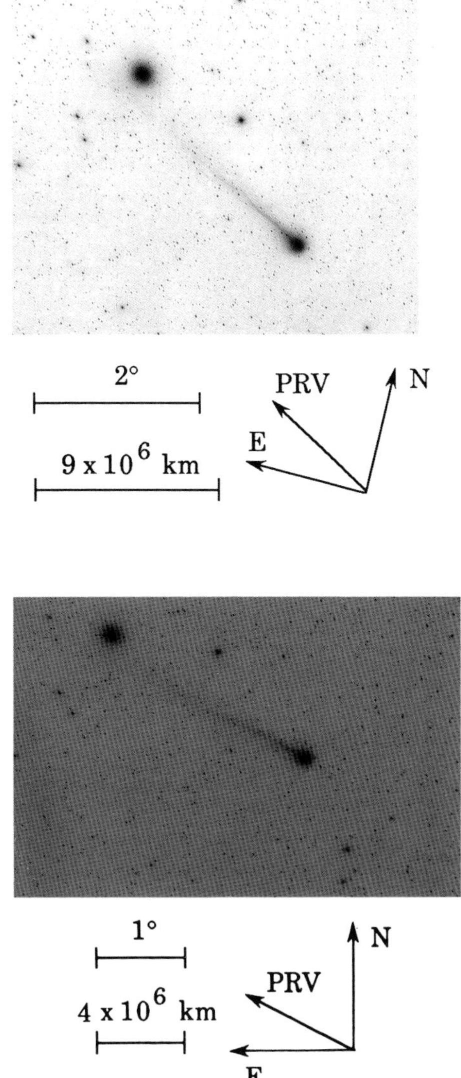

(top) Fig. 151: 1986 Jan 3.23055 UT; exposure 30.0 minutes on IIIa-J emulsion with no filter; r=0.97, Δ=1.20, β=127.5°; LSPN-1872.
Photograph by D. Cruikshank, Mauna Kea Observatory.

(bottom) Fig. 152: 1986 Jan 3.26493 UT; exposure 30.0 minutes on IIIa-J emulsion with no filter; r=0.97, Δ=1.20, β=127.5°; LSPN-1873.
Photograph by D. Cruikshank, Mauna Kea Observatory.

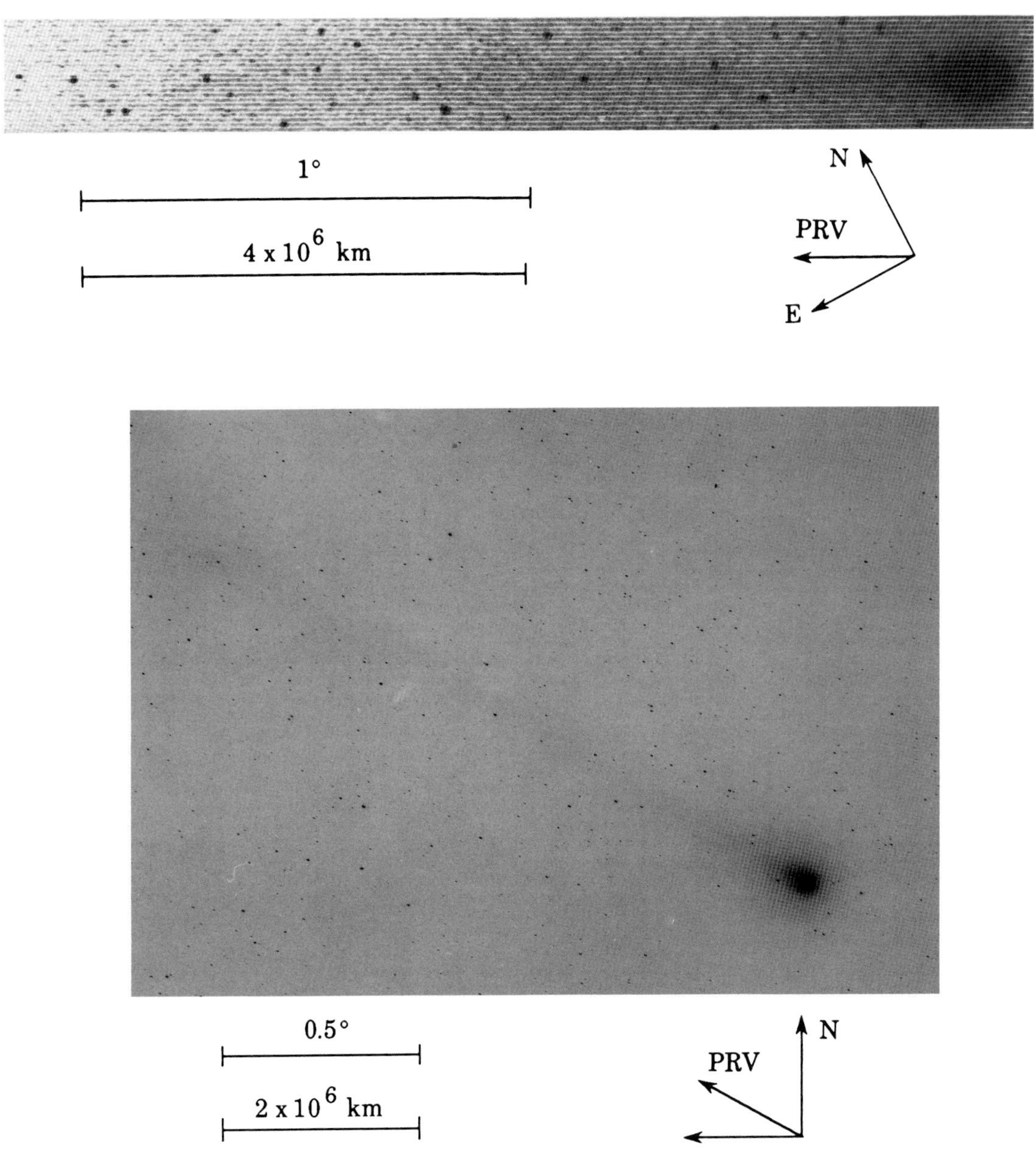

(top) Fig. 153: 1986 Jan 3.62015 UT; exposure 15.0 minutes on ORWO ZU-21 emulsion with no filter; r=0.97, Δ=1.21, β=127.7°; LSPN-3288.
Photograph by N. Kiselev, Gissar Observatory.

(bottom) Fig. 154: 1986 Jan 3.71528 UT; exposure 10.0 minutes on ORWO ZU-21 emulsion with GG-13 filter; r=0.97, Δ=1.21, β=127.8°; LSPN-3463.
Photograph by F. Börngen/F. Ludwig, Karl Schwarzschild Observatory.

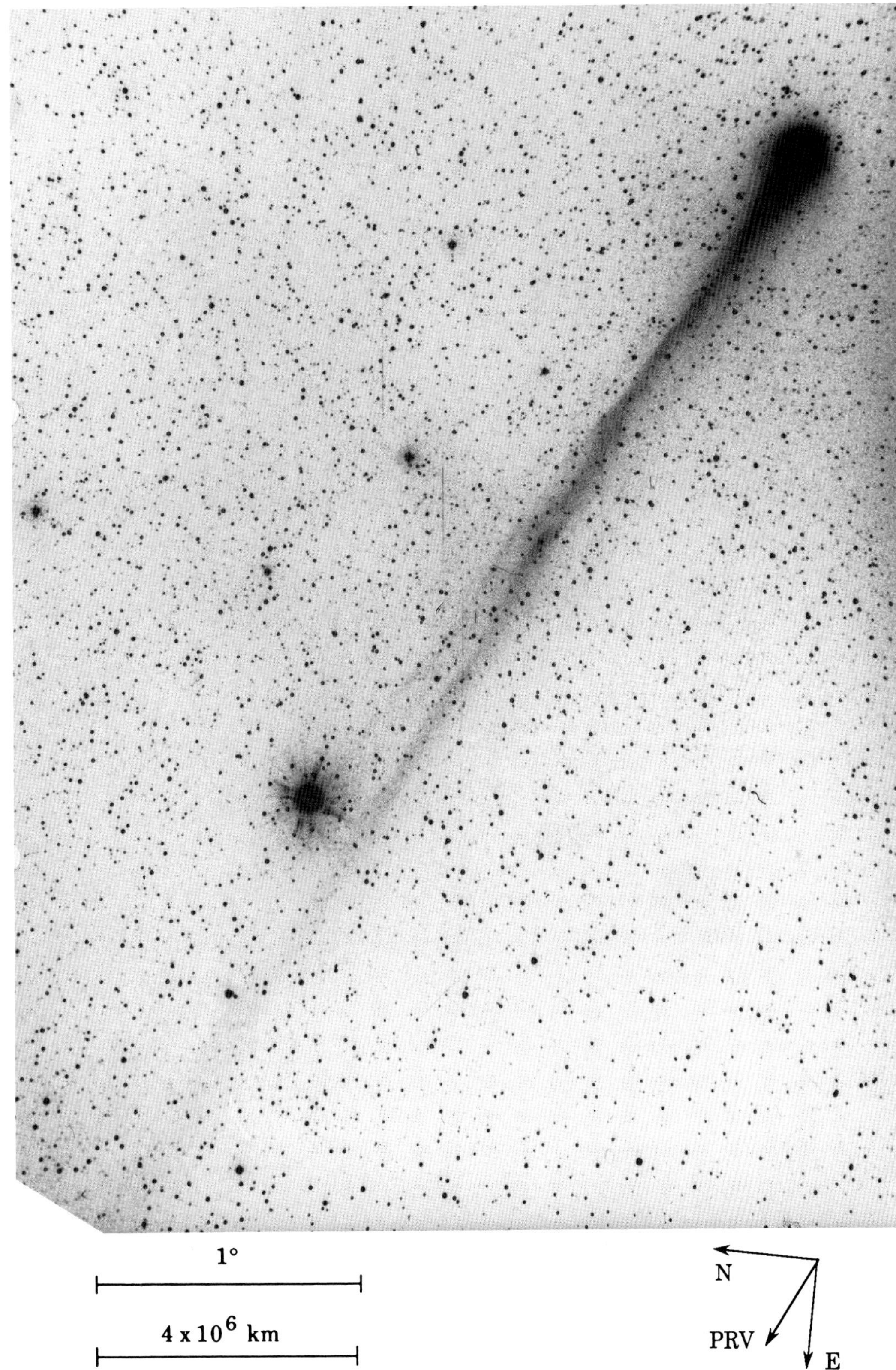

Fig. 155: 1986 Jan 3.71790 UT; two exposures 3.0 minutes each on Tech. Pan 2415 emulsion; r=0.97, Δ=1.21, β=127.8°; AON-850595.
Photograph by M. Jäger, Austria.

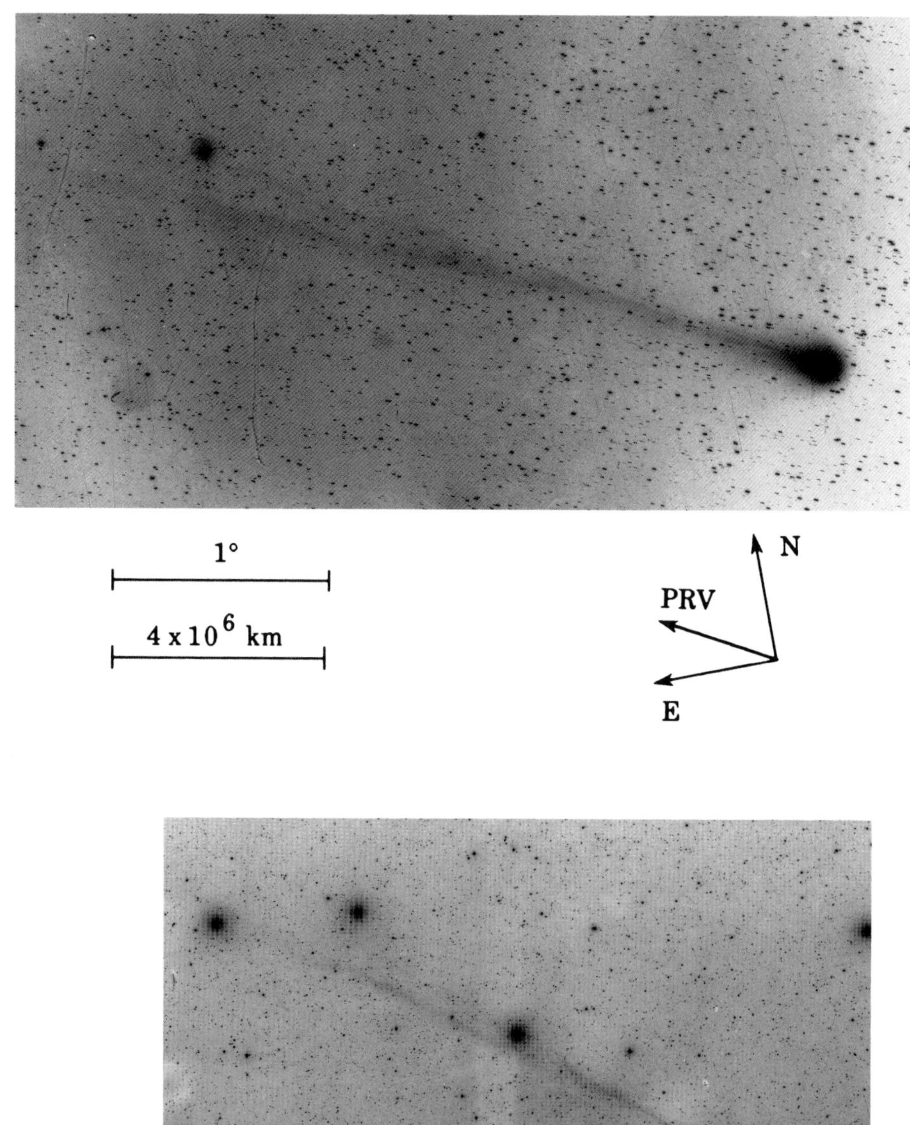

(top) Fig. 156: 1986 Jan 3.75069 UT; exposure 30.0 minutes on IIIa-J emulsion with no filter; r=0.96, Δ=1.21, β=127.8°; LSPN-2491.
Photograph by V. Ivanova et al., Bulgarian National Observatory.

(bottom) Fig. 157: 1986 Jan 4.09375 UT; exposure 10.0 minutes on IIa-O emulsion with no filter; r=0.96, Δ=1.22, β=128.0°; LSPN-855.
Photograph by E. Moore et al., Joint Observatory for Cometary Research.

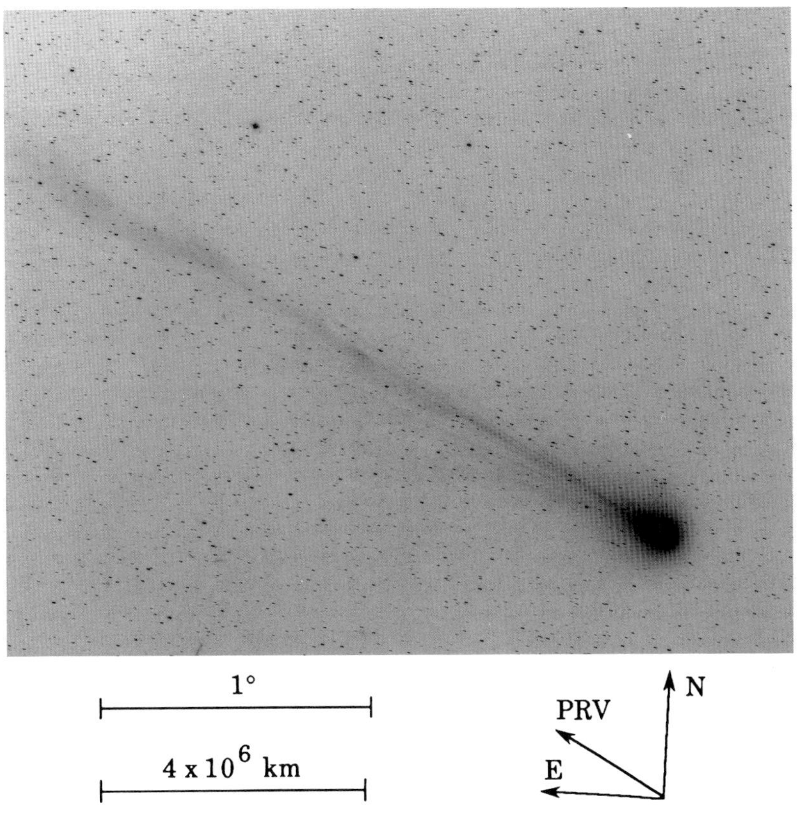

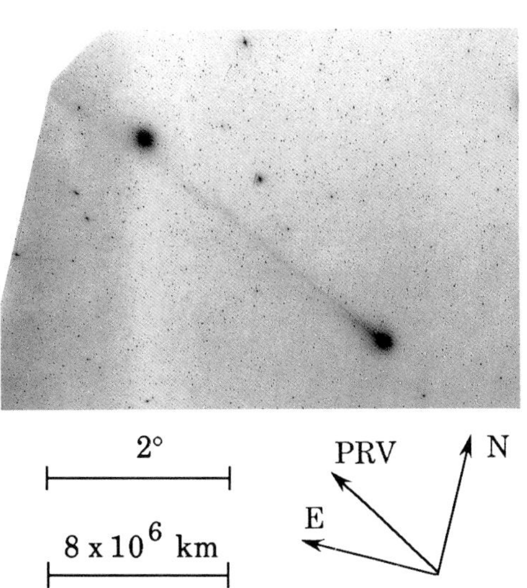

(top) Fig. 158: 1986 Jan 4.11250 UT; exposure 20.0 minutes on IIa-O emulsion with no filter; r=0.96, Δ=1.22, β=128.0°; LSPN-2584.
Photograph by R. Hill, Warner and Swasey Observatory.

(bottom) Fig. 159: 1986 Jan 4.21215 UT; exposure 15.0 minutes on IIIa-J emulsion with no filter; r=0.96, Δ=1.22, β=128.0°; LSPN-1875.
Photograph by D. Cruikshank, Mauna Kea Observatory.

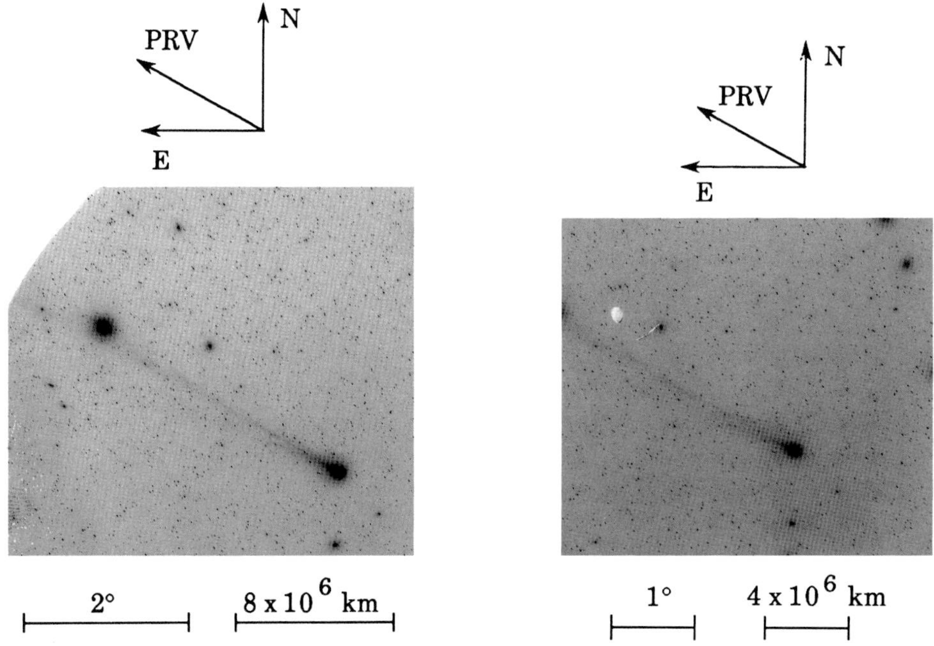

(left) Fig. 160: 1986 Jan 4.23351 UT; exposure 30.0 minutes on IIIa-J emulsion with no filter; r=0.96, Δ=1.22, β=128.0°; LSPN-1874.
Photograph by D. Cruikshank, Mauna Kea Observatory.

(right) Fig. 161: 1986 Jan 4.26112 UT; exposure 30.0 minutes on IIIa-J emulsion with no filter; r=0.96, Δ=1.22, β=128.1°; LSPN-1876.
Photograph by D. Cruikshank, Mauna Kea Observatory.

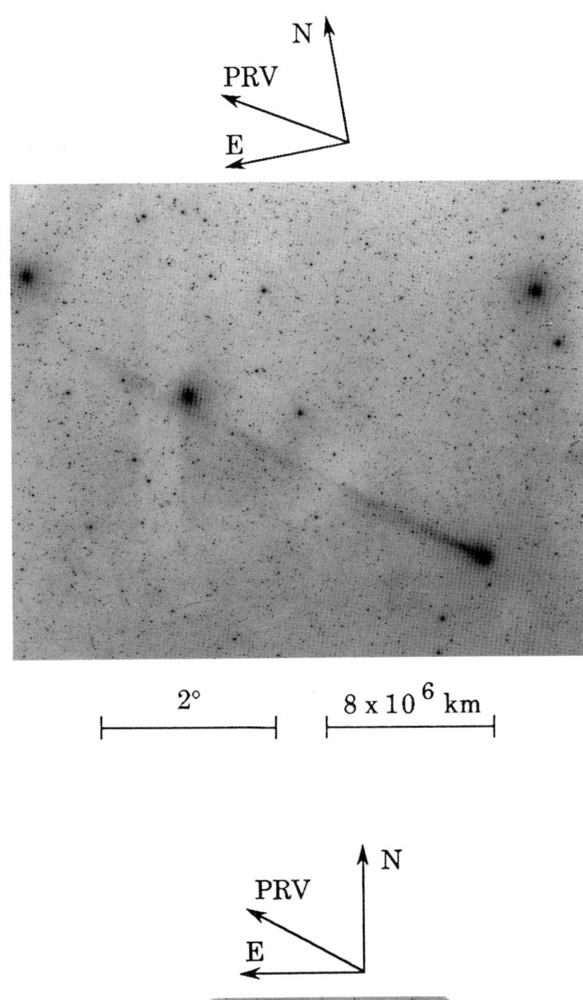

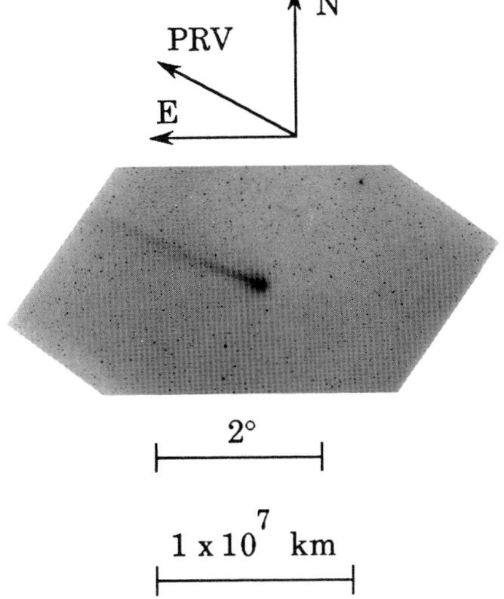

(top) Fig. 162: 1986 Jan 5.09306 UT; exposure 10.0 minutes on IIa-O emulsion with no filter; r=0.94, Δ=1.24, β=128.5°; LSPN-863.
Photograph by E. Moore et al., Joint Observatory for Cometary Research.

(bottom) Fig. 163: 1986 Jan 5.09583 UT; exposure 10.0 minutes on IIa-O emulsion with no filter; r=0.94, Δ=1.24, β=128.5°; LSPN-129.
Photograph by G. Emerson, E. E. Barnard Observatory.

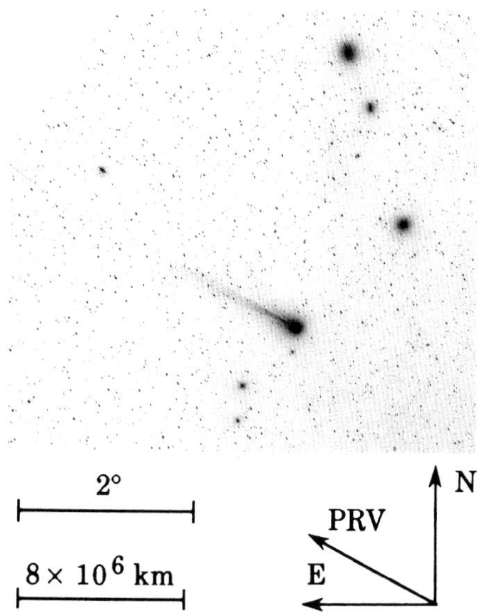

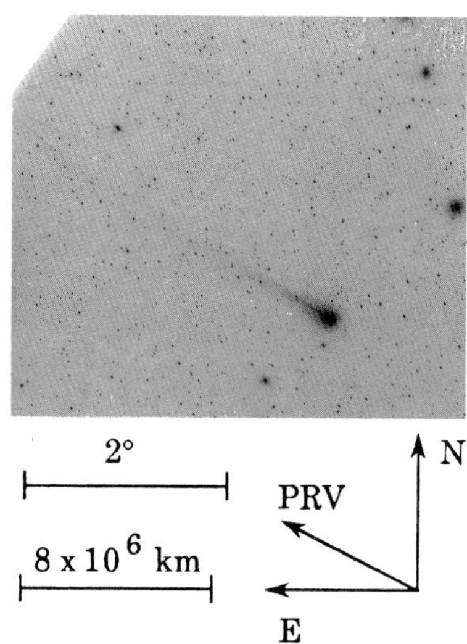

(top) Fig. 164: 1986 Jan 5.23594 UT; exposure 30.0 minutes on IIIa-J emulsion with no filter; r=0.94, Δ=1.24, β=128.6°; LSPN-1877.
Photograph by D. Cruikshank, Mauna Kea Observatory.

(bottom) Fig. 165: 1986 Jan 5.26146 UT; exposure 30.0 minutes on IIIa-J emulsion with no filter; r=0.94, Δ=1.24, β=128.6°; LSPN-1878.
Photograph by D. Cruikshank, Mauna Kea Observatory.

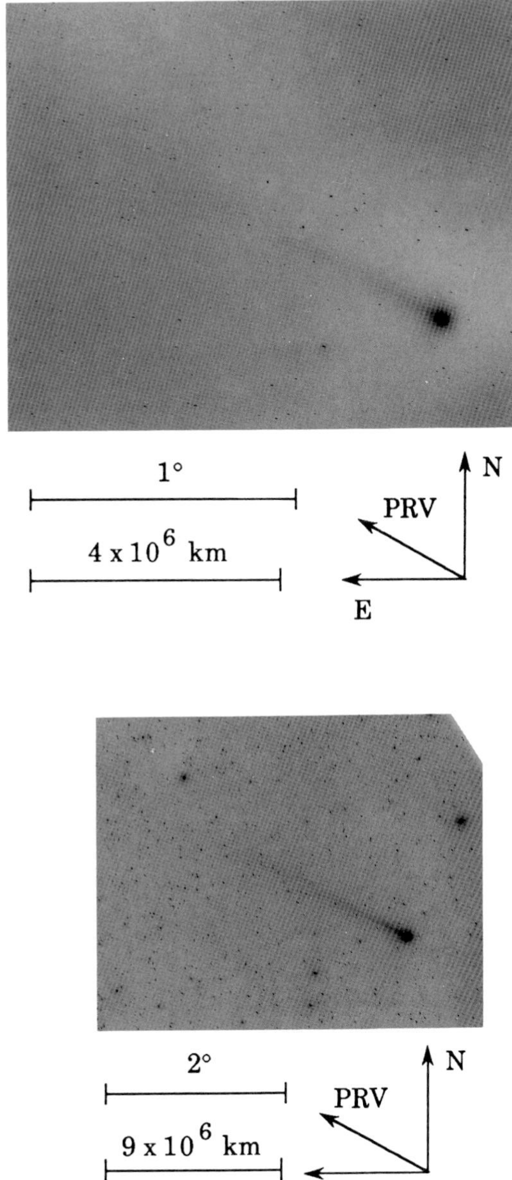

(top) Fig. 166: 1986 Jan 6.00907 UT; exposure 10.0 minutes on IIa-O emulsion with GG-13 filter; r=0.93, Δ=1.25, β=129.1°; LSPN-137.
Photograph by A. Heiser, A. J. Dyer Observatory.

(bottom) Fig. 167: 1986 Jan 6.25417 UT; exposure 30.0 minutes on IIIa-J emulsion with no filter; r=0.93, Δ=1.26, β=129.3°; LSPN-1879.
Photograph by D. Cruikshank, Mauna Kea Observatory.

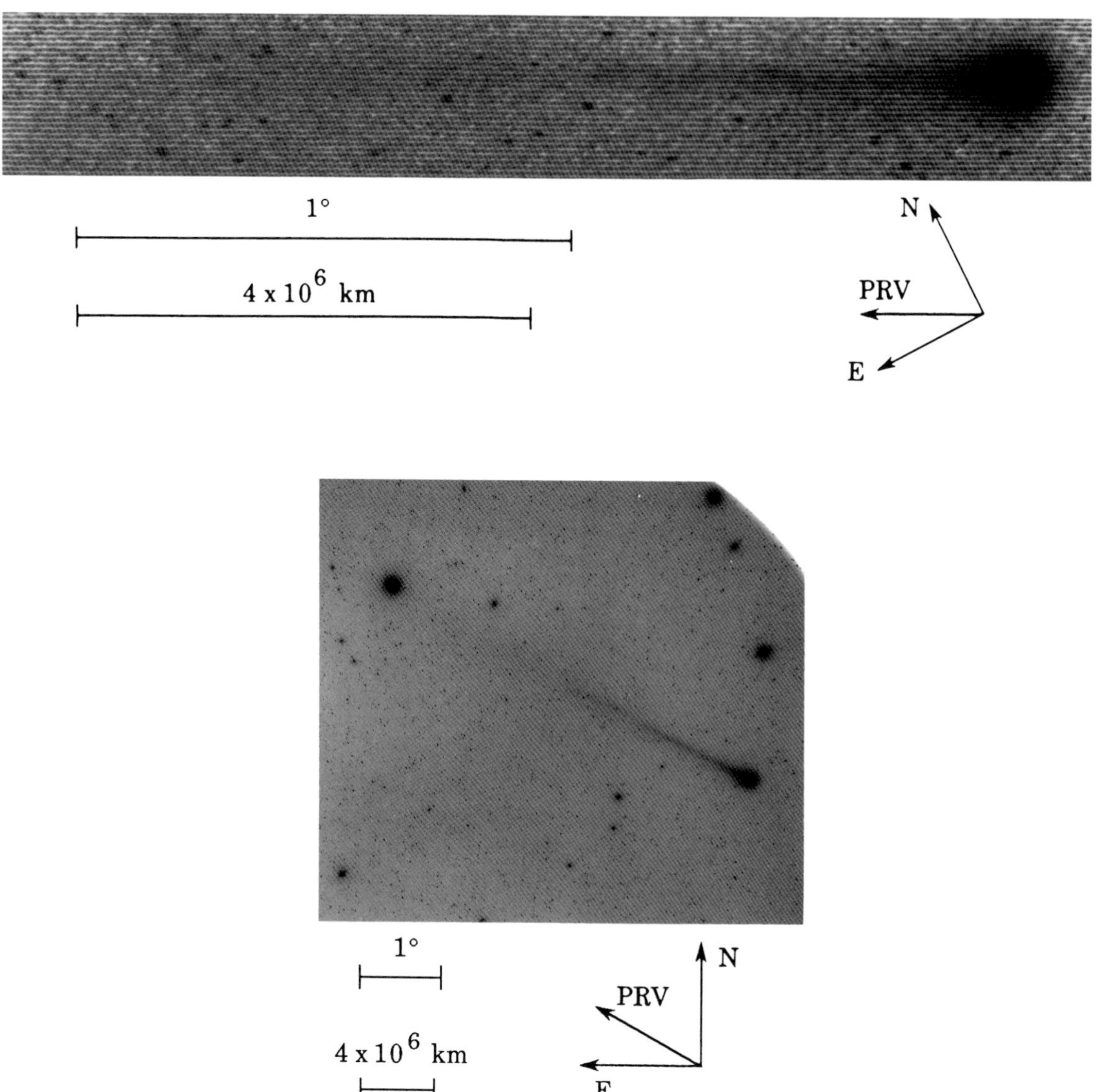

(top) Fig. 168: 1986 Jan 6.58683 UT; exposure 40.0 minutes on ORWO ZU-21 emulsion with no filter; r=0.92, Δ=1.26, β=129.5°; LSPN-3290.
Photograph by N. Kiselev, Gissar Observatory.

(bottom) Fig. 169: 1986 Jan 7.22899 UT; exposure 30.0 minutes on IIIa-J emulsion with no filter; r=0.91, Δ=1.28, β=129.9°; LSPN-1881.
Photograph by D. Cruikshank, Mauna Kea Observatory.

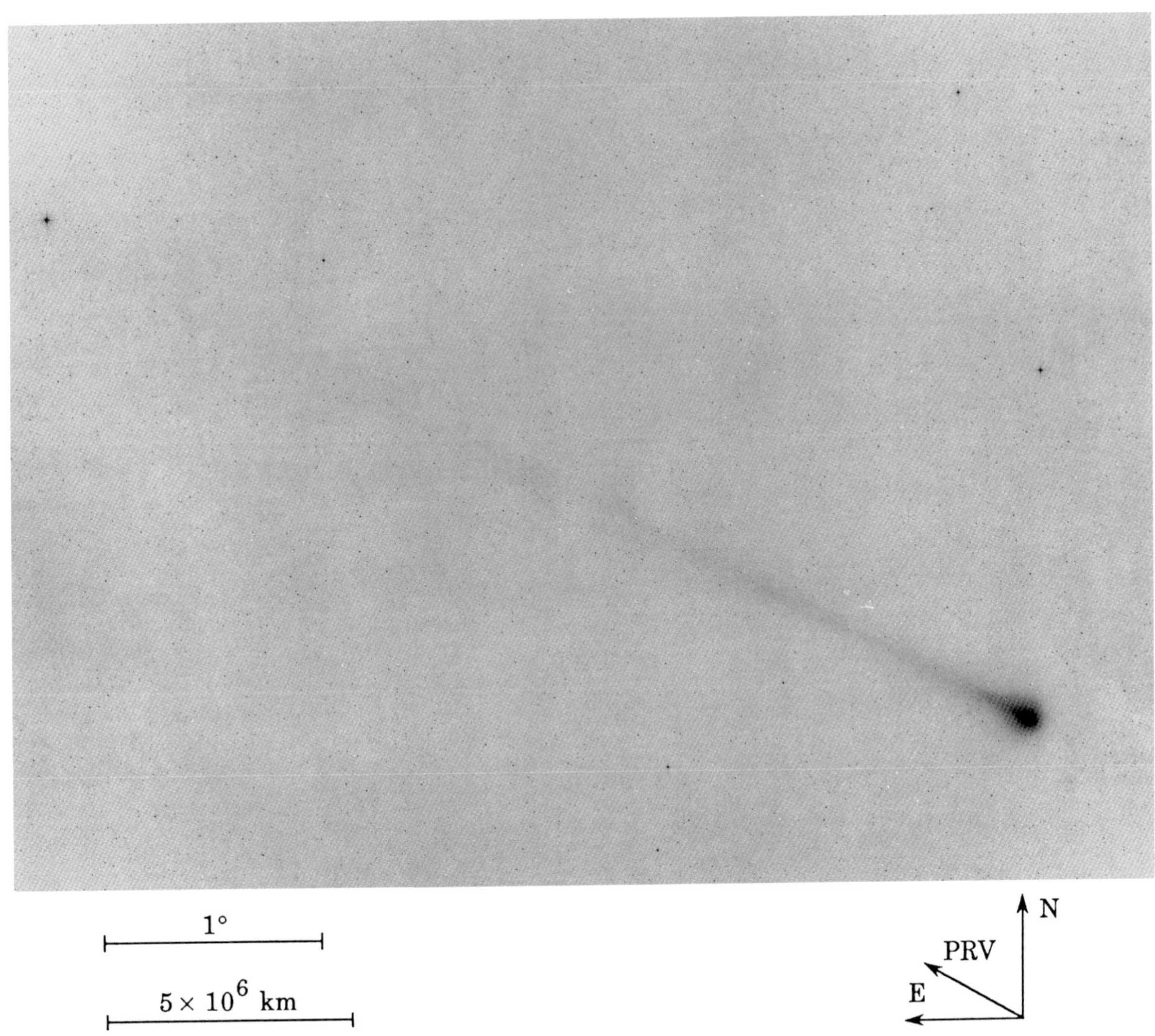

Fig. 170: 1986 Jan 7.38760 UT; exposure 10.0 minutes on IIa-O emulsion with no filter; r=0.91, Δ=1.28, β=130.0°; LSPN-1624.
Photograph by H. Maehara, Kiso Observatory.

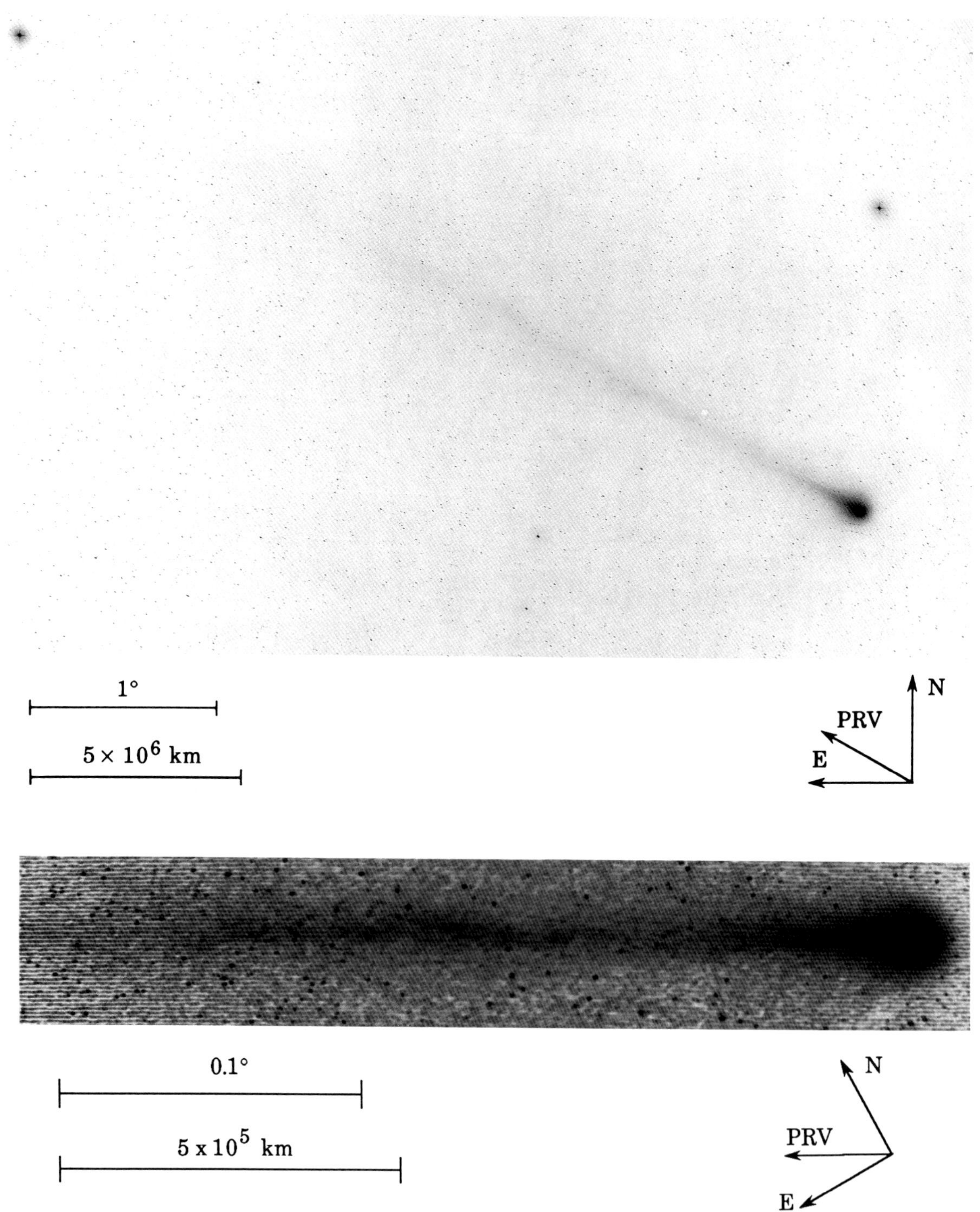

(top) Fig. 171: 1986 Jan 7.40412 UT; exposure 20.0 minutes on IIa-O emulsion with no filter; r=0.91, Δ=1.28, β=130.0°; LSPN-1625.
Photograph by H. Maehara, Kiso Observatory.

(bottom) Fig. 172: 1986 Jan 7.55127 UT; exposure 10.0 minutes on ORWO ZU-21 emulsion with no filter; r=0.91, Δ=1.28, β=130.1°; LSPN-3369.
Photograph by D. Gorodetskij, Mountain Observatory.

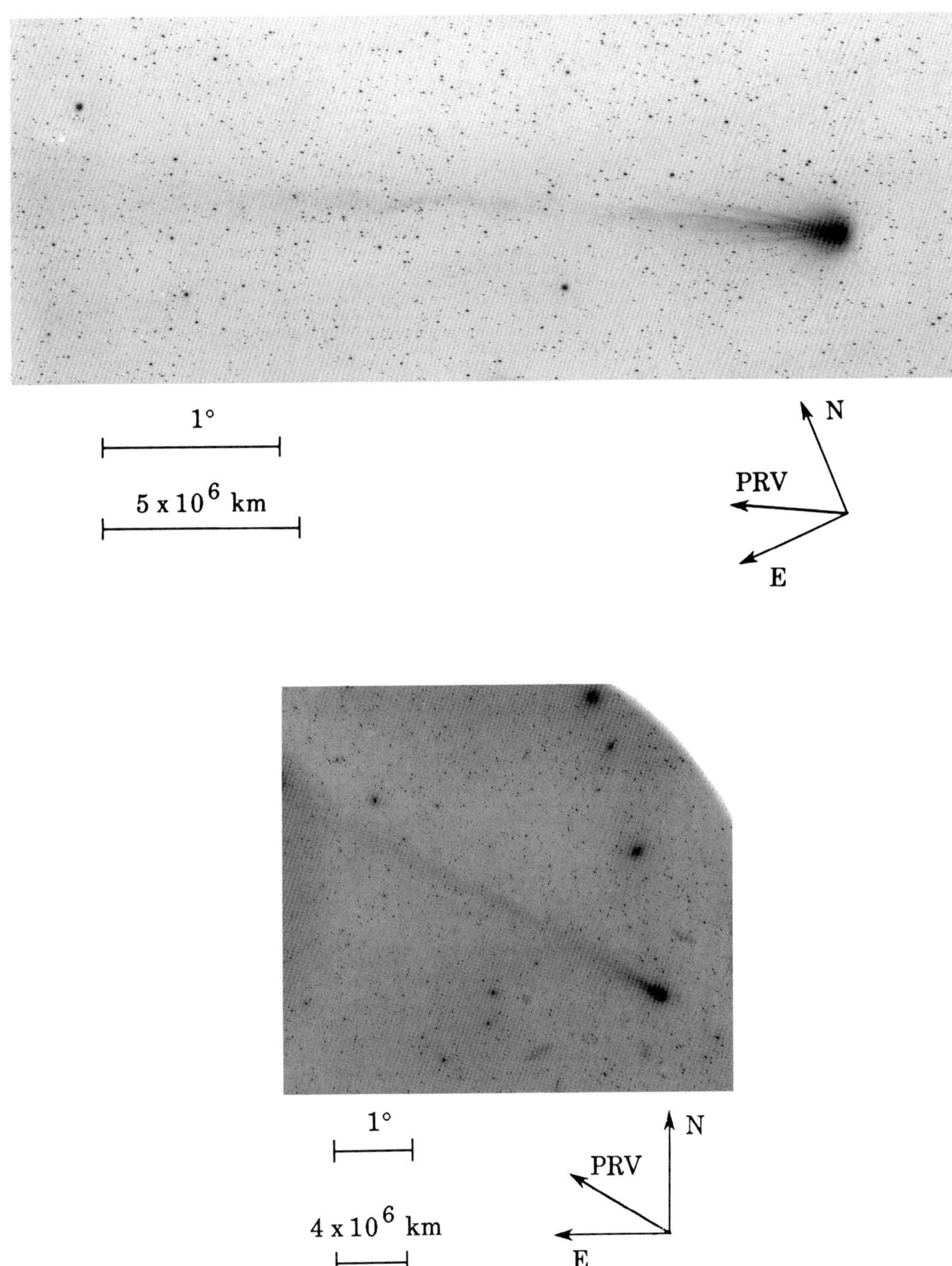

(top) Fig. 173: 1986 Jan 8.10764 UT; exposure 30.0 minutes on 103a-O emulsion with no filter; r=0.90, Δ=1.29, β=130.5°; LSPN-3571.
Photograph by H. Giclas, Lowell Observatory.

(bottom) Fig. 174: 1986 Jan 8.23023 UT; exposure 30.0 minutes on IIIa-J emulsion with no filter; r=0.90, Δ=1.29, β=130.6°; LSPN-1883.
Photograph by D. Cruikshank, Mauna Kea Observatory.

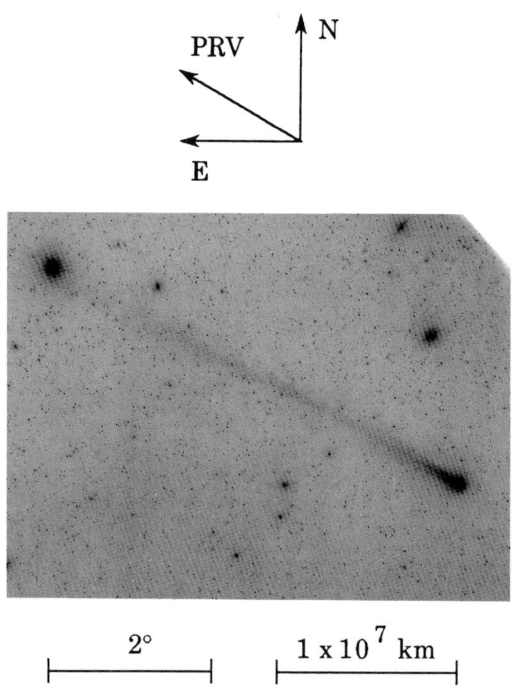

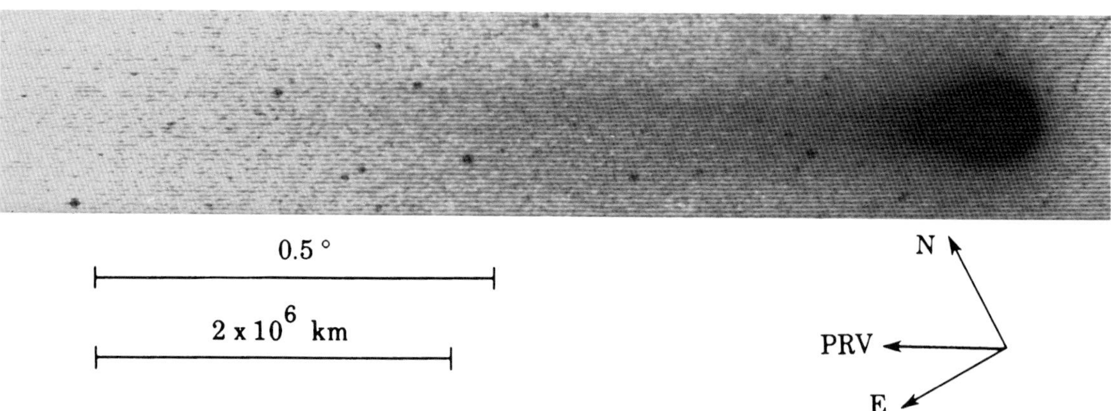

(top) Fig. 175: 1986 Jan 8.26235 UT; exposure 30.0 minutes on IIIa-J emulsion with no filter; r=0.90, Δ=1.29, β=130.7°; LSPN-1882.
Photograph by D. Cruikshank, Mauna Kea Observatory.

(bottom) Fig. 176: 1986 Jan 8.58333 UT; exposure 20.0 minutes on ORWO ZU-21 emulsion with no filter; r=0.89, Δ=1.30, β=130.9°; LSPN-3292.
Photograph by N. Kiselev, Gissar Observatory.

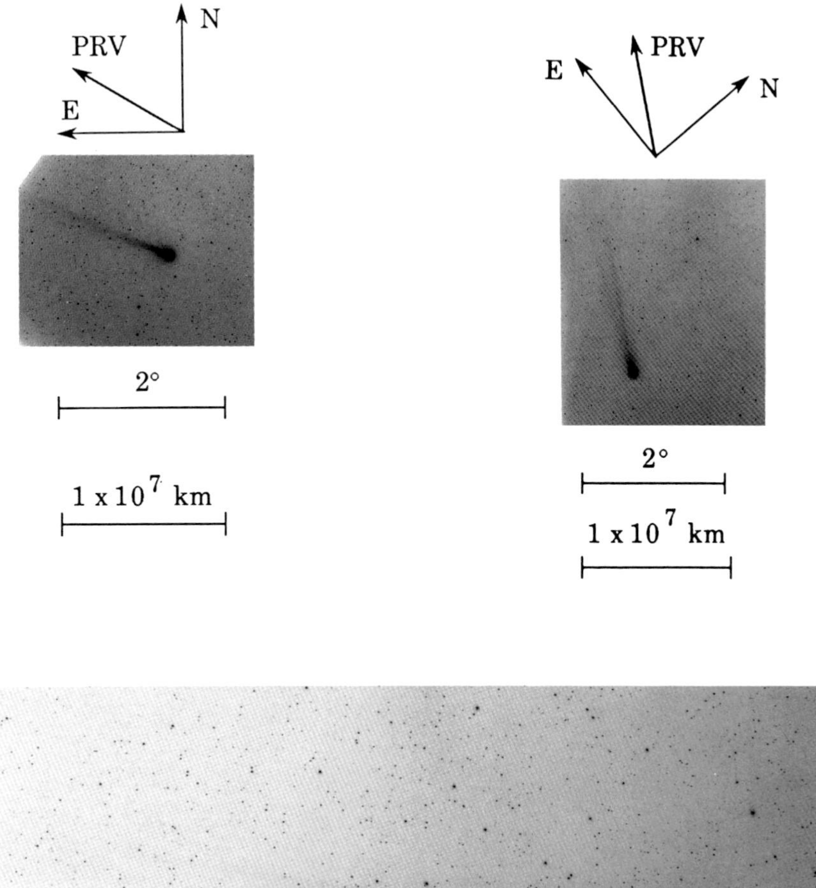

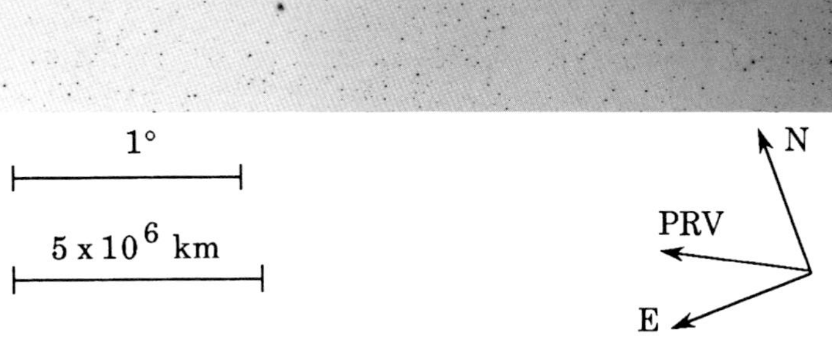

(left) Fig. 177: 1986 Jan 9.04896 UT; exposure 7.0 minutes on IIa-O emulsion; r=0.89, Δ=1.31, β=131.3°; LSPN-130.
Photograph by G. Emerson, E. E. Barnard Observatory.

(right) Fig. 178: 1986 Jan 9.06250 UT; exposure 7.0 minutes on IIa-O emulsion with no filter; r=0.89, Δ=1.31, β=131.3°; LSPN-131.
Photograph by G. Emerson, E. E. Barnard Observatory.

(bottom) Fig. 179: 1986 Jan 9.08194 UT; exposure 20.0 minutes on 103a-O emulsion with no filter; r=0.89, Δ=1.31, β=131.3°; LSPN-3572.
Photograph by H. Giclas, Lowell Observatory.

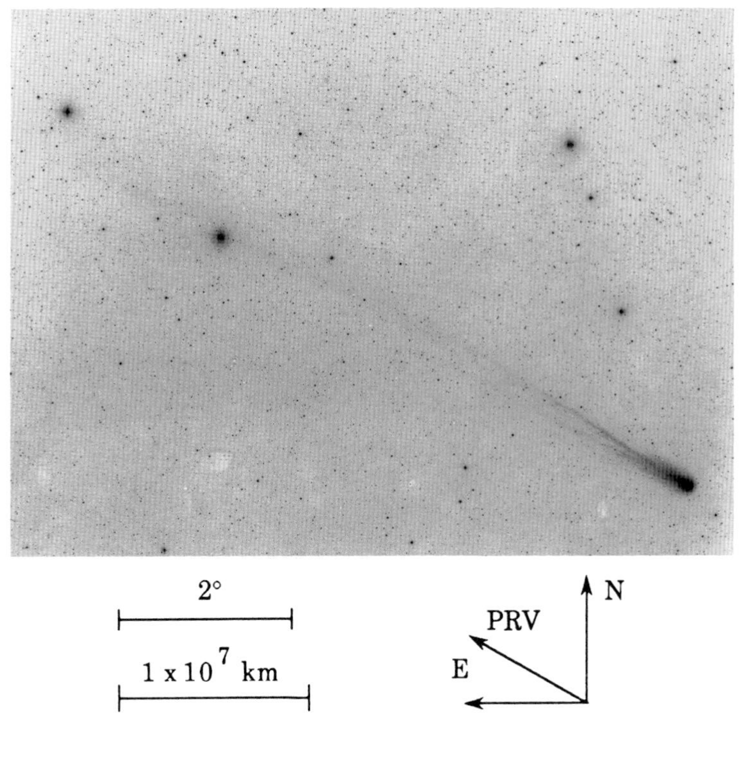

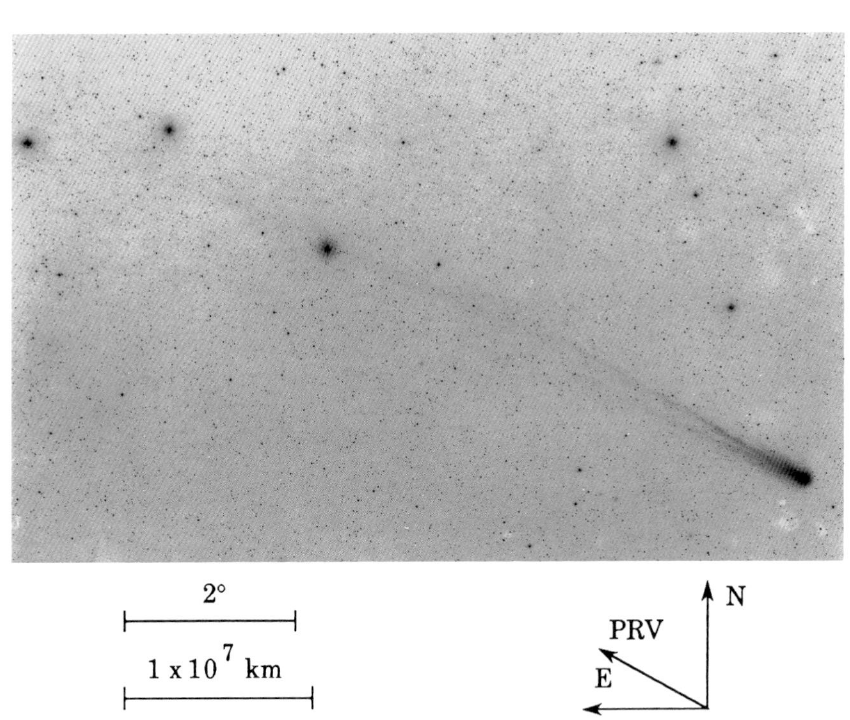

(top) Fig. 180: 1986 Jan 9.10625 UT; exposure 10.0 minutes on IIa-O emulsion with no filter; r=0.88, Δ=1.31, β=131.3°; LSPN-2088.
Photograph by E. Moore/E. Marr, Joint Observatory for Cometary Research.

(bottom) Fig. 181: 1986 Jan 9.11458 UT; exposure 10.0 minutes on IIa-O emulsion with no filter; r=0.88, Δ=1.31, β=131.3°; LSPN-2089.
Photograph by E. Moore/E. Marr, Joint Observatory for Cometary Research.

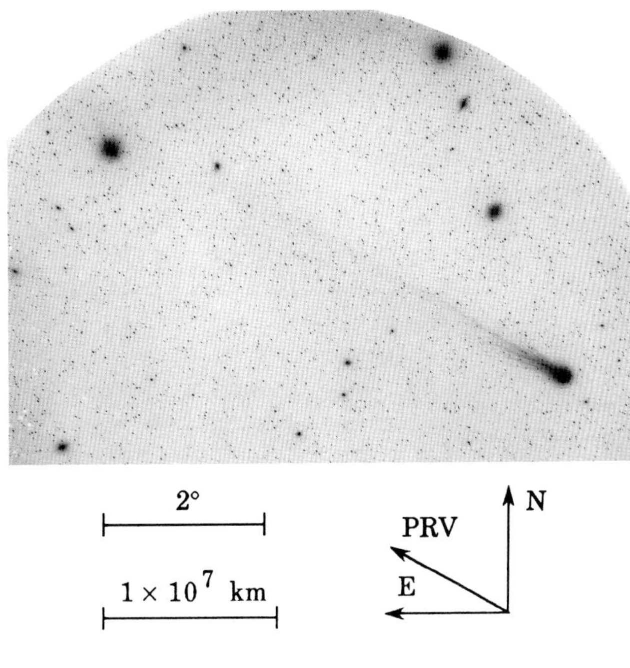

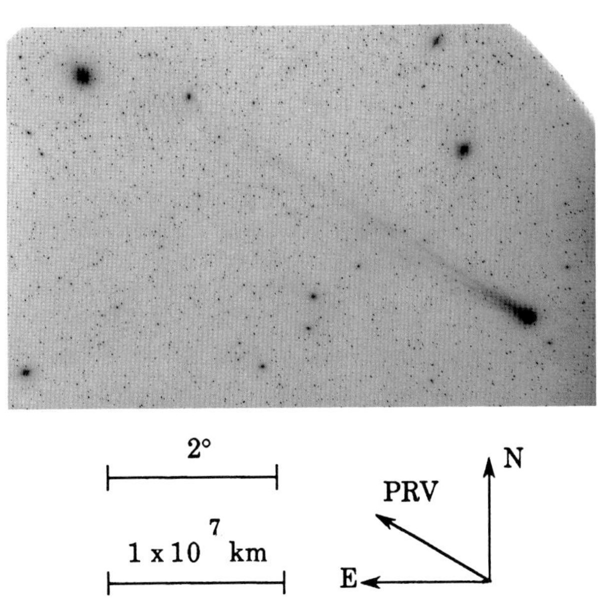

(top) Fig. 182: 1986 Jan 9.22708 UT; exposure 30.0 minutes on IIIa-J emulsion with no filter; r=0.88, Δ=1.31, β=131.4°; LSPN-1884.
Photograph by M. Buie/A. Storrs, Mauna Kea Observatory.

(bottom) Fig. 183: 1986 Jan 9.25208 UT; exposure 30.0 minutes on IIIa-J emulsion with no filter; r=0.88, Δ=1.31, β=131.4°; LSPN-1885.
Photograph by M. Buie/A. Storrs, Mauna Kea Observatory.

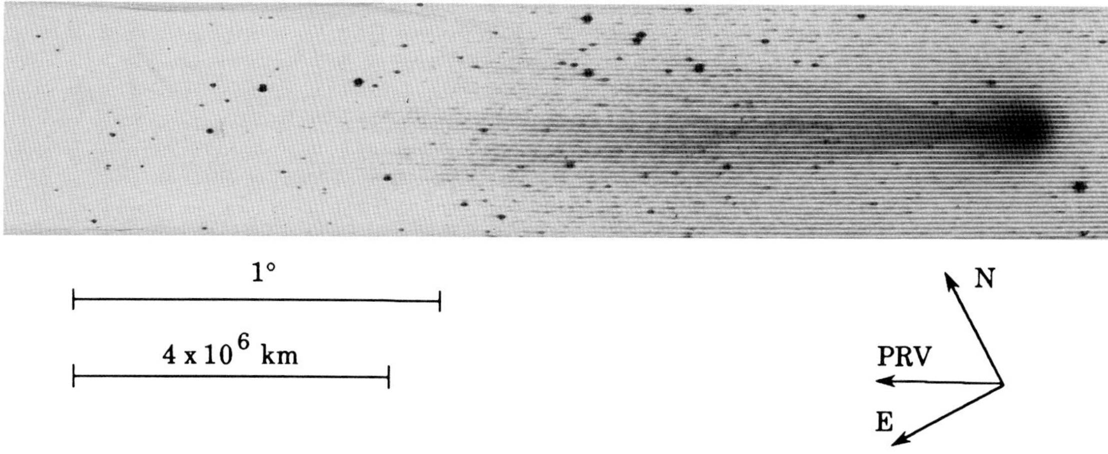

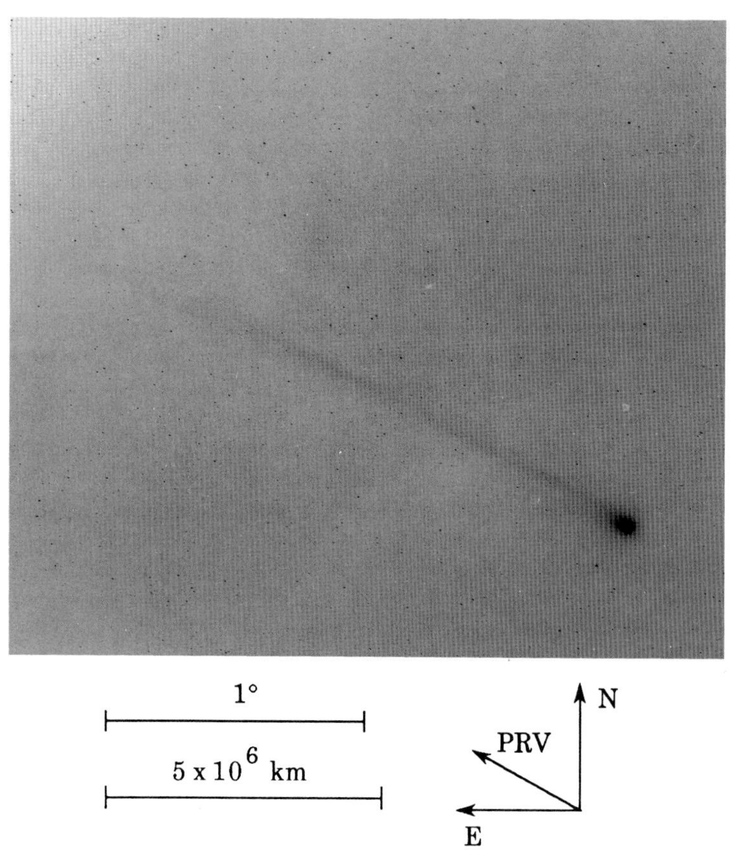

(top) Fig. 184: 1986 Jan 9.66928 UT; exposure 28.0 minutes on ORWO ZU-21 emulsion with no filter; r=0.88, Δ=1.32, β=131.8°; LSPN-3439.
Photograph by N. Chernykh, Crimean Astrophysical Observatory.

(bottom) Fig. 185: 1986 Jan 10.01919 UT; exposure 10.0 minutes on IIa-O emulsion with GG-13 filter; r=0.87, Δ=1.32, β=132.1°; LSPN-140.
Photograph by A. Heiser, A. J. Dyer Observatory.

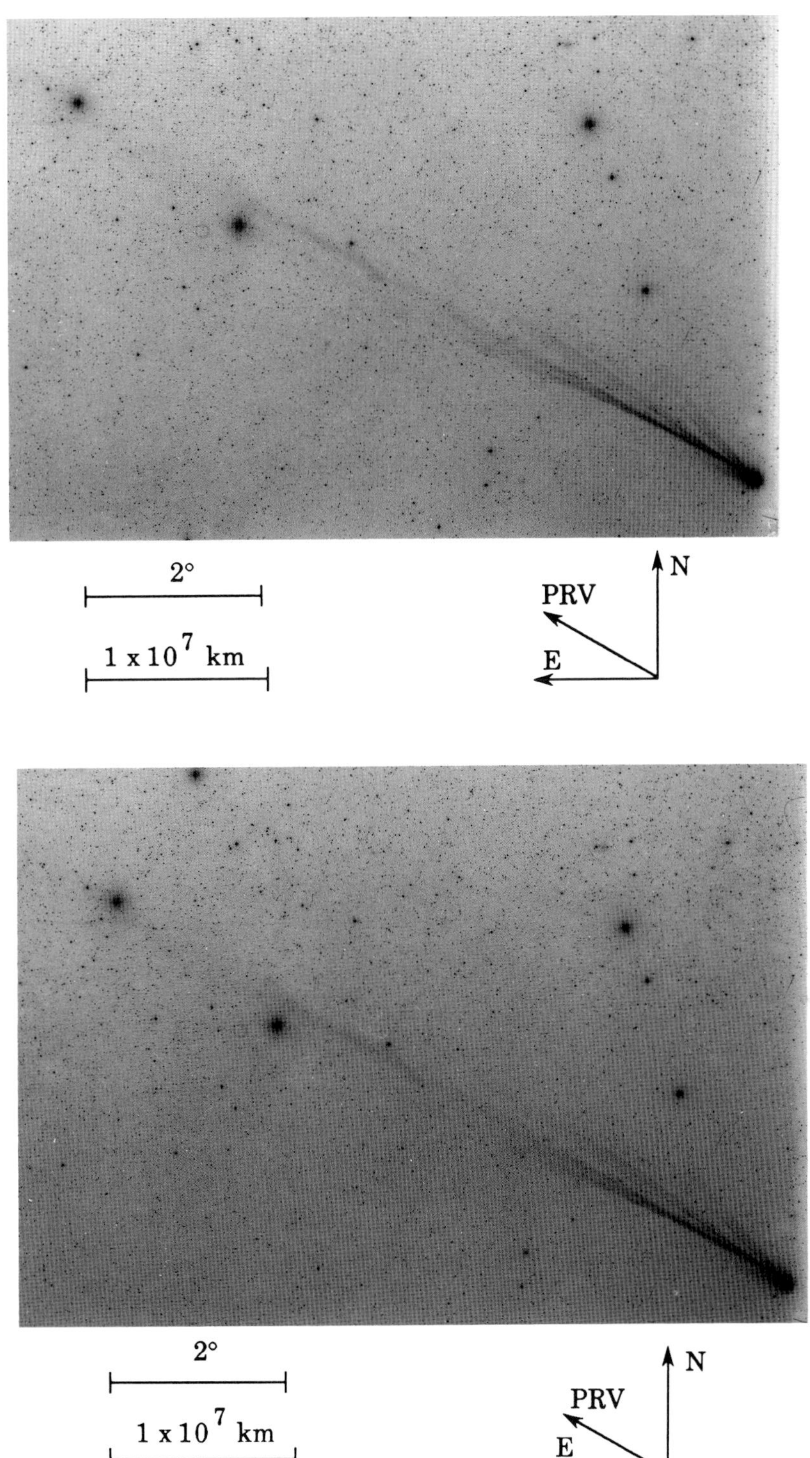

(top) Fig. 186: 1986 Jan 10.08056 UT; exposure 10.0 minutes on IIa-O emulsion with no filter; r=0.87, Δ=1.33, β=132.1°; LSPN-2091.
Photograph by E. Moore/E. Marr, Joint Observatory for Cometary Research.

(bottom) Fig. 187: 1986 Jan 10.08889 UT; exposure 10.0 minutes on IIa-O emulsion with no filter; r=0.87, Δ=1.33, β=132.1°; LSPN-2092.
Photograph by E. Moore/E. Marr, Joint Observatory for Cometary Research.

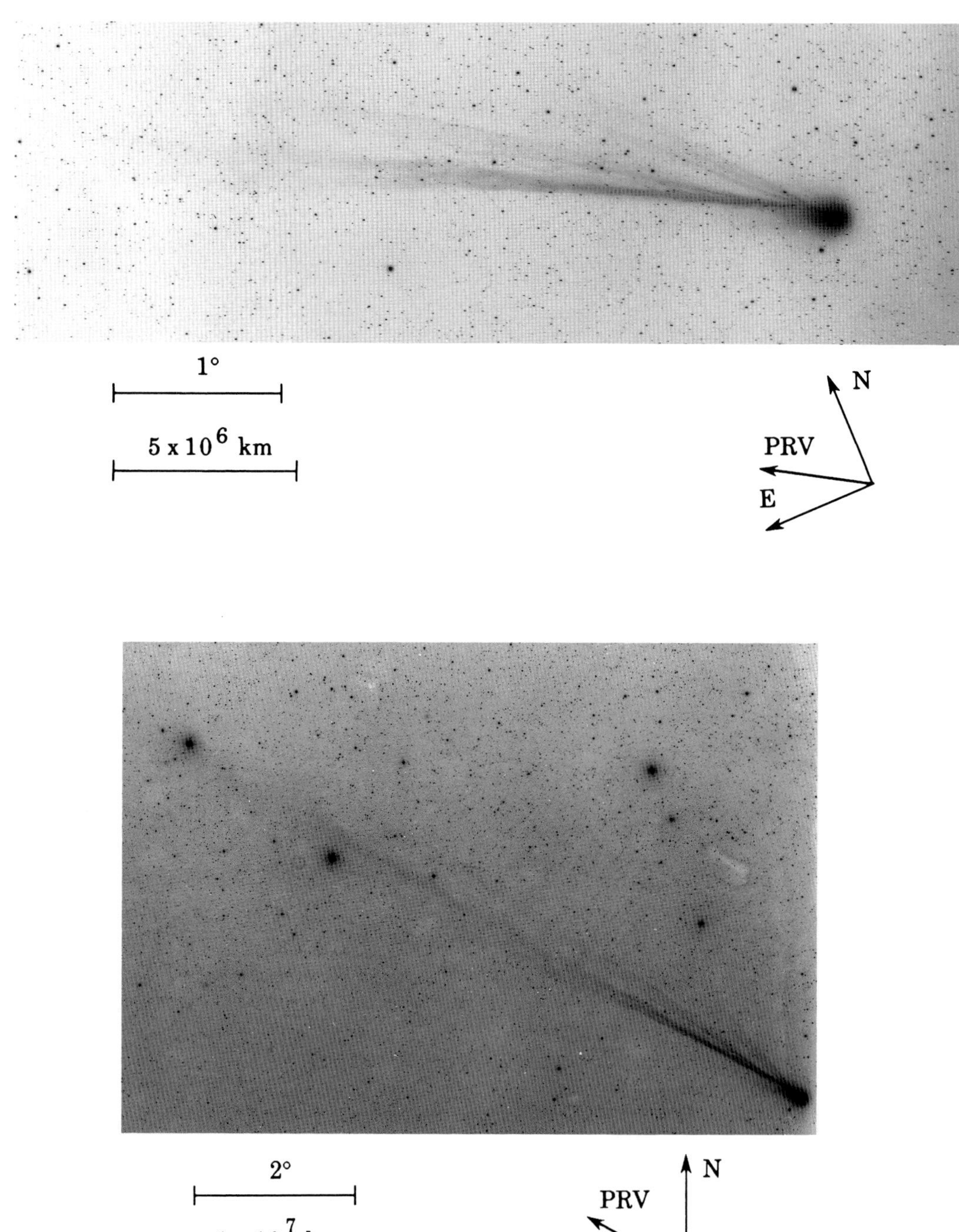

(top) Fig. 188: 1986 Jan 10.09028 UT; exposure 30.0 minutes on 103a-O emulsion with no filter; r=0.87, Δ=1.33, β=132.1°; LSPN-3569.
Photograph by H. Giclas, Lowell Observatory.

(bottom) Fig. 189: 1986 Jan 10.09722 UT; exposure 10.0 minutes on IIa-O emulsion with no filter; r=0.87, Δ=1.33, β=132.1°; LSPN-2093.
Photograph by E. Moore/E. Marr, Joint Observatory for Cometary Research.

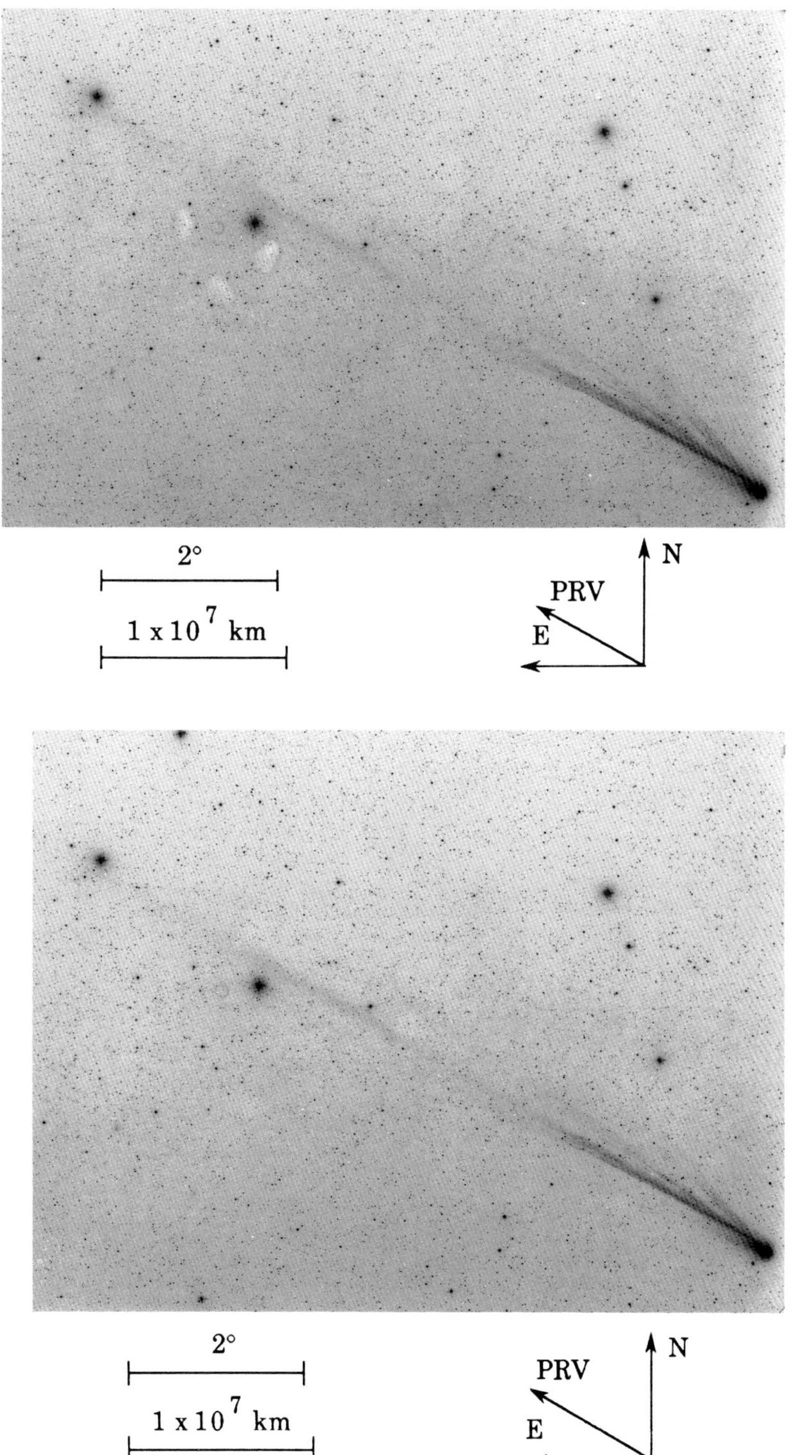

(top) Fig. 190: 1986 Jan 10.10556 UT; exposure 10.0 minutes on IIa-O emulsion with no filter; r=0.87, Δ=1.33, β=132.1°; LSPN-2094.
Photograph by E. Moore/E. Marr, Joint Observatory for Cometary Research.

(bottom) Fig. 191: 1986 Jan 10.11389 UT; exposure 10.0 minutes on IIa-O emulsion with no filter; r=0.87, Δ=1.33, β=132.1°; LSPN-2095.
Photograph by E. Moore/E. Marr, Joint Observatory for Cometary Research.

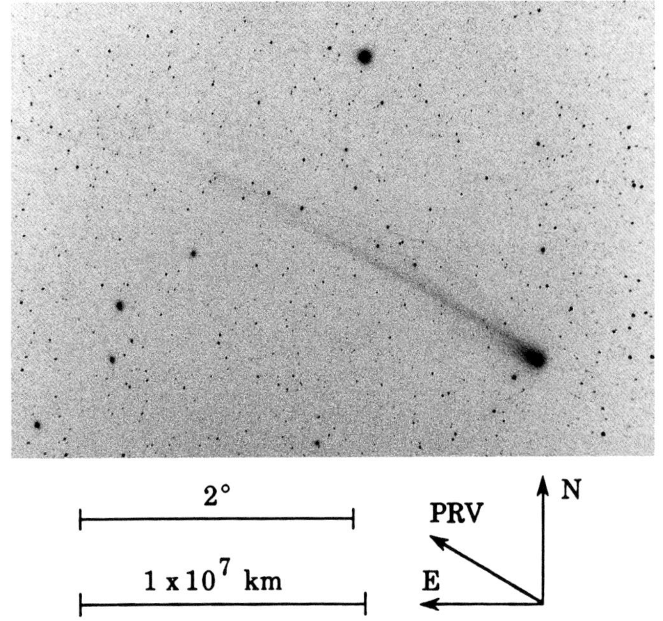

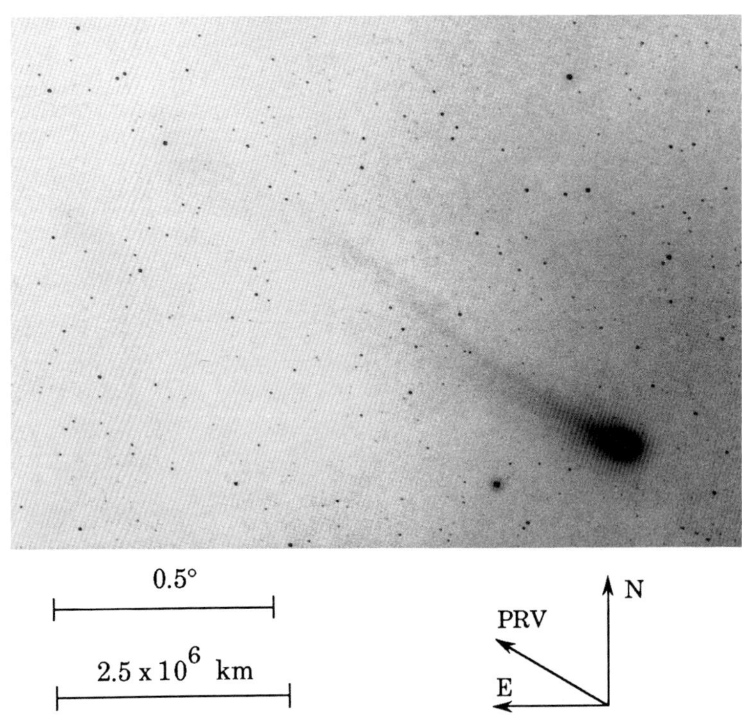

(top) Fig. 192: 1986 Jan 10.12080 UT; exposure 20.0 minutes on 3M 1000 emulsion; r=0.87, Δ=1.33, β=132.1°; AON-850728.
Photograph by S. Edberg, USA.

(bottom) Fig. 193: 1986 Jan 10.38750 UT; exposure 10.0 minutes on Tech. Pan 2415 emulsion; r=0.87, Δ=1.33, β=132.4°; AON-850739.
Photograph by T. Kojima, Japan.

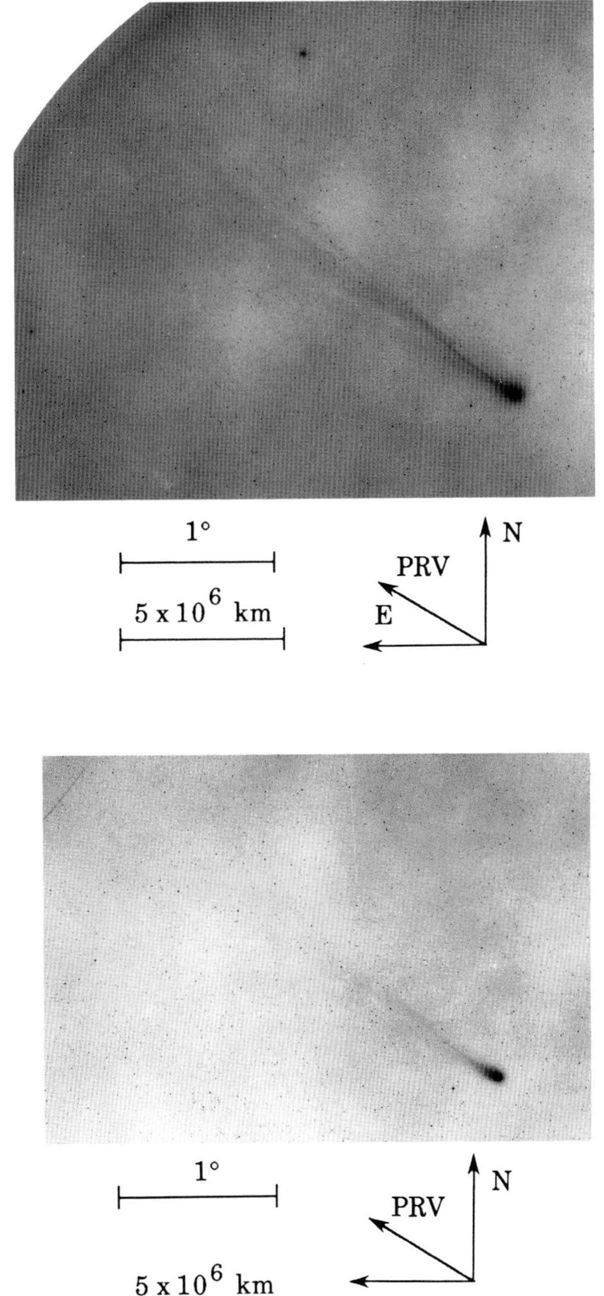

(top) Fig. 194: 1986 Jan 10.39063 UT; exposure 5.0 minutes on IIa-O emulsion with no filter; r=0.87, Δ=1.33, β=132.4°; LSPN-1341.
Photograph by T. Tsujimura, University Kyoto Observatory.

(bottom) Fig. 195: 1986 Jan 10.40104 UT; exposure 5.0 minutes on 103a-F emulsion with Y-50 filter; r=0.87, Δ=1.33, β=132.4°; LSPN-1342.
Photograph by T. Tsujimura, University Kyoto Observatory.

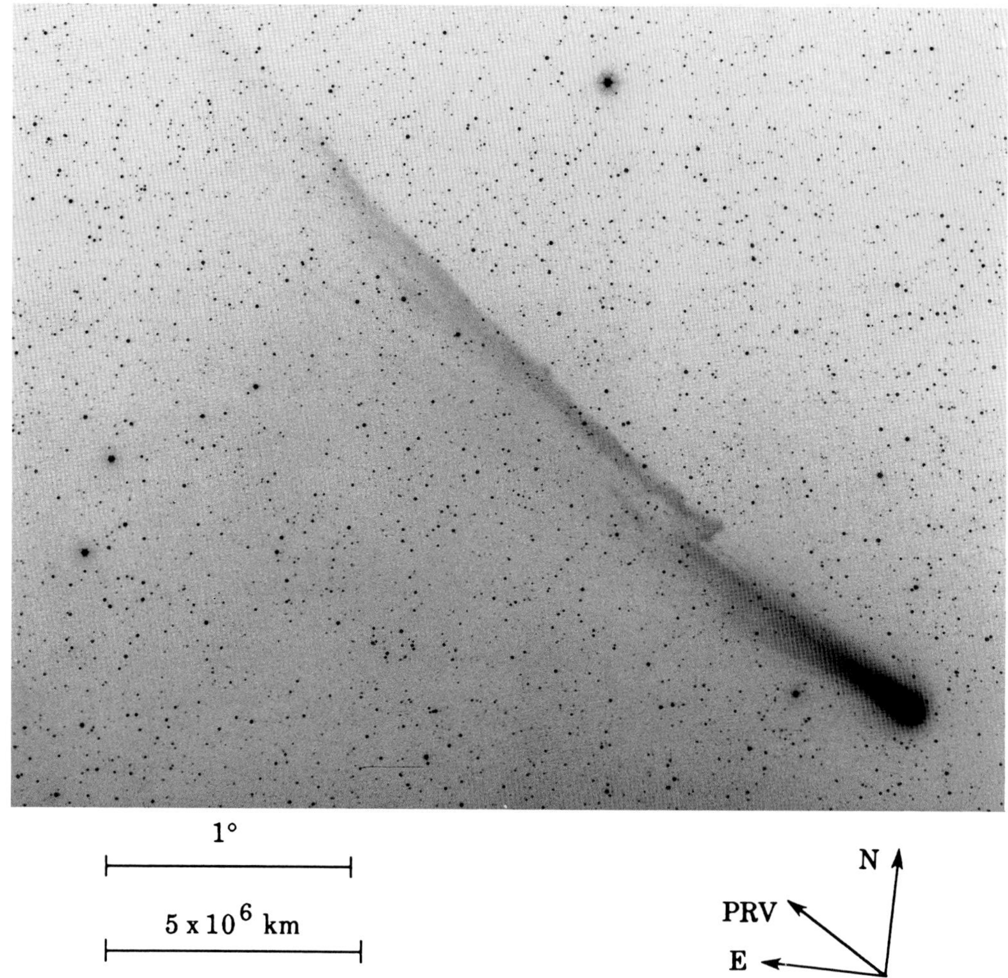

Fig. 196: 1986 Jan 10.73900 UT; two exposures 4.0 minutes each on Tech. Pan 2415 emulsion; r=0.86, Δ=1.34, β=132.7°; AON-850749.
Photograph by M. Jäger, Austria.

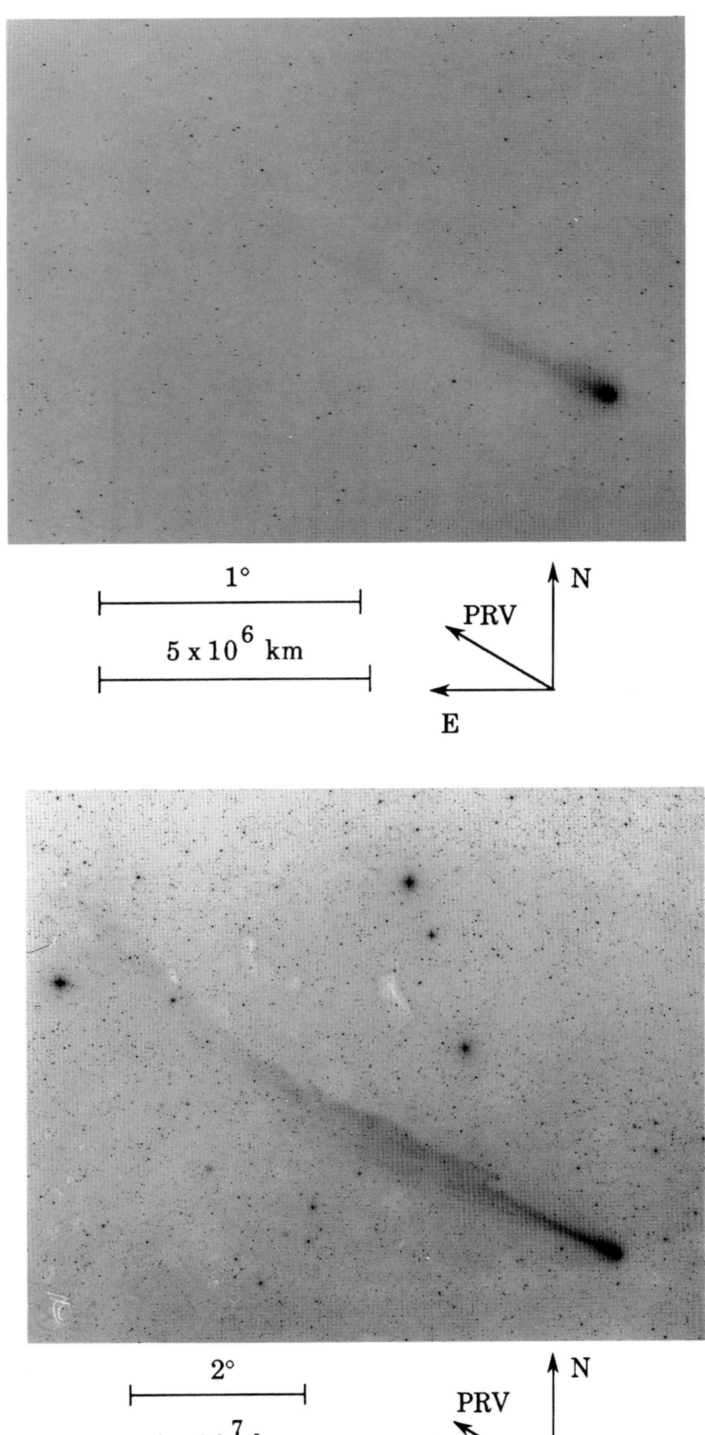

(top) Fig. 197: 1986 Jan 11.01535 UT; exposure 12.0 minutes on IIa-O emulsion with GG-13 filter; r=0.86, Δ=1.34, β=132.9°; LSPN-141.
Photograph by A. Heiser, A. J. Dyer Observatory.

(bottom) Fig. 198: 1986 Jan 11.08889 UT; exposure 10.0 minutes on IIa-O emulsion with no filter; r=0.86, Δ=1.34, β=133.0°; LSPN-2097.
Photograph by E. Moore/E. Marr, Joint Observatory for Cometary Research.

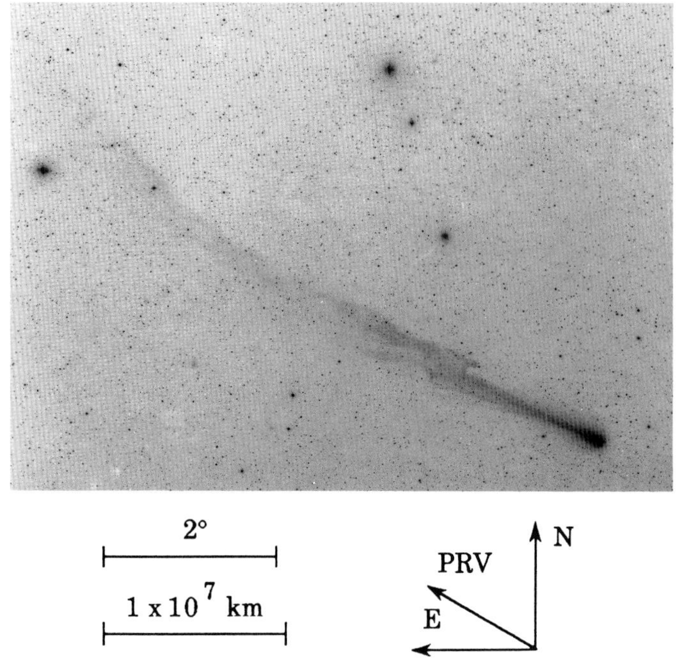

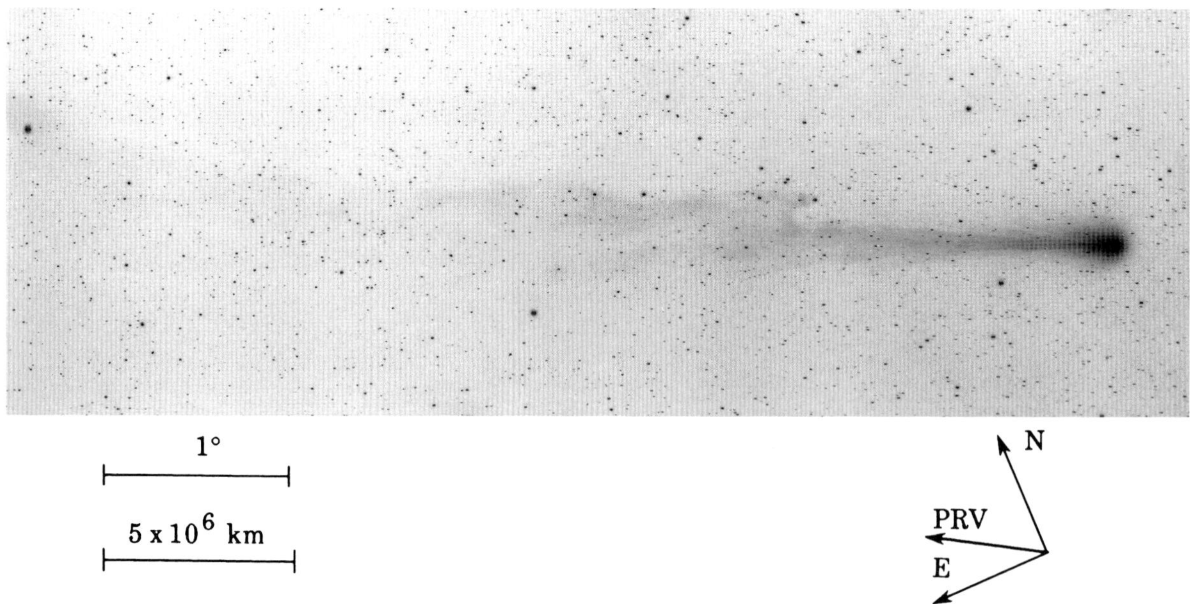

(top) Fig. 199: 1986 Jan 11.09722 UT; exposure 10.0 minutes on IIa-O emulsion with no filter; r=0.86, Δ=1.34, β=133.0°; LSPN-2098.
Photograph by E. Moore/E. Marr, Joint Observatory for Cometary Research.

(bottom) Fig. 200: 1986 Jan 11.10069 UT; exposure 30.0 minutes on 103a-O emulsion with no filter; r=0.86, Δ=1.34, β=133.0°; LSPN-3573.
Photograph by H. Giclas, Lowell Observatory.

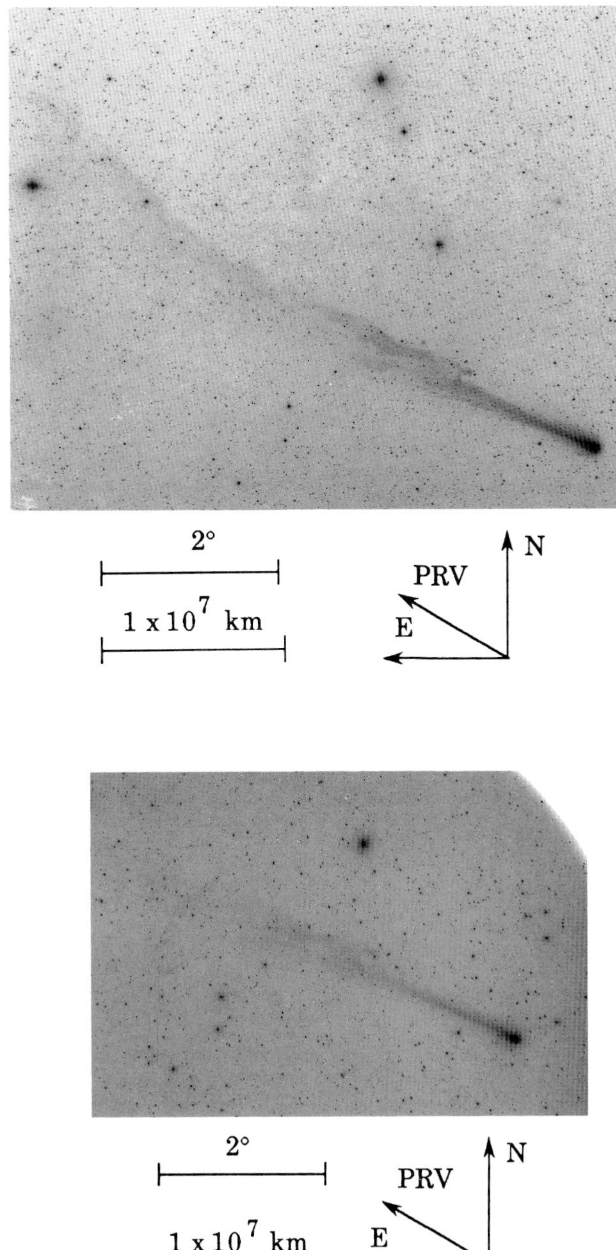

(top) Fig. 201: 1986 Jan 11.10556 UT; exposure 10.0 minutes on IIa-O emulsion with no filter; r=0.86, Δ=1.34, β=133.0°; LSPN-2099.
Photograph by E. Moore/E. Marr, Joint Observatory for Cometary Research.

(bottom) Fig. 202: 1986 Jan 11.24236 UT; exposure 30.0 minutes on IIIa-J emulsion with no filter; r=0.85, Δ=1.34, β=133.1°; LSPN-1889.
Photograph by A. Storrs, Mauna Kea Observatory.

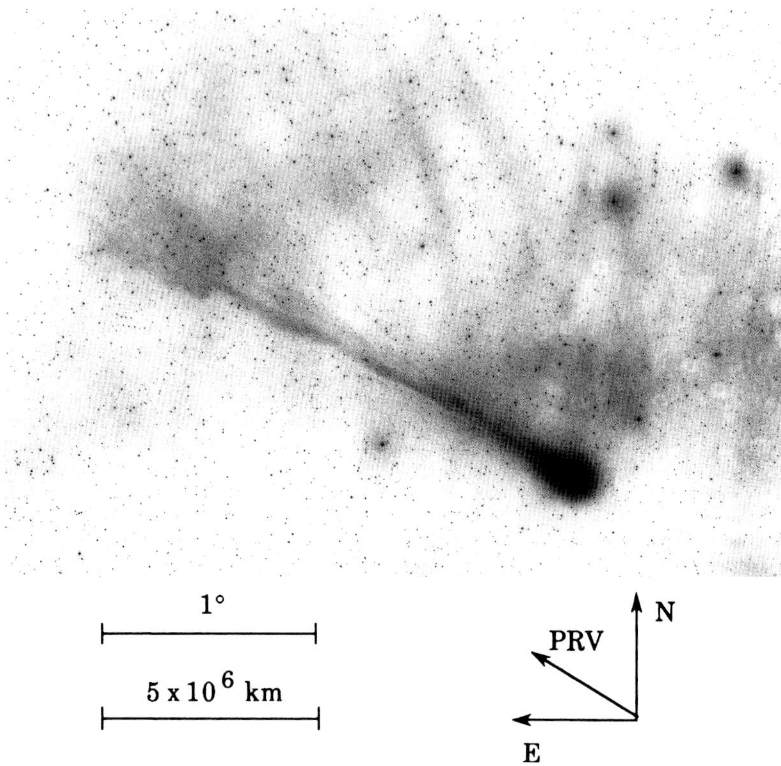

Fig. 203: 1986 Jan 11.73229 UT; exposure 10.0 minutes on IIa-O emulsion with no filter; r=0.85, Δ=1.35, β=133.6°; LSPN-2492.
Photograph by V. Shkodrov et al., Bulgarian National Observatory.

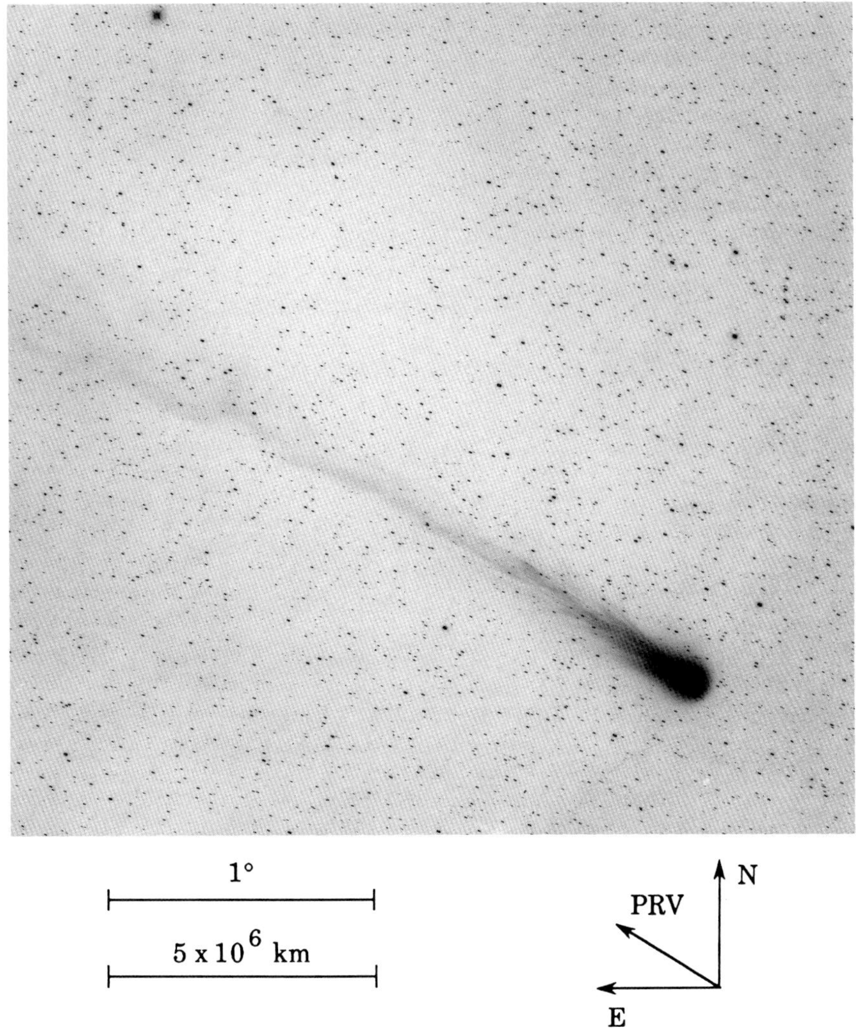

Fig. 204: 1986 Jan 11.75694 UT; exposure 24.0 minutes on IIIa-J emulsion with no filter; r=0.85, Δ=1.35, β=133.6°; LSPN-3196.
Photograph by G. Sause/A. Laugier, Observatory of Haute-Provence.

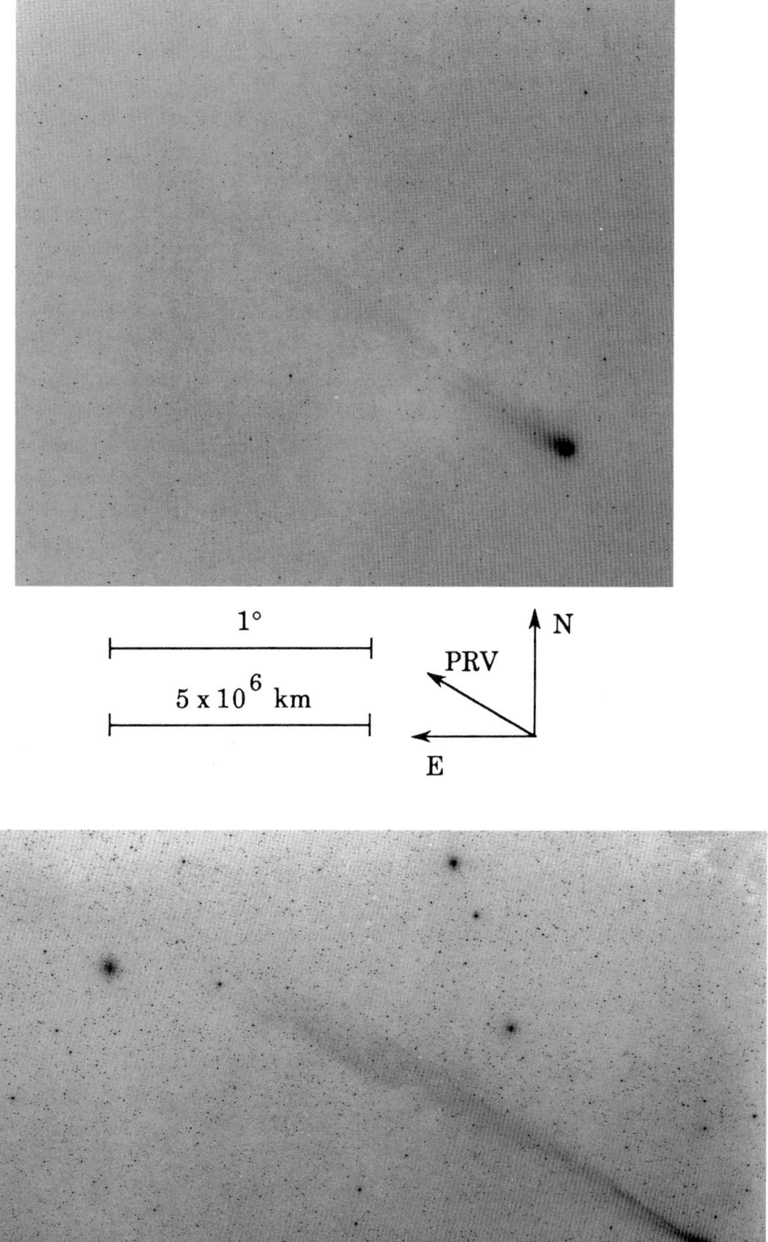

(top) Fig. 205: 1986 Jan 12.01422 UT; exposure 10.0 minutes on IIa-O emulsion with GG-13 filter; r=0.84, Δ=1.36, β=133.9°; LSPN-142.
Photograph by A. Heiser, A. J. Dyer Observatory.

(bottom) Fig. 206: 1986 Jan 12.08472 UT; exposure 10.0 minutes on IIa-O emulsion with no filter; r=0.84, Δ=1.36, β=133.9°; LSPN-2101.
Photograph by E. Moore/E. Marr, Joint Observatory for Cometary Research.

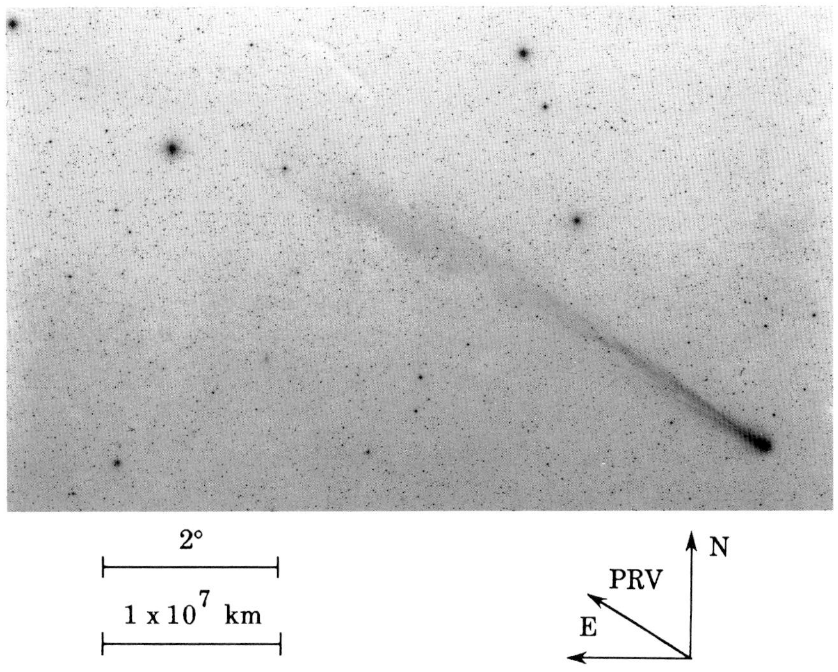

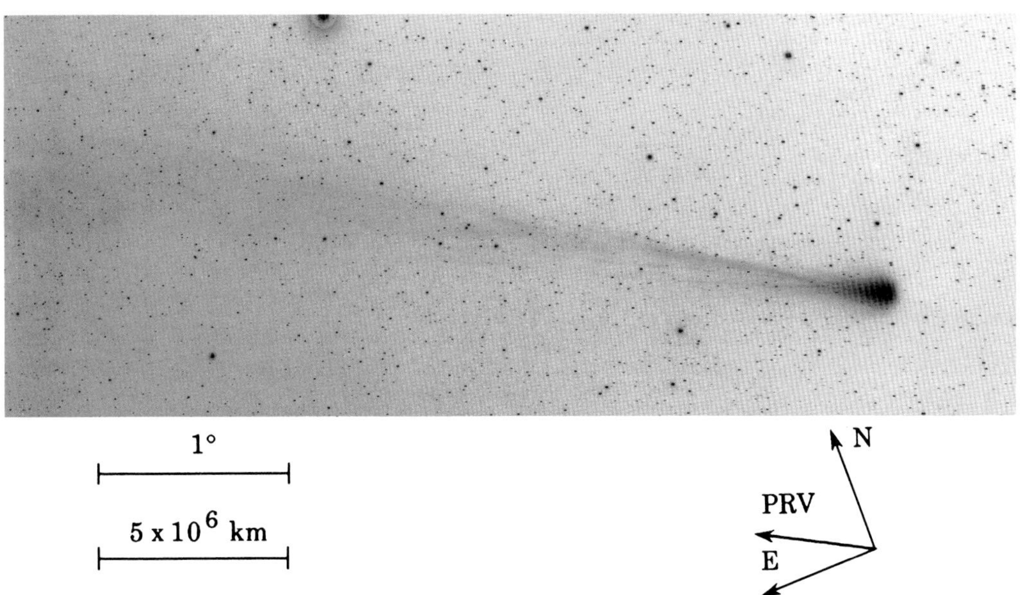

(top) Fig. 207: 1986 Jan 12.09306 UT; exposure 10.0 minutes on IIa-O emulsion with no filter; r=0.84, Δ=1.36, β=133.9°; LSPN-2102.
Photograph by E. Moore/E. Marr, Joint Observatory for Cometary Research.

(bottom) Fig. 208: 1986 Jan 12.09861 UT; exposure 28.0 minutes on 103a-O emulsion with no filter; r=0.84, Δ=1.36, β=133.9°; LSPN-3575.
Photograph by H. Giclas, Lowell Observatory.

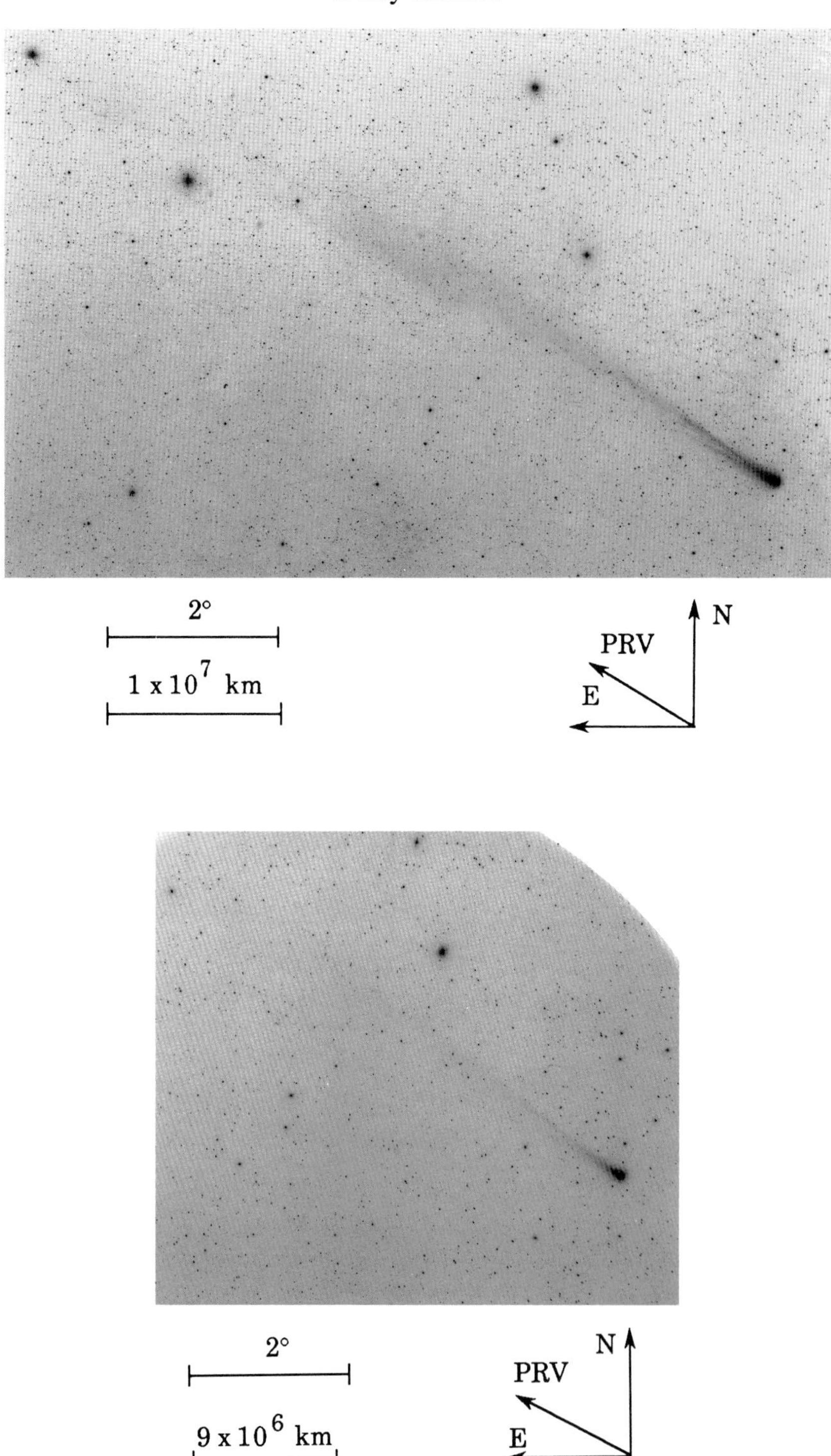

(top) Fig. 209: 1986 Jan 12.10139 UT; exposure 10.0 minutes on IIa-O emulsion with no filter; r=0.84, Δ=1.36, β=133.9°; LSPN-2103.
Photograph by E. Moore/E. Marr, Joint Observatory for Cometary Research.

(bottom) Fig. 210: 1986 Jan 12.24236 UT; exposure 30.0 minutes on IIIa-J emulsion with no filter; r=0.84, Δ=1.36, β=134.1°; LSPN-1892.
Photograph by A. Storrs, Mauna Kea Observatory.

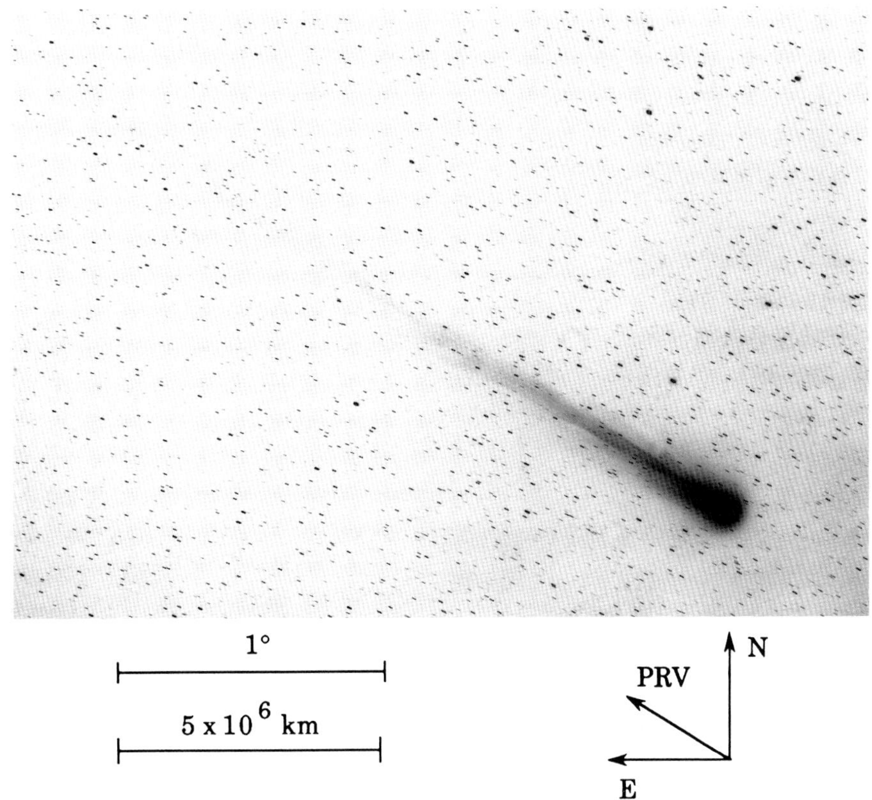

Fig. 211: 1986 Jan 12.76111 UT; exposure 40.0 minutes on IIIa-J emulsion with no filter; r=0.83, Δ=1.37, β=134.6°; LSPN-3197.
Photograph by G. Sause/A. Laugier, Observatory of Haute-Provence.

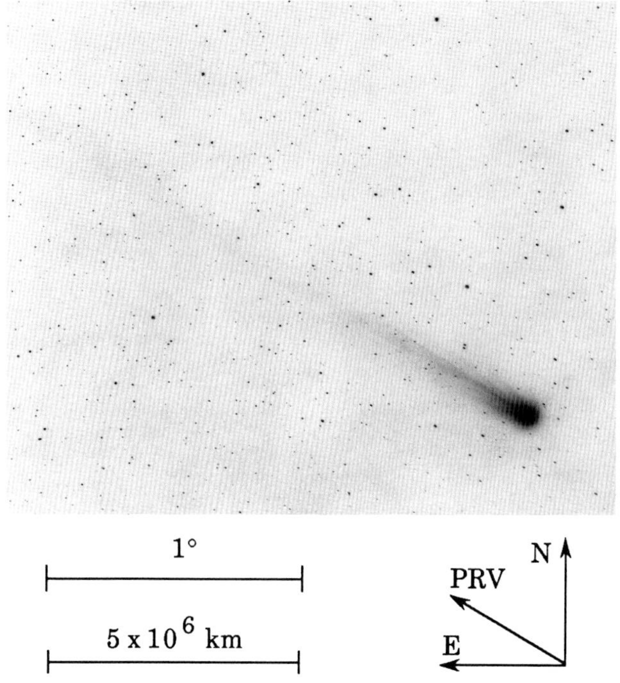

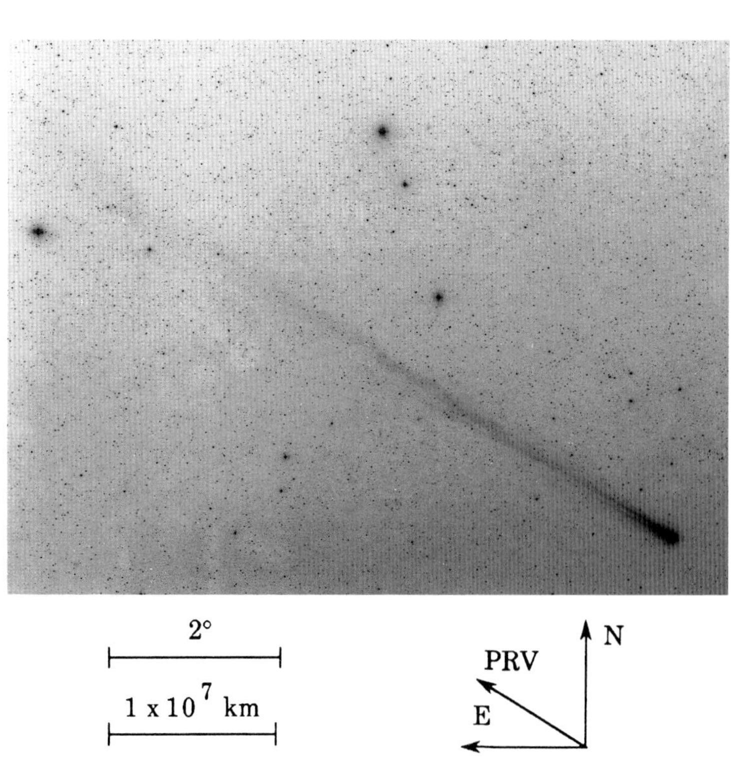

(top) Fig. 212: 1986 Jan 12.78438 UT; exposure 15.0 minutes on IIa-O emulsion with no filter; $r=0.83$, $\Delta=1.37$, $\beta=134.6°$; LSPN-3198.
Photograph by G. Sause/A. Laugier, Observatory of Haute-Provence.

(bottom) Fig. 213: 1986 Jan 13.08194 UT; exposure 10.0 minutes on IIa-O emulsion with no filter; $r=0.83$, $\Delta=1.37$, $\beta=134.9°$; LSPN-2105.
Photograph by E. Moore/E. Marr, Joint Observatory for Cometary Research.

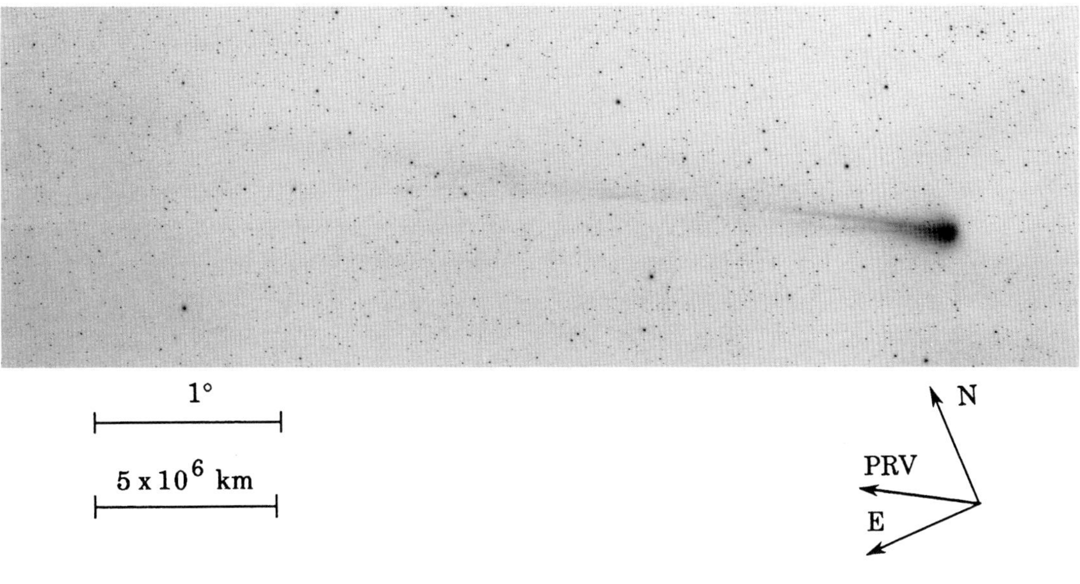

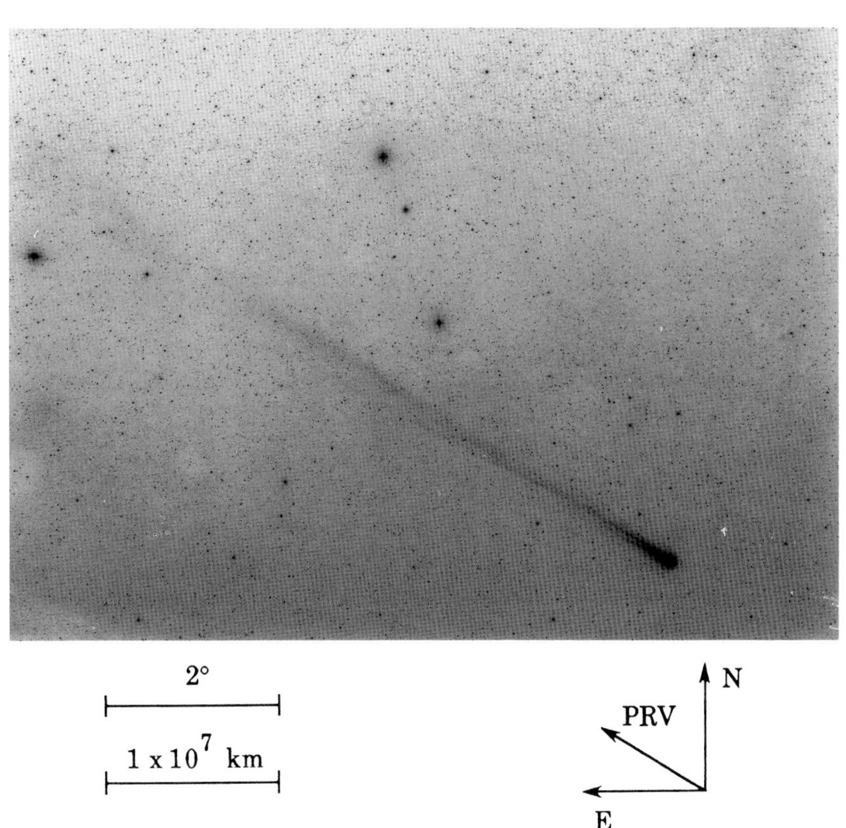

(top) Fig. 214: 1986 Jan 13.08866 UT; exposure 15.0 minutes on 103a-O emulsion with no filter; $r=0.83$, $\Delta=1.37$, $\beta=134.9^{\circ}$; LSPN-3576.
Photograph by H. Giclas, Lowell Observatory.

(bottom) Fig. 215: 1986 Jan 13.08958 UT; exposure 10.0 minutes on IIa-O emulsion with no filter; $r=0.83$, $\Delta=1.37$, $\beta=134.9^{\circ}$; LSPN-869.
Photograph by E. Moore/E. Marr, Joint Observatory for Cometary Research.

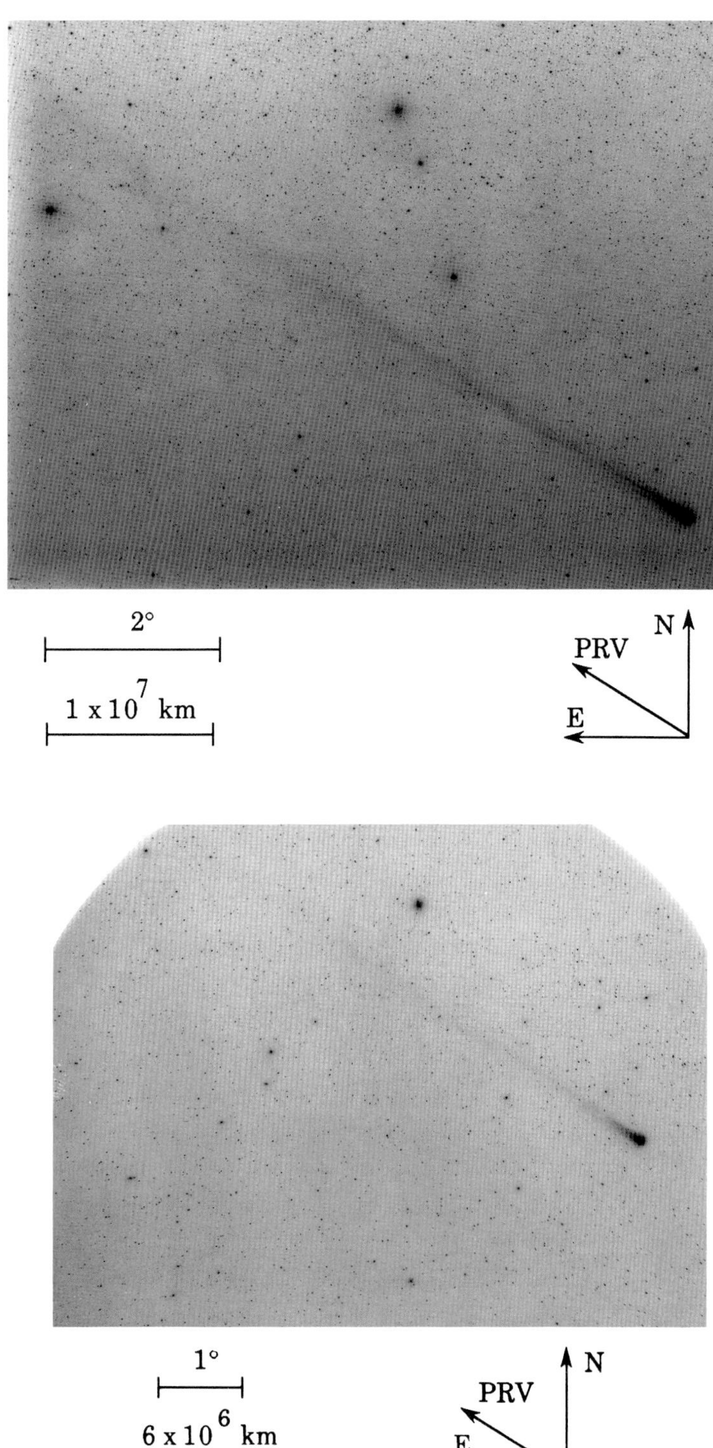

(top) Fig. 216: 1986 Jan 13.09792 UT; exposure 10.0 minutes on IIa-O emulsion with no filter; r=0.83, Δ=1.37, β=134.9°; LSPN-870.
Photograph by E. Moore/E. Marr, Joint Observatory for Cometary Research.

(bottom) Fig. 217: 1986 Jan 13.24844 UT; exposure 27.0 minutes on IIIa-J emulsion with no filter; r=0.83, Δ=1.38, β=135.1°; LSPN-1894.
Photograph by A. Storrs, Mauna Kea Observatory.

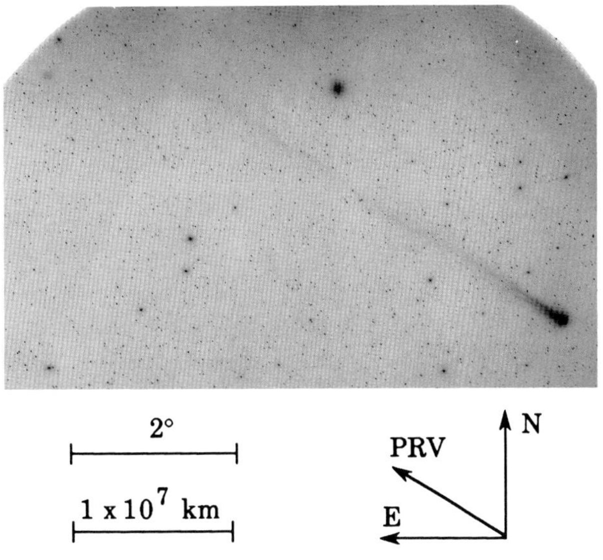

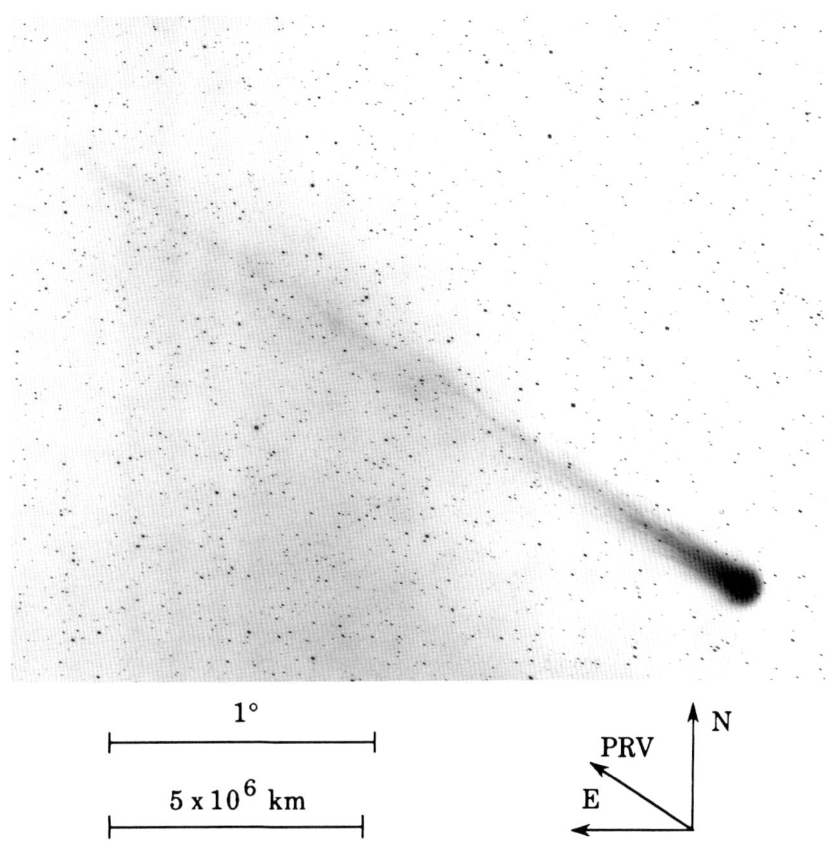

(top) Fig. 218: 1986 Jan 13.30937 UT; exposure 30.0 minutes on IIIa-J emulsion with no filter; r=0.82, Δ=1.38, β=135.1°; LSPN-1893.
Photograph by A. Storrs, Mauna Kea Observatory.

(bottom) Fig. 219: 1986 Jan 13.75833 UT; exposure 20.0 minutes on IIIa-J emulsion with no filter; r=0.82, Δ=1.38, β=135.6°; LSPN-3199.
Photograph by G. Sause, Observatory of Haute-Provence.

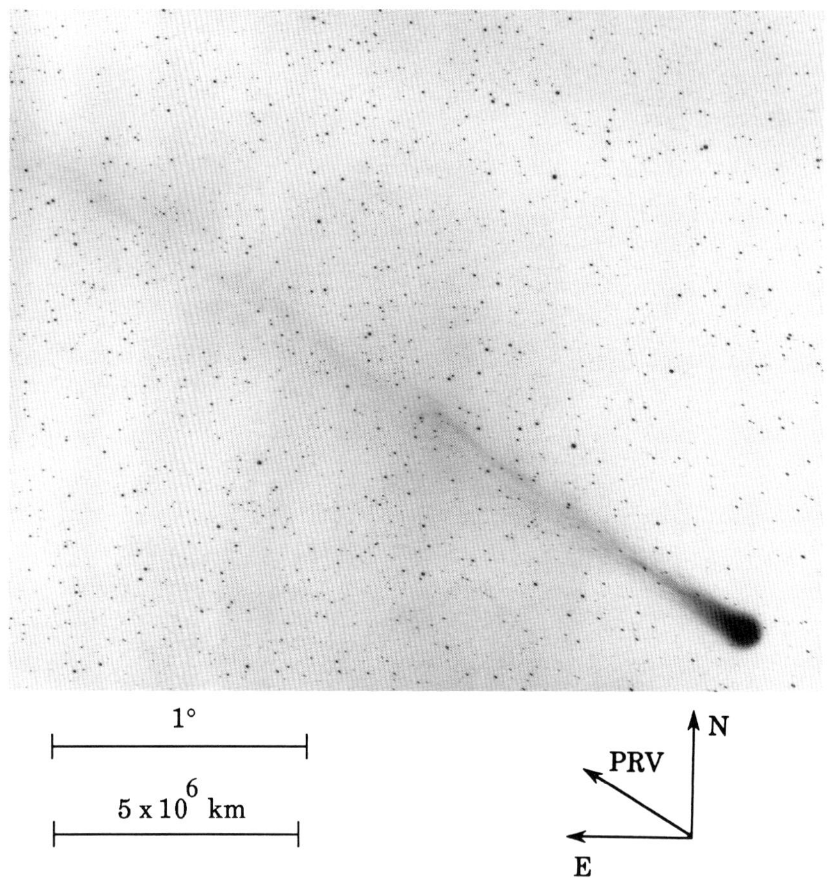

Fig. 220: 1986 Jan 13.77326 UT; exposure 15.0 minutes on IIIa-J emulsion with no filter; r=0.82, Δ=1.38, β=135.6°; LSPN-3200.
Photograph by G. Sause, Observatory of Haute-Provence.

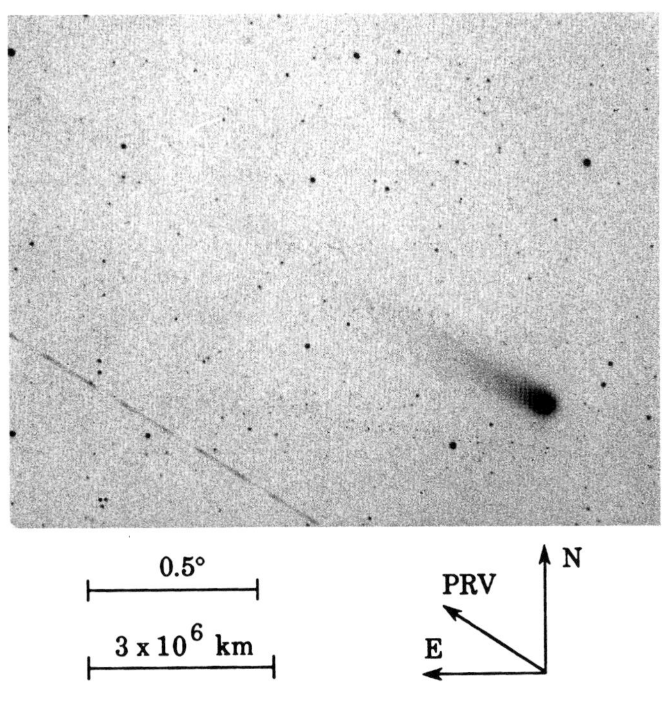

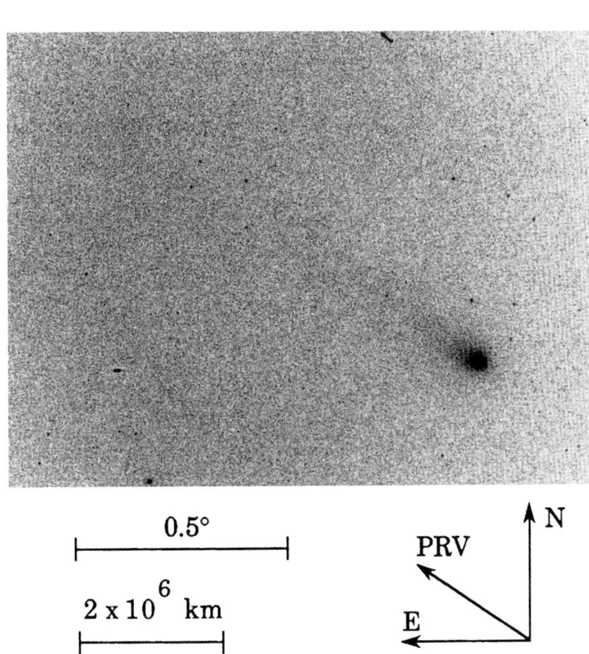

(top) Fig. 221: 1986 Jan 15.75880 UT; exposure 8.0 minutes on IIa-F emulsion; r=0.79, Δ=1.41, β=137.8°; AON-850871.
Photograph by H. B. Ridley, UK.

(bottom) Fig. 222: 1986 Jan 17.72900 UT; exposure 3.0 minutes on Ilford HP 5 emulsion; r=0.76, Δ=1.44, β=140.2°; AON-850886.
Photograph by H. Mikuz, Yugoslavia.

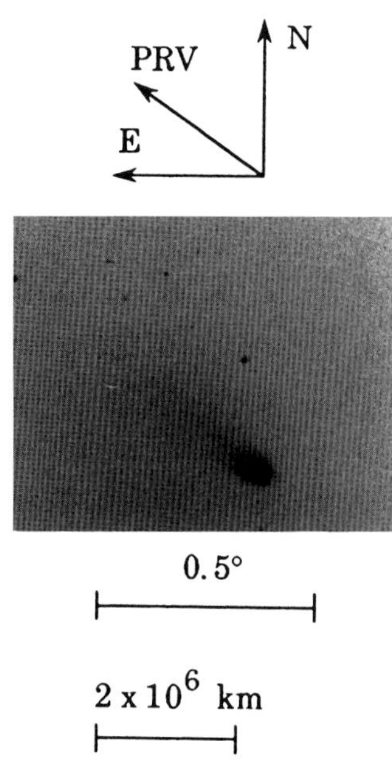

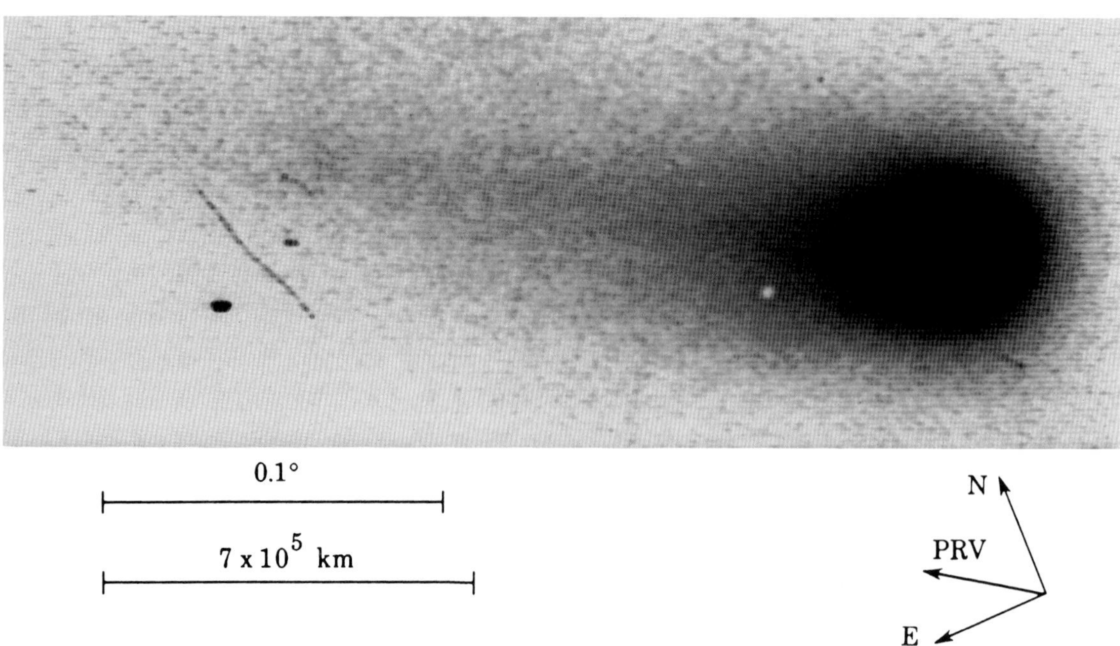

(top) Fig. 223: 1986 Jan 19.05270 UT; exposure 12.0 minutes on Fujichrome 400 emulsion; r=0.75, Δ=1.46, β=142.0°; AON-850893.
Photograph by A. Levy, USA.

(bottom) Fig. 224: 1986 Jan 19.81807 UT; exposure 10.0 minutes on ORWO ZU-21 emulsion with BS-8 filter; r=0.74, Δ=1.47, β=143.0°; LSPN-3349.
Photograph by K. Tarasov, Sanglok Observatory.

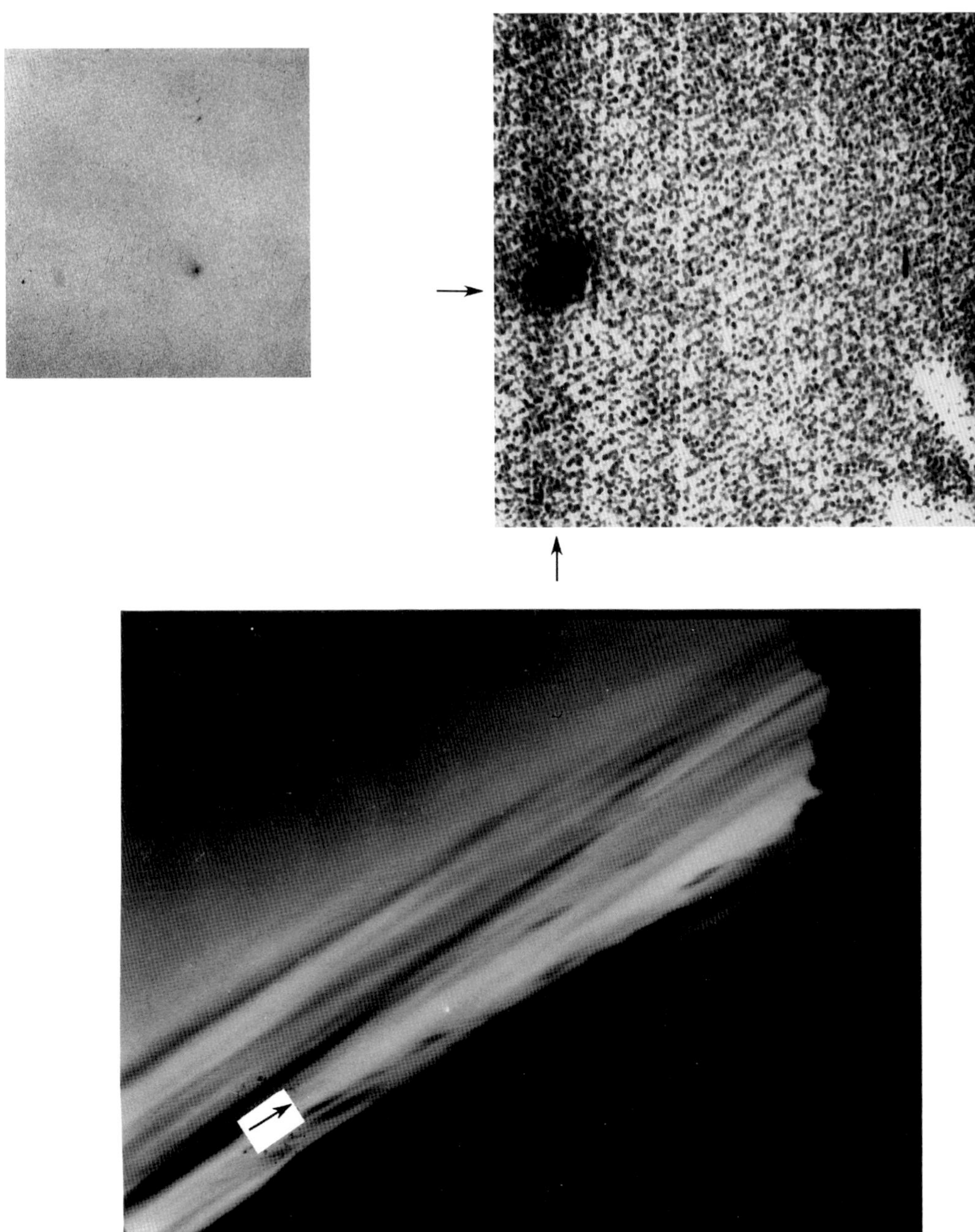

(left) Fig. 225: 1986 Jan 25.74480 UT; exposure 0.2 minute on Sakuracolor SR 1600 emulsion; r=0.67, Δ=1.53, β=152.3°; AON-850911.
Photograph by T. Niijima, Japan.

(right) Fig. 226: 1986 Jan 28.08450 UT; exposure 1.0 minute on RCA CCD with RG-665 filter; r=0.65, Δ=1.54, β=156.5°.
Photograph by D. Levy, USA.

(bottom) Fig. 227: 1986 Jan 28.70800 UT; exposure 0.2 minute on Tech. Pan 2415 emulsion; r=0.64, Δ=1.55, β=157.6°; AON-850912.
Photograph by M. Jäger, Austria.

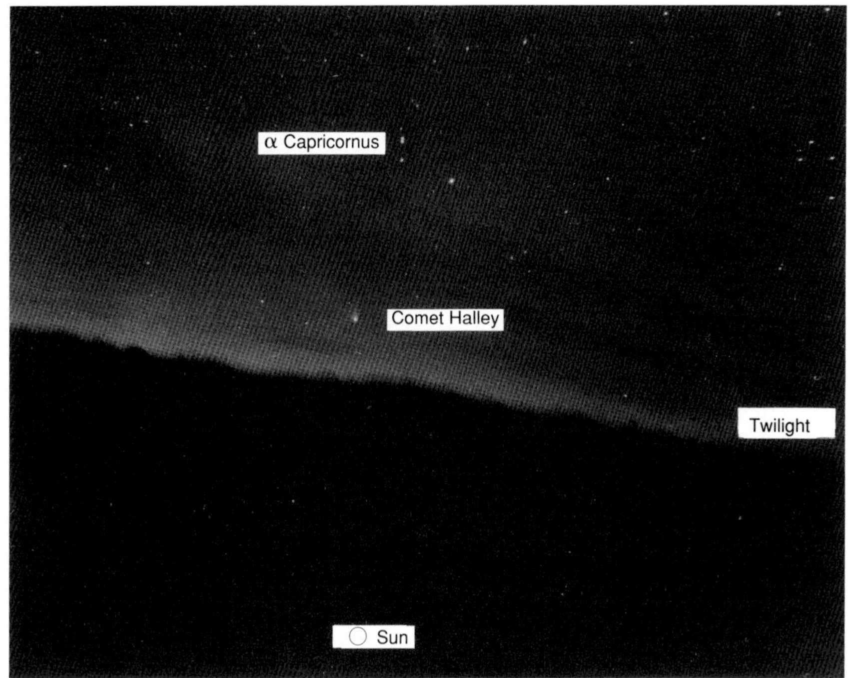

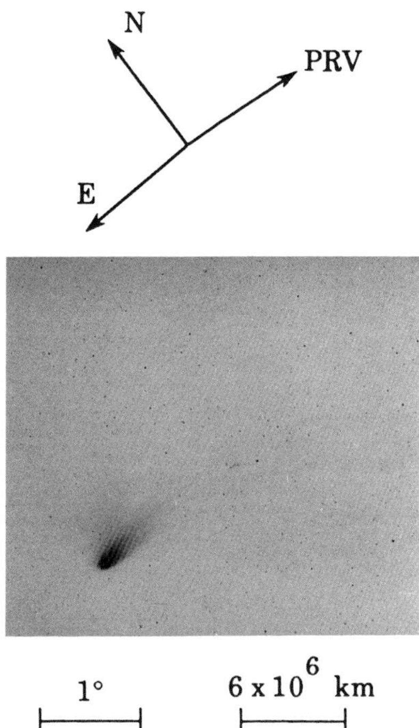

(top) Fig. 228: 1986 Feb 19.41667 UT; exposure 2.0 minutes on Tri-X emulsion; r=0.63, Δ=1.44, β=145.8°; AON-850913.
Photograph by I. Ferrin, Venezuela.

(bottom) Fig. 229: 1986 Feb 21.76700 UT; exposure 3.0 minutes on Tech. Pan 2415 emulsion; r=0.65, Δ=1.40, β=141.1°; AON-850914.
Photograph by G. Garradd, Australia.

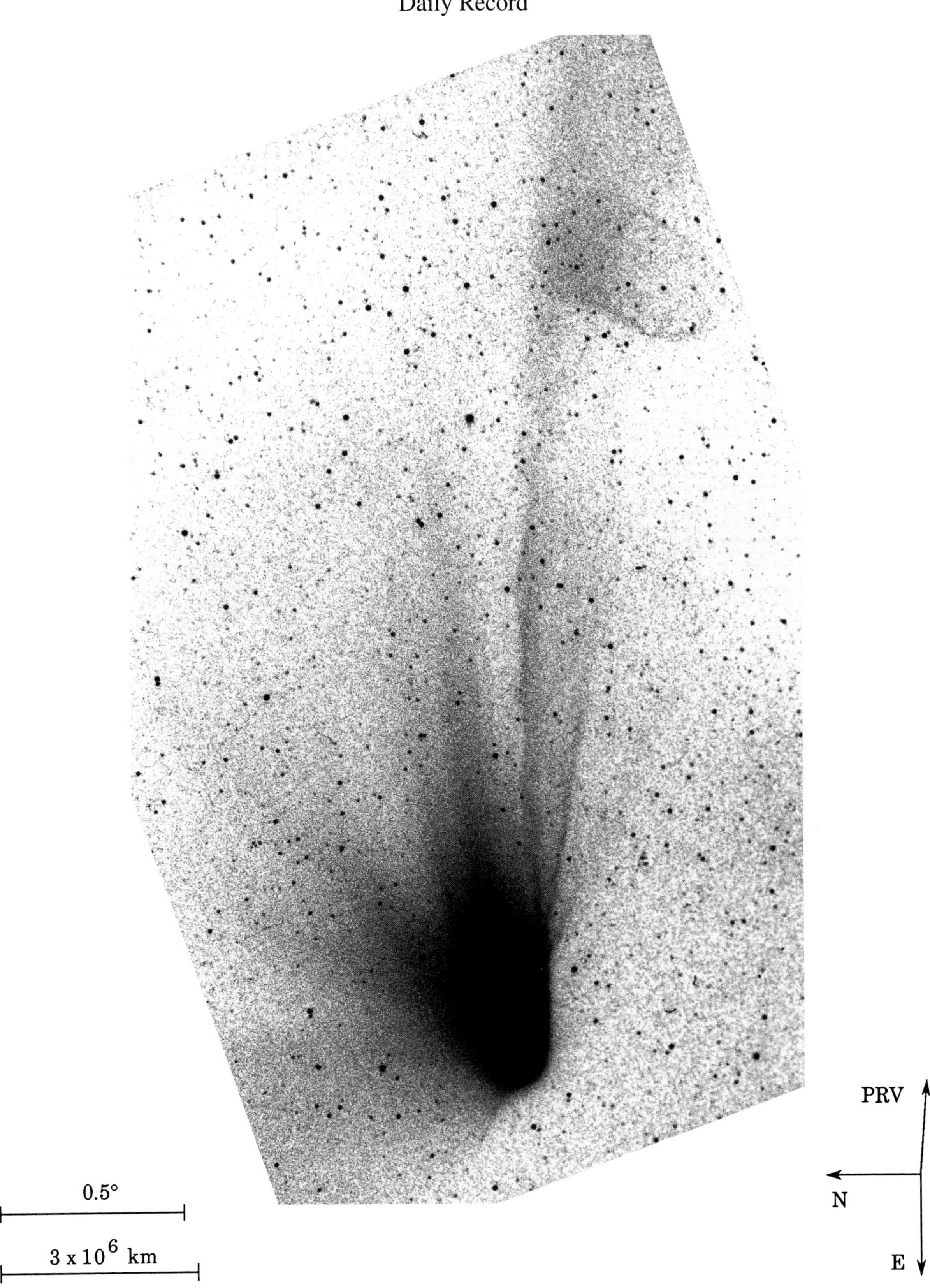

Fig. 230: 1986 Feb 22.12569 UT; exposure 2.0 minutes on Tech. Pan 2415 emulsion with no filter; r=0.65, Δ=1.40, β=140.4°; LSPN-1781.
Photograph by F. Van Wyk, Sutherland, LSPN Island Network.

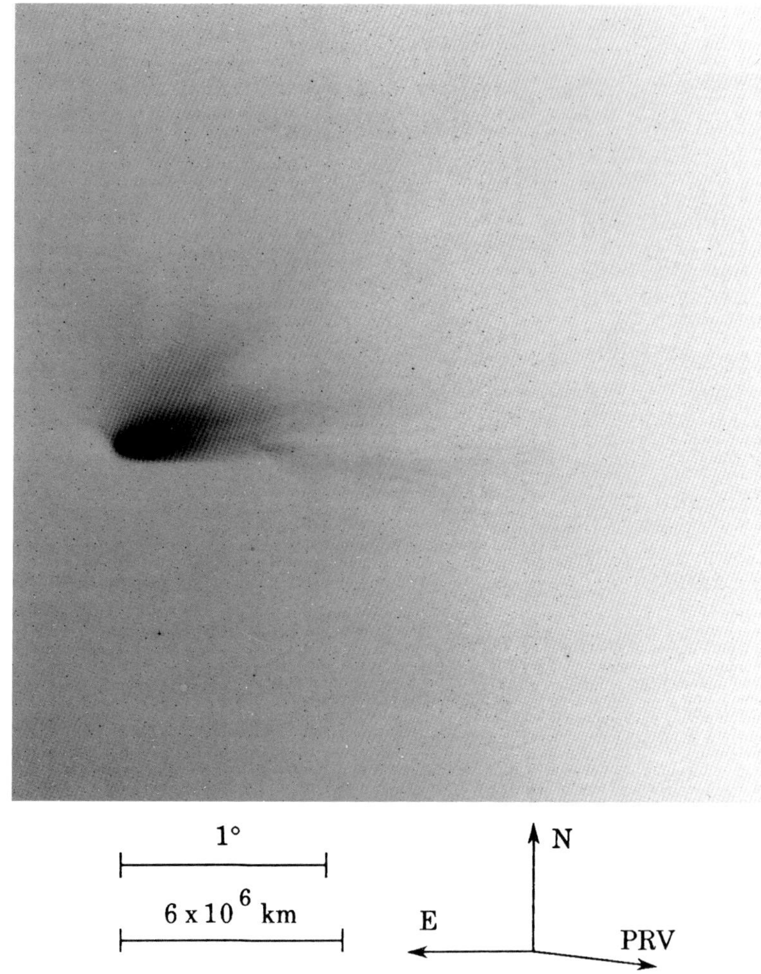

Fig. 231: 1986 Feb 22.77933 UT; exposure 2.0 minutes on IIIa-F emulsion with no filter; r=0.66, Δ=1.39, β=139.1°; LSPN-2376.
Photograph from Royal Observatory/UK Schmidt Telescope Unit.

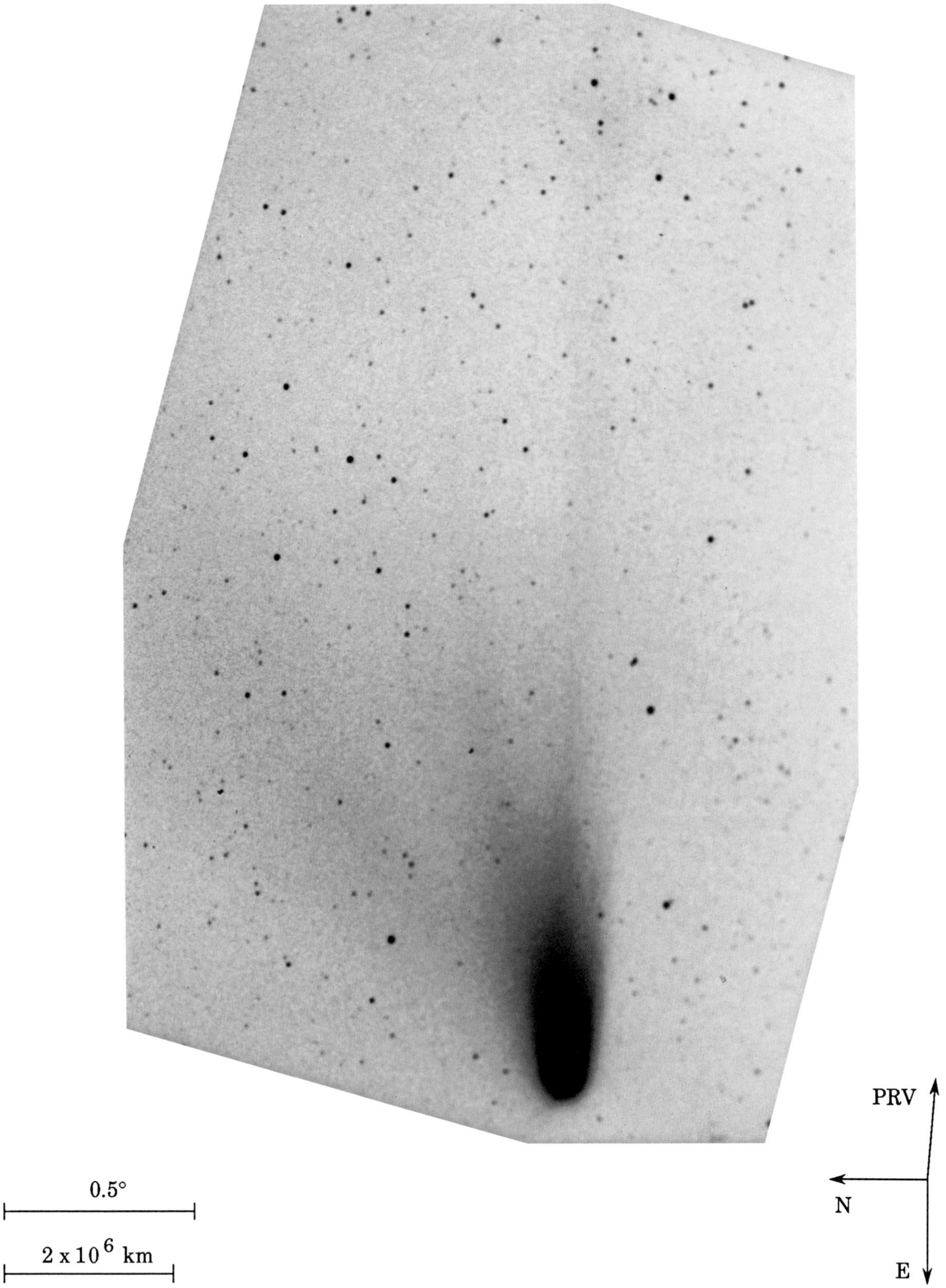

Fig. 232: 1986 Feb 27.12188 UT; exposure 5.0 minutes on Tech. Pan 2415 emulsion with no filter; r=0.70, Δ=1.31, β=131.6°; LSPN-1786.
Photograph by F. Marang, Sutherland, LSPN Island Network.

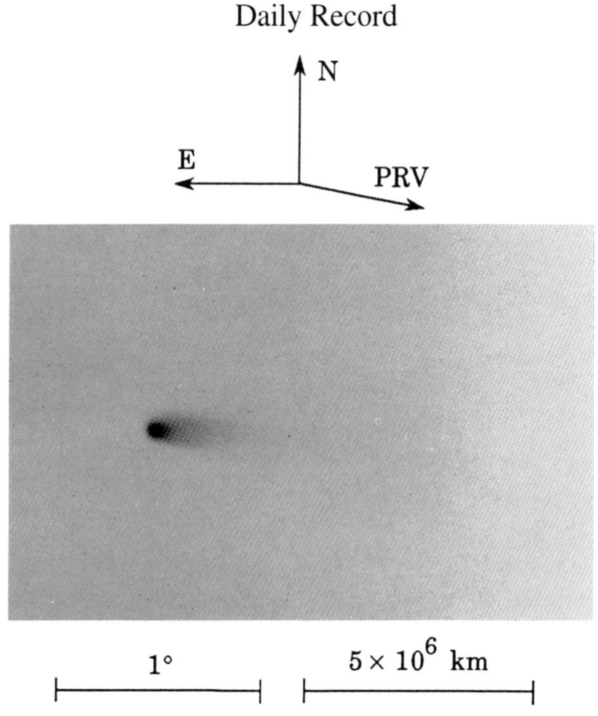

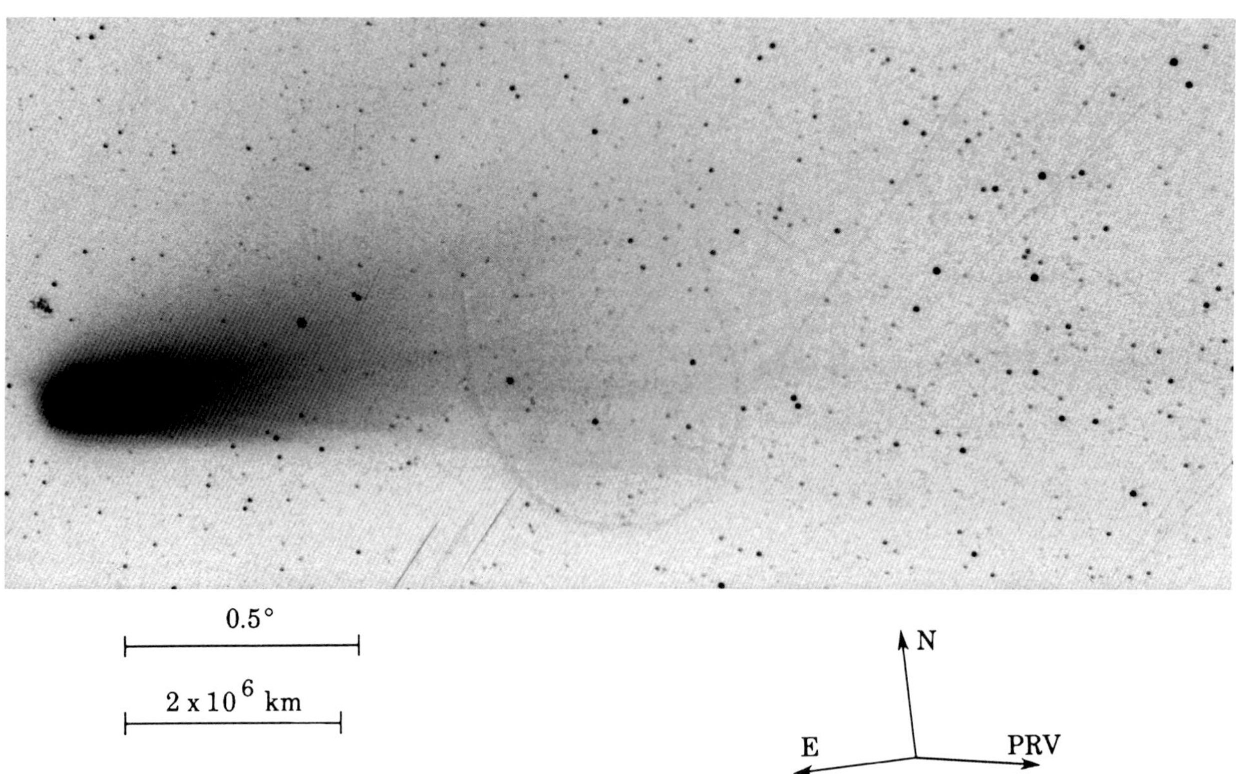

(top) Fig. 233: 1986 Feb 27.78472 UT; exposure 3.0 minutes on IIIa-J emulsion with no filter; r=0.71, Δ=1.29, β=130.6°; LSPN-2377.
Photograph from Royal Observatory/UK Schmidt Telescope Unit.

(bottom) Fig. 234: 1986 Feb 28.12569 UT; exposure 2.0 minutes on Tech. Pan 2415 emulsion with no filter; r=0.71, Δ=1.29, β=130.0°; LSPN-1789.
Photograph by F. Marang, Sutherland, LSPN Island Network.

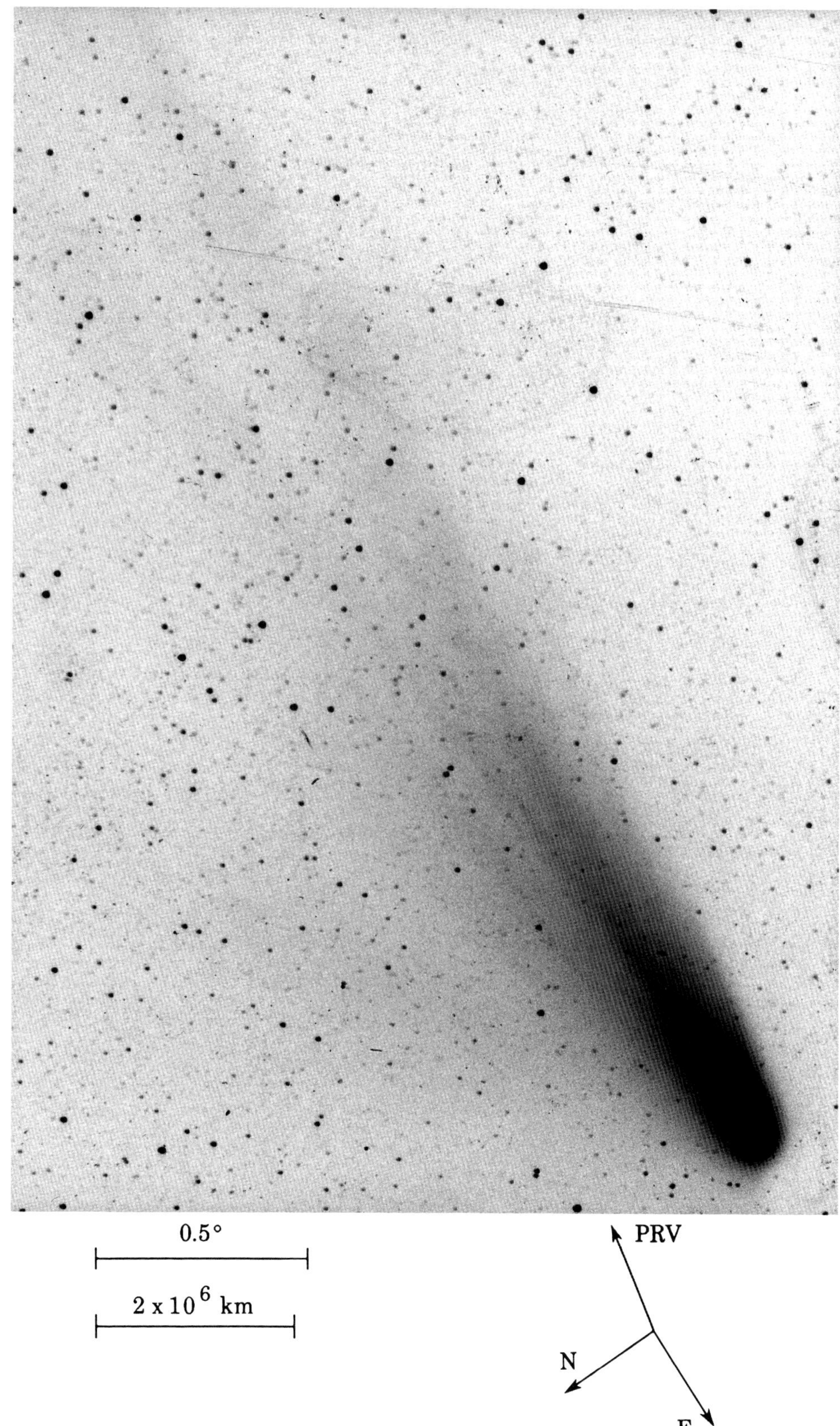

Fig. 235: 1986 Mar 1.12500 UT; exposure 2.0 minutes on Tech. Pan 2415 emulsion with no filter; r=0.73, Δ=1.26, β=128.6°; LSPN-1790.
Photograph by F. Marang, Sutherland, LSPN Island Network.

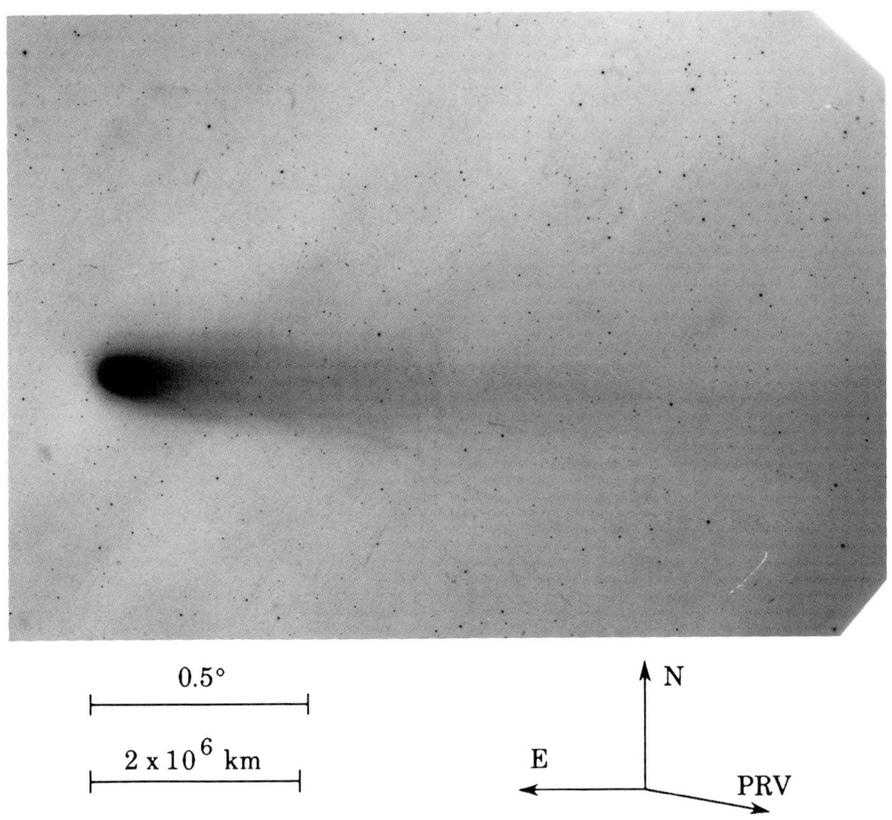

Fig. 236: 1986 Mar 1.85278 UT; exposure 10.0 minutes on 103a-O emulsion with no filter; r=0.73, Δ=1.25, β=127.5°; LSPN-3066.
Photograph by M. Candy, Perth Observatory.

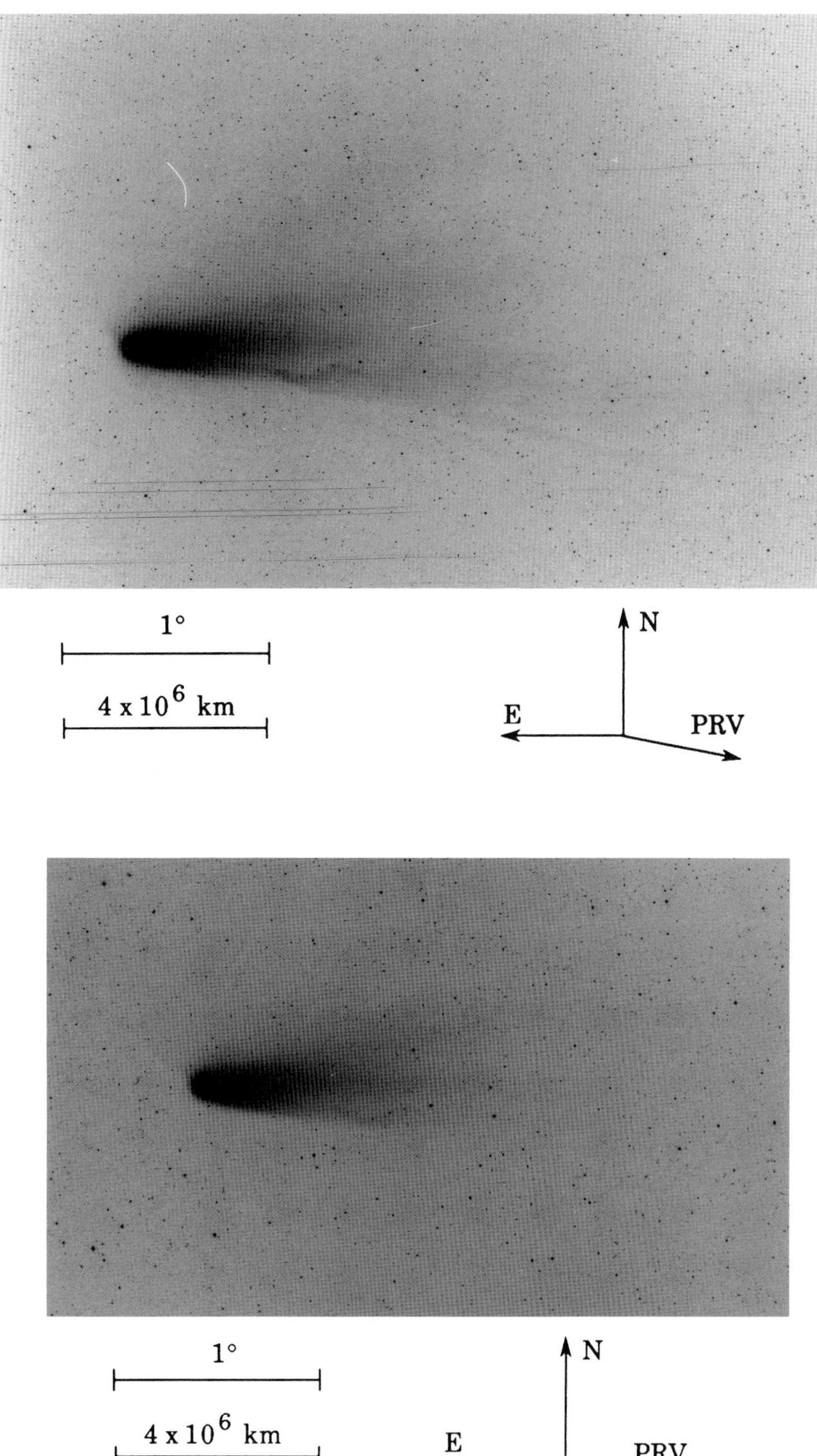

(top) Fig. 237: 1986 Mar 2.00000 UT; exposure 2.0 minutes on Tech. Pan 2415 emulsion with no filter; r=0.74, Δ=1.25, β=127.3°; LSPN-1357.
Photograph by W. Liller,Easter Island, LSPN Island Network.

(bottom) Fig. 238: 1986 Mar 2.47847 UT; exposure 6.0 minutes on Tech. Pan 2415 emulsion with Wratten 25 filter; r=0.74, Δ=1.24, β=126.7°; LSPN-1354.
Photograph by W. Liller,Easter Island, LSPN Island Network.

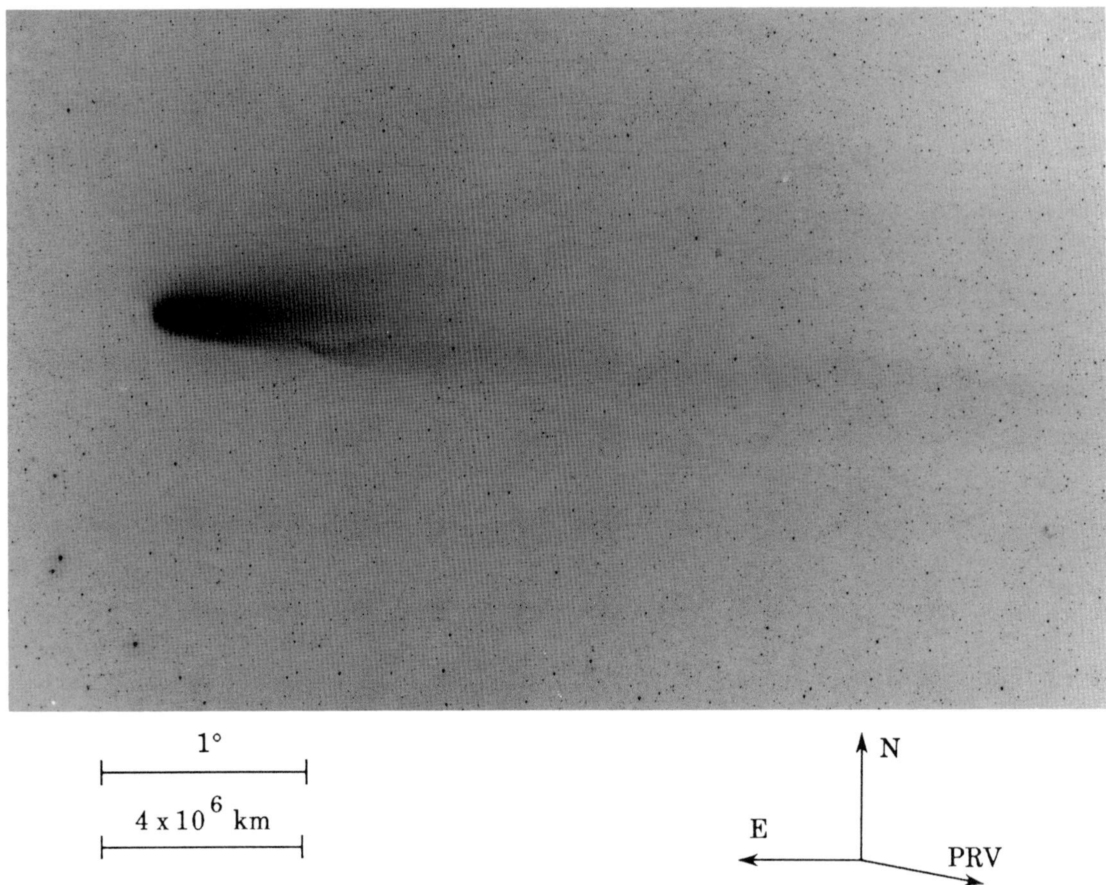

Fig. 239: 1986 Mar 2.49010 UT; exposure 2.5 minutes on Tech. Pan 2415 emulsion with no filter; r=0.74, Δ=1.24, β=126.6°; LSPN-1356.
Photograph by W. Liller, Easter Island, LSPN Island Network.

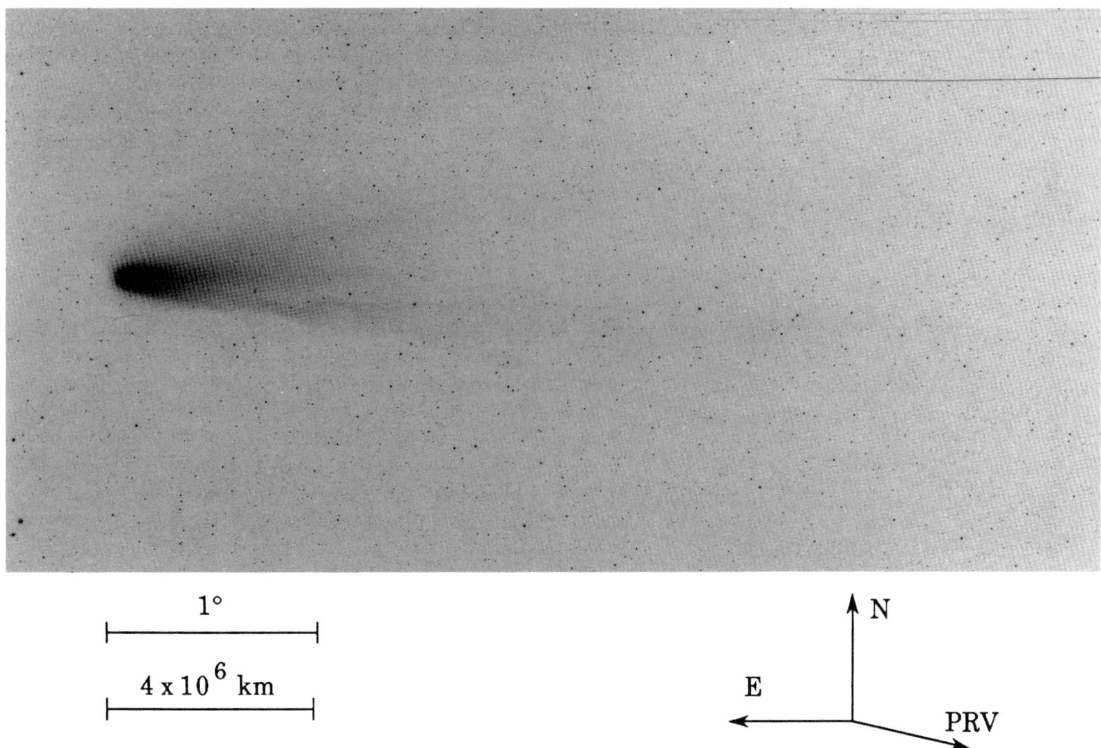

Fig. 240: 1986 Mar 2.50148 UT; exposure 0.2 minute on Tech. Pan 2415 emulsion with no filter; r=0.74, Δ=1.23, β=126.6°; LSPN-1358.
Photograph by W. Liller, Easter Island, LSPN Island Network.

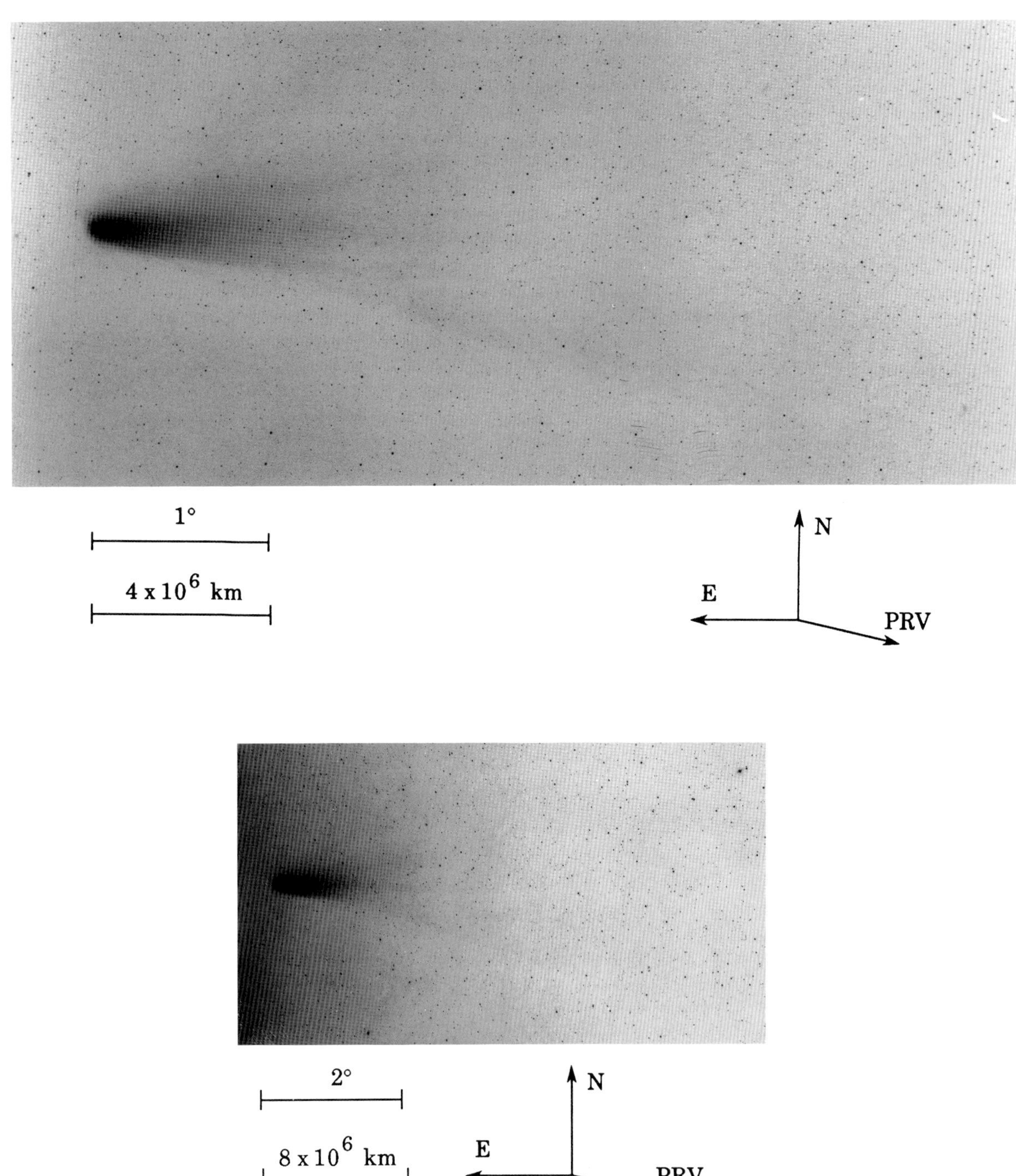

(top) Fig. 241: 1986 Mar 3.47951 UT; exposure 2.0 minutes on Tech. Pan 2415 emulsion with no filter; r=0.75, Δ=1.21, β=125.3°; LSPN-1363.
Photograph by W. Liller, Easter Island, LSPN Island Network.

(bottom) Fig. 242: 1986 Mar 3.64514 UT; exposure 20.0 minutes on IIIa-J emulsion with no filter; r=0.76, Δ=1.21, β=125.1°; LSPN-1898.
Photograph by A. Storrs, Mauna Kea Observatory.

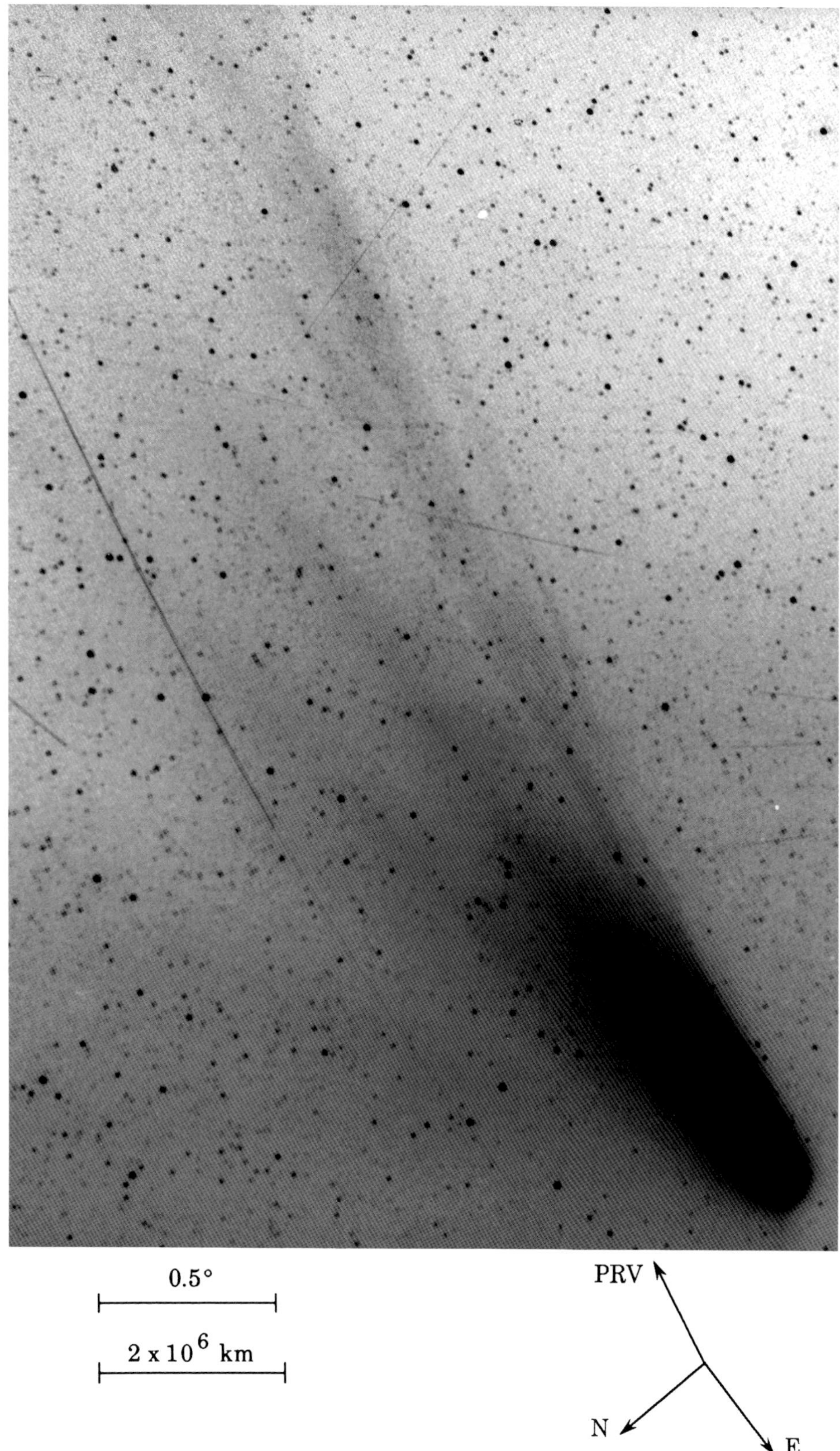

Fig. 243: 1986 Mar 4.10694 UT; exposure 10.0 minutes on Tech. Pan 2415 emulsion with no filter; r=0.76, Δ=1.20, β=124.5°; LSPN-1794.
Photograph by F. Marang, Sutherland, LSPN Island Network.

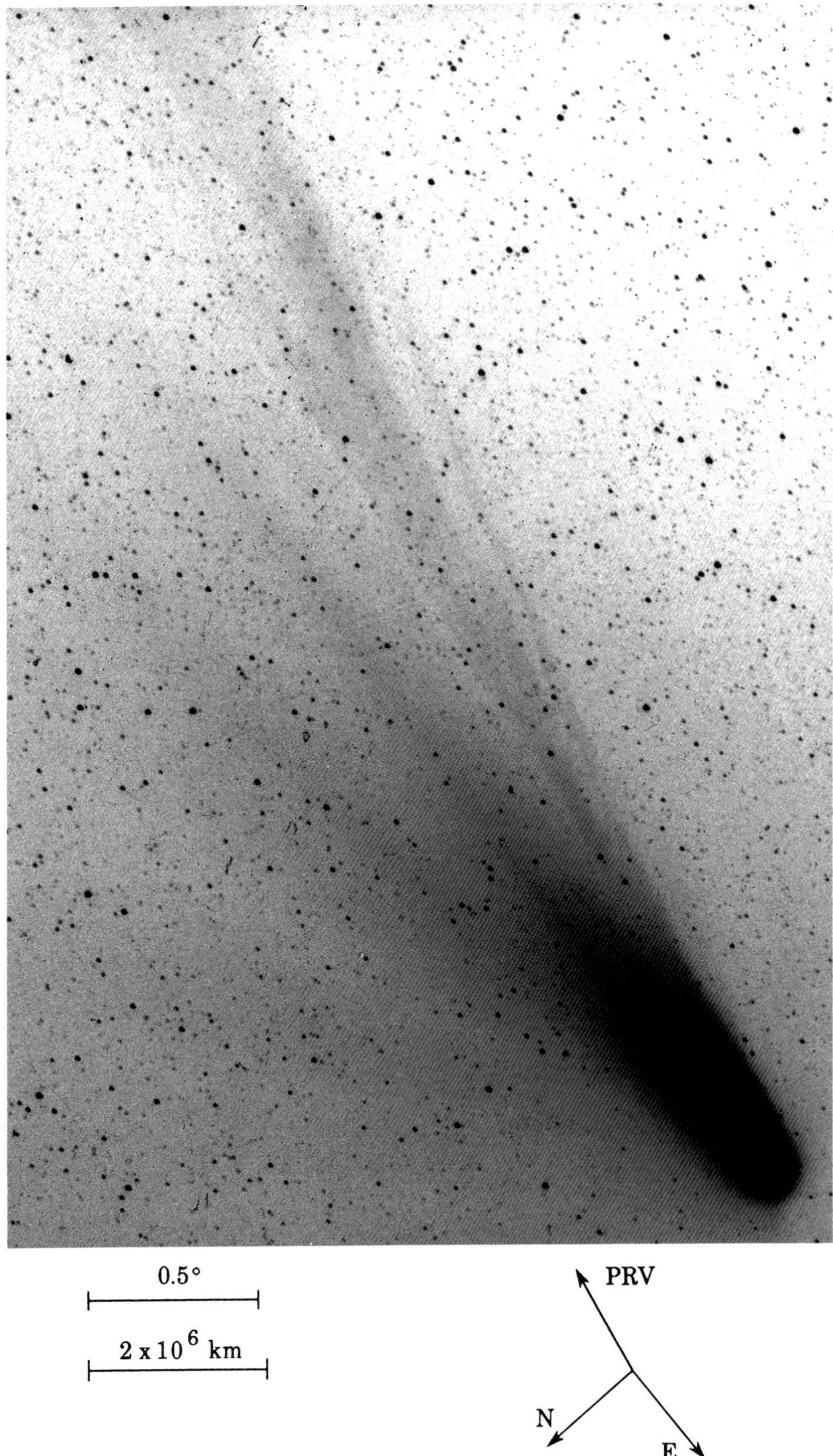

Fig. 244: 1986 Mar 4.11354 UT; exposure 5.0 minutes on Tech. Pan 2415 emulsion with no filter; r=0.76, Δ=1.20, β=124.5°; LSPN-1795.
Photograph by F. Marang, Sutherland, LSPN Island Network.

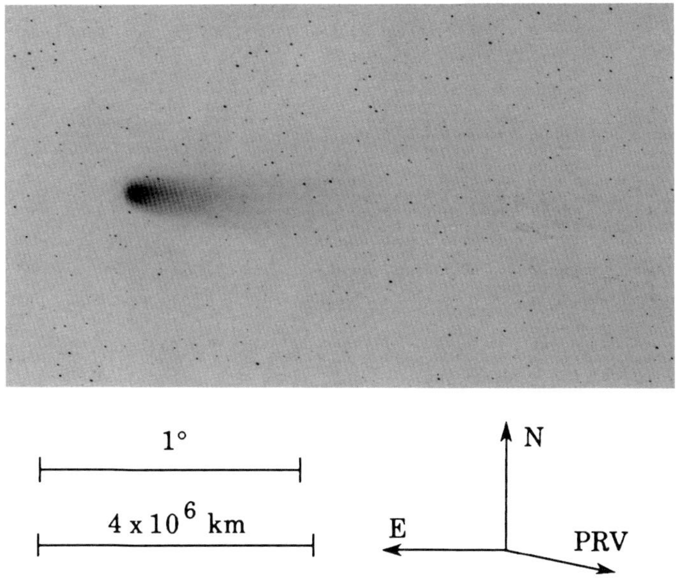

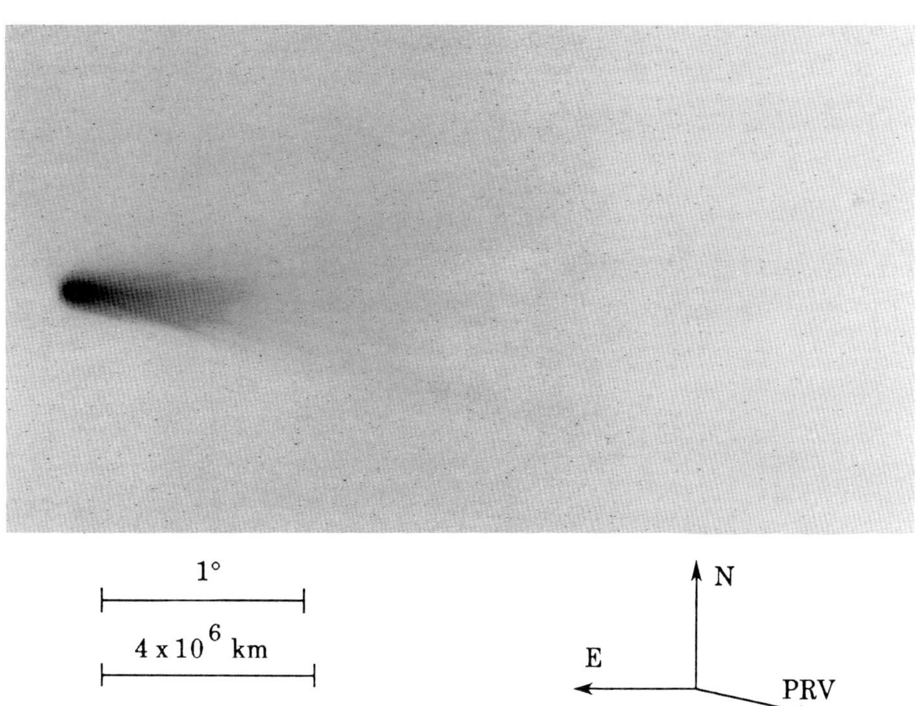

(top) Fig. 245: 1986 Mar 4.69502 UT; exposure 10.0 minutes on 103a-G emulsion with GG-495 filter; r=0.77, Δ=1.19, β=123.8°; LSPN-380.
Photograph by G. Douglass, U.S. Naval Observatory Station.

(bottom) Fig. 246: 1986 Mar 4.77419 UT; exposure 2.0 minutes on IIa-0 emulsion with GG-385 filter; r=0.77, Δ=1.18, β=123.7°; LSPN-2378.
Photograph from Royal Observatory/UK Schmidt Telescope Unit.

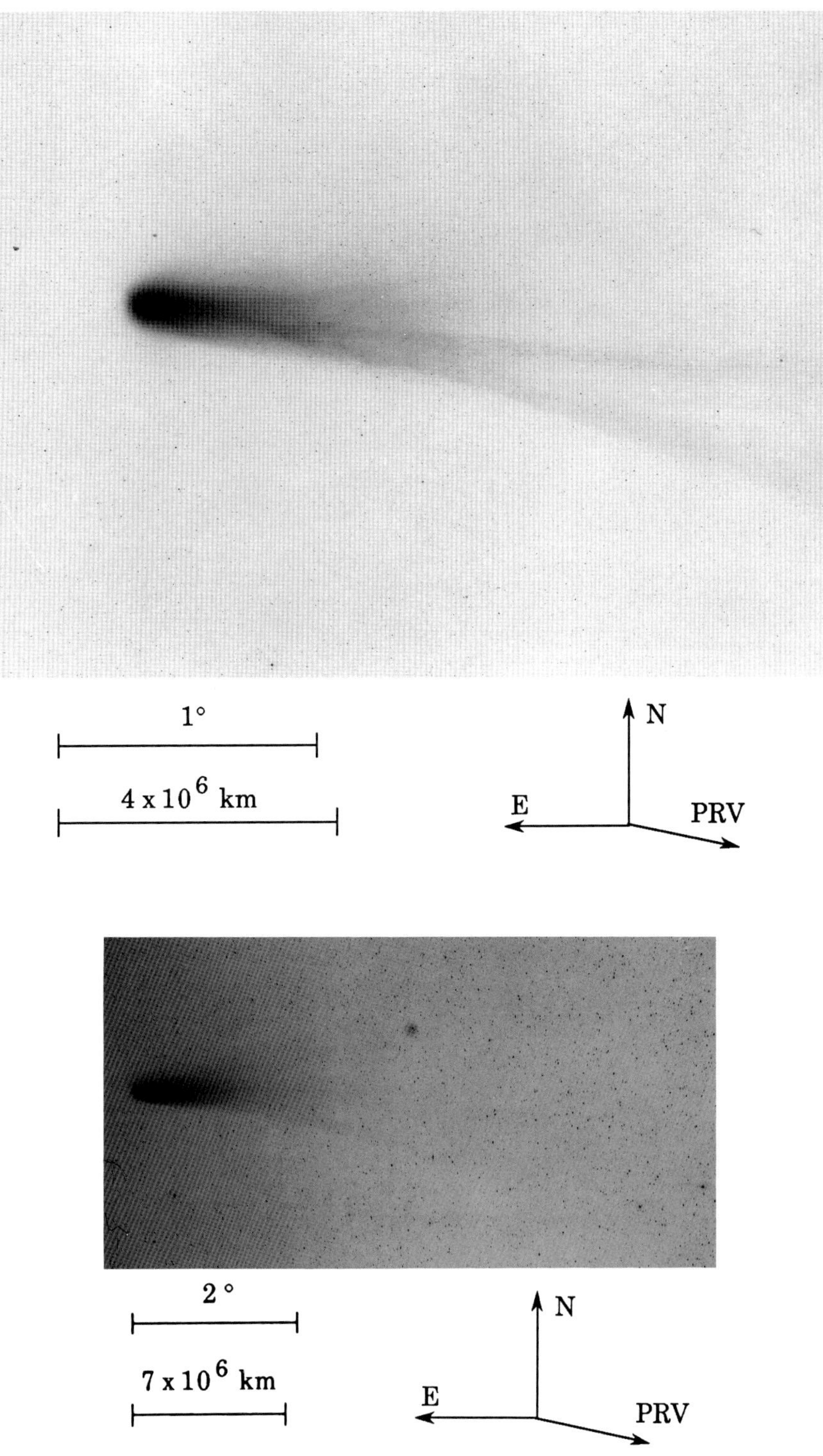

(top) Fig. 247: 1986 Mar 5.37674 UT; exposure 5.0 minutes on 103a-O emulsion with no filter; r=0.78, Δ=1.17, β=123.0°; LSPN-724.
Photograph by C. Torres/H. Wroblewski, Cerro el Roble Astronomical Observatory.

(bottom) Fig. 248: 1986 Mar 5.62677 UT; exposure 20.0 minutes on IIIa-F emulsion with Wratten 2B filter; r=0.78, Δ=1.17, β=122.7°; LSPN-1900.
Photograph by D. Cruikshank, Mauna Kea Observatory.

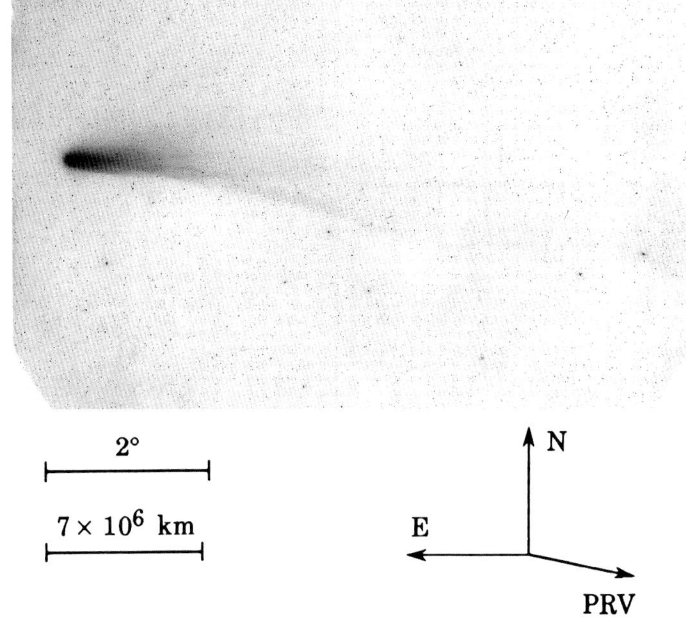

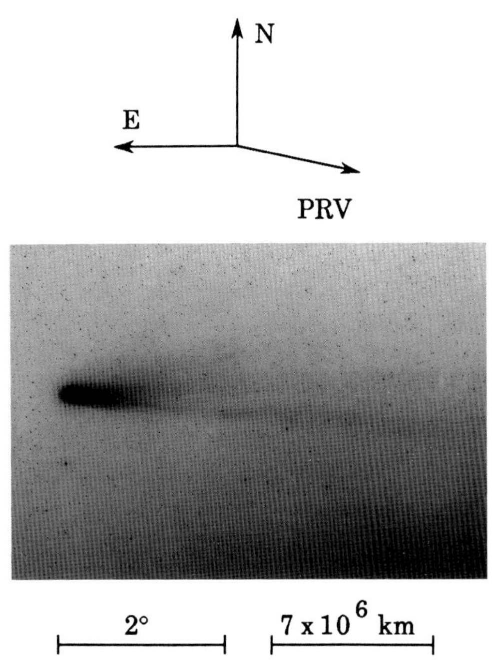

(top) Fig. 249: 1986 Mar 5.64190 UT; exposure 10.0 minutes on IIIa-J emulsion with no filter; r=0.78, Δ=1.16, β=122.7°; LSPN-1901.
Photograph by D. Cruikshank, Mauna Kea Observatory.

(bottom) Fig. 250: 1986 Mar 6.63985 UT; exposure 20.0 minutes on IIIa-J emulsion with no filter; r=0.80, Δ=1.14, β=121.6°; LSPN-1902.
Photograph by D. Cruikshank/M. Buie, Mauna Kea Observatory.

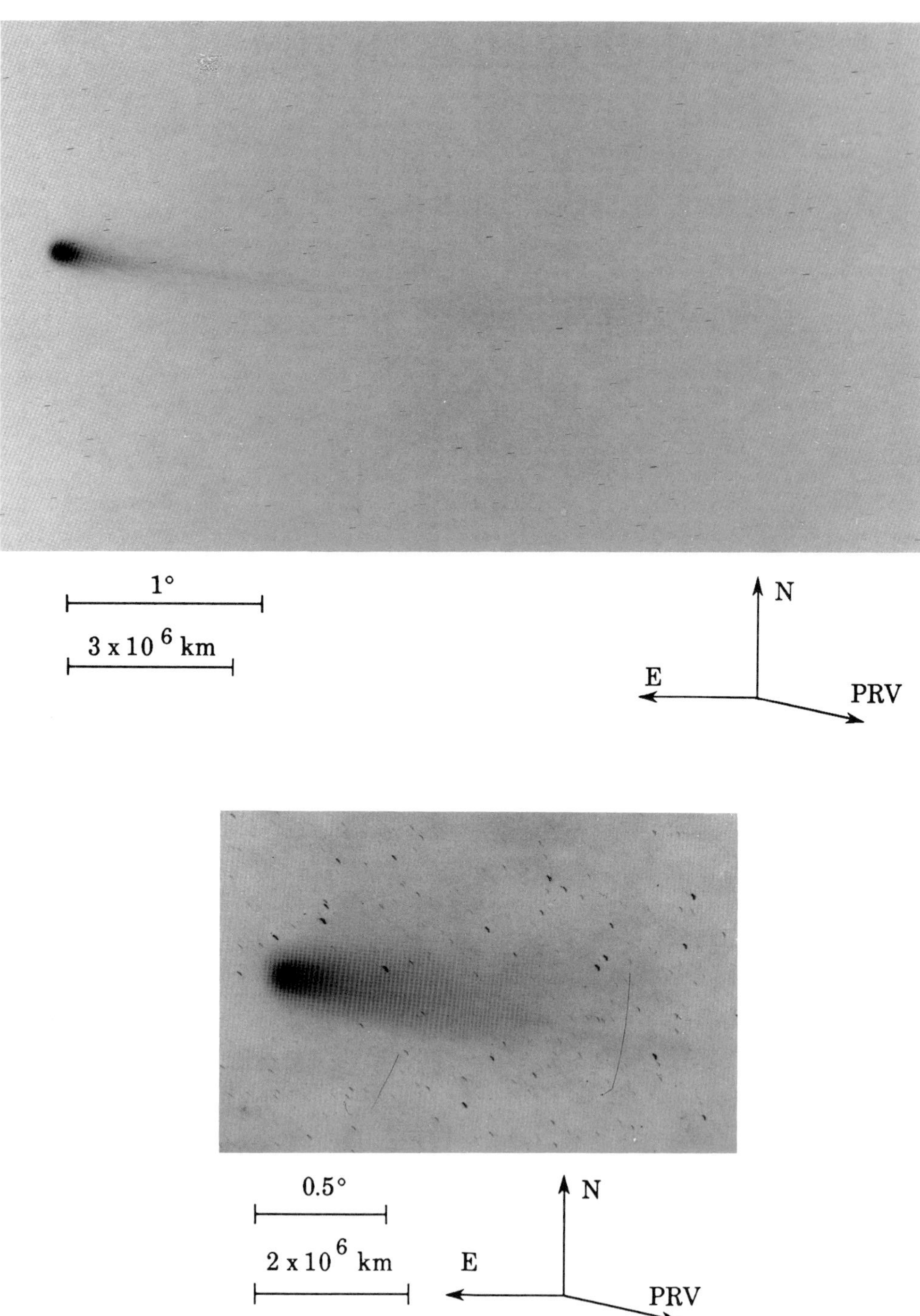

(top) Fig. 251: 1986 Mar 6.74623 UT; exposure 7.0 minutes on IIIa-J emulsion with UG-1 filter; r=0.80, Δ=1.14, β=121.5°; LSPN-2381.
Photograph from Royal Observatory/UK Schmidt Telescope Unit.

(bottom) Fig. 252: 1986 Mar 6.97396 UT; exposure 45.0 minutes on 103a-O emulsion with no filter; r=0.80, Δ=1.13, β=121.3°; LSPN-3619.
Photograph submitted by K. Sivaraman, Indian Institute for Astrophysics, Kodaikanal Station.

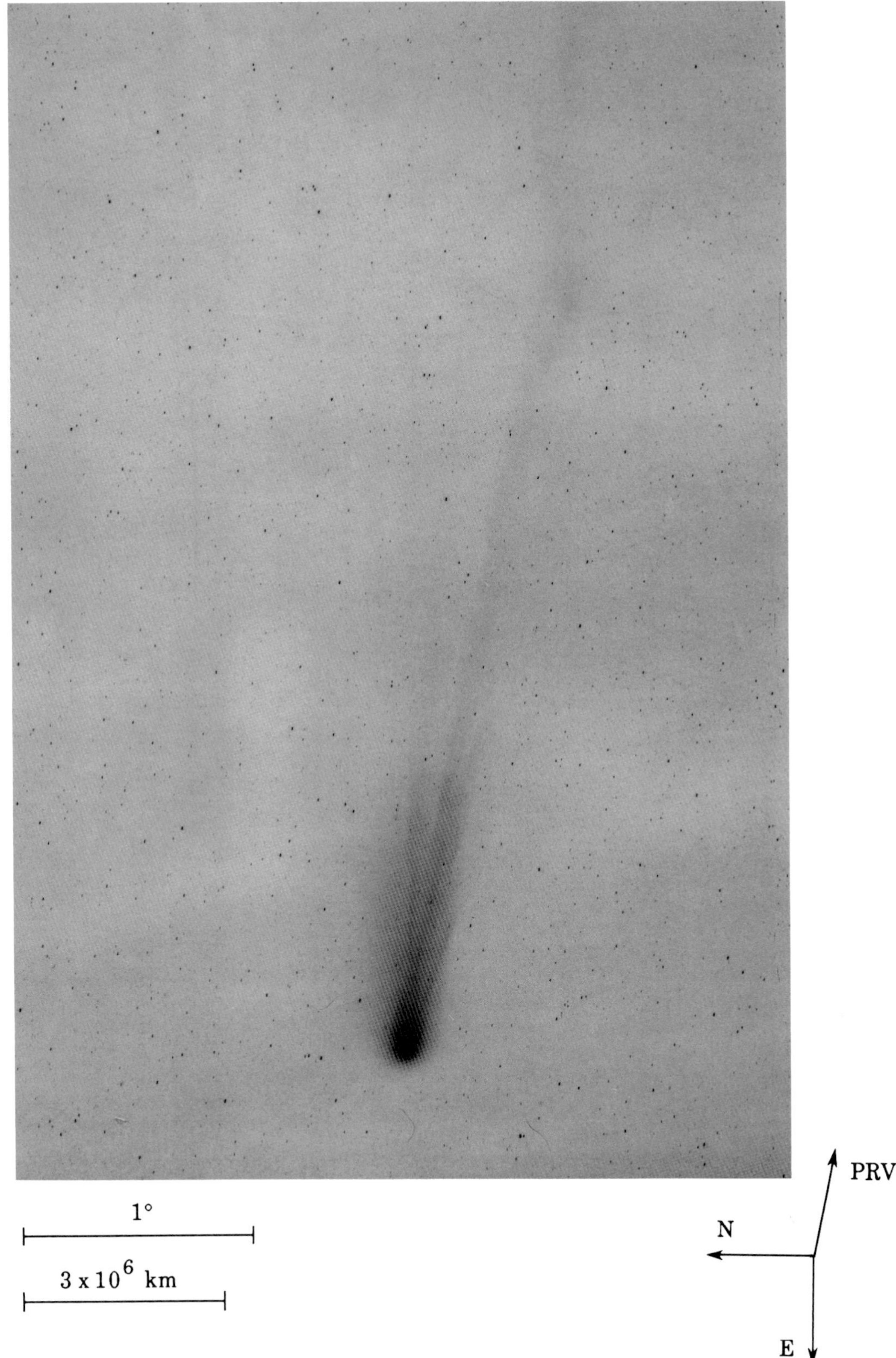

Fig. 253: 1986 Mar 7.36319 UT; exposure 10.0 minutes on IIa-O emulsion with no filter; r=0.81, Δ=1.12, β=120.9°; LSPN-1266.
Photograph by F. Miller, University of Michigan/CTIO.

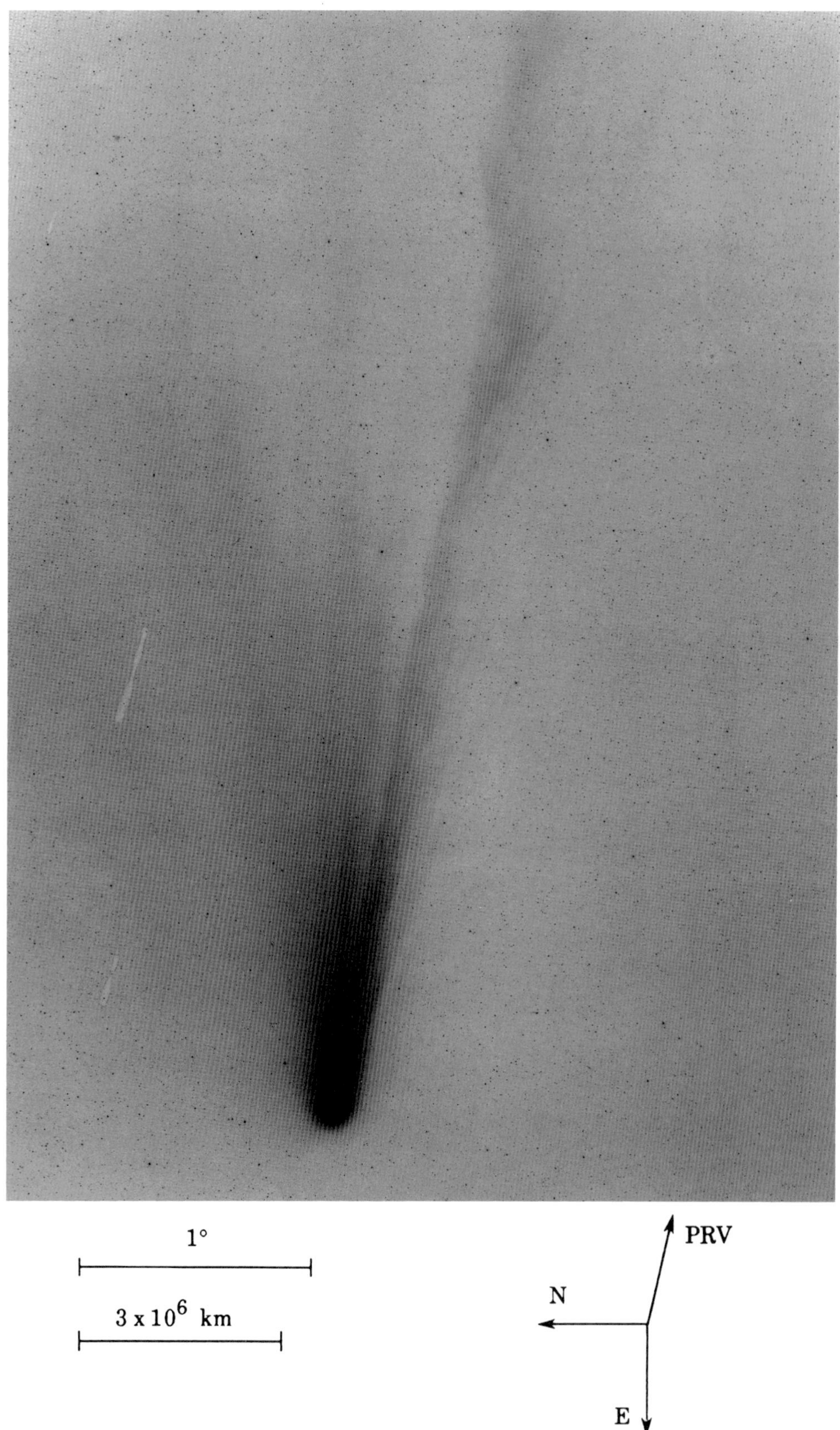

Fig. 254: 1986 Mar 7.38125 UT; exposure 10.0 minutes on IIa-O emulsion with no filter; r=0.81, Δ=1.12, β=120.8°; LSPN-1267.
Photograph by F. Miller, University of Michigan/CTIO.

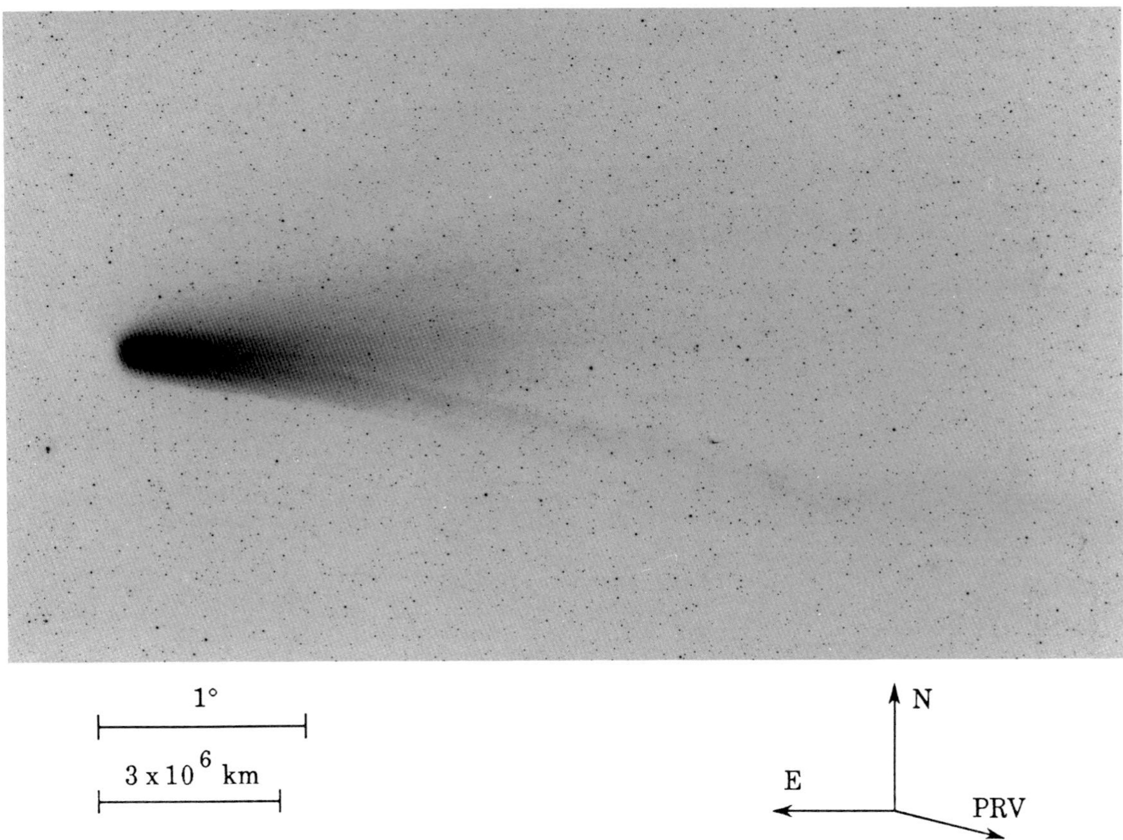

Fig. 255: 1986 Mar 7.49062 UT; exposure 3.0 minutes on Tech. Pan 2415 emulsion with no filter; r=0.81, Δ=1.12, β=120.7°; LSPN-1375.
Photograph by W. Liller, Easter Island, LSPN Island Network.

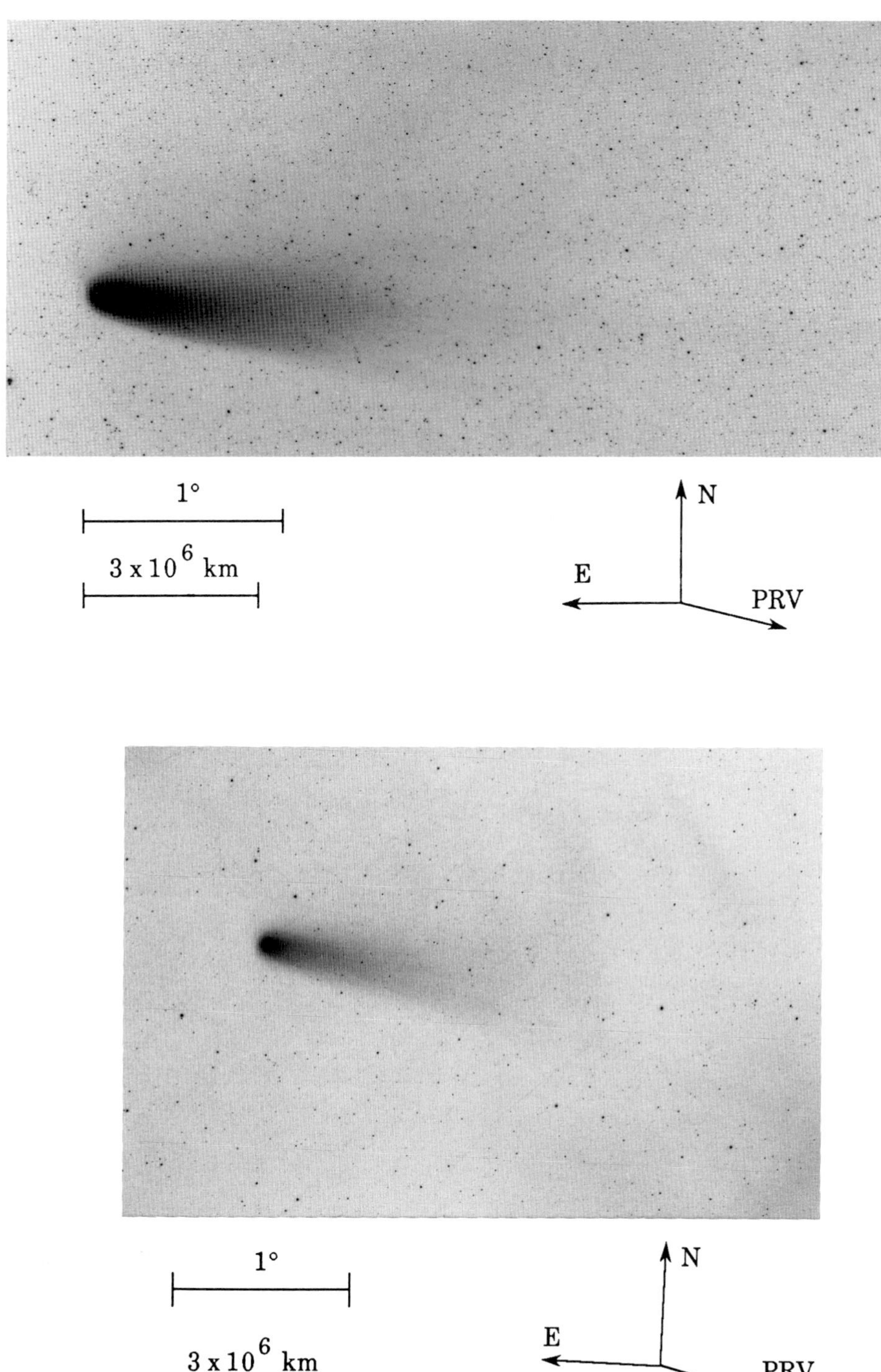

(top) Fig. 256: 1986 Mar 7.50000 UT; exposure 8.0 minutes on Tech. Pan 2415 emulsion with Wratten 25 filter; r=0.81, Δ=1.12, β=120.7°; LSPN-1376.
Photograph by W. Liller, Easter Island, LSPN Island Network.

(bottom) Fig. 257: 1986 Mar 7.52500 UT; exposure 14.0 minutes on IIa-O emulsion with no filter; r=0.81, Δ=1.12, β=120.7°; LSPN-3577.
Photograph by H. Giclas, Lowell Observatory.

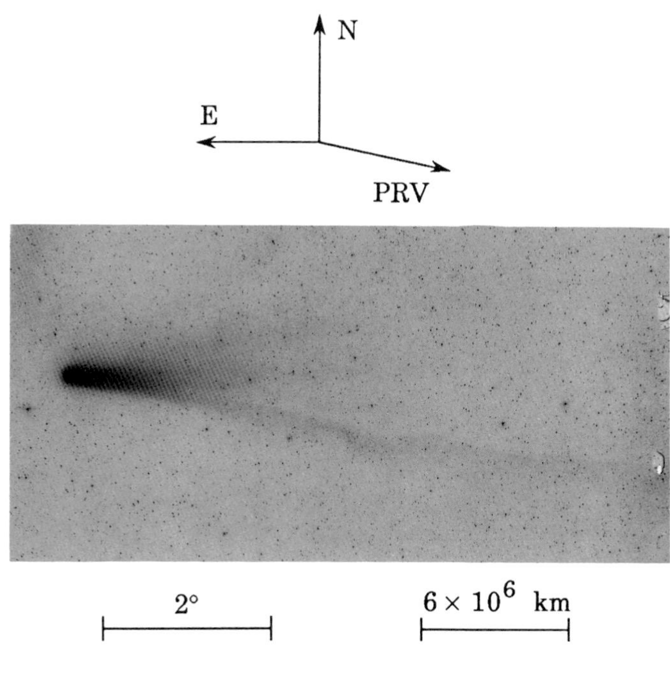

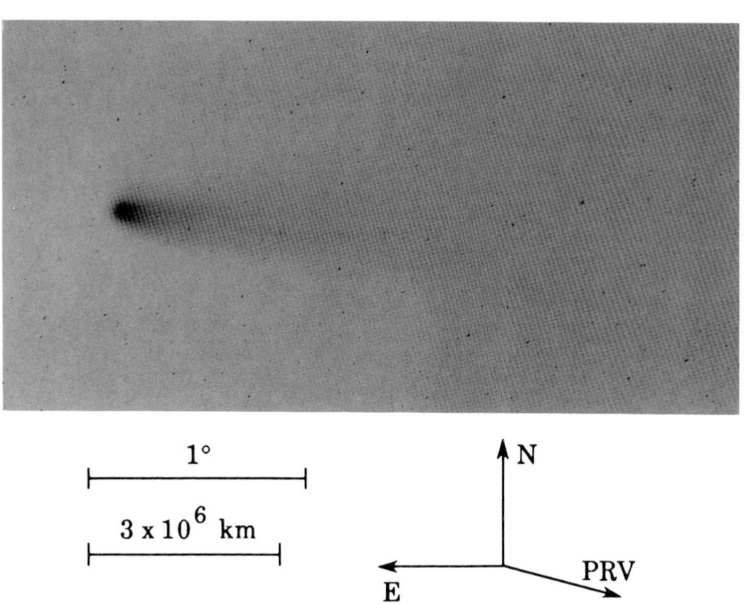

(top) Fig. 258: 1986 Mar 7.62425 UT; exposure 20.0 minutes on IIIa-J emulsion with no filter; r=0.81, Δ=1.12, β=120.6°; LSPN-1905.
Photograph by D. Cruikshank/M. Buie, Mauna Kea Observatory.

(bottom) Fig. 259: 1986 Mar 7.82486 UT; exposure 5.0 minutes on IIa-O emulsion with no filter; r=0.81, Δ=1.11, β=120.4°; LSPN-1627.
Photograph by H. Maehara, Kiso Observatory.

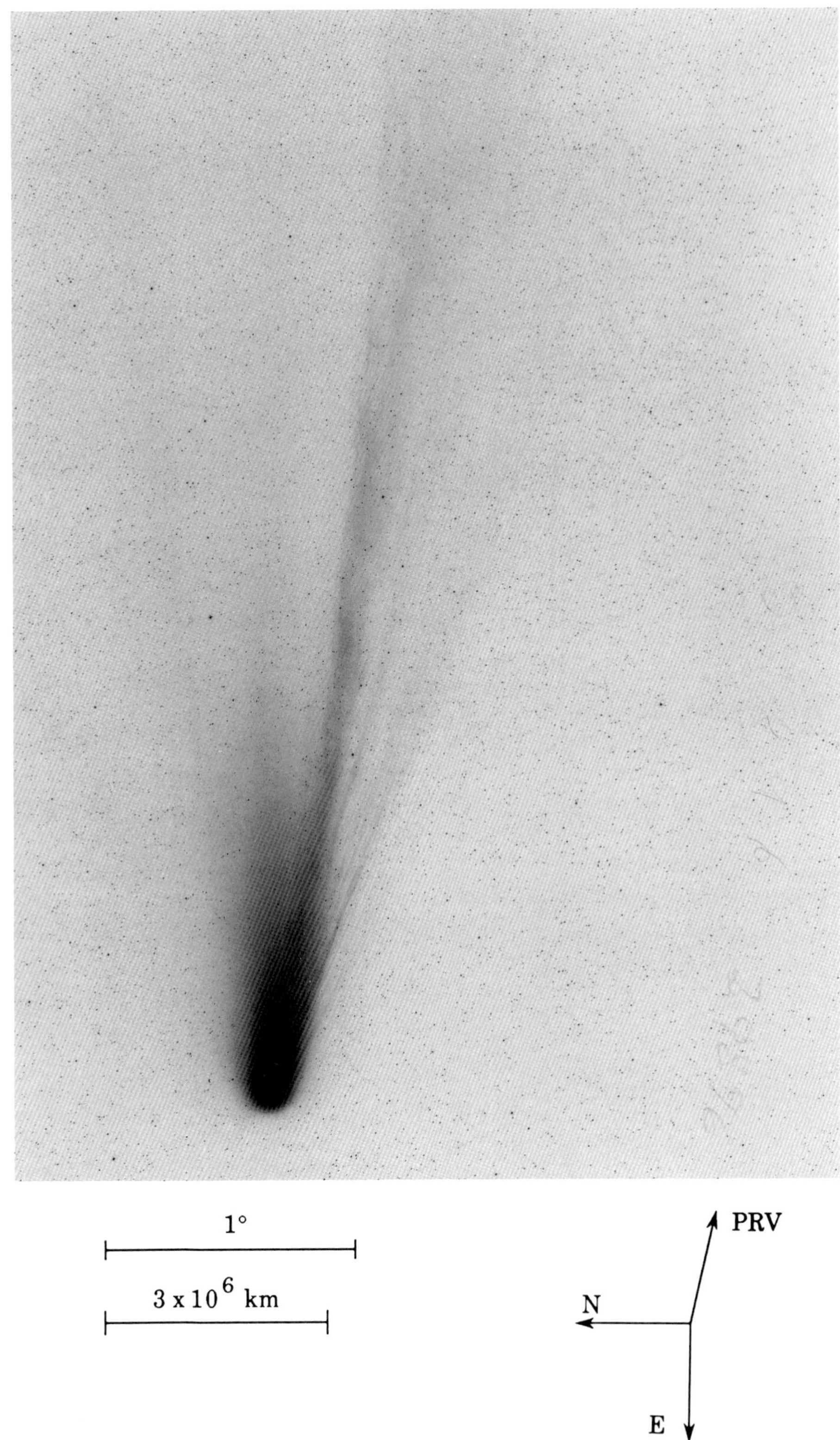

Fig. 260: 1986 Mar 8.35486 UT; exposure 10.0 minutes on IIa-O emulsion with no filter; r=0.82, Δ=1.10, β=119.9°; LSPN-1269.
Photograph by F. Miller, University of Michigan/CTIO.

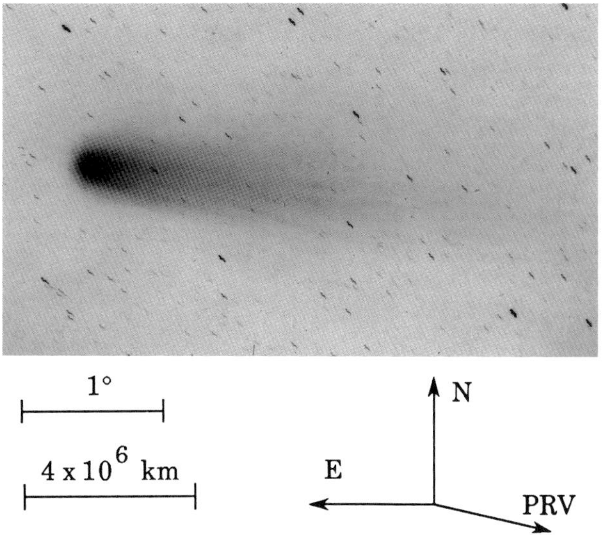

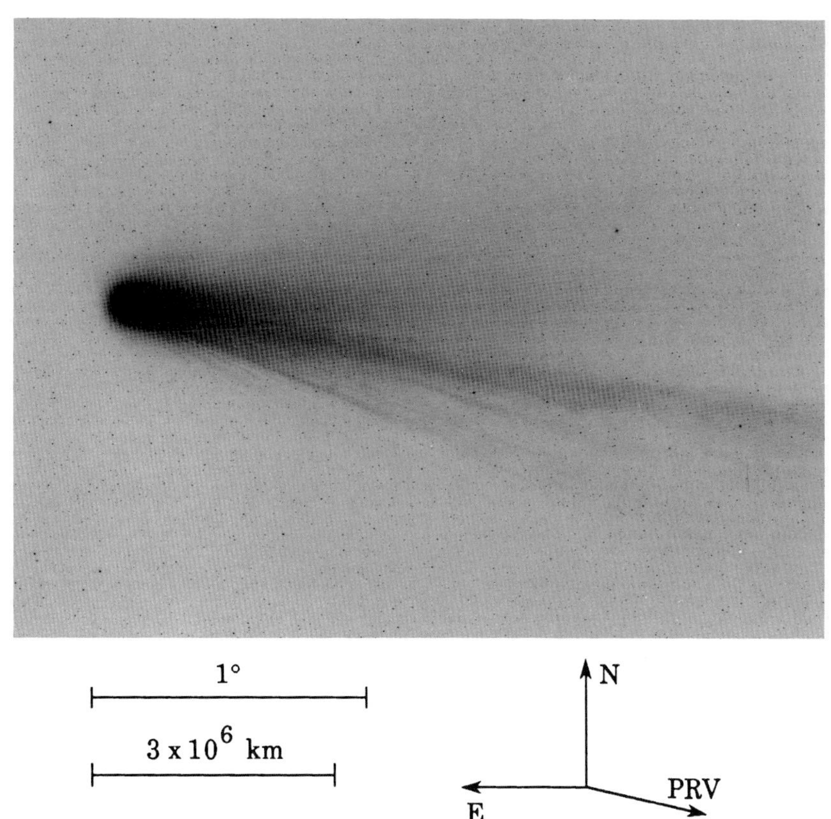

(top) Fig. 261*: 1986 Mar 7.97639 UT; exposure 60.0 minutes on 103a-O emulsion with no filter; r=0.82, Δ=1.11, β=120.3°; LSPN-3621.
Photograph submitted by K. Sivaraman, Indian Institute for Astrophysics, Kodaikanal Station.

(bottom) Fig. 262: 1986 Mar 8.38507 UT; exposure 3.0 minutes on IIa-O emulsion with no filter; r=0.82, Δ=1.10, β=119.9°; LSPN-1271.
Photograph by F. Miller, University of Michigan/CTIO.

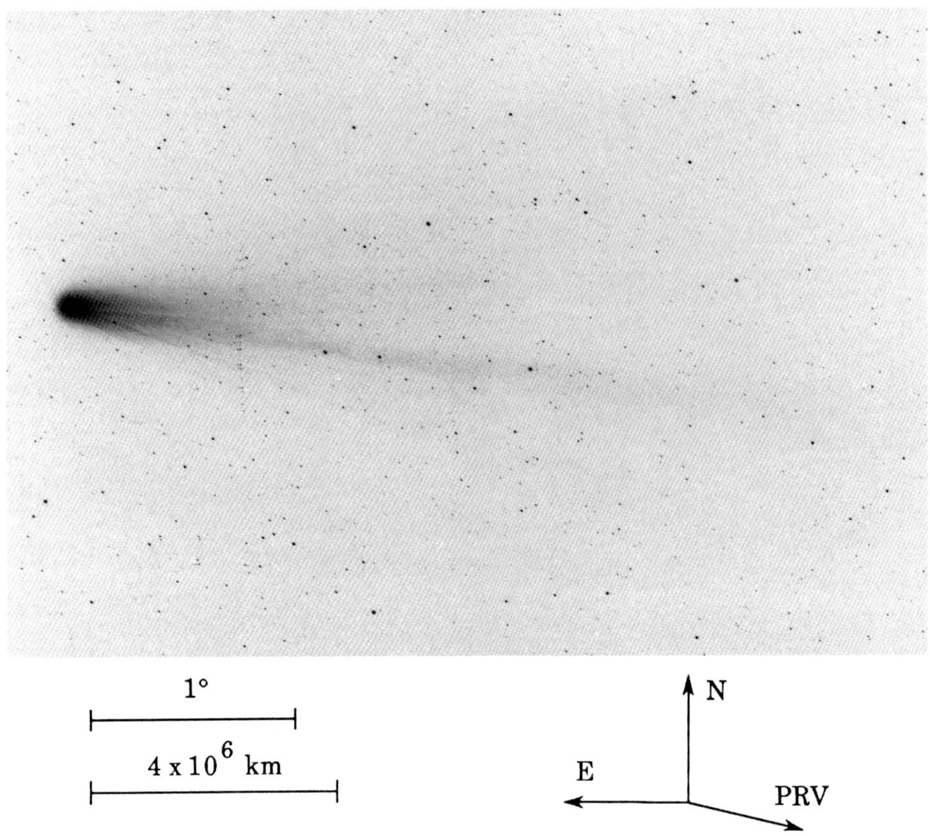

Fig. 263: 1986 Mar 8.48958 UT; exposure 32.0 minutes on Tech. Pan 2415 emulsion with Wratten 47B filter; r=0.82, Δ=1.10, β=119.8°; LSPN-1379.
Photograph by W. Liller, Easter Island, LSPN Island Network.

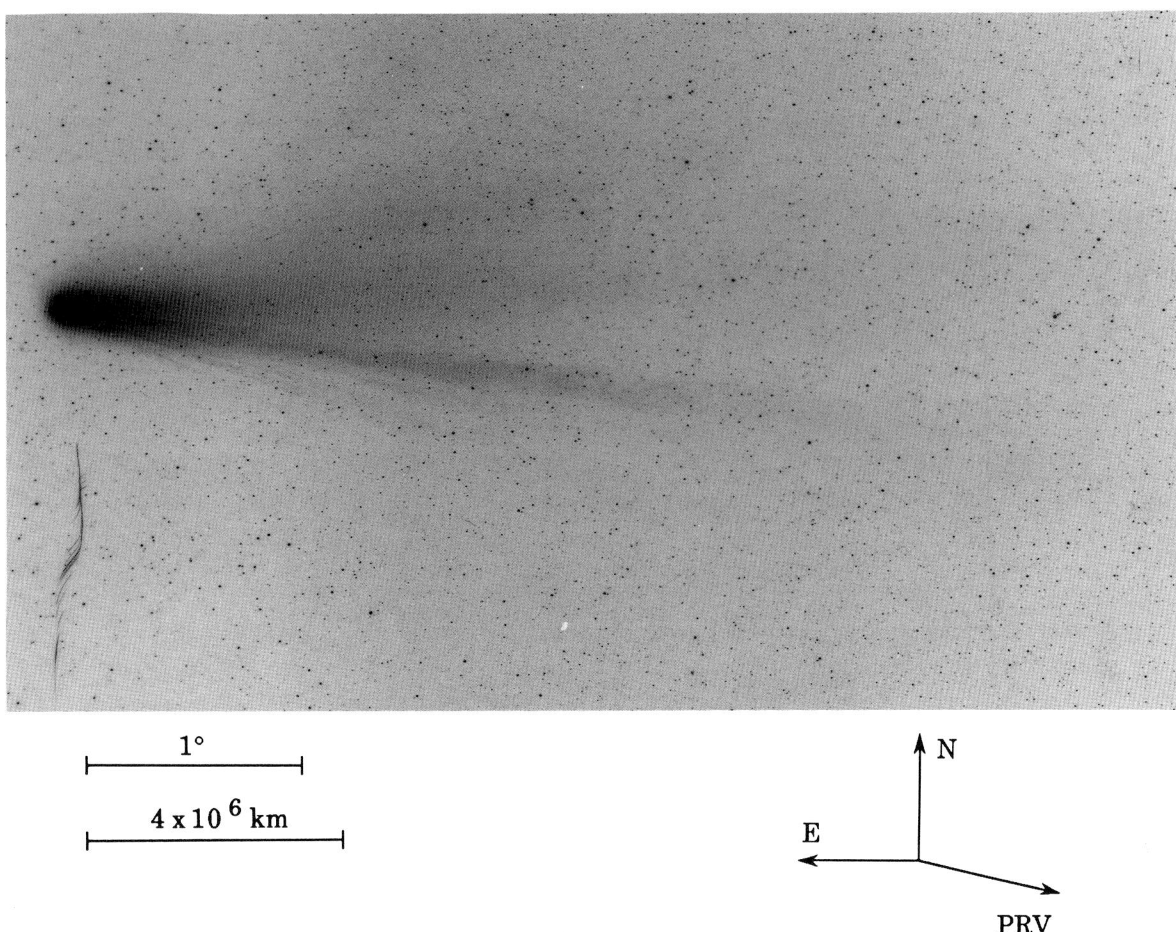

Fig. 264: 1986 Mar 8.50567 UT; exposure 3.0 minutes on Tech. Pan 2415 emulsion with no filter; r=0.82, Δ=1.10, β=119.7°; LSPN-1380.
Photograph by W. Liller, Easter Island, LSPN Island Network.

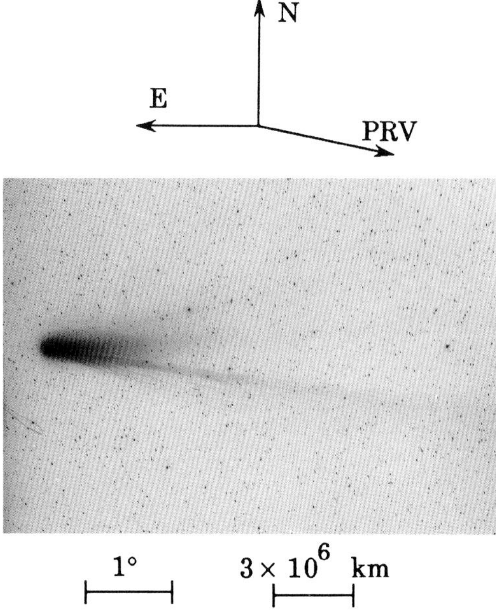

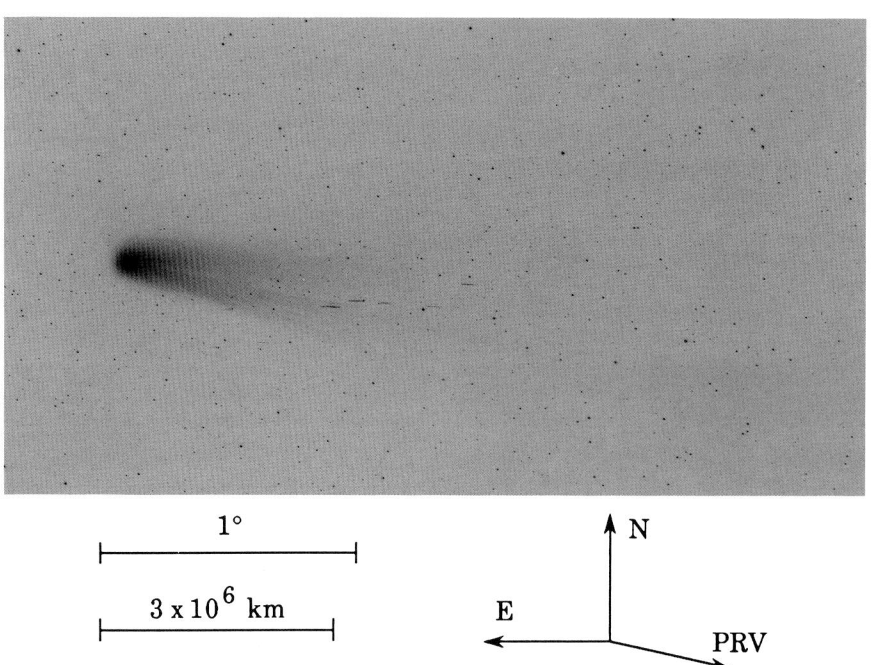

(top) Fig. 265: 1986 Mar 8.61840 UT; exposure 20.0 minutes on IIIa-J emulsion with no filter; r=0.82, Δ=1.09, β=119.6°; LSPN-1907.
Photograph by D. Cruikshank, Mauna Kea Observatory.

(bottom) Fig. 266: 1986 Mar 8.68264 UT; exposure 10.0 minutes on 103a-O emulsion with GG-400 filter; r=0.83, Δ=1.09, β=119.6°; LSPN-382.
Photograph by G. Douglass, U.S. Naval Observatory Station.

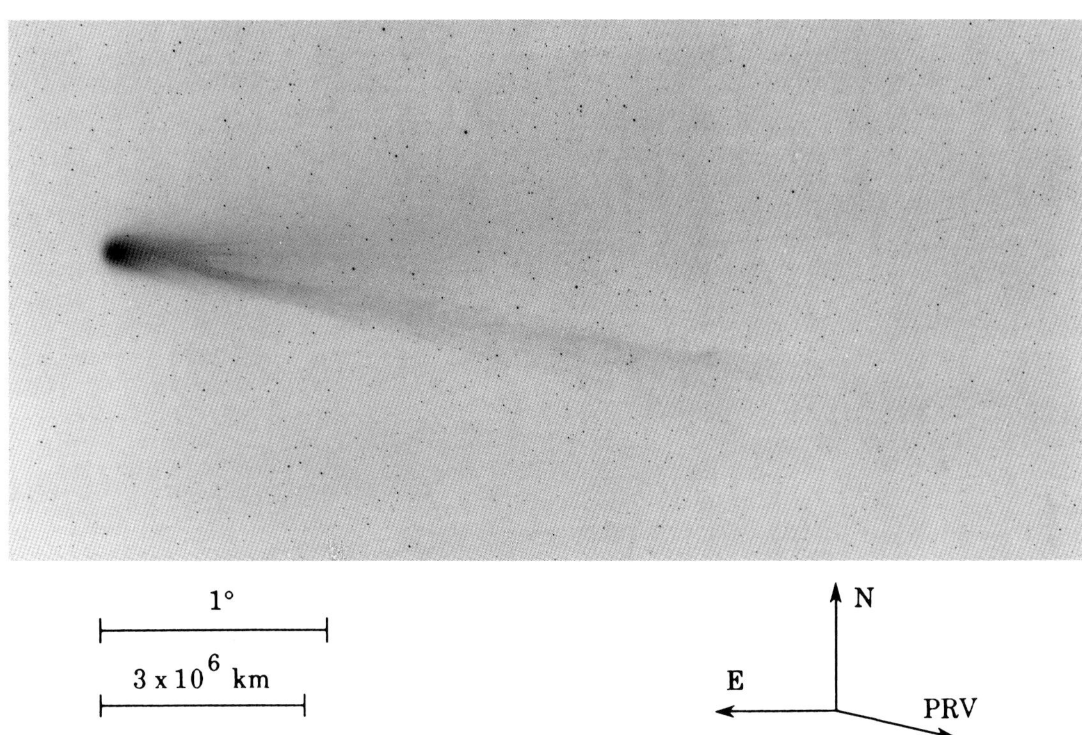

Fig. 267: 1986 Mar 8.82484 UT; exposure 7.0 minutes on IIa-O emulsion with no filter; r=0.83, Δ=1.09, β=119.4°; LSPN-1628.
Photograph by H. Maehara, Kiso Observatory.

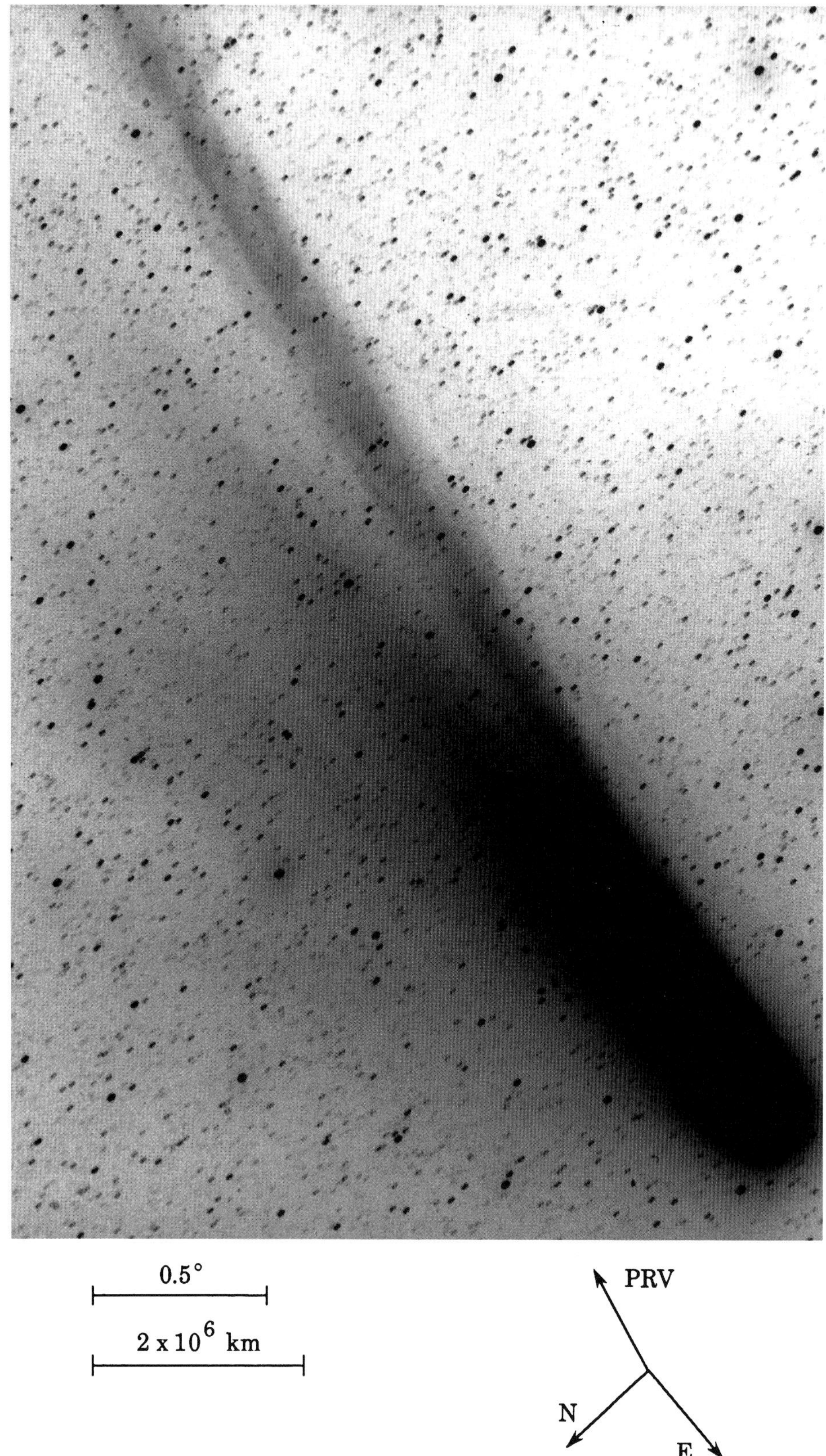

Fig. 268: 1986 Mar 9.11944 UT; exposure 12.0 minutes on Tech. Pan 2415 emulsion with no filter; r=0.83, Δ=1.08, β=119.2°; LSPN-1800.
Photograph by F. Marang, Sutherland, LSPN Island Network.

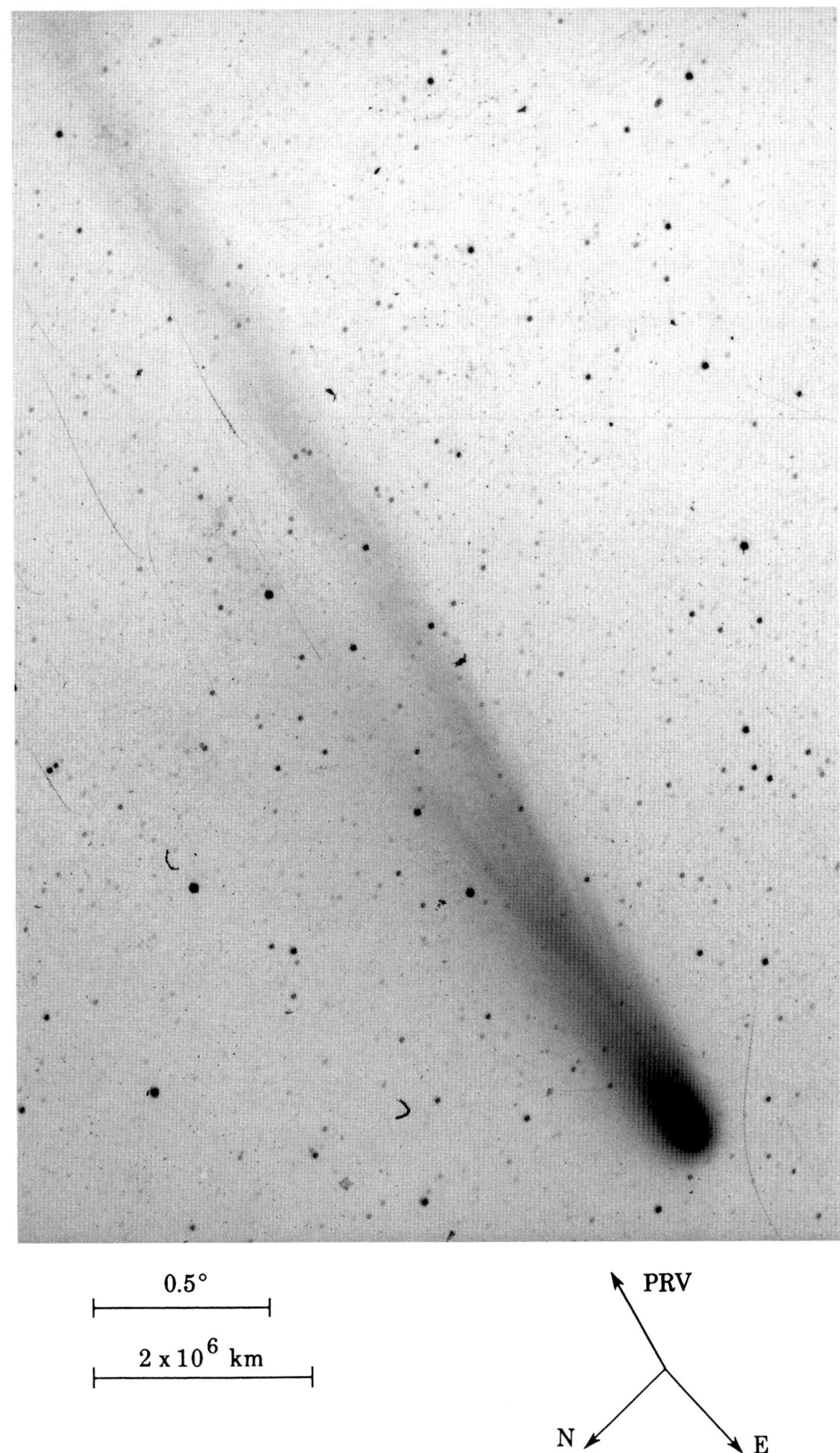

Fig. 269: 1986 Mar 9.14236 UT; exposure 10.0 minutes on Tech. Pan 2415 emulsion with Wratten 47B filter; r=0.83, Δ=1.08, β=119.2°; LSPN-1802.
Photograph by F. Marang, Sutherland, LSPN Island Network.

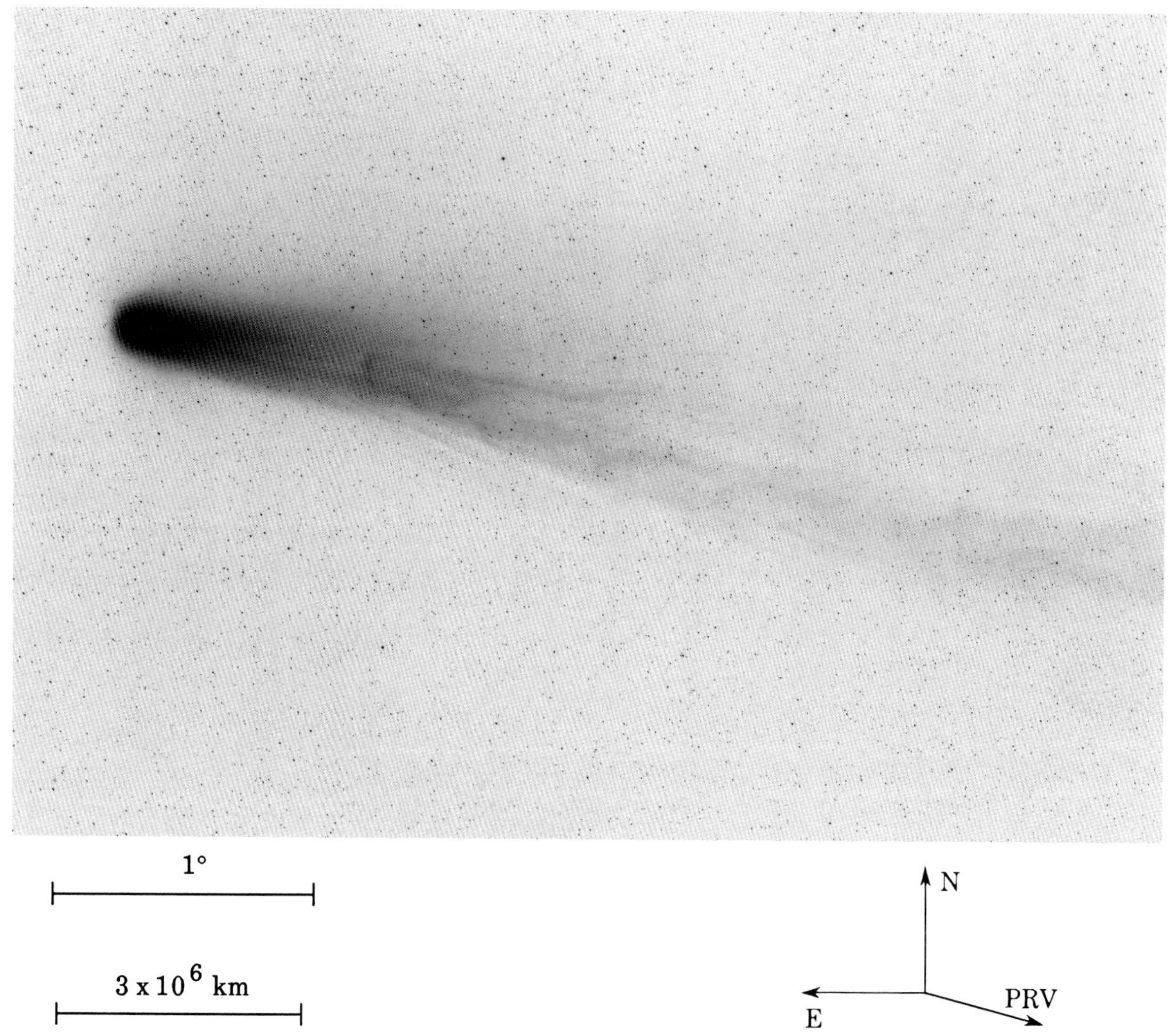

Fig. 270: 1986 Mar 9.34444 UT; exposure 10.0 minutes on IIa-O emulsion with no filter; r=0.83, Δ=1.08, β=119.0°; LSPN-1272.
Photograph by F. Miller, University of Michigan/CTIO.

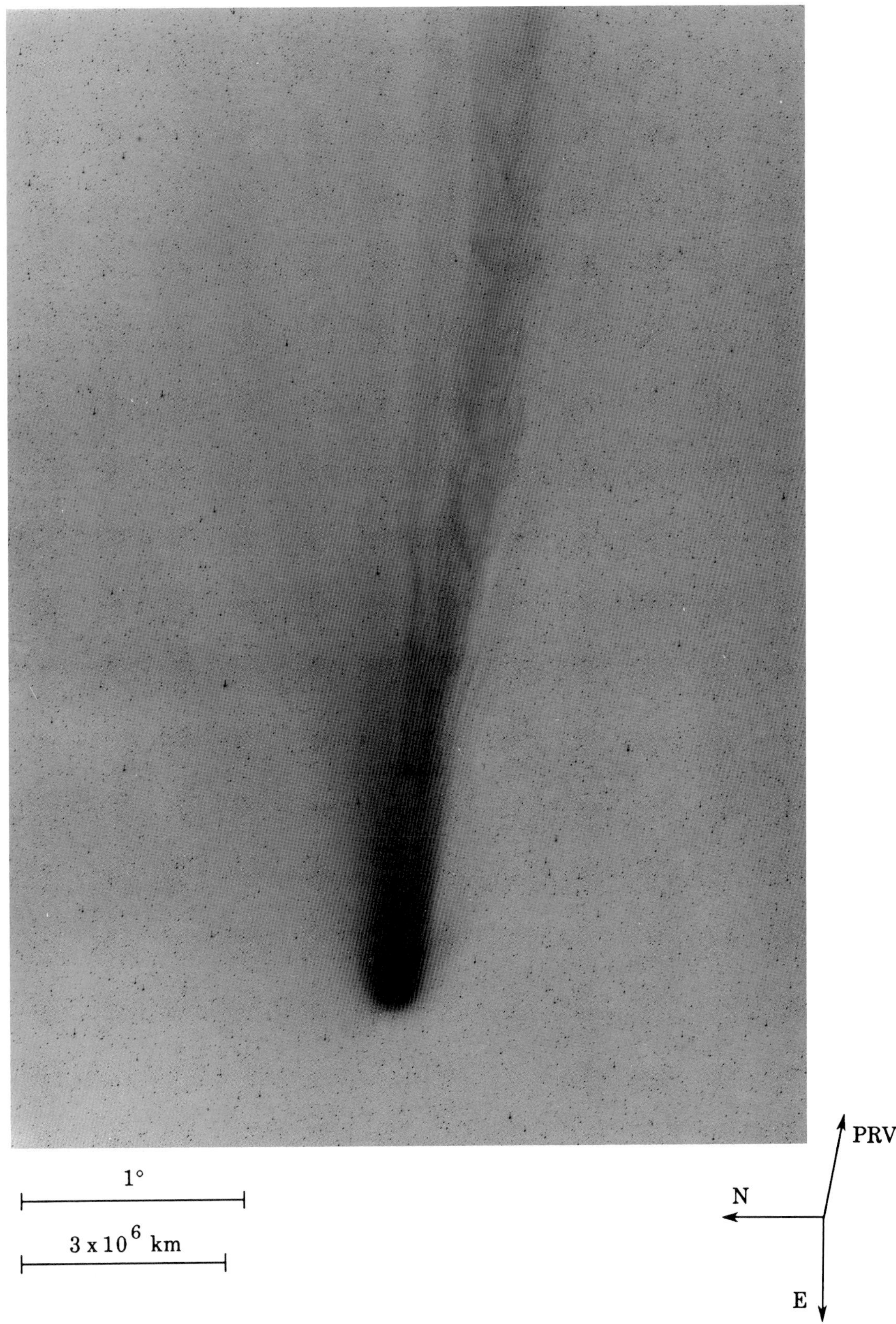

Fig. 271: 1986 Mar 9.37083 UT; exposure 10.0 minutes on IIa-O emulsion with no filter; r=0.84, Δ=1.08, β=119.0°; LSPN-1274.
Photograph by F. Miller, University of Michigan/CTIO.

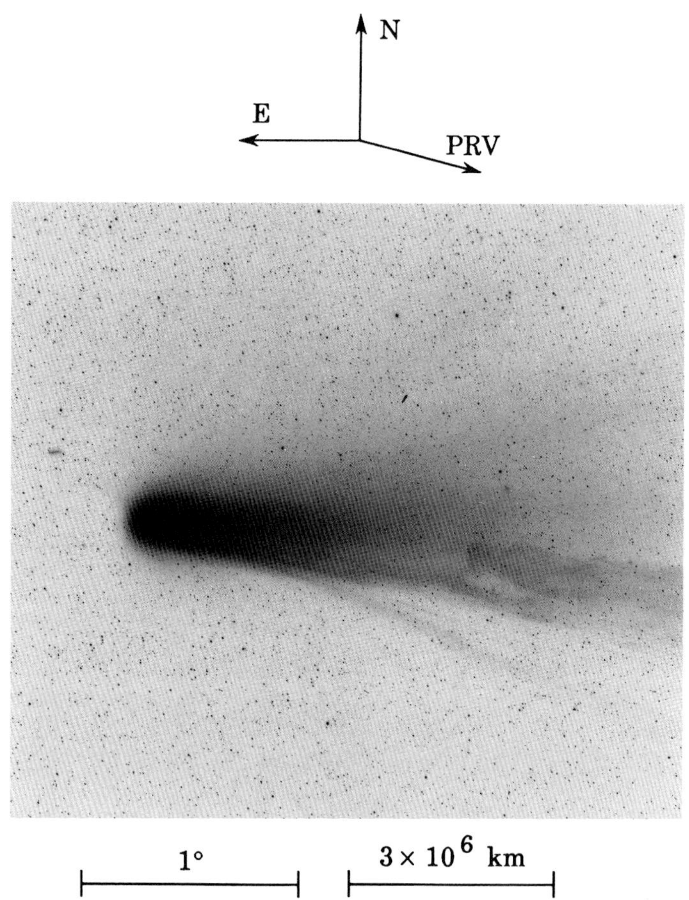

Fig. 272: 1986 Mar 9.75708 UT; exposure 12.0 minutes on IIa-O emulsion with GG-385 filter; r=0.84, Δ=1.07, β=118.6°; LSPN-2116.
Photograph by P. Magnusson, Uppsala Southern Station.

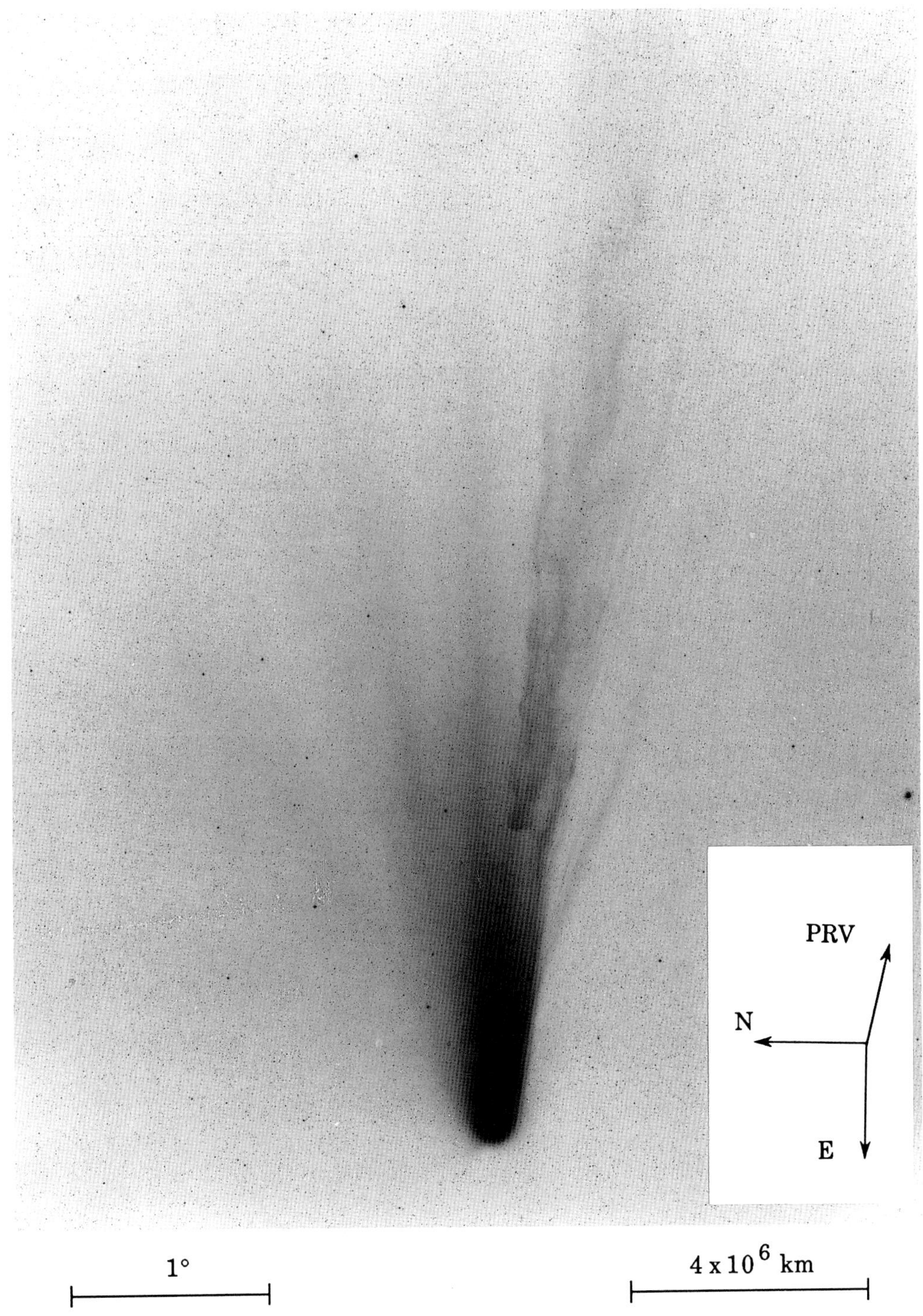

Fig. 273: 1986 Mar 9.77716 UT; exposure 8.0 minutes on IIIa-J emulsion with GG-395 filter; r=0.84, Δ=1.07, β=118.6°; LSPN-2382.
Photograph from Royal Observatory/UK Schmidt Telescope Unit.

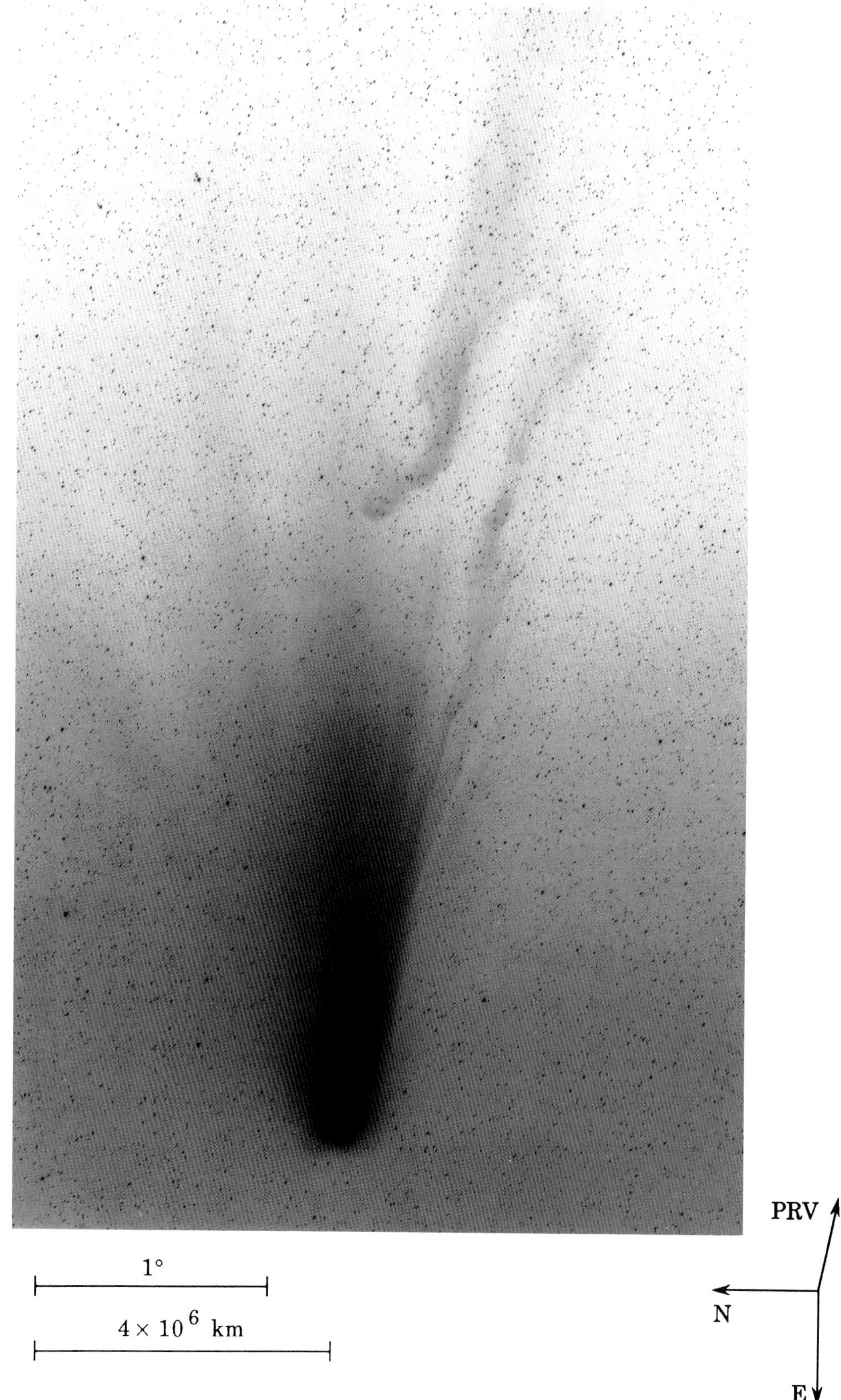

Fig. 274: 1986 Mar 10.37604 UT; exposure 15.0 minutes on IIa-O emulsion with no filter; r=0.85, Δ=1.05, β=118.1°; LSPN-1277.
Photograph by F. Miller, University of Michigan/CTIO.

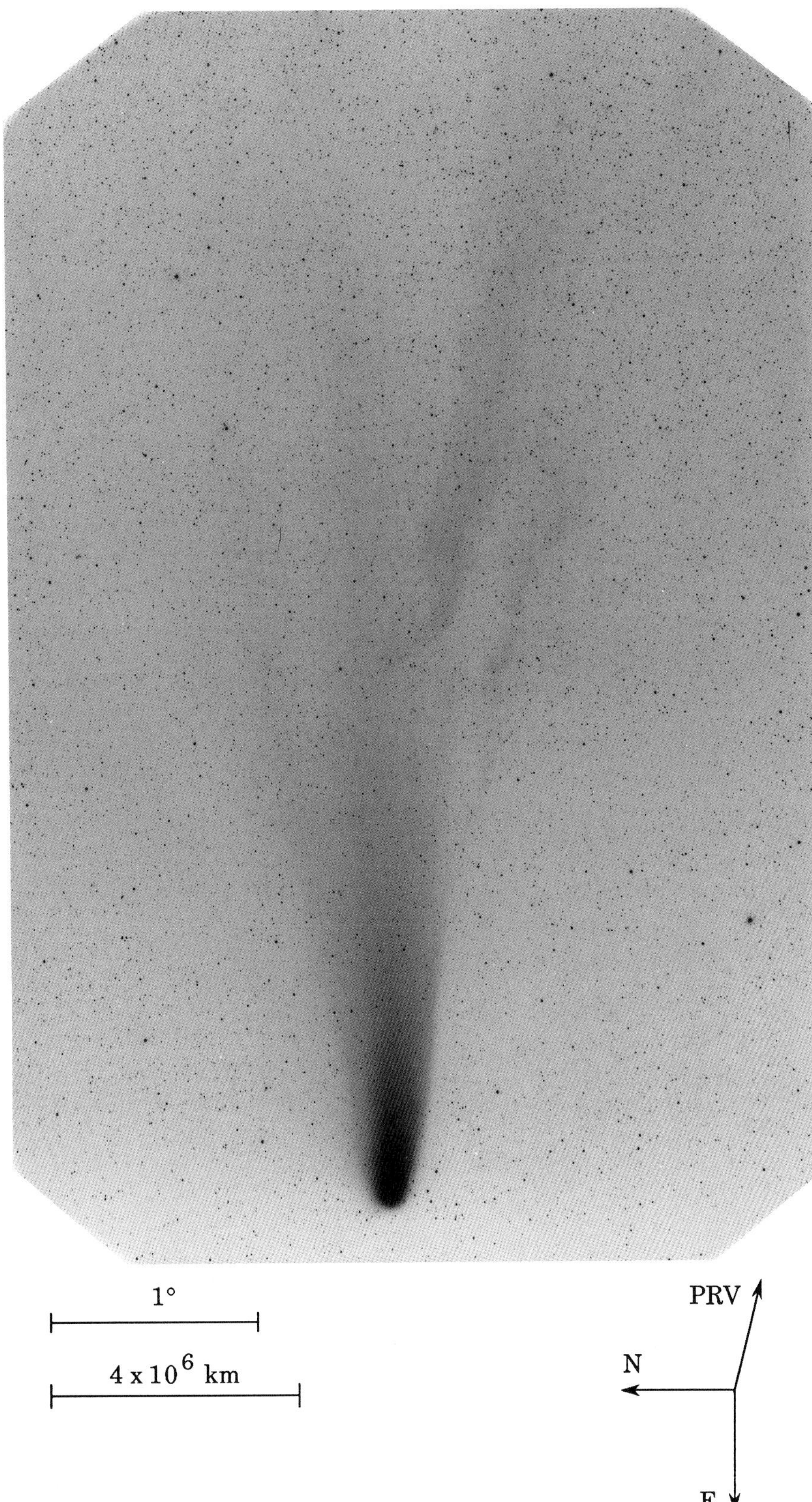

Fig. 275: 1986 Mar 10.43958 UT; exposure 10.0 minutes on Tech. Pan 2415 emulsion with no filter; r=0.85, Δ=1.05, β=118.1°; LSPN-1384.
Photograph by W. Liller, Easter Island, LSPN Island Network.

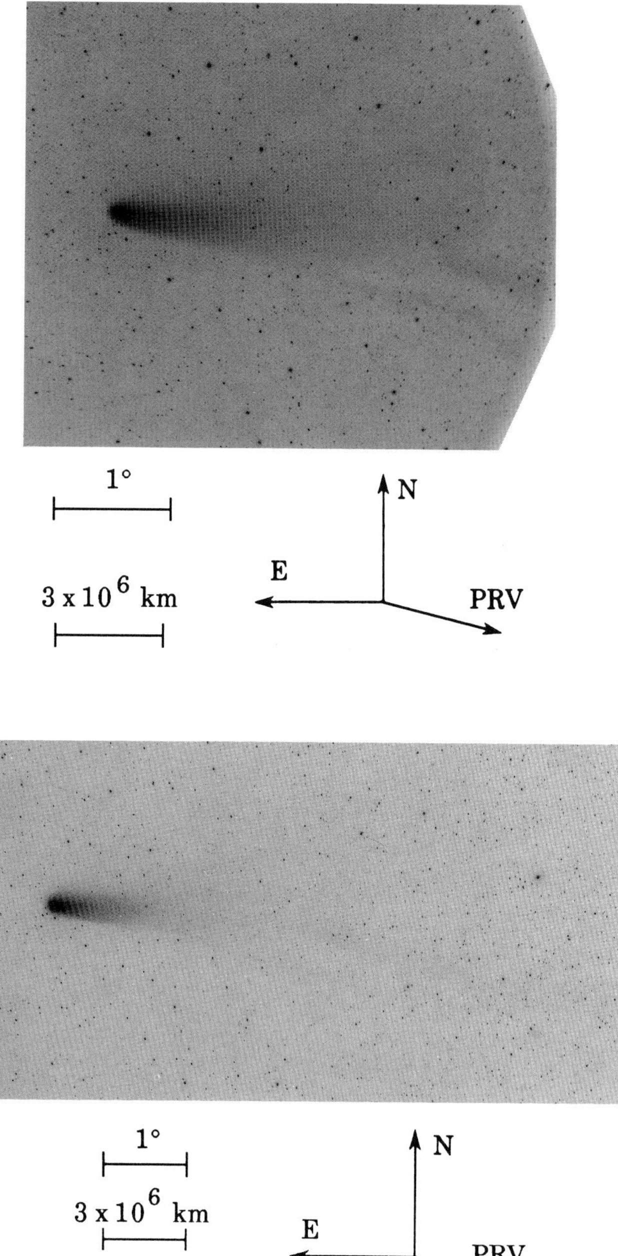

(top) Fig. 276: 1986 Mar 10.47570 UT; exposure 25.0 minutes on IIa-O emulsion with no filter; r=0.85, Δ=1.05, β=118.0°; LSPN-3415.
Photograph by G. Emerson, E. E. Barnard Observatory Station.

(bottom) Fig. 277: 1986 Mar 10.62014 UT; exposure 20.0 minutes on IIIa-J emulsion with no filter; r=0.85, Δ=1.05, β=117.9°; LSPN-1908.
Photograph by A. Storrs, Mauna Kea Observatory.

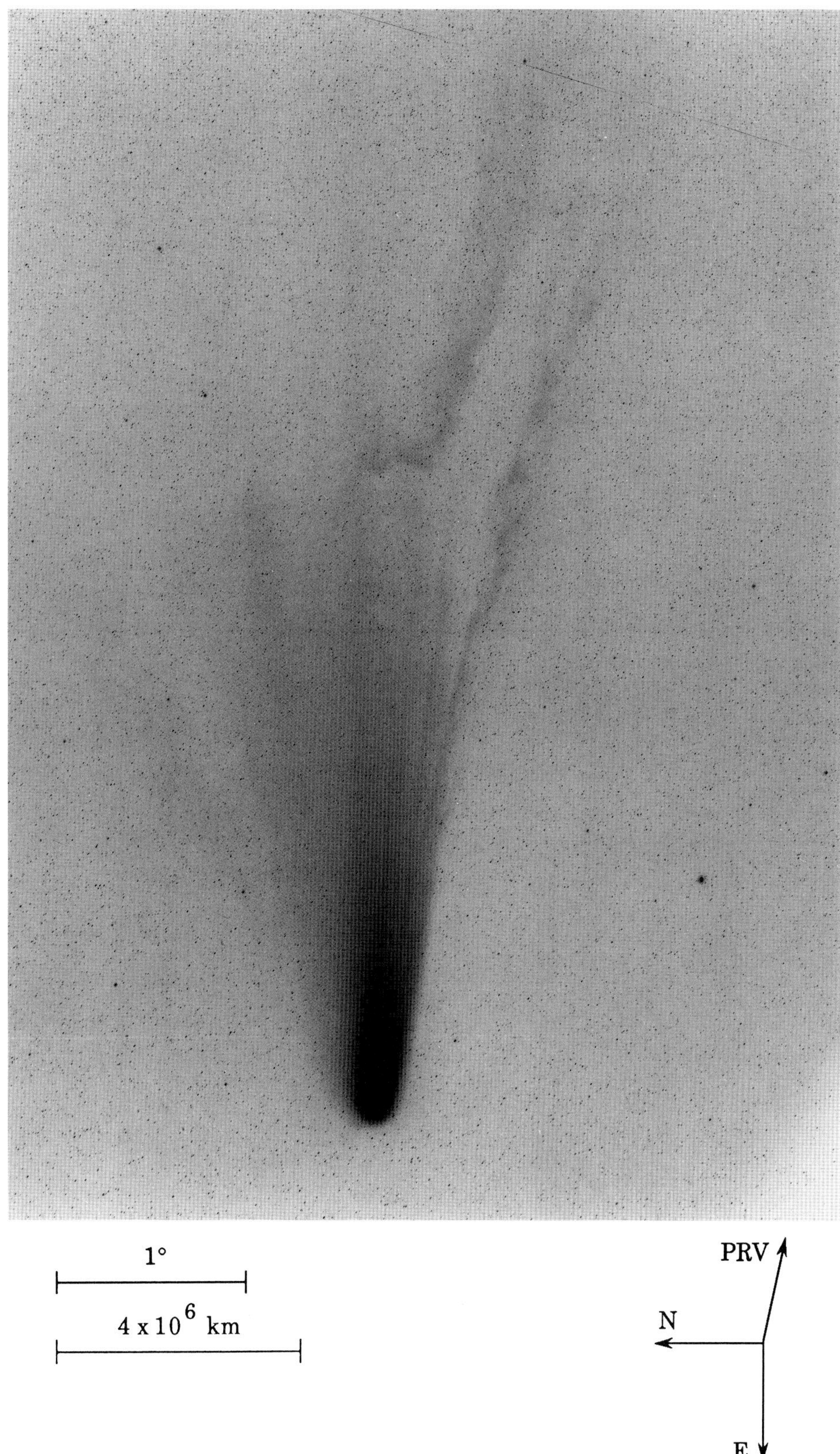

Fig. 278: 1986 Mar 10.73808 UT; exposure 15.0 minutes on IIa-O emulsion with GG-395 filter; r=0.85, Δ=1.04, β=117.8°; LSPN-2383.
Photograph from Royal Observatory/UK Schmidt Telescope Unit.

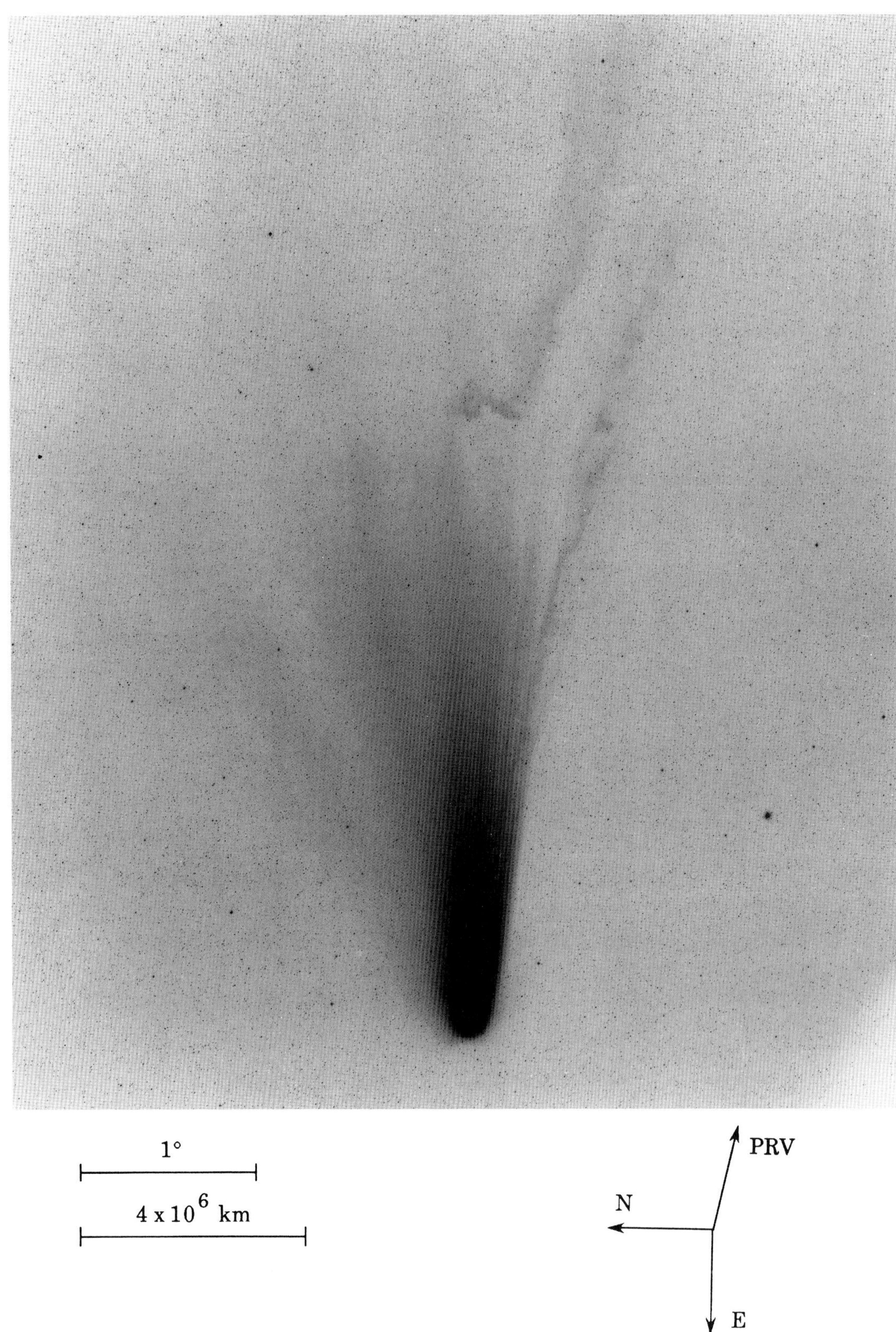

Fig. 279: 1986 Mar 10.77582 UT; exposure 10.0 minutes on IIIa-J emulsion with GG-395 filter; r=0.86, Δ=1.04, β=117.8°; LSPN-2385.
Photograph from Royal Observatory/UK Schmidt Telescope Unit.

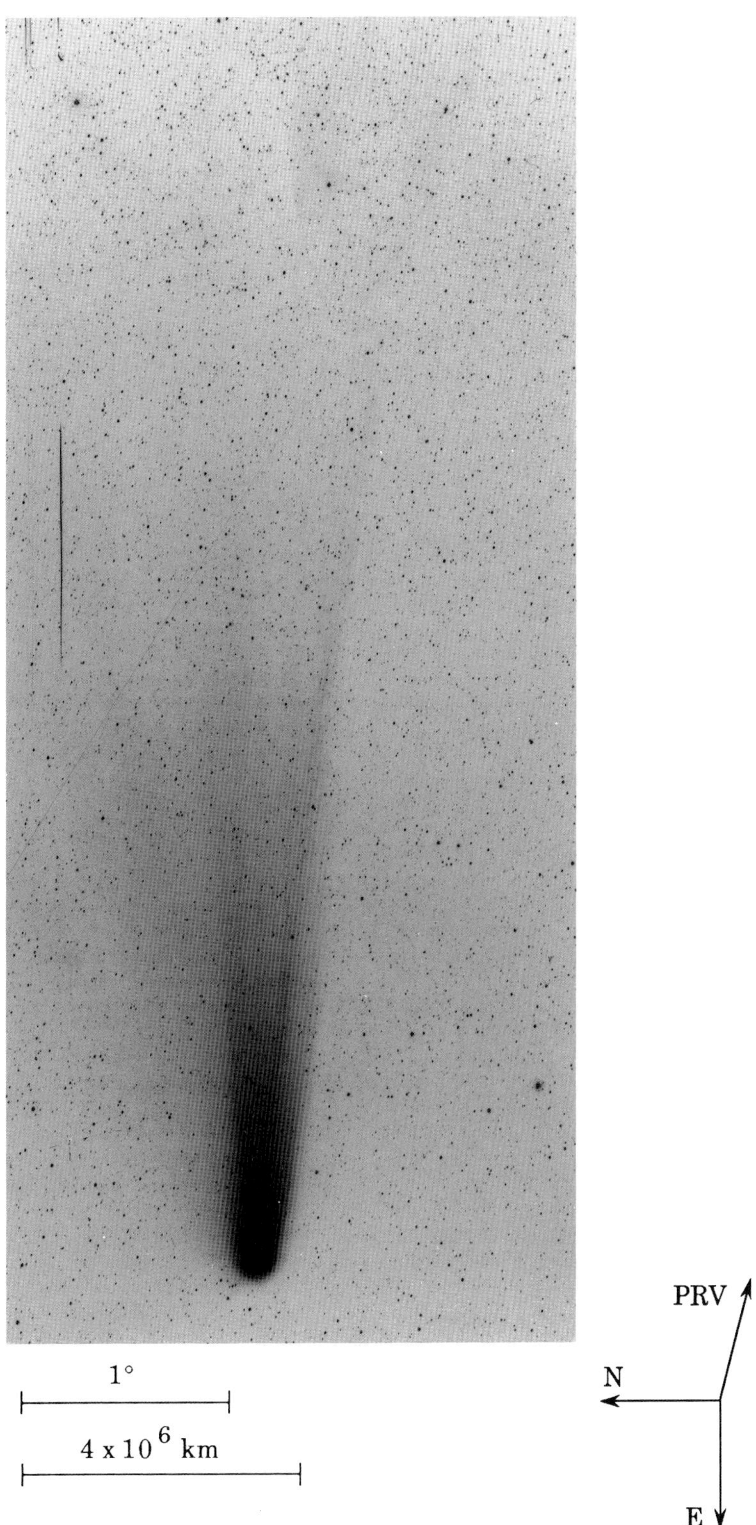

Fig. 280: 1986 Mar 11.45660 UT; exposure 15.0 minutes on Tech. Pan 2415 emulsion with no filter; r=0.87, Δ=1.03, β=117.3°; LSPN-1386.
Photograph by W. Liller, Easter Island, LSPN Island Network.

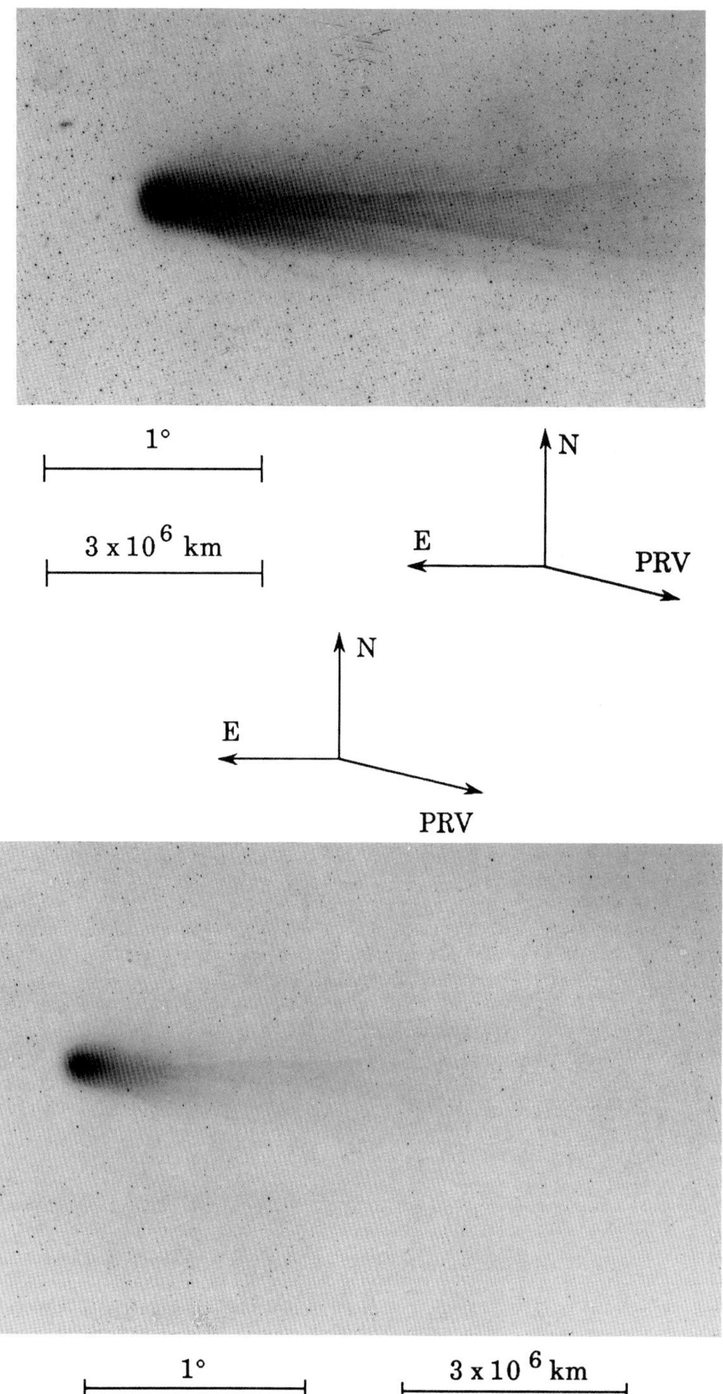

(top) Fig. 281: 1986 Mar 11.78590 UT; exposure 5.0 minutes on IIIa-J emulsion with GG-385 filter; r=0.87, Δ=1.02, β=117.0°; LSPN-2117.
Photograph by P. Magnusson, Uppsala Southern Station.

(bottom) Fig. 282: 1986 Mar 11.82273 UT; exposure 10.0 minutes on IIa-O emulsion with no filter; r=0.87, Δ=1.02, β=117.0°; LSPN-1629.
Photograph by H. Maehara, Kiso Observatory.

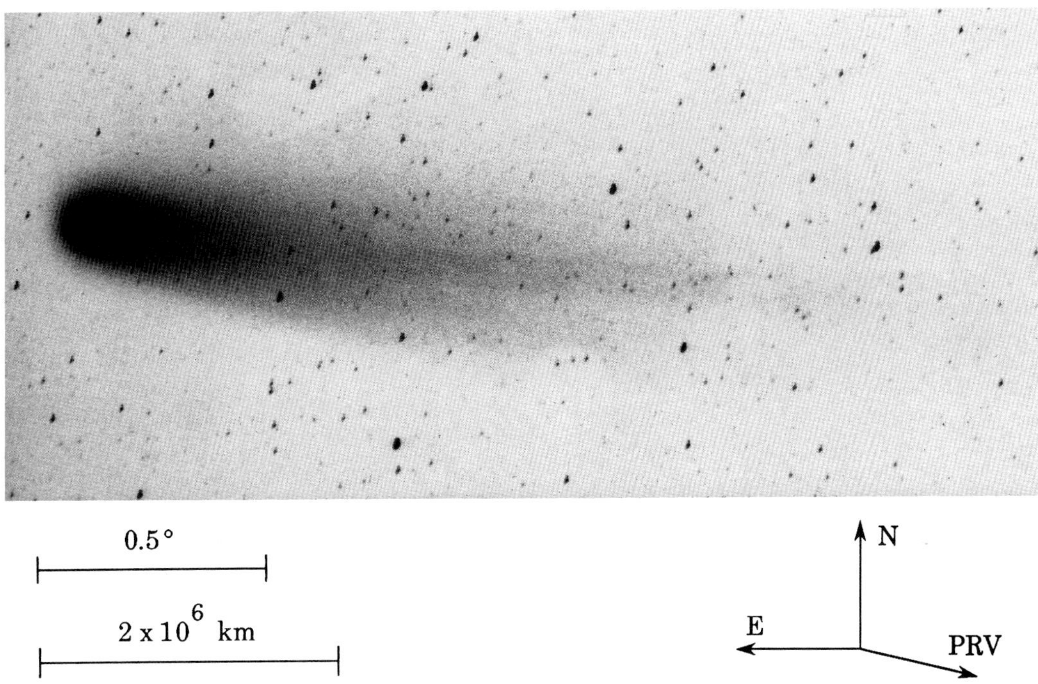

Fig. 283: 1986 Mar 12.04510 UT; exposure 5.0 minutes on Tech. Pan 2415 emulsion; r=0.87, Δ=1.01, β=116.9°. *Photograph by A. Marafie, Wafra Observatory.*

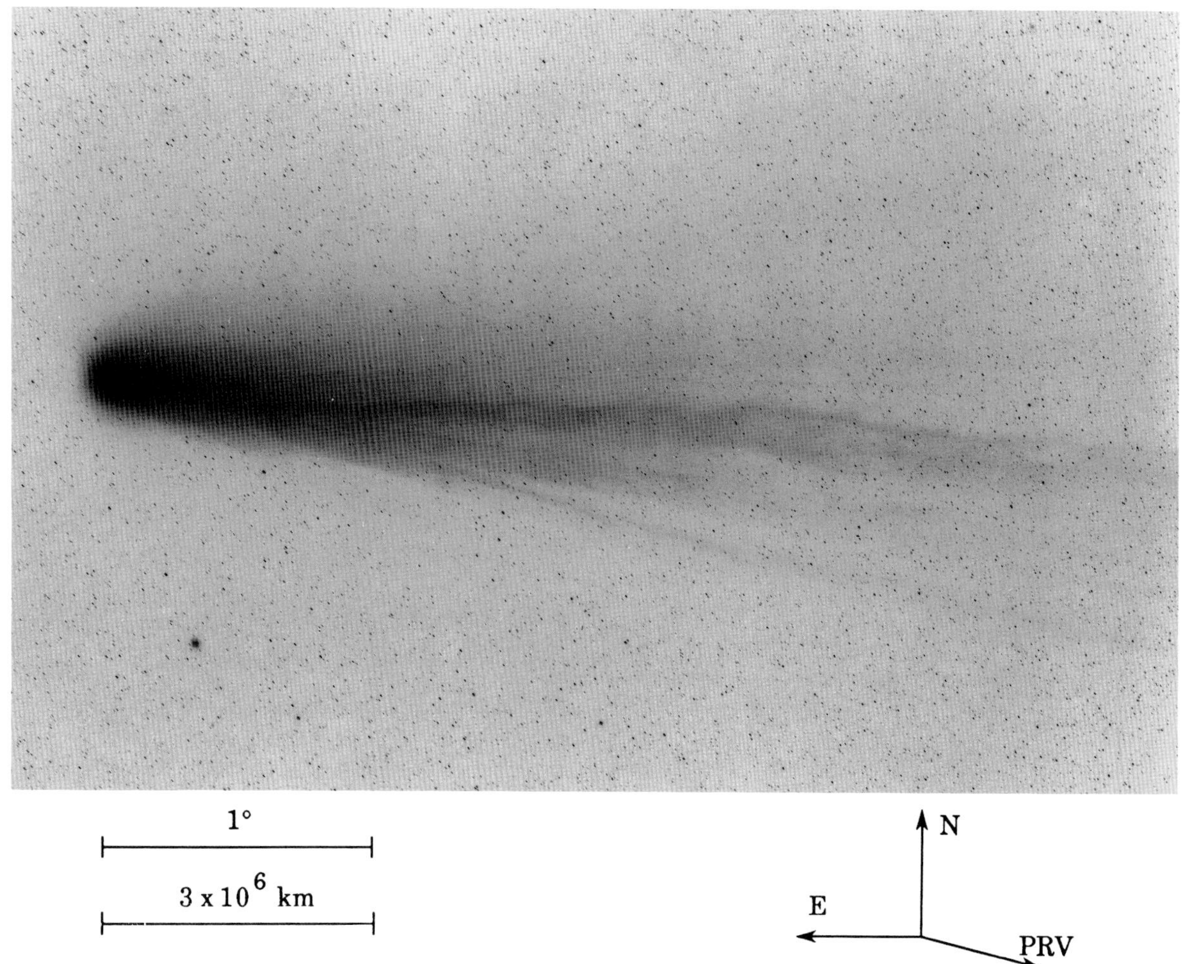

Fig. 284: 1986 Mar 12.32917 UT; exposure 20.0 minutes on IIa-O emulsion with no filter; r=0.88, Δ=1.00, β=116.7°; LSPN-1280.
Photograph by F. Miller, University of Michigan/CTIO.

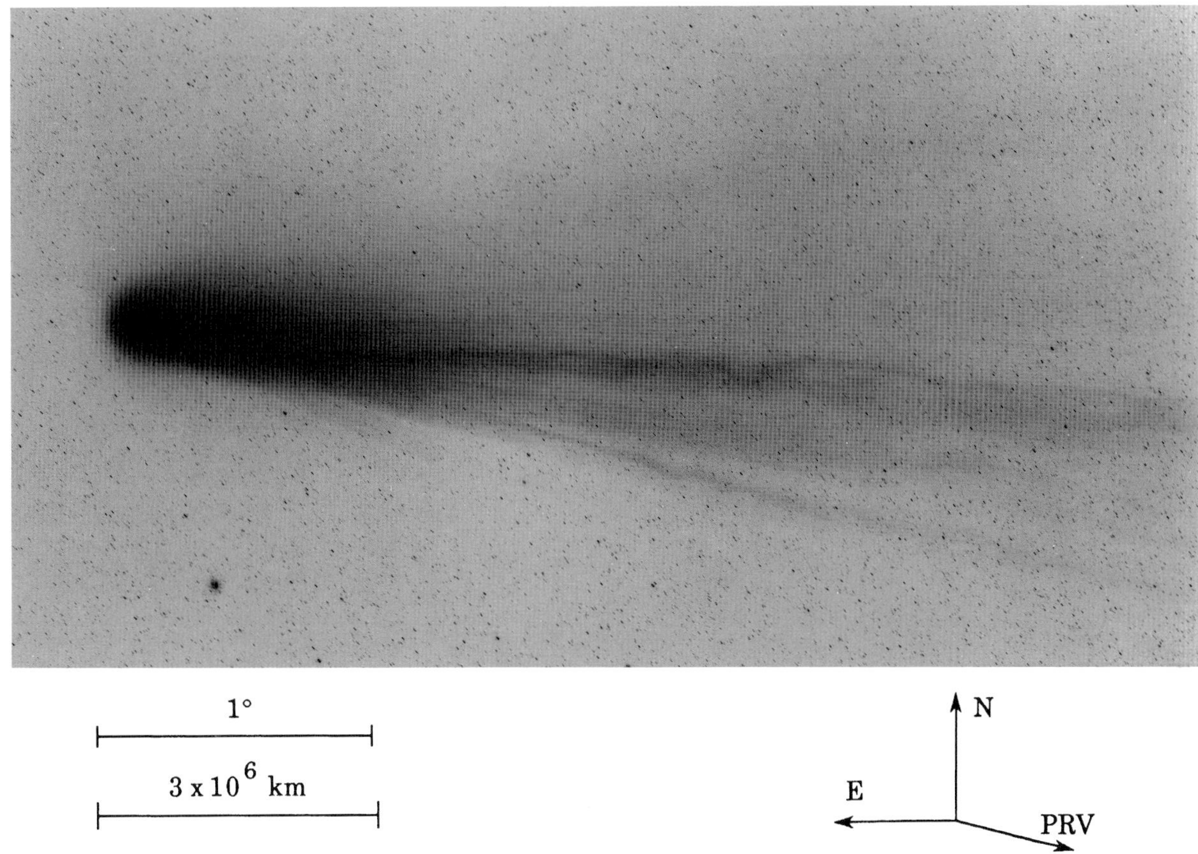

Fig. 285: 1986 Mar 12.36319 UT; exposure 20.0 minutes on IIa-O emulsion with no filter; r=0.88, Δ=1.00, β=116.7°; LSPN-1281.
Photograph by F. Miller, University of Michigan/CTIO.

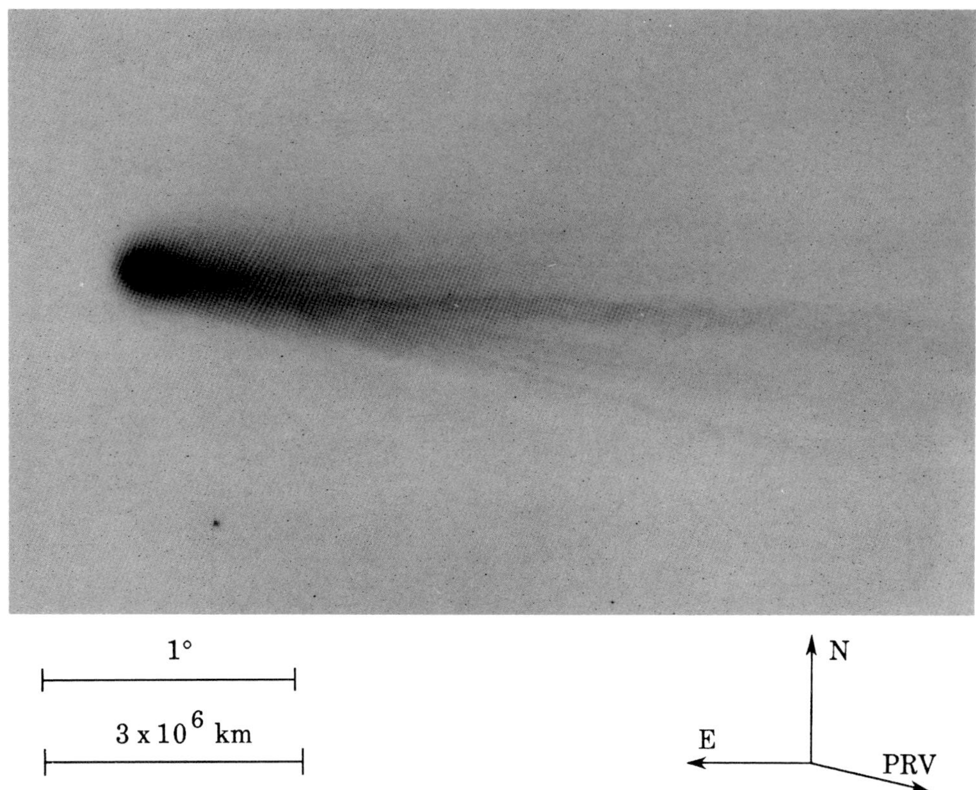

Fig. 286: 1986 Mar 12.36771 UT; exposure 5.0 minutes on 103a-O emulsion with no filter; r=0.88, Δ=1.00, β=116.7°; LSPN-726.
Photograph by C. Torres/H. Wroblewski, Cerro el Roble Astronomical Observatory.

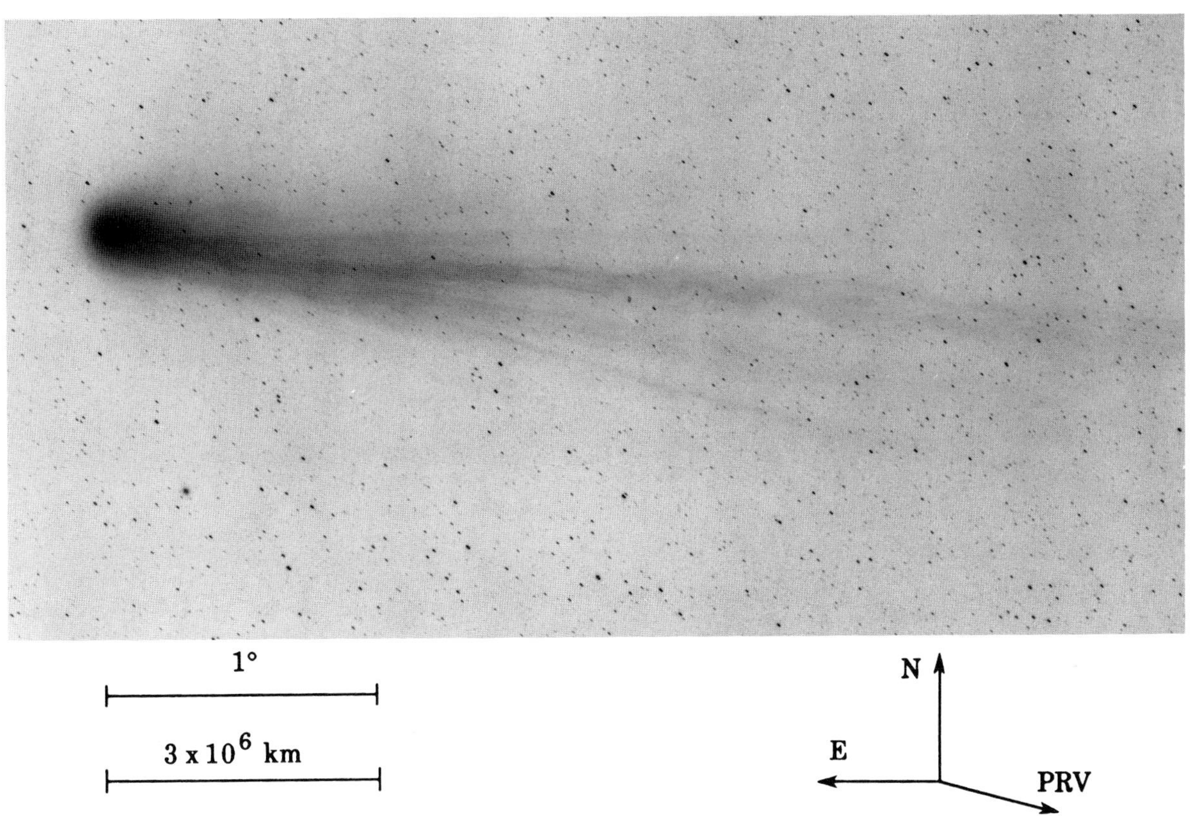

Fig. 287: 1986 Mar 12.38750 UT; exposure 20.0 minutes on IIa-O emulsion with no filter; r=0.88, Δ=1.00, β=116.6°; LSPN-1282.
Photograph by F. Miller, University of Michigan/CTIO.

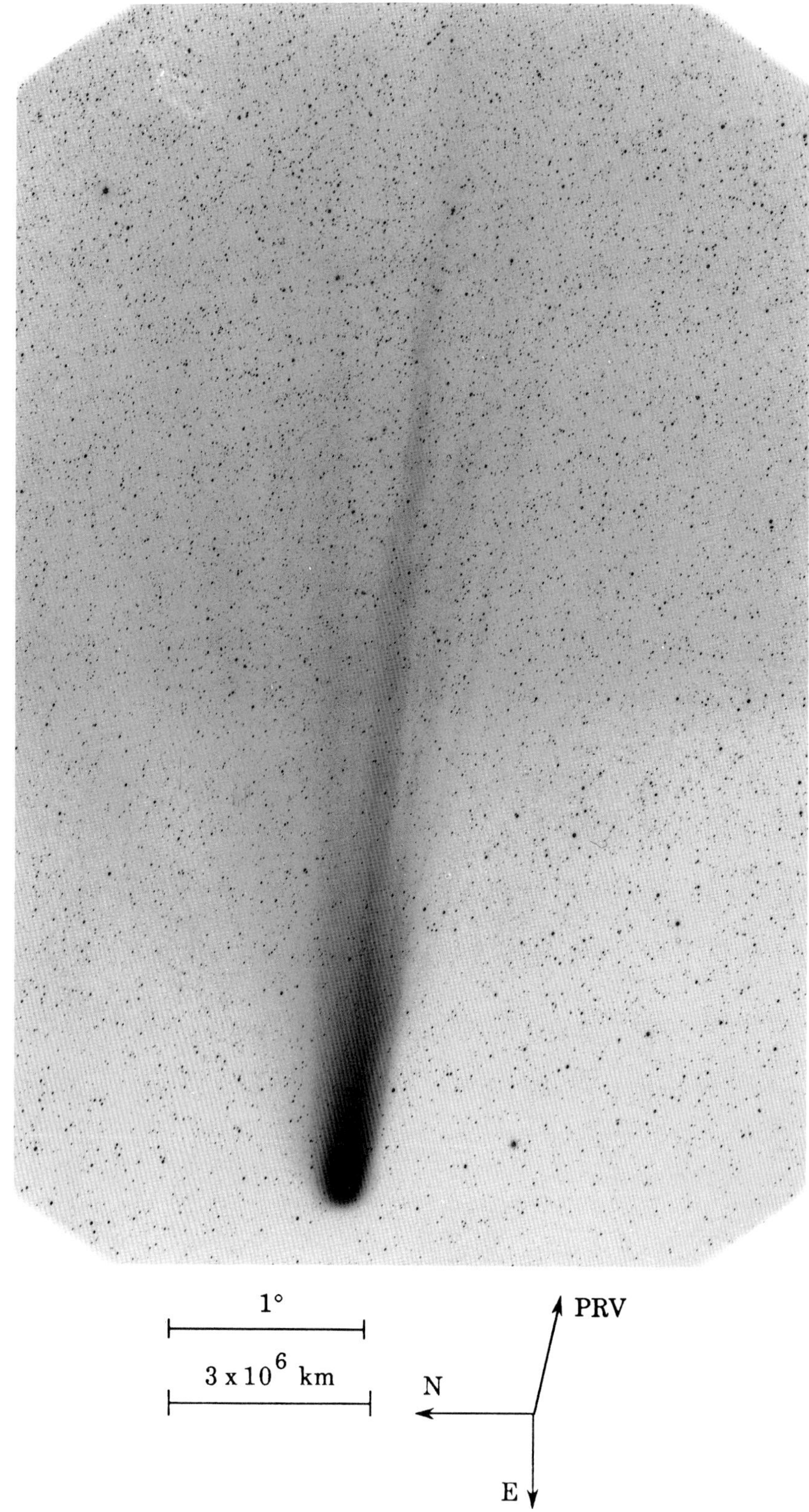

Fig. 288: 1986 Mar 12.49583 UT; exposure 16.0 minutes on Tech. Pan 2415 emulsion with no filter; r=0.88, Δ=1.00, β=116.6°; LSPN-1390.
Photograph by W. Liller, Easter Island, LSPN Island Network.

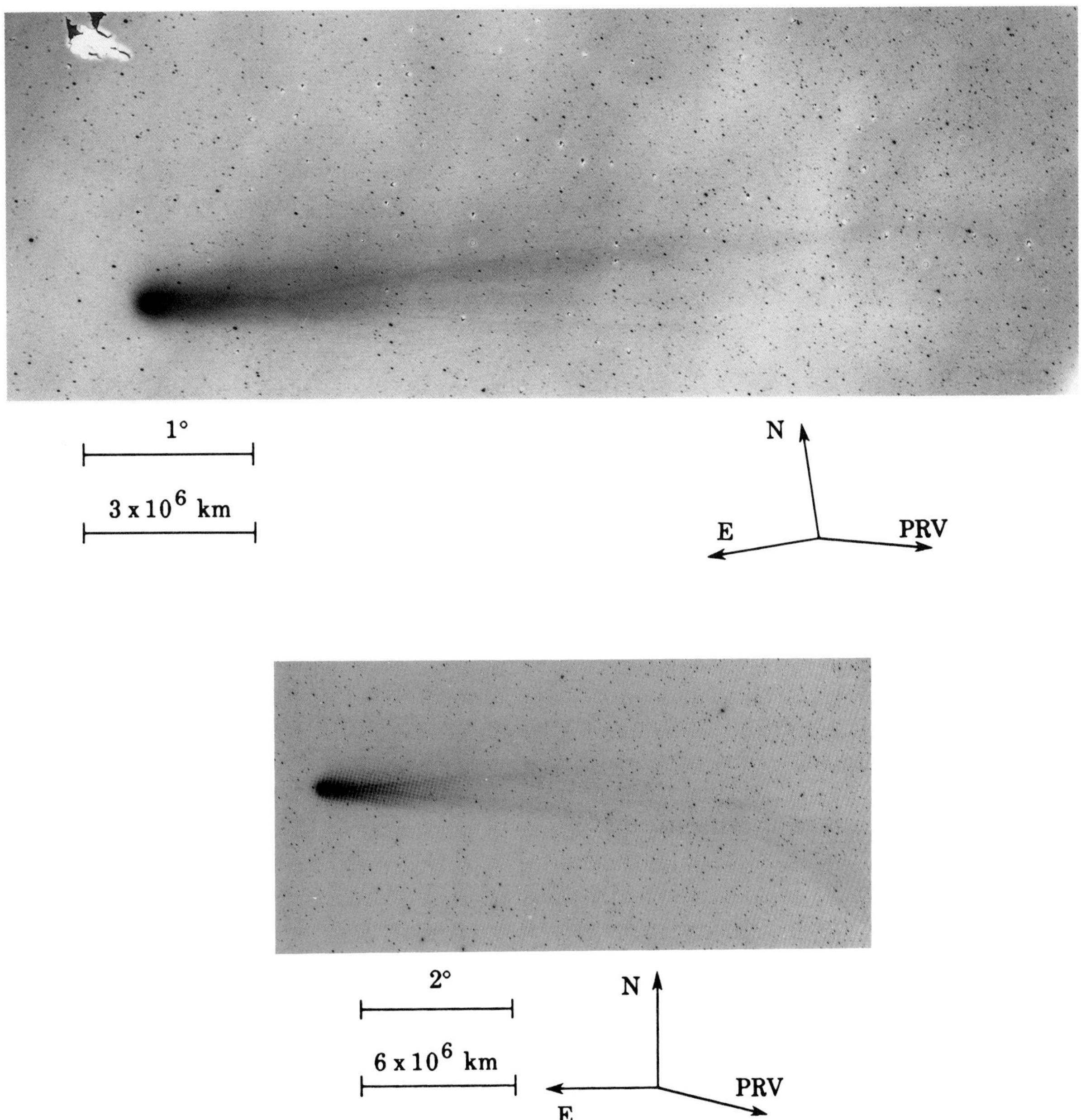

(top) Fig. 289: 1986 Mar 12.51927 UT; exposure 21.0 minutes on 103a-O emulsion with no filter; r=0.88, Δ=1.00, β=116.6°; LSPN-3578.
Photograph by H. Giclas, Lowell Observatory.

(bottom) Fig. 290: 1986 Mar 12.62708 UT; exposure 30.0 minutes on IIIa-J emulsion with no filter; r=0.88, Δ=1.00, β=116.5°; LSPN-1910.
Photograph by A. Storrs, Mauna Kea Observatory.

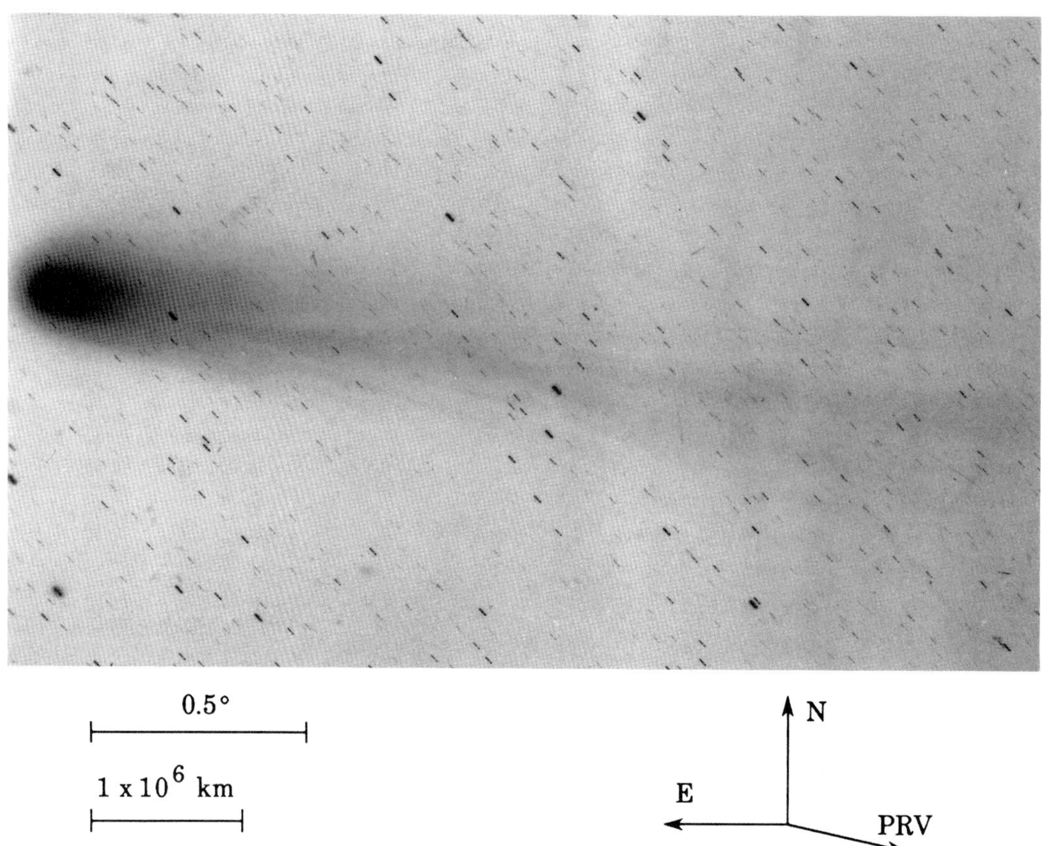

Fig. 291: 1986 Mar 12.86667 UT; exposure 36.0 minutes on 103a-O emulsion with no filter; r=0.89, Δ=0.99, β=116.3°; LSPN-3075.
Photograph by P. Jekabsons, Perth Observatory.

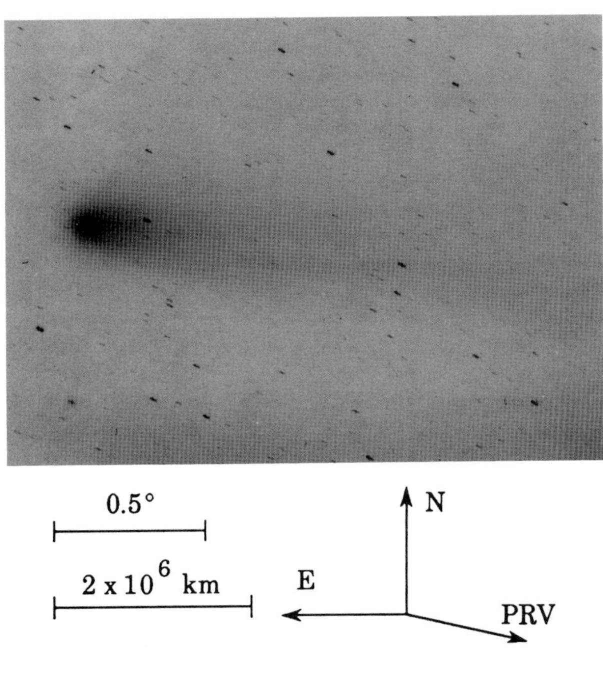

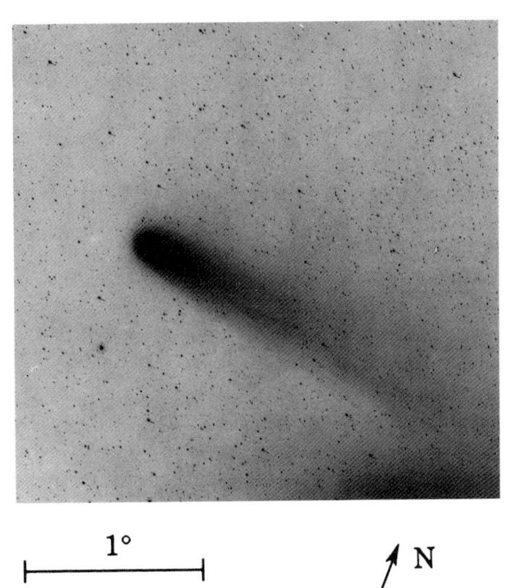

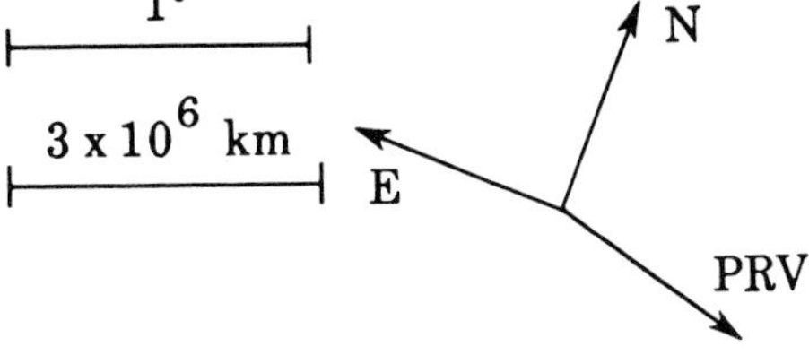

(top) Fig. 292: 1986 Mar 12.95903 UT; exposure 60.0 minutes on 103a-O emulsion with no filter; r=0.89, Δ=0.99, β=116.3°; LSPN-3622.
Photograph submitted by K. Sivaraman, Indian Institute for Astrophysics, Kodaikanal Station.

(bottom) Fig. 293: 1986 Mar 12.98194 UT; exposure 8.0 minutes on 103a-F emulsion with no filter; r=0.89, Δ=0.99, β=116.3°; LSPN-3663.
Photograph submitted by K. Sivaraman, Indian Institute for Astrophysics, Kavalur Station.

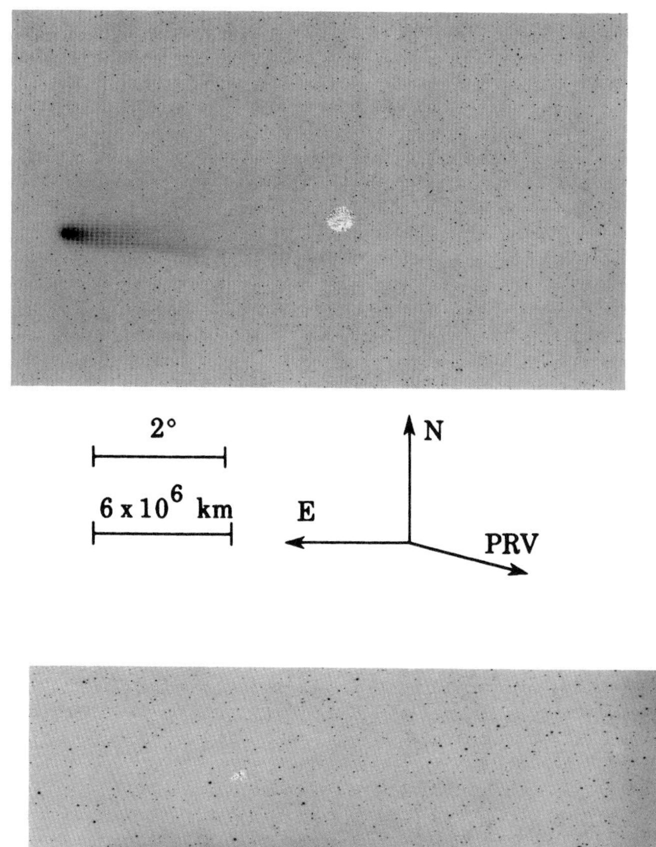

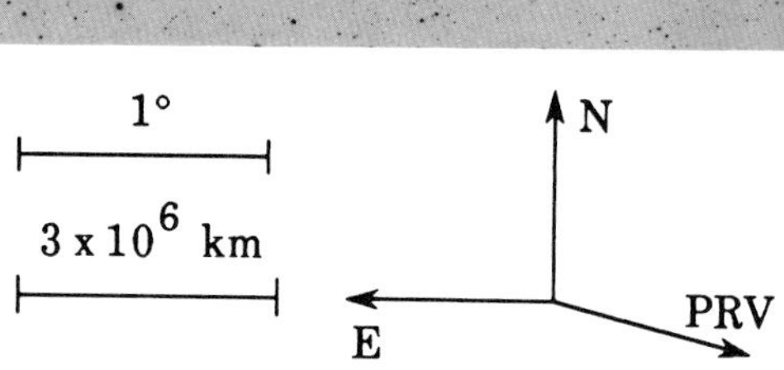

(top) Fig. 294: 1986 Mar 13.09618 UT; exposure 15.0 minutes on IIa-O emulsion with no filter; r=0.89, Δ=0.99, β=116.2°; LSPN-2843.
Photograph by G. Malcolm/A. Jarrett, Boyden Observatory.

(bottom) Fig. 295: 1986 Mar 13.09618 UT; exposure 15.0 minutes on IIa-O emulsion with no filter; r=0.89, Δ=0.99, β=116.2°; LSPN-2844.
Photograph by G. Malcolm/A. Jarrett, Boyden Observatory.

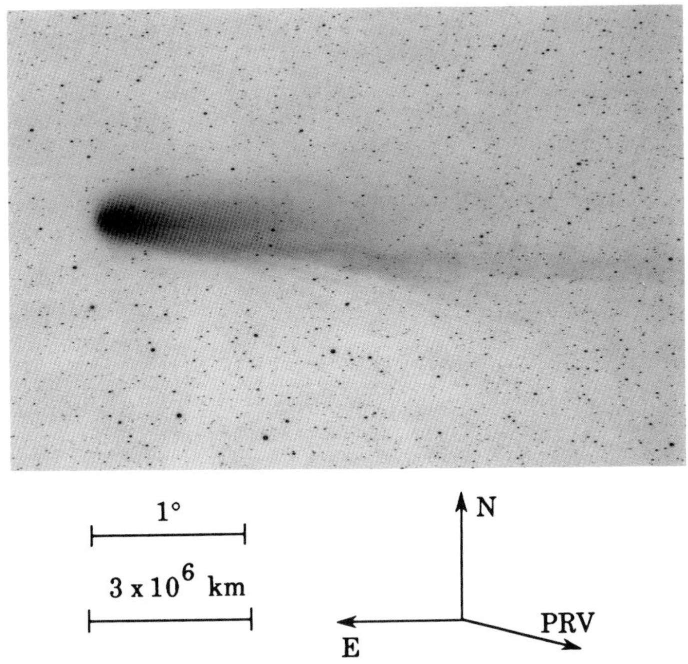

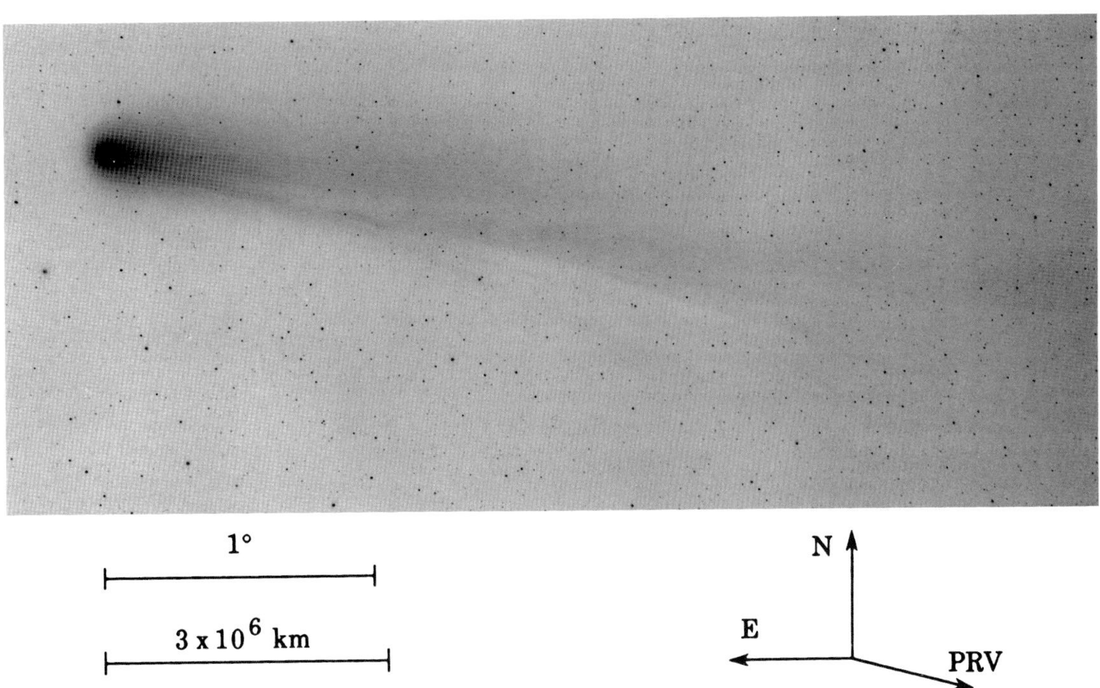

(top) Fig. 296: 1986 Mar 13.11493 UT; exposure 25.0 minutes on IIa-O emulsion with no filter; r=0.89, Δ=0.98, β=116.2°; LSPN-2845.
Photograph by G. Malcolm/A. Jarrett, Boyden Observatory.

(bottom) Fig. 297: 1986 Mar 13.30972 UT; exposure 20.0 minutes on IIa-O emulsion with no filter; r=0.89, Δ=0.98, β=116.1°; LSPN-1283.
Photograph by F. Miller, University of Michigan/CTIO.

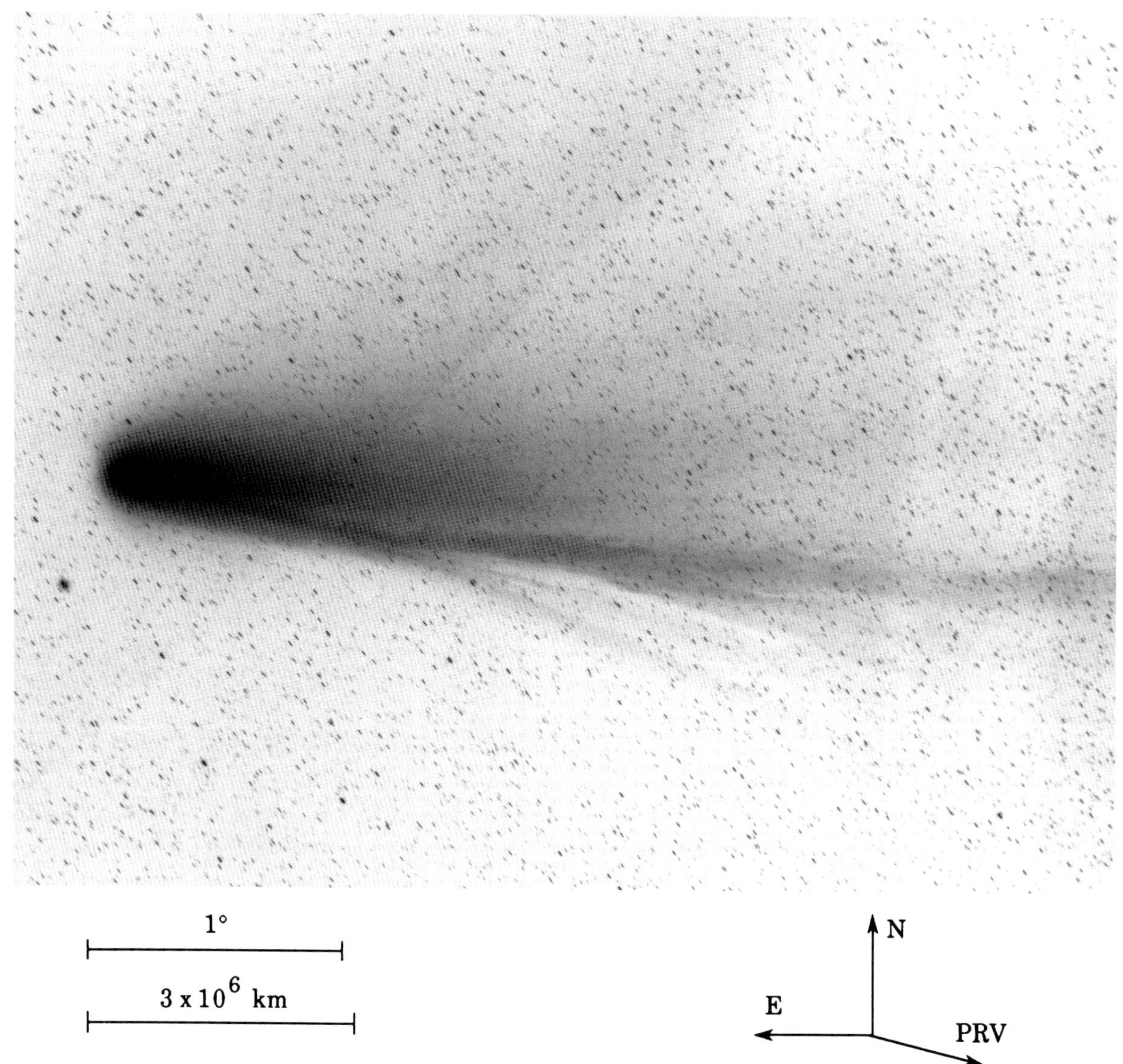

Fig. 298: 1986 Mar 13.34792 UT; exposure 30.0 minutes on IIa-O emulsion with no filter; r=0.89, Δ=0.98, β=116.0°; LSPN-1285.
Photograph by F. Miller, University of Michigan/CTIO.

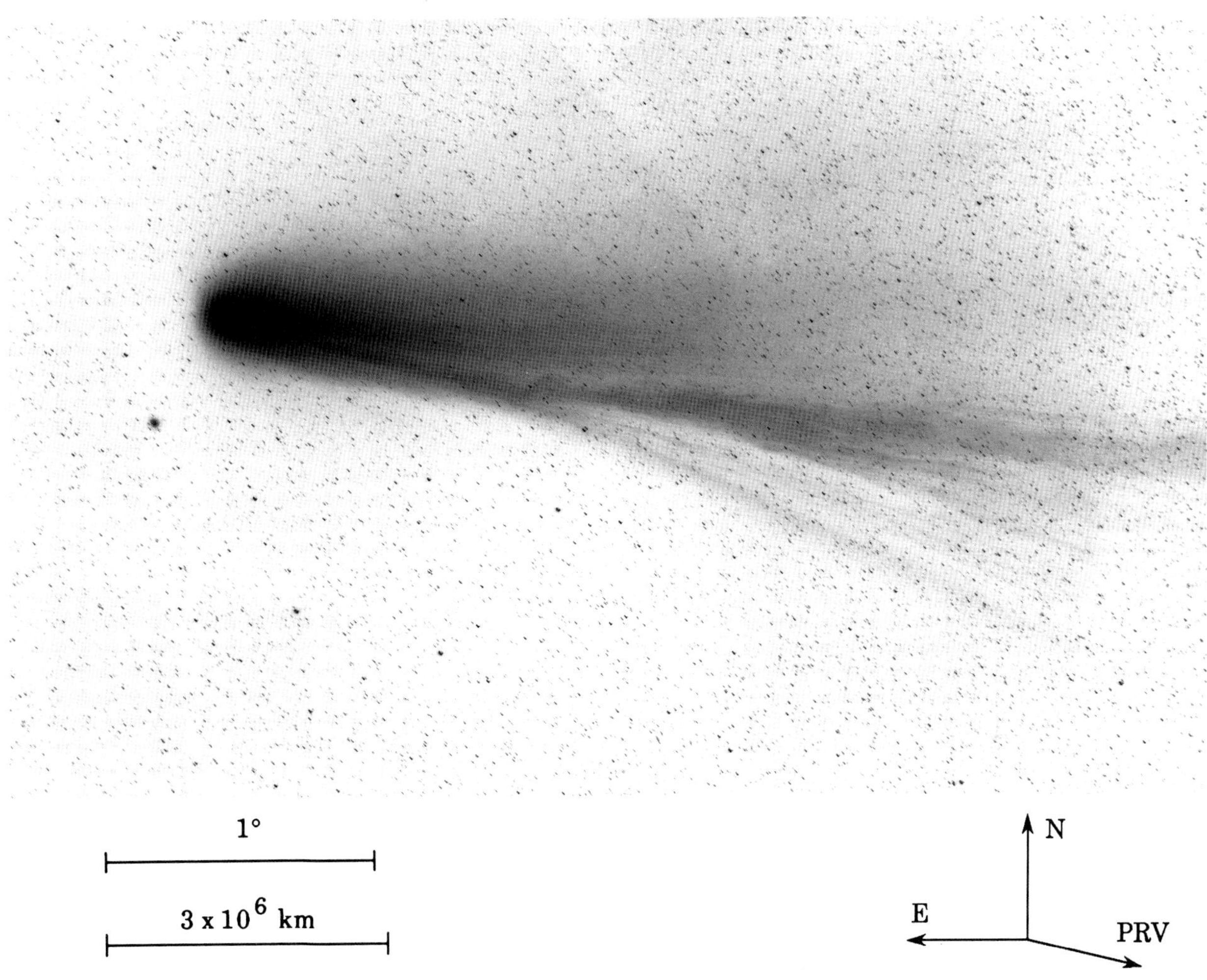

Fig. 299: 1986 Mar 13.37569 UT; exposure 20.0 minutes on IIa-O emulsion with no filter; r=0.89, Δ=0.98, β=116.0°; LSPN-1286.
Photograph by F. Miller, University of Michigan/CTIO.

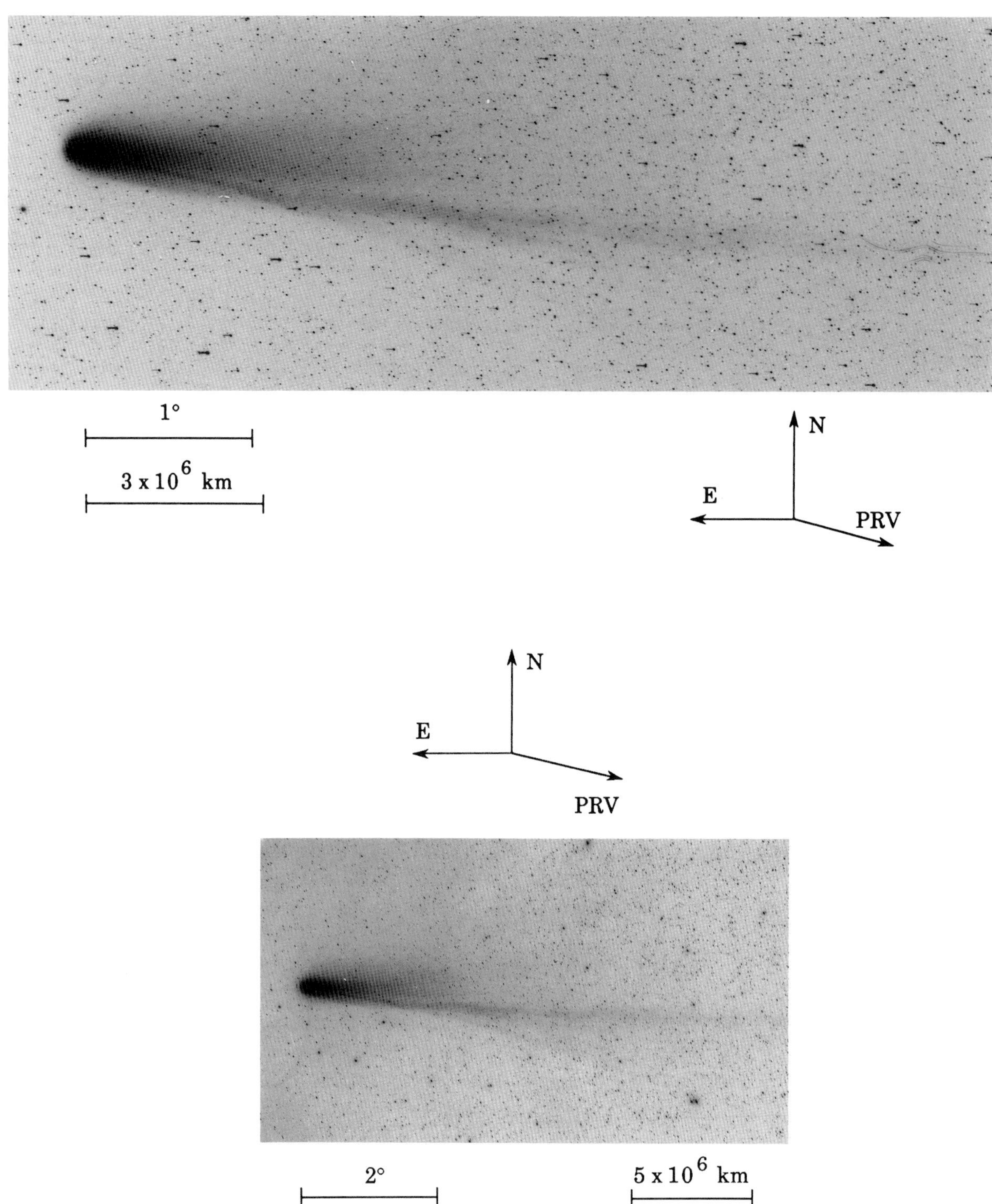

(top) Fig. 300: 1986 Mar 13.48438 UT; exposure 13.0 minutes on Tech. Pan 2415 emulsion with no filter; r=0.89, Δ=0.98, β=116.0°; LSPN-1392.
Photograph by W. Liller, Easter Island, LSPN Island Network.

(bottom) Fig. 301: 1986 Mar 13.62014 UT; exposure 30.0 minutes on IIIa-J emulsion with no filter; r=0.90, Δ=0.97, β=115.9°; LSPN-1911.
Photograph by A. Storrs, Mauna Kea Observatory.

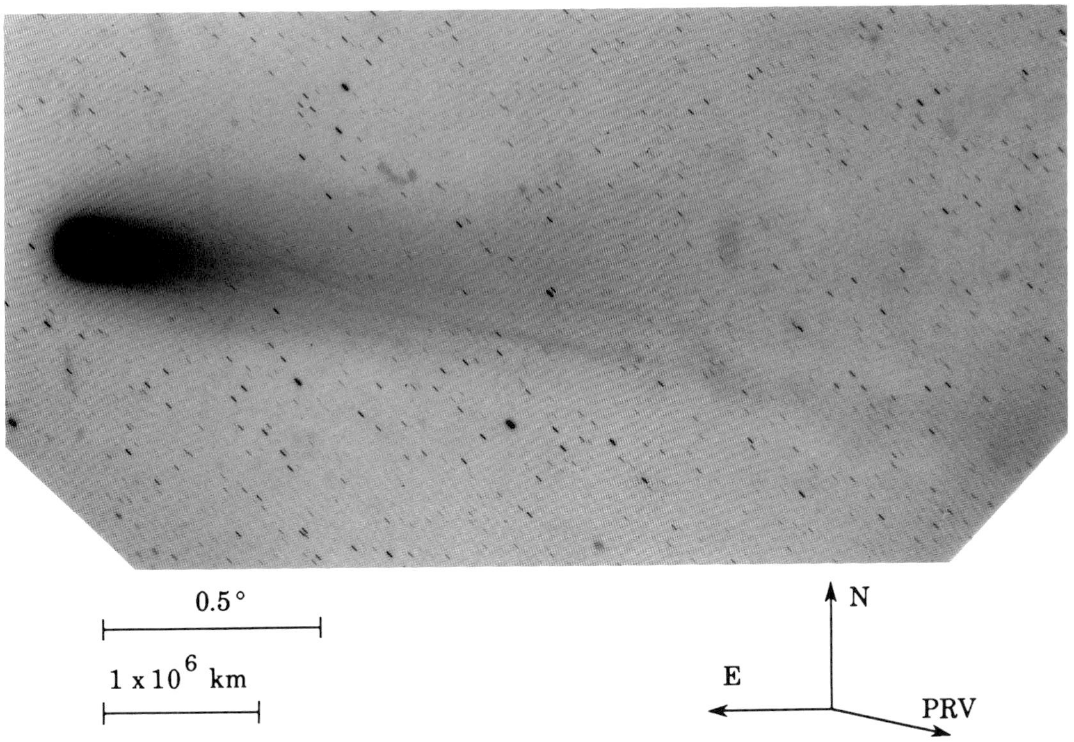

0.5°

1 x 10^6 km

N

E

PRV

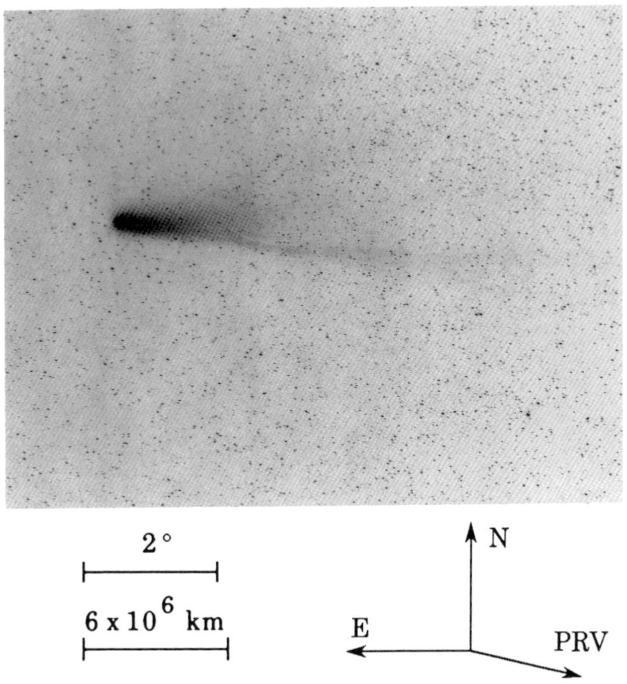

2°

6 x 10^6 km

N

E

PRV

(top) Fig. 302: 1986 Mar 13.87153 UT; exposure 28.0 minutes on 103a-O emulsion with no filter; r=0.90, Δ=0.97, β=115.7°; LSPN-3077.
Photograph by P. Jekabsons, Perth Observatory.

(bottom) Fig. 303: 1986 Mar 14.10104 UT; exposure 45.0 minutes on IIa-O emulsion with no filter; r=0.90, Δ=0.96, β=115.6°; LSPN-2847.
Photograph by G. Malcolm, Boyden Observatory.

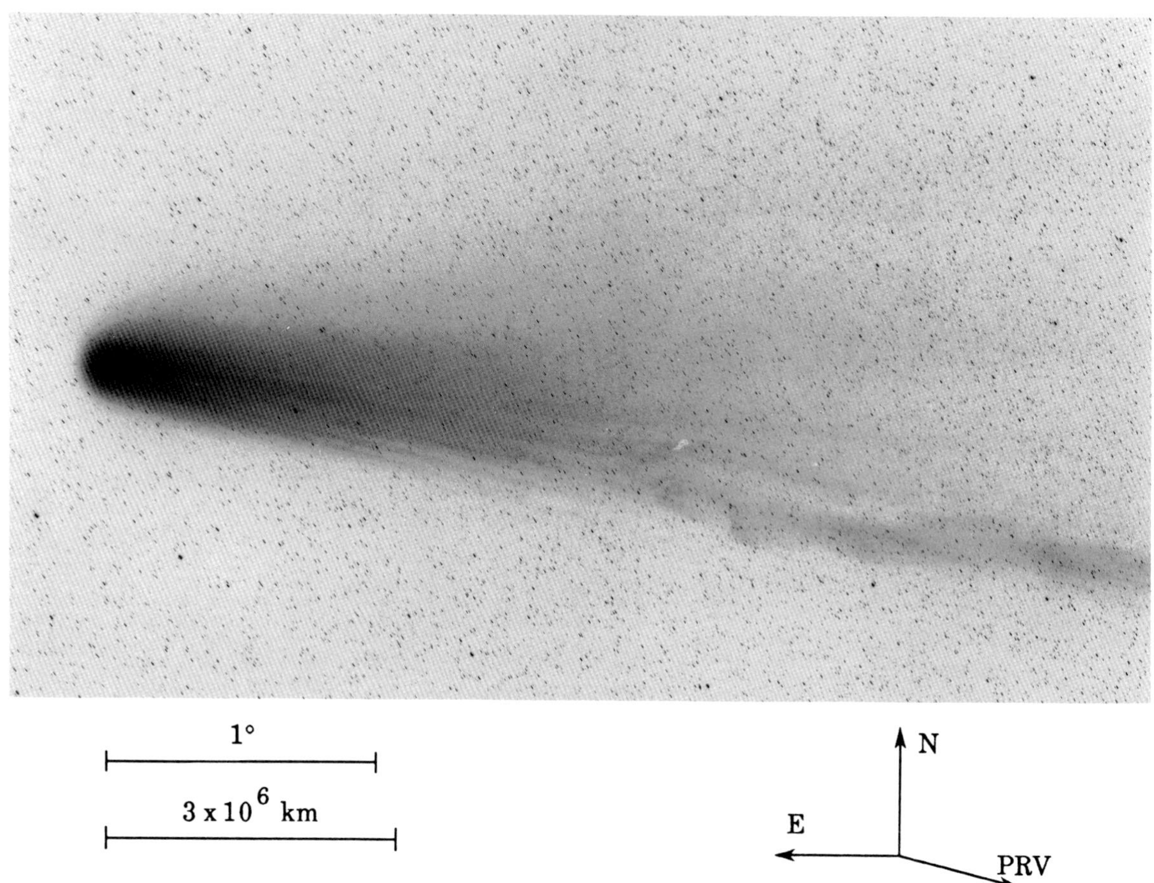

Fig. 304: 1986 Mar 14.31667 UT; exposure 20.0 minutes on IIa-O emulsion with no filter; r=0.91, Δ=0.95, β=115.5°; LSPN-1287.
Photograph by F. Miller, University of Michigan/CTIO.

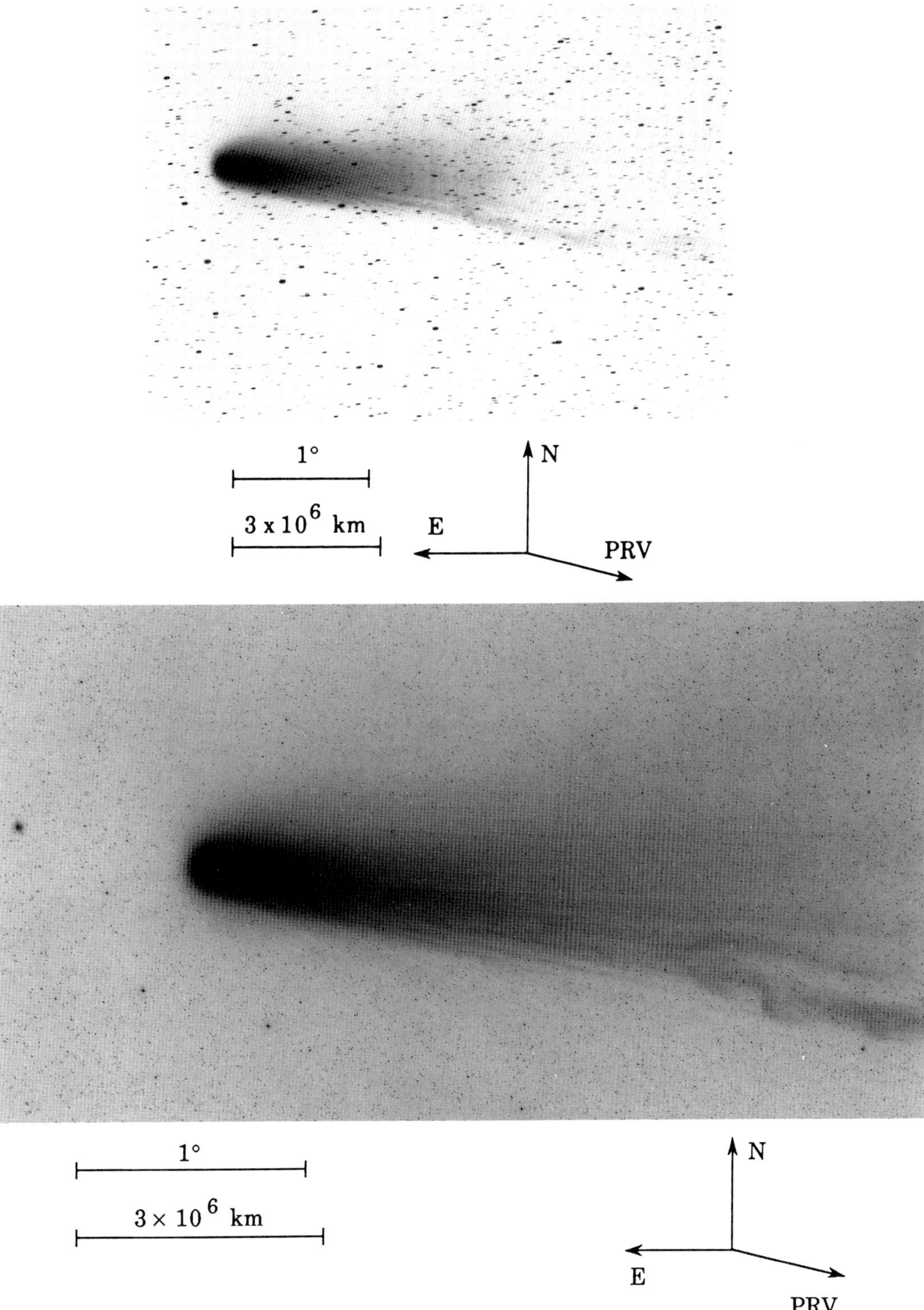

(top) Fig. 305*: 1986 Mar 14.10104 UT; exposure 45.0 minutes on IIa-O emulsion with no filter; r=0.90, Δ=0.96, β=115.6°; LSPN-2846.
Photograph by G. Malcolm, Boyden Observatory.

(bottom) Fig. 306: 1986 Mar 14.35278 UT; exposure 10.0 minutes on 103a-O emulsion with no filter; r=0.91, Δ=0.95, β=115.5°; LSPN-727.
Photograph by C. Torres/H. Wroblewski, Cerro el Roble Astronomical Observatory.

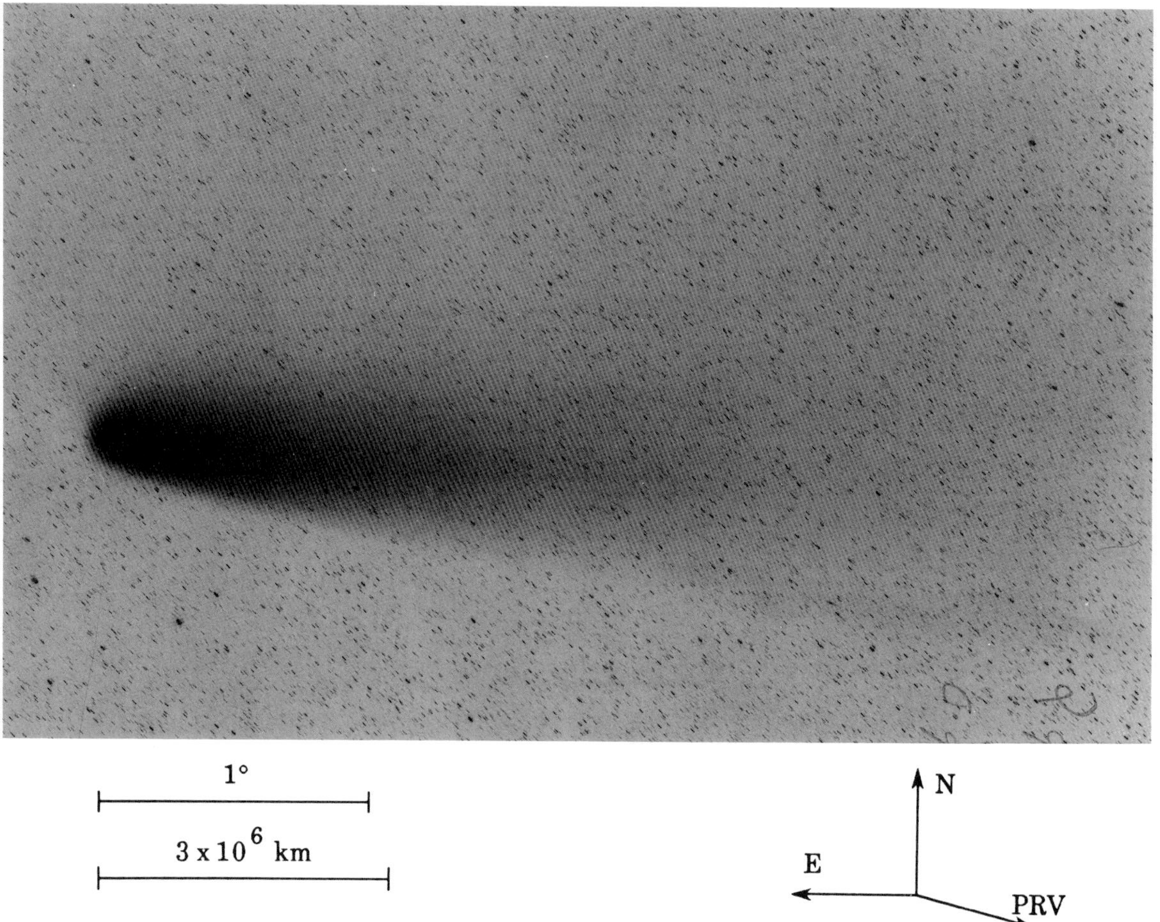

Fig. 307: 1986 Mar 14.35556 UT; exposure 30.0 minutes on 098-04 emulsion with RG-610 filter; r=0.91, Δ=0.95, β=115.5°; LSPN-1289.
Photograph by F. Miller, University of Michigan/CTIO.

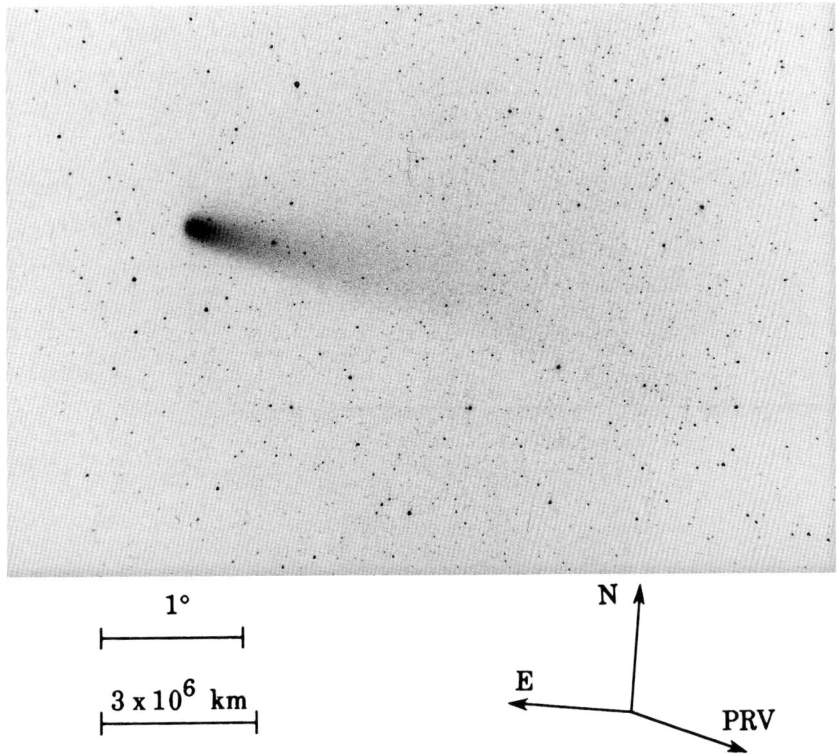

Fig. 308: 1986 Mar 14.52670 UT; exposure 5.0 minutes on Tri-X emulsion; r=0.91, Δ=0.95, β=115.4°; AON-851064.
Photograph by S. Edberg, USA.

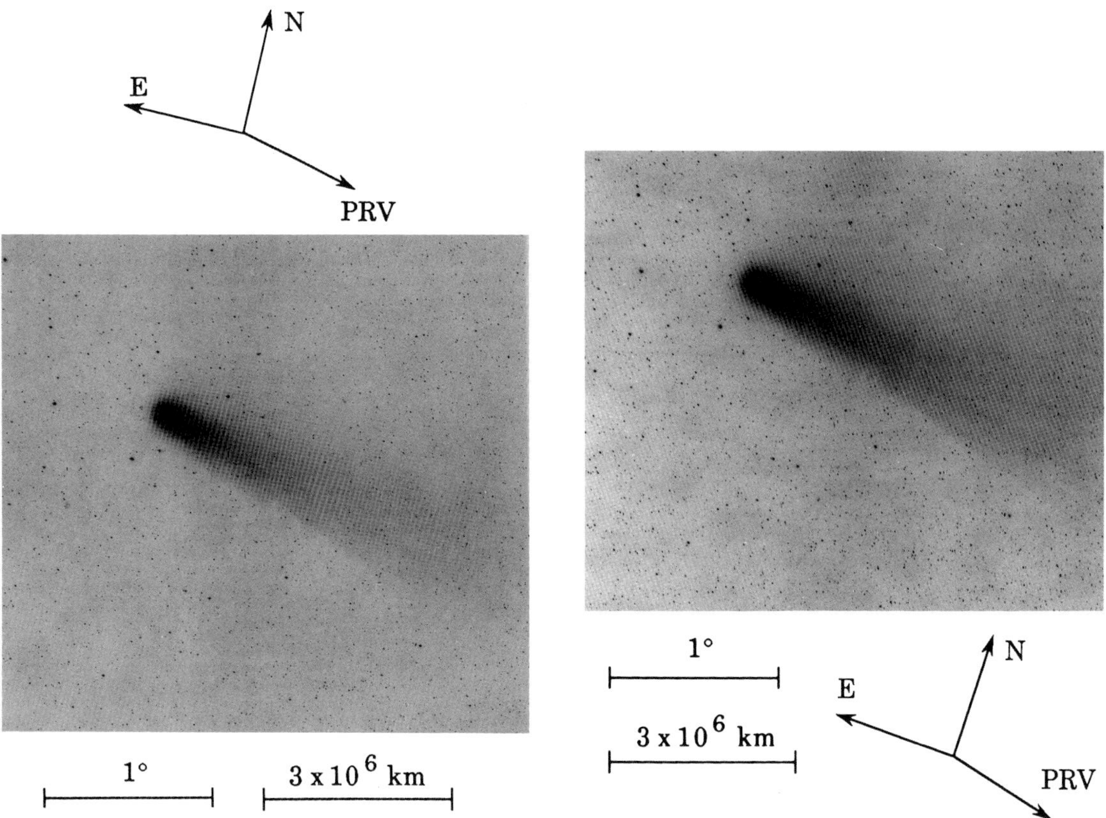

(left) Fig. 309: 1986 Mar 14.95660 UT; exposure 5.0 minutes on 103a-D emulsion with no filter; r=0.92, Δ=0.94, β=115.2°; LSPN-3665.
Photograph submitted by K. Sivaraman, Indian Institute for Astrophysics, Kavalur Station.

(right) Fig. 310: 1986 Mar 14.95660 UT; exposure 5.0 minutes on 103a-D emulsion with no filter; r=0.92, Δ=0.94, β=115.2°; LSPN-3666.
Photograph submitted by K. Sivaraman, Indian Institute for Astrophysics, Kavalur Station.

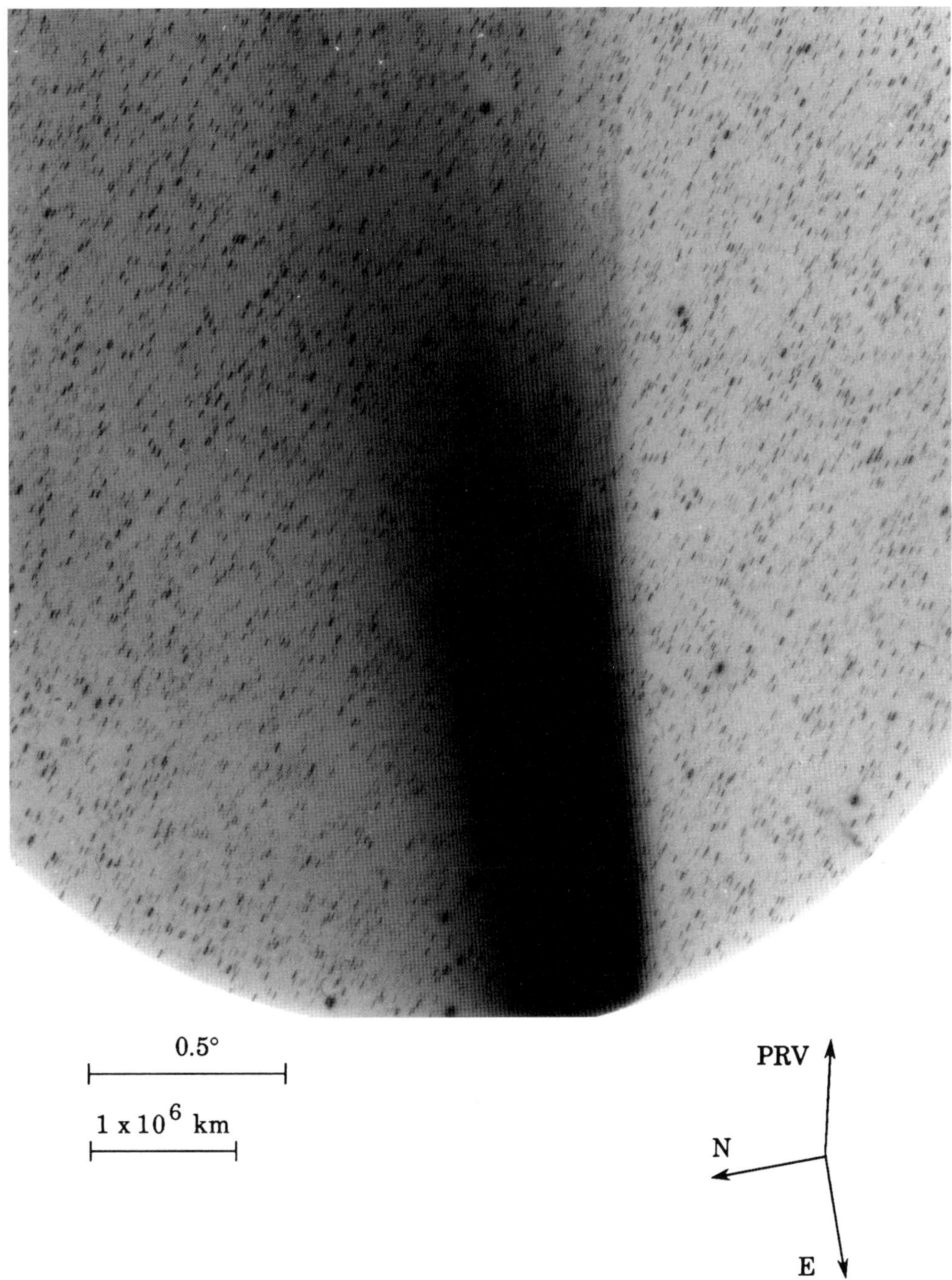

Fig. 311: 1986 Mar 15.11649 UT; exposure 30.0 minutes on Tech. Pan 2415 emulsion with no filter; r=0.92, Δ=0.93, β=115.1°; LSPN-3479.
Photograph by B. Carter, University of Perugia/South African Astronomical Observatory.

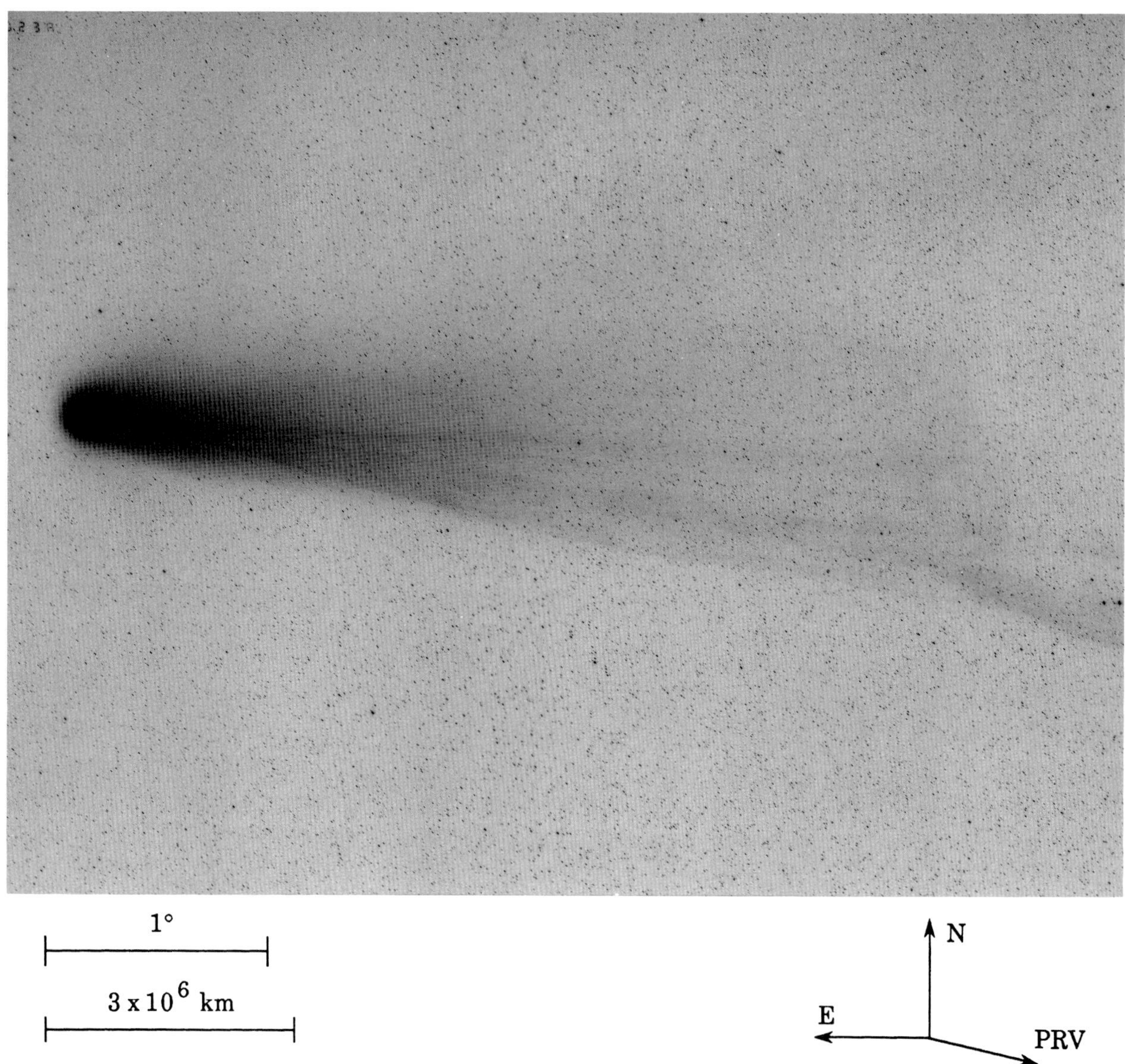

Fig. 312: 1986 Mar 15.33924 UT; exposure 15.0 minutes on IIa-O emulsion with GG-385 filter; r=0.92, Δ=0.93, β=115.0°; LSPN-2891.
Photograph by G. Pizarro, European Southern Observatory.

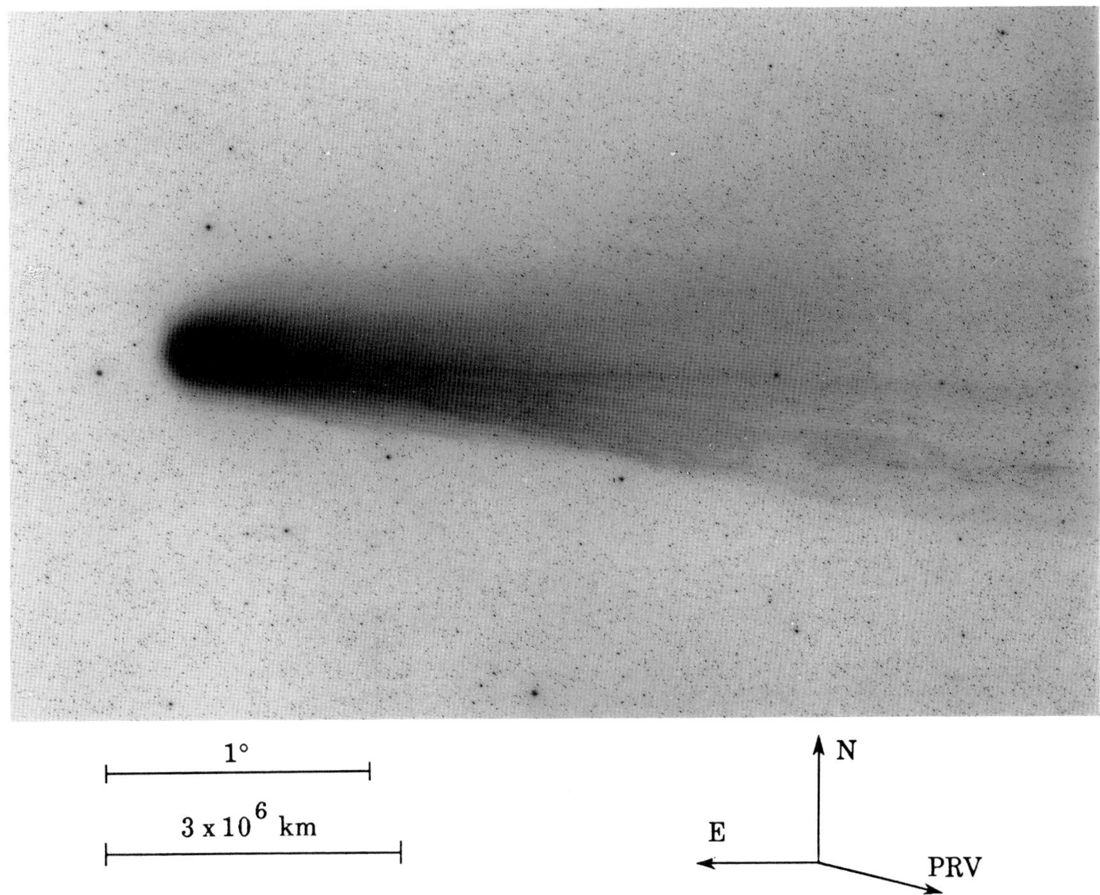

Fig. 313: 1986 Mar 15.35903 UT; exposure 10.0 minutes on 103a-O emulsion with GG-385 filter; r=0.92, Δ=0.93, β=115.0°; LSPN-729.
Photograph by C. Torres/H. Wroblewski, Cerro el Roble Astronomical Observatory.

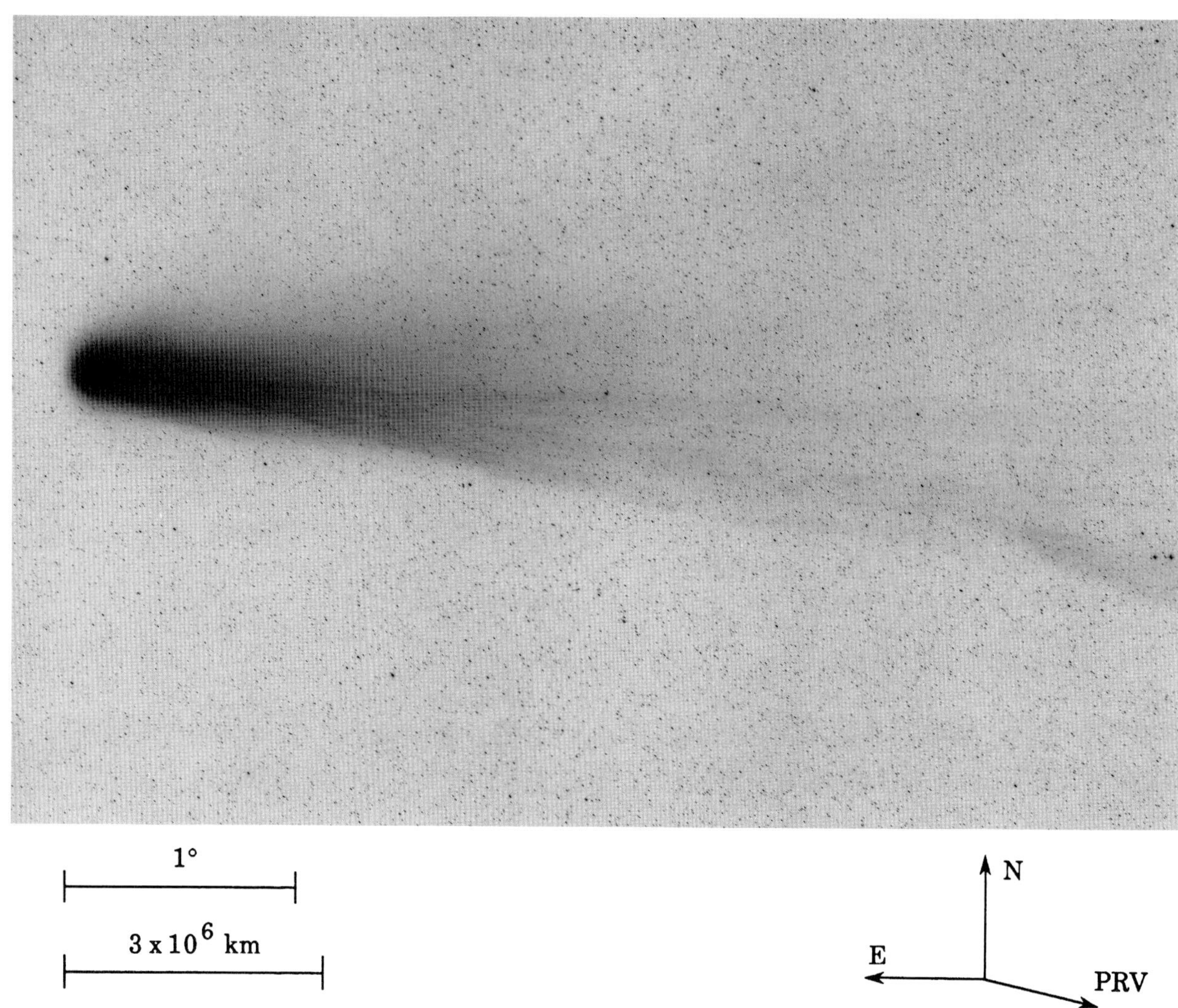

Fig. 314: 1986 Mar 15.37066 UT; exposure 15.0 minutes on IIa-O emulsion with GG-385 filter; r=0.92, Δ=0.93, β=115.0°; LSPN-2892.
Photograph by G. Pizarro, European Southern Observatory.

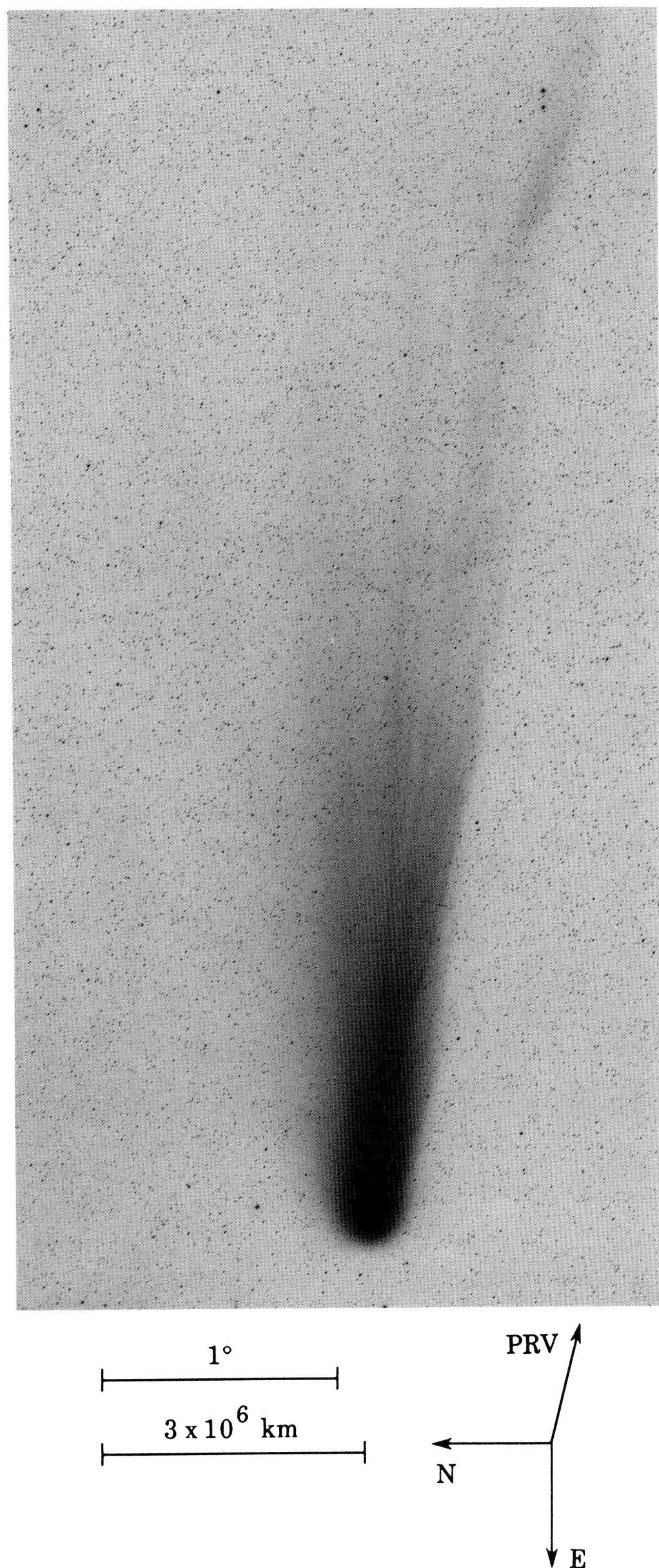

Fig. 315: 1986 Mar 15.38924 UT; exposure 15.0 minutes on IIa-O emulsion with GG-385 filter; r=0.92, Δ=0.93, β=115.0°; LSPN-2893.
Photograph by G. Pizarro, European Southern Observatory.

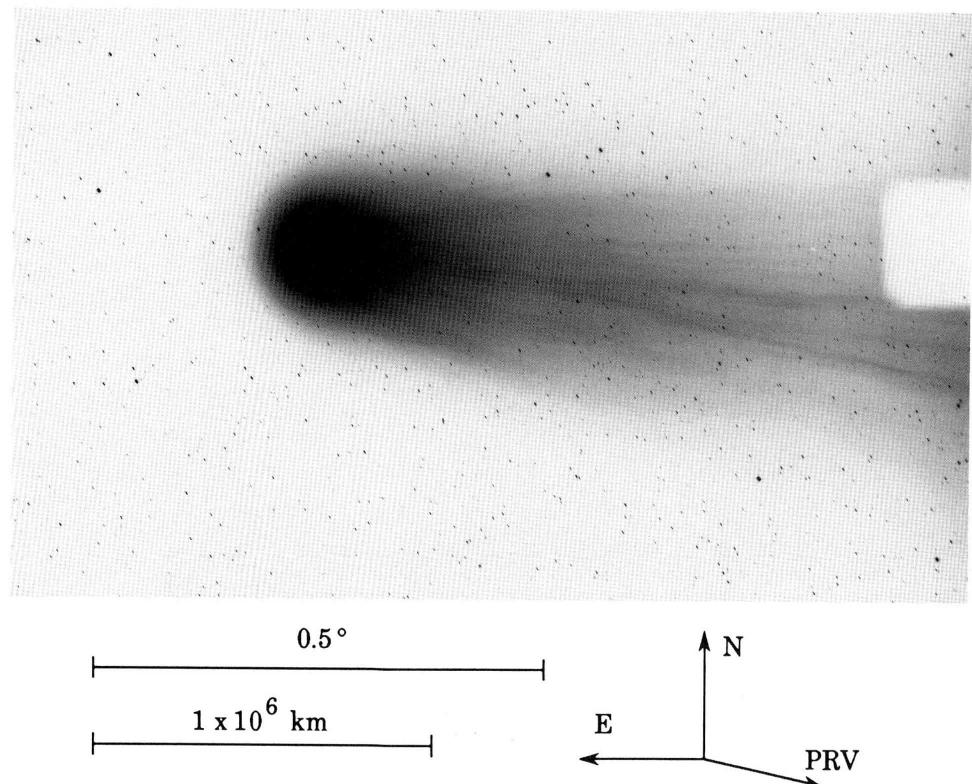

Fig. 316: 1986 Mar 15.40521 UT; exposure 5.0 minutes on 103a-O emulsion with Wratten 2C filter; r=0.92, Δ=0.93, β=115.0°; LSPN-2564.
Photograph by A. Dressler/R. Windhorst, Mount Wilson/Las Campanas Observatories.

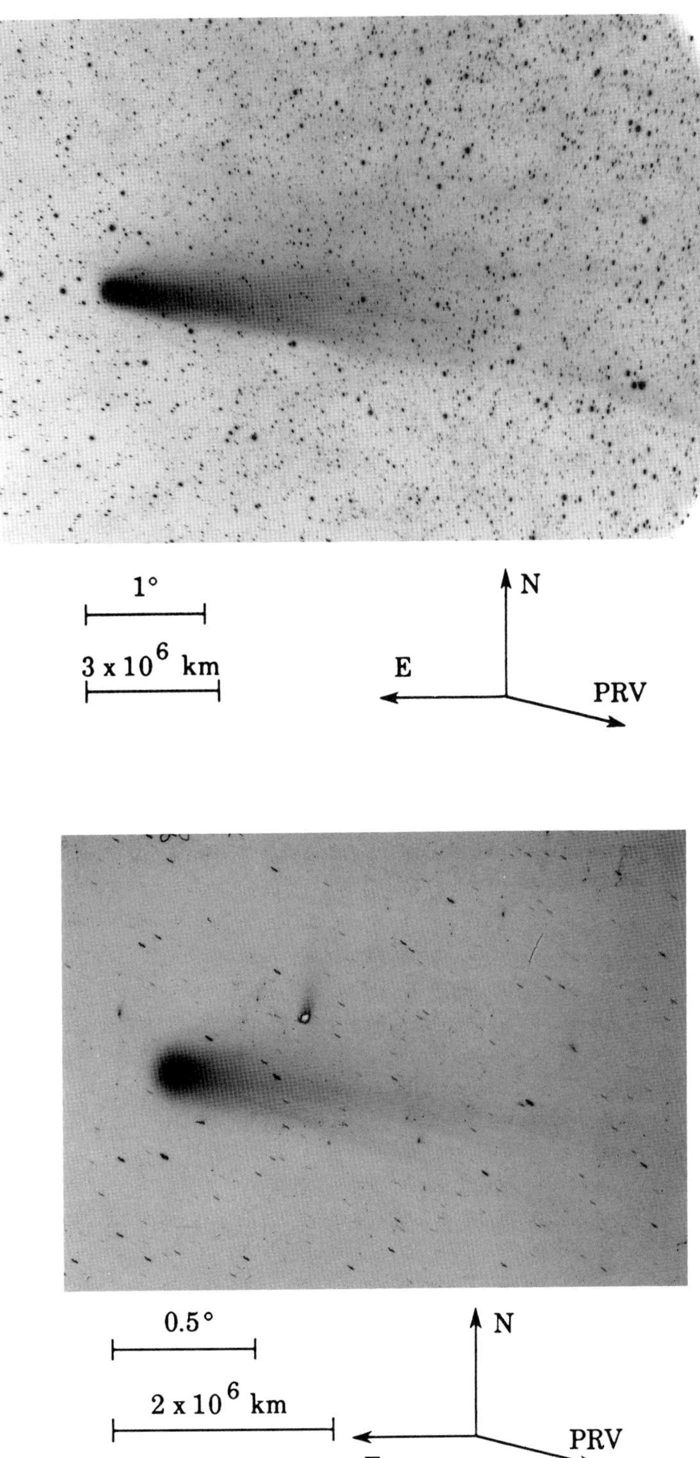

(top) Fig. 317: 1986 Mar 15.46810 UT; exposure 35.0 minutes on IIa-O emulsion with no filter; r=0.92, Δ=0.93, β=115.0°; LSPN-3416.
Photograph by G. Emerson, E. E. Barnard Observatory Station.

(bottom) Fig. 318*: 1986 Mar 15.97292 UT; exposure 30.0 minutes on 103a-O emulsion with no filter; r=0.93, Δ=0.91, β=114.7°; LSPN-3625.
Photograph submitted by K. Sivaraman, Indian Institute for Astrophysics, Kodaikanal Station.

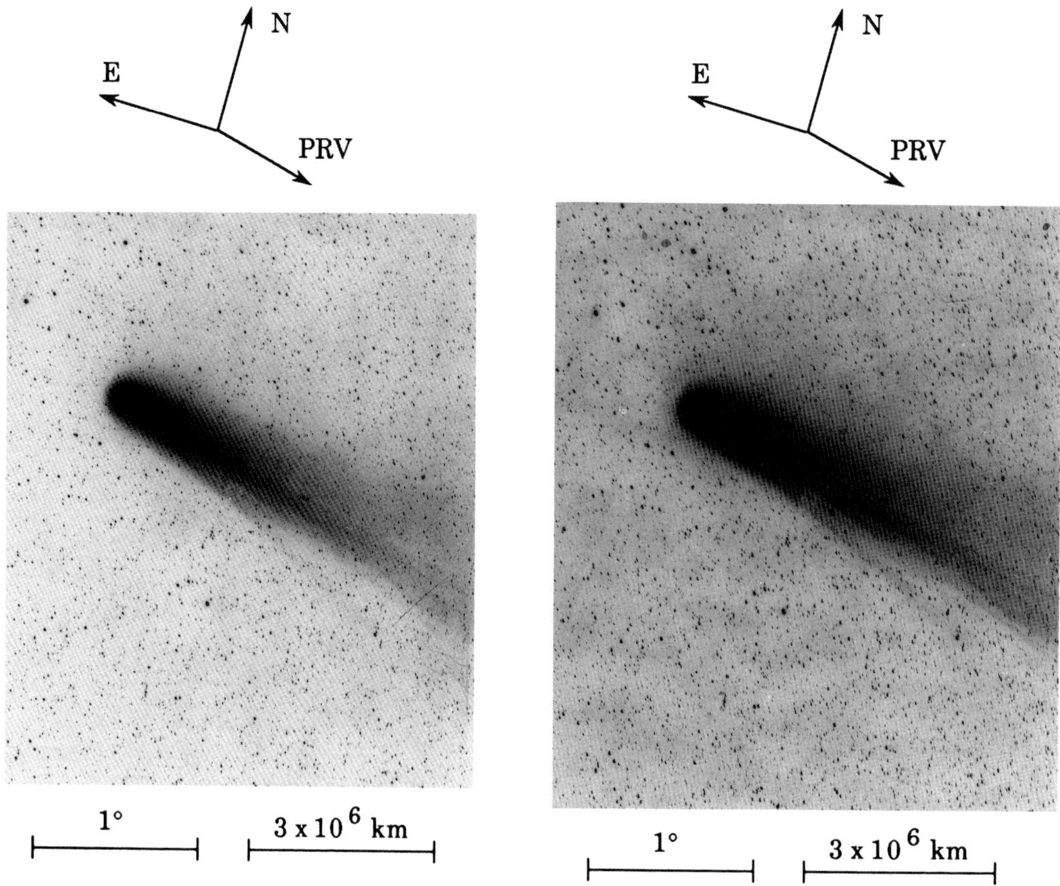

(left) Fig. 319: 1986 Mar 15.96562 UT; exposure 20.0 minutes on 103a-D emulsion with no filter; r=0.93, Δ=0.91, β=114.7°; LSPN-3667.
Photograph submitted by K. Sivaraman, Indian Institute for Astrophysics, Kavalur Station.

(right) Fig. 320: 1986 Mar 15.98125 UT; exposure 20.0 minutes on 103a-D emulsion with no filter; r=0.93, Δ=0.91, β=114.7°; LSPN-3668.
Photograph submitted by K. Sivaraman, Indian Institute for Astrophysics, Kavalur Station.

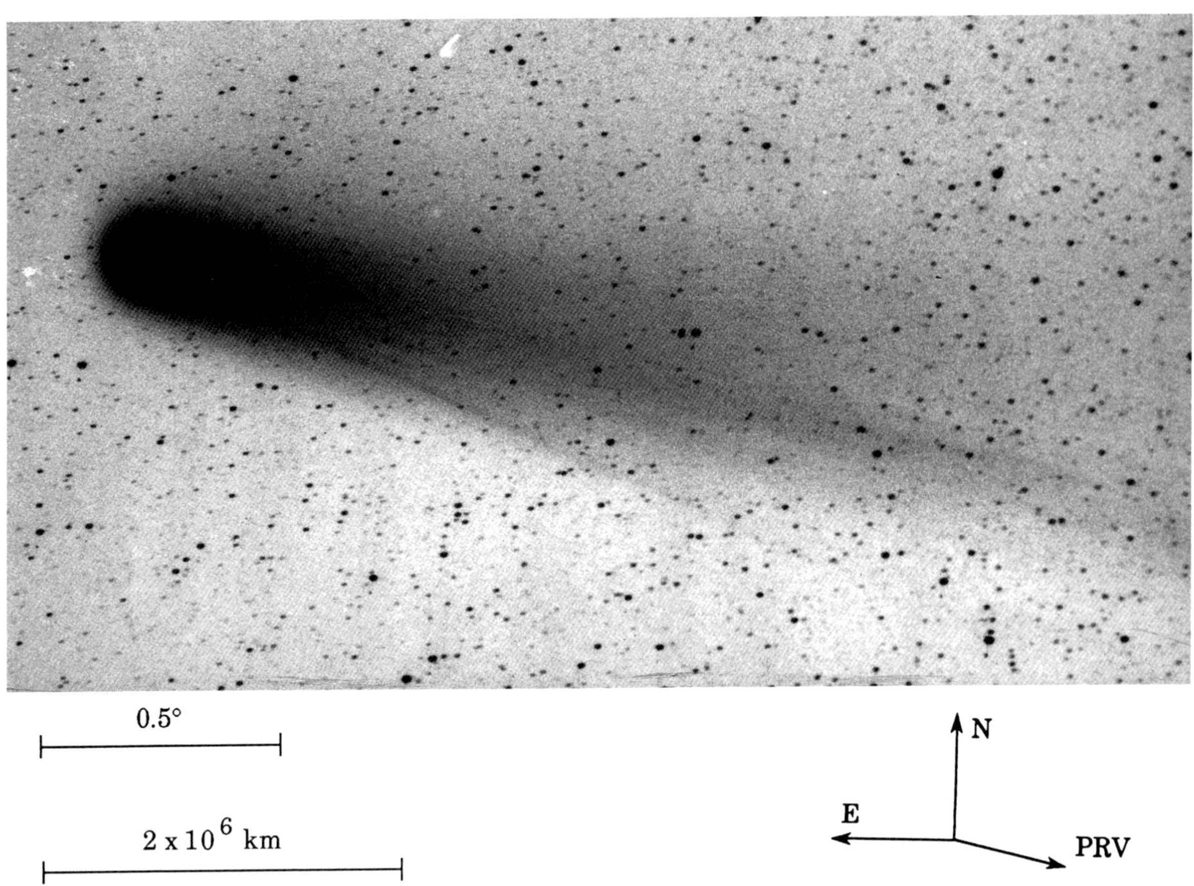

Fig. 321: 1986 Mar 16.05970 UT; exposure 12.0 minutes on Tech. Pan 2415 emulsion; r=0.93, Δ=0.91, β=114.7°. *Photograph by A. Marafie, Wafra Observatory.*

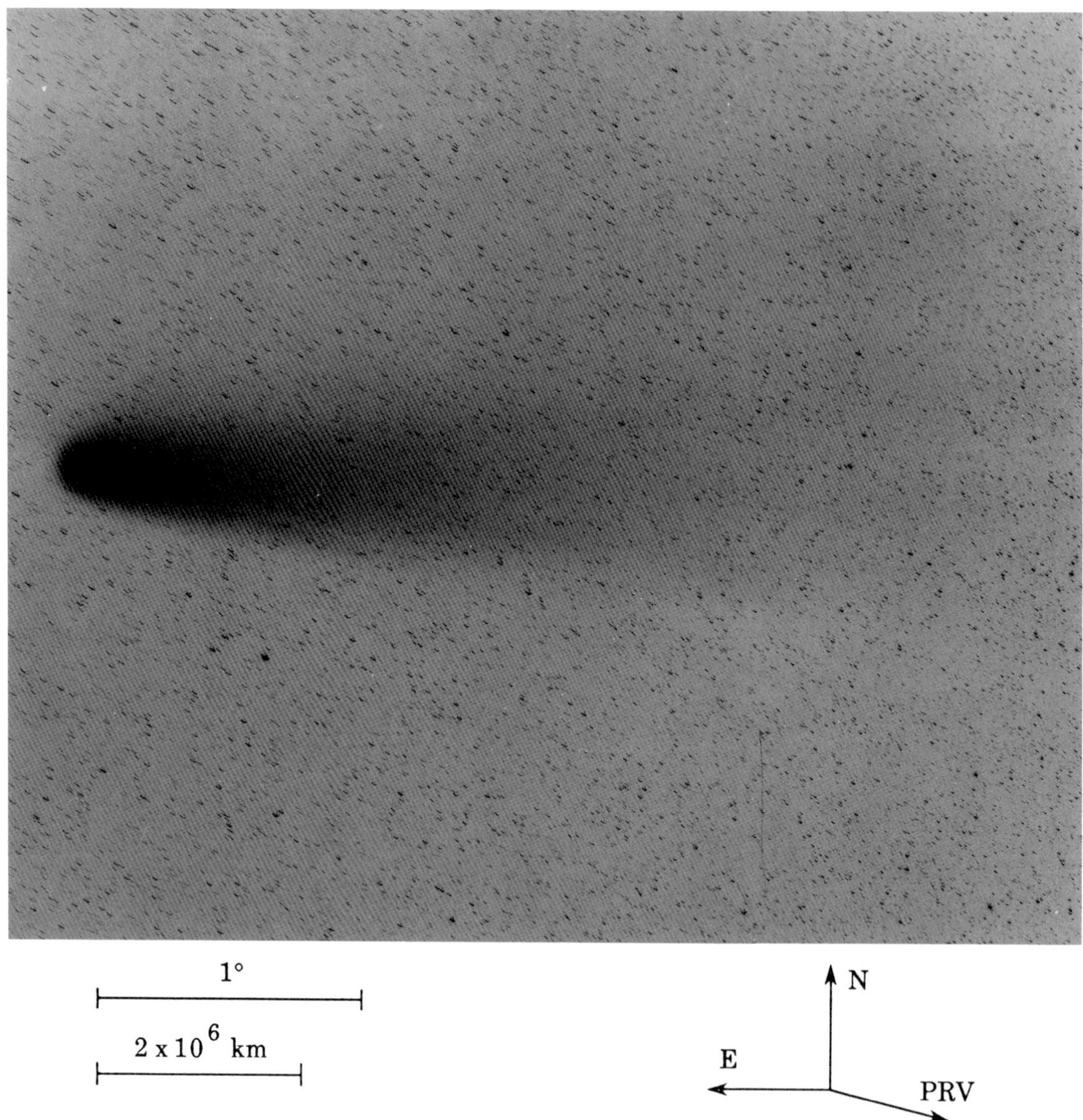

Fig. 322: 1986 Mar 16.29236 UT; exposure 30.0 minutes on 098-04 emulsion with RG-610 filter; r=0.94, Δ=0.90, β=114.6°; LSPN-1294.
Photograph by F. Miller, University of Michigan/CTIO.

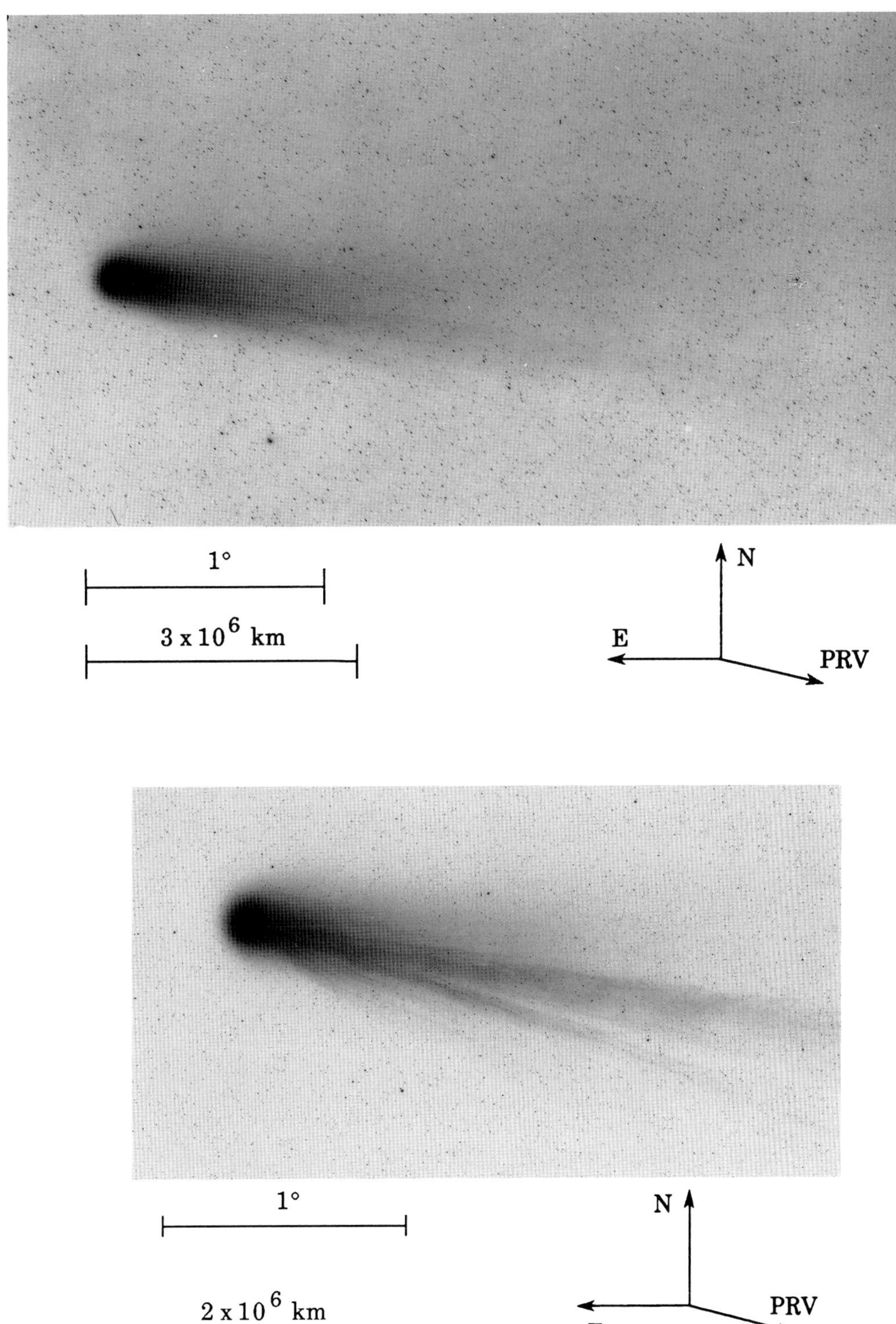

(top) Fig. 323: 1986 Mar 16.34409 UT; exposure 15.0 minutes on IIa-D emulsion with GG-495 filter; r=0.94, Δ=0.90, β=114.6°; LSPN-730.
Photograph by C. Torres/H. Wroblewski, Cerro el Roble Astronomical Observatory.

(bottom) Fig. 324: 1986 Mar 16.36215 UT; exposure 3.0 minutes on IIa-O emulsion with no filter; r=0.94, Δ=0.90, β=114.6°; LSPN-1297.
Photograph by F. Miller, University of Michigan/CTIO.

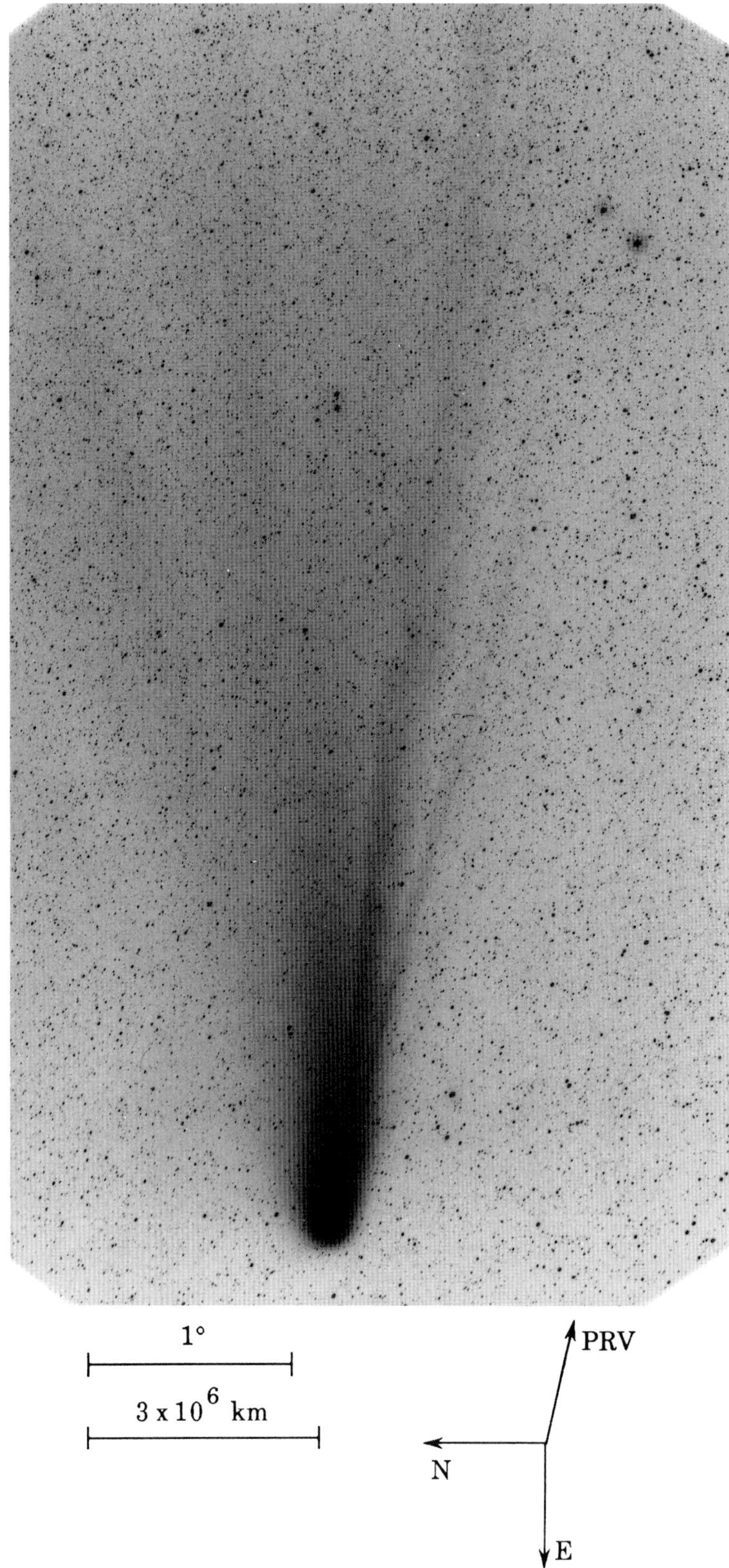

Fig. 325: 1986 Mar 16.41597 UT; exposure 16.0 minutes on Tech. Pan 2415 emulsion with no filter; r=0.94, Δ=0.90, β=114.6°; LSPN-1397.
Photograph by W. Liller, Easter Island, LSPN Island Network.

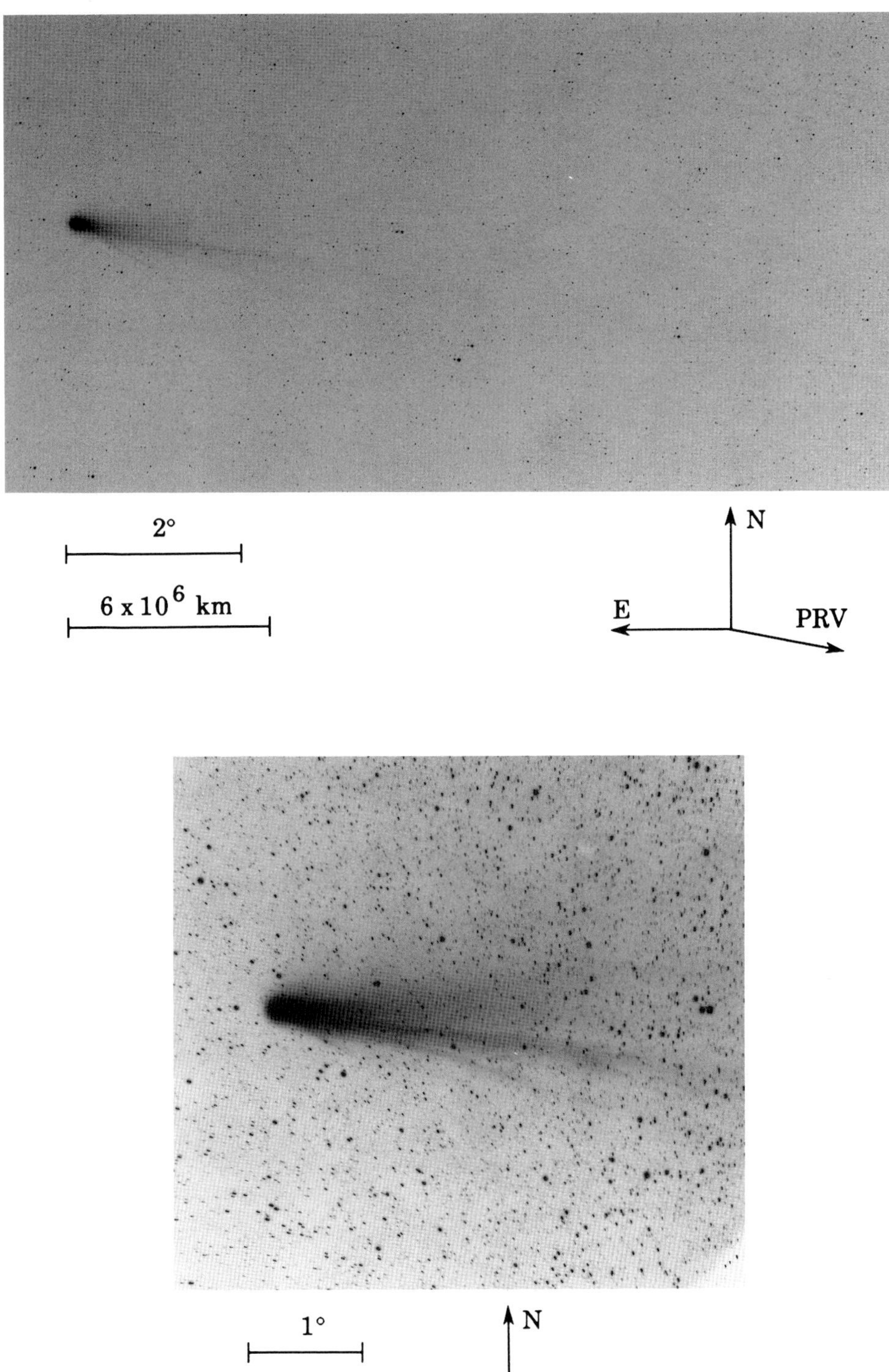

(top) Fig. 326: 1986 Mar 16.44900 UT; exposure 10.0 minutes on IIa-O emulsion; r=0.94, Δ=0.90, β=114.6°.
Photograph by A. Butcher/D. Doss, McDonald Observatory.

(bottom) Fig. 327: 1986 Mar 16.46150 UT; exposure 35.0 minutes on IIa-O emulsion with no filter; r=0.94, Δ=0.90, β=114.6°; LSPN-3417.
Photograph by G. Emerson, E. E. Barnard Observatory Station.

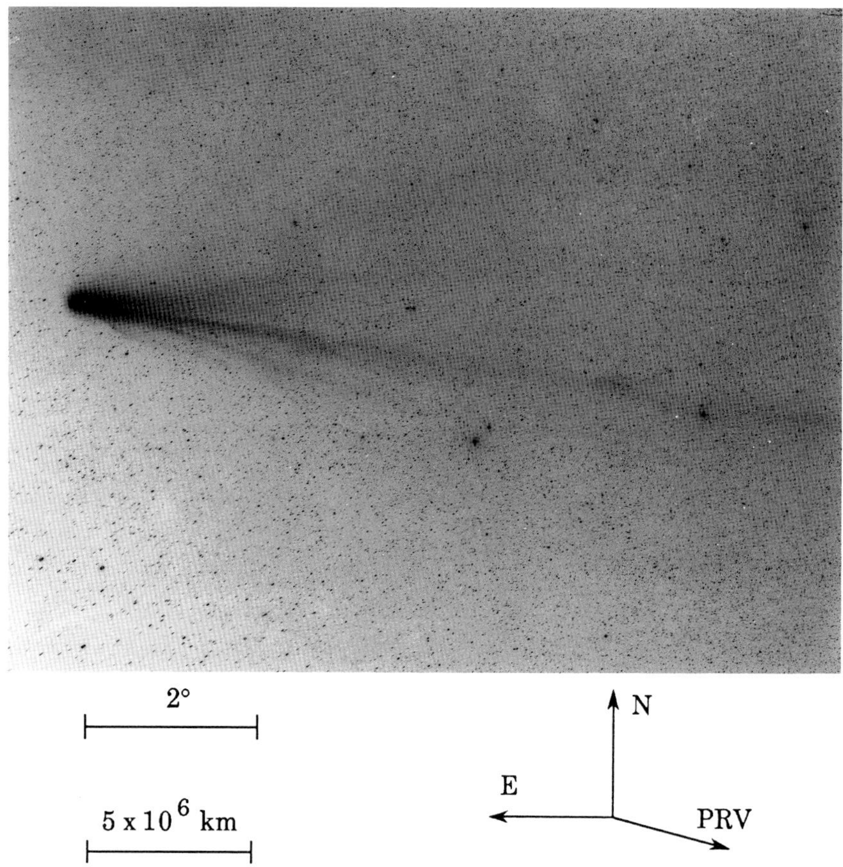

Fig. 328: 1986 Mar 16.46806 UT; exposure 10.0 minutes on IIa-O emulsion with no filter; r=0.94, Δ=0.90, β=114.6°; LSPN-893.
Photograph by E. Moore, Joint Observatory for Cometary Research.

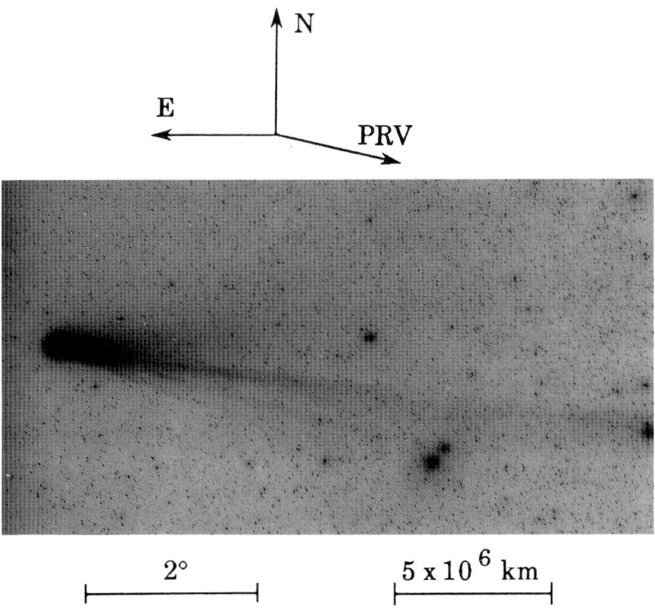

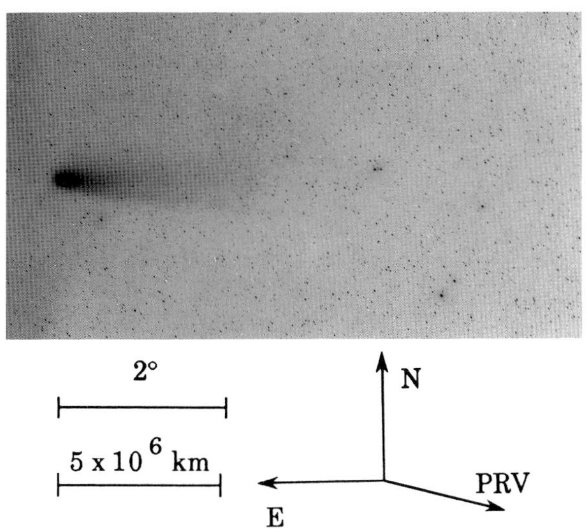

(top) Fig. 329: 1986 Mar 16.61296 UT; exposure 30.0 minutes on IIIa-J emulsion with no filter; r=0.94, Δ=0.90, β=114.5°; LSPN-1912.
Photograph by D. Cruikshank, Mauna Kea Observatory.

(bottom) Fig. 330: 1986 Mar 16.63646 UT; exposure 22.0 minutes on IIIa-F emulsion with Wratten 25 filter; r=0.94, Δ=0.90, β=114.5°; LSPN-1913.
Photograph by D. Cruikshank, Mauna Kea Observatory.

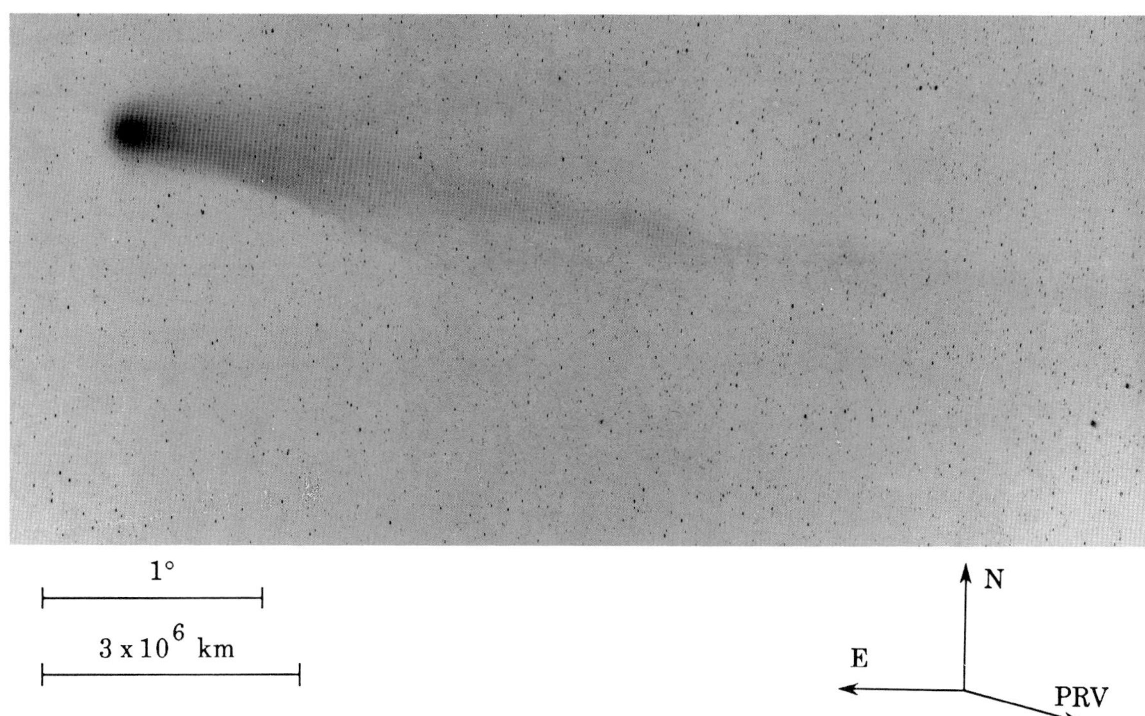

Fig. 331: 1986 Mar 16.82516 UT; exposure 10.0 minutes on IIa-O emulsion with no filter; r=0.94, Δ=0.89, β=114.4°; LSPN-1631.
Photograph by H. Maehara, Kiso Observatory.

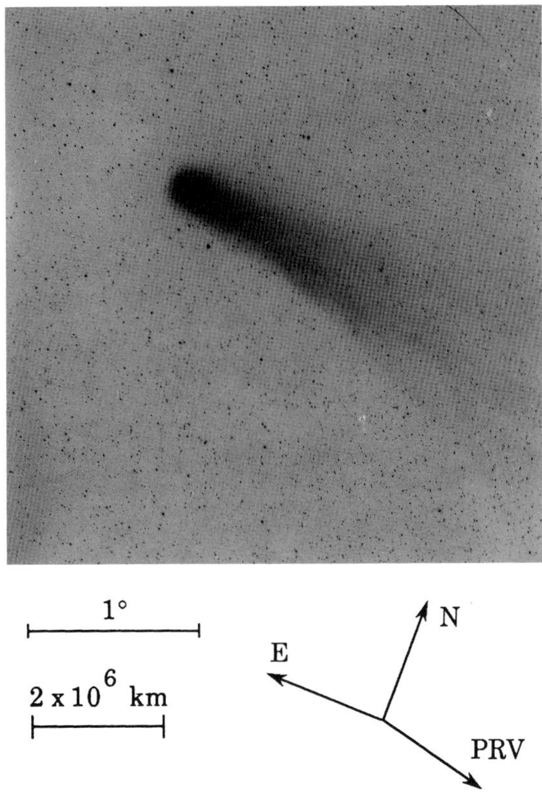

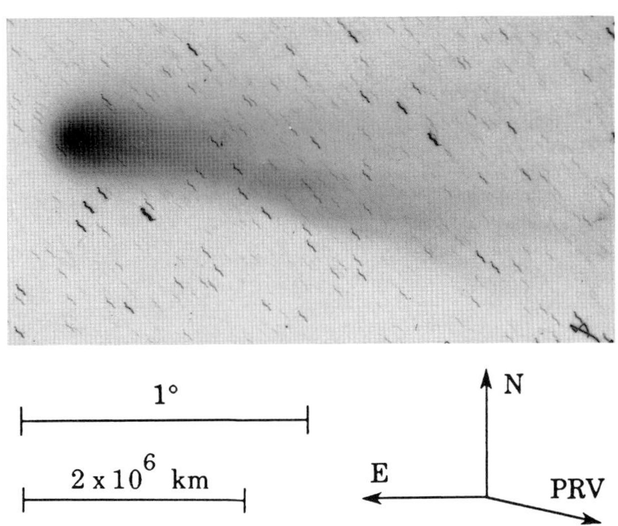

(top) Fig. 332: 1986 Mar 16.96111 UT; exposure 15.0 minutes on 103a-D emulsion with no filter; r=0.95, Δ=0.89, β=114.4°; LSPN-3669.
Photograph submitted by K. Sivaraman, Indian Institute for Astrophysics, Kavalur Station.

(bottom) Fig. 333: 1986 Mar 16.96667 UT; exposure 60.0 minutes on 103a-O emulsion with no filter; r=0.95, Δ=0.89, β=114.4°; LSPN-3626.
Photograph submitted by K. Sivaraman, Indian Institute for Astrophysics, Kodaikanal Station.

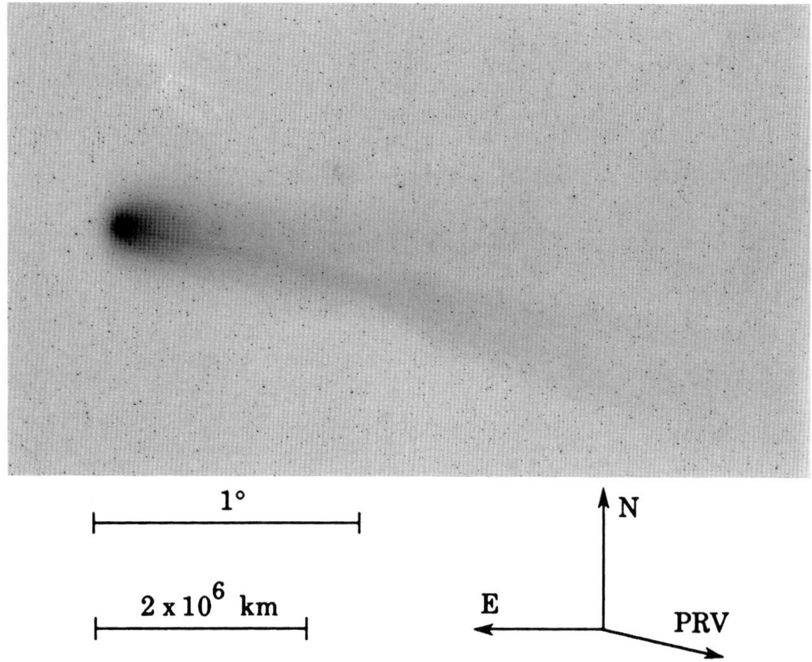

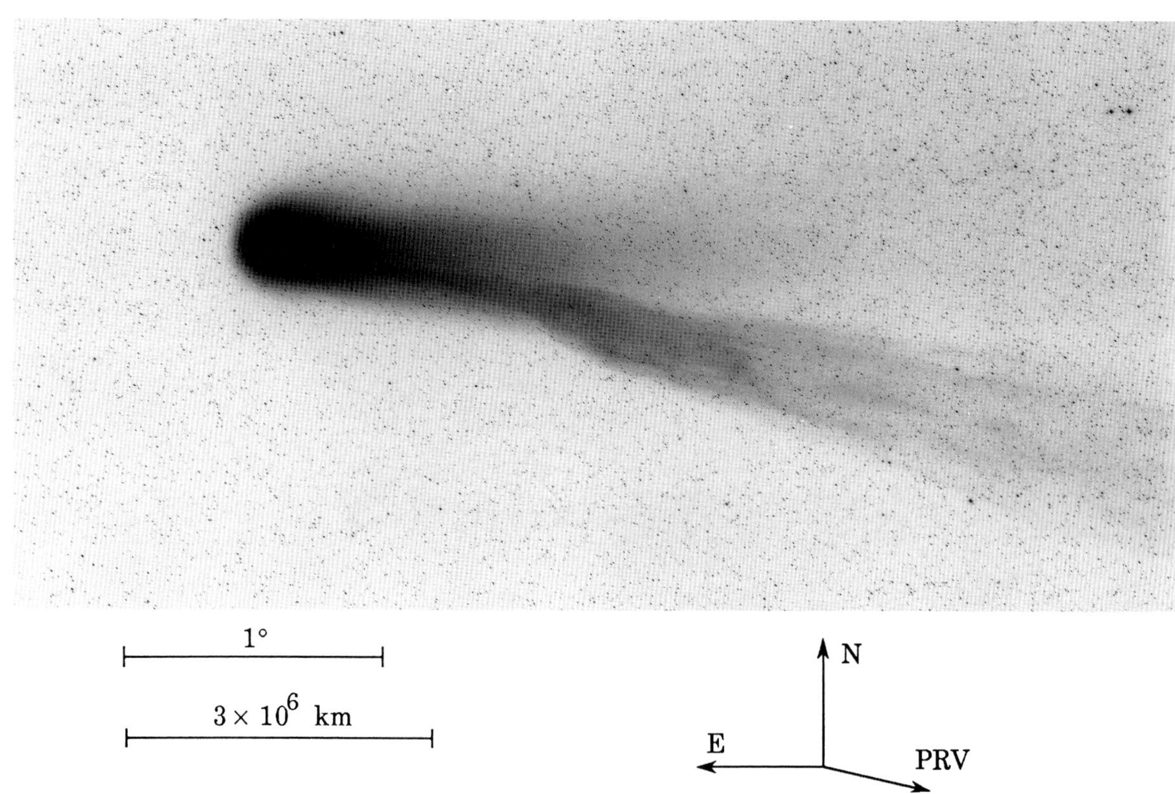

(top) Fig. 334: 1986 Mar 17.28229 UT; exposure 3.0 minutes on IIa-O emulsion with no filter; r=0.95, Δ=0.88, β=114.3°; LSPN-1298.
Photograph by F. Miller, University of Michigan/CTIO.

(bottom) Fig. 335: 1986 Mar 17.35208 UT; exposure 10.0 minutes on 103a-O emulsion with no filter; r=0.95, Δ=0.88, β=114.3°; LSPN-731.
Photograph by C. Torres/H. Wroblewski, Cerro el Roble Astronomical Observatory.

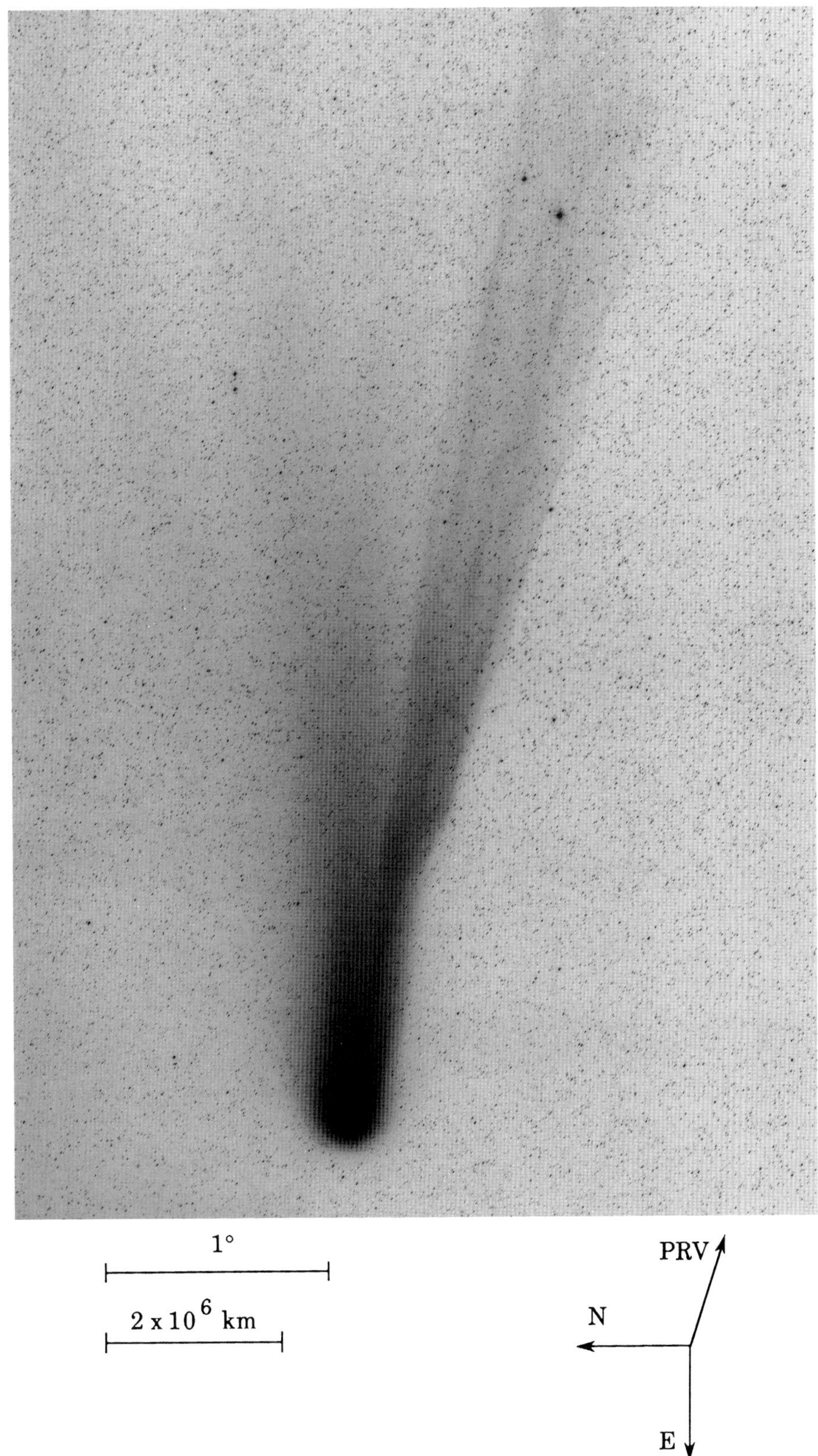

Fig. 336: 1986 Mar 17.36910 UT; exposure 15.0 minutes on IIa-O emulsion with GG-385 filter; r=0.95, Δ=0.88, β=114.3°; LSPN-2899.
Photograph by G. Pizarro, European Southern Observatory.

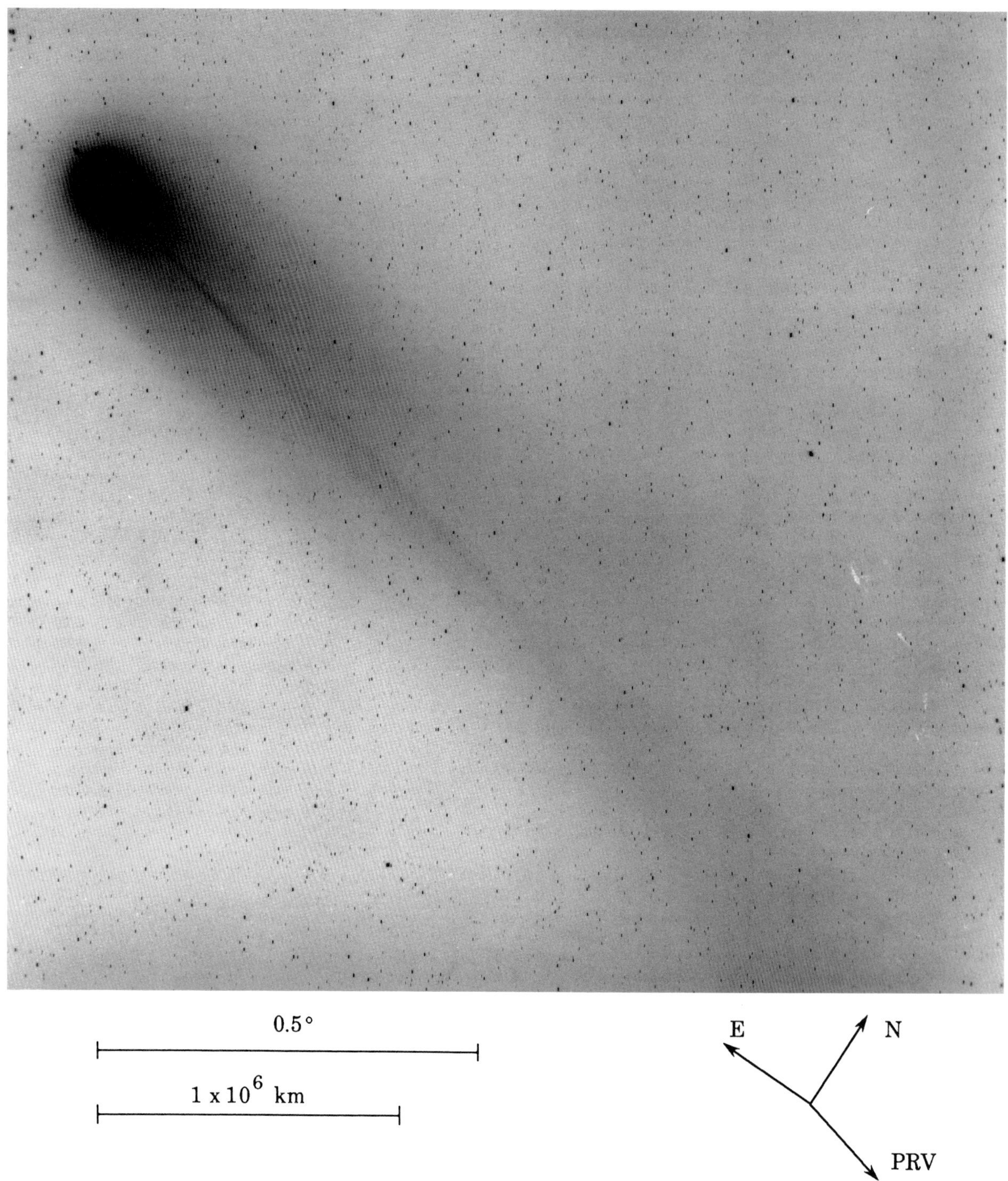

Fig. 337: 1986 Mar 17.39410 UT; exposure 5.0 minutes on 103a-O emulsion with Wratten 2C filter; r=0.95, Δ=0.88, β=114.3°; LSPN-2566.
Photograph by A. Dressler/R. Windhorst, Mount Wilson/Las Campanas Observatories.

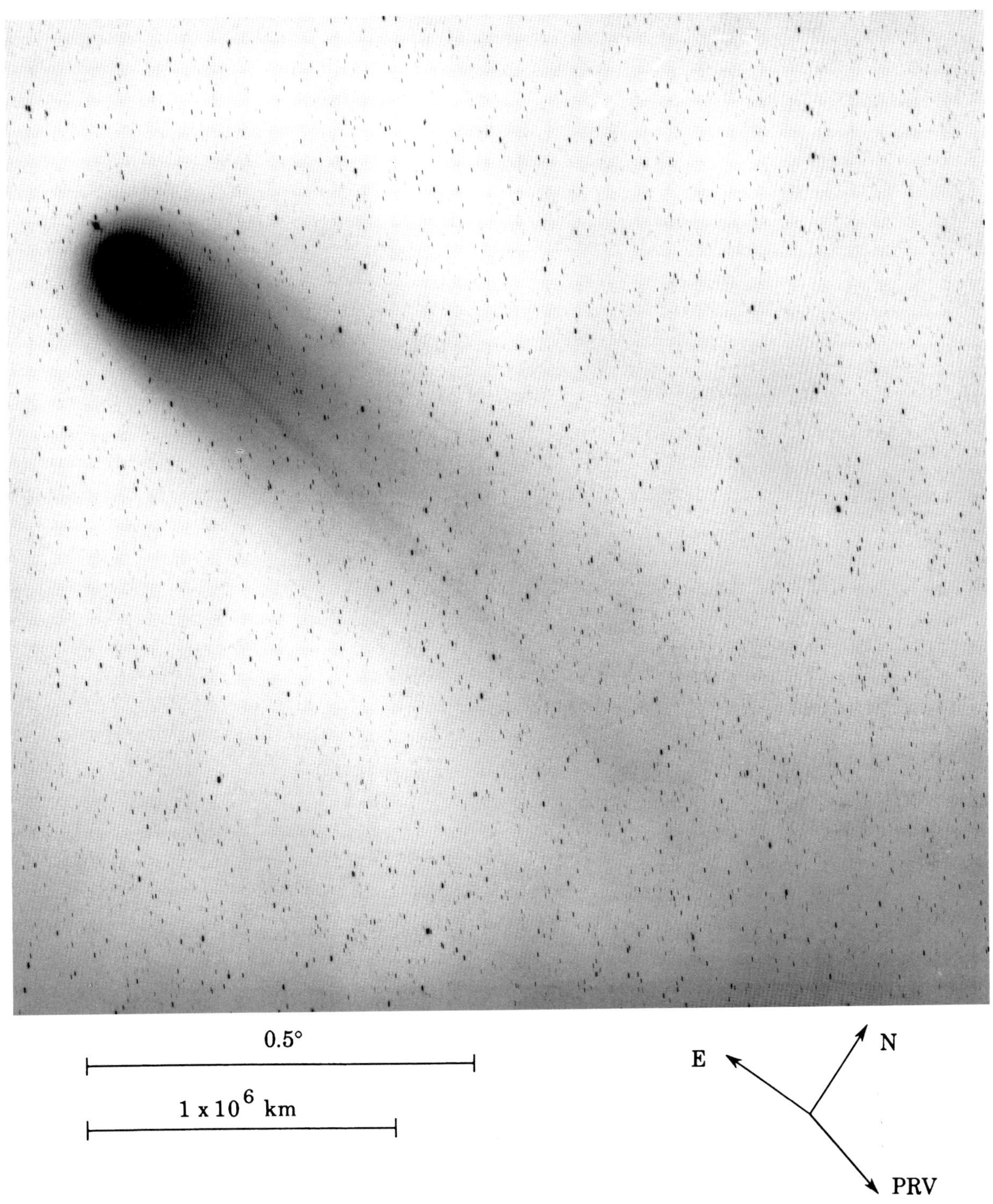

Fig. 338: 1986 Mar 17.40747 UT; exposure 7.5 minutes on 103a-D emulsion with Wratten 12 filter; r=0.95, Δ=0.88, β=114.3°; LSPN-2567.
Photograph by A. Dressler/R. Windhorst, Mount Wilson/Las Campanas Observatories.

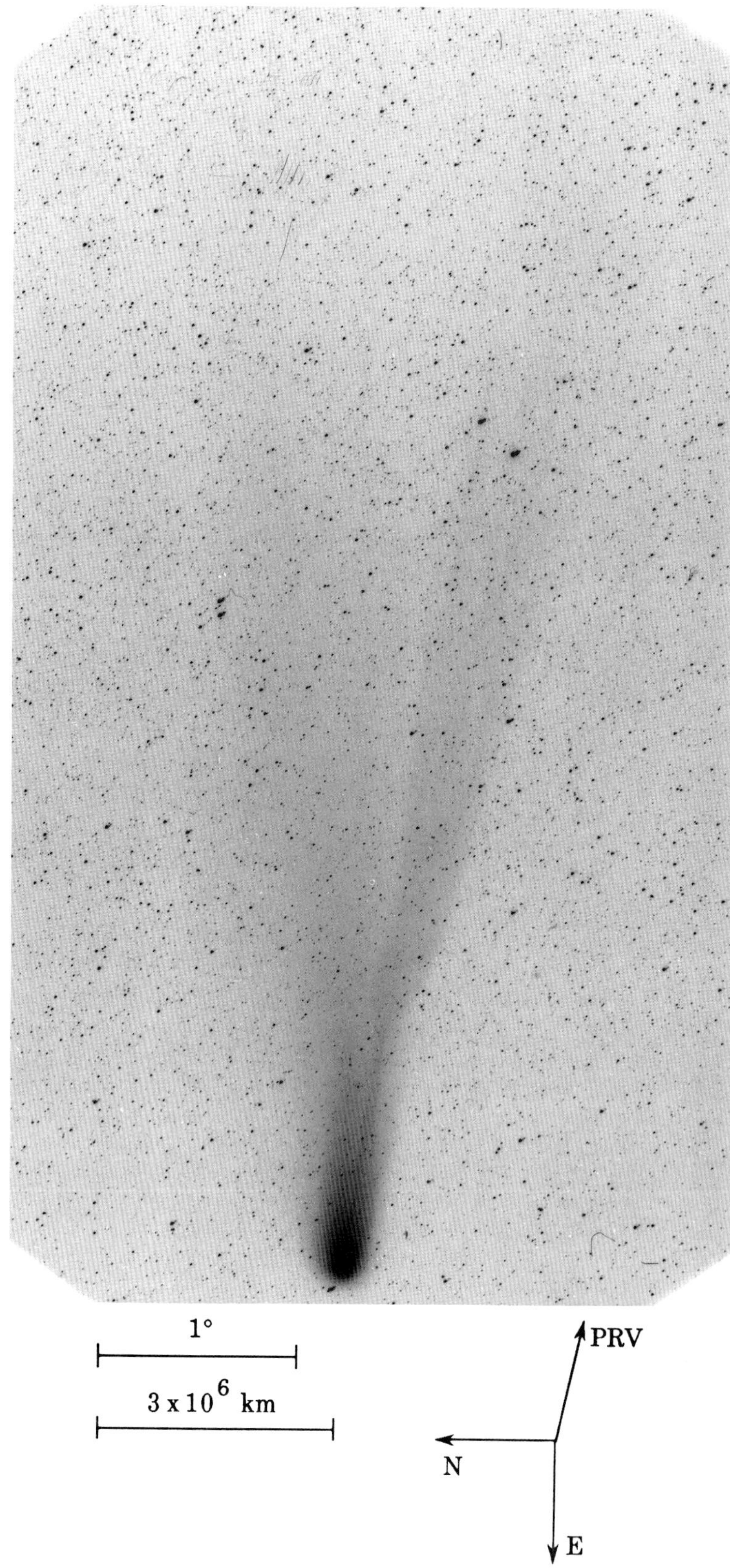

Fig. 339: 1986 Mar 17.45417 UT; exposure 8.0 minutes on Tech. Pan 2415 emulsion with no filter; r=0.95, Δ=0.88, β=114.3°; LSPN-1401.
Photograph by W. Liller, Easter Island, LSPN Island Network.

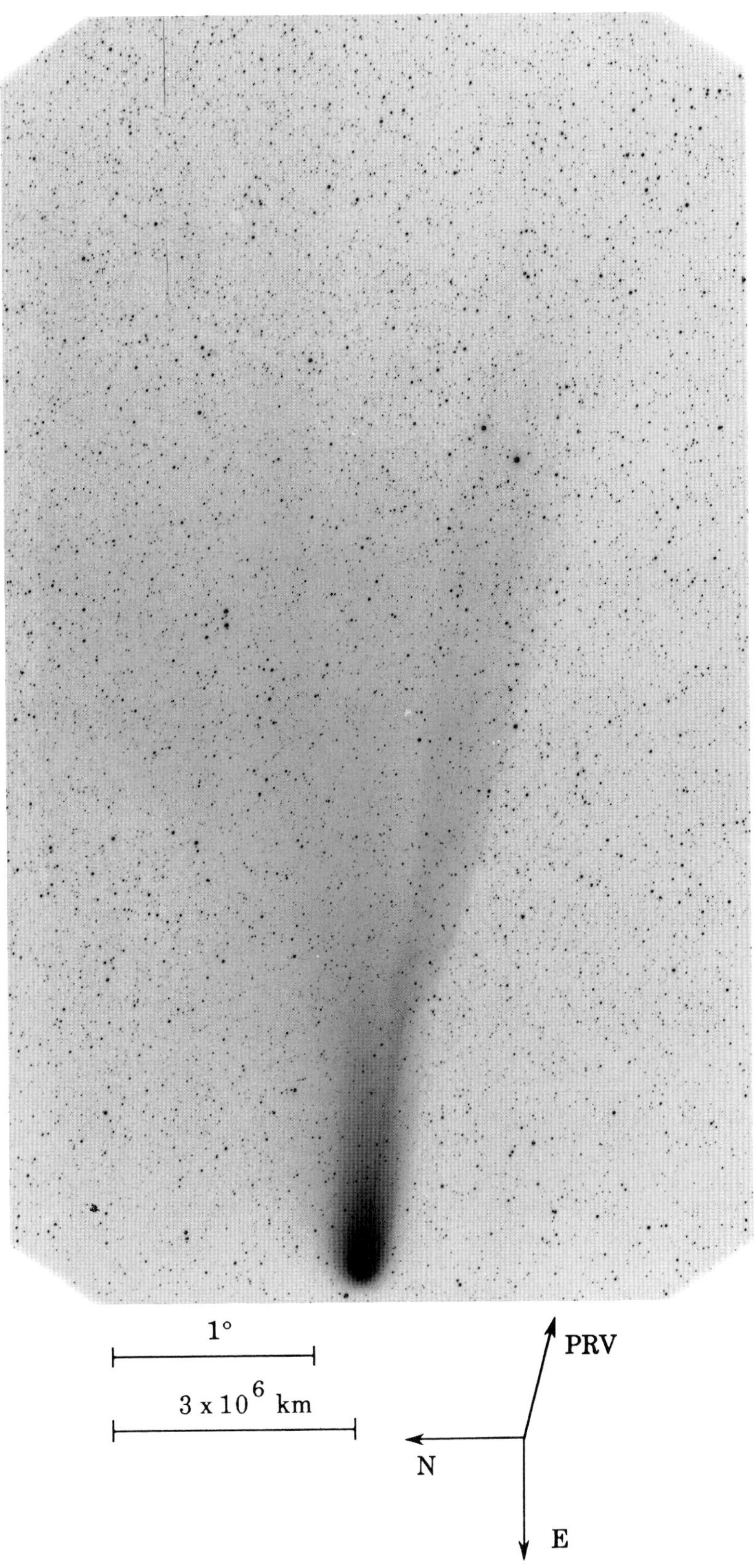

Fig. 340: 1986 Mar 17.49896 UT; exposure 7.0 minutes on Tech. Pan 2415 emulsion with no filter; r=0.95, Δ=0.87, β=114.2°; LSPN-1402.
Photograph by W. Liller, Easter Island, LSPN Island Network.

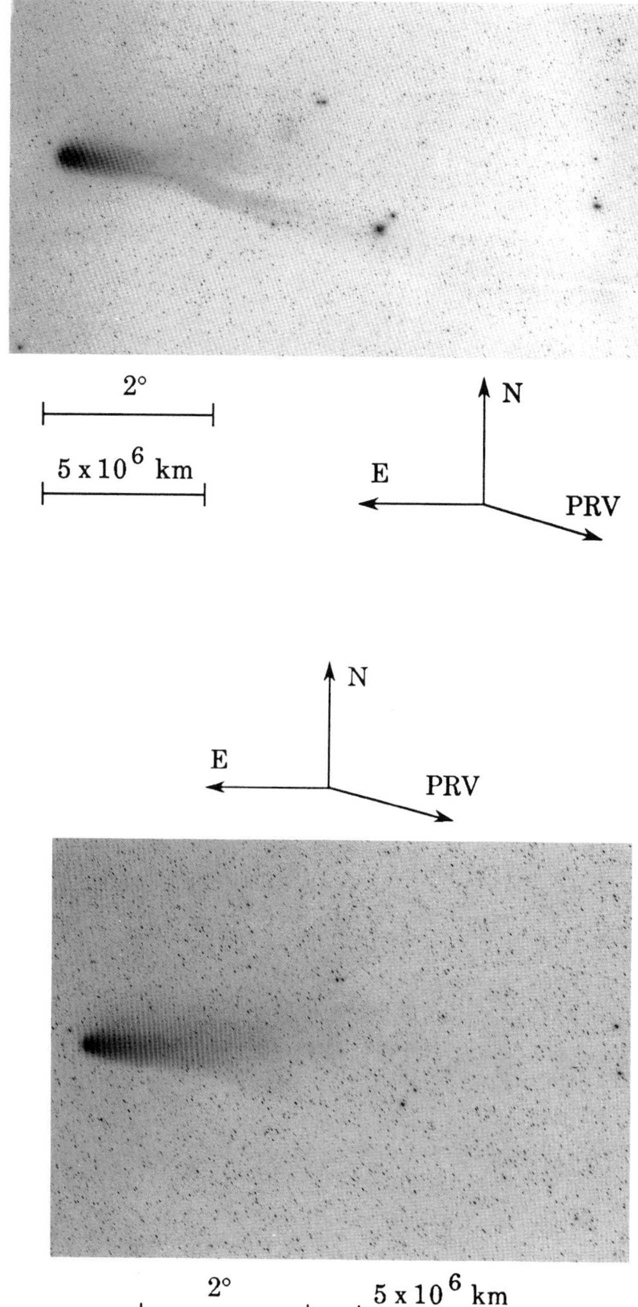

(top) Fig. 341: 1986 Mar 17.60841 UT; exposure 25.0 minutes on IIIa-J emulsion with no filter; r=0.96, Δ=0.87, β=114.2°; LSPN-1915.
Photograph by D. Cruikshank, Mauna Kea Observatory.

(bottom) Fig. 342: 1986 Mar 17.63251 UT; exposure 30.0 minutes on IIIa-F emulsion with Wratten 25 filter; r=0.96, Δ=0.87, β=114.2°; LSPN-1914.
Photograph by D. Cruikshank, Mauna Kea Observatory.

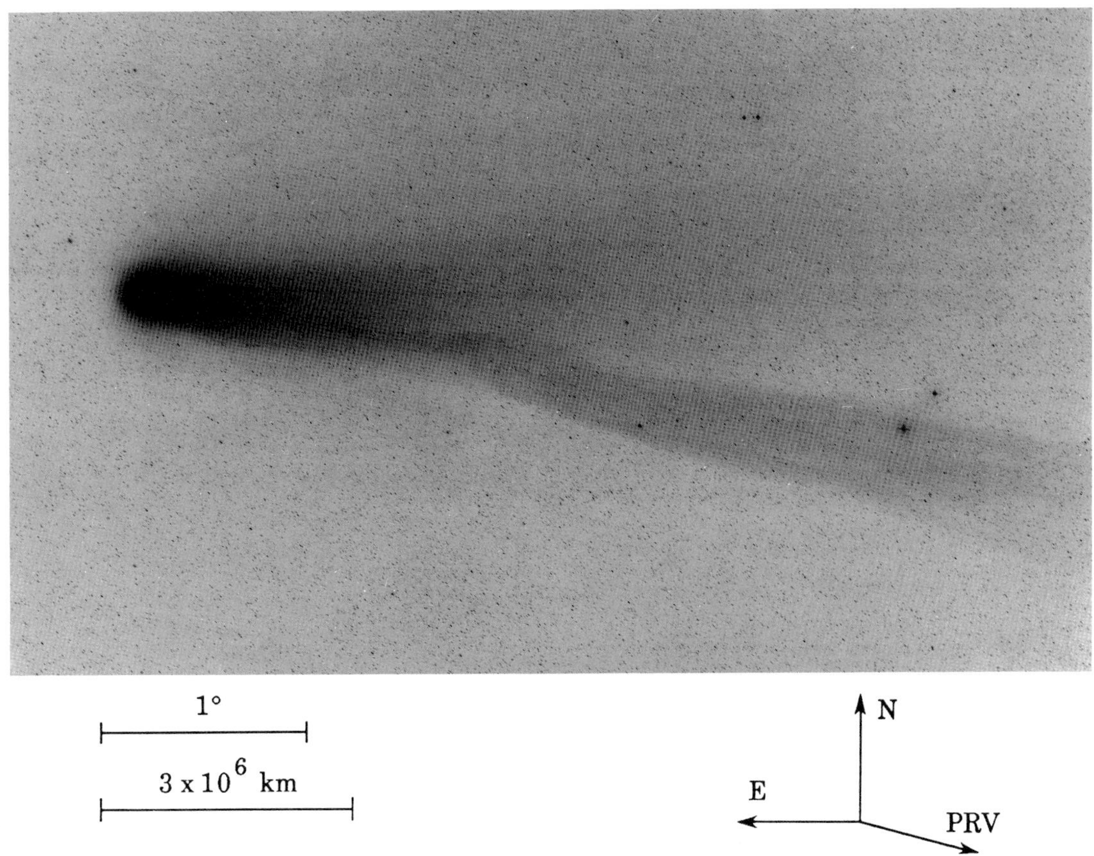

Fig. 343: 1986 Mar 17.77091 UT; exposure 15.0 minutes on IIa-O emulsion with GG-395 filter; r=0.96, Δ=0.87, β=114.2°; LSPN-2386.
Photograph from Royal Observatory/UK Schmidt Telescope Unit.

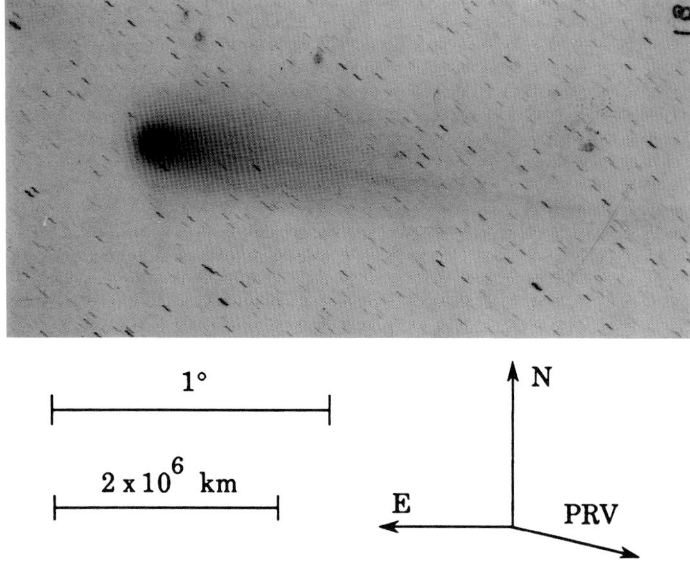

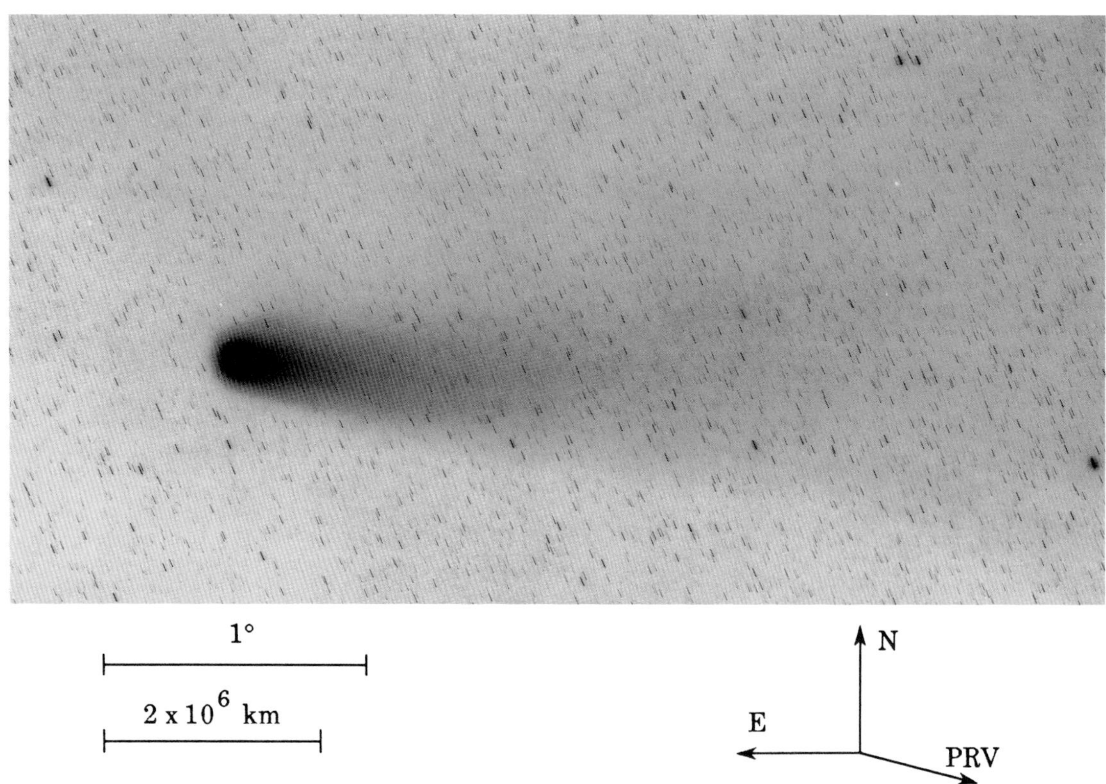

(top) Fig. 344: 1986 Mar 17.99132 UT; exposure 45.0 minutes on 103a-O emulsion with no filter; r=0.96, Δ=0.86, β=114.1°; LSPN-3628.
Photograph submitted by K. Sivaraman, Indian Institute for Astrophysics, Kodaikanal Station.

(bottom) Fig. 345: 1986 Mar 18.32396 UT; exposure 15.0 minutes on IIa-D emulsion with GG-495 filter; r=0.97, Δ=0.85, β=114.1°; LSPN-732.
Photograph by C. Torres/H. Wroblewski, Cerro el Roble Astronomical Observatory.

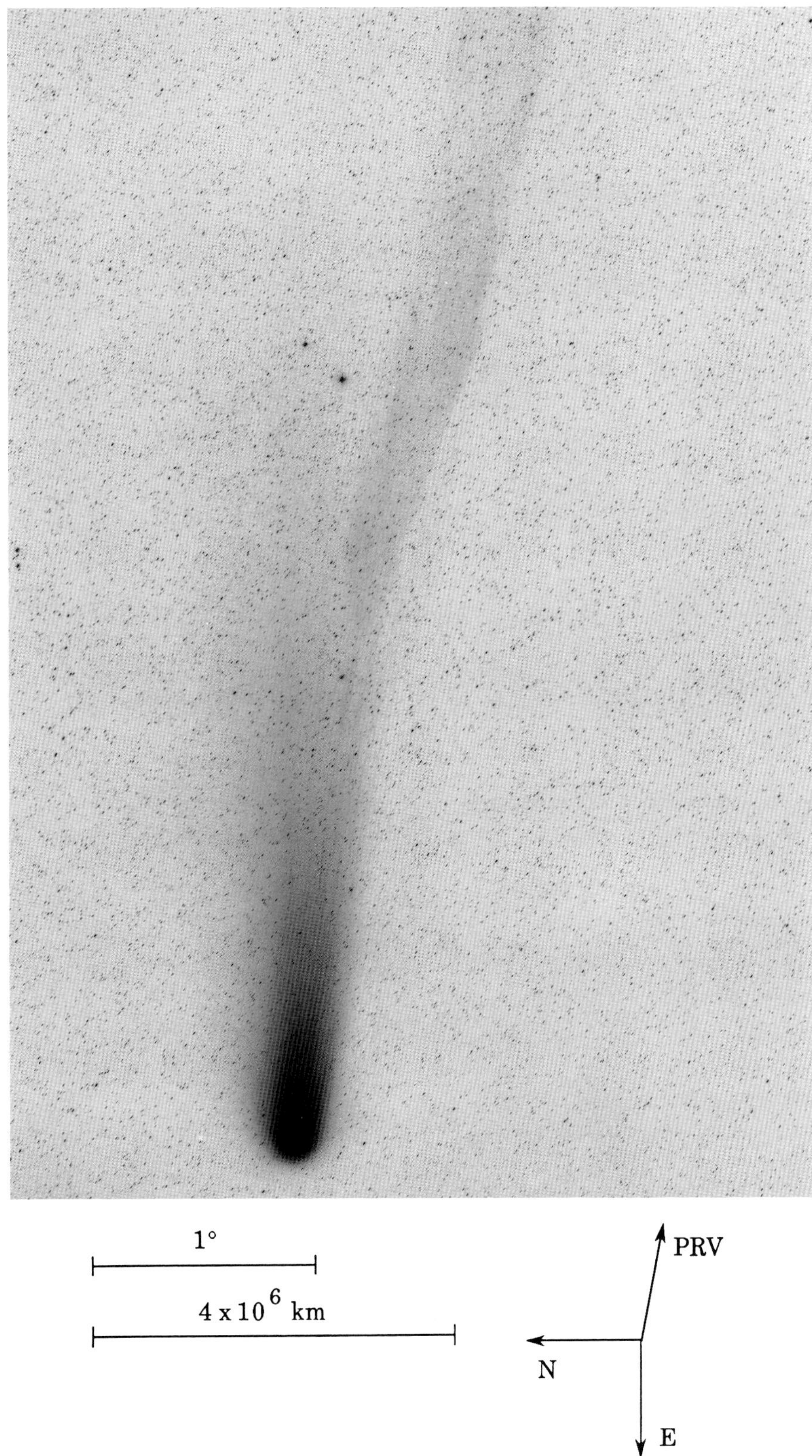

Fig. 346: 1986 Mar 18.32813 UT; exposure 15.0 minutes on IIa-O emulsion with GG-385 filter; r=0.97, Δ=0.85, β=114.1°; LSPN-2900.
Photograph by G. Pizarro, European Southern Observatory.

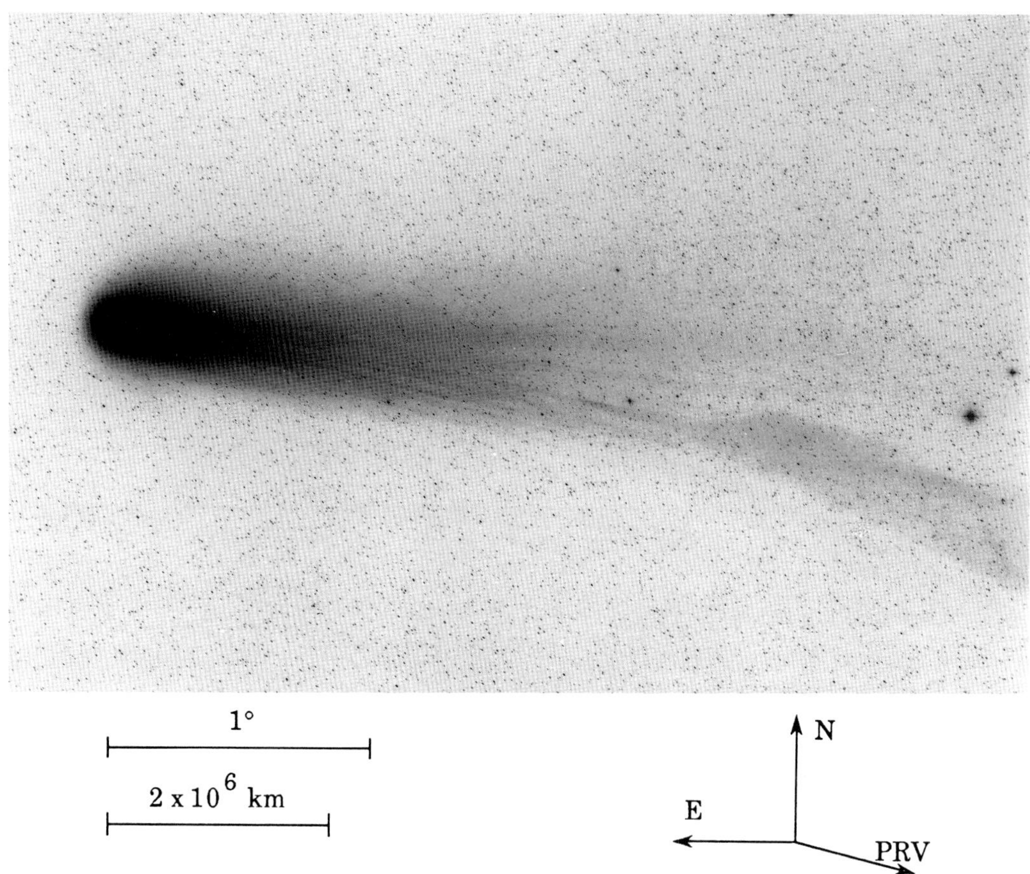

Fig. 347: 1986 Mar 18.34375 UT; exposure 10.0 minutes on 103a-O emulsion with no filter; r=0.97, Δ=0.85, β=114.1°; LSPN-733.
Photograph by C. Torres/H. Wroblewski, Cerro el Roble Astronomical Observatory.

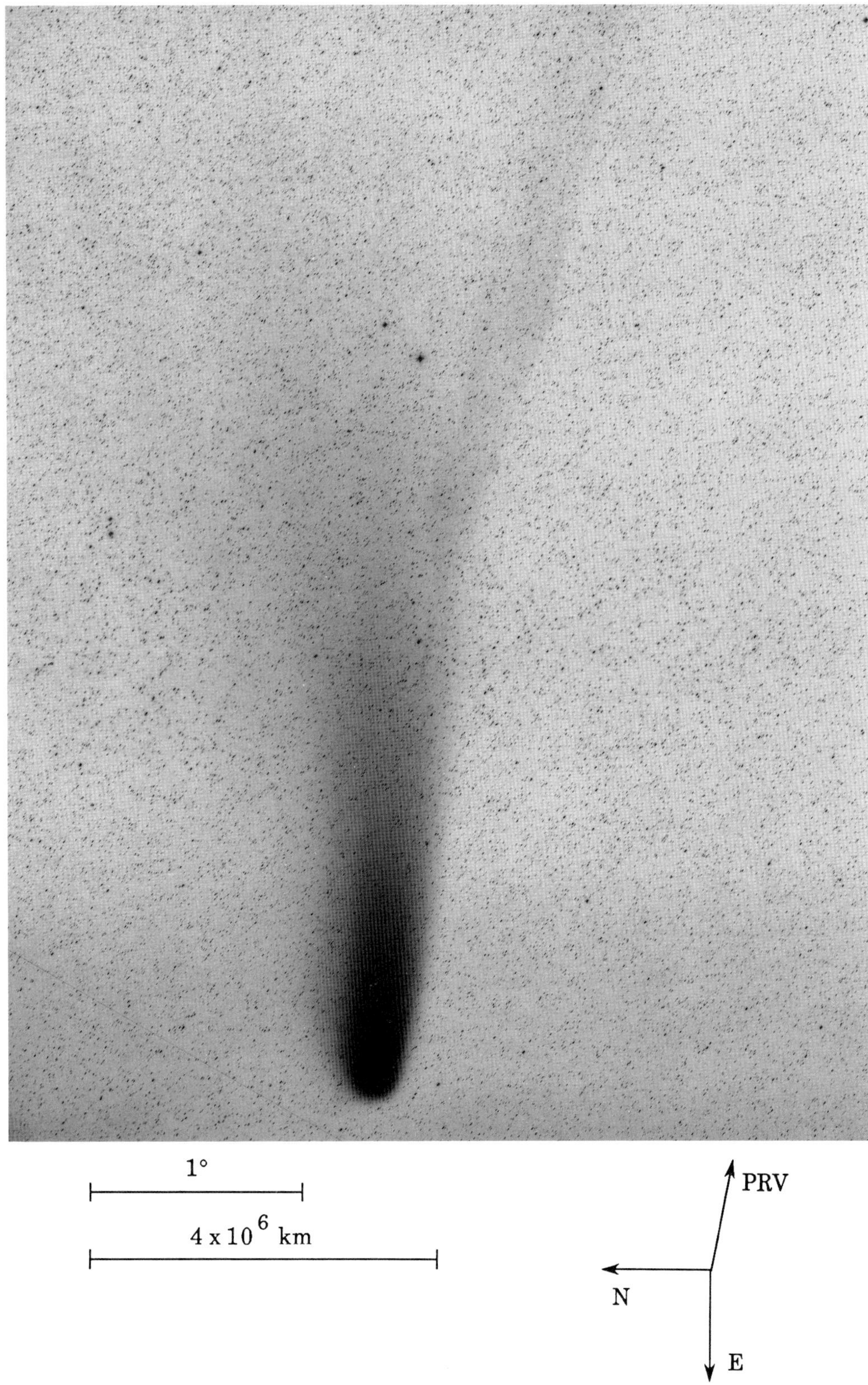

Fig. 348: 1986 Mar 18.35000 UT; exposure 15.0 minutes on 103a-D emulsion with GG-495 filter; r=0.97, Δ=0.85, β=114.1°; LSPN-2901.
Photograph by G. Pizarro, European Southern Observatory.

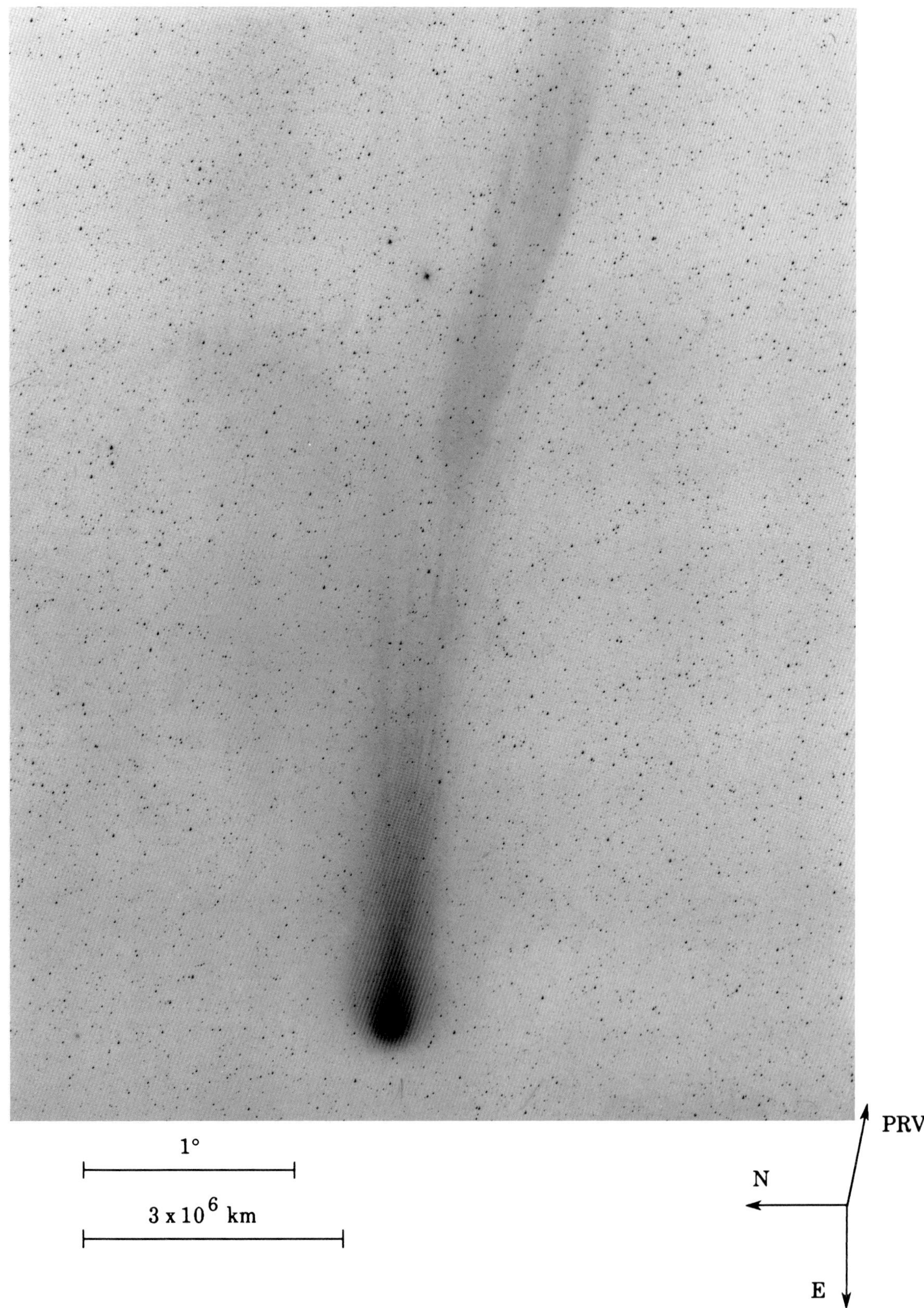

Fig. 349: 1986 Mar 18.36111 UT; exposure 20.0 minutes on IIa-O emulsion with no filter; r=0.97, Δ=0.85, β=114.1°; LSPN-1307.
Photograph by F. Miller, University of Michigan/CTIO.

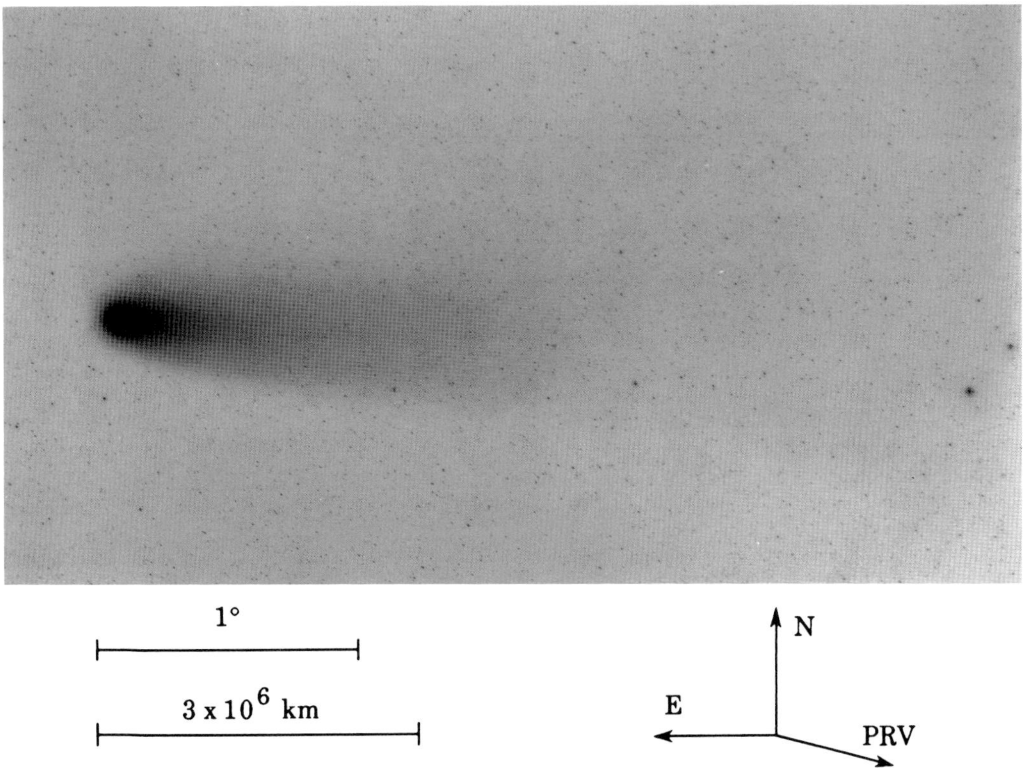

Fig. 350: 1986 Mar 18.36181 UT; exposure 10.0 minutes on 098-04 emulsion with RG-610 filter; r=0.97, Δ=0.85, β=114.1°; LSPN-734.
Photograph by C. Torres/H. Wroblewski, Cerro el Roble Astronomical Observatory.

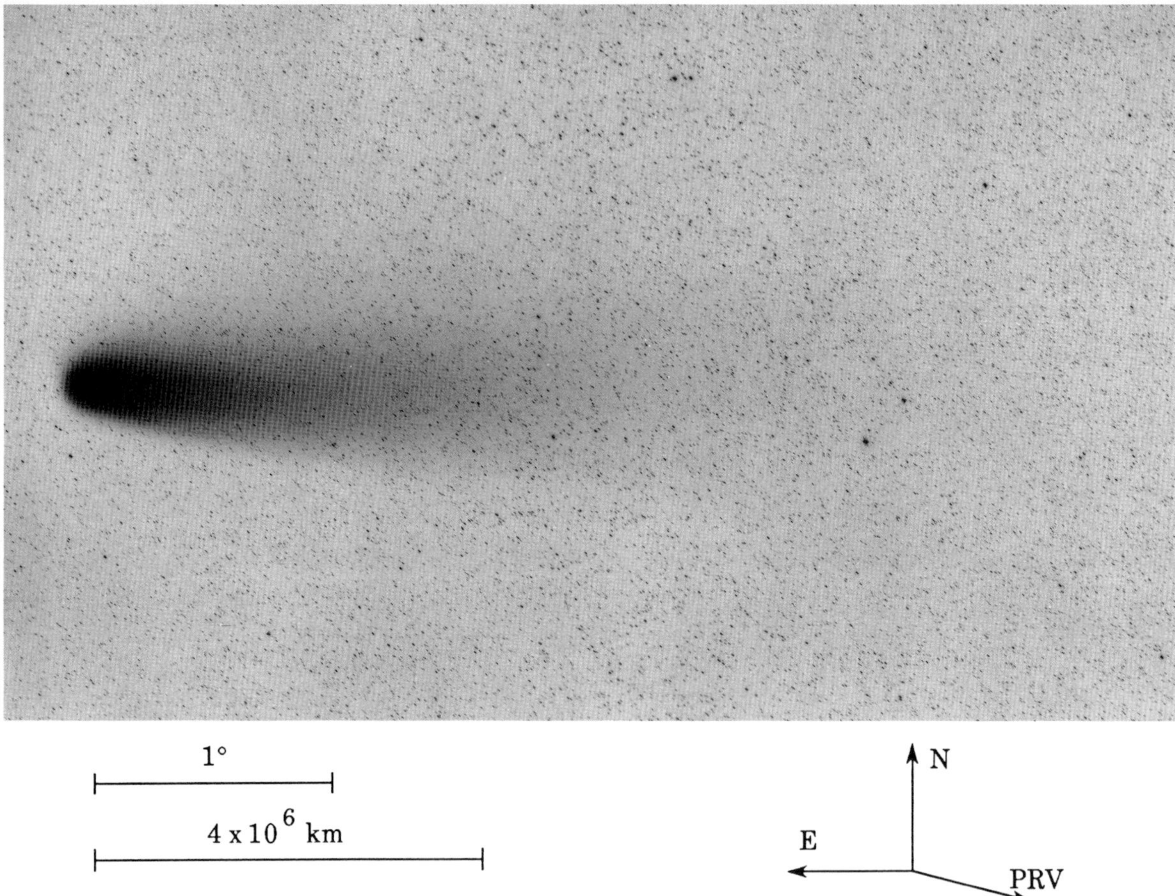

Fig. 351: 1986 Mar 18.37049 UT; exposure 15.0 minutes on 098-04 emulsion with RG-630 filter; r=0.97, Δ=0.85, β=114.0°; LSPN-2902.
Photograph by G. Pizarro, European Southern Observatory.

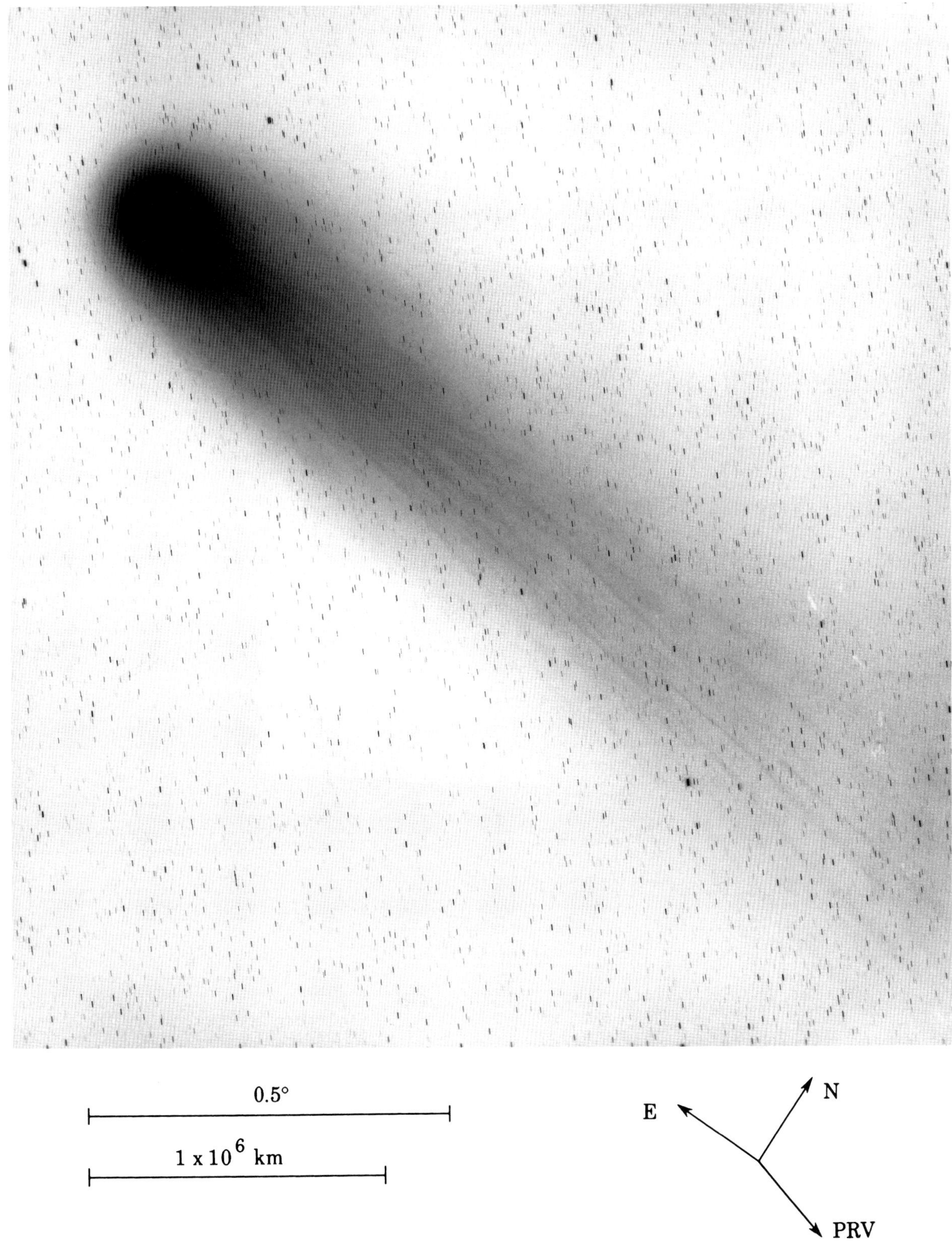

Fig. 352: 1986 Mar 18.40208 UT; exposure 10.0 minutes on 103a-O emulsion with Wratten 2C filter; r=0.97, Δ=0.85, β=114.0°; LSPN-2568.
Photograph by A. Dressler/R. Windhorst, Mount Wilson/Las Campanas Observatories.

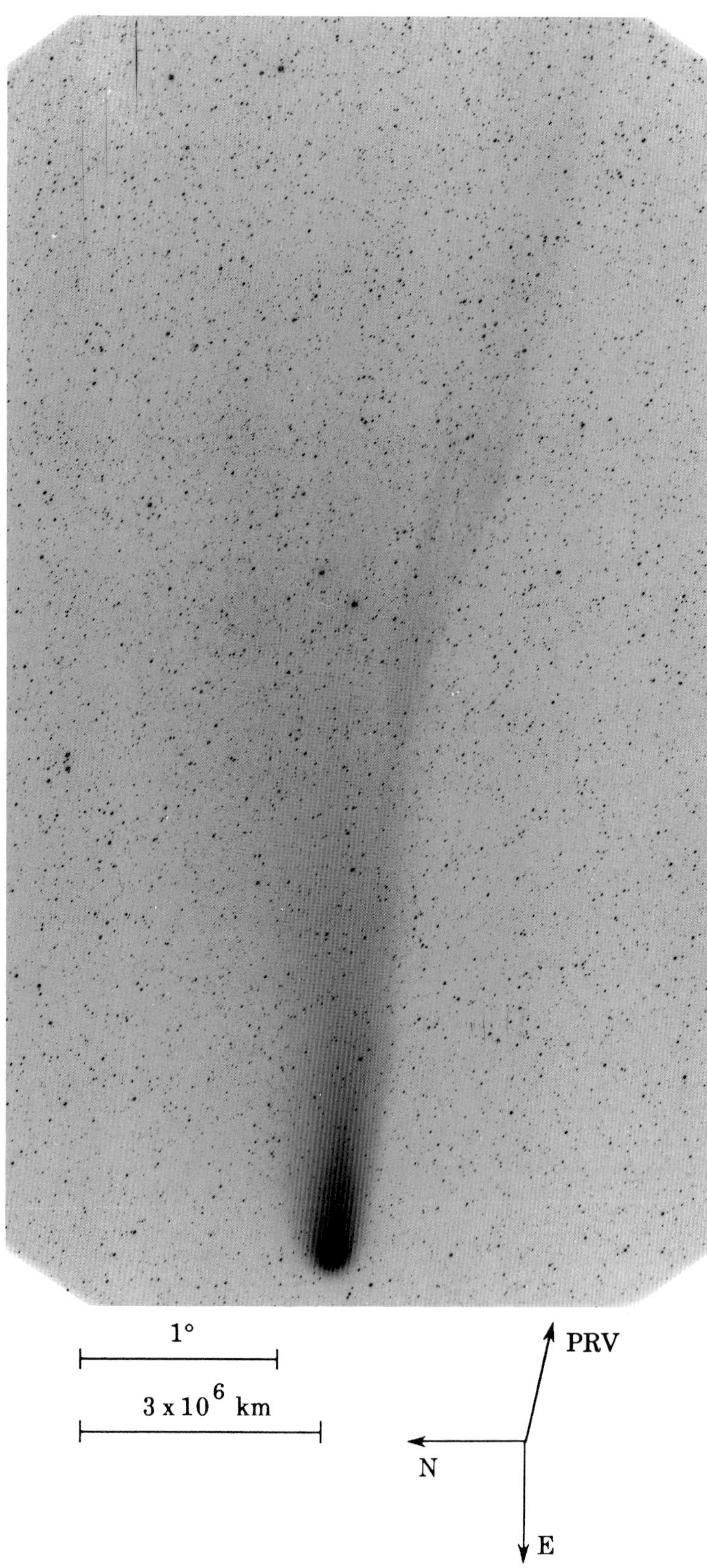

Fig. 353: 1986 Mar 18.41042 UT; exposure 16.0 minutes on Tech. Pan 2415 emulsion with no filter; r=0.97, Δ=0.85, β=114.0°; LSPN-1404.
Photograph by W. Liller, Easter Island, LSPN Island Network.

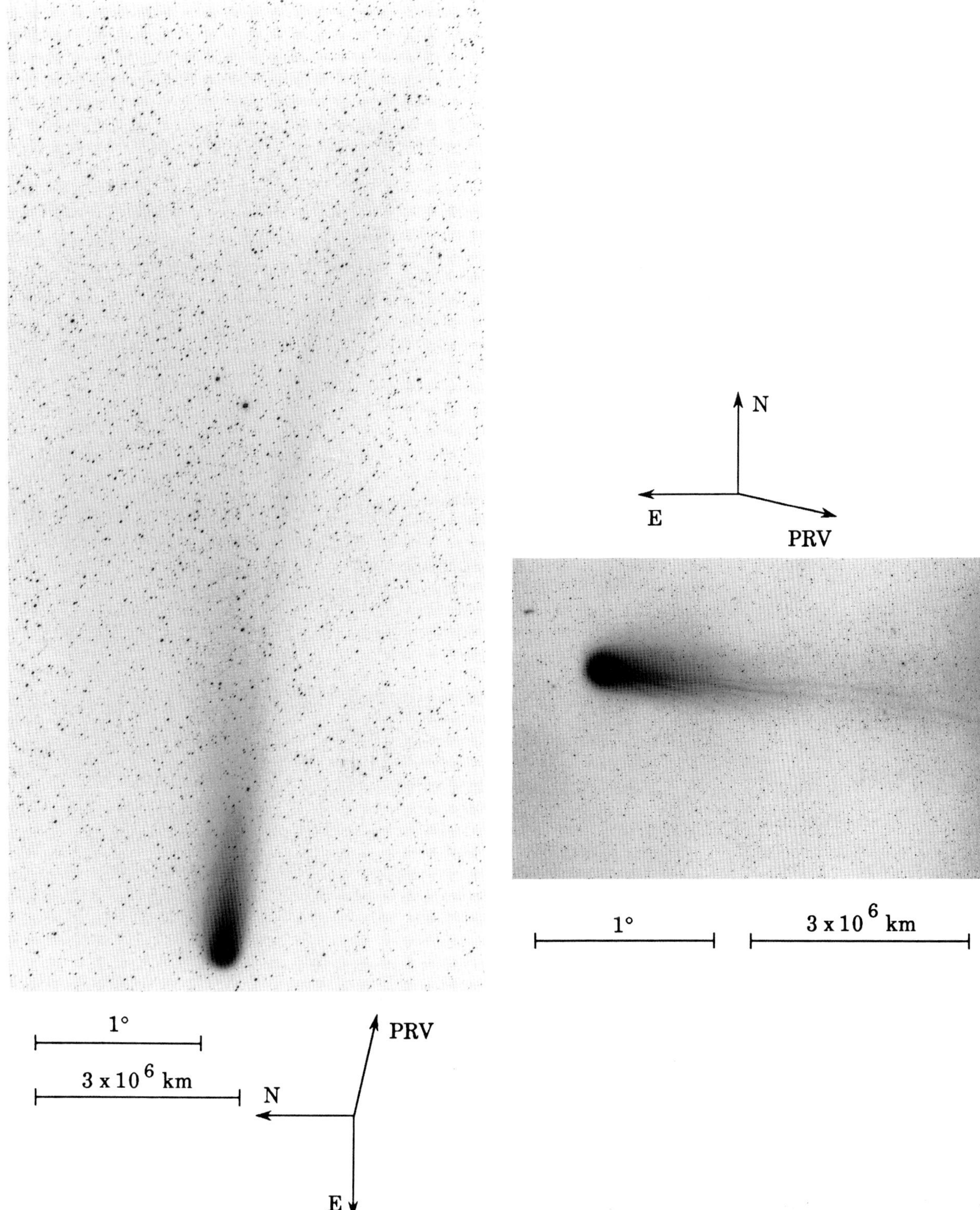

(left) Fig. 354: 1986 Mar 18.50139 UT; exposure 14.0 minutes on Tech. Pan 2415 emulsion with no filter; r=0.97, Δ=0.85, β=114.0°; LSPN-1403.
Photograph by W. Liller, Easter Island, LSPN Island Network.

(right) Fig. 355*: 1986 Mar 18.69245 UT; exposure 5.0 minutes on IIIa-J emulsion with no filter; r=0.97, Δ=0.84, β=114.0°; LSPN-2118.
Photograph by P. Magnusson, Uppsala Southern Station.

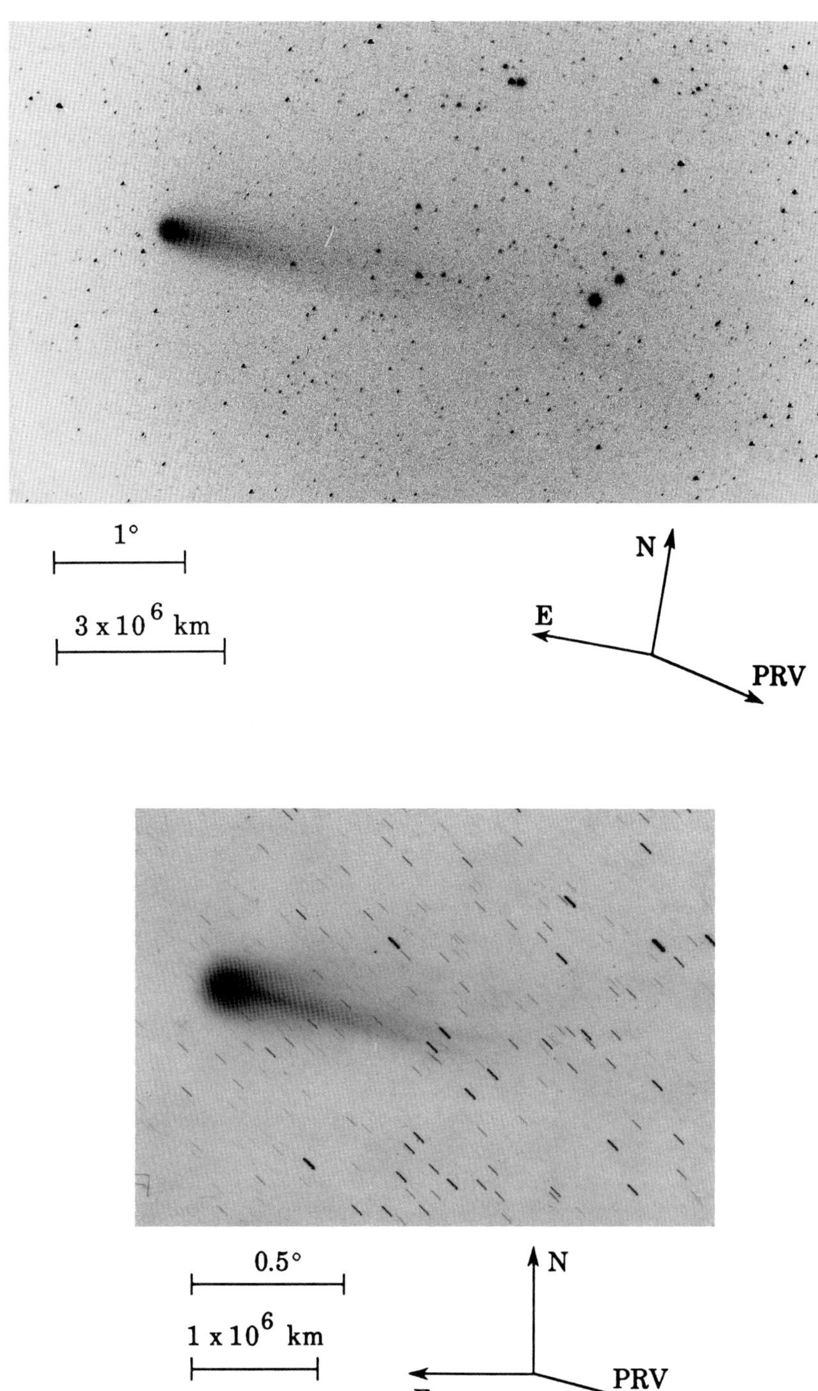

(top) Fig. 356: 1986 Mar 18.52920 UT; exposure 6.8 minutes on Tri-X emulsion; r=0.97, Δ=0.85, β=114.0°; AON-851117.
Photograph by S. Edberg, USA.

(bottom) Fig. 357*: 1986 Mar 18.96042 UT; exposure 60.0 minutes on 103a-O emulsion with no filter; r=0.98, Δ=0.84, β=114.0°; LSPN-3629.
Photograph submitted by K. Sivaraman, Indian Institute for Astrophysics, Kodaikanal Station.

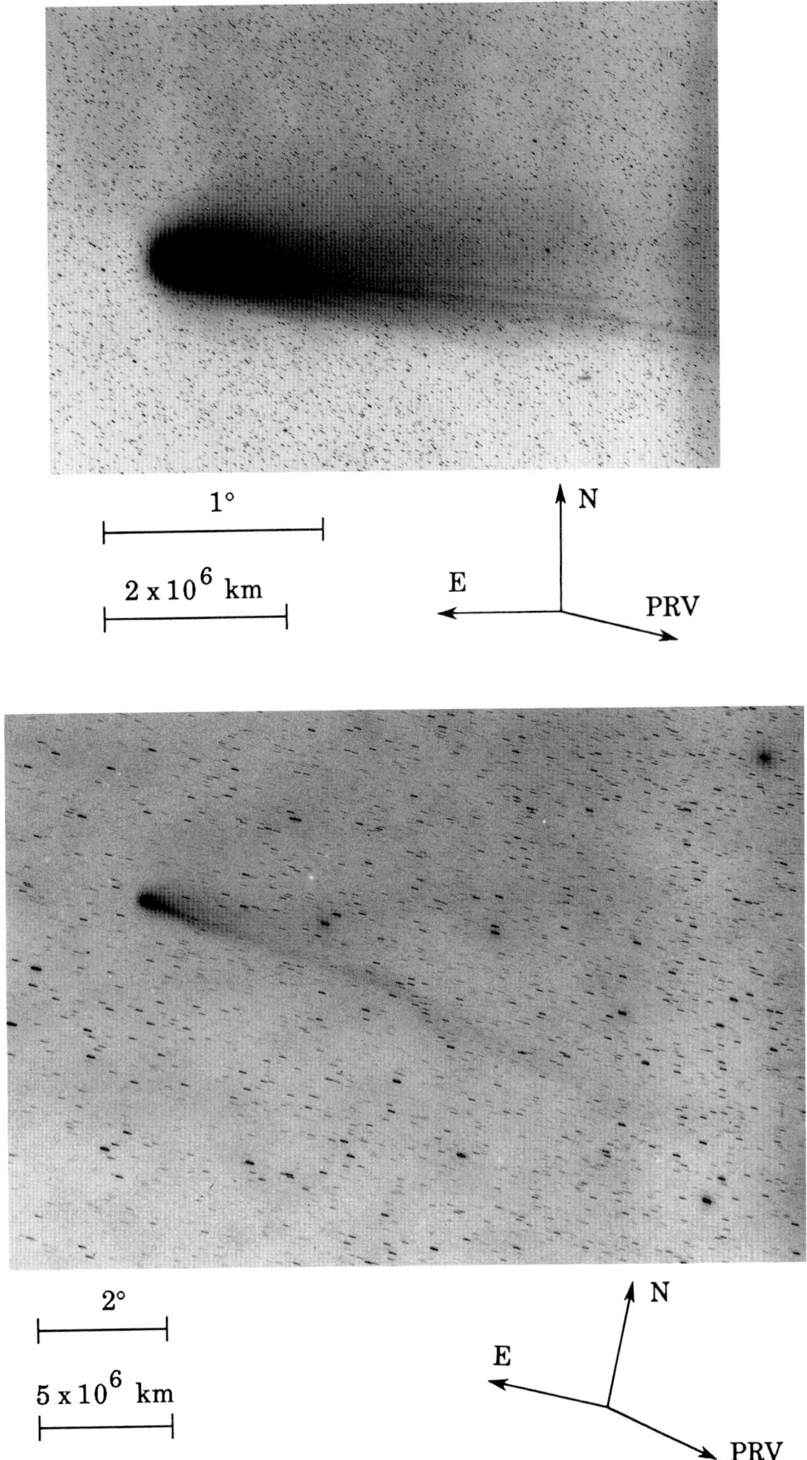

(top) Fig. 358: 1986 Mar 18.72281 UT; exposure 20.0 minutes on IIIa-J emulsion with no filter; r=0.97, Δ=0.84, β=114.0°; LSPN-2120.
Photograph by P. Magnusson, Uppsala Southern Station.

(bottom) Fig. 359: 1986 Mar 18.93750 UT; exposure 60.0 minutes on 103a-O emulsion with no filter; r=0.98, Δ=0.84, β=114.0°; LSPN-3652.
Photograph submitted by K. Sivaraman, Indian Institute for Astrophysics, Kodaikanal Station.

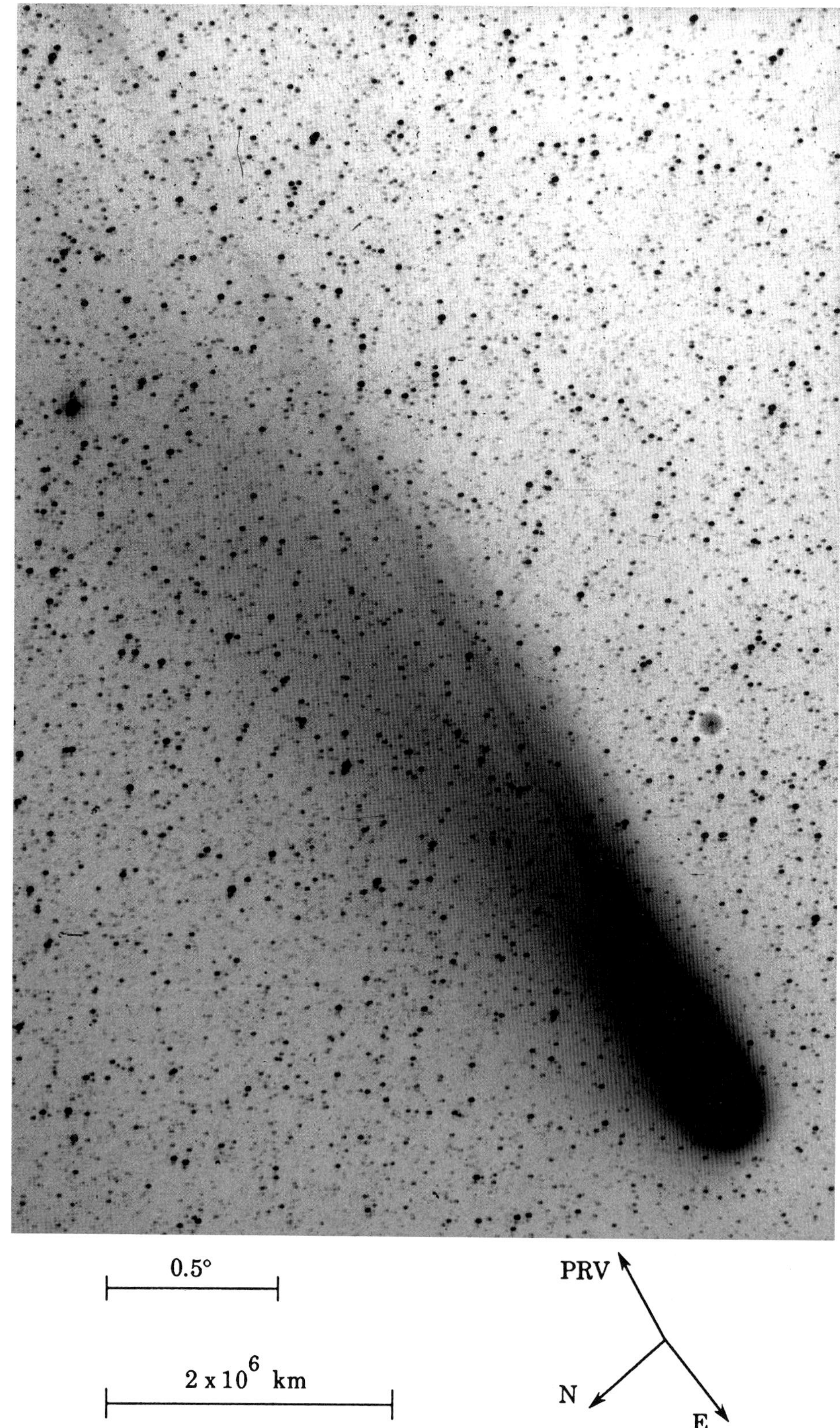

Fig. 360: 1986 Mar 19.02778 UT; exposure 6.0 minutes on Tech. Pan 2415 emulsion with no filter; r=0.98, Δ=0.84, β=113.9°; LSPN-1803.
Photograph by F. Marang, Sutherland, LSPN Island Network.

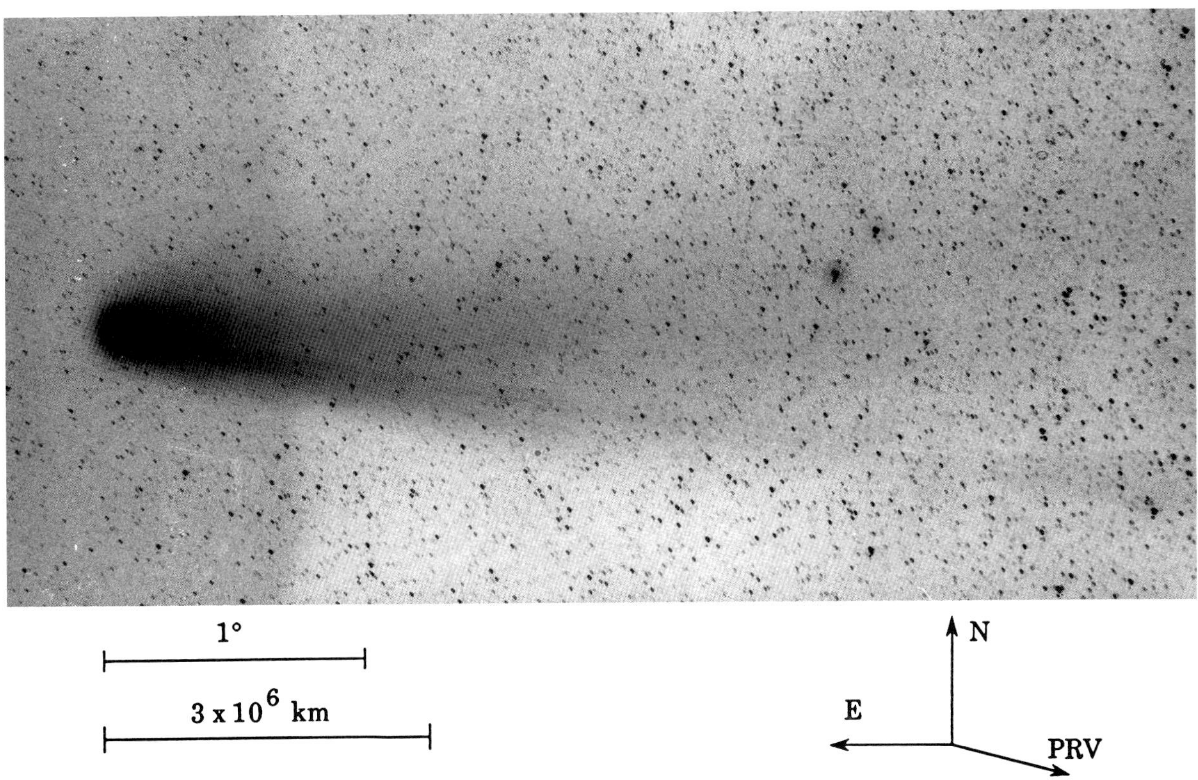

Fig. 361: 1986 Mar 19.03958 UT; exposure 12.0 minutes on Tech. Pan 2415 emulsion; r=0.98, Δ=0.83, β=113.9°. *Photograph by A. Marafie, Wafra Observatory.*

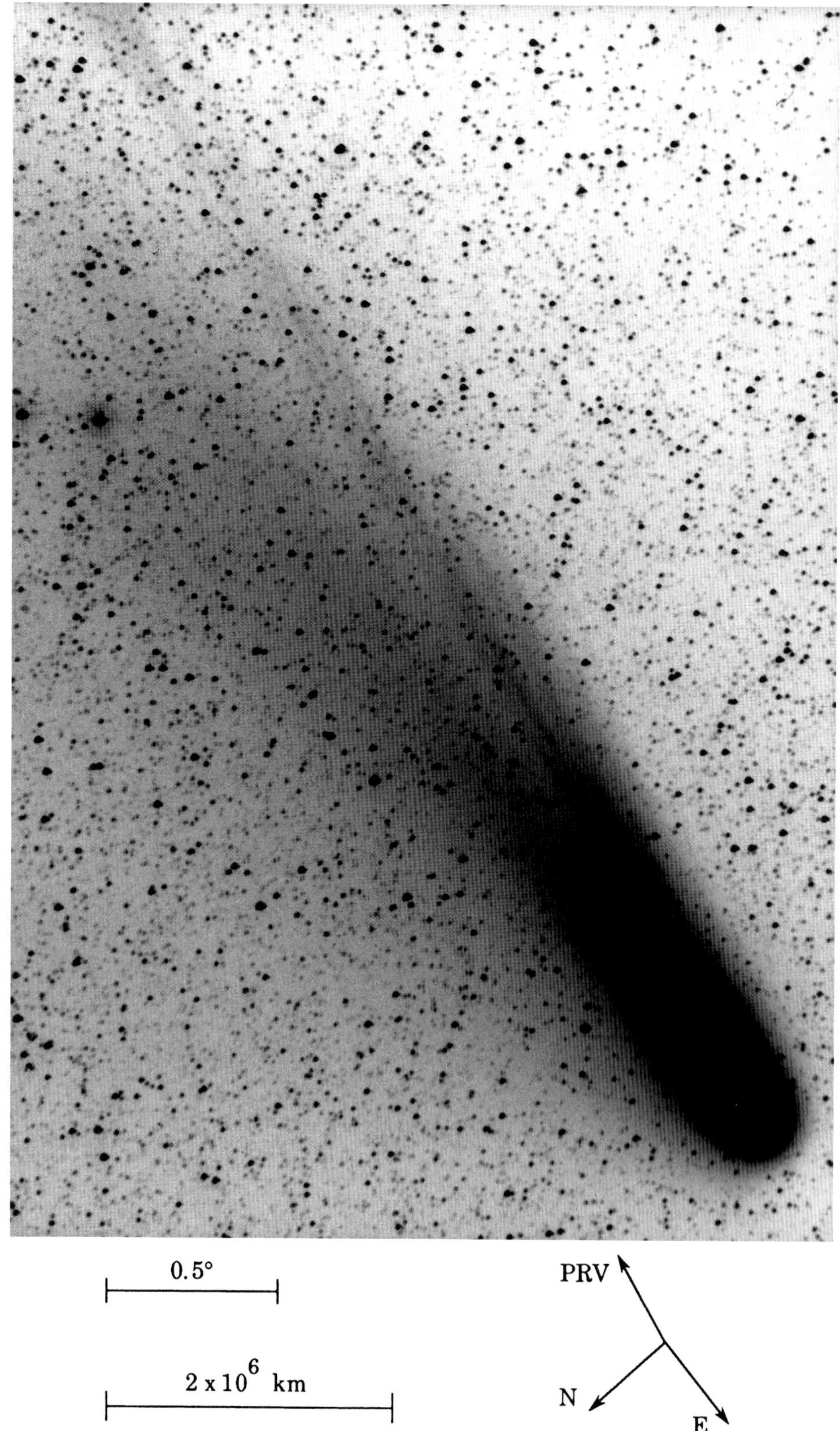

Fig. 362: 1986 Mar 19.04271 UT; exposure 25.0 minutes on Tech. Pan 2415 emulsion with no filter; r=0.98, Δ=0.83, β=113.9°; LSPN-1804.
Photograph by F. Marang, Sutherland, LSPN Island Network.

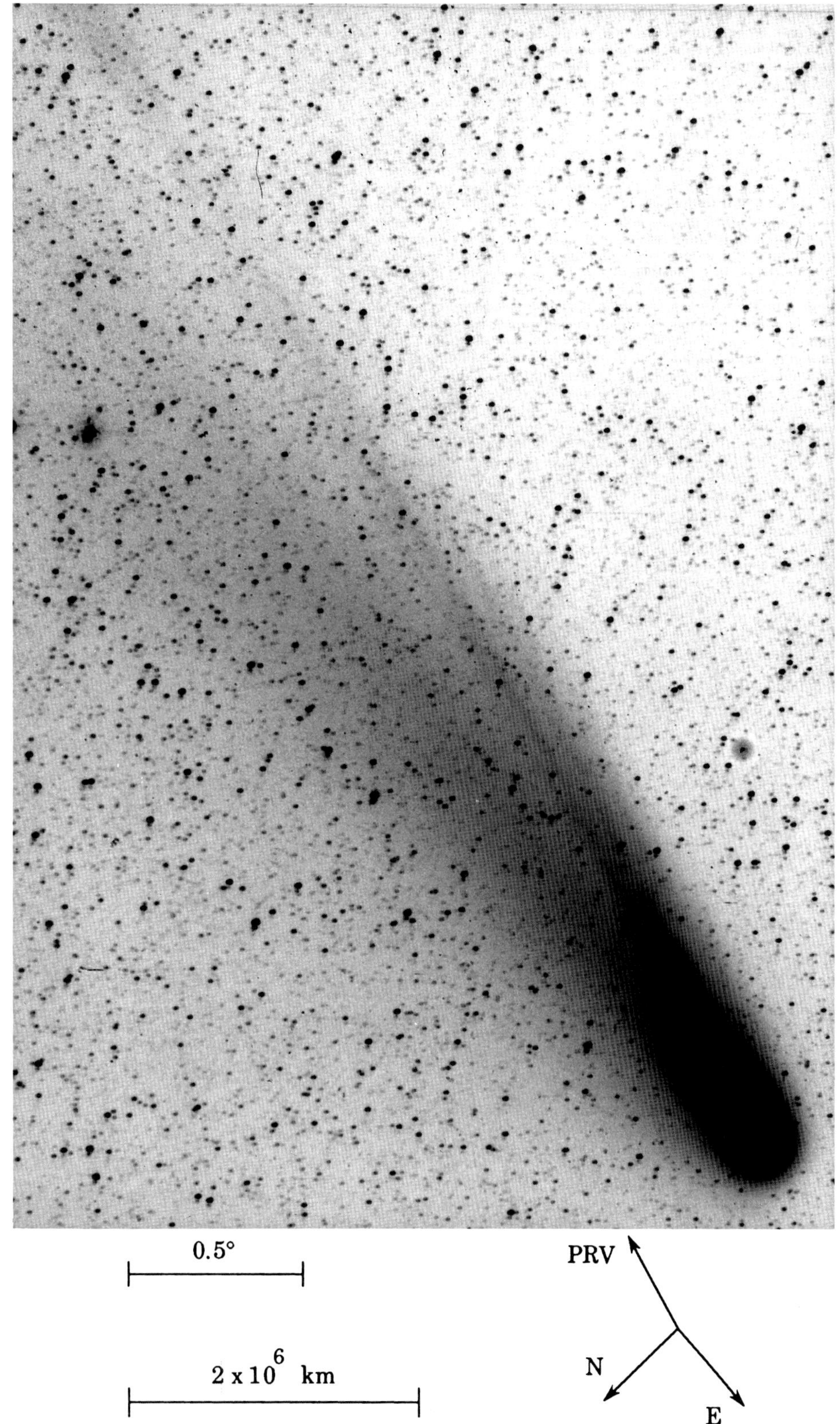

Fig. 363: 1986 Mar 19.06285 UT; exposure 5.0 minutes on Tech. Pan 2415 emulsion with no filter; r=0.98, Δ=0.83, β=113.9°; LSPN-1805.
Photograph by F. Marang, Sutherland, LSPN Island Network.

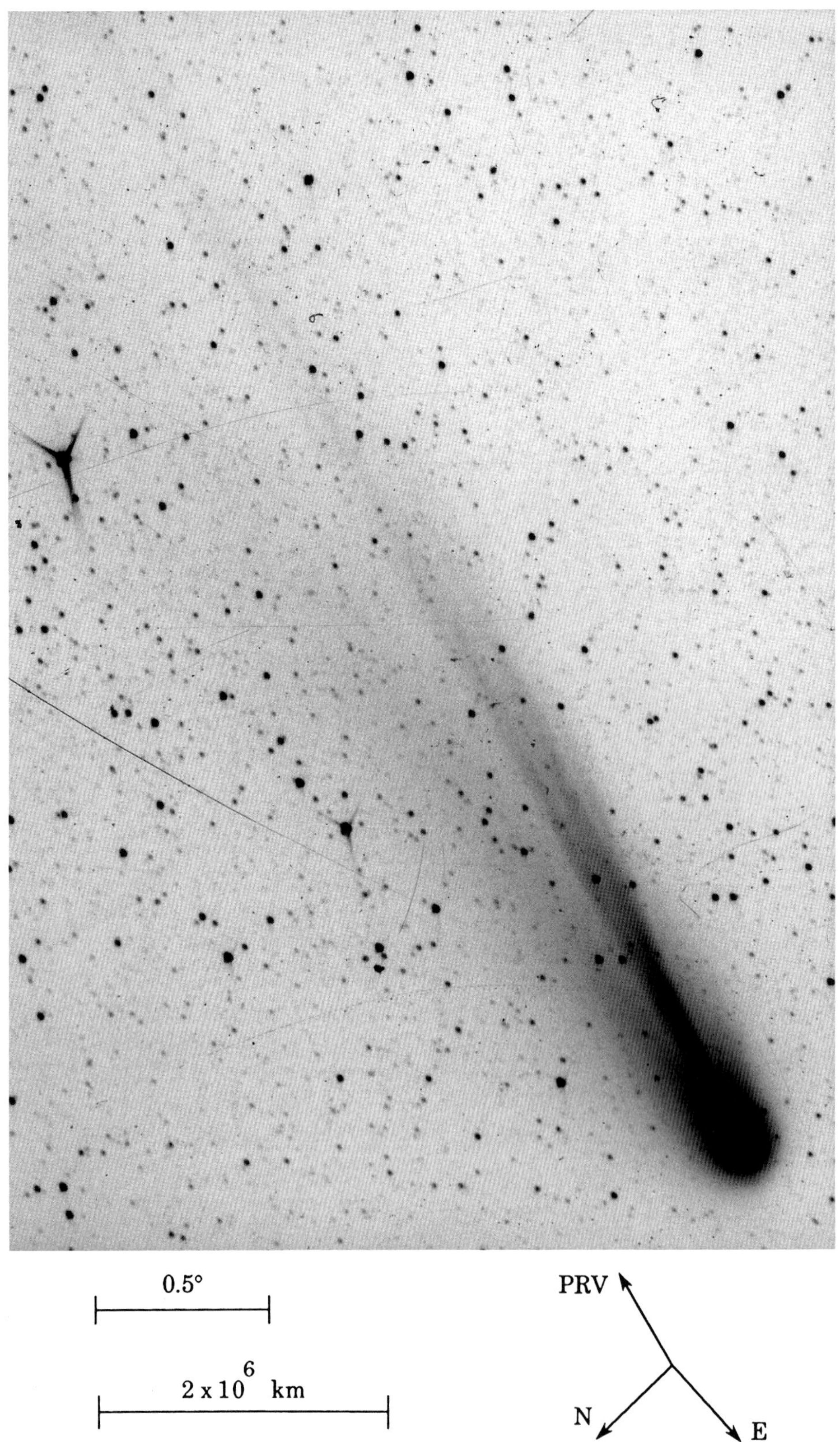

Fig. 364: 1986 Mar 19.11875 UT; exposure 30.0 minutes on Tech. Pan 2415 emulsion with Wratten 47B filter; r=0.98, Δ=0.83, β=113.9°; LSPN-1806.
Photograph by F. Marang, Sutherland, LSPN Island Network.

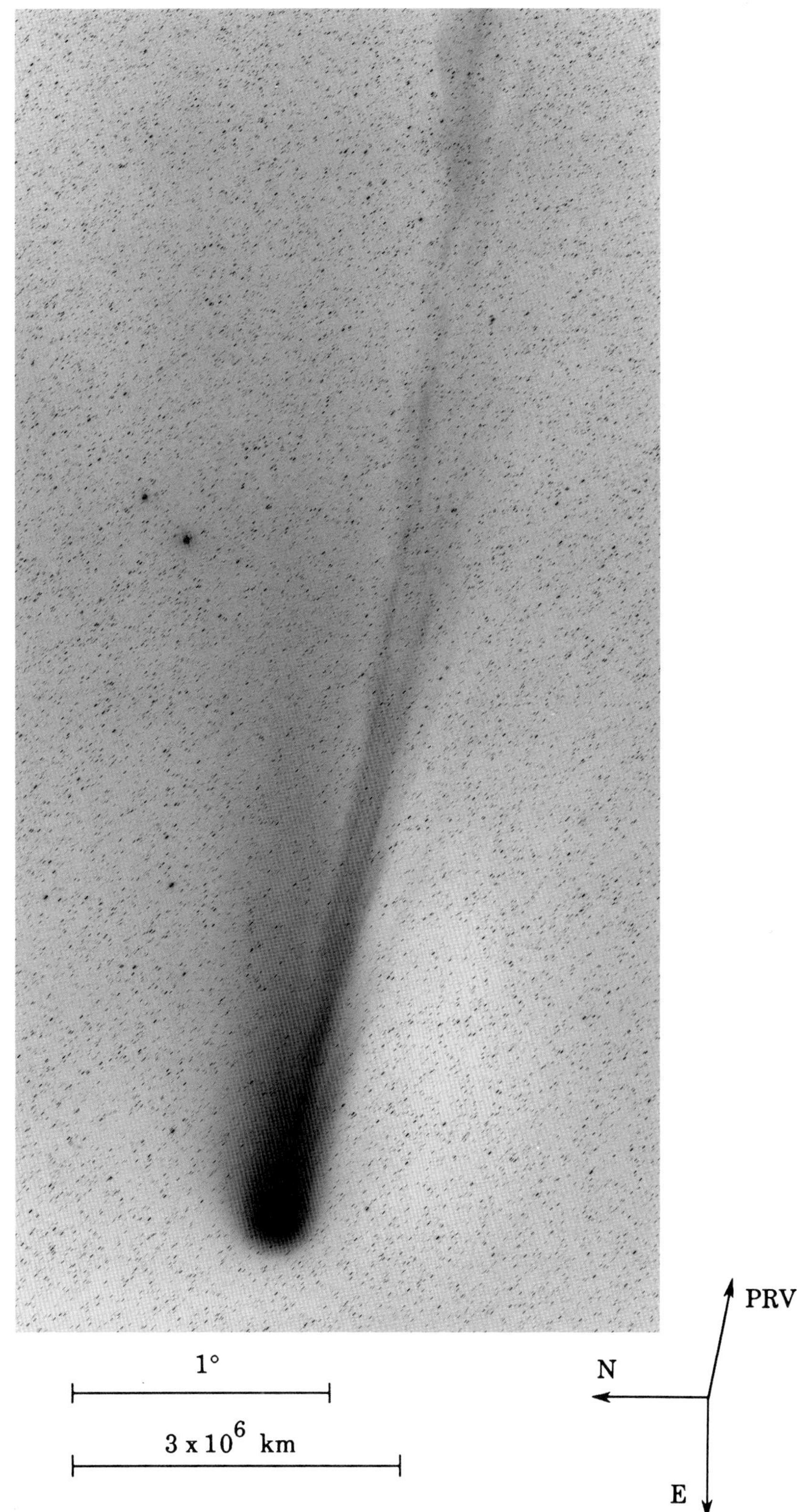

Fig. 365: 1986 Mar 19.30000 UT; exposure 20.0 minutes on IIa-O emulsion with no filter; r=0.98, Δ=0.83, β=113.9°; LSPN-1310.
Photograph by F. Miller, University of Michigan/CTIO.

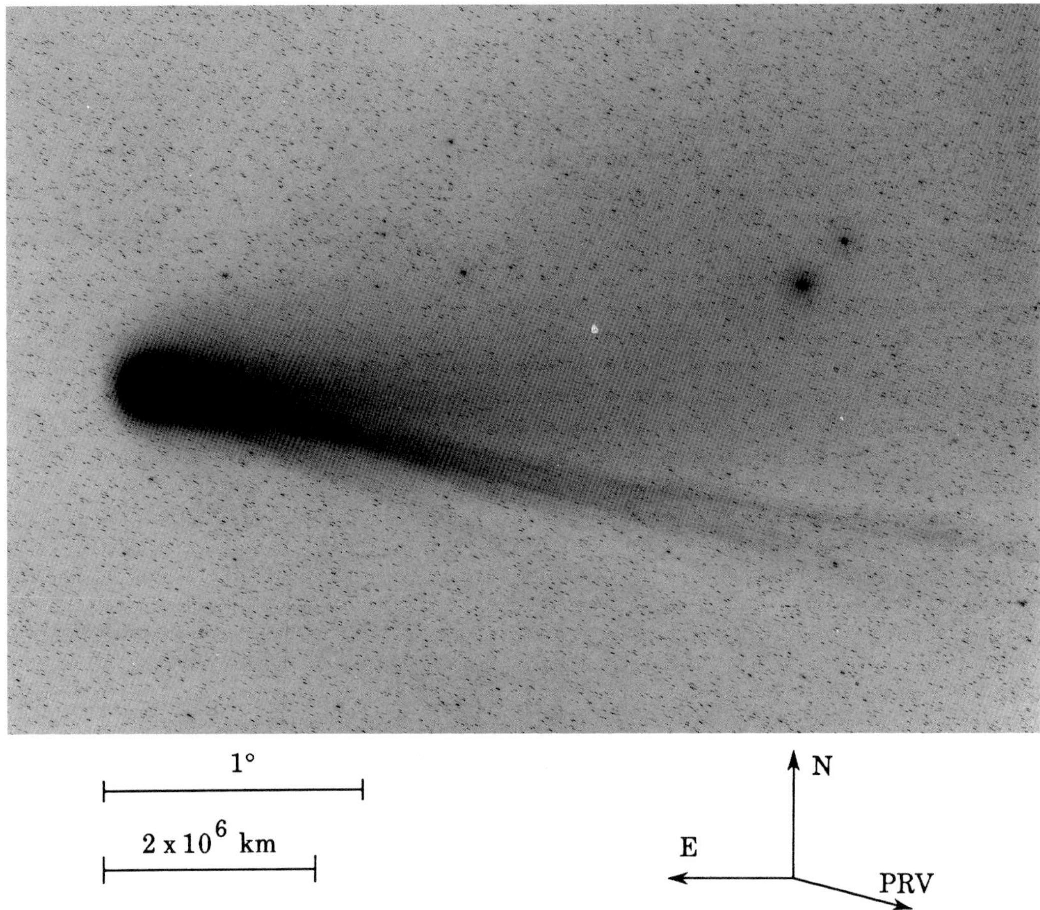

Fig. 366: 1986 Mar 19.31042 UT; exposure 10.0 minutes on 103a-O emulsion with GG-385 filter; r=0.98, Δ=0.83, β=113.9°; LSPN-735.
Photograph by C. Torres/H. Wroblewski, Cerro el Roble Astronomical Observatory.

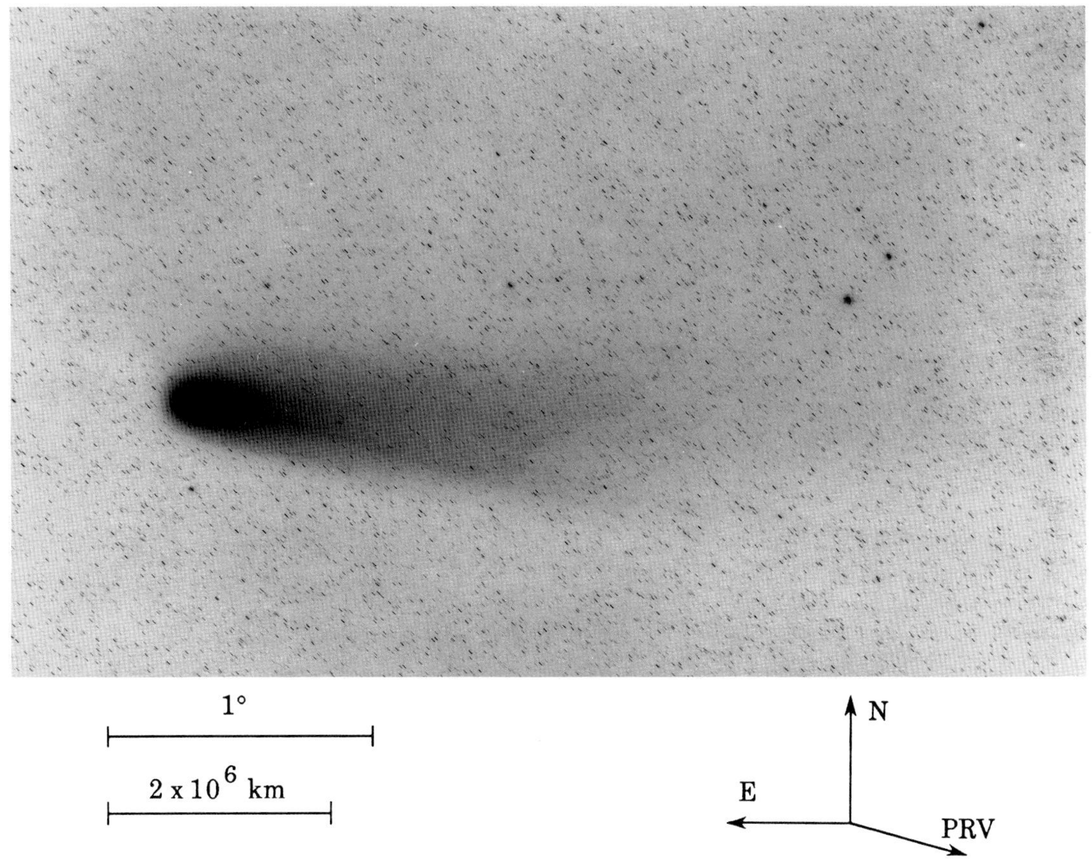

Fig. 367: 1986 Mar 19.33125 UT; exposure 20.0 minutes on IIIa-F emulsion with RG-610 filter; r=0.98, Δ=0.83, β=113.9°; LSPN-736.
Photograph by C. Torres/H. Wroblewski, Cerro el Roble Astronomical Observatory.

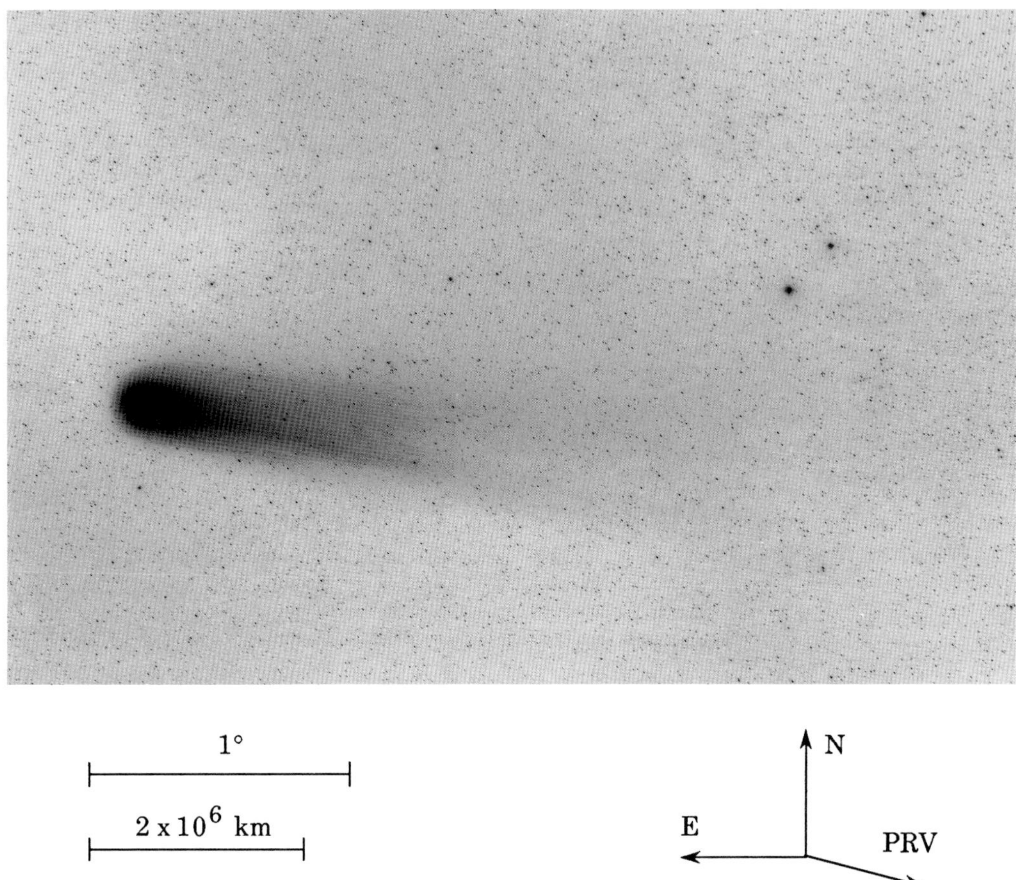

Fig. 368: 1986 Mar 19.35208 UT; exposure 10.0 minutes on 098-04 emulsion with GG-495 filter; r=0.98, Δ=0.83, β=113.9°; LSPN-737.
Photograph by C. Torres/H. Wroblewski, Cerro el Roble Astronomical Observatory.

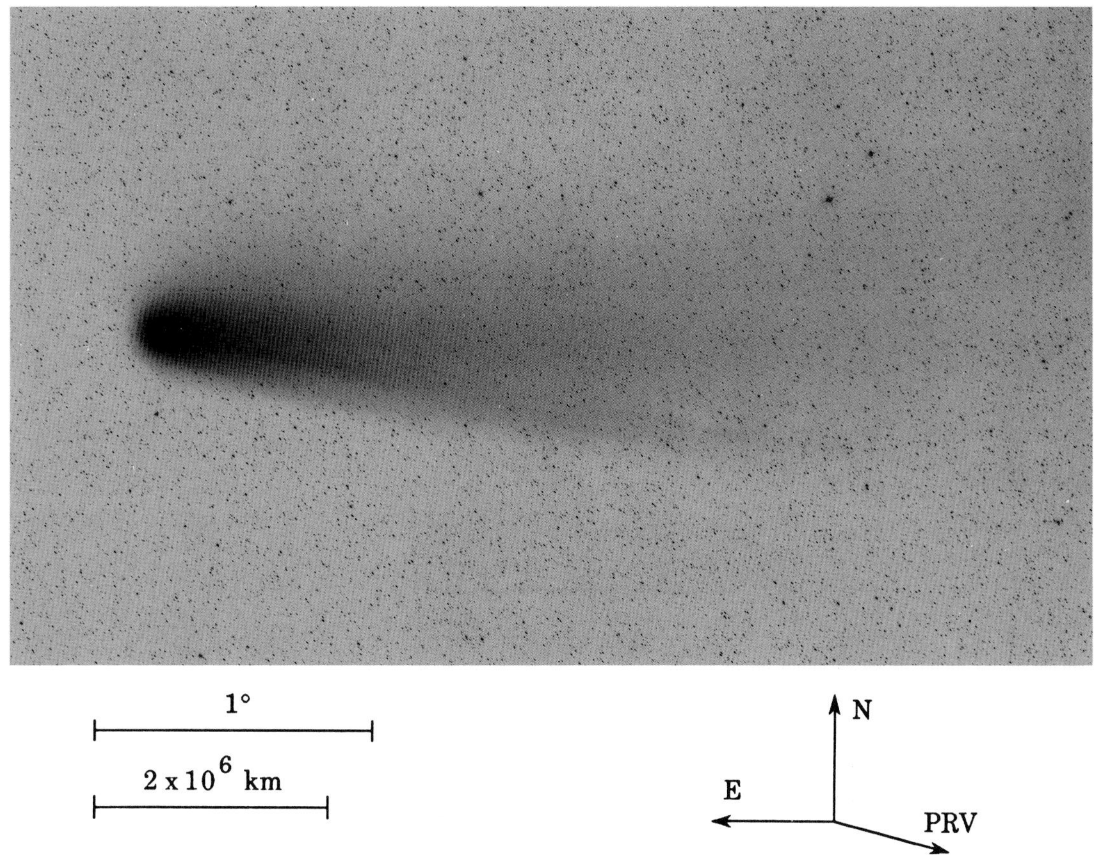

Fig. 369: 1986 Mar 19.37431 UT; exposure 30.0 minutes on 098-04 emulsion with RG-610 filter; r=0.98, Δ=0.83, β=113.9°; LSPN-1313.
Photograph by F. Miller, University of Michigan/CTIO.

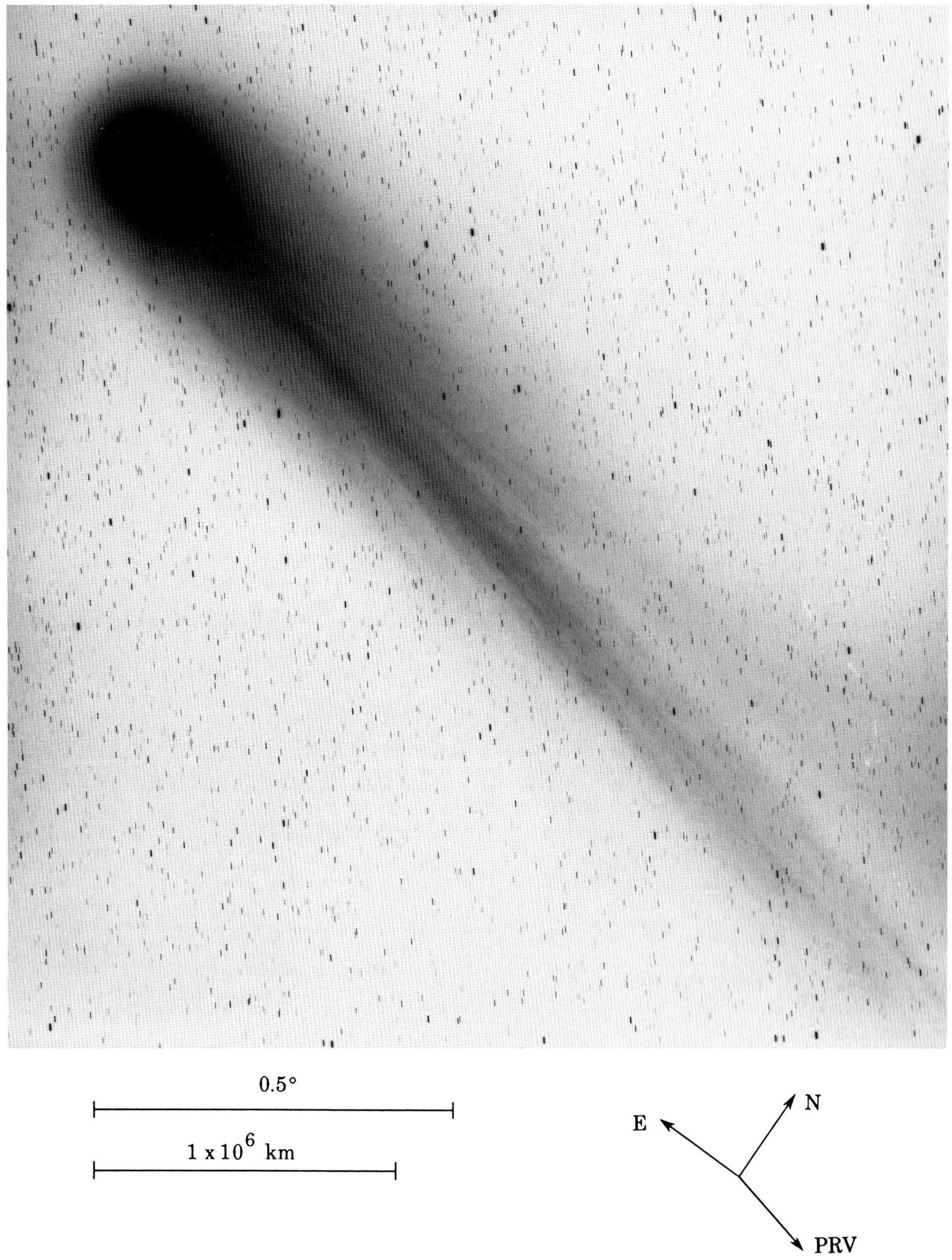

Fig. 370: 1986 Mar 19.40000 UT; exposure 10.0 minutes on 103a-O emulsion with Wratten 2C filter; r=0.98, Δ=0.83, β=113.9°; LSPN-2569.
Photograph by A. Dressler/R. Windhorst, Mount Wilson/Las Campanas Observatories.

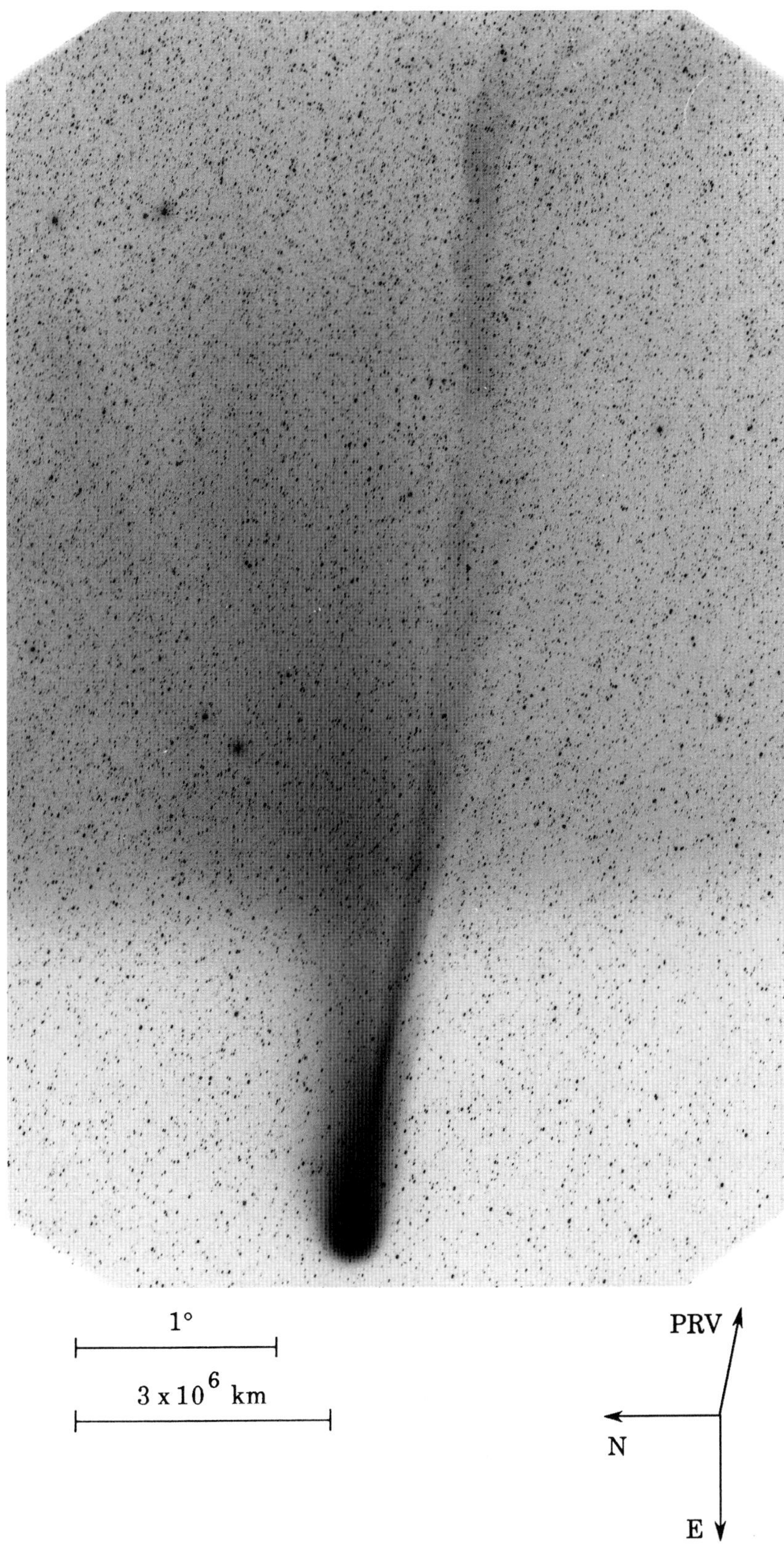

Fig. 371: 1986 Mar 19.47292 UT; exposure 15.0 minutes on Tech. Pan 2415 emulsion with no filter; r=0.98, Δ=0.82, β=113.9°; LSPN-1406.
Photograph by W. Liller, Easter Island, LSPN Island Network.

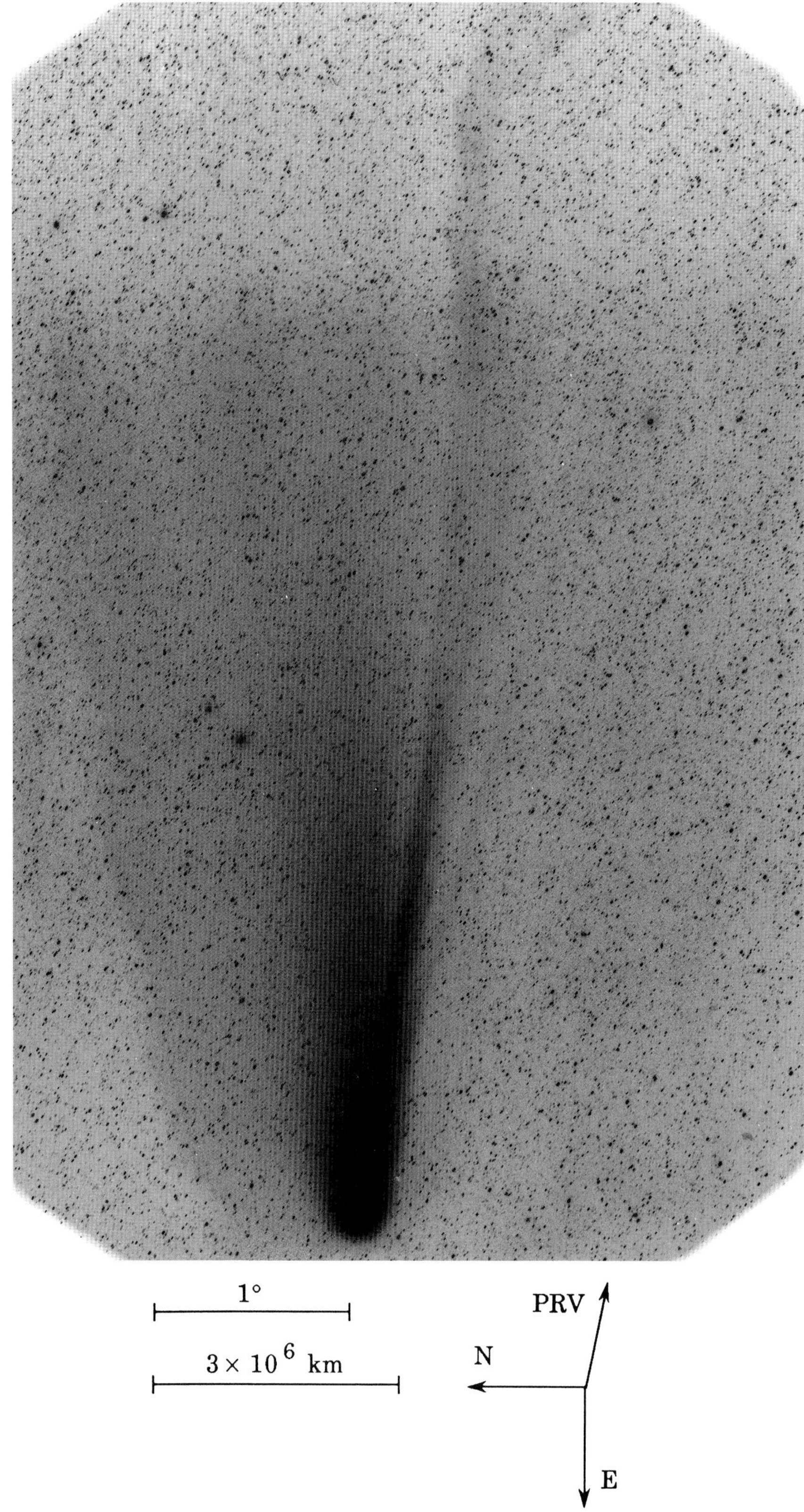

Fig. 372: 1986 Mar 19.49757 UT; exposure 16.0 minutes on Tech. Pan 2415 emulsion with no filter; r=0.99, Δ=0.82, β=113.9°; LSPN-1407.
Photograph by W. Liller, Easter Island, LSPN Island Network.

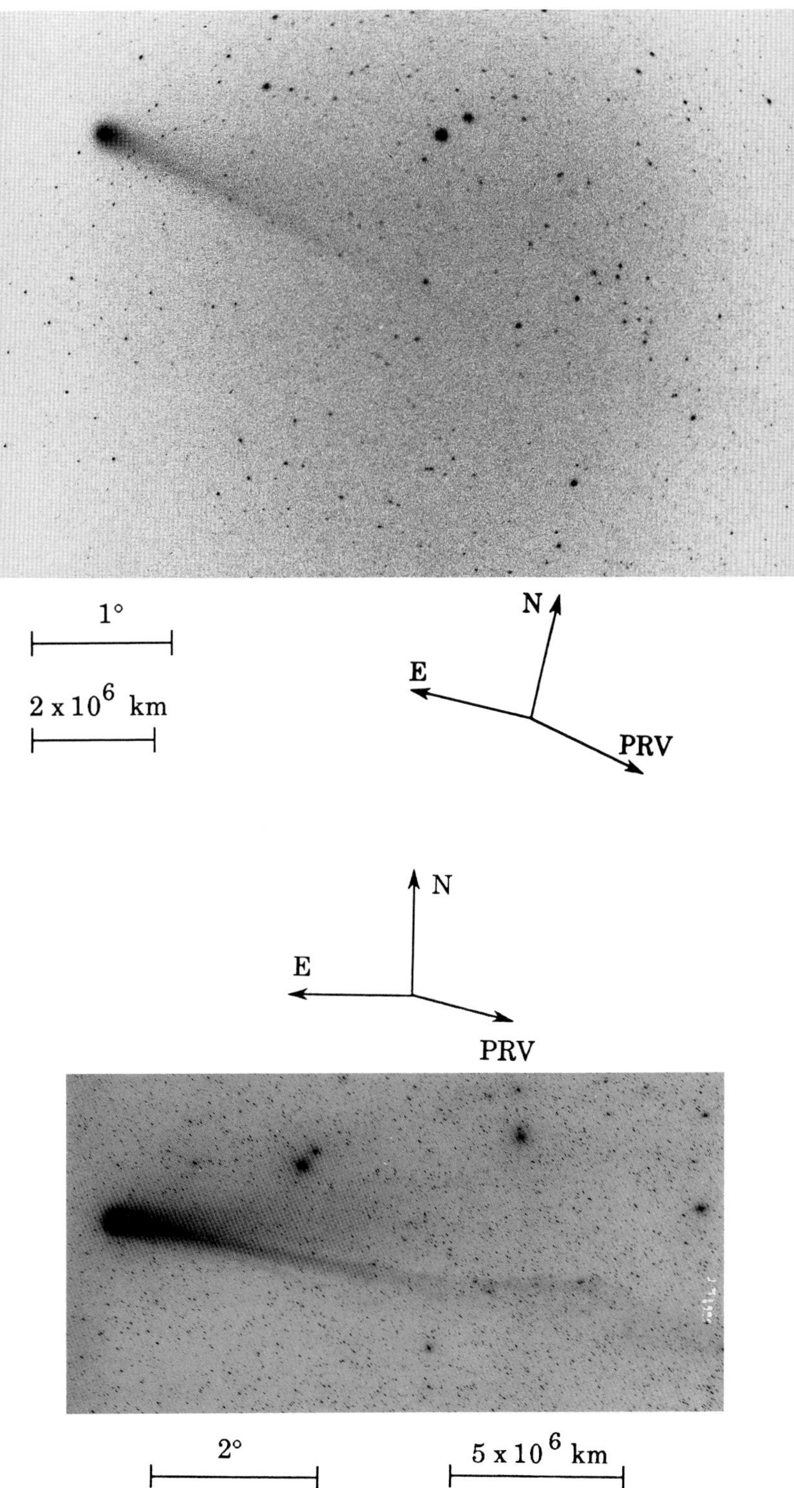

(top) Fig. 373: 1986 Mar 19.53670 UT; exposure 4.6 minutes on 3M 1000 emulsion; r=0.99, Δ=0.82, β=113.9°; AON-851146.
Photograph by S. Edberg, USA.

(bottom) Fig. 374: 1986 Mar 19.63542 UT; exposure 30.0 minutes on IIIa-J emulsion with no filter; r=0.99, Δ=0.82, β=113.9°; LSPN-1917.
Photograph by M. Buie, Mauna Kea Observatory.

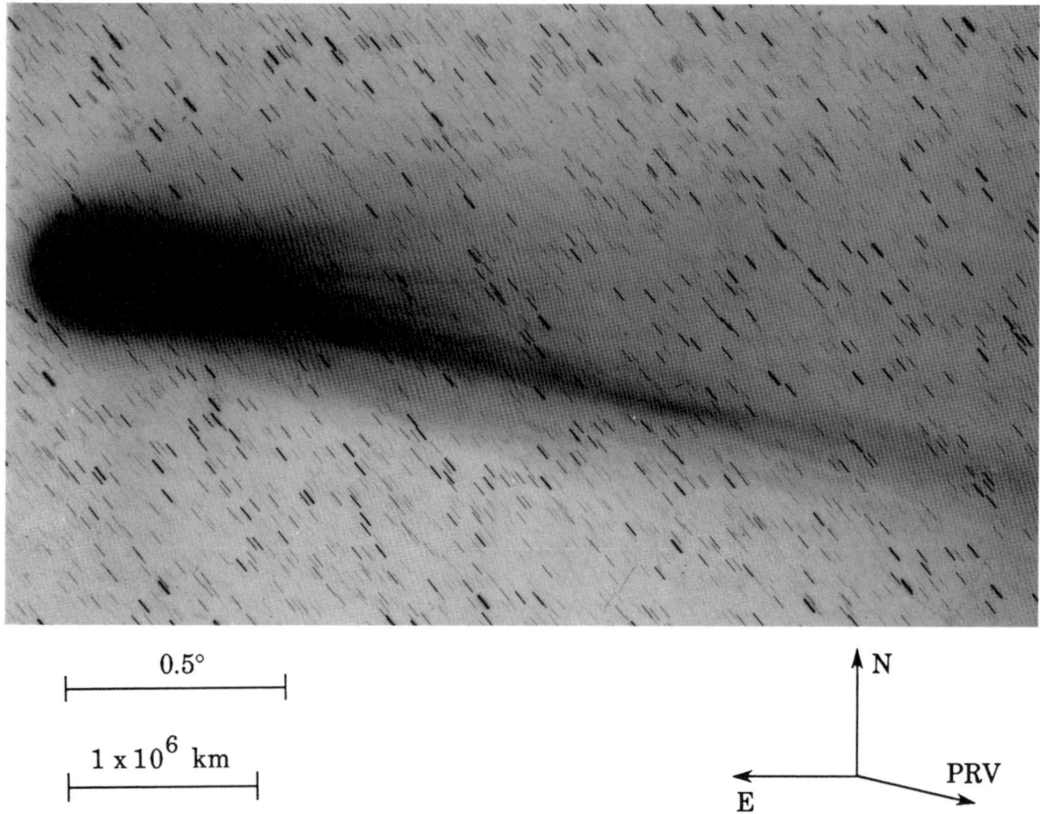

Fig. 375: 1986 Mar 19.82604 UT; exposure 45.0 minutes on 103a-O emulsion with no filter; r=0.99, Δ=0.81, β=113.9°; LSPN-3080.
Photograph by P. Jekabsons, Perth Observatory.

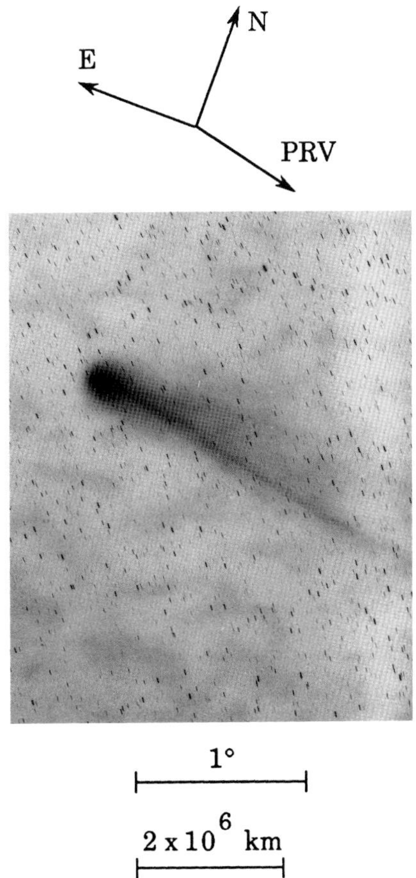

Fig. 376: 1986 Mar 19.96771 UT; exposure 15.0 minutes on 103a-O emulsion with Blue filter; r=0.99, Δ=0.81, β=113.9°; LSPN-3671.
Photograph submitted by K. Sivaraman, Indian Institute for Astrophysics, Kavalur Station.

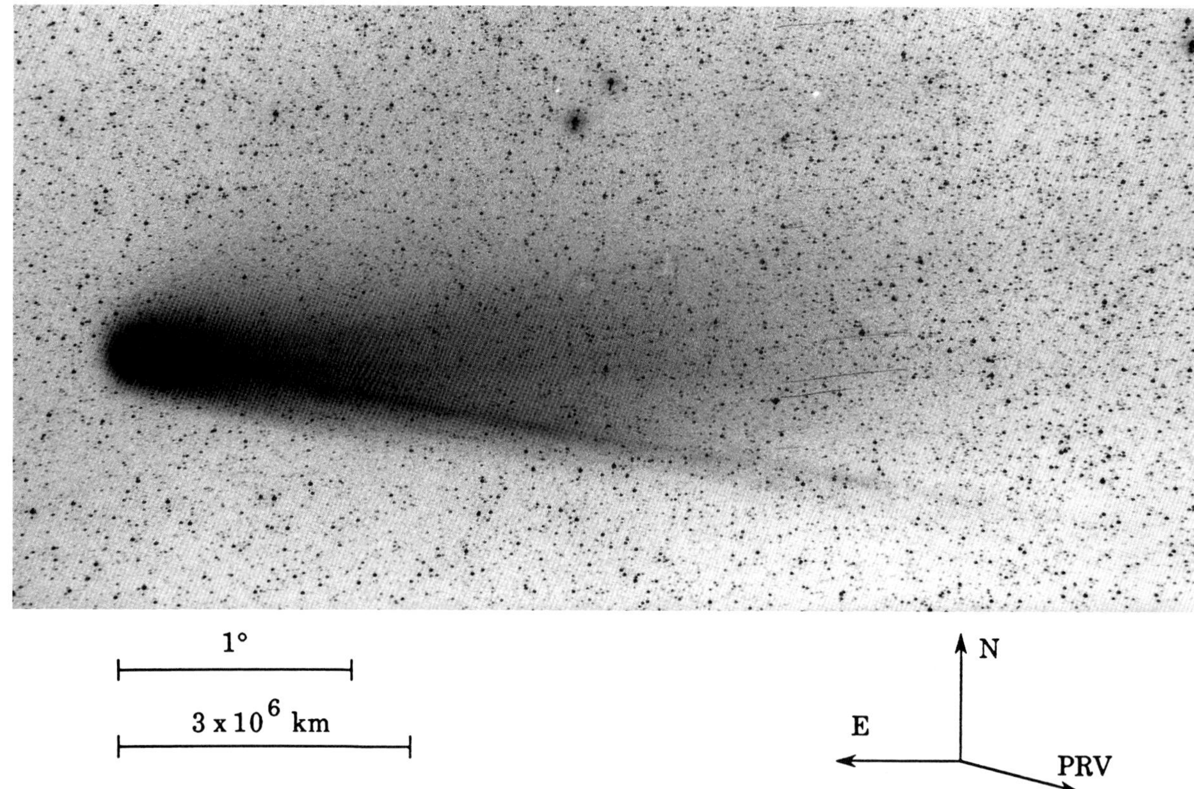

Fig. 377: 1986 Mar 20.05069 UT; exposure 8.0 minutes on Tech. Pan 2415 emulsion; r=0.99, Δ=0.81, β=113.9°. *Photograph by A. Marafie, Wafra Observatory.*

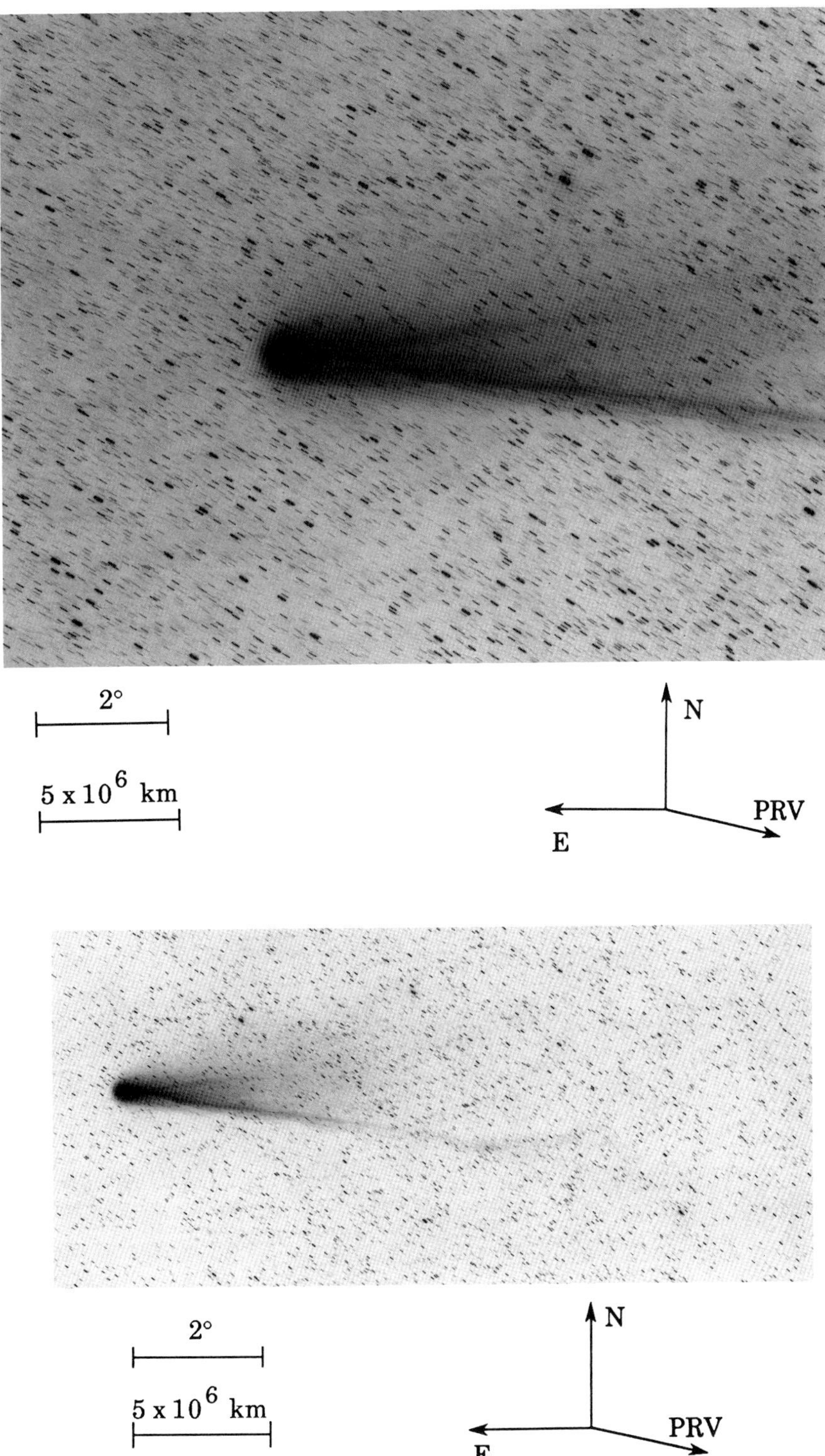

(top) Fig. 378: 1986 Mar 20.10417 UT; exposure 60.0 minutes on IIa-O emulsion with no filter; r=0.99, Δ=0.81, β=113.9°; LSPN-2848.
Photograph by G. Malcolm, Boyden Observatory.

(bottom) Fig. 379: 1986 Mar 20.10417 UT; exposure 60.0 minutes on IIa-O emulsion with no filter; r=0.99, Δ=0.81, β=113.9°; LSPN-2849.
Photograph by G. Malcolm/A. Jarrett, Boyden Observatory.

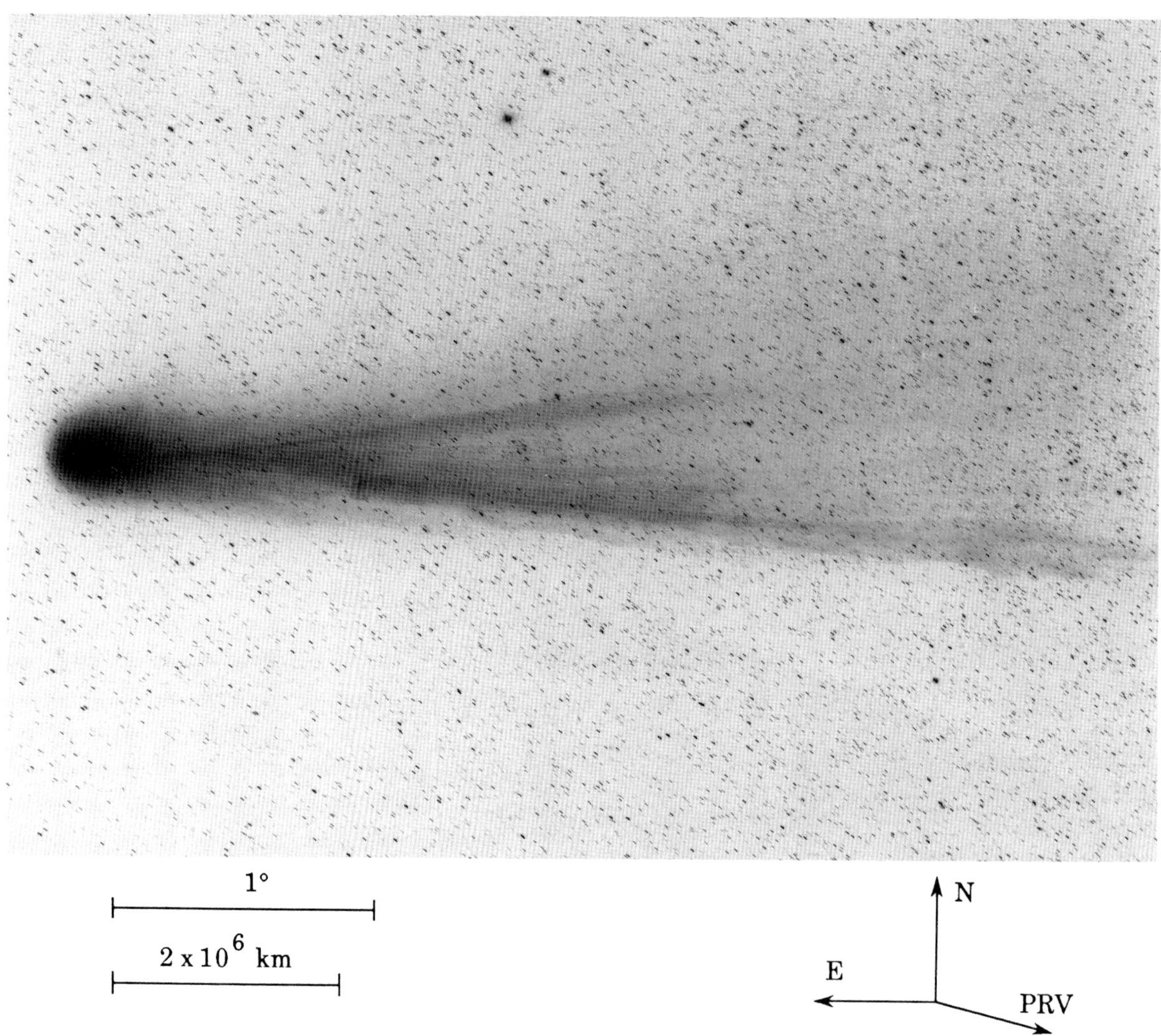

Fig. 380: 1986 Mar 20.29028 UT; exposure 20.0 minutes on IIa-O emulsion with no filter; r=1.00, Δ=0.80, β=113.9°; LSPN-1103.
Photograph by F. Miller, University of Michigan/CTIO.

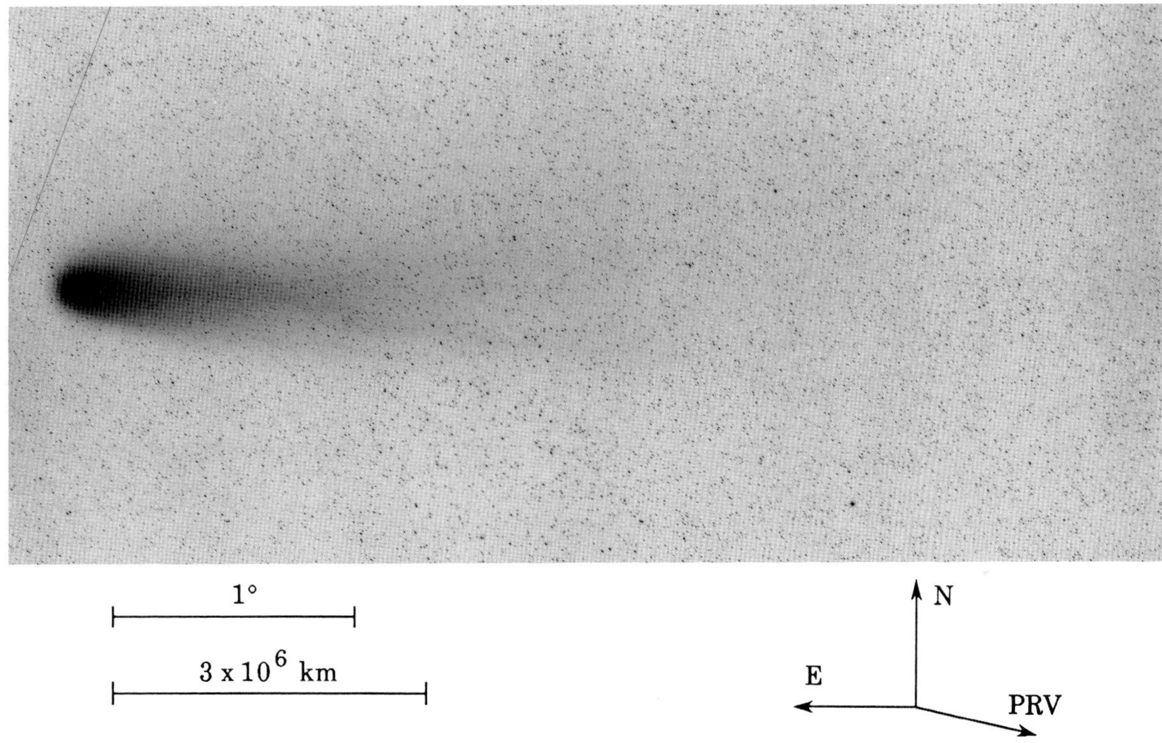

Fig. 381: 1986 Mar 20.32500 UT; exposure 10.0 minutes on 098-04 emulsion with RG-630 filter; r=1.00, Δ=0.80, β=113.9°; LSPN-2903.
Photograph by H. Schuster/O. Pizarro, European Southern Observatory.

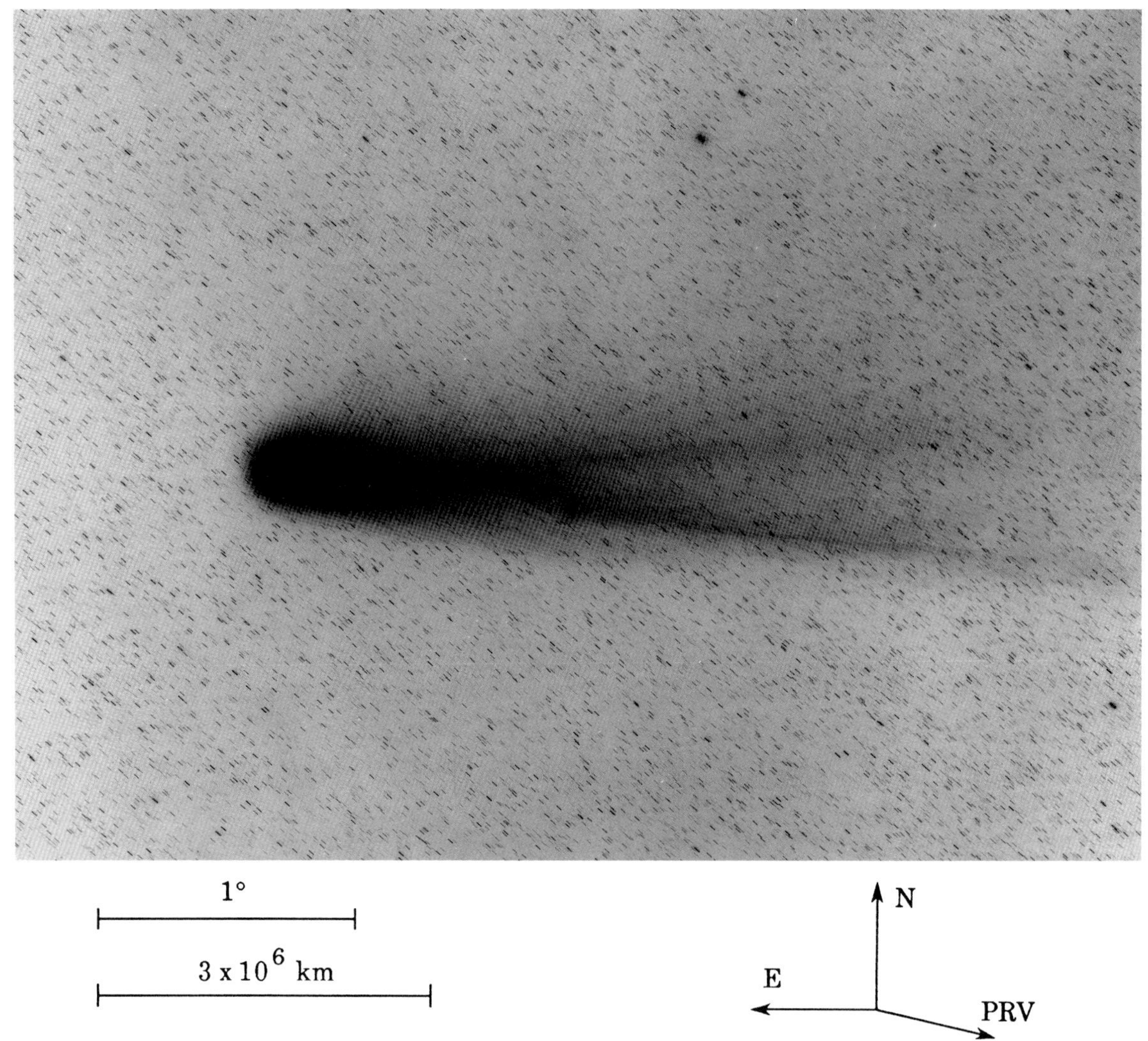

Fig. 382: 1986 Mar 20.34028 UT; exposure 30.0 minutes on IIIa-F emulsion with no filter; r=1.00, Δ=0.80, β=113.9°; LSPN-738.
Photograph by C. Torres/H. Wroblewski, Cerro el Roble Astronomical Observatory.

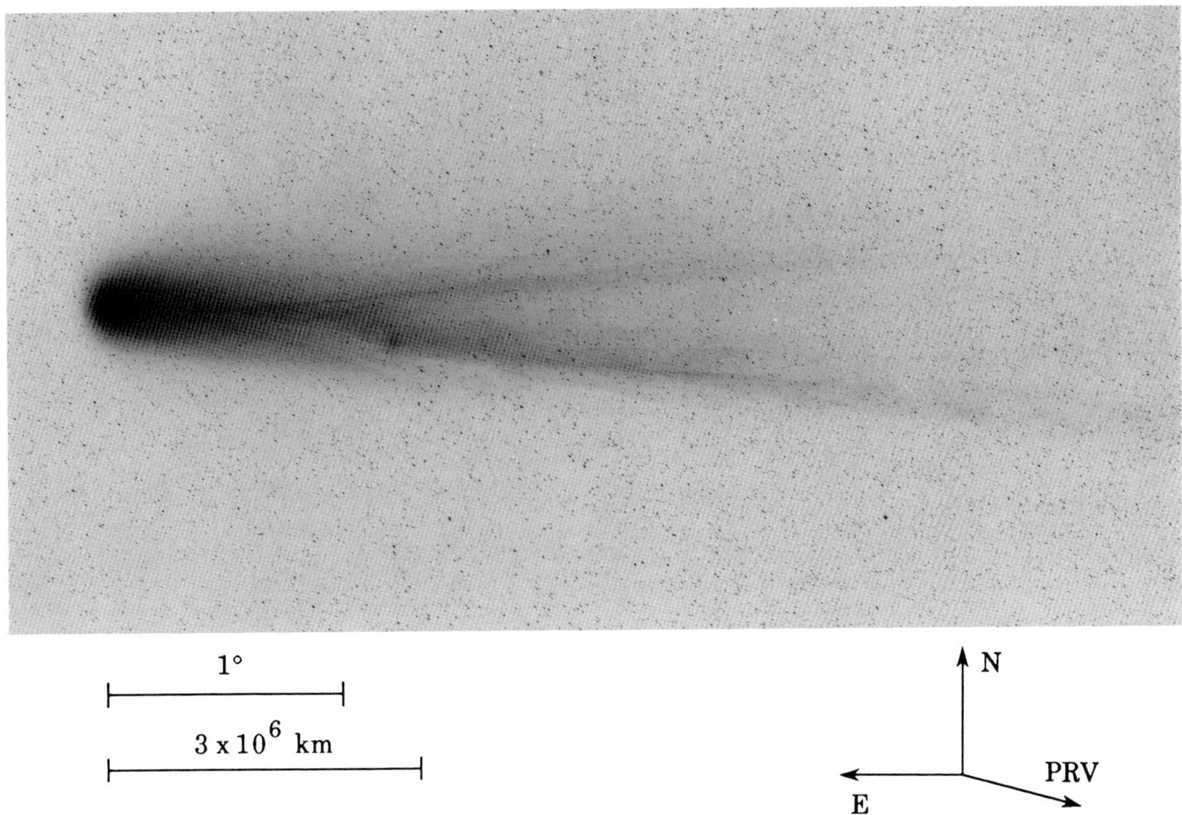

Fig. 383: 1986 Mar 20.34861 UT; exposure 10.0 minutes on IIa-O emulsion with GG-385 filter; r=1.00, Δ=0.80, β=113.9°; LSPN-2904.
Photograph by H. Schuster/O. Pizarro, European Southern Observatory.

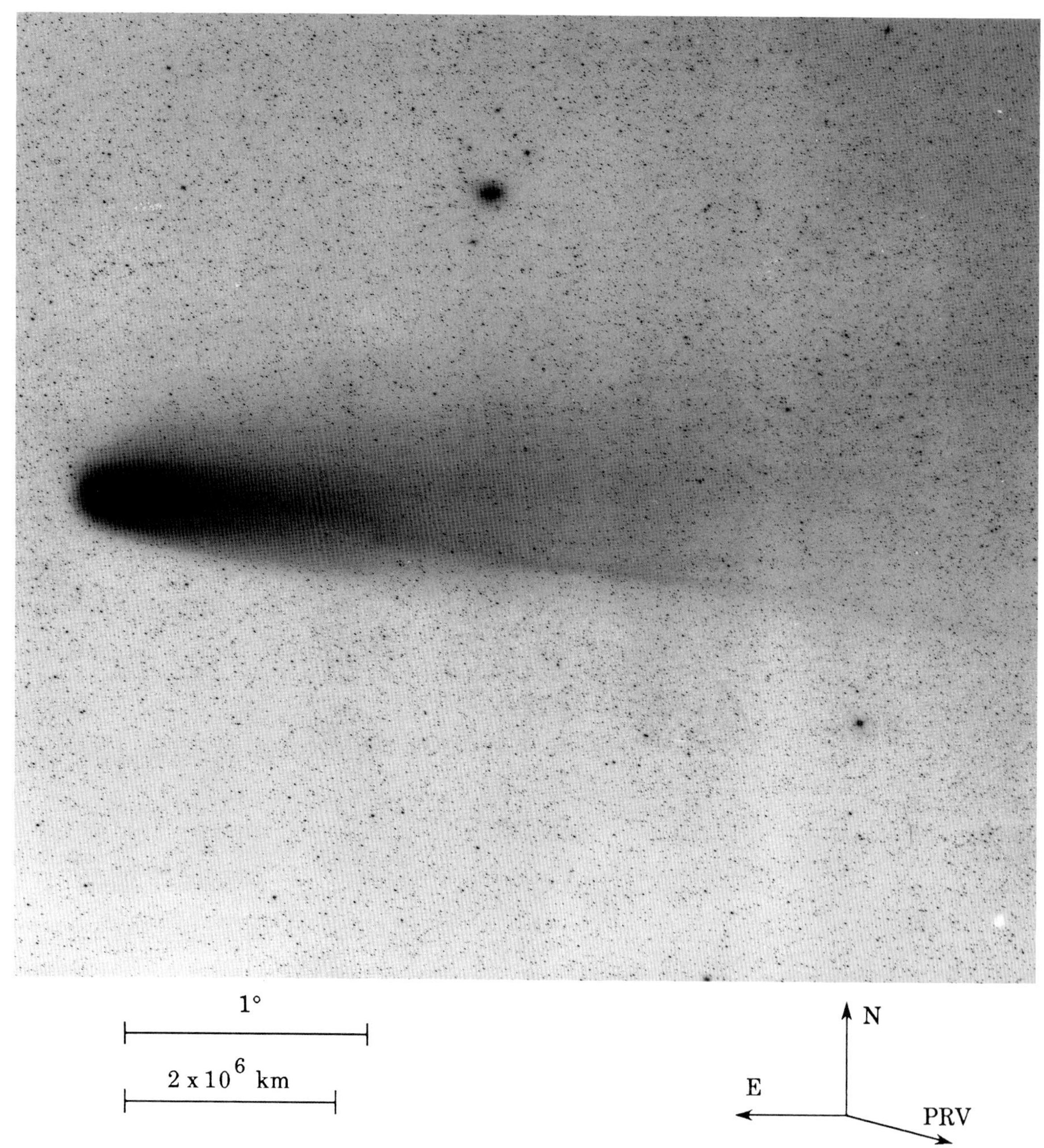

Fig. 384: 1986 Mar 20.36458 UT; exposure 30.0 minutes on 098-04 emulsion with RG-610 filter; r=1.00, Δ=0.80, β=113.9°; LSPN-1106.
Photograph by F. Miller, University of Michigan/CTIO.

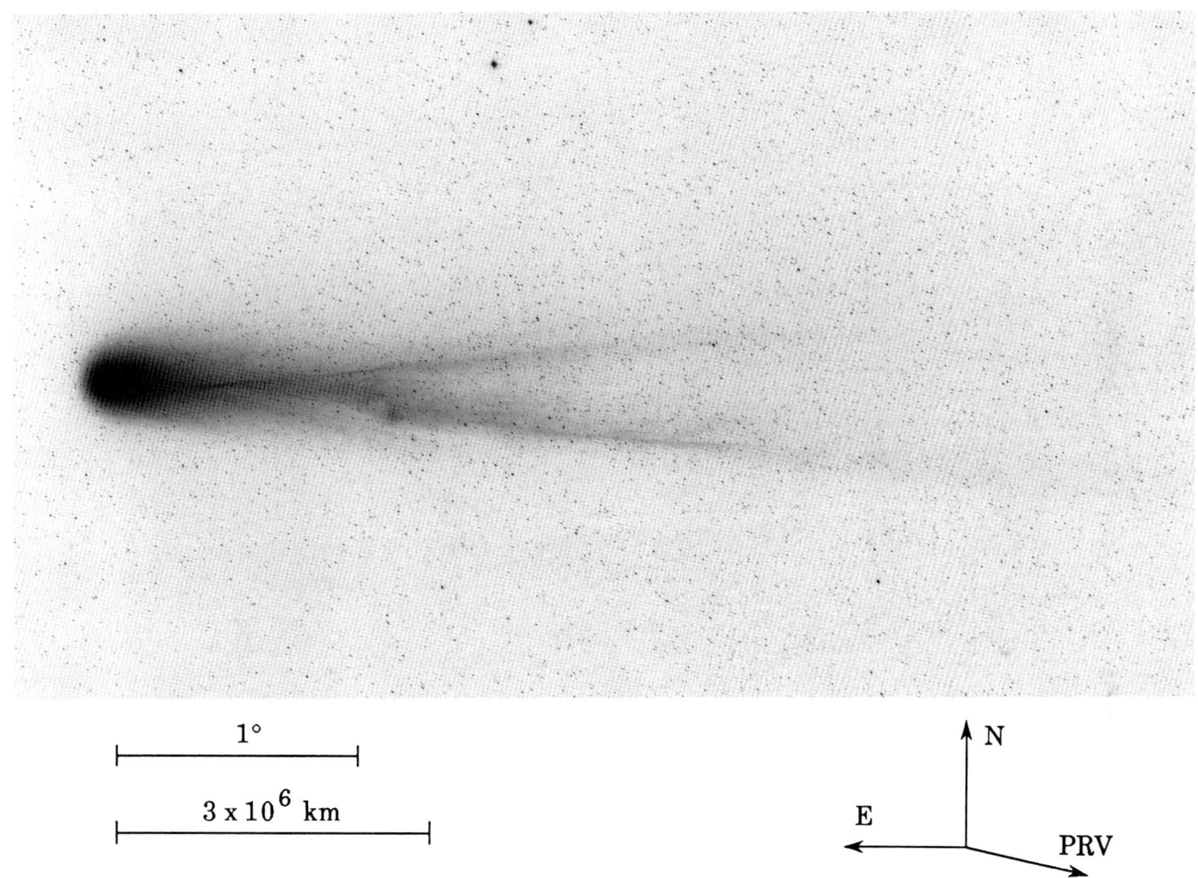

Fig. 385: 1986 Mar 20.36667 UT; exposure 10.0 minutes on IIa-O emulsion with GG-385 filter; r=1.00, Δ=0.80, β=113.9°; LSPN-2905.
Photograph by H. Schuster/O. Pizarro, European Southern Observatory.

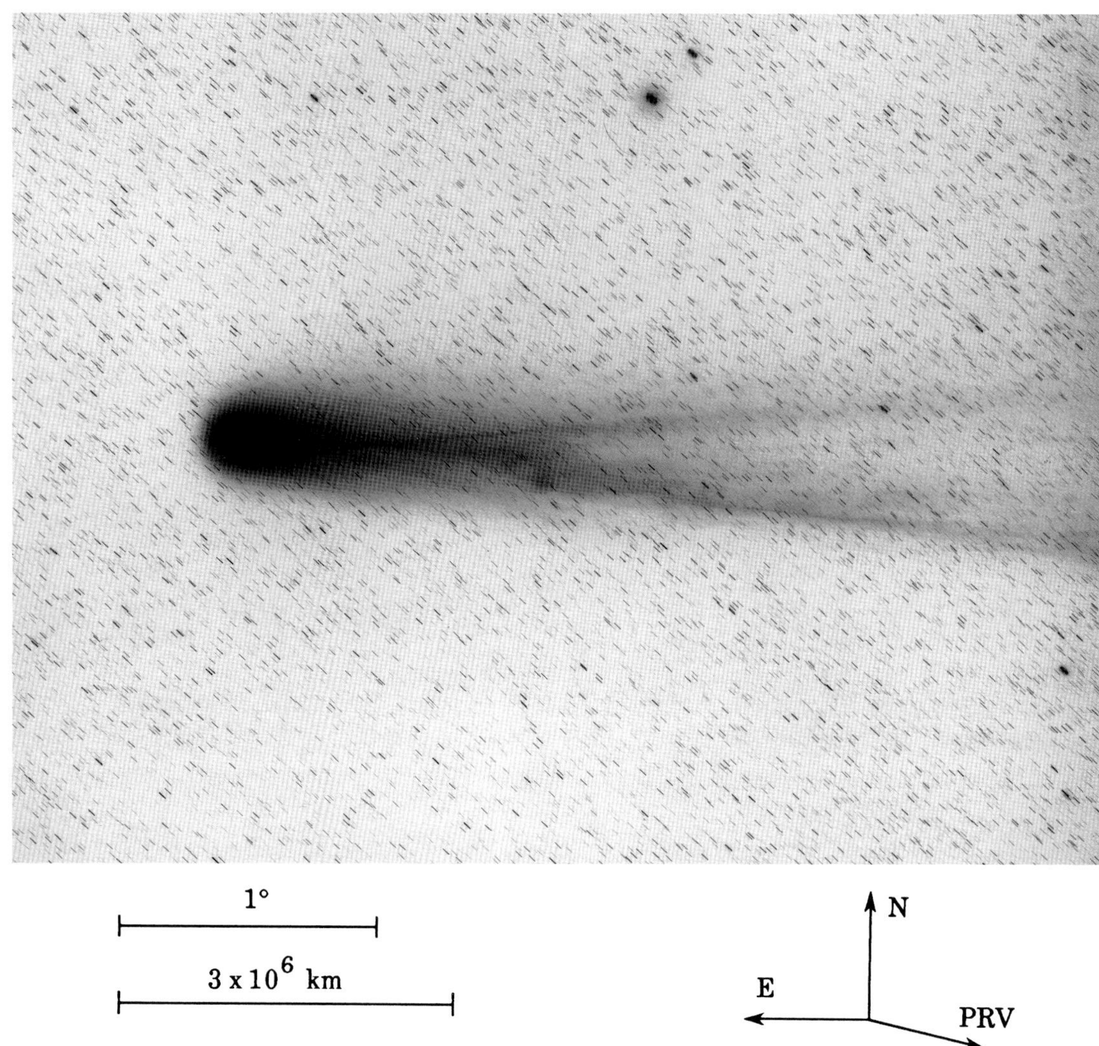

Fig. 386: 1986 Mar 20.37986 UT; exposure 40.0 minutes on IIa-O emulsion with GG-385 filter; r=1.00, Δ=0.80, β=113.9°; LSPN-739.
Photograph by C. Torres/H. Wroblewski, Cerro el Roble Astronomical Observatory.

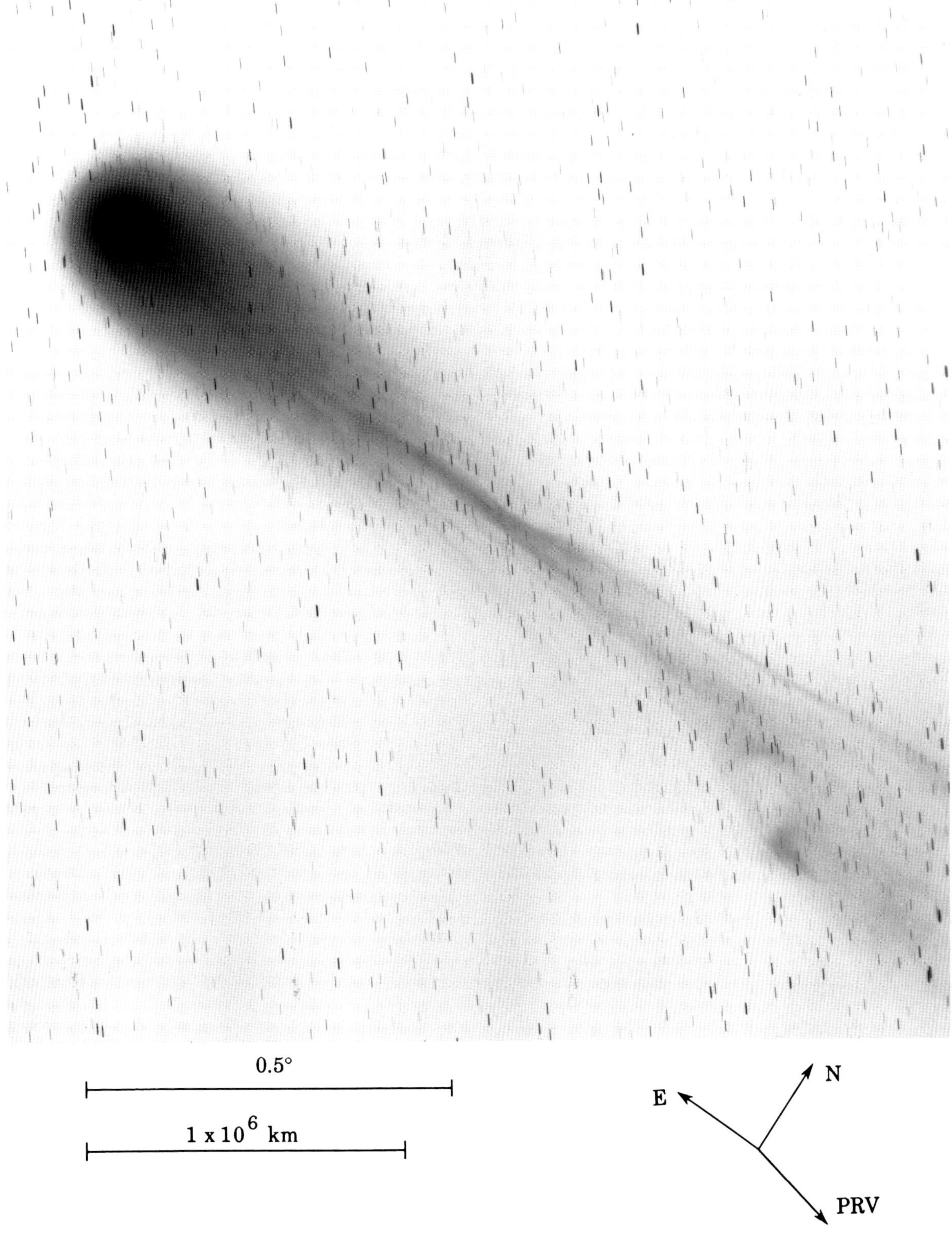

Fig. 387: 1986 Mar 20.38403 UT; exposure 18.0 minutes on 103a-O emulsion with Wratten 2C filter; r=1.00, Δ=0.80, β=113.9°; LSPN-2570.
Photograph by A. Dressler/R. Windhorst, Mount Wilson/Las Campanas Observatories.

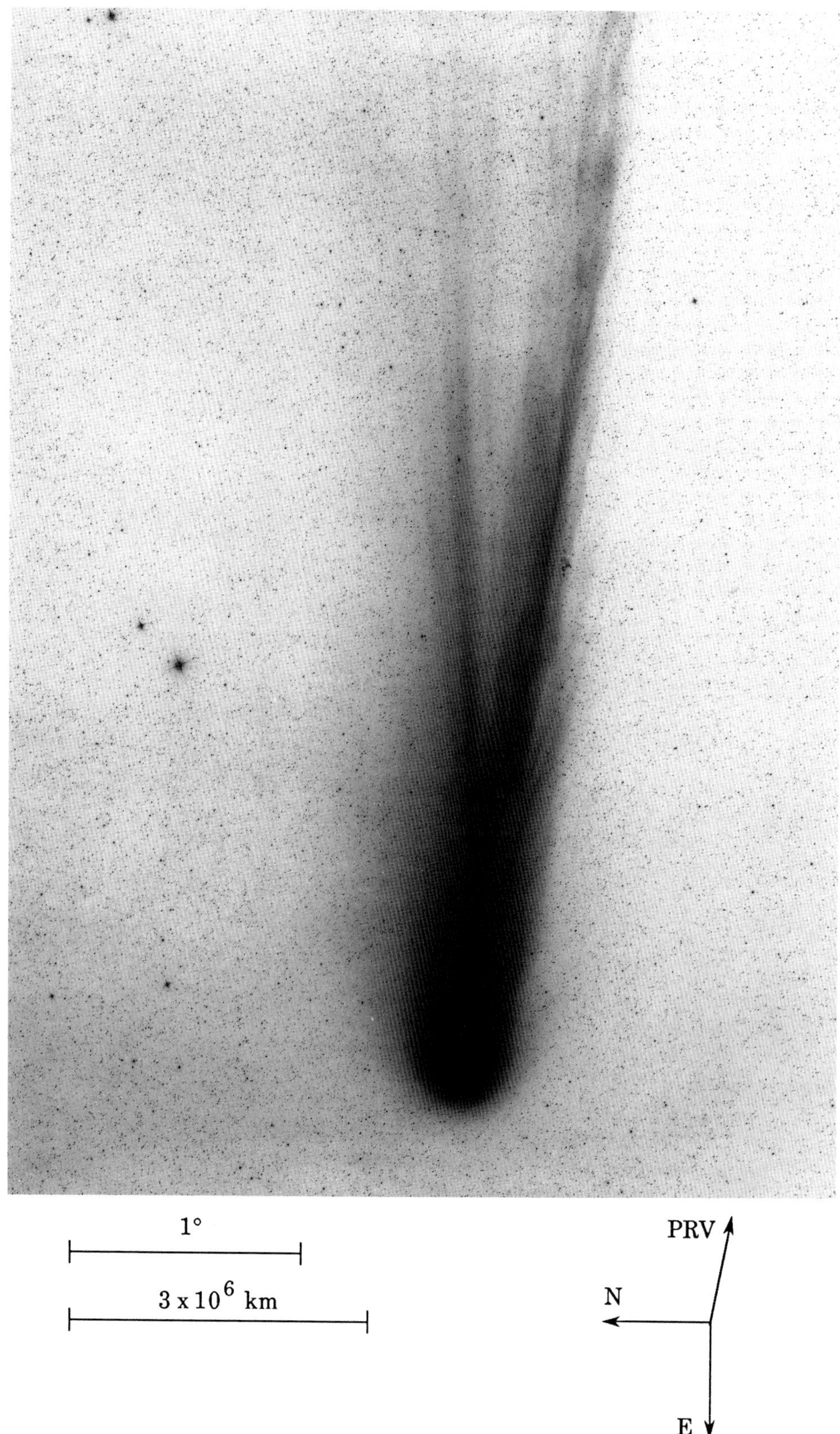

Fig. 388: 1986 Mar 20.39097 UT; exposure 20.0 minutes on IIa-O emulsion with no filter; r=1.00, Δ=0.80, β=113.9°; LSPN-1107.
Photograph by F. Miller, University of Michigan/CTIO.

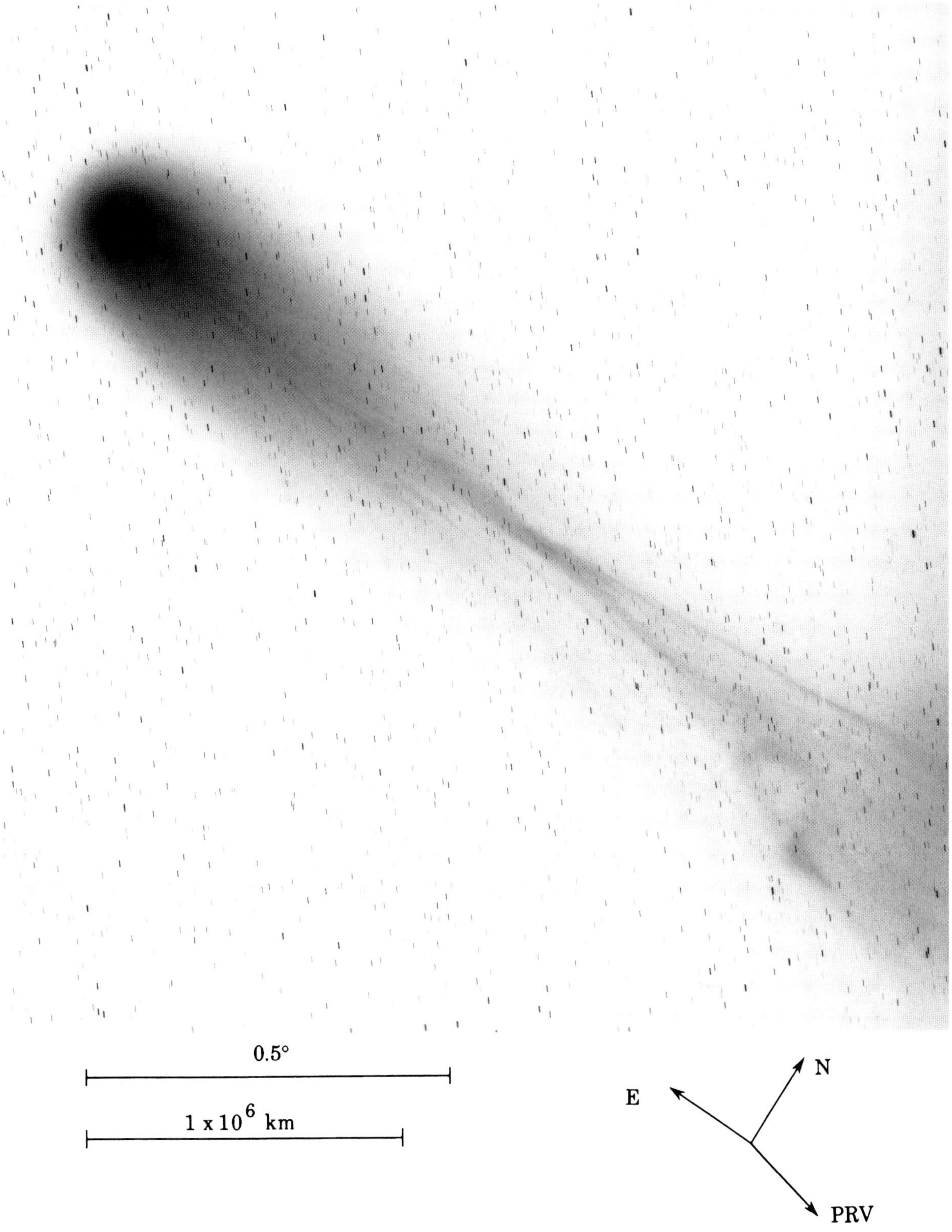

Fig. 389: 1986 Mar 20.40139 UT; exposure 10.0 minutes on 103a-O emulsion with Wratten 2C filter; r=1.00, Δ=0.80, β=113.9°; LSPN-2571.
Photograph by A. Dressler/R. Windhorst, Mount Wilson/Las Campanas Observatories.

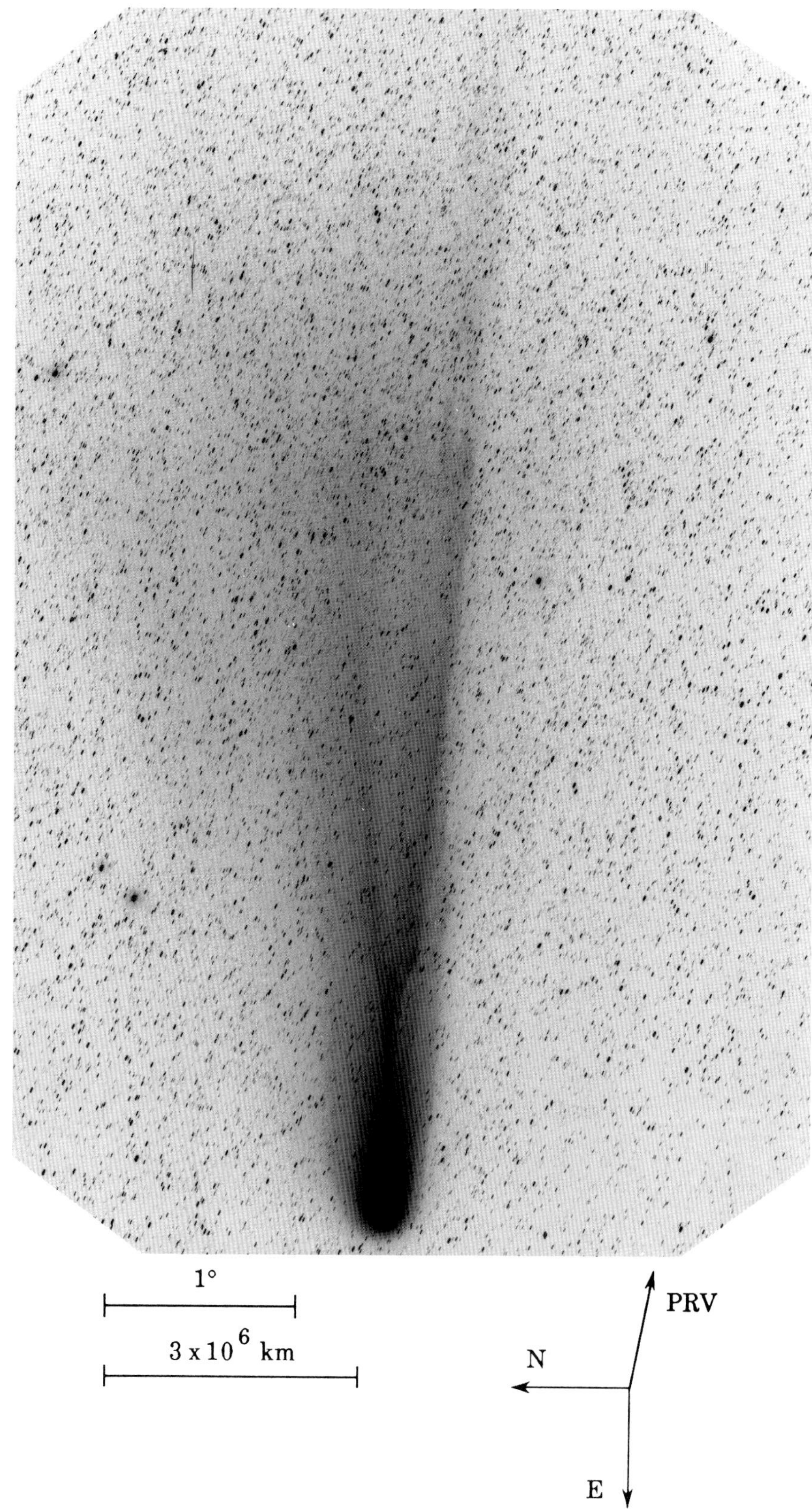

Fig. 390: 1986 Mar 20.41806 UT; exposure 21.0 minutes on Tech. Pan 2415 emulsion with no filter; r=1.00, Δ=0.80, β=113.9°; LSPN-1408.
Photograph by W. Liller, Easter Island, LSPN Island Network.

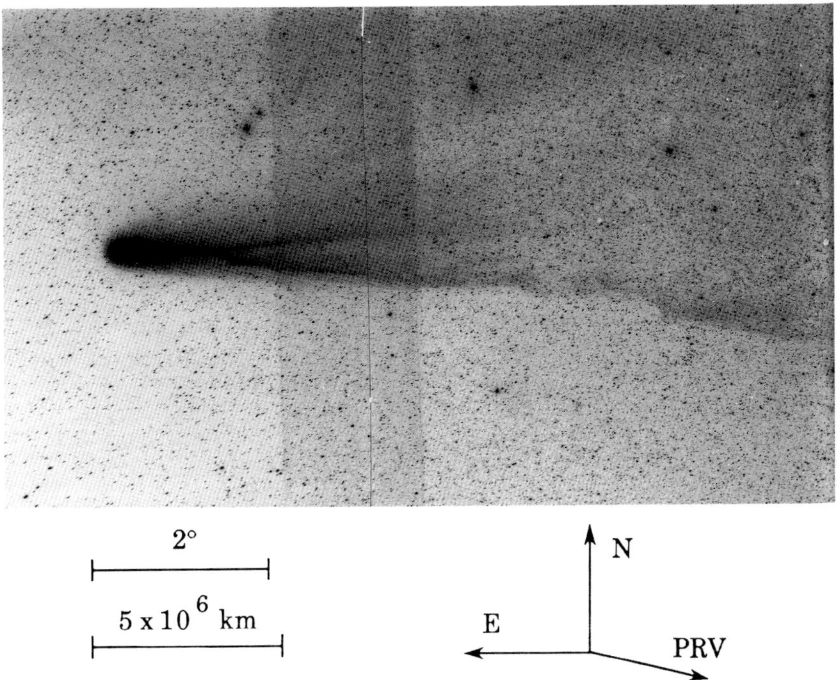

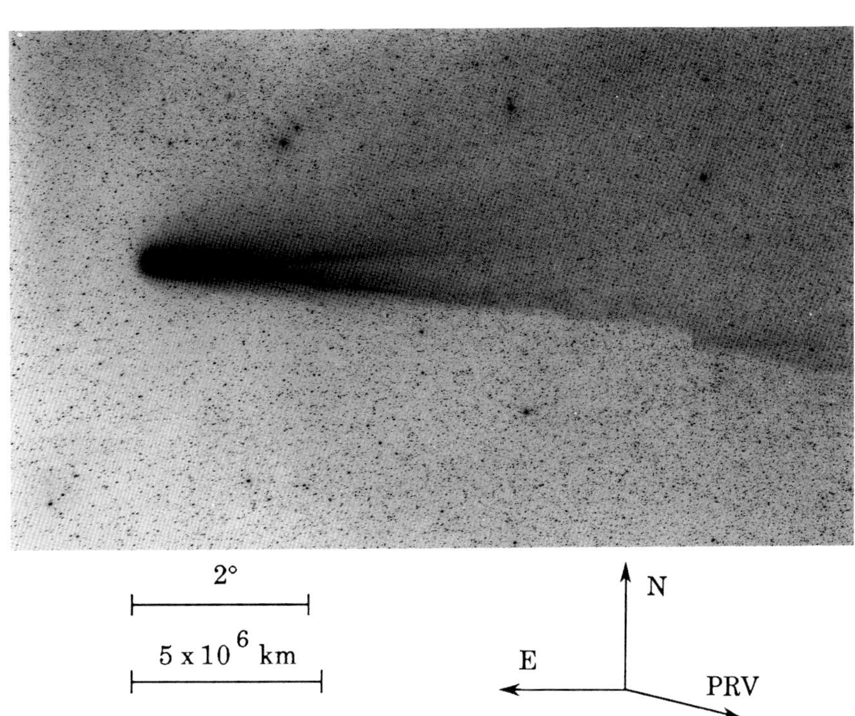

(top) Fig. 391: 1986 Mar 20.45903 UT; exposure 10.0 minutes on IIa-O emulsion with no filter; r=1.00, Δ=0.80, β=113.9°; LSPN-897.
Photograph by E. Moore, Joint Observatory for Cometary Research.

(bottom) Fig. 392: 1986 Mar 20.46806 UT; exposure 10.0 minutes on IIa-O emulsion with no filter; r=1.00, Δ=0.80, β=113.9°; LSPN-898.
Photograph by E. Moore, Joint Observatory for Cometary Research.

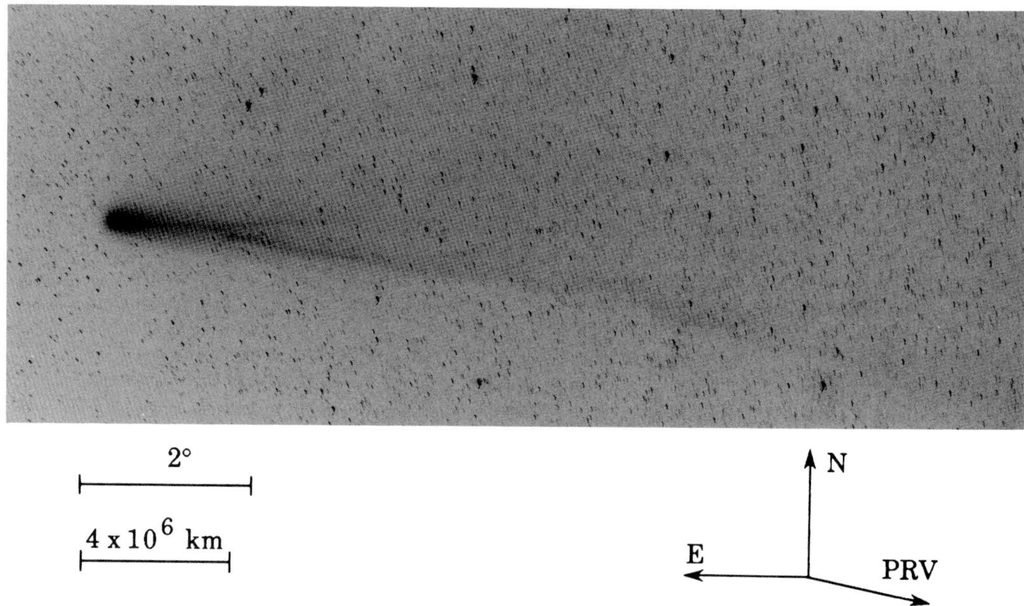

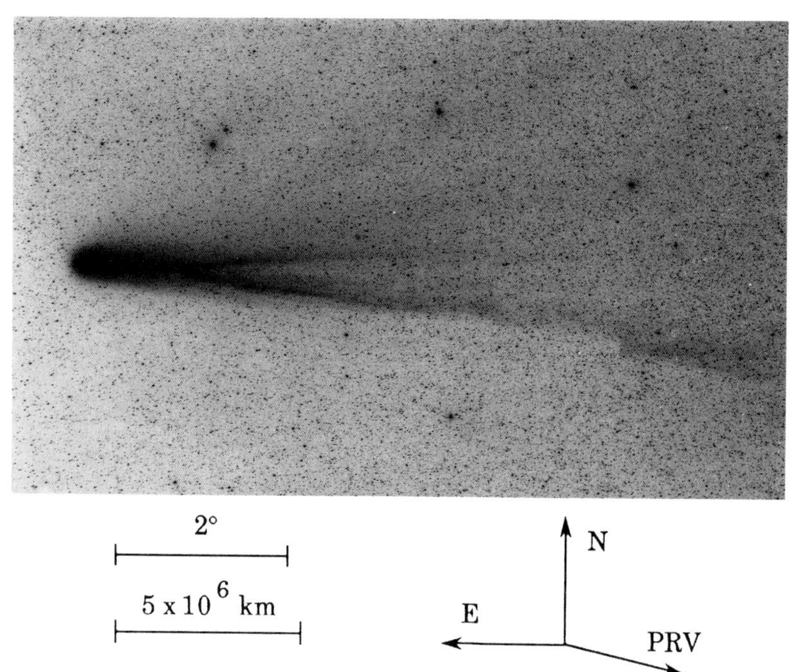

(top) Fig. 393: 1986 Mar 20.47430 UT; exposure 60.0 minutes on IIa-O emulsion; r=1.00, Δ=0.80, β=113.9°.
Photograph by A. Butcher/D. Doss, McDonald Observatory.

(bottom) Fig. 394: 1986 Mar 20.47708 UT; exposure 10.0 minutes on IIa-O emulsion with no filter; r=1.00, Δ=0.80, β=113.9°; LSPN-899.
Photograph by E. Moore, Joint Observatory for Cometary Research.

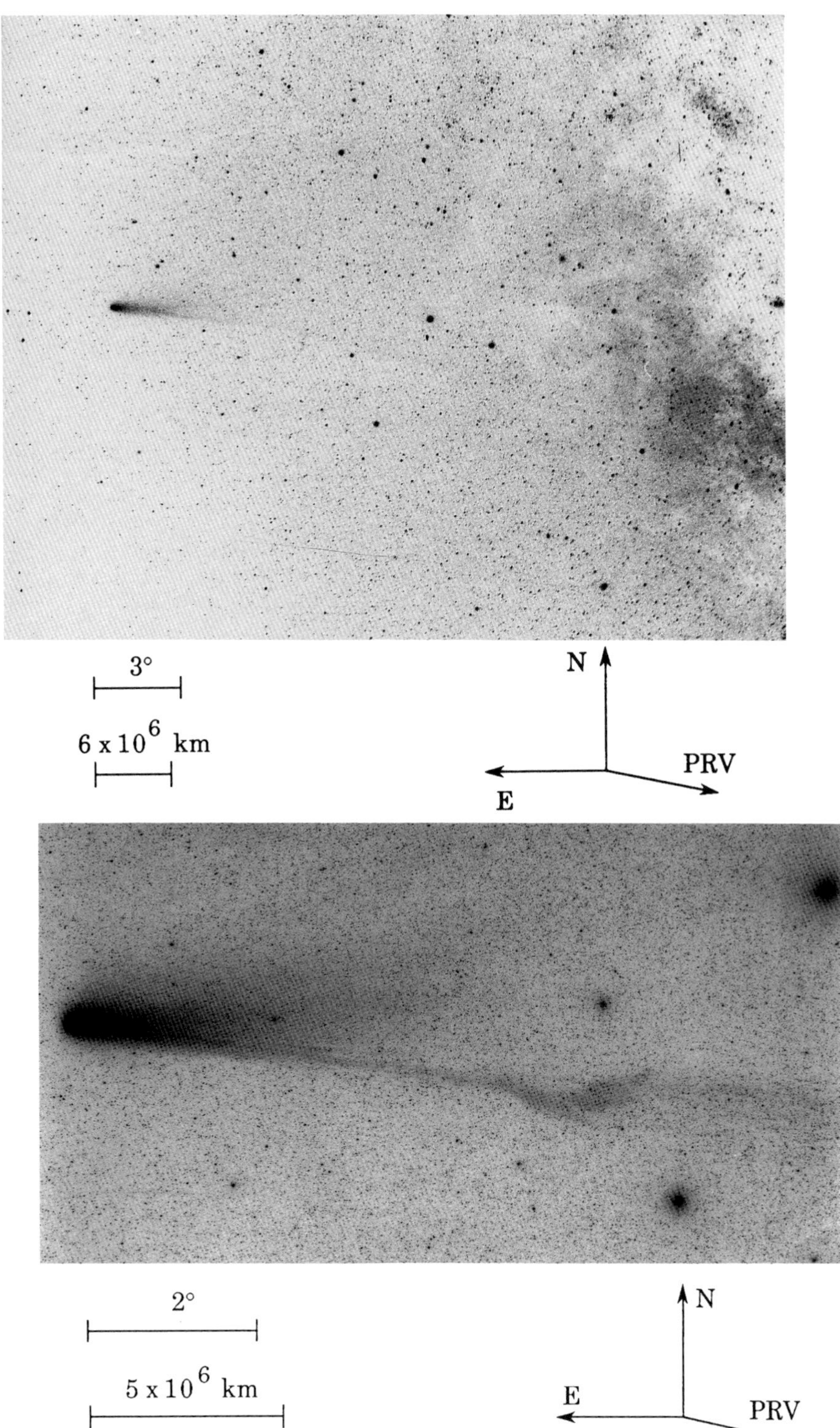

(top) Fig. 395: 1986 Mar 20.48500 UT; exposure 20.0 minutes on Tri-X emulsion; r=1.00, Δ=0.80, β=113.9°; AON-851187.
Photograph by S. Edberg, USA.

(bottom) Fig. 396*: 1986 Mar 22.49306 UT; exposure 8.0 minutes on IIa-O emulsion with no filter; r=1.03, Δ=0.75, β=114.1°; LSPN-908.
Photograph by E. Moore, Joint Observatory for Cometary Research.

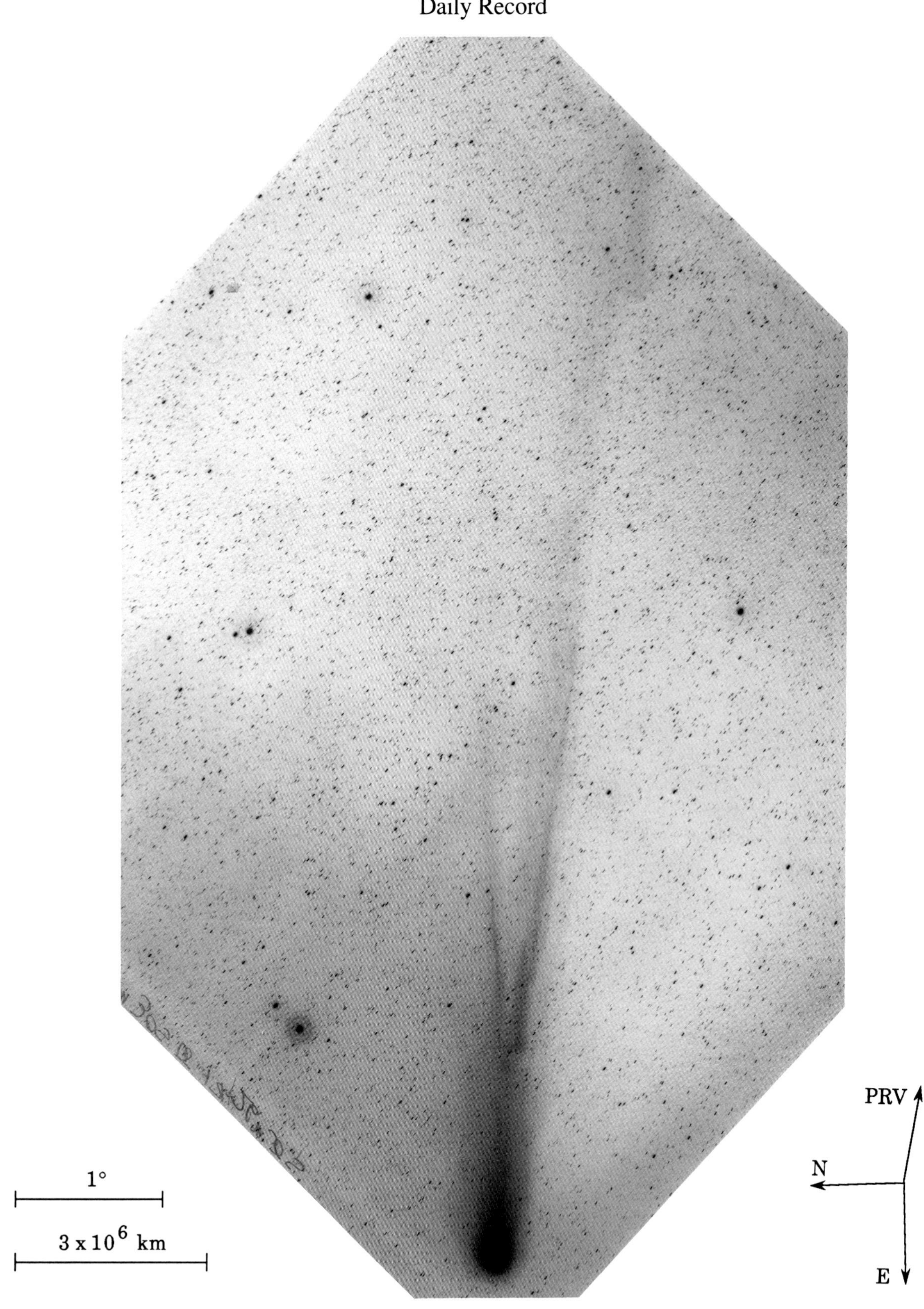

Fig. 397: 1986 Mar 20.51007 UT; exposure 24.0 minutes on 103a-O emulsion with no filter; r=1.00, Δ=0.80, β=113.9°; LSPN-3580.
Photograph by H. Giclas, Lowell Observatory.

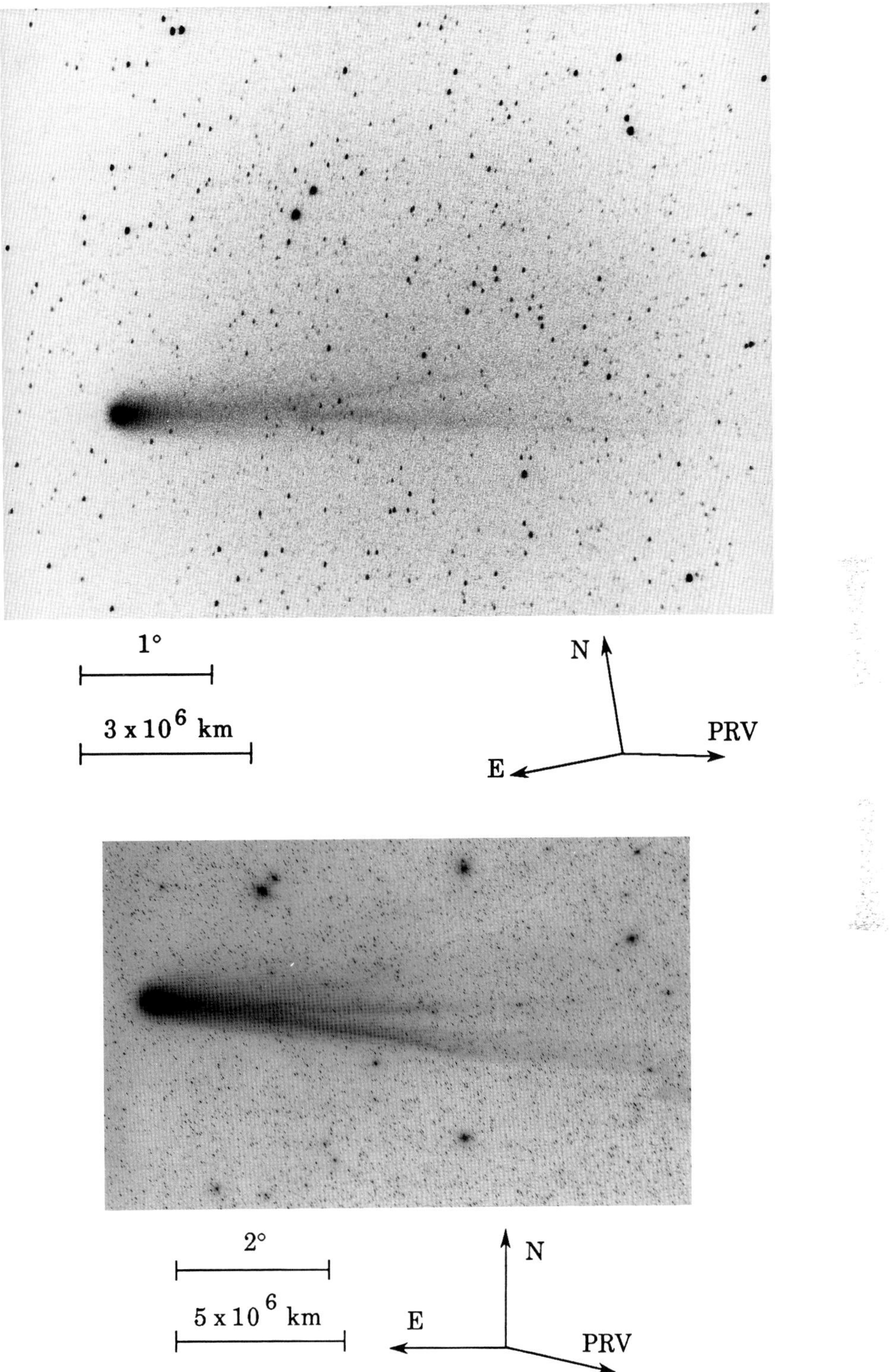

(top) Fig. 398: 1986 Mar 20.52590 UT; exposure 10.0 minutes on 3M 1000 emulsion; r=1.00, Δ=0.80, β=113.9°; AON-851186.
Photograph by S. Edberg, USA.

(bottom) Fig. 399: 1986 Mar 20.62570 UT; exposure 30.0 minutes on IIIa-J emulsion with no filter; r=1.00, Δ=0.79, β=113.9°; LSPN-1918.
Photograph by M. Buie, Mauna Kea Observatory.

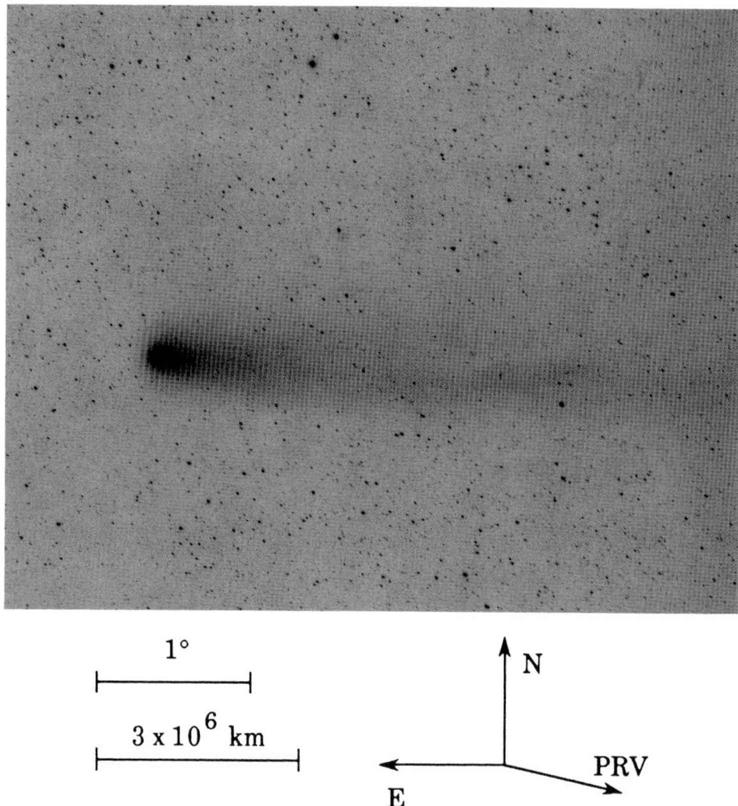

Fig. 400: 1986 Mar 21.04826 UT; exposure 15.0 minutes on IIa-O emulsion with no filter; r=1.01, Δ=0.78, β=113.9°; LSPN-2850.
Photograph by G. Malcolm, Boyden Observatory.

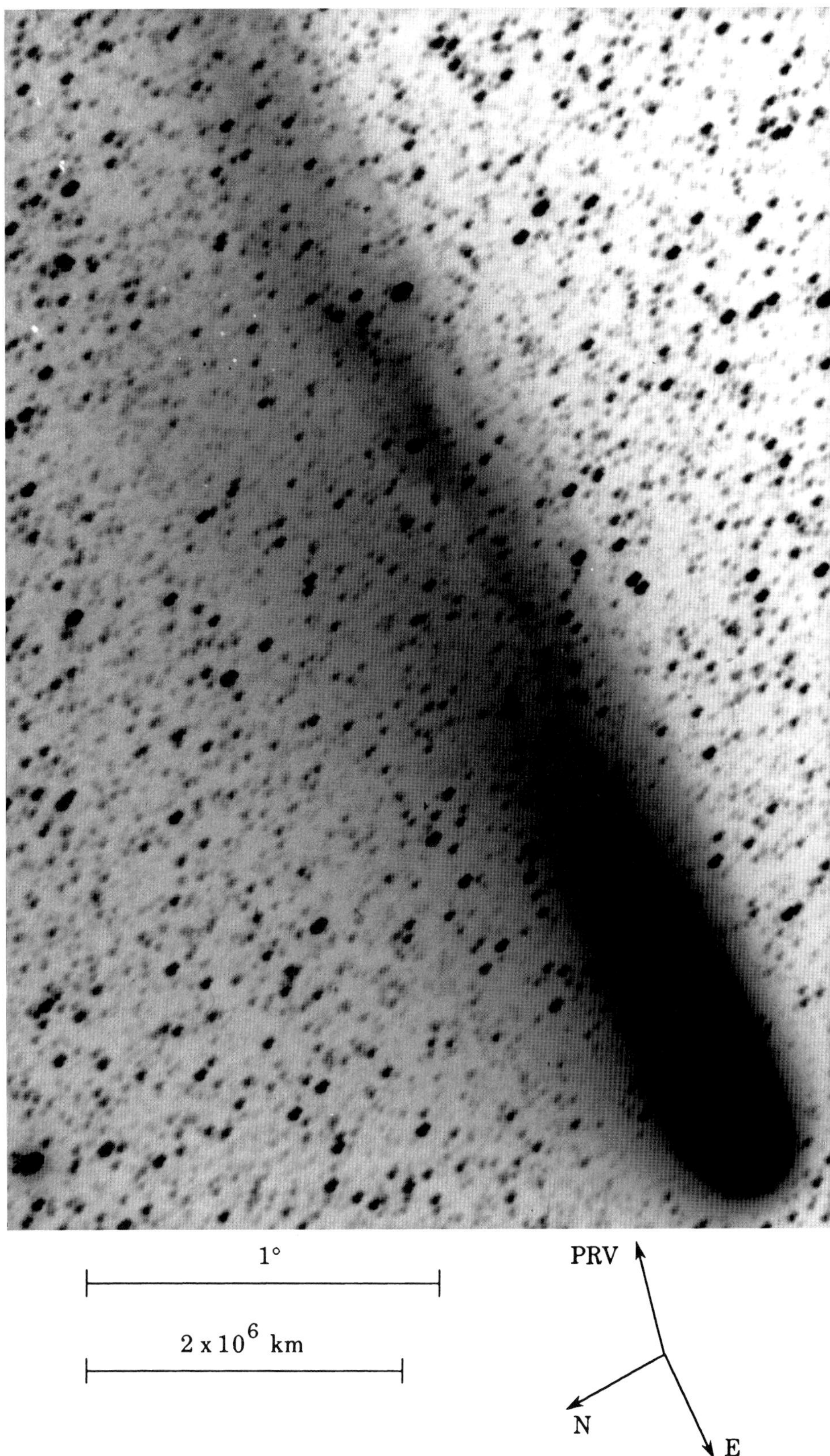

Fig. 401: 1986 Mar 21.06771 UT; exposure 25.0 minutes on Tech. Pan 2415 emulsion with no filter; r=1.01, Δ=0.78, β=113.9°; LSPN-1807.
Photograph by F. Marang, Sutherland, LSPN Island Network.

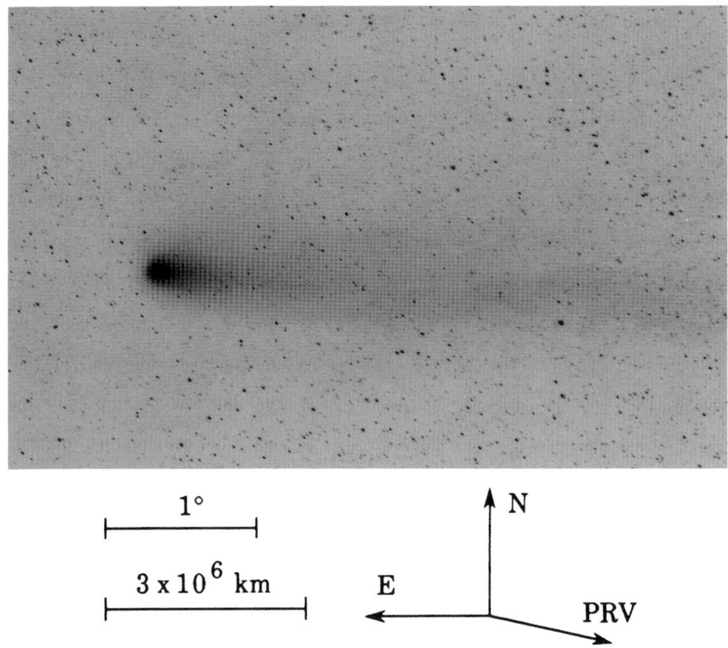

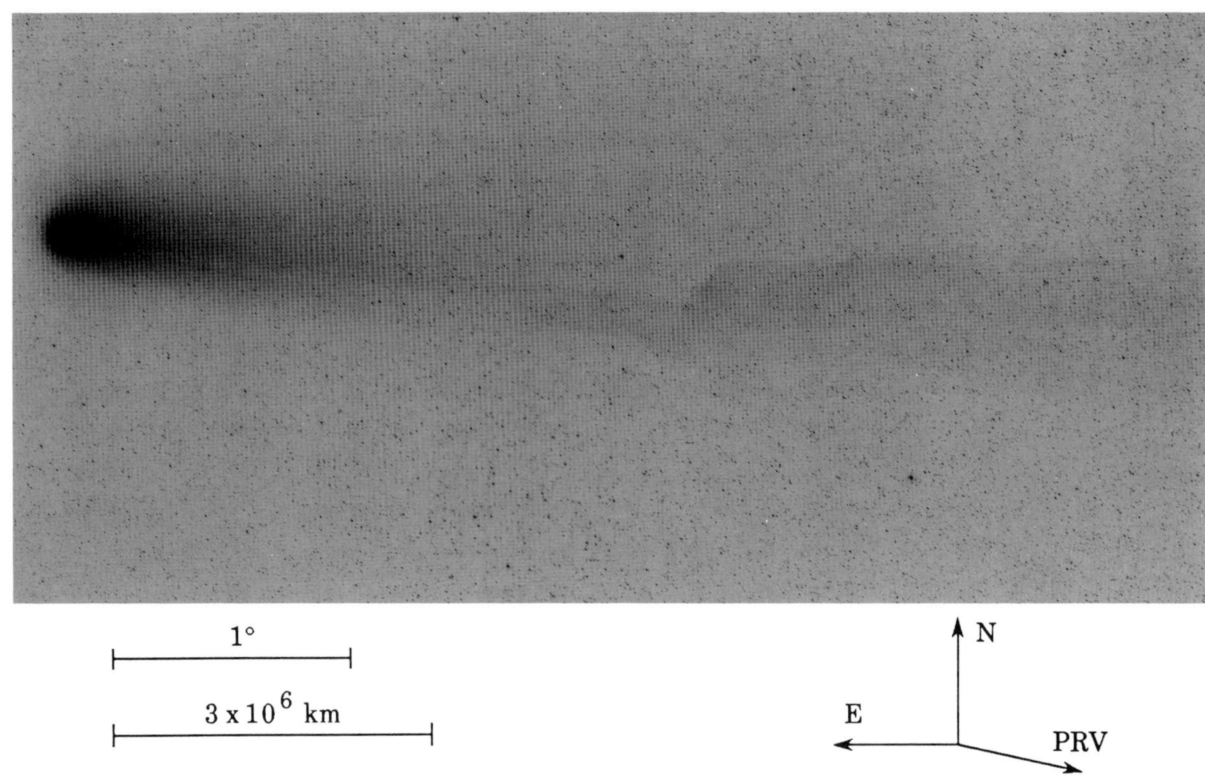

(top) Fig. 402: 1986 Mar 21.06771 UT; exposure 15.0 minutes on IIa-O emulsion with no filter; r=1.01, Δ=0.78, β=113.9°; LSPN-2853.
Photograph by G. Malcolm, Boyden Observatory.

(bottom) Fig. 403: 1986 Mar 21.32986 UT; exposure 10.0 minutes on IIa-O emulsion with GG-385 filter; r=1.01, Δ=0.78, β=113.9°; LSPN-2906.
Photograph by H. Schuster/O. Pizarro, European Southern Observatory.

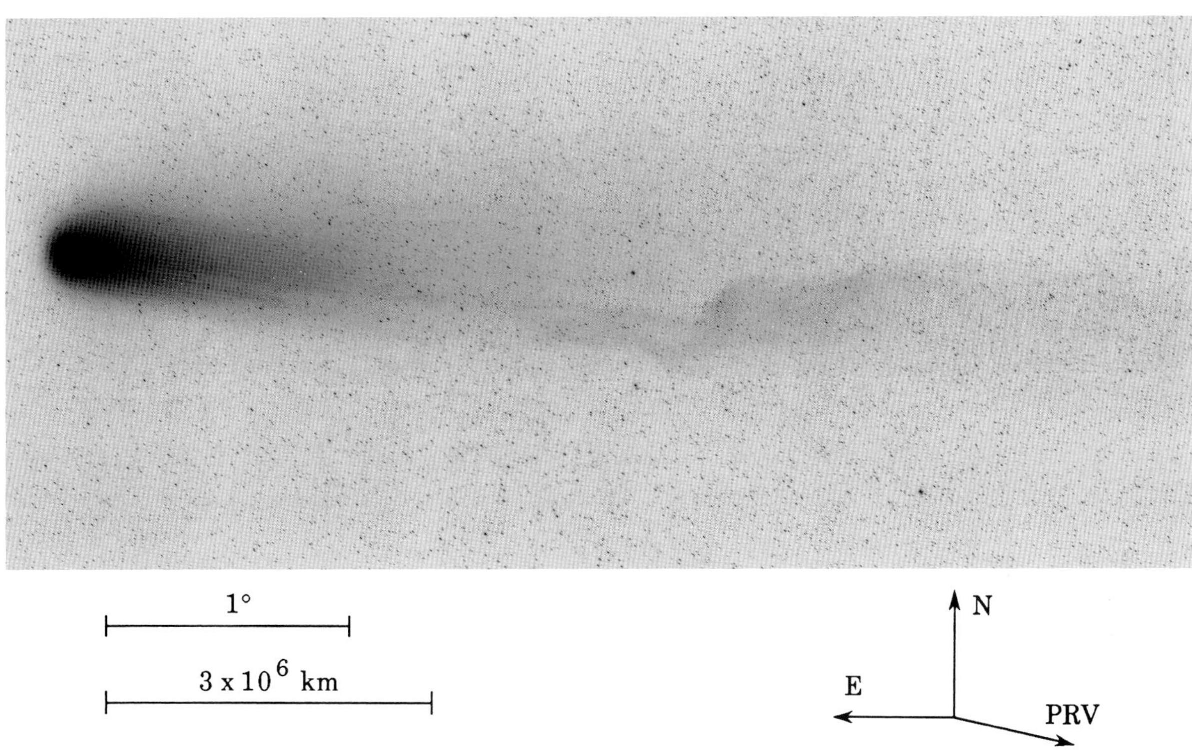

Fig. 404: 1986 Mar 21.34167 UT; exposure 10.0 minutes on IIa-O emulsion with GG-385 filter; r=1.01, Δ=0.78, β=113.9°; LSPN-2907.
Photograph by H. Schuster/O. Pizarro, European Southern Observatory.

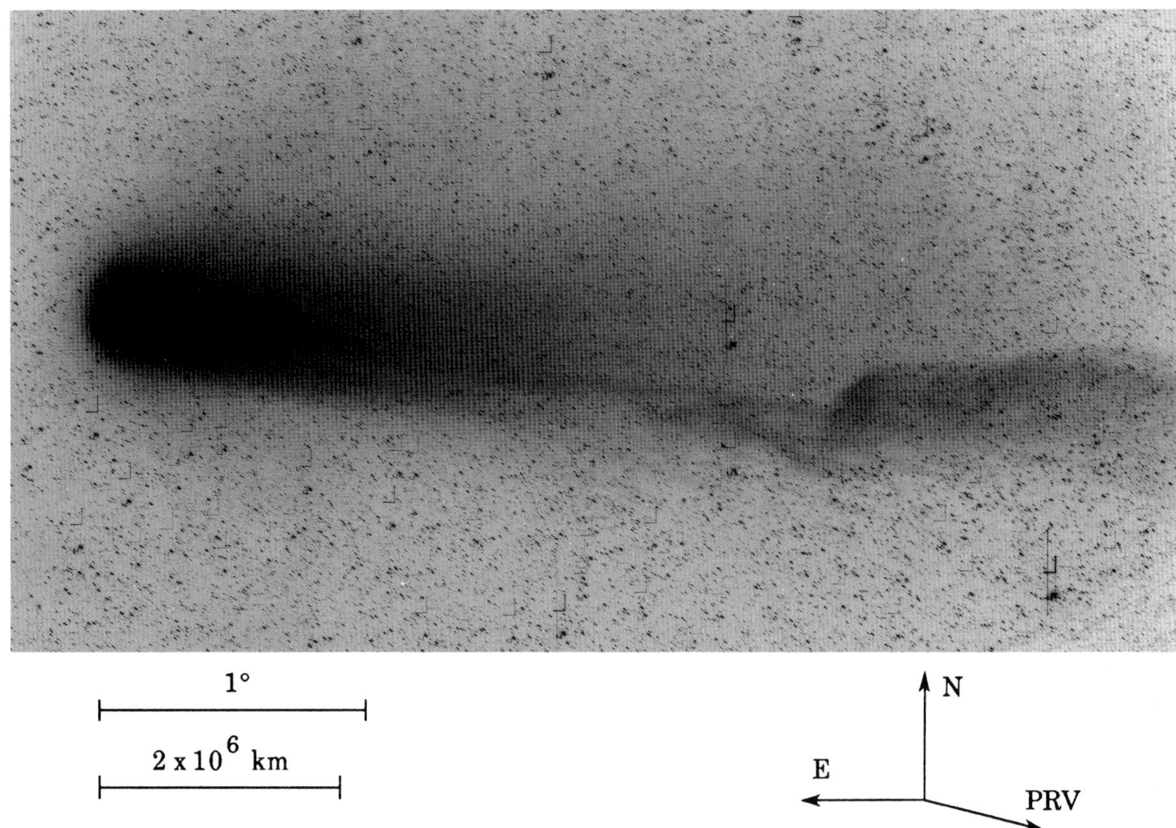

Fig. 405: 1986 Mar 21.36181 UT; exposure 20.0 minutes on IIa-O emulsion with no filter; r=1.01, Δ=0.78, β=113.9°; LSPN-1112.
Photograph by F. Miller, University of Michigan/CTIO.

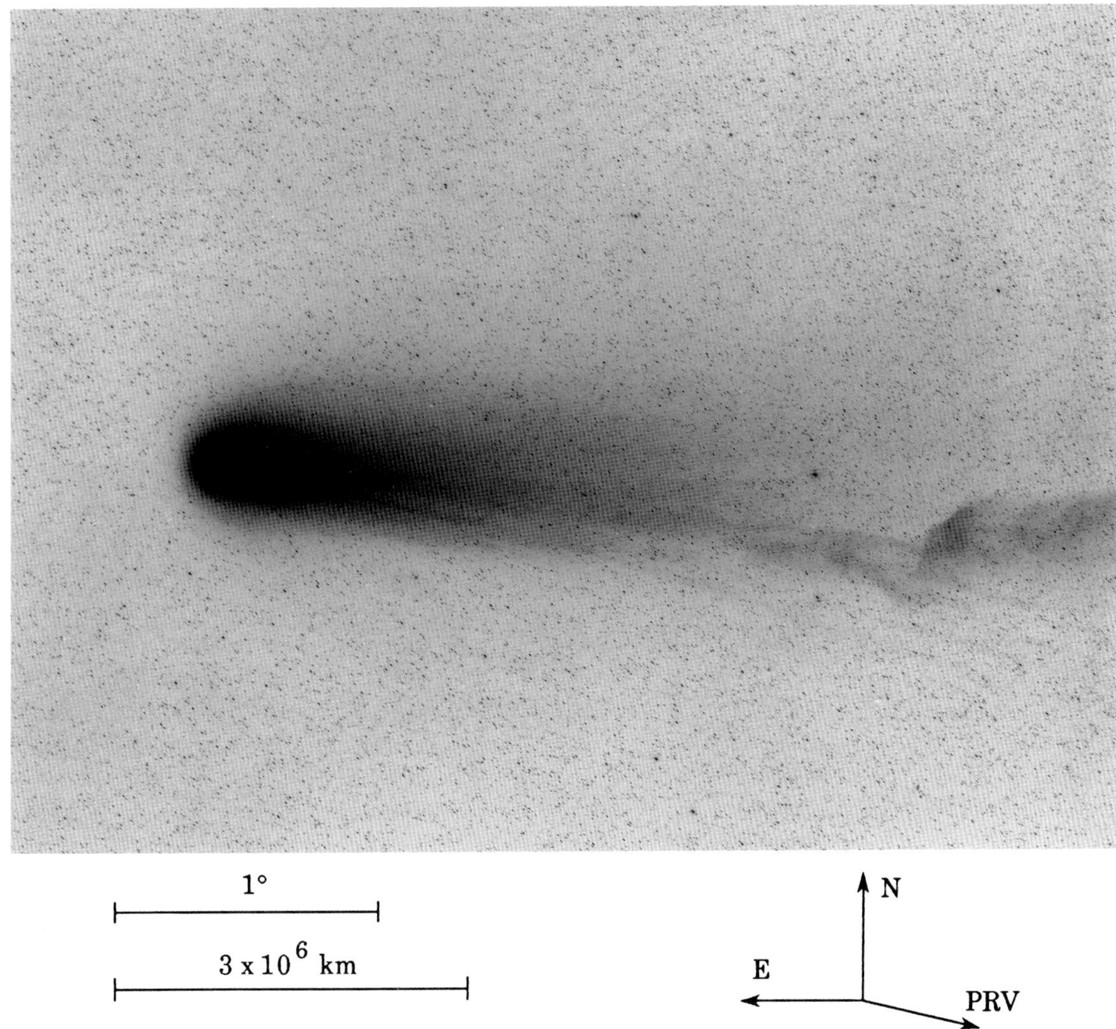

Fig. 406: 1986 Mar 21.38264 UT; exposure 10.0 minutes on 103a-O emulsion with no filter; r=1.01, Δ=0.78, β=113.9°; LSPN-741.
Photograph by C. Torres/H. Wroblewski, Cerro el Roble Astronomical Observatory.

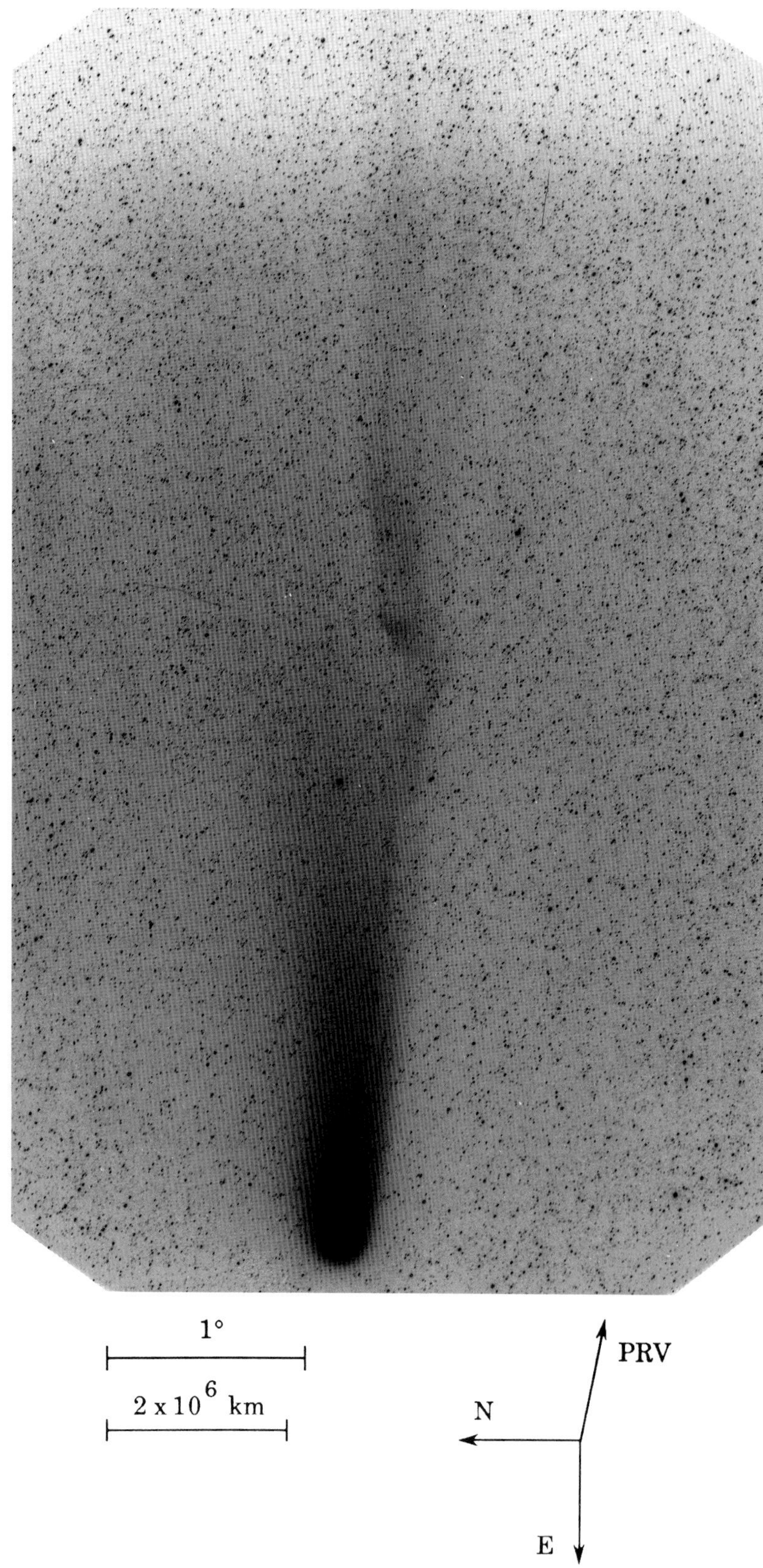

Fig. 407: 1986 Mar 21.44201 UT; exposure 15.0 minutes on Tech. Pan 2415 emulsion with no filter; r=1.01, Δ=0.77, β=113.9°; LSPN-1410.
Photograph by W. Liller, Easter Island, LSPN Island Network.

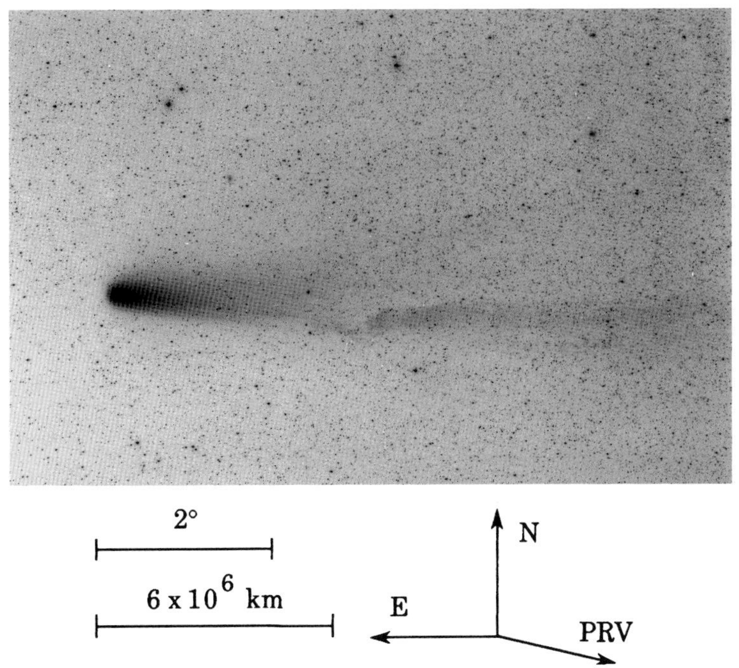

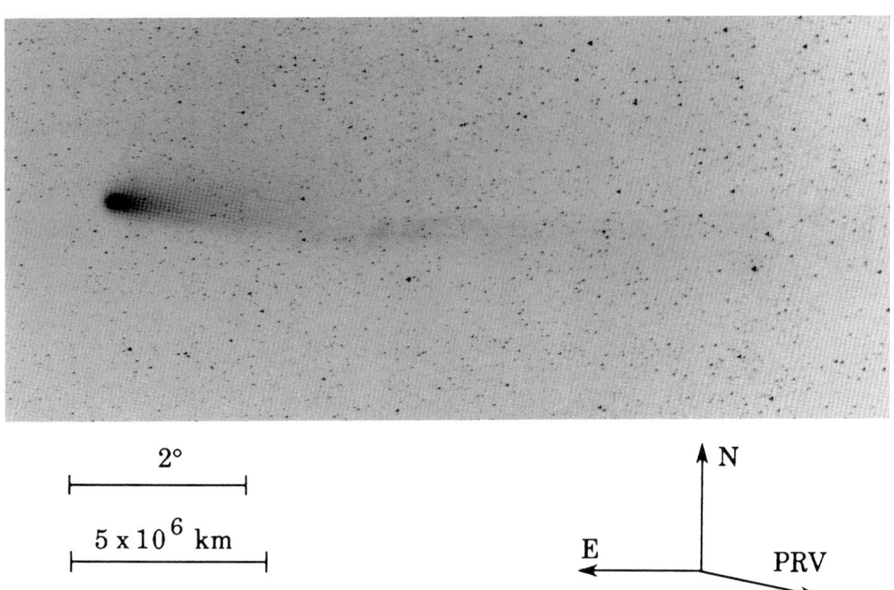

(top) Fig. 408: 1986 Mar 21.45868 UT; exposure 5.0 minutes on IIa-O emulsion with no filter; $r=1.02$, $\Delta=0.77$, $\beta=113.9°$; LSPN-902.
Photograph by E. Moore, Joint Observatory for Cometary Research.

(bottom) Fig. 409*: 1986 Mar 21.50000 UT; exposure on 103a-O emulsion; $r=1.02$, $\Delta=0.77$, $\beta=113.9°$.
Photograph by A. Butcher/D. Doss, McDonald Observatory.

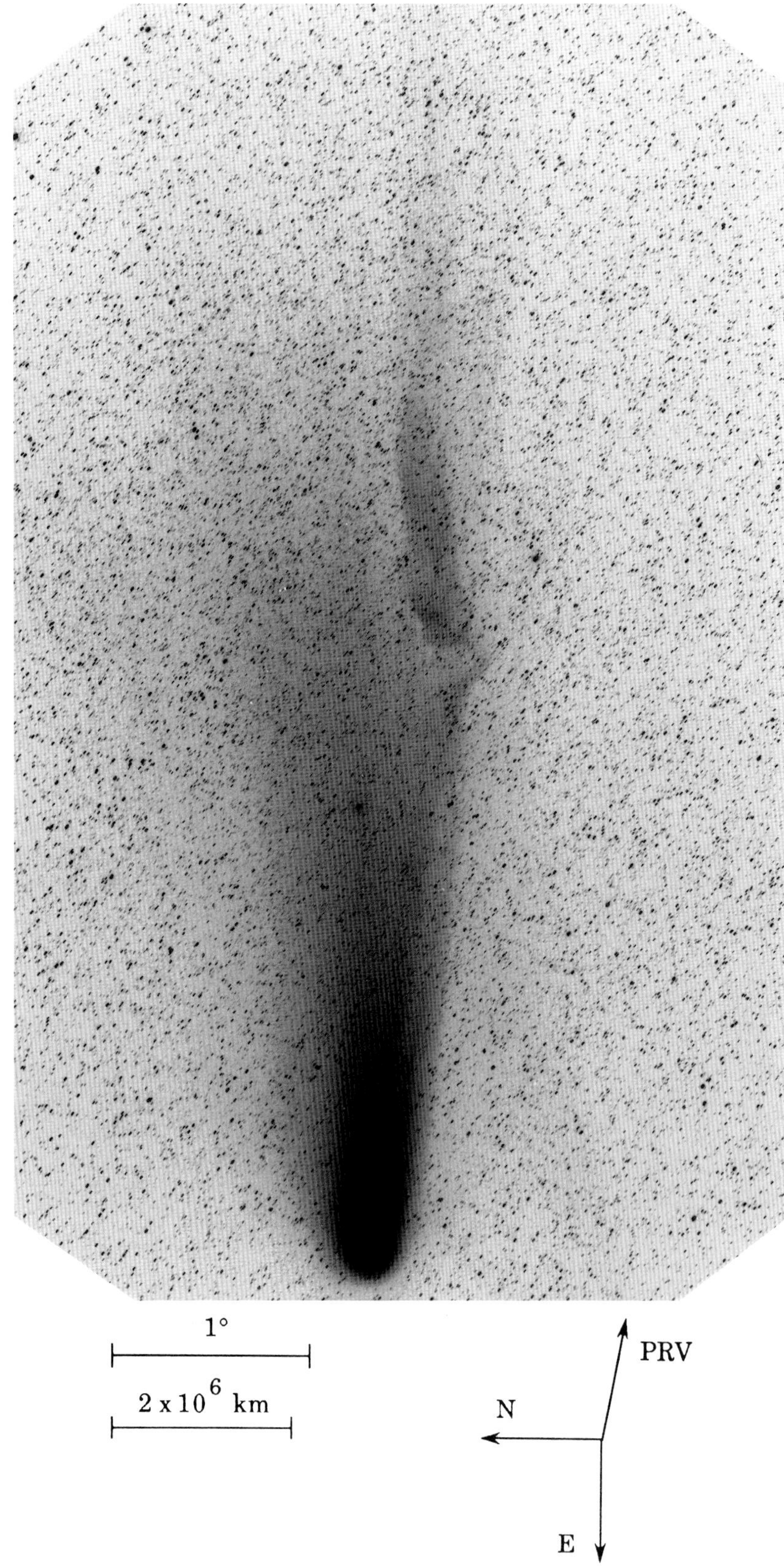

Fig. 410: 1986 Mar 21.49896 UT; exposure 16.0 minutes on Tech. Pan 2415 emulsion with no filter; r=1.02, Δ=0.77, β=113.9°; LSPN-1412.
Photograph by W. Liller, Easter Island, LSPN Island Network.

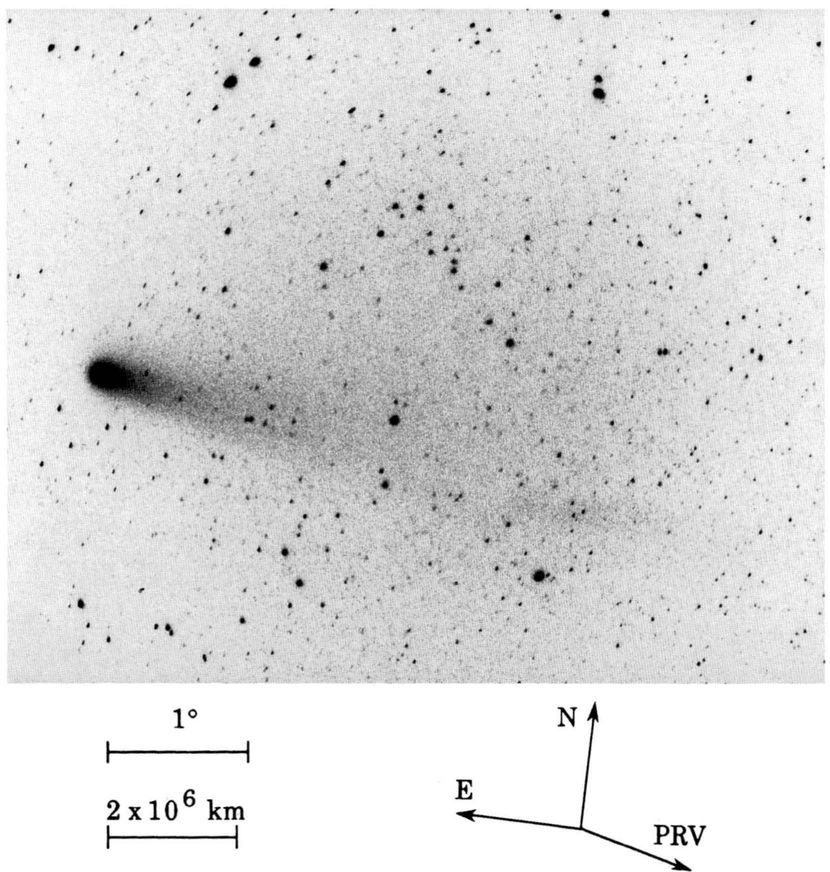

Fig. 411: 1986 Mar 21.52220 UT; exposure 10.0 minutes on 3M 1000 emulsion; r=1.02, Δ=0.77, β=113.9°; AON-851222.
Photograph by S. Edberg, USA.

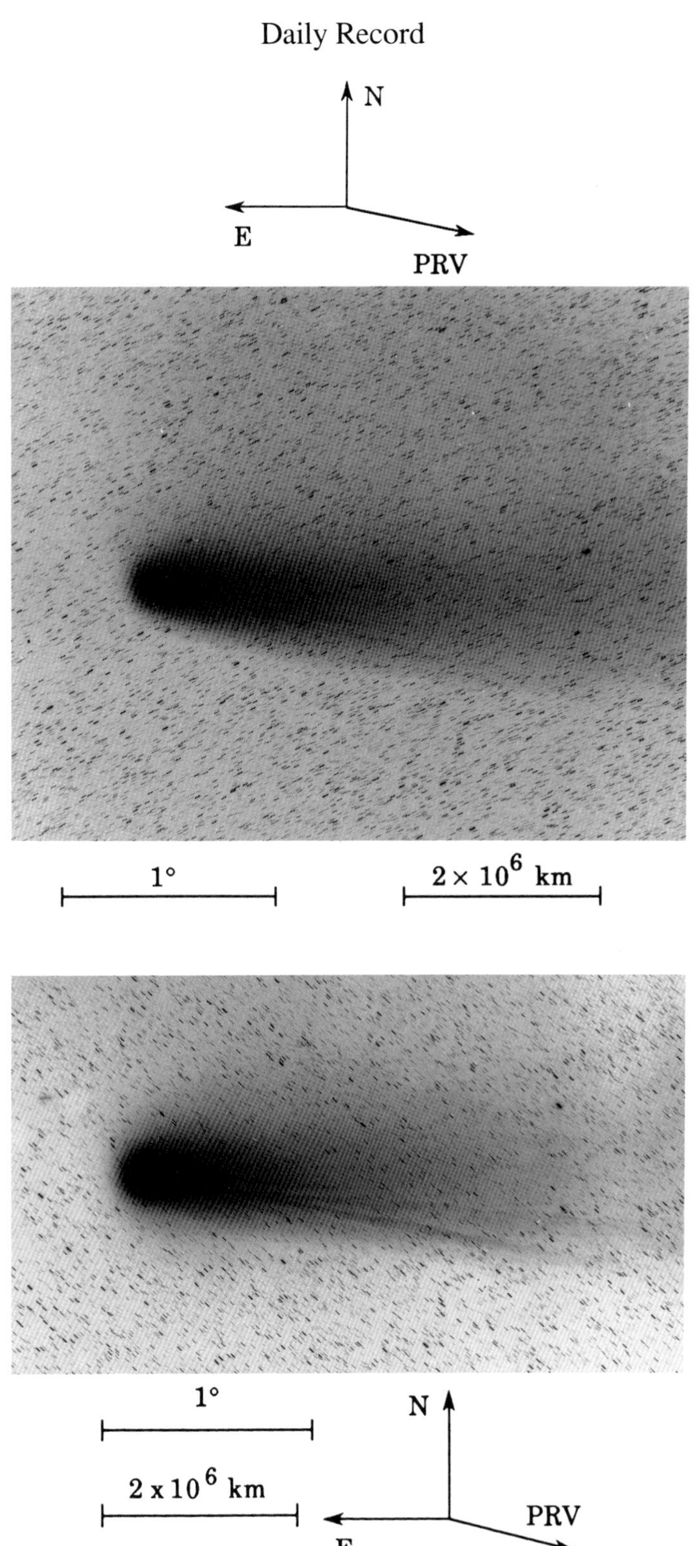

(top) Fig. 412: 1986 Mar 21.69315 UT; exposure 30.0 minutes on 098-04 emulsion with RG-610 filter; r=1.02, Δ=0.77, β=113.9°; LSPN-2125.
Photograph by P. Magnusson, Uppsala Southern Station.

(bottom) Fig. 413*: 1986 Mar 21.77083 UT; exposure 25.0 minutes on 103a-O emulsion with GG-385 filter; r=1.02, Δ=0.77, β=113.9°; LSPN-2128.
Photograph by P. Magnusson, Uppsala Southern Station.

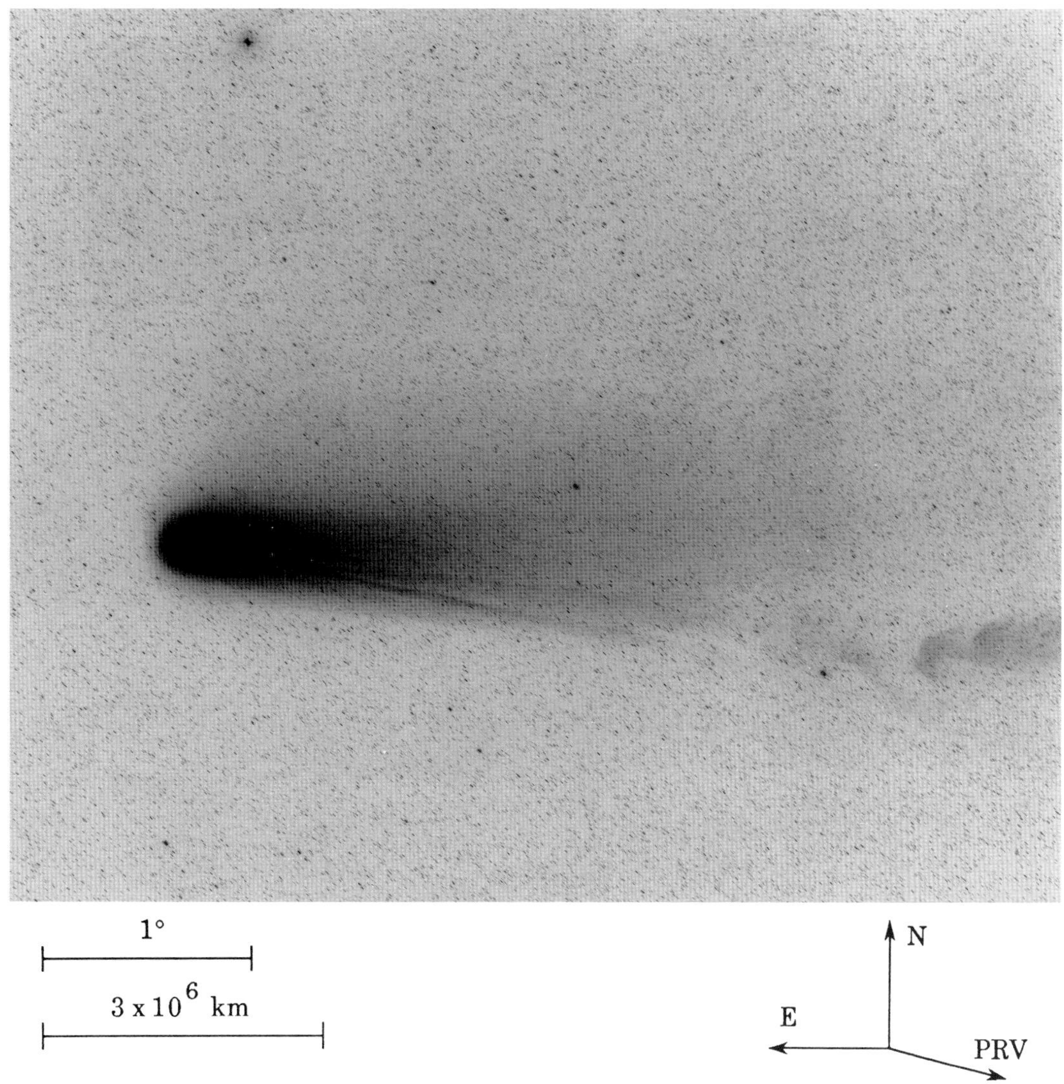

Fig. 414: 1986 Mar 21.76934 UT; exposure 20.0 minutes on IIa-O emulsion with GG-395 filter; r=1.02, Δ=0.77, β=113.9°; LSPN-2395.
Photograph from Royal Observatory/UK Schmidt Telescope Unit.

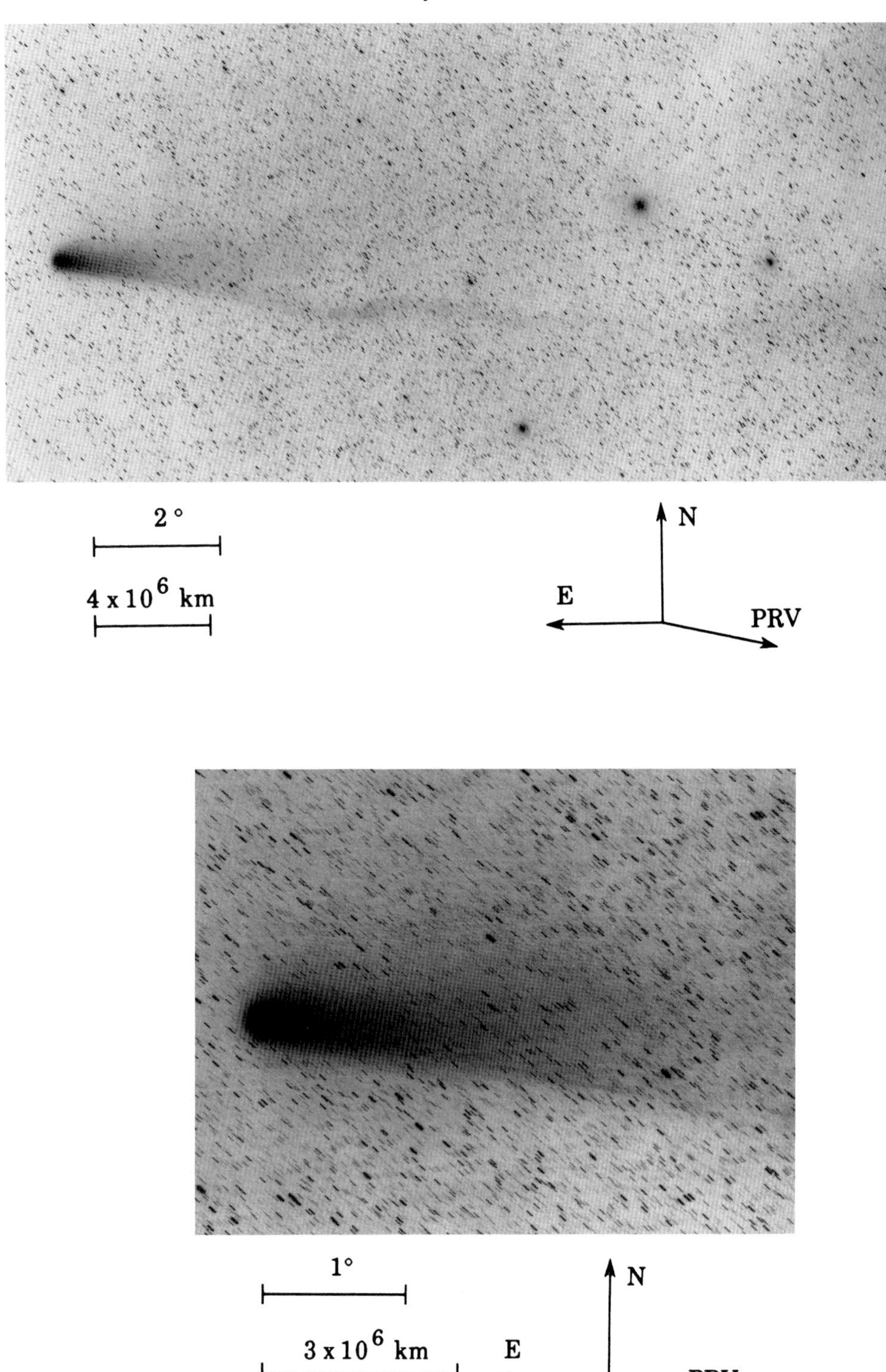

(top) Fig. 415: 1986 Mar 22.06424 UT; exposure 60.0 minutes on IIa-O emulsion with no filter; r=1.02, Δ=0.76, β=114.0°; LSPN-2854.
Photograph by G. Malcolm, Boyden Observatory.

(bottom) Fig. 416: 1986 Mar 22.06424 UT; exposure 60.0 minutes on IIa-O emulsion with no filter; r=1.02, Δ=0.76, β=114.0°; LSPN-2855.
Photograph by G. Malcolm, Boyden Observatory.

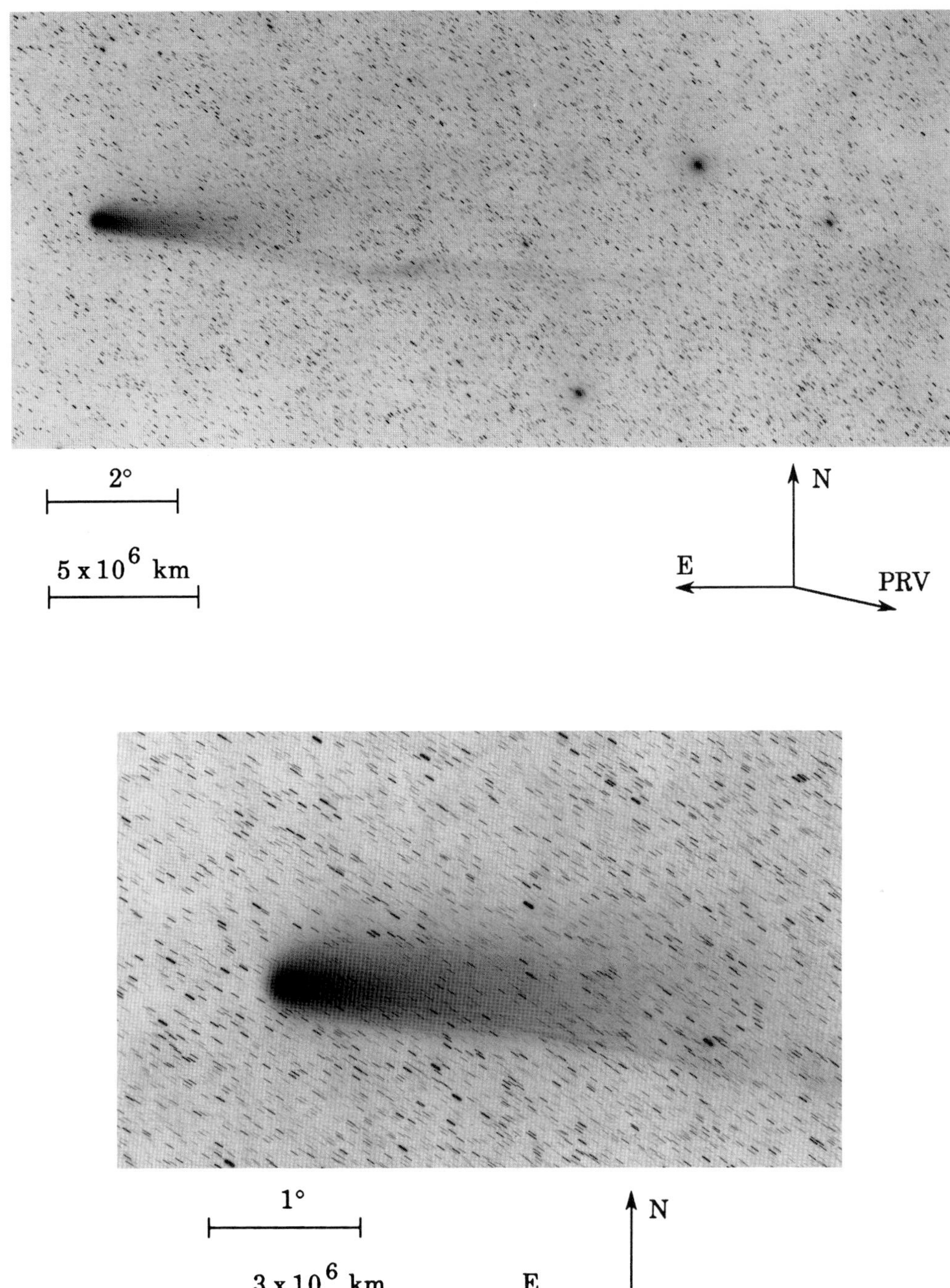

(top) Fig. 417: 1986 Mar 22.10868 UT; exposure 60.0 minutes on IIa-O emulsion with no filter; r=1.03, Δ=0.76, β=114.0°; LSPN-2856.
Photograph by G. Malcolm, Boyden Observatory.

(bottom) Fig. 418: 1986 Mar 22.10868 UT; exposure 60.0 minutes on IIa-O emulsion with no filter; r=1.03, Δ=0.76, β=114.0°; LSPN-2857.
Photograph by G. Malcolm, Boyden Observatory.

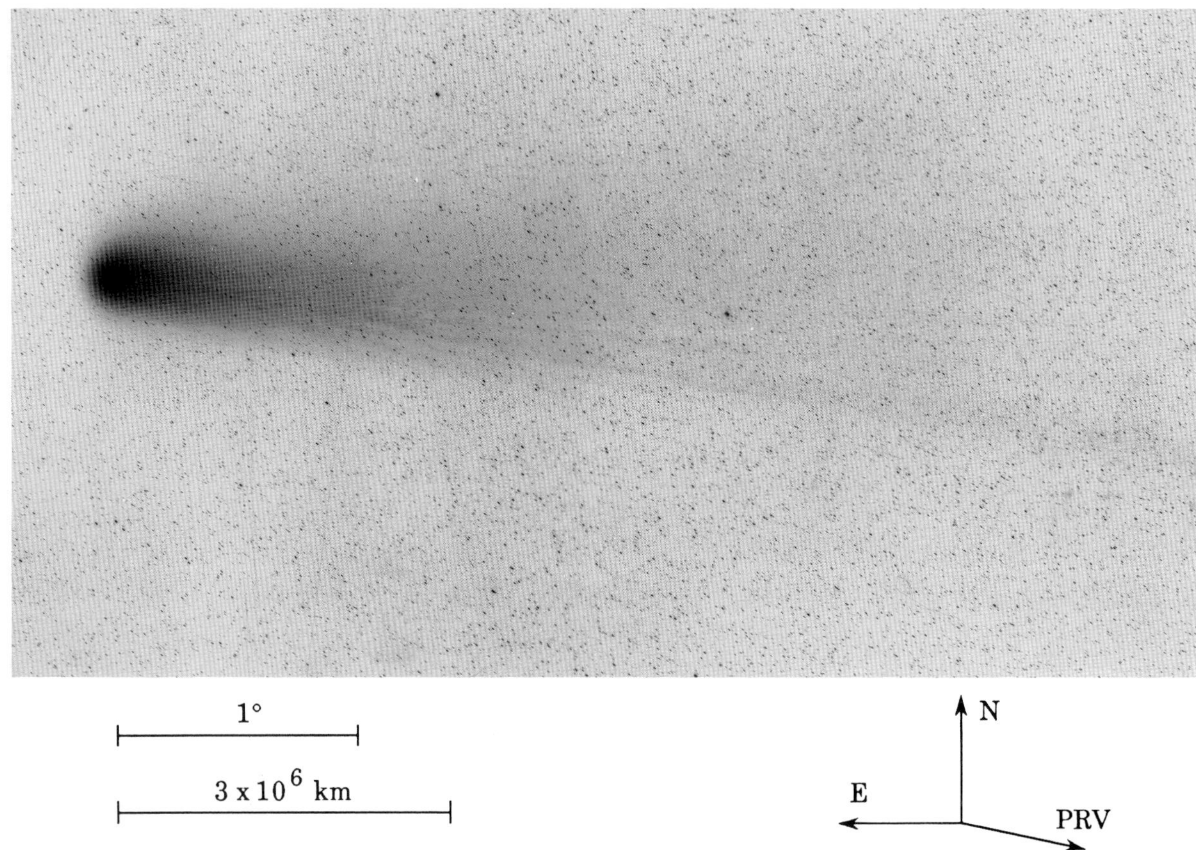

Fig. 419: 1986 Mar 22.32292 UT; exposure 10.0 minutes on IIa-O emulsion with GG-385 filter; r=1.03, Δ=0.75, β=114.0°; LSPN-2908.
Photograph by H. Schuster/O. Pizarro, European Southern Observatory.

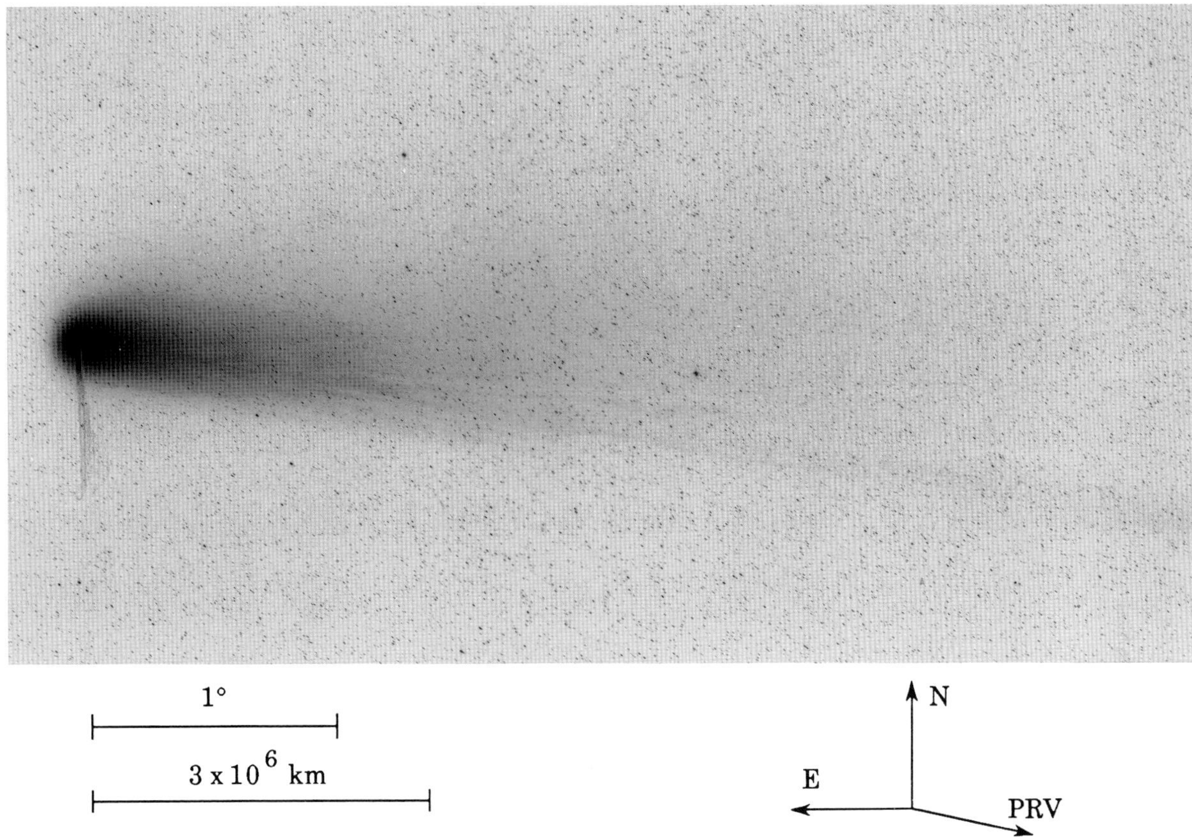

Fig. 420: 1986 Mar 22.34236 UT; exposure 10.0 minutes on IIa-O emulsion with GG-385 filter; r=1.03, Δ=0.75, β=114.0°; LSPN-2909.
Photograph by H. Schuster/O. Pizarro, European Southern Observatory.

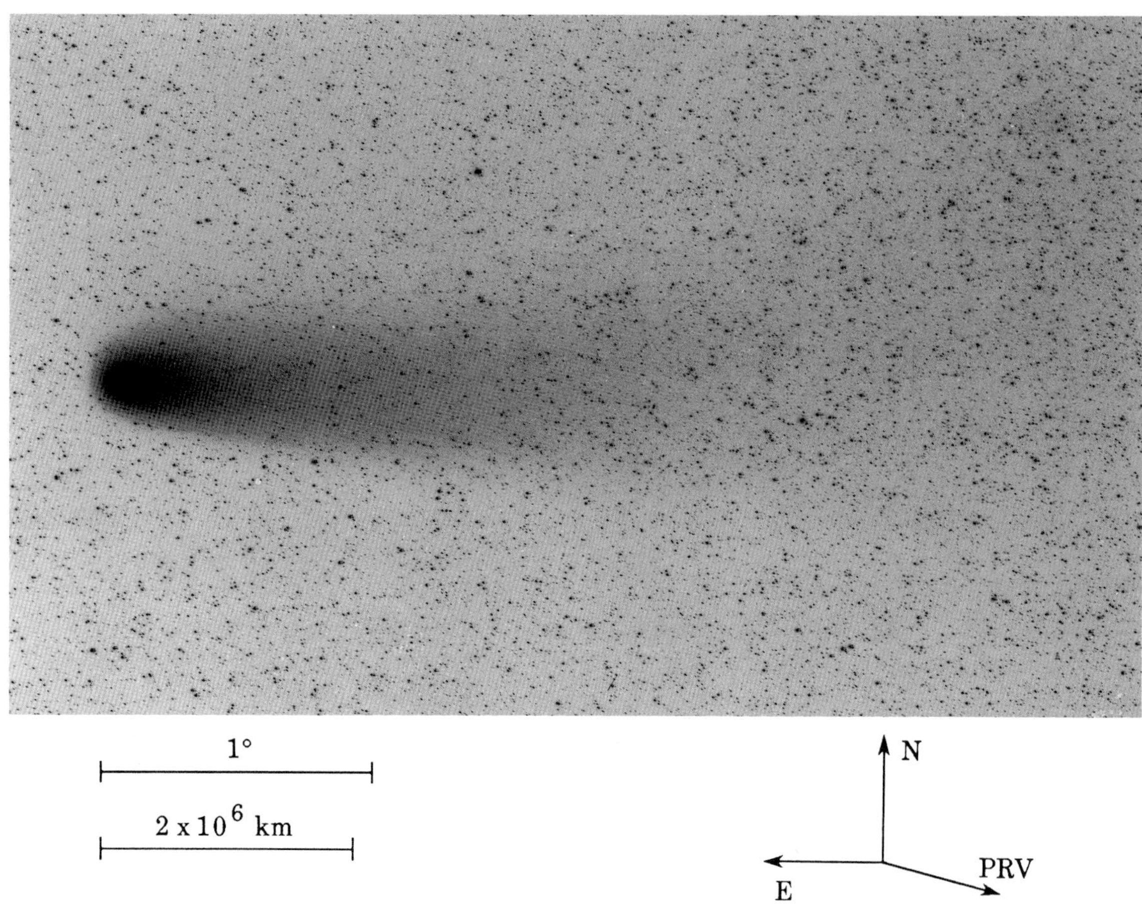

Fig. 421: 1986 Mar 22.34722 UT; exposure 20.0 minutes on 098-04 emulsion with RG-610 filter; r=1.03, Δ=0.75, β=114.0°; LSPN-1118.
Photograph by F. Miller, University of Michigan/CTIO.

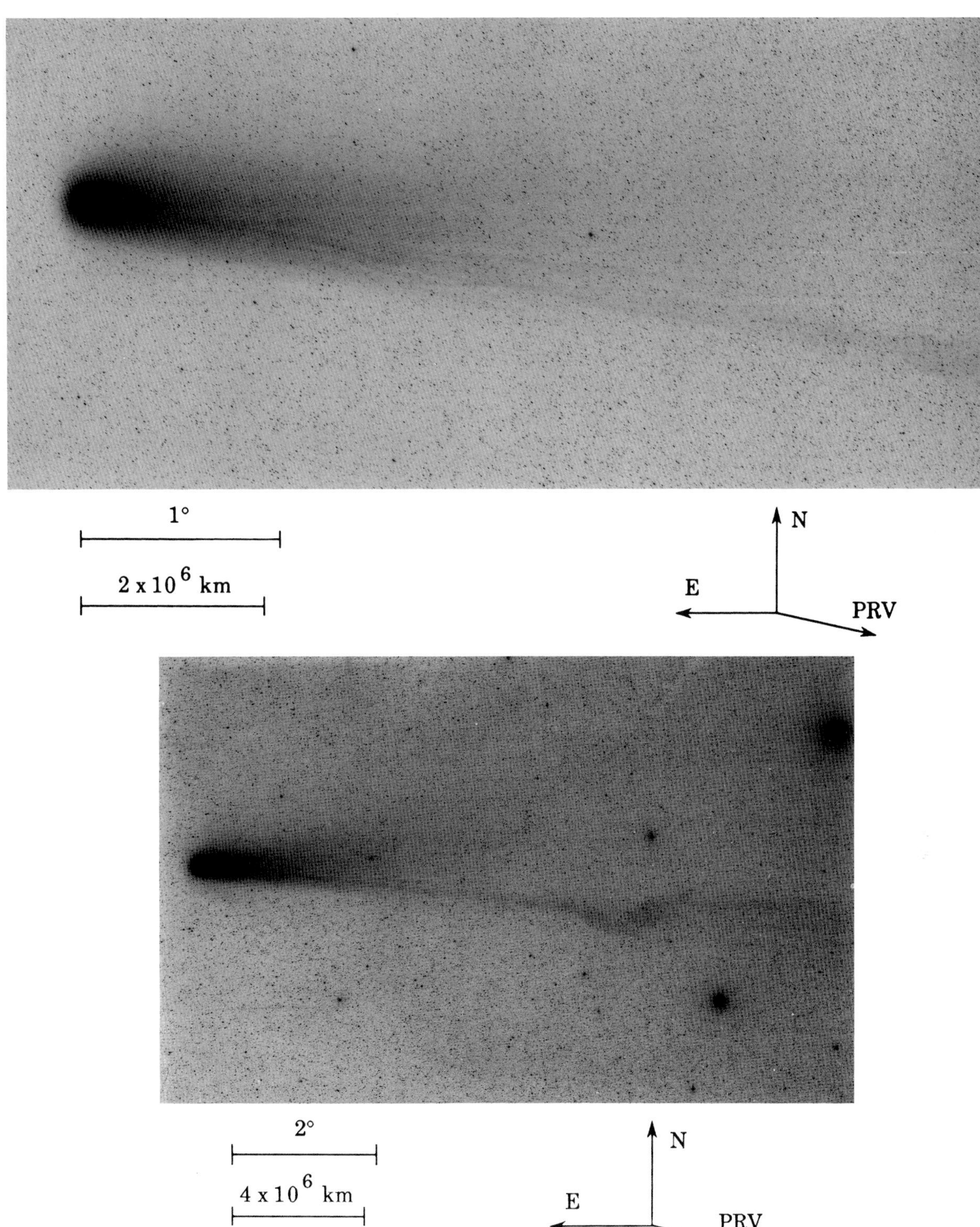

(top) Fig. 422: 1986 Mar 22.36042 UT; exposure 10.0 minutes on IIa-O emulsion with GG-385 filter; r=1.03, Δ=0.75, β=114.1°; LSPN-2910.
Photograph by H. Schuster/O. Pizarro, European Southern Observatory.

(bottom) Fig. 423*: 1986 Mar 22.48542 UT; exposure 10.0 minutes on IIa-O emulsion with no filter; r=1.03, Δ=0.75, β=114.1°; LSPN-907.
Photograph by E. Moore, Joint Observatory for Cometary Research.

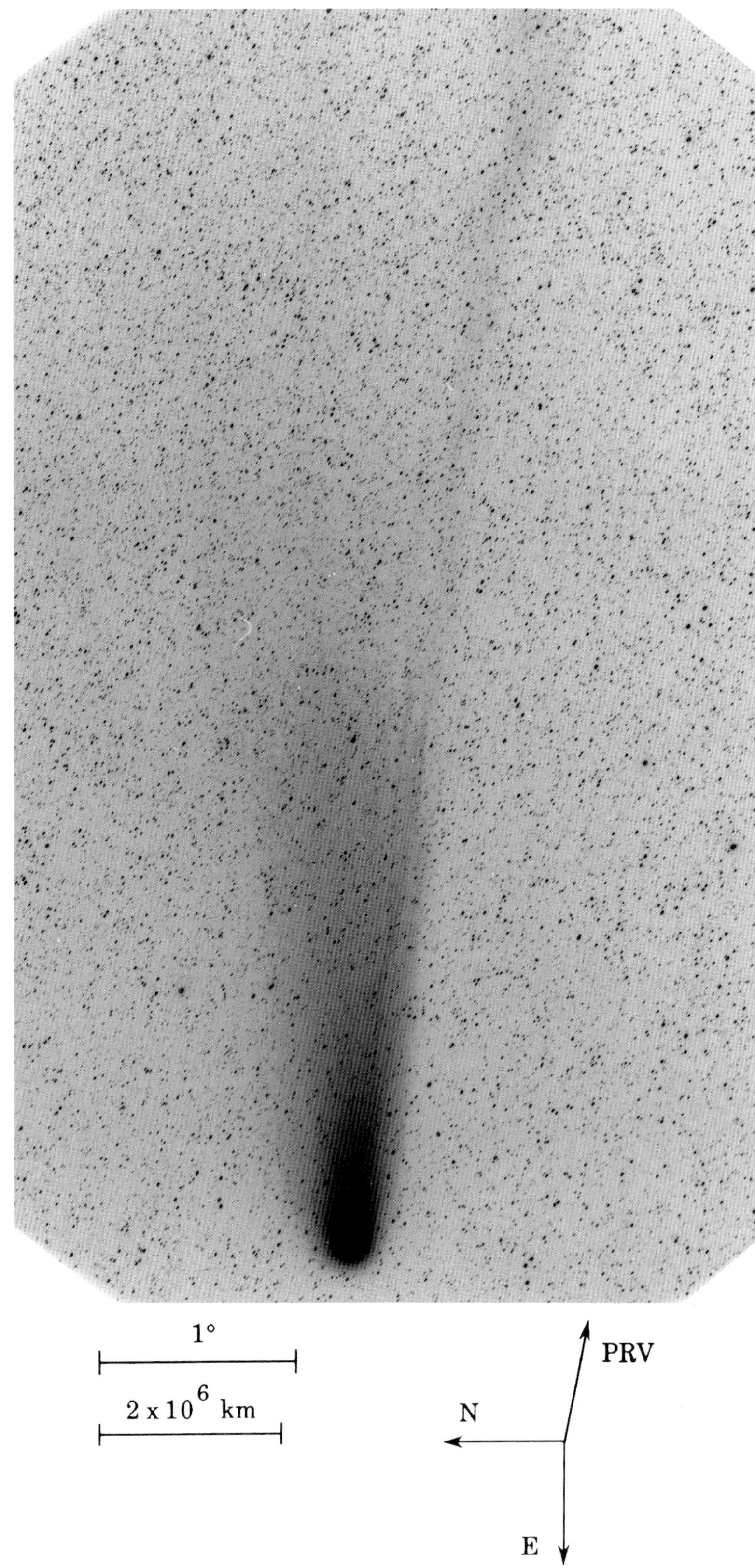

Fig. 424: 1986 Mar 22.44410 UT; exposure 15.0 minutes on Tech. Pan 2415 emulsion with no filter; r=1.03, Δ=0.75, β=114.1°; LSPN-1413.
Photograph by W. Liller, Easter Island, LSPN Island Network.

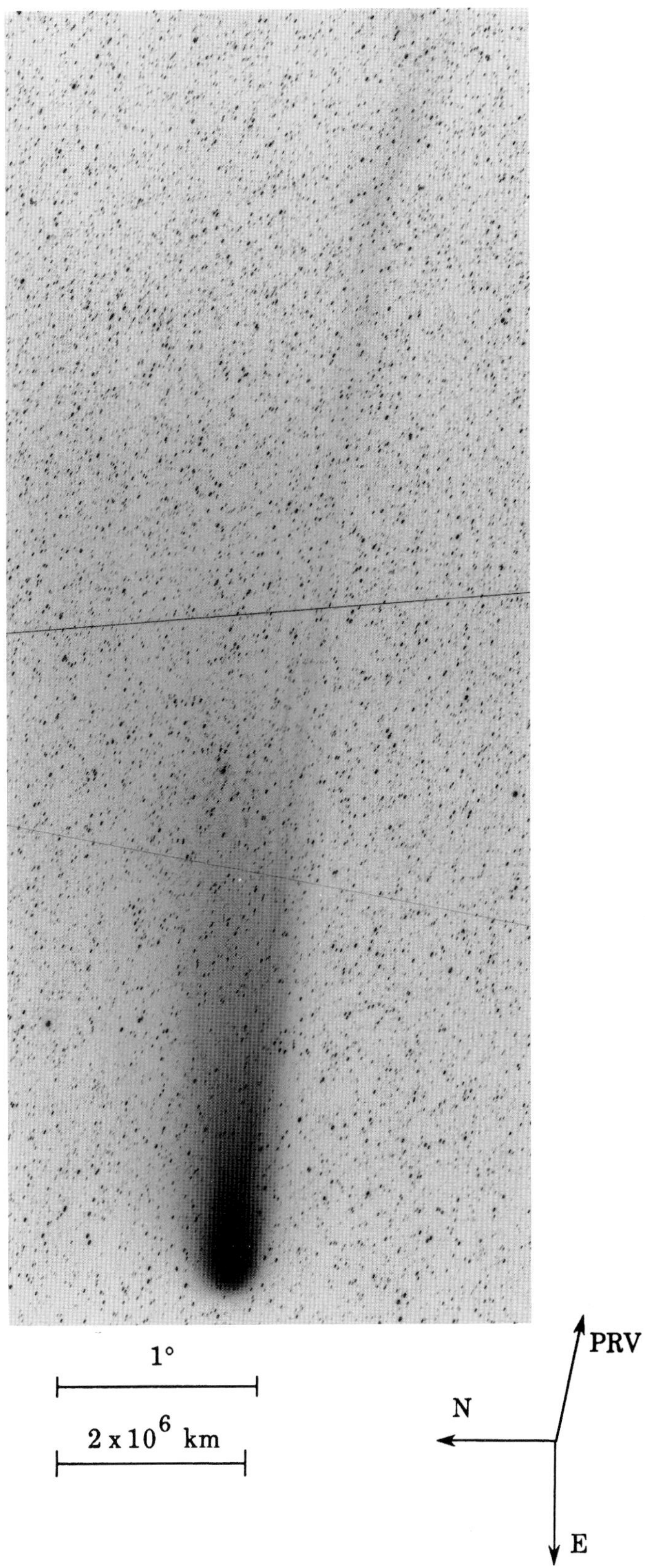

Fig. 425: 1986 Mar 22.50035 UT; exposure 15.0 minutes on Tech. Pan 2415 emulsion with no filter; r=1.03, Δ=0.75, β=114.1°; LSPN-1416.
Photograph by W. Liller,Easter Island, LSPN Island Network.

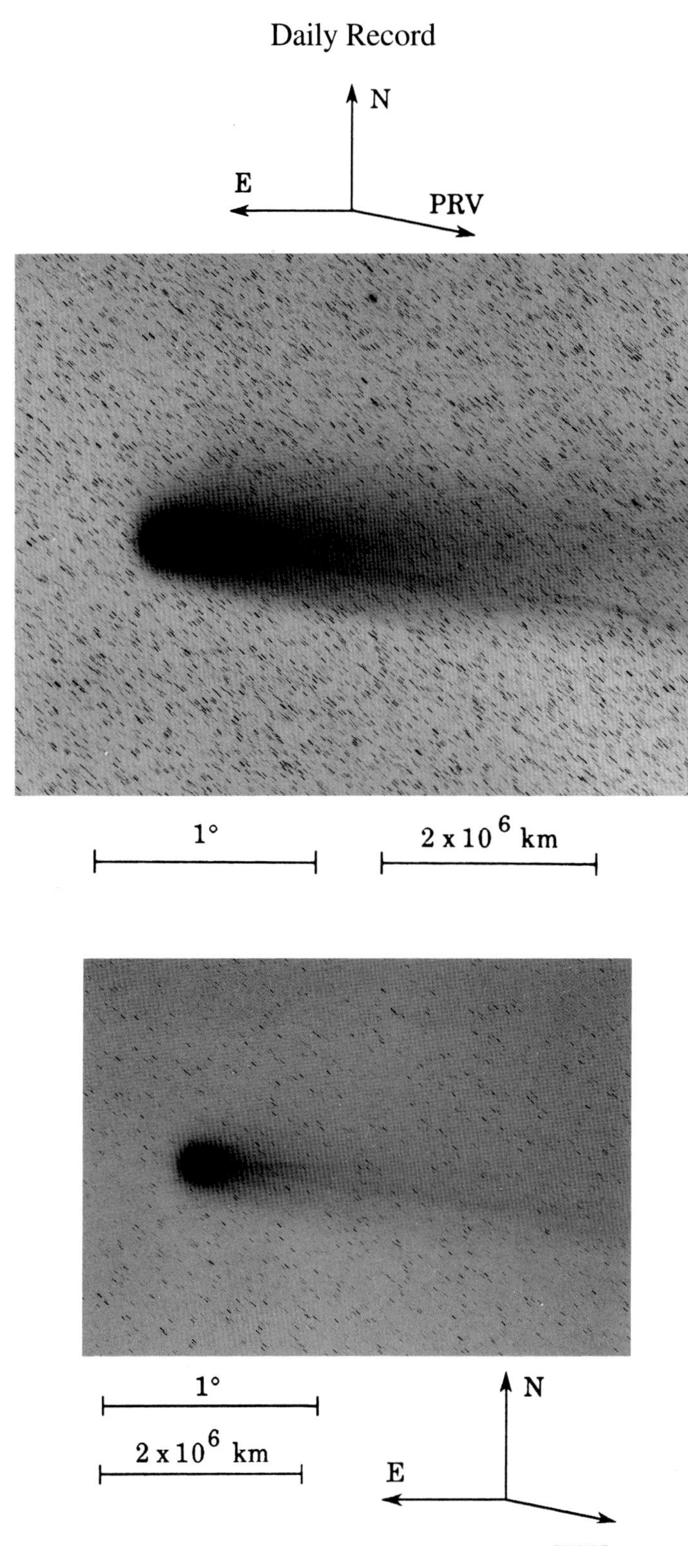

(top) Fig. 426: 1986 Mar 22.72851 UT; exposure 30.0 minutes on IIa-D emulsion with GG-495 filter; r=1.03, Δ=0.74, β=114.1°; LSPN-2130.
Photograph by P. Magnusson, Uppsala Southern Station.

(bottom) Fig. 427*: 1986 Mar 22.78080 UT; exposure 25.0 minutes on IIIa-J emulsion with GG-385 filter; r=1.04, Δ=0.74, β=114.1°; LSPN-2122.
Photograph by P. Magnusson, Uppsala Southern Station.

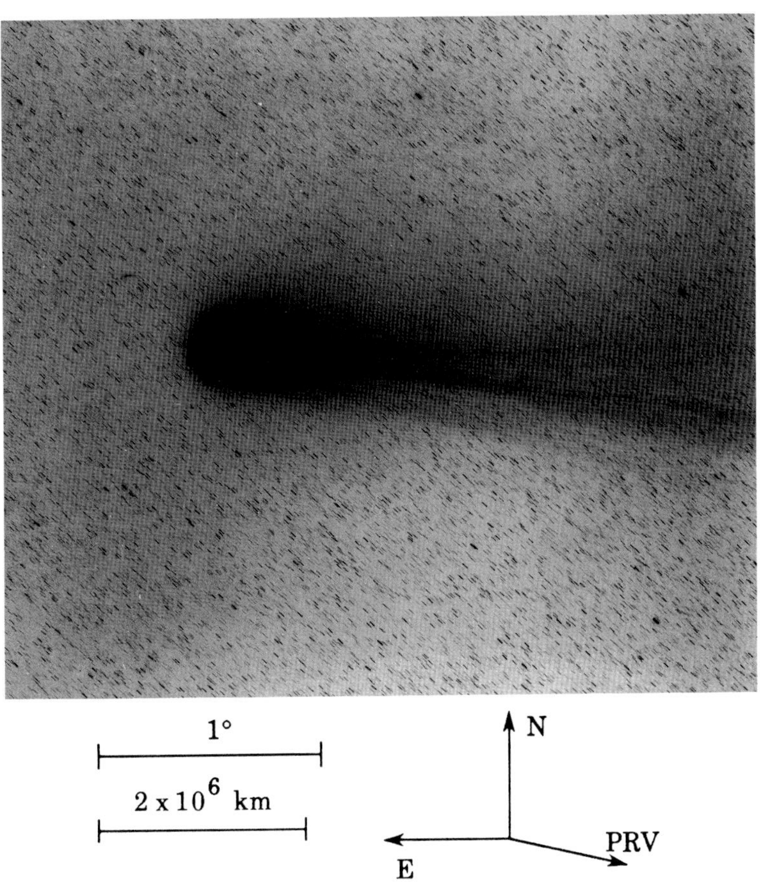

Fig. 428: 1986 Mar 22.75760 UT; exposure 30.0 minutes on IIa-O emulsion with GG-385 filter; r=1.04, Δ=0.74, β=114.1°; LSPN-2121.
Photograph by P. Magnusson, Uppsala Southern Station.

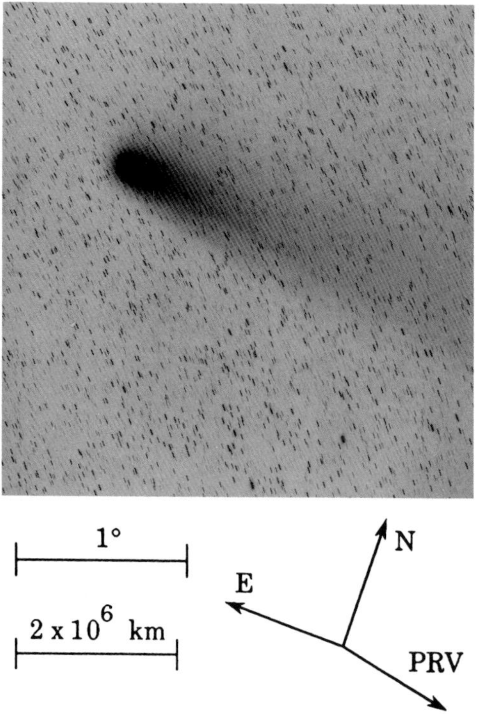

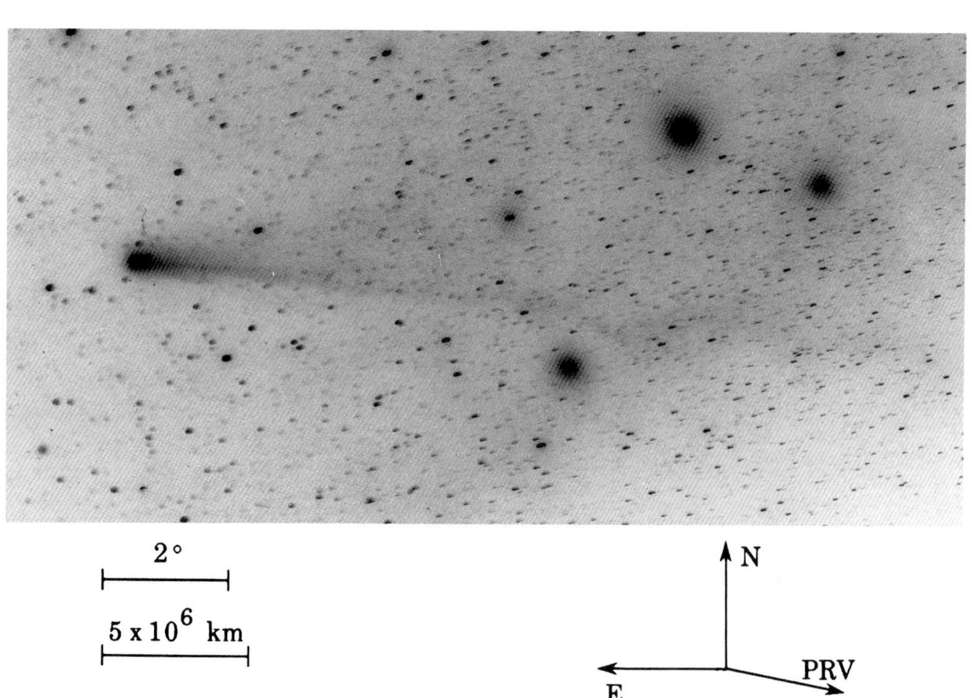

(top) Fig. 429: 1986 Mar 22.99028 UT; exposure 34.0 minutes on 103a-F emulsion with Red filter; r=1.04, Δ=0.74, β=114.2°; LSPN-3674.
Photograph submitted by K. Sivaraman, Indian Institute for Astrophysics, Kavalur Station.

(bottom) Fig. 430: 1986 Mar 23.00000 UT; exposure 75.0 minutes on 103a-O emulsion with no filter; r=1.04, Δ=0.74, β=114.2°; LSPN-3653.
Photograph submitted by K. Sivaraman, Indian Institute for Astrophysics, Kodaikanal Station.

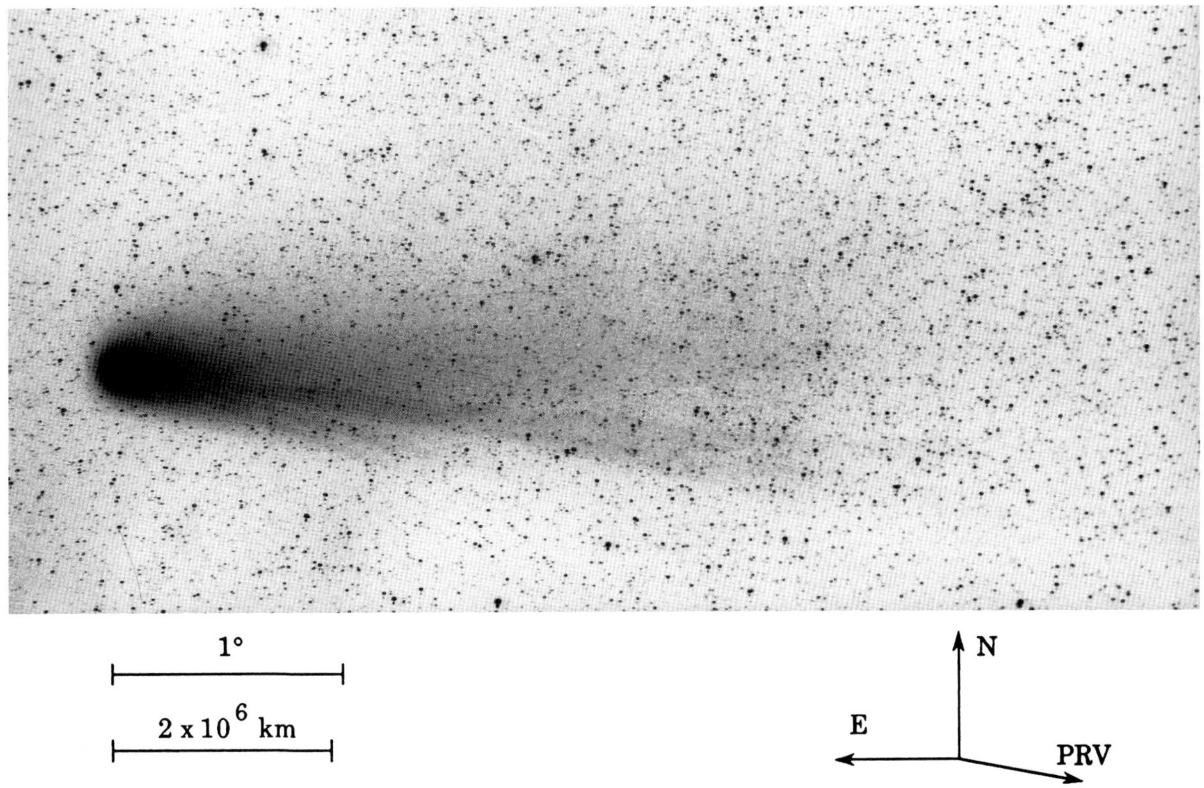

Fig. 431: 1986 Mar 23.02431 UT; exposure 5.0 minutes on Tech. Pan 2415 emulsion; r=1.04, Δ=0.73, β=114.2°. *Photograph by A. Marafie, Wafra Observatory.*

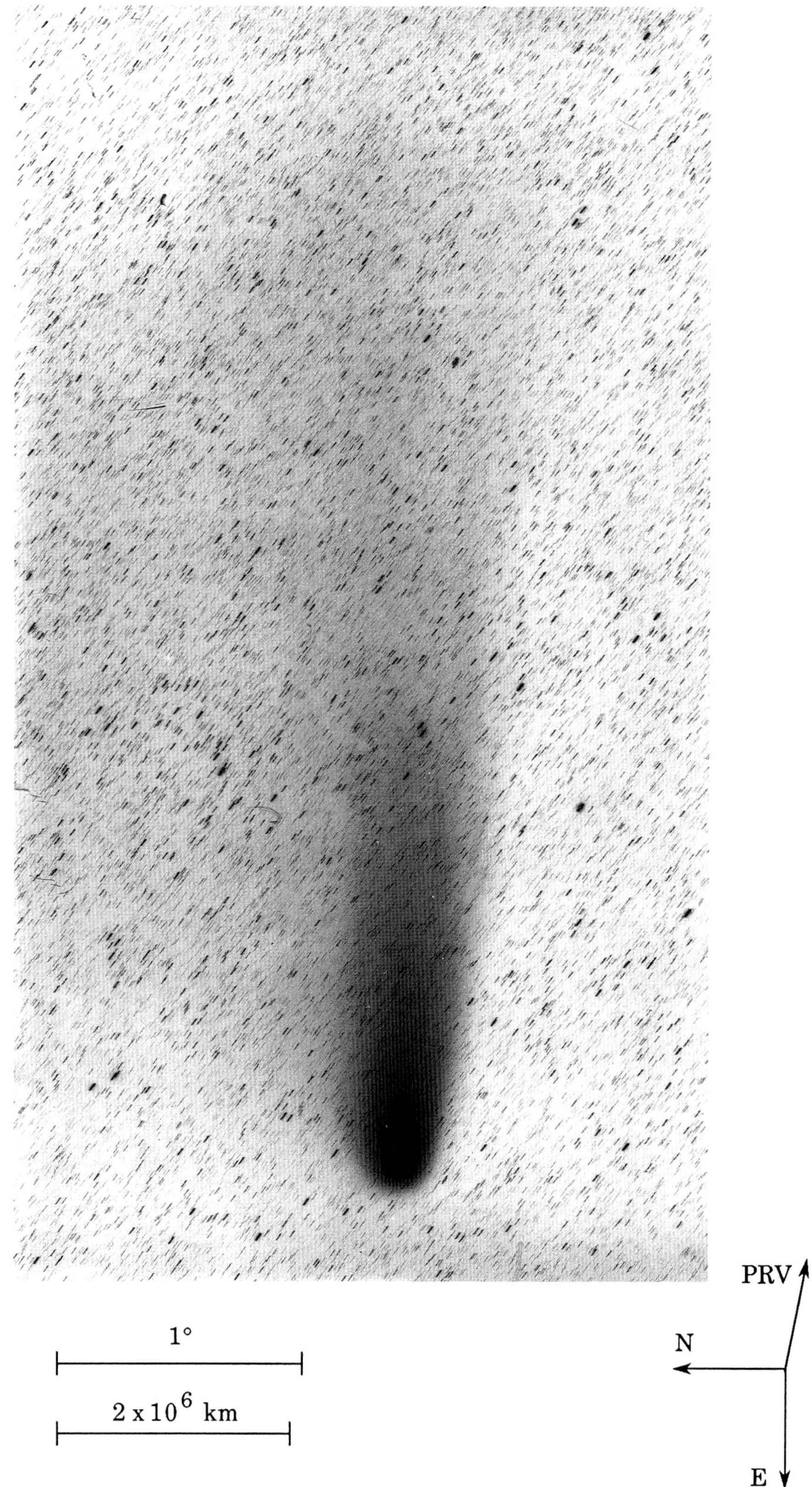

Fig. 432: 1986 Mar 23.32639 UT; exposure 30.0 minutes on 098-04 emulsion with RG-610 filter; r=1.04, Δ=0.73, β=114.3°; LSPN-1121.
Photograph by F. Miller, University of Michigan/CTIO.

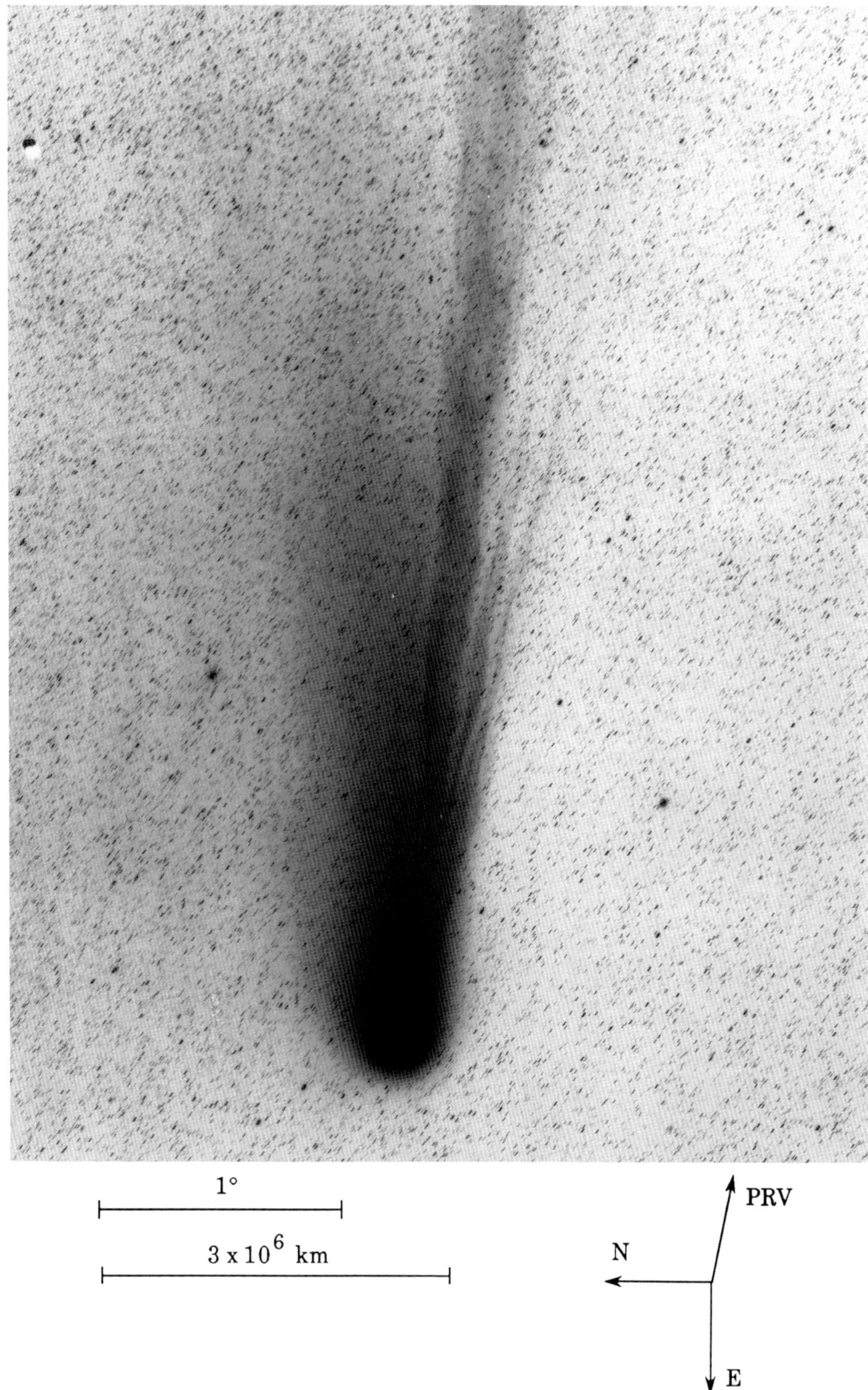

Fig. 433: 1986 Mar 23.35347 UT; exposure 20.0 minutes on IIa-O emulsion with no filter; r=1.04, Δ=0.73, β=114.3°; LSPN-1122.
Photograph by F. Miller, University of Michigan/CTIO.

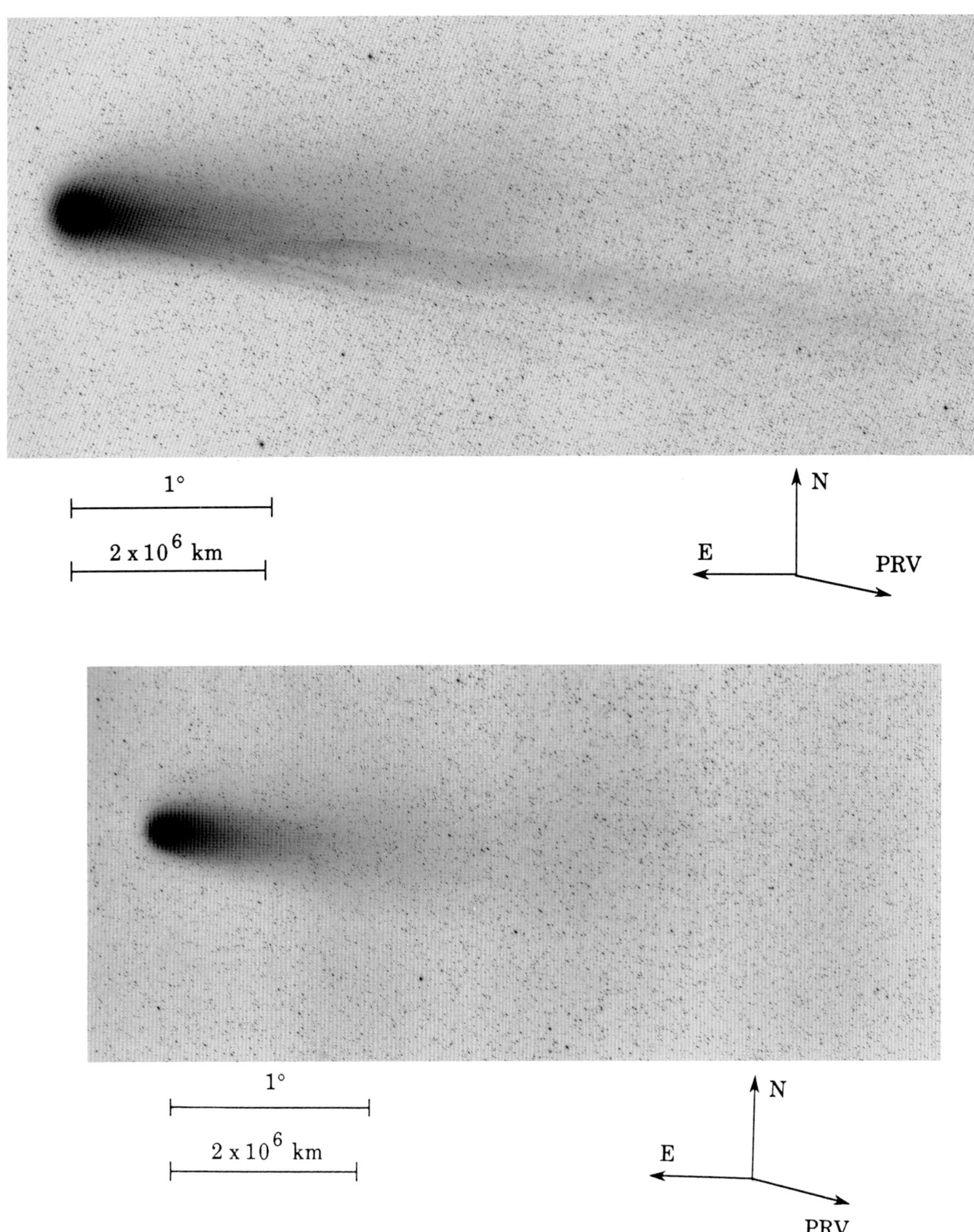

(top) Fig. 434: 1986 Mar 23.35833 UT; exposure 10.0 minutes on IIa-O emulsion with GG-385 filter; r=1.04, Δ=0.73, β=114.3°; LSPN-2911.
Photograph by H. Schuster/O. Pizarro, European Southern Observatory.

(bottom) Fig. 435: 1986 Mar 23.37708 UT; exposure 10.0 minutes on 098-04 emulsion with RG-630 filter; r=1.04, Δ=0.73, β=114.3°; LSPN-2912.
Photograph by H. Schuster/O. Pizarro, European Southern Observatory.

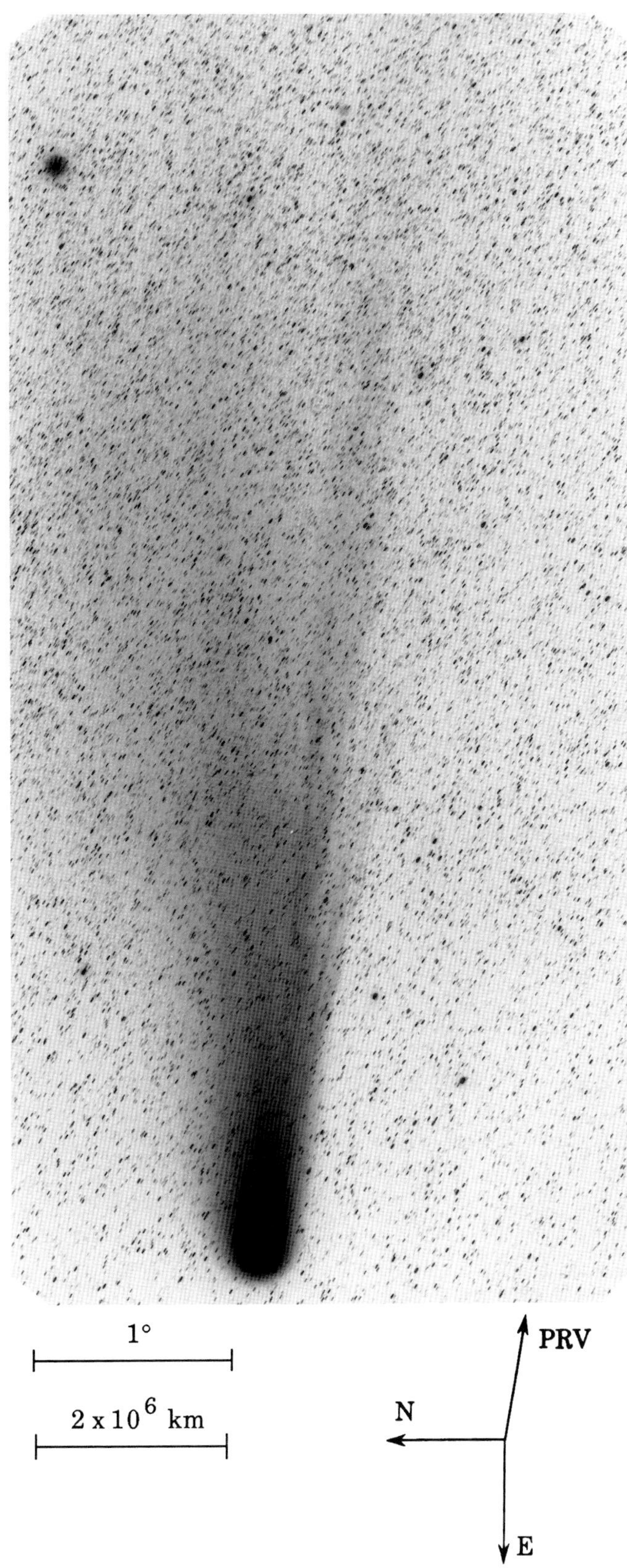

Fig. 436: 1986 Mar 23.47049 UT; exposure 20.0 minutes on Tech. Pan 2415 emulsion with no filter; r=1.05, Δ=0.72, β=114.3°; LSPN-1418.
Photograph by W. Liller, Easter Island, LSPN Island Network.

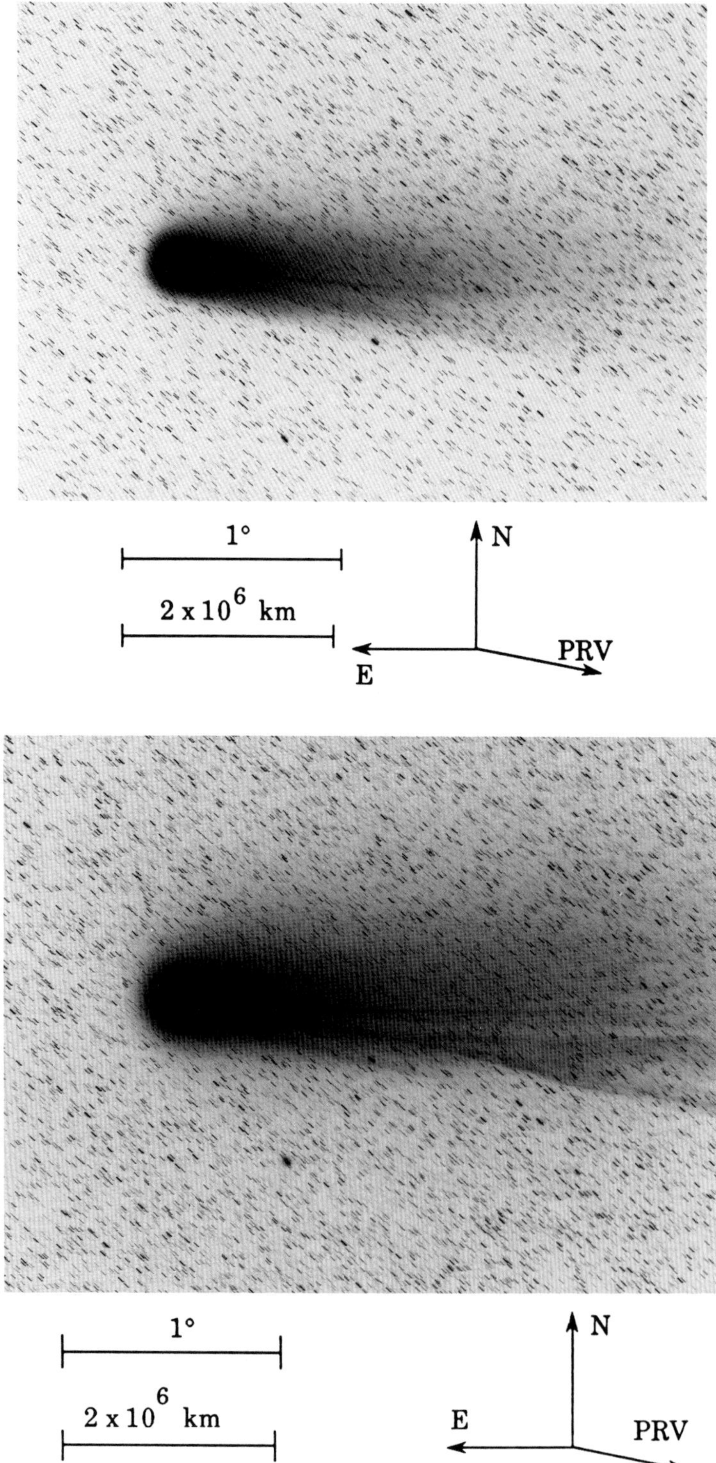

(top) Fig. 437: 1986 Mar 23.74575 UT; exposure 30.0 minutes on IIa-D emulsion with GG-495 filter; r=1.05, Δ=0.72, β=114.4°; LSPN-2131.
Photograph by P. Magnusson, Uppsala Southern Station.

(bottom) Fig. 438: 1986 Mar 23.77304 UT; exposure 30.0 minutes on IIa-O emulsion with GG-385 filter; r=1.05, Δ=0.72, β=114.5°; LSPN-2123.
Photograph by P. Magnusson, Uppsala Southern Station.

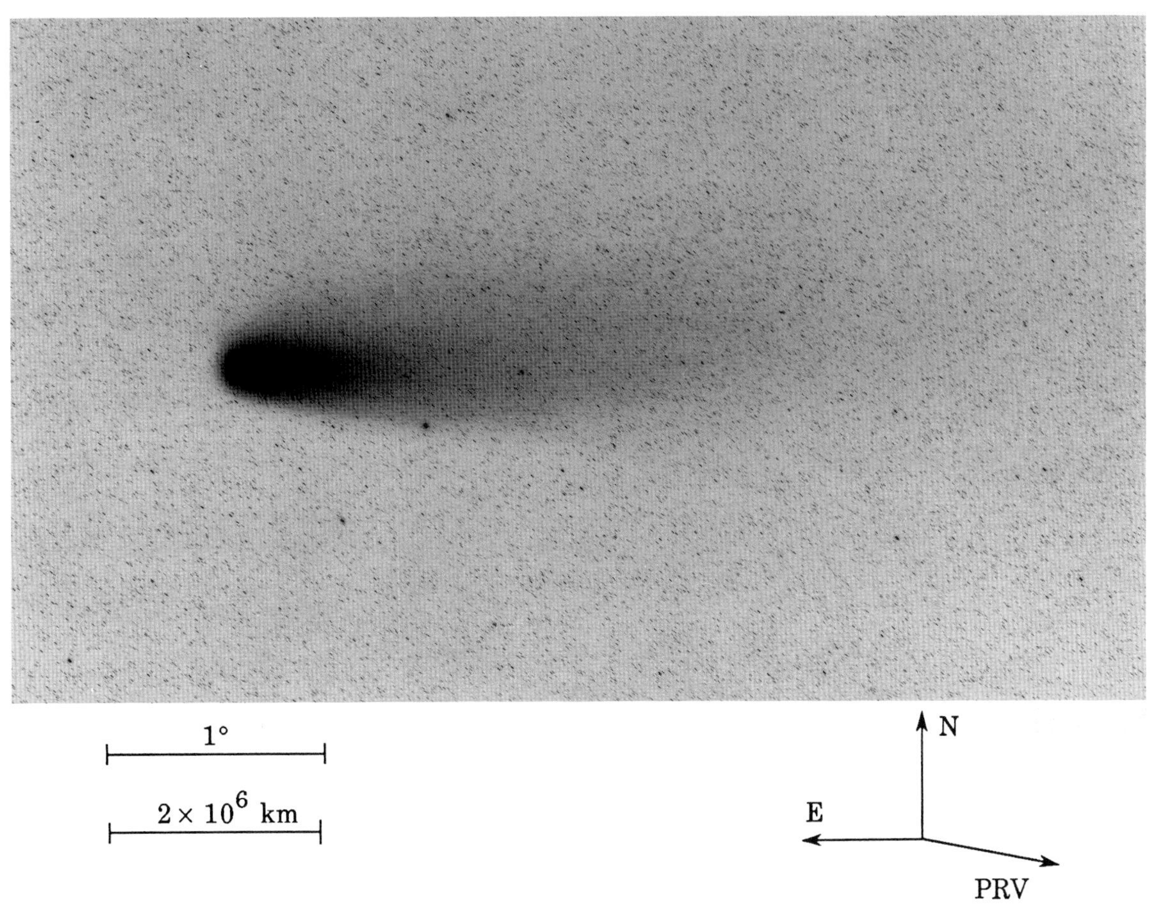

Fig. 439: 1986 Mar 23.79123 UT; exposure 15.0 minutes on IIIa-F emulsion with RG-630 filter; r=1.05, Δ=0.72, β=114.5°; LSPN-2398.
Photograph from Royal Observatory/UK Schmidt Telescope Unit.

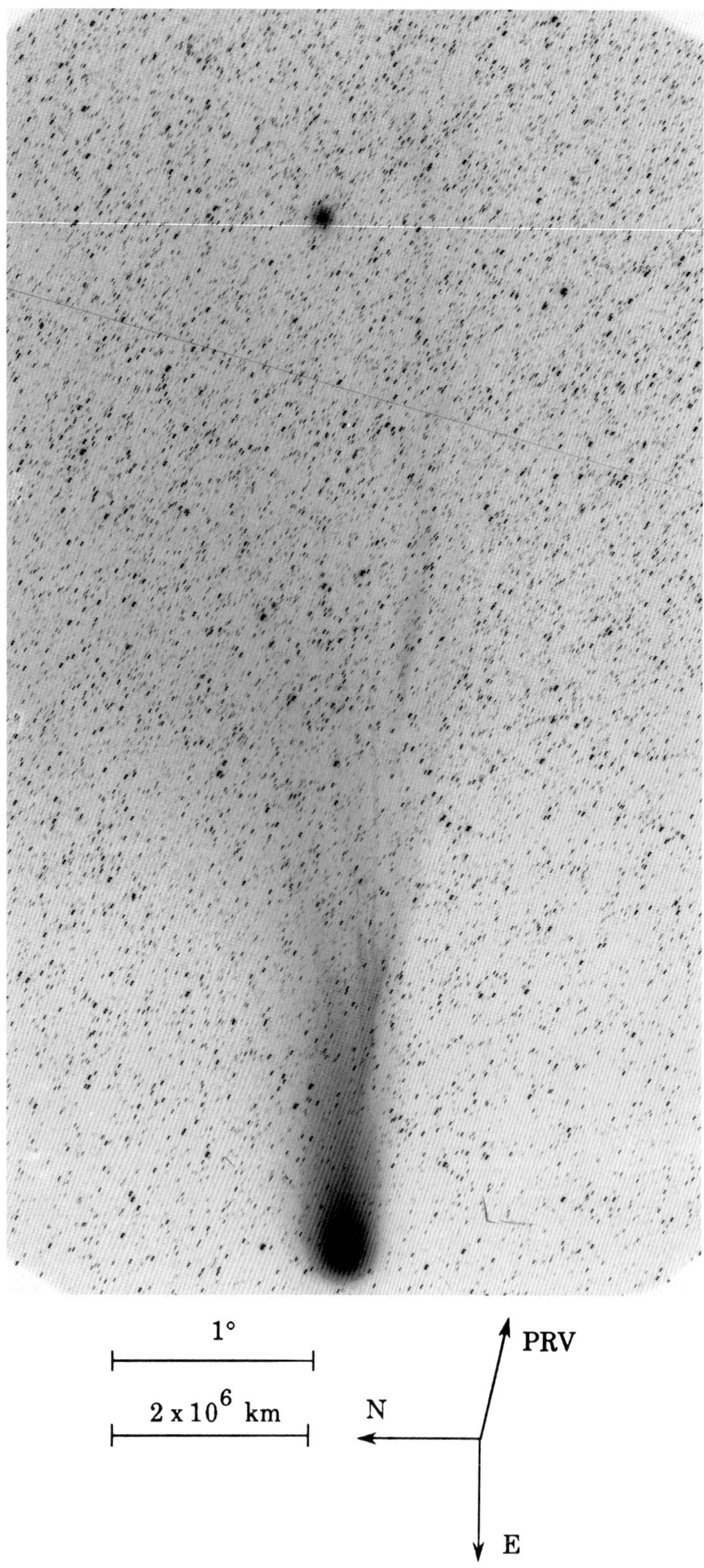

Fig. 440: 1986 Mar 24.50174 UT; exposure 19.0 minutes on Tech. Pan 2415 emulsion with no filter; r=1.06, Δ=0.70, β=114.8°; LSPN-1421.
Photograph by W. Liller, Easter Island, LSPN Island Network.

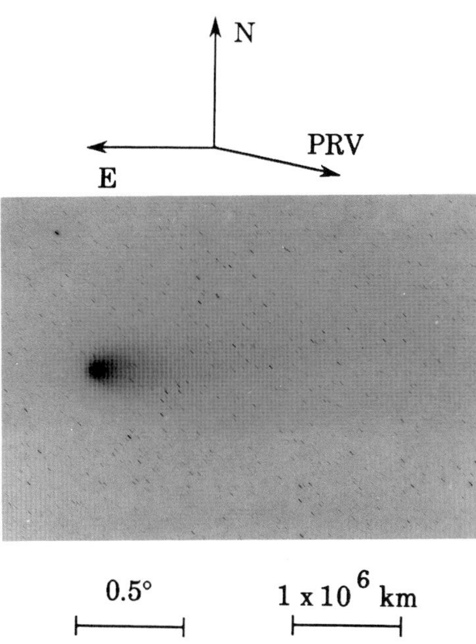

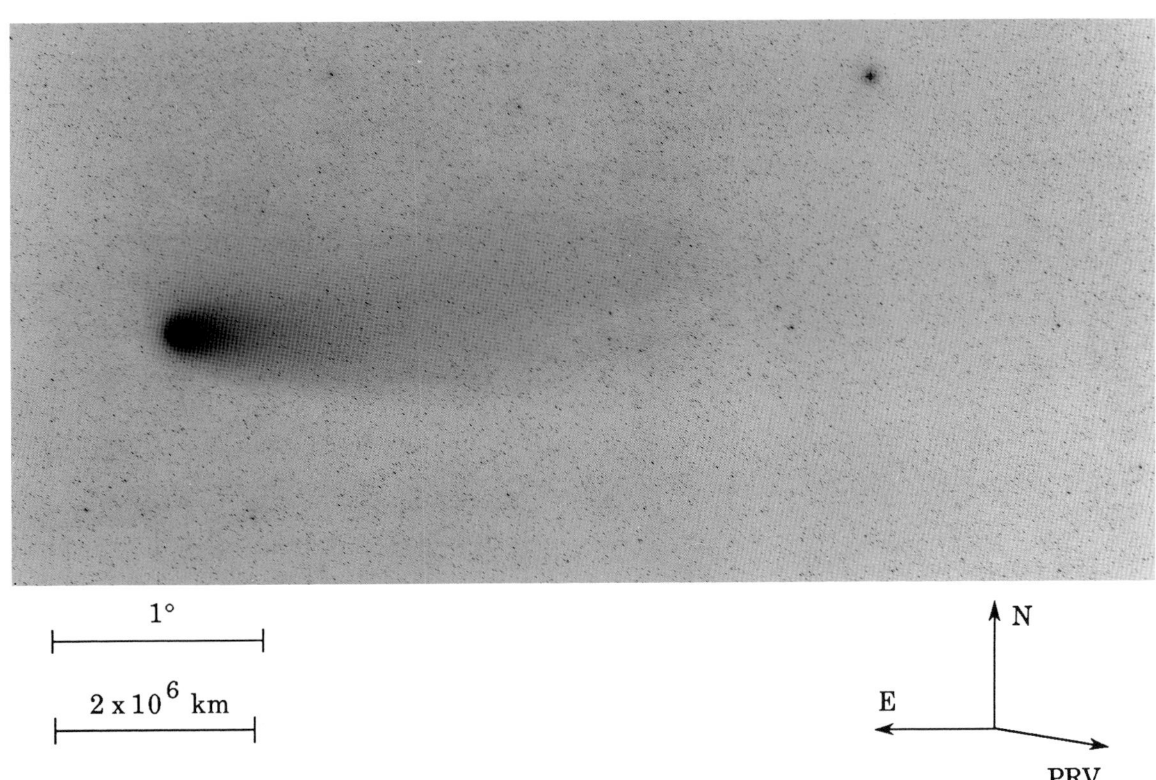

(top) Fig. 441: 1986 Mar 24.72306 UT; exposure 20.0 minutes on IIIa-F emulsion with AAO-643 filter; $r=1.07$, $\Delta=0.69$, $\beta=114.9°$; LSPN-2399.
Photograph from Royal Observatory/UK Schmidt Telescope Unit.

(bottom) Fig. 442*: 1986 Mar 25.77711 UT; exposure 10.0 minutes on IIIa-F emulsion with RG-630 filter; $r=1.08$, $\Delta=0.67$, $\beta=115.4°$; LSPN-2402.
Photograph from Royal Observatory/UK Schmidt Telescope Unit.

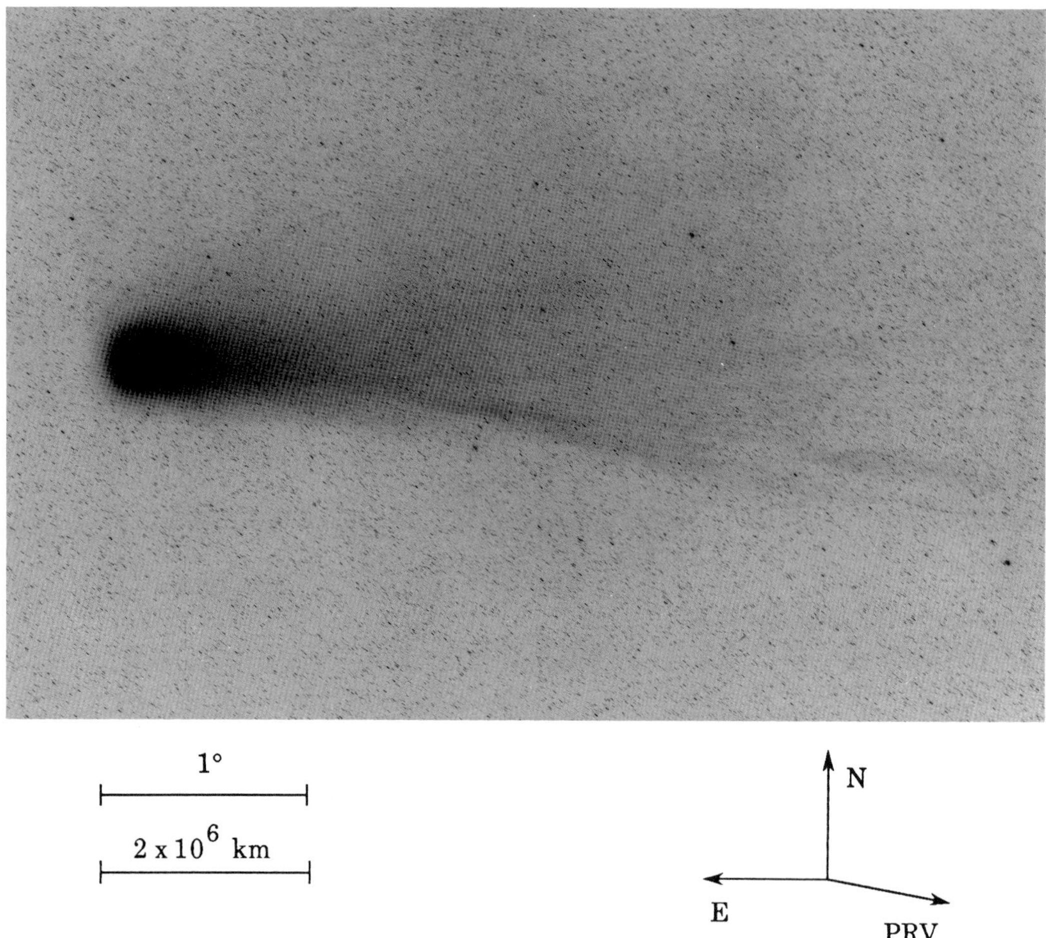

Fig. 443: 1986 Mar 24.78296 UT; exposure 15.0 minutes on IIa-O emulsion with GG-395 filter; r=1.07, Δ=0.69, β=114.9°; LSPN-2400.
Photograph from Royal Observatory/UK Schmidt Telescope Unit.

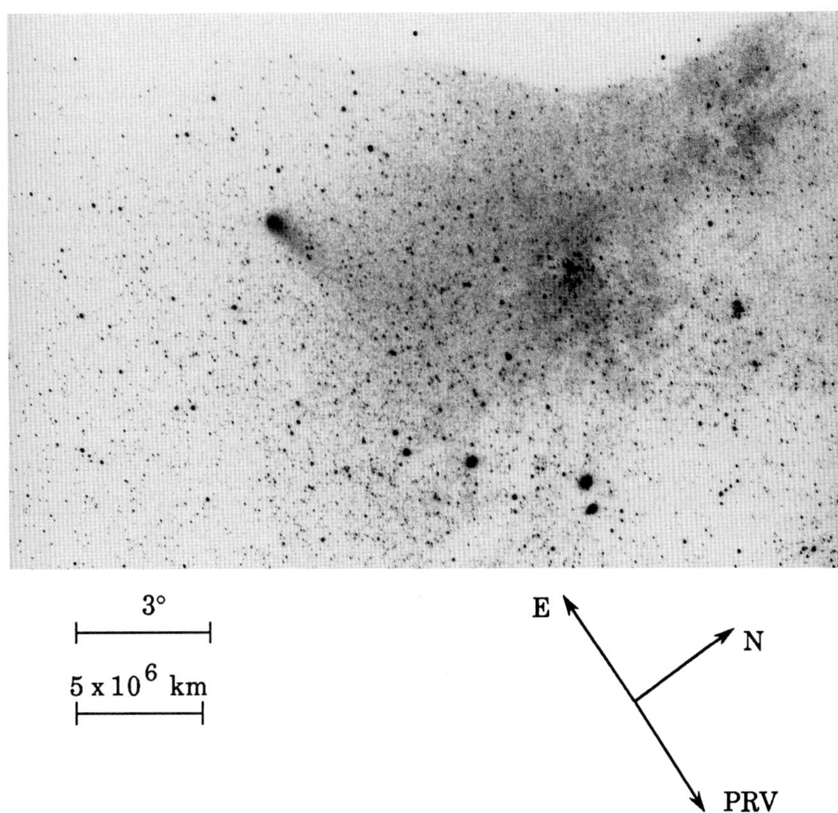

Fig. 444: 1986 Apr 1.51830 UT; exposure 10.0 minutes on Tri-X emulsion; r=1.19, Δ=0.52, β=123.4°; AON-851295.
Photograph by A. Ward, UK.

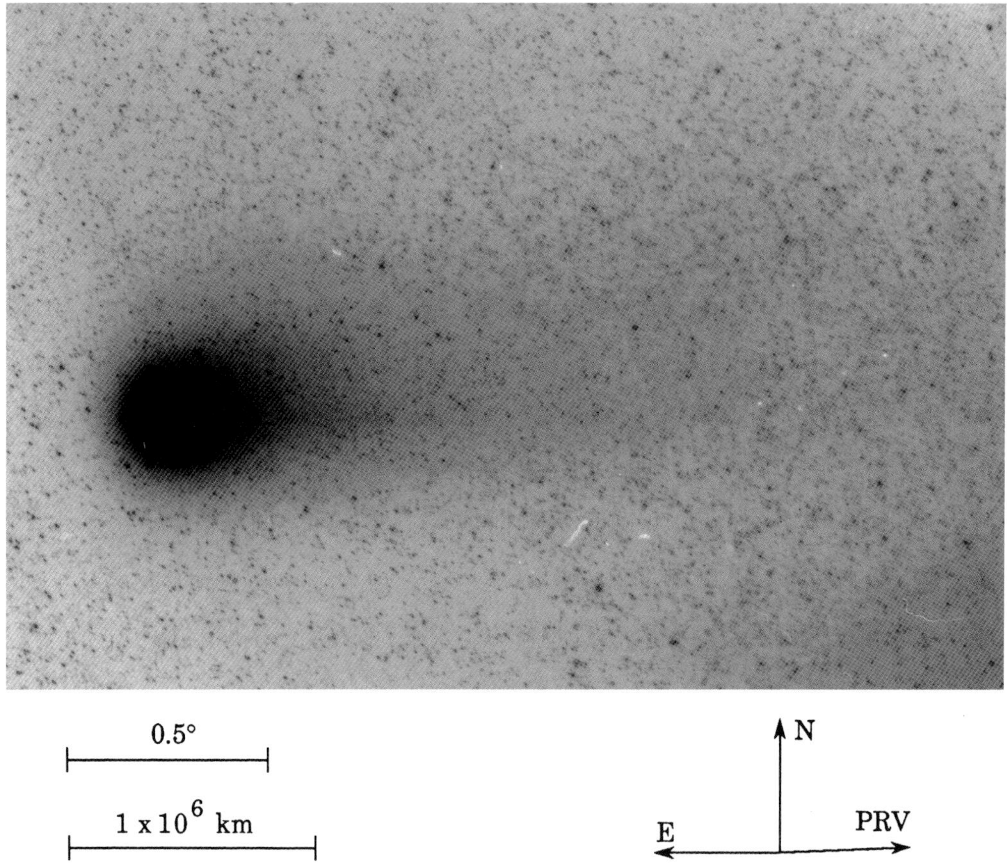

Fig. 445: 1986 Apr 2.03090 UT; exposure 3.0 minutes on Tech. Pan 2415 emulsion with no filter; $r=1.19$, $\Delta=0.51$, $\beta=124.3°$; LSPN-3491.
Photograph by C. Butler, University of Perugia/South African Astronomical Observatory.

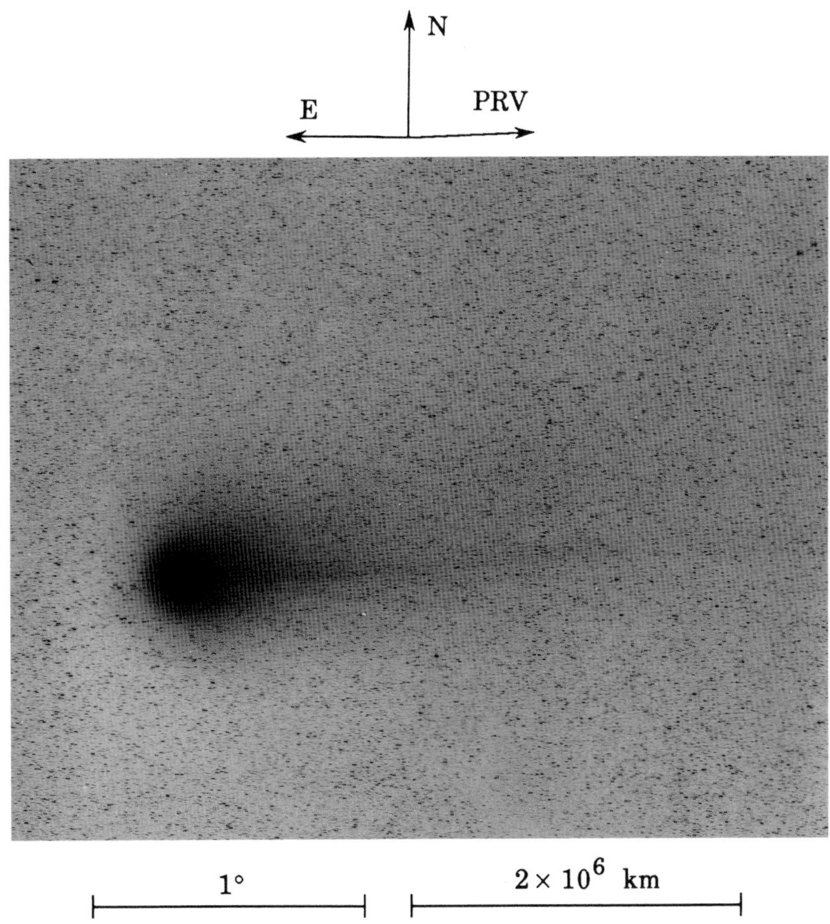

Fig. 446: 1986 Apr 2.17674 UT; exposure 5.0 minutes on IIa-O emulsion with no filter; r=1.20, Δ=0.51, β=124.6°; LSPN-1125.
Photograph by F. Miller, University of Michigan/CTIO.

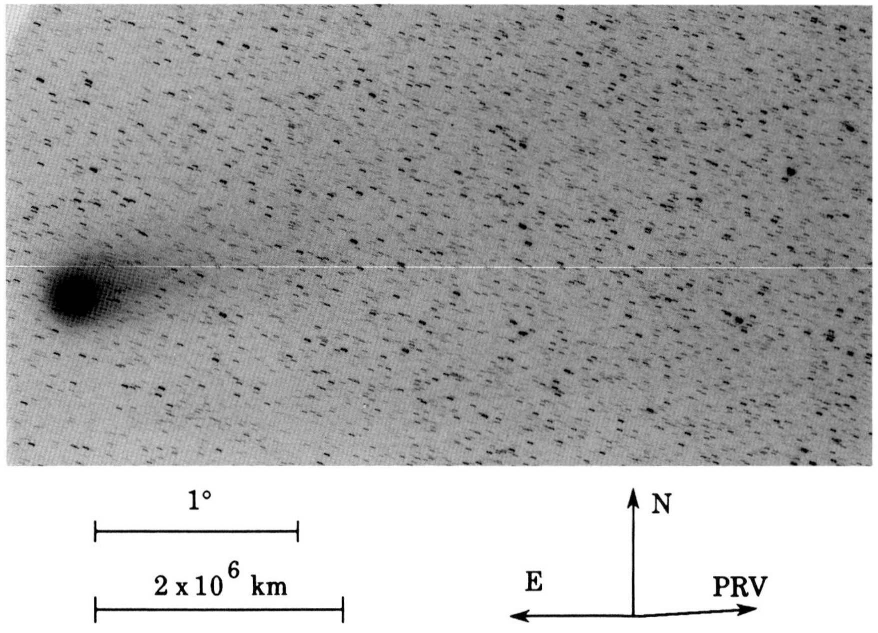

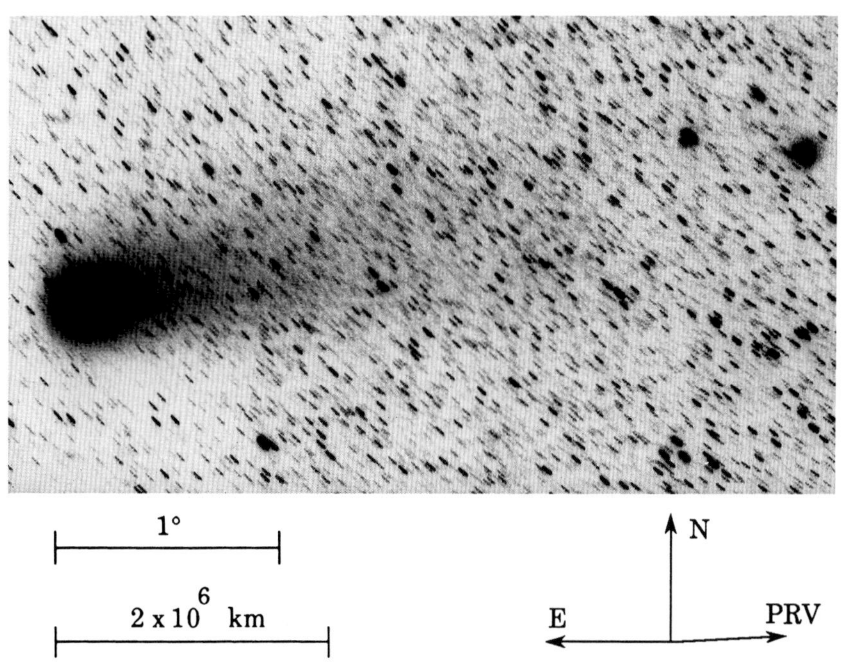

(top) Fig. 447: 1986 Apr 2.25590 UT; exposure 15.0 minutes on Tech. Pan 2415 emulsion with no filter; r=1.20, Δ=0.50, β=124.8°; LSPN-1440.
Photograph by W. Liller, Easter Island, LSPN Island Network.

(bottom) Fig. 448: 1986 Apr 2.51840 UT; exposure 15.0 minutes on Tri-X emulsion; r=1.20, Δ=0.50, β=125.3°; AON-851301.
Photograph by A. Ward, UK.

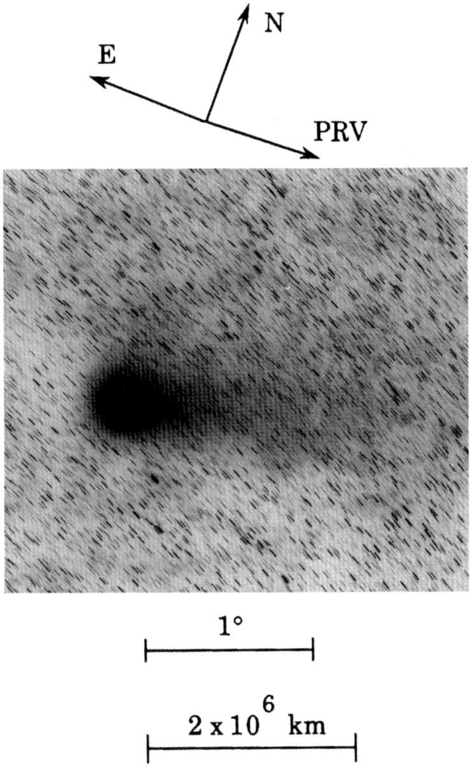

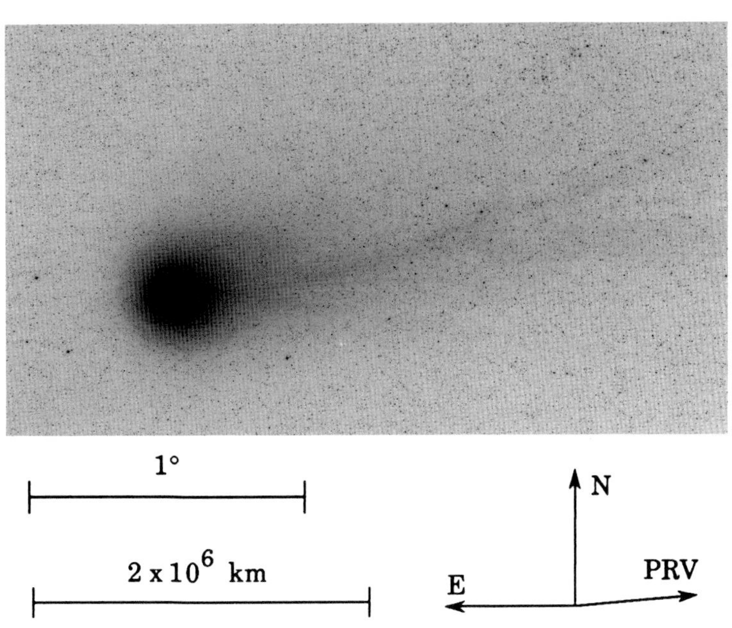

(top) Fig. 449: 1986 Apr 2.86944 UT; exposure 20.0 minutes on 103a-F emulsion with UV filter; r=1.21, Δ=0.49, β=126.0°; LSPN-3676.
Photograph submitted by K. Sivaraman, Indian Institute for Astrophysics, Kavalur Station.

(bottom) Fig. 450: 1986 Apr 3.16146 UT; exposure 5.0 minutes on IIa-O emulsion with no filter; r=1.21, Δ=0.49, β=126.6°; LSPN-1127.
Photograph by F. Miller, University of Michigan/CTIO.

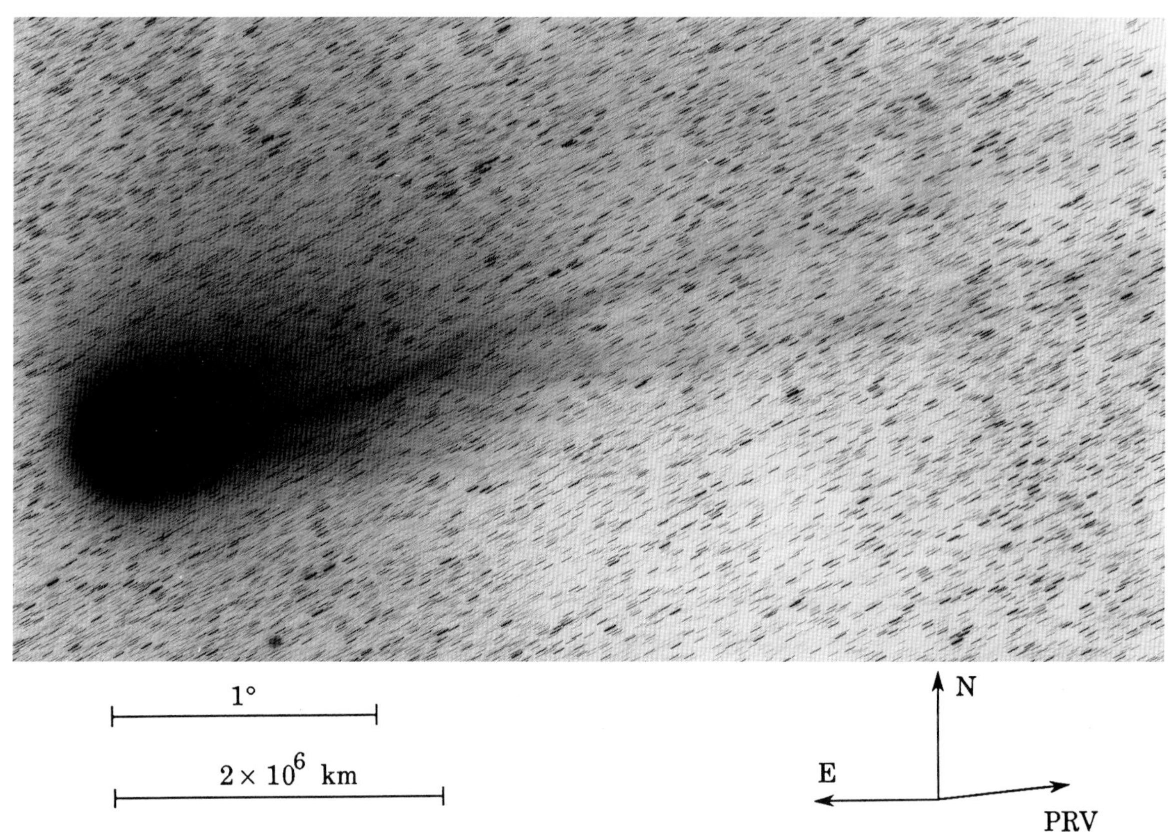

Fig. 451: 1986 Apr 3.20486 UT; exposure 20.0 minutes on IIa-O emulsion with no filter; r=1.21, Δ=0.49, β=126.7°; LSPN-1128.
Photograph by F. Miller, University of Michigan/CTIO.

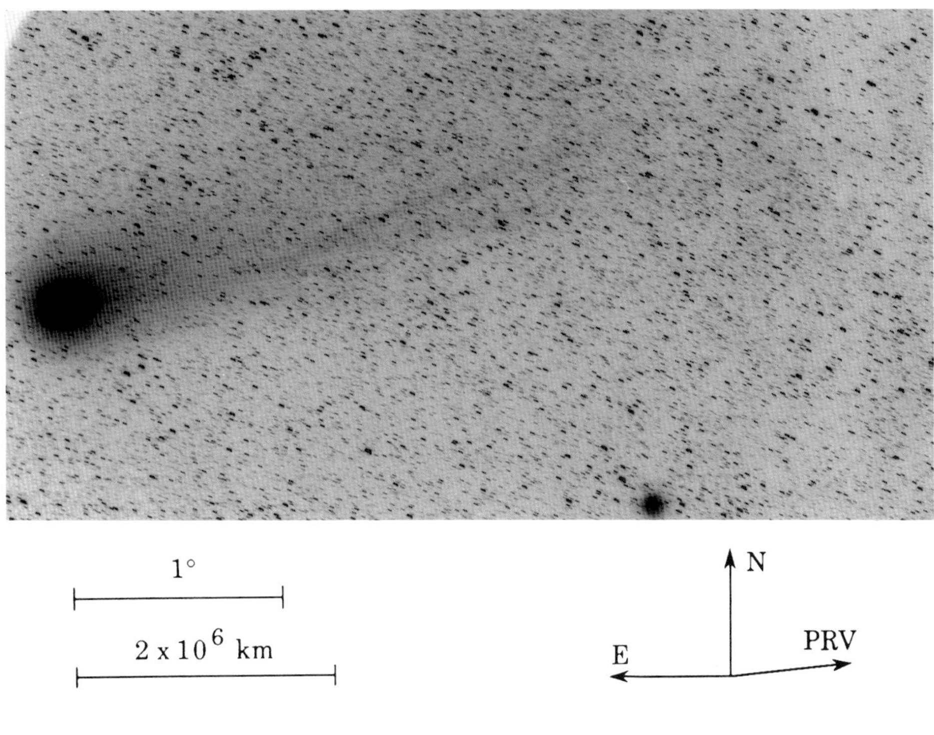

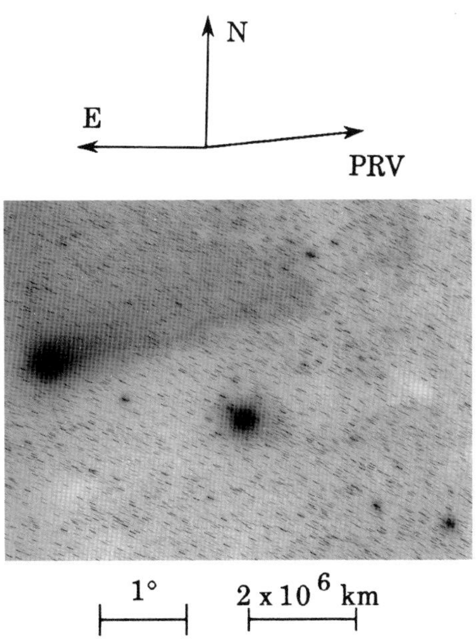

(top) Fig. 452: 1986 Apr 3.32917 UT; exposure 8.0 minutes on Tech. Pan 2415 emulsion with no filter; r=1.21, Δ=0.49, β=127.0°; LSPN-1444.
Photograph by W. Liller, Easter Island, LSPN Island Network.

(bottom) Fig. 453: 1986 Apr 3.52153 UT; exposure 30.0 minutes on IIIa-J emulsion with no filter; r=1.22, Δ=0.48, β=127.4°; LSPN-1922.
Photograph by M. Buie, Mauna Kea Observatory.

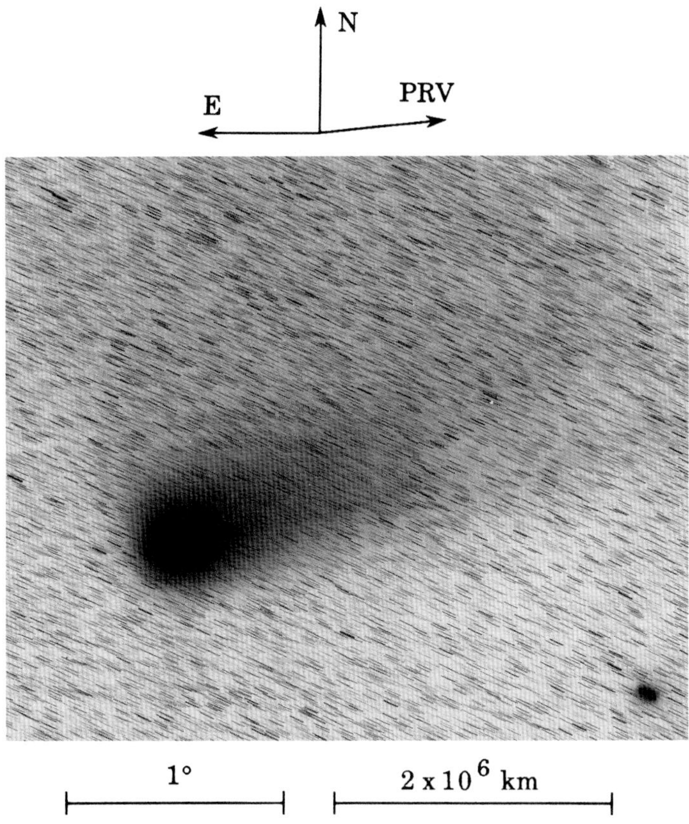

Fig. 454: 1986 Apr 3.55065 UT; exposure 30.0 minutes on IIa-D emulsion with GG-495 filter; r=1.22, Δ=0.48, β=127.5°; LSPN-2132.
Photograph by P. Magnusson, Uppsala Southern Station.

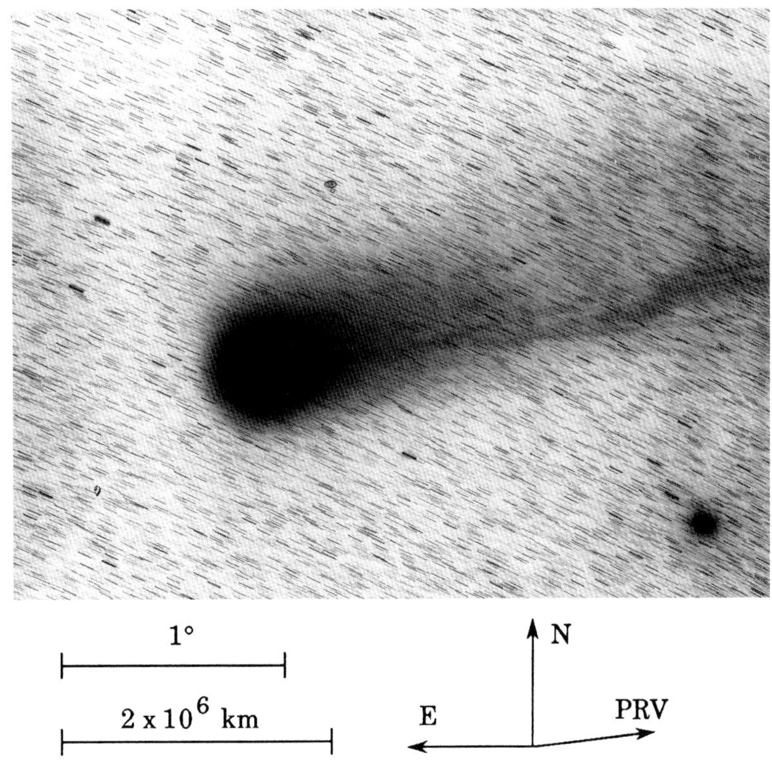

Fig. 455: 1986 Apr 3.58130 UT; exposure 30.0 minutes on IIIa-J emulsion with no filter; r=1.22, Δ=0.48, β=127.6°; LSPN-2124.
Photograph by P. Magnusson, Uppsala Southern Station.

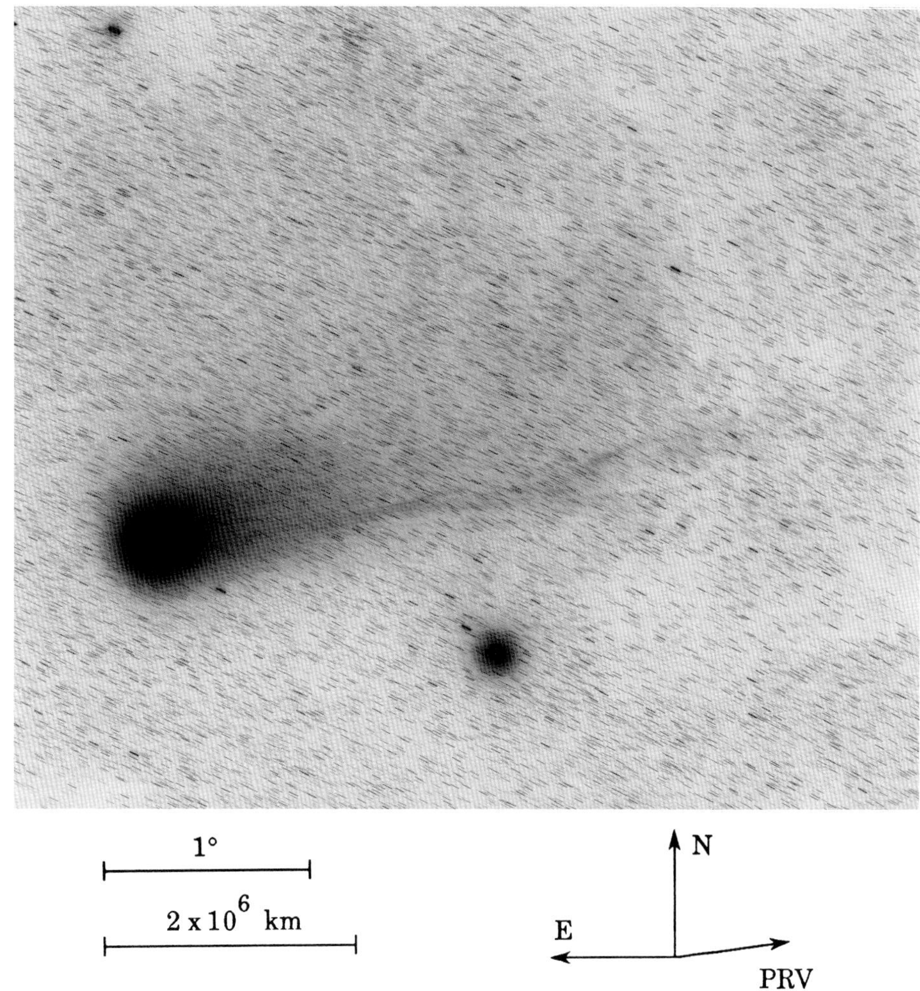

Fig. 456: 1986 Apr 3.69853 UT; exposure 20.0 minutes on IIa-O emulsion with GG-395 filter; $r=1.22$, $\Delta=0.48$, $\beta=127.8°$; LSPN-2409.
Photograph from Royal Observatory/UK Schmidt Telescope Unit.

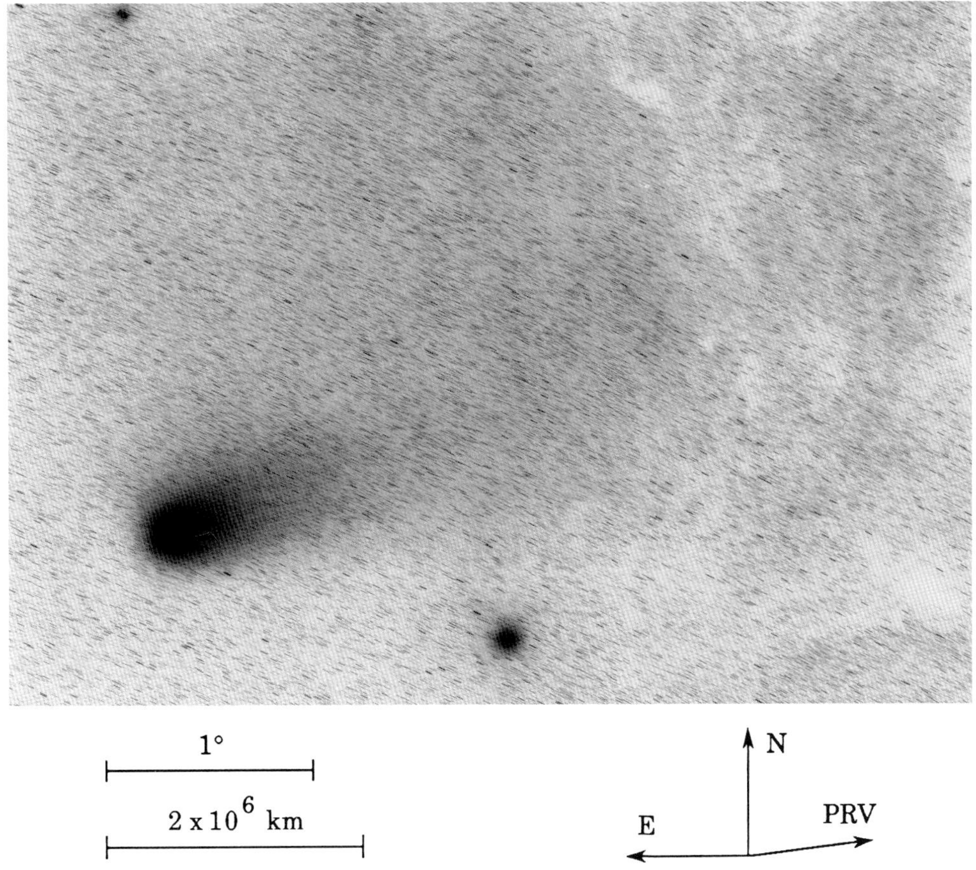

Fig. 457: 1986 Apr 3.72103 UT; exposure 15.0 minutes on IIIa-F emulsion with RG-630 filter; r=1.22, Δ=0.48, β=127.9°; LSPN-2410.
Photograph from Royal Observatory/UK Schmidt Telescope Unit.

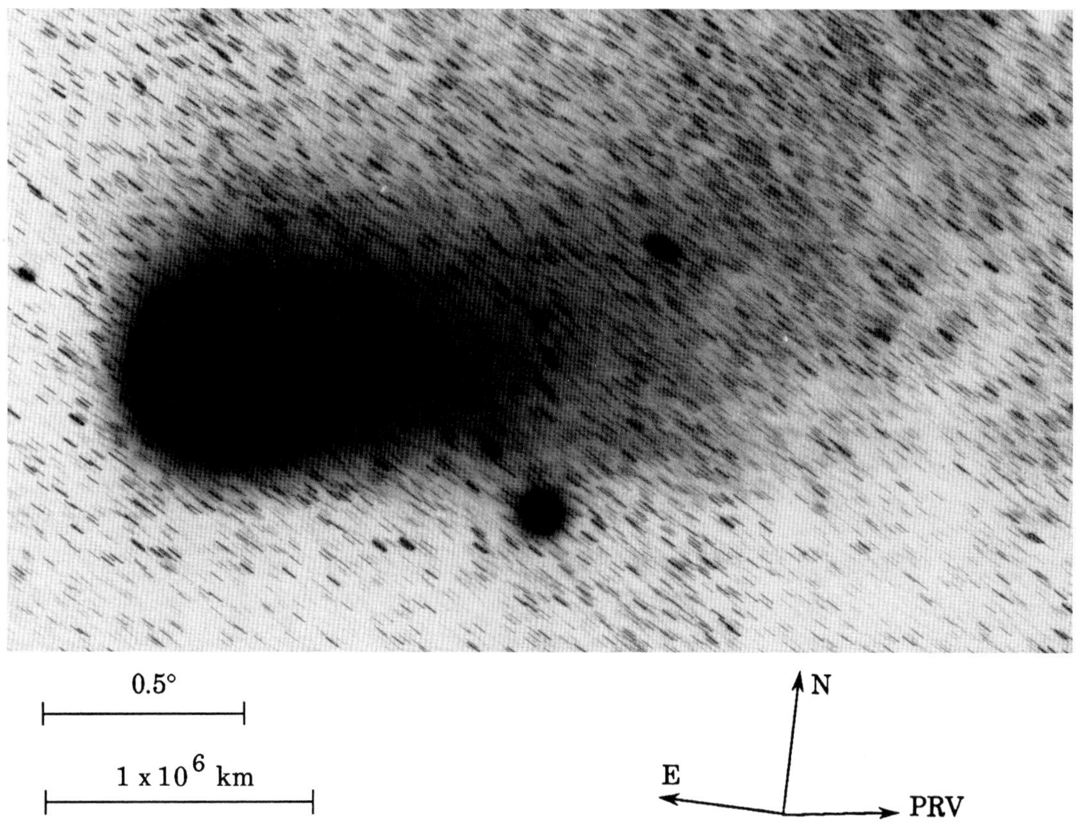

Fig. 458: 1986 Apr 4.03646 UT; exposure 15.0 minutes on Tech. Pan 2415 emulsion with no filter; r=1.22, Δ=0.47, β=128.6°; LSPN-3493.
Photograph by C. Butler, University of Perugia/South African Astronomical Observatory.

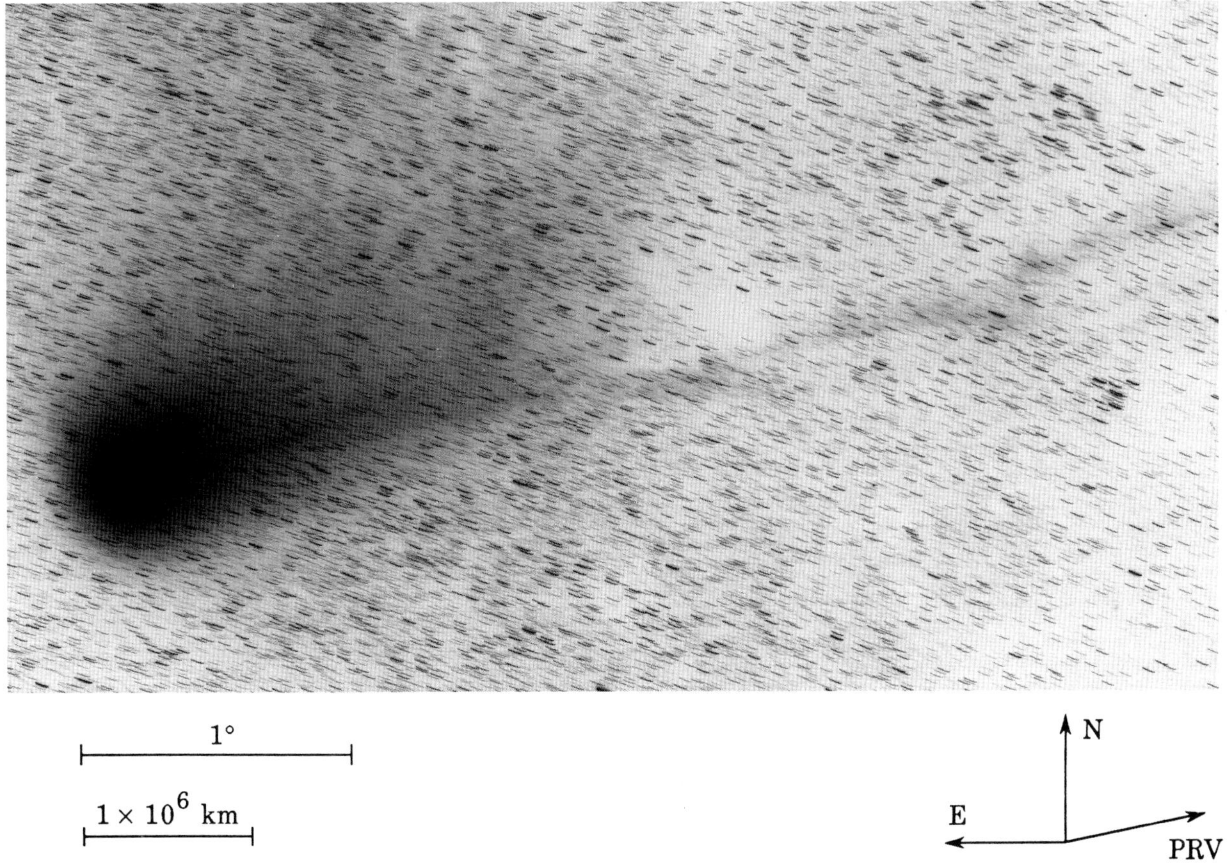

Fig. 459: 1986 Apr 4.17500 UT; exposure 20.0 minutes on IIa-O emulsion with no filter; r=1.23, Δ=0.47, β=128.9°; LSPN-1130.
Photograph by F. Miller, University of Michigan/CTIO.

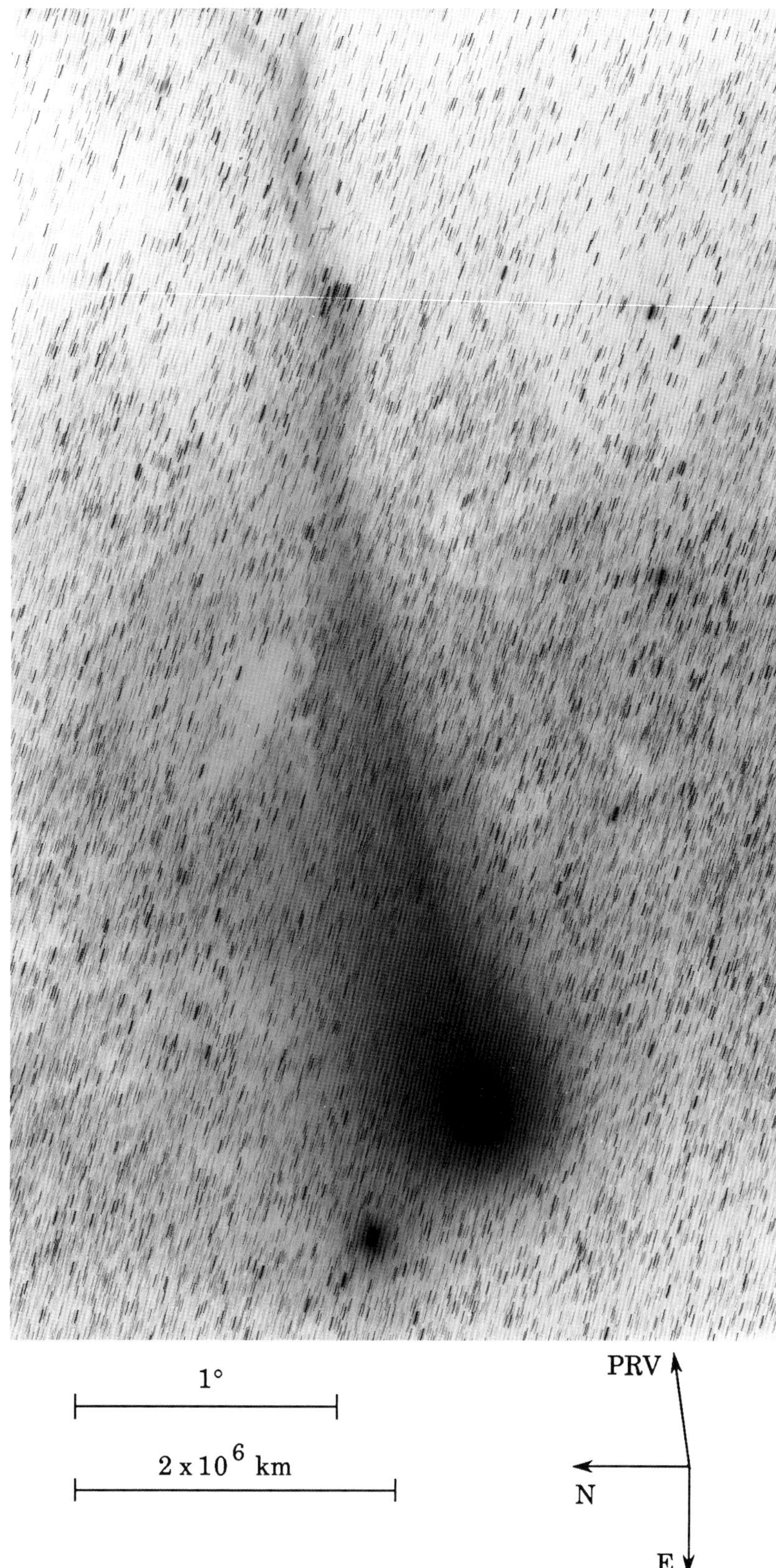

Fig. 460: 1986 Apr 4.37292 UT; exposure 20.0 minutes on IIa-O emulsion with no filter; r=1.23, Δ=0.47, β=129.4°; LSPN-1134.
Photograph by F. Miller, University of Michigan/CTIO.

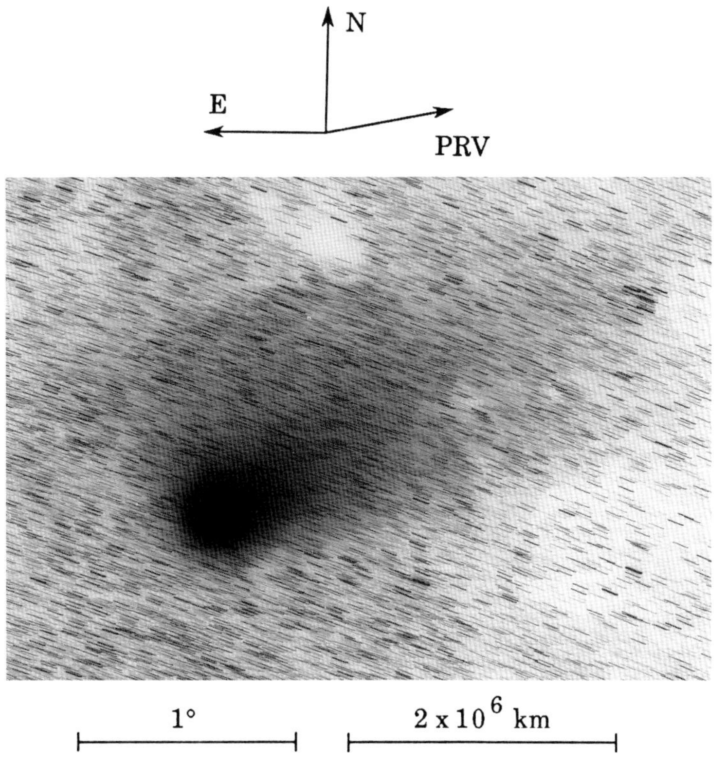

Fig. 461: 1986 Apr 4.68412 UT; exposure 30.0 minutes on IIa-D emulsion with GG-495 filter; $r=1.23$, $\Delta=0.46$, $\beta=130.2°$; LSPN-2134.
Photograph by P. Magnusson, Uppsala Southern Station.

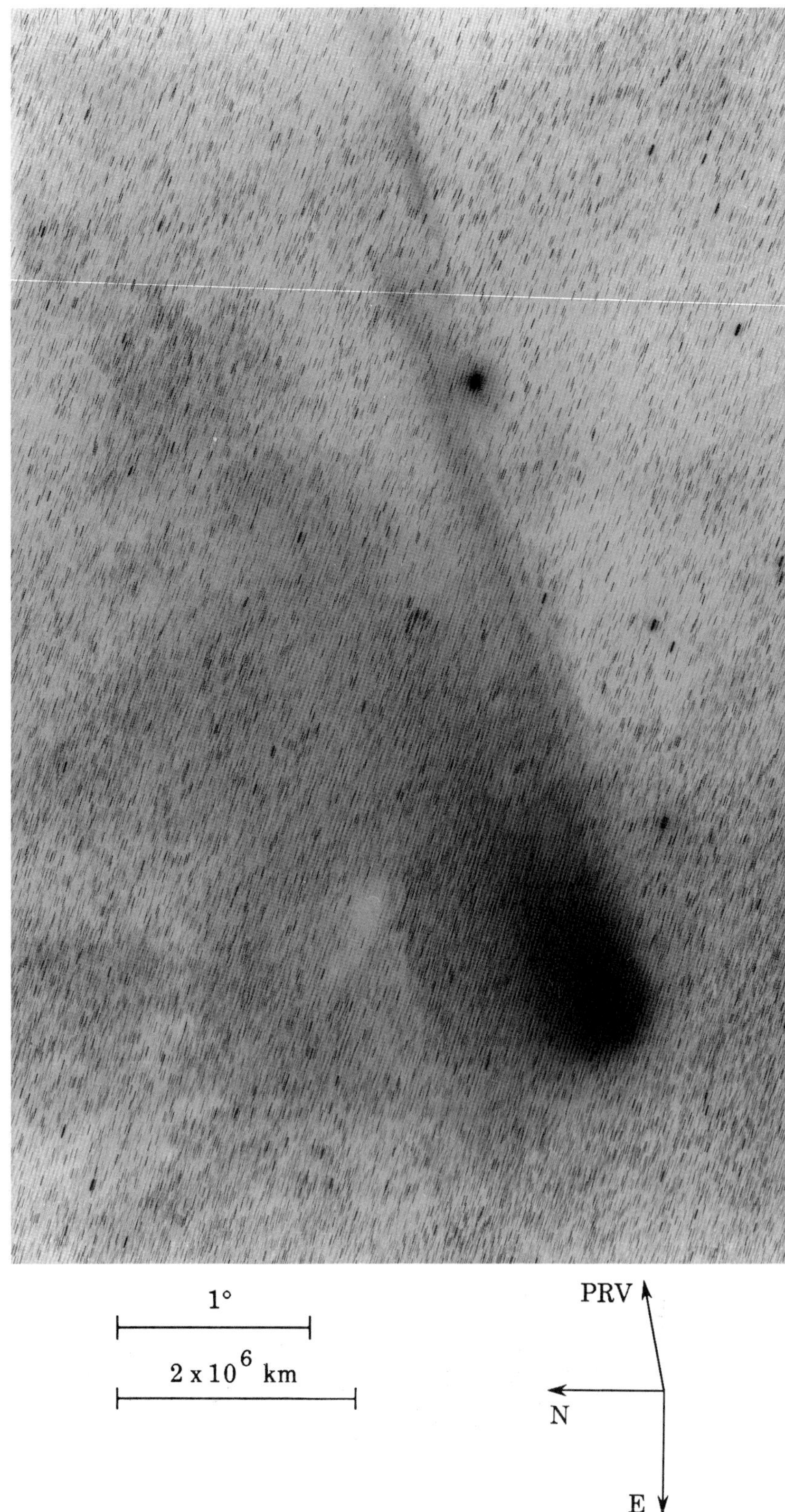

Fig. 462: 1986 Apr 4.68679 UT; exposure 20.0 minutes on IIa-O emulsion with GG-395 filter; r=1.23, Δ=0.46, β=130.2°; LSPN-2411.
Photograph from Royal Observatory/UK Schmidt Telescope Unit.

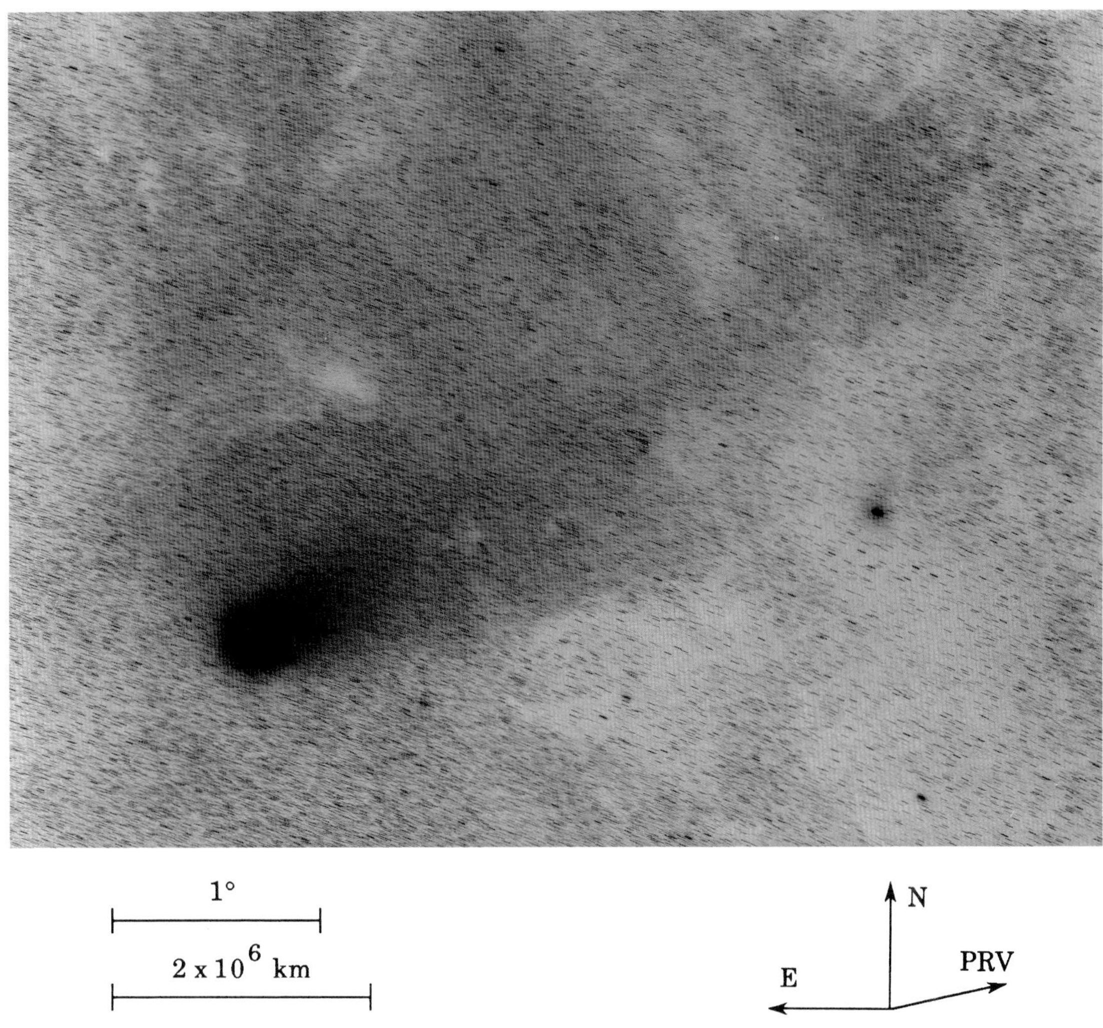

Fig. 463: 1986 Apr 4.70929 UT; exposure 15.0 minutes on IIIa-F emulsion with RG-630 filter; r=1.24, Δ=0.46, β=130.2°; LSPN-2412.
Photograph from Royal Observatory/UK Schmidt Telescope Unit.

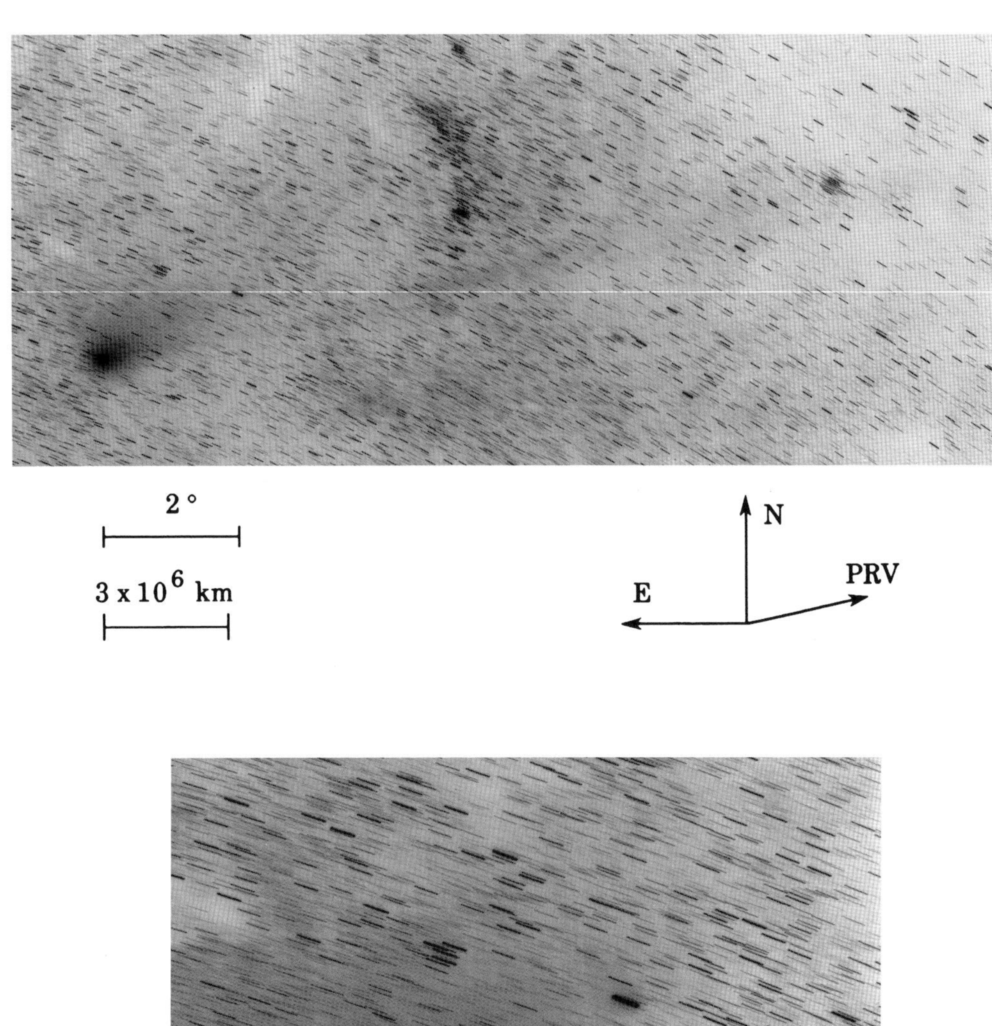

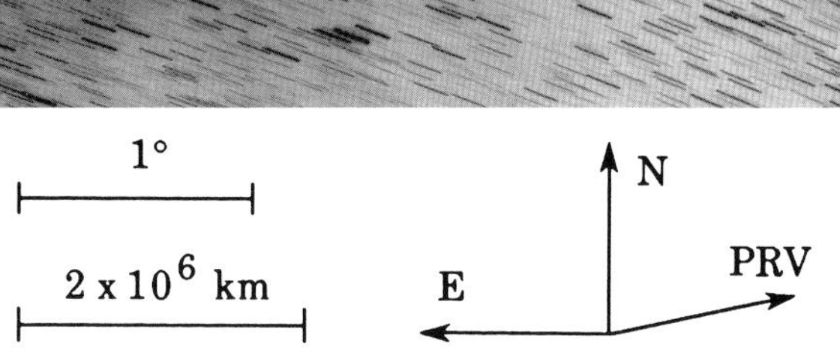

(top) Fig. 464: 1986 Apr 5.00625 UT; exposure 60.0 minutes on IIa-O emulsion with no filter; r=1.24, Δ=0.46, β=131.0°; LSPN-2859.
Photograph by G. Malcolm, Boyden Observatory.

(bottom) Fig. 465: 1986 Apr 5.00625 UT; exposure 60.0 minutes on IIa-O emulsion with no filter; r=1.24, Δ=0.46, β=131.0°; LSPN-2860.
Photograph by G. Malcolm, Boyden Observatory.

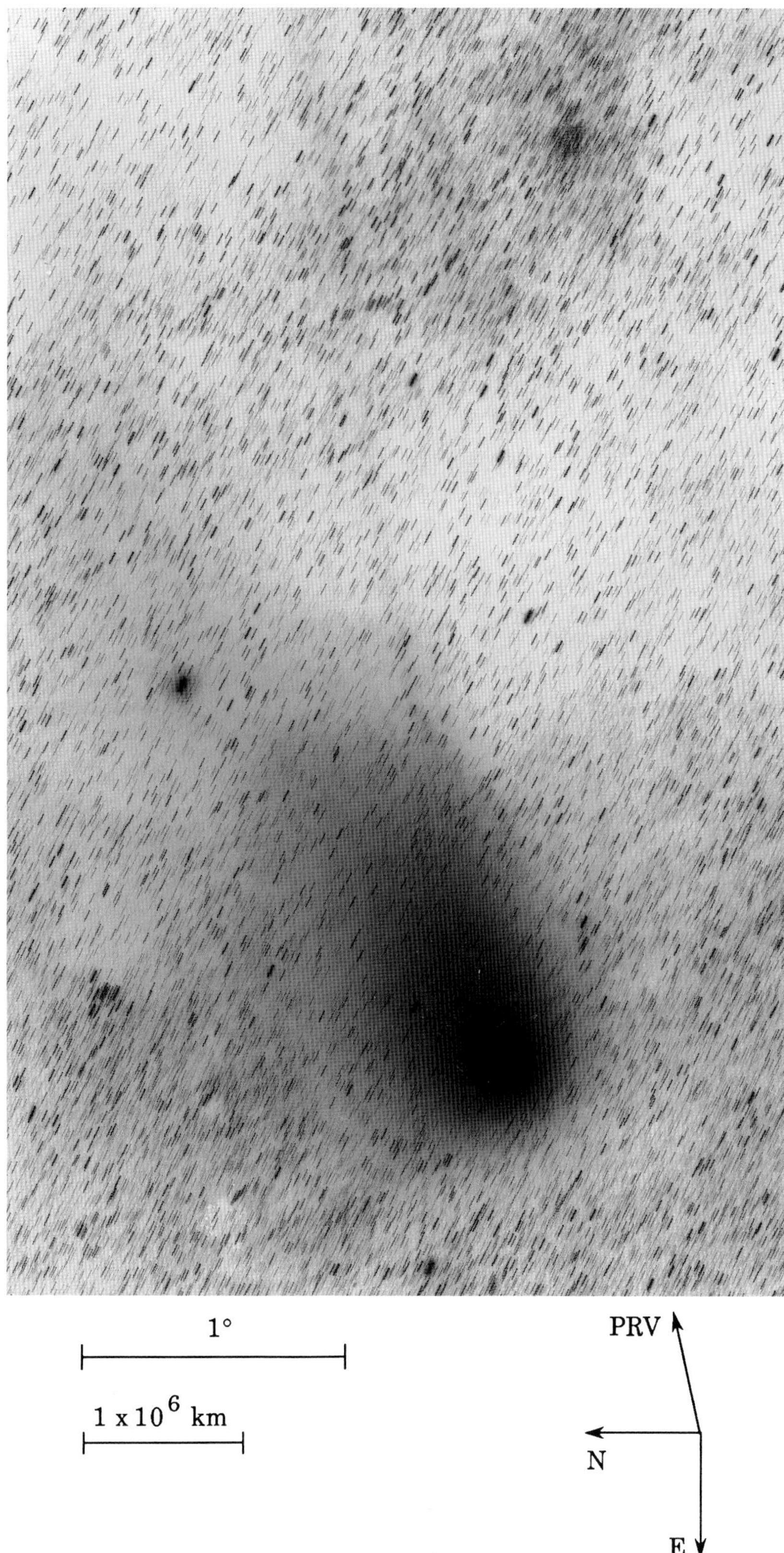

Fig. 466: 1986 Apr 5.17639 UT; exposure 20.0 minutes on IIa-O emulsion with no filter; r=1.24, Δ=0.46, β=131.4°; LSPN-1136.
Photograph by F. Miller, University of Michigan/CTIO.

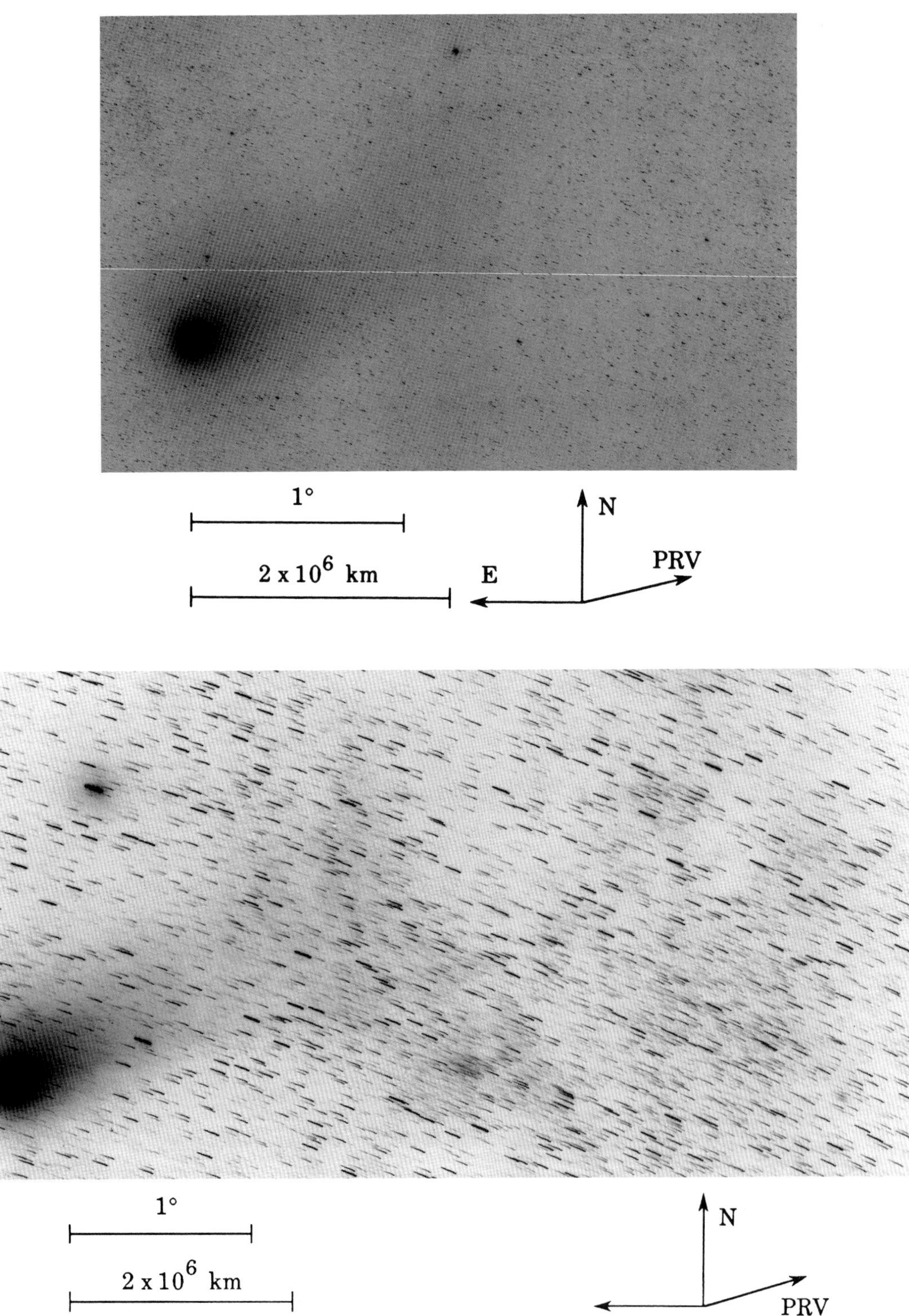

(top) Fig. 467: 1986 Apr 5.21736 UT; exposure 4.0 minutes on IIa-O emulsion with GG-385 filter; r=1.24, Δ=0.46, β=131.5°; LSPN-2913.
Photograph by H. Schuster/O. Pizarro, European Southern Observatory.

(bottom) Fig. 468: 1986 Apr 5.43889 UT; exposure 27.0 minutes on Tech. Pan 2415 emulsion with no filter; r=1.25, Δ=0.45, β=132.1°; LSPN-1450.
Photograph by W. Liller, Easter Island, LSPN Island Network.

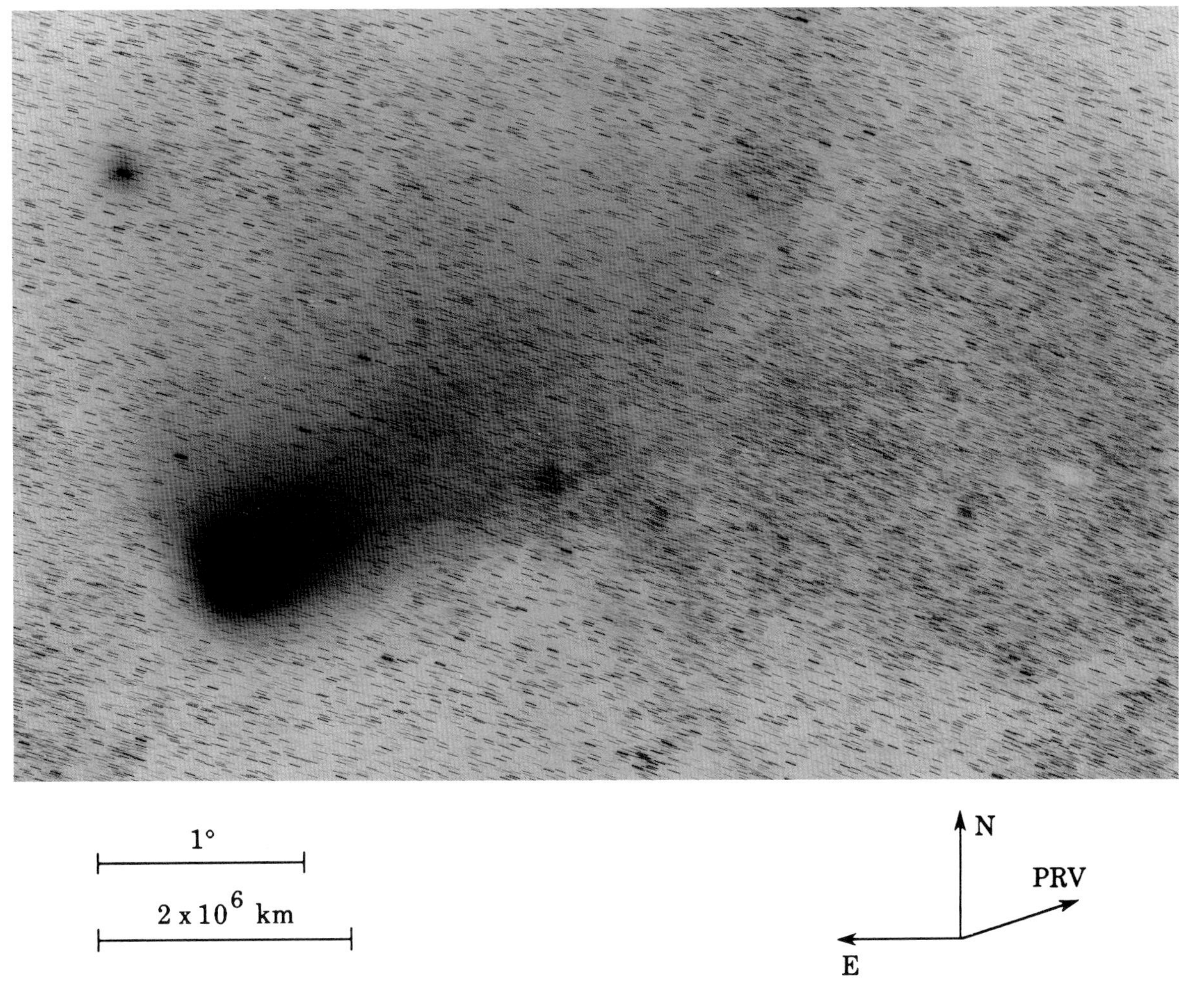

Fig. 469: 1986 Apr 5.72007 UT; exposure 20.0 minutes on IIa-O emulsion with GG-395 filter; r=1.25, Δ=0.45, β=132.8°; LSPN-2414.
Photograph from Royal Observatory/UK Schmidt Telescope Unit.

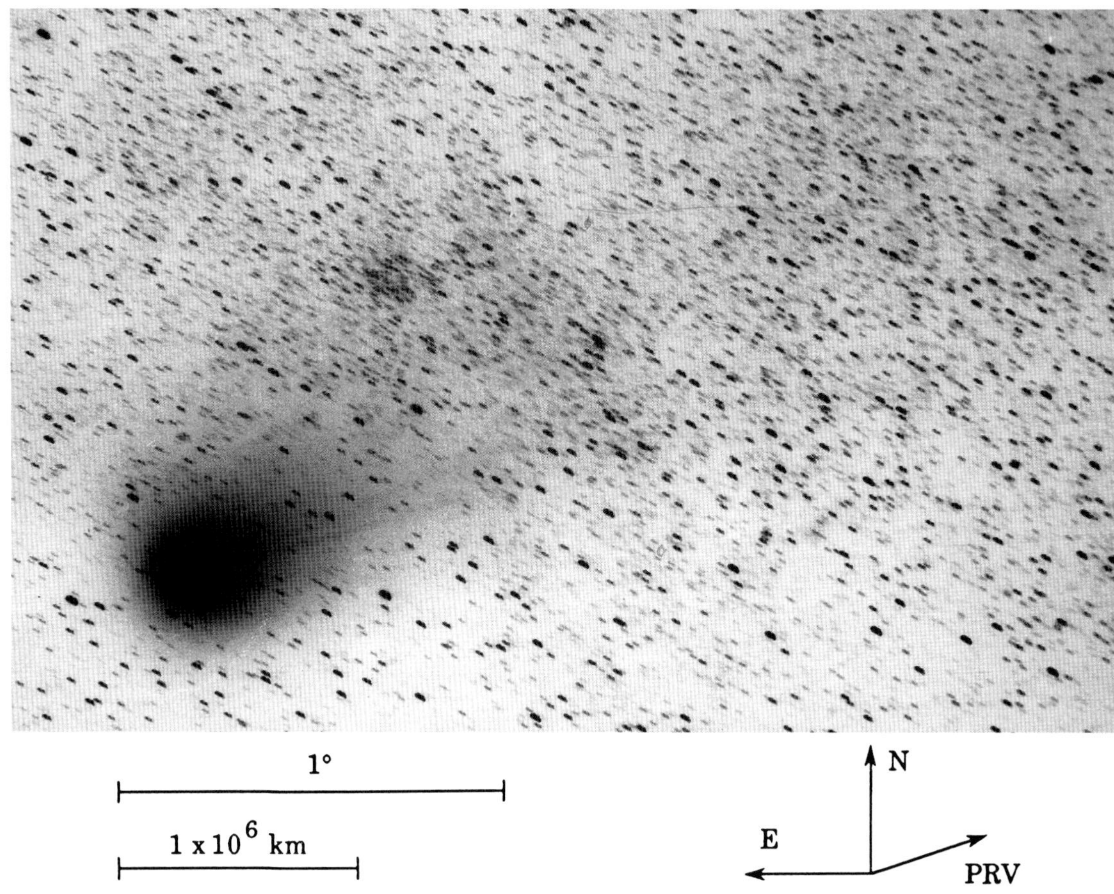

Fig. 470: 1986 Apr 5.99306 UT; exposure 8.0 minutes on Tech. Pan 2415 emulsion; r=1.26, Δ=0.45, β=133.6°. *Photograph by A. Marafie, Wafra Observatory.*

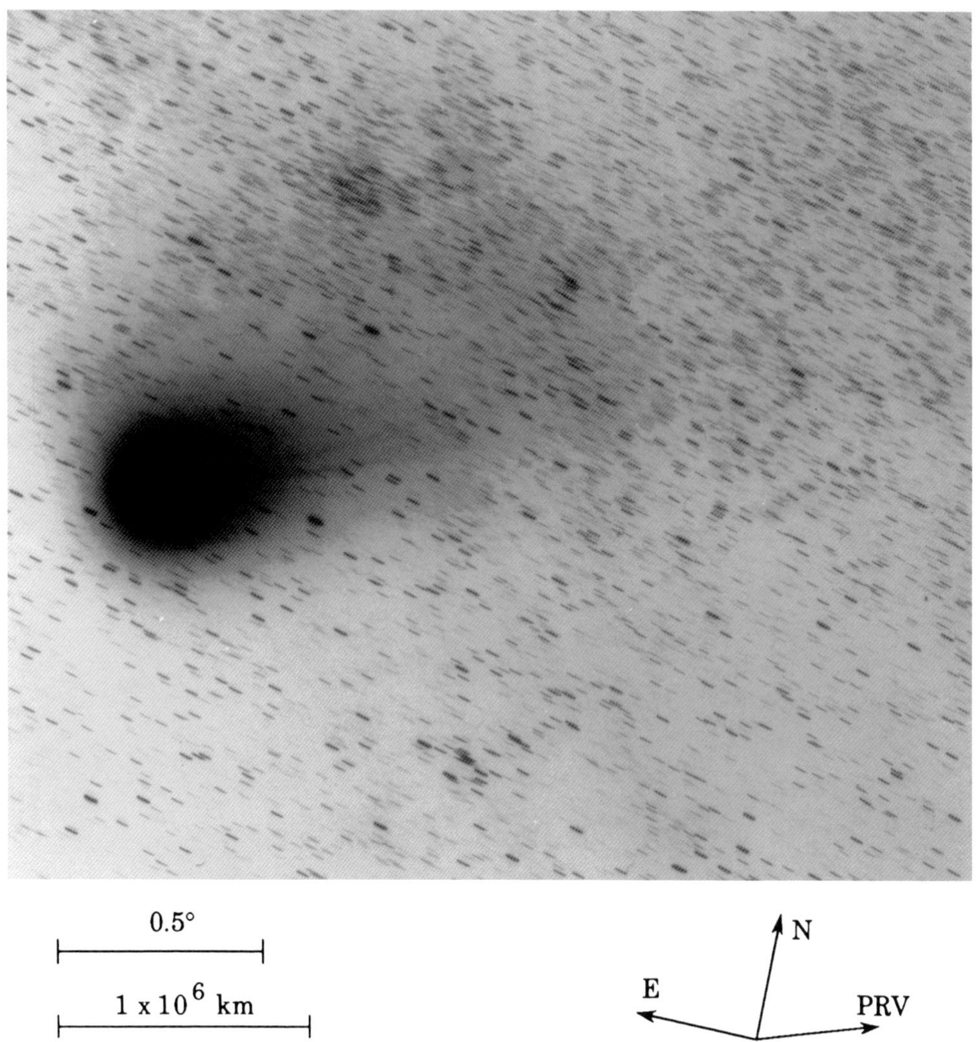

Fig. 471: 1986 Apr 6.02083 UT; exposure 10.0 minutes on Tech. Pan 2415 emulsion with no filter; r=1.26, Δ=0.45, β=133.6°; LSPN-3495.
Photograph by C. Butler, University of Perugia/South African Astronomical Observatory.

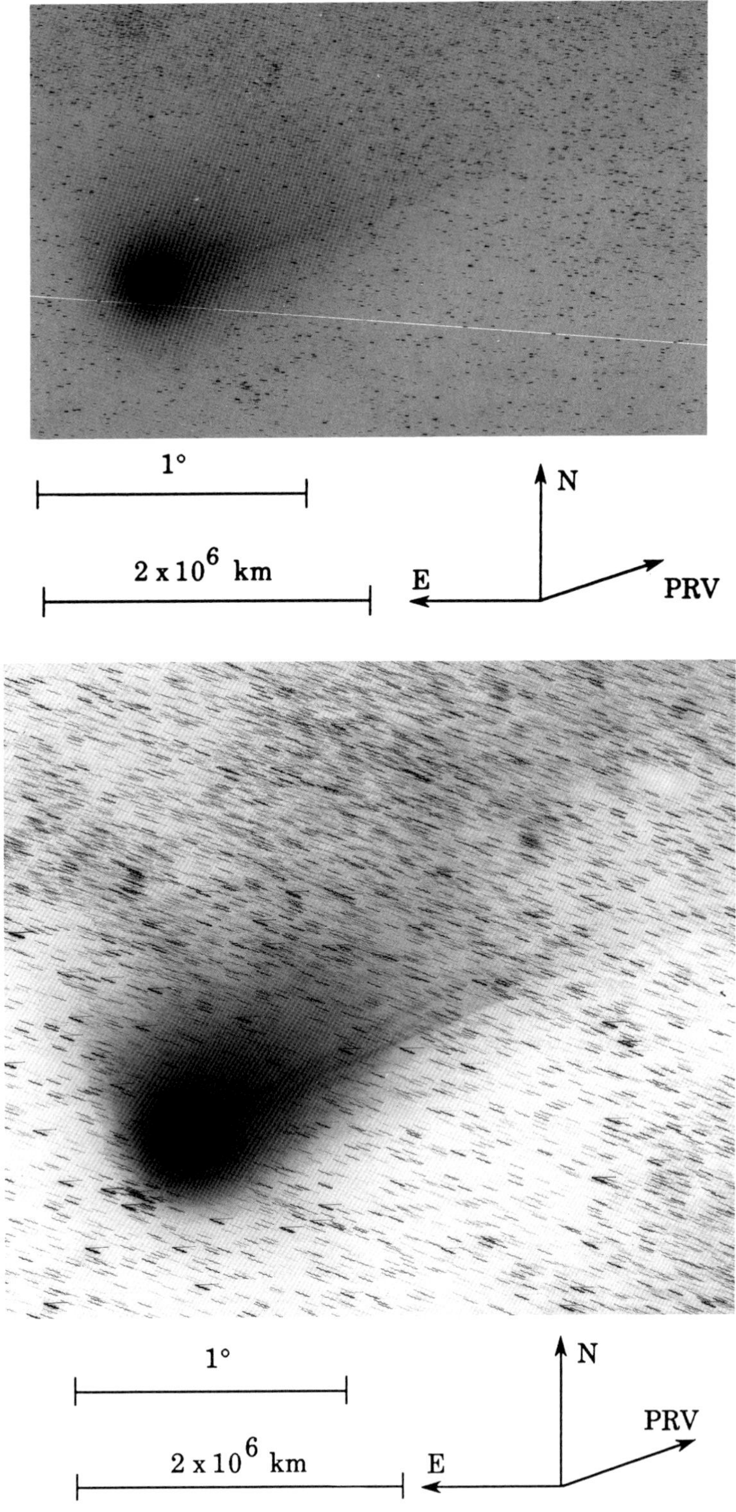

(top) Fig. 472: 1986 Apr 6.15243 UT; exposure 5.0 minutes on IIa-O emulsion with no filter; r=1.26, Δ=0.45, β=134.0°; LSPN-1143.
Photograph by F. Miller, University of Michigan/CTIO.

(bottom) Fig. 473*: 1986 Apr 6.30417 UT; exposure 20.0 minutes on IIa-O emulsion with no filter; r=1.26, Δ=0.44, β=134.4°; LSPN-1148.
Photograph by F. Miller, University of Michigan/CTIO.

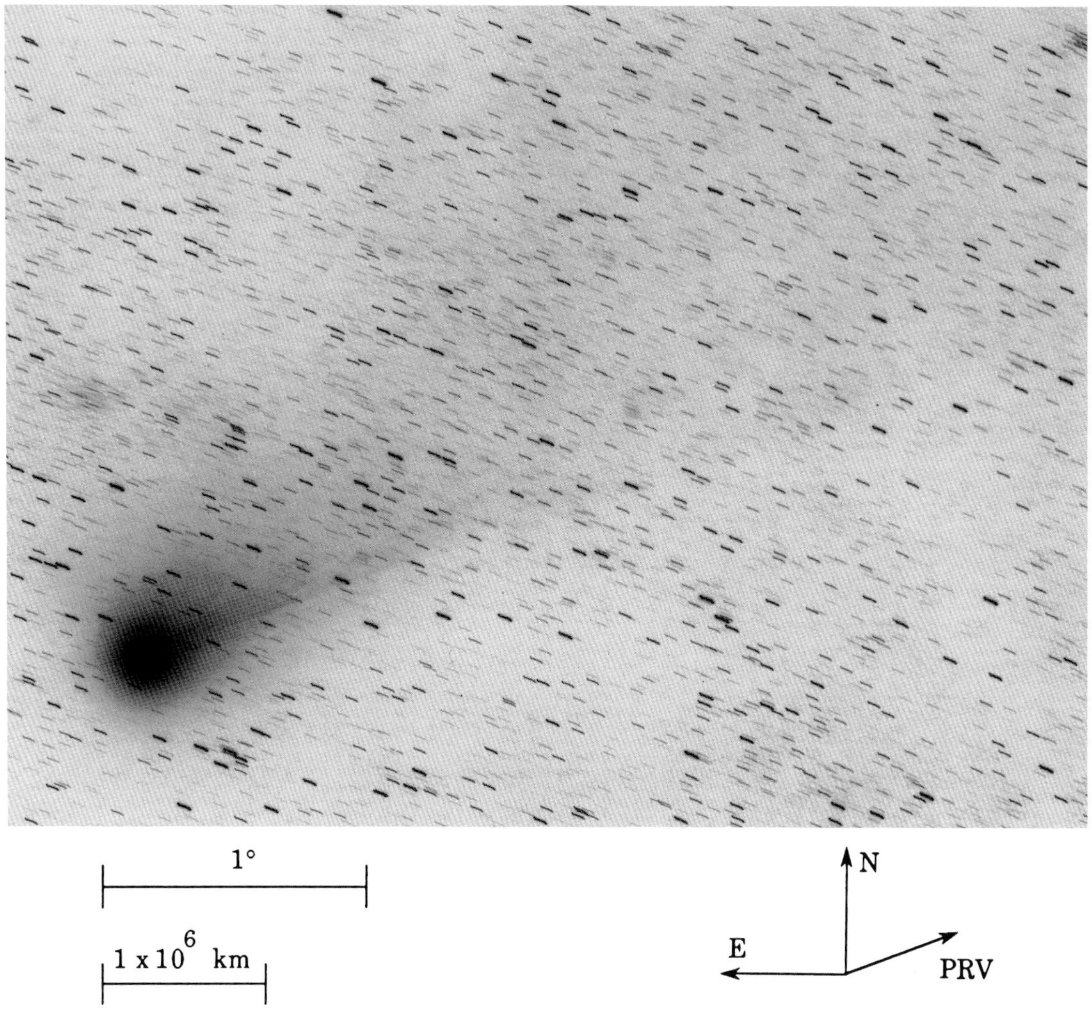

Fig. 474: 1986 Apr 6.17014 UT; exposure 20.0 minutes on IIa-O emulsion with no filter; r=1.26, Δ=0.45, β=134.0°; LSPN-1144.
Photograph by F. Miller, University of Michigan/CTIO.

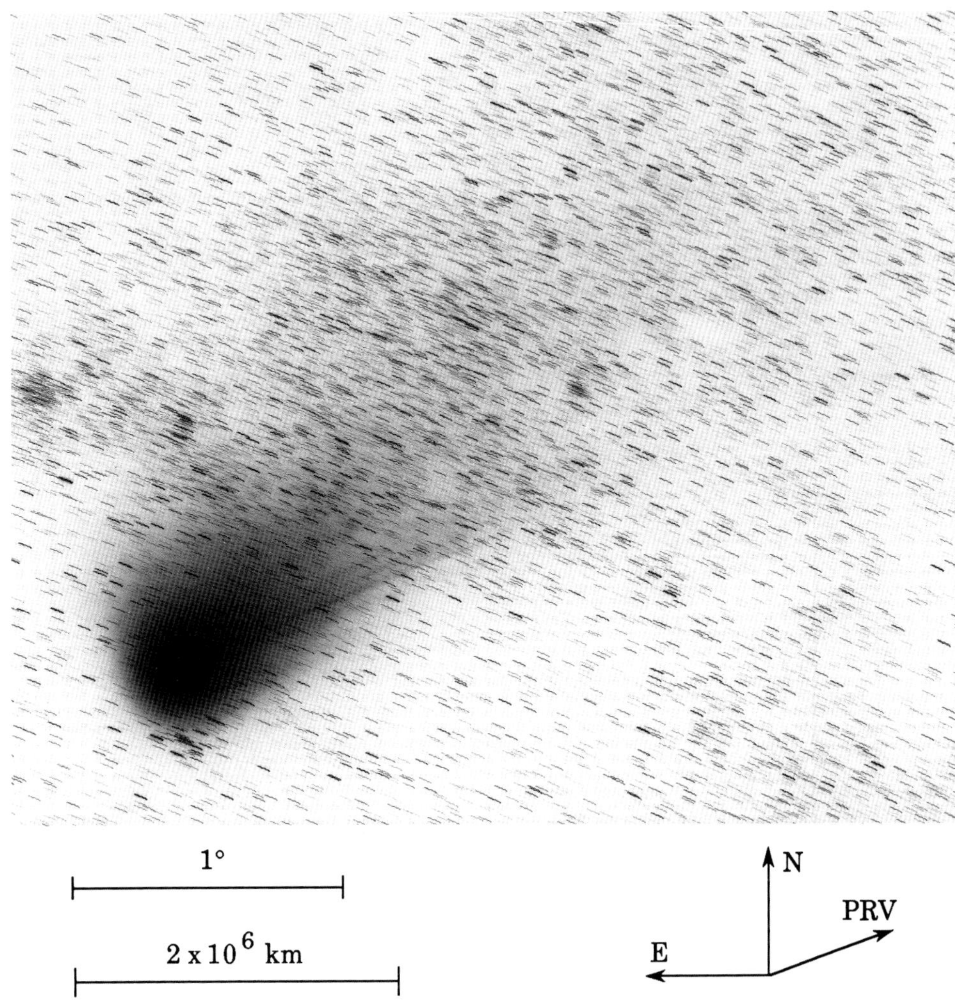

Fig. 475: 1986 Apr 6.24931 UT; exposure 20.0 minutes on IIa-O emulsion with no filter; r=1.26, Δ=0.44, β=134.3°; LSPN-1147.
Photograph by F. Miller, University of Michigan/CTIO.

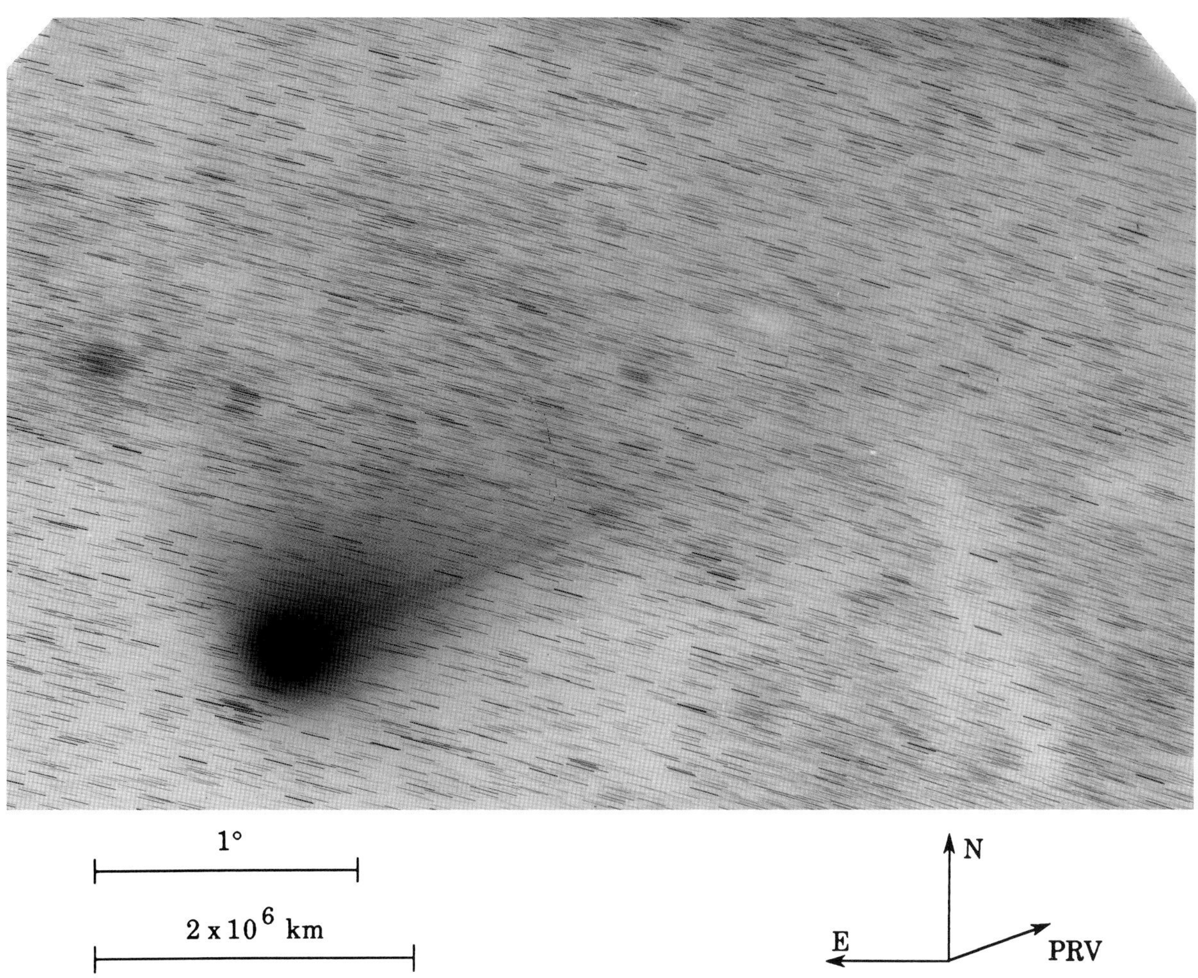

Fig. 476: 1986 Apr 6.30208 UT; exposure 40.0 minutes on IIa-O emulsion with no filter; r=1.26, Δ=0.44, β=134.4°; LSPN-742.
Photograph by C. Torres/H. Wroblewski, Cerro el Roble Astronomical Observatory.

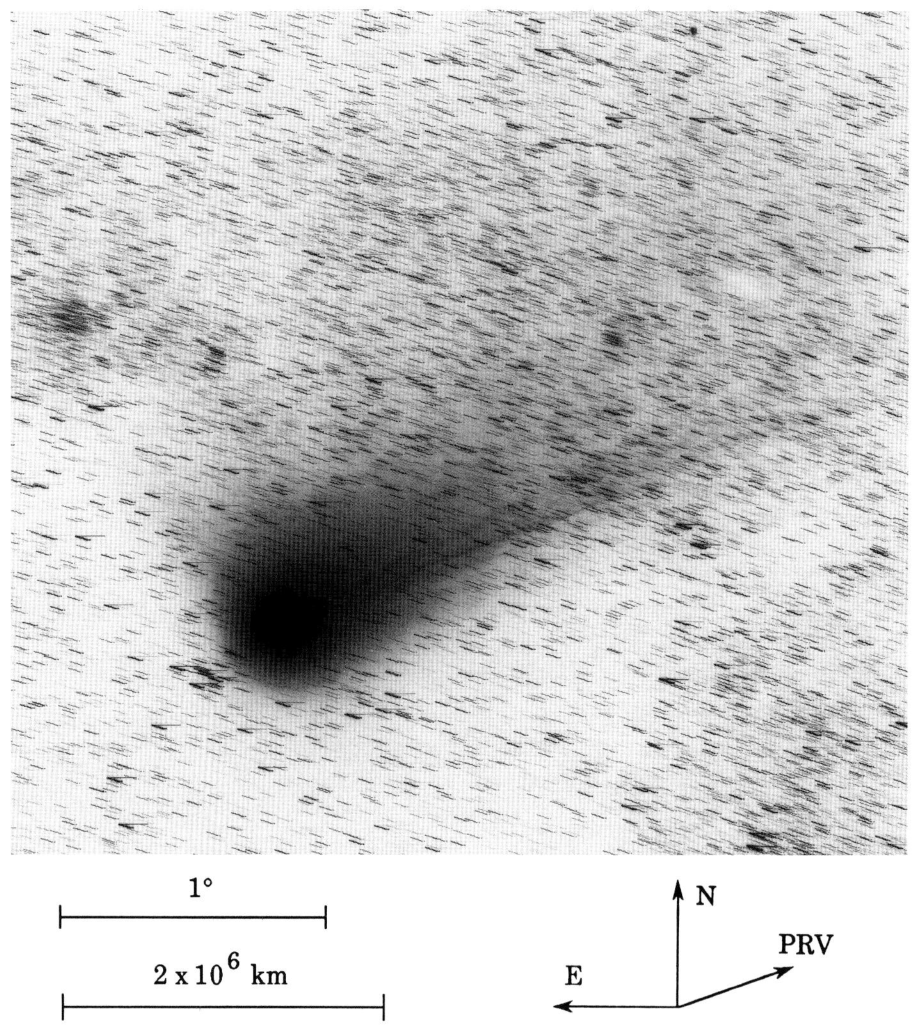

Fig. 477: 1986 Apr 6.32708 UT; exposure 20.0 minutes on IIa-O emulsion with no filter; r=1.26, Δ=0.44, β=134.5°; LSPN-1149.
Photograph by F. Miller, University of Michigan/CTIO.

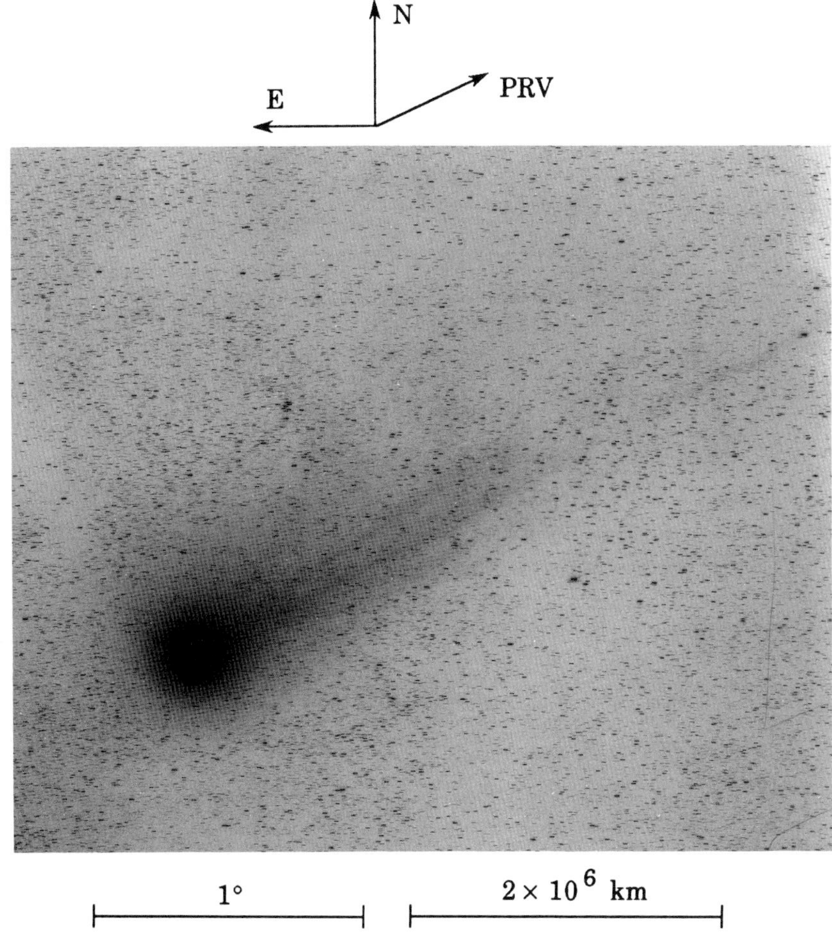

Fig. 478: 1986 Apr 7.14618 UT; exposure 5.0 minutes on IIa-O emulsion with no filter; r=1.27, Δ=0.44, β=136.8°; LSPN-1150.
Photograph by F. Miller, University of Michigan/CTIO.

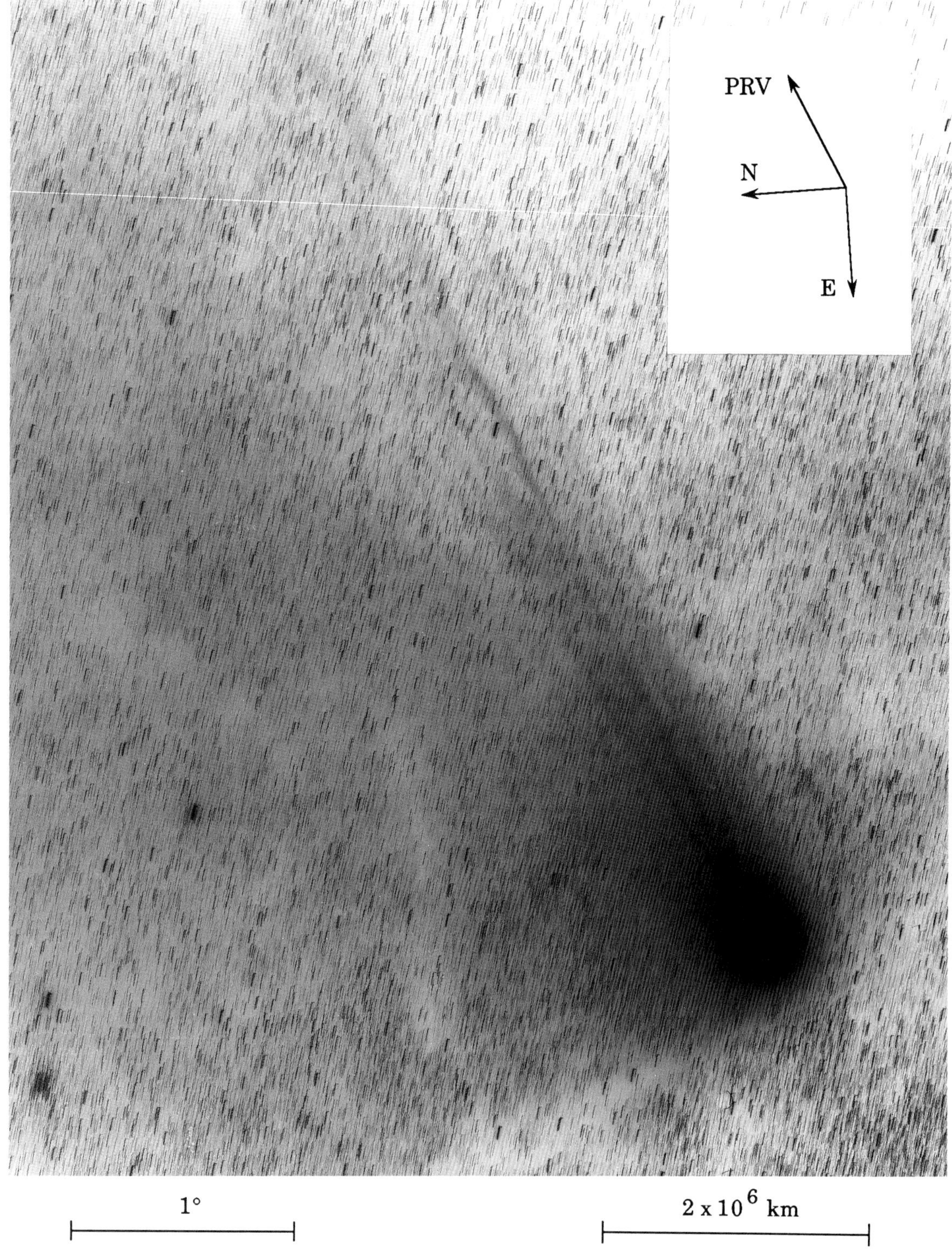

Fig. 479: 1986 Apr 7.16319 UT; exposure 20.0 minutes on IIa-O emulsion with no filter; r=1.27, Δ=0.44, β=136.8°; LSPN-1151.
Photograph by F. Miller, University of Michigan/CTIO.

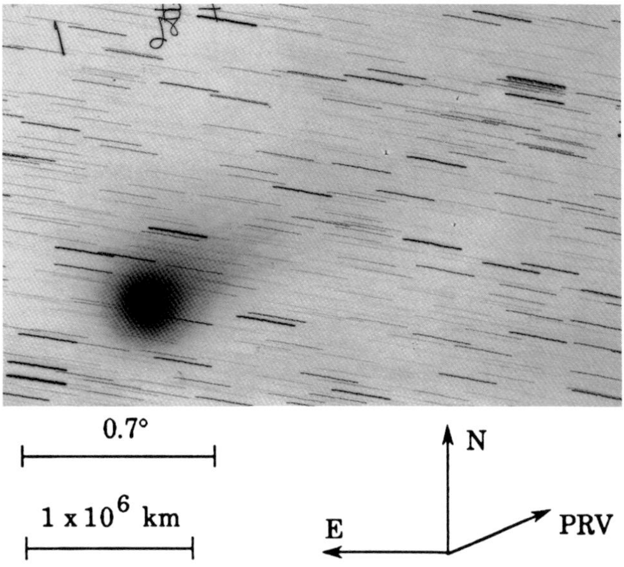

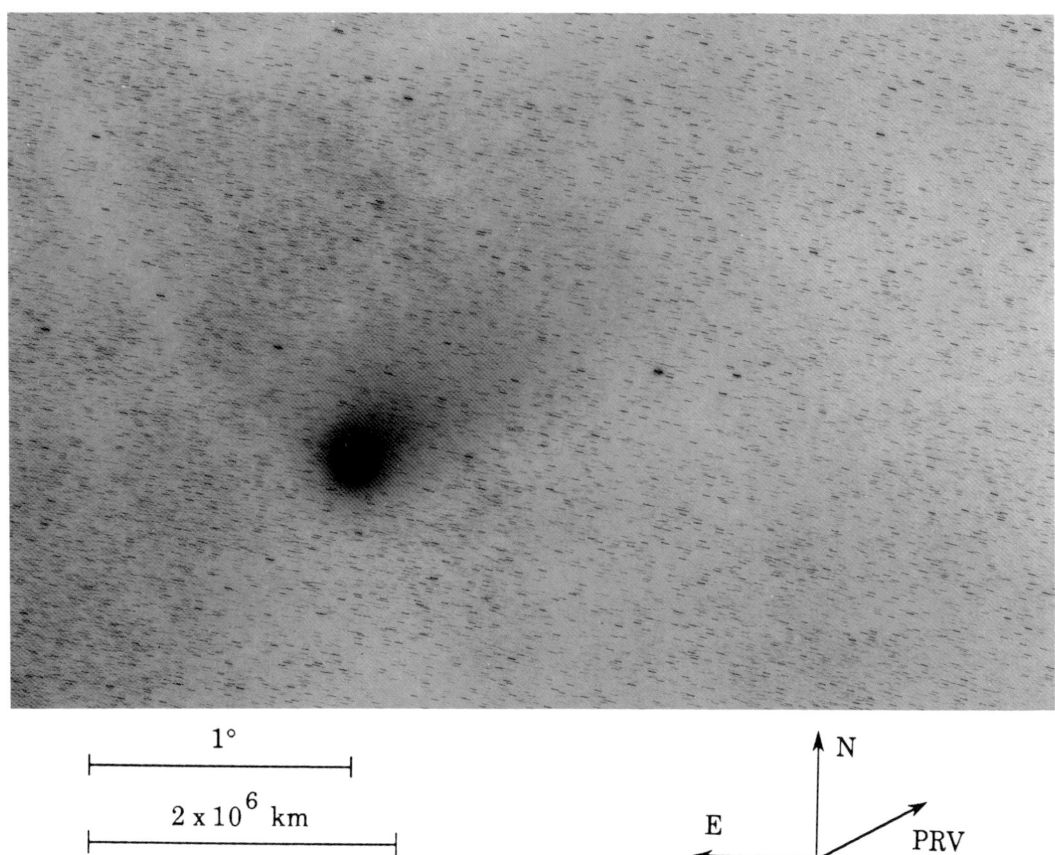

(top) Fig. 480*: 1986 Apr 6.89271 UT; exposure 69.0 minutes on 103a-O emulsion with no filter; r=1.27, Δ=0.44, β=136.1°; LSPN-3637.
Photograph submitted by K. Sivaraman, Indian Institute for Astrophysics, Kodaikanal Station.

(bottom) Fig. 481: 1986 Apr 7.20556 UT; exposure 10.0 minutes on 098-04 emulsion with RG-610 filter; r=1.27, Δ=0.43, β=137.0°; LSPN-743.
Photograph by C. Torres/H. Wroblewski, Cerro el Roble Astronomical Observatory.

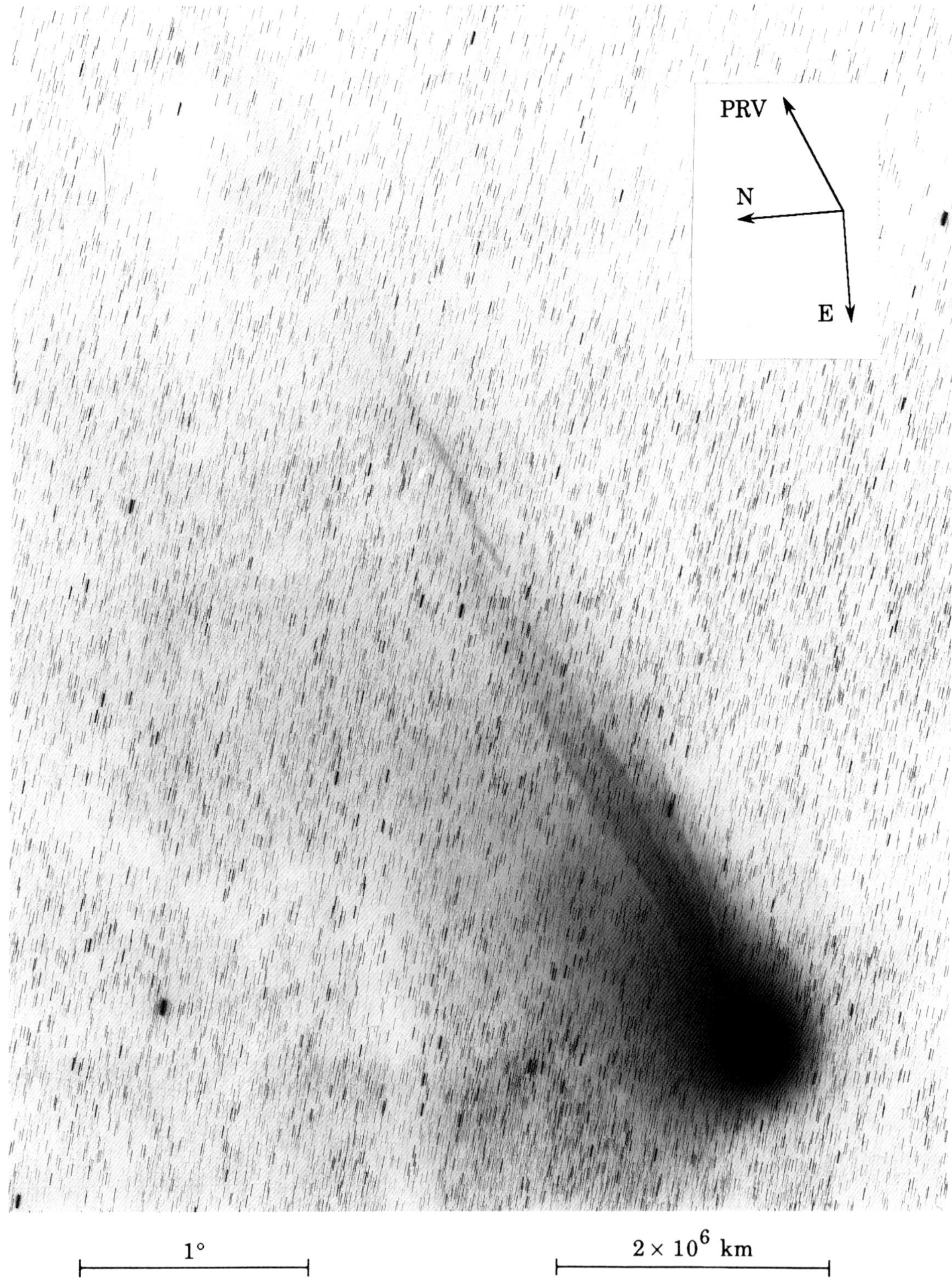

1°

2×10^6 km

Fig. 482: 1986 Apr 7.24236 UT; exposure 20.0 minutes on IIa-O emulsion with no filter; r=1.27, Δ=0.43, β=137.1°; LSPN-1154.
Photograph by F. Miller, University of Michigan/CTIO.

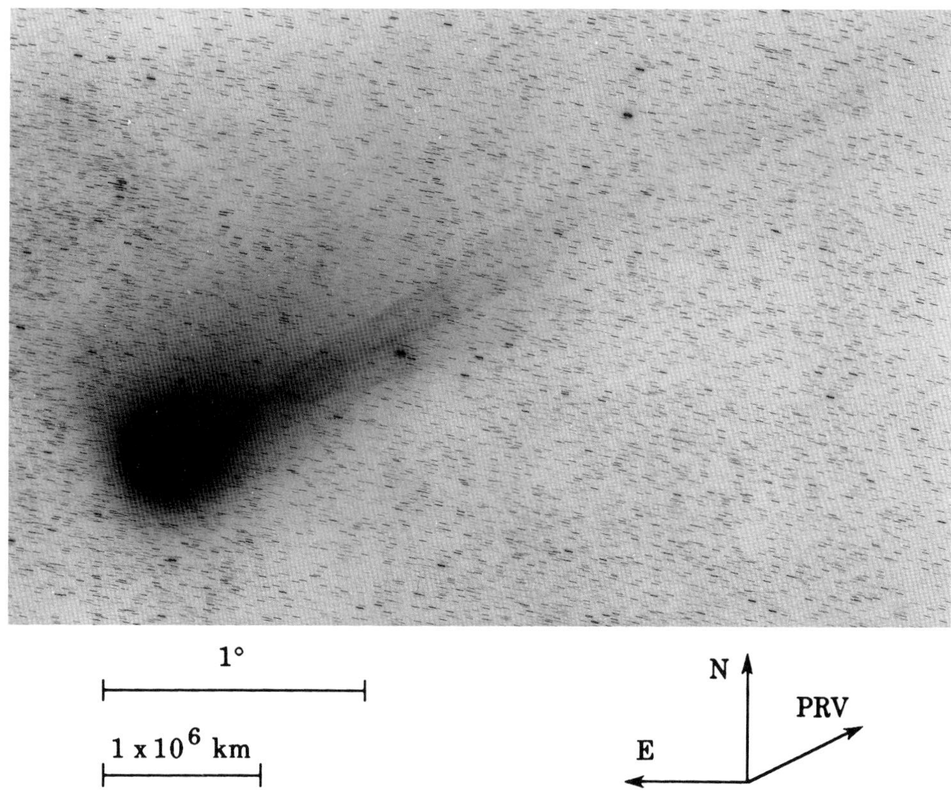

Fig. 483: 1986 Apr 7.27222 UT; exposure 10.0 minutes on 103a-O emulsion with GG-385 filter; r=1.27, Δ=0.43, β=137.1°; LSPN-744.
Photograph by C. Torres/H. Wroblewski, Cerro el Roble Astronomical Observatory.

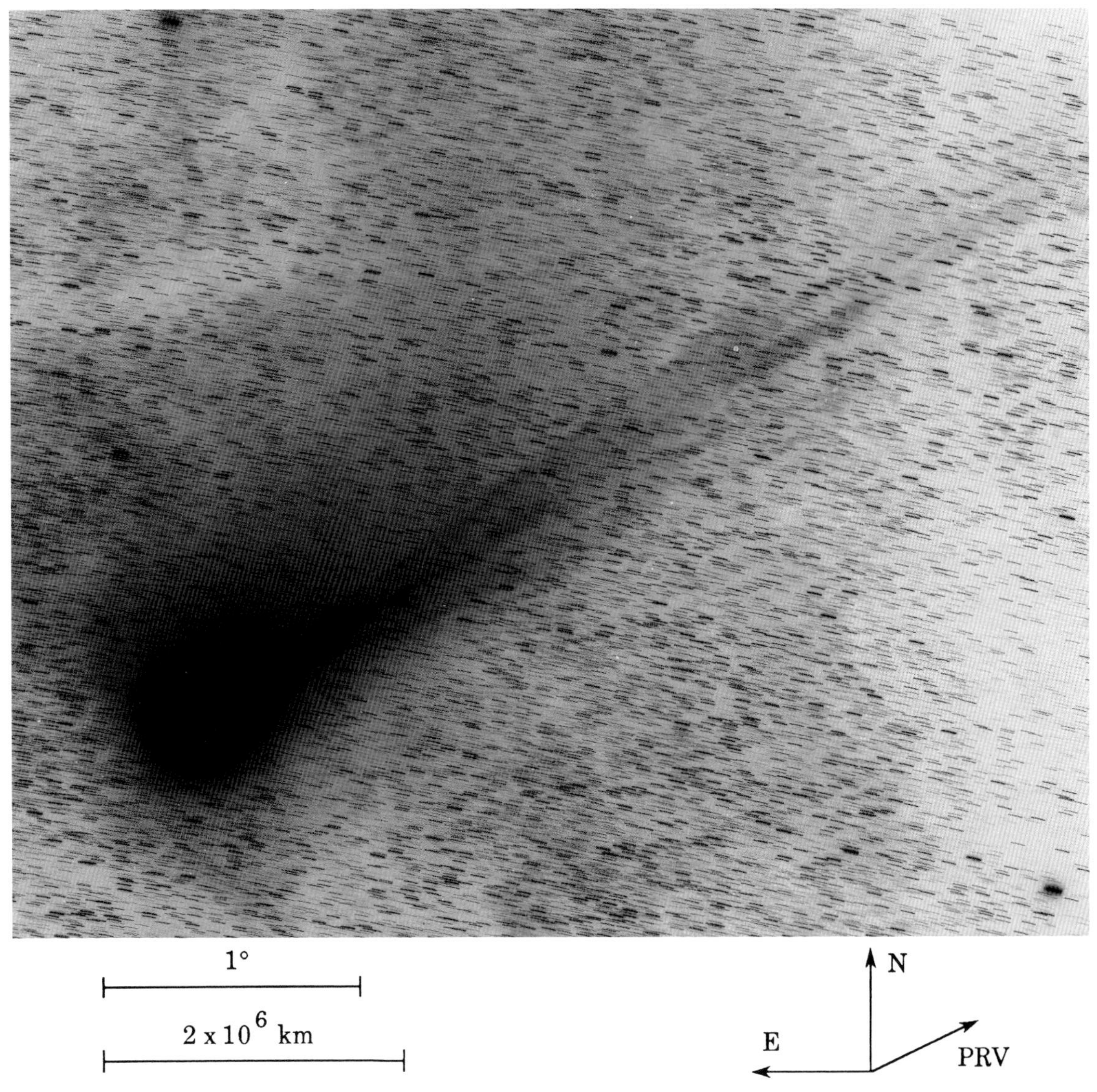

Fig. 484: 1986 Apr 7.28472 UT; exposure 20.0 minutes on IIa-O emulsion with no filter; r=1.28, Δ=0.43, β=137.2°; LSPN-1155.
Photograph by F. Miller, University of Michigan/CTIO.

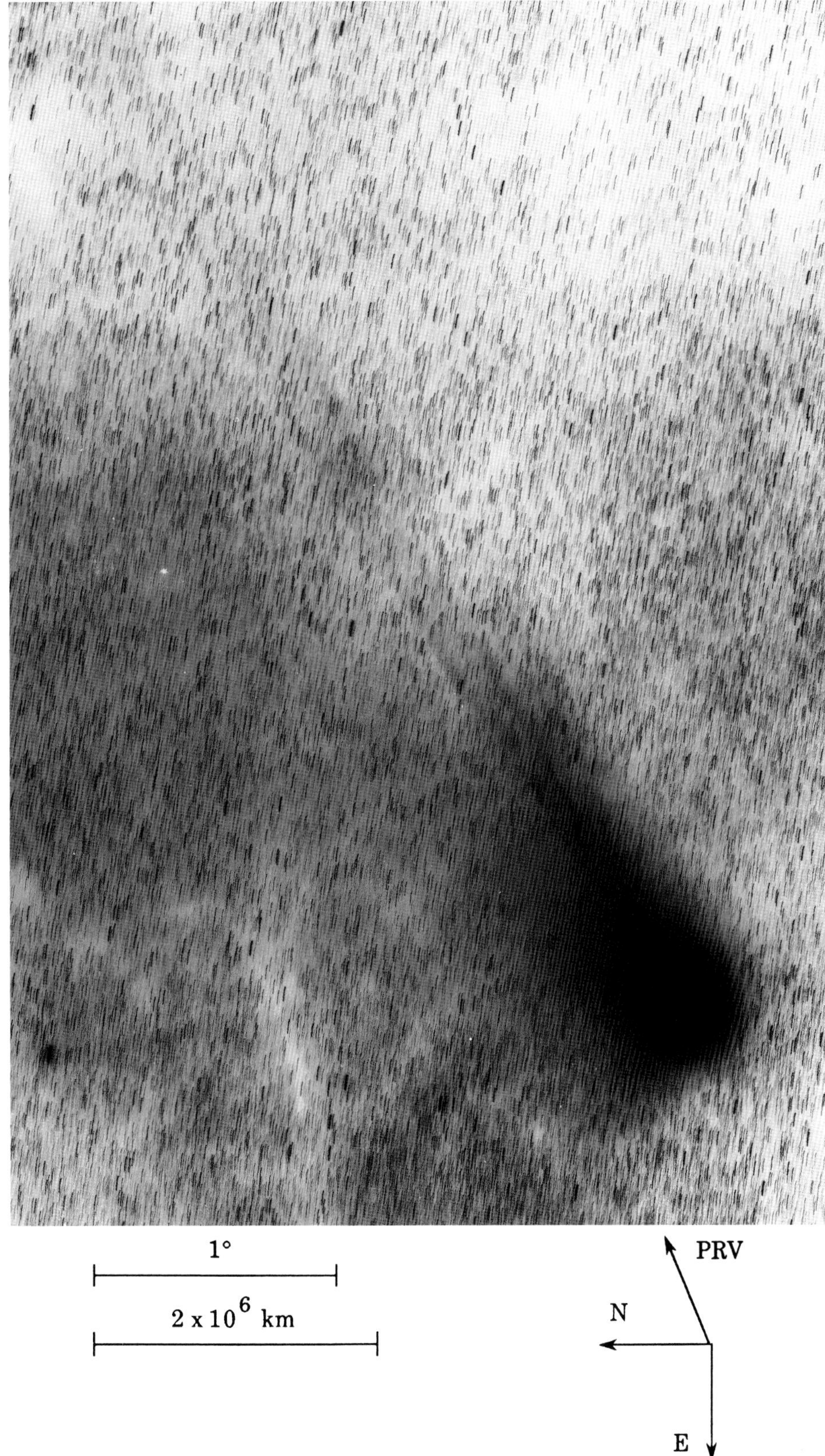

Fig. 485: 1986 Apr 7.30556 UT; exposure 20.0 minutes on 098-04 emulsion with no filter; r=1.28, Δ=0.43, β=137.2°; LSPN-1156.
Photograph by F. Miller, University of Michigan/CTIO.

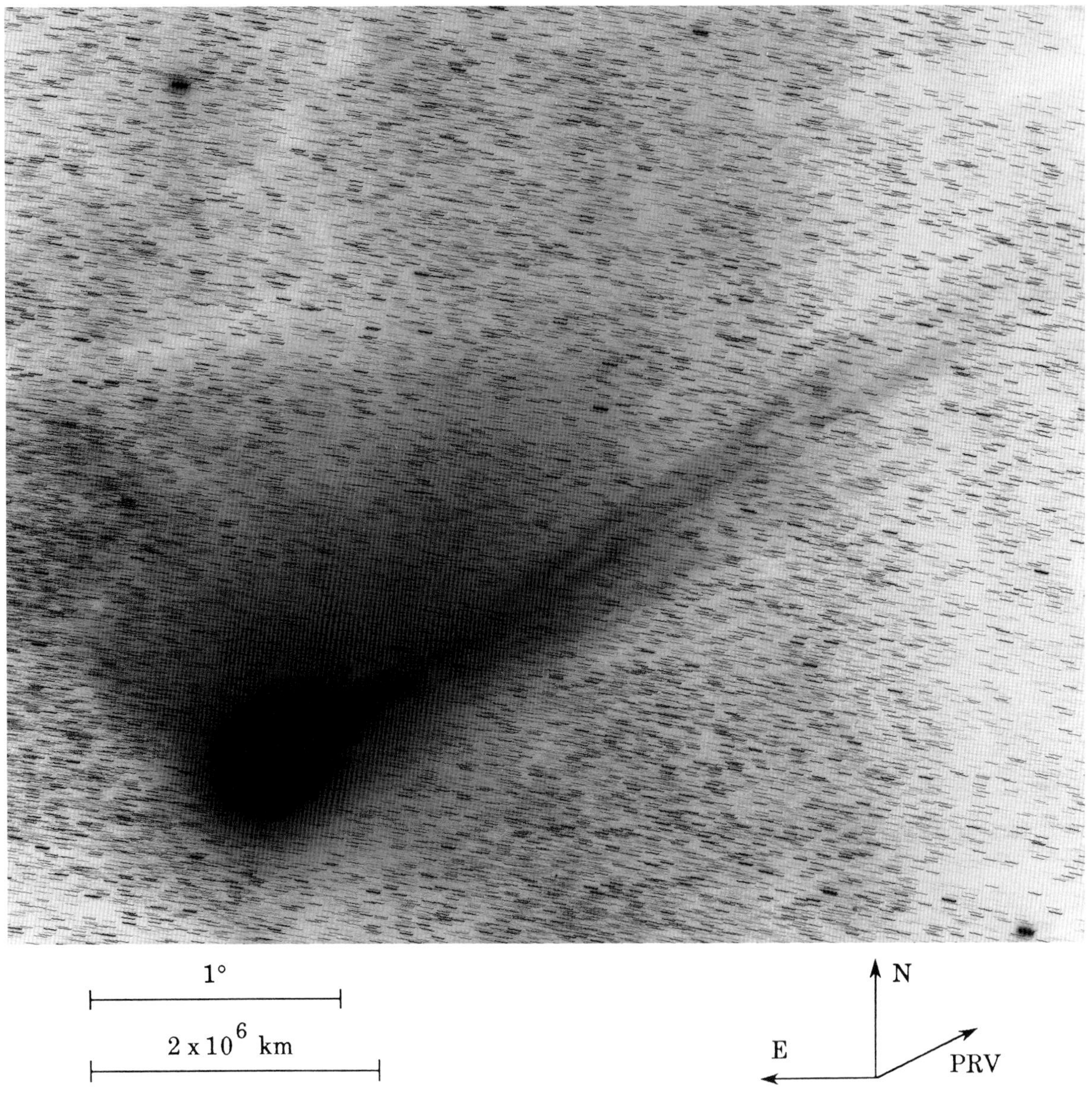

Fig. 486: 1986 Apr 7.34653 UT; exposure 20.0 minutes on IIa-O emulsion with no filter; r=1.28, Δ=0.43, β=137.4°; LSPN-1157.
Photograph by F. Miller, University of Michigan/CTIO.

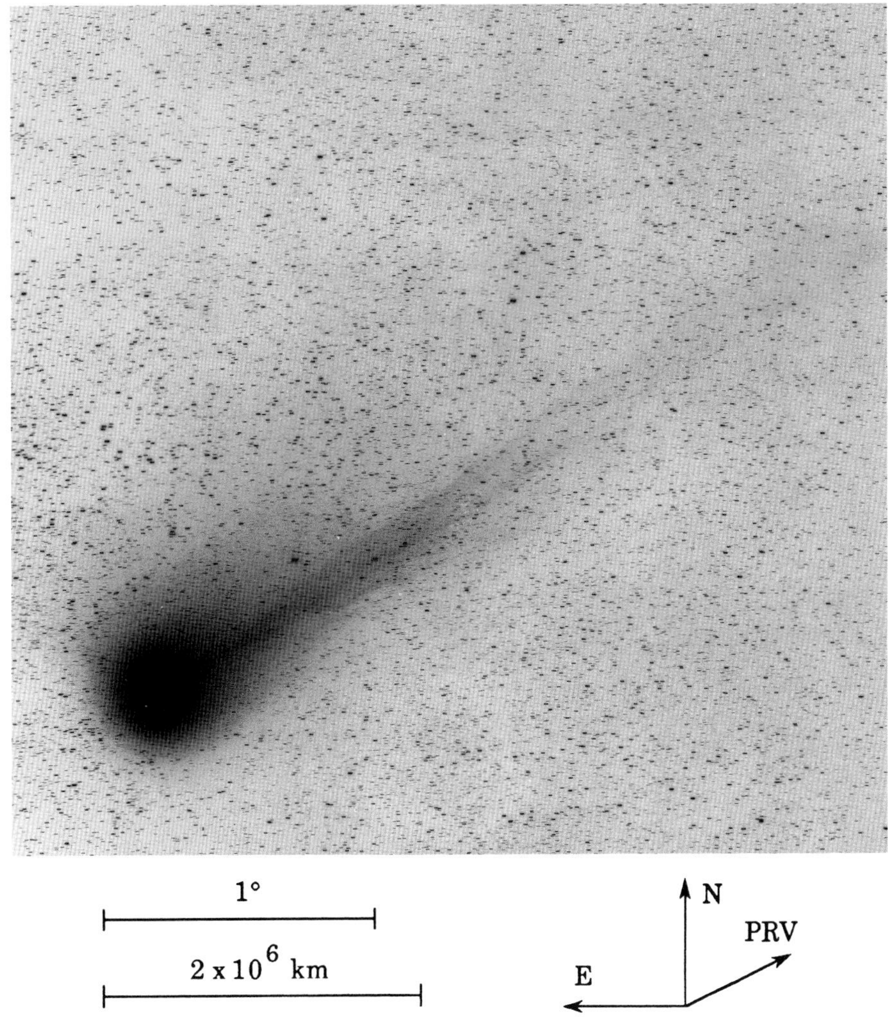

Fig. 487: 1986 Apr 7.35938 UT; exposure 5.0 minutes on IIa-O emulsion with no filter; r=1.28, Δ=0.43, β=137.4°; LSPN-1158.
Photograph by F. Miller, University of Michigan/CTIO.

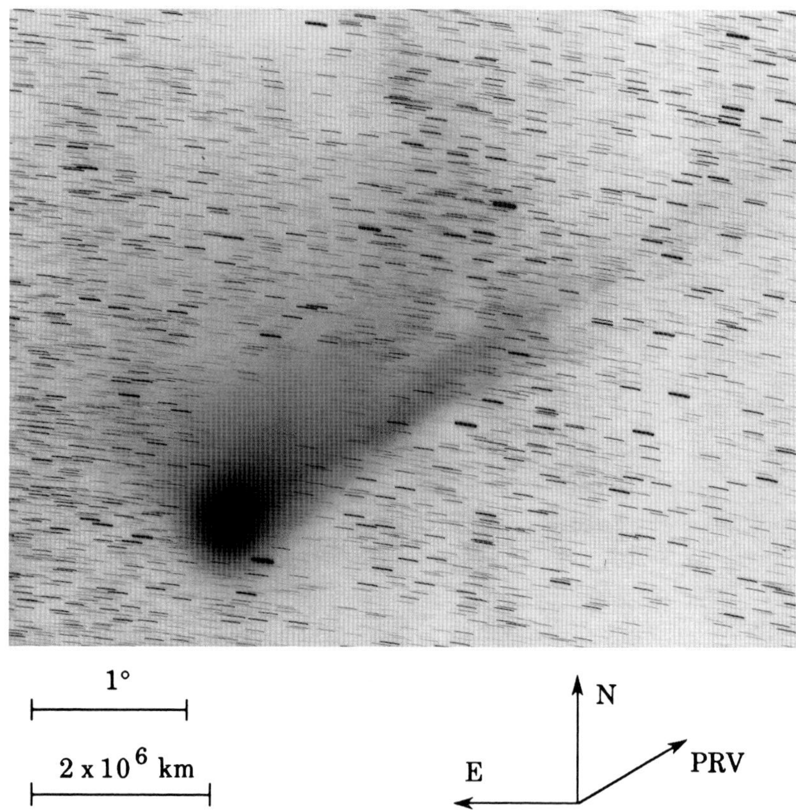

Fig. 488: 1986 Apr 8.02292 UT; exposure 43.0 minutes on IIa-O emulsion with no filter; r=1.29, Δ=0.43, β=139.3°; LSPN-2863.
Photograph by G. Malcolm, Boyden Observatory.

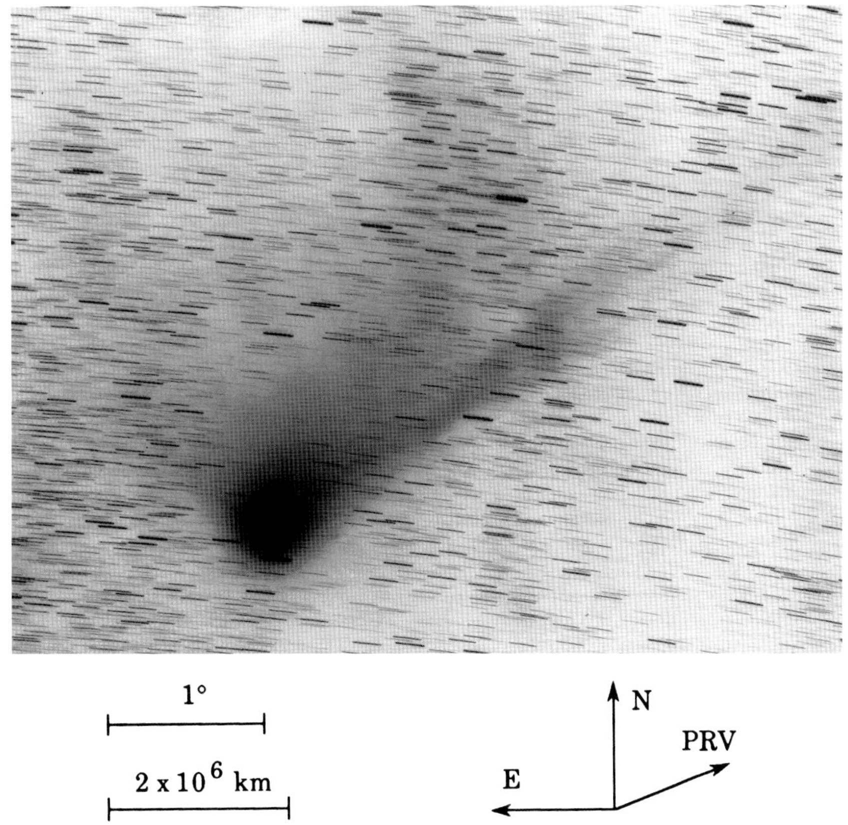

Fig. 489: 1986 Apr 8.06736 UT; exposure 60.0 minutes on IIa-O emulsion with no filter; r=1.29, Δ=0.43, β=139.5°; LSPN-2864.
Photograph by G. Malcolm, Boyden Observatory.

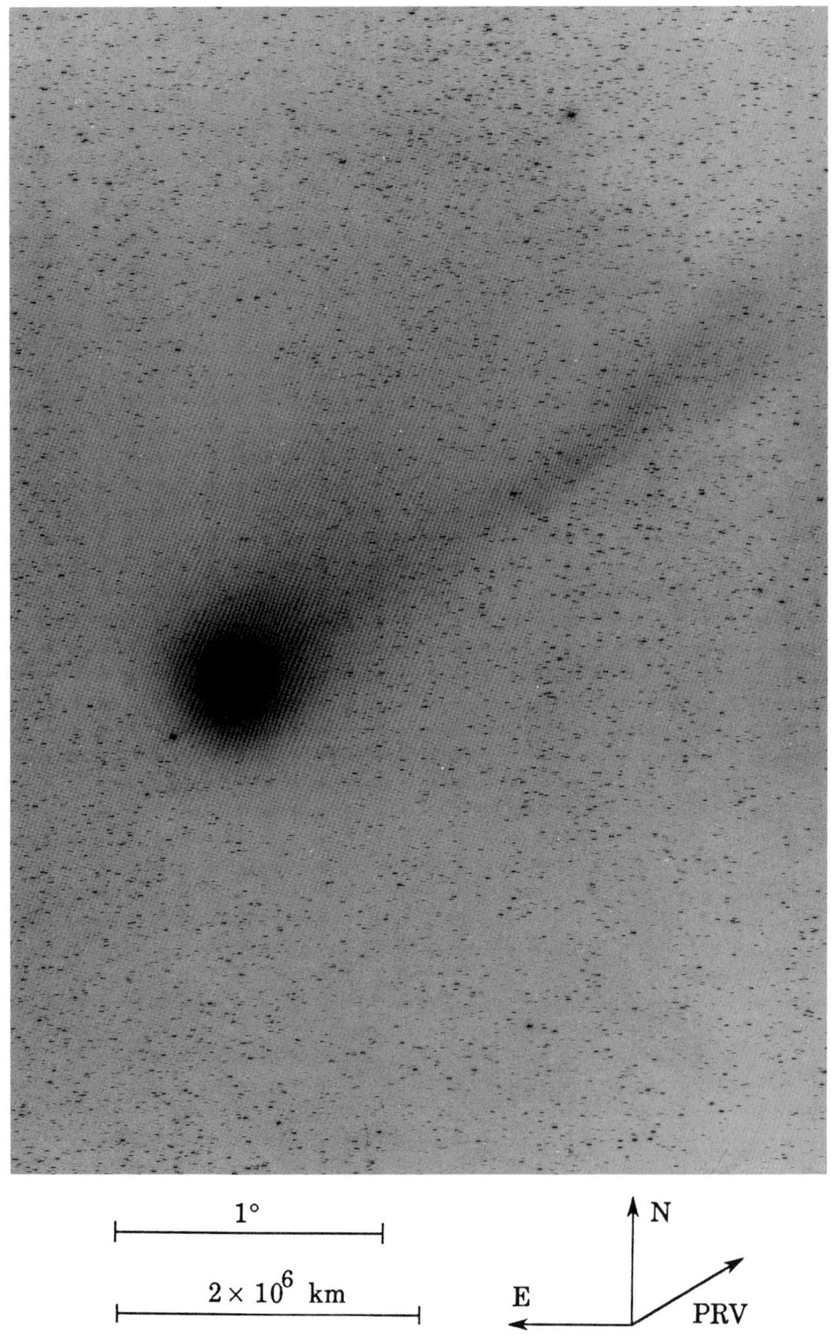

Fig. 490: 1986 Apr 8.12951 UT; exposure 5.0 minutes on 103a-O emulsion with no filter; r=1.29, Δ=0.43, β=139.7°; LSPN-1159.
Photograph by F. Miller, University of Michigan/CTIO.

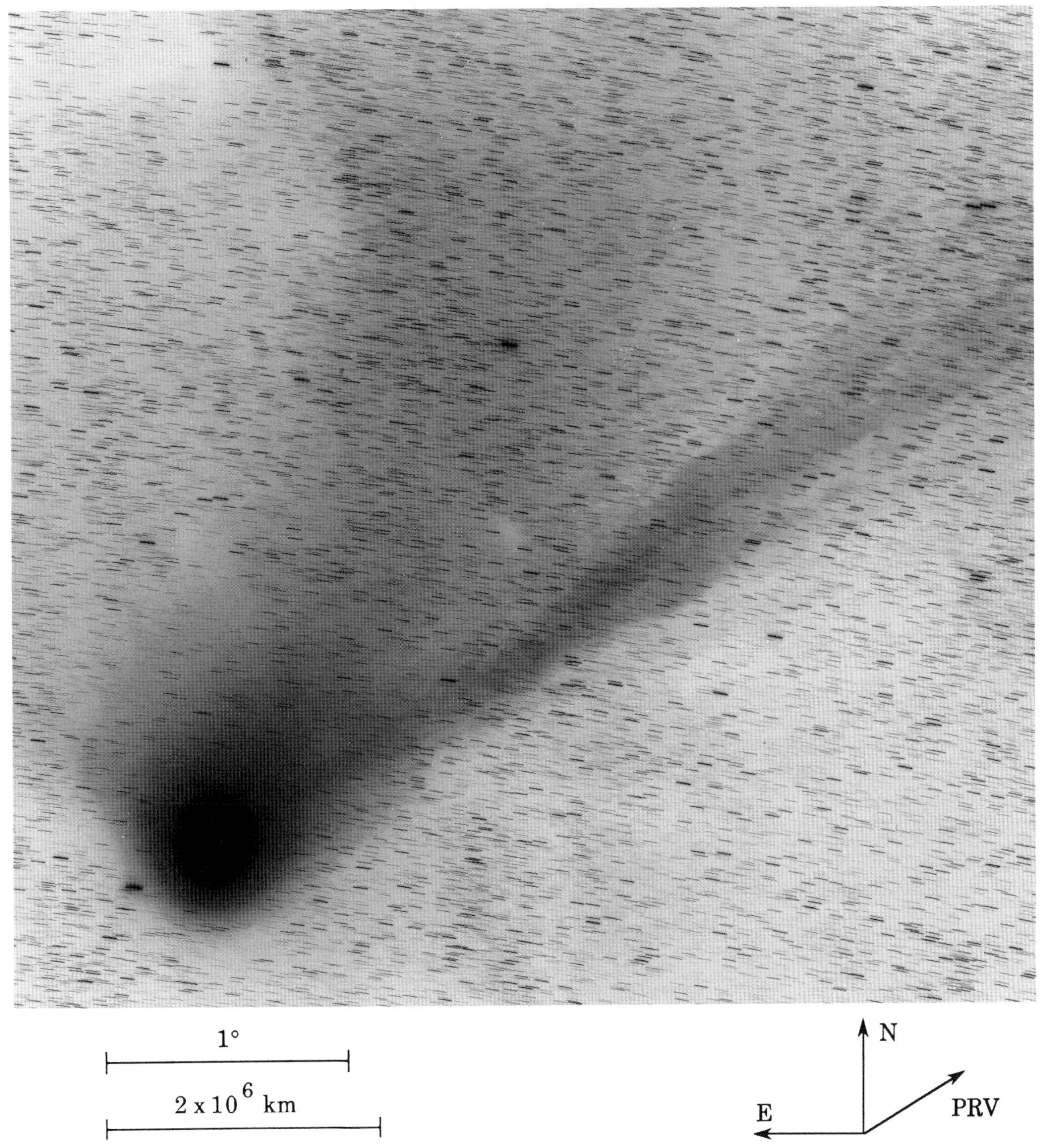

Fig. 491: 1986 Apr 8.15208 UT; exposure 20.0 minutes on IIa-O emulsion with no filter; r=1.29, Δ=0.43, β=139.7°; LSPN-1160.
Photograph by F. Miller, University of Michigan/CTIO.

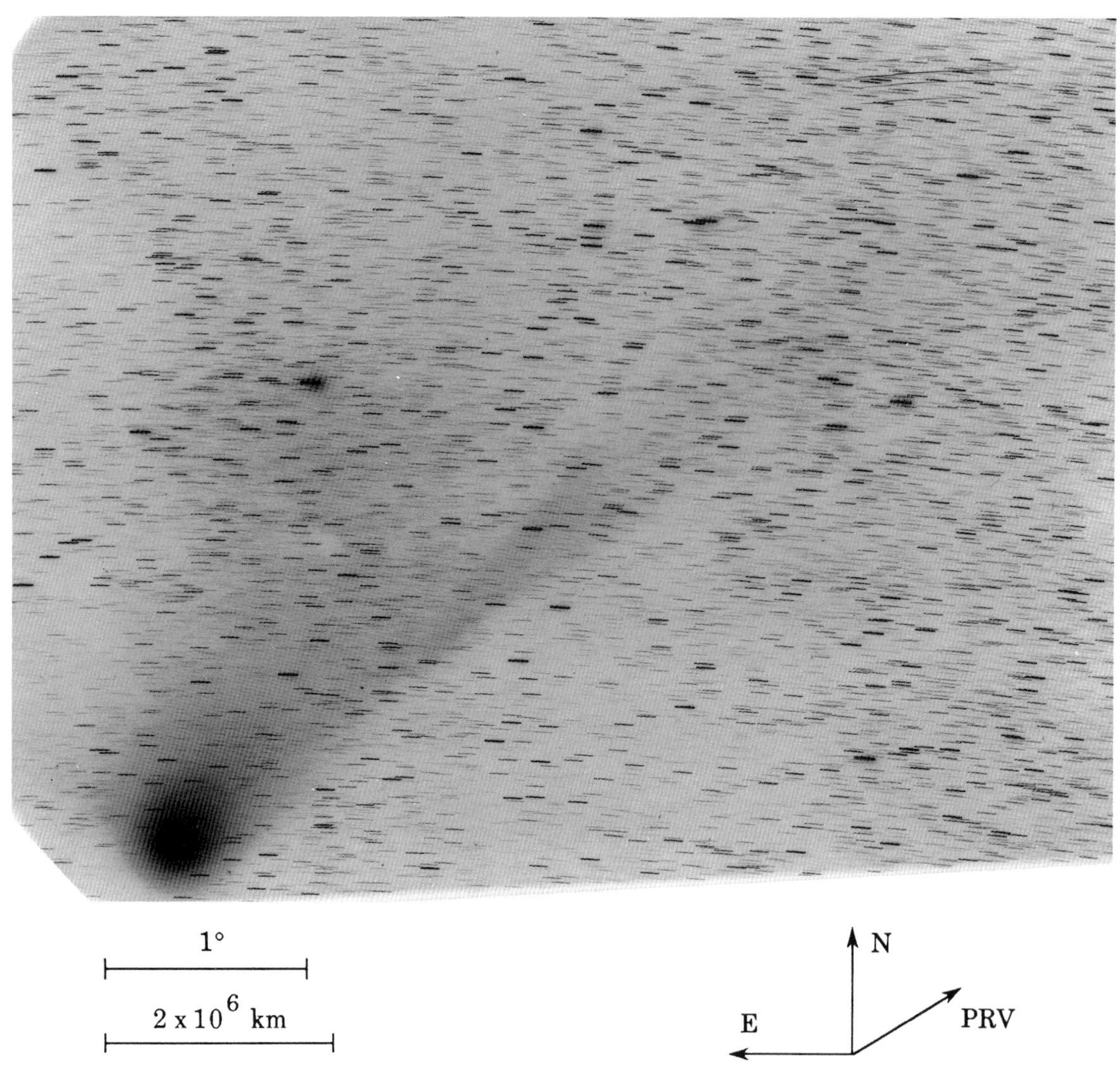

Fig. 492: 1986 Apr 8.23264 UT; exposure 28.0 minutes on Tech. Pan 2415 emulsion with no filter; r=1.29, Δ=0.43, β=140.0°; LSPN-1454.
Photograph by W. Liller, Easter Island, LSPN Island Network.

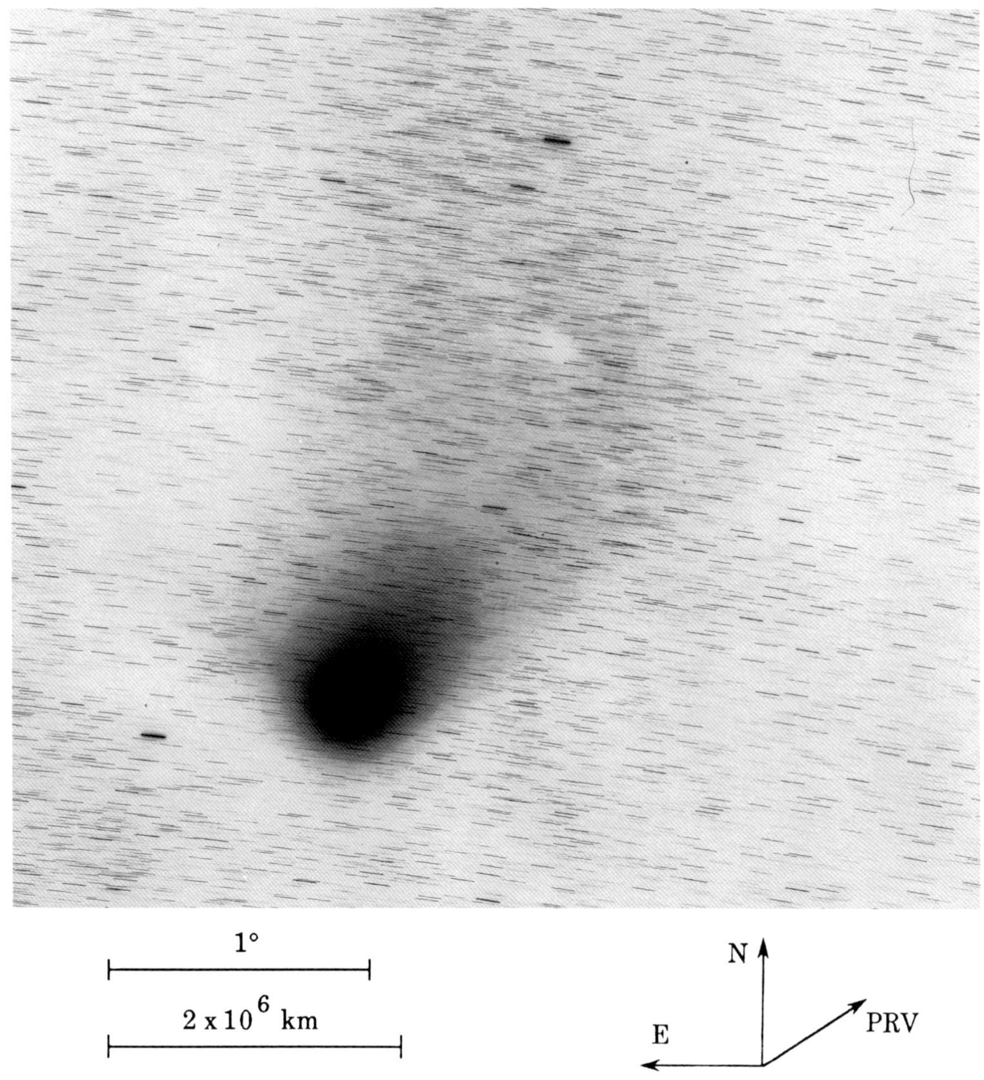

Fig. 493: 1986 Apr 8.24722 UT; exposure 30.0 minutes on IIIa-F emulsion with RG-610 filter; r=1.29, Δ=0.43, β=140.0°; LSPN-745.
Photograph by C. Torres/H. Wroblewski, Cerro el Roble Astronomical Observatory.

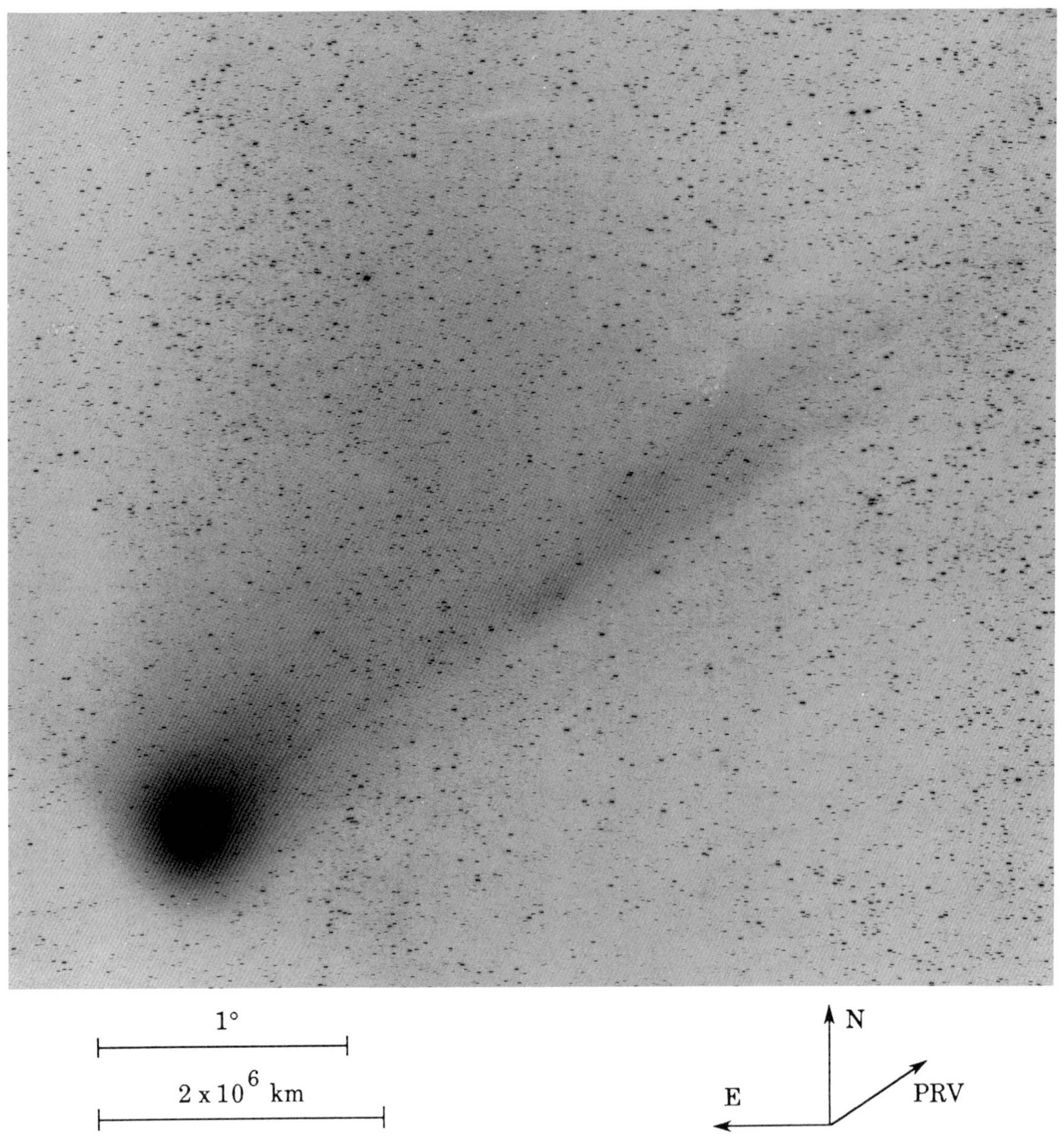

Fig. 494: 1986 Apr 8.25451 UT; exposure 5.0 minutes on 103a-O emulsion with no filter; r=1.29, Δ=0.43, β=140.0°; LSPN-1163.
Photograph by F. Miller, University of Michigan/CTIO.

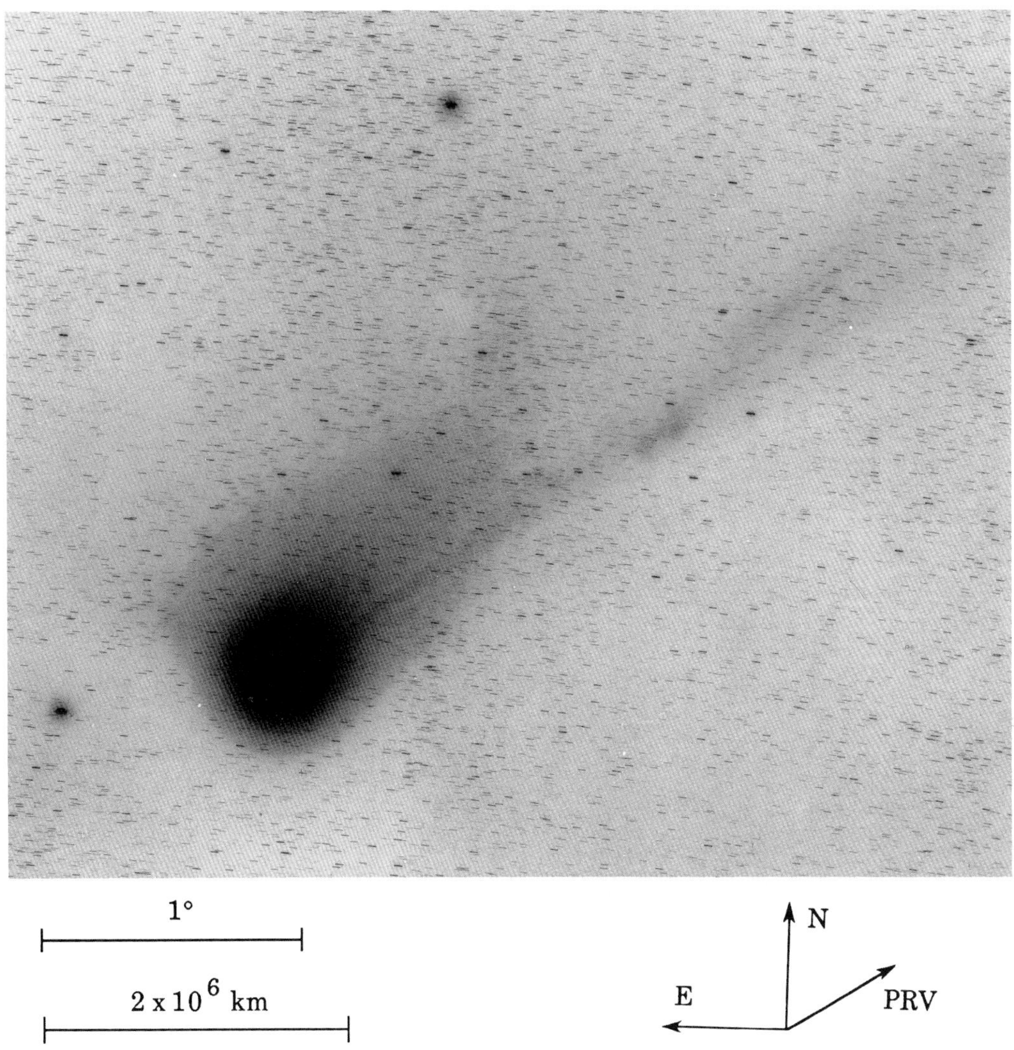

Fig. 495: 1986 Apr 8.27153 UT; exposure 10.0 minutes on 103a-O emulsion with GG-385 filter; r=1.29, Δ=0.43, β=140.1°; LSPN-746.
Photograph by C. Torres/H. Wroblewski, Cerro el Roble Astronomical Observatory.

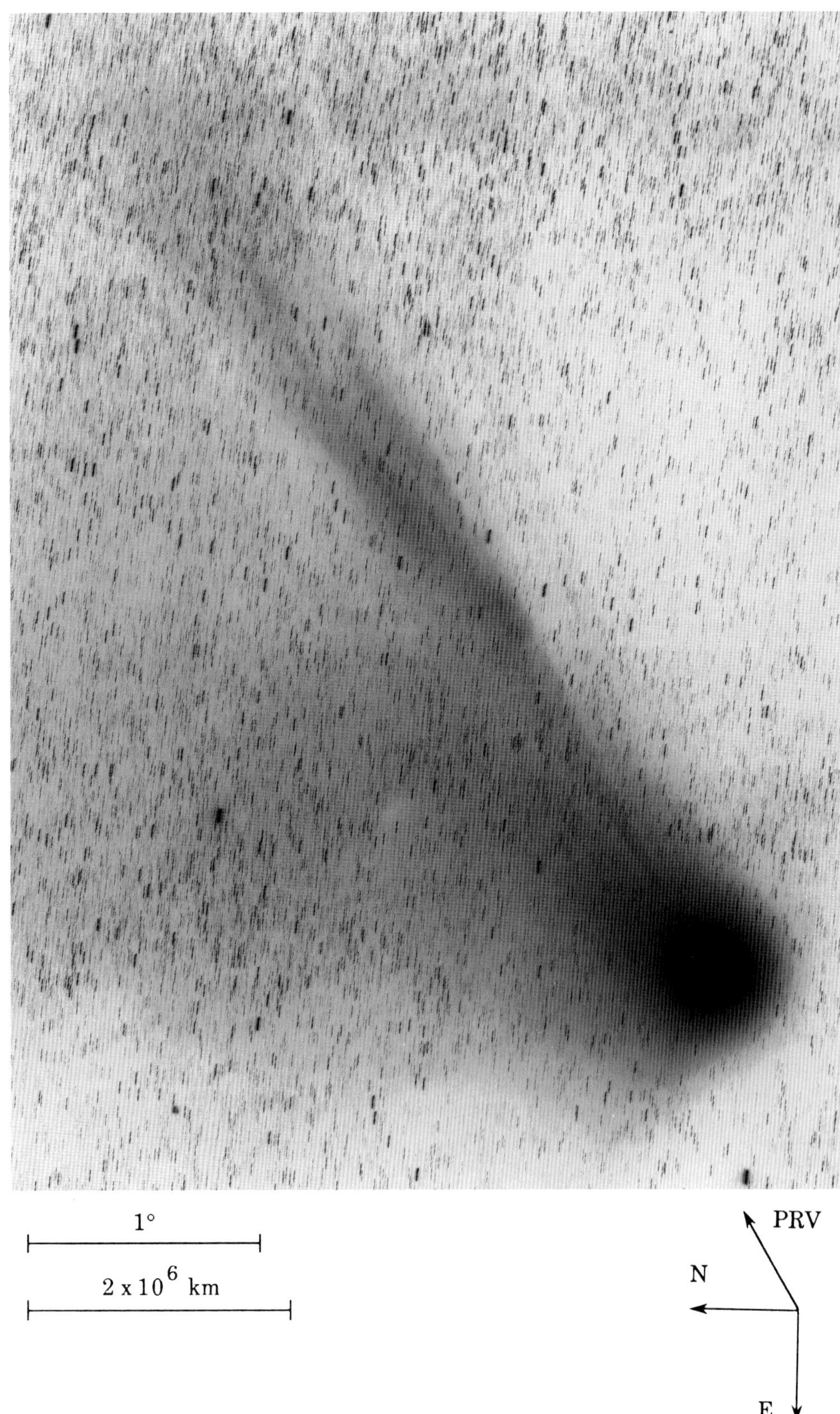

Fig. 496: 1986 Apr 8.27361 UT; exposure 20.0 minutes on IIa-O emulsion with no filter; r=1.29, Δ=0.43, β=140.1°; LSPN-1164.
Photograph by F. Miller, University of Michigan/CTIO.

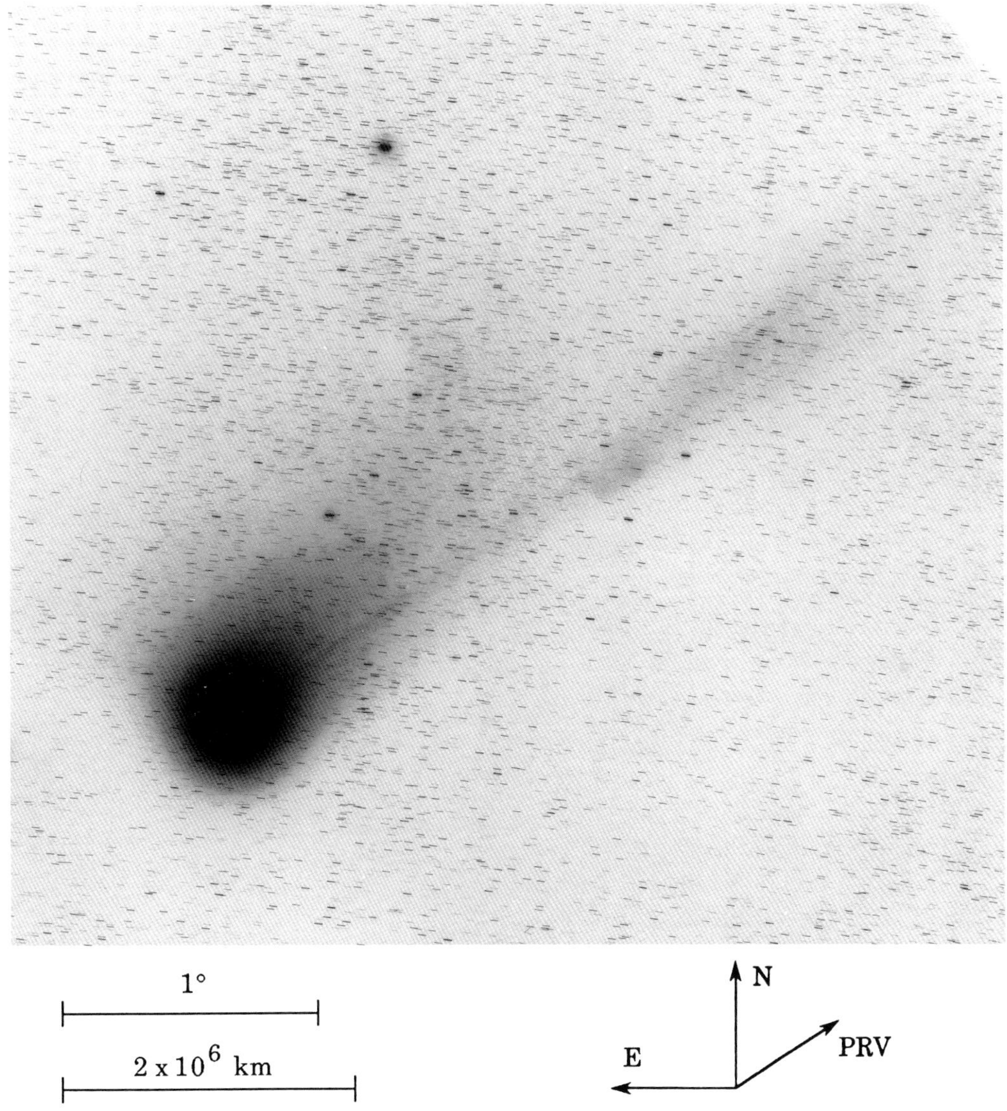

Fig. 497: 1986 Apr 8.28681 UT; exposure 10.0 minutes on 103a-O emulsion with GG-385 filter; $r=1.29$, $\Delta=0.43$, $\beta=140.1°$; LSPN-747.
Photograph by C. Torres/H. Wroblewski, Cerro el Roble Astronomical Observatory.

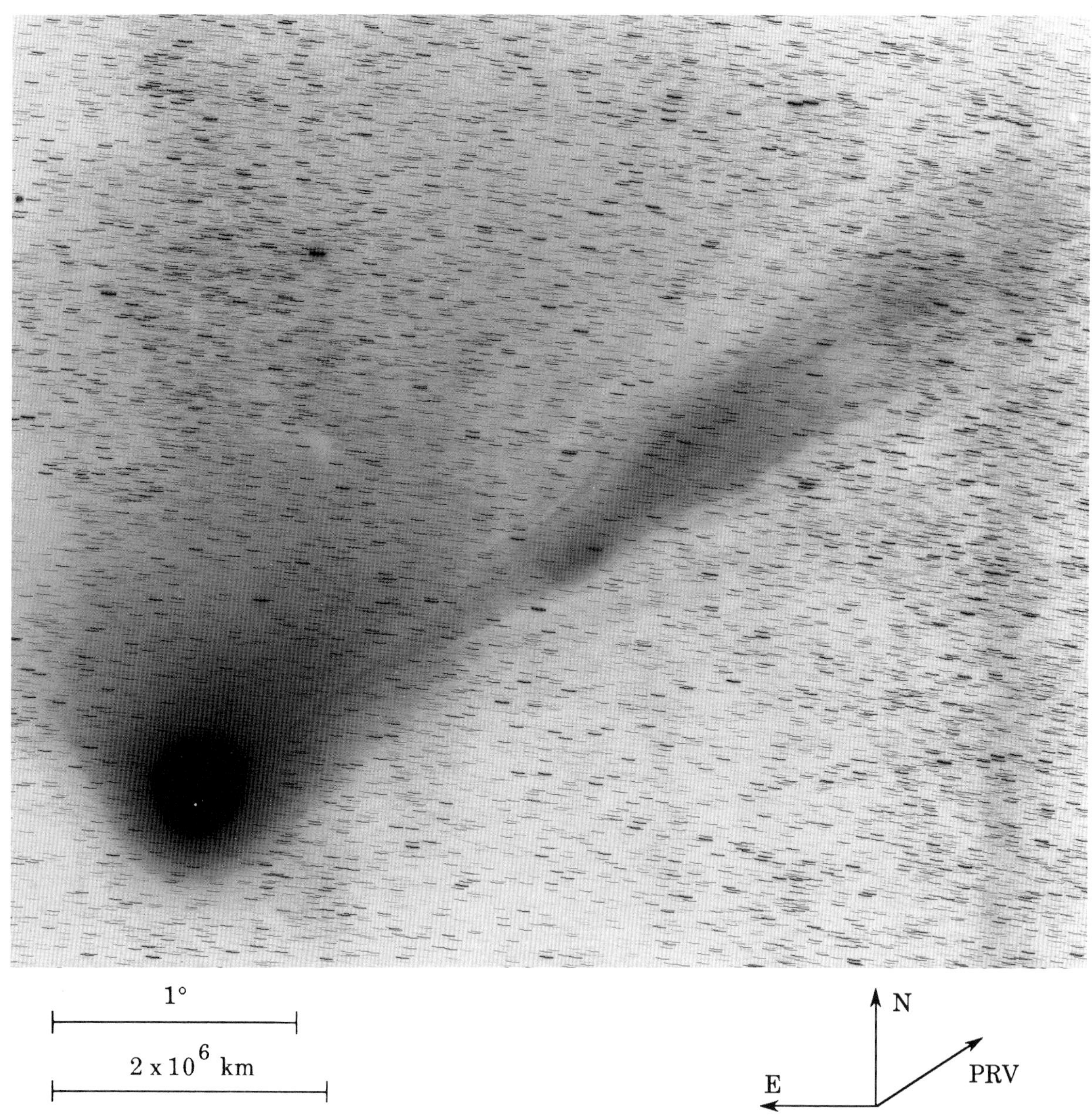

Fig. 498: 1986 Apr 8.31319 UT; exposure 20.0 minutes on IIa-O emulsion with no filter; r=1.29, Δ=0.43, β=140.2°; LSPN-1165.
Photograph by F. Miller, University of Michigan/CTIO.

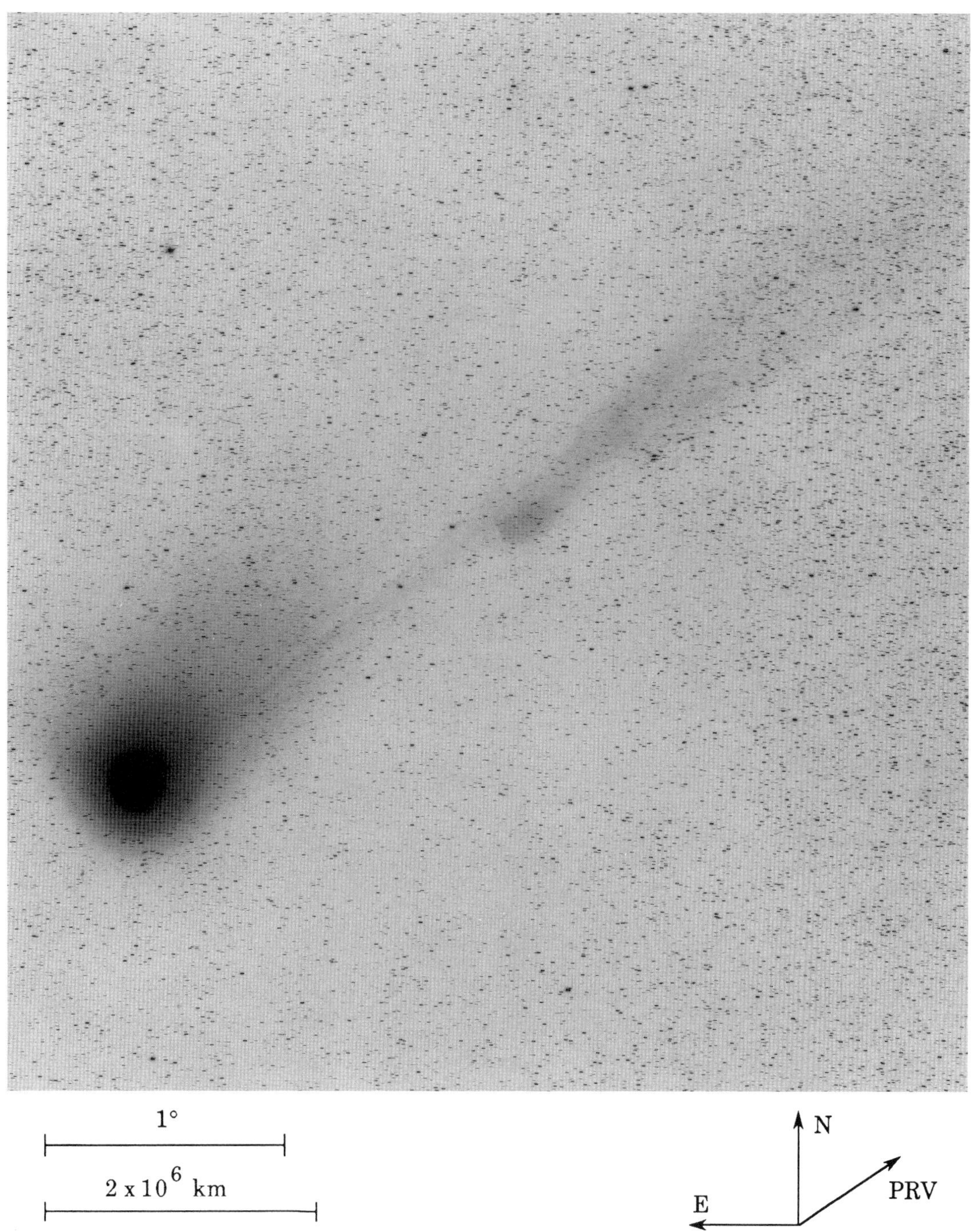

Fig. 499: 1986 Apr 8.37743 UT; exposure 5.0 minutes on 103a-O emulsion with no filter; r=1.29, Δ=0.43, β=140.4°; LSPN-1167.
Photograph by F. Miller, University of Michigan/CTIO.

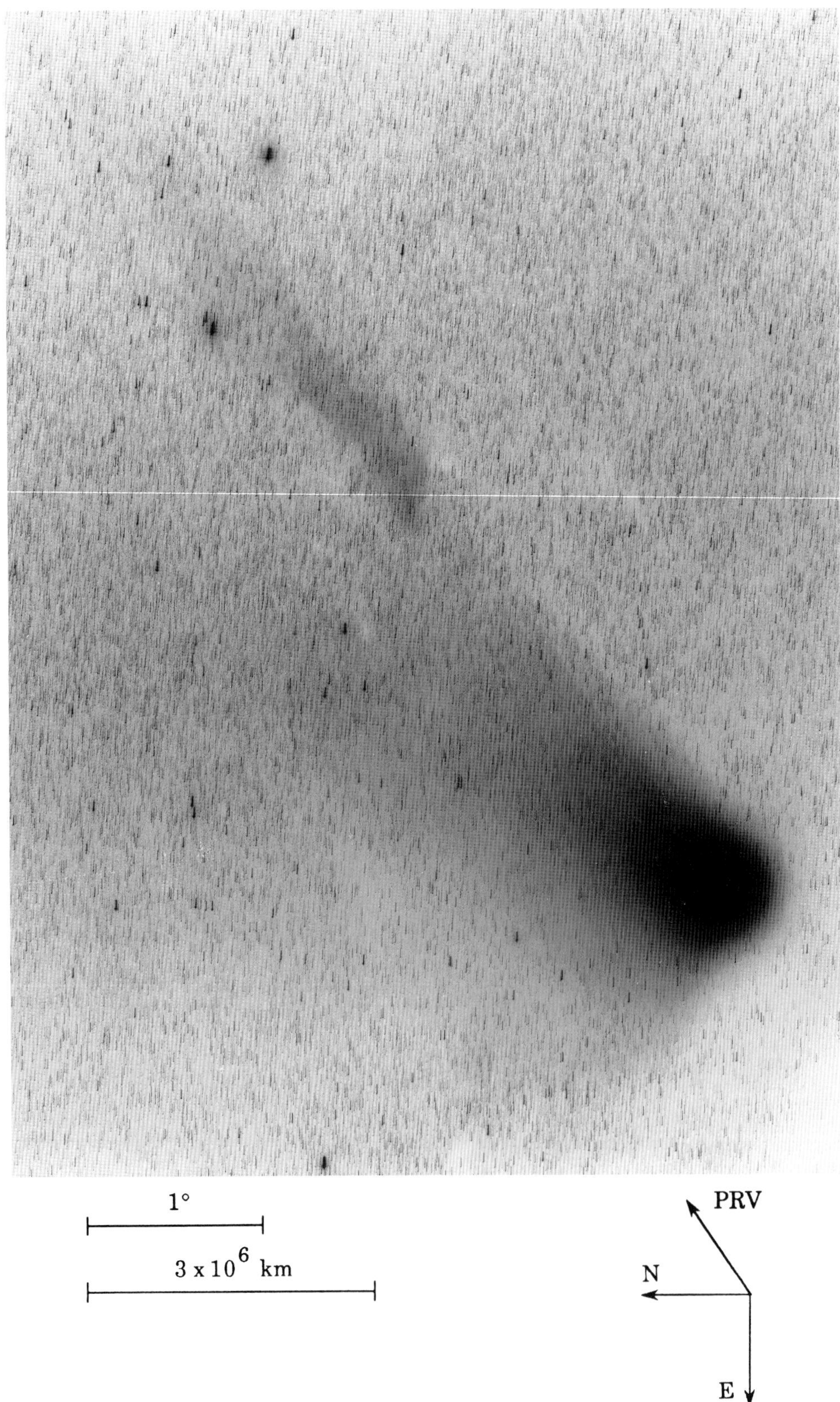

Fig. 500: 1986 Apr 8.72921 UT; exposure 30.0 minutes on IIa-O emulsion with GG-395 filter; r=1.30, Δ=0.42, β=141.5°; LSPN-2416.
Photograph from Royal Observatory/UK Schmidt Telescope Unit.

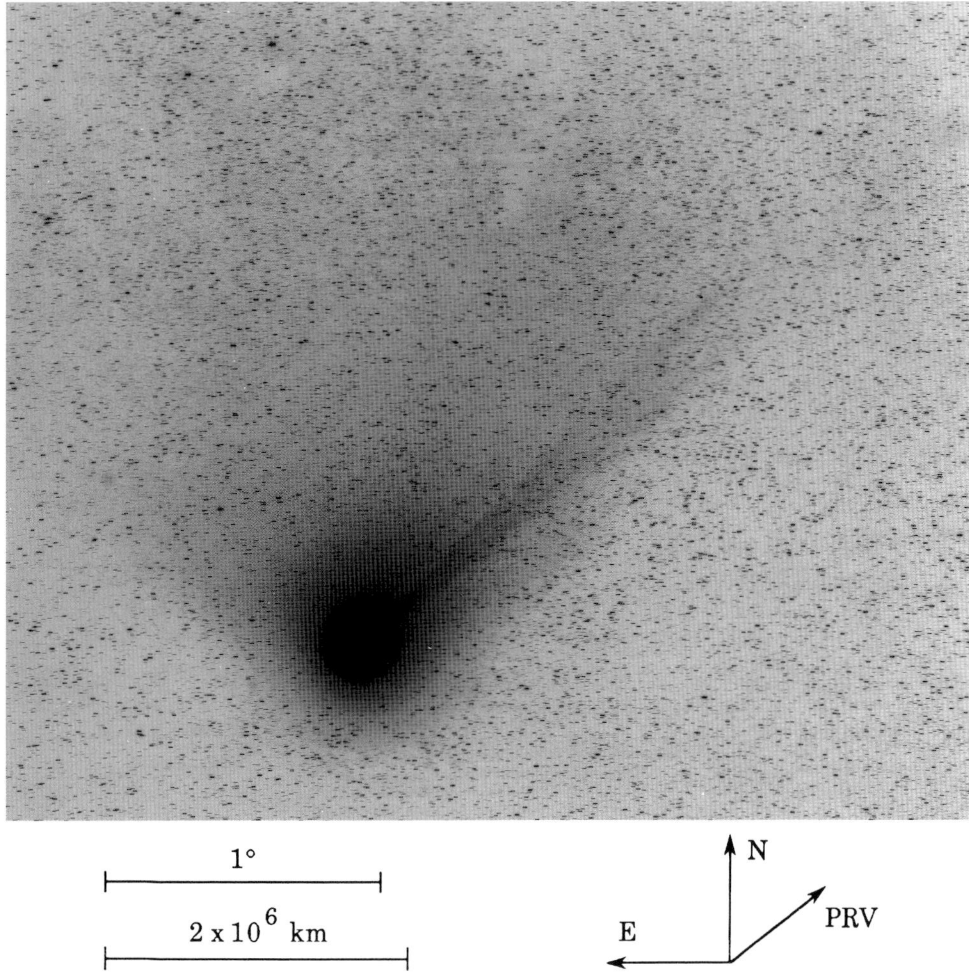

Fig. 501: 1986 Apr 9.12118 UT; exposure 5.0 minutes on IIa-O emulsion with no filter; r=1.30, Δ=0.42, β=142.6°; LSPN-1168.
Photograph by F. Miller, University of Michigan/CTIO.

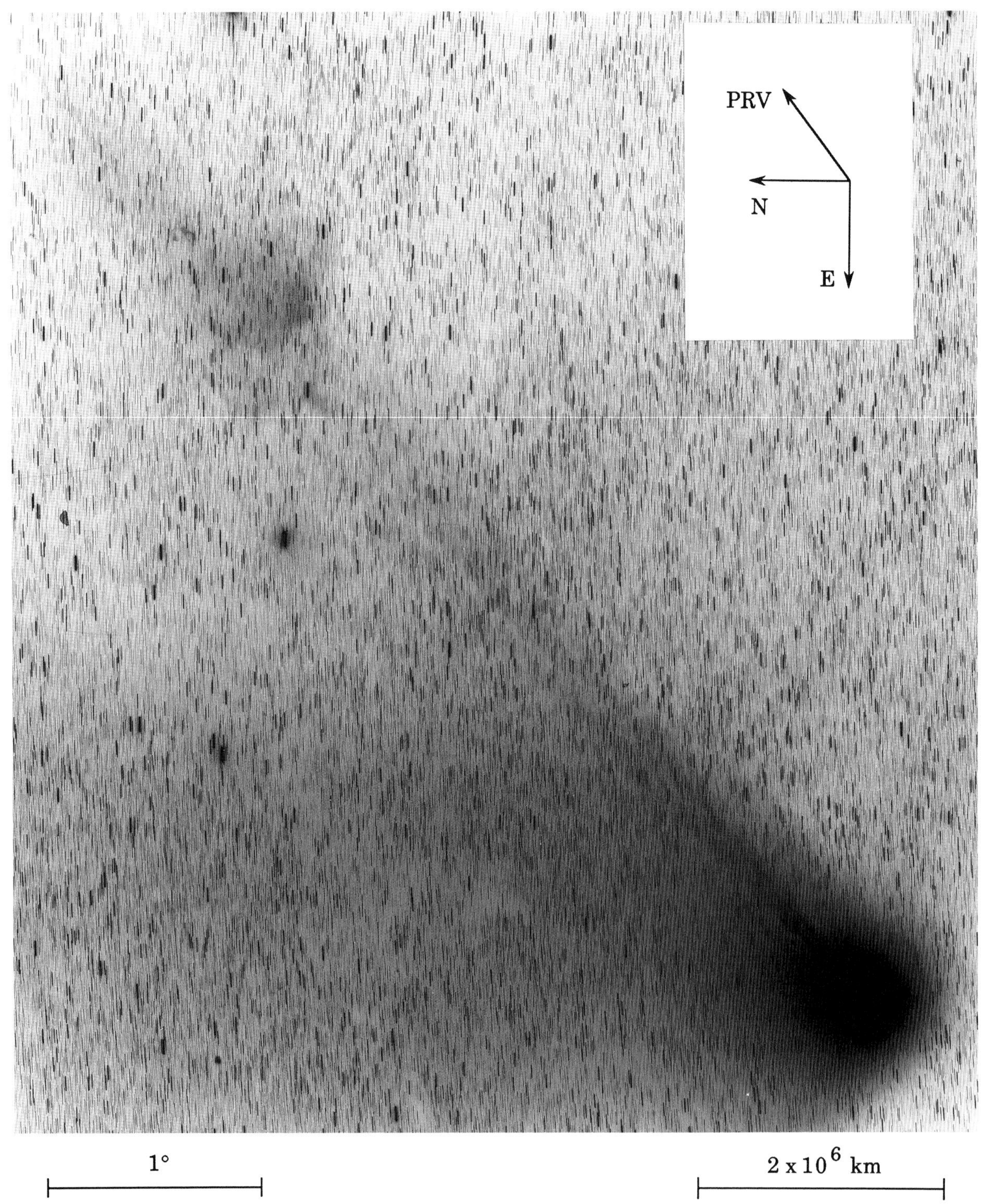

1°

2×10^6 km

Fig. 502: 1986 Apr 9.13958 UT; exposure 20.0 minutes on IIa-O emulsion with no filter; r=1.30, Δ=0.42, β=142.7°; LSPN-1169.
Photograph by F. Miller, University of Michigan/CTIO.

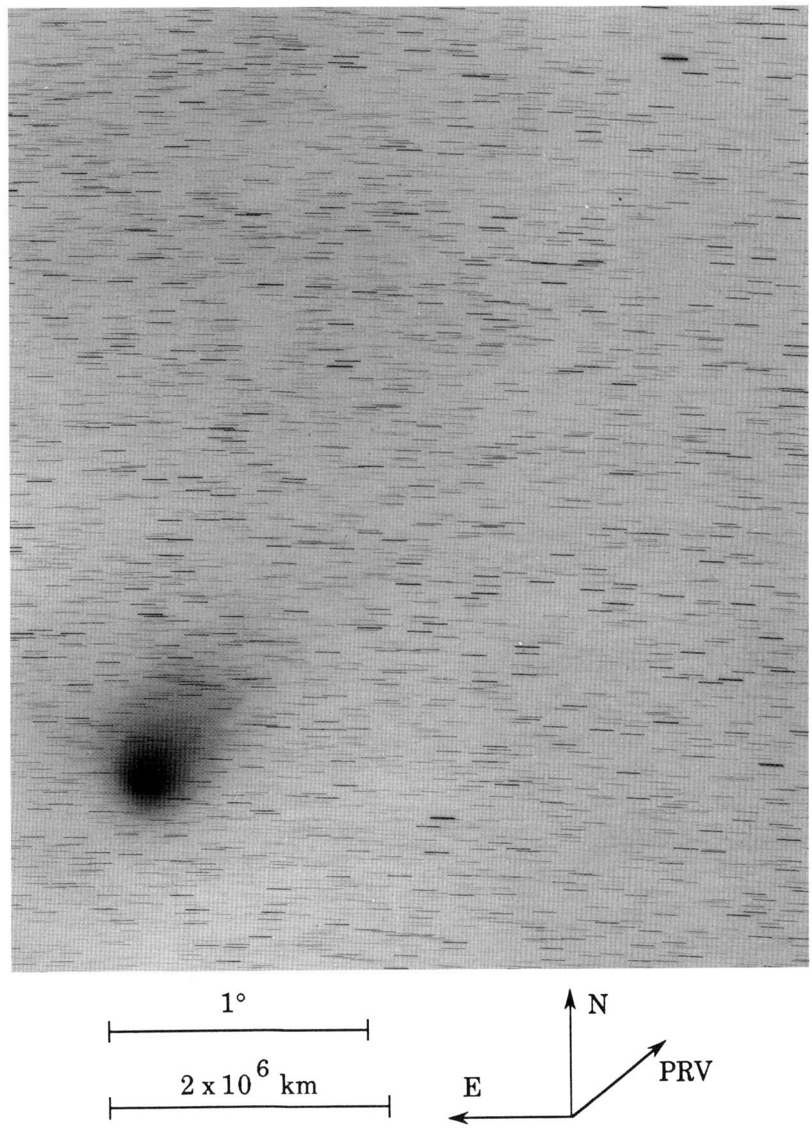

Fig. 503: 1986 Apr 9.16111 UT; exposure 30.0 minutes on IIIa-F emulsion with RG-610 filter; r=1.30, Δ=0.42, β=142.8°; LSPN-748.
Photograph by C. Torres/H. Wroblewski, Cerro el Roble Astronomical Observatory.

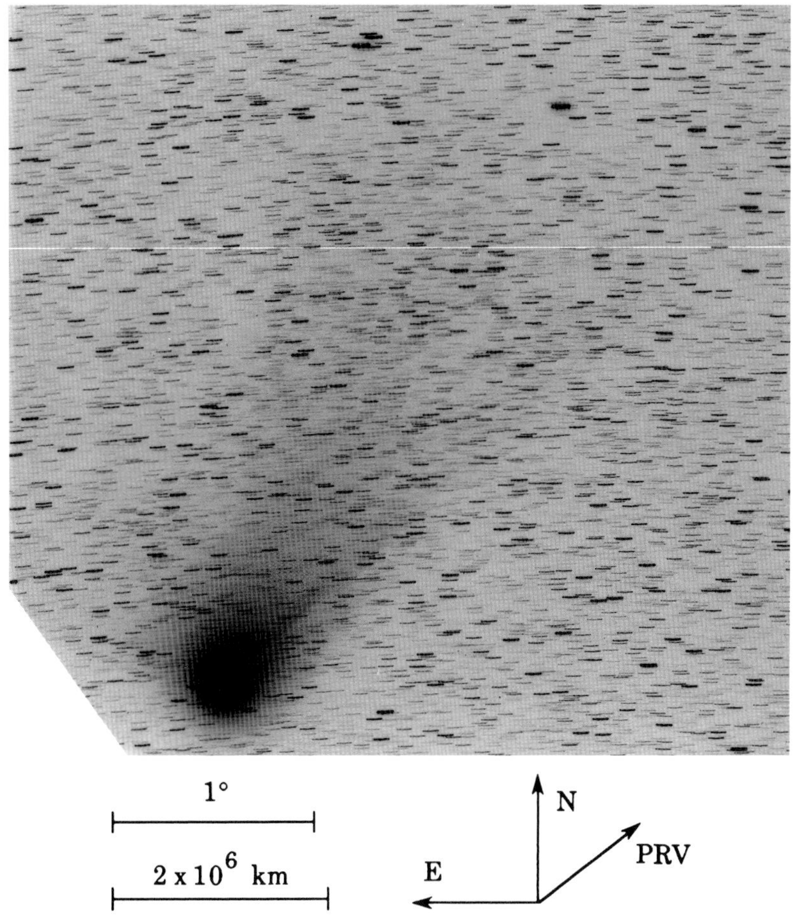

Fig. 504: 1986 Apr 9.23542 UT; exposure 25.0 minutes on Tech. Pan 2415 emulsion with no filter; r = 1.31, Δ = 0.42, β = 143.0°; LSPN-1459.
Photograph by W. Liller, Easter Island, LSPN Island Network.

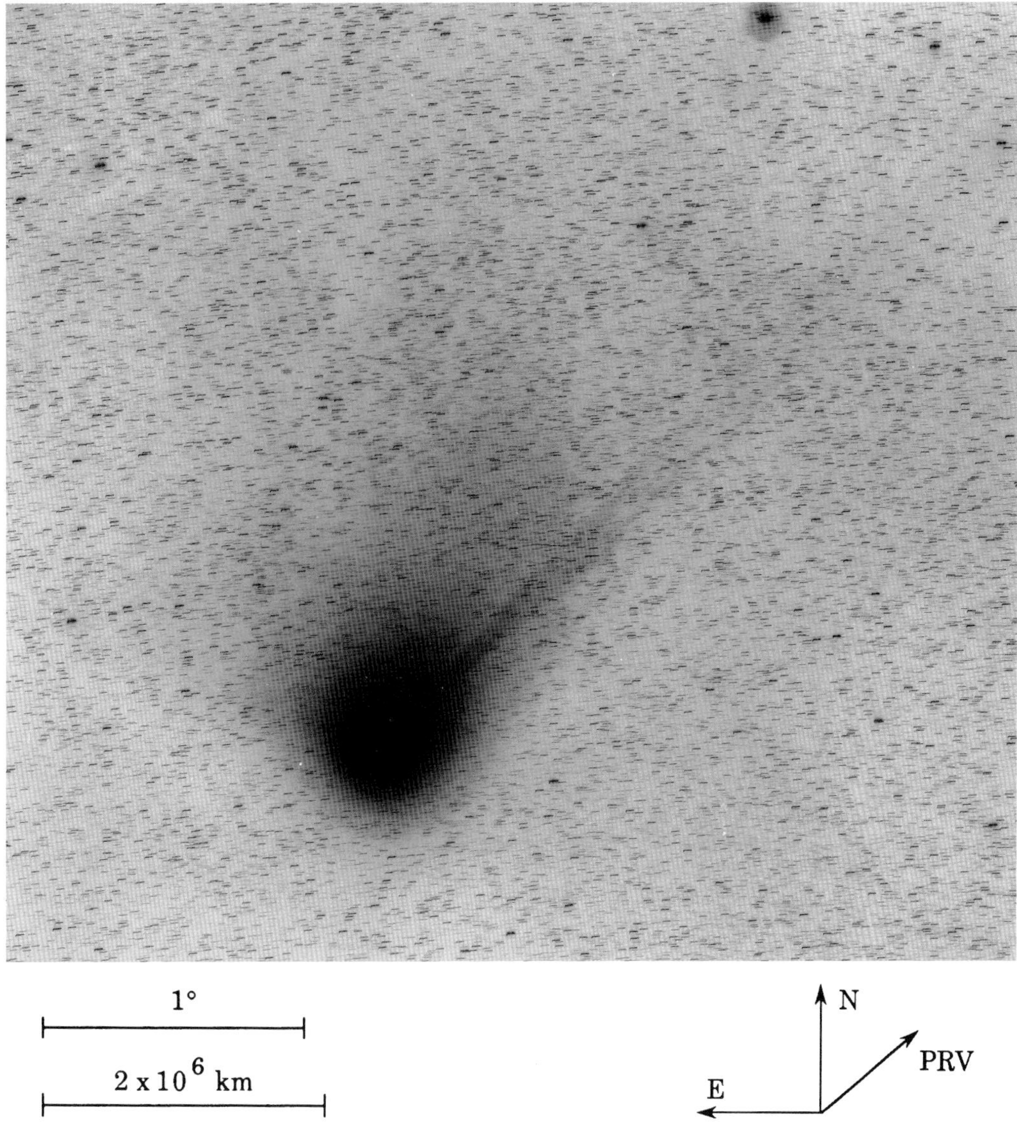

Fig. 505: 1986 Apr 9.26319 UT; exposure 10.0 minutes on 103a-O emulsion with GG-385 filter; r=1.31, Δ=0.42, β=143.1°; LSPN-749.
Photograph by C. Torres/H. Wroblewski, Cerro el Roble Astronomical Observatory.

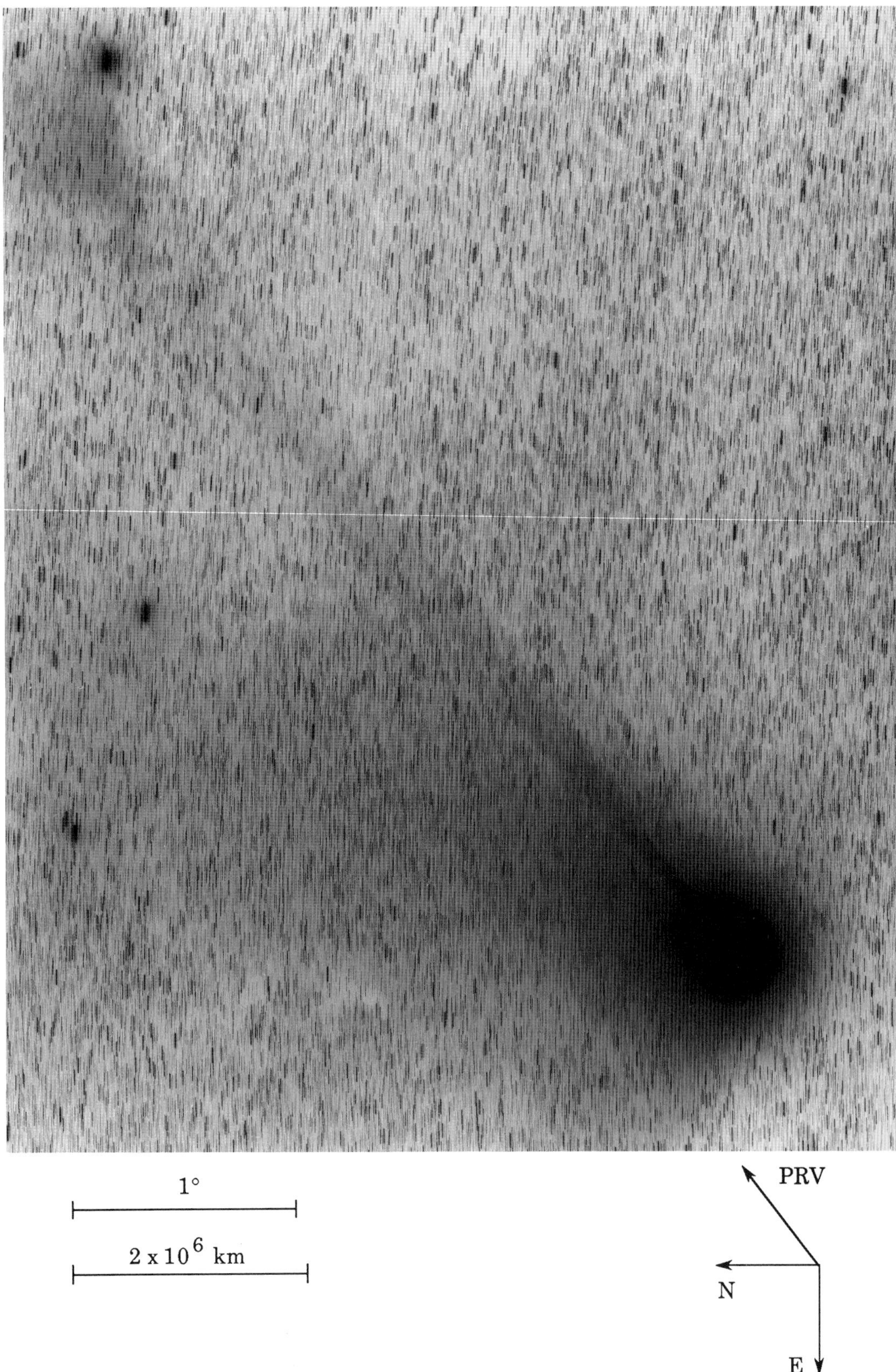

Fig. 506: 1986 Apr 9.28403 UT; exposure 20.0 minutes on IIa-O emulsion with no filter; r=1.31, Δ=0.42, β=143.1°; LSPN-1173.
Photograph by F. Miller, University of Michigan/CTIO.

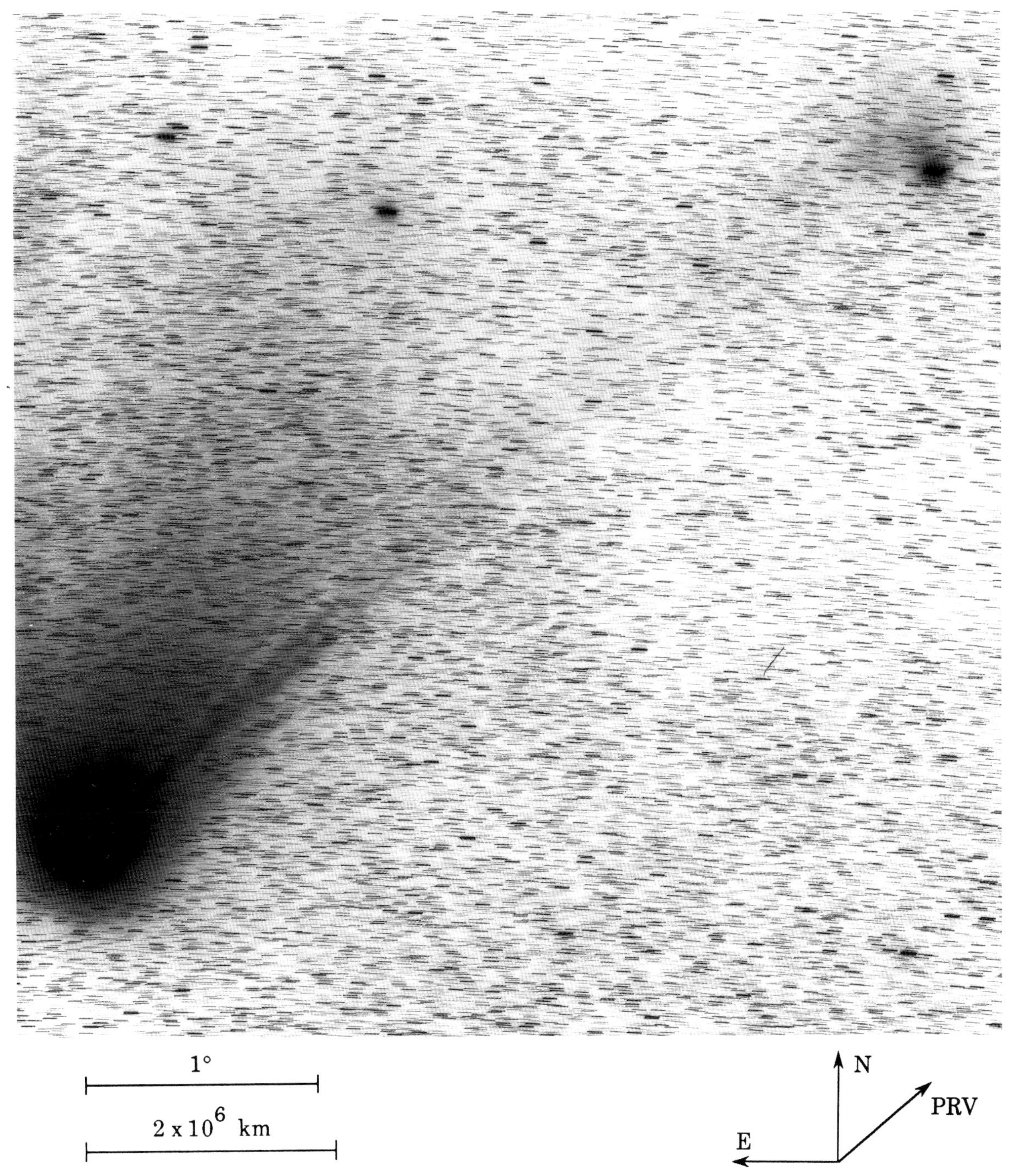

Fig. 507: 1986 Apr 9.32361 UT; exposure 20.0 minutes on 103a-O emulsion with no filter; r=1.31, Δ=0.42, β=143.2°; LSPN-1174.
Photograph by F. Miller, University of Michigan/CTIO.

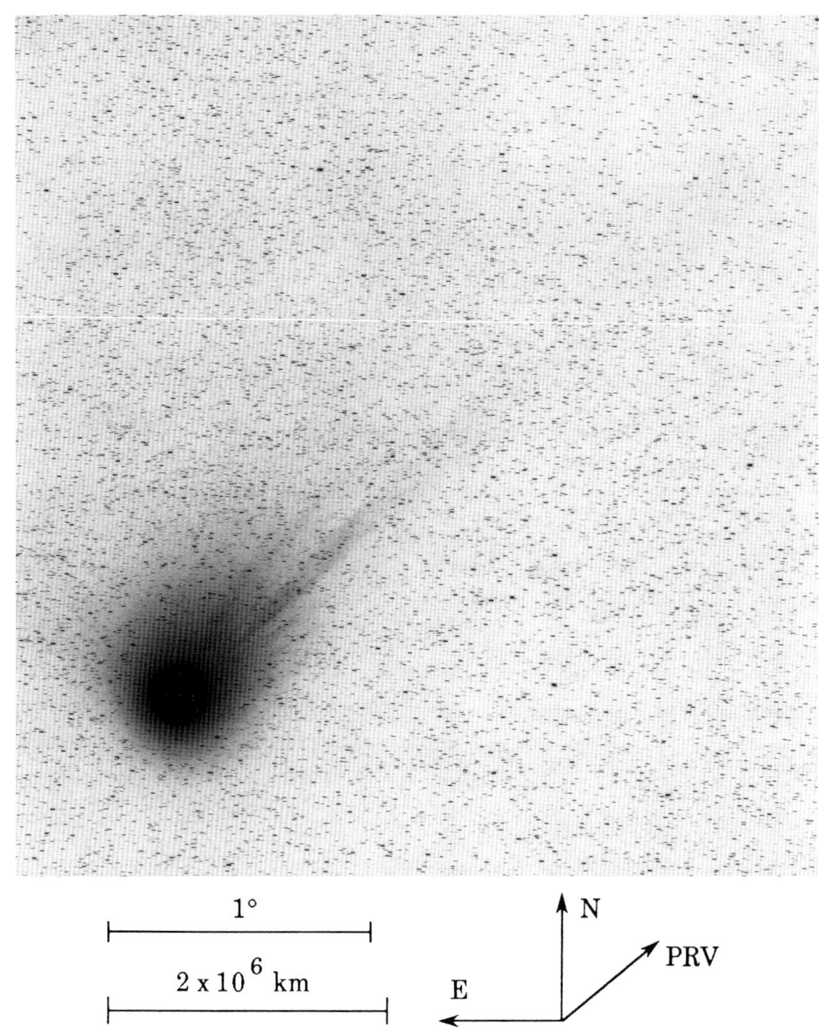

Fig. 508: 1986 Apr 9.37604 UT; exposure 5.0 minutes on 103a-O emulsion with no filter; r=1.31, Δ=0.42, β=143.4°; LSPN-1177.
Photograph by F. Miller, University of Michigan/CTIO.

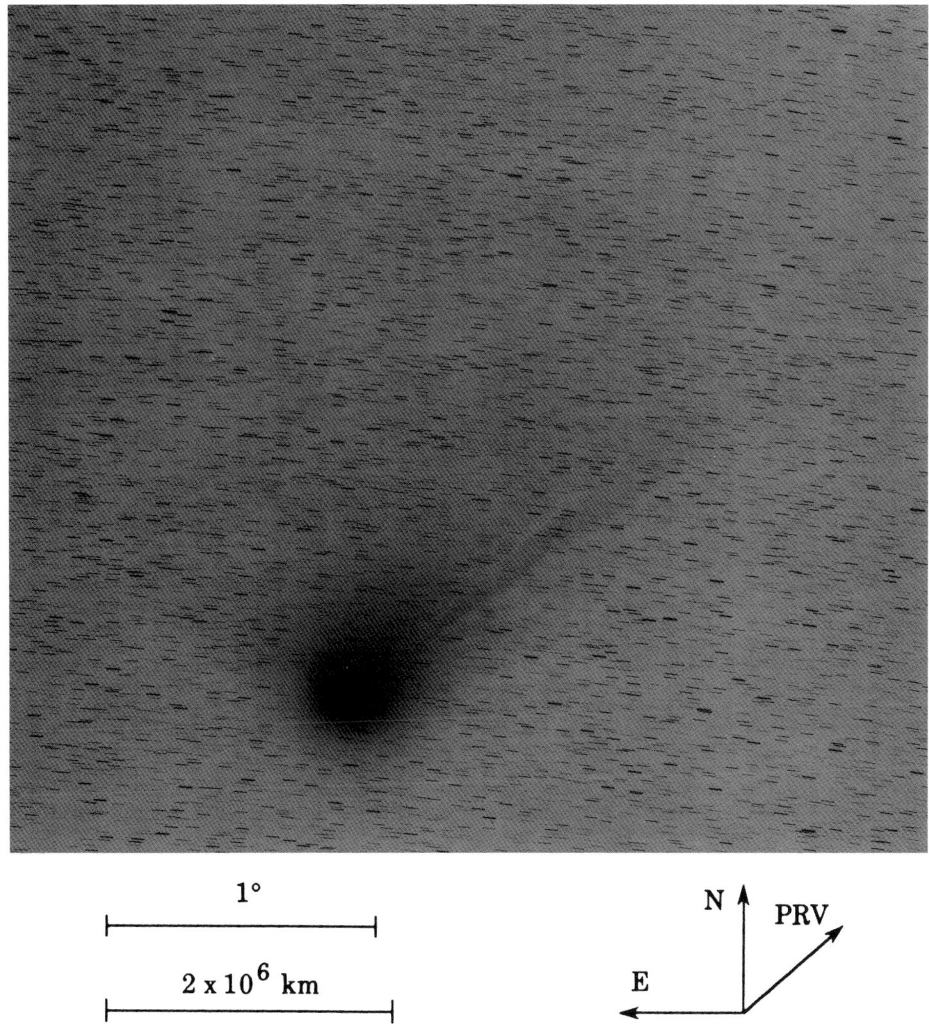

Fig. 509: 1986 Apr 9.40104 UT; exposure 15.0 minutes on 103a-O emulsion with no filter; r=1.31, Δ=0.42, β=143.5°; LSPN-2591.
Photograph by R. Hill, Warner and Swasey Observatory.

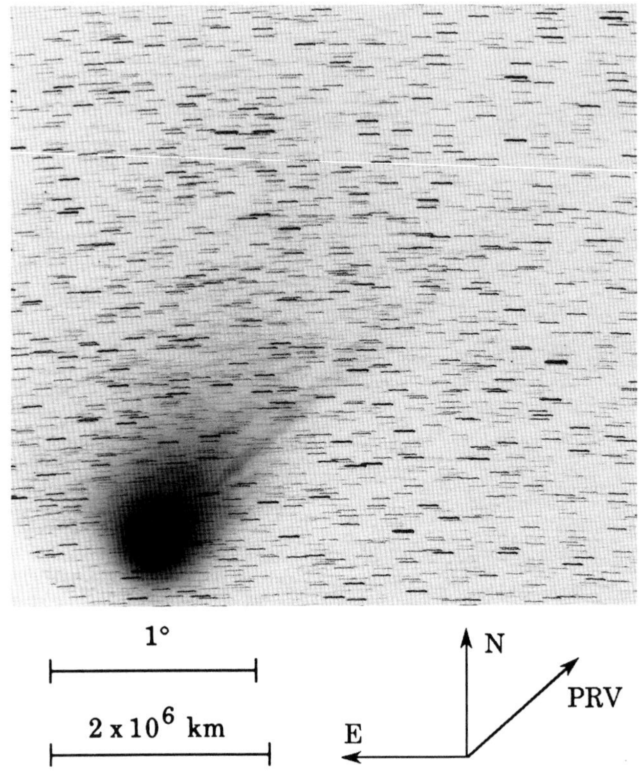

Fig. 510: 1986 Apr 9.42153 UT; exposure 28.0 minutes on Tech. Pan 2415 emulsion with no filter; r=1.31, Δ=0.42, β=143.5°; LSPN-1463.
Photograph by W. Liller, Easter Island, LSPN Island Network.

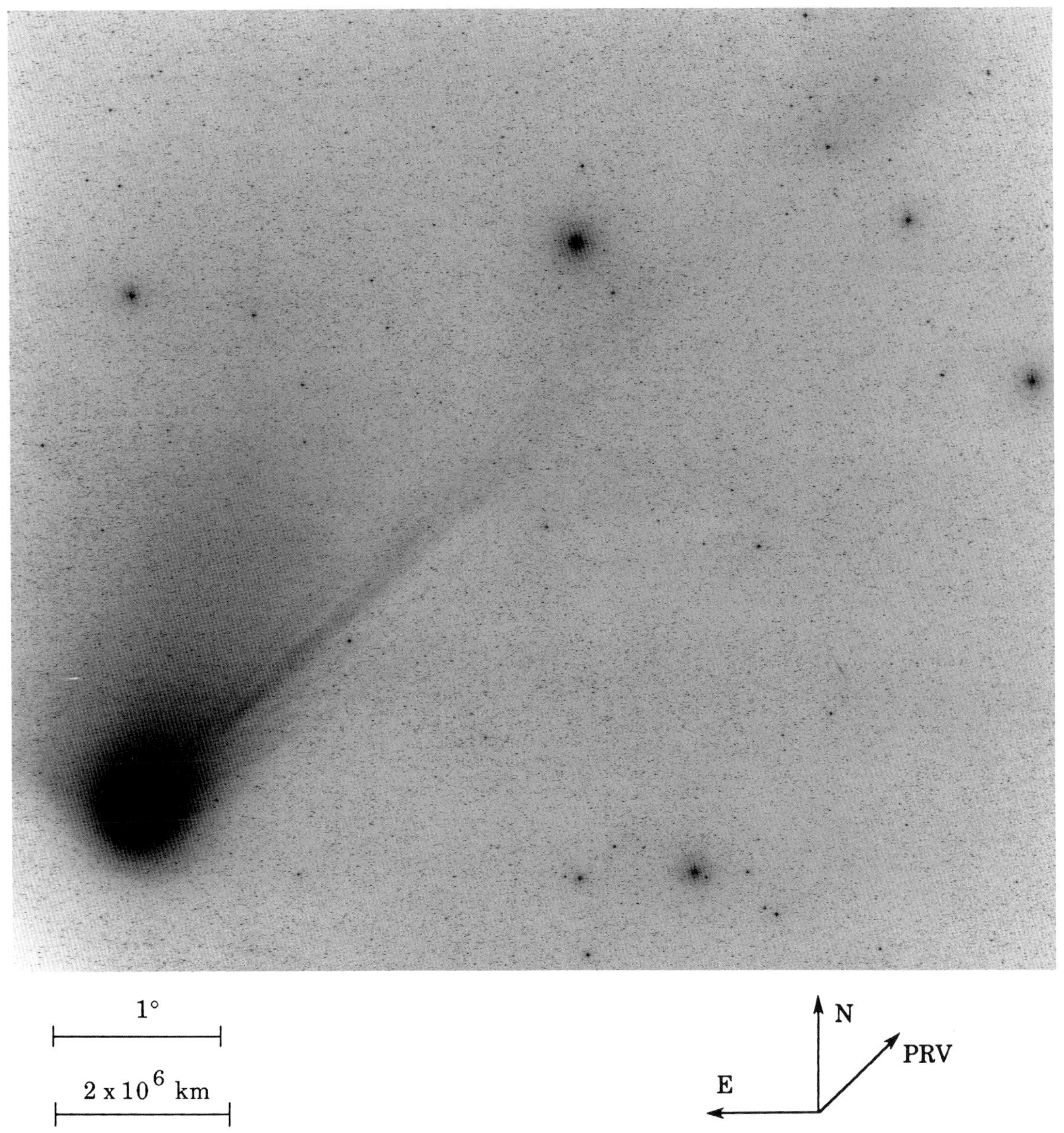

Fig. 511: 1986 Apr 9.59696 UT; exposure 20.0 minutes on IIa-O emulsion with GG-385 filter; r=1.31, Δ=0.42, β=144.1°; LSPN-2417.
Photograph from Royal Observatory/UK Schmidt Telescope Unit.

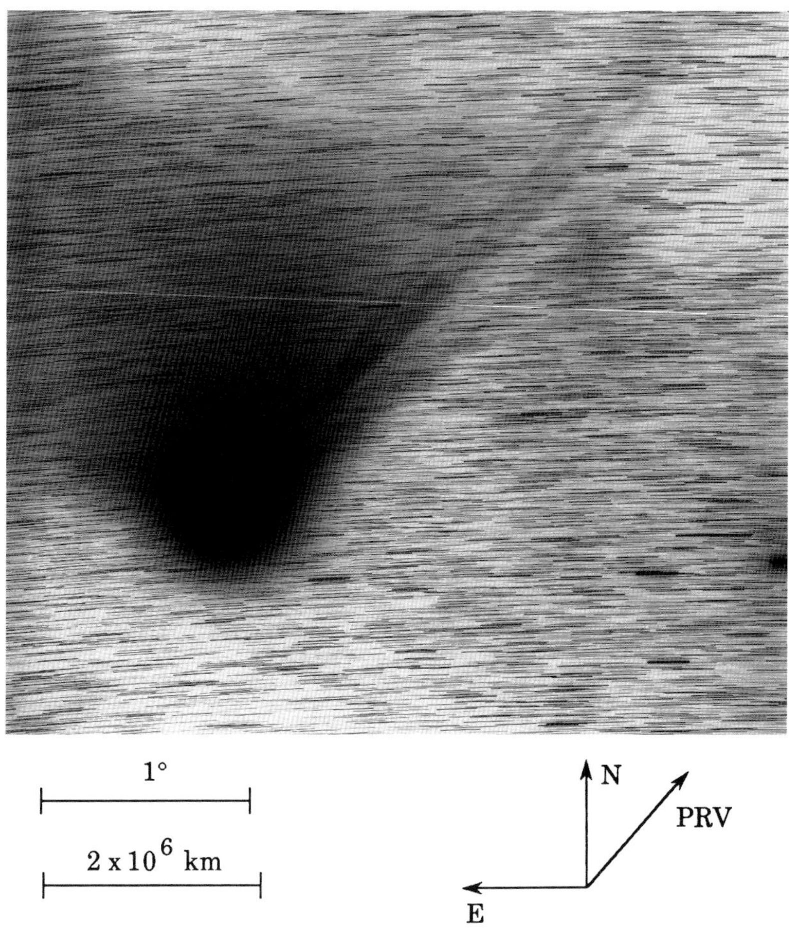

Fig. 512: 1986 Apr 9.68306 UT; exposure 60.0 minutes on IIIa-J emulsion with GG-385 filter; r=1.31, Δ=0.42, β=144.3°; LSPN-2135.
Photograph by P. Magnusson, Uppsala Southern Station.

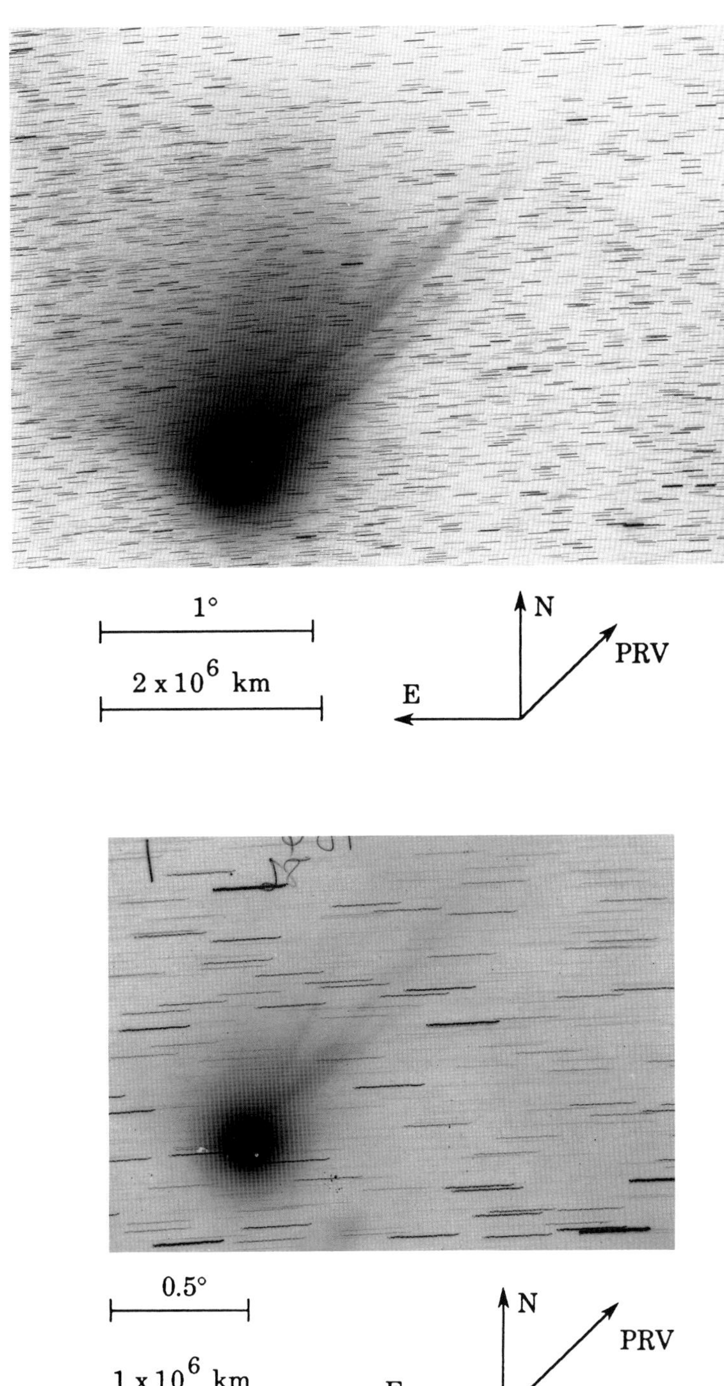

(top) Fig. 513: 1986 Apr 9.72091 UT; exposure 30.0 minutes on IIa-O emulsion with GG-385 filter; r=1.31, Δ=0.42, β=144.4°; LSPN-2136.
Photograph by P. Magnusson, Uppsala Southern Station.

(bottom) Fig. 514*: 1986 Apr 9.83854 UT; exposure 75.0 minutes on 103a-O emulsion with no filter; r=1.31, Δ=0.42, β=144.8°; LSPN-3642.
Photograph submitted by K. Sivaraman, Indian Institute for Astrophysics, Kodaikanal Station.

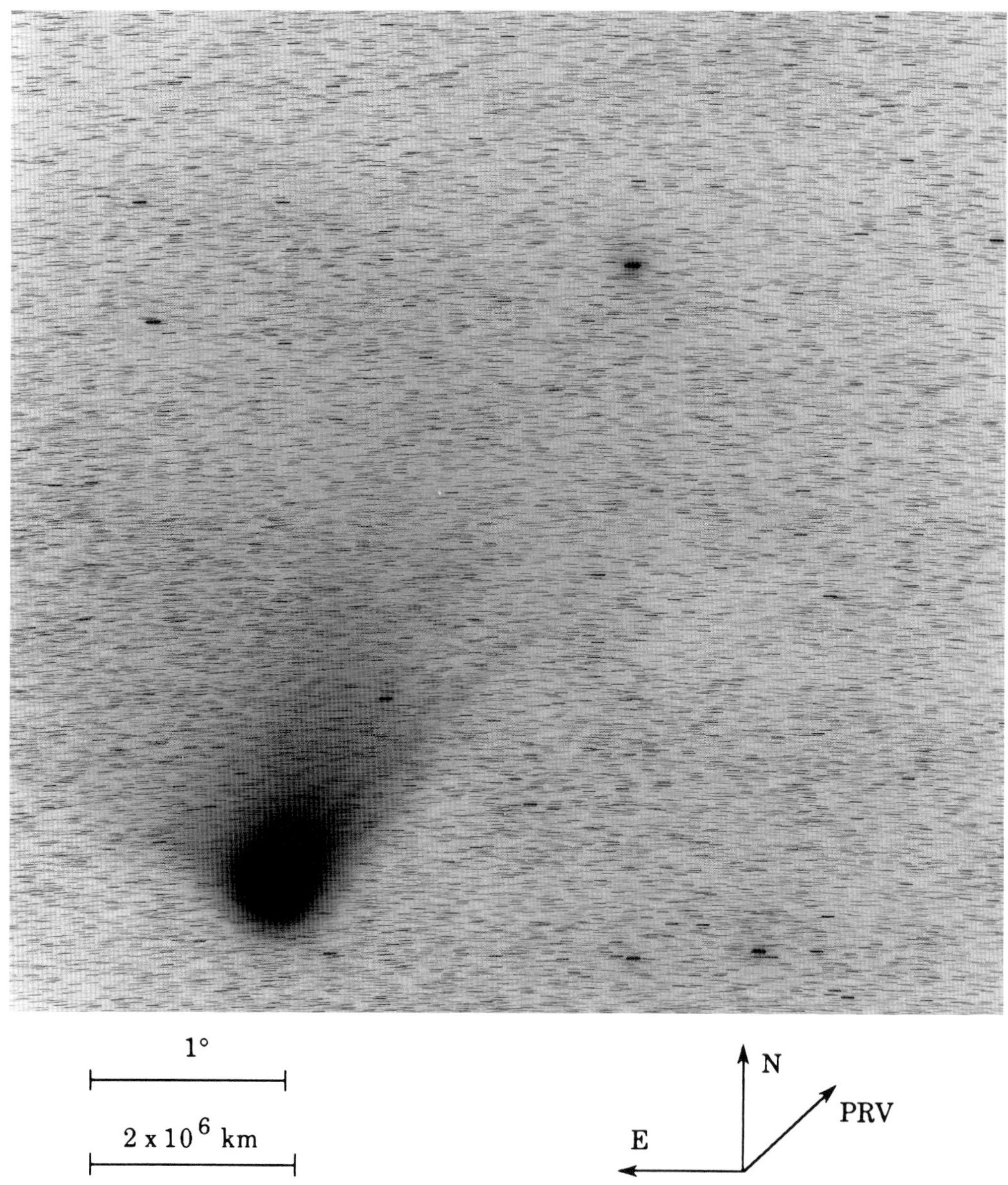

Fig. 515: 1986 Apr 9.72508 UT; exposure 20.0 minutes on IIa-D emulsion with GG-495 filter; r=1.31, Δ=0.42, β=144.4°; LSPN-2420.
Photograph from Royal Observatory/UK Schmidt Telescope Unit.

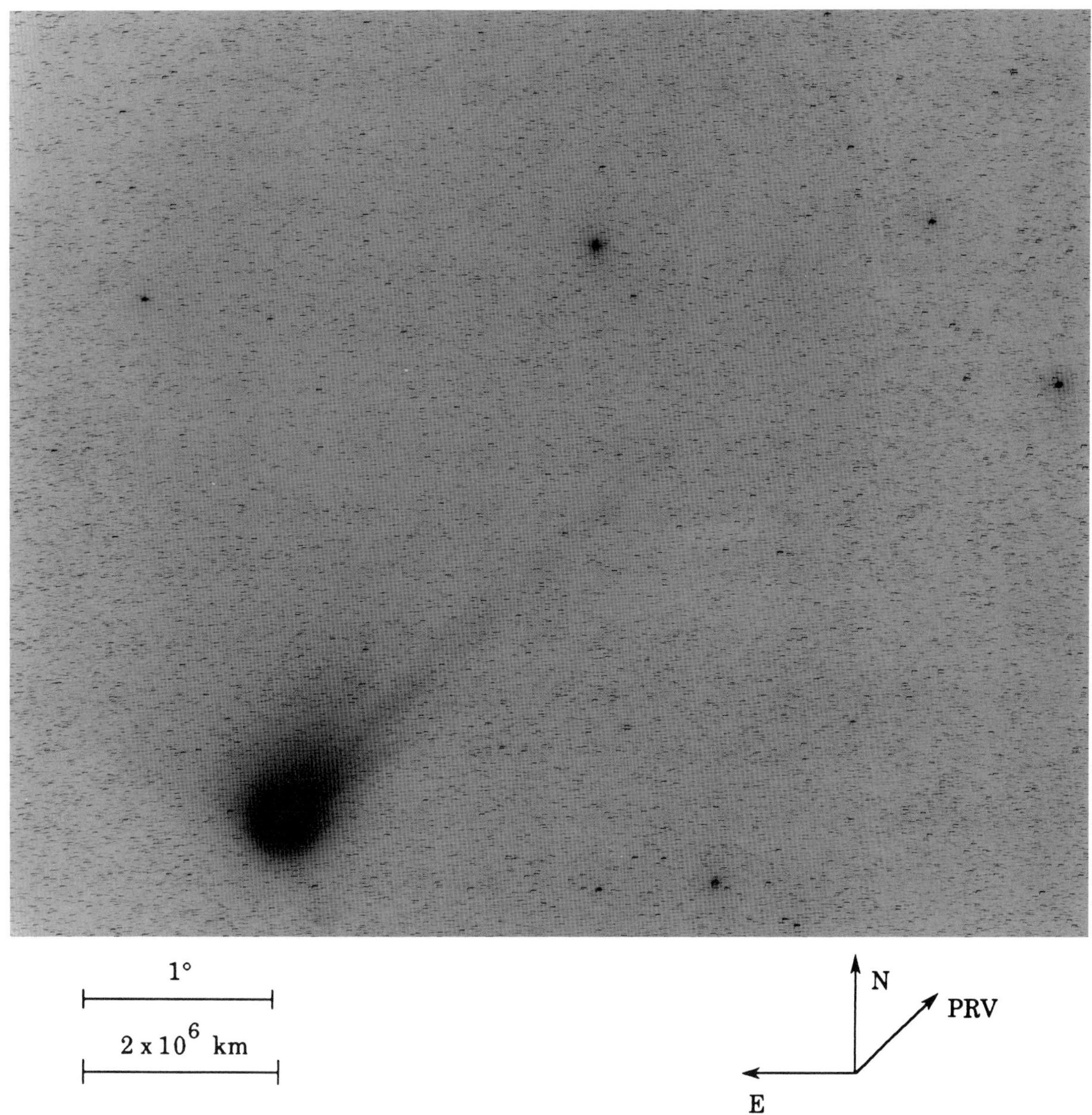

Fig. 516: 1986 Apr 9.75049 UT; exposure 9.4 minutes on IIa-O emulsion with GG-385 filter; r=1.31, Δ=0.42, β=144.5°; LSPN-2421.
Photograph from Royal Observatory/UK Schmidt Telescope Unit.

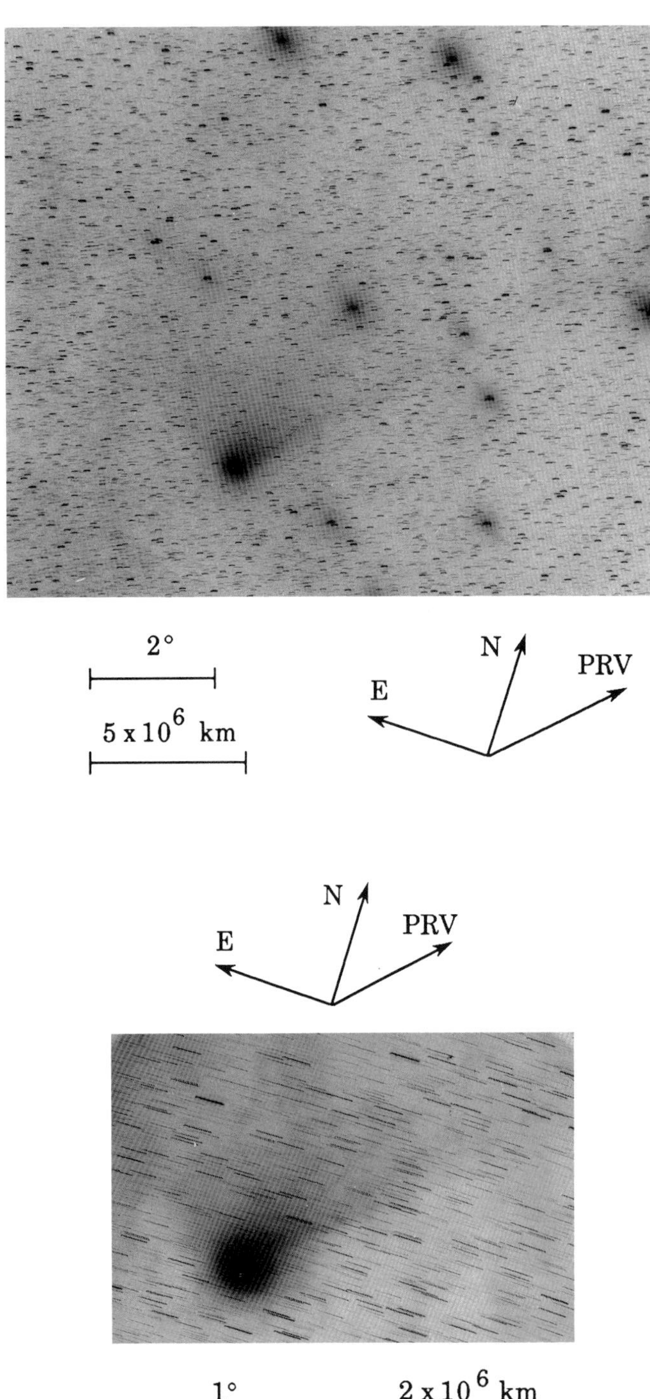

(top) Fig. 517: 1986 Apr 9.87674 UT; exposure 75.0 minutes on 103a-O emulsion with no filter; r=1.31, Δ=0.42, β=144.9°; LSPN-3657.
Photograph submitted by K. Sivaraman, Indian Institute for Astrophysics, Kodaikanal Station.

(bottom) Fig. 518: 1986 Apr 9.95139 UT; exposure 60.0 minutes on 103a-O emulsion with Blue filter; r=1.32, Δ=0.42, β=145.1°; LSPN-3688.
Photograph submitted by K. Sivaraman, Indian Institute for Astrophysics, Kavalur Station.

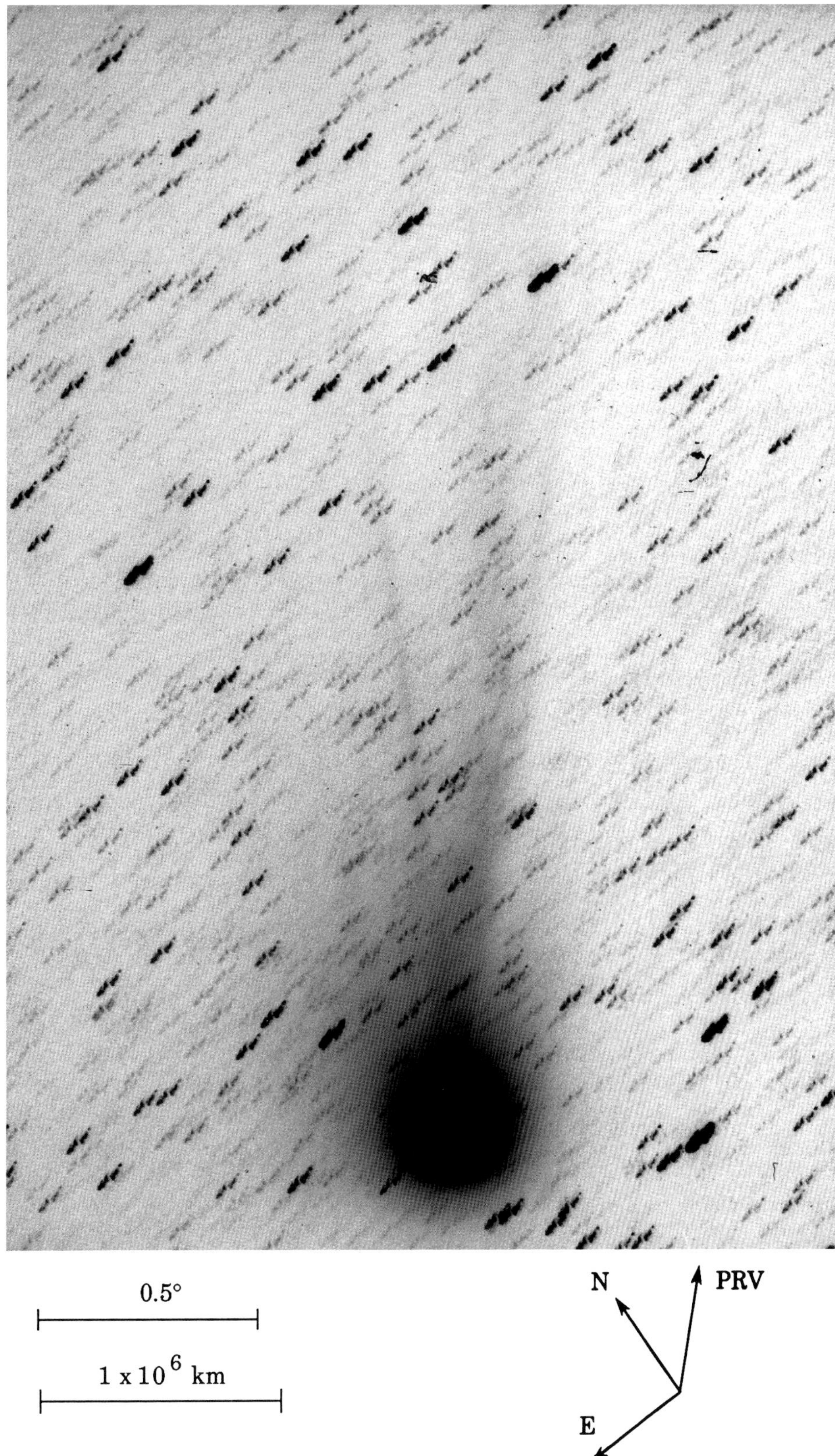

Fig. 519: 1986 Apr 10.02570 UT; exposure 20.0 minutes on Tech. Pan 2415 emulsion with Wratten 47B filter; r=1.32, Δ=0.42, β=145.3°; LSPN-1953.
Photograph by M. Clilverd/M. Dowson, British Antarctic Survey.

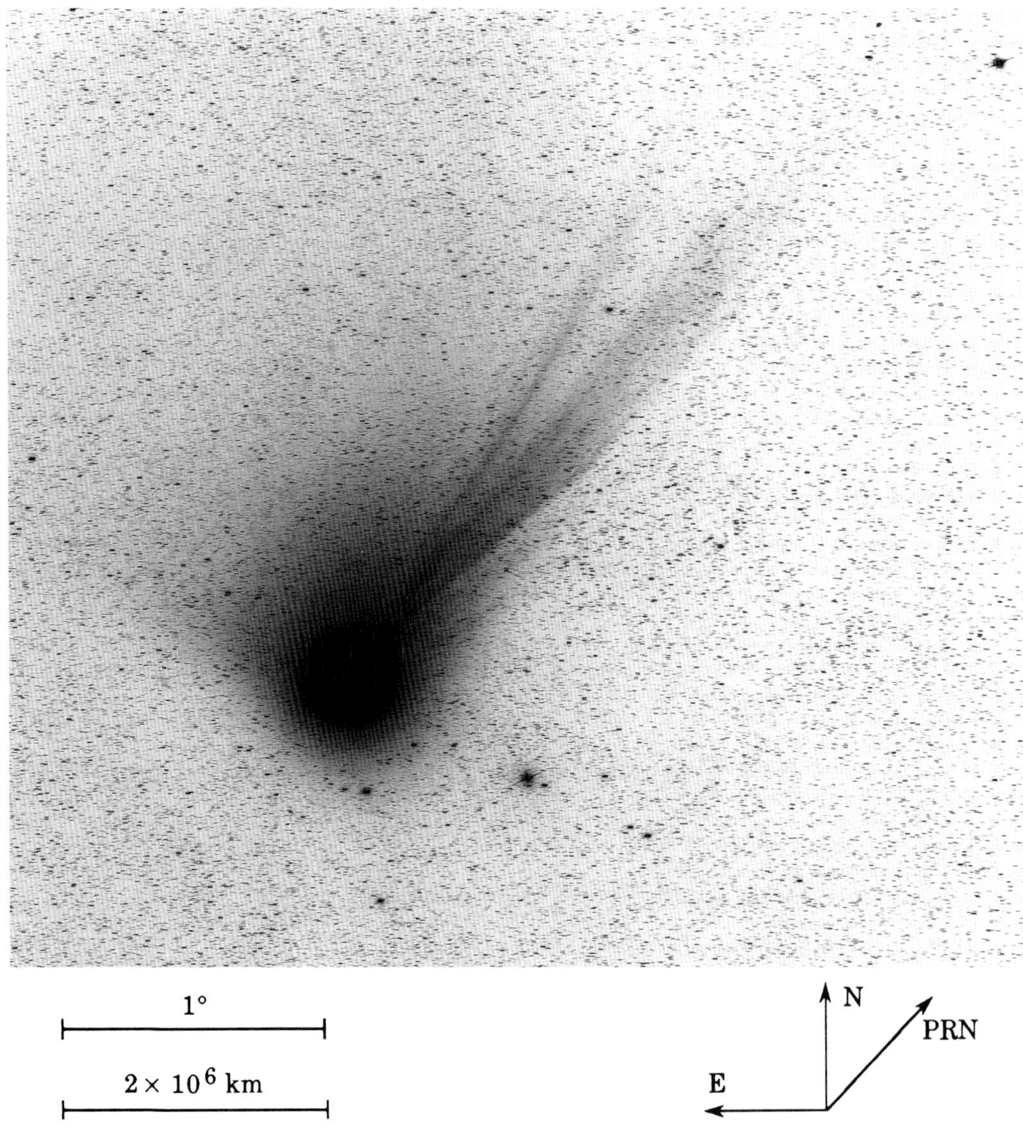

Fig. 520: 1986 Apr 10.10521 UT; exposure 5.0 minutes on 103a-O emulsion with no filter; r=1.32, Δ=0.42, β=145.6°; LSPN-1178.
Photograph by F. Miller, University of Michigan/CTIO.

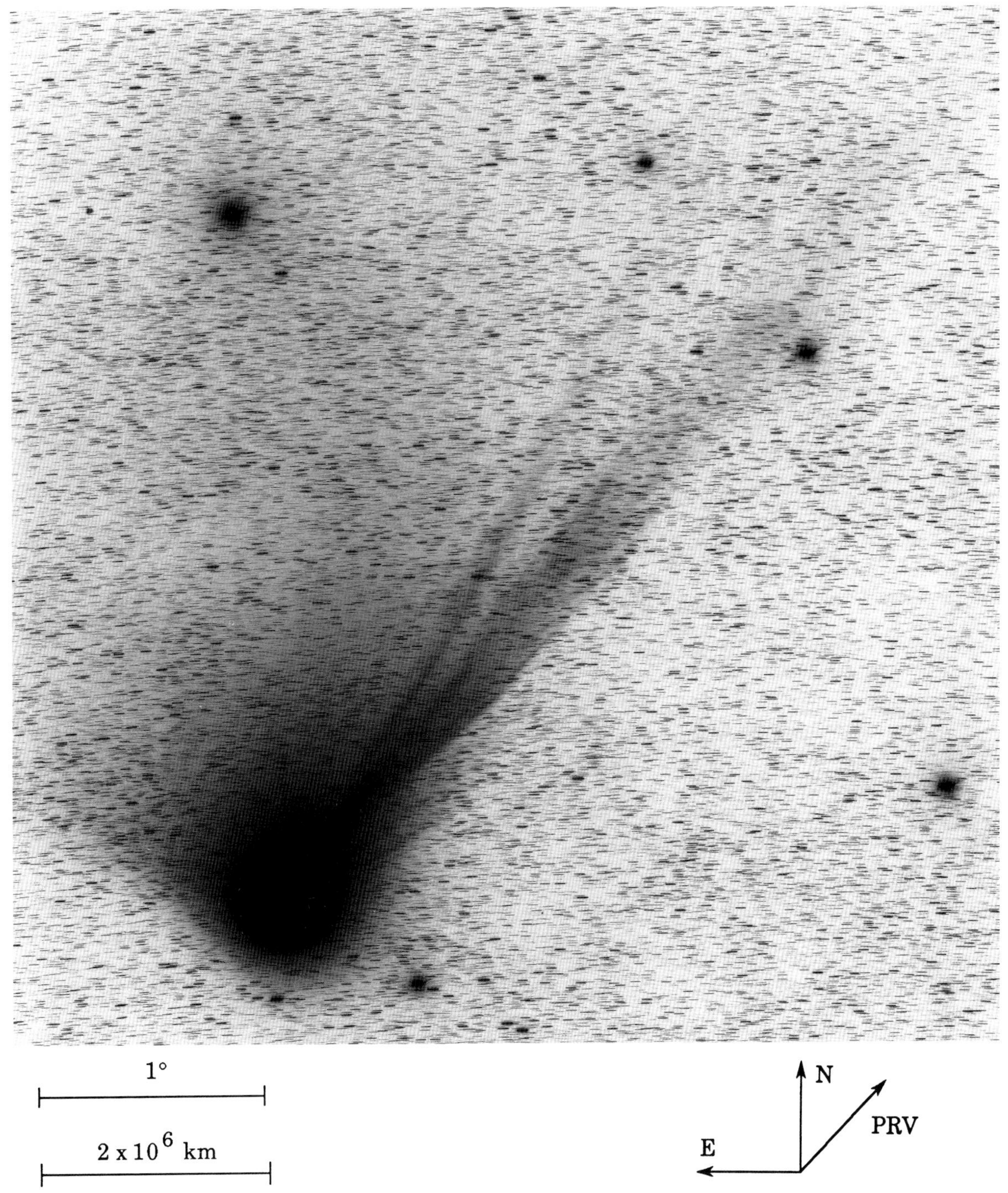

Fig. 521: 1986 Apr 10.12917 UT; exposure 20.0 minutes on 103a-O emulsion with no filter; r=1.32, Δ=0.42, β=145.6°; LSPN-1179.
Photograph by F. Miller, University of Michigan/CTIO.

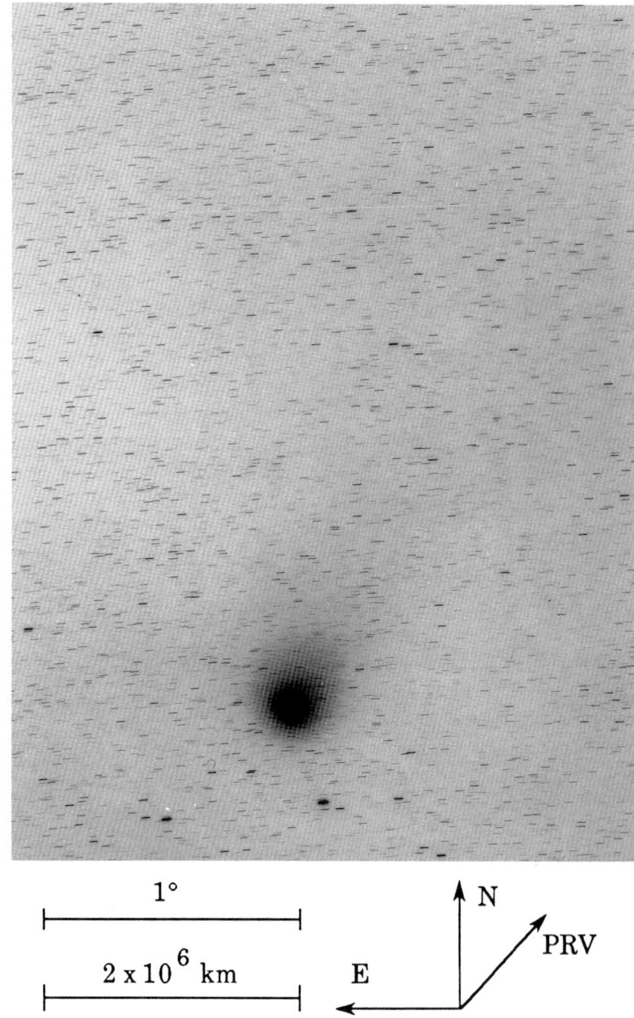

Fig. 522: 1986 Apr 10.23125 UT; exposure 10.0 minutes on 098-04 emulsion with RG-610 filter; r=1.32, Δ=0.42, β=145.9°; LSPN-750.
Photograph by C. Torres/H. Wroblewski, Cerro el Roble Astronomical Observatory.

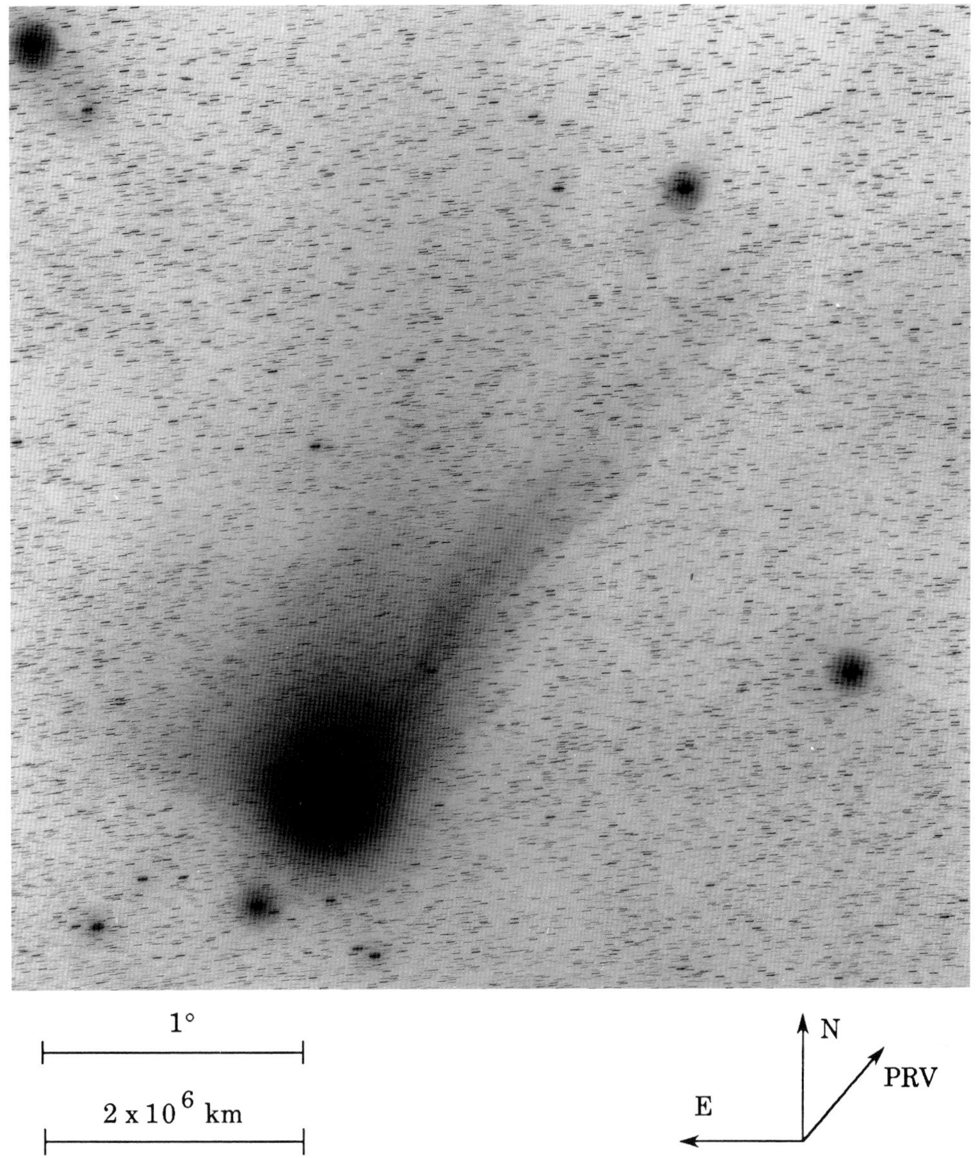

Fig. 523: 1986 Apr 10.31597 UT; exposure 10.0 minutes on 103a-O emulsion with GG-385 filter; r=1.32, Δ=0.42, β=146.2°; LSPN-751.
Photograph by C. Torres/H. Wroblewski, Cerro el Roble Astronomical Observatory.

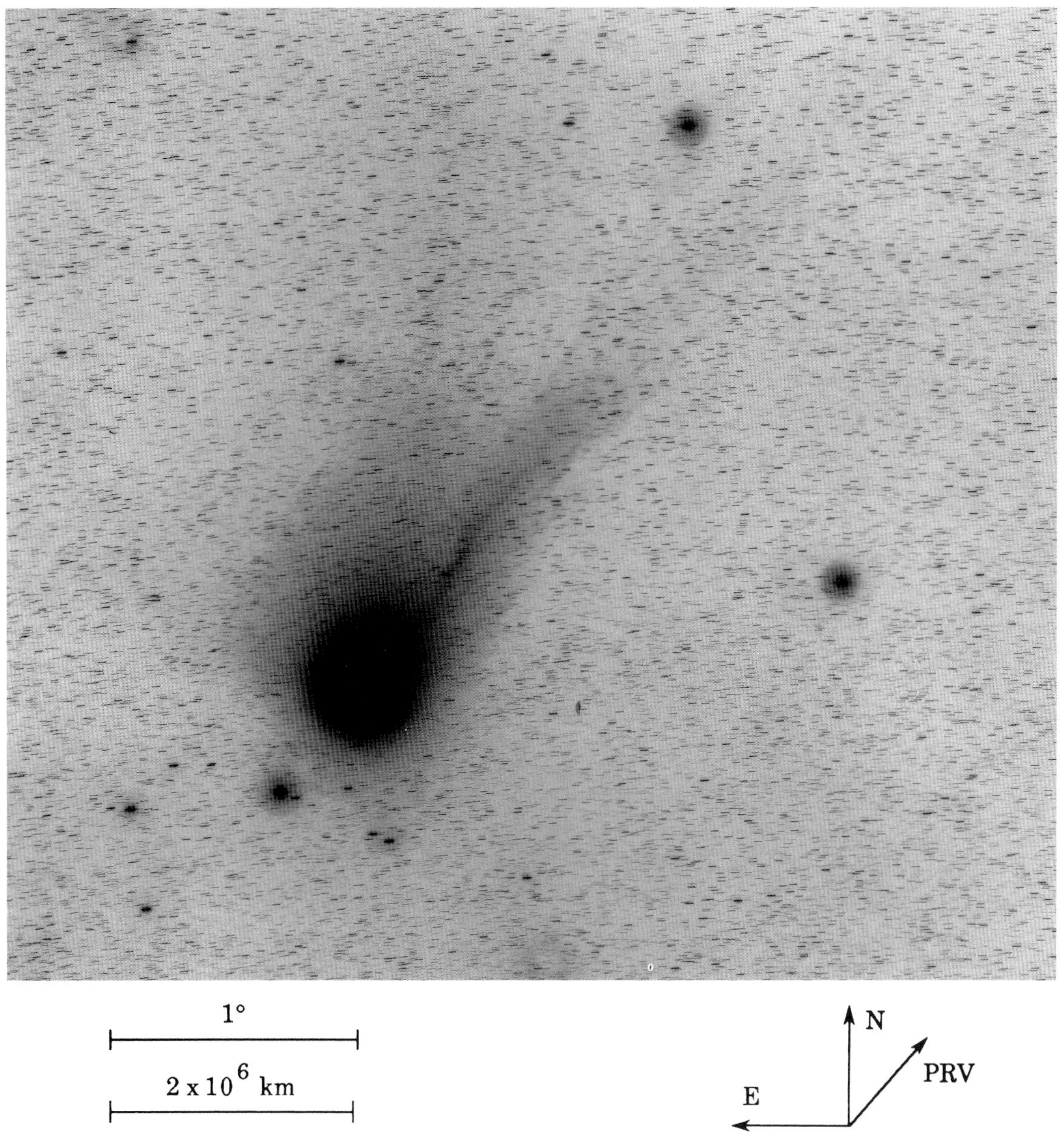

Fig. 524: 1986 Apr 10.33194 UT; exposure 10.0 minutes on 103a-O emulsion with GG-385 filter; r=1.32, Δ=0.42, β=146.2°; LSPN-752.
Photograph by C. Torres/H. Wroblewski, Cerro el Roble Astronomical Observatory.

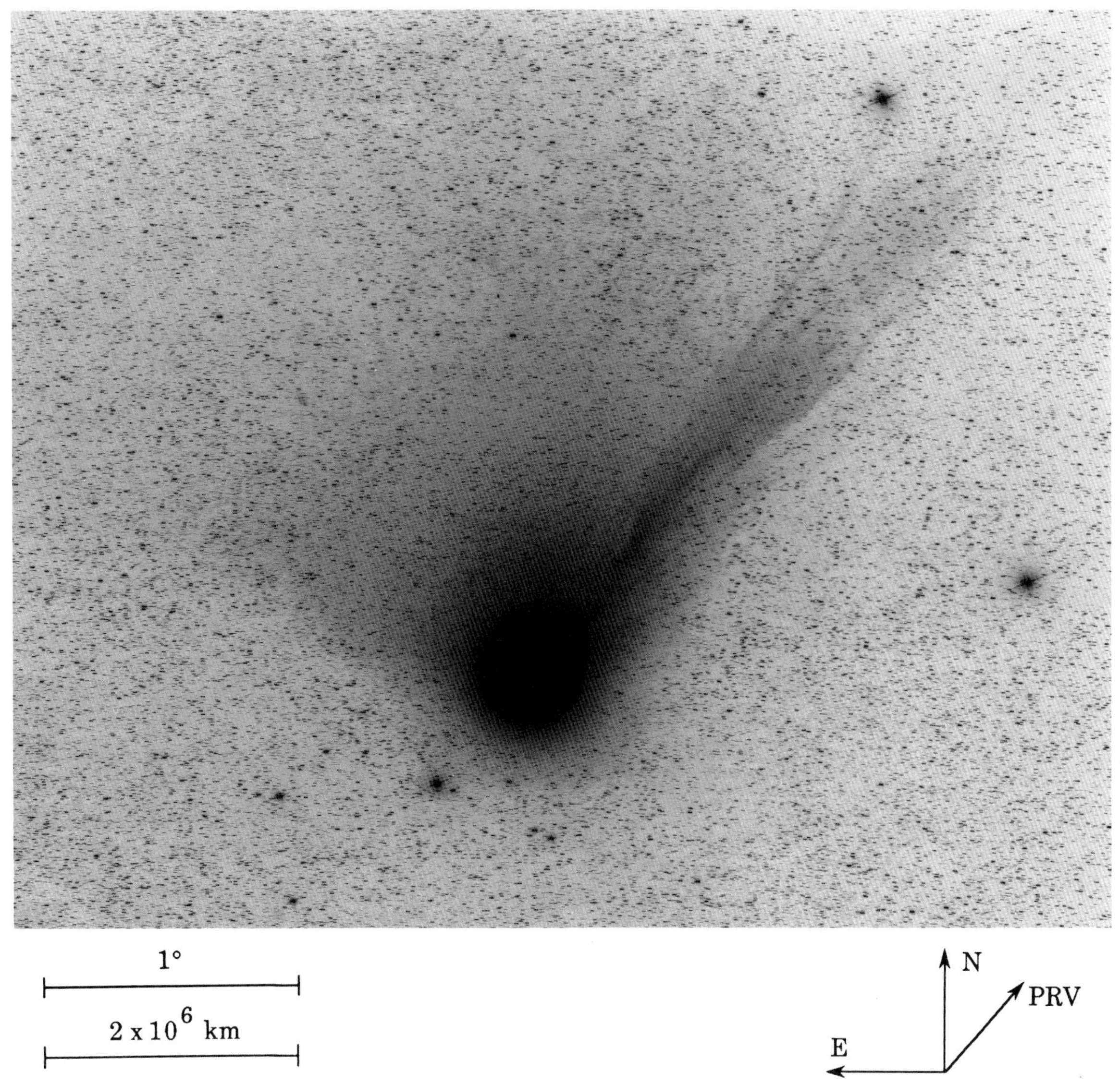

Fig. 525: 1986 Apr 10.33299 UT; exposure 5.0 minutes on 103a-O emulsion with no filter; r=1.32, Δ=0.42, β=146.2°; LSPN-1180.
Photograph by F. Miller, University of Michigan/CTIO.

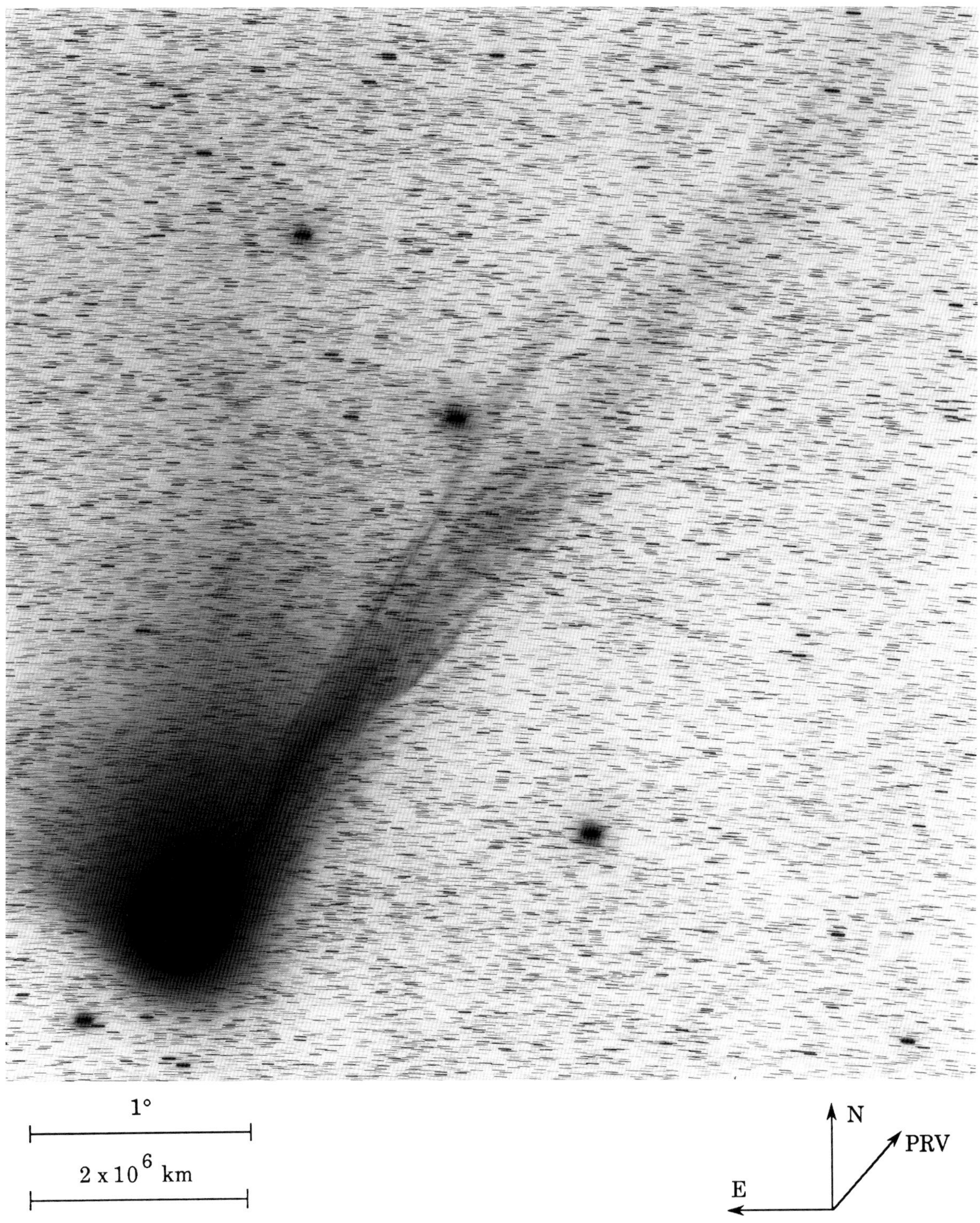

Fig. 526: 1986 Apr 10.34792 UT; exposure 20.0 minutes on 103a-O emulsion with no filter; r=1.32, Δ=0.42, β=146.3°; LSPN-1181.
Photograph by F. Miller, University of Michigan/CTIO.

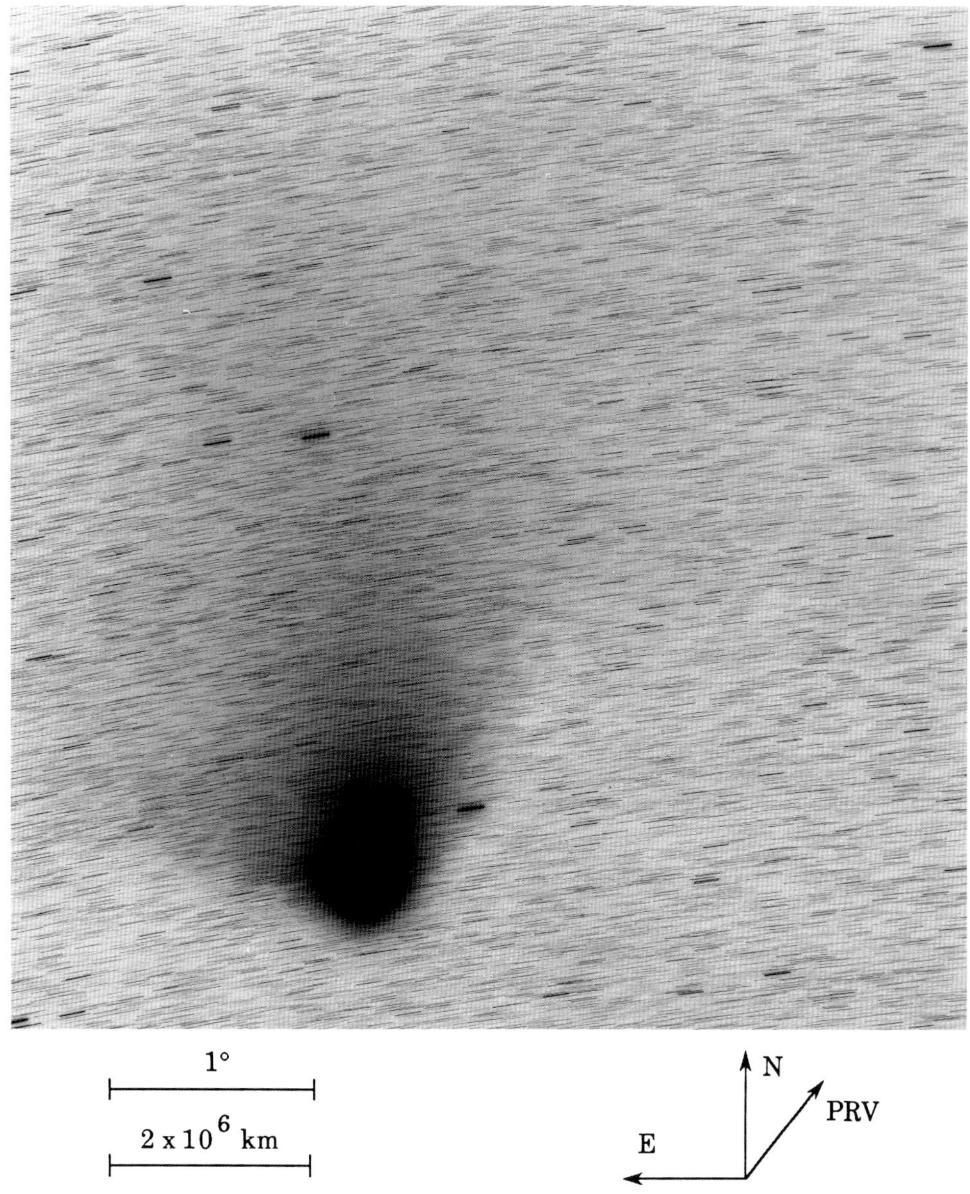

Fig. 527: 1986 Apr 10.64134 UT; exposure 40.0 minutes on IIIa-F emulsion with RG-630 filter; $r=1.33$, $\Delta=0.42$, $\beta=147.1^\circ$; LSPN-2422.
Photograph from Royal Observatory/UK Schmidt Telescope Unit.

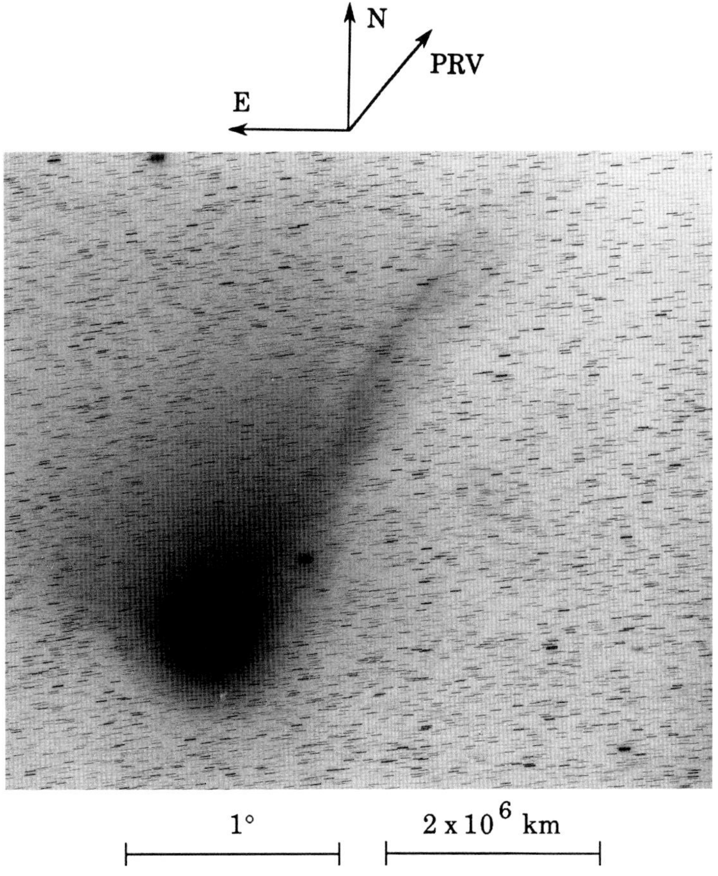

Fig. 528: 1986 Apr 10.66289 UT; exposure 15.0 minutes on IIa-O emulsion with GG-385 filter; r=1.33, Δ=0.42, β=147.2°; LSPN-2140.
Photograph by P. Magnusson, Uppsala Southern Station.

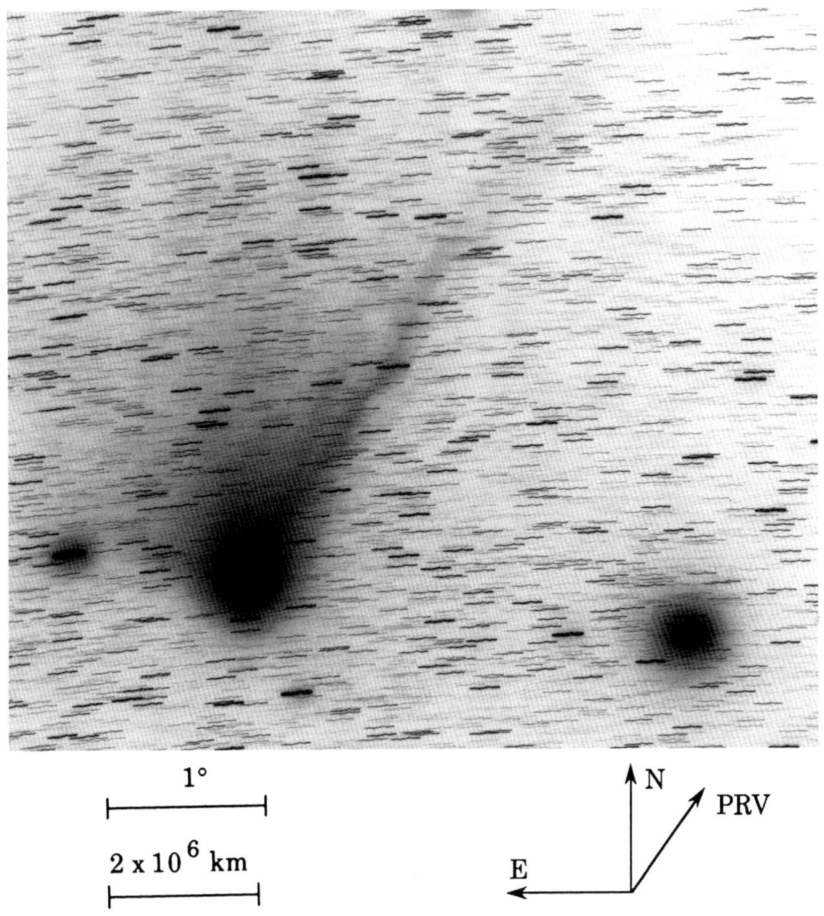

Fig. 529: 1986 Apr 11.00764 UT; exposure 60.0 minutes on IIa-O emulsion with no filter; r=1.33, Δ=0.42, β=148.2°; LSPN-2865.
Photograph by G. Malcolm, Boyden Observatory.

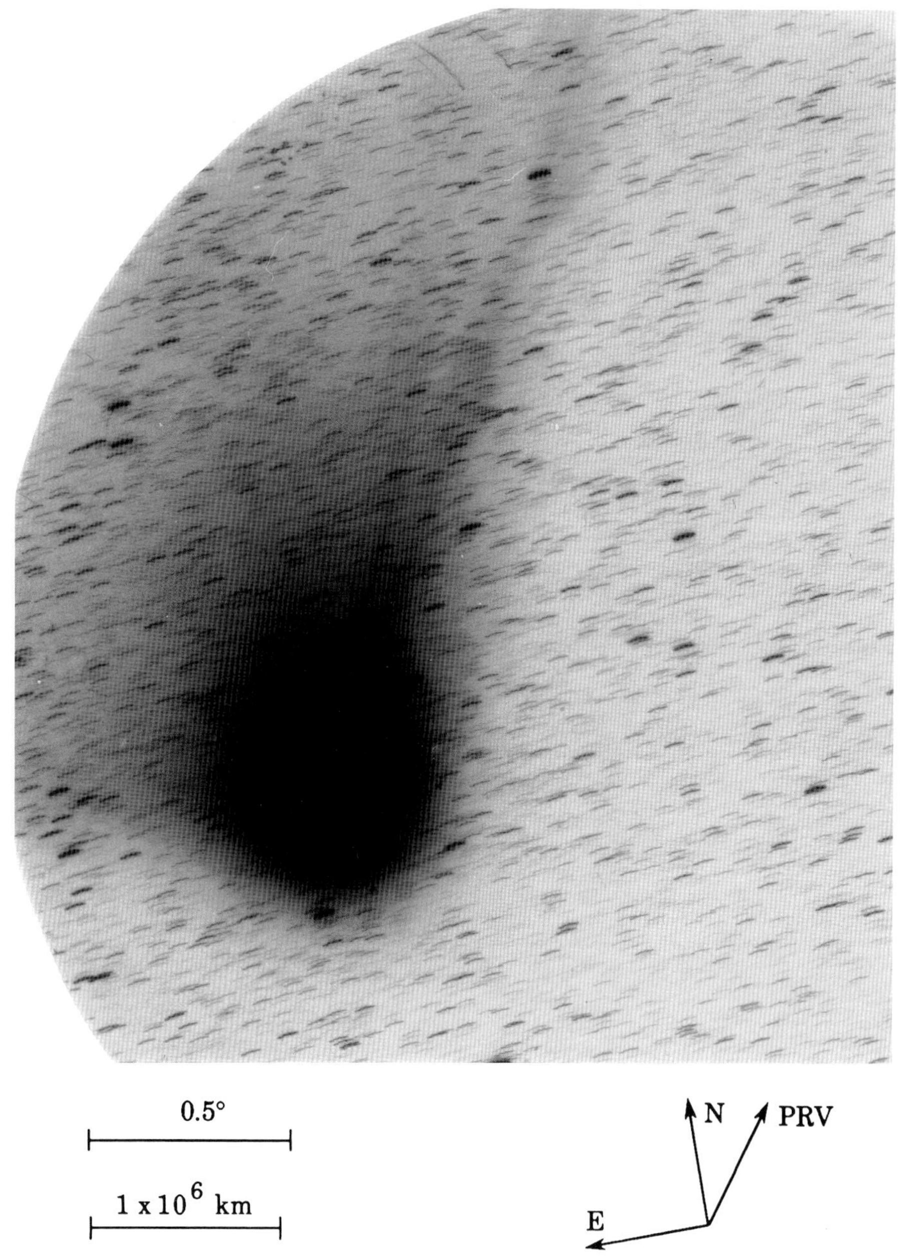

Fig. 530: 1986 Apr 11.02465 UT; exposure 15.0 minutes on Tech. Pan 2415 emulsion with no filter; r=1.33, Δ=0.42, β=148.2°; LSPN-3497.
Photograph by D. O'Donoghue, University of Perugia/South African Astronomical Observatory.

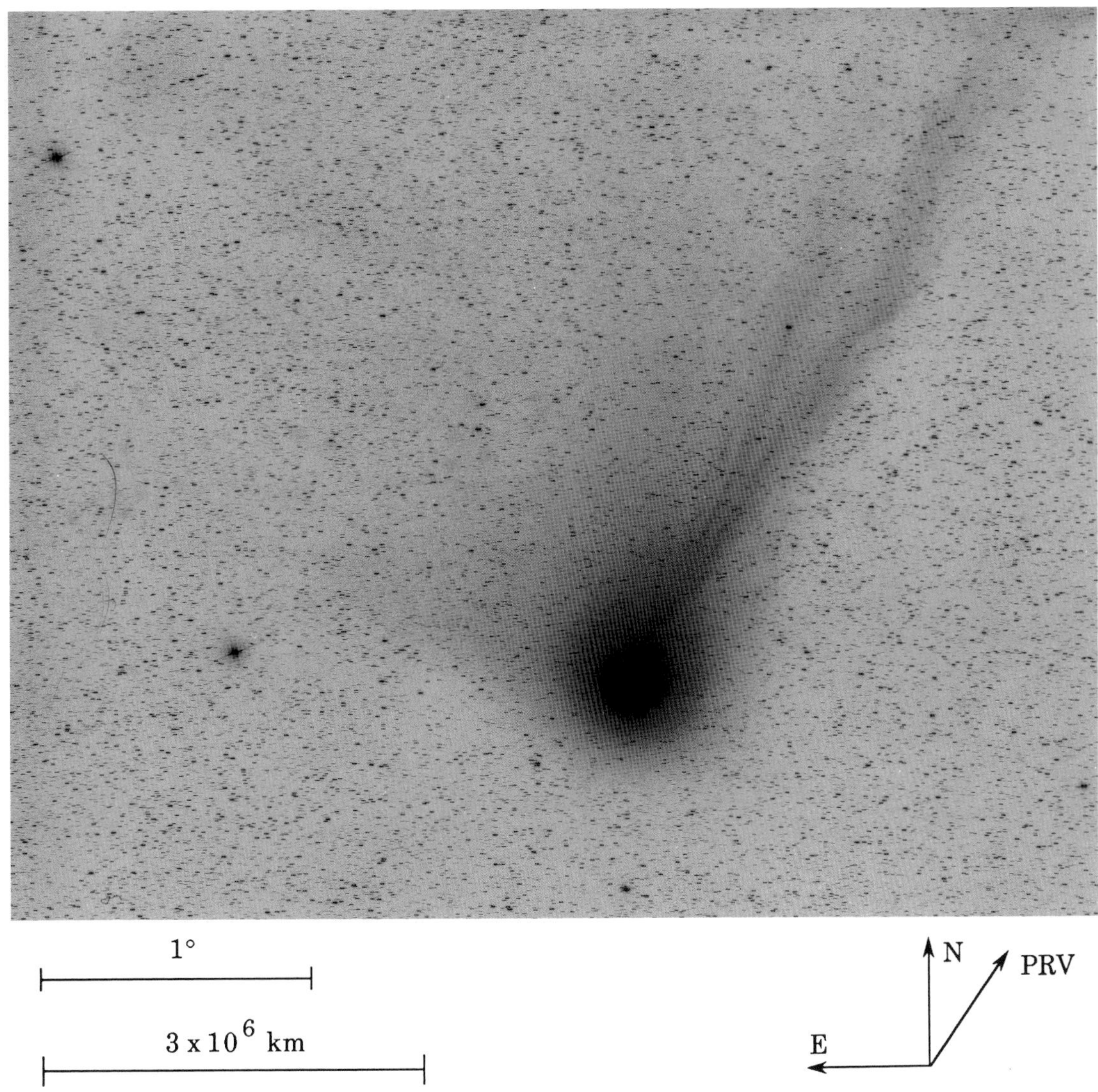

Fig. 531: 1986 Apr 11.08715 UT; exposure 5.0 minutes on 103a-O emulsion with no filter; r=1.33, Δ=0.42, β=148.4°; LSPN-1183.
Photograph by F. Miller, University of Michigan/CTIO.

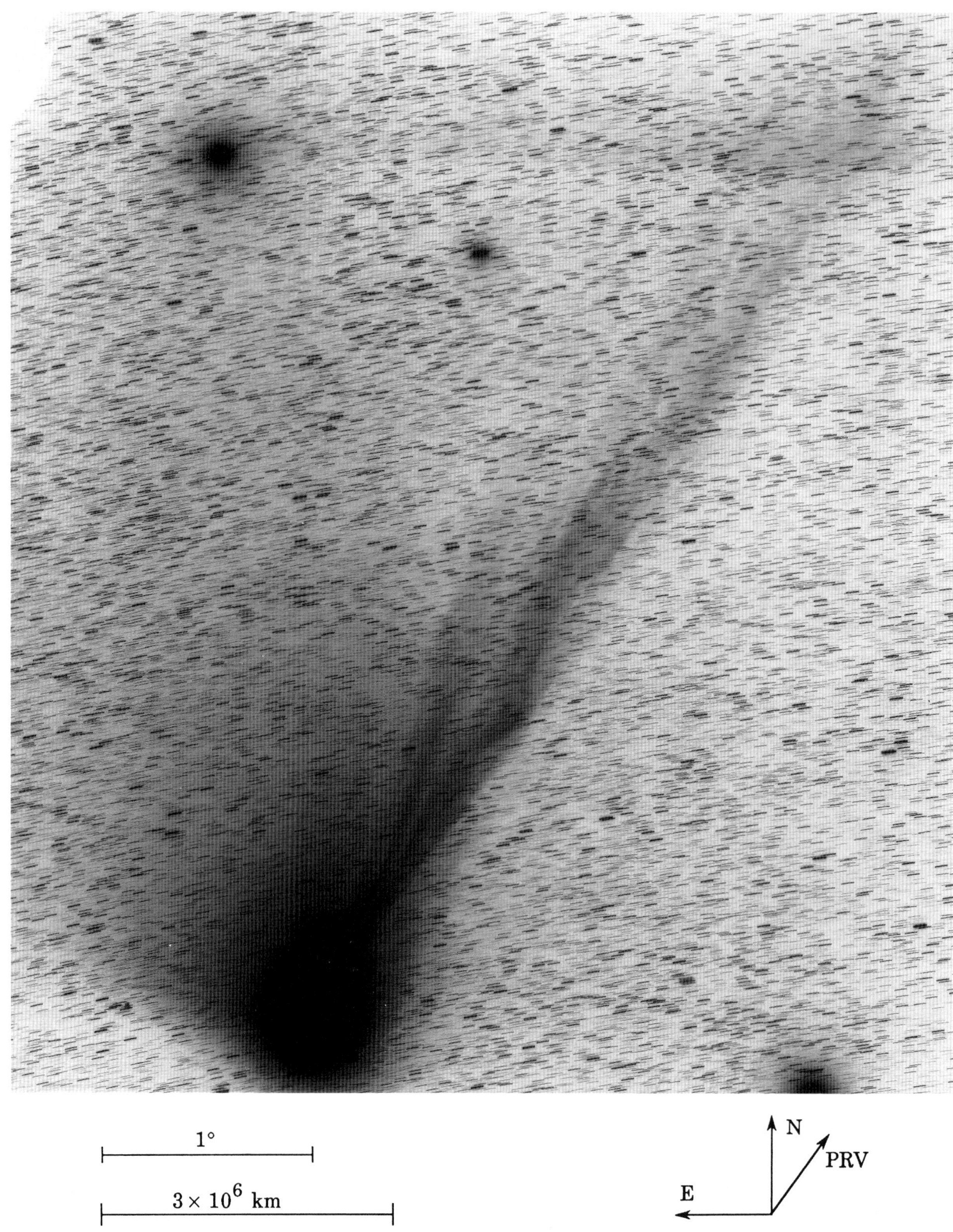

Fig. 532: 1986 Apr 11.10625 UT; exposure 20.0 minutes on 103a-O emulsion with no filter; r=1.33, Δ=0.42, β=148.5°; LSPN-1184.
Photograph by F. Miller, University of Michigan/CTIO.

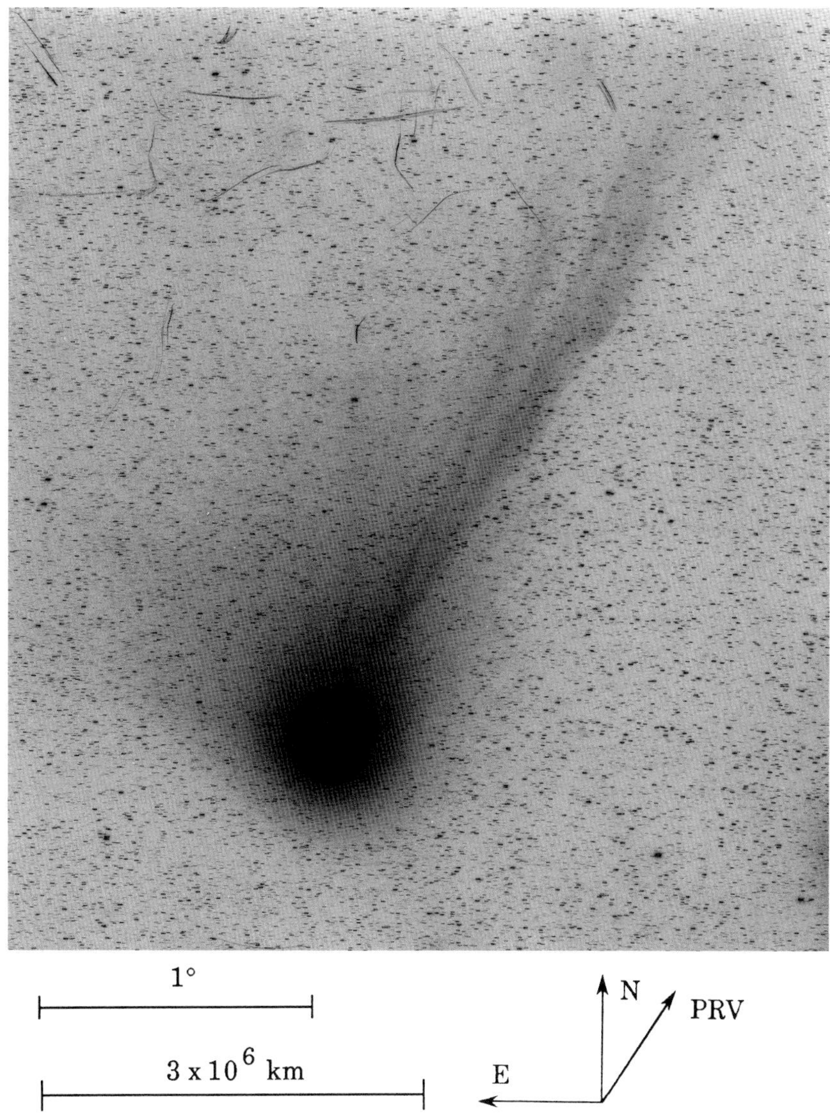

Fig. 533: 1986 Apr 11.19201 UT; exposure 5.0 minutes on 103a-O emulsion with no filter; r=1.34, Δ=0.42, β=148.7°; LSPN-1187.
Photograph by F. Miller, University of Michigan/CTIO.

Fig. 534: 1986 Apr 11.19410 UT; exposure 20.0 minutes on Tech. Pan 2415 emulsion with Wratten 47B filter; r=1.34, Δ=0.42, β=148.7°; LSPN-1961.
Photograph by M. Clilverd/M. Dowson, British Antarctic Survey.

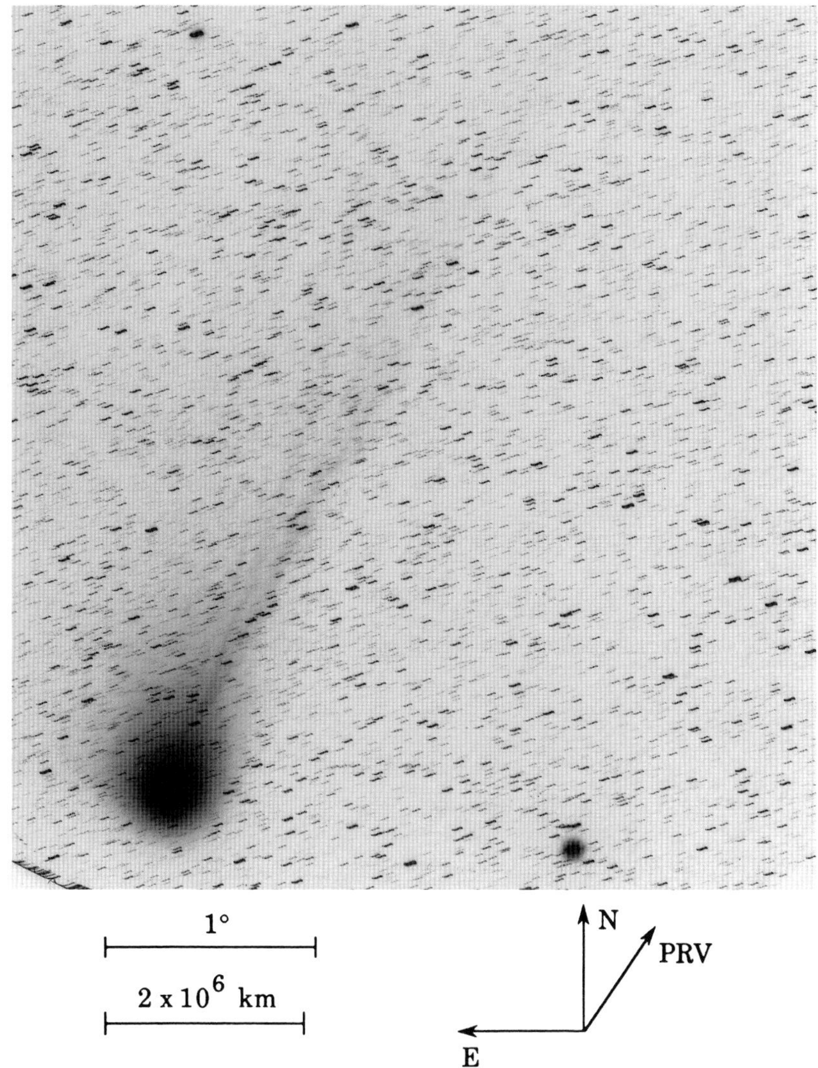

Fig. 535: 1986 Apr 11.21181 UT; exposure 14.0 minutes on Tech. Pan 2415 emulsion with no filter; r=1.34, Δ=0.42, β=148.8°; LSPN-1467.
Photograph by W. Liller, Easter Island, LSPN Island Network.

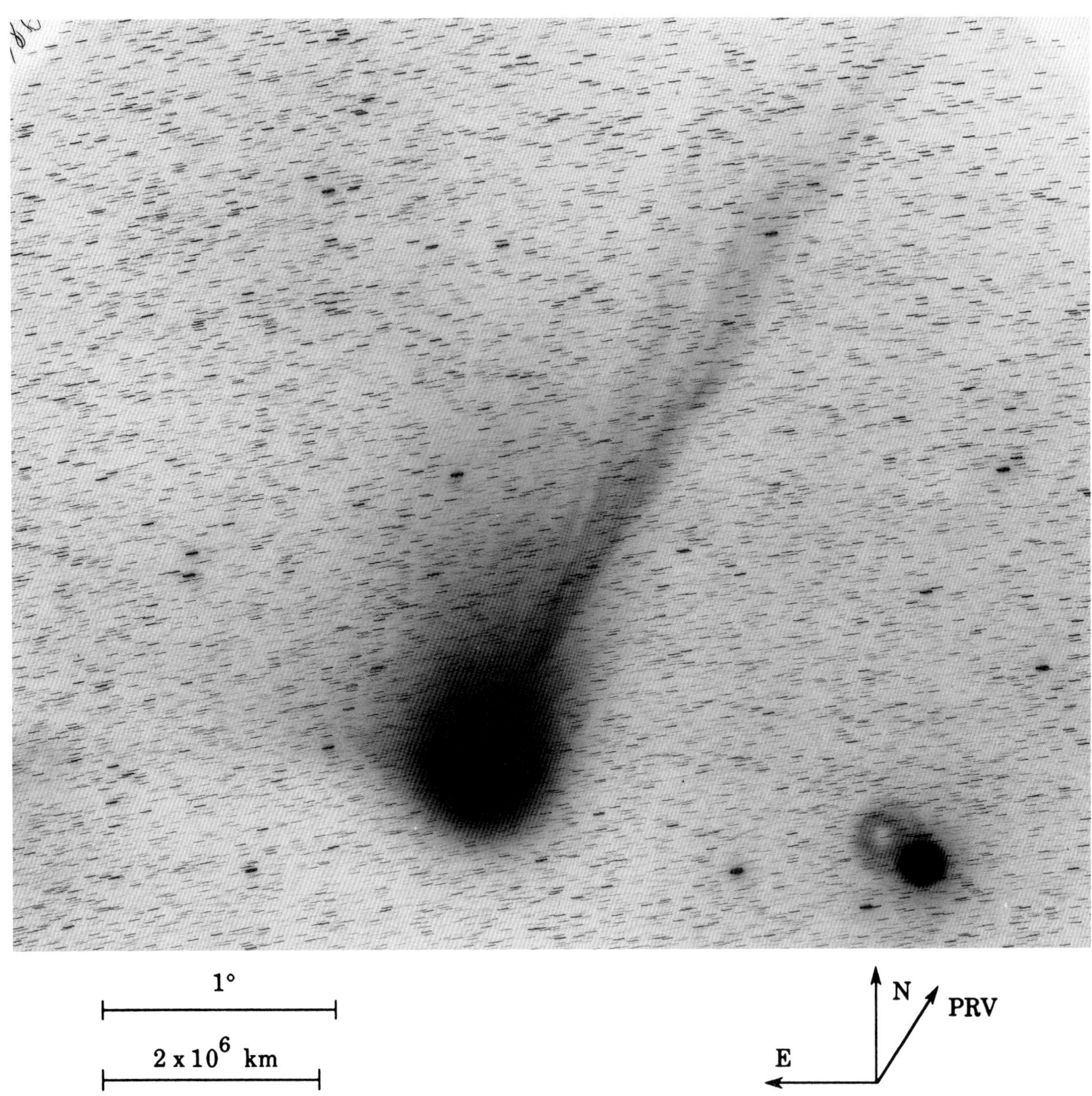

Fig. 536: 1986 Apr 11.22812 UT; exposure 15.0 minutes on 103a-O emulsion with GG-385 filter; $r = 1.34$, $\Delta = 0.42$, $\beta = 148.8°$; LSPN-753.
Photograph by C. Torres/H. Wroblewski, Cerro el Roble Astronomical Observatory.

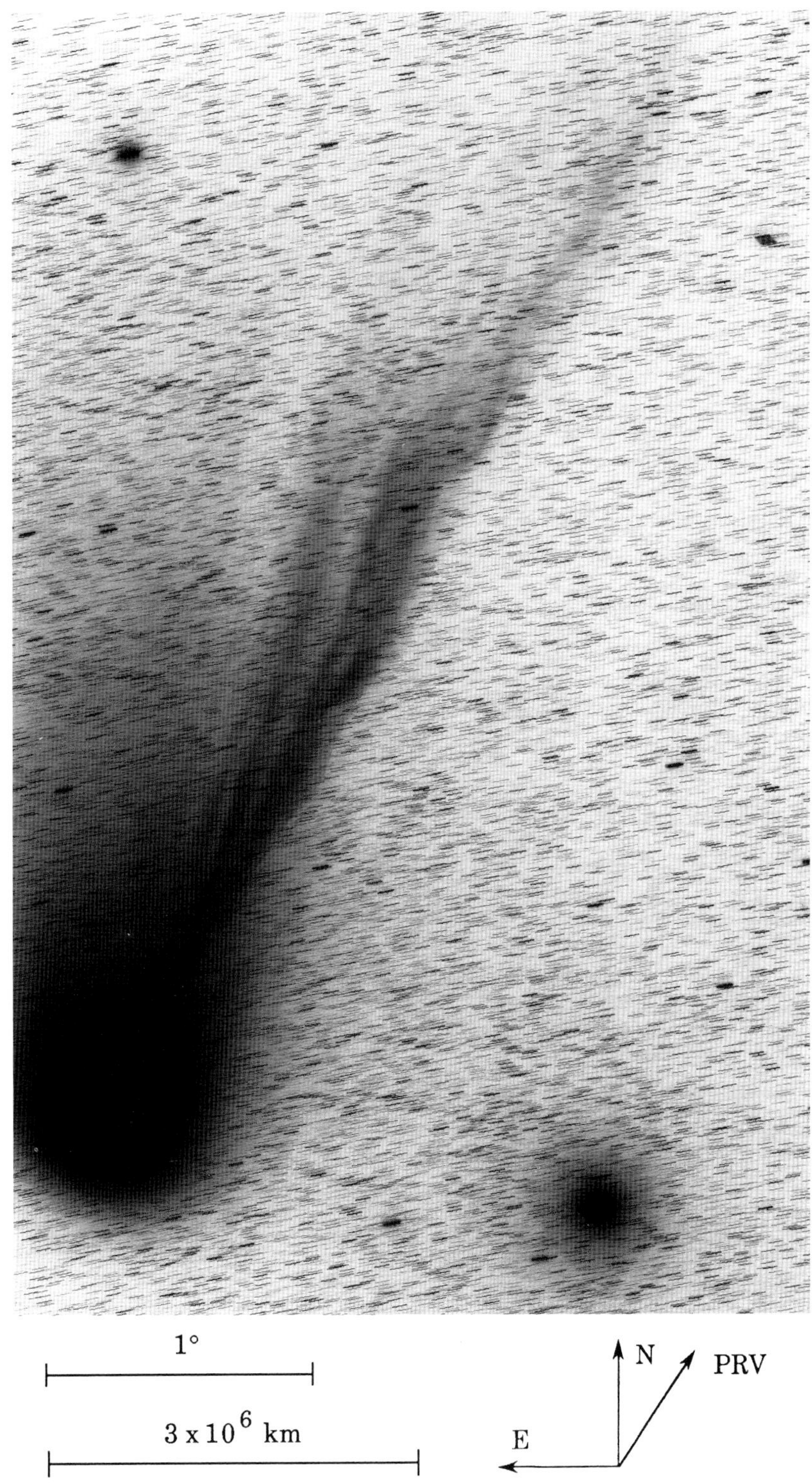

Fig. 537: 1986 Apr 11.23056 UT; exposure 20.0 minutes on 103a-O emulsion with no filter; r=1.34, Δ=0.42, β=148.8°; LSPN-1189.
Photograph by F. Miller, University of Michigan/CTIO.

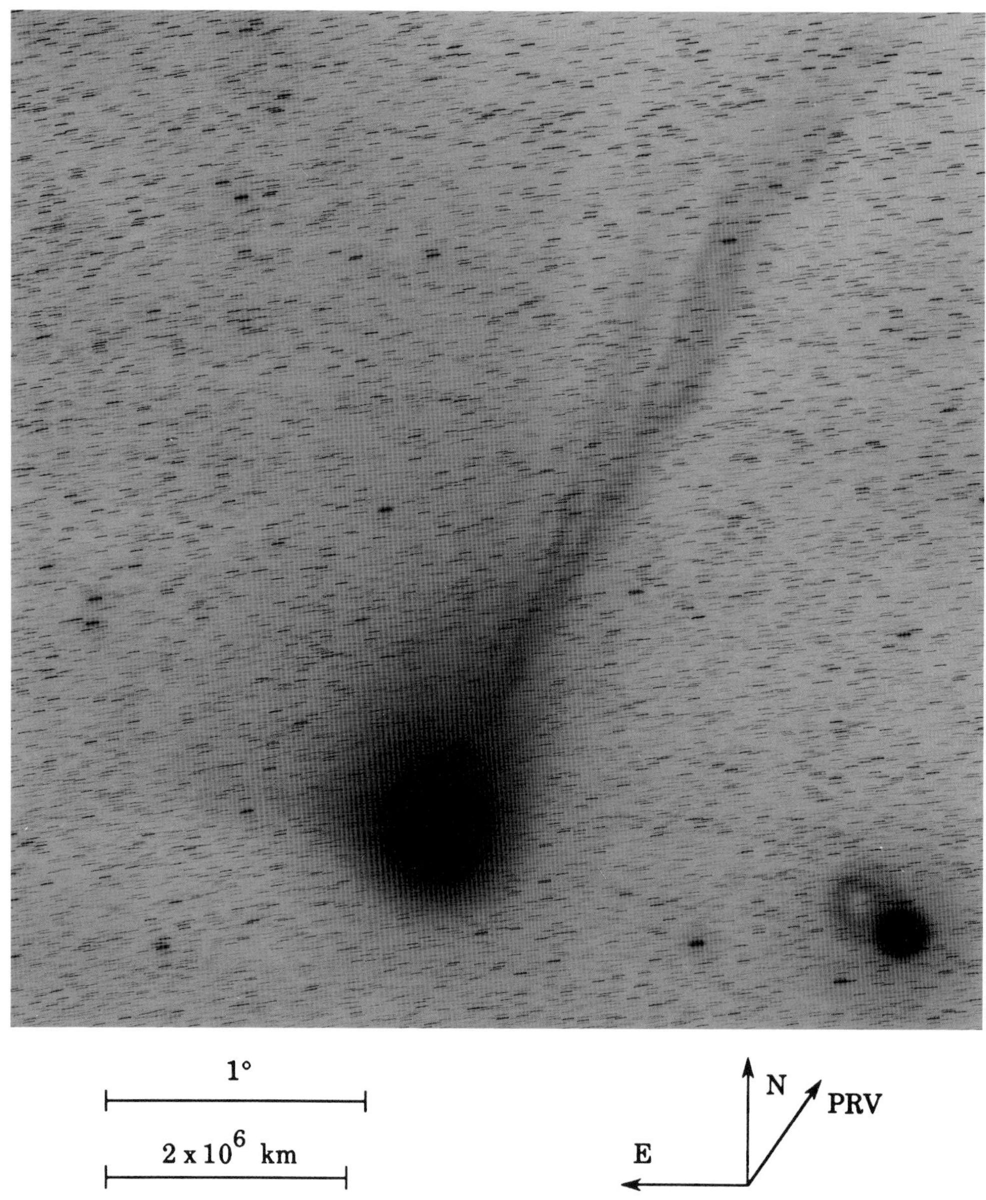

Fig. 538: 1986 Apr 11.24479 UT; exposure 15.0 minutes on 103a-O emulsion with GG-385 filter; r=1.34, Δ=0.42, β=148.9°; LSPN-754.
Photograph by C. Torres/H. Wroblewski, Cerro el Roble Astronomical Observatory.

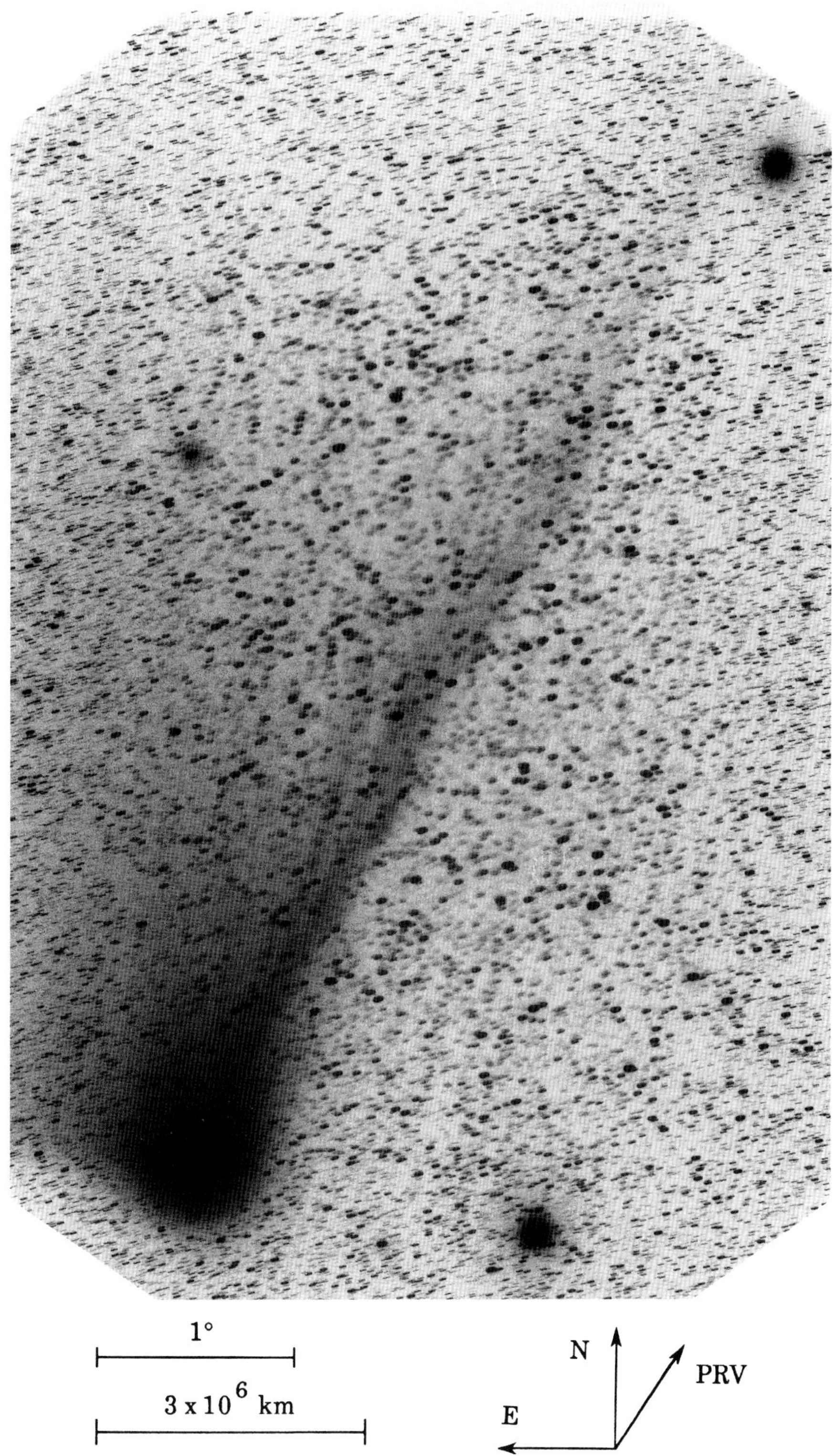

Fig. 539: 1986 Apr 11.25972 UT; exposure 10.0 minutes on Tri-X emulsion with no filter; r=1.34, Δ=0.42, β=148.9°; LSPN-1468.
Photograph by W. Liller, Easter Island, LSPN Island Network.

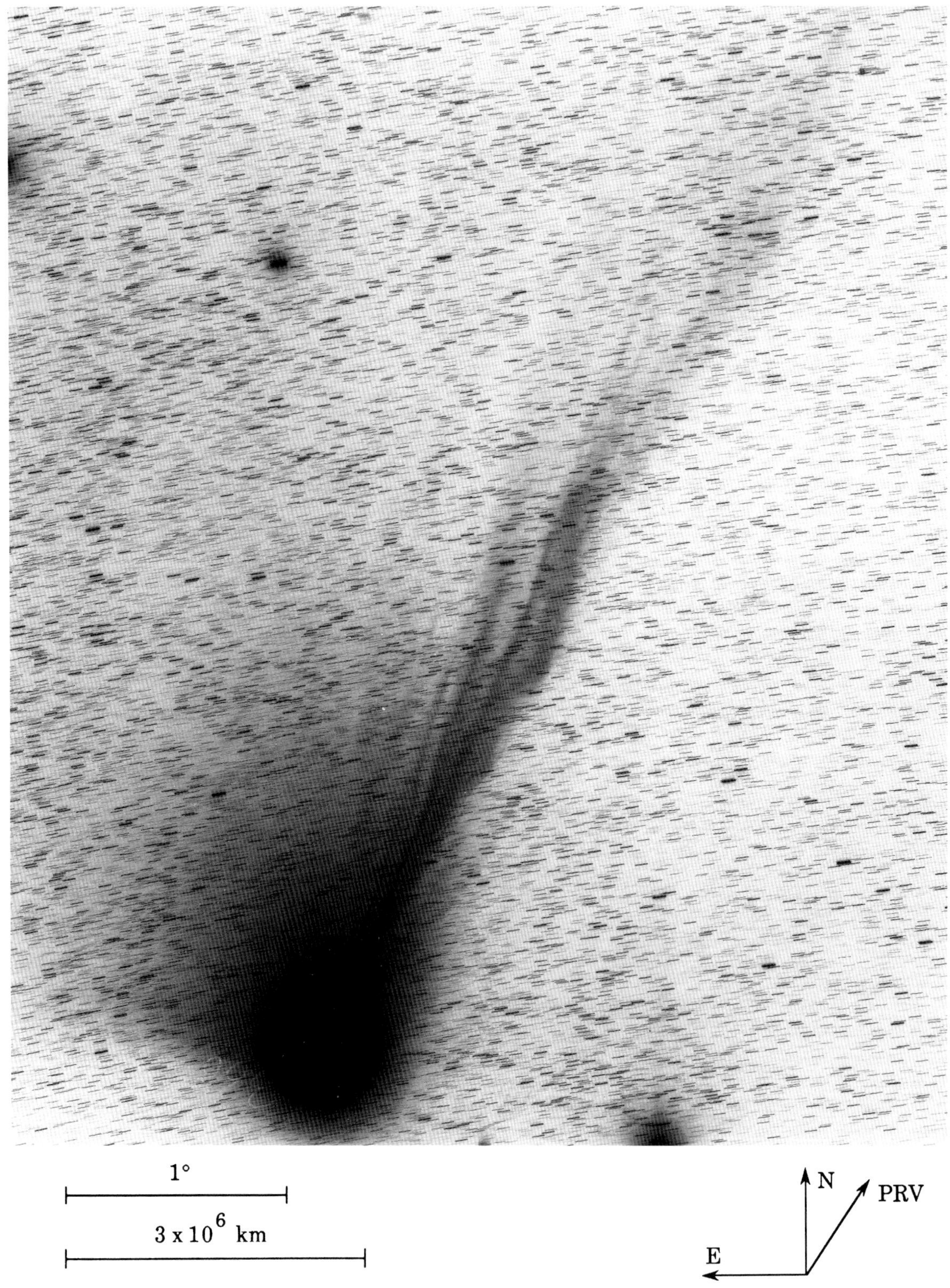

Fig. 540: 1986 Apr 11.30208 UT; exposure 20.0 minutes on 103a-O emulsion with no filter; r=1.34, Δ=0.42, β=149.0°; LSPN-1190.
Photograph by F. Miller, University of Michigan/CTIO.

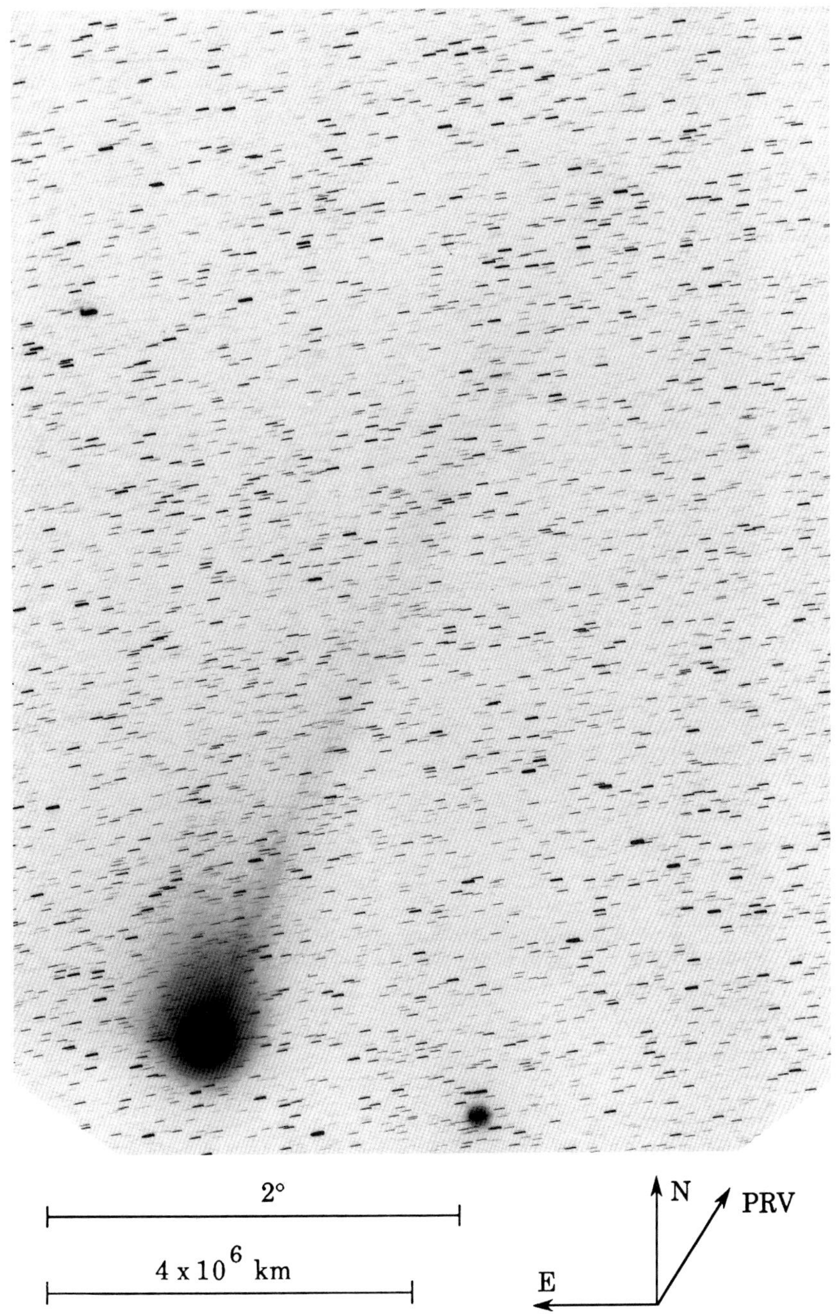

Fig. 541: 1986 Apr 11.35035 UT; exposure 15.0 minutes on Tech. Pan 2415 emulsion with no filter; r=1.34, Δ=0.42, β=149.2°; LSPN-1470.
Photograph by W. Liller, Easter Island, LSPN Island Network.

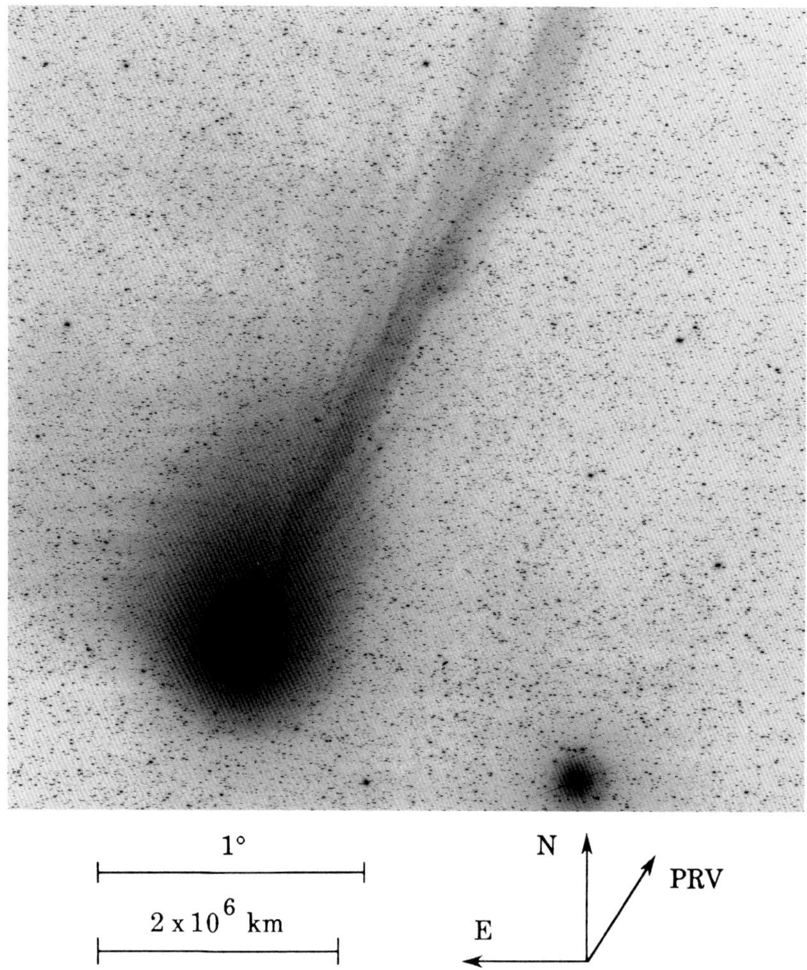

Fig. 542: 1986 Apr 11.35312 UT; exposure 3.0 minutes on 103a-O emulsion with no filter; r=1.34, Δ=0.42, β=149.2°; LSPN-1192.
Photograph by F. Miller, University of Michigan/CTIO.

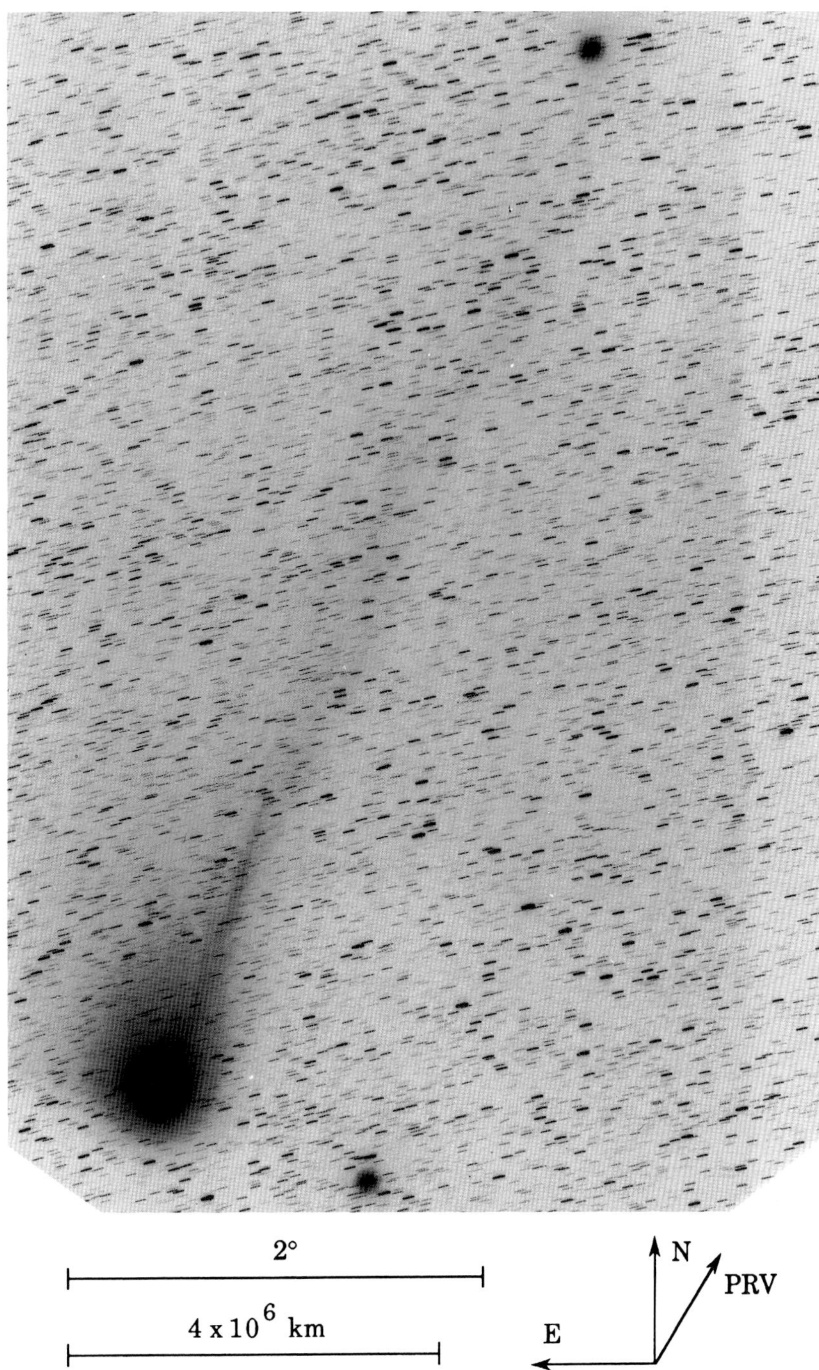

Fig. 543: 1986 Apr 11.41944 UT; exposure 14.0 minutes on Tech. Pan 2415 emulsion with no filter; r=1.34, Δ=0.42, β=149.3°; LSPN-1471.
Photograph by W. Liller, Easter Island, LSPN Island Network.

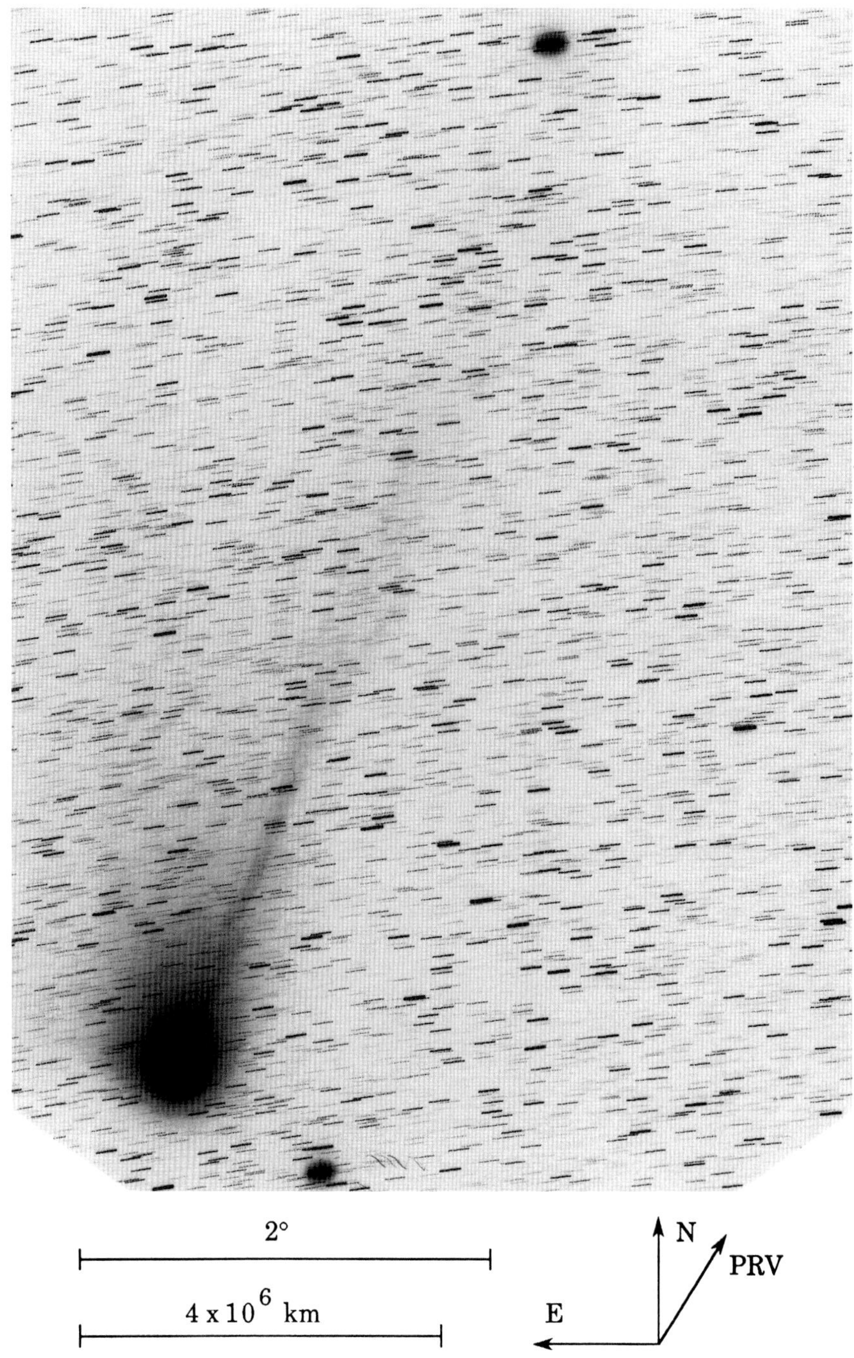

Fig. 544: 1986 Apr 11.48646 UT; exposure 30.0 minutes on Tech. Pan 2415 emulsion with no filter; r=1.34, Δ=0.42, β=149.5°; LSPN-1473.
Photograph by W. Liller, Easter Island, LSPN Island Network.

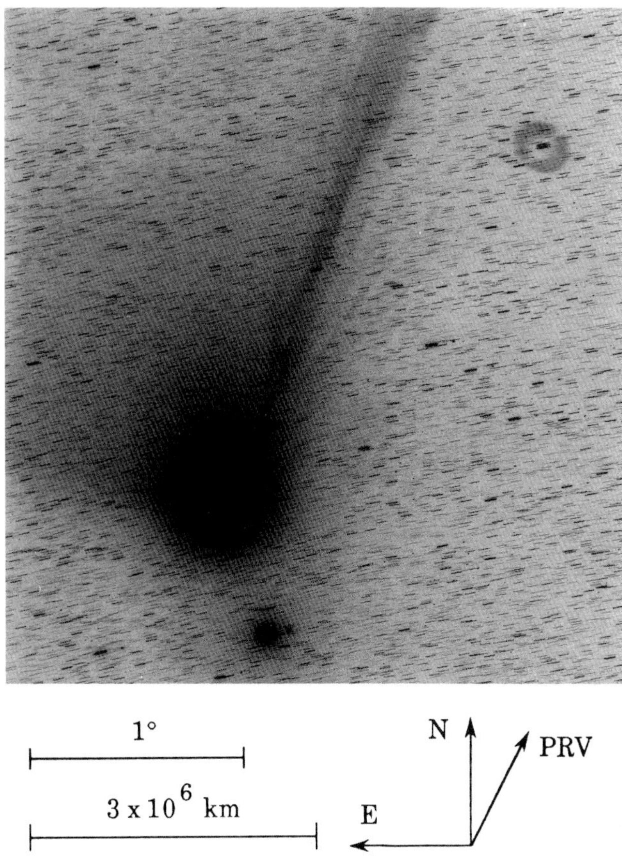

Fig. 545: 1986 Apr 11.57833 UT; exposure 6.0 minutes on IIa-O emulsion with GG-385 filter; r=1.34, Δ=0.42, β=149.8°; LSPN-2146.
Photograph by P. Magnusson, Uppsala Southern Station.

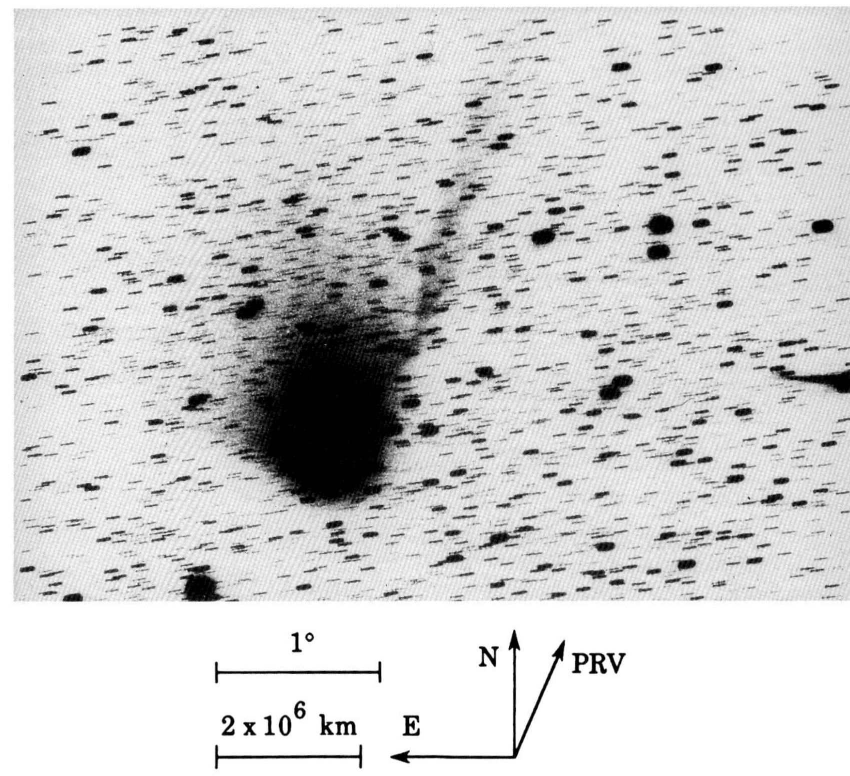

Fig. 546: 1986 Apr 11.82190 UT; exposure 20.0 minutes on AGFA 1000 RS emulsion; r=1.34, Δ=0.42, β=150.4°. *Photograph from Volkssternwarte Frankfurt, Germany.*

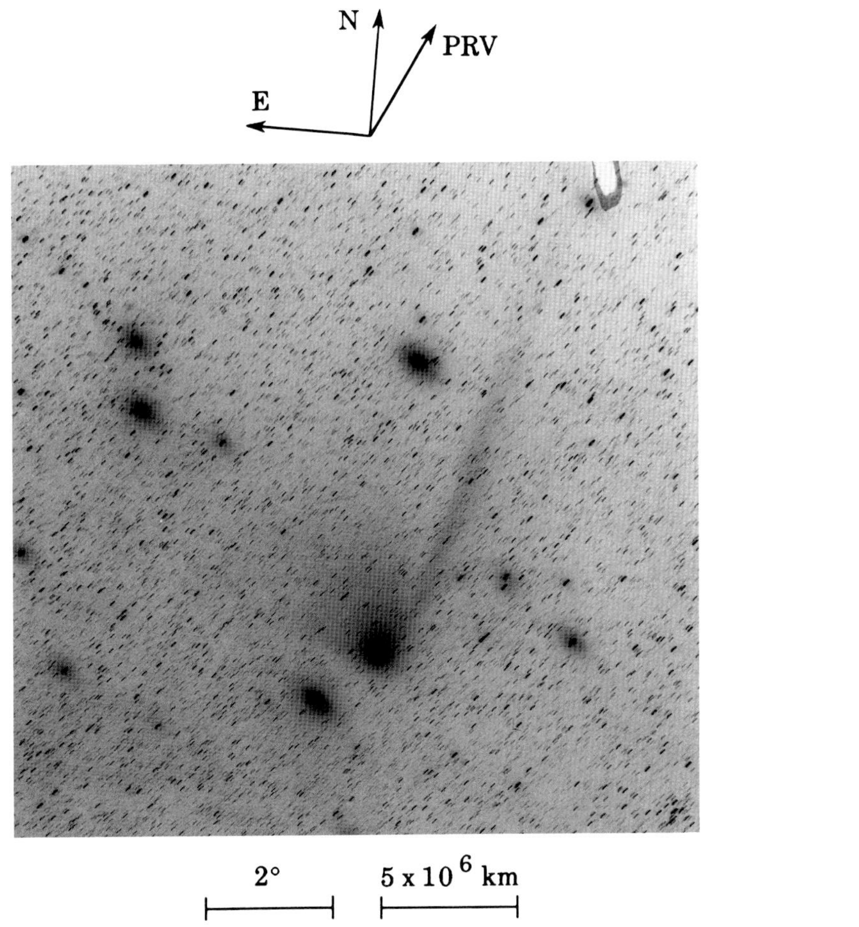

Fig. 547: 1986 Apr 11.84549 UT; exposure 75.0 minutes on 103a-O emulsion with no filter; r=1.35, Δ=0.42, β=150.5°; LSPN-3659.
Photograph submitted by K. Sivaraman, Indian Institute for Astrophysics, Kodaikanal Station.

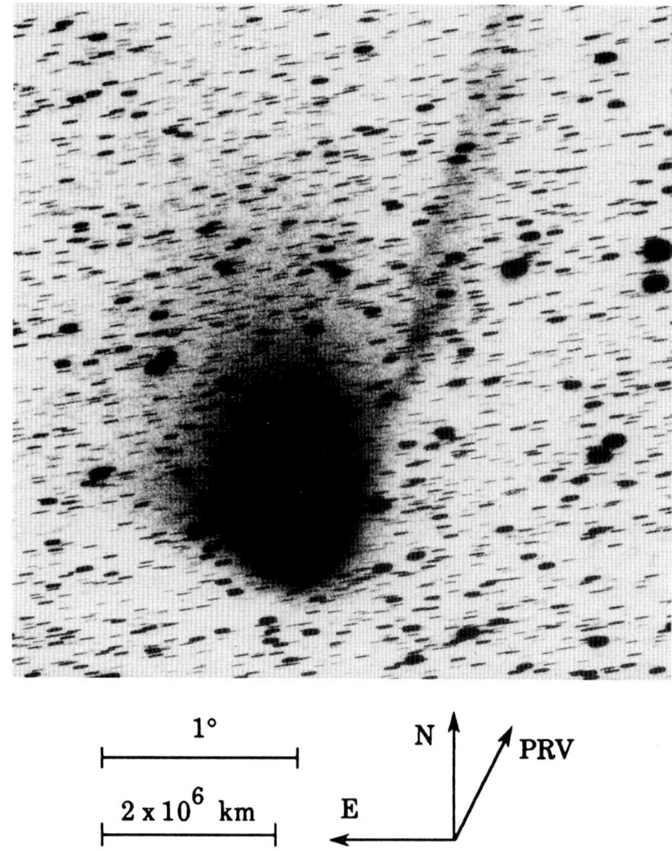

Fig. 548: 1986 Apr 11.86111 UT; exposure 20.0 minutes on AGFA 1000 RS emulsion; r=1.35, Δ=0.42, β=150.5°; AON-851697.
Photograph from Volkssternwarte Frankfurt, Germany.

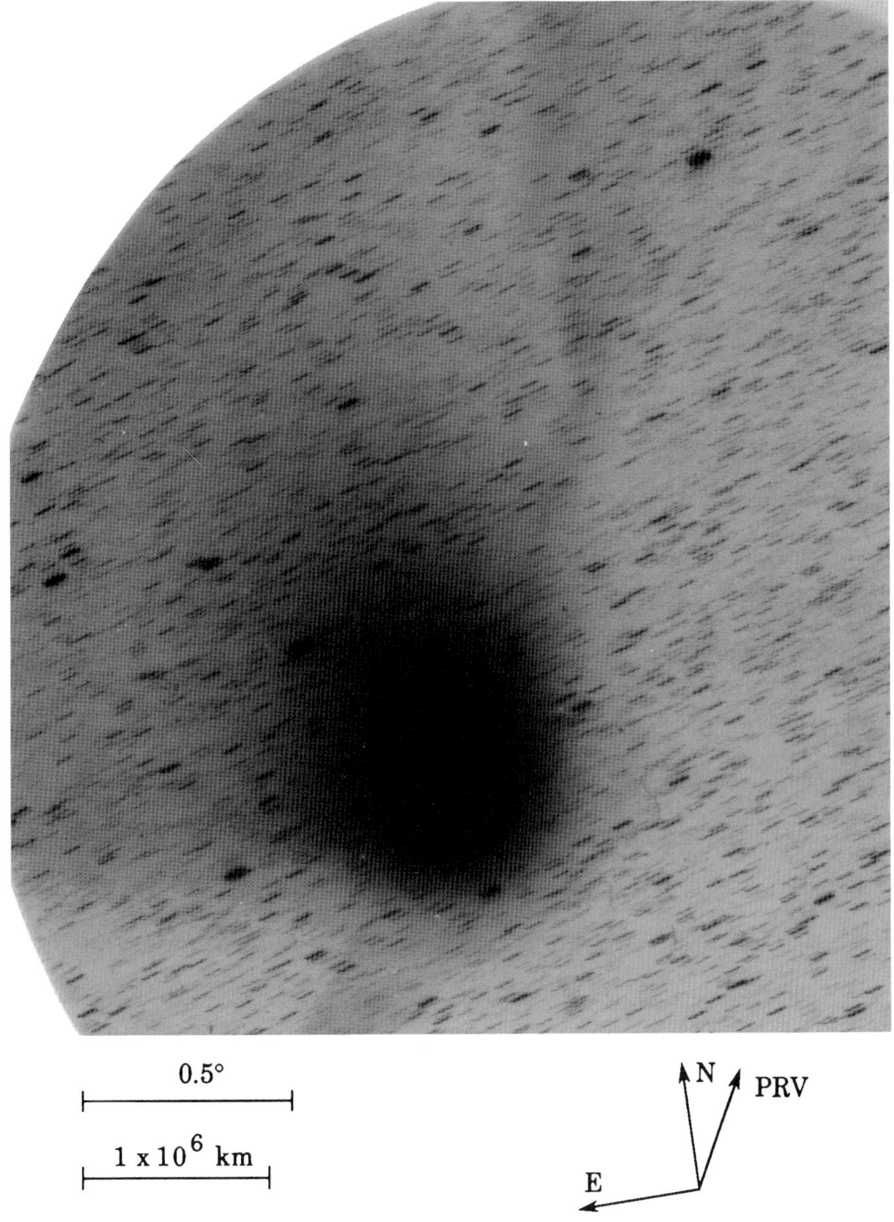

Fig. 549: 1986 Apr 11.87396 UT; exposure 17.0 minutes on Tech. Pan 2415 emulsion with no filter; r=1.35, Δ=0.42, β=150.6°; LSPN-3498.
Photograph by D. O'Donoghue, University of Perugia/South African Astronomical Observatory.

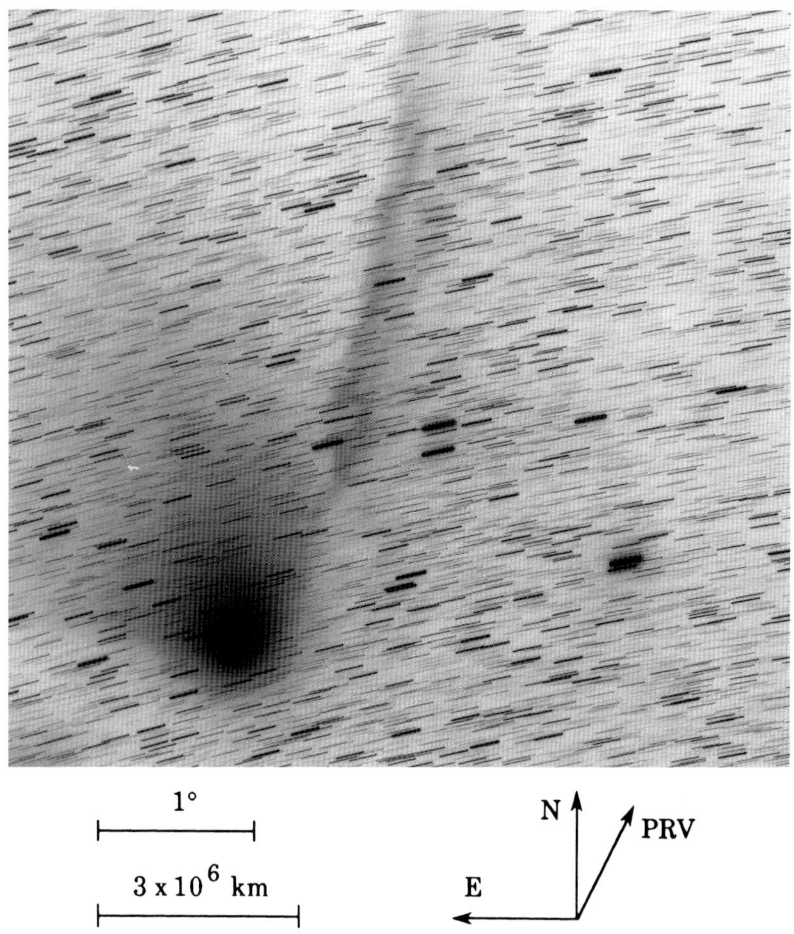

Fig. 550: 1986 Apr 11.95556 UT; exposure 60.0 minutes on IIa-O emulsion with no filter; r=1.35, Δ=0.42, β=150.8°; LSPN-2867.
Photograph by G. Malcolm/A. Jarrett, Boyden Observatory.

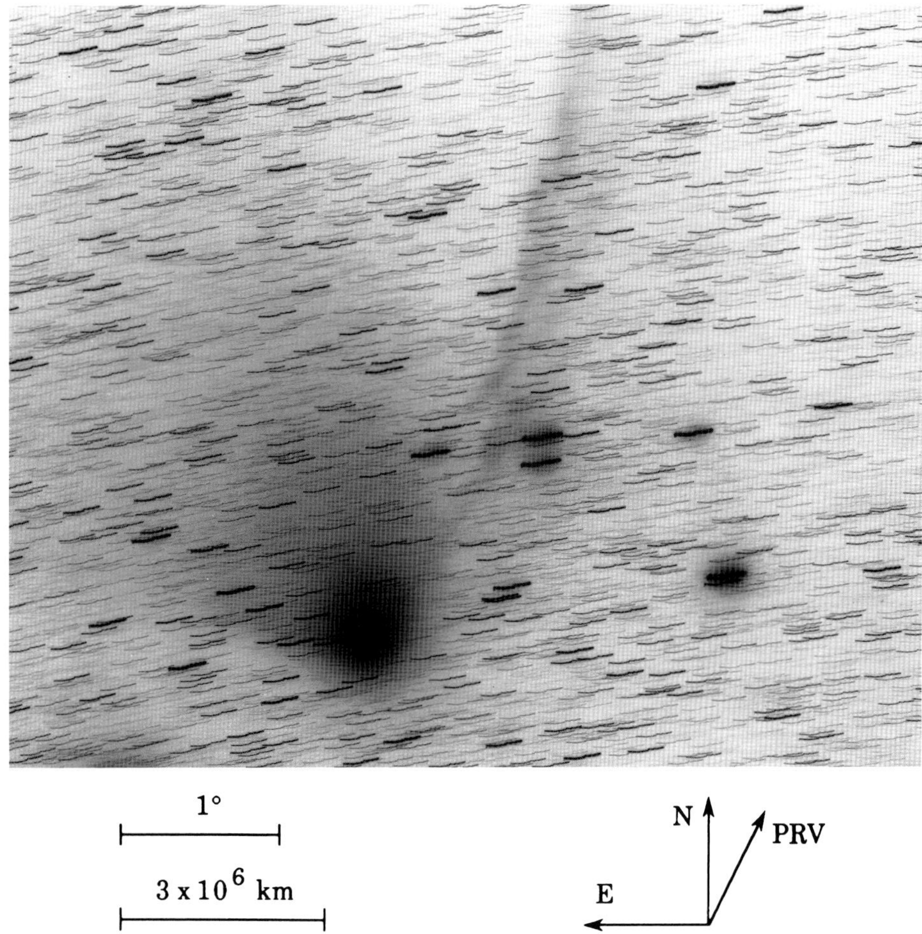

Fig. 551: 1986 Apr 12.00521 UT; exposure 65.0 minutes on IIa-O emulsion with no filter; r=1.35, Δ=0.42, β=150.9°; LSPN-2868.
Photograph by G. Malcolm, Boyden Observatory.

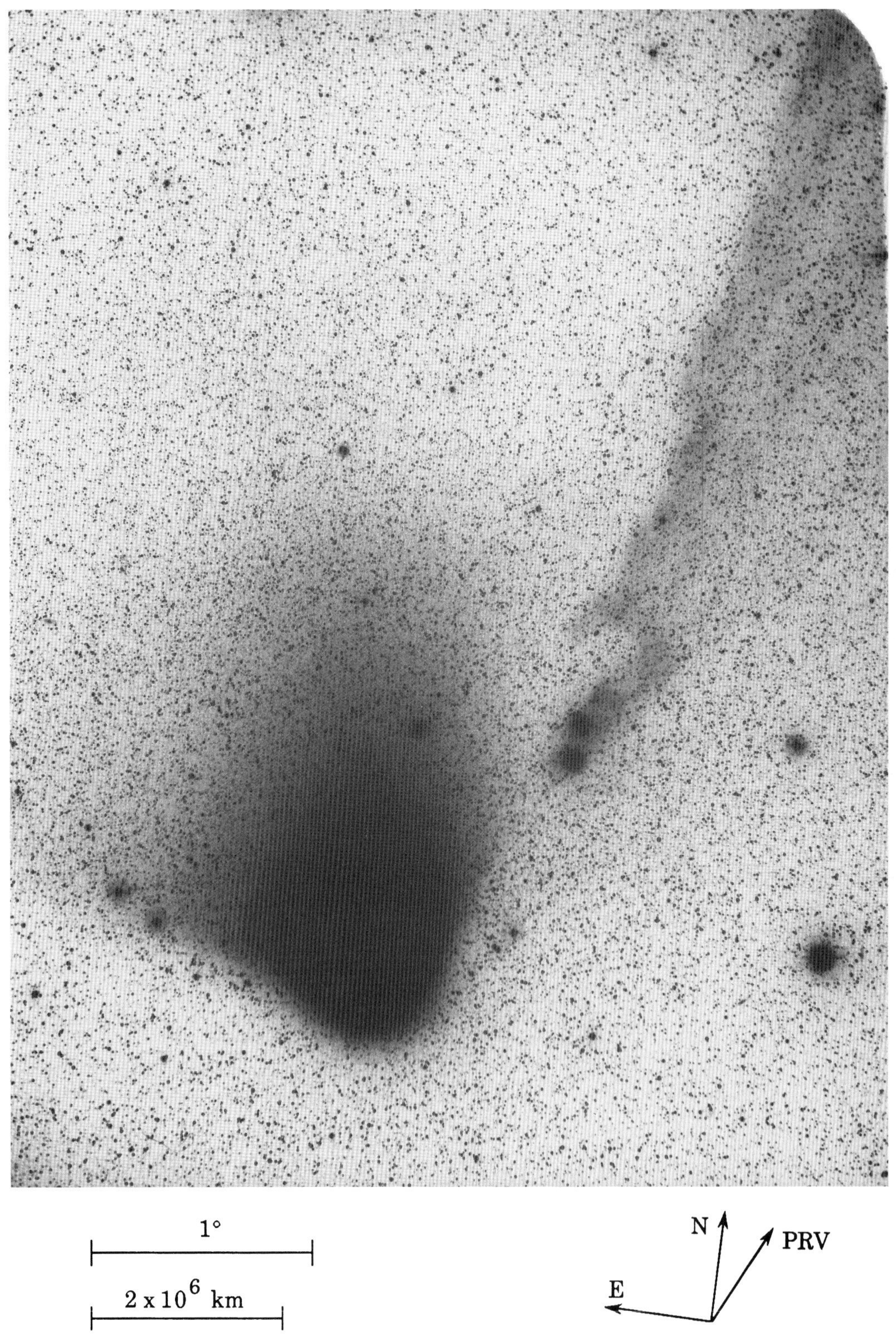

Fig. 552: 1986 Apr 12.03299 UT; exposure 5.0 minutes on Tech. Pan 2415 emulsion; r=1.35, Δ=0.42, β=151.0°; AON-851705.
Photograph by M. Jäger, Austria.

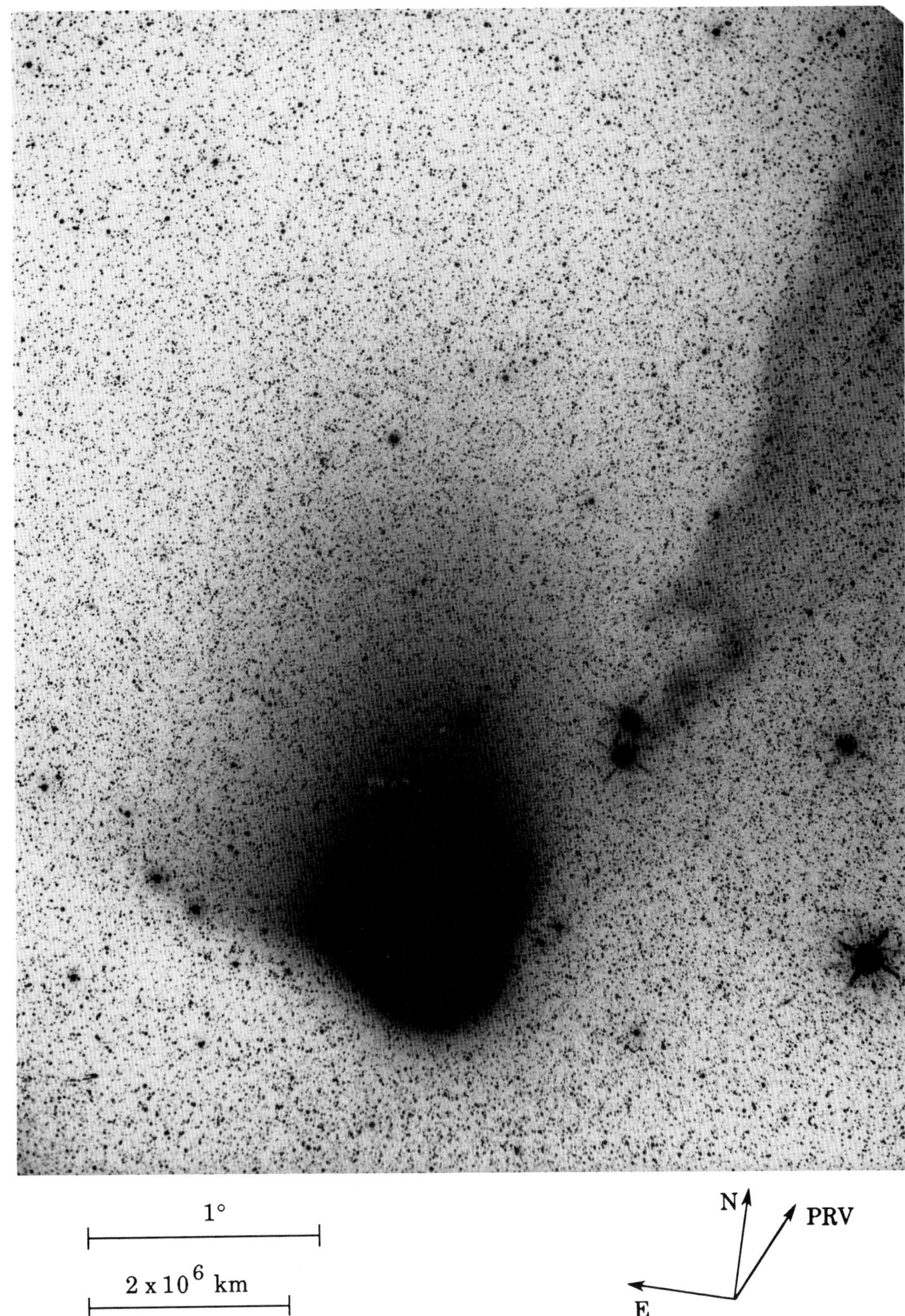

Fig. 553: 1986 Apr 12.04583 UT; exposure 6.0 minutes on Tech. Pan 2415 emulsion; r=1.35, Δ=0.42, β=151.0°; AON-851707.
Photograph by M. Jäger, Austria.

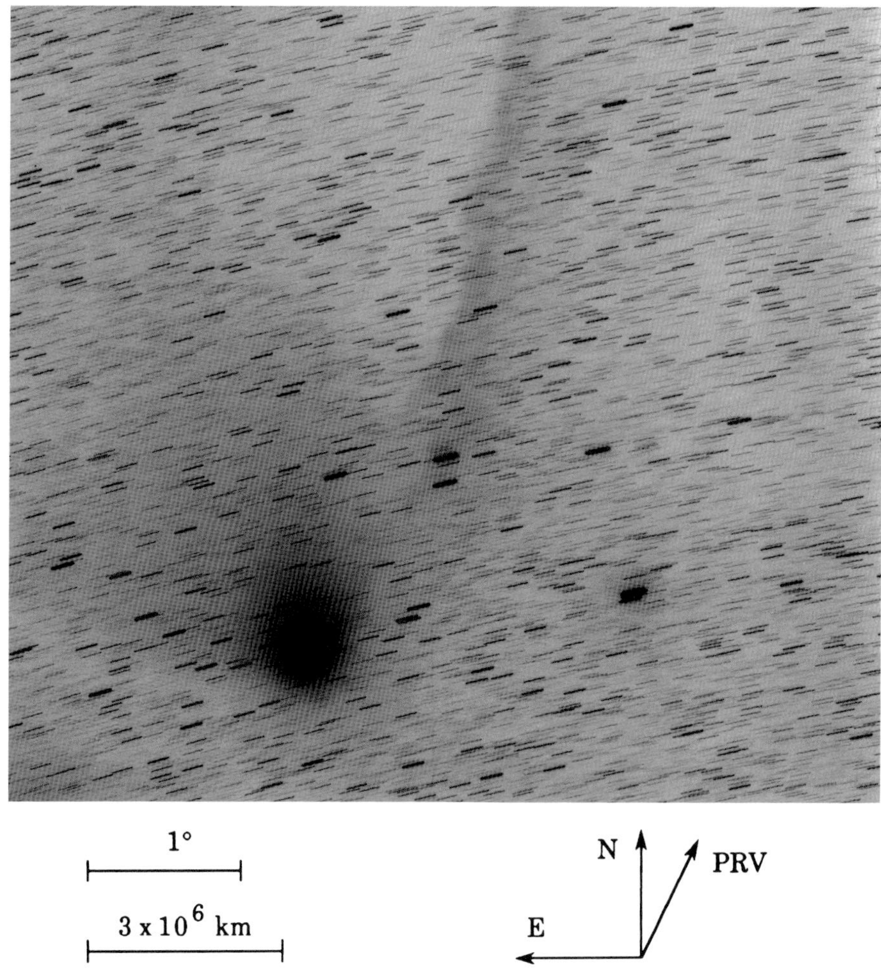

Fig. 554: 1986 Apr 12.05035 UT; exposure 45.0 minutes on IIa-O emulsion with no filter; r=1.35, Δ=0.42, β=151.0°; LSPN-2869.
Photograph by G. Malcolm, Boyden Observatory.

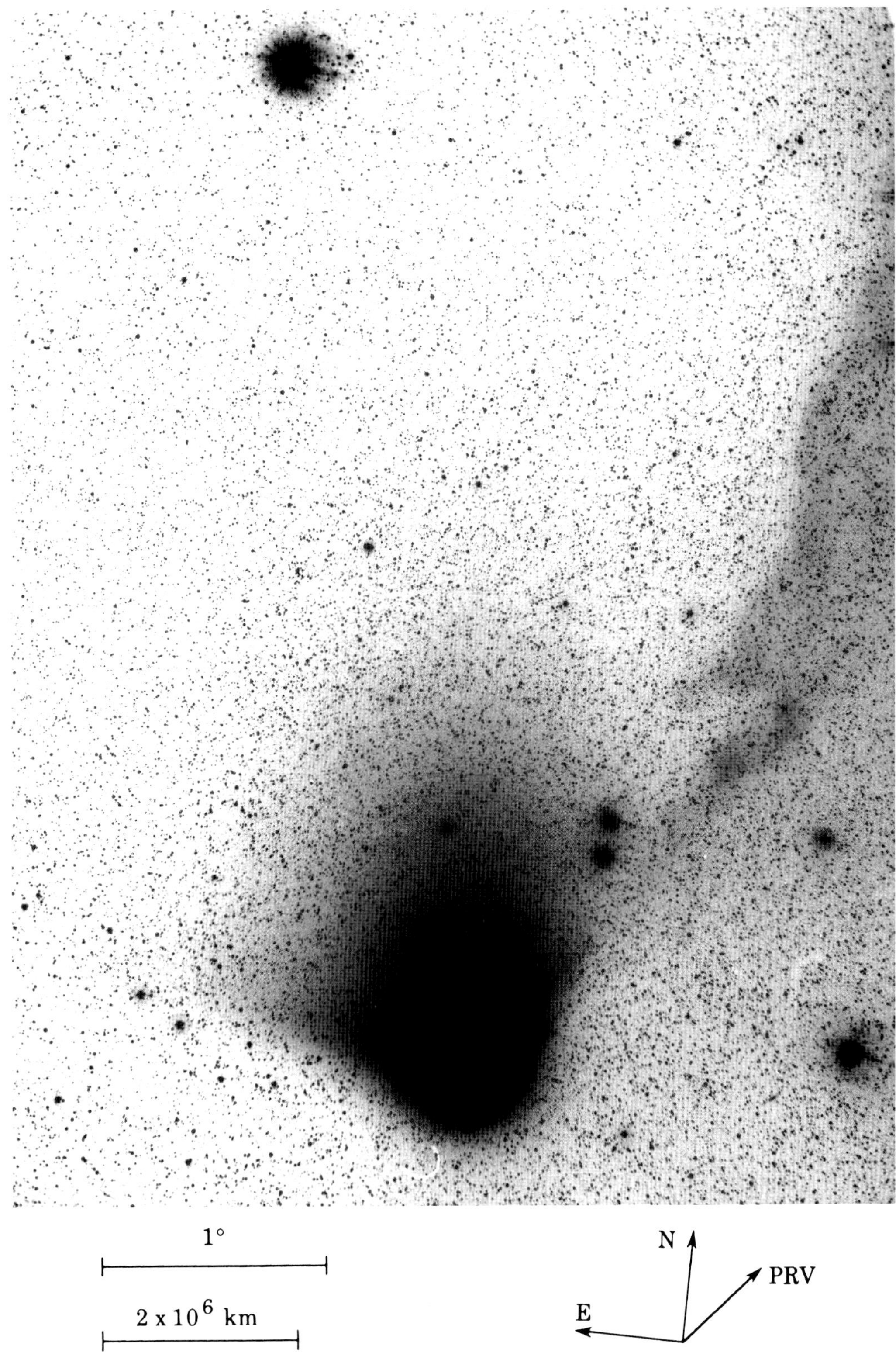

Fig. 555: 1986 Apr 12.09722 UT; exposure 6.0 minutes on Tech. Pan 2415 emulsion; r=1.35, Δ=0.42, β=151.2°; AON-851708.
Photograph by M. Jäger, Austria.

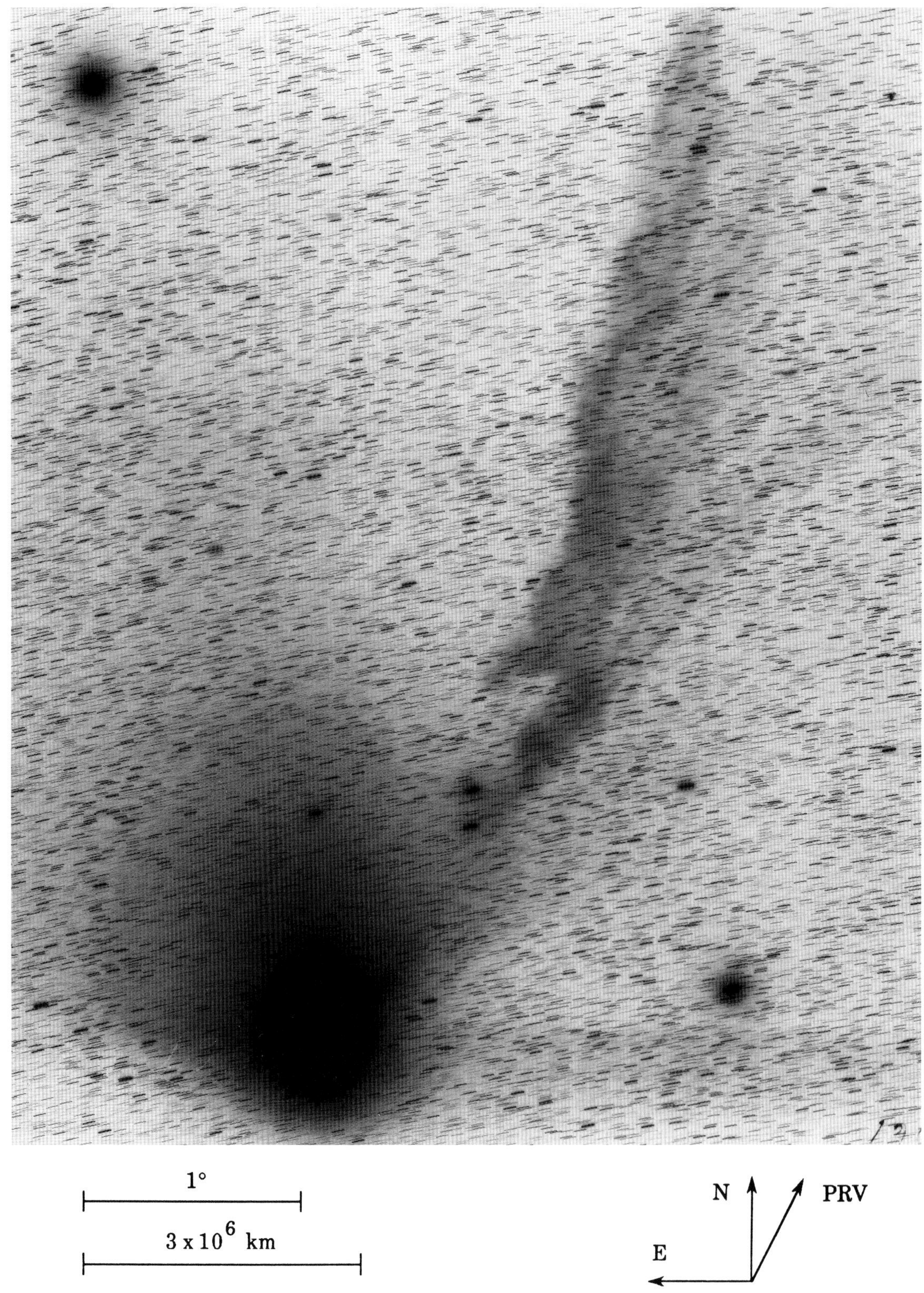

Fig. 556: 1986 Apr 12.10417 UT; exposure 20.0 minutes on 103a-O emulsion with no filter; r=1.35, Δ=0.42, β=151.2°; LSPN-1194.
Photograph by F. Miller, University of Michigan/CTIO.

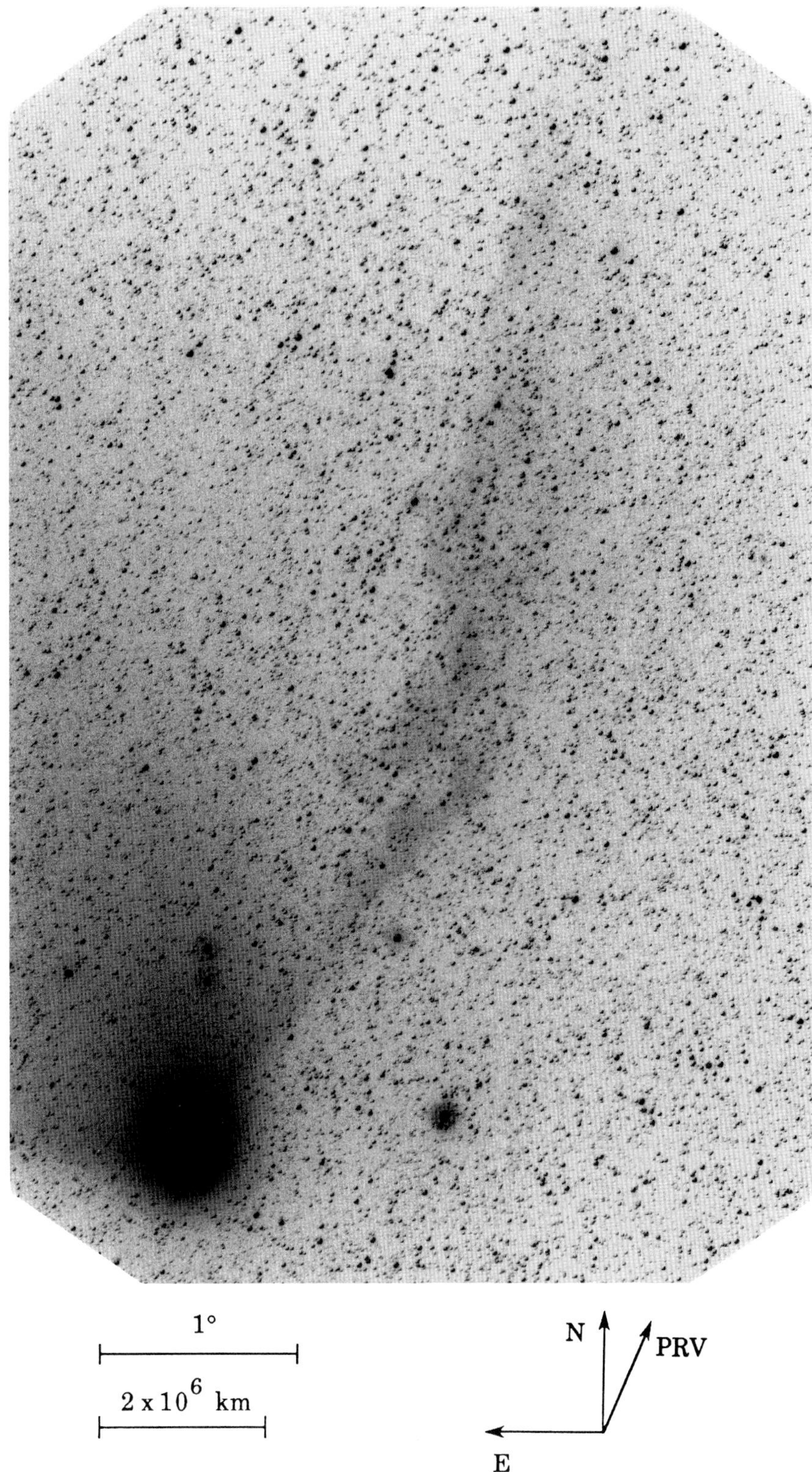

Fig. 557: 1986 Apr 12.23160 UT; exposure 7.0 minutes on Tri-X emulsion with no filter; r=1.35, Δ=0.42, β=151.5°; LSPN-1474.
Photograph by W. Liller, Easter Island, LSPN Island Network.

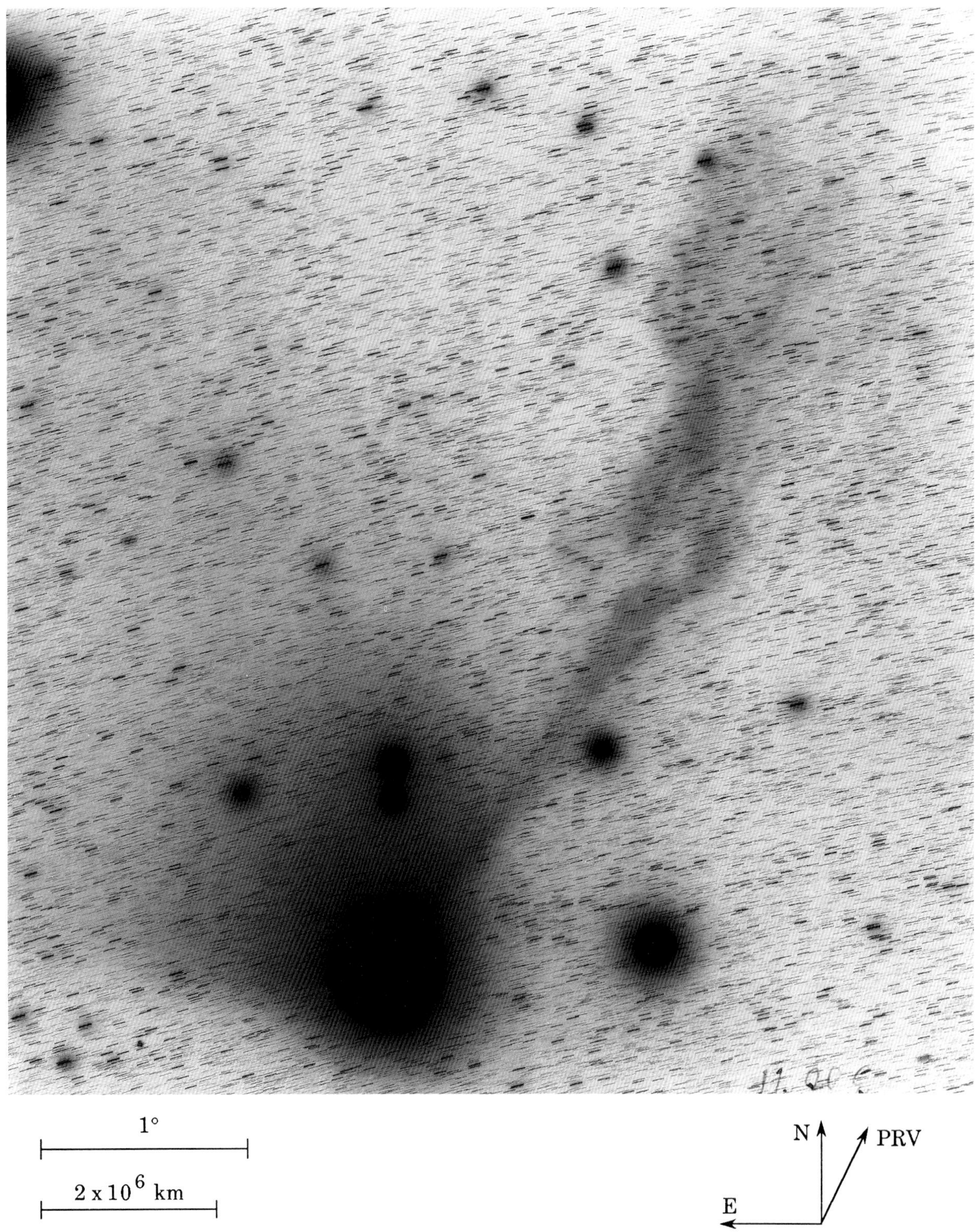

Fig. 558: 1986 Apr 12.25208 UT; exposure 20.0 minutes on 103a-O emulsion with no filter; r=1.35, Δ=0.42, β=151.6°; LSPN-1197.
Photograph by F. Miller, University of Michigan/CTIO.

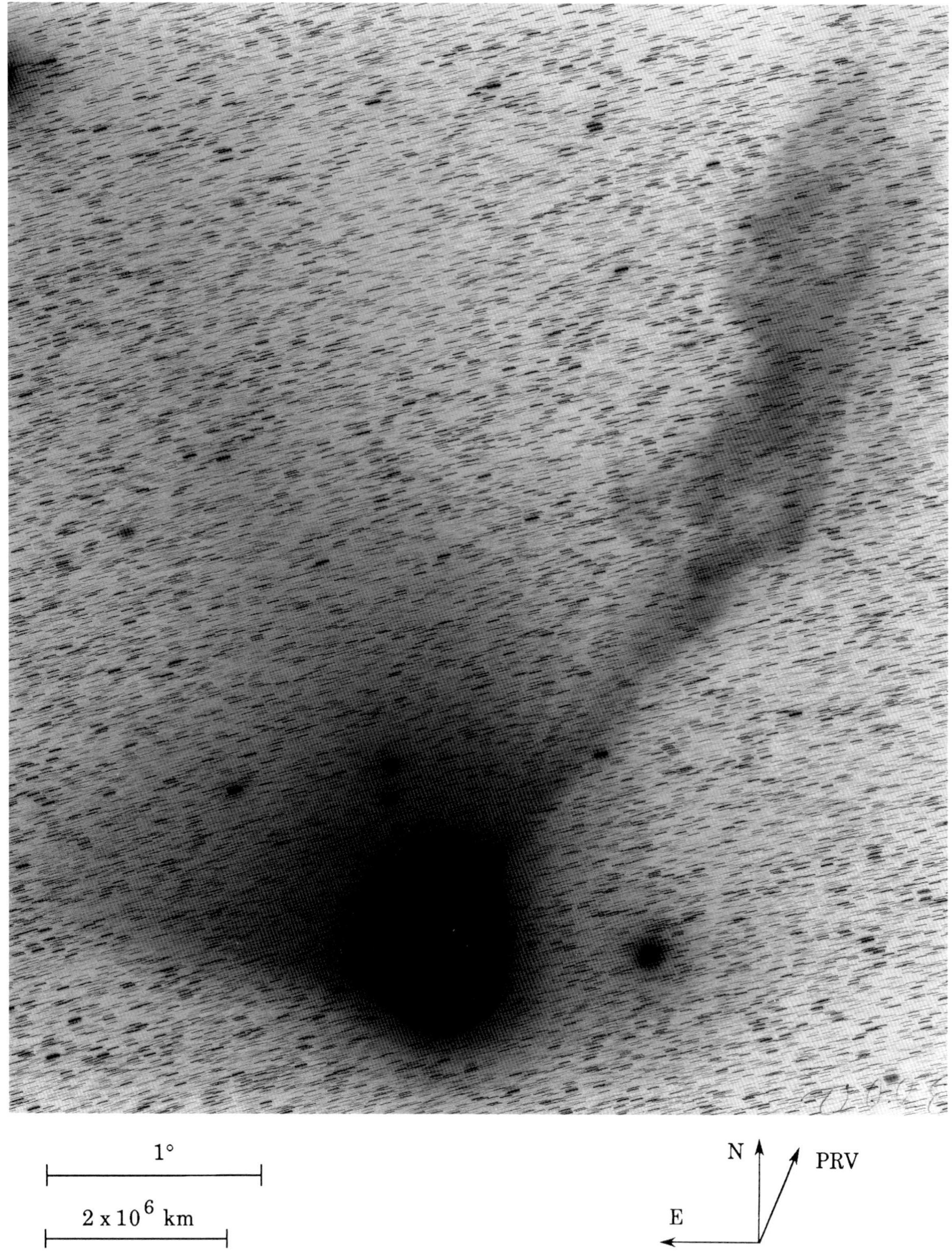

Fig. 559: 1986 Apr 12.30903 UT; exposure 20.0 minutes on 103a-O emulsion with no filter; r=1.35, Δ=0.42, β=151.7°; LSPN-1198.
Photograph by F. Miller, University of Michigan/CTIO.

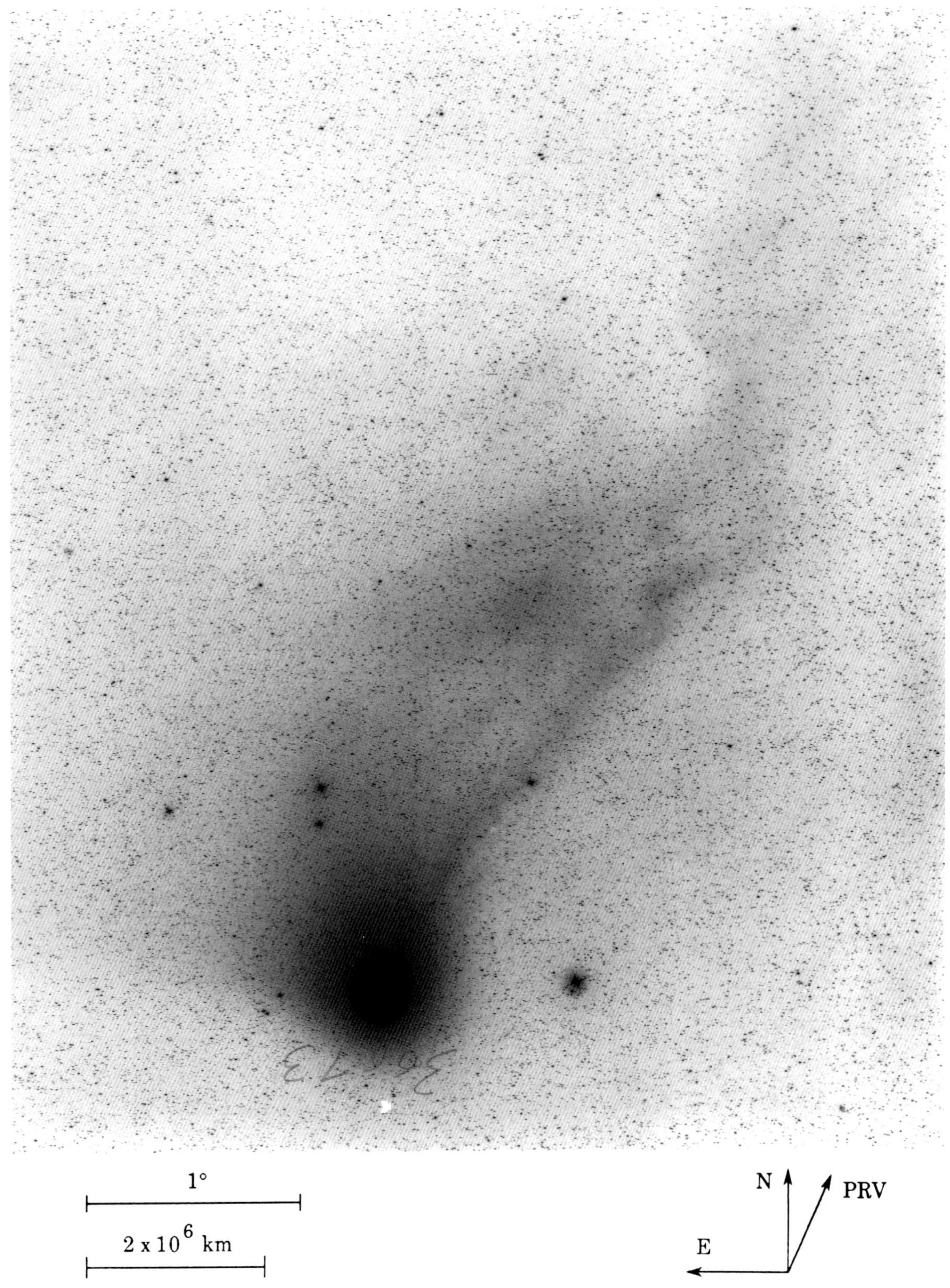

Fig. 560: 1986 Apr 12.32951 UT; exposure 3.0 minutes on 103a-O emulsion with no filter; r=1.35, Δ=0.42, β=151.7°; LSPN-1199.
Photograph by F. Miller, University of Michigan/CTIO.

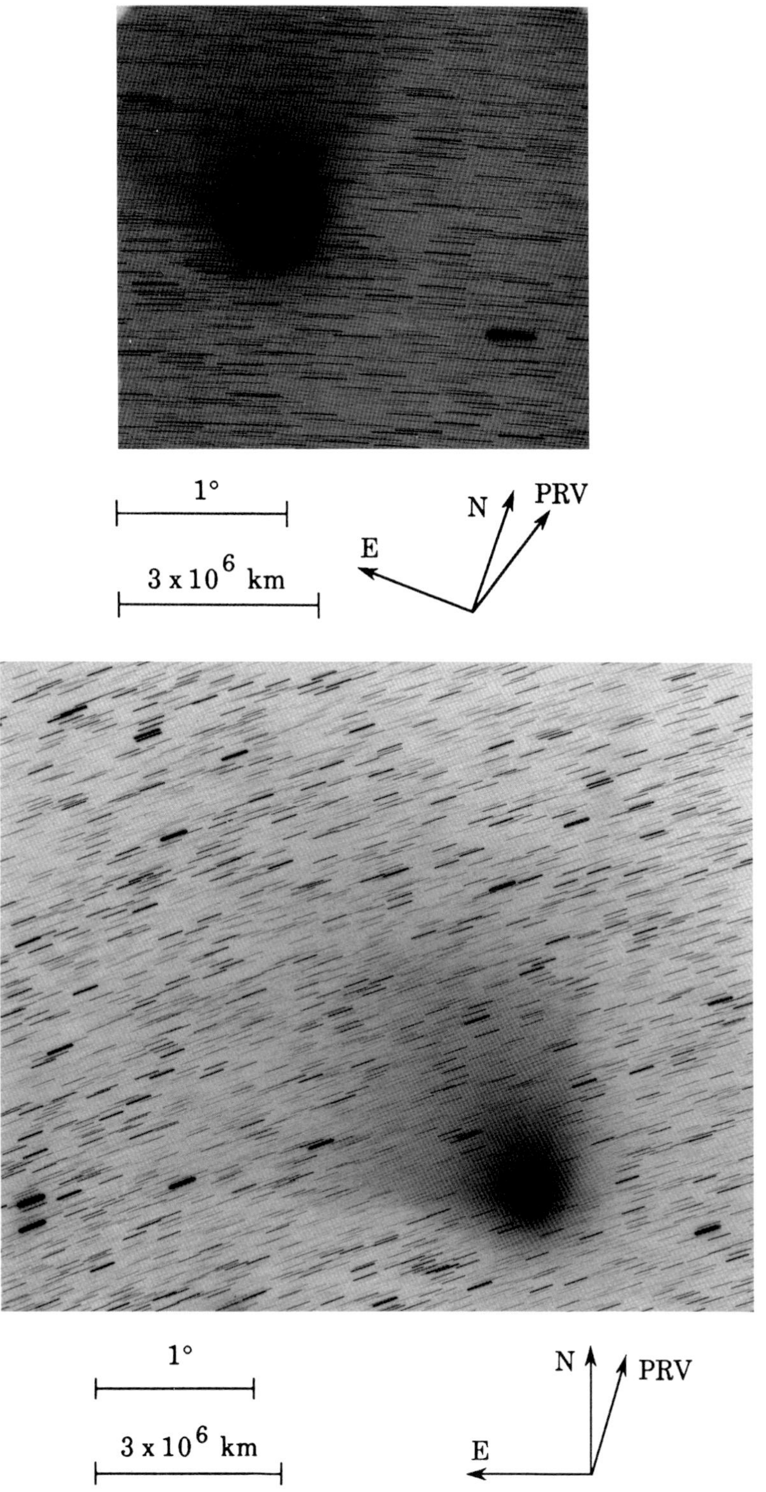

(top) Fig. 561: 1986 Apr 12.89236 UT; exposure 90.0 minutes on 103a-O emulsion with UV filter; r=1.36, Δ=0.42, β=153.1°; LSPN-3692.
Photograph submitted by K. Sivaraman, Indian Institute for Astrophysics, Kavalur Station.

(bottom) Fig. 562: 1986 Apr 12.98715 UT; exposure 50.0 minutes on IIa-O emulsion with no filter; r=1.36, Δ=0.42, β=153.3°; LSPN-2870.
Photograph by G. Malcolm/A. Jarrett, Boyden Observatory.

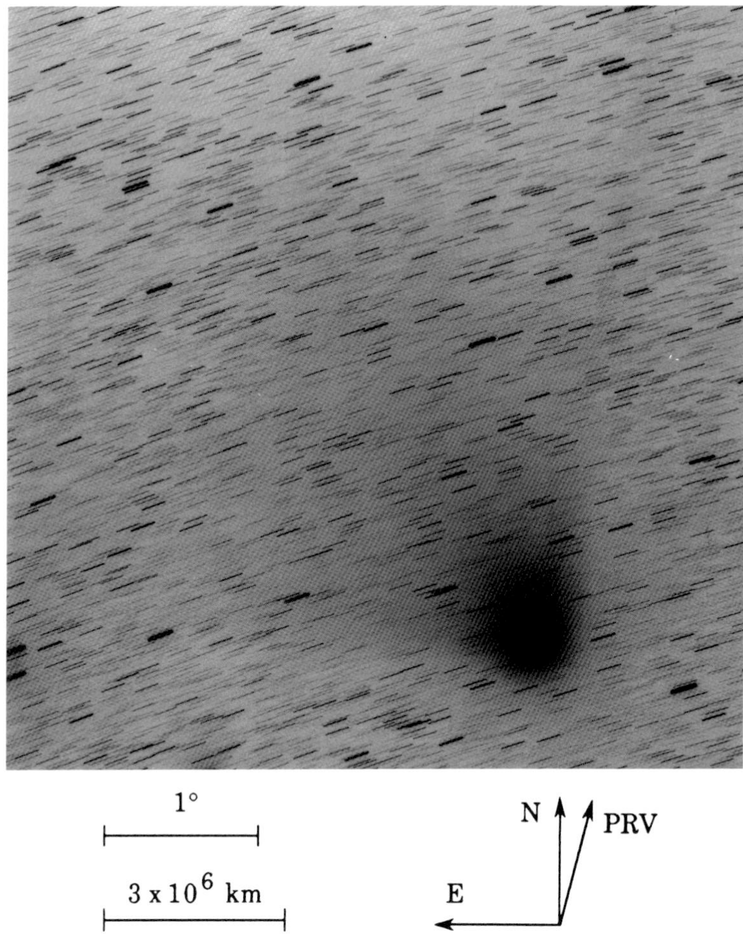

Fig. 563: 1986 Apr 13.02708 UT; exposure 50.0 minutes on IIa-O emulsion with no filter; r=1.36, Δ=0.42, β=153.4°; LSPN-2871.
Photograph by G. Malcolm, Boyden Observatory.

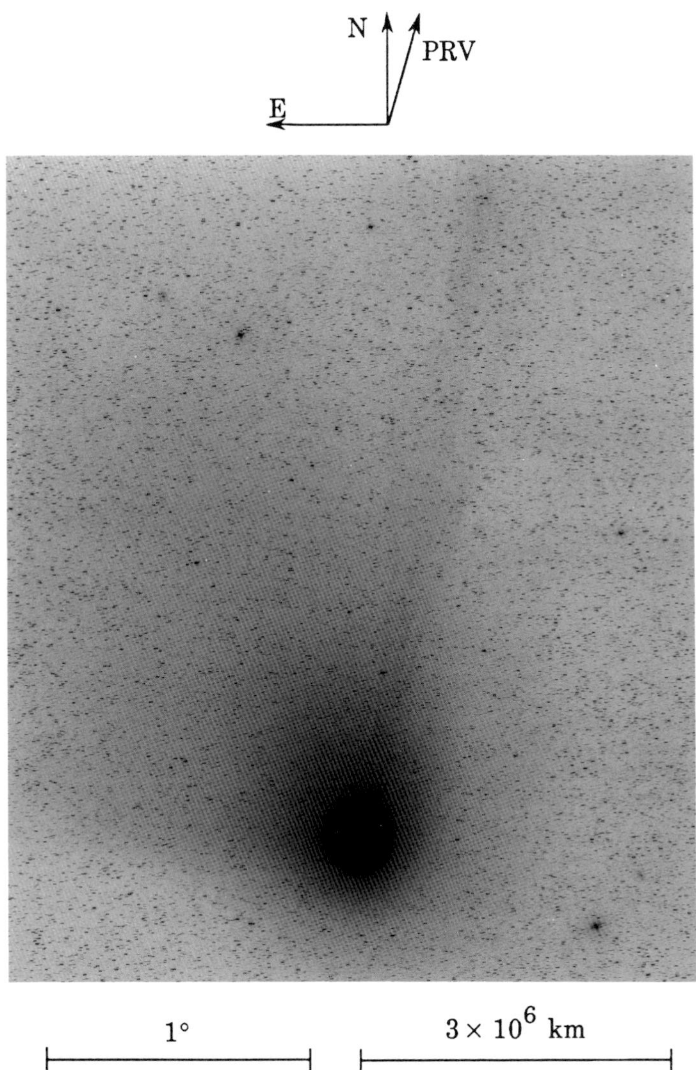

Fig. 564: 1986 Apr 13.04931 UT; exposure 4.0 minutes on 103a-O emulsion with no filter; r=1.36, Δ=0.42, β=153.5°; LSPN-1200.
Photograph by F. Miller, University of Michigan/CTIO.

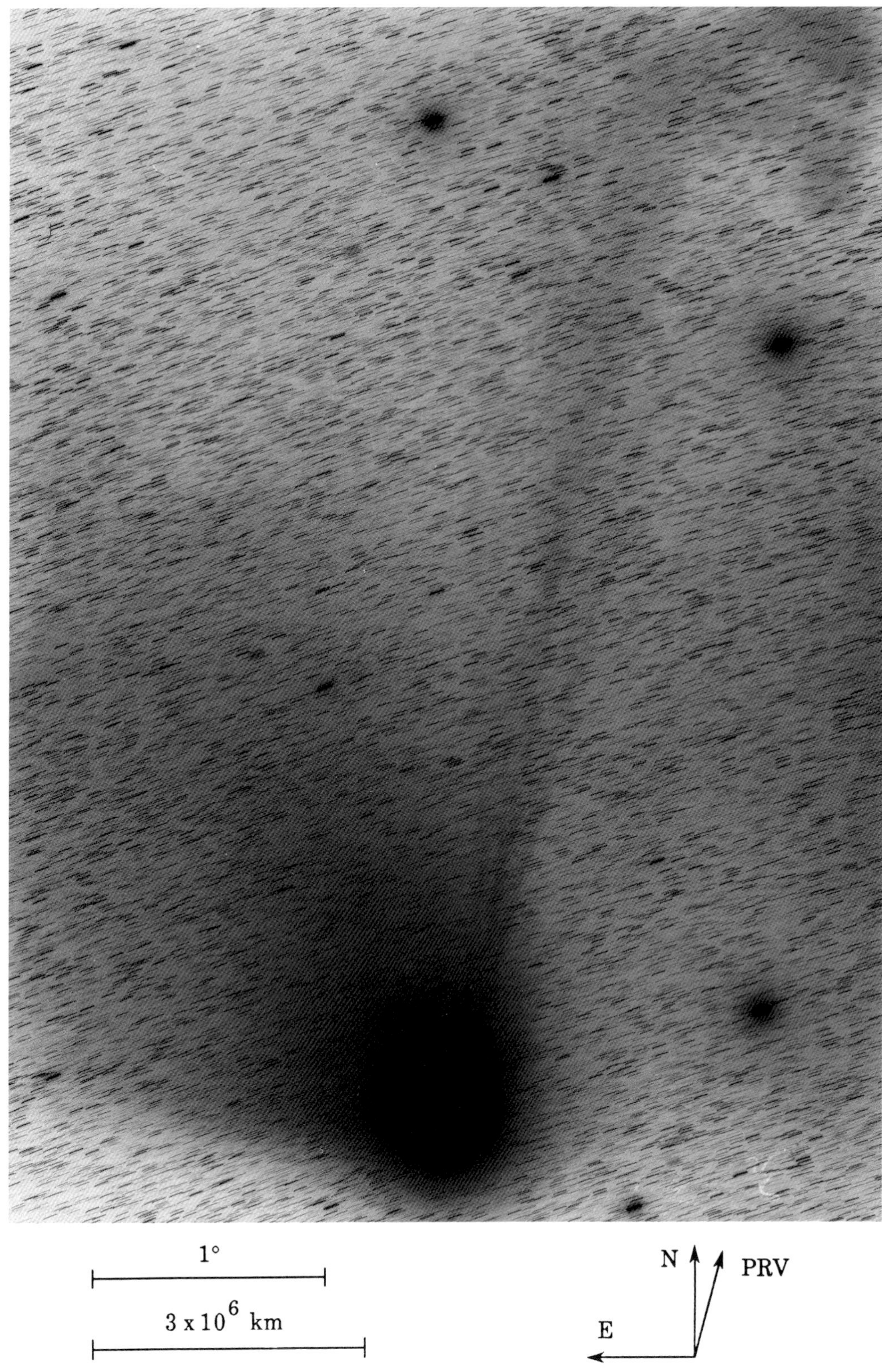

Fig. 565: 1986 Apr 13.06319 UT; exposure 20.0 minutes on 103a-O emulsion with no filter; r=1.36, Δ=0.42, β=153.5°; LSPN-1201.
Photograph by F. Miller, University of Michigan/CTIO.

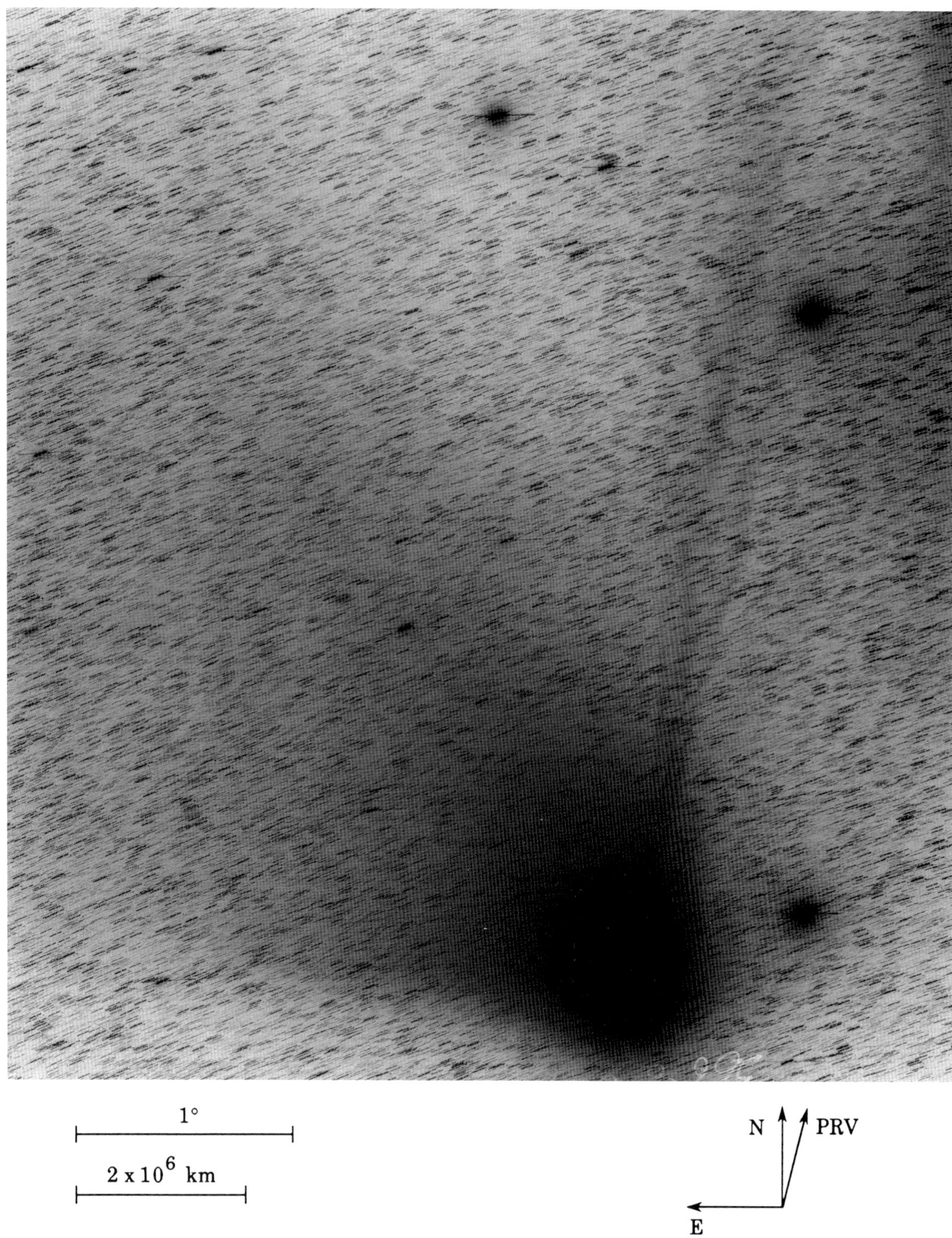

Fig. 566: 1986 Apr 13.18542 UT; exposure 20.0 minutes on 103a-O emulsion with no filter; r=1.37, Δ=0.42, β=153.8°; LSPN-1204.
Photograph by F. Miller, University of Michigan/CTIO.

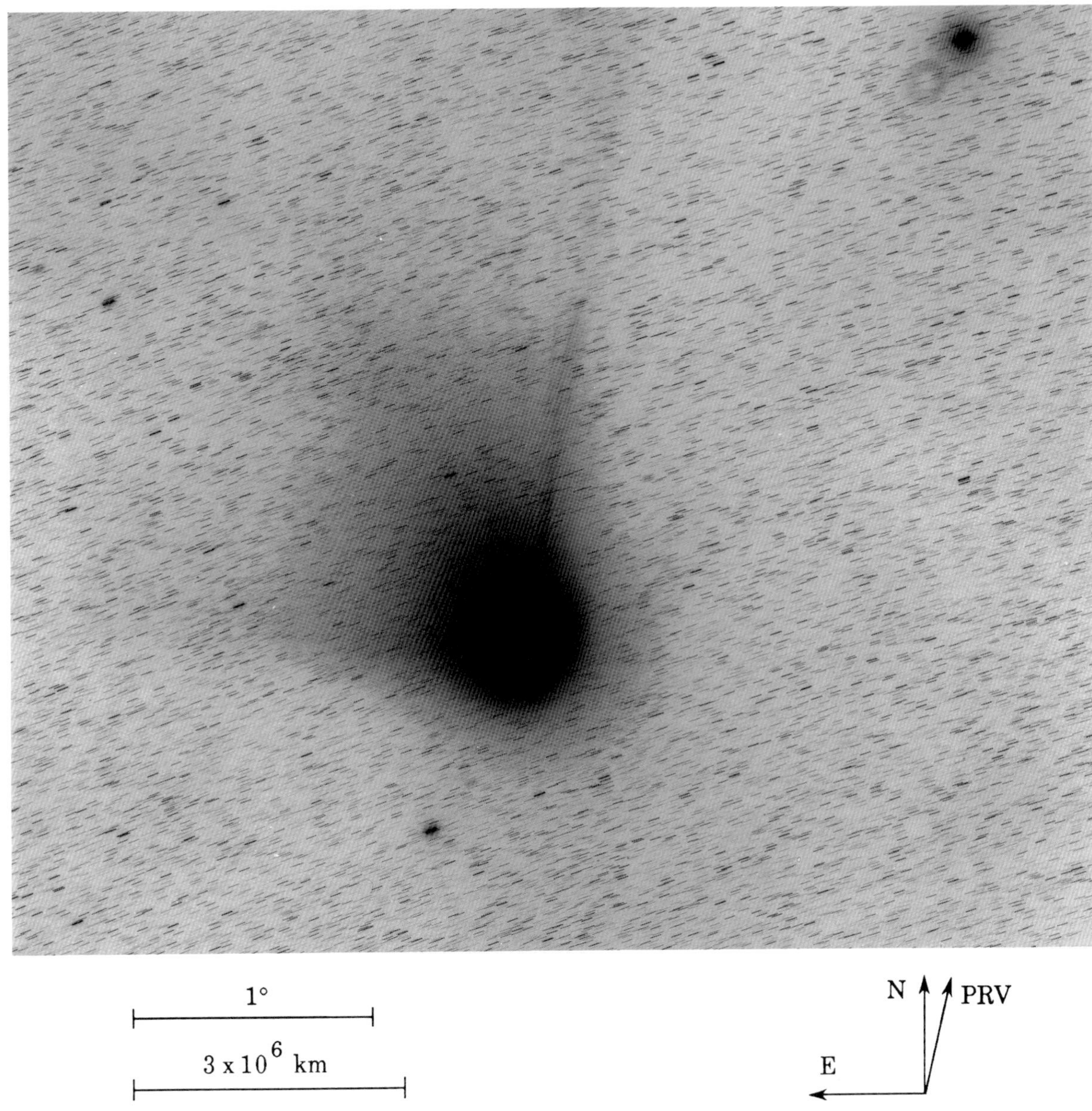

Fig. 567: 1986 Apr 13.34549 UT; exposure 15.0 minutes on 103a-O emulsion with no filter; r=1.37, Δ=0.43, β=154.1°; LSPN-755.
Photograph by C. Torres/H. Wroblewski, Cerro el Roble Astronomical Observatory.

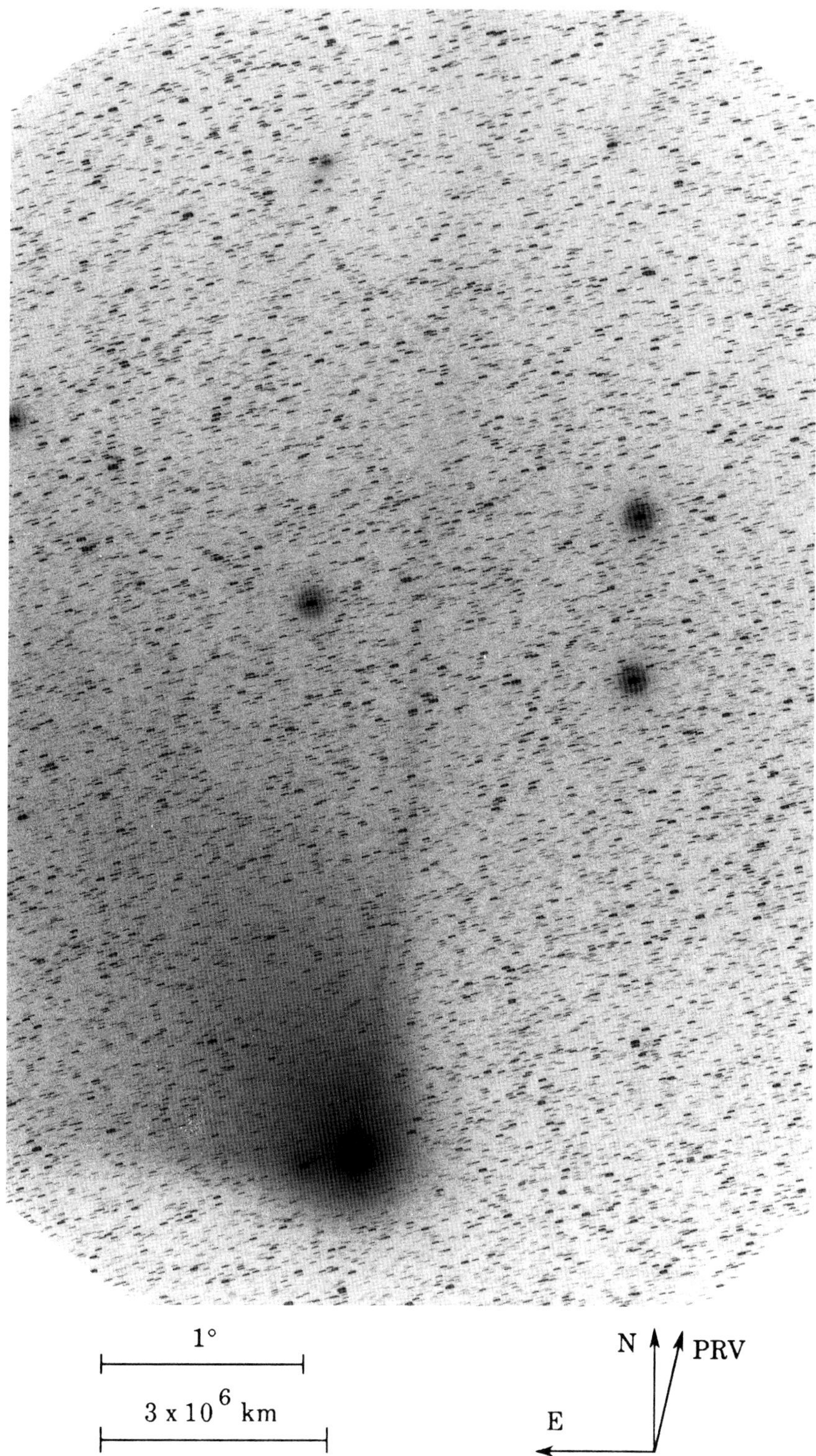

Fig. 568: 1986 Apr 13.44514 UT; exposure 10.0 minutes on Tri-X emulsion with no filter; r=1.37, Δ=0.43, β=154.3°; LSPN-1479.
Photograph by W. Liller, Easter Island, LSPN Island Network.

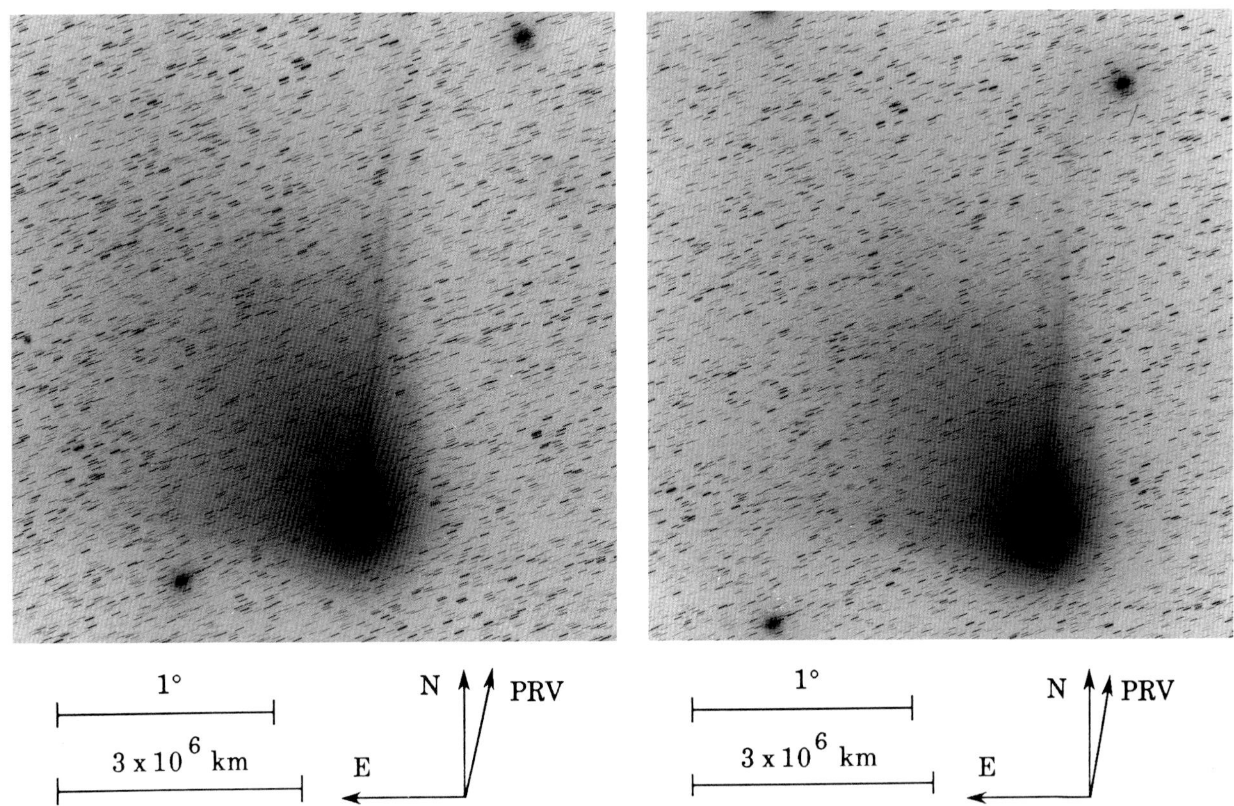

(left) Fig. 569: 1986 Apr 13.58683 UT; exposure 14.3 minutes on IIa-O emulsion with GG-385 filter; $r=1.37$, $\Delta=0.43$, $\beta=154.6°$; LSPN-2153.
Photograph by P. Magnusson, Uppsala Southern Station.

(right) Fig. 570: 1986 Apr 13.68297 UT; exposure 12.0 minutes on IIa-O emulsion with GG-385 filter; $r=1.37$, $\Delta=0.43$, $\beta=154.8°$; LSPN-2160.
Photograph by P. Magnusson, Uppsala Southern Station.

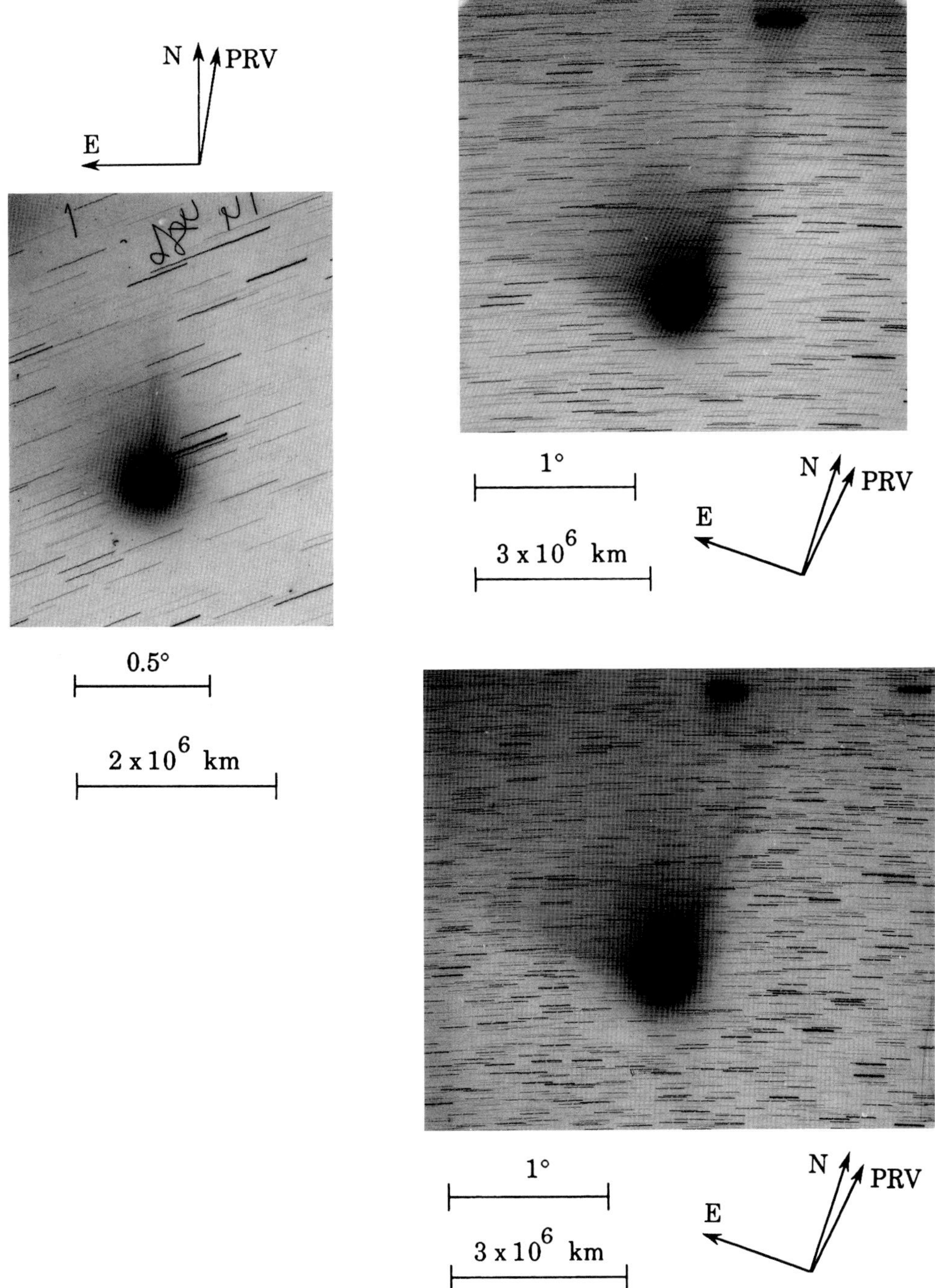

(left) Fig. 571: 1986 Apr 13.74618 UT; exposure 75.0 minutes on 103a-O emulsion with no filter; r=1.37, Δ=0.43, β=155.0°; LSPN-3649.
Photograph submitted by K. Sivaraman, Indian Institue for Astrophysics, Kodaikanal Station.

(top right) Fig. 572: 1986 Apr 13.76736 UT; exposure 90.0 minutes on 103a-O emulsion with Blue filter; r=1.37, Δ=0.43, β=155.0°; LSPN-3693.
Photograph submitted by K. Sivaraman, Indian Institute for Astrophysics, Kavalur Station.

(bottom right) Fig. 573: 1986 Apr 13.82639 UT; exposure 60.0 minutes on 103a-O emulsion with UV filter; r=1.38, Δ=0.43, β=155.1°; LSPN-3694.
Photograph submitted by K. Sivaraman, Indian Institute for Astrophysics, Kavalur Station.

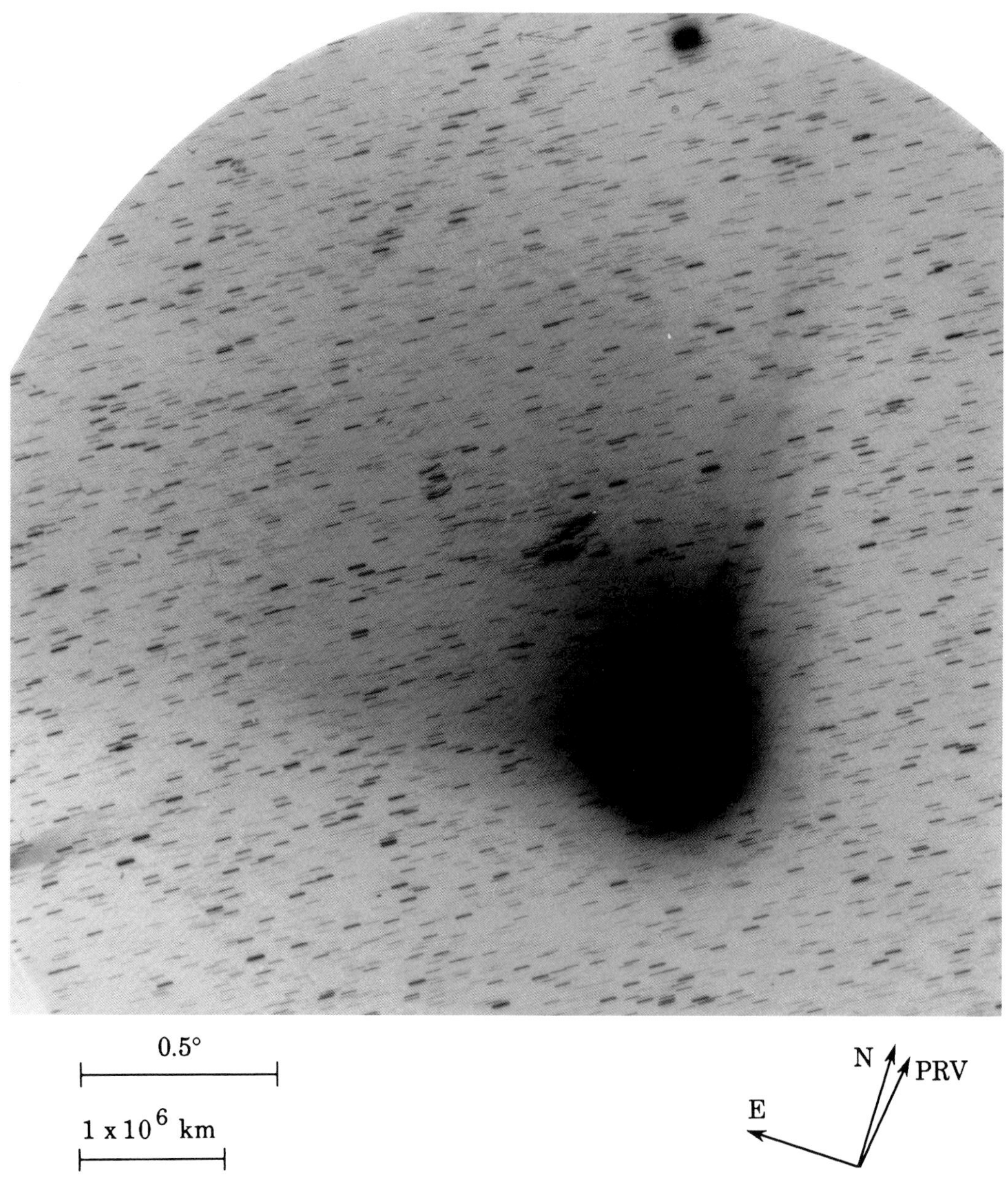

Fig. 574: 1986 Apr 13.84132 UT; exposure 15.0 minutes on Tech. Pan 2415 emulsion with no filter; r=1.38, Δ=0.43, β=155.1°; LSPN-3500.
Photograph by D. O'Donoghue, University of Perugia/South African Astronomical Observatory.

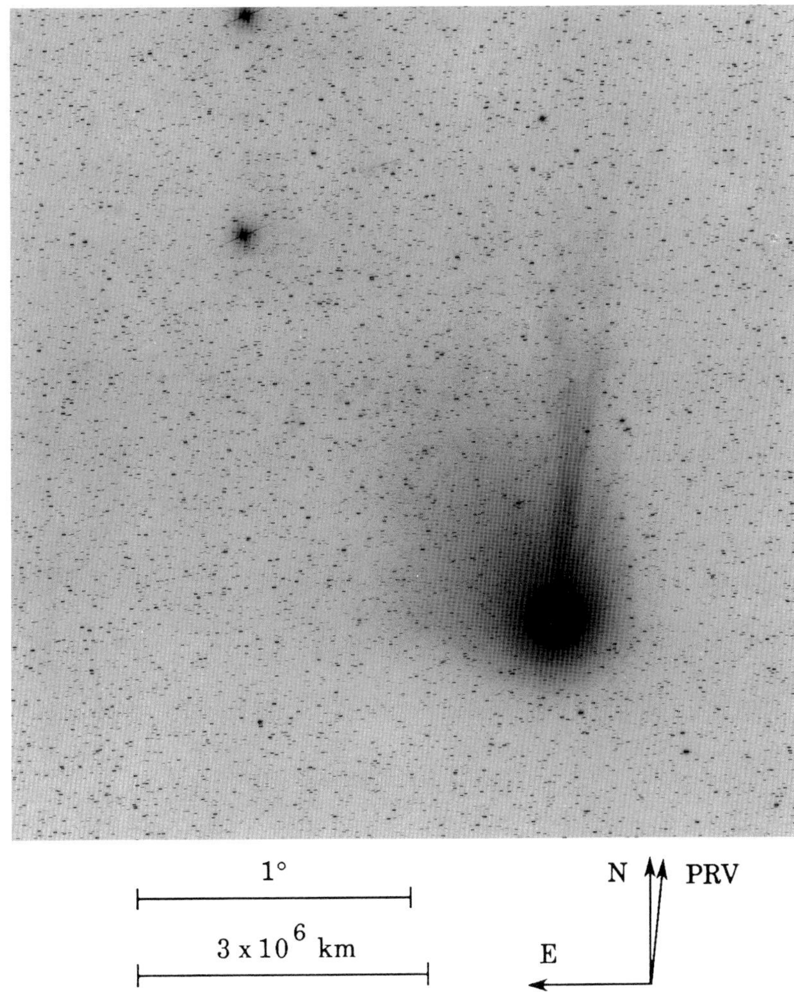

Fig. 575: 1986 Apr 14.05868 UT; exposure 5.0 minutes on 103a-O emulsion with no filter; r=1.38, Δ=0.43, β=155.6°; LSPN-1208.
Photograph by F. Miller, University of Michigan/CTIO.

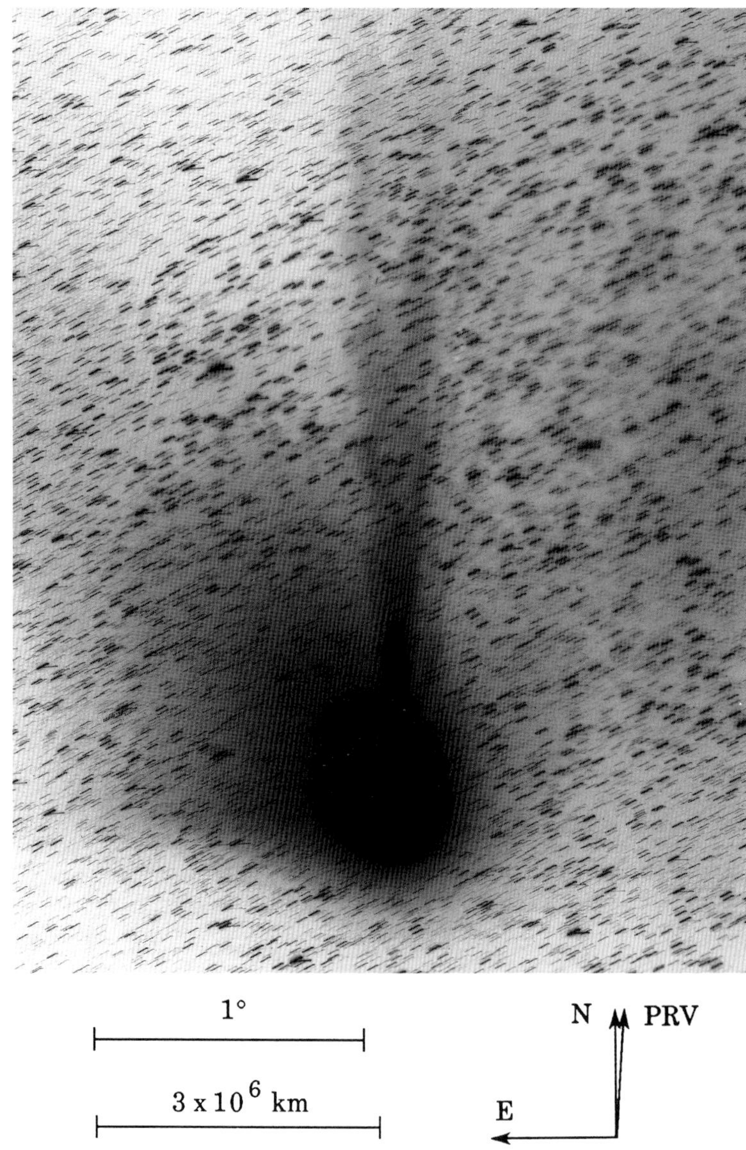

Fig. 576: 1986 Apr 14.18333 UT; exposure 20.0 minutes on 103a-O emulsion with no filter; r=1.38, Δ=0.43, β=155.8°; LSPN-1212.
Photograph by F. Miller, University of Michigan/CTIO.

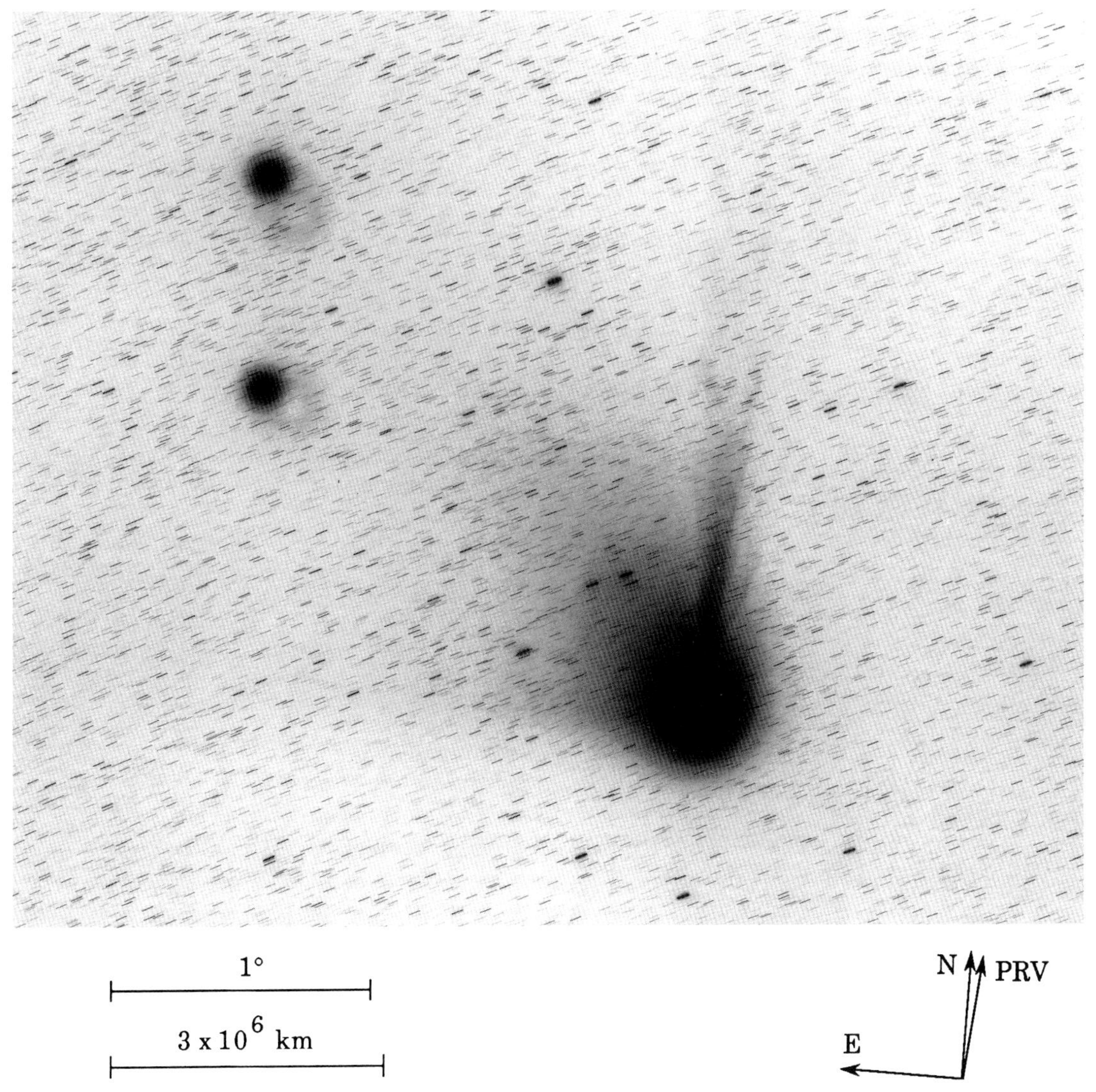

Fig. 577: 1986 Apr 14.19410 UT; exposure 15.0 minutes on 103a-O emulsion with GG-385 filter; r=1.38, Δ=0.43, β=155.8°; LSPN-756.
Photograph by C. Torres/H. Wroblewski, Cerro el Roble Astronomical Observatory.

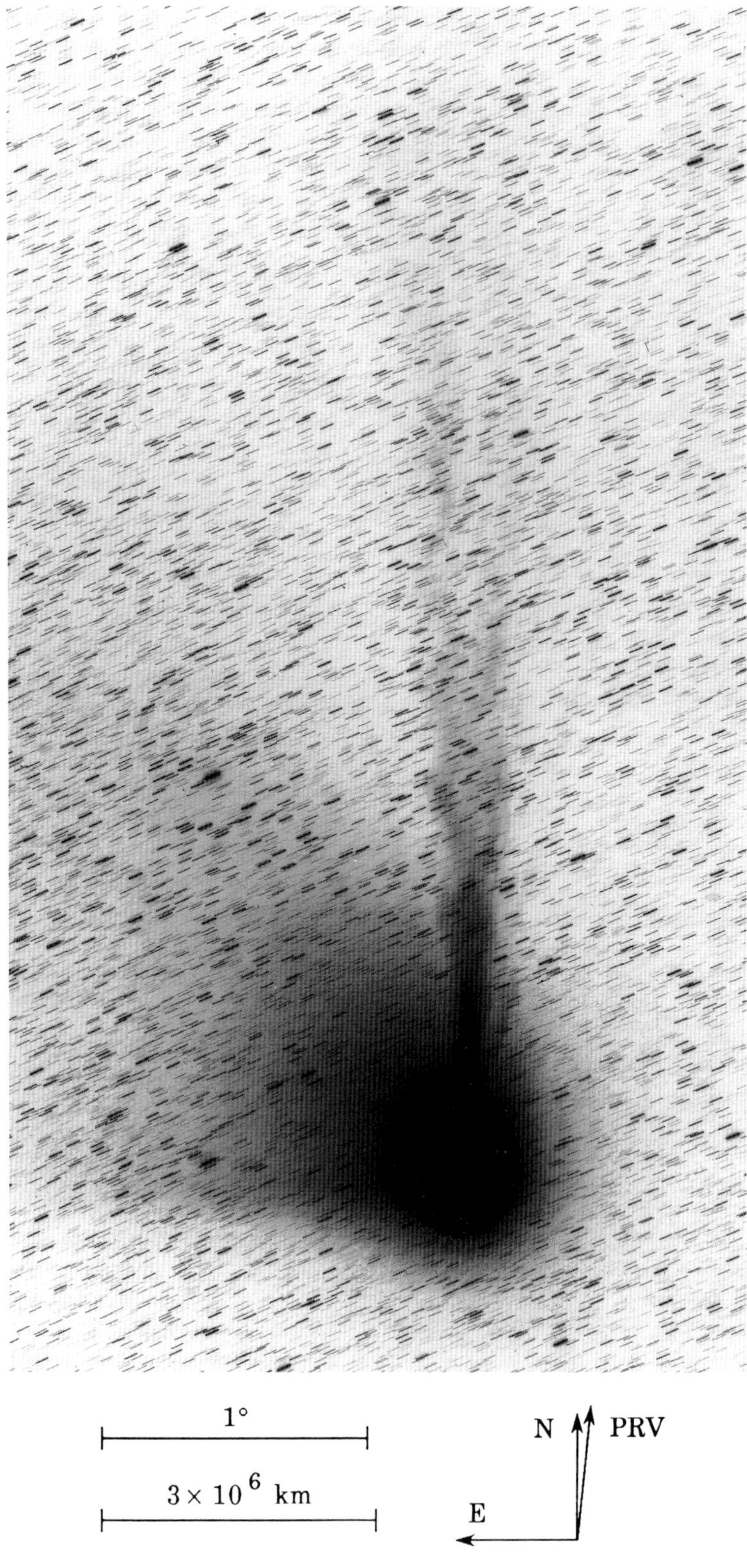

Fig. 578: 1986 Apr 14.25972 UT; exposure 20.0 minutes on 103a-O emulsion with no filter; r=1.38, Δ=0.43, β=155.9°; LSPN-1215.
Photograph by F. Miller, University of Michigan/CTIO.

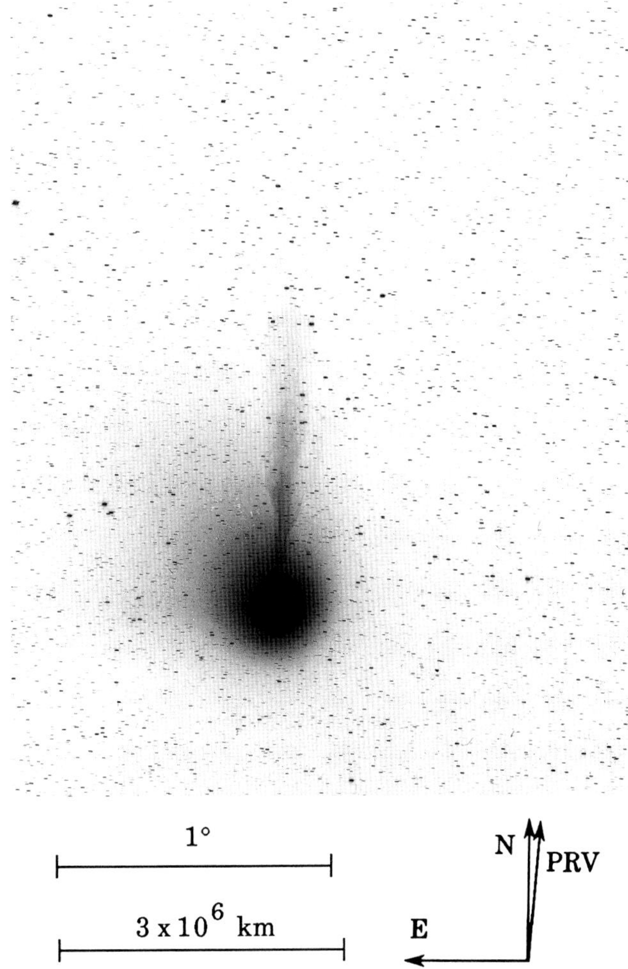

Fig. 579: 1986 Apr 14.27465 UT; exposure 5.0 minutes on 103a-O emulsion with no filter; r=1.38, Δ=0.43, β=156.0°; LSPN-1216.
Photograph by F. Miller, University of Michigan/CTIO.

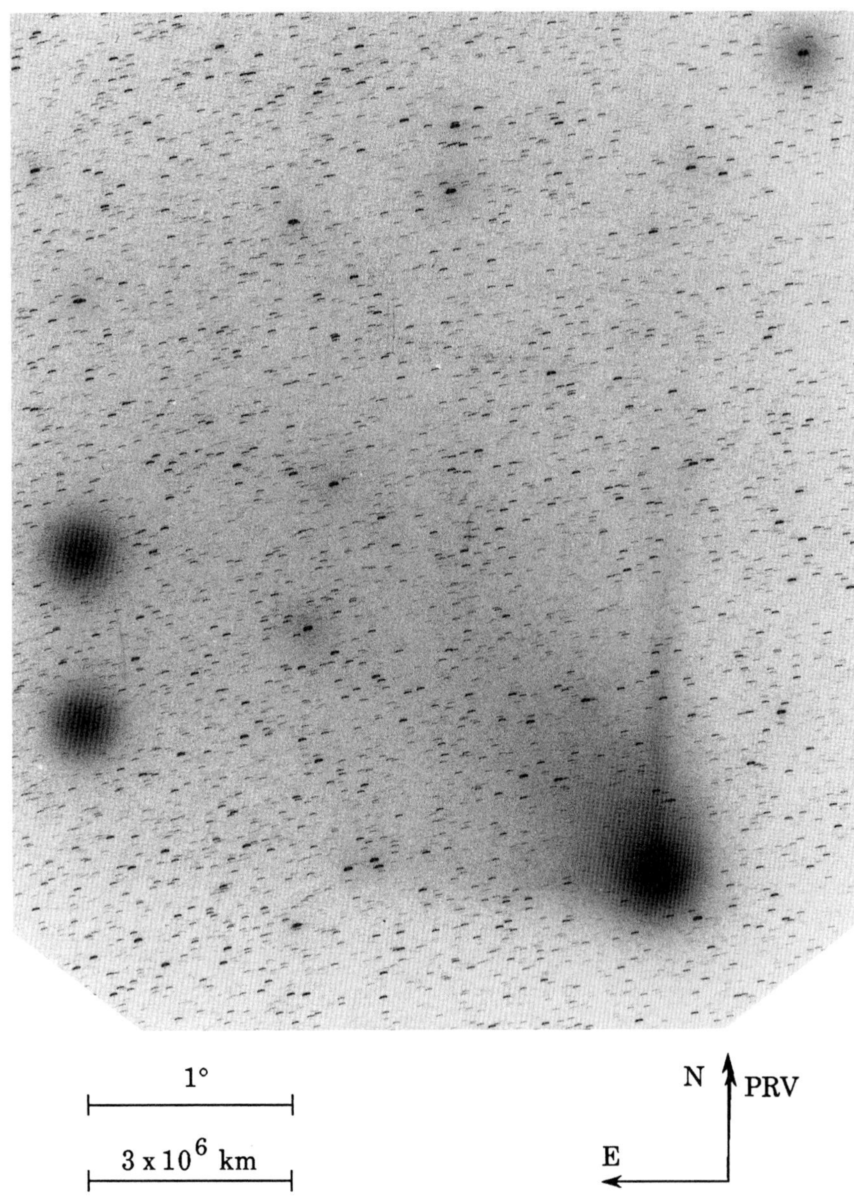

Fig. 580: 1986 Apr 14.46285 UT; exposure 7.0 minutes on Tri-X emulsion with no filter; r=1.39, Δ=0.43, β=156.3°; LSPN-1482.
Photograph by W. Liller, Easter Island, LSPN Island Network.

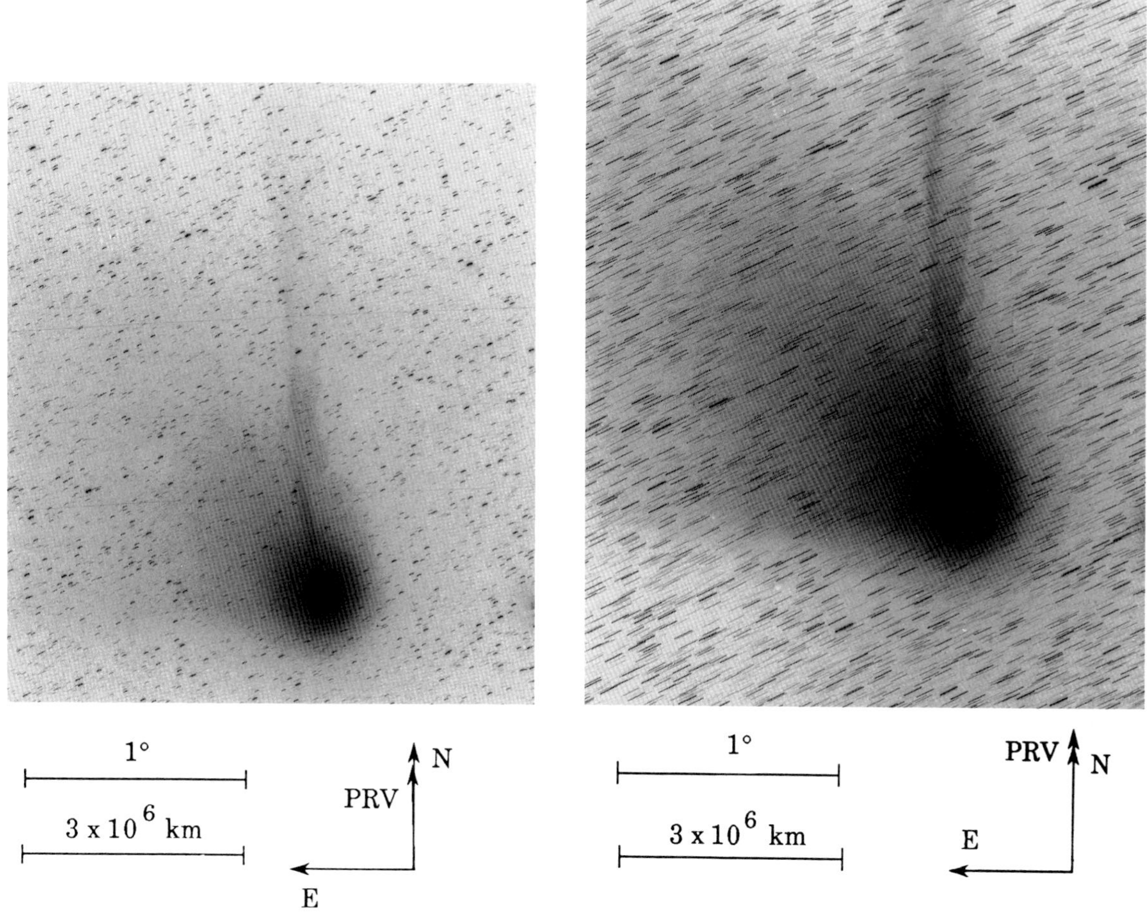

(left) Fig. 581: 1986 Apr 14.64285 UT; exposure 8.0 minutes on IIa-O emulsion with GG-385 filter; r=1.39, Δ=0.44, β=156.6°; LSPN-2162.
Photograph by P. Magnusson, Uppsala Southern Station.

(right) Fig. 582: 1986 Apr 14.66155 UT; exposure 30.0 minutes on IIa-O emulsion with GG-385 filter; r=1.39, Δ=0.44, β=156.6°; LSPN-2163.
Photograph by P. Magnusson, Uppsala Southern Station.

Fig. 583: 1986 Apr 14.67090 UT; exposure 5.0 minutes on IIIa-F emulsion with no filter; r=1.39, Δ=0.44, β=156.6°; LSPN-2426.
Photograph from Royal Observatory/UK Schmidt Telescope Unit.

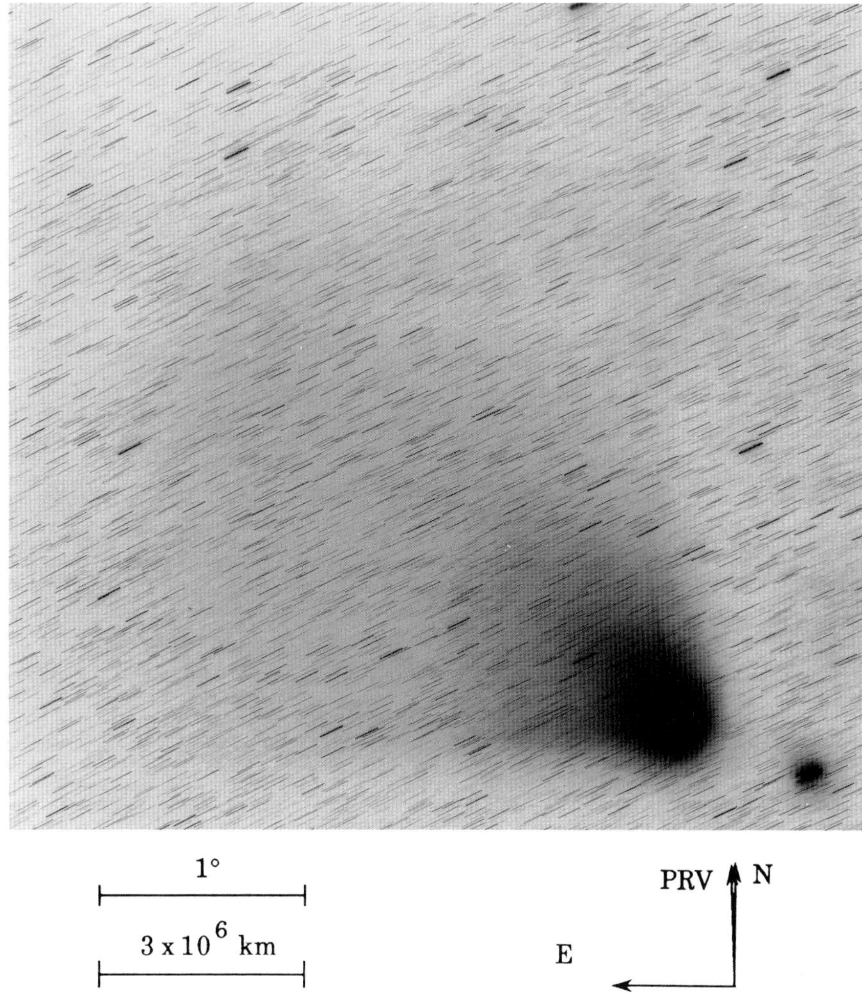

Fig. 584: 1986 Apr 14.72876 UT; exposure 40.0 minutes on IIIa-F emulsion with RG-630 filter; r=1.39, Δ=0.44, β=156.7°; LSPN-2428.
Photograph from Royal Observatory/UK Schmidt Telescope Unit.

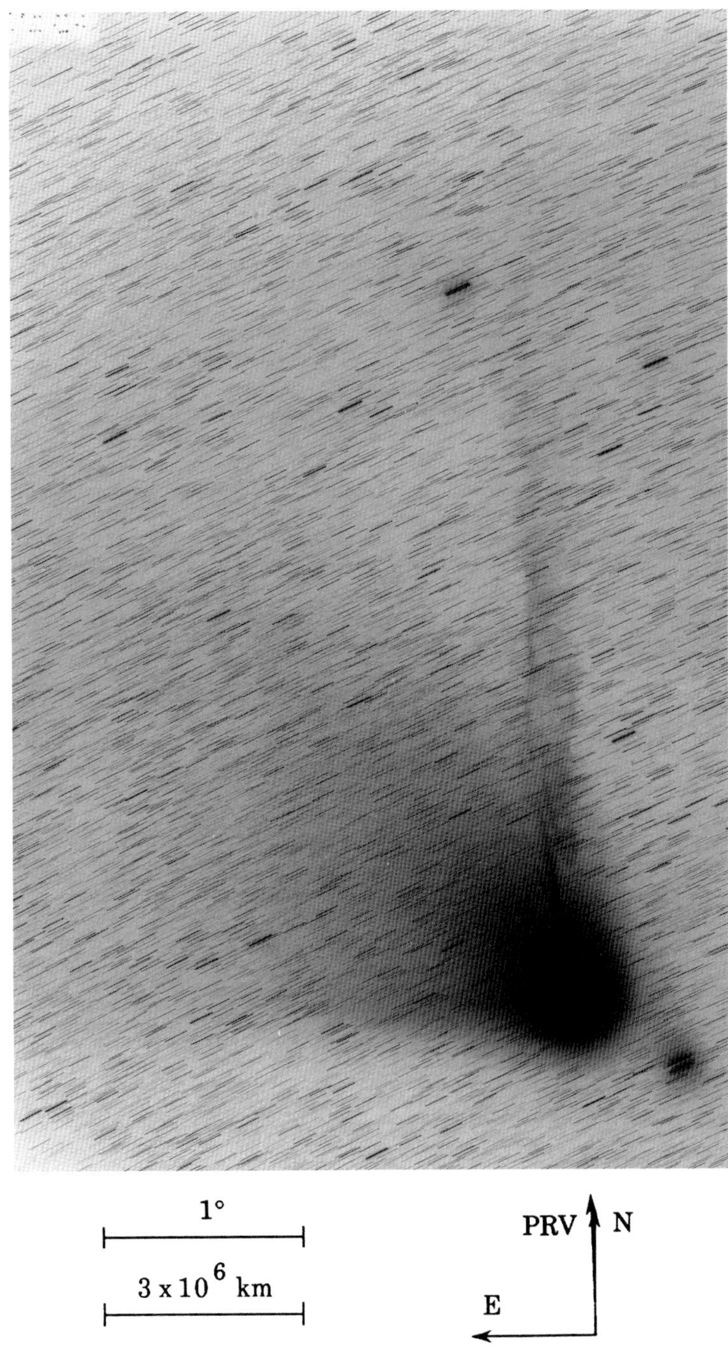

Fig. 585: 1986 Apr 14.76685 UT; exposure 40.0 minutes on IIa-O emulsion with GG-395 filter; r=1.39, Δ=0.44, β=156.8°; LSPN-2429.
Photograph from Royal Observatory/UK Schmidt Telescope Unit.

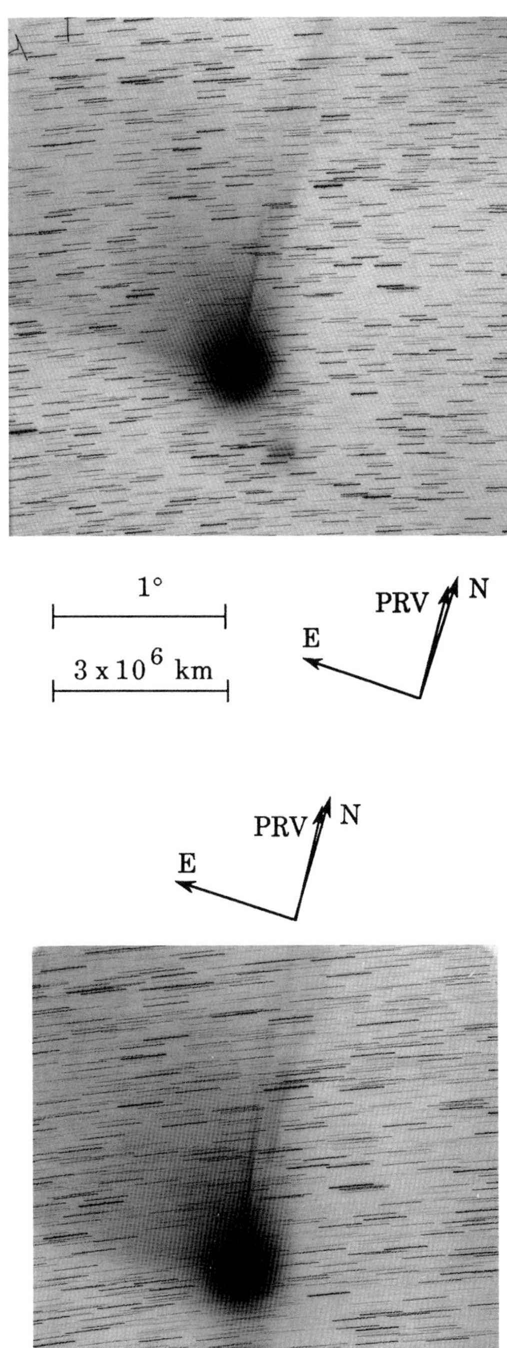

(top) Fig. 586: 1986 Apr 14.79479 UT; exposure 63.0 minutes on 103a-O emulsion with UV filter; r=1.39, Δ=0.44, β=156.8°; LSPN-3695.
Photograph submitted by K. Sivaraman, Indian Institute for Astrophysics, Kavalur Station.

(bottom) Fig. 587: 1986 Apr 14.85069 UT; exposure 90.0 minutes on 103a-O emulsion with Blue filter; r=1.39, Δ=0.44, β=156.9°; LSPN-3696.
Photograph submitted by K. Sivaraman, Indian Institute for Astrophysics, Kavalur Station.

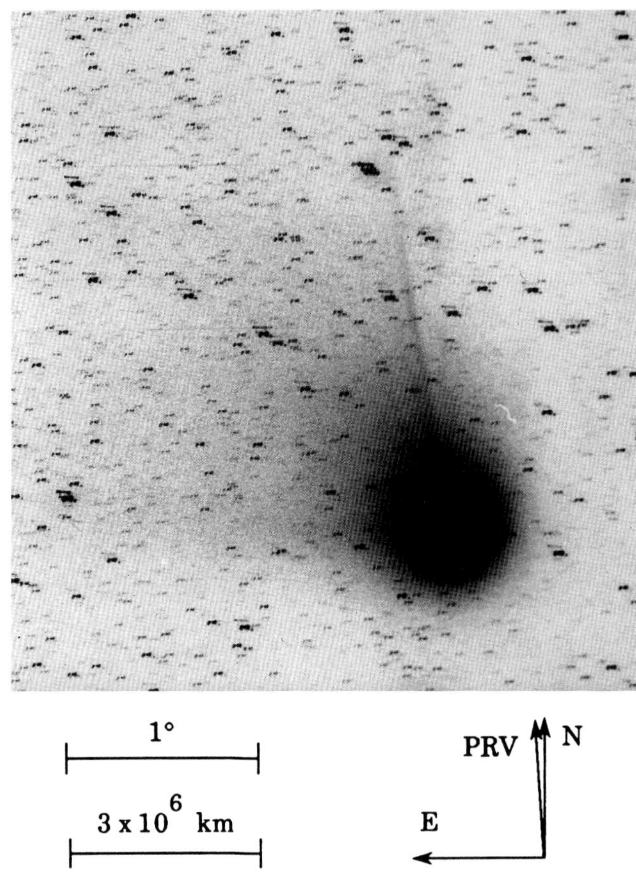

Fig. 588: 1986 Apr 14.90278 UT; exposure 10.0 minutes on Tech. Pan 2415 emulsion; r=1.39, Δ=0.44, β=157.0°. *Photograph by A. Marafie, Wafra Observatory.*

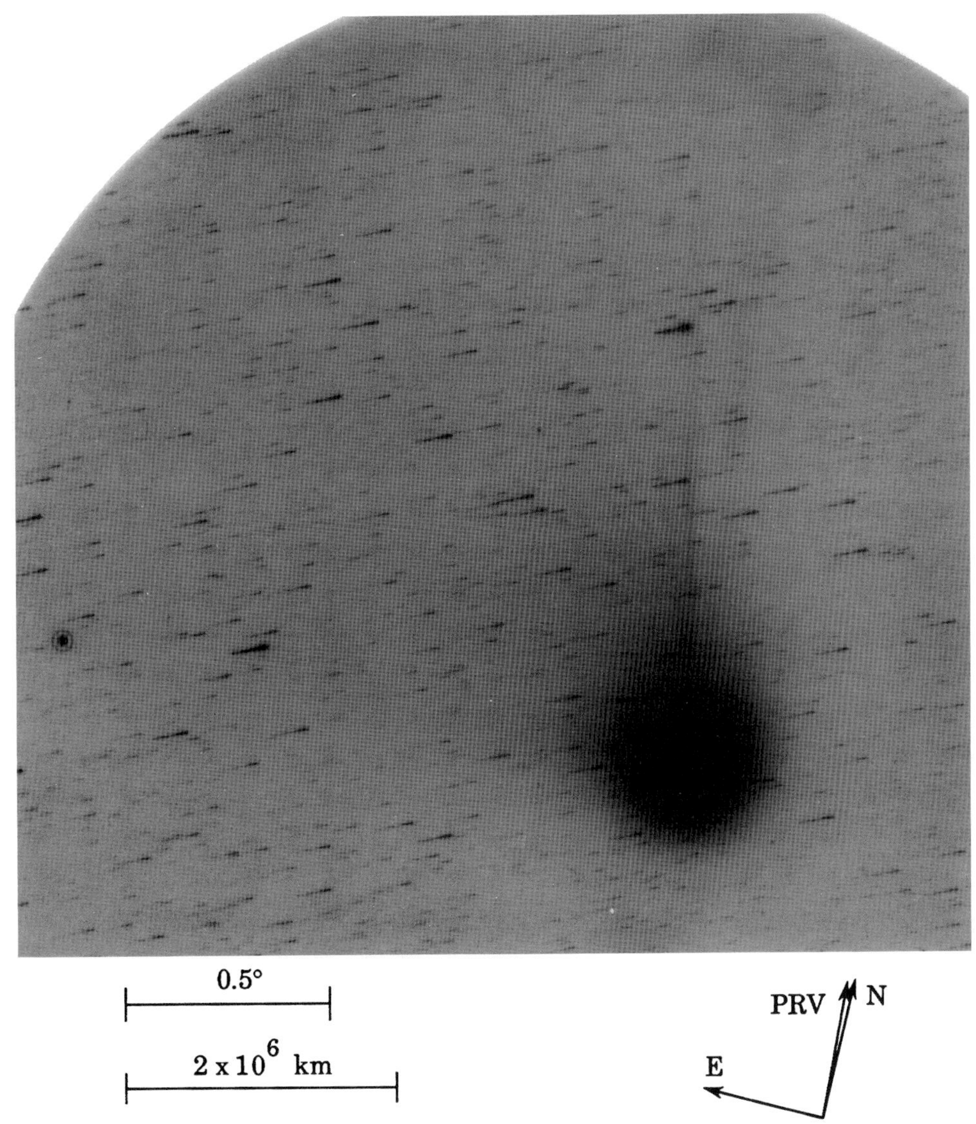

Fig. 589: 1986 Apr 14.90417 UT; exposure 30.0 minutes on Tech. Pan 2415 emulsion with no filter; r=1.39, Δ=0.44, β=157.0°; LSPN-3501.
Photograph by D. O'Donoghue, University of Perugia/South African Astronomical Observatory.

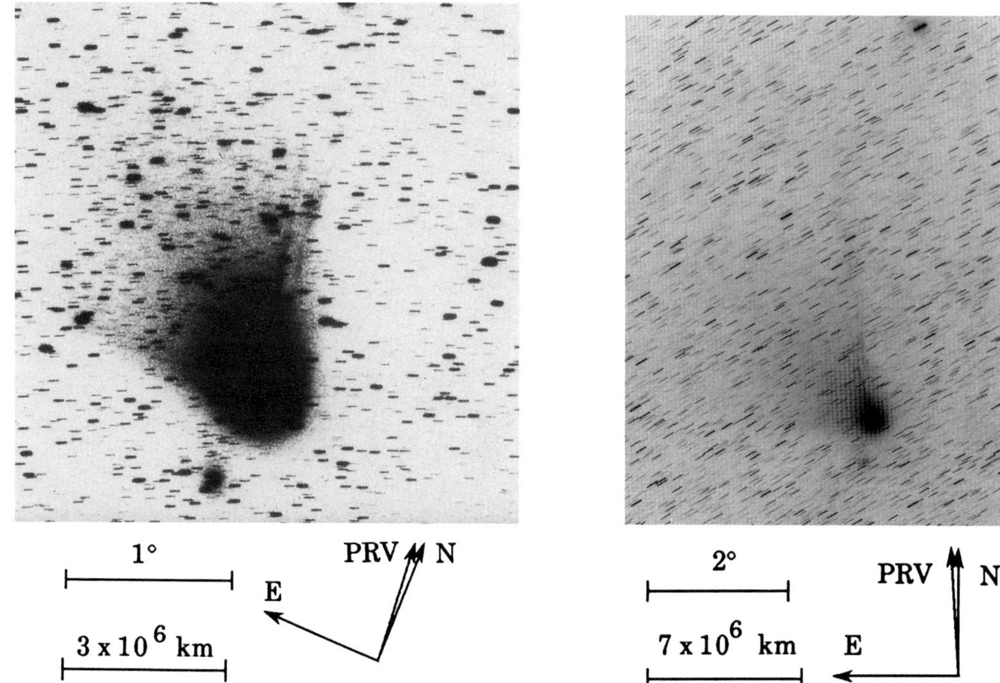

(left) Fig. 590: 1986 Apr 14.92360 UT; exposure 20.0 minutes on AGFA 1000 RS emulsion; r=1.39, Δ=0.44, β=157.0°; AON-851830.
Photograph from Volkssternwarte Frankfurt, Germany.

(right) Fig. 591: 1986 Apr 14.94410 UT; exposure 60.0 minutes on IIa-O emulsion with no filter; r=1.39, Δ=0.44, β=157.0°; LSPN-2872.
Photograph by G. Malcolm/A. Jarrett, Boyden Observatory.

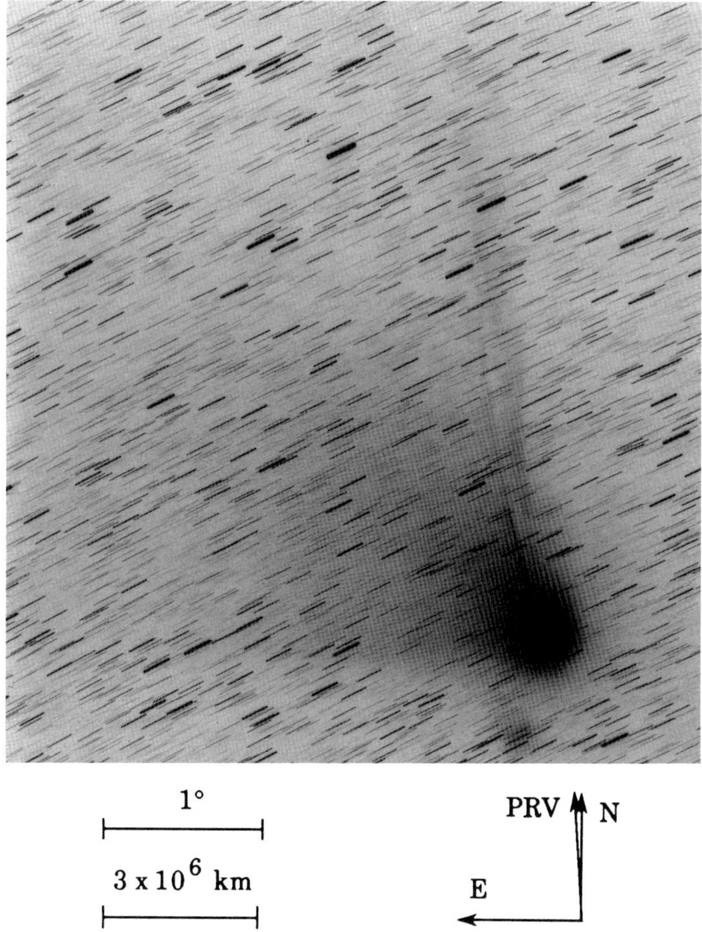

Fig. 592: 1986 Apr 14.94410 UT; exposure 60.0 minutes on IIa-O emulsion with no filter; r=1.39, Δ=0.44, β=157.0°; LSPN-2873.
Photograph by G. Malcolm/A. Jarrett, Boyden Observatory.

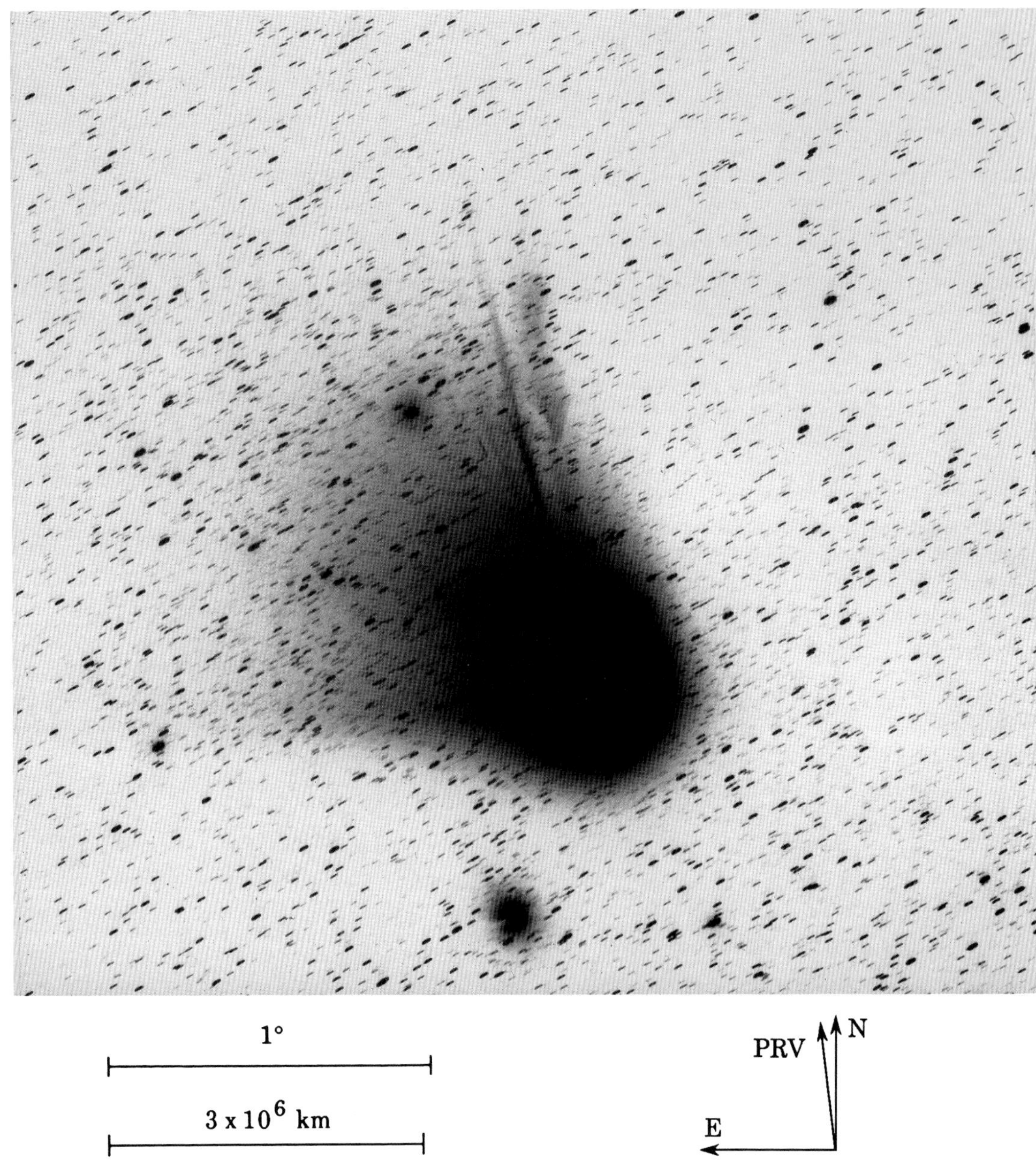

Fig. 593: 1986 Apr 14.97500 UT; exposure 10.0 minutes on Tech. Pan 2415 emulsion; $r=1.39$, $\Delta=0.44$, $\beta=157.1°$; AON-851833.
Photograph by B. Koch, Germany.

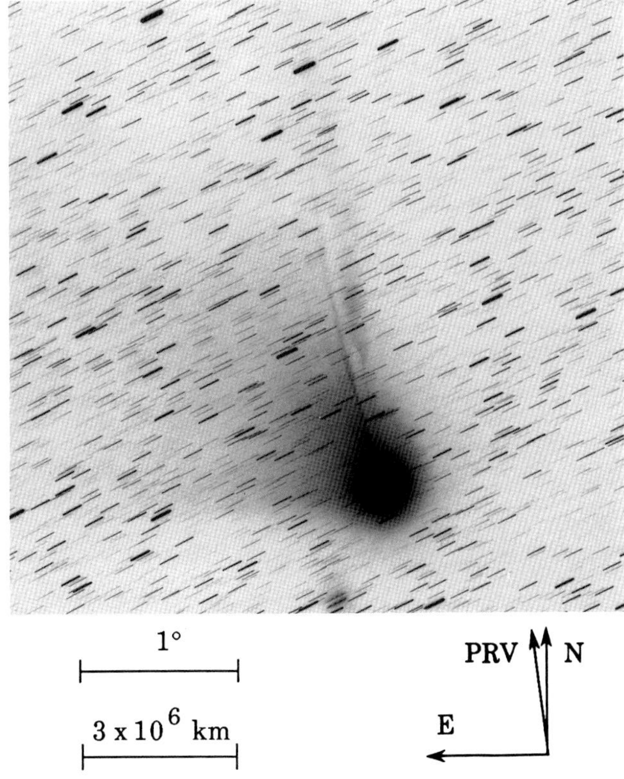

Fig. 594: 1986 Apr 14.98368 UT; exposure 45.0 minutes on IIa-O emulsion with no filter; r=1.39, Δ=0.44, β=157.1°; LSPN-2874.
Photograph by G. Malcolm, Boyden Observatory.

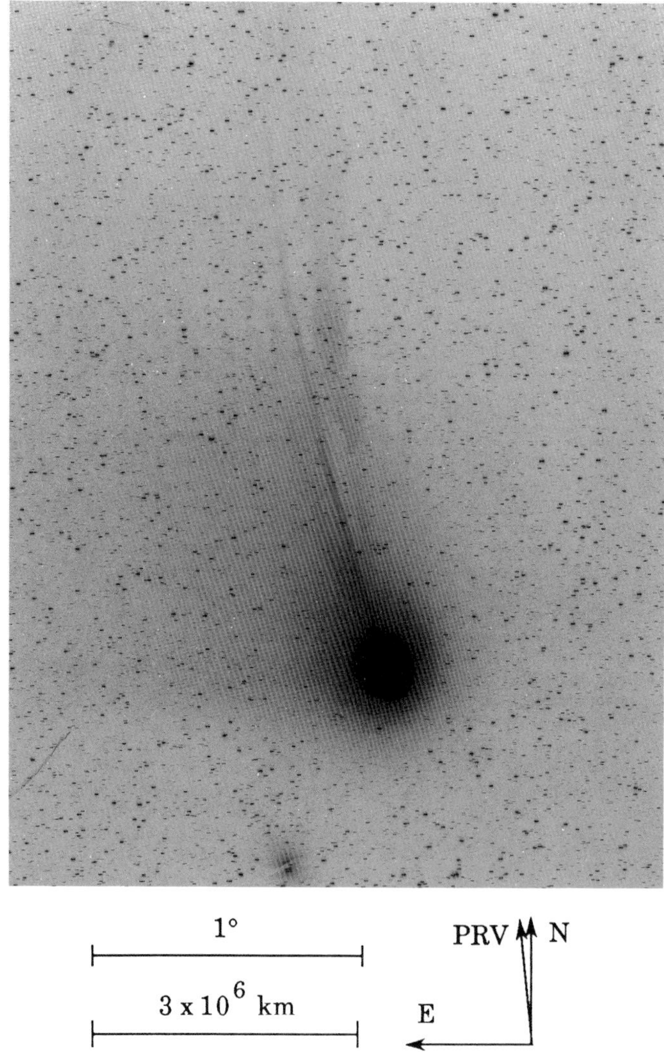

Fig. 595: 1986 Apr 14.99826 UT; exposure 5.0 minutes on 103a-O emulsion with no filter; r=1.39, Δ=0.44, β=157.1°; LSPN-1217.
Photograph by F. Miller, University of Michigan/CTIO.

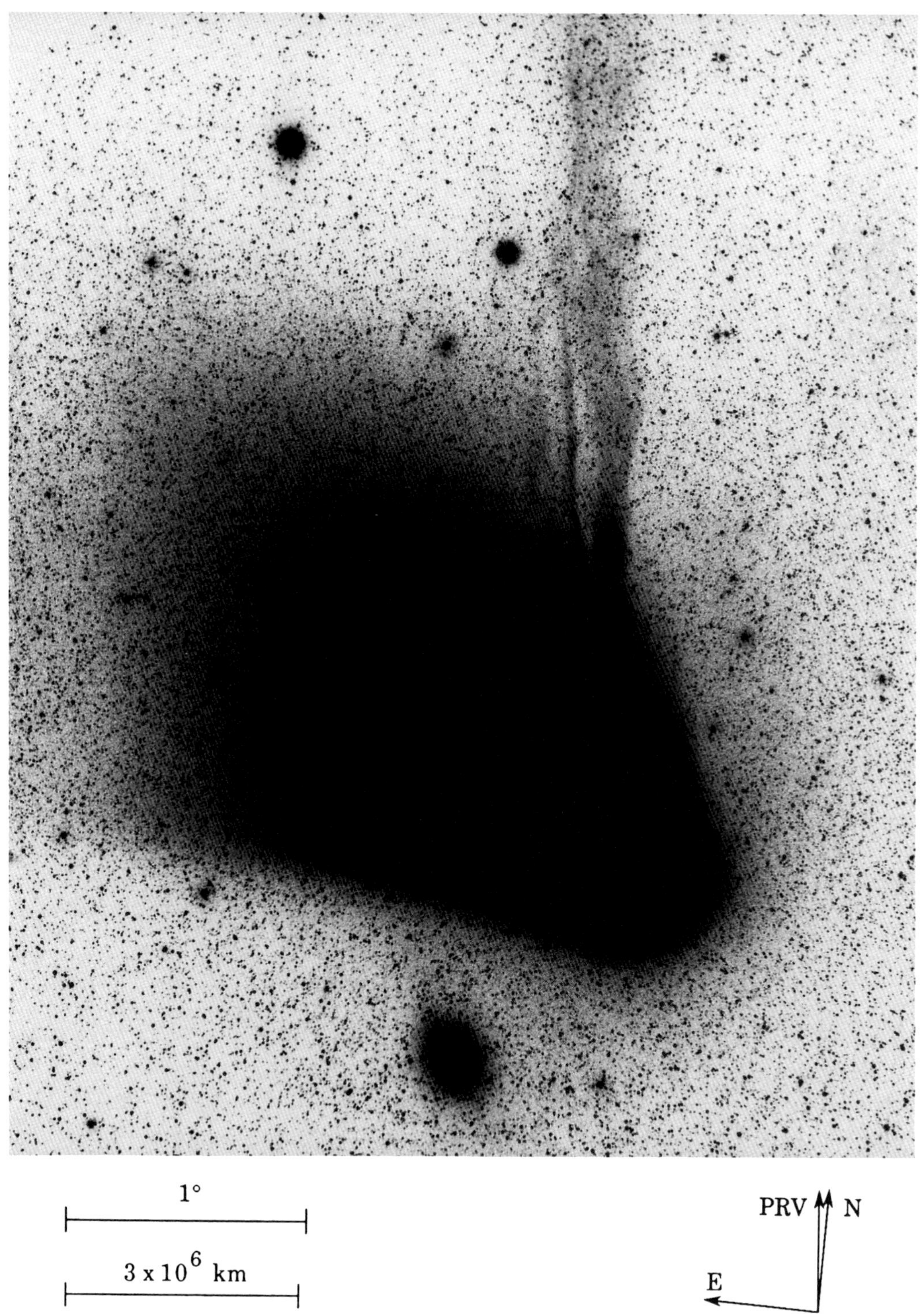

Fig. 596: 1986 Apr 15.03715 UT; exposure 7.0 minutes on Tech. Pan 2415 emulsion; r=1.39, Δ=0.44, β=157.2°; AON-851839.
Photograph by M. Jäger, Austria.

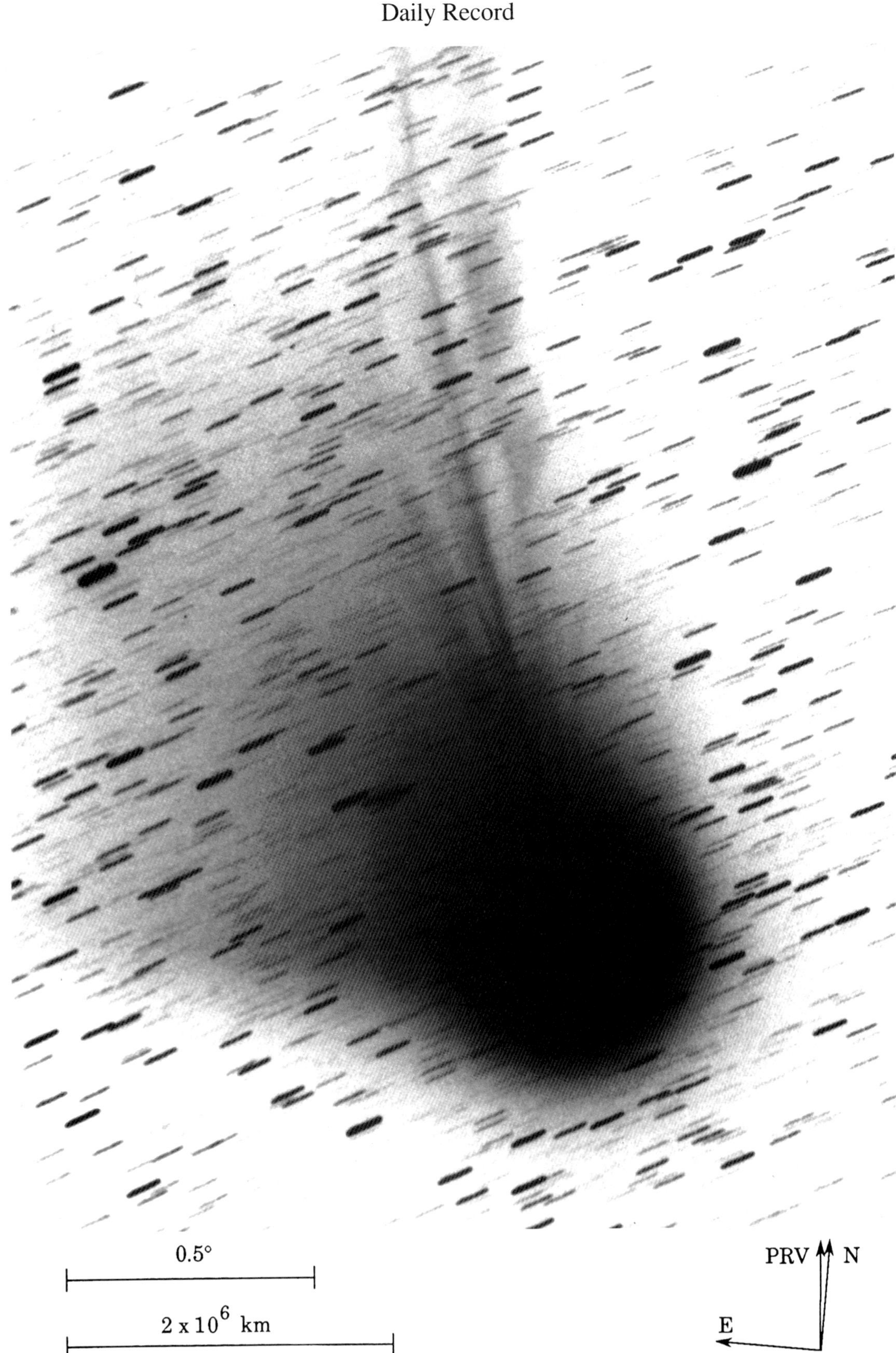

Fig. 597: 1986 Apr 15.05382 UT; exposure 25.0 minutes on AGFA Ortho emulsion; r=1.39, Δ=0.44, β=157.2°; AON-851841.
Photograph by M. Jäger, Austria.

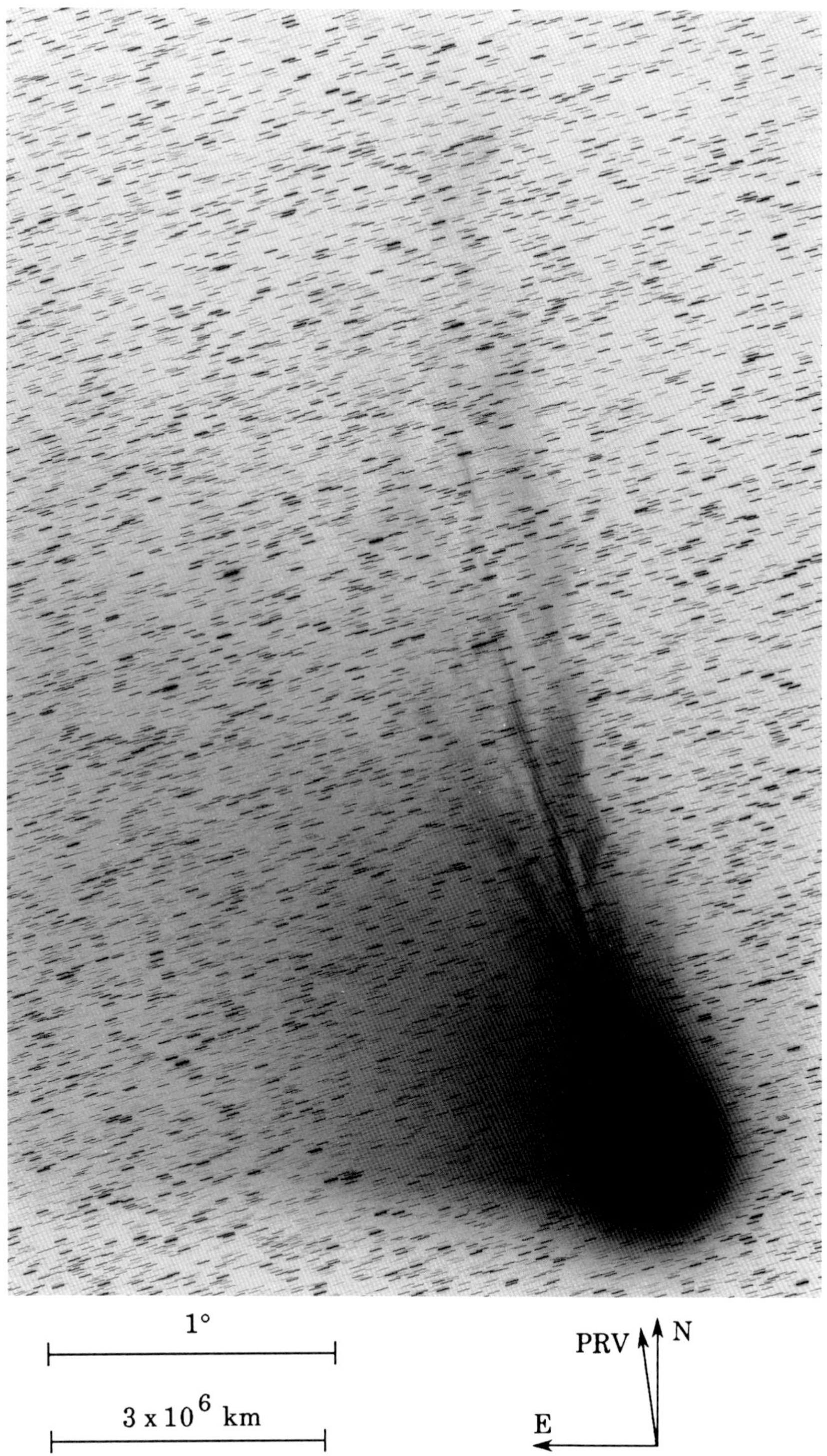

Fig. 598: 1986 Apr 15.29375 UT; exposure 20.0 minutes on 103a-O emulsion with no filter; r=1.40, Δ=0.44, β=157.5°; LSPN-1227.
Photograph by F. Miller, University of Michigan/CTIO.

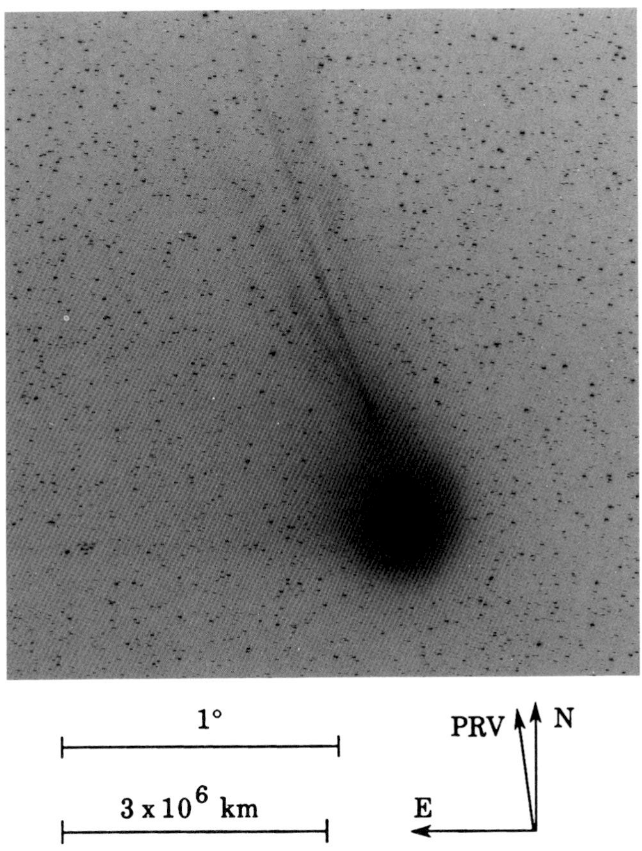

Fig. 599: 1986 Apr 15.30660 UT; exposure 5.0 minutes on 103a-O emulsion with no filter; r=1.40, Δ=0.44, β=157.5°; LSPN-1228.
Photograph by F. Miller, University of Michigan/CTIO.

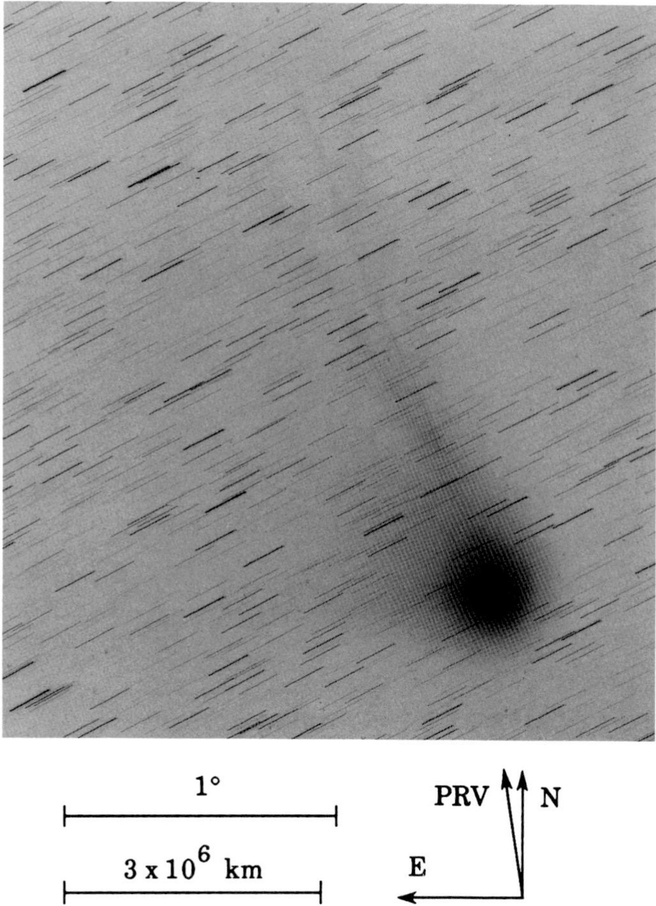

Fig. 600: 1986 Apr 15.44861 UT; exposure 60.0 minutes on 103a-O emulsion with GG-400 filter; r=1.40, Δ=0.44, β=157.7°; LSPN-424.
Photograph by G. Douglass, U.S. Naval Observatory Station.

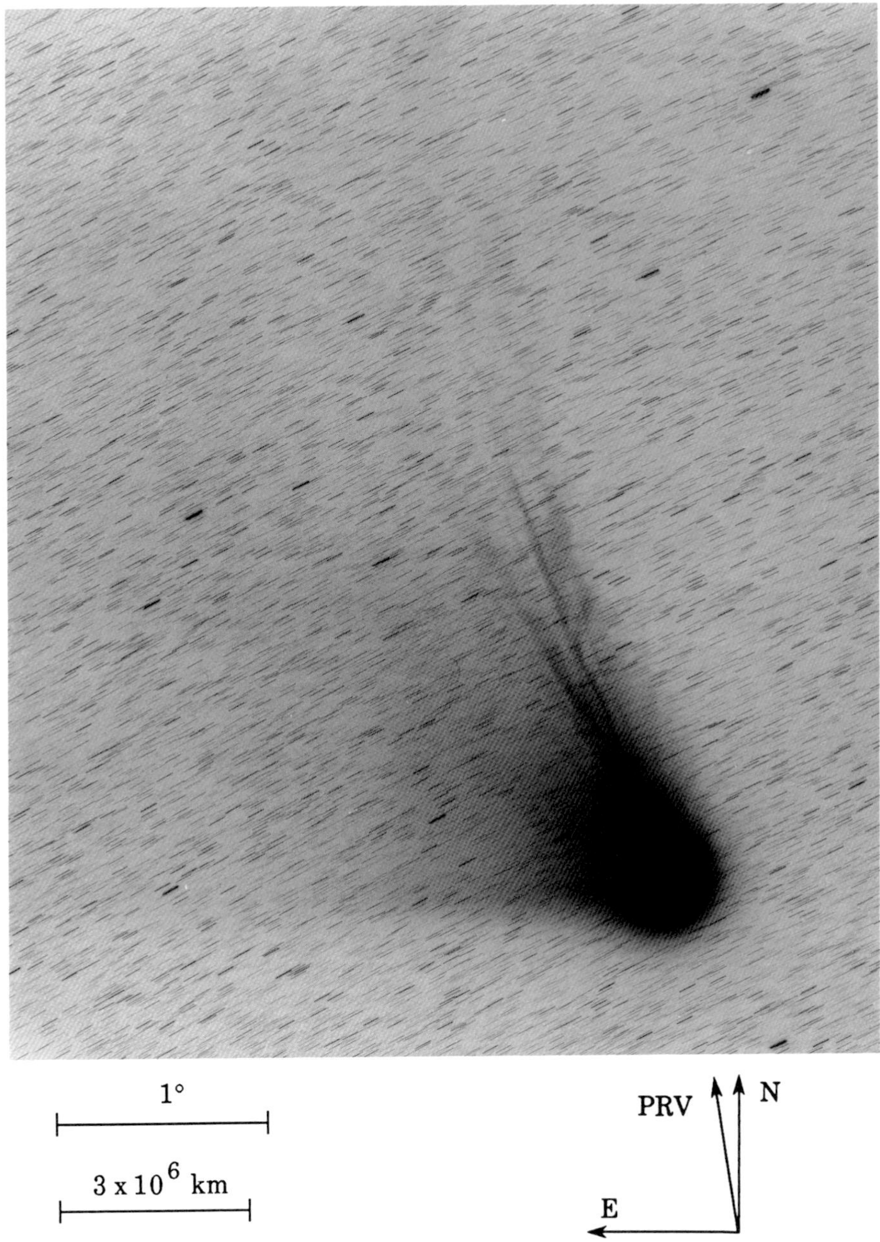

Fig. 601: 1986 Apr 15.47463 UT; exposure 30.0 minutes on IIa-O emulsion with GG-395 filter; r=1.40, Δ=0.44, β=157.7°; LSPN-2430.
Photograph from Royal Observatory/UK Schmidt Telescope Unit.

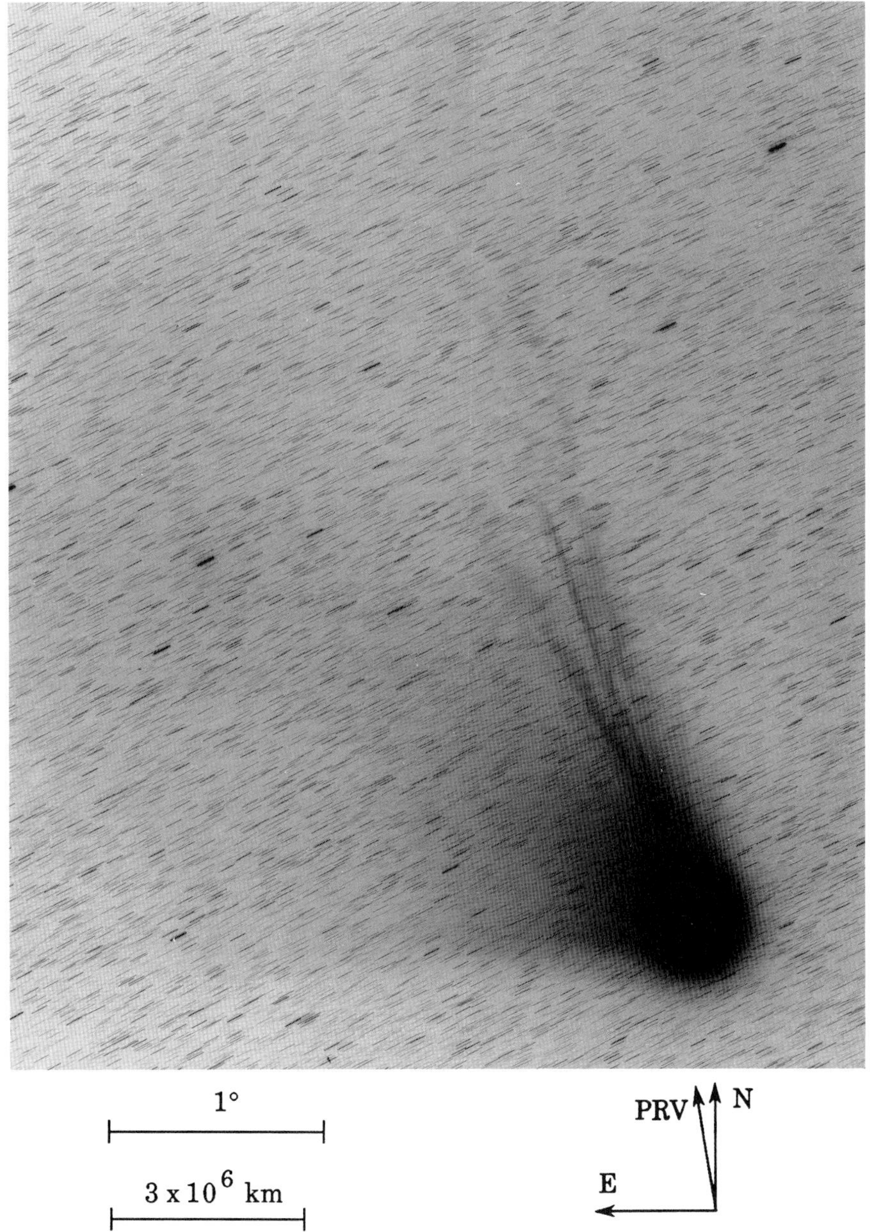

Fig. 602: 1986 Apr 15.50198 UT; exposure 29.0 minutes on IIa-O emulsion with GG-395 filter; r=1.40, Δ=0.44, β=157.7°; LSPN-2431.
Photograph from Royal Observatory/UK Schmidt Telescope Unit.

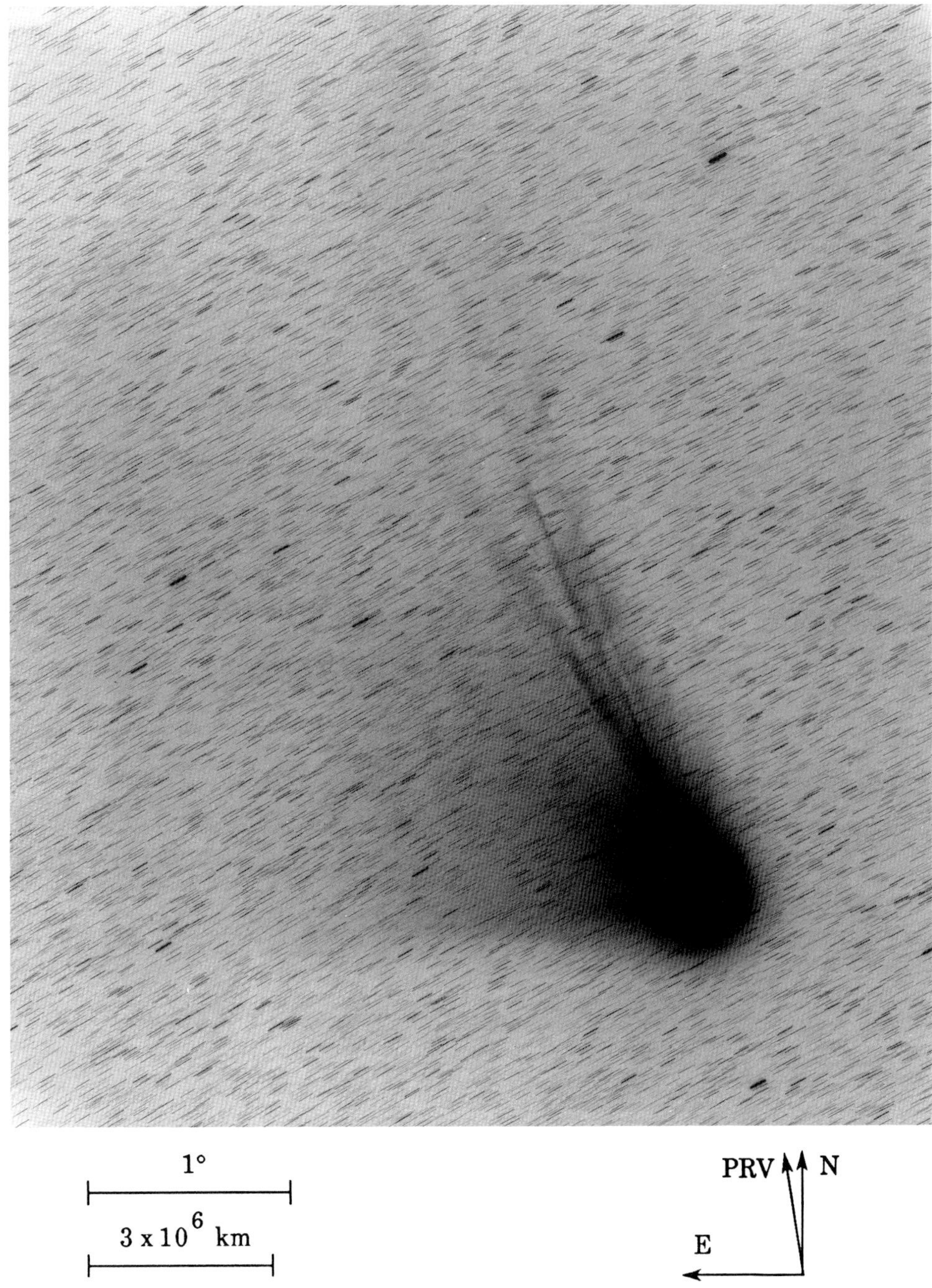

Fig. 603: 1986 Apr 15.55012 UT; exposure 30.0 minutes on IIa-O emulsion with GG-395 filter; r=1.40, Δ=0.45, β=157.8°; LSPN-2432.
Photograph from Royal Observatory/UK Schmidt Telescope Unit.

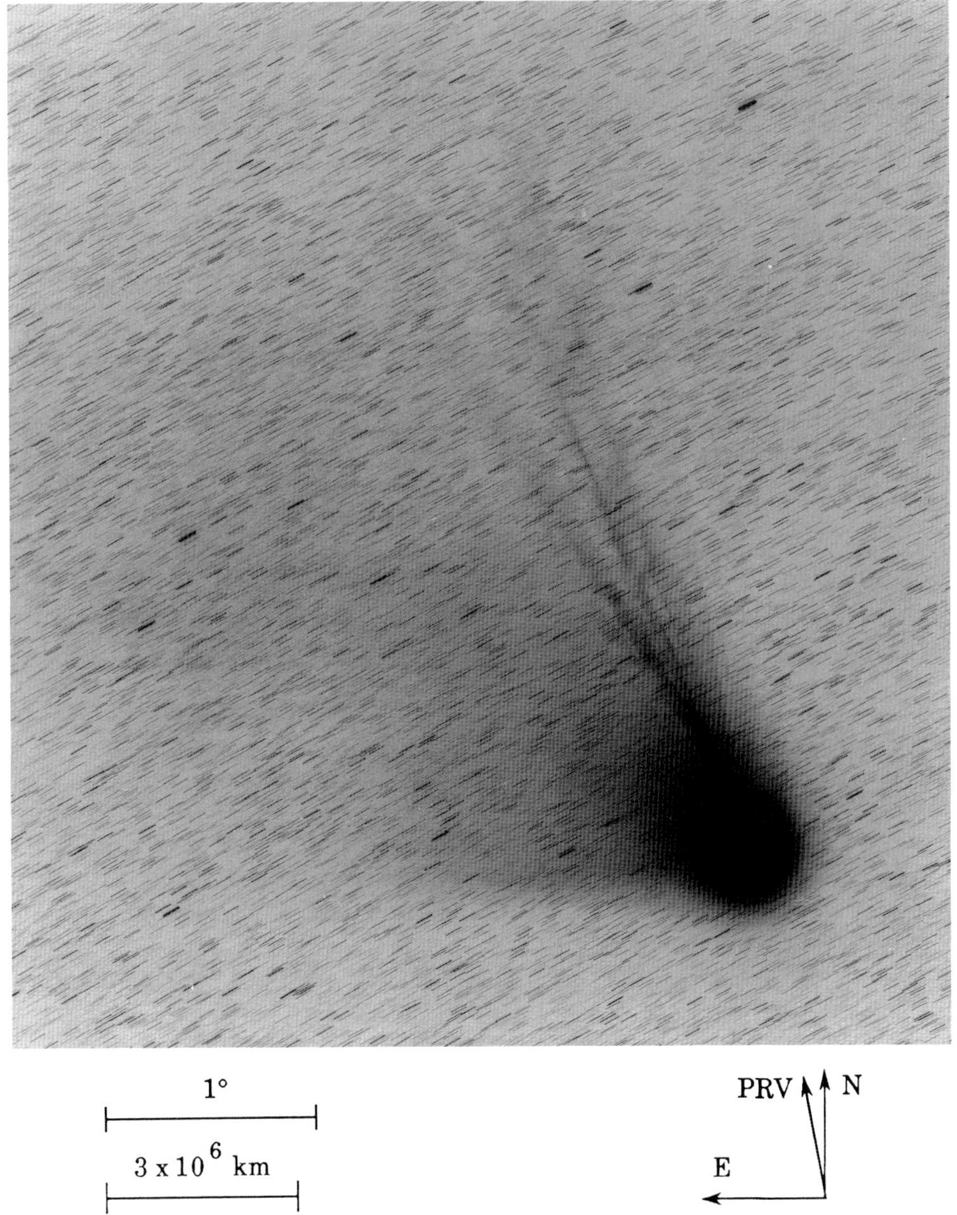

Fig. 604: 1986 Apr 15.57851 UT; exposure 30.0 minutes on IIa-O emulsion with GG-395 filter; r=1.40, Δ=0.45, β=157.8°; LSPN-2433.
Photograph from Royal Observatory/UK Schmidt Telescope Unit.

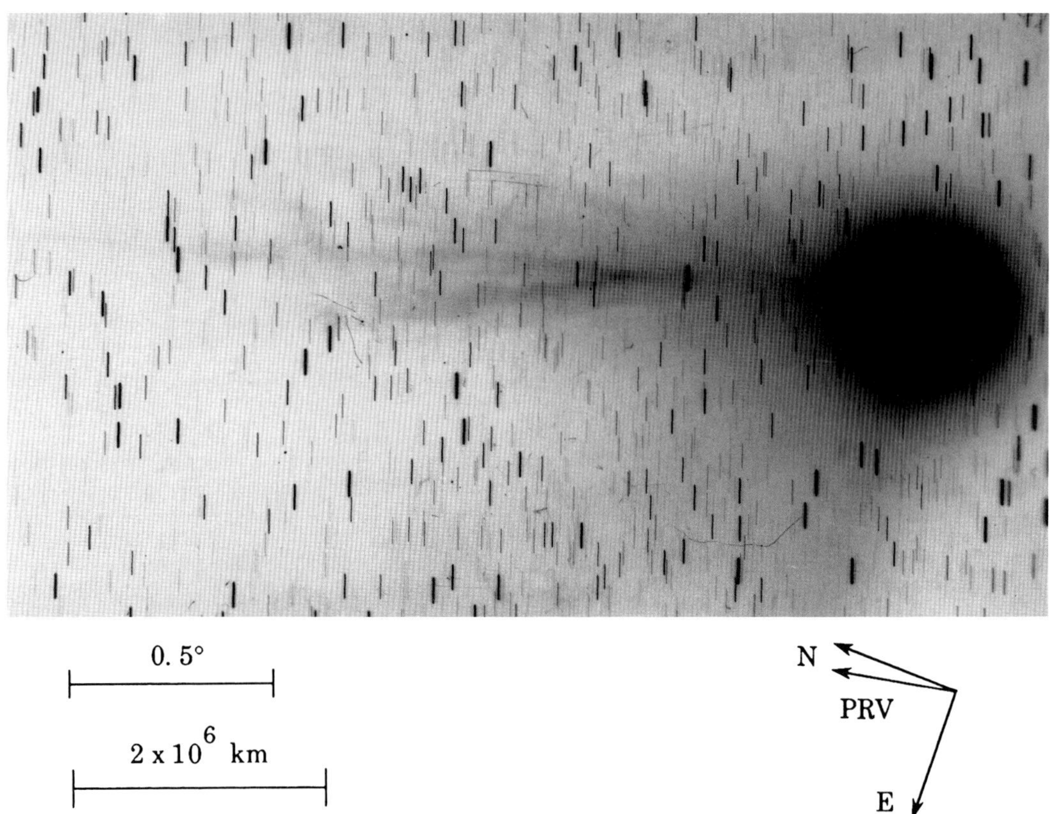

Fig. 605: 1986 Apr 15.59514 UT; exposure 20.0 minutes on 103a-O emulsion with no filter; r=1.40, Δ=0.45, β=157.9°; LSPN-3093.
Photograph by A. McGrath, Perth Observatory.

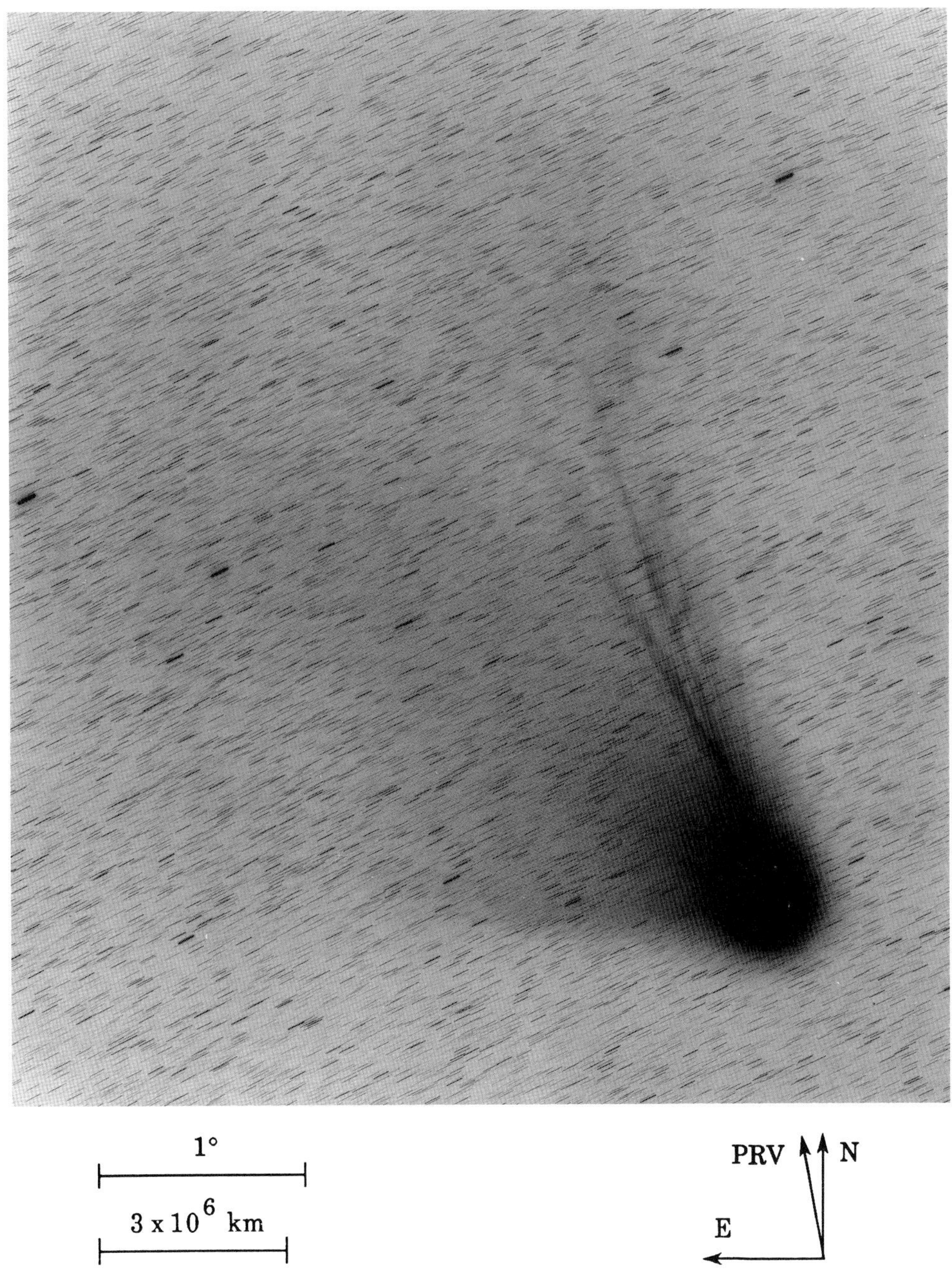

Fig. 606: 1986 Apr 15.60829 UT; exposure 30.0 minutes on IIa-O emulsion with GG-395 filter; r=1.40, Δ=0.45, β=157.9°; LSPN-2434.
Photograph from Royal Observatory/UK Schmidt Telescope Unit.

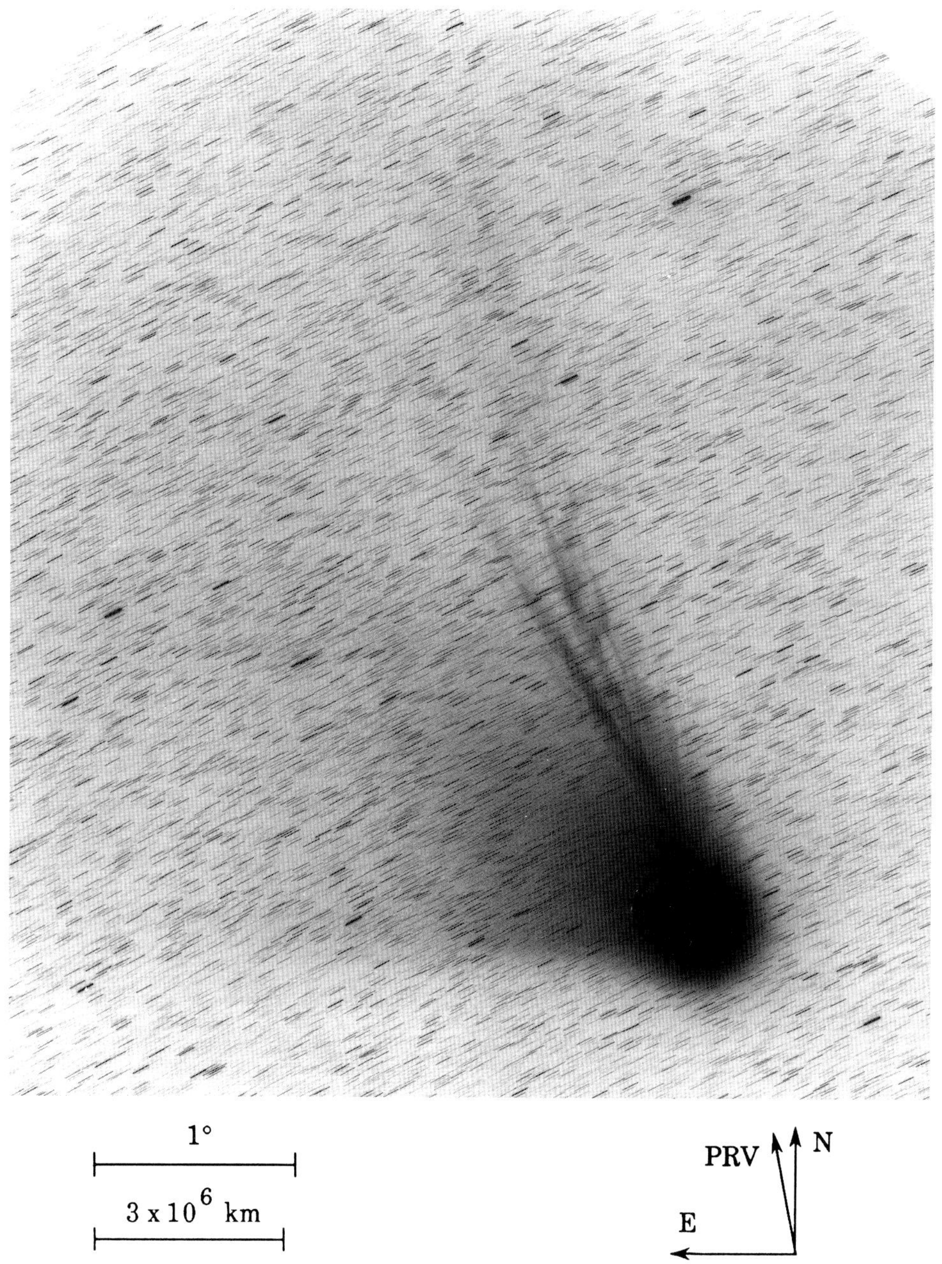

Fig. 607: 1986 Apr 15.63737 UT; exposure 30.0 minutes on IIa-O emulsion with GG-395 filter; r=1.40, Δ=0.45, β=157.9°; LSPN-2435.
Photograph from Royal Observatory/UK Schmidt Telescope Unit.

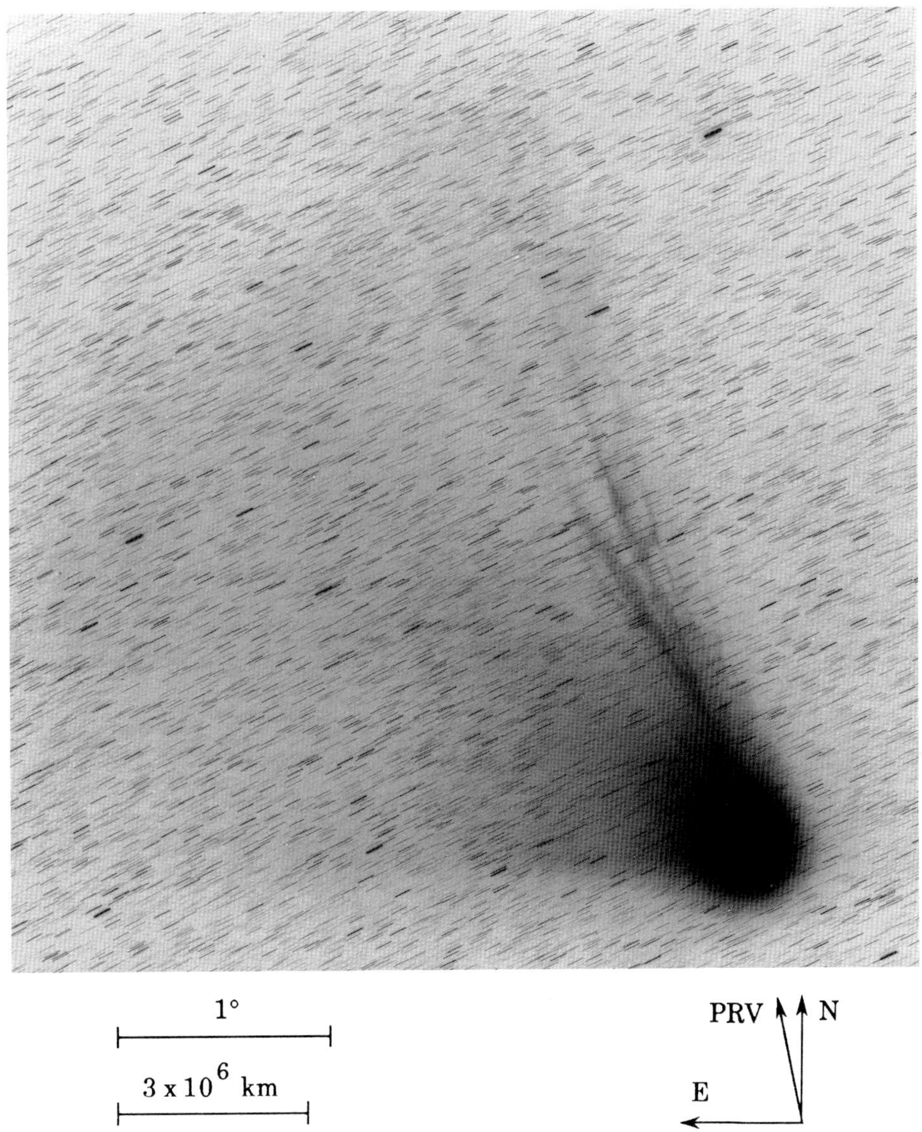

Fig. 608: 1986 Apr 15.67061 UT; exposure 30.0 minutes on IIa-O emulsion with GG-395 filter; r=1.40, Δ=0.45, β=157.9°; LSPN-2436.
Photograph from Royal Observatory/UK Schmidt Telescope Unit.

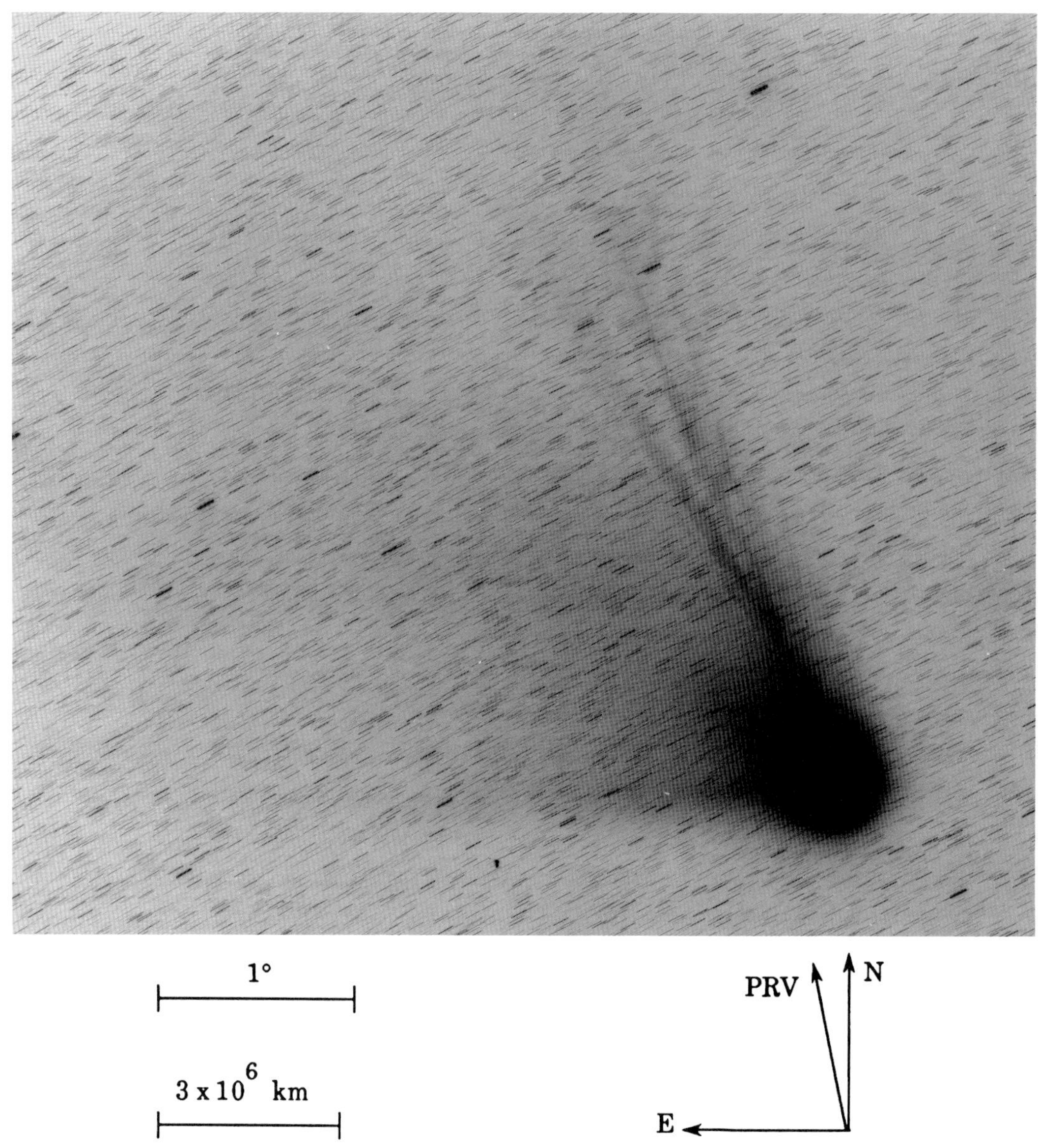

Fig. 609: 1986 Apr 15.69900 UT; exposure 30.0 minutes on IIa-O emulsion with GG-395 filter; r=1.40, Δ=0.45, β=158.0°; LSPN-2437.
Photograph from Royal Observatory/UK Schmidt Telescope Unit.

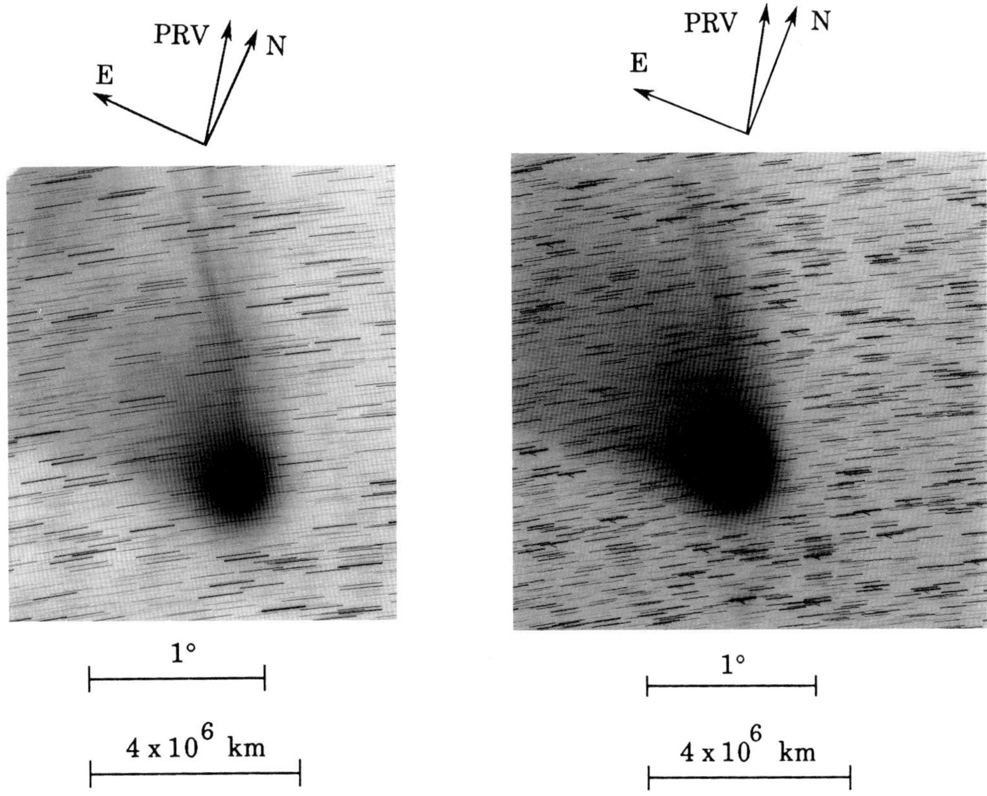

(left) Fig. 610: 1986 Apr 15.72743 UT; exposure 85.0 minutes on 103a-O emulsion with Blue filter; r=1.40, Δ=0.45, β=158.0°; LSPN-3698.
Photograph submitted by K. Sivaraman, Indian Institute for Astrophysics, Kavalur Station.

(right) Fig. 611*: 1986 Apr 15.78819 UT; exposure 60.0 minutes on 103a-O emulsion with UV filter; r=1.41, Δ=0.45, β=158.1°; LSPN-3699.
Photograph submitted by K. Sivaraman, Indian Institute for Astrophysics, Kavalur Station.

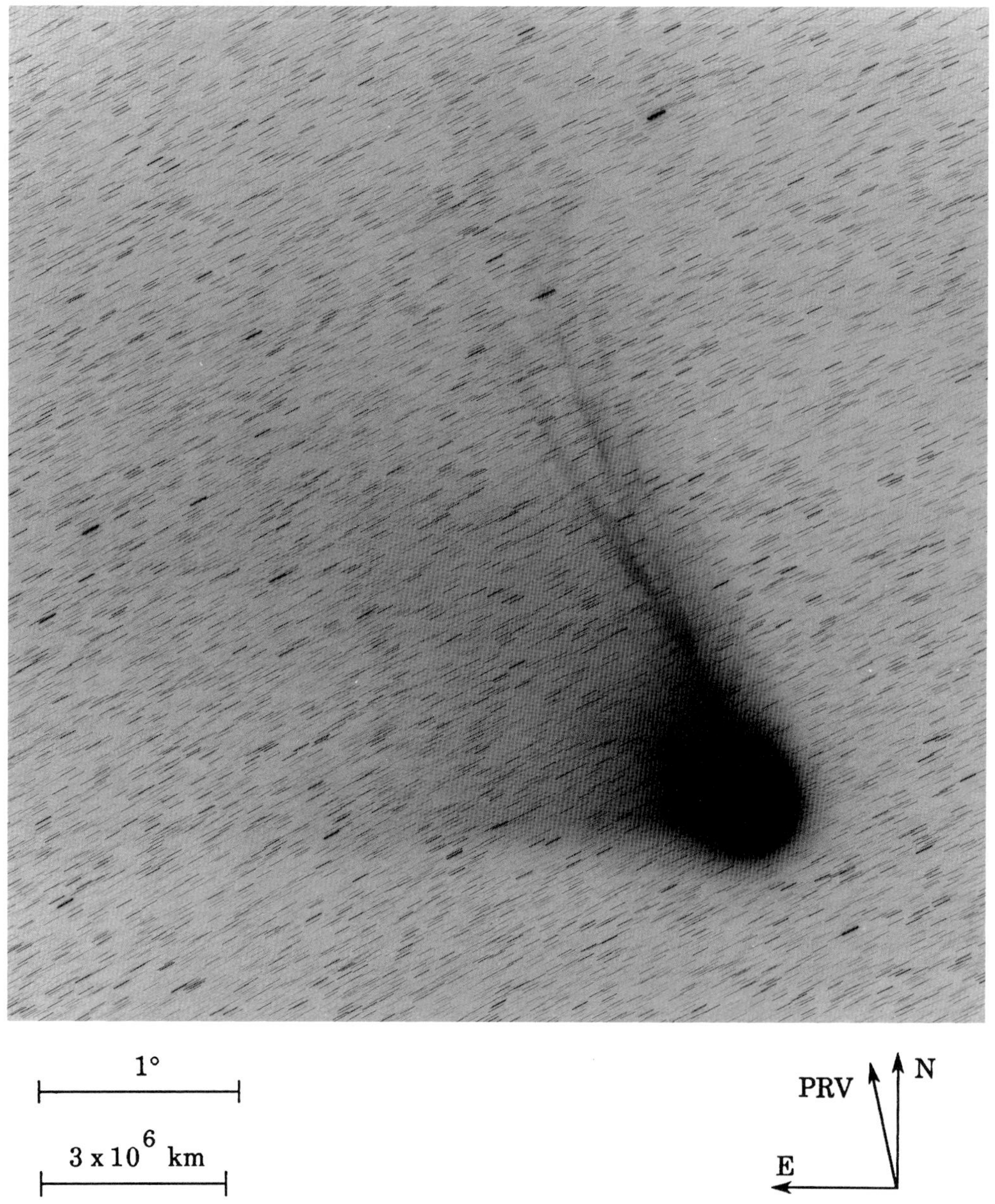

Fig. 612: 1986 Apr 15.72880 UT; exposure 30.0 minutes on IIa-O emulsion with GG-395 filter; r=1.40, Δ=0.45, β=158.0°; LSPN-2438.
Photograph from Royal Observatory/UK Schmidt Telescope Unit.

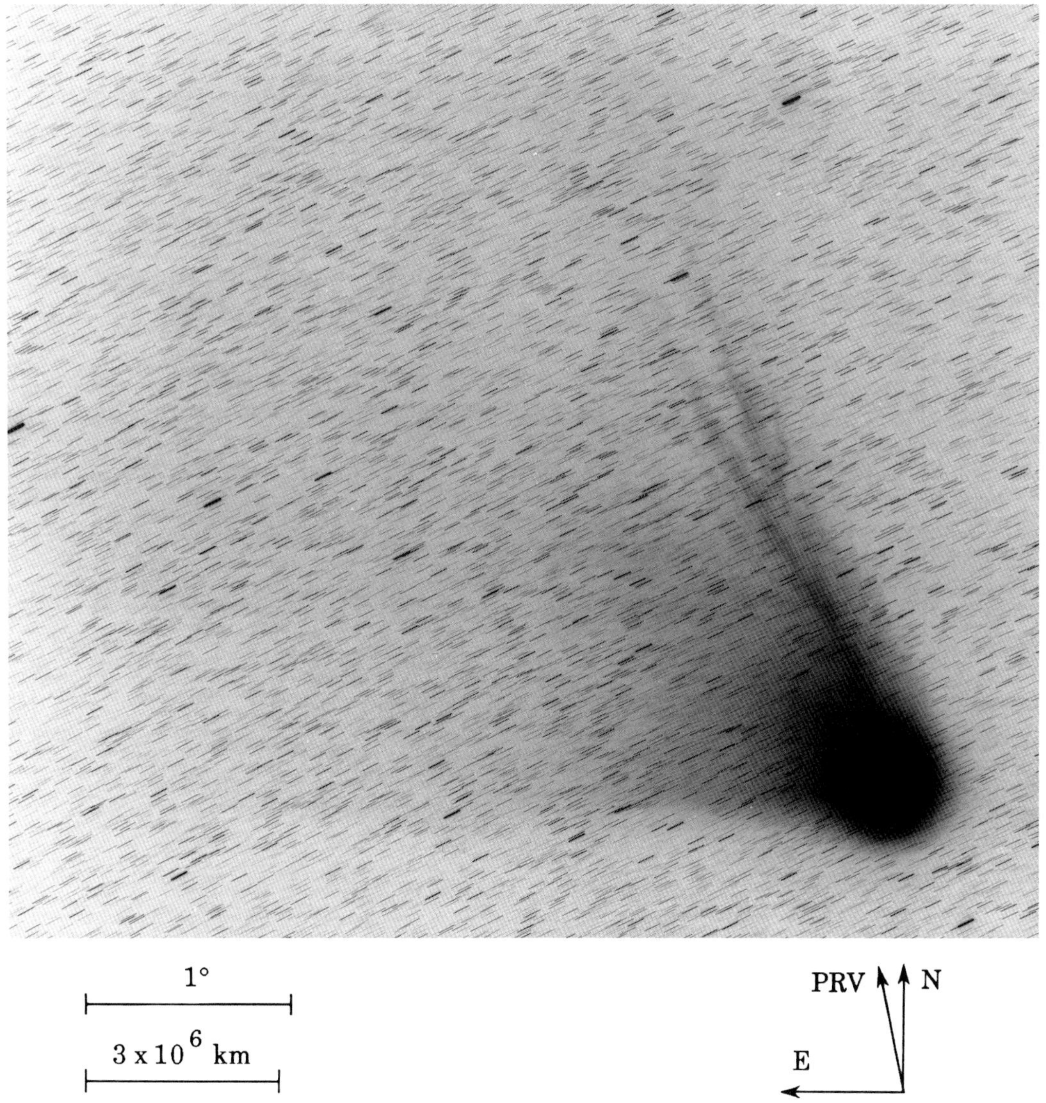

Fig. 613: 1986 Apr 15.75788 UT; exposure 30.0 minutes on IIa-O emulsion with GG-395 filter; r=1.40, Δ=0.45, β=158.0°; LSPN-2439.
Photograph from Royal Observatory/UK Schmidt Telescope Unit.

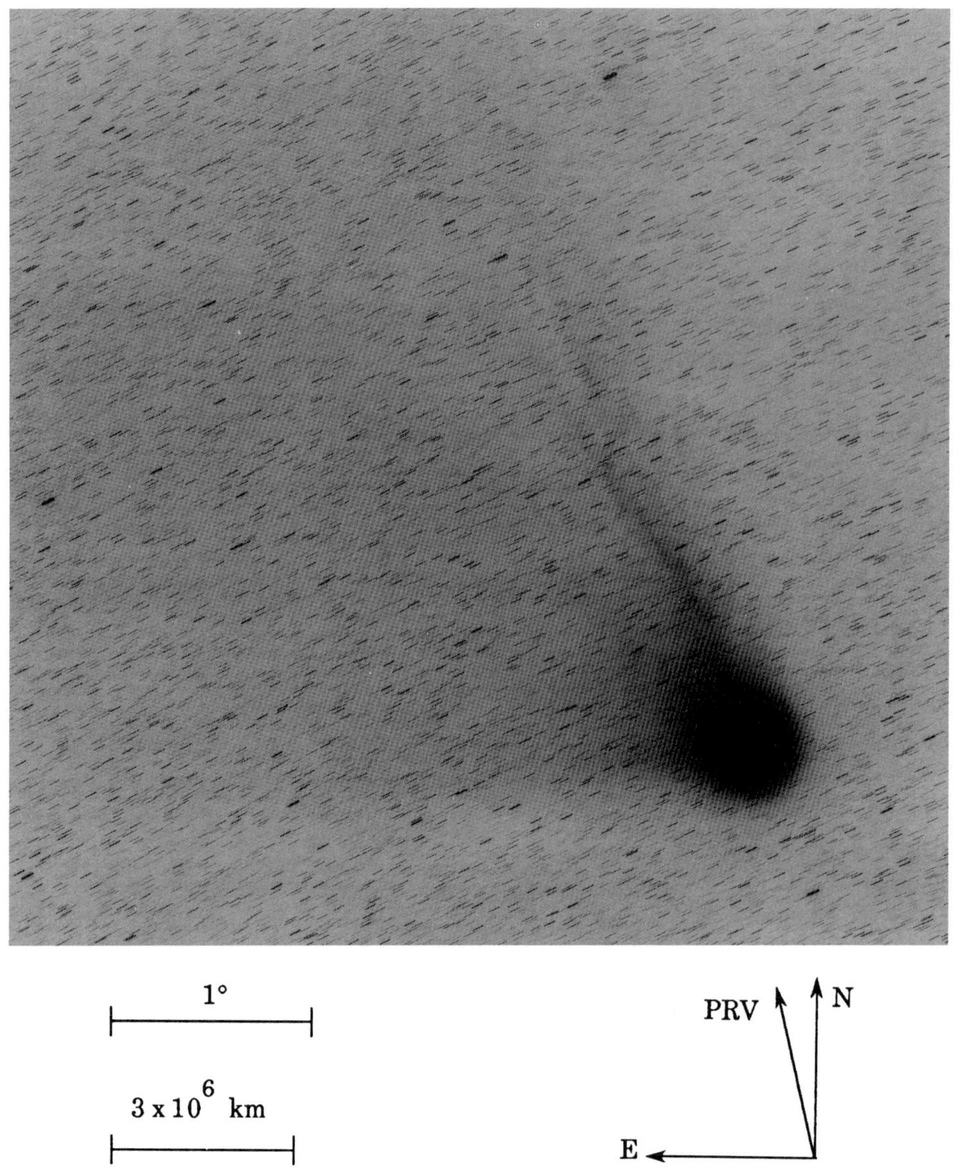

Fig. 614: 1986 Apr 15.78315 UT; exposure 21.0 minutes on IIa-O emulsion with GG-395 filter; r=1.41, Δ=0.45, β=158.1°; LSPN-2440.
Photograph from Royal Observatory/UK Schmidt Telescope Unit.

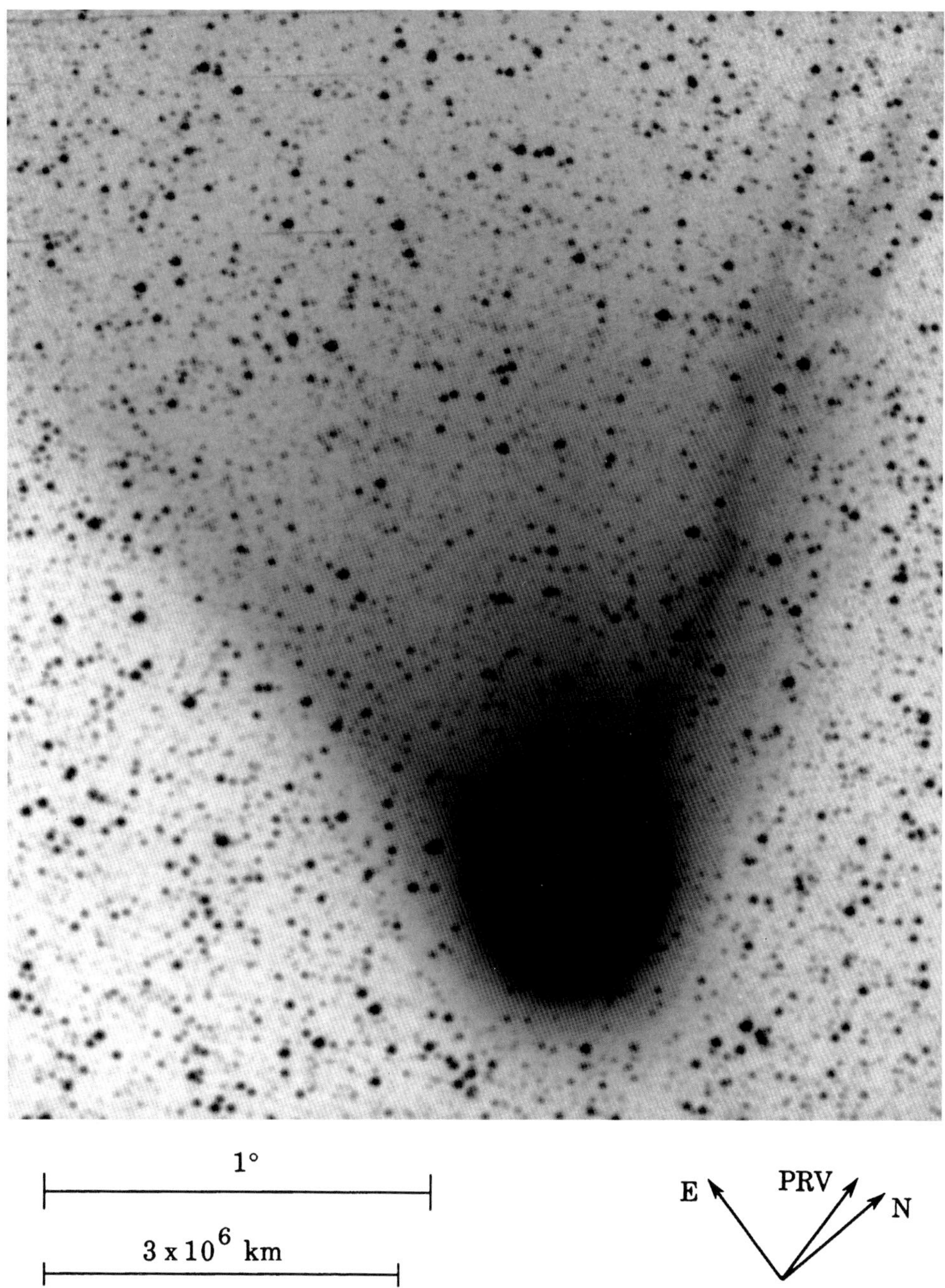

Fig. 615: 1986 Apr 15.83403 UT; exposure 10.0 minutes on Tech. Pan 2415 emulsion with no filter; r=1.41, Δ=0.45, β=158.1°; LSPN-1828.
Photograph by F. Marang/F. Van Wyk, Sutherland, LSPN Island Network.

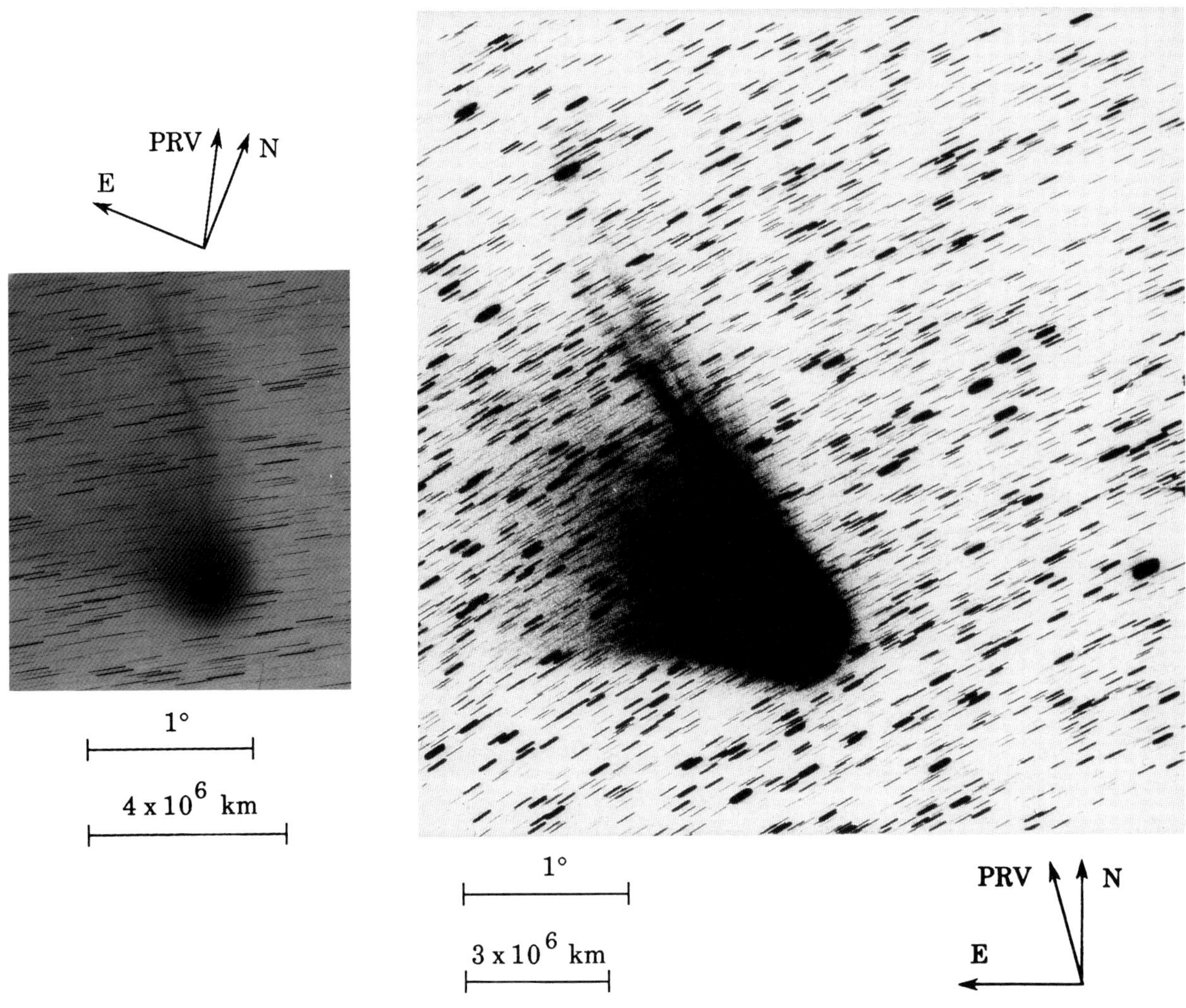

(left) Fig. 616: 1986 Apr 15.89410 UT; exposure 85.0 minutes on 103a-O emulsion with Blue filter; r=1.41, Δ=0.45, β=158.2°; LSPN-3700.
Photograph submitted by K. Sivaraman, Indian Institute for Astrophysics, Kavalur Station.

(right) Fig. 617: 1986 Apr 15.97010 UT; exposure 40.0 minutes on AGFA 1000 RS emulsion; r=1.41, Δ=0.45, β=158.2°; AON-851884.
Photograph from Volkssternwarte Frankfurt, Germany.

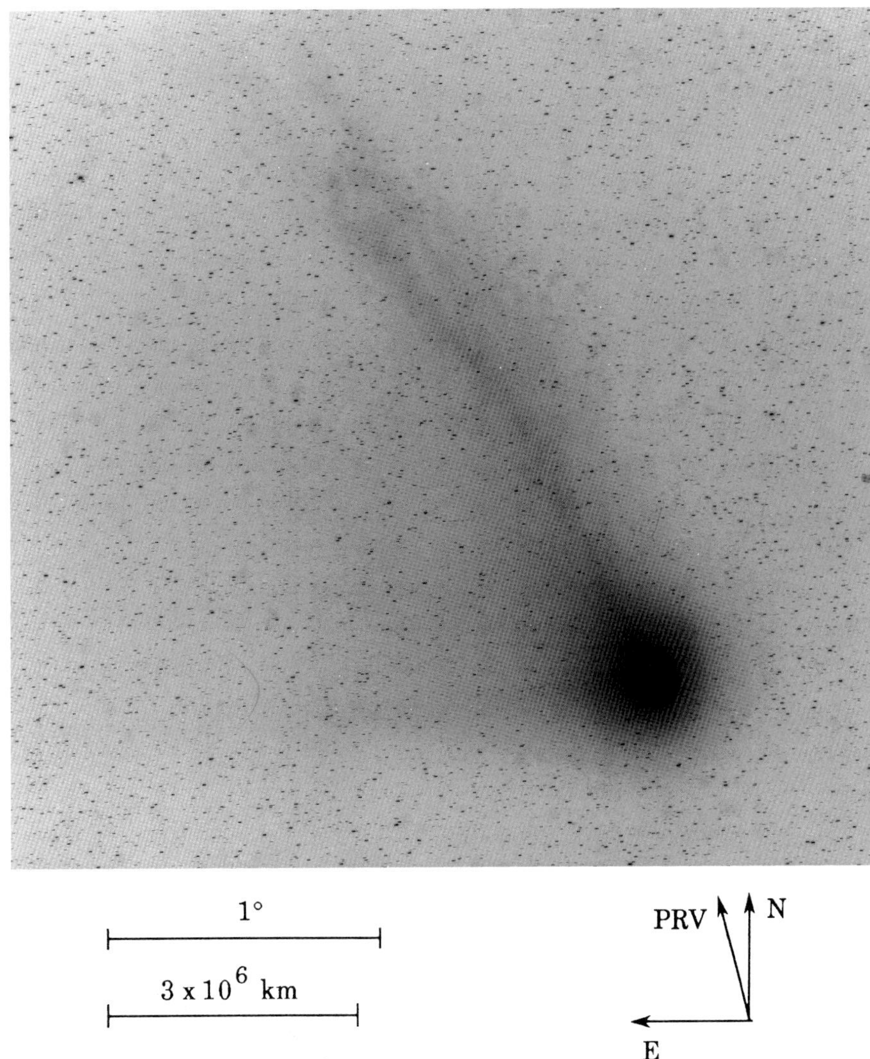

Fig. 618: 1986 Apr 16.00104 UT; exposure 5.0 minutes on 103a-O emulsion with no filter; r=1.41, Δ=0.45, β=158.3°; LSPN-1229.
Photograph by F. Miller, University of Michigan/CTIO.

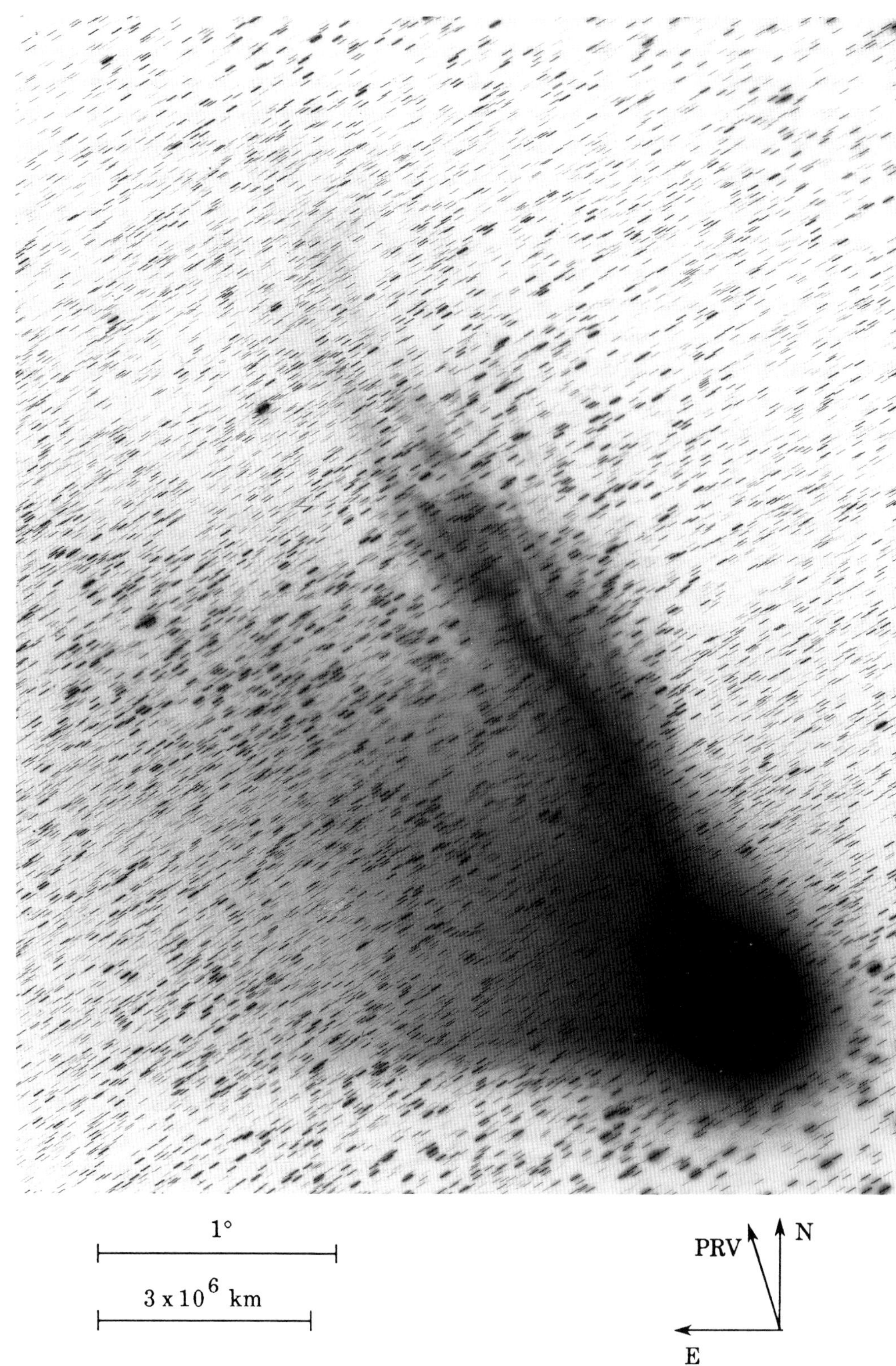

Fig. 619: 1986 Apr 16.13958 UT; exposure 20.0 minutes on 103a-O emulsion with no filter; r=1.41, Δ=0.45, β=158.4°; LSPN-1233.
Photograph by F. Miller, University of Michigan/CTIO.

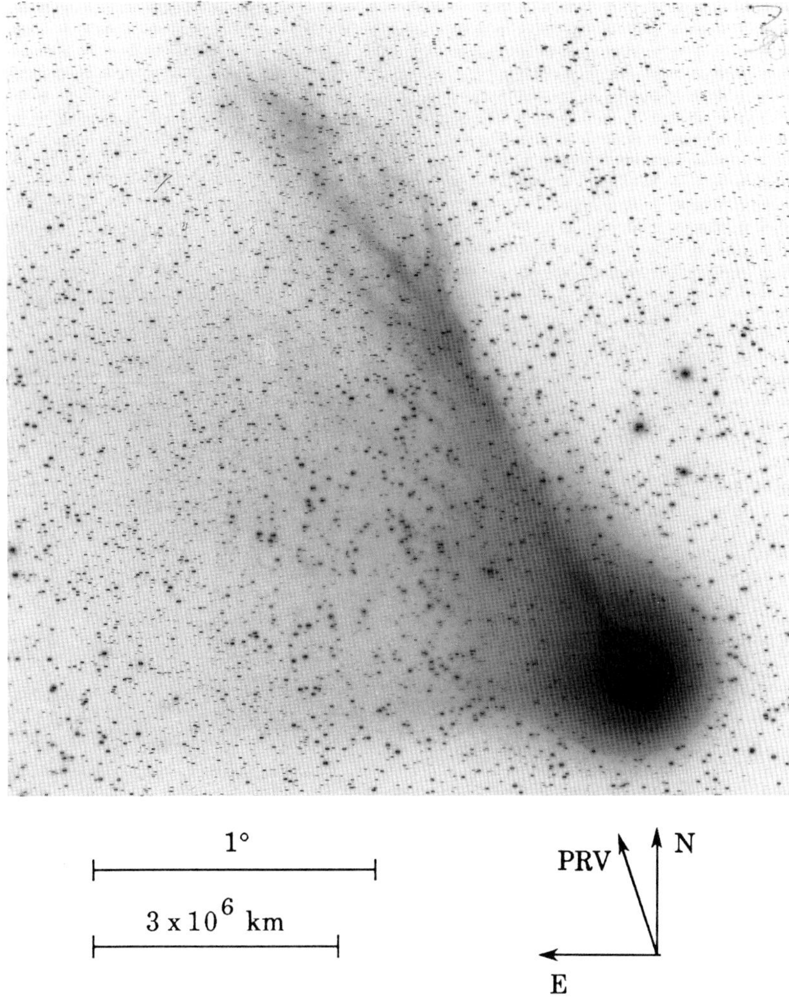

Fig. 620: 1986 Apr 16.29965 UT; exposure 5.0 minutes on 103a-O emulsion with no filter; r=1.41, Δ=0.46, β=158.5°; LSPN-1236.
Photograph by F. Miller, University of Michigan/CTIO.

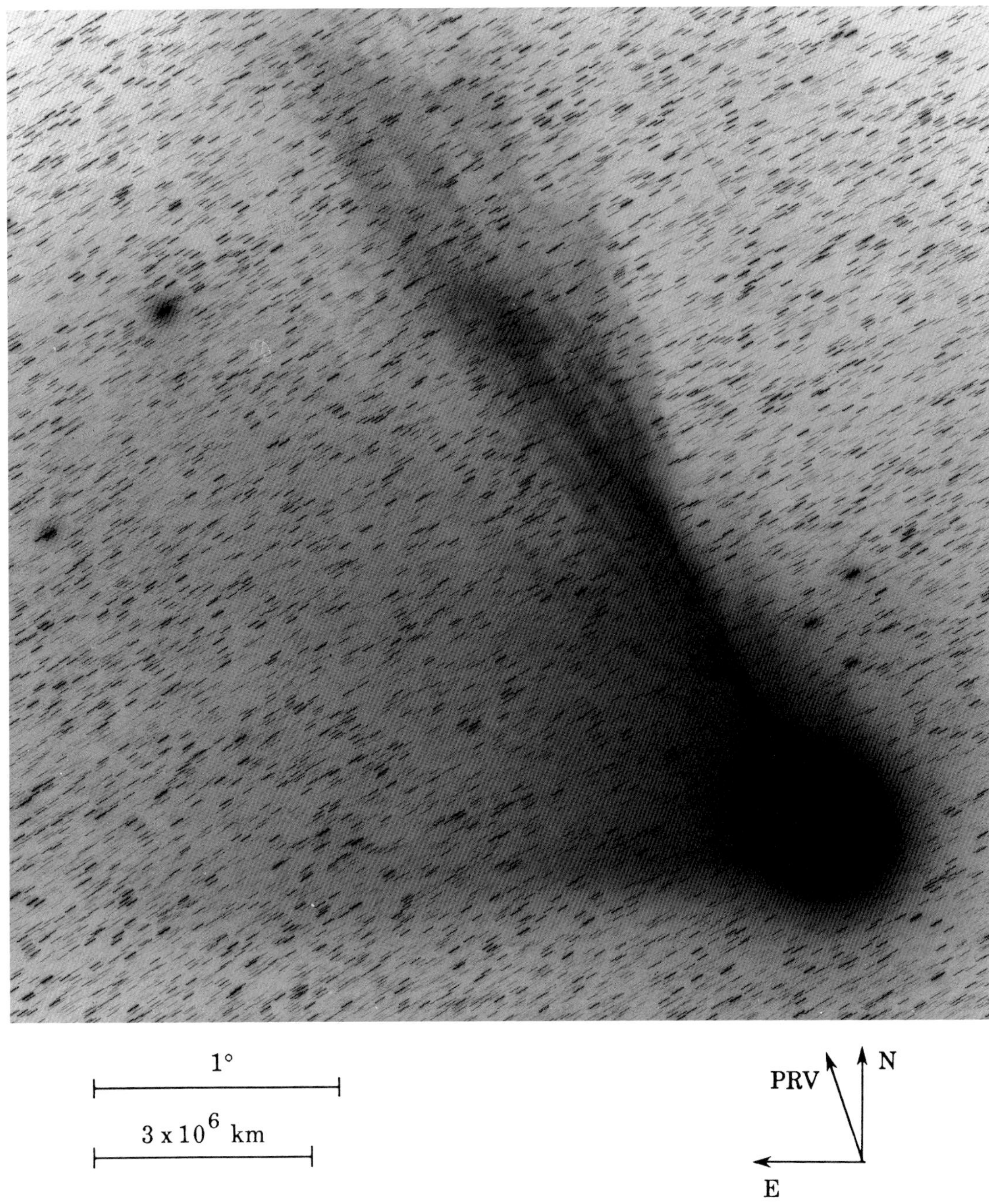

Fig. 621: 1986 Apr 16.33264 UT; exposure 20.0 minutes on 103a-O emulsion with no filter; r=1.41, Δ=0.46, β=158.5°; LSPN-1237.
Photograph by F. Miller, University of Michigan/CTIO.

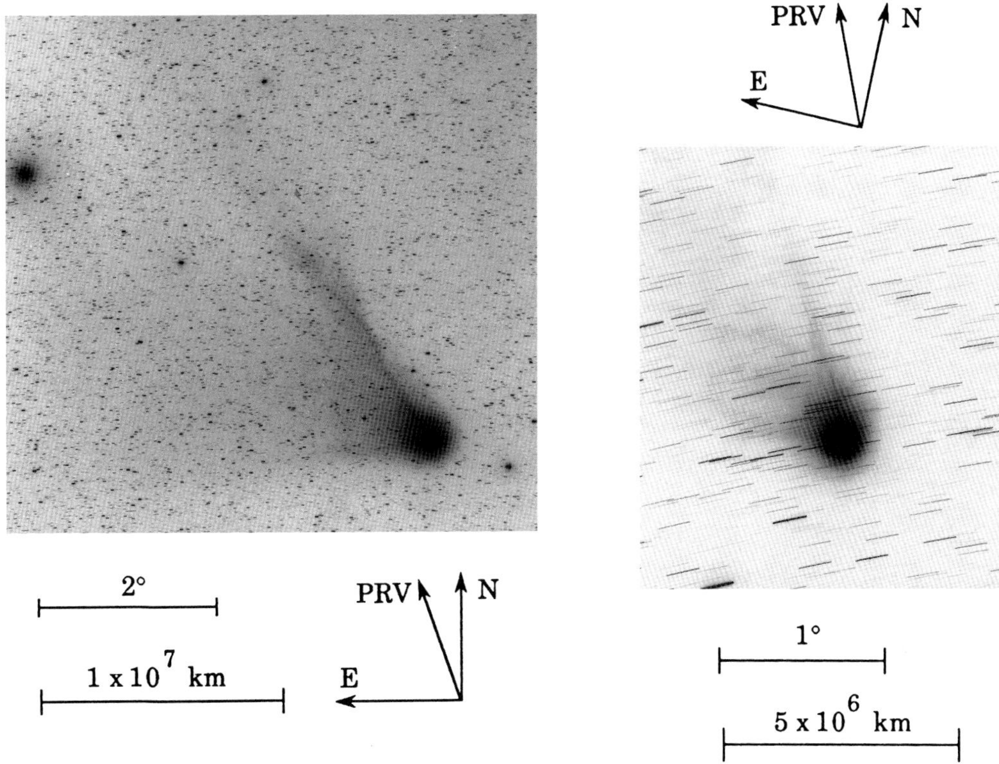

(left) Fig. 622: 1986 Apr 16.35833 UT; exposure 10.0 minutes on IIa-O emulsion with no filter; r=1.41, Δ=0.46, β=158.6°; LSPN-918.
Photograph by E. Moore/J. Brandt, Joint Observatory for Cometary Research.

(right) Fig. 623*: 1986 Apr 16.74826 UT; exposure 75.0 minutes on 103a-O emulsion with Blue filter; r=1.42, Δ=0.46, β=158.8°; LSPN-3701.
Photograph submitted by K. Sivaraman, Indian Institute for Astrophysics, Kavalur Station.

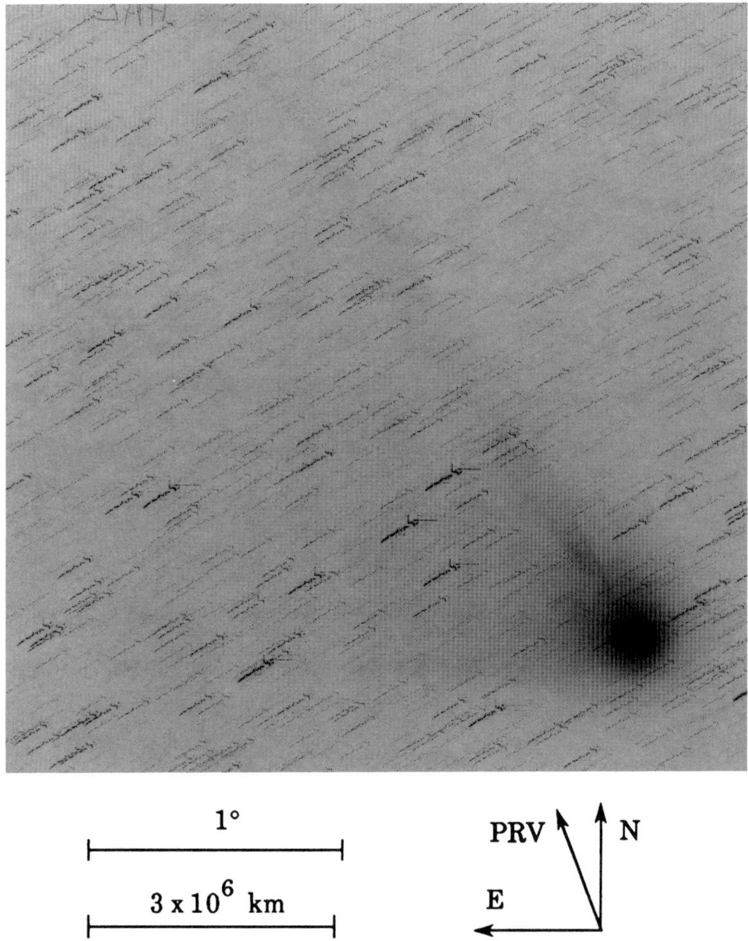

Fig. 624: 1986 Apr 16.56771 UT; exposure 57.0 minutes on 103a-O emulsion with GG-400 filter; r=1.42, Δ=0.46, β=158.7°; LSPN-426.
Photograph by S. Dick, U.S. Naval Observatory Station.

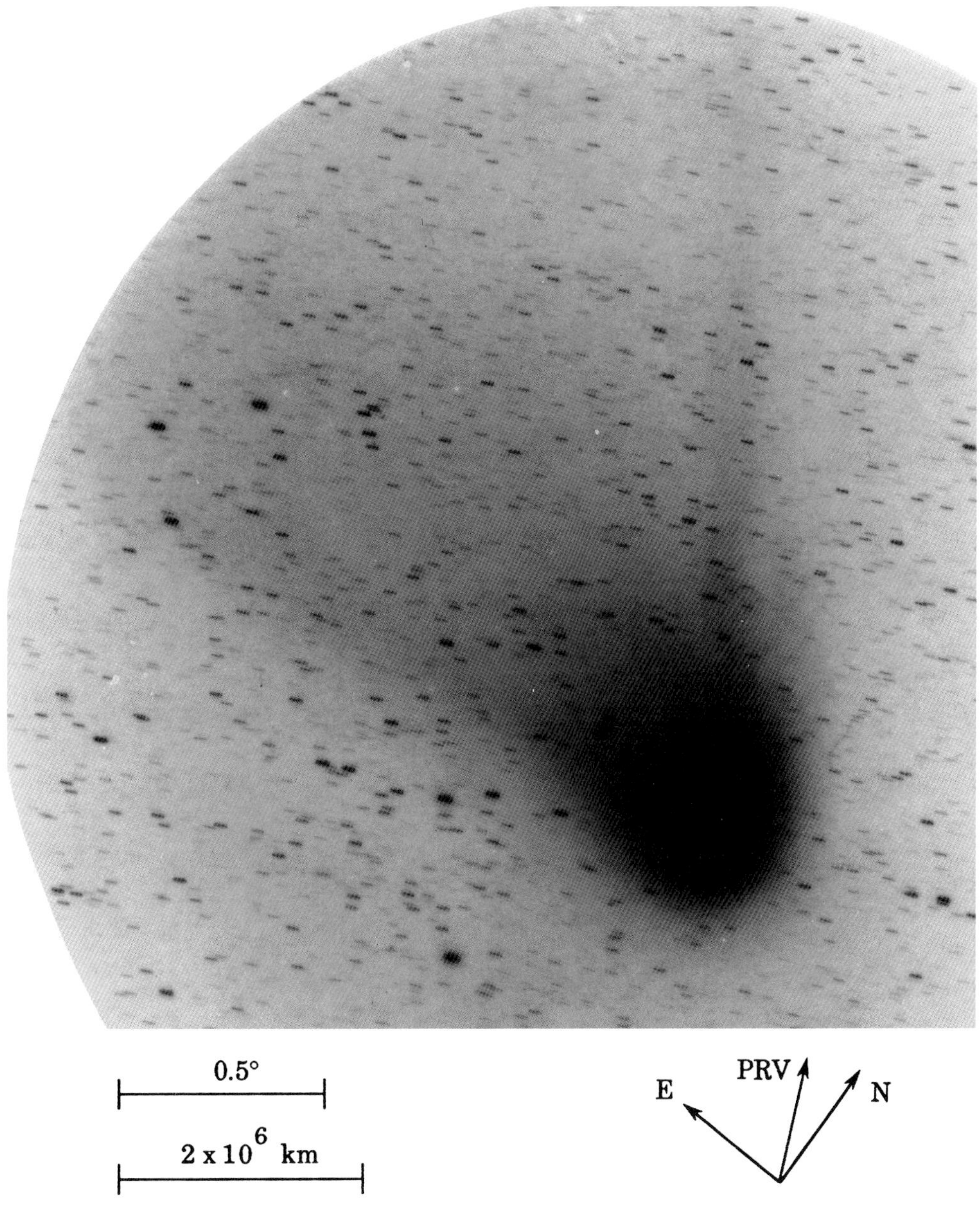

Fig. 625: 1986 Apr 16.80208 UT; exposure 10.0 minutes on Tech. Pan 2415 emulsion with no filter; r=1.42, Δ=0.46, β=158.8°; LSPN-3502.
Photograph by G. Roberts, University of Perugia/South African Astronomical Observatory.

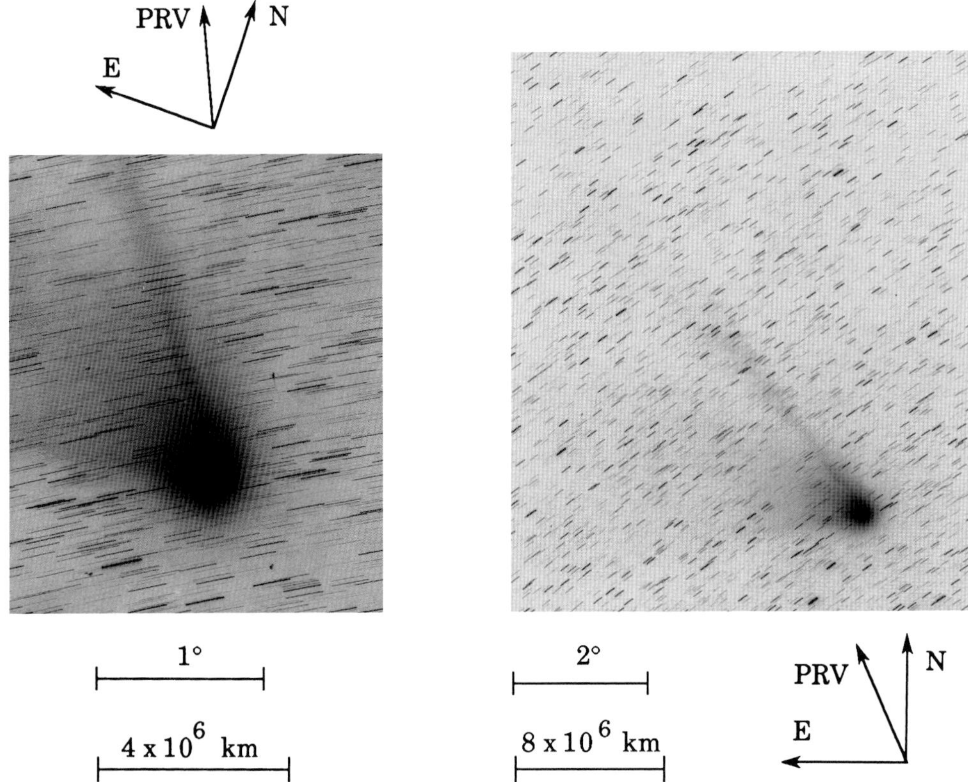

(left) Fig. 626: 1986 Apr 16.81944 UT; exposure 60.0 minutes on 103a-O emulsion with Blue filter; r=1.42, Δ=0.46, β=158.8°; LSPN-3702.
Photograph submitted by K. Sivaraman, Indian Institute for Astrophysics, Kavalur Station.

(right) Fig. 627: 1986 Apr 16.85590 UT; exposure 60.0 minutes on IIa-O emulsion with no filter; r=1.42, Δ=0.46, β=158.9°; LSPN-2875.
Photograph by G. Malcolm/A. Jarrett, Boyden Observatory.

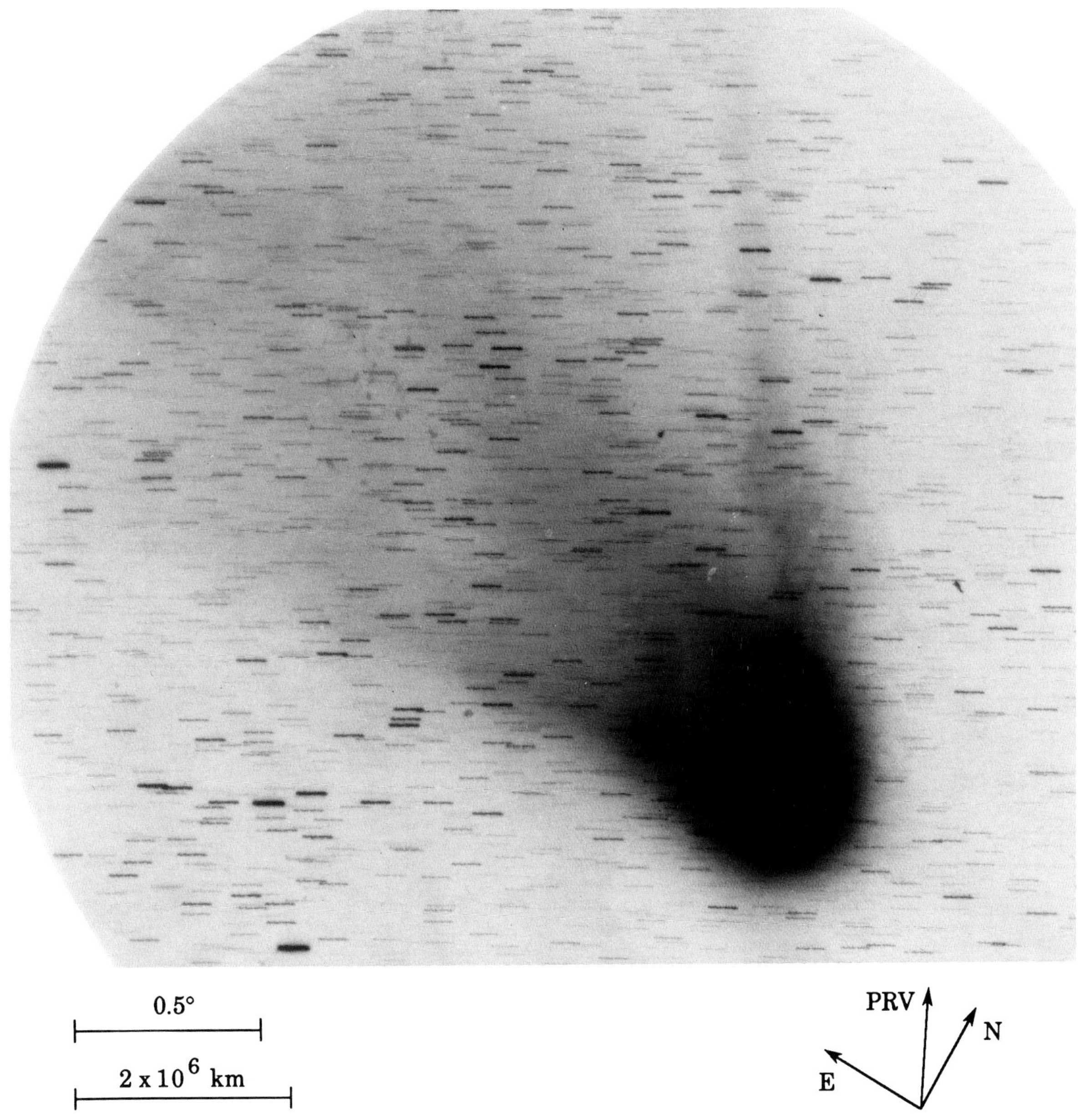

Fig. 628: 1986 Apr 16.99653 UT; exposure 30.0 minutes on Tech. Pan 2415 emulsion with no filter; r=1.42, Δ=0.46, β=158.9°; LSPN-3503.
Photograph by G. Roberts, University of Perugia/South African Astronomical Observatory.

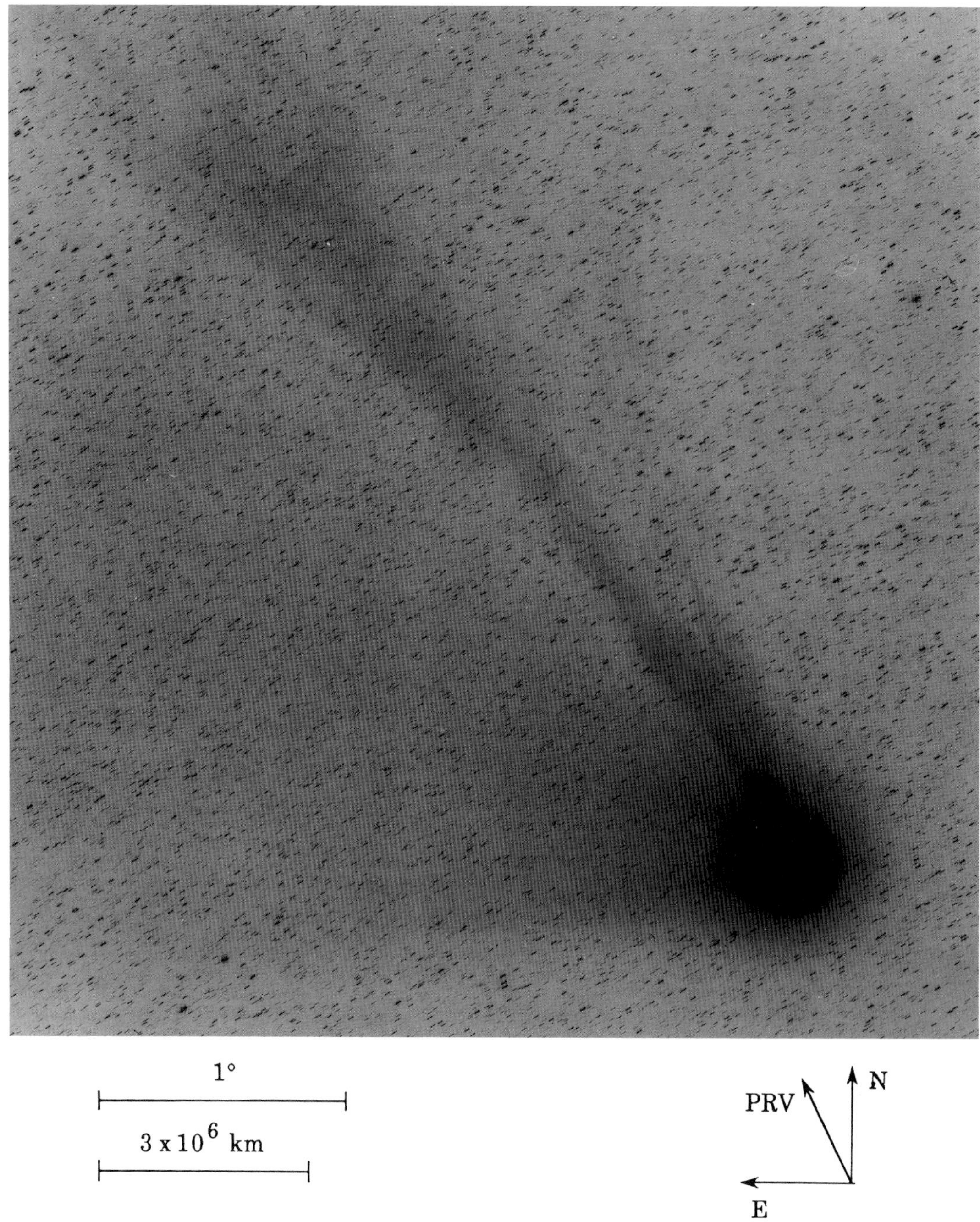

Fig. 629: 1986 Apr 17.05764 UT; exposure 10.0 minutes on 103a-O emulsion with no filter; r=1.42, Δ=0.47, β=159.0°; LSPN-1240.
Photograph by F. Miller, University of Michigan/CTIO.

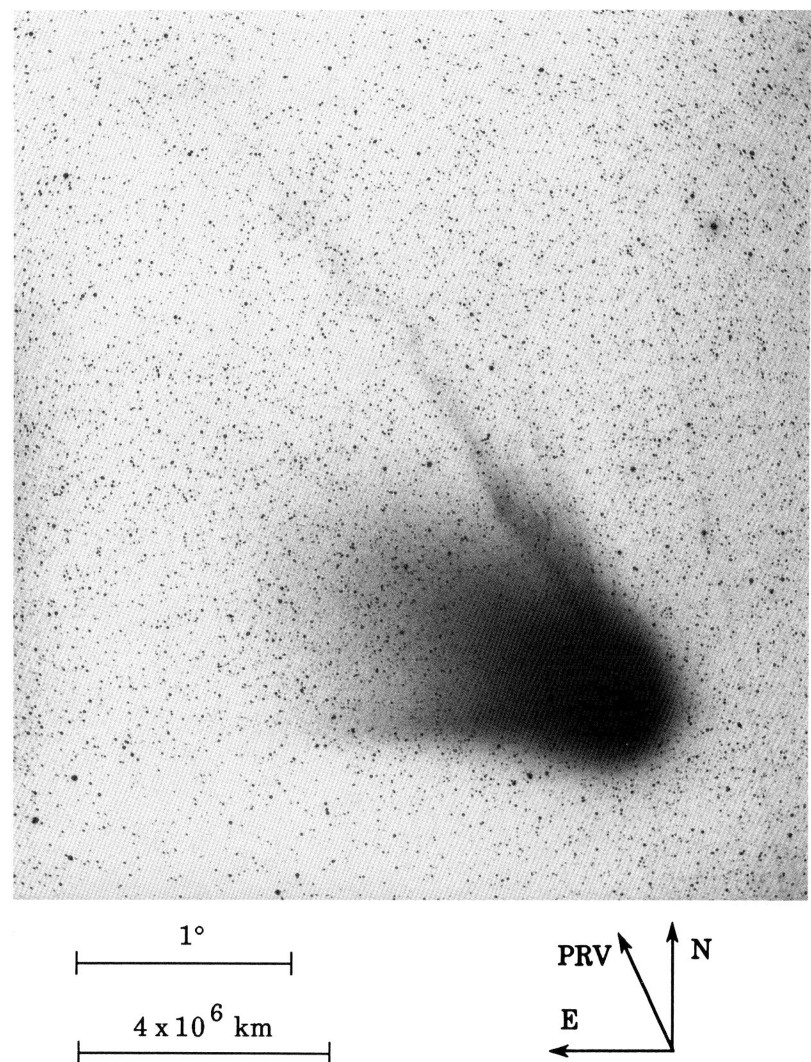

Fig. 630: 1986 Apr 17.06007 UT; exposure 5.0 minutes on Tech. Pan 2415 emulsion; r=1.42, Δ=0.47, β=159.0°; AON-851892.
Photograph by M. Jäger, Austria.

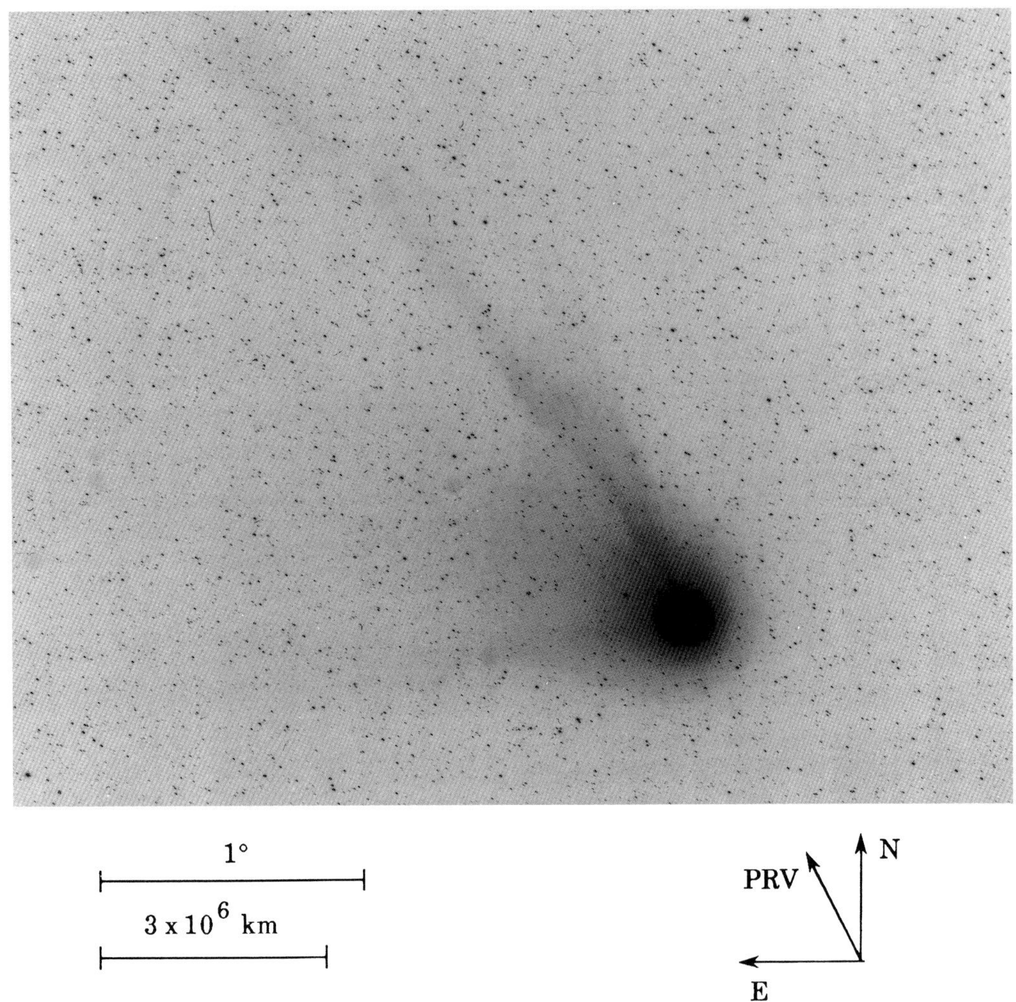

Fig. 631: 1986 Apr 17.10833 UT; exposure 4.0 minutes on 103a-O emulsion with no filter; r=1.43, Δ=0.47, β=159.0°; LSPN-1241.
Photograph by F. Miller, University of Michigan/CTIO.

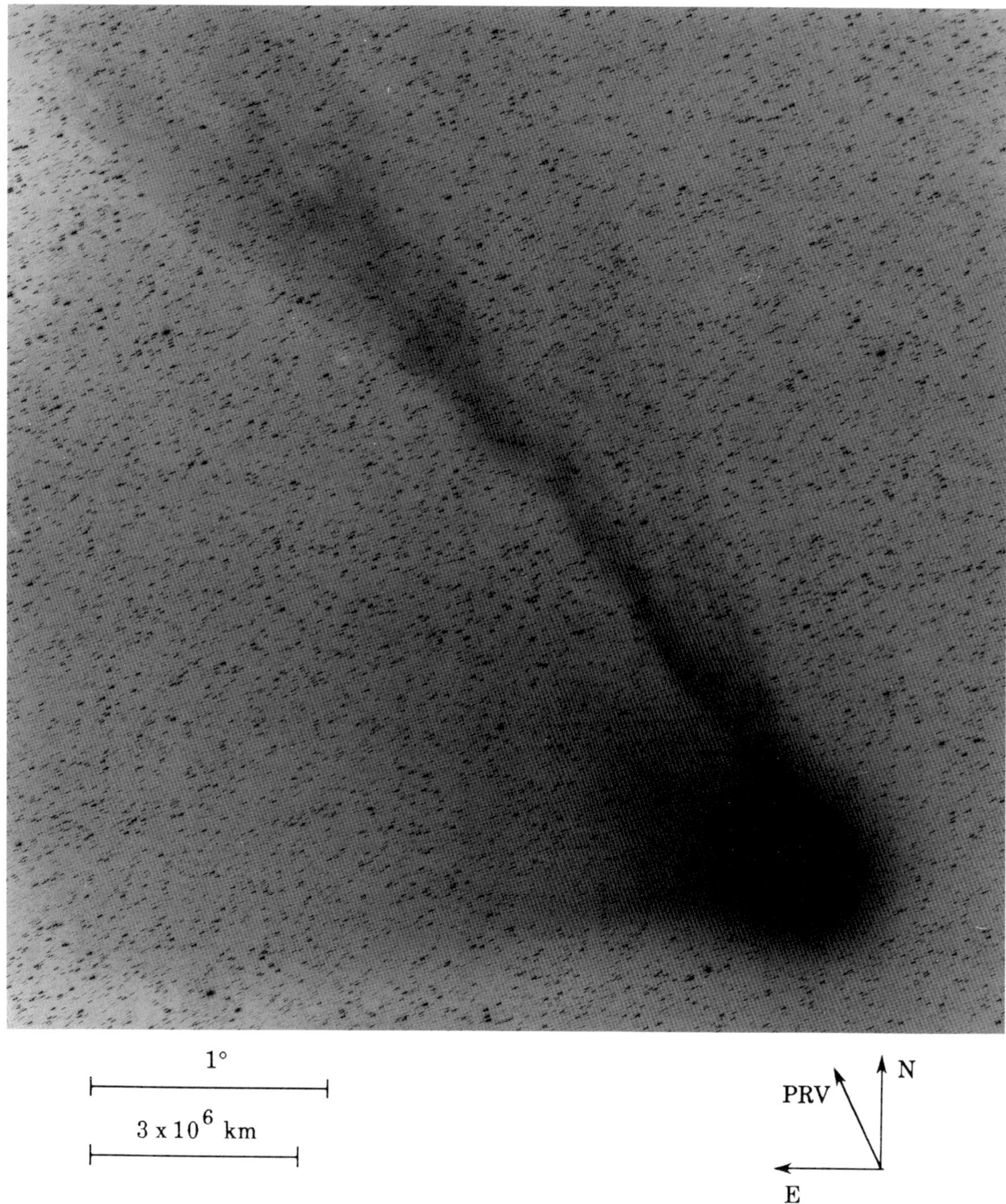

Fig. 632: 1986 Apr 17.11875 UT; exposure 10.0 minutes on 103a-O emulsion with no filter; r=1.43, Δ=0.47, β=159.0°; LSPN-1242.
Photograph by F. Miller, University of Michigan/CTIO.

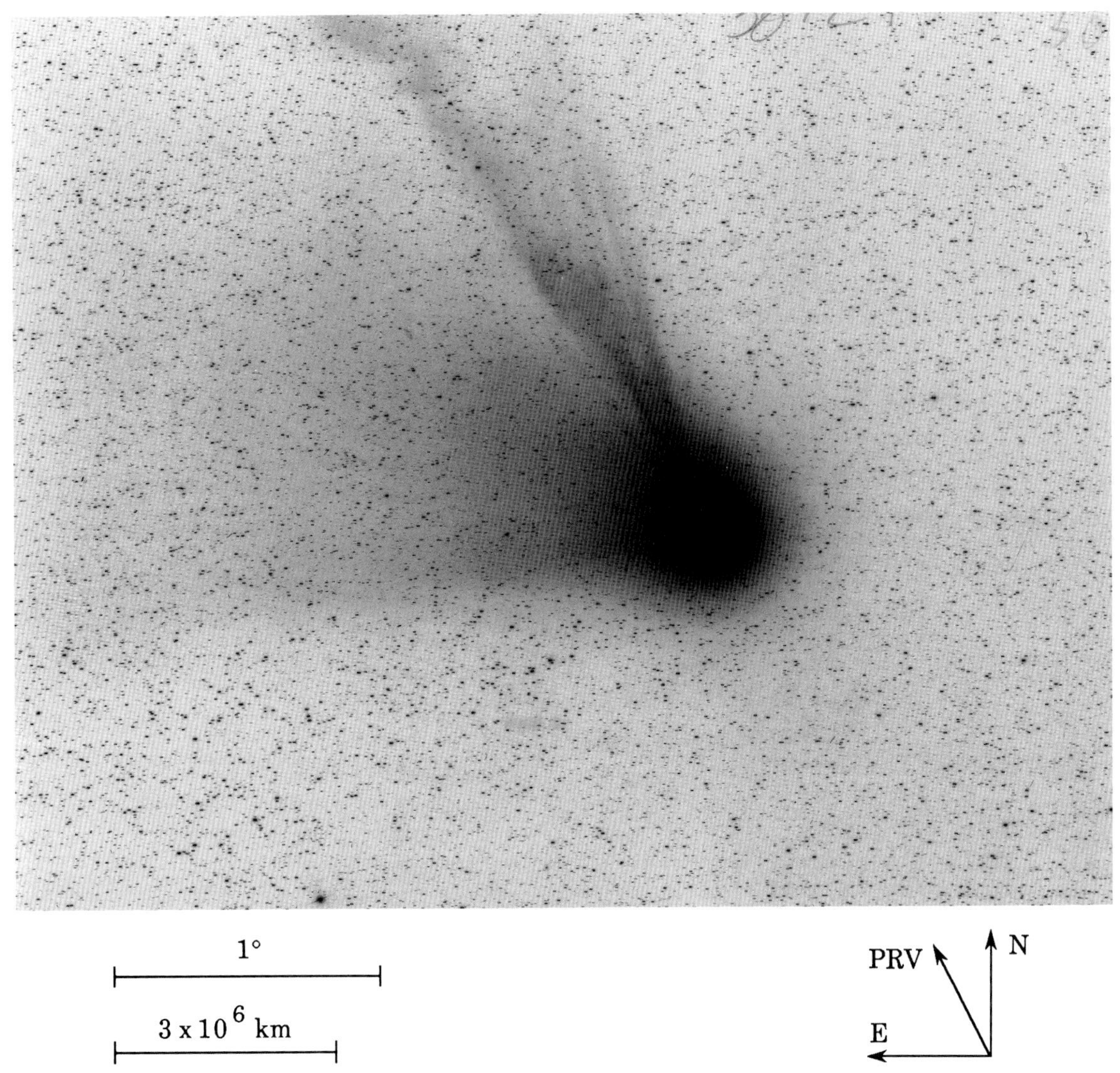

Fig. 633: 1986 Apr 17.16910 UT; exposure 5.0 minutes on 103a-O emulsion with no filter; r=1.43, Δ=0.47, β=159.0°; LSPN-1243.
Photograph by F. Miller, University of Michigan/CTIO.

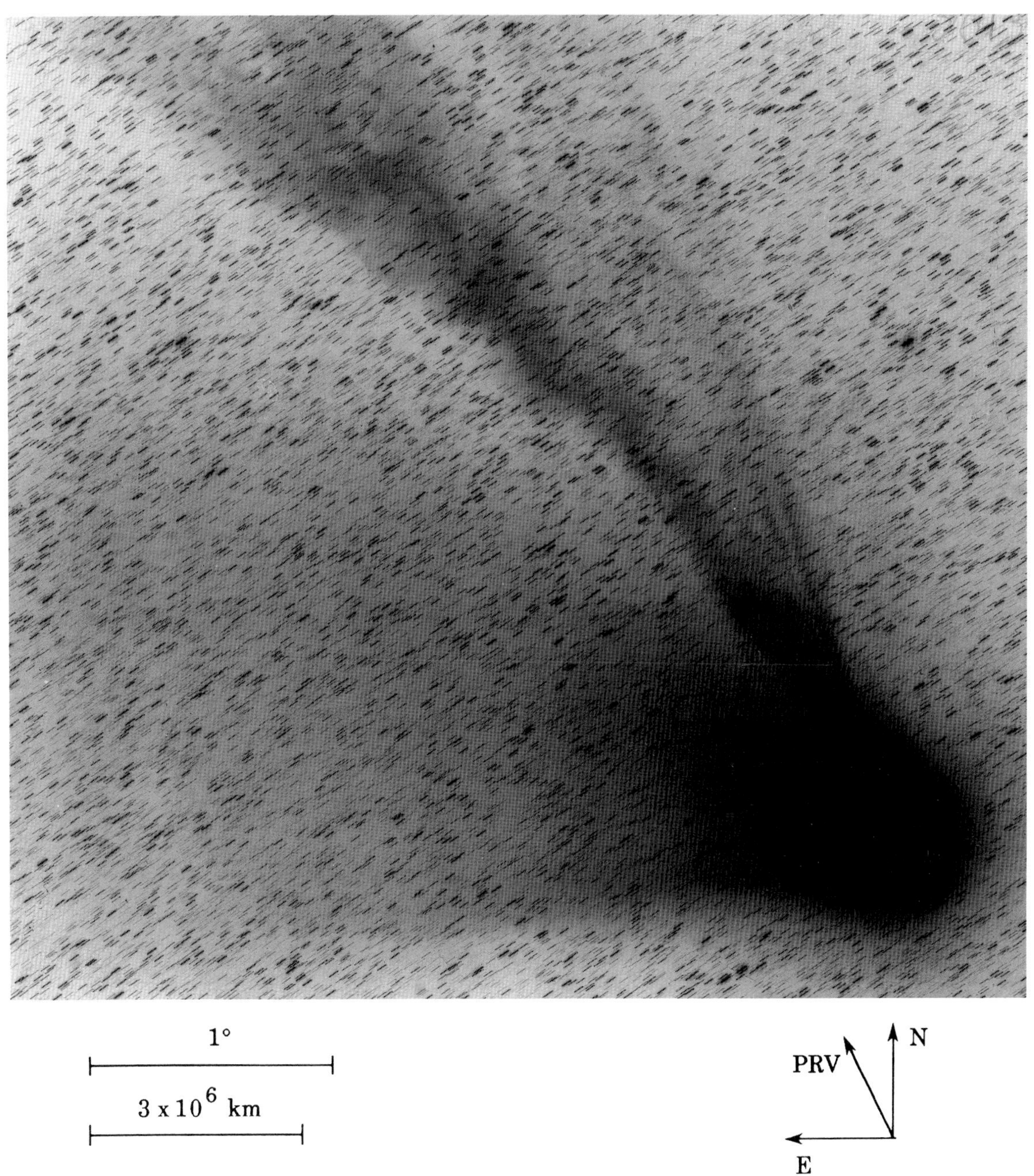

Fig. 634: 1986 Apr 17.18403 UT; exposure 20.0 minutes on 103a-O emulsion with no filter; r=1.43, Δ=0.47, β=159.0°; LSPN-1244.
Photograph by F. Miller, University of Michigan/CTIO.

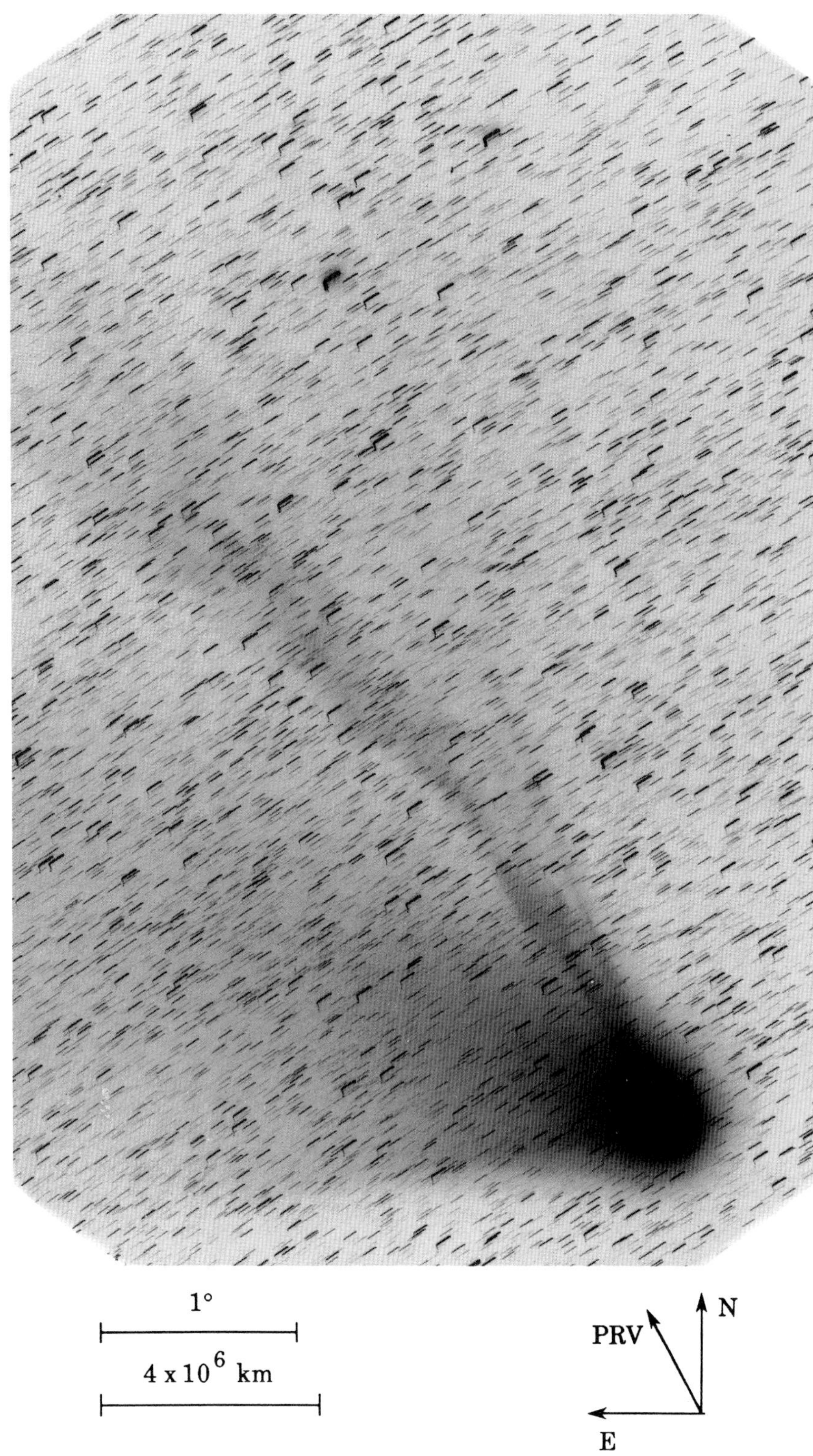

Fig. 635: 1986 Apr 17.28889 UT; exposure 28.0 minutes on Tech. Pan 2415 emulsion with no filter; r=1.43, Δ=0.47, β=159.0°; LSPN-1495.
Photograph by W. Liller, Easter Island, LSPN Island Network.

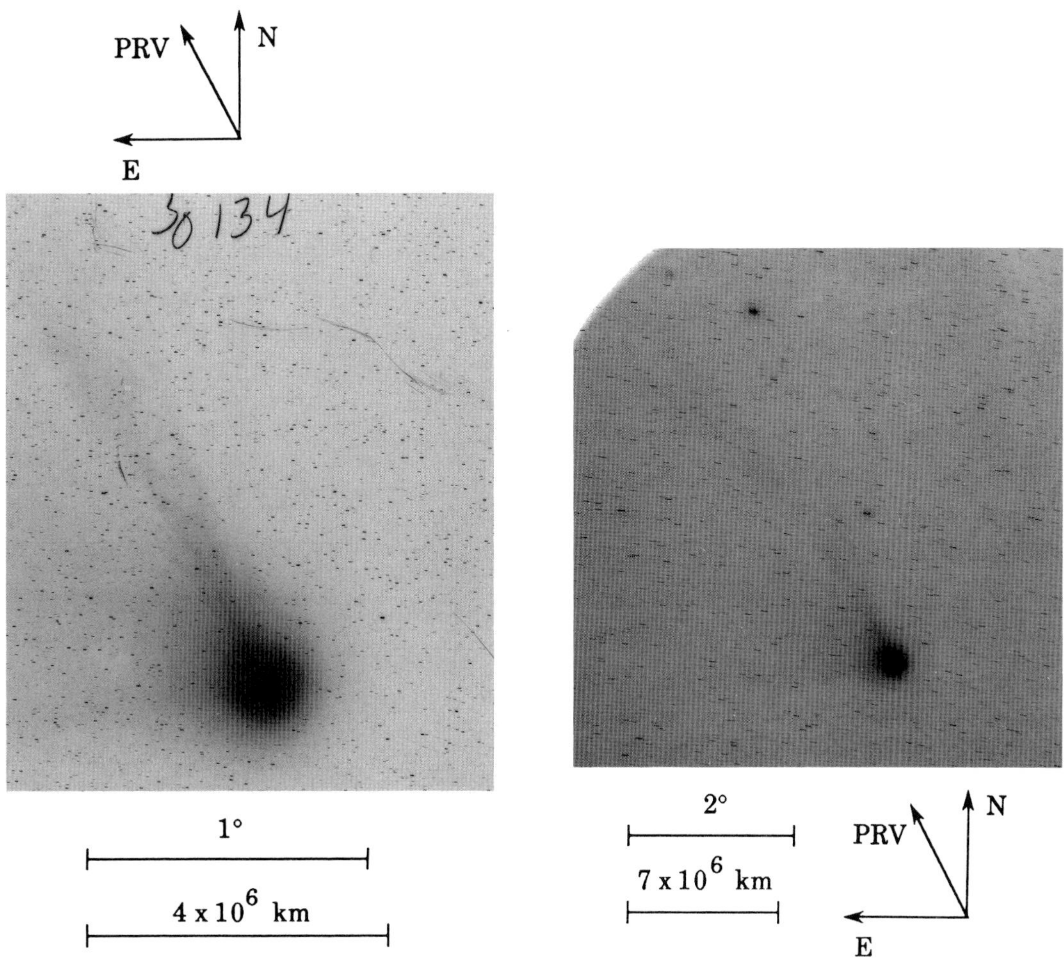

(left) Fig. 636: 1986 Apr 17.29688 UT; exposure 5.0 minutes on 103a-O emulsion with no filter; r=1.43, Δ=0.47, β=159.0°; LSPN-1248.
Photograph by F. Miller, University of Michigan/CTIO.

(right) Fig. 637: 1986 Apr 17.31250 UT; exposure 30.0 minutes on IIIa-J emulsion with no filter; r=1.43, Δ=0.47, β=159.0°; LSPN-1925.
Photograph by A. Storrs, Mauna Kea Observatory.

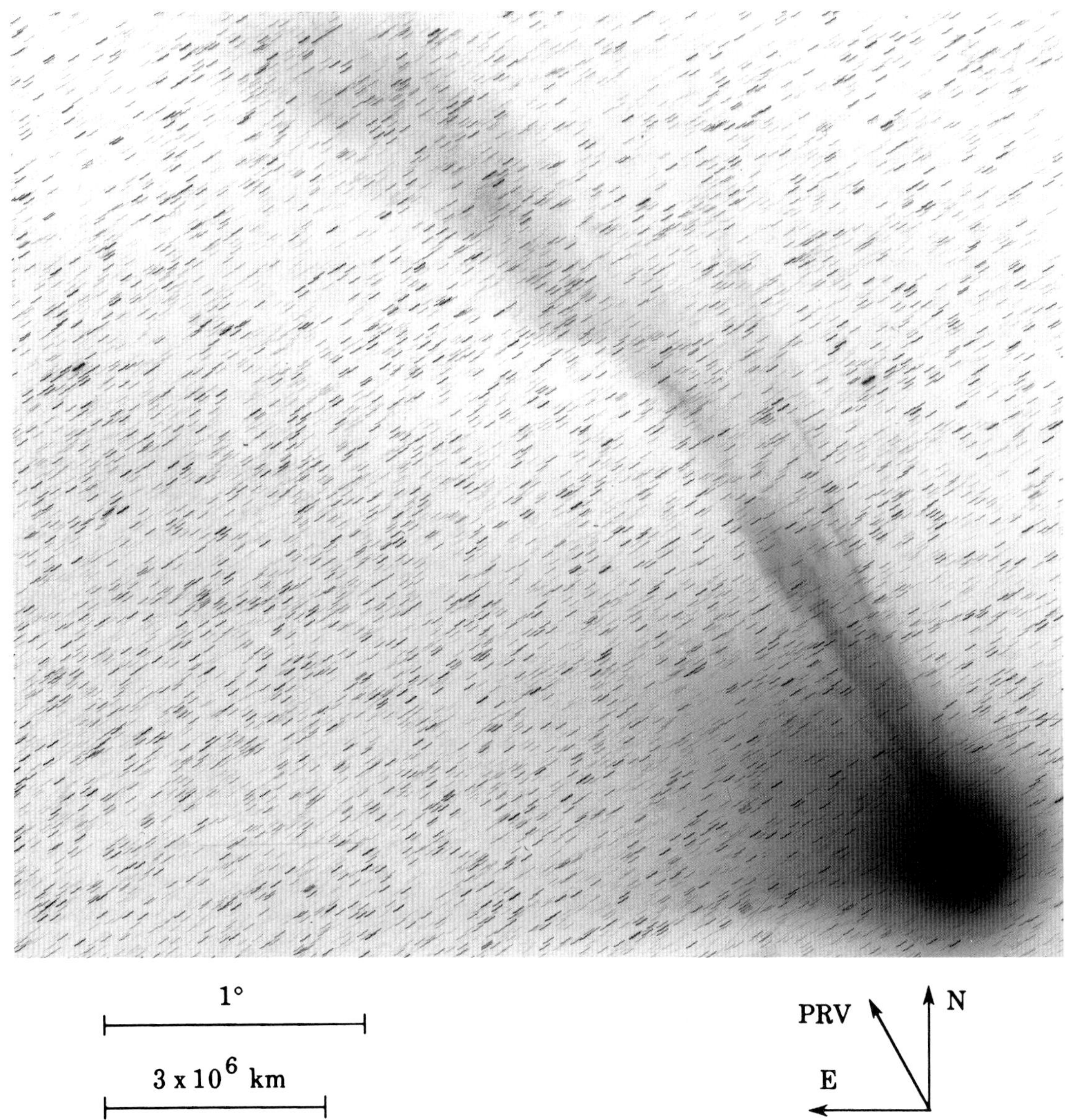

Fig. 638: 1986 Apr 17.31597 UT; exposure 20.0 minutes on 103a-O emulsion with no filter; r=1.43, Δ=0.47, β=159.0°; LSPN-1249.
Photograph by F. Miller, University of Michigan/CTIO.

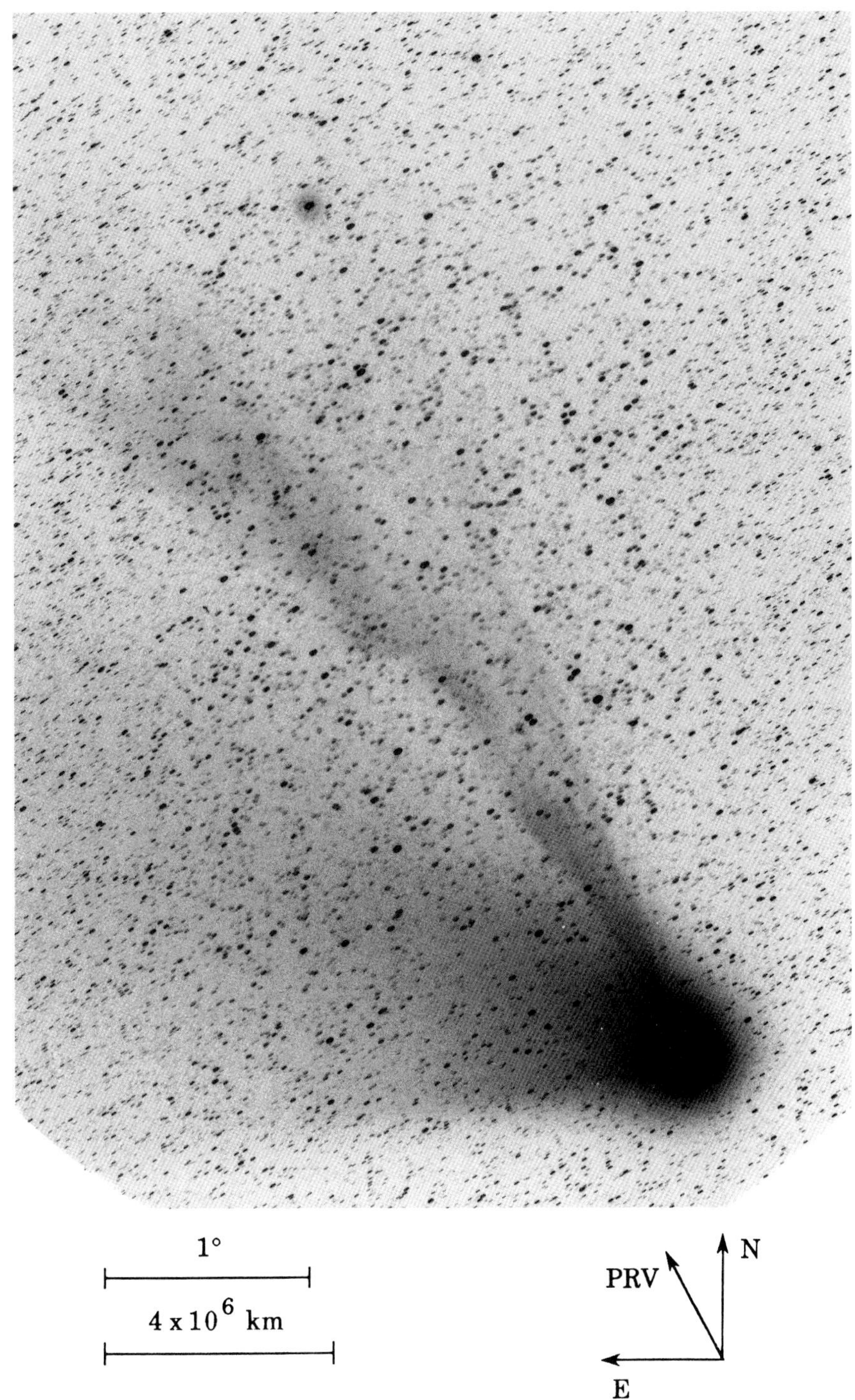

Fig. 639: 1986 Apr 17.34097 UT; exposure 8.0 minutes on Tri-X emulsion with no filter; r=1.43, Δ=0.47, β=159.1°; LSPN-1496.
Photograph by W. Liller, Easter Island, LSPN Island Network.

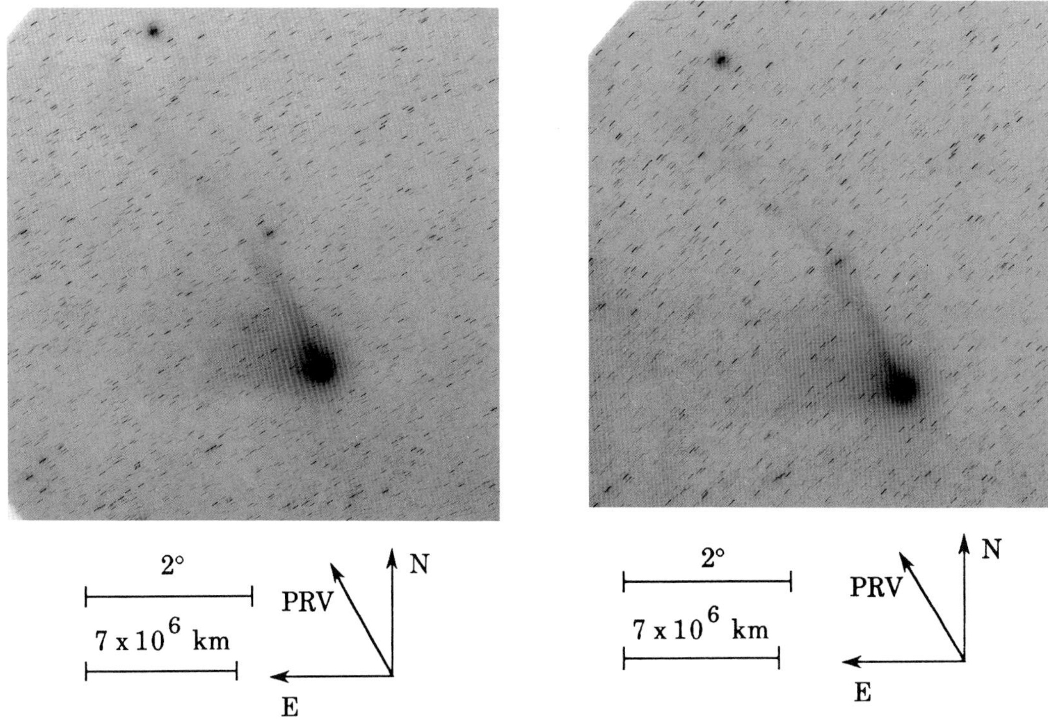

(left) Fig. 640: 1986 Apr 17.40347 UT; exposure 30.0 minutes on IIIa-J emulsion with no filter; r=1.43, Δ=0.47, β=159.1°; LSPN-1926.
Photograph by A. Storrs, Mauna Kea Observatory.

(right) Fig. 641: 1986 Apr 17.45278 UT; exposure 30.0 minutes on IIIa-J emulsion with no filter; r=1.43, Δ=0.47, β=159.1°; LSPN-1927.
Photograph by A. Storrs, Mauna Kea Observatory.

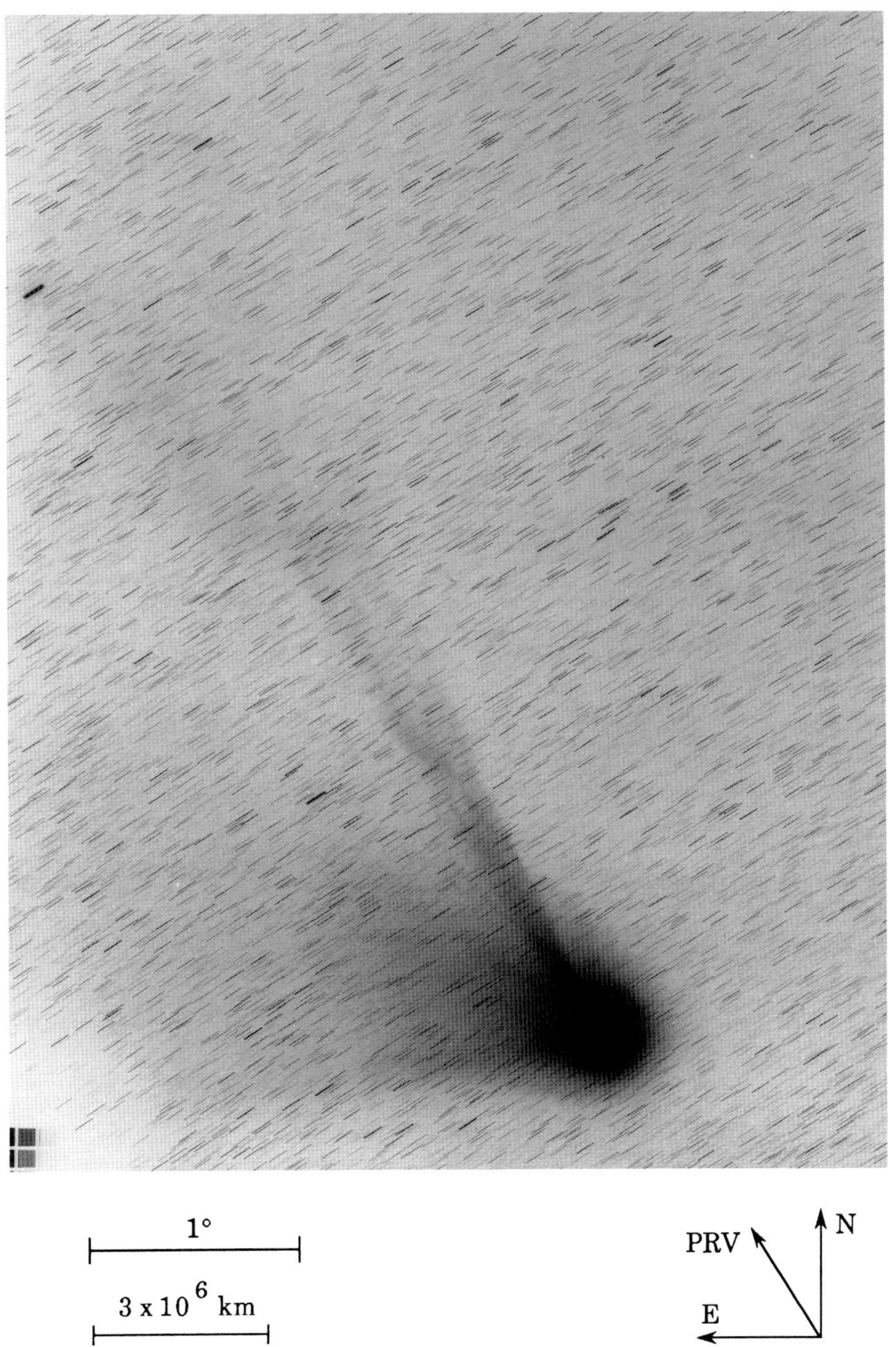

Fig. 642: 1986 Apr 17.65617 UT; exposure 40.0 minutes on IIa-O emulsion with GG-395 filter; r=1.43, Δ=0.48, β=159.1°; LSPN-2441.
Photograph from Royal Observatory/UK Schmidt Telescope Unit.

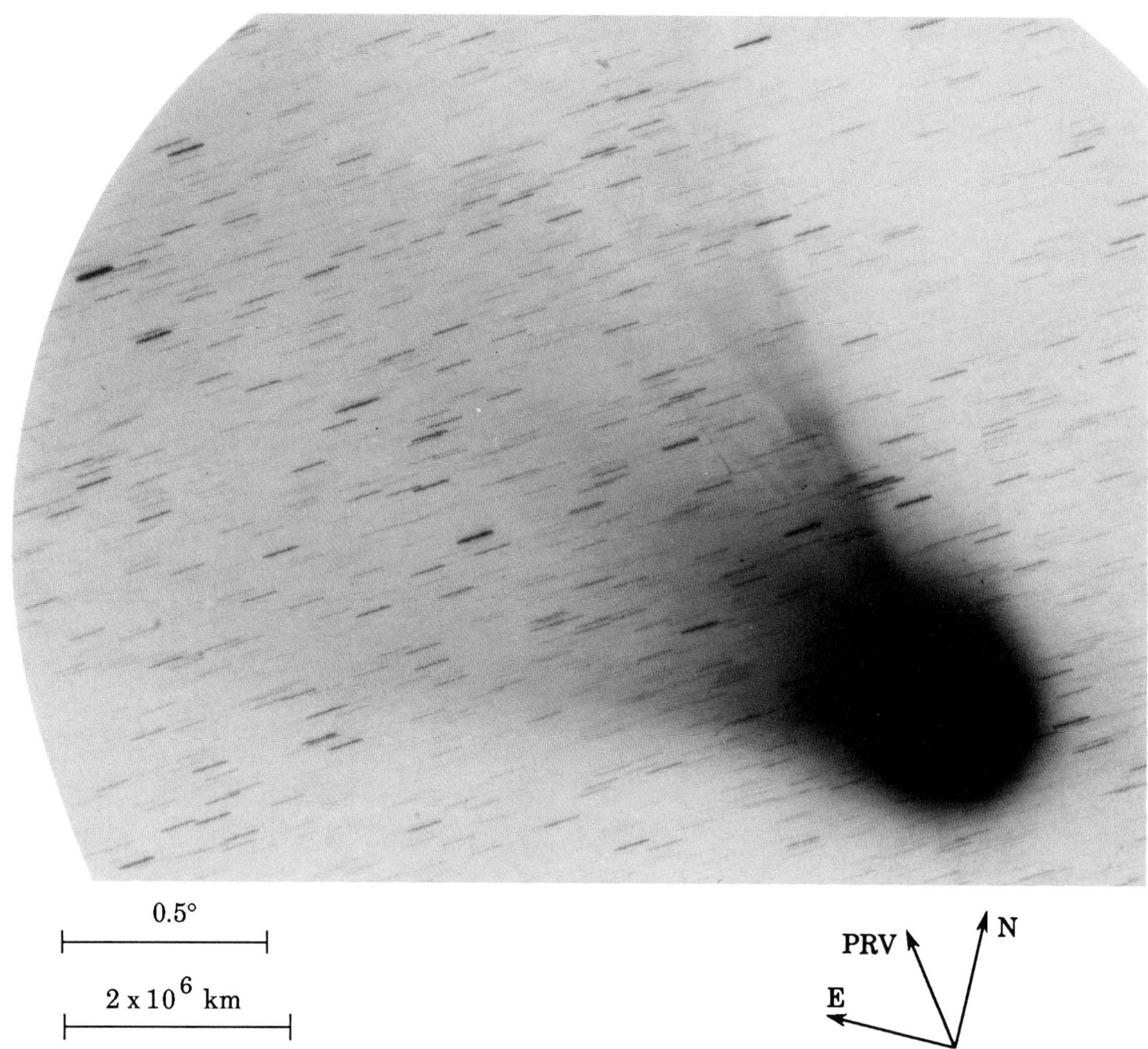

Fig. 643: 1986 Apr 17.96007 UT; exposure 35.0 minutes on Tech. Pan 2415 emulsion with no filter; r=1.44, Δ=0.48, β=159.2°; LSPN-3504.
Photograph by G. Roberts, University of Perugia/South African Astronomical Observatory.

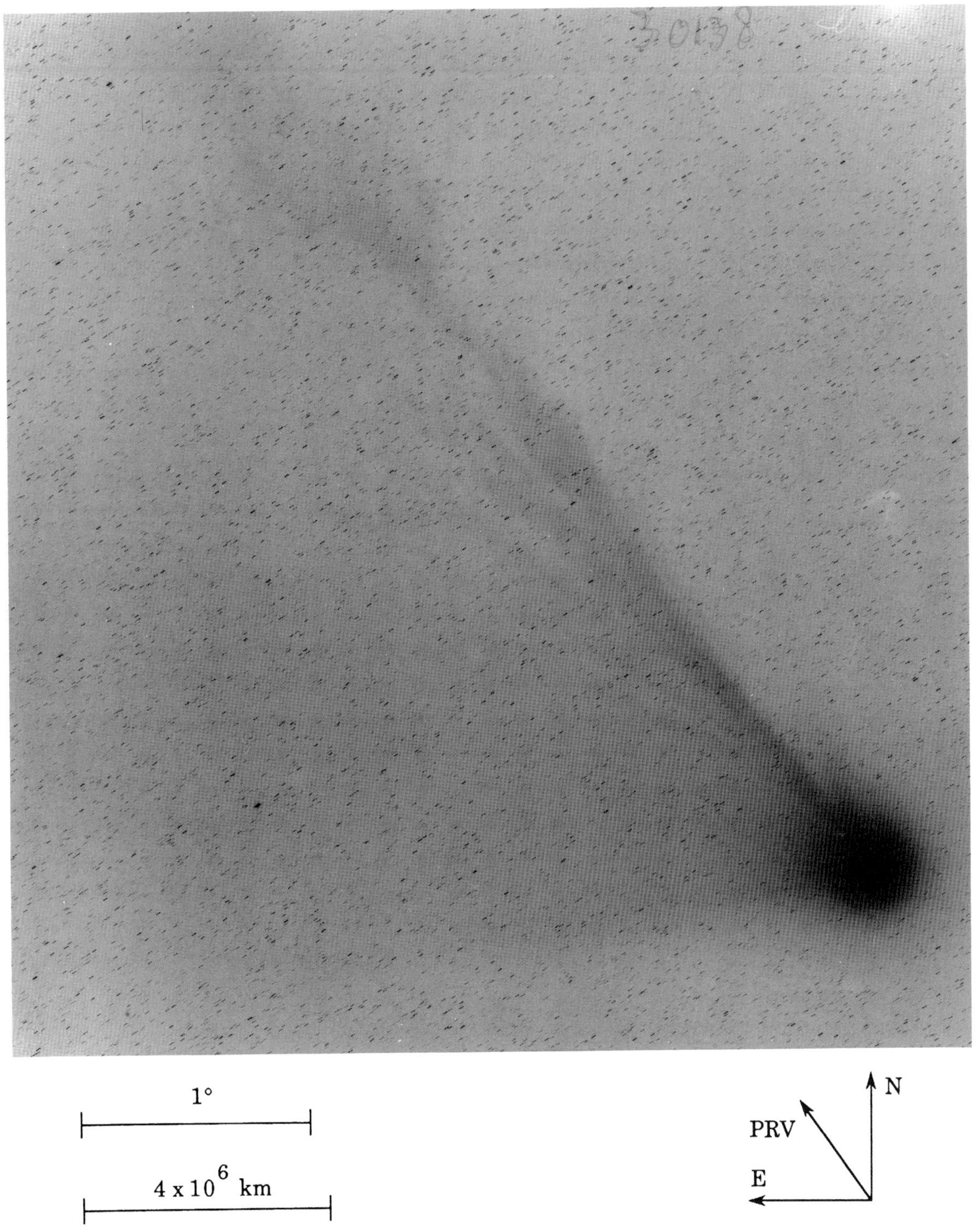

Fig. 644: 1986 Apr 18.08611 UT; exposure 10.0 minutes on 103a-O emulsion with no filter; r=1.44, Δ=0.48, β=159.2°; LSPN-1252.
Photograph by F. Miller, University of Michigan/CTIO.

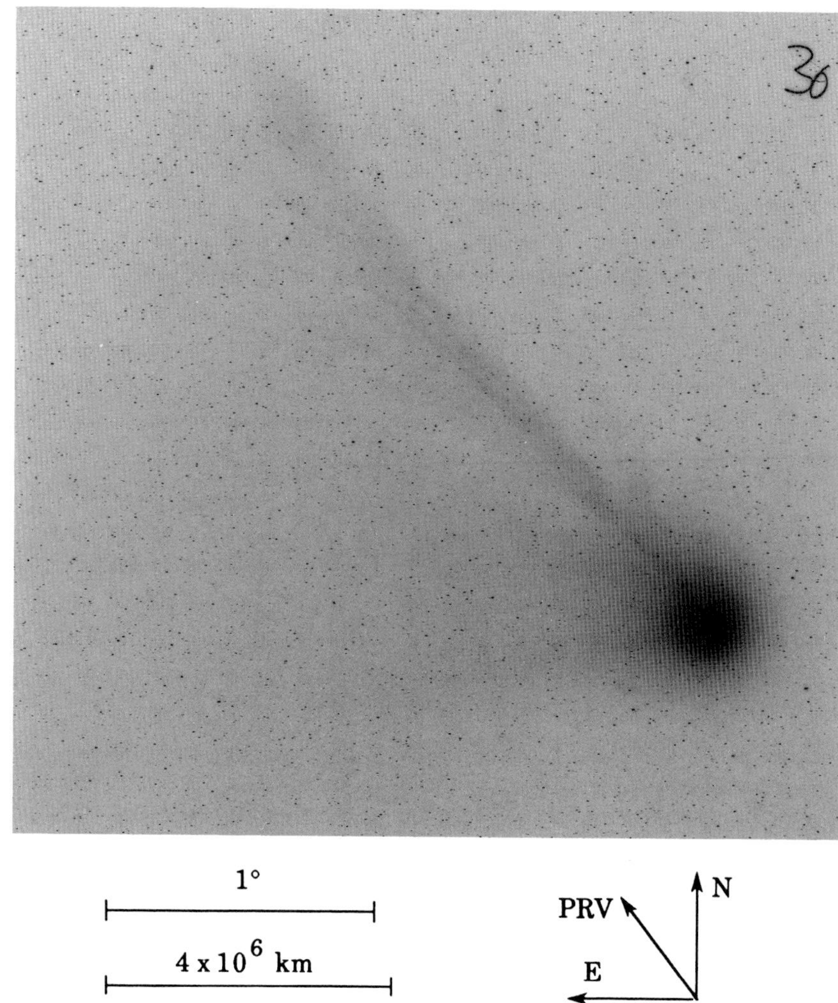

Fig. 645: 1986 Apr 18.15625 UT; exposure 4.0 minutes on 103a-O emulsion with no filter; r=1.44, Δ=0.48, β=159.2°; LSPN-1253.
Photograph by F. Miller, University of Michigan/CTIO.

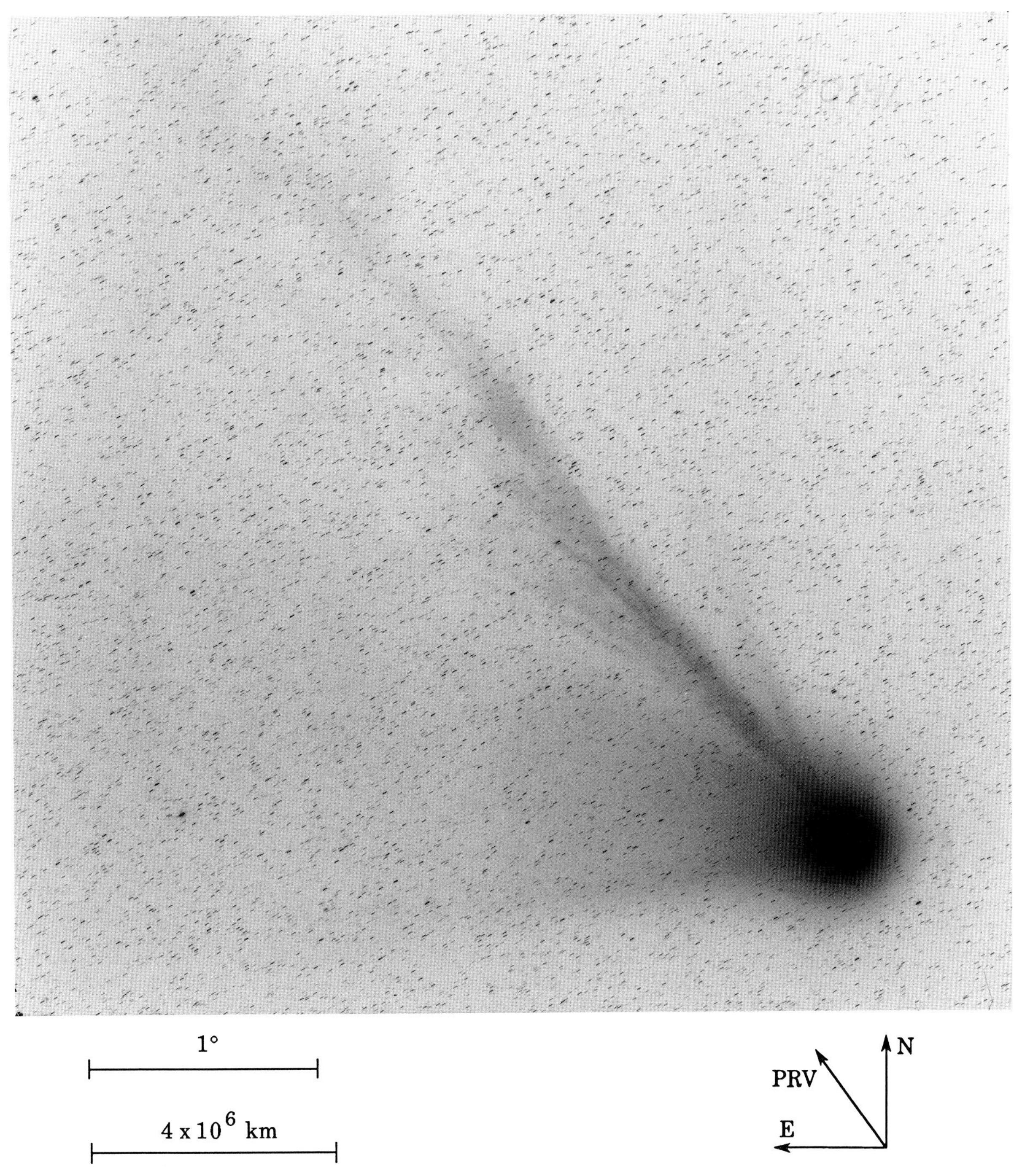

Fig. 646: 1986 Apr 18.16667 UT; exposure 10.0 minutes on 103a-O emulsion with no filter; r=1.44, Δ=0.48, β=159.1°; LSPN-1254.
Photograph by F. Miller, University of Michigan/CTIO.

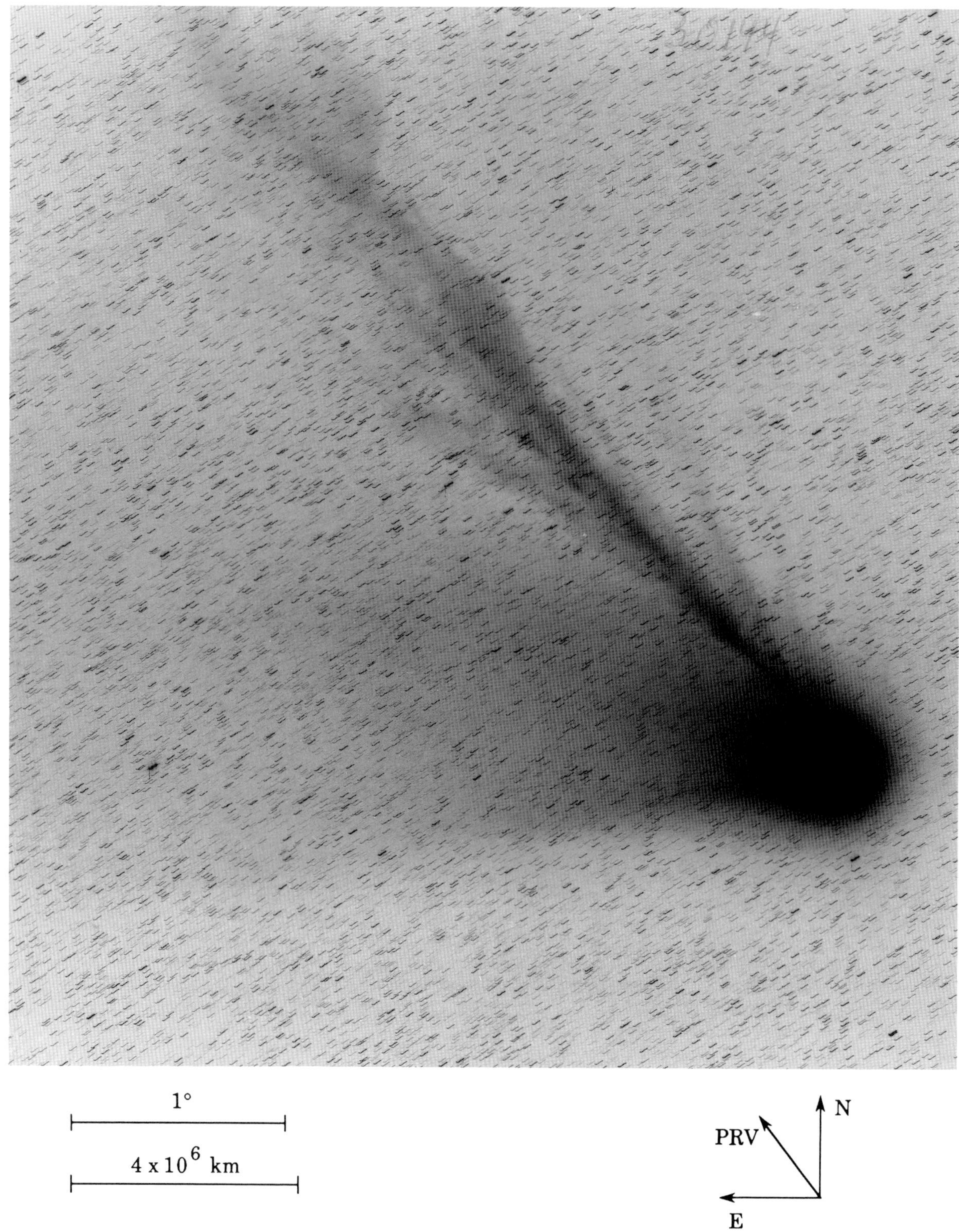

Fig. 647: 1986 Apr 18.24792 UT; exposure 20.0 minutes on 103a-O emulsion with no filter; r=1.44, Δ=0.49, β=159.1°; LSPN-1256.
Photograph by F. Miller, University of Michigan/CTIO.

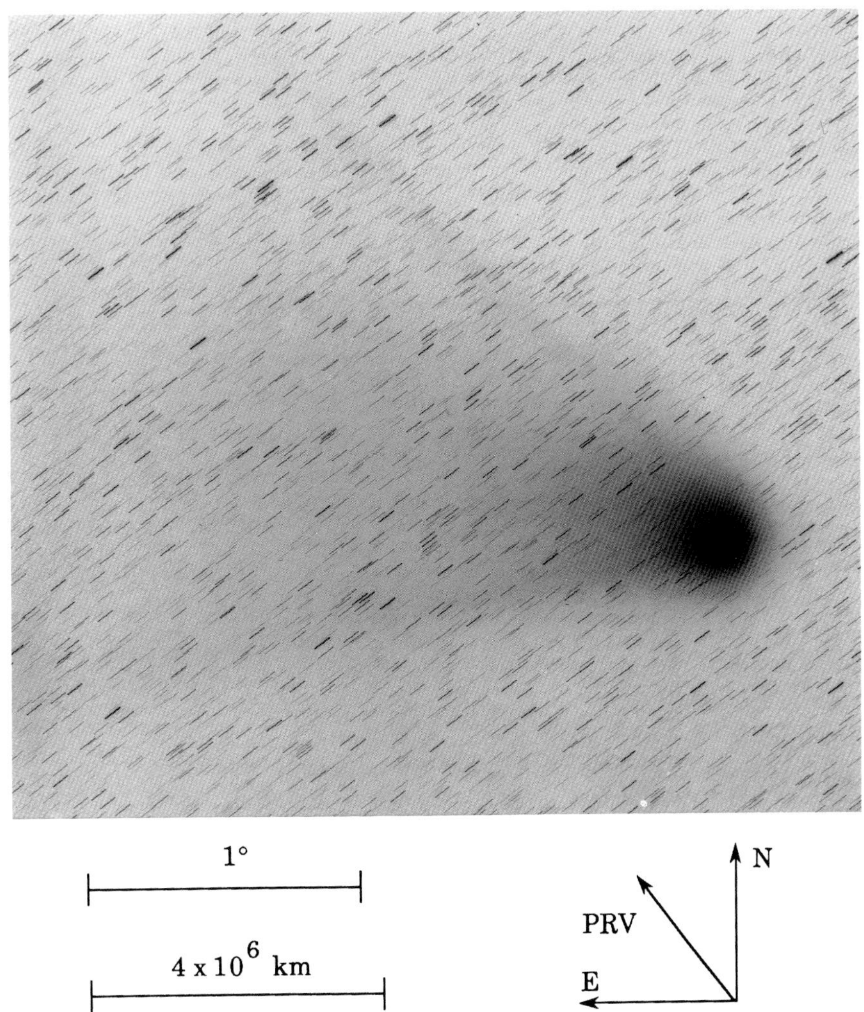

Fig. 648: 1986 Apr 18.31389 UT; exposure 30.0 minutes on 098-04 emulsion with RG-610 filter; r=1.44, Δ=0.49, β=159.1°; LSPN-1258.
Photograph by F. Miller, University of Michigan/CTIO.

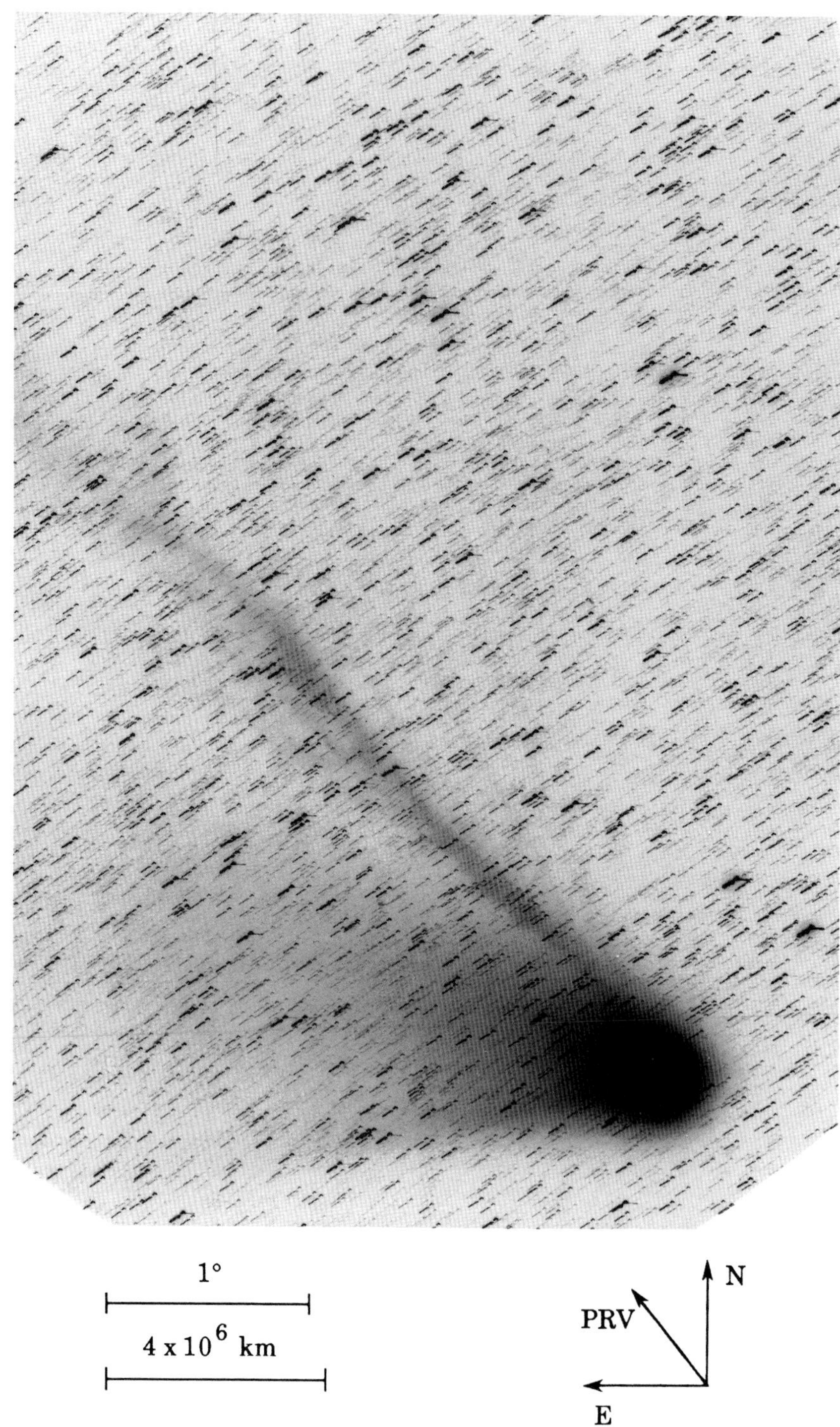

Fig. 649: 1986 Apr 18.35278 UT; exposure 30.0 minutes on Tech. Pan 2415 emulsion with no filter; r=1.44, Δ=0.49, β=159.1°; LSPN-1500.
Photograph by W. Liller, Easter Island, LSPN Island Network.

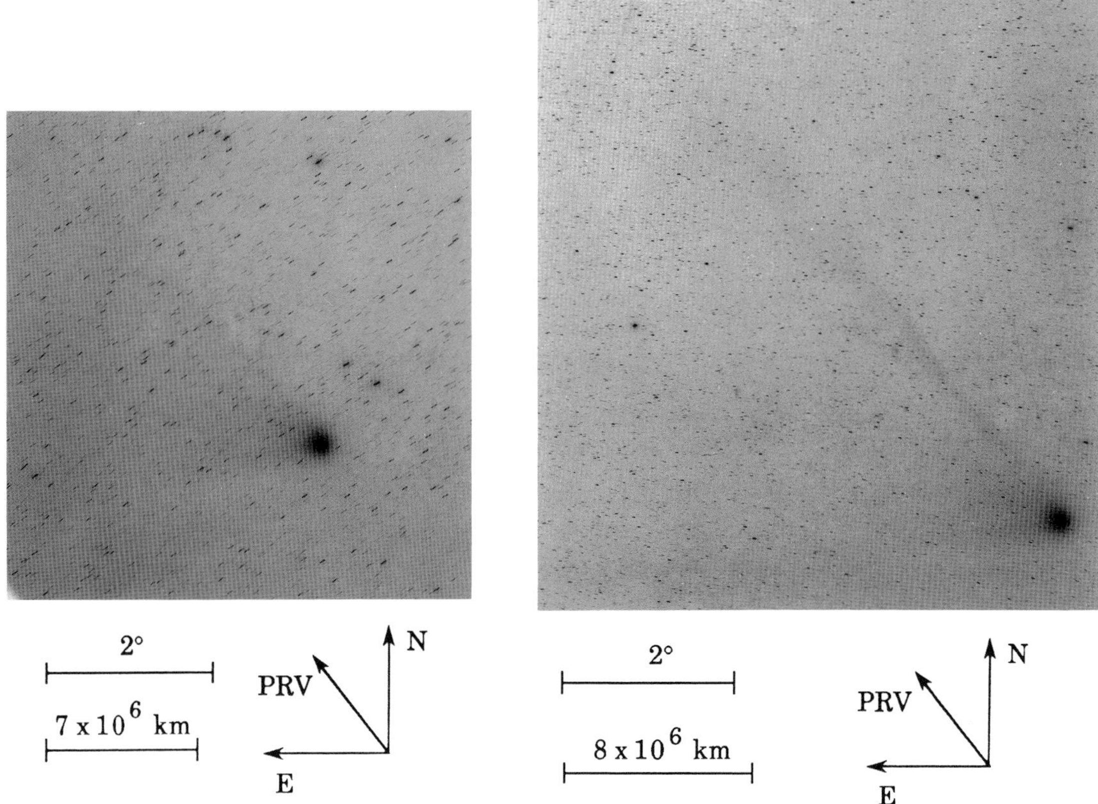

(left) Fig. 650: 1986 Apr 18.35625 UT; exposure 30.0 minutes on IIIa-F emulsion with Wratten 2B filter; r=1.44, Δ=0.49, β=159.1°; LSPN-1933.
Photograph by A. Storrs, Mauna Kea Observatory.

(right) Fig. 651*: 1986 Apr 18.37917 UT; exposure 10.0 minutes on IIa-O emulsion with no filter; r=1.44, Δ=0.49, β=159.1°; LSPN-927.
Photograph by E. Moore/J. Brandt, Joint Observatory for Cometary Research.

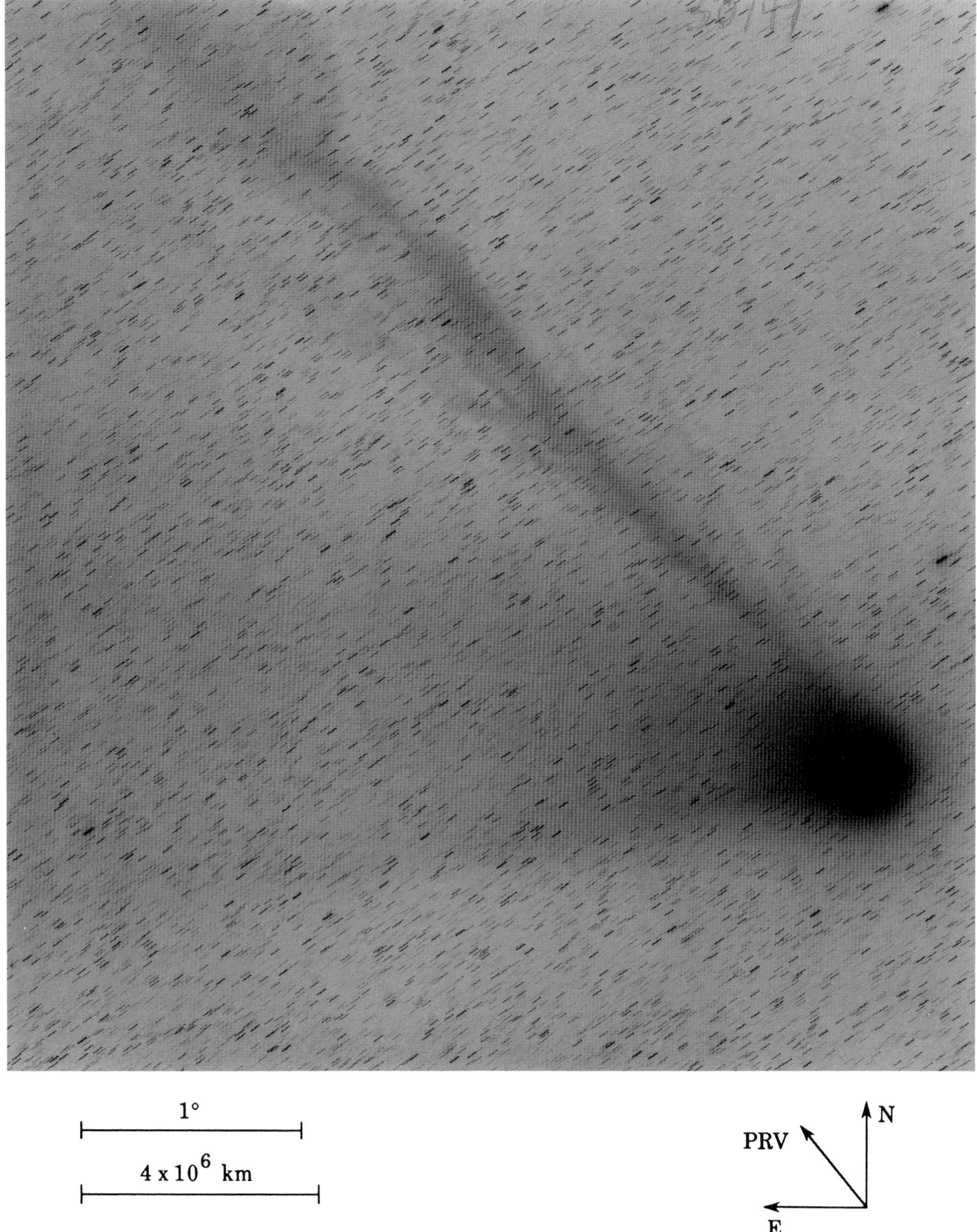

Fig. 652: 1986 Apr 18.35764 UT; exposure 20.0 minutes on 103a-O emulsion with no filter; r=1.44, Δ=0.49, β=159.1°; LSPN-1259.
Photograph by F. Miller, University of Michigan/CTIO.

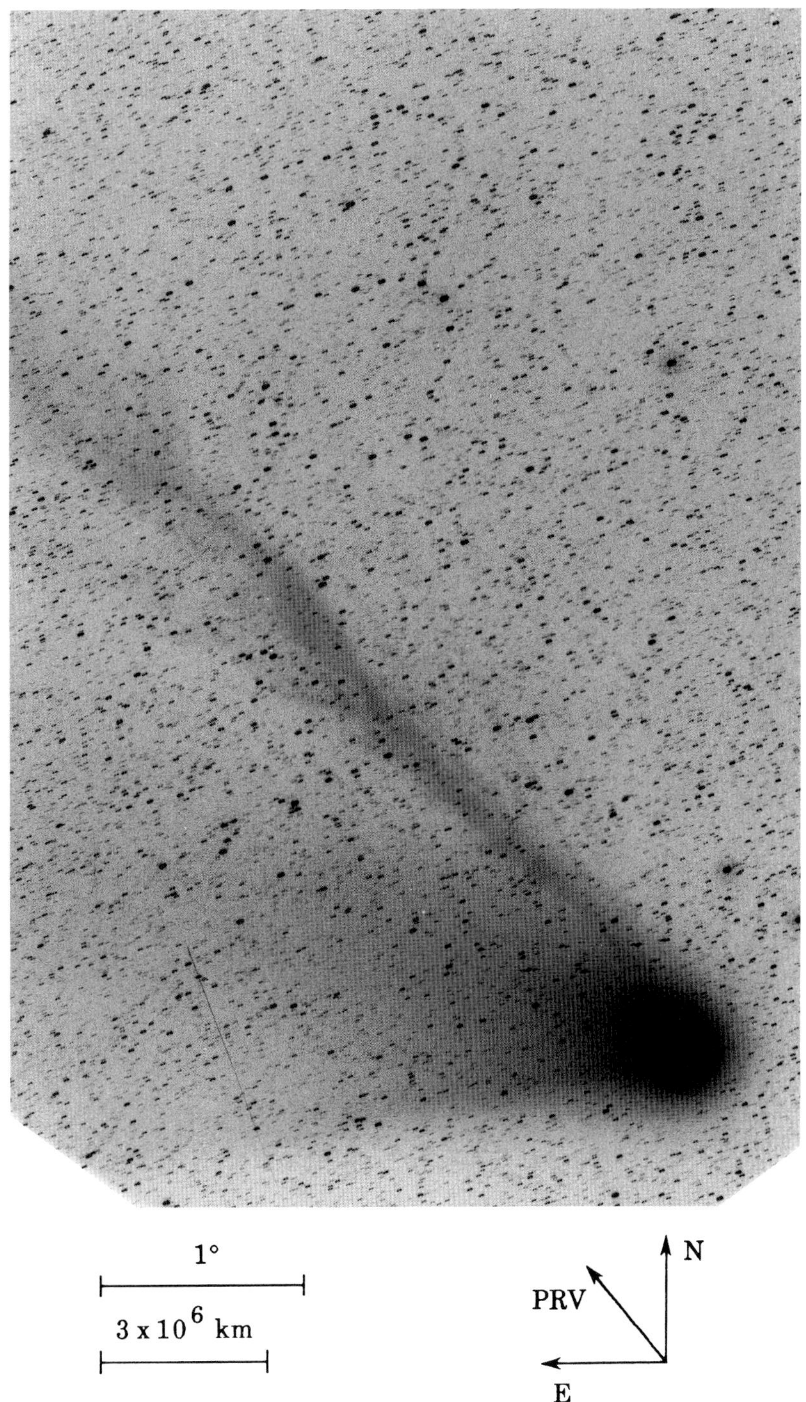

Fig. 653: 1986 Apr 18.39549 UT; exposure 7.0 minutes on Tri-X emulsion with no filter; r=1.44, Δ=0.49, β=159.1°; LSPN-1501.
Photograph by W. Liller, Easter Island, LSPN Island Network.

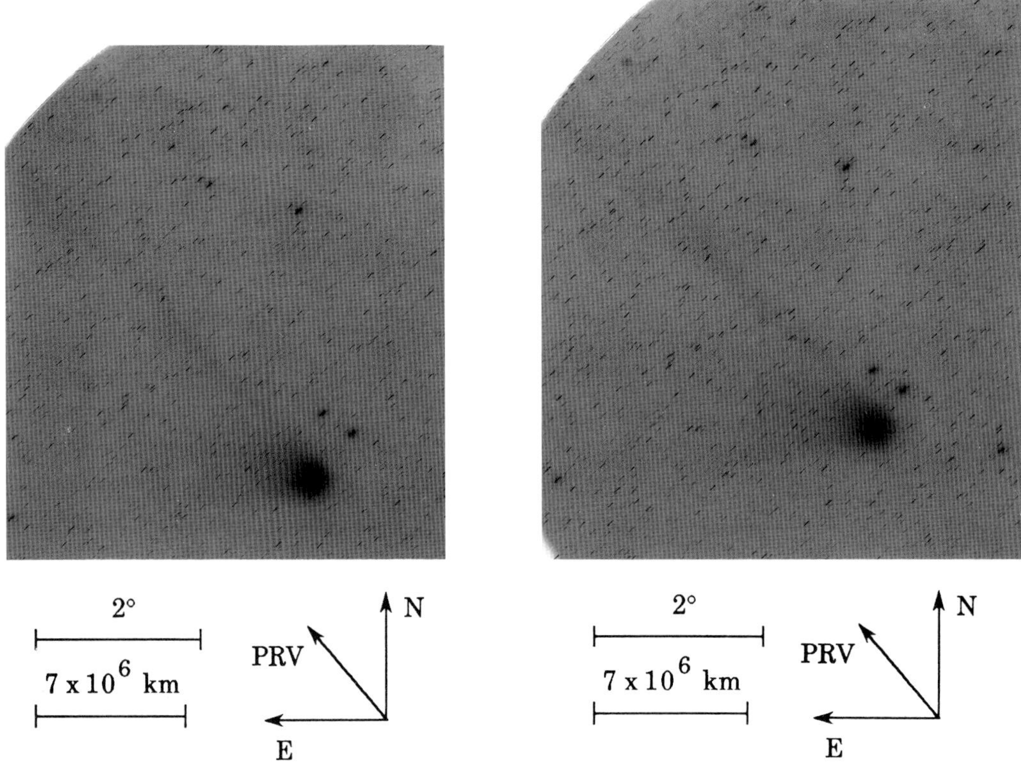

(left) Fig. 654: 1986 Apr 18.43472 UT; exposure 30.0 minutes on IIIa-J emulsion with no filter; $r=1.45$, $\Delta=0.49$, $\beta=159.1°$; LSPN-1929.
Photograph by A. Storrs, Mauna Kea Observatory.

(right) Fig. 655*: 1986 Apr 18.48565 UT; exposure 30.0 minutes on IIIa-J emulsion with no filter; $r=1.45$, $\Delta=0.49$, $\beta=159.1°$; LSPN-1931.
Photograph by A. Storrs, Mauna Kea Observatory.

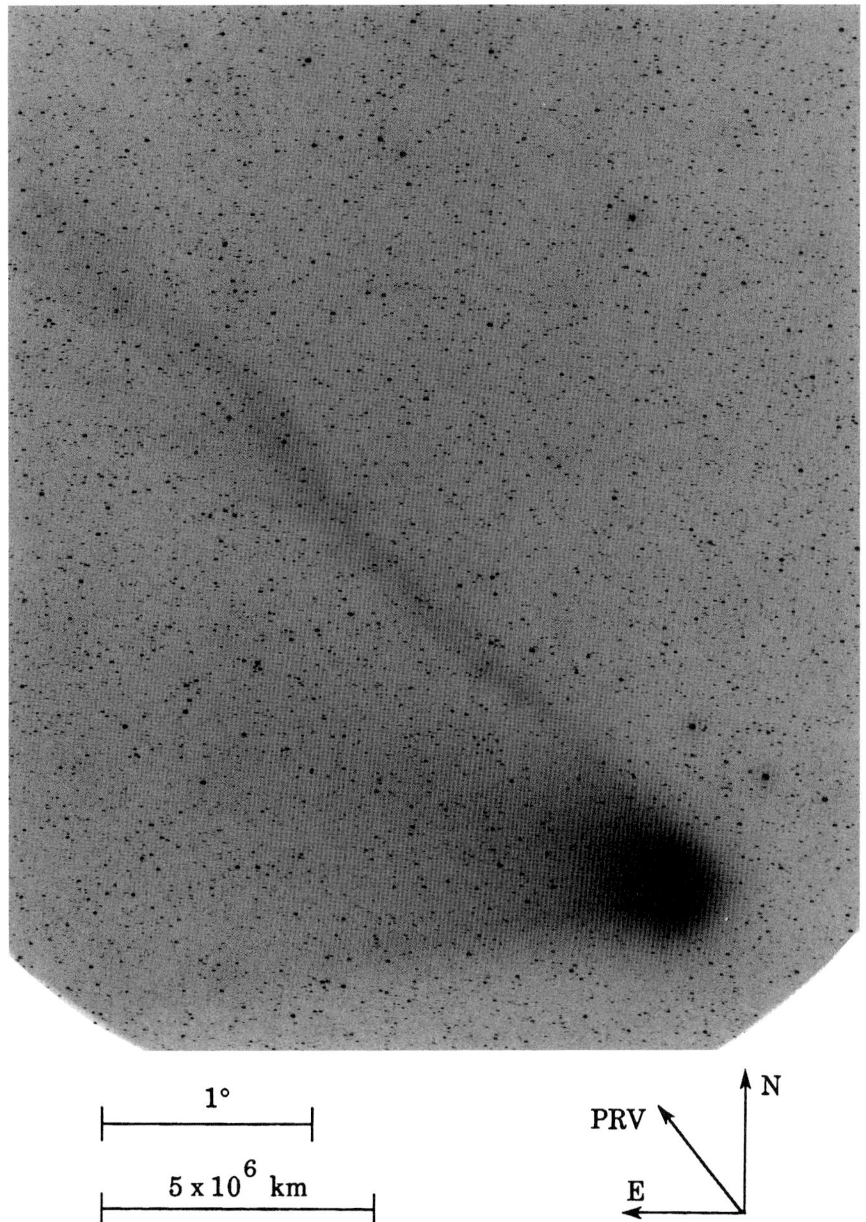

Fig. 656: 1986 Apr 18.44444 UT; exposure 6.0 minutes on Tri-X emulsion with no filter; r=1.45, Δ=0.49, β=159.1°; LSPN-1502.
Photograph by W. Liller, Easter Island, LSPN Island Network.

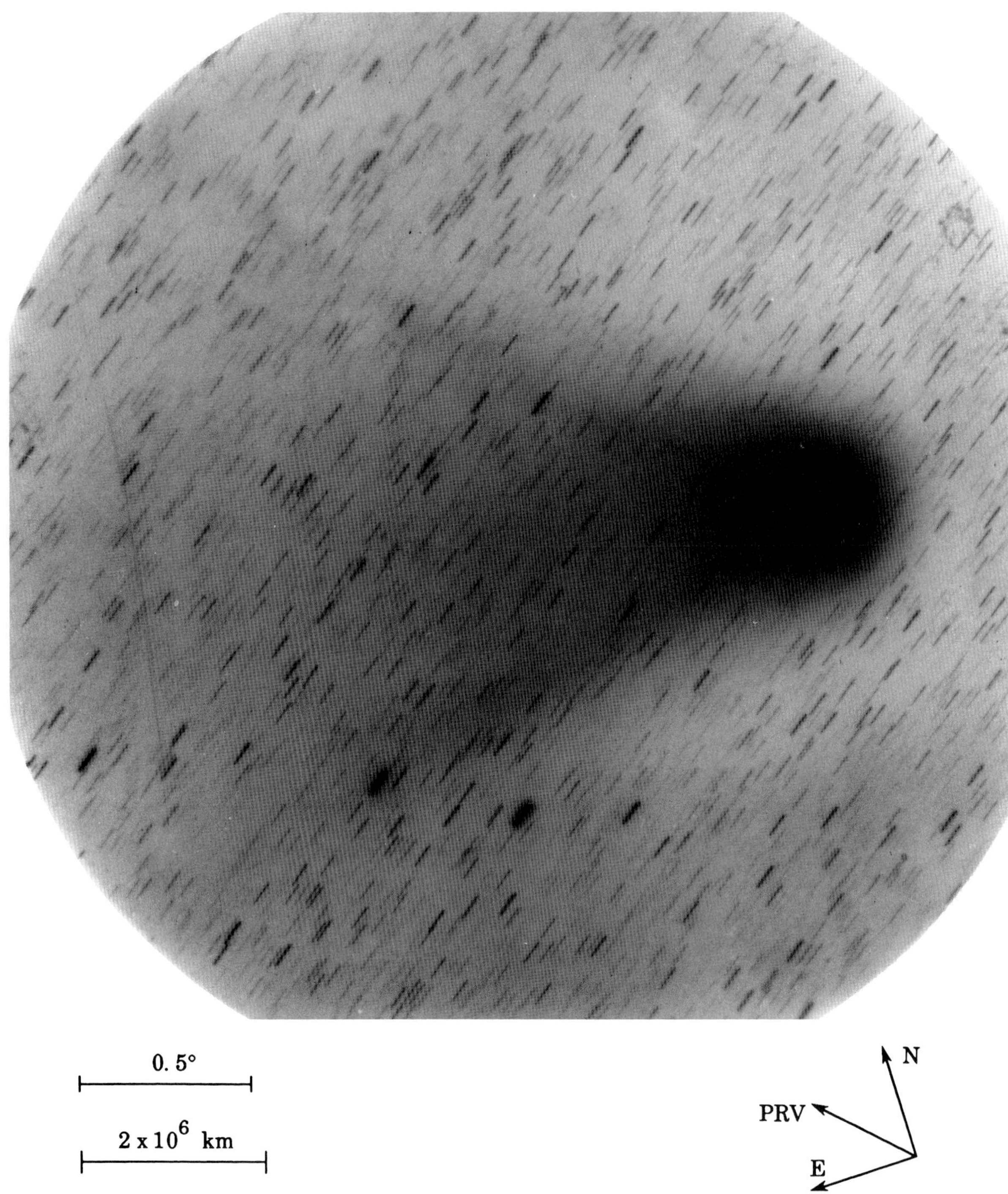

Fig. 657: 1986 Apr 19.00995 UT; exposure 20.6 minutes on Tech. Pan 2415 emulsion with no filter; r=1.45, Δ=0.50, β=159.0°; LSPN-3505.
Photograph by G. Roberts, University of Perugia/South African Astronomical Observatory.

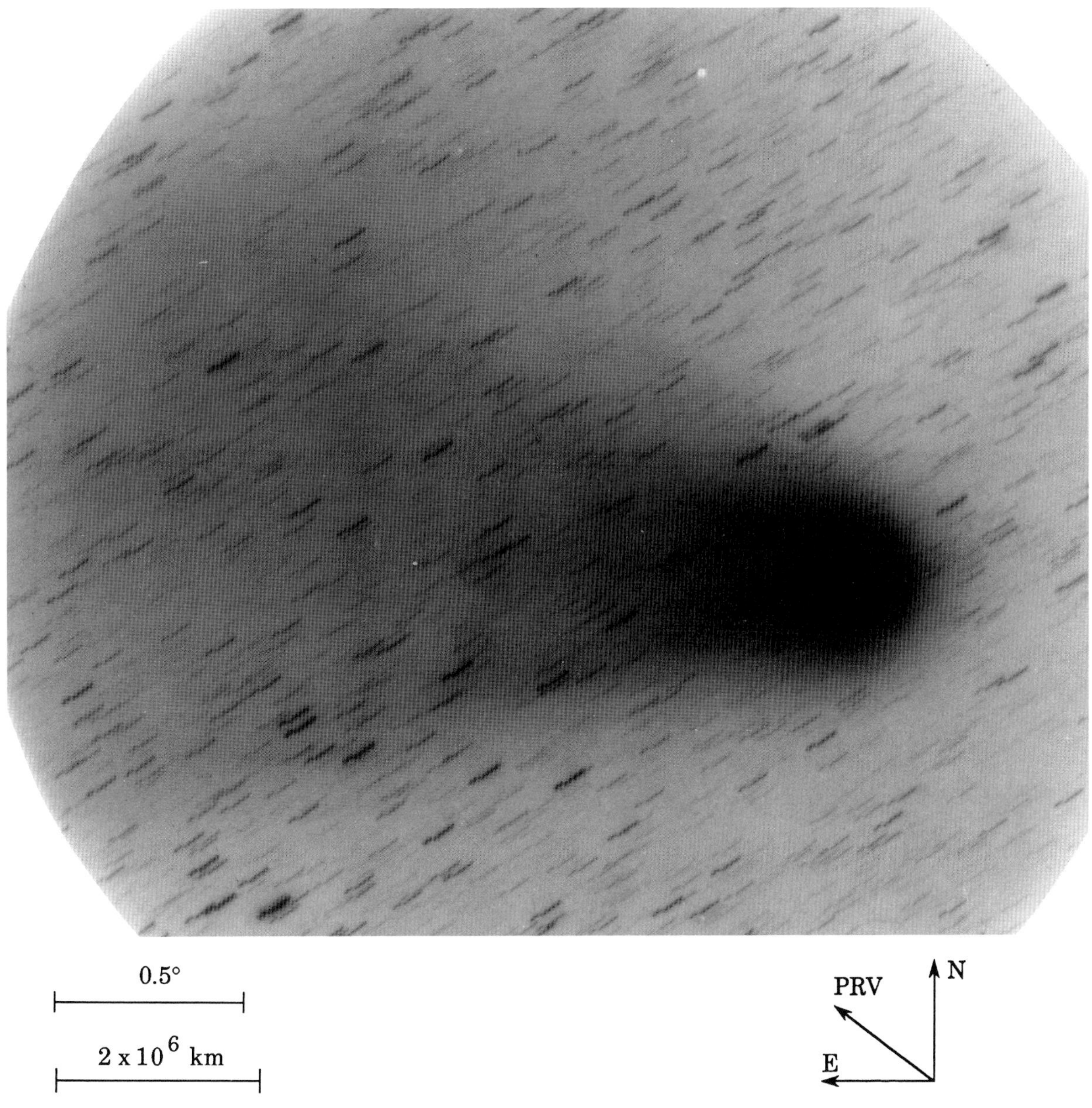

Fig. 658: 1986 Apr 20.04167 UT; exposure 40.0 minutes on Tech. Pan 2415 emulsion with no filter; r=1.47, Δ=0.52, β=158.6°; LSPN-3506.
Photograph by G. Roberts, University of Perugia/South African Astronomical Observatory.

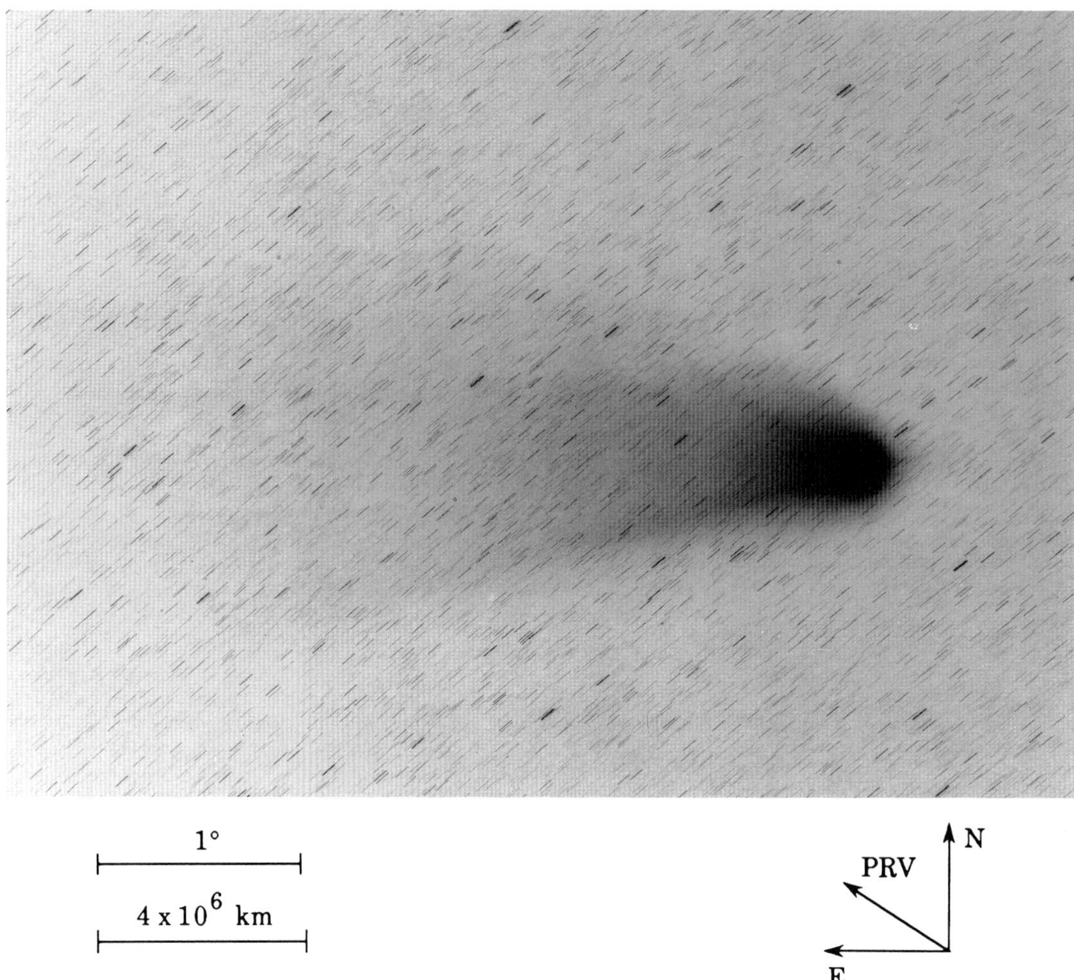

Fig. 659: 1986 Apr 20.70891 UT; exposure 40.0 minutes on IIIa-F emulsion with RG-630 filter; r=1.48, Δ=0.53, β=158.2°; LSPN-2445.
Photograph from Royal Observatory/UK Schmidt Telescope Unit.

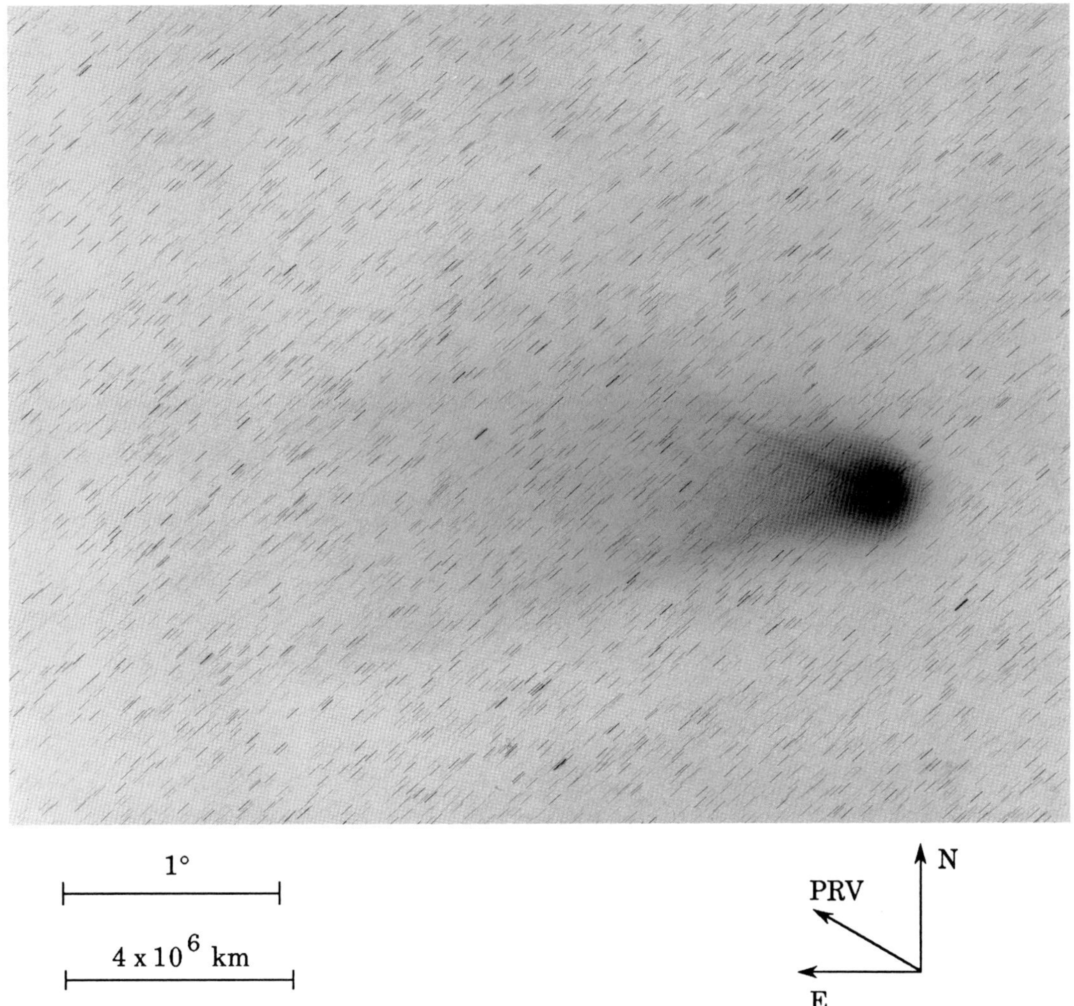

Fig. 660: 1986 Apr 20.75116 UT; exposure 40.0 minutes on IIa-O emulsion with GG-395 filter; r=1.48, Δ=0.53, β=158.1°; LSPN-2446.
Photograph from Royal Observatory/UK Schmidt Telescope Unit.

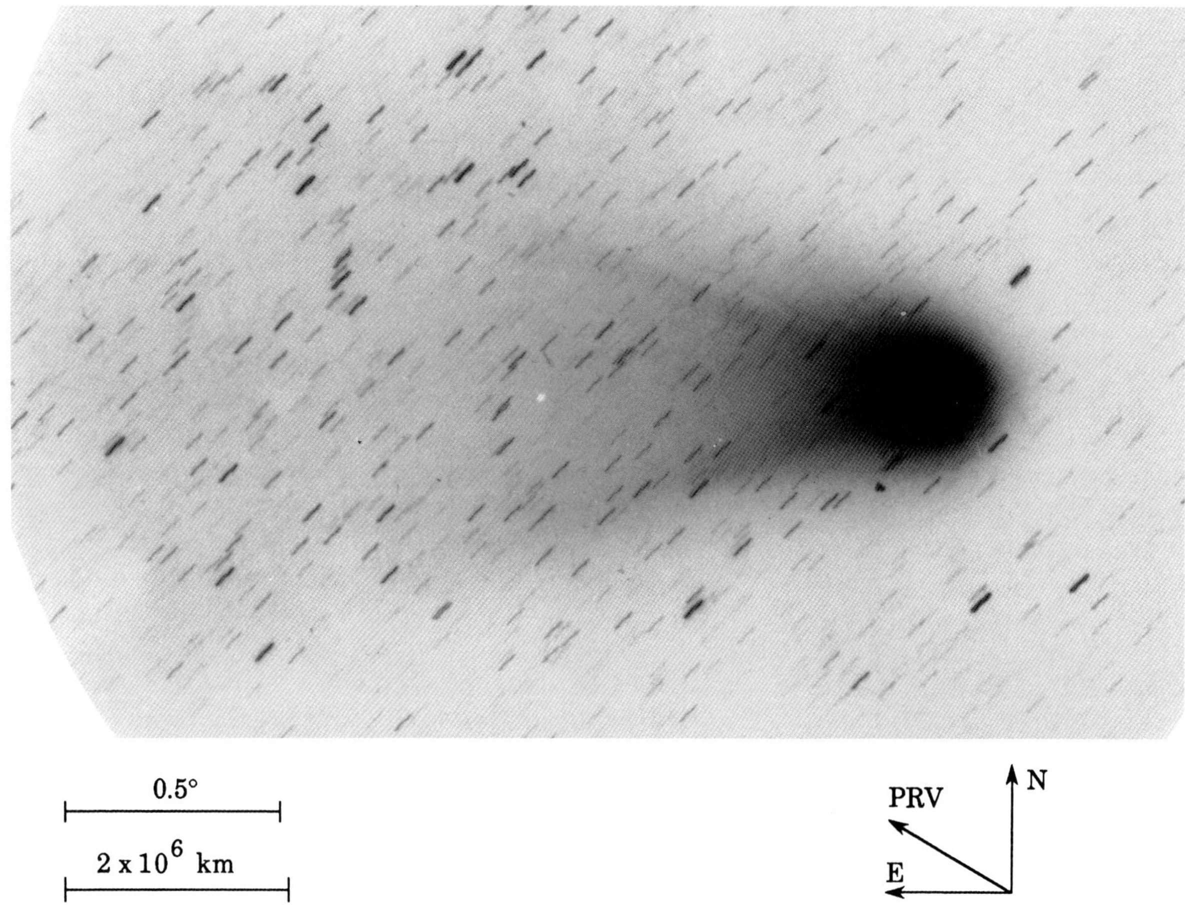

Fig. 661: 1986 Apr 21.08125 UT; exposure 30.0 minutes on Tech. Pan 2415 emulsion with no filter; $r = 1.49$, $\Delta = 0.54$, $\beta = 157.9°$; LSPN-3507.
Photograph by G. Roberts, University of Perugia/South African Astronomical Observatory.

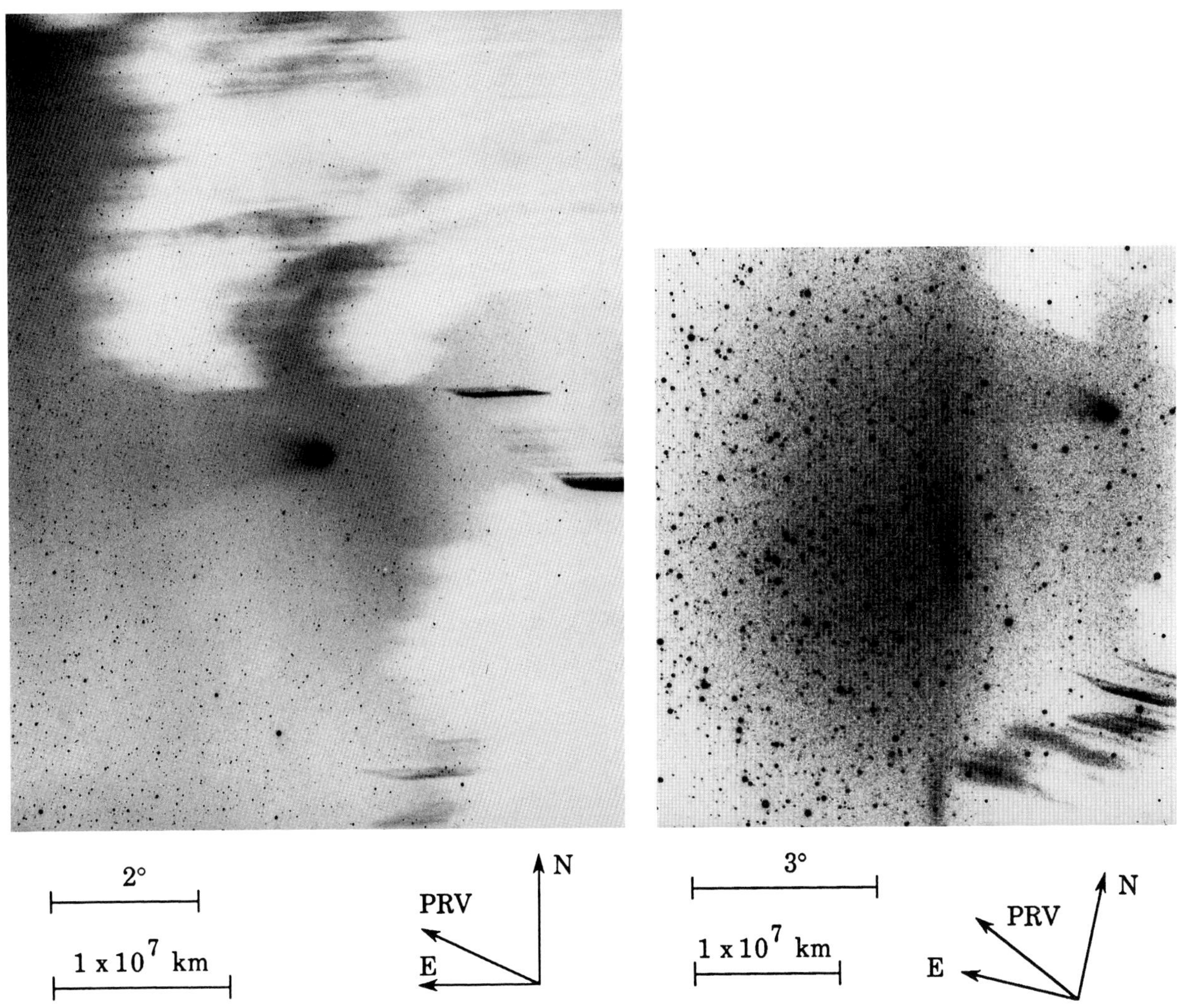

(left) Fig. 662: 1986 Apr 21.74980 UT; exposure 6.0 minutes on Fujichrome emulsion; r=1.50, Δ=0.56, β=157.4°; AON-851910.
Photograph by R. Royer, USA

(right) Fig. 663: 1986 Apr 21.74980 UT; exposure 6.0 minutes on Kodak 098 emulsion; r=1.50, Δ=0.56, β=157.4°; AON-851908.
Photograph by R. Royer, USA.

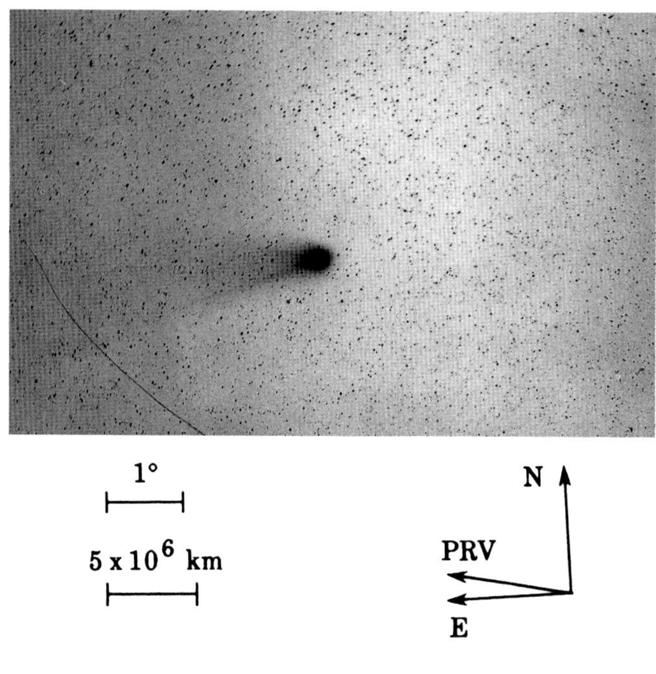

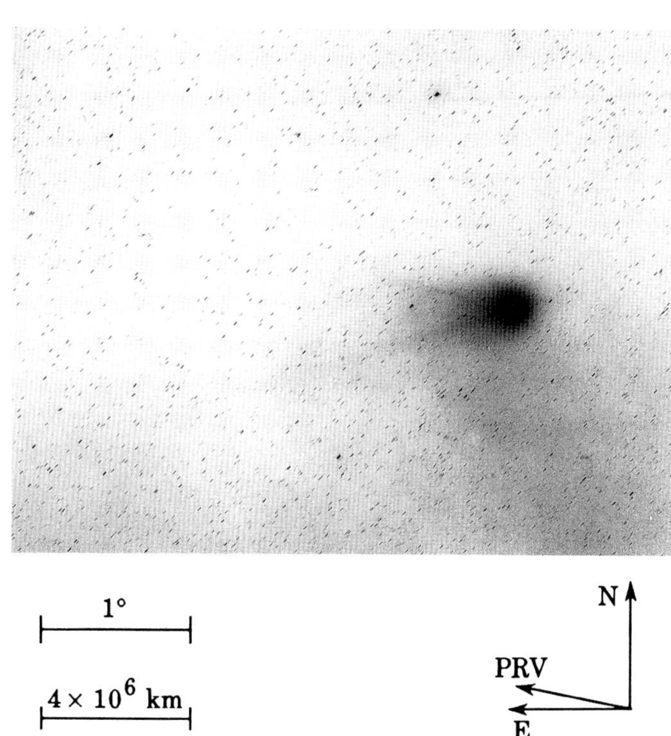

(top) Fig. 664: 1986 Apr 24.53640 UT; exposure 15.0 minutes on IIa-D emulsion; r=1.54, Δ=0.62, β=155.2°; AON-851915.
Photograph by R. Royer, USA.

(bottom) Fig. 665: 1986 Apr 24.54340 UT; exposure 15.0 minutes on 103a-F emulsion with no filter; r=1.54, Δ=0.62, β=155.2°; LSPN-1345.
Photograph by T. Tsujimura, University Kyoto Observatory.

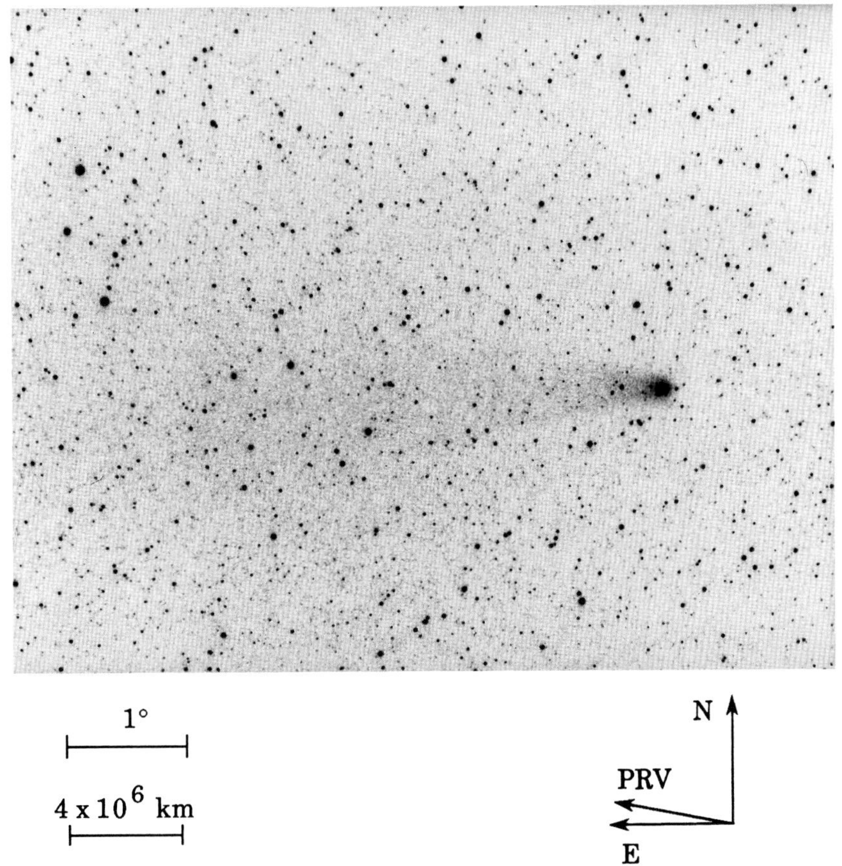

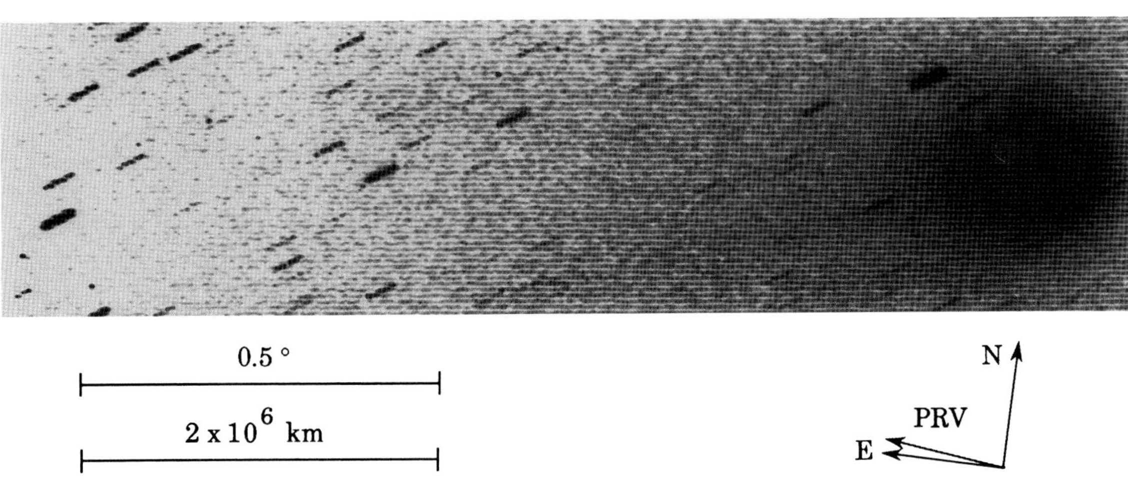

(top) Fig. 666: 1986 Apr 24.55555 UT; exposure 10.0 minutes on 3M 1000 emulsion; r=1.54, Δ=0.62, β=155.2°; AON-851921 (taken during lunar eclipse).
Photograph by G. Robertson, USA.

(bottom) Fig. 667: 1986 Apr 26.68056 UT; exposure 40.0 minutes on ORWO ZU-21 emulsion with no filter; r=1.57, Δ=0.68, β=153.6°; LSPN-3296.
Photograph by S. Gerasimenko, Gissar Observatory.

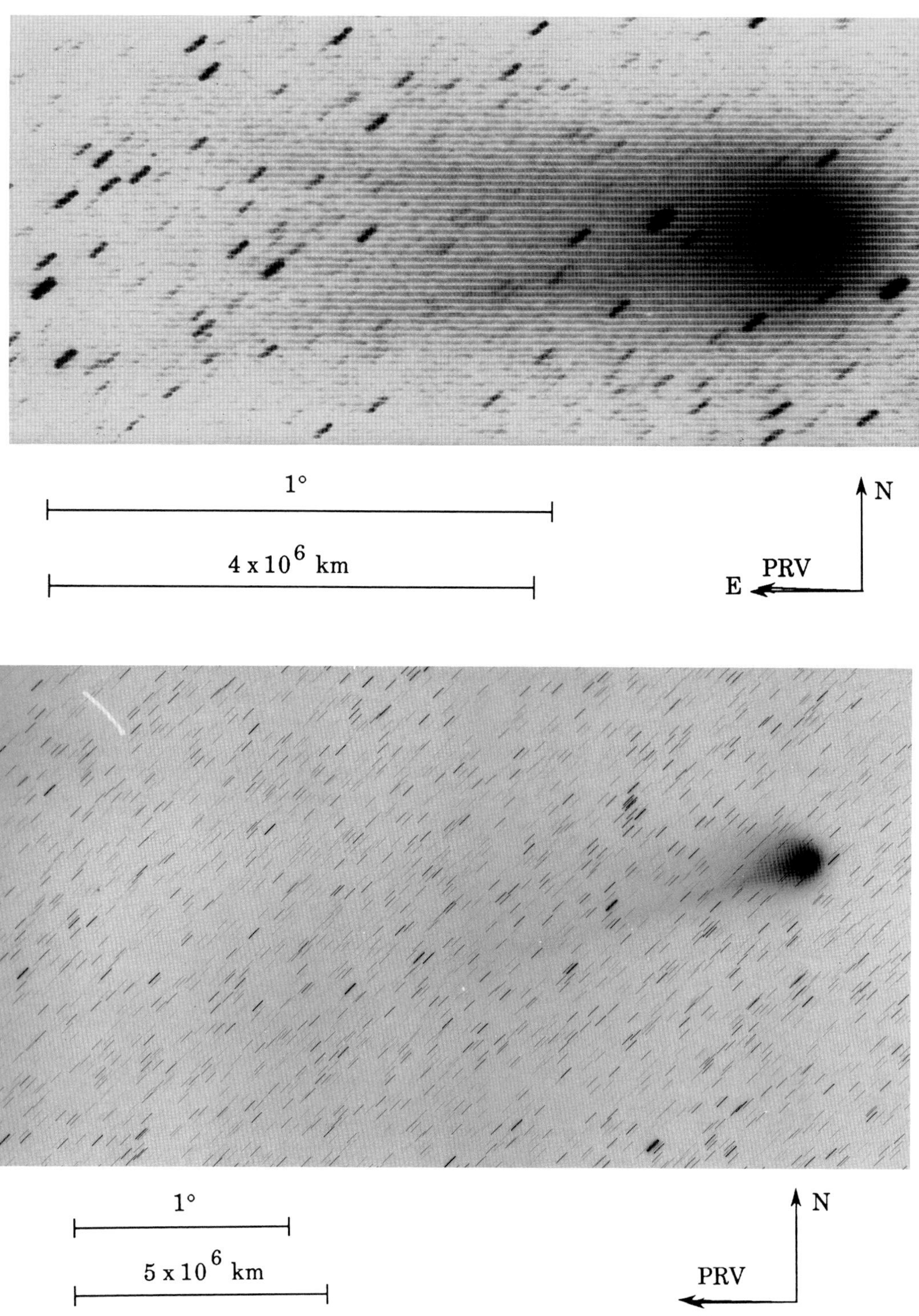

(top) Fig. 668: 1986 Apr 26.78514 UT; exposure 40.0 minutes on ORWO ZU-21 emulsion with no filter; r=1.57, Δ=0.68, β=153.6°; LSPN-3440.
Photograph by N. Chernykh, Crimean Astrophysical Observatory.

(bottom) Fig. 669*: 1986 Apr 27.99306 UT; exposure 20.0 minutes on 098-04 emulsion with RG-630 filter; r=1.59, Δ=0.72, β=152.8°; LSPN-2914.
Photograph by G. Pizarro, European Southern Observatory.

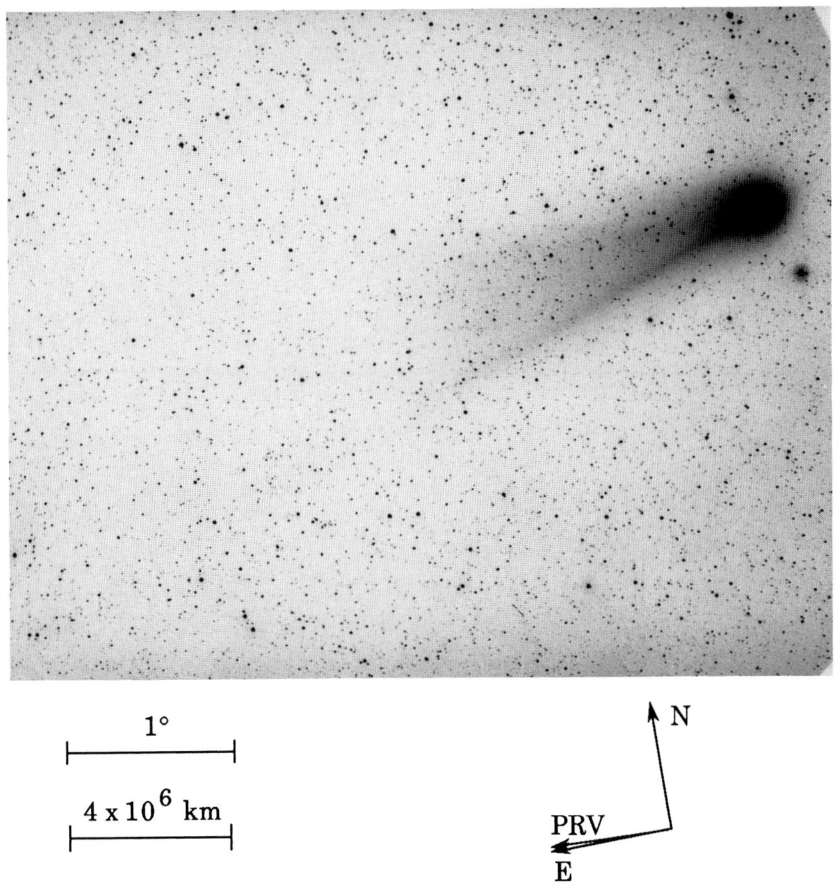

Fig. 670: 1986 Apr 27.22410 UT; exposure 6.0 minutes on Tech. Pan 2415 emulsion; r=1.58, Δ=0.69, β=153.3°; AON-851925.
Photograph by B. Yen, USA.

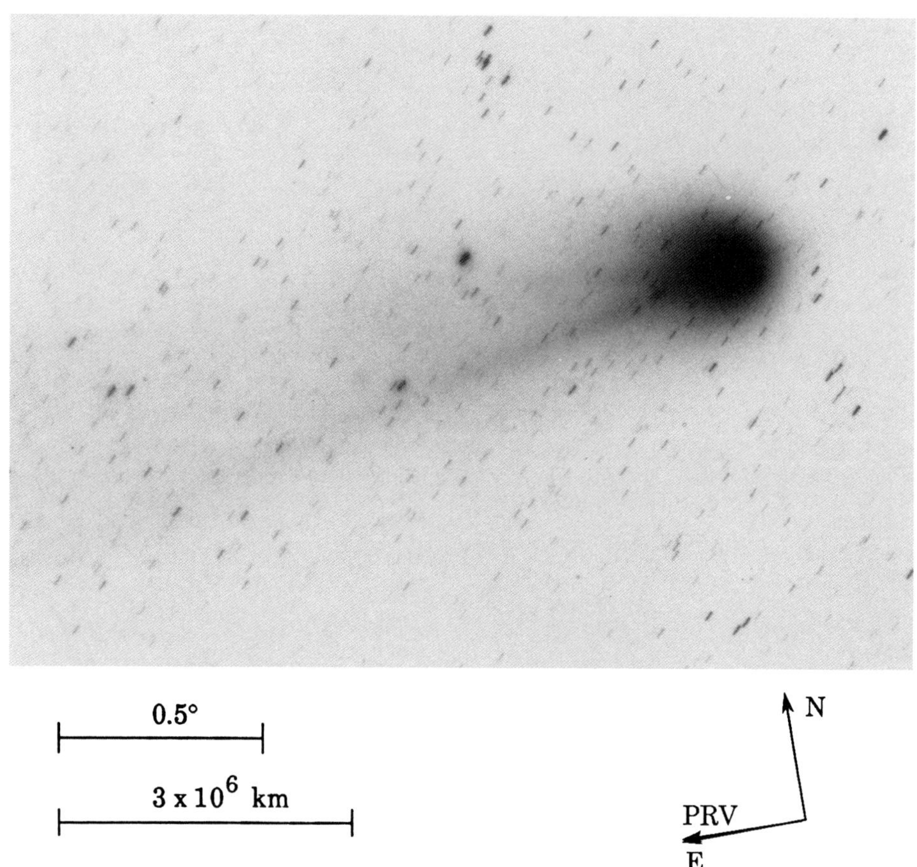

Fig. 671: 1986 Apr 27.74896 UT; exposure 25.0 minutes on Tech. Pan 2415 emulsion with no filter; r=1.59, Δ=0.71, β=152.9°; LSPN-3508.
Photograph by B. Carter, University of Perugia/South African Astronomical Observatory.

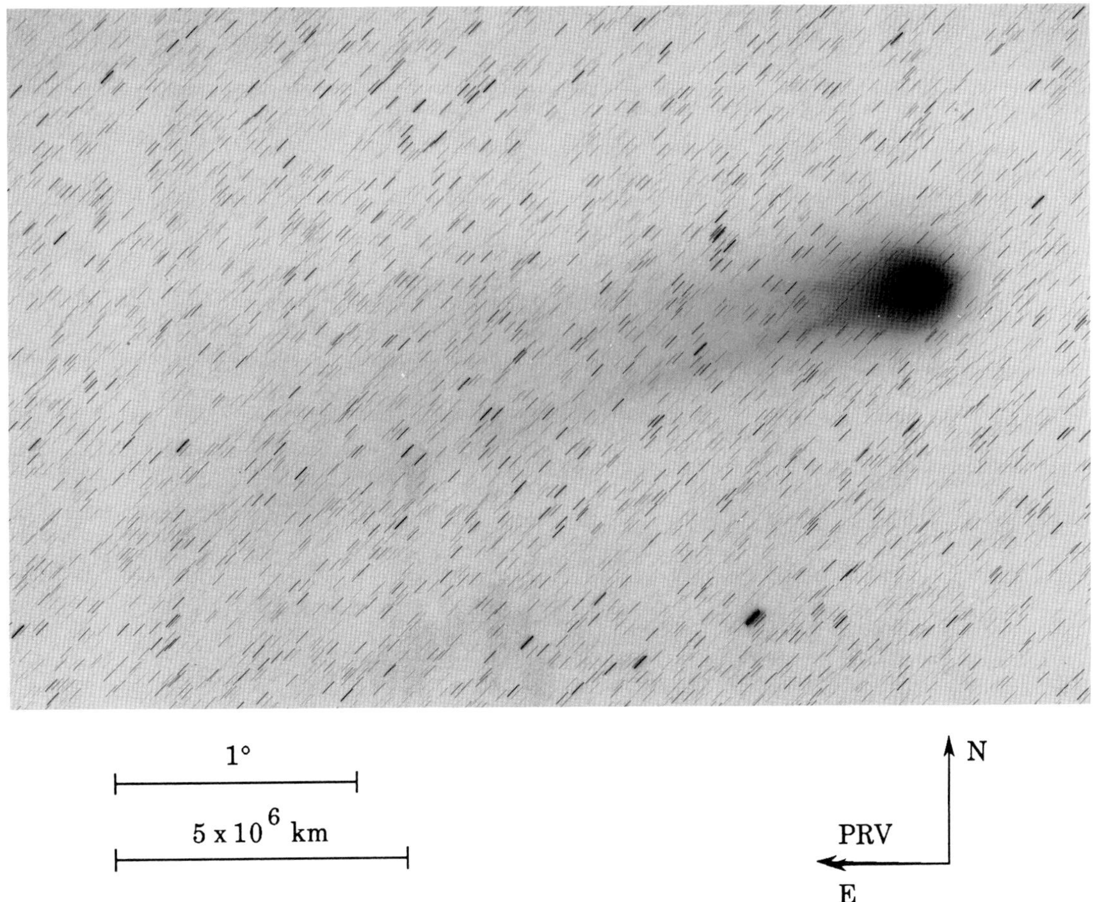

Fig. 672: 1986 Apr 28.01736 UT; exposure 20.0 minutes on IIa-O emulsion with GG-385 filter; r=1.59, Δ=0.72, β=152.7°; LSPN-2915.
Photograph by G. Pizarro, European Southern Observatory.

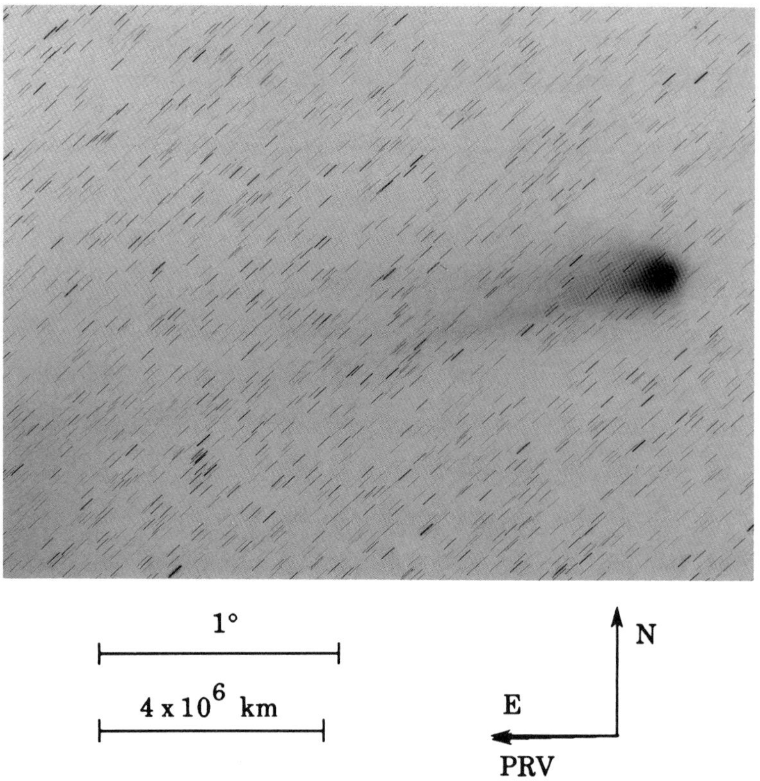

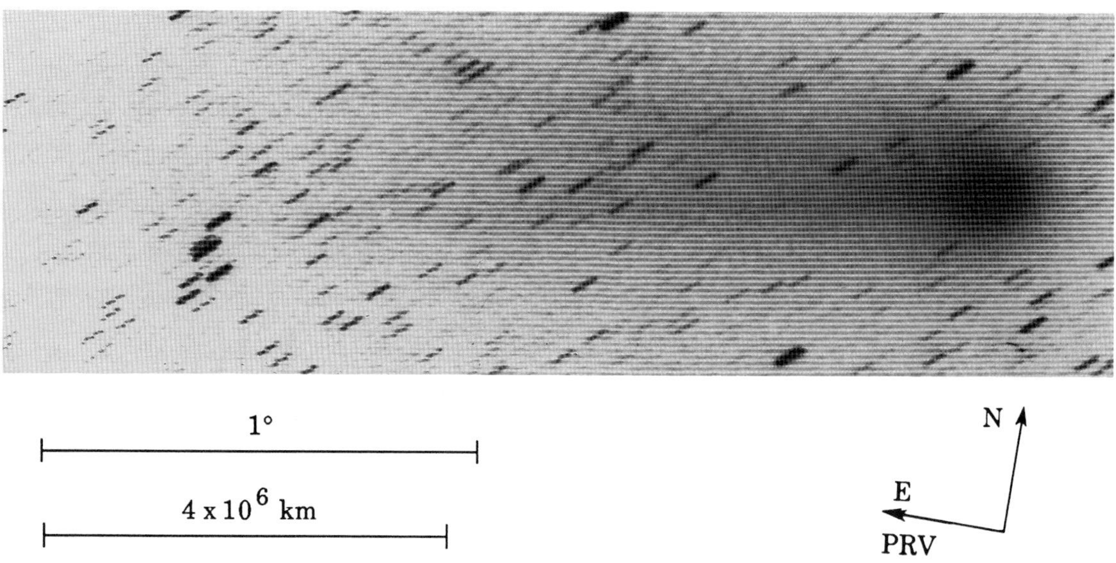

(top) Fig. 673: 1986 Apr 28.02292 UT; exposure 22.0 minutes on 098-04 emulsion with RG-630 filter; r=1.59, Δ=0.72, β=152.7°; LSPN-2917.
Photograph by G. Pizarro, European Southern Observatory.

(bottom) Fig. 674: 1986 Apr 28.80566 UT; exposure 60.0 minutes on ORWO ZU-21 emulsion with no filter; r=1.60, Δ=0.74, β=152.3°; LSPN-3442.
Photograph by N. Chernykh, Crimean Astrophysical Observatory.

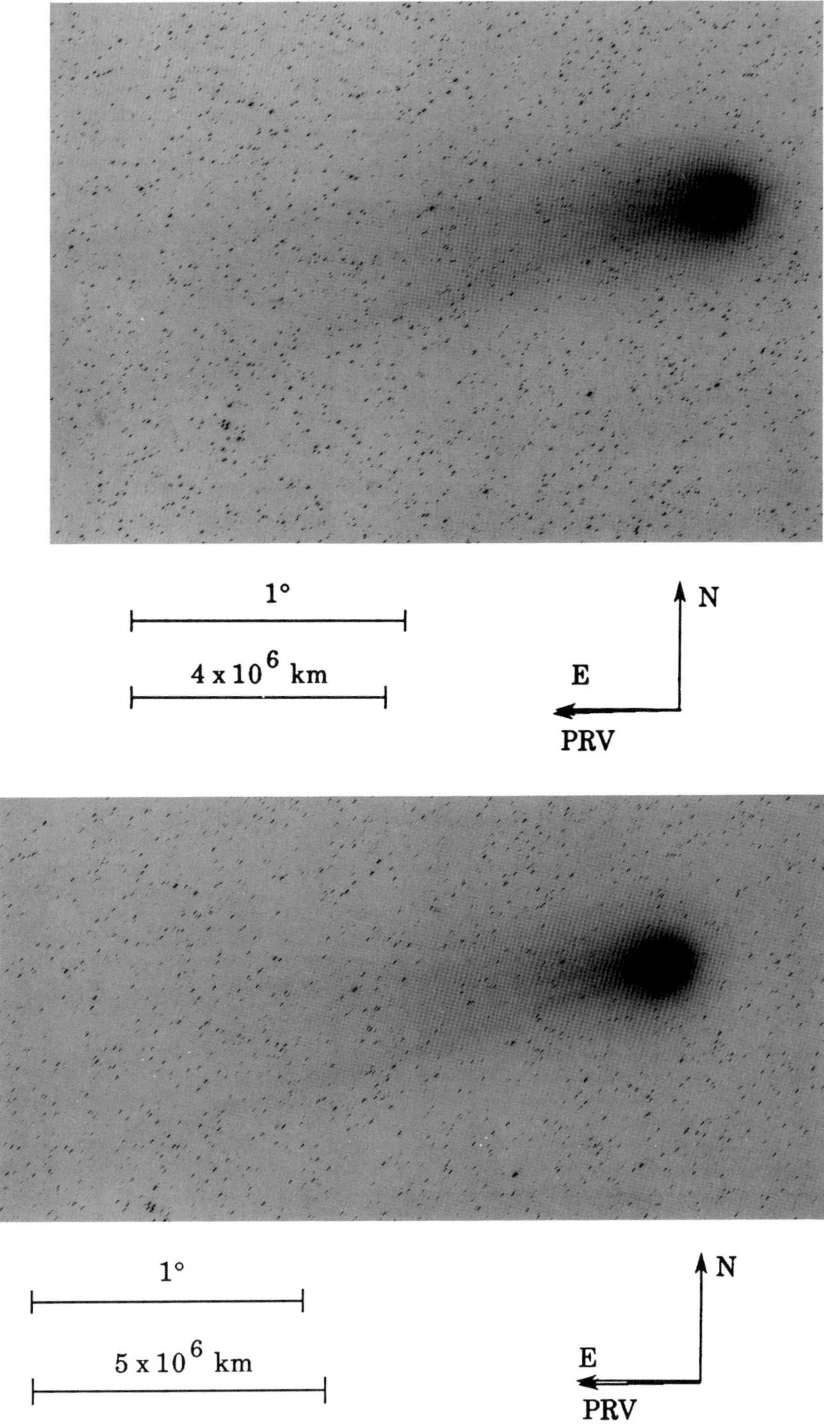

(top) Fig. 675: 1986 Apr 28.99340 UT; exposure 16.0 minutes on IIIa-J emulsion with no filter; r=1.60, Δ=0.74, β=152.2°; LSPN-2989.
Photograph by K. Meech, University of Michigan/CTIO.

(bottom) Fig. 676*: 1986 Apr 29.05932 UT; exposure 15.0 minutes on IIIa-J emulsion with no filter; r=1.60, Δ=0.75, β=152.1°; LSPN-2993.
Photograph by K. Meech, University of Michigan/CTIO.

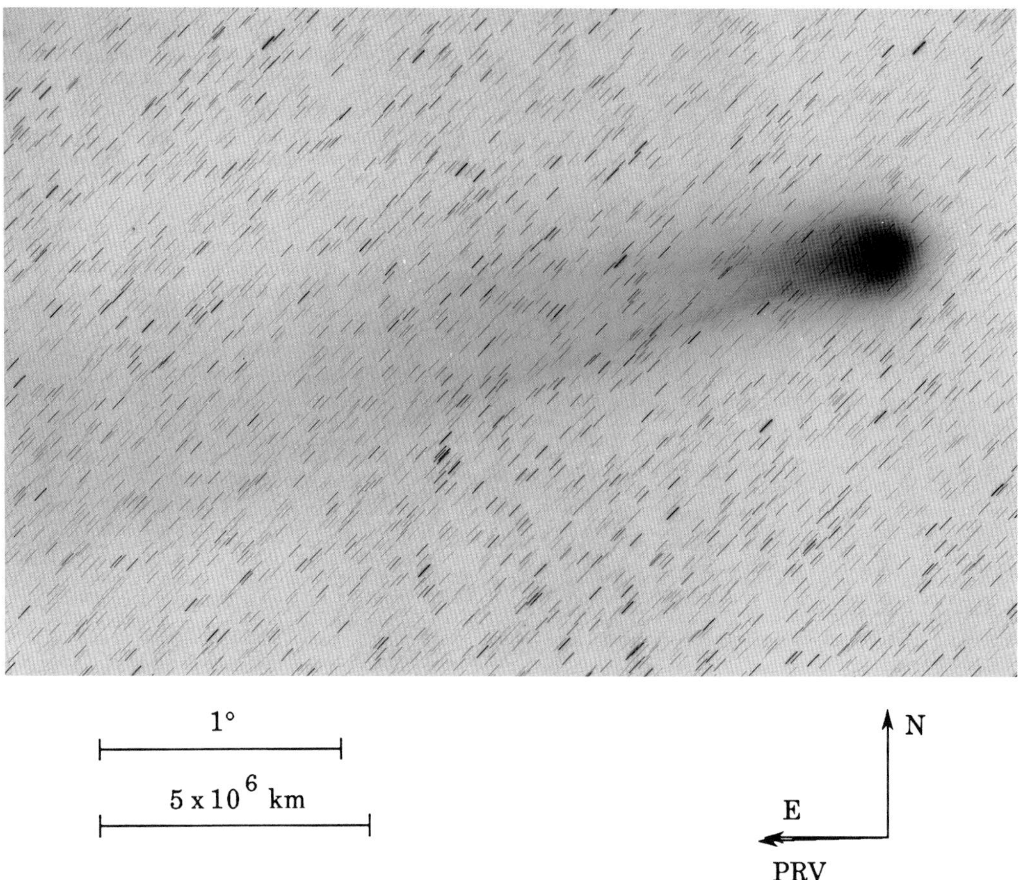

Fig. 677: 1986 Apr 28.99514 UT; exposure 22.0 minutes on IIa-O emulsion with GG-385 filter; r=1.60, Δ=0.74, β=152.2°; LSPN-2916.
Photograph by G. Pizarro, European Southern Observatory.

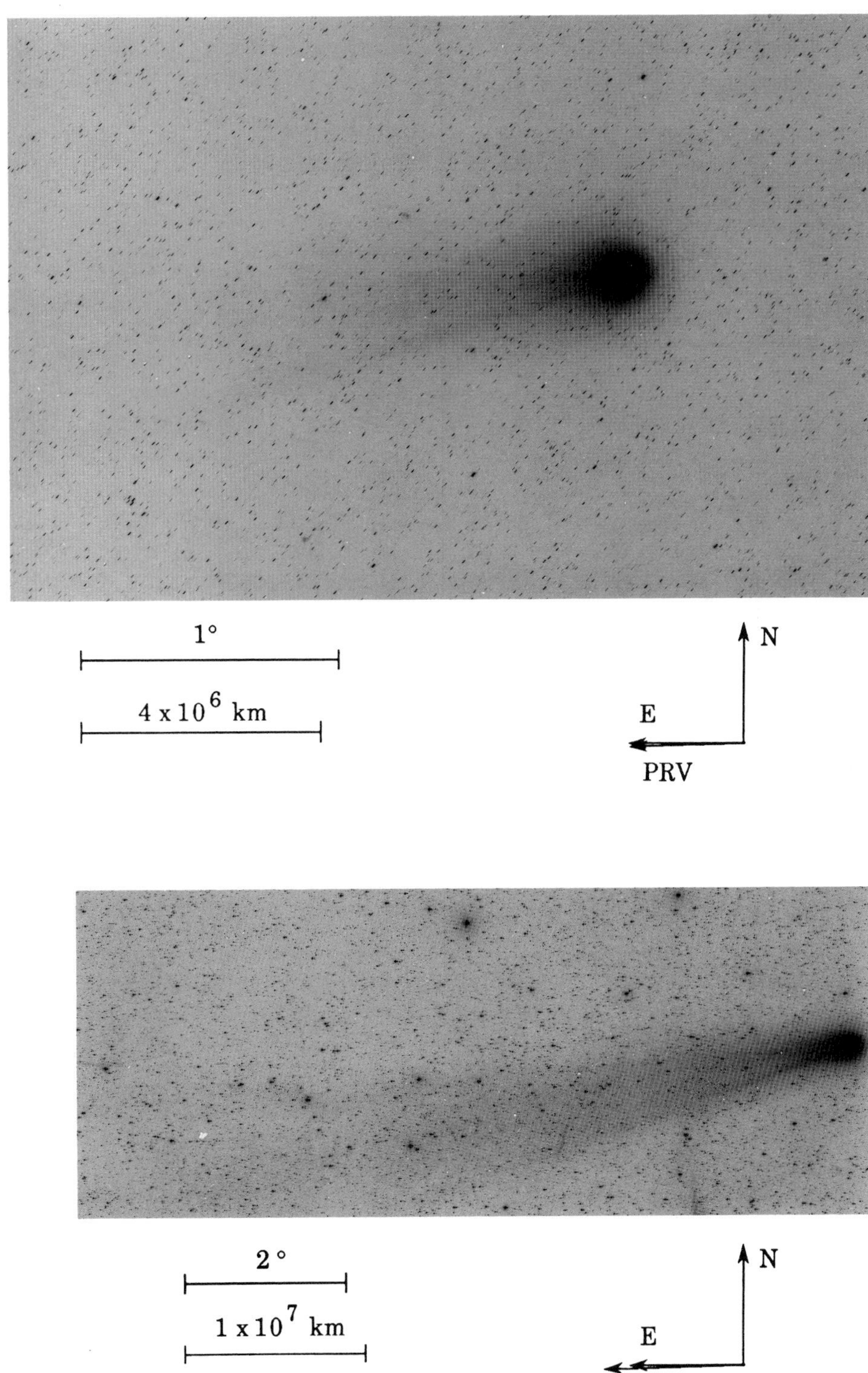

(top) Fig. 678: 1986 Apr 29.07535 UT; exposure 15.0 minutes on IIIa-J emulsion with no filter; r=1.60, Δ=0.75, β=152.1°; LSPN-2994.
Photograph by K. Meech, University of Michigan/CTIO.

(bottom) Fig. 679: 1986 Apr 29.21458 UT; exposure 10.0 minutes on IIa-O emulsion with no filter; r=1.61, Δ=0.75, β=152.0°; LSPN-936.
Photograph by E. Moore et al., Joint Observatory for Cometary Research.

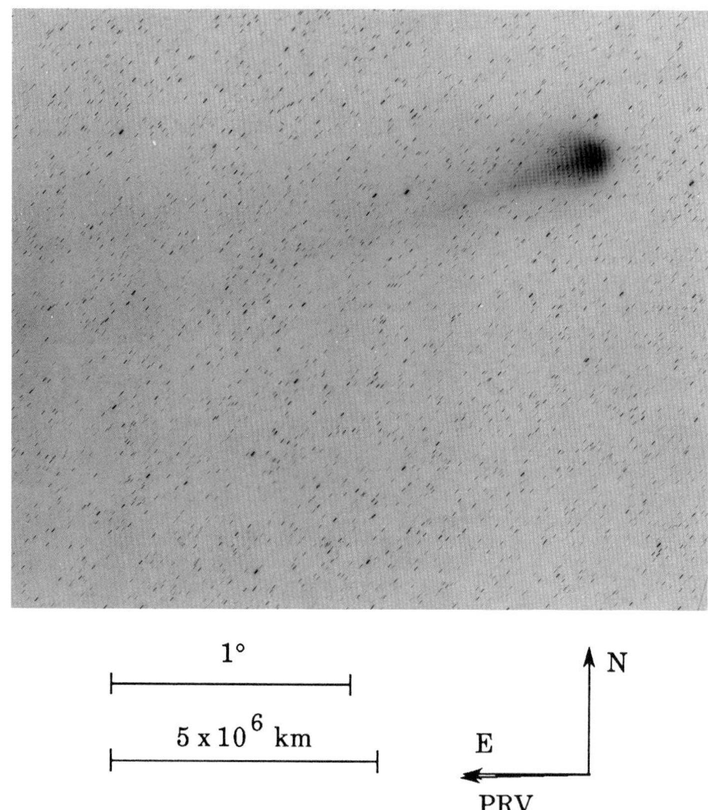

Fig. 680: 1986 Apr 29.98264 UT; exposure 20.0 minutes on 098-04 emulsion with RG-630 filter; r=1.62, Δ=0.77, β=151.6°; LSPN-2918.
Photograph by O. Pizarro, European Southern Observatory.

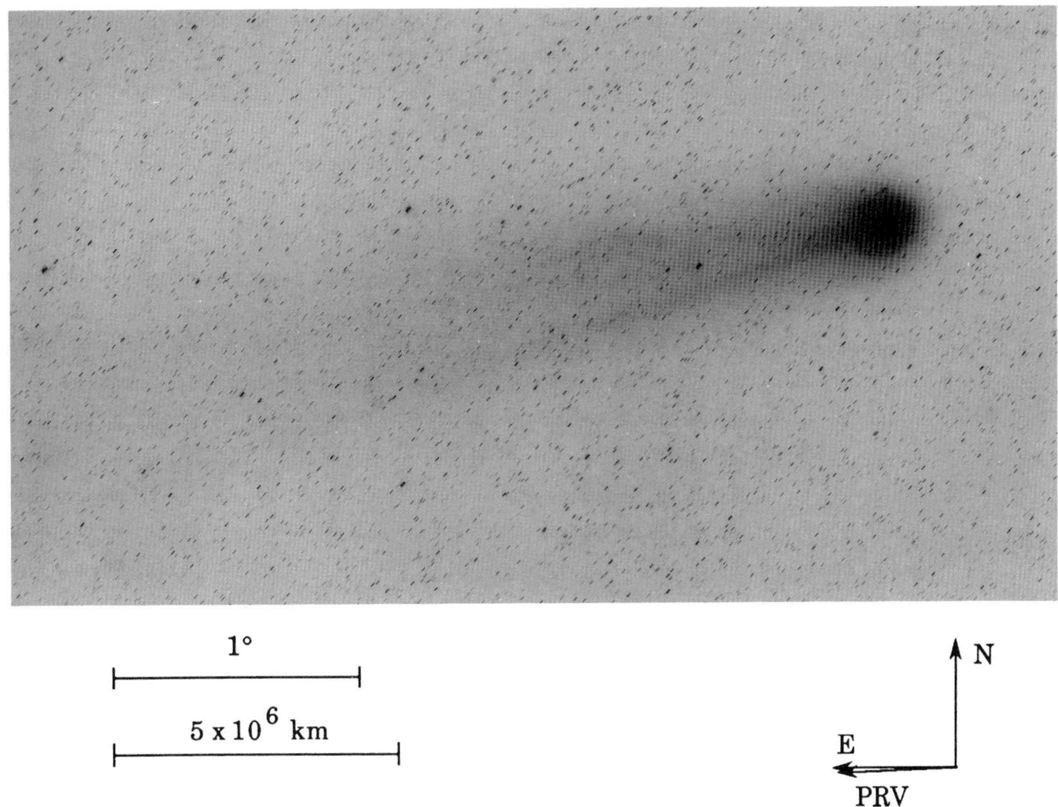

Fig. 681: 1986 Apr 30.01597 UT; exposure 20.0 minutes on IIa-O emulsion with GG-385 filter; r=1.62, Δ=0.77, β=151.6°; LSPN-2919.
Photograph by O. Pizarro, European Southern Observatory.

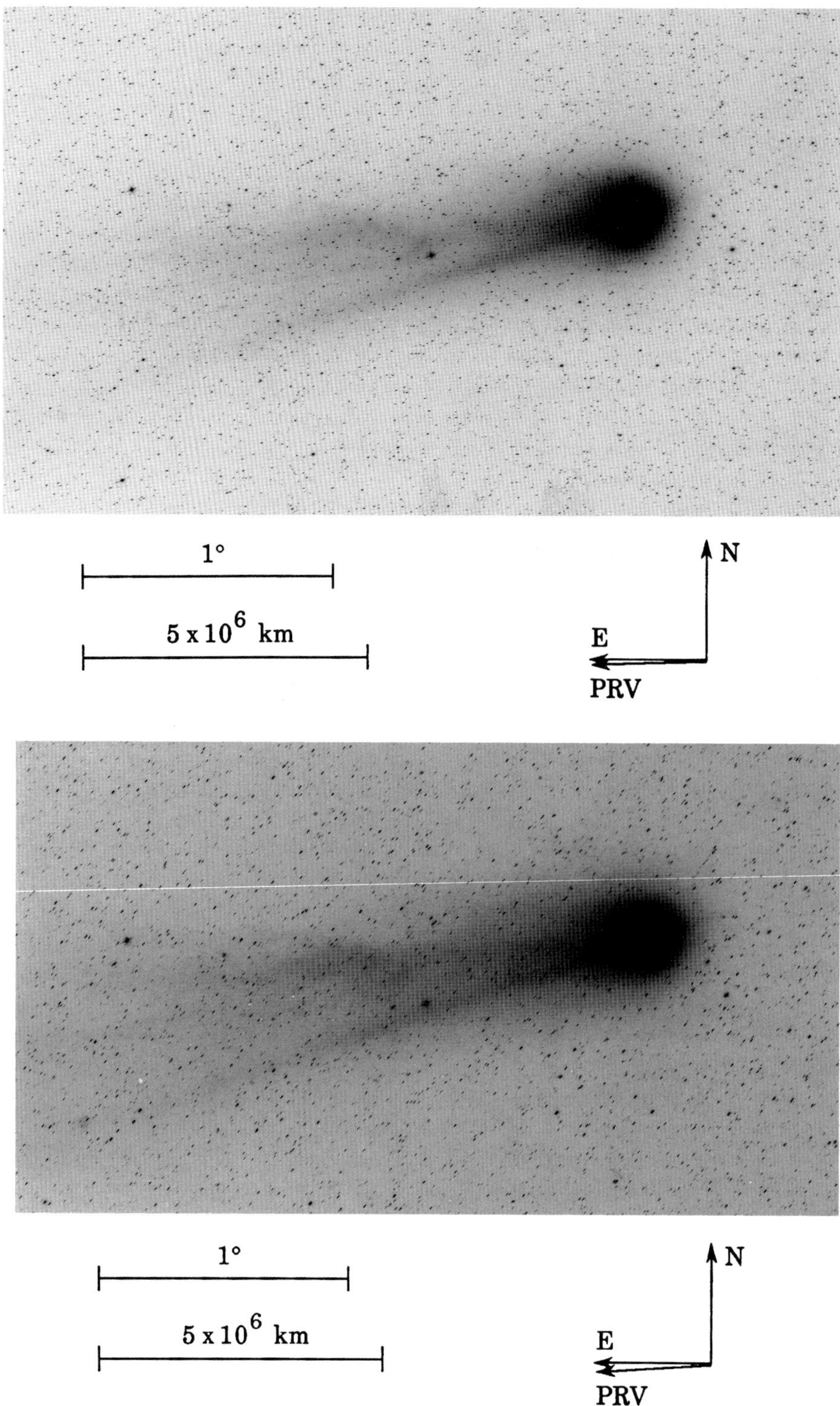

(top) Fig. 682*: 1986 Apr 30.01580 UT; exposure 15.0 minutes on IIa-O emulsion with no filter; r=1.62, Δ=0.77, β=151.6°; LSPN-3000.
Photograph by K. Meech, University of Michigan/CTIO.

(bottom) Fig. 683*: 1986 Apr 30.11123 UT; exposure 15.0 minutes on IIa-O emulsion with no filter; r=1.62, Δ=0.78, β=151.6°; LSPN-3005.
Photograph by K. Meech, University of Michigan/CTIO.

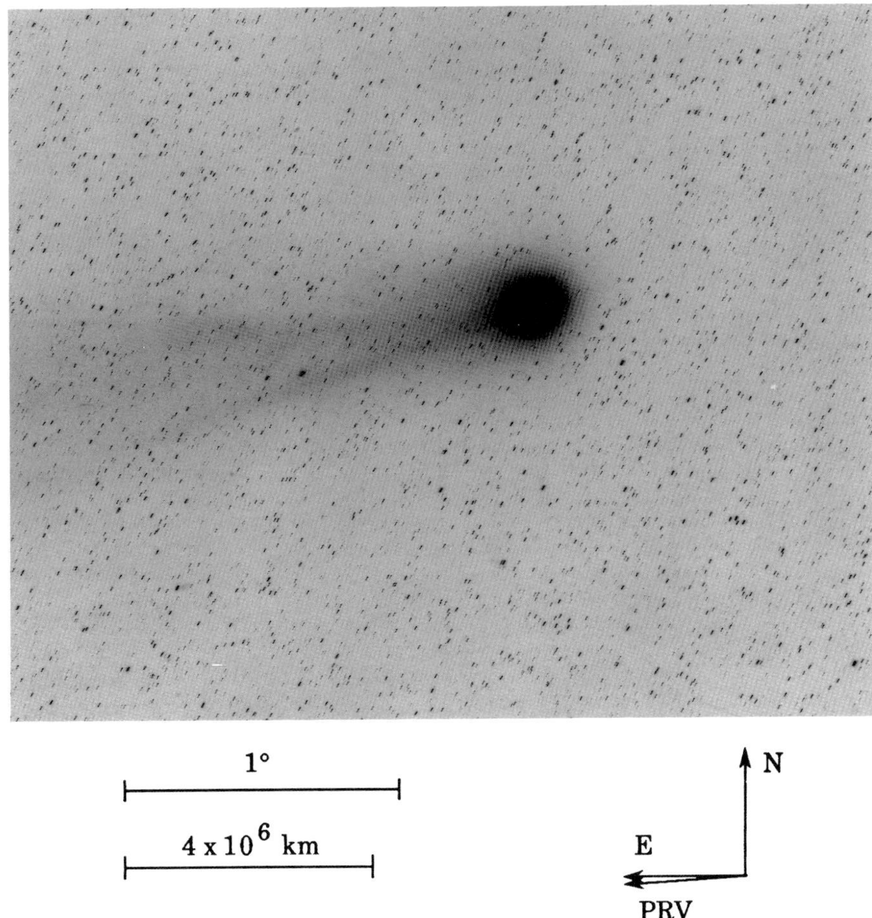

Fig. 684: 1986 Apr 30.09225 UT; exposure 15.0 minutes on IIIa-J emulsion with no filter; r=1.62, Δ=0.78, β=151.6°; LSPN-3004.
Photograph by K. Meech, University of Michigan/CTIO.

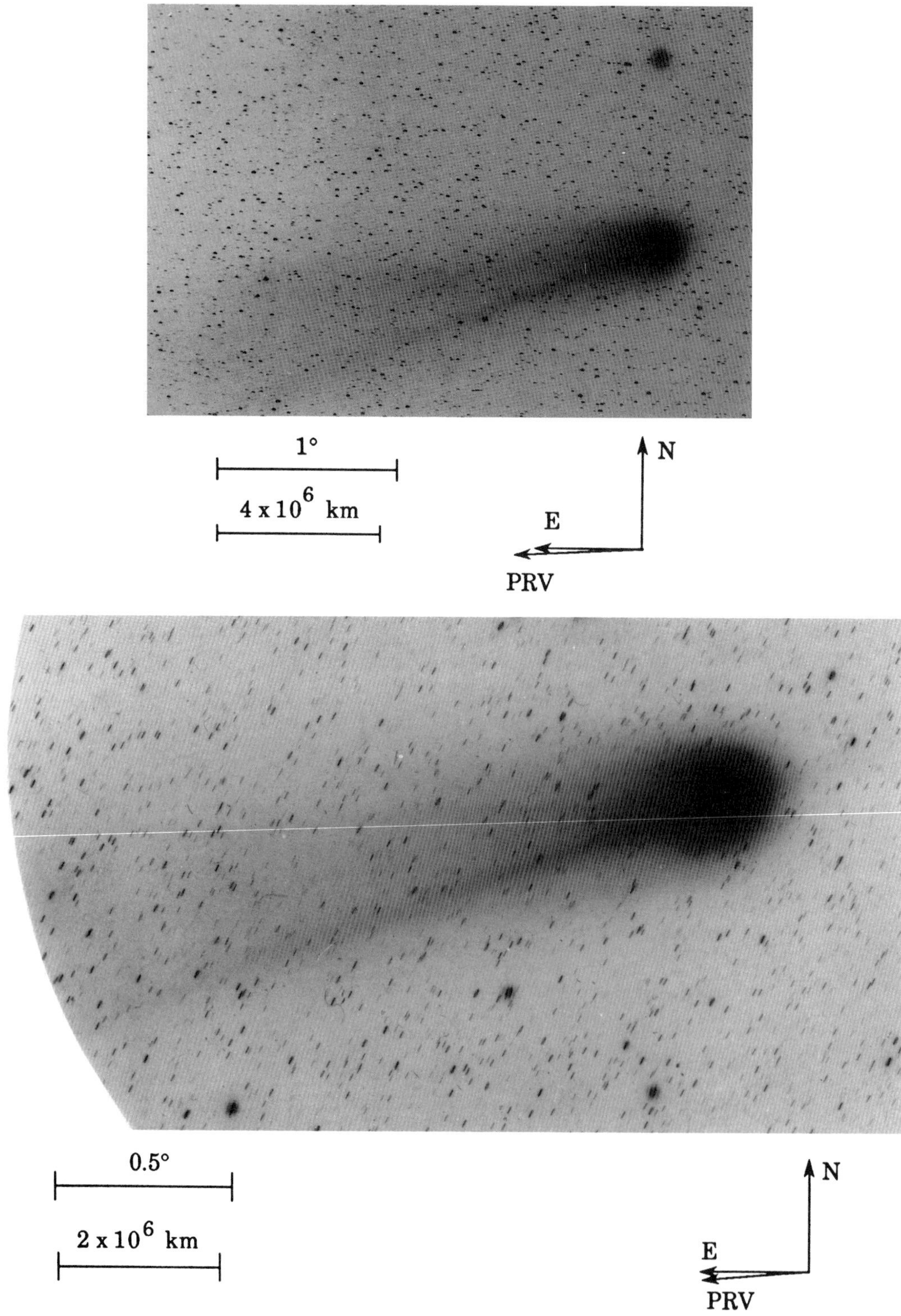

(top) Fig. 685: 1986 Apr 30.22187 UT; exposure 15.0 minutes on Tech. Pan 2415 emulsion with no filter; r=1.62, Δ=0.78, β=151.5°; LSPN-1503.
Photograph by W. Liller, Easter Island, LSPN Island Network.

(bottom) Fig. 686: 1986 Apr 30.73229 UT; exposure 25.0 minutes on Tech. Pan 2415 emulsion with no filter; r=1.63, Δ=0.80, β=151.3°; LSPN-3509.
Photograph by D. O'Donoghue, University of Perugia/South African Astronomical Observatory.

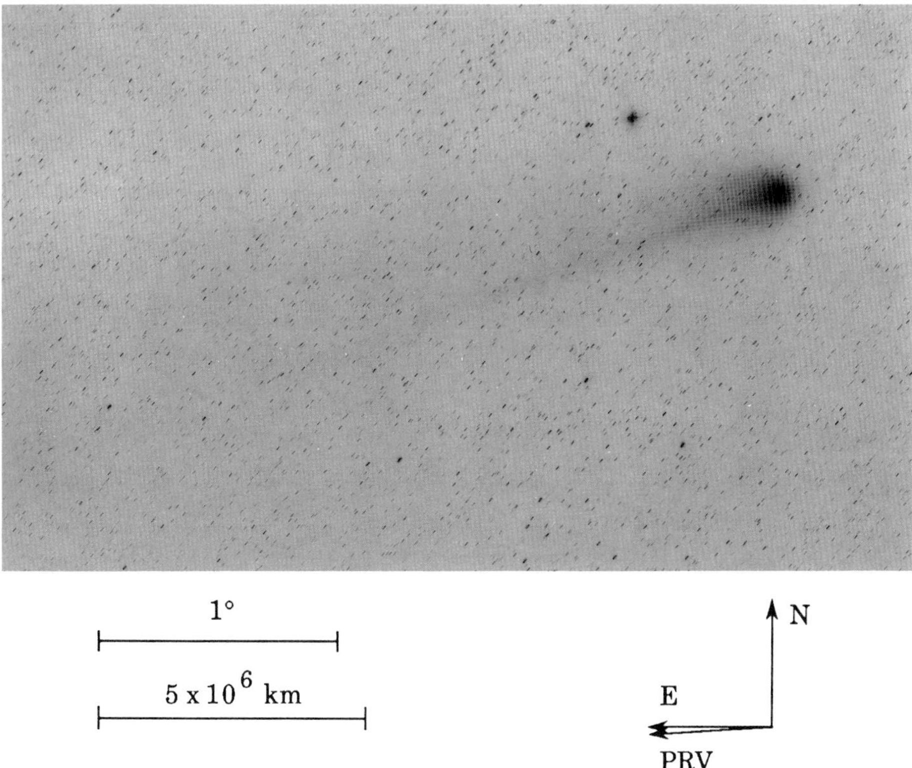

Fig. 687: 1986 Apr 30.97569 UT; exposure 20.0 minutes on 098-04 emulsion with RG-630 filter; r=1.63, Δ=0.80, β=151.1°; LSPN-2920.
Photograph by O. Pizarro, European Southern Observatory.

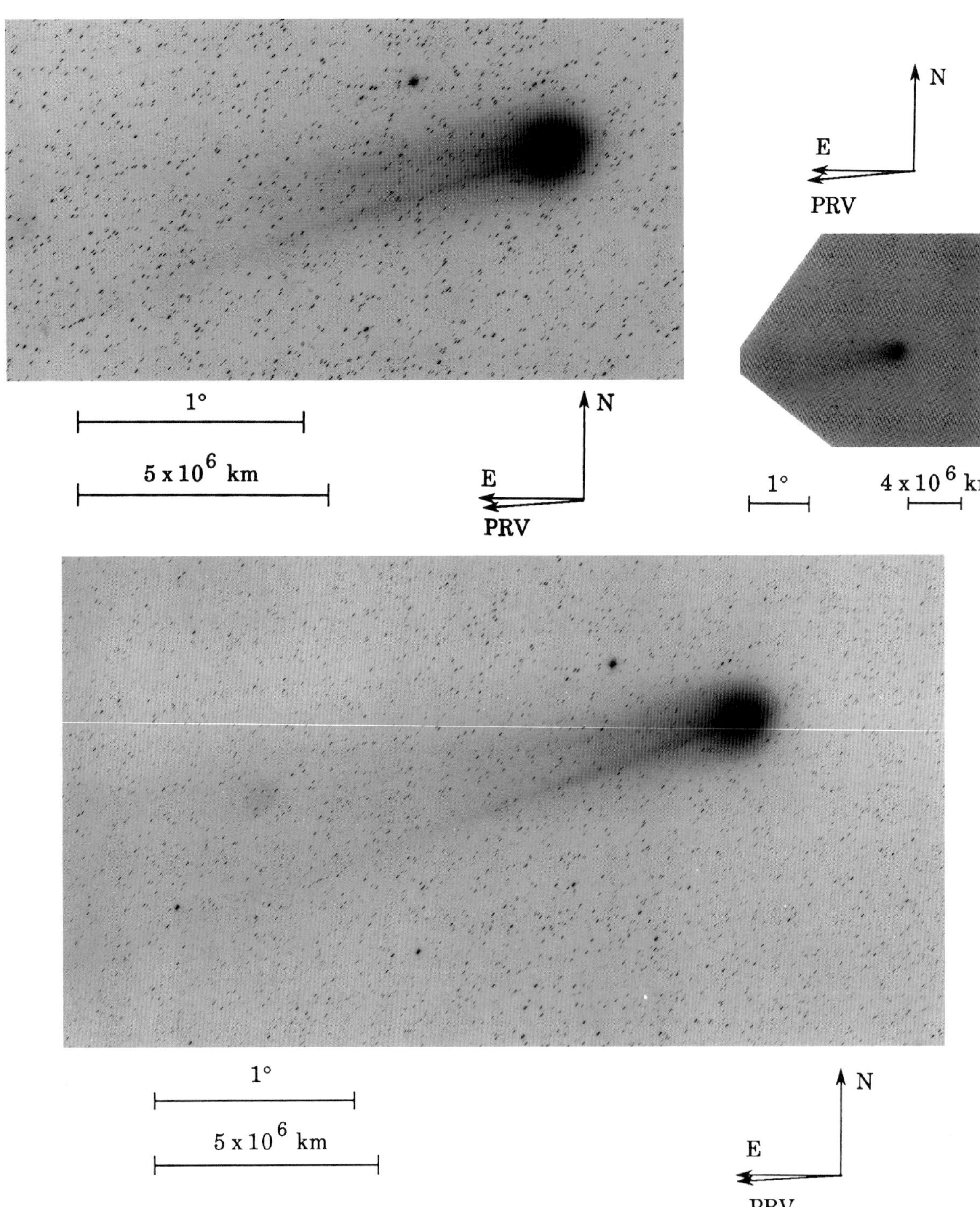

(left) Fig. 688: 1986 May 1.01128 UT; exposure 20.0 minutes on IIIa-J emulsion with no filter; r=1.63, Δ=0.80, β=151.1°; LSPN-3010.
Photograph by K. Meech, University of Michigan/CTIO.

(bottom) Fig. 689: 1986 May 1.02014 UT; exposure 20.0 minutes on IIa-O emulsion with GG-385 filter; r=1.63, Δ=0.80, β=151.1°; LSPN-2921.
Photograph by O. Pizarro, European Southern Observatory.

(right) Fig. 690: 1986 May 1.16458 UT; exposure 32.0 minutes on Tech. Pan 2415 emulsion with no filter; r=1.64, Δ=0.81, β=151.1°; LSPN-981.
Photograph by G. Emerson, E. E. Barnard Observatory.

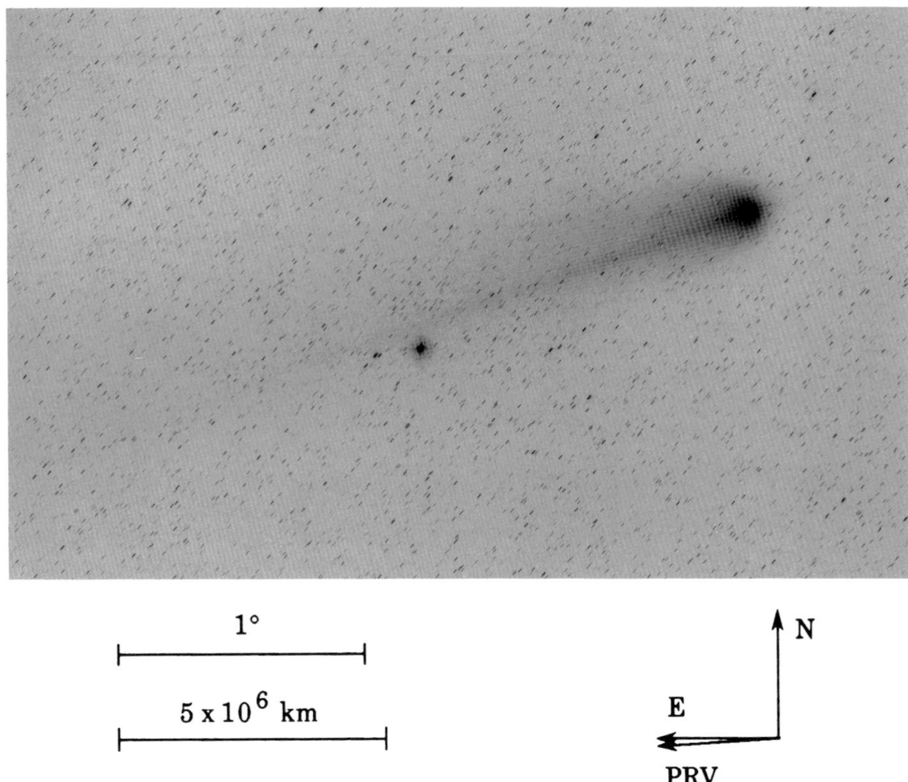

Fig. 691: 1986 May 1.97396 UT; exposure 25.0 minutes on 098-04 emulsion with RG-630 filter; r=1.65, Δ=0.83, β=150.7°; LSPN-2922.
Photograph by O. Pizarro, European Southern Observatory.

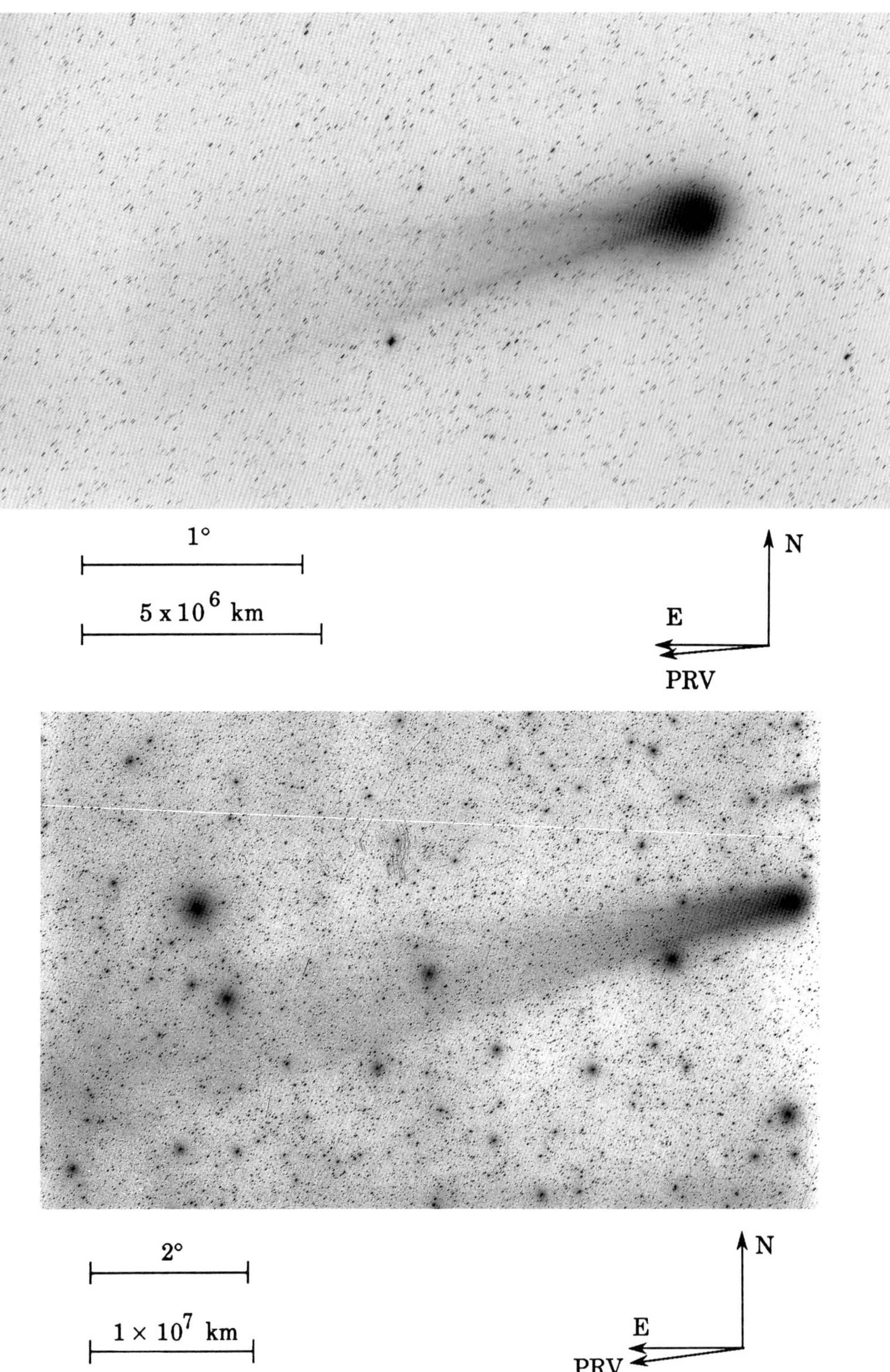

(top) Fig. 692: 1986 May 2.00660 UT; exposure 25.0 minutes on IIa-O emulsion with GG-385 filter; r=1.65, Δ=0.83, β=150.7°; LSPN-2923.
Photograph by O. Pizarro, European Southern Observatory.

(bottom) Fig. 693*: 1986 May 2.17222 UT; exposure 30.0 minutes on IIa-O emulsion with no filter; r=1.65, Δ=0.84, β=150.6°; LSPN-941.
Photograph by E. Moore, Joint Observatory for Cometary Research.

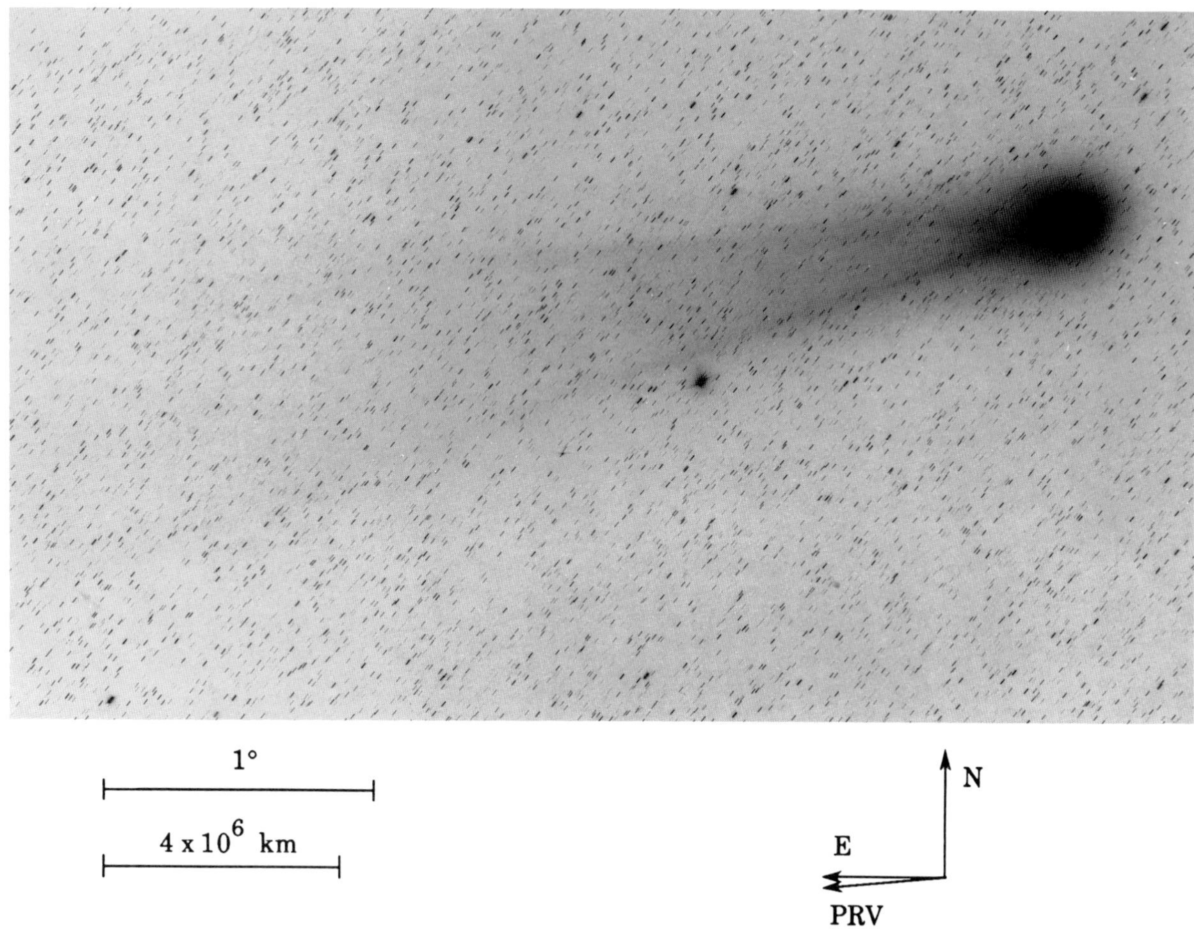

Fig. 694: 1986 May 2.01389 UT; exposure 30.0 minutes on IIIa-J emulsion with no filter; r=1.65, Δ=0.83, β=150.7°; LSPN-3020.
Photograph by K. Meech, University of Michigan/CTIO.

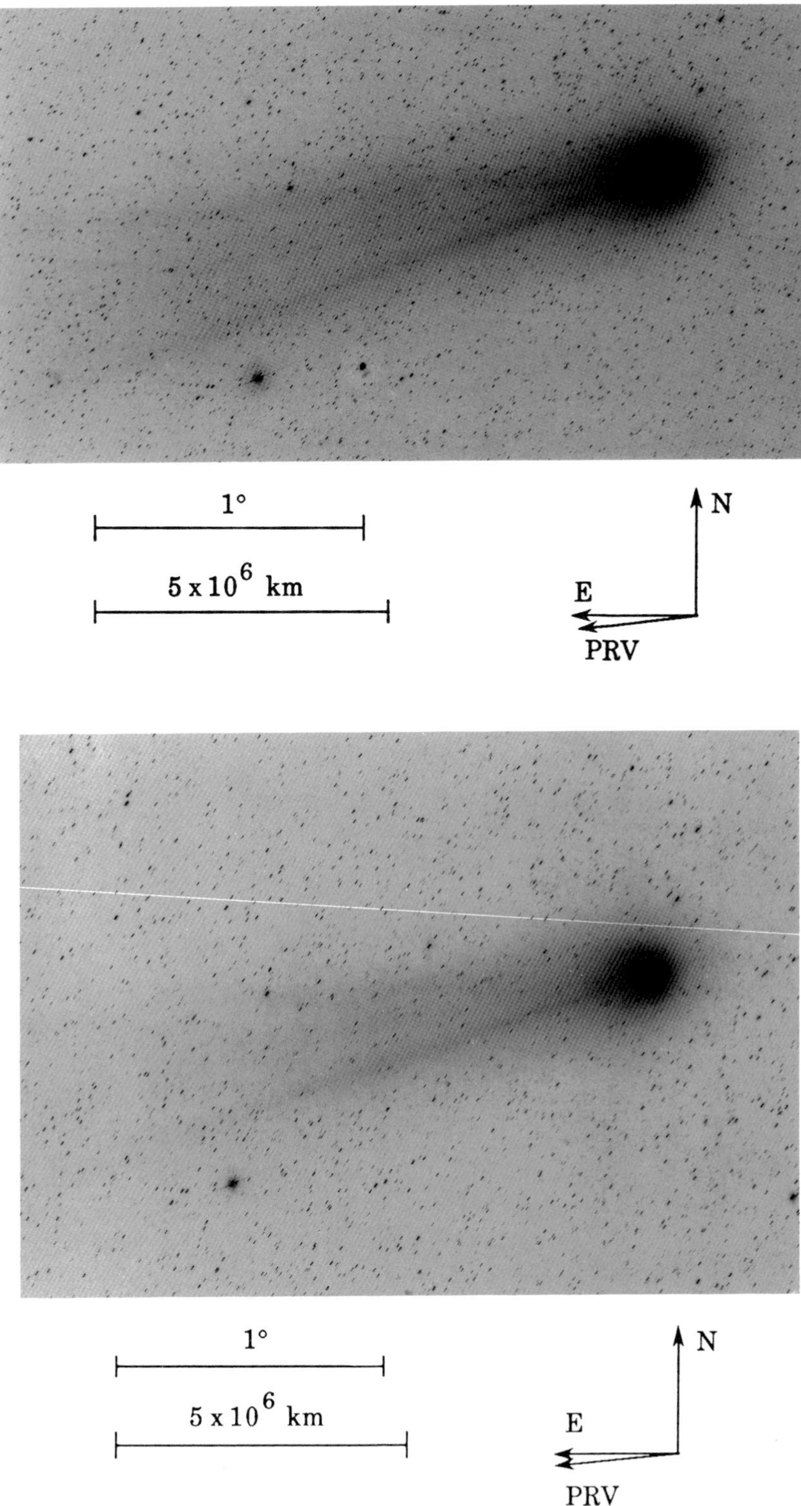

(top) Fig. 695: 1986 May 2.20703 UT; exposure 15.0 minutes on 103a-O emulsion with no filter; r=1.65, Δ=0.84, β=150.6°; LSPN-3023.
Photograph by K. Meech, University of Michigan/CTIO.

(bottom) Fig. 696: 1986 May 2.22766 UT; exposure 20.0 minutes on IIIa-J emulsion with no filter; r=1.65, Δ=0.84, β=150.6°; LSPN-3024.
Photograph by K. Meech, University of Michigan/CTIO.

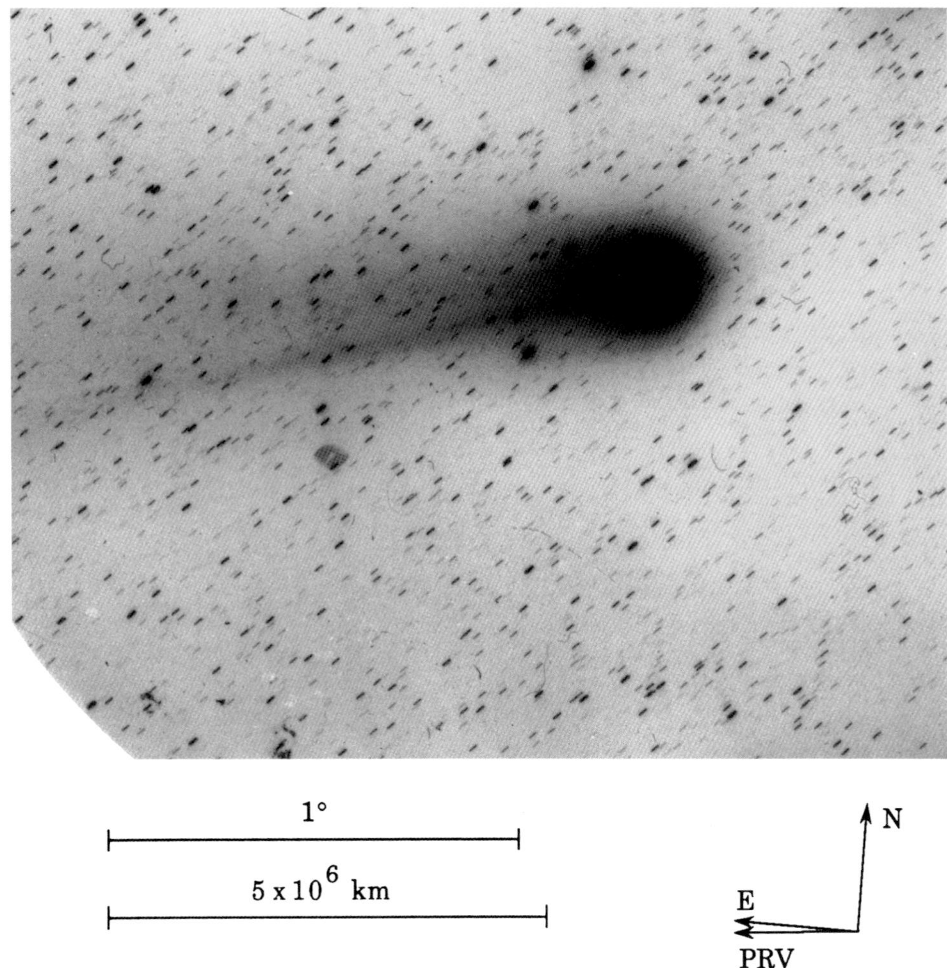

Fig. 697: 1986 May 2.82118 UT; exposure 25.0 minutes on Tech. Pan 2415 emulsion with no filter; r=1.66, Δ=0.86, β=150.4°; LSPN-3510.
Photograph by D. O'Donoghue, University of Perugia/South African Astronomical Observatory.

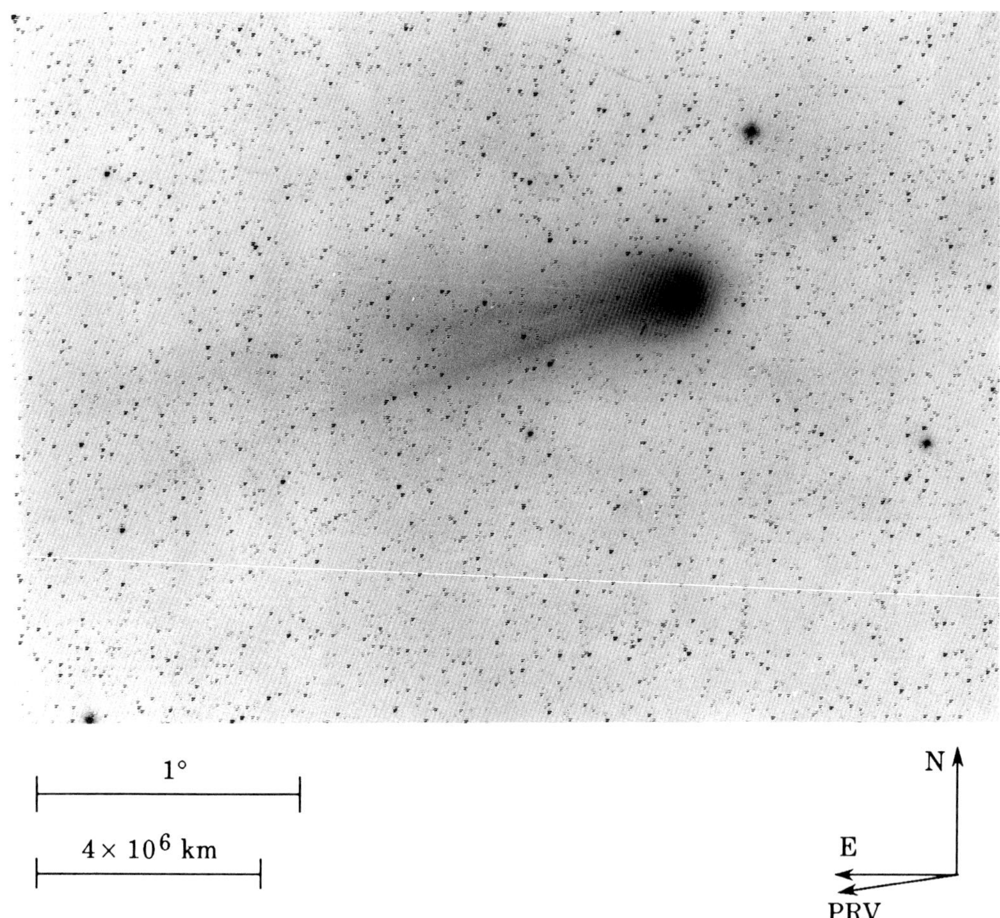

Fig. 698: 1986 May 3.21667 UT; exposure 20.0 minutes on IIa-O emulsion with no filter; $r=1.67$, $\Delta=0.87$, $\beta=150.2°$; LSPN-2592.
Photograph by R. Hill, Warner and Swasey Observatory.

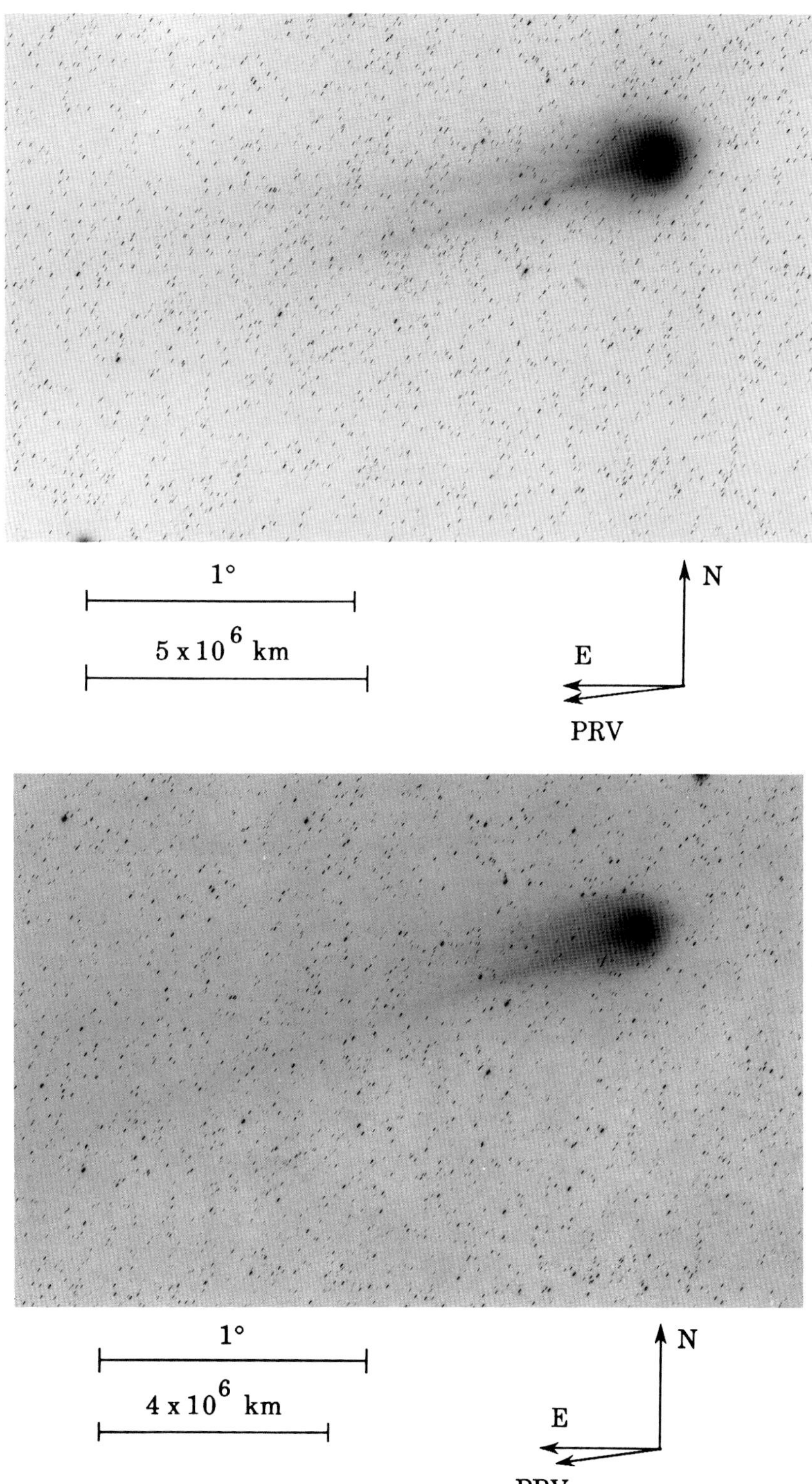

(top) Fig. 699*: 1986 May 3.11018 UT; exposure 20.0 minutes on IIIa-J emulsion with no filter; r=1.66, Δ=0.87, β=150.3°; LSPN-3028.
Photograph by K. Meech, University of Michigan/CTIO.

(bottom) Fig. 700: 1986 May 3.23472 UT; exposure 20.0 minutes on IIa-F emulsion with no filter; r=1.67, Δ=0.87, β=150.2°; LSPN-2593.
Photograph by R. Hill, Warner and Swasey Observatory.

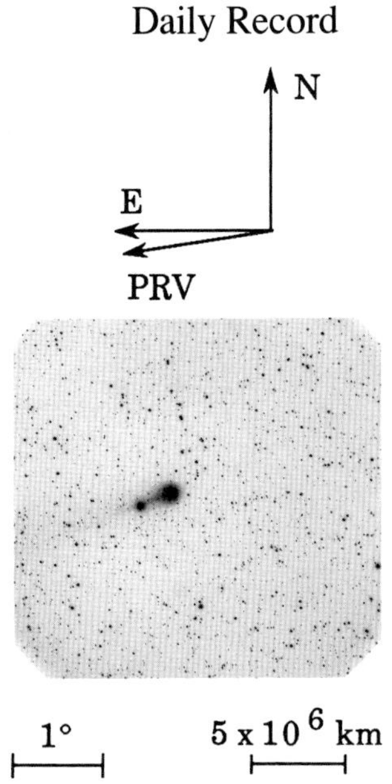

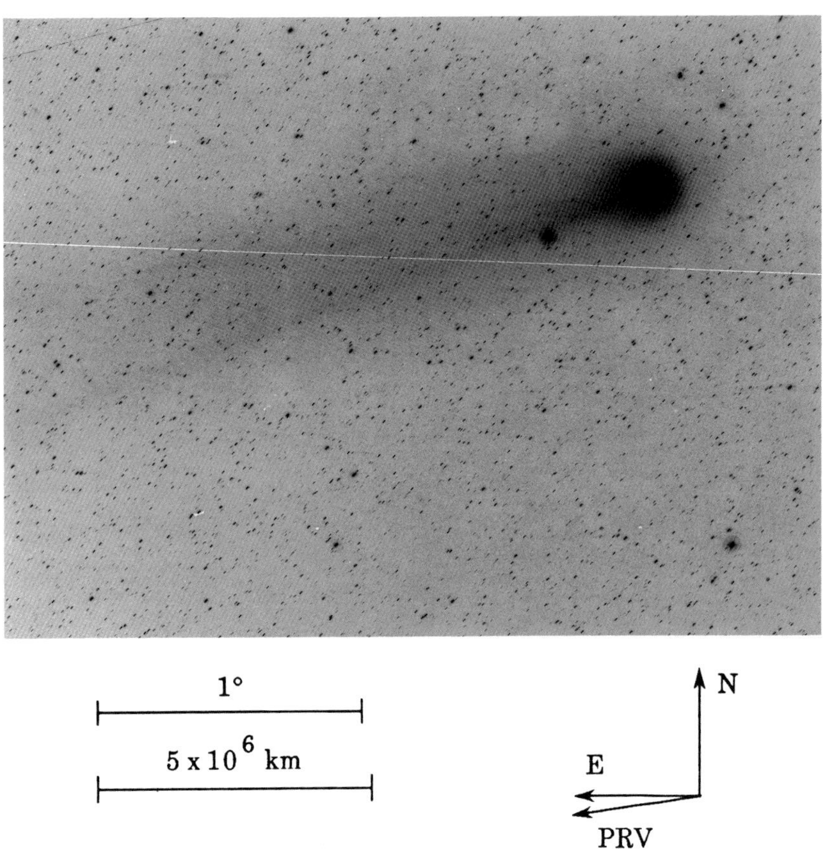

(top) Fig. 701: 1986 May 4.19444 UT; exposure 30.0 minutes on Tech. Pan 2415 emulsion with no filter; r=1.68, Δ=0.90, β=149.9°; LSPN-983.
Photograph by G. Emerson, E. E. Barnard Observatory.

(bottom) Fig. 702: 1986 May 4.24514 UT; exposure 20.0 minutes on IIa-O emulsion with no filter; r=1.68, Δ=0.90, β=149.9°; LSPN-2595.
Photograph by R. Hill, Warner and Swasey Observatory.

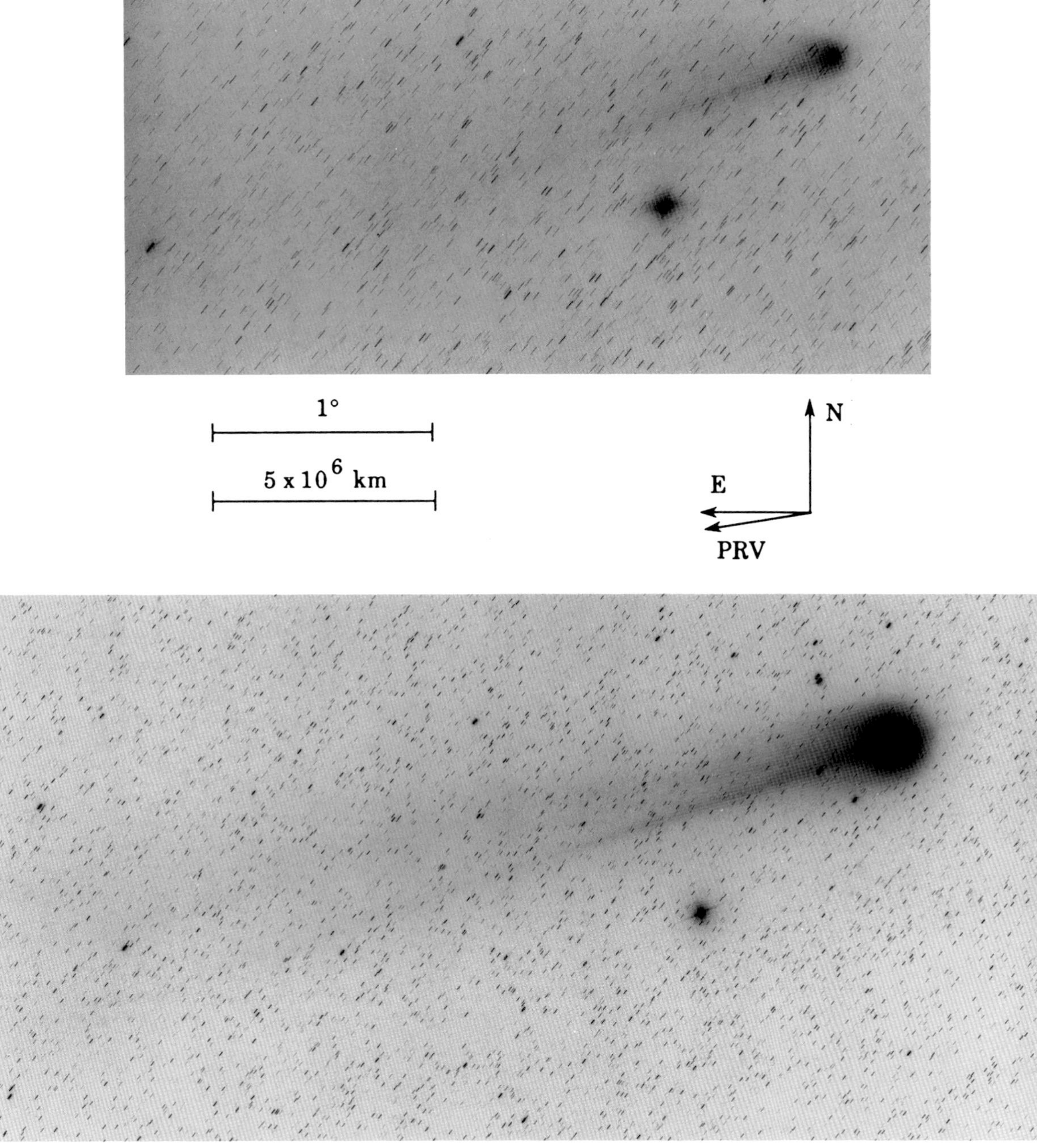

(top) Fig. 703: 1986 May 4.96910 UT; exposure 25.0 minutes on 098-04 emulsion with RG-630 filter; r=1.69, Δ=0.92, β=149.7°; LSPN-2924.
Photograph by G. Pizarro/O. Pizarro, European Southern Observatory.

(bottom) Fig. 704: 1986 May 5.01042 UT; exposure 30.0 minutes on IIIa-J emulsion with no filter; r=1.69, Δ=0.92, β=149.7°; LSPN-3034.
Photograph by K. Meech, University of Michigan/CTIO.

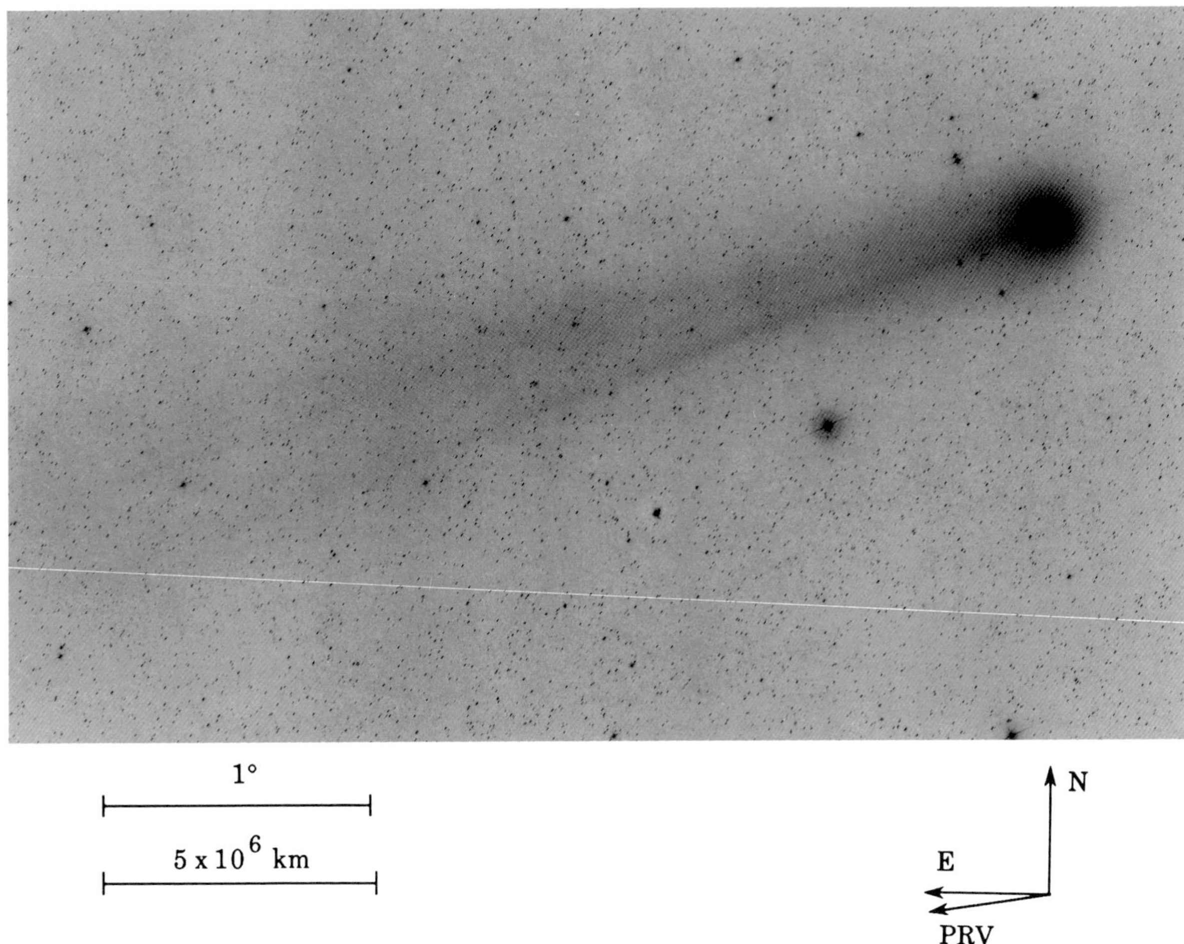

Fig. 705: 1986 May 5.03438 UT; exposure 15.0 minutes on 103a-O emulsion with no filter; r=1.69, Δ=0.93, β=149.7°; LSPN-3035.
Photograph by K. Meech, University of Michigan/CTIO.

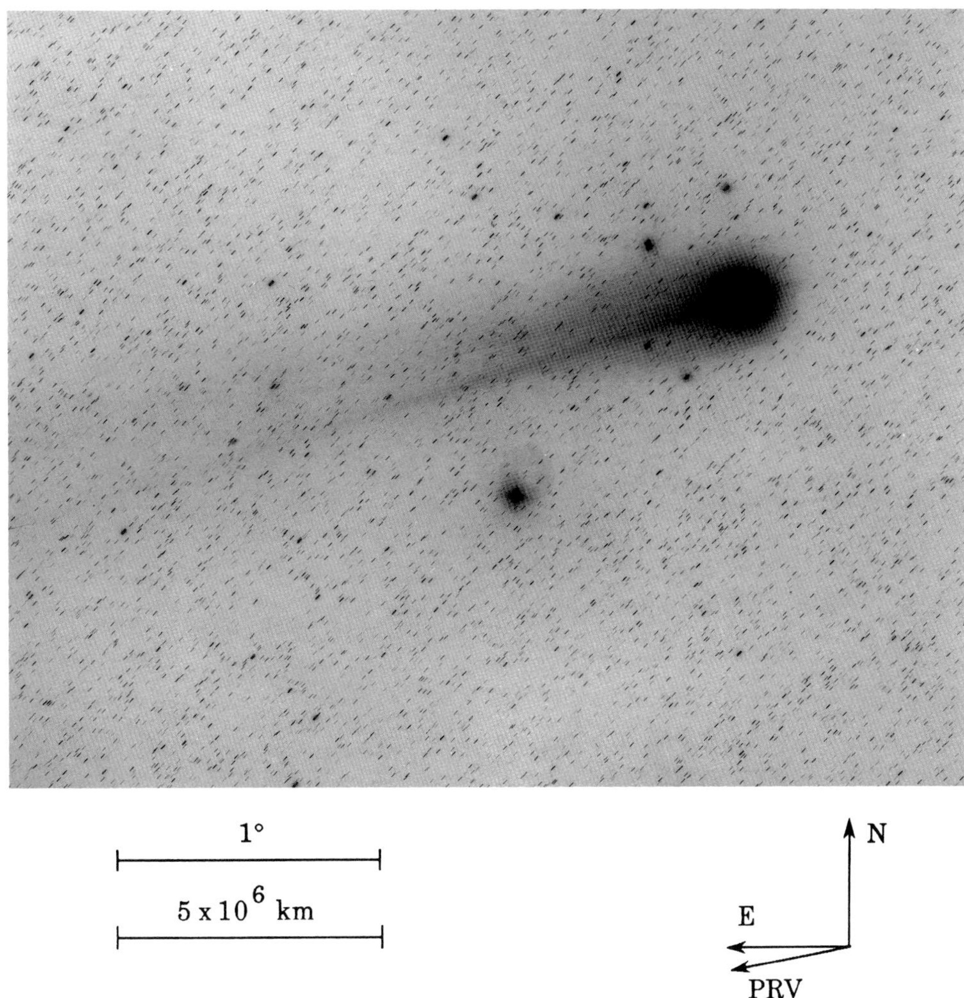

Fig. 706: 1986 May 5.11111 UT; exposure 30.0 minutes on 103a-O emulsion with no filter; r=1.69, Δ=0.93, β=149.6°; LSPN-757.
Photograph by C. Torres/H. Wroblewski, Cerro el Roble Astronomical Observatory.

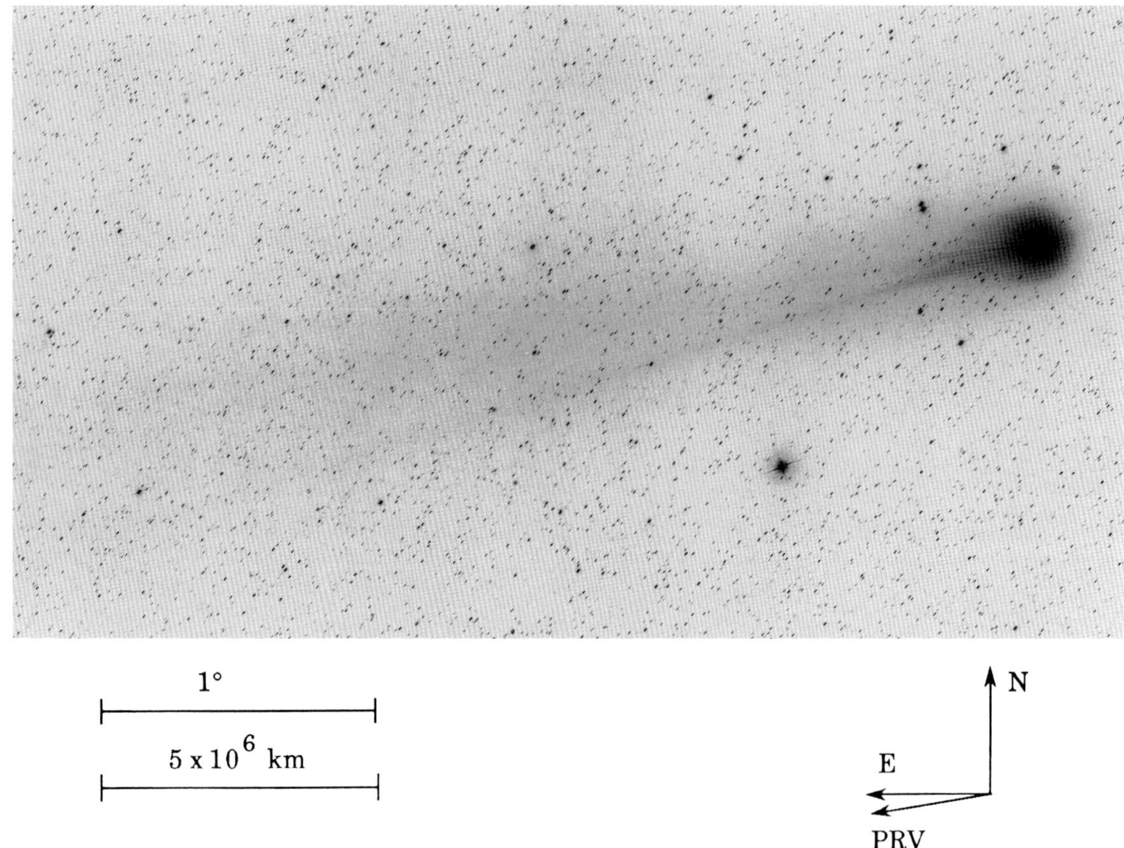

Fig. 707: 1986 May 5.21146 UT; exposure 15.0 minutes on 103a-O emulsion with no filter; r = 1.70, Δ = 0.93, β = 149.6°; LSPN-3040.
Photograph by K. Meech, University of Michigan/CTIO.

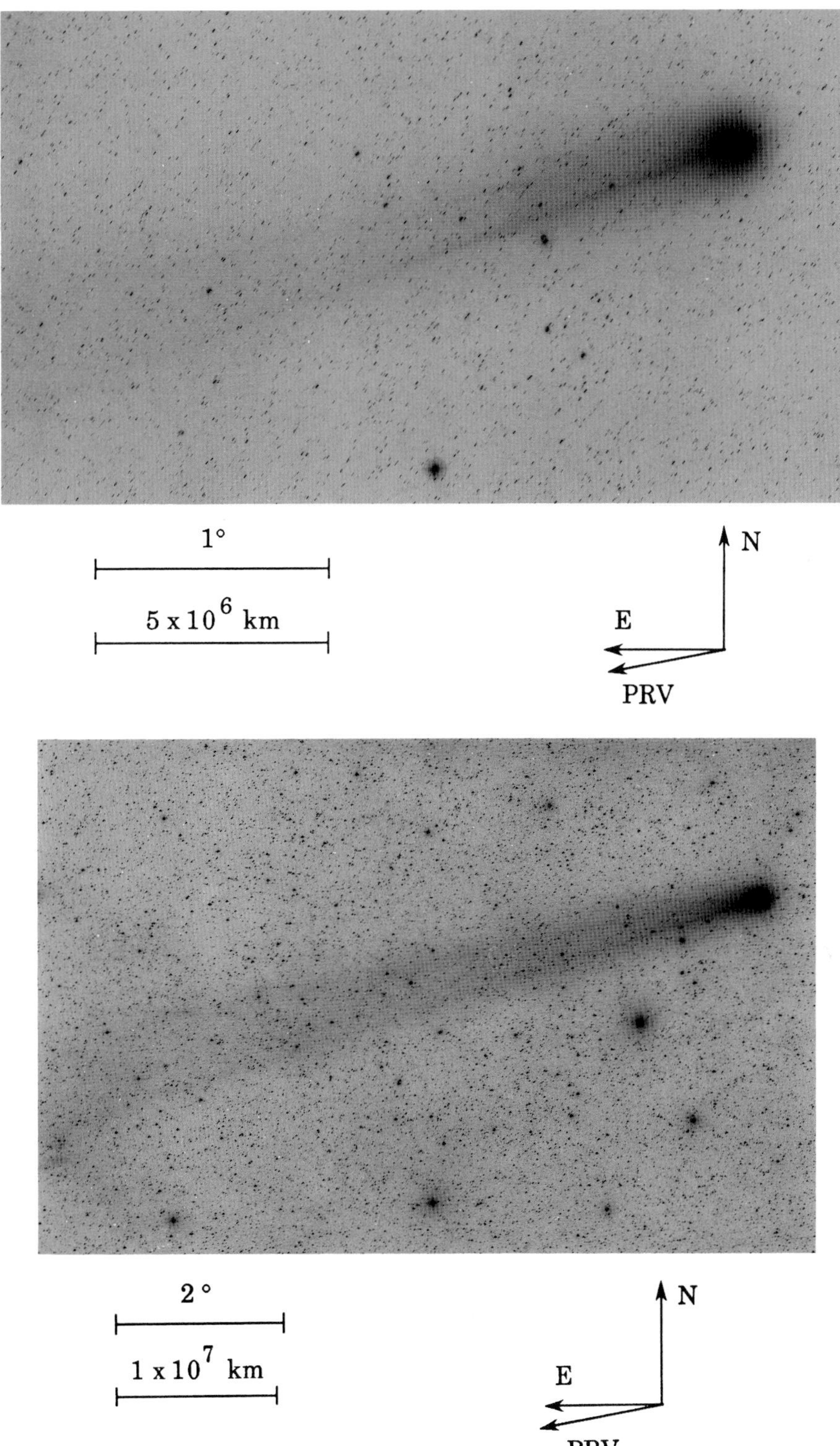

(top) Fig. 708: 1986 May 5.98229 UT; exposure 25.0 minutes on IIa-O emulsion with GG-385 filter; r=1.71, Δ=0.96, β=149.4°; LSPN-2926.
Photograph by G. Pizarro/O. Pizarro, European Southern Observatory.

(bottom) Fig. 709*: 1986 May 6.17569 UT; exposure 30.0 minutes on IIa-O emulsion with no filter; r=1.71, Δ=0.96, β=149.4°; LSPN-944.
Photograph by E. Moore, Joint Observatory for Cometary Research.

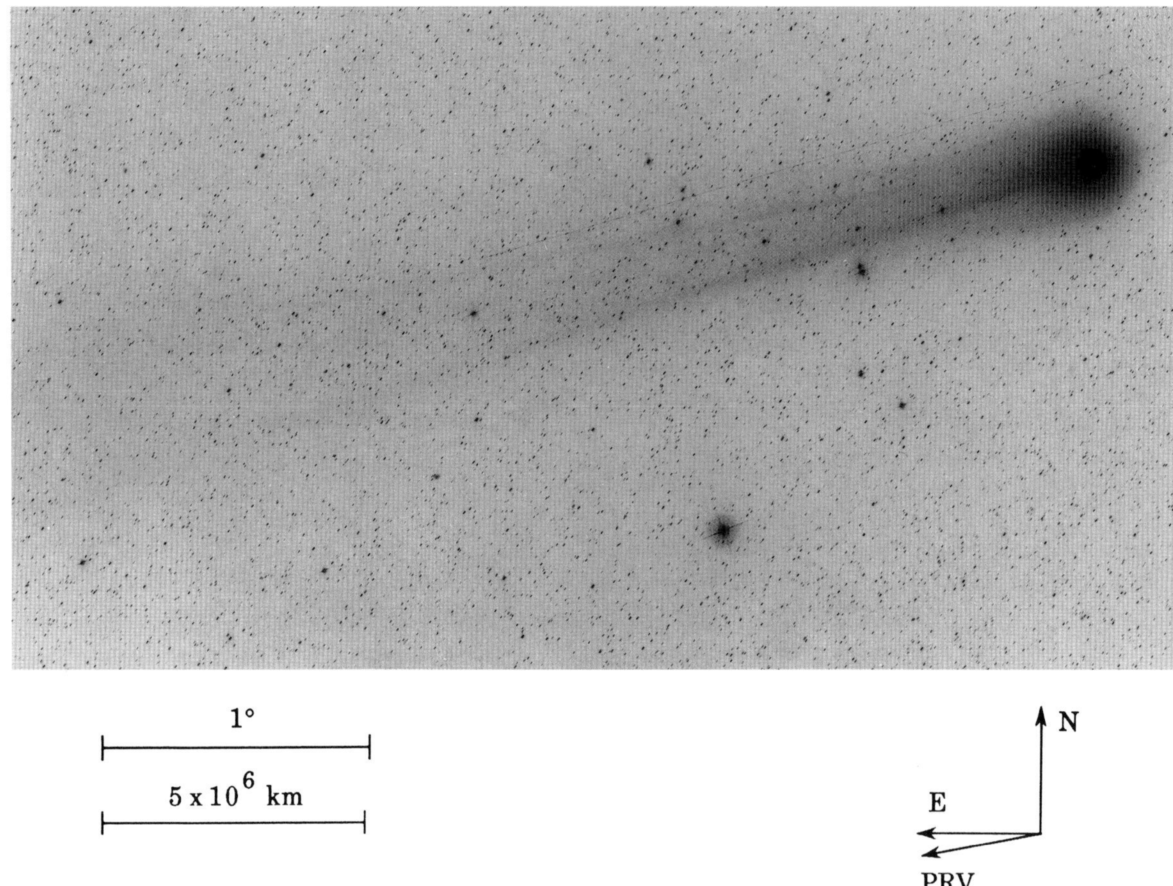

Fig. 710: 1986 May 6.00689 UT; exposure 15.0 minutes on 103a-O emulsion with no filter; r=1.71, Δ=0.96, β=149.4°; LSPN-3043.
Photograph by K. Meech, University of Michigan/CTIO.

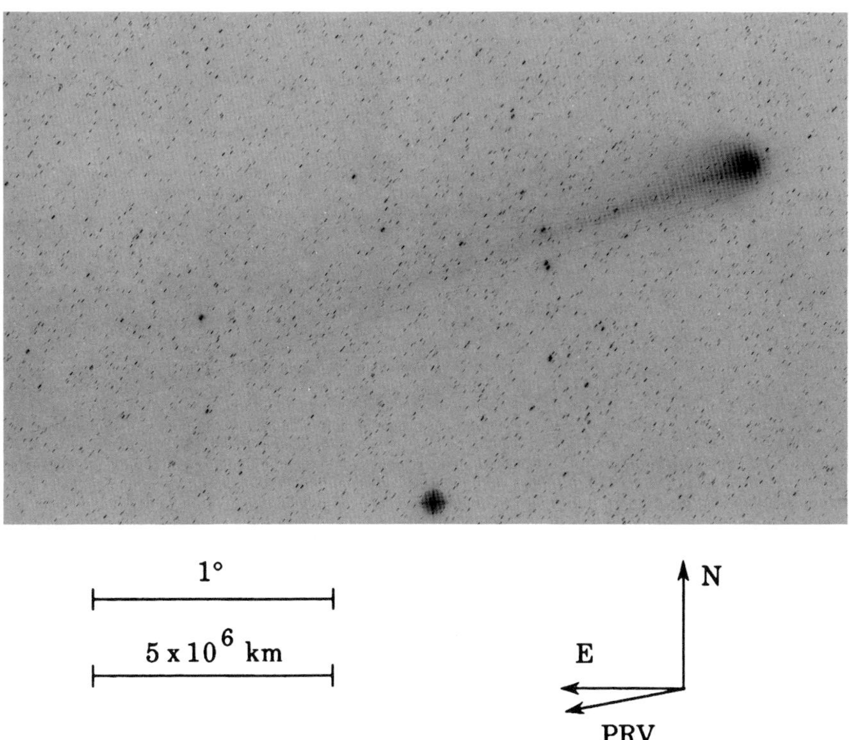

Fig. 711: 1986 May 6.00938 UT; exposure 25.0 minutes on 098-04 emulsion with RG-630 filter; r=1.71, Δ=0.96, β=149.4°; LSPN-2927.
Photograph by G. Pizarro/O. Pizarro, European Southern Observatory.

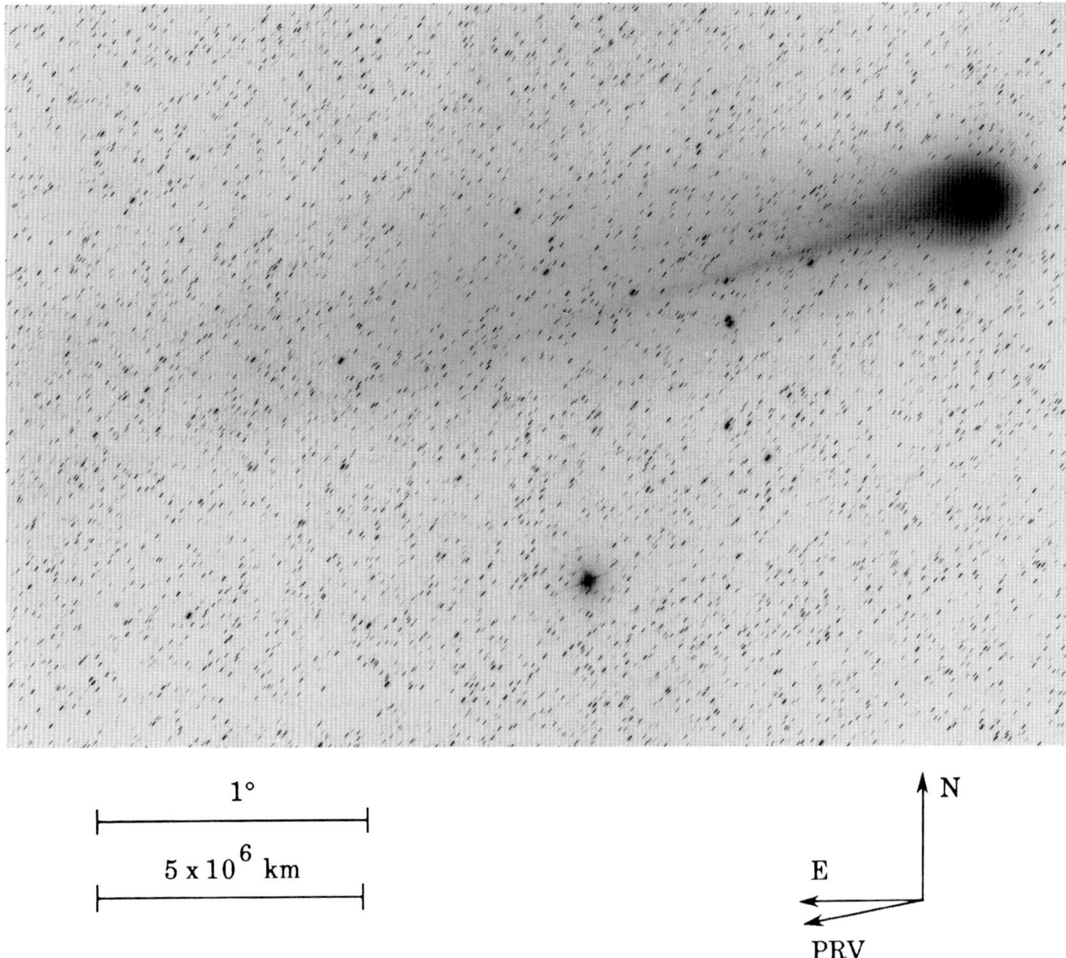

Fig. 712: 1986 May 6.13056 UT; exposure 30.0 minutes on IIIa-J emulsion with no filter; r=1.71, Δ=0.96, β=149.4°; LSPN-3045.
Photograph by K. Meech, University of Michigan/CTIO.

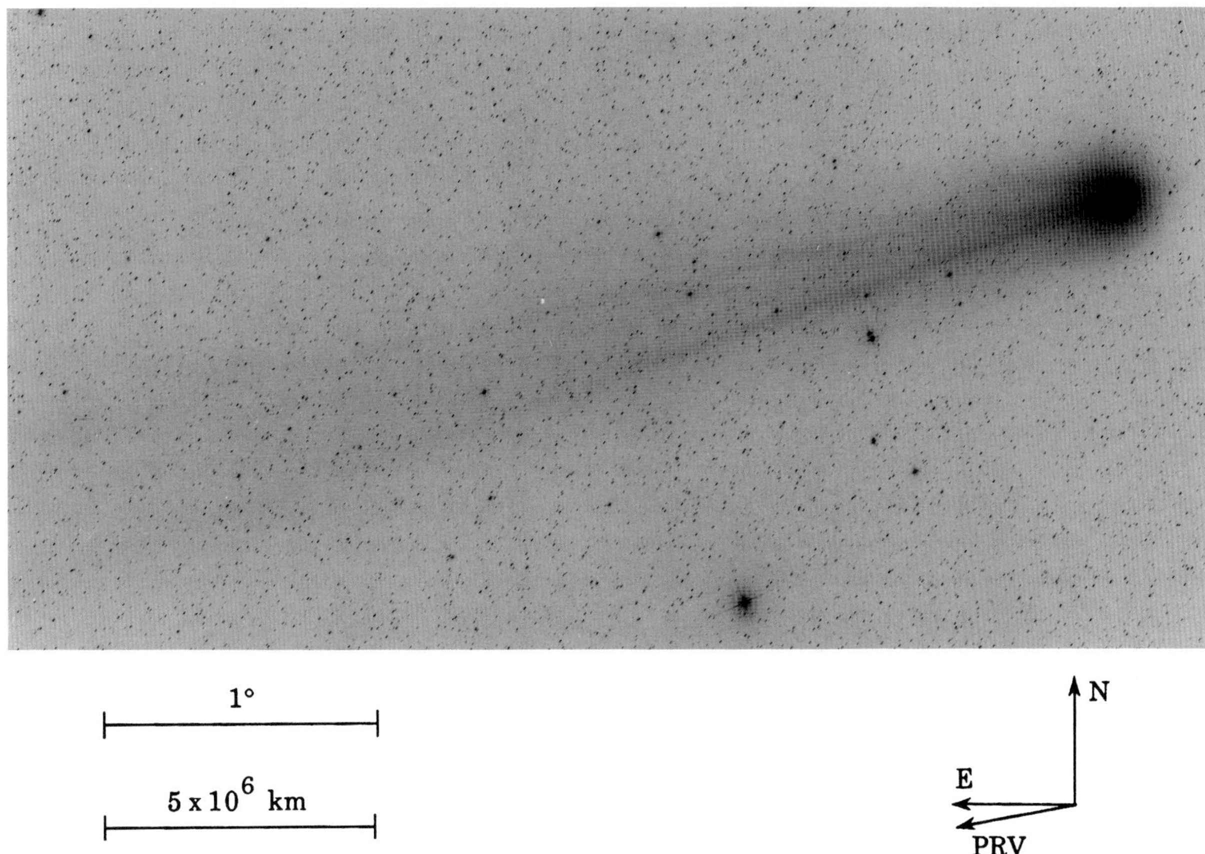

Fig. 713: 1986 May 6.15660 UT; exposure 15.0 minutes on 103a-O emulsion with no filter; r=1.71, Δ=0.96, β=149.4°; LSPN-3046.
Photograph by K. Meech, University of Michigan/CTIO.

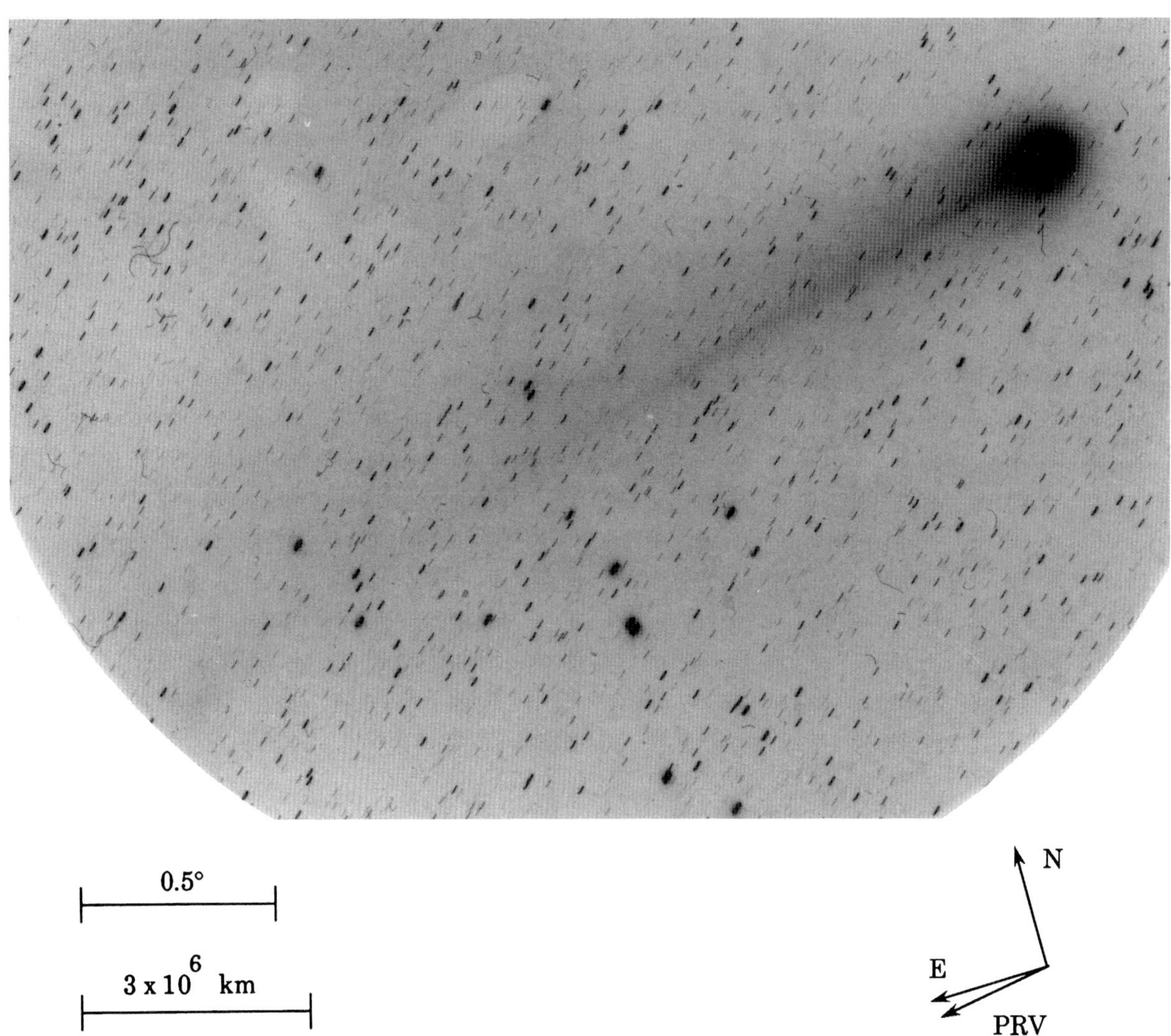

Fig. 714: 1986 May 6.92465 UT; exposure 41.0 minutes on Tech. Pan 2415 emulsion with no filter; r=1.72, Δ=0.98, β=149.2°; LSPN-3511.
Photograph by R. Catchpole, University of Perugia/South African Astronomical Observatory.

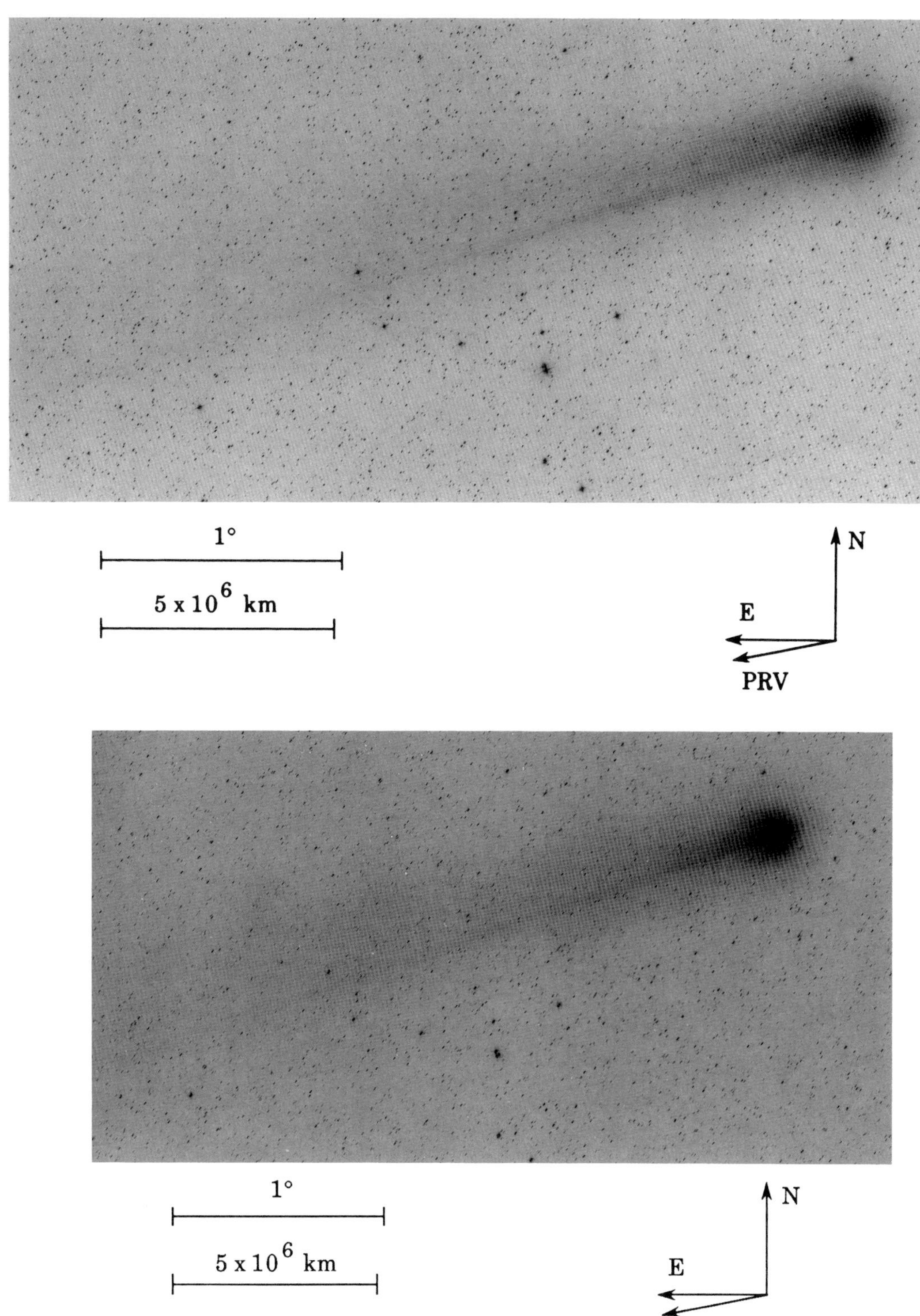

(top) Fig. 715: 1986 May 7.00132 UT; exposure 15.0 minutes on 103a-O emulsion with no filter; r=1.72, Δ=0.99, β=149.2°; LSPN-3053.
Photograph by K. Meech, University of Michigan/CTIO.

(bottom) Fig. 716: 1986 May 7.00660 UT; exposure 25.0 minutes on IIa-O emulsion with GG-385 filter; r=1.72, Δ=0.99, β=149.2°; LSPN-2929.
Photograph by G. Pizarro/O. Pizarro, European Southern Observatory.

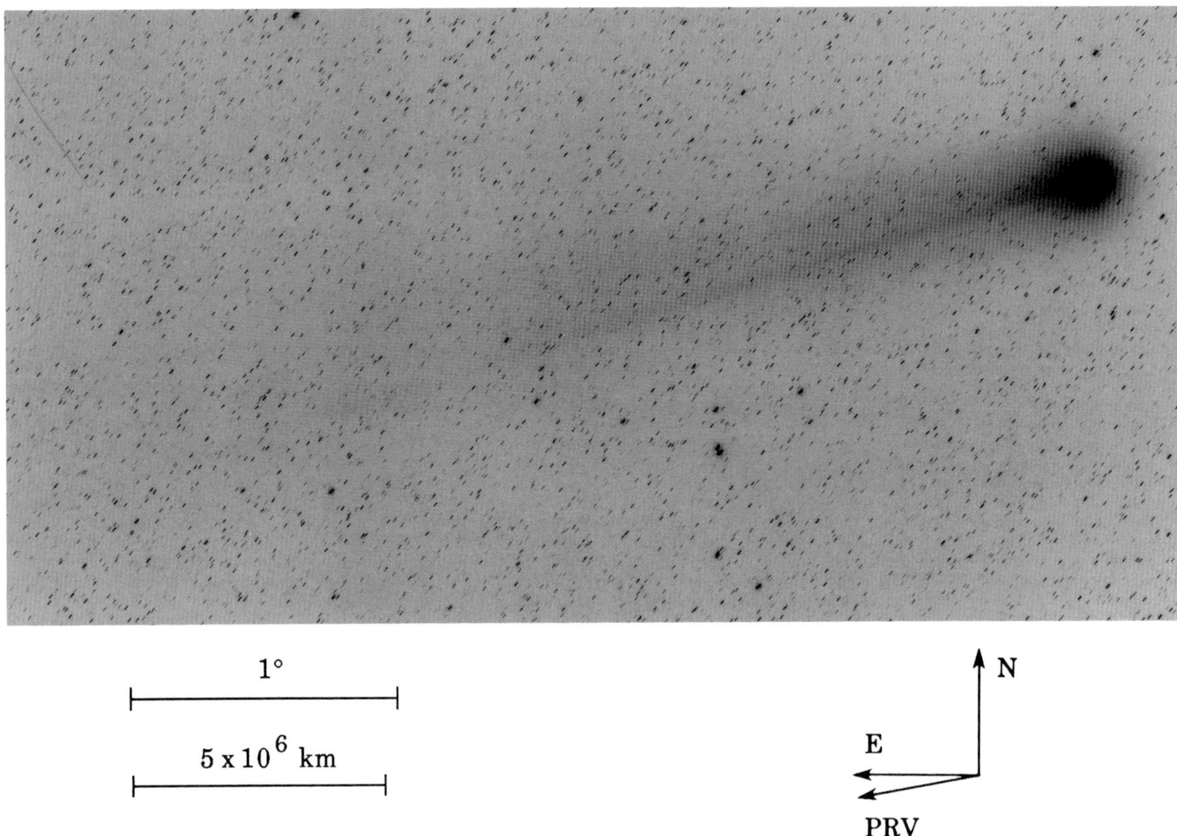

Fig. 717: 1986 May 7.02292 UT; exposure 30.0 minutes on IIIa-J emulsion with no filter; r=1.72, Δ=0.99, β=149.2°; LSPN-3054.
Photograph by K. Meech, University of Michigan/CTIO.

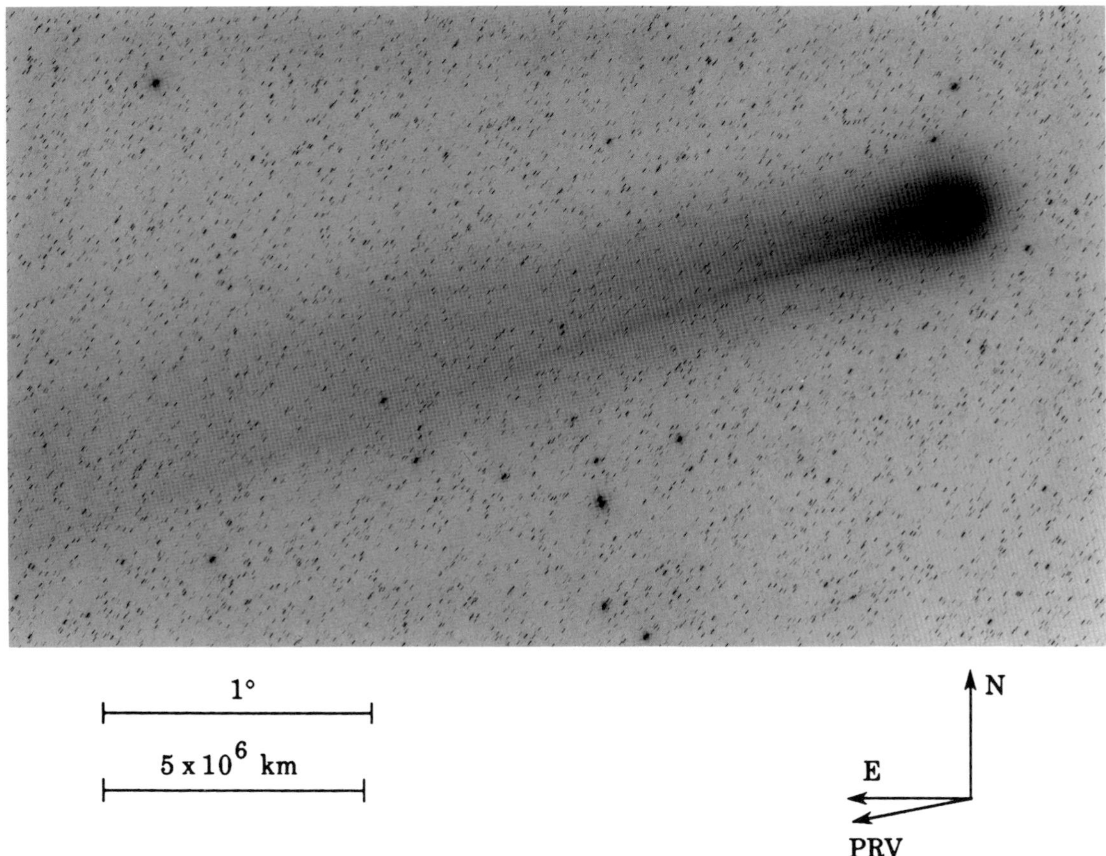

Fig. 718: 1986 May 7.05306 UT; exposure 30.0 minutes on 103a-O emulsion with no filter; r=1.72, Δ=0.99, β=149.2°; LSPN-3055.
Photograph by K. Meech, University of Michigan/CTIO.

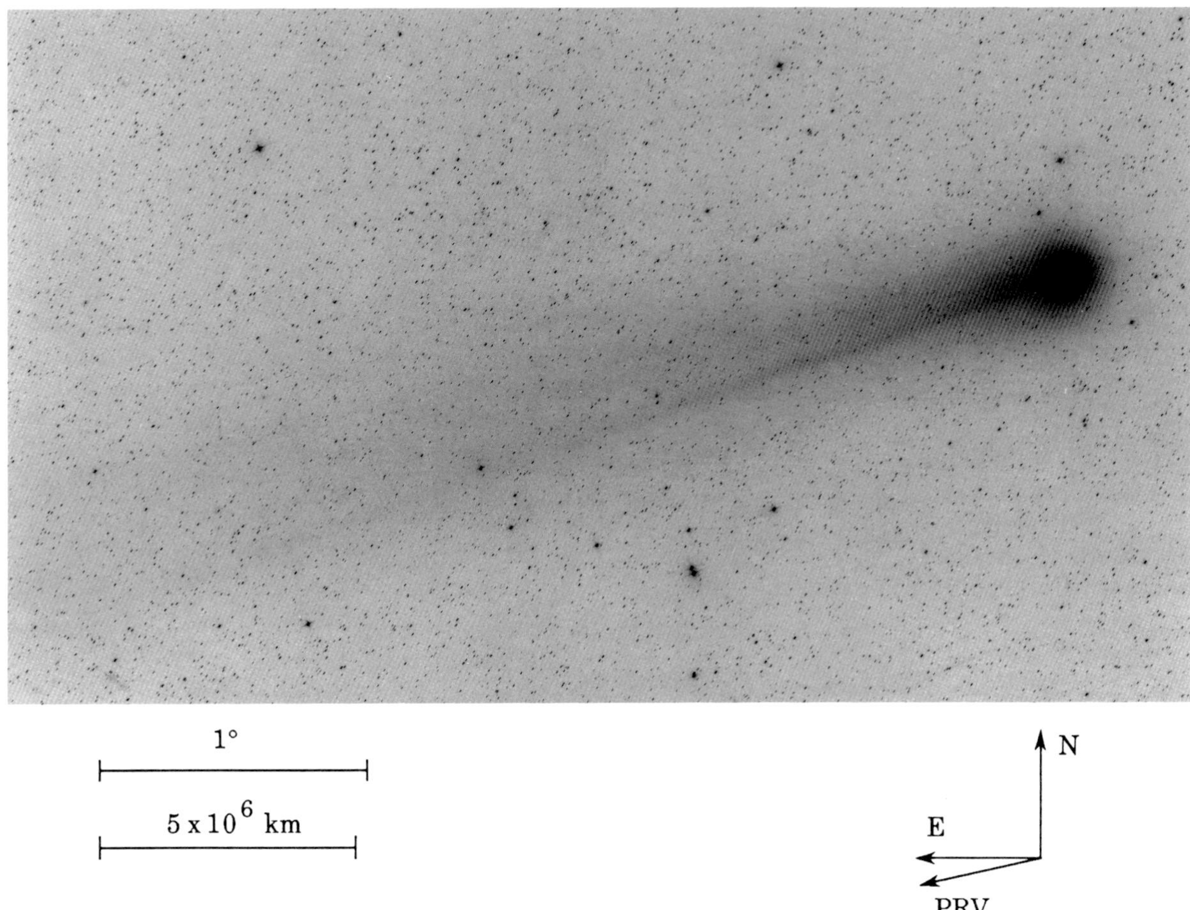

Fig. 719: 1986 May 7.10625 UT; exposure 15.0 minutes on 103a-O emulsion with no filter; r=1.72, Δ=0.99, β=149.2°; LSPN-3057.
Photograph by K. Meech, University of Michigan/CTIO.

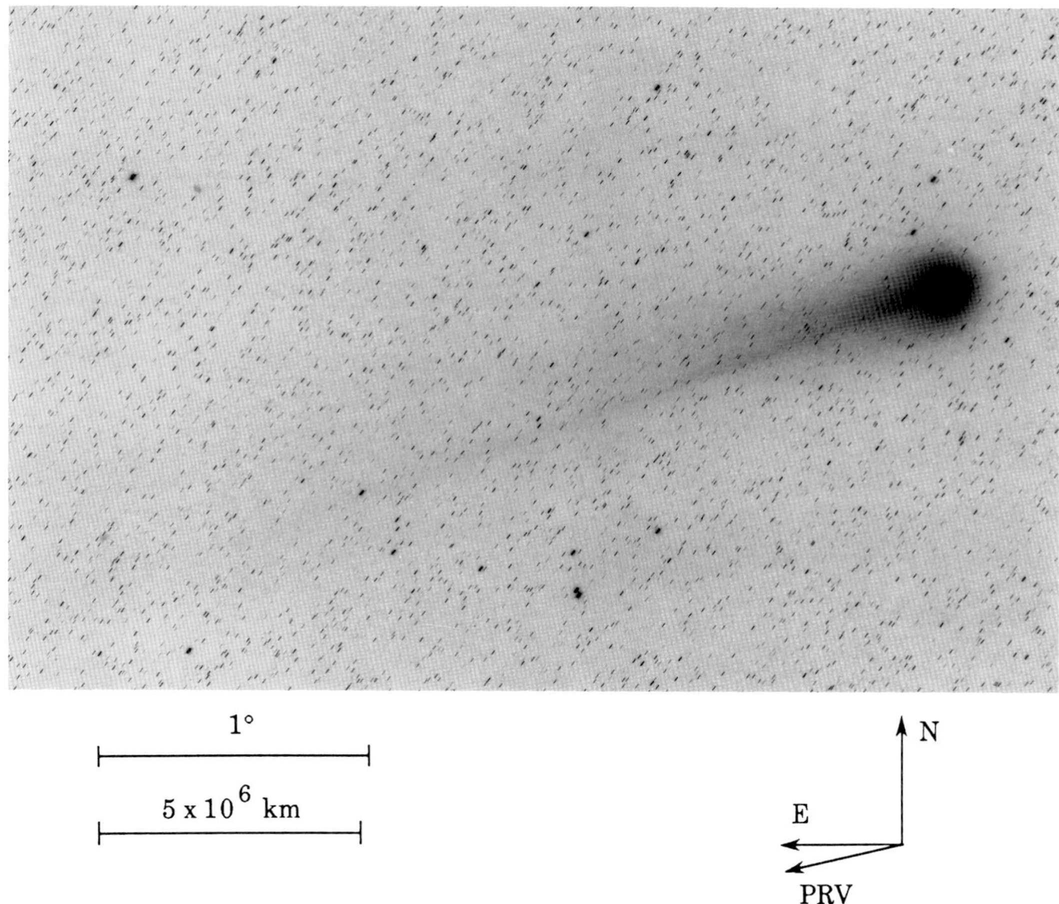

Fig. 720: 1986 May 7.13264 UT; exposure 30.0 minutes on IIIa-J emulsion with no filter; r=1.72, Δ=0.99, β=149.2°; LSPN-3058.
Photograph by K. Meech, University of Michigan/CTIO.

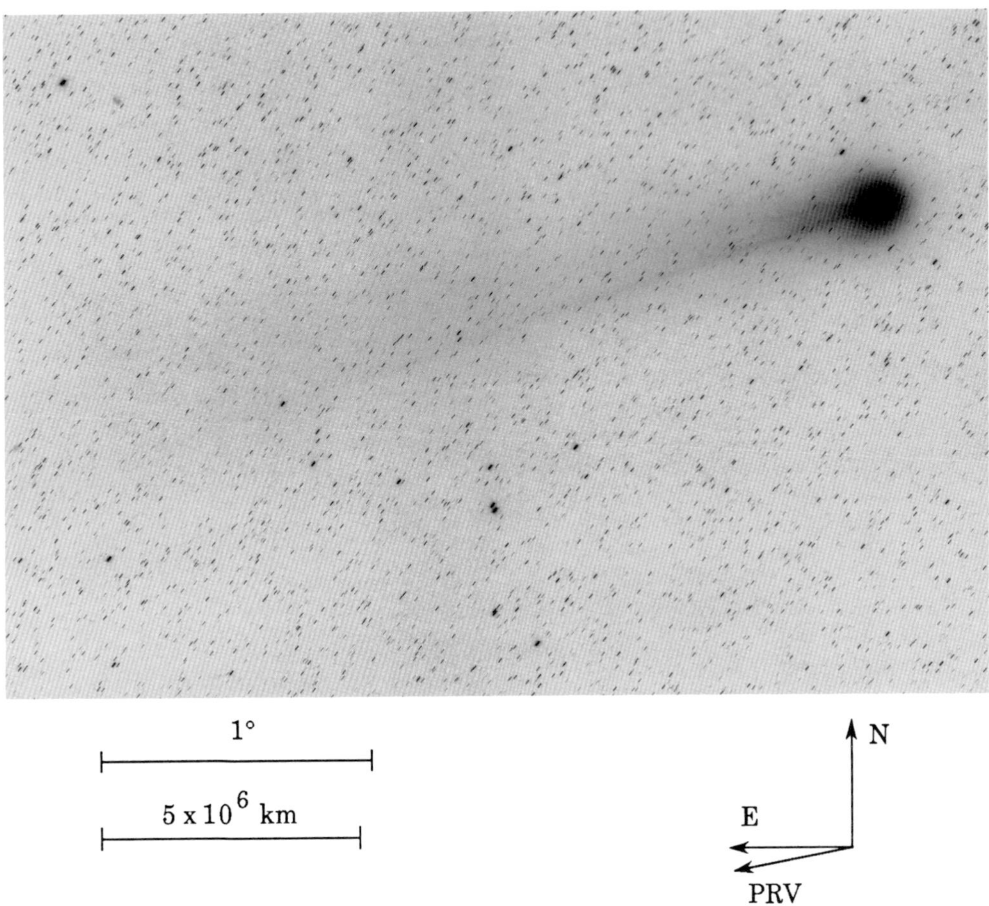

Fig. 721: 1986 May 7.16944 UT; exposure 30.0 minutes on IIIa-J emulsion with no filter; r=1.72, Δ=0.99, β=149.2°; LSPN-3059.
Photograph by K. Meech, University of Michigan/CTIO.

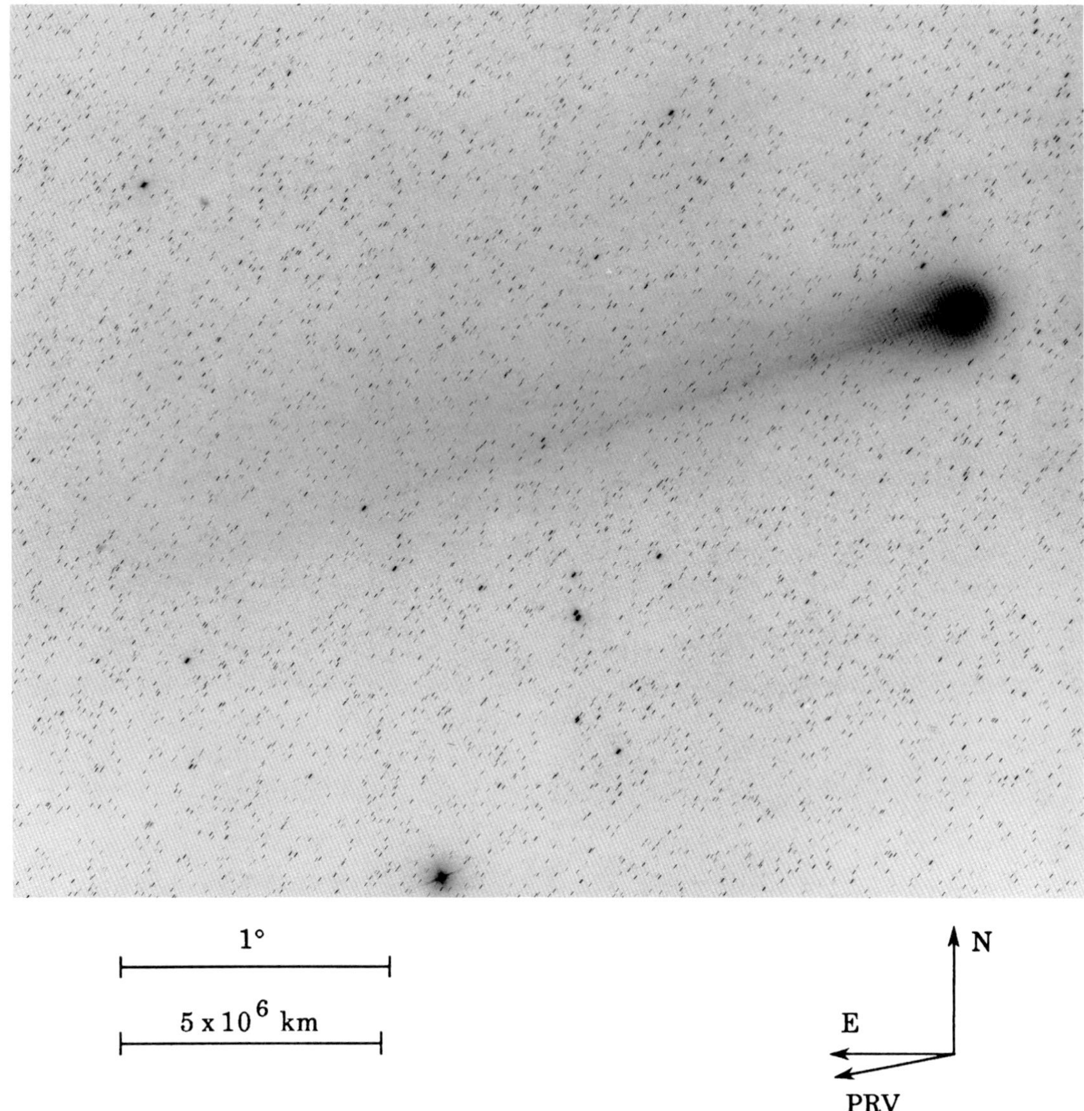

Fig. 722: 1986 May 7.20747 UT; exposure 30.0 minutes on IIIa-J emulsion with no filter; r=1.72, Δ=0.99, β=149.2°; LSPN-3060.
Photograph by K. Meech, University of Michigan/CTIO.

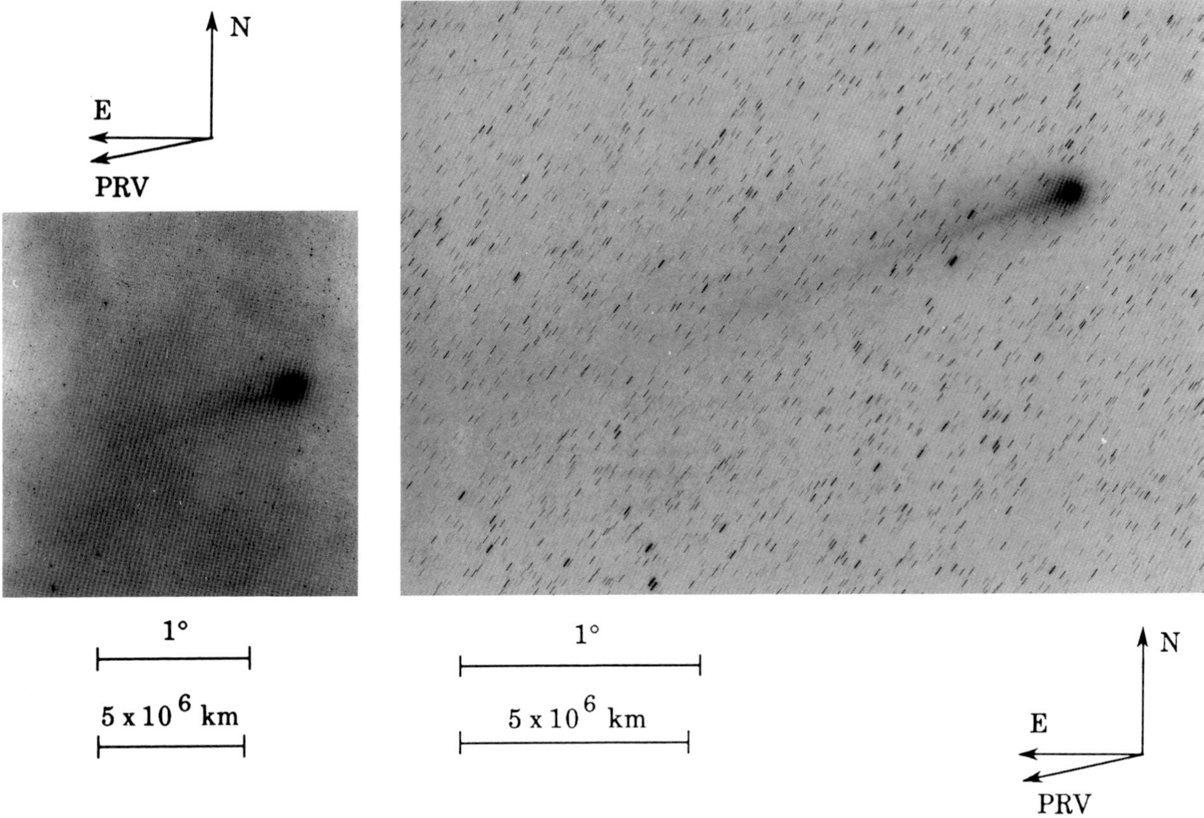

(left) Fig. 723: 1986 May 7.51701 UT; exposure 15.0 minutes on IIa-O emulsion with no filter; r=1.73, Δ=1.00, β=149.1°; LSPN-1349.
Photograph by T. Tsujimura, University Kyoto Observatory.

(right) Fig. 724: 1986 May 7.98368 UT; exposure 25.0 minutes on 098-04 emulsion with RG-630 filter; r=1.74, Δ=1.02, β=149.0°; LSPN-2930.
Photograph by G. Pizarro, European Southern Observatory.

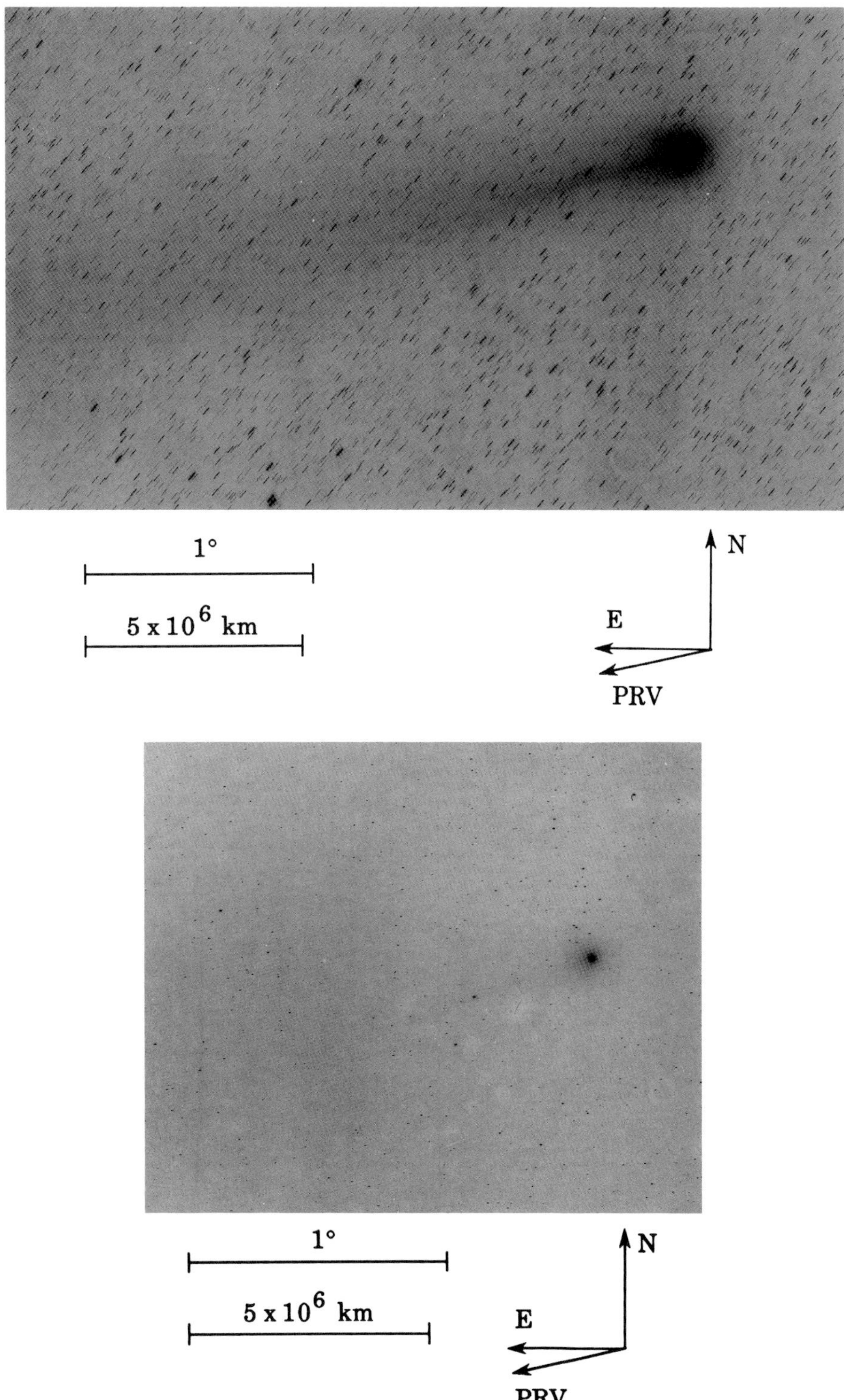

(top) Fig. 725: 1986 May 8.01632 UT; exposure 25.0 minutes on IIa-O emulsion with GG-385 filter; r=1.74, Δ=1.02, β=149.0°; LSPN-2931.
Photograph by G. Pizarro, European Southern Observatory.

(bottom) Fig. 726: 1986 May 8.11250 UT; exposure 10.0 minutes on IIa-O emulsion with GG-13 filter; r=1.74, Δ=1.02, β=149.0°; LSPN-660.
Photograph by A. Heiser, A. J. Dyer Observatory.

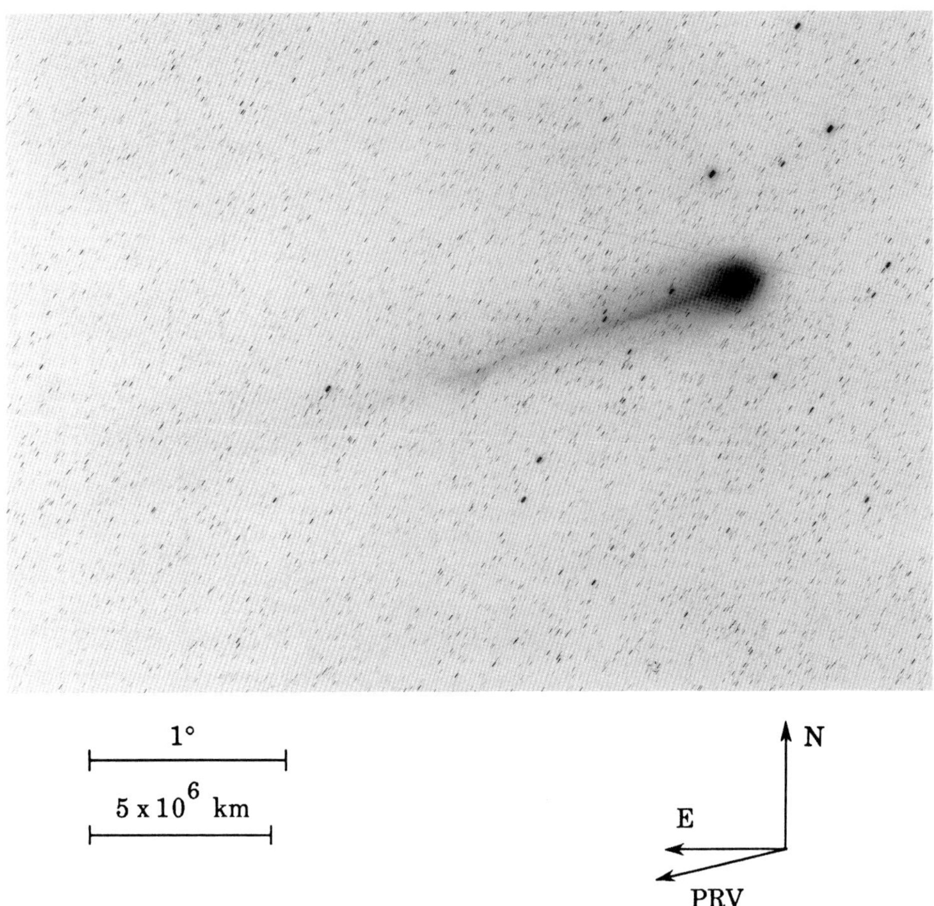

Fig. 727: 1986 May 9.42227 UT; exposure 60.0 minutes on IIa-O emulsion with GG-395 filter; $r=1.76$, $\Delta=1.06$, $\beta=148.8°$; LSPN-2447.
Photograph from Royal Observatory/UK Schmidt Telescope Unit.

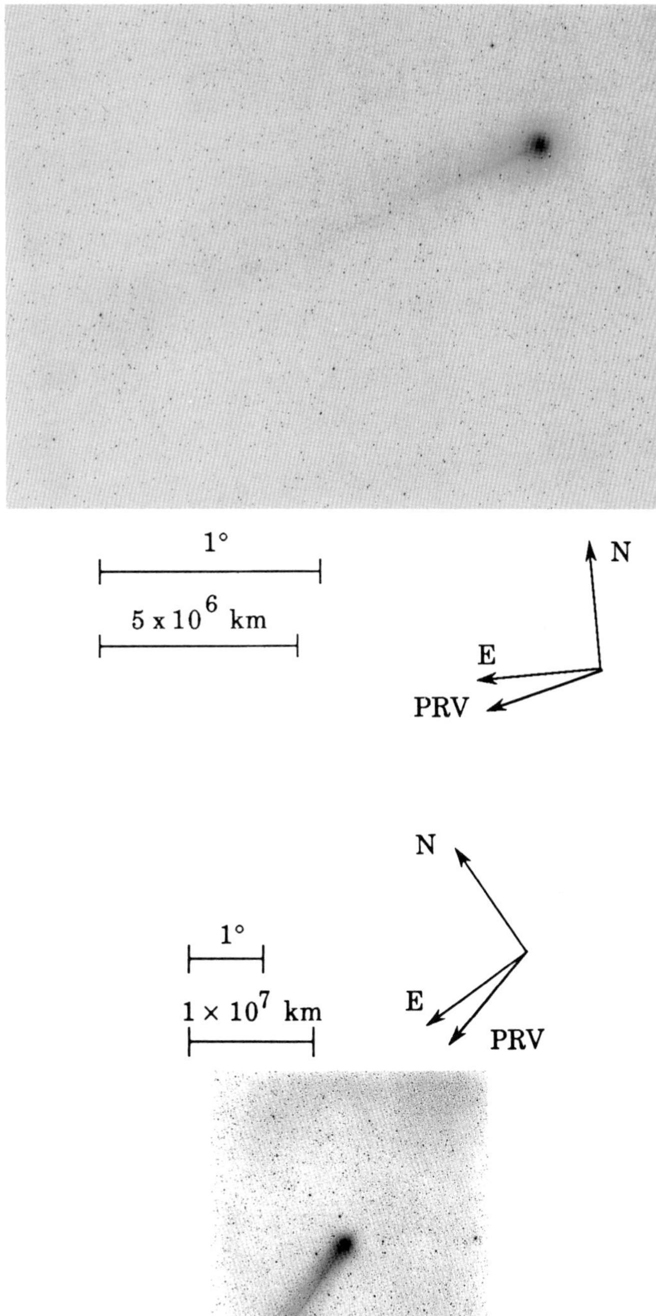

(top) Fig. 728: 1986 May 9.53598 UT; exposure 15.0 minutes on IIa-O emulsion with no filter; r=1.76, Δ=1.07, β=148.8°; LSPN-1641.
Photograph by H. Maehara, Kiso Observatory.

(bottom) Fig. 729: 1986 May 10.18472 UT; exposure 8.0 minutes on IIa-O emulsion with no filter; r=1.77, Δ=1.09, β=148.7°; LSPN-984.
Photograph by G. Emerson, E. E. Barnard Observatory.

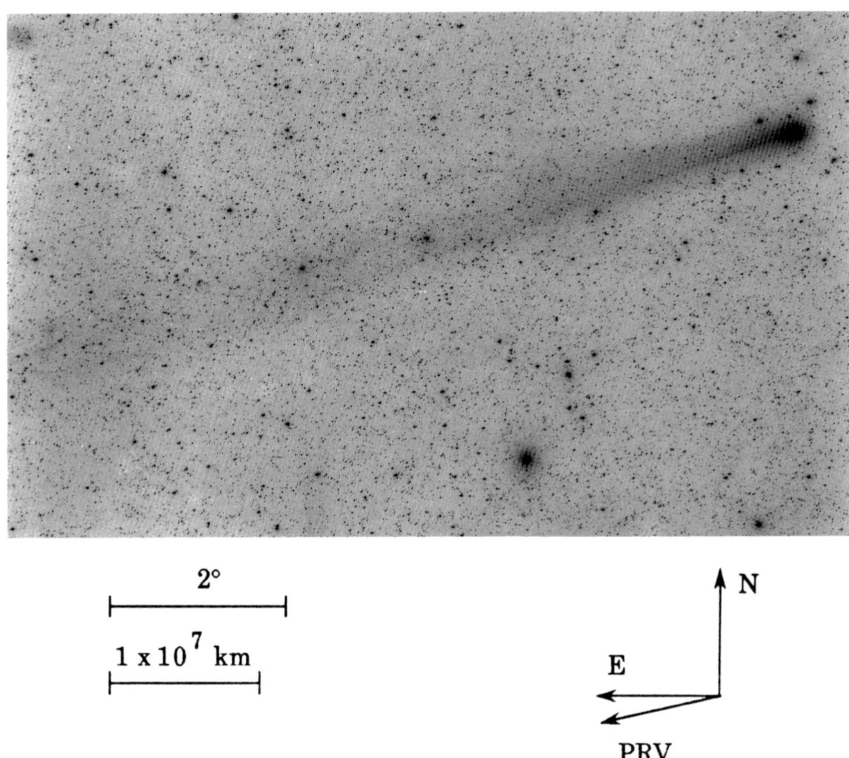

Fig. 730: 1986 May 10.21944 UT; exposure 30.0 minutes on IIa-O emulsion with no filter; r = 1.77, Δ = 1.09, β = 148.7°; LSPN-946.
Photograph by E. Moore/E. Marr, Joint Observatory for Cometary Research.

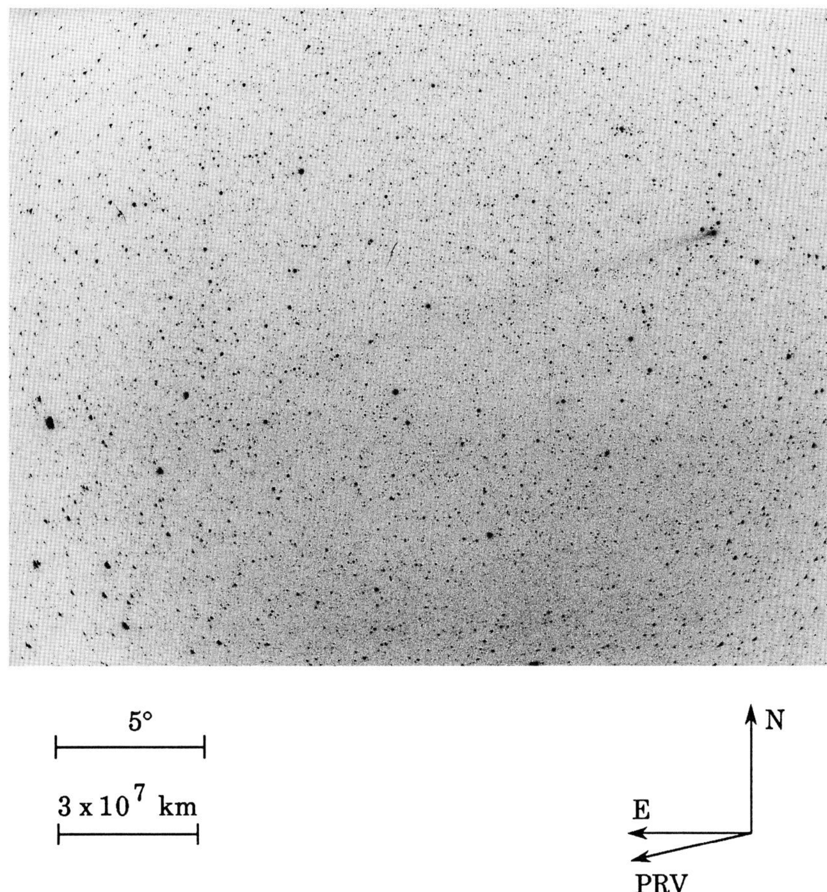

Fig. 731: 1986 May 10.22920 UT; exposure 10.0 minutes on Tri-X emulsion; r=1.77, Δ=1.09, β=148.7°; AON-852104.
Photograph by S. Edberg, USA.

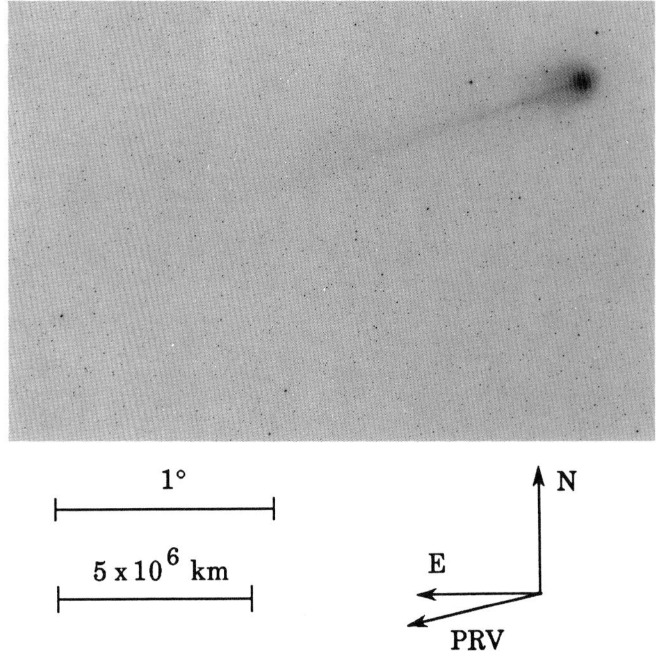

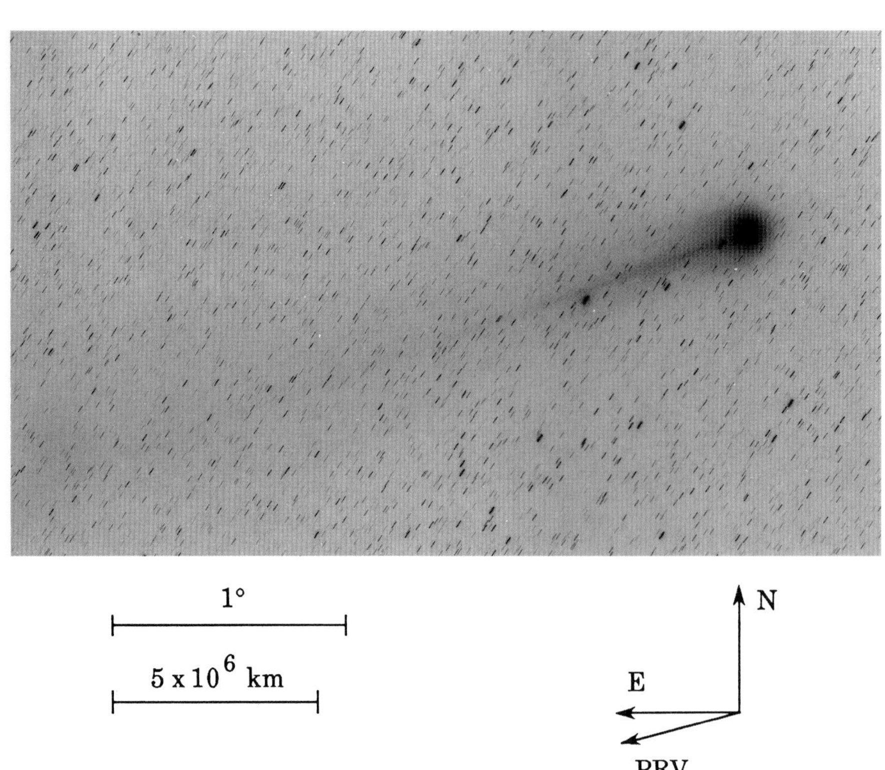

(top) Fig. 732: 1986 May 10.48160 UT; exposure 15.0 minutes on IIa-O emulsion with no filter; $r=1.77$, $\Delta=1.10$, $\beta=148.7°$; LSPN-1642.
Photograph by H. Maehara, Kiso Observatory.

(bottom) Fig. 733: 1986 May 11.04896 UT; exposure 25.0 minutes on IIa-O emulsion with GG-385 filter; $r=1.78$, $\Delta=1.12$, $\beta=148.7°$; LSPN-2932.
Photograph by G. Pizarro, European Southern Observatory.

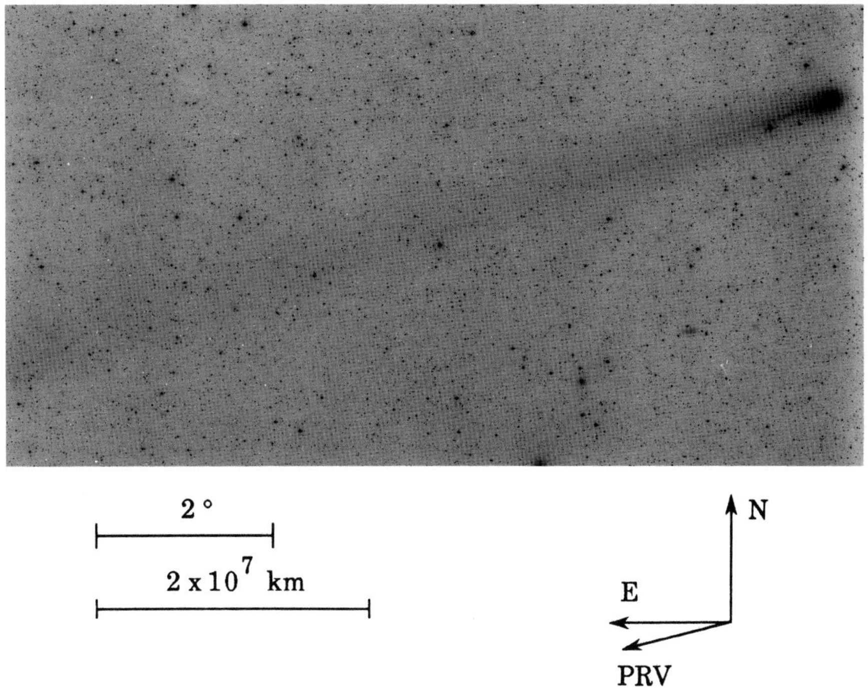

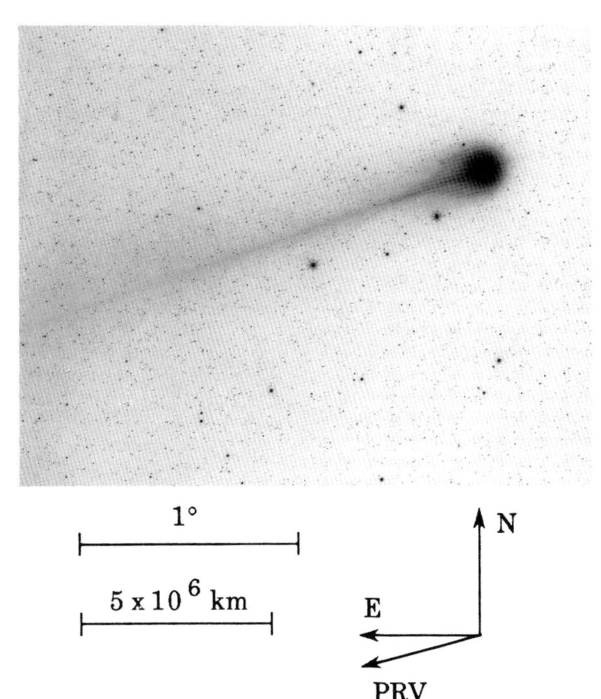

(top) Fig. 734: 1986 May 11.18194 UT; exposure 10.0 minutes on IIa-O emulsion with no filter; r=1.78, Δ=1.12, β=148.7°; LSPN-947.
Photograph by E. Moore/E. Marr, Joint Observatory for Cometary Research.

(bottom) Fig. 735: 1986 May 11.40465 UT; exposure 27.0 minutes on IIIa-J emulsion with GG-385 filter; r=1.79, Δ=1.13, β=148.7°; LSPN-2165.
Photograph by P. Magnusson, Uppsala Southern Station.

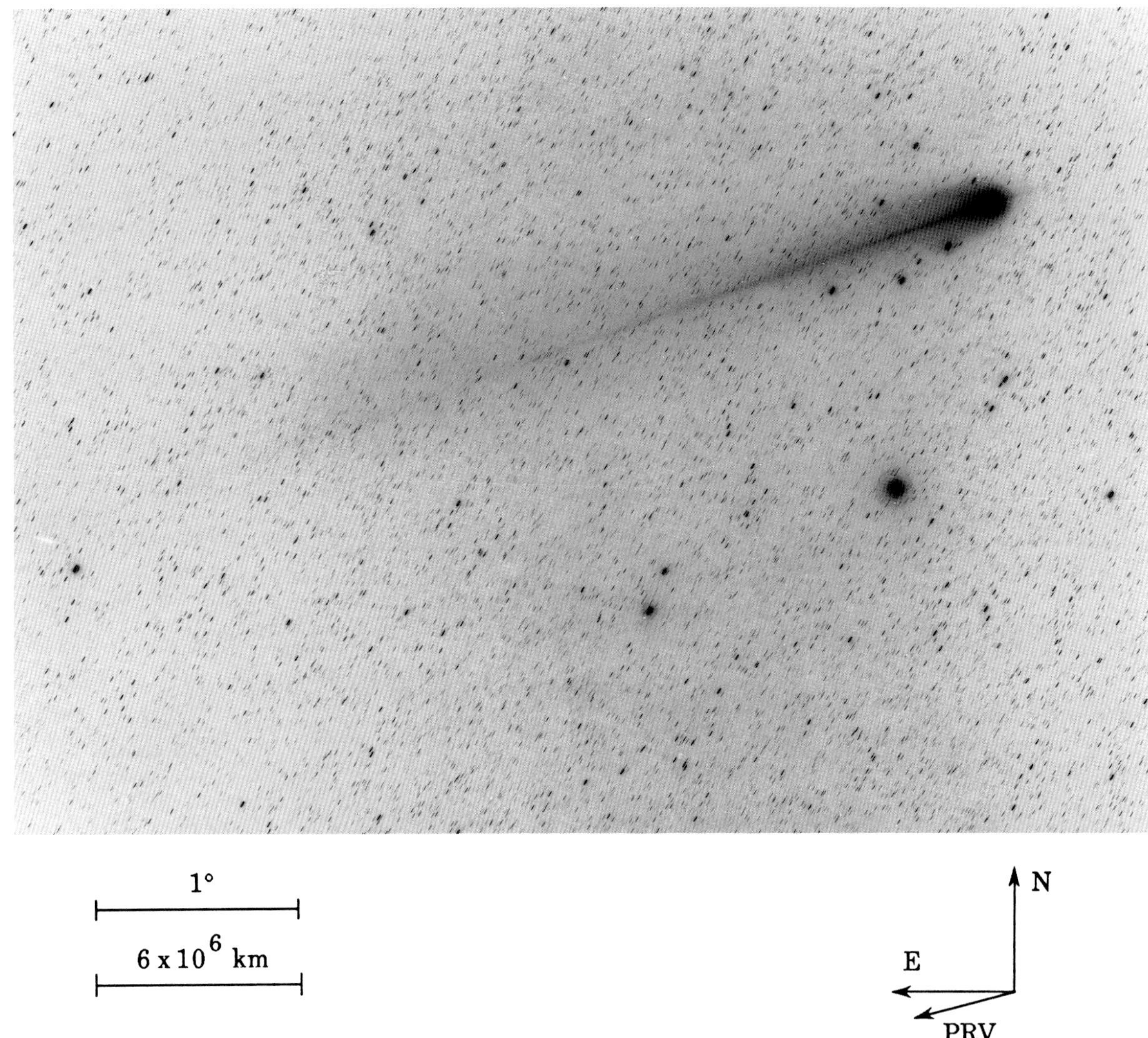

Fig. 736: 1986 May 11.41163 UT; exposure 75.0 minutes on IIIa-F emulsion with RG-630 filter; r=1.79, Δ=1.13, β=148.7°; LSPN-2448.
Photograph from Royal Observatory/UK Schmidt Telescope Unit.

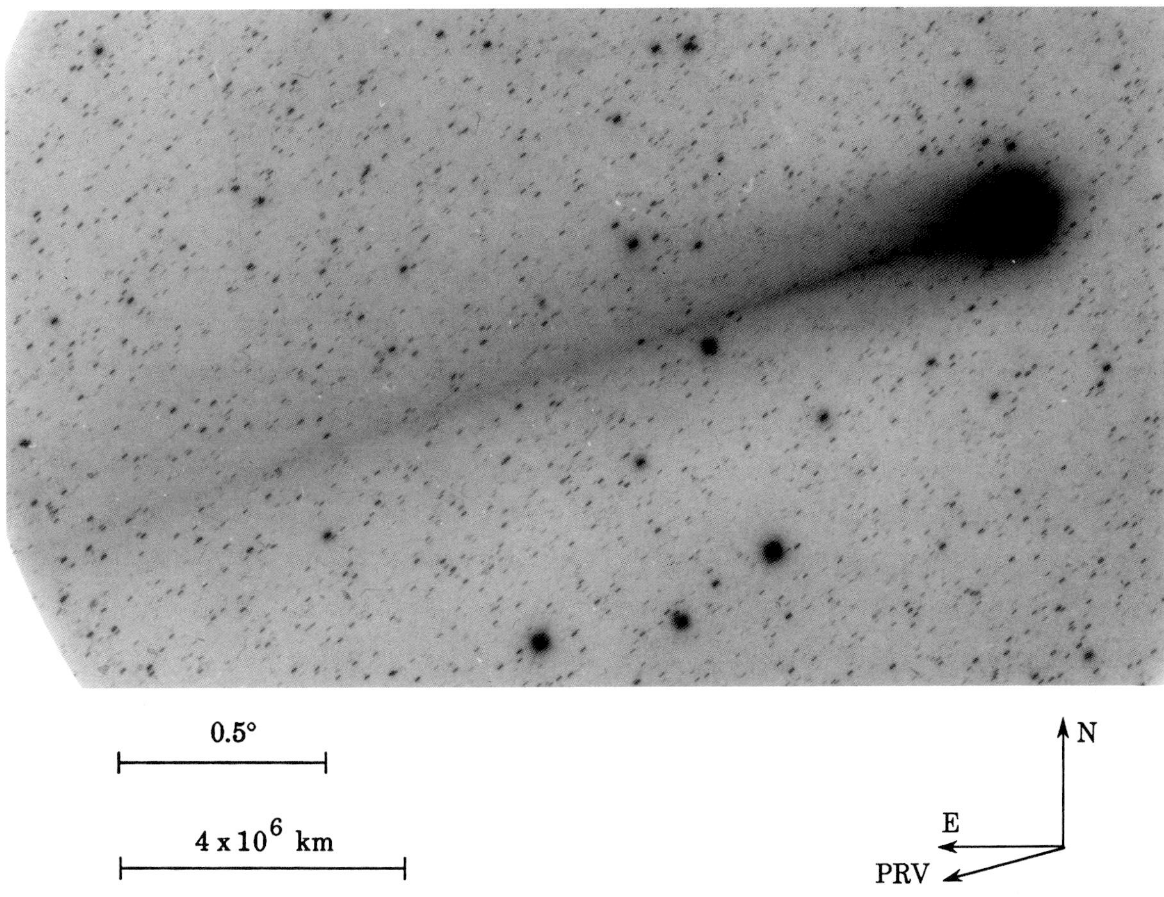

Fig. 737: 1986 May 11.86875 UT; exposure 54.0 minutes on Tech. Pan 2415 emulsion with no filter; r=1.79, Δ=1.14, β=148.6°; LSPN-3512.
Photograph by R. Catchpole, University of Perugia/South African Astronomical Observatory.

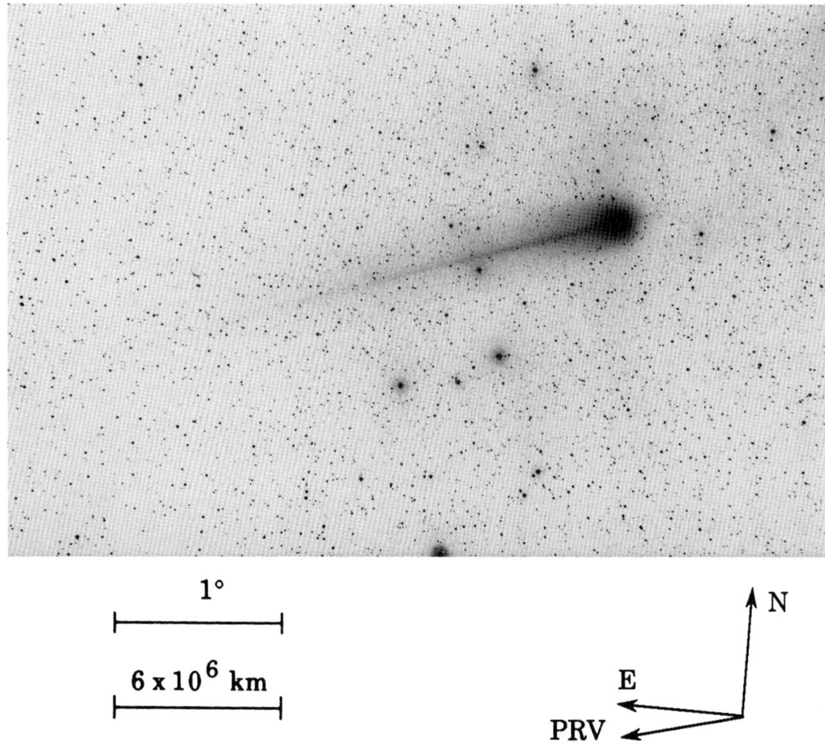

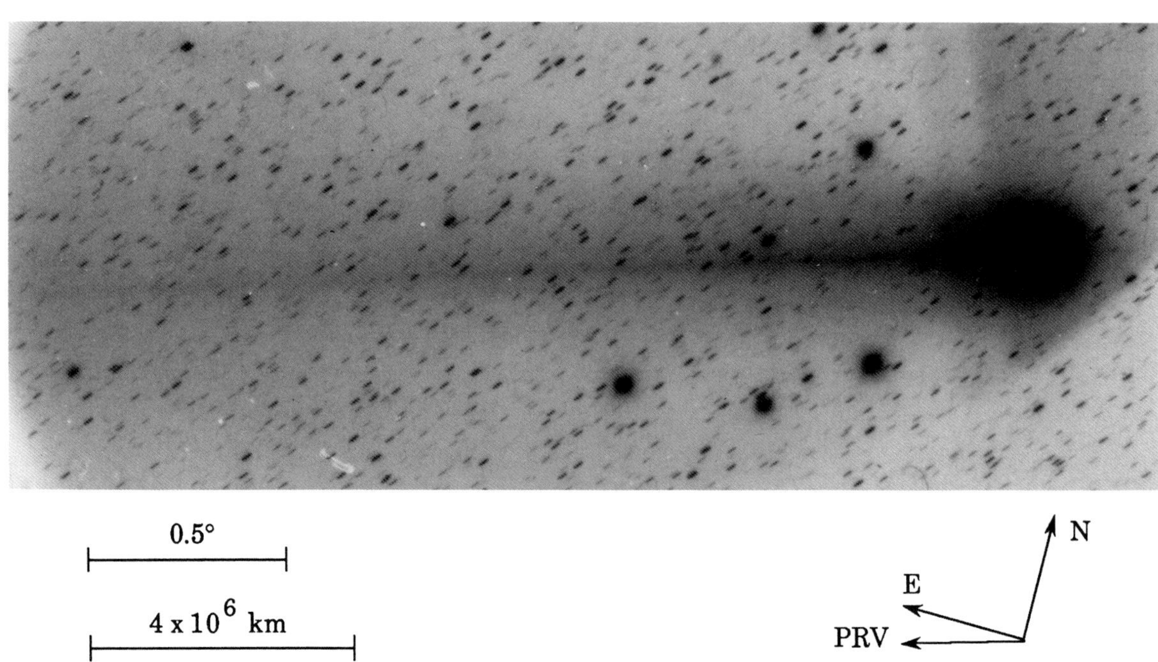

(top) Fig. 738: 1986 May 12.84310 UT; exposure 9.5 minutes on Tech. Pan 2415 emulsion; r=1.81, Δ=1.17, β=148.6°; AON-852130.
Photograph by M. Jäger, Austria.

(bottom) Fig. 739: 1986 May 12.89375 UT; exposure 40.0 minutes on Tech. Pan 2415 emulsion with no filter; r=1.81, Δ=1.18, β=148.6°; LSPN-3513.
Photograph by R. Catchpole, University of Perugia/South African Astronomical Observatory.

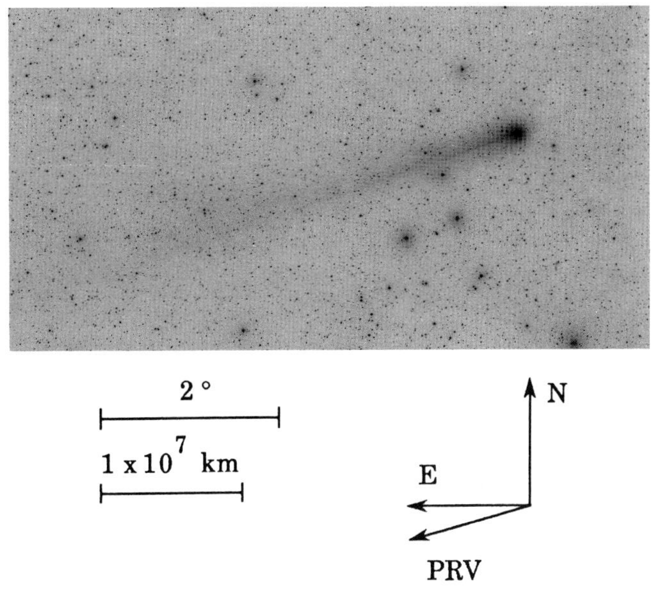

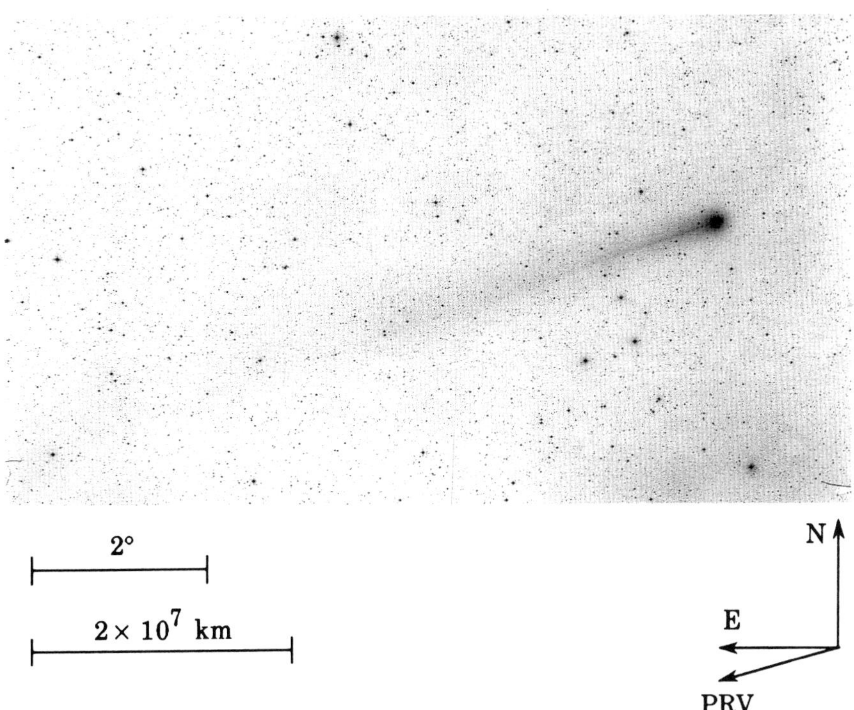

(top) Fig. 740: 1986 May 13.15833 UT; exposure 30.0 minutes on IIa-O emulsion with no filter; $r=1.81$, $\Delta=1.18$, $\beta=148.6°$; LSPN-953.
Photograph by E. Moore/E. Marr, Joint Observatory for Cometary Research.

(bottom) Fig. 741: 1986 May 14.17292 UT; exposure 10.0 minutes on IIa-O emulsion with no filter; $r=1.82$, $\Delta=1.22$, $\beta=148.6°$; LSPN-955.
Photograph by E. Moore/E. Marr, Joint Observatory for Cometary Research.

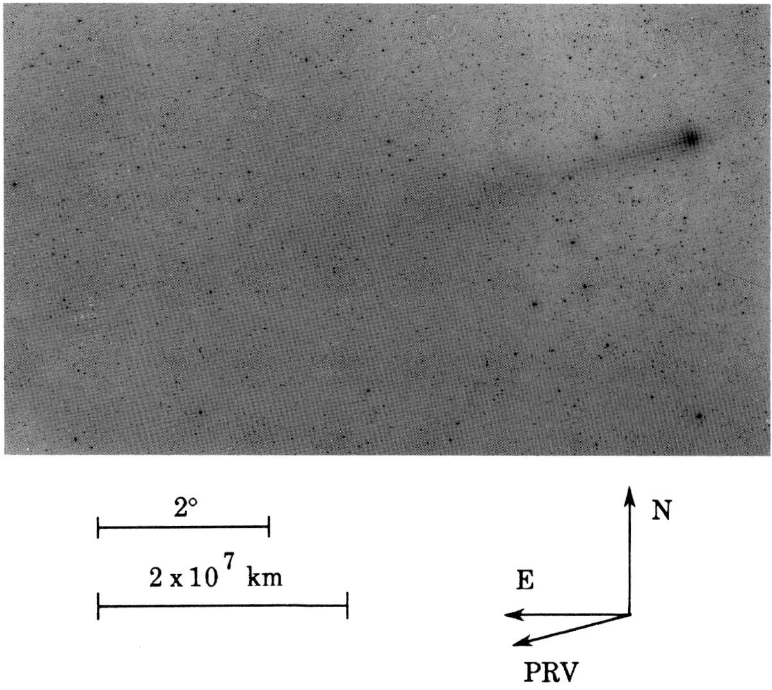

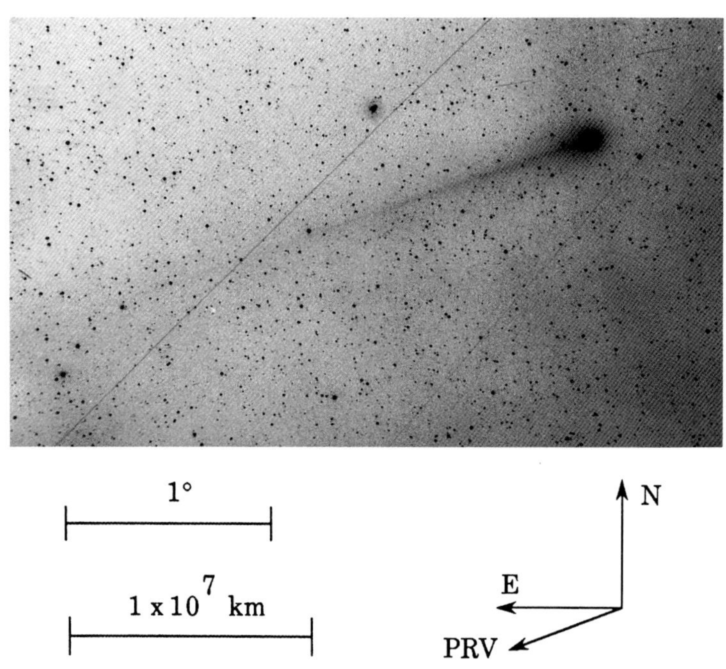

(top) Fig. 742: 1986 May 15.15104 UT; exposure 7.0 minutes on IIa-O emulsion with no filter; r=1.84, Δ=1.25, β=148.6°; LSPN-957.
Photograph by E. Moore/E. Marr, Joint Observatory for Cometary Research.

(bottom) Fig. 743: 1986 May 25.83130 UT; exposure 10.0 minutes on Tech. Pan 2415 emulsion; r=1.99, Δ=1.60, β=149.6°; AON-852143.
Photograph by M. Jäger, Austria.

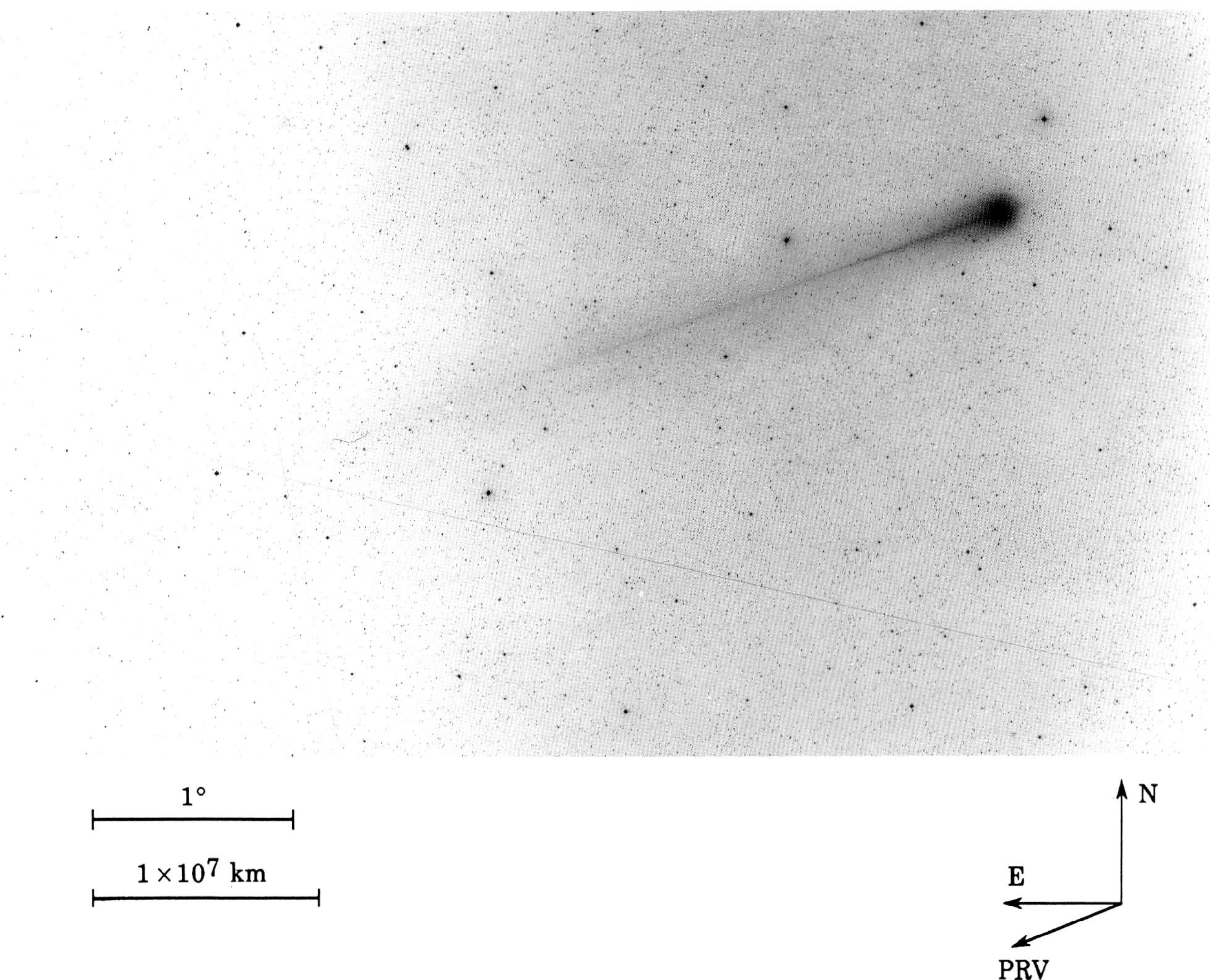

Fig. 744: 1986 May 27.37936 UT; exposure 60.0 minutes on IIa-O emulsion with GG-395 filter; r=2.01, Δ=1.65, β=149.9°; LSPN-2449.
Photograph from Royal Observatory/UK Schmidt Telescope Unit.

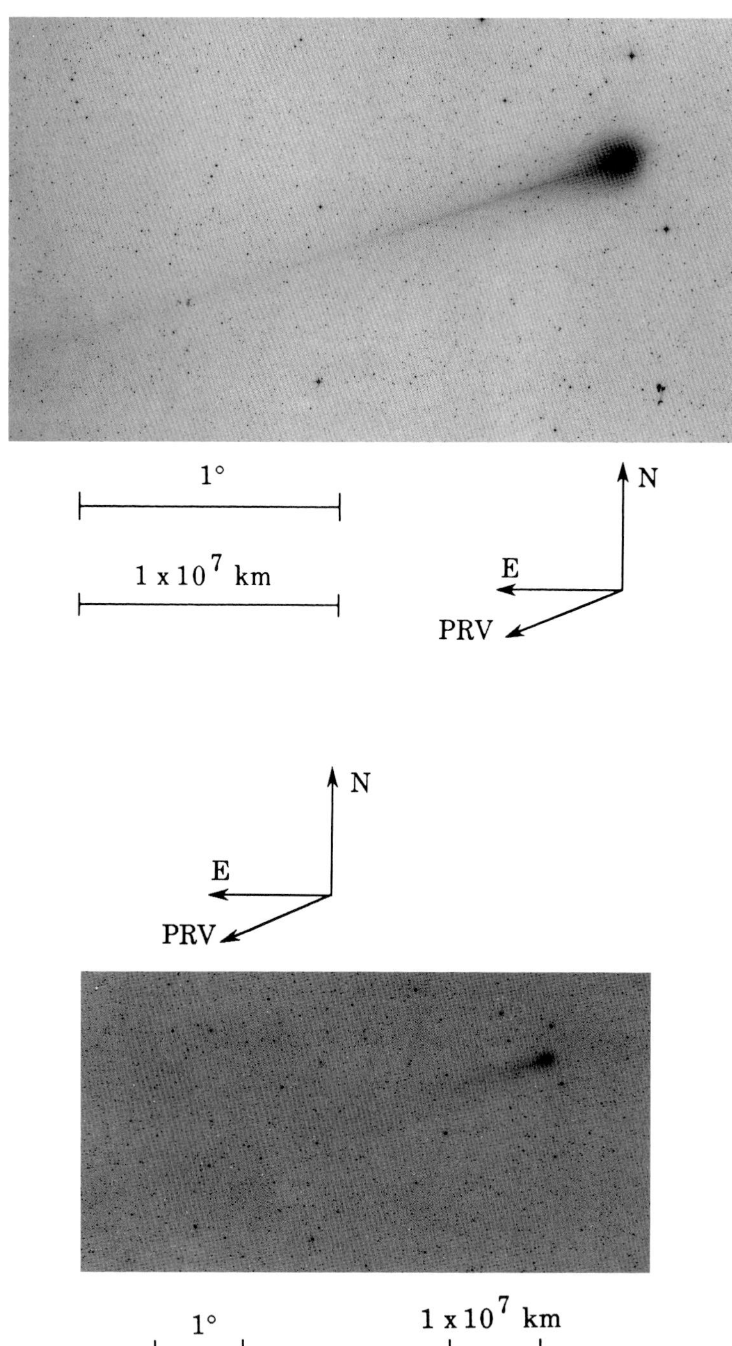

(top) Fig. 745: 1986 Jun 1.06250 UT; exposure 30.0 minutes on 103a-O emulsion with no filter; r=2.08, Δ=1.80, β=150.8°; LSPN-759.
Photograph by C. Torres/H. Wroblewski, Cerro el Roble Astronomical Observatory.

(bottom) Fig. 746: 1986 Jun 1.21493 UT; exposure 11.0 minutes on IIa-O emulsion with no filter; r=2.08, Δ=1.81, β=150.8°; LSPN-964.
Photograph by E. Moore, Joint Observatory for Cometary Research.

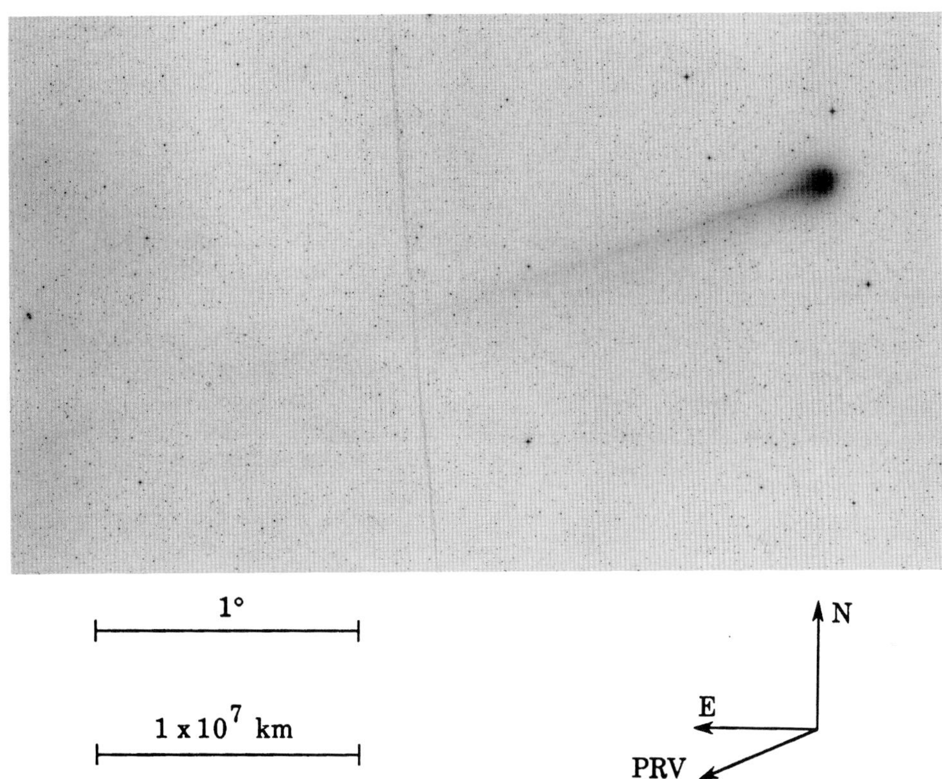

Fig. 747: 1986 Jun 2.03125 UT; exposure 30.0 minutes on 103a-O emulsion with no filter; r=2.09, Δ=1.84, β=151.0°; LSPN-760.
Photograph by C. Torres/H. Wroblewski, Cerro el Roble Astronomical Observatory.

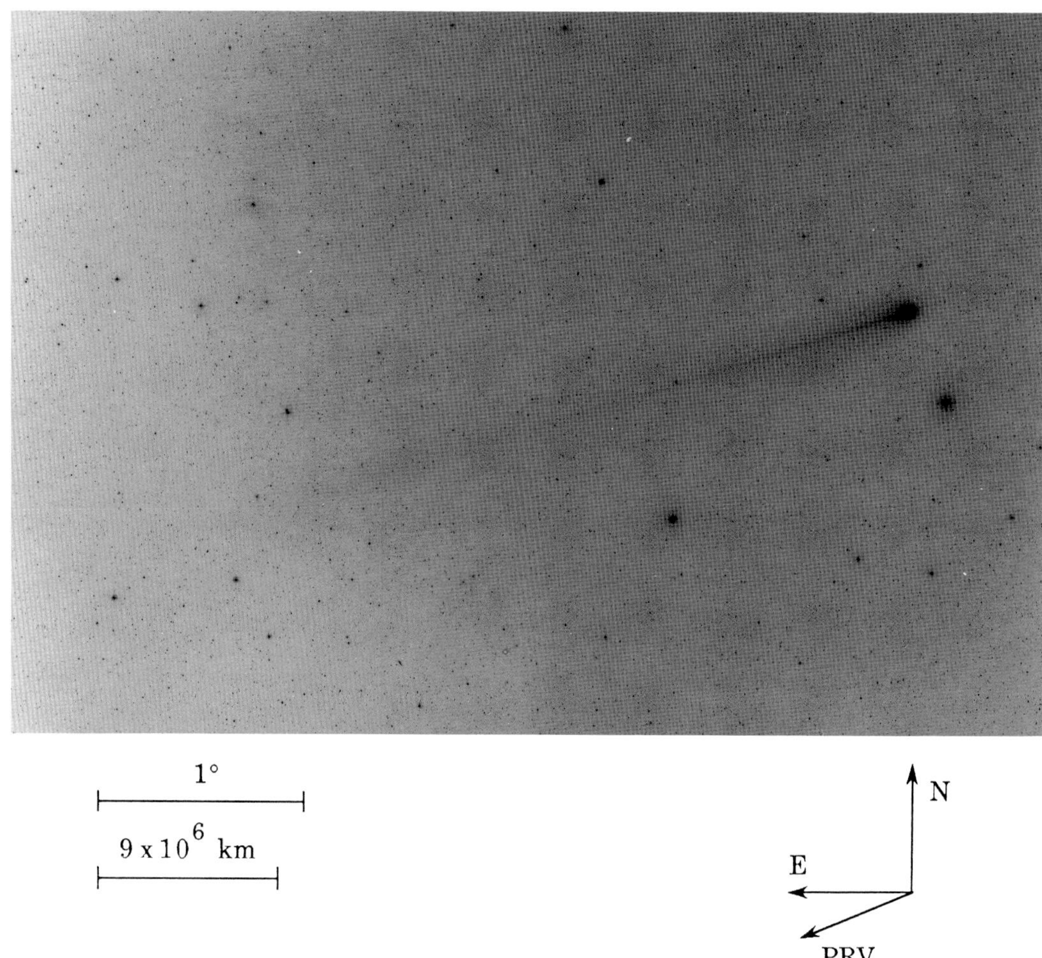

Fig. 748: 1986 Jun 2.41043 UT; exposure 75.0 minutes on IIIa-F emulsion with RG-630 filter; r=2.09, Δ=1.85, β=151.0°; LSPN-2450.
Photograph from Royal Observatory/UK Schmidt Telescope Unit.

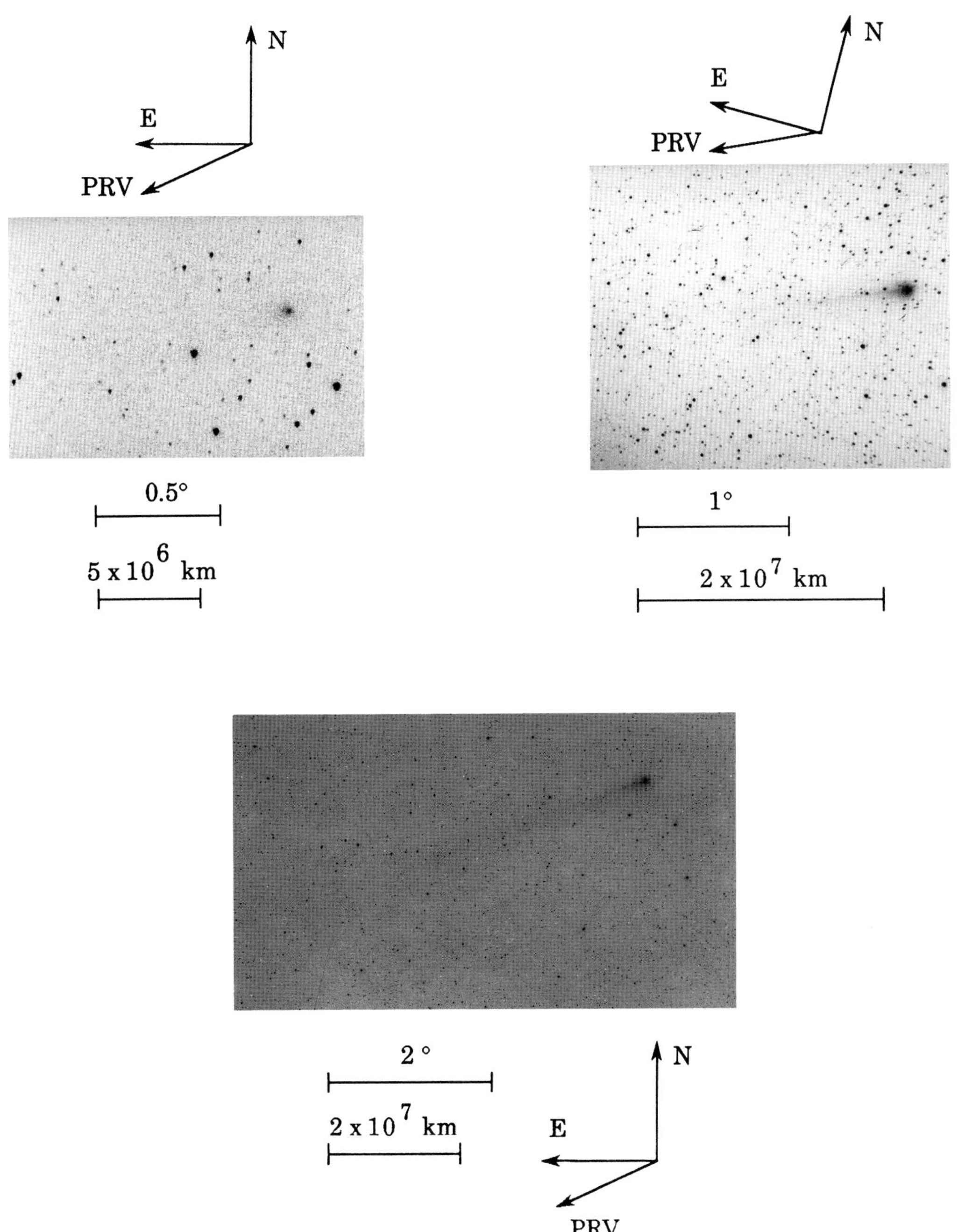

(left) Fig. 749: 1986 Jun 7.61220 UT; exposure 10.0 minutes on 3M 1000 emulstion; r=2.17, Δ=2.02, β=152.2°; AON-852153.
Photograph by S. Edberg, USA.

(right) Fig. 750: 1986 Jun 9.27050 UT; exposure 5.0 minutes on Tech. Pan 2415 emulsion; r=2.19, Δ=2.07, β=152.6°; AON-852164.
Photograph by B. Yen, USA.

(bottom) Fig. 751: 1986 Jun 10.16076 UT; exposure 3.0 minutes on IIa-O emulsion with no filter; r=2.20, Δ=2.10, β=152.8°; LSPN-971.
Photograph by E. Moore/E. Marr, Joint Observatory for Cometary Research.

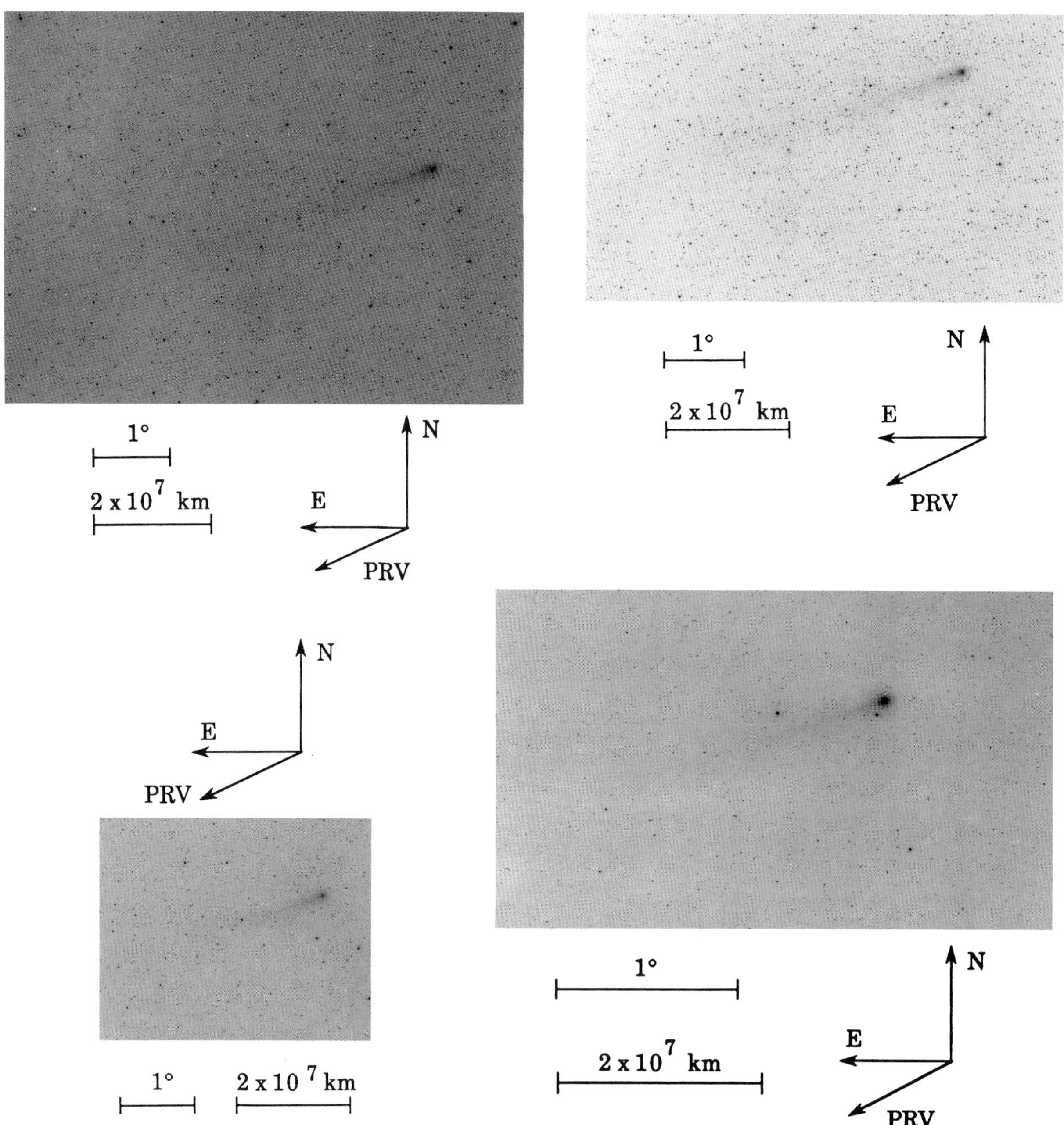

(top left) Fig. 752: 1986 Jun 10.17500 UT; exposure 30.0 minutes on IIa-O emulsion with no filter; r=2.20, Δ=2.10, β=152.8°; LSPN-965.
Photograph by E. Moore, Joint Observatory for Cometary Research.

(top right) Fig. 753: 1986 Jun 11.16945 UT; exposure 20.0 minutes on IIa-O emulsion with no filter; r=2.21, Δ=2.13, β=153.1°; LSPN-967.
Photograph by E. Moore/E. Marr, Joint Observatory for Cometary Research.

(bottom left) Fig. 754: 1986 Jun 12.16528 UT; exposure 10.0 minutes on IIa-O emulsion with no filter; r=2.23, Δ=2.16, β=153.3°; LSPN-969.
Photograph by E. Moore/E. Marr, Joint Observatory for Cometary Research.

(bottom right) Fig. 755: 1986 Jun 25.37428 UT; exposure 60.0 minutes on IIa-O emulsion with GG-395 filter; r=2.40, Δ=2.57, β=156.8°; LSPN-2451.
Photograph from Royal Observatory/UK Schmidt Telescope Unit.

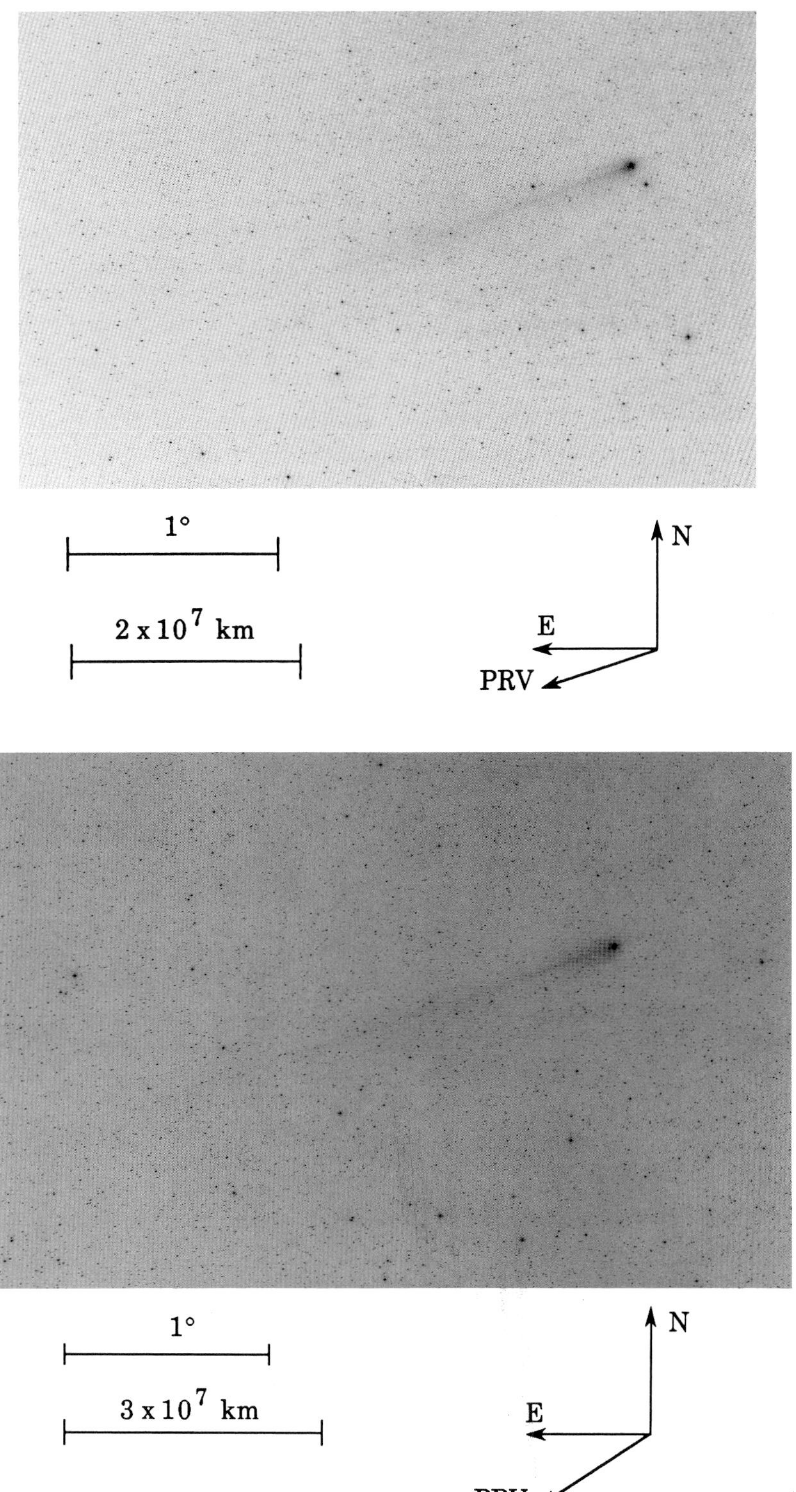

(top) Fig. 756: 1986 Jun 26.38197 UT; exposure 90.0 minutes on IIIa-F emulsion with RG-630 filter; r=2.42, Δ=2.60, β=157.0°; LSPN-2452.
Photograph from Royal Observatory/UK Schmidt Telescope Unit.

(bottom) Fig. 757: 1986 Jul 6.39137 UT; exposure 90.0 minutes on IIIa-F emulsion with RG-630 filter; r=2.54, Δ=2.90, β=159.8°; LSPN-2453.
Photograph from Royal Observatory/UK Schmidt Telescope Unit.

Supplemental

Daily

Record

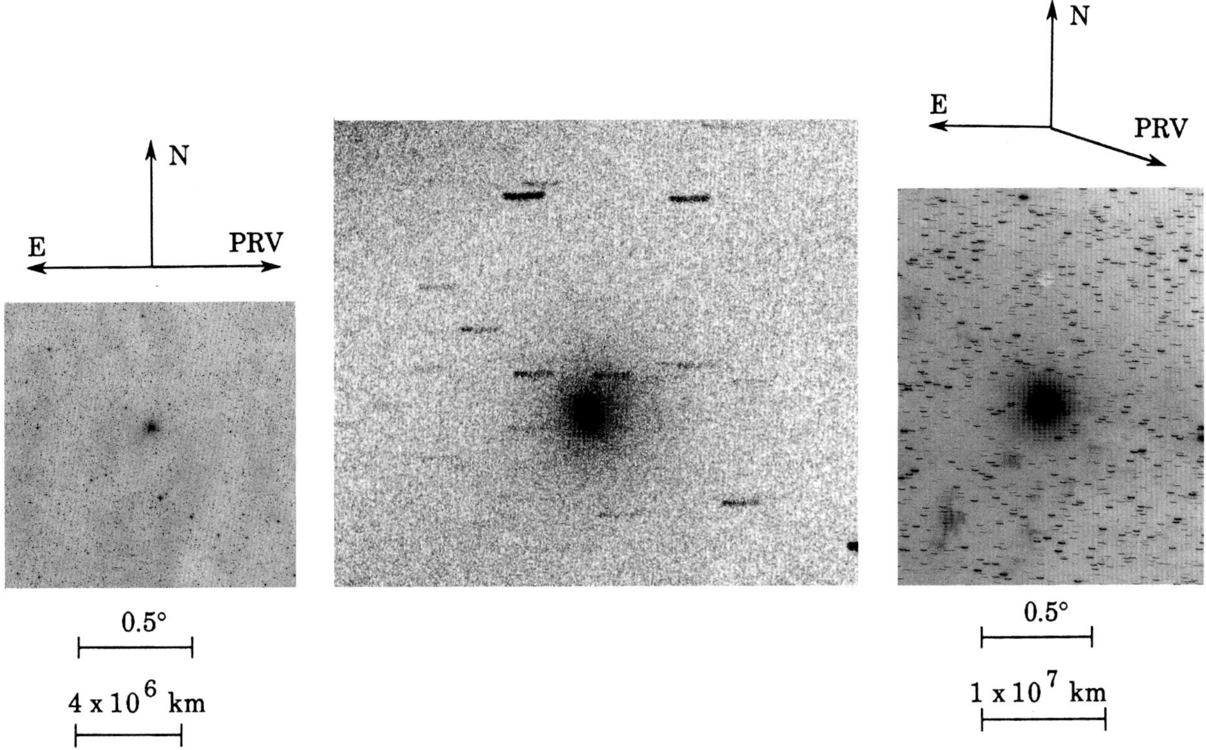

(left) Fig. 758: 1985 Oct 22.04418 UT; exposure 15.0 minutes on 103a-D emulsion with GG-14 filter; r=2.06, Δ=1.37, β=155.4°; LSPN-3432.
Photograph by I. Toth, Konkoly Observatory, Piszkesteto Station.

(center) Fig. 759: 1985 Nov 11.01736 UT; exposure 18.4 minutes on Fuji 1600 emulsion; r=1.78, Δ=0.82, β=168.5°; AON-850133.
Photograph by M. Mobberley, UK.

(right) Fig. 760: 1985 Nov 14.08958 UT; exposure 24.0 minutes on ORWO ZU-21 emulsion with no filter; r=1.73, Δ=0.75, β=172.9°; LSPN-2313.
Photograph by V. Ivanova et al., Bulgarian National Observatory.

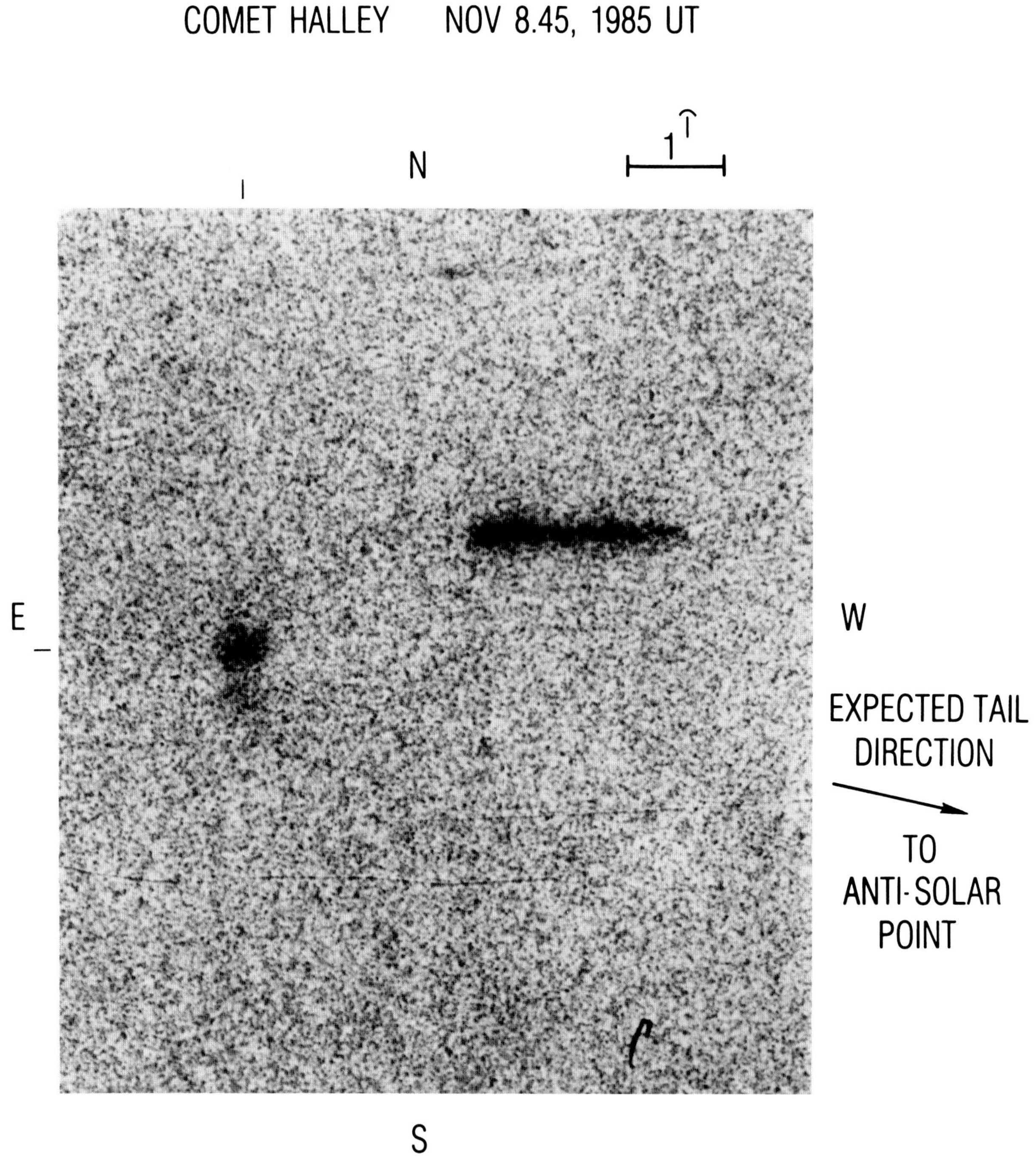

COMET HALLEY NOV 8.45, 1985 UT

N

1′

E

W

EXPECTED TAIL DIRECTION

TO ANTI-SOLAR POINT

S

Fig. 761*: 1985 Nov. 8.45000 UT; exposure 38.0 minutes on Kodacolor 100 with no filter; r=1.82, Δ= 0.88, β=165.0°.
Photograph by D. Lynch/R. Russell, Steward Observatory.

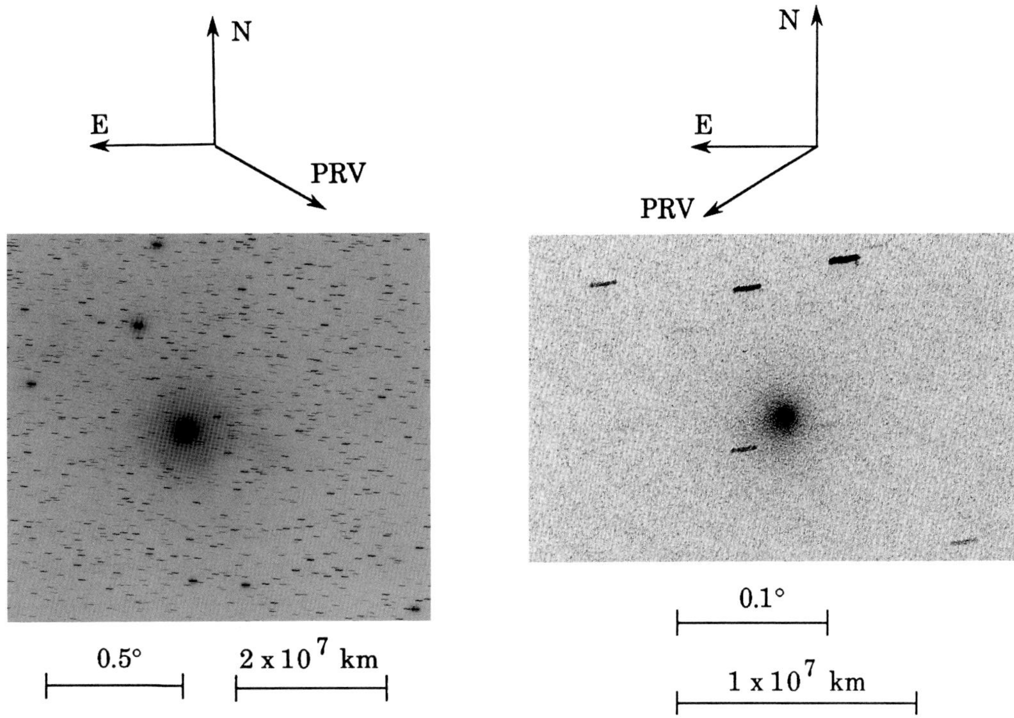

(left) Fig. 762: 1985 Nov 15.28090 UT; exposure 15.0 minutes on IIa-O emulsion with no filter; r=1.72, Δ=0.73, β=174.8°; LSPN-2576.
Photograph by R. Hill, Warner and Swasey Observatory.

(right) Fig. 763: 1985 Nov 18.55208 UT; exposure 10.0 minutes on Kodacolor 1000 emulsion with no filter; r=1.67, Δ=0.68, β=178.2°.
Photograph by I. S. Nha/H. I. Kim, Yonsei University Observatory.

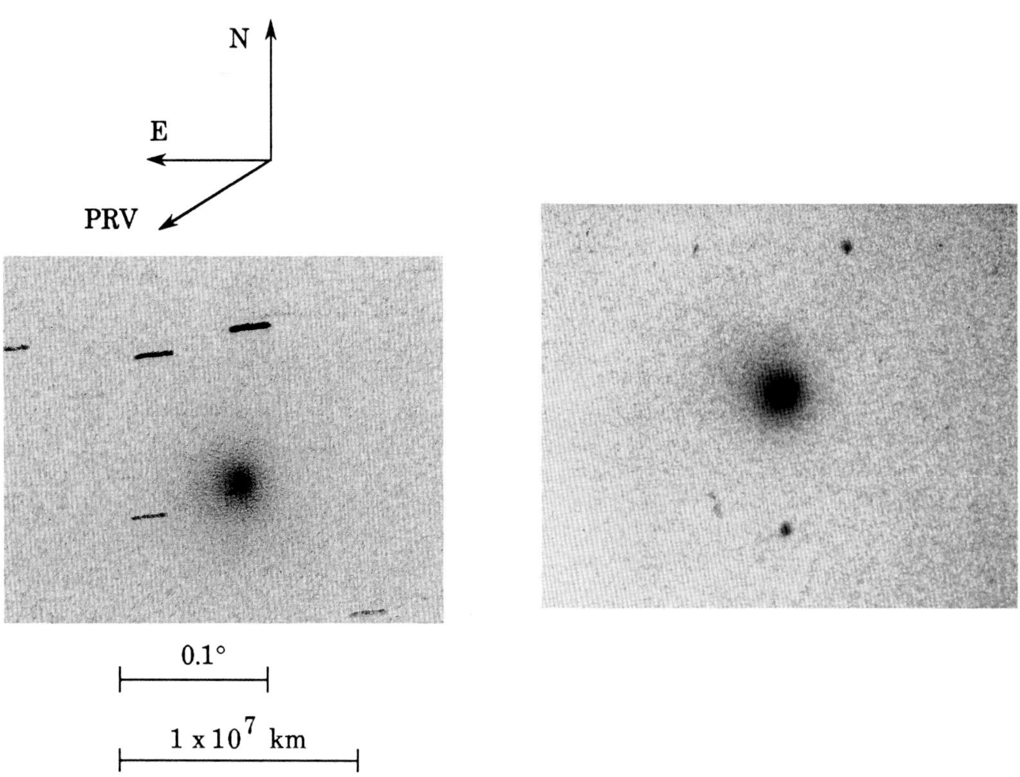

(left) Fig. 764: 1985 Nov 18.56597 UT; exposure 15.0 minutes on Kodacolor 1000 emulsion with no filter; r=1.67, Δ=0.68, β=178.2°.
Photograph by Y. S. Lee/Y. W. Chun, Yonsei University Observatory.

(right) Fig. 765: 1985 Dec 27.68100 UT; exposure 1.0 minute on Tech. Pan 2415 emulsion; r=1.07, Δ=1.07, β=125.5°; AON-850539.
Photograph by R. Conrad, Austria

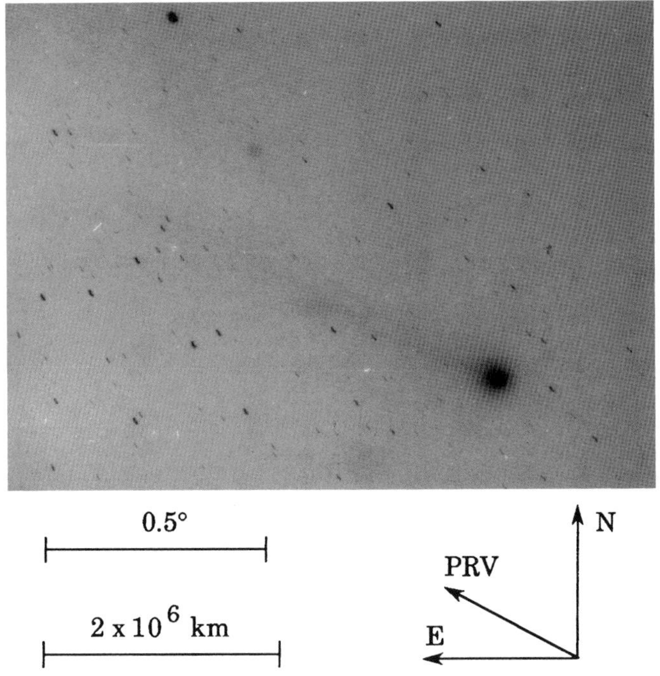

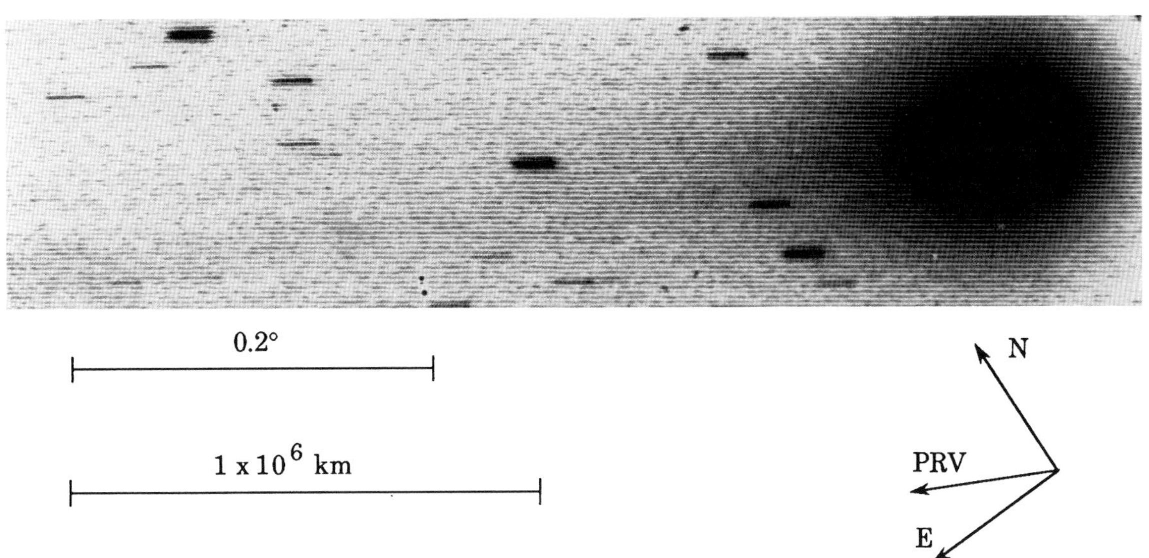

(top) Fig. 766: 1985 Dec 31.53924 UT; exposure 33.0 minutes on 103a-O emulsion with no filter; r=1.01, Δ=1.15, β=126.4°; LSPN-3063.
Photograph by J. Johnston, Perth Observatory.

(bottom) Fig. 767: 1986 Jan 1.58681 UT; exposure 50.0 minutes on ORWO ZU-21 emulsion with no filter; r=1.00, Δ=1.17, β=126.8°; LSPN-3269.
Photograph by S. Gerasimenko, Gissar Observatory.

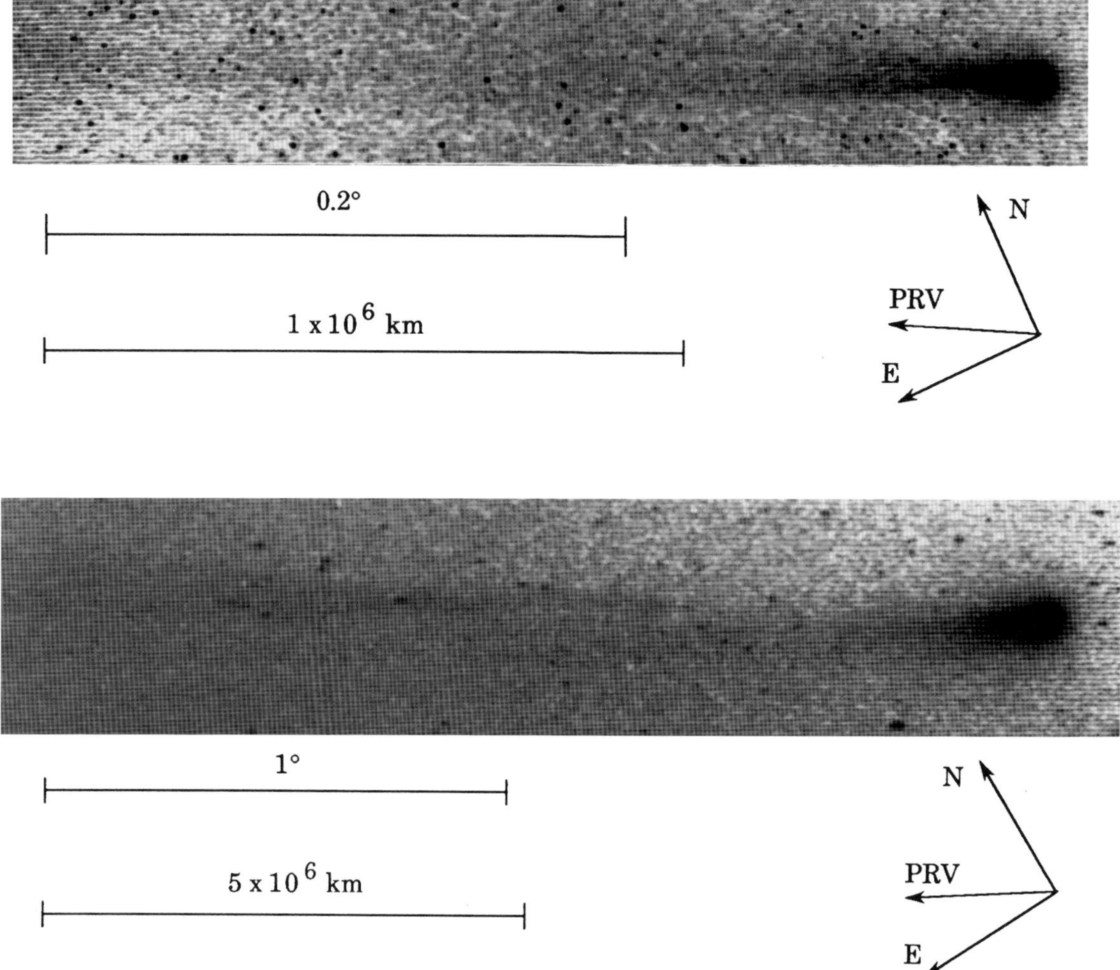

(top) Fig. 768: 1986 Jan 8.57965 UT; exposure 8.8 minutes on ORWO ZU-21 emulsion with no filter; r=0.89, Δ=1.30, β=130.9°; LSPN-3372.
Photograph by D. Gorodetskij, Mountain Observatory.

(bottom) Fig. 769: 1986 Jan 10.58718 UT; exposure 40.0 minutes on ORWO ZU-21 emulsion with no filter; r=0.86, Δ=1.33, β=132.5°; LSPN-3293.
Photograph by N. Kiselev, Gissar Observatory.

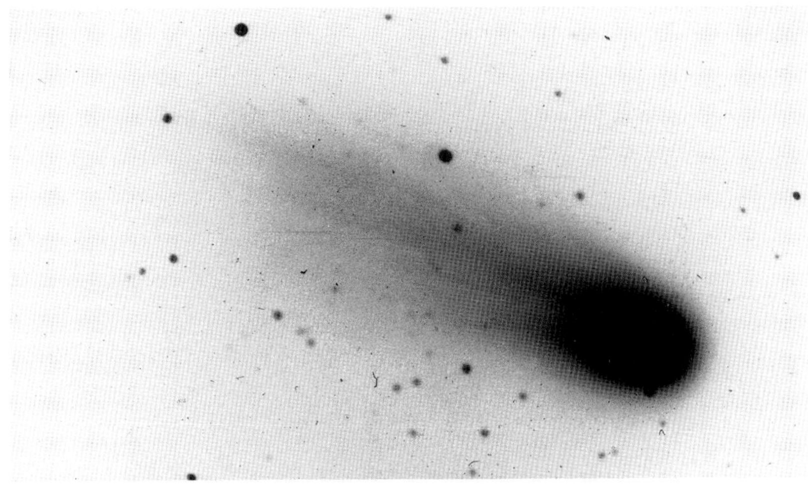

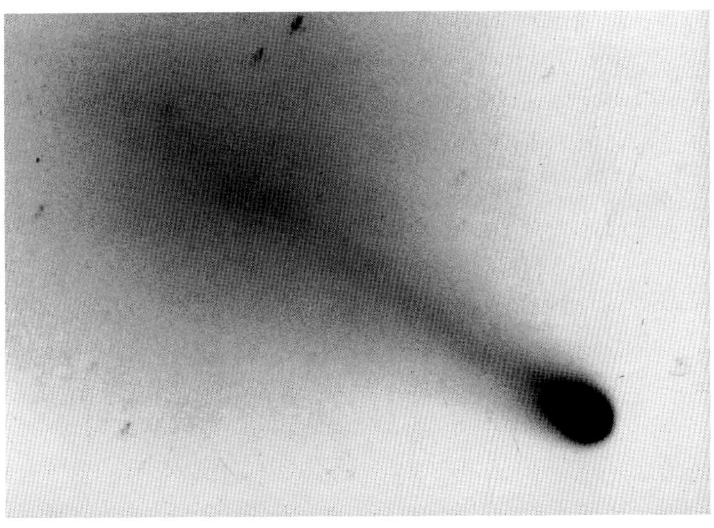

(top) Fig. 770: 1986 Jan 16.75000 UT; exposure 4.0 minutes on Tech. Pan 2415 emulsion; r=0.78, Δ=1.43, β=139.0°; AON-850884.
Photograph by V. Belli, Italy

(center) Fig. 771: 1986 Jan 18.97104 UT; exposure 3.0 minutes on Tri-X emulsion; r=0.75, Δ=1.46, β=141.9°; AON-850891.
Photograph by H. Lazerson, USA

(bottom) Fig. 772: 1986 Jan 21.05347 UT; exposure 10.0 minutes on Tech. Pan 2415 emulsion; r=0.72, Δ=1.48, β=144.8°; AON-850909.
Photograph by R. Minton, USA

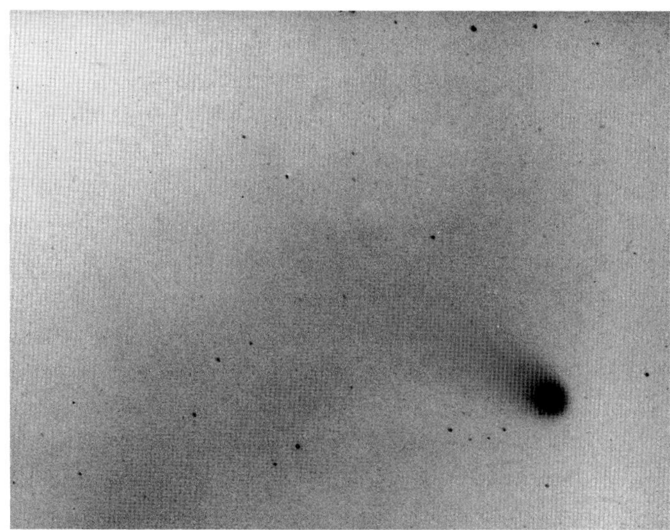

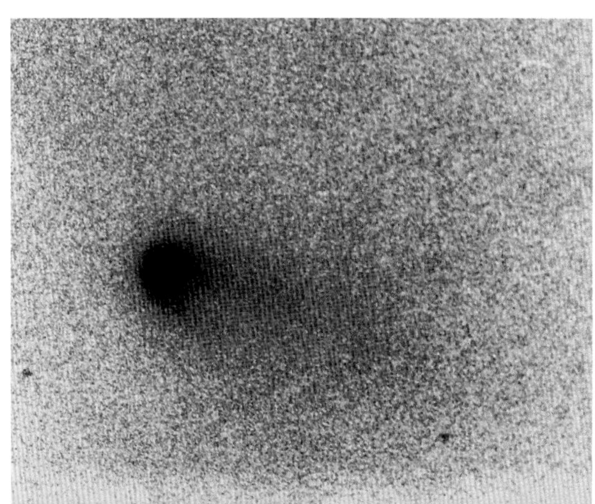

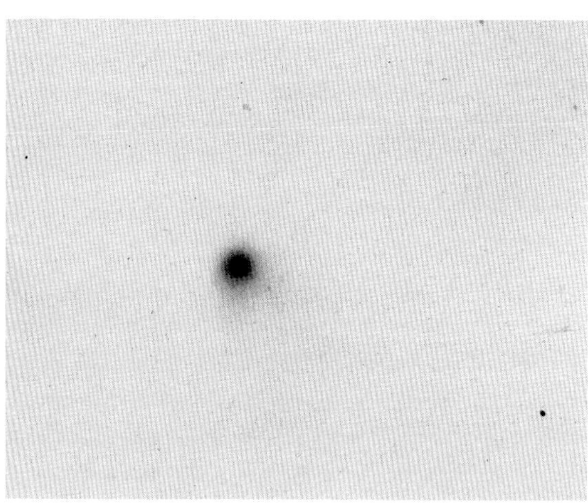

(top) Fig. 773: 1986 Jan 22.37708 UT; exposure 1.0 minute on Tech. Pan 2415 emulsion; r=0.71, Δ=1.50, β=146.8°; AON-850910.
Photograph by T. Niijima, Japan

(left) Fig. 774: 1986 Feb 24.55934 UT; exposure 0.8 minutes on Fujichrome 1600 emulsion; r=0.67, Δ=1.36, β=135.9°; AON-850923.
Photograph by R. Royer, USA

(right) Fig. 775: 1986 Feb 25.55934 UT; exposure 1.0 minute on Tech. Pan 2415 emulsion; r=0.68, Δ=1.34, β=134.2°; AON-850925.
Photograph by R. Royer, USA

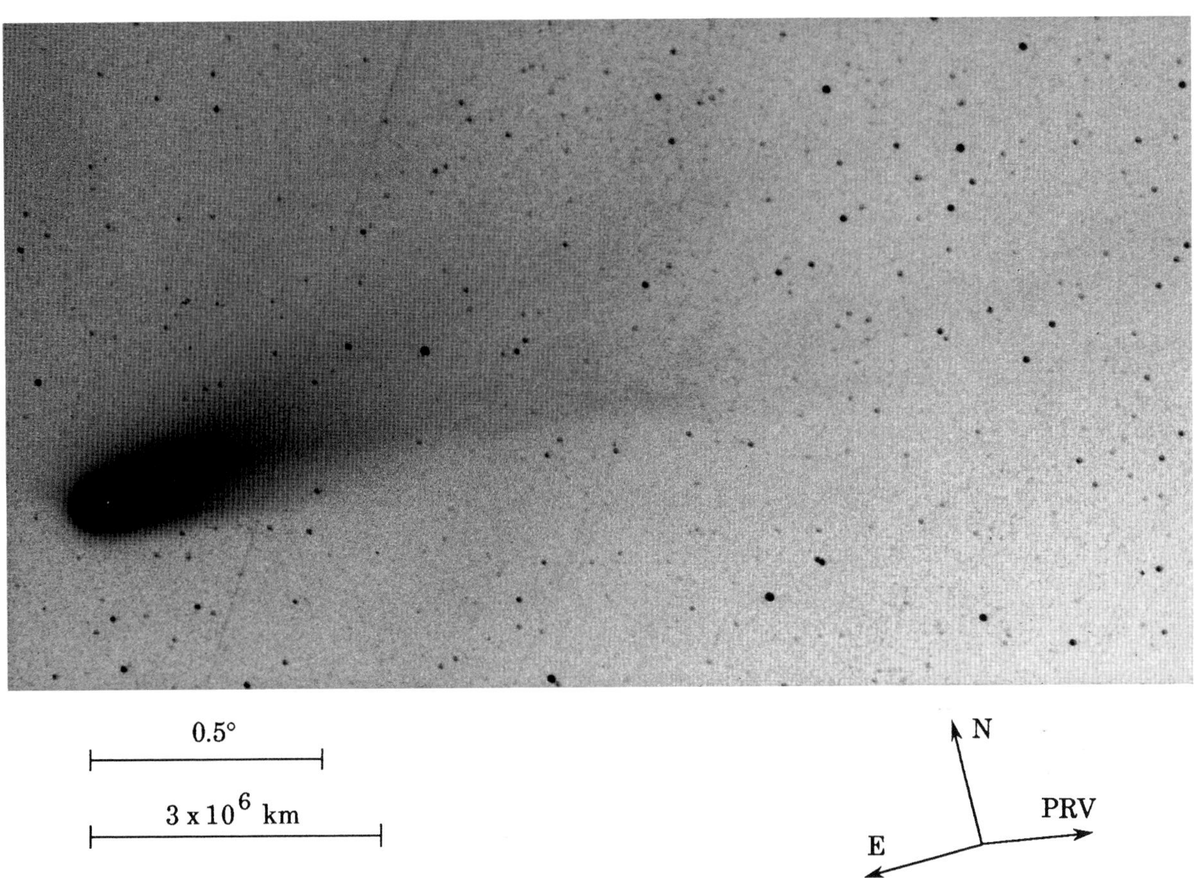

Fig. 776: 1986 Feb 26.13194 UT; exposure 2.0 minutes on Tech. Pan 2415 emulsion with no filter; r=0.69, Δ=1.33, β=133.2°; LSPN-1783.
Photograph by F. Marang, Sutherland, LSPN Island Network

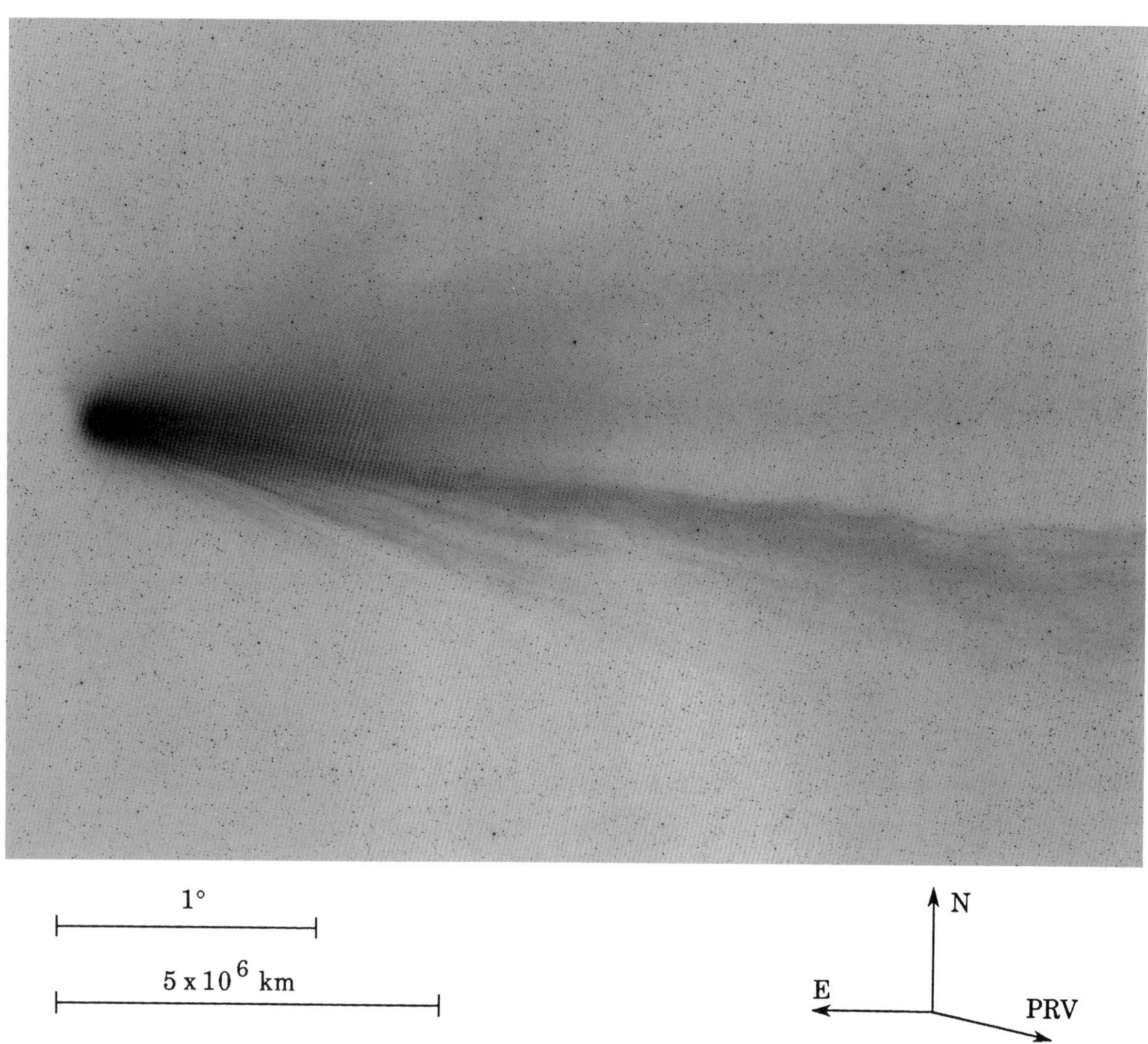

Fig. 777: 1986 Mar 8.37292 UT; exposure 10.0 minutes on IIa-O emulsion with no filter; r=0.82, Δ=1.10, β=119.9°; LSPN-1270.
Photograph by F. Miller, University of Michigan/CTIO.

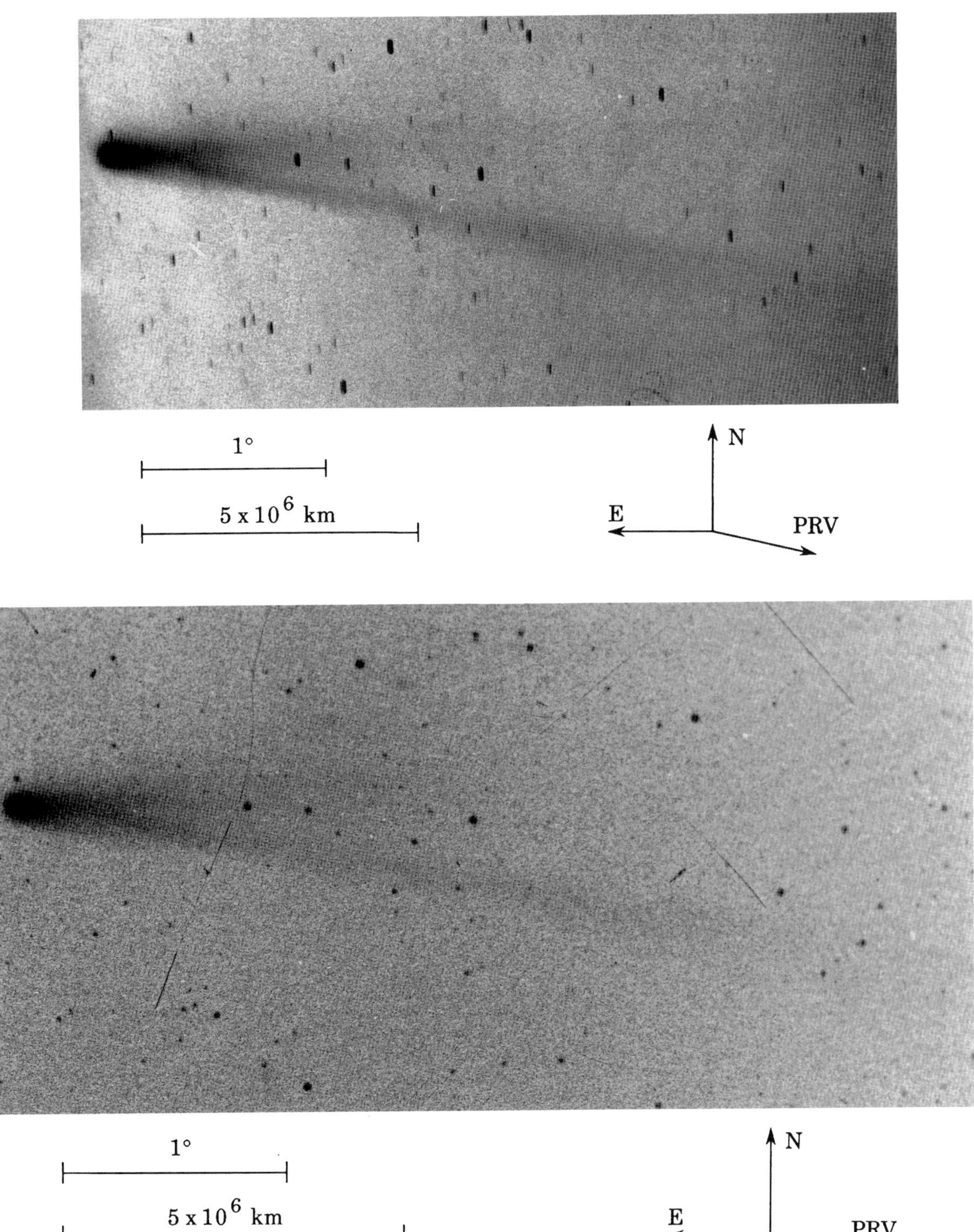

(top) Fig. 778: 1986 Mar 8.90532 UT; exposure 25.0 minutes on 103a-O emulsion with GG-13 filter; r=0.83, Δ=1.09, β=119.4°; LSPN-2729.
Photograph by W. Lai, Shanghai Observatory.

(bottom) Fig. 779: 1986 Mar 8.92654 UT; exposure 13.0 minutes on IIa-O emulsion with no filter; r=0.83, Δ=1.09, β=119.4°; LSPN-2608.
Photograph by Y. Zhang et al., Yunnan Observatory.

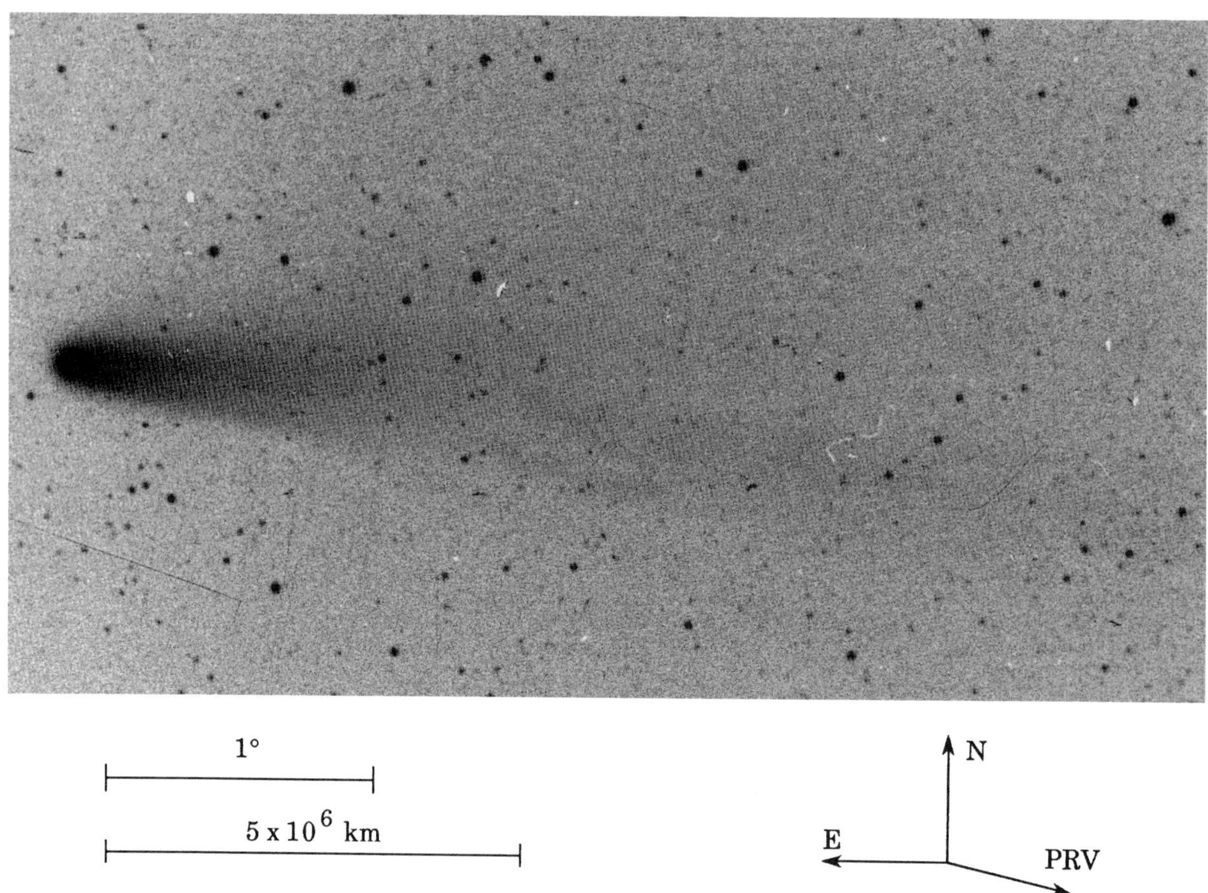

Fig. 780: 1986 Mar 9.90397 UT; exposure 20.0 minutes on IIa-O emulsion with no filter; r=0.84, Δ=1.06, β=118.5°; LSPN-2610.
Photograph by Y. Zhang et al., Yunnan Observatory .

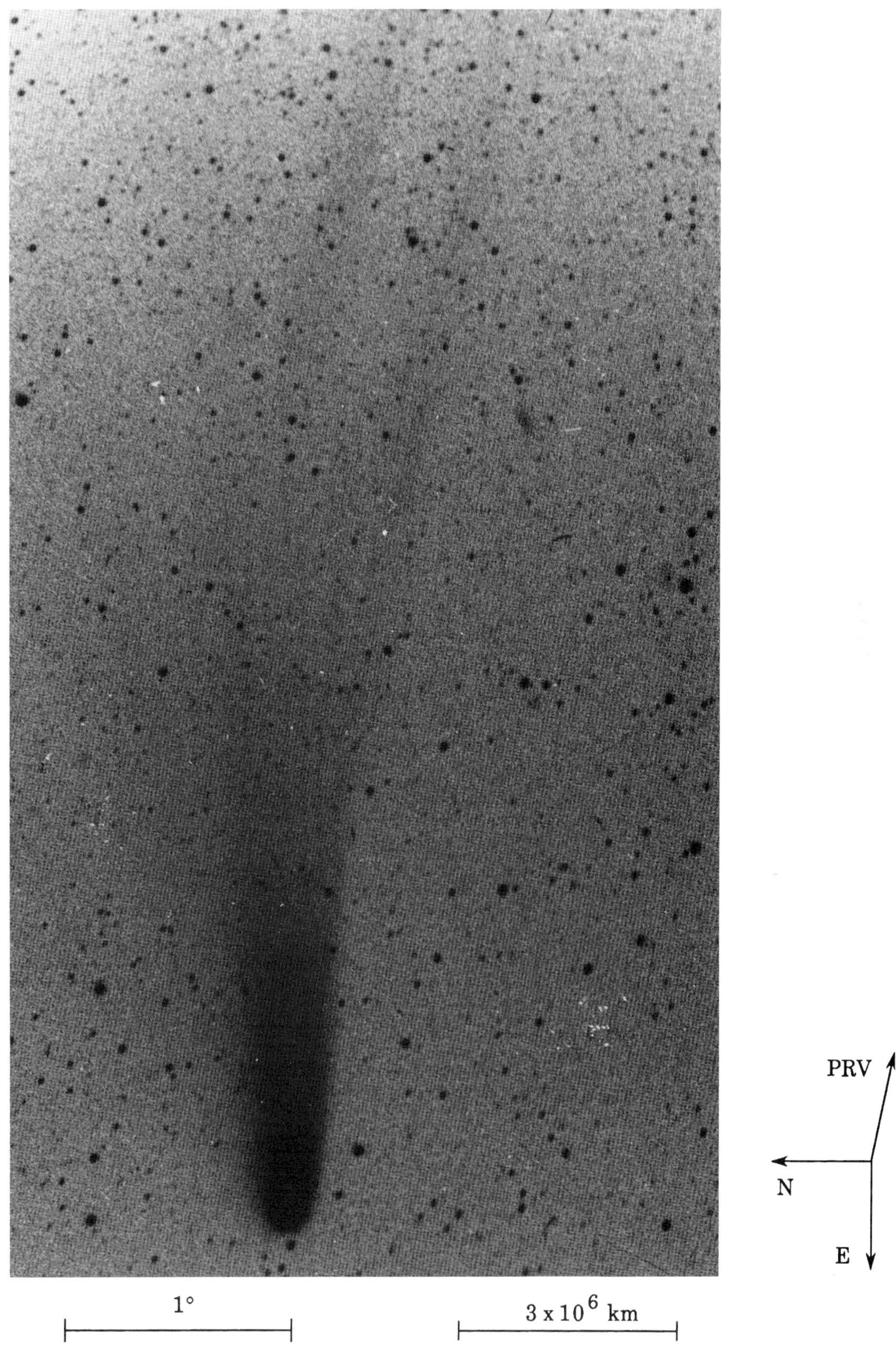

Fig. 781: 1986 Mar 10.91308 UT; exposure 40.0 minutes on IIa-O emulsion with GG-13 filter; r=0.86, Δ=1.04, β=117.7°; LSPN-2612.
Photograph by Y. Zhang et al., Yunnan Observatory .

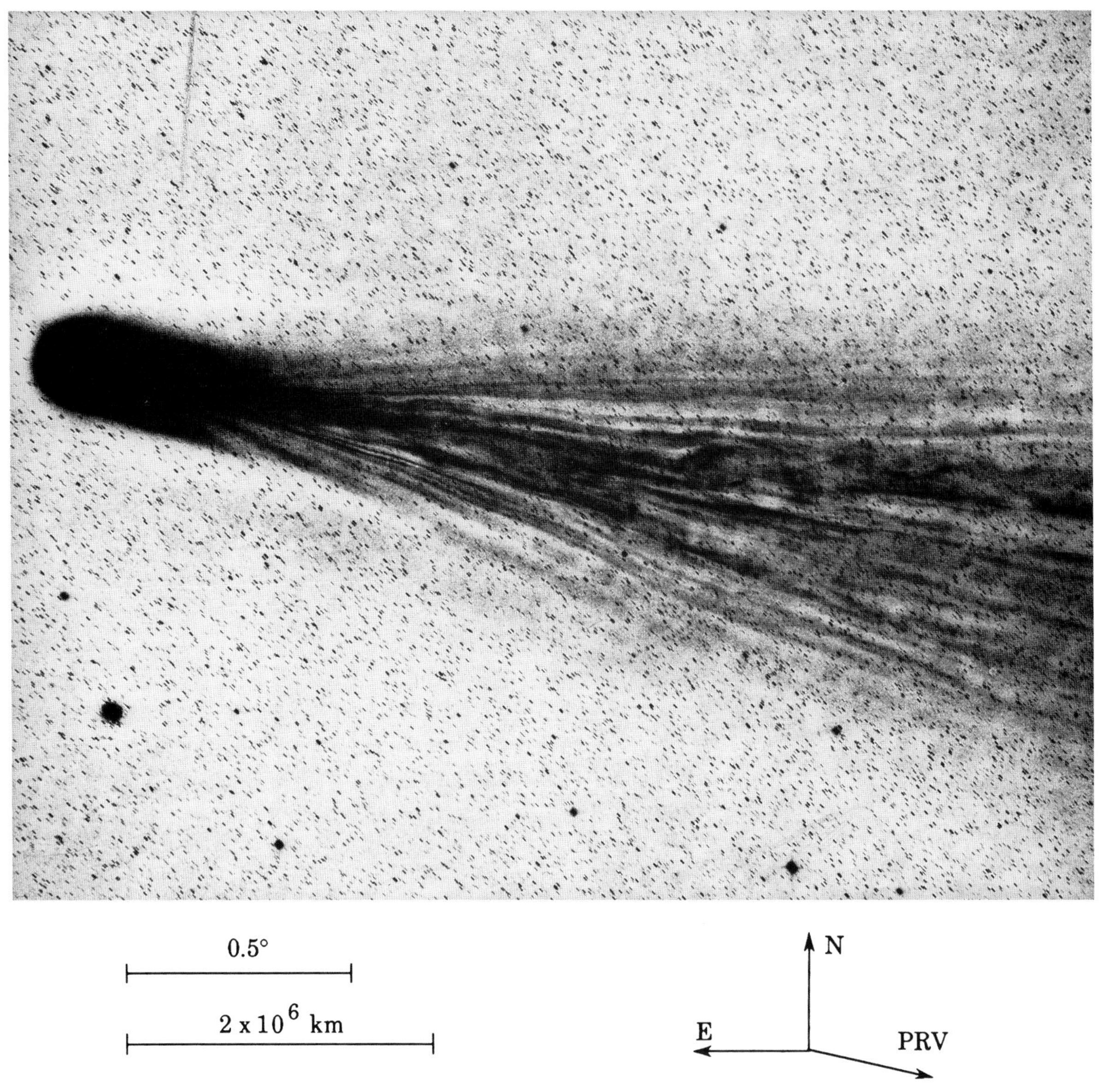

Fig. 782: 1986 Mar 12.21680 UT; exposure 15.0 minutes on IIa-O emulsion with GG-385 filter; r=0.88, Δ=1.01, β=116.8°.
Photograph by D. Malin, Anglo-Australian Observatory.

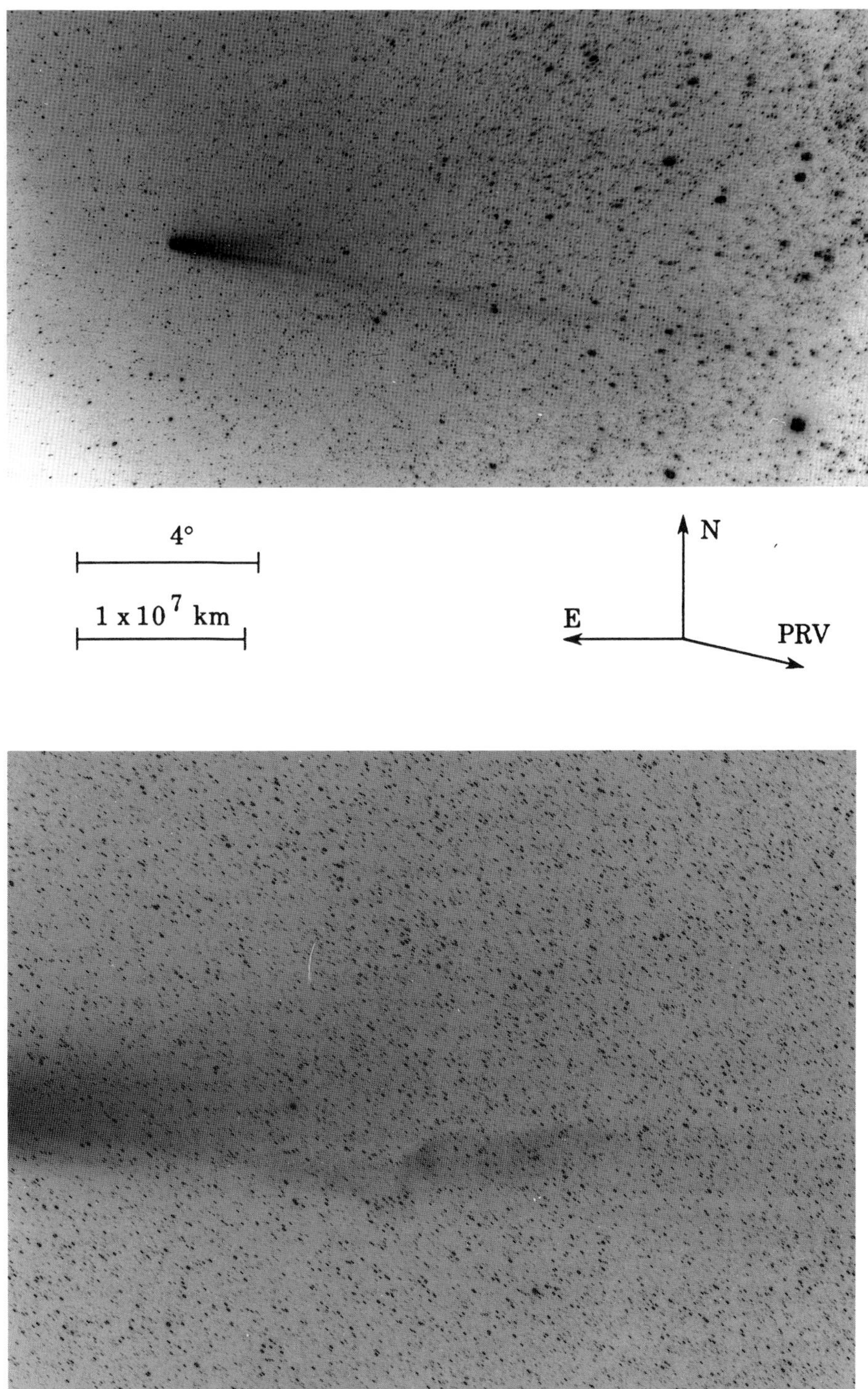

(top) Fig. 783: 1986 Mar 16.46150 UT; exposure 35.0 minutes on IIa-O emulsion with no filter; r=0.94, Δ=0.90, β=114.6°; LSPN-3421.
Photograph by G. Emerson, E. E. Barnard Observatory Station.

(bottom) Fig. 784: 1986 Mar 21.41319 UT; exposure 16.0 minutes on Tech. Pan 2415 emulsion with no filter; r=1.01, Δ=0.77, β=113.9°; LSPN-1409.
Photograph by W. Liller, Easter Island, LSPN Island Network

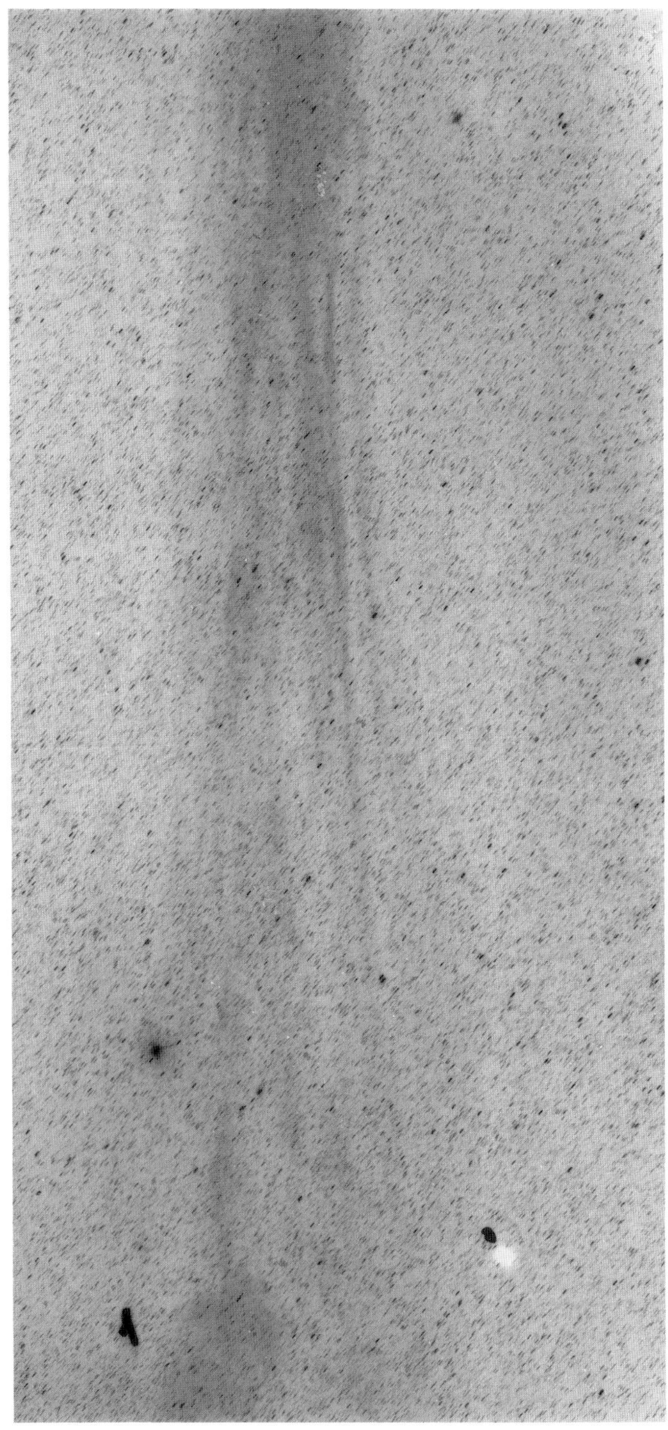

Fig. 785*: 1986 Mar 21.30694 UT; exposure 20.0 minutes on IIa-O emulsion with no filter; r=1.01, Δ=0.78, β=113.9°; LSPN-1110.
Photograph by F. Miller, University of Michigan/CTIO.

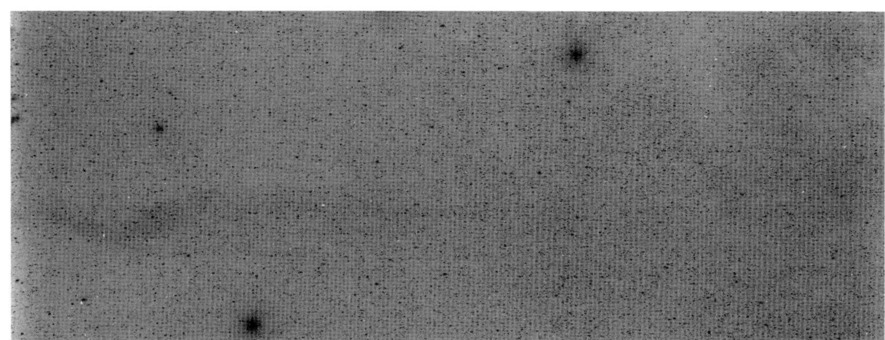

(top) Fig. 786: 1986 Mar 22.47500 UT; exposure 10.0 minutes on IIa-O emulsion with no filter; r=1.03, Δ=0.75, β=114.1°; LSPN-906.
Photograph by E. Moore, Joint Observatory for Cometary Research.

(bottom) Fig. 787: 1986 Mar 22.90383 UT; exposure 30.0 minutes on IIa-O emulsion with GG-13 filter; r=1.04, Δ=0.74, β=114.2°; LSPN-2618.
Photograph by Y. Zhang et al., Yunnan Observatory .

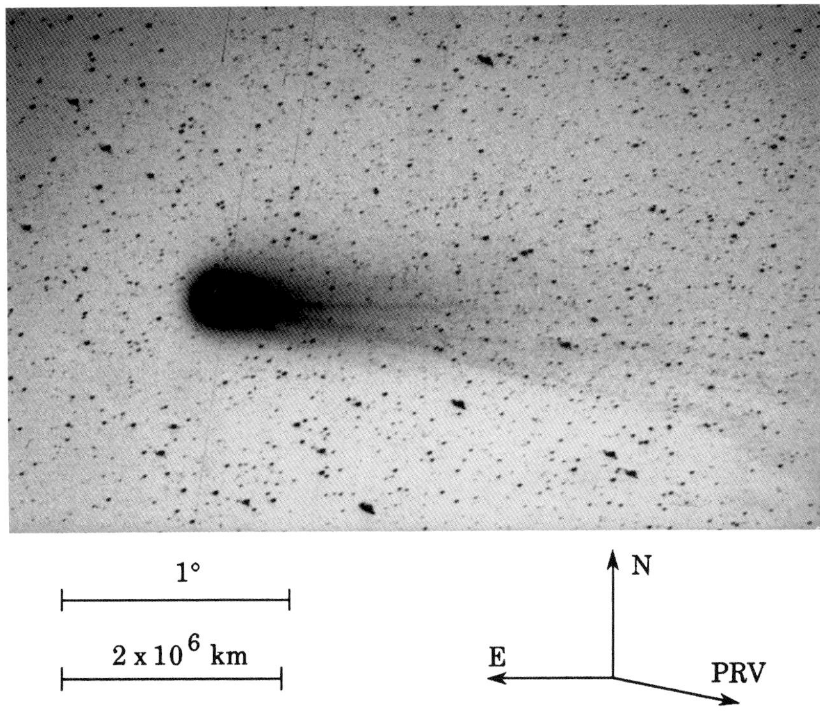

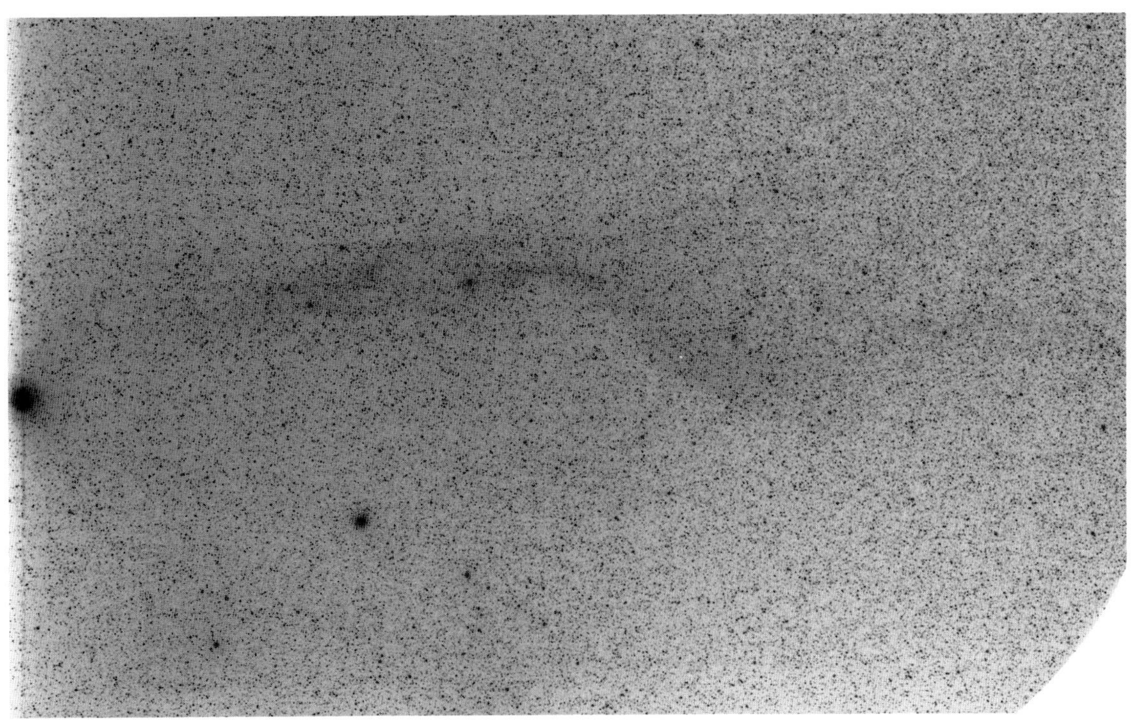

(top) Fig. 788: 1986 Mar 23.48785 UT; exposure 20.0 minutes on Tech. Pan 2415 emulsion with no filter; r=1.05, Δ=0.72, β=114.4°; LSPN-1419.
Photograph by W. Liller, Easter Island, LSPN Island Network

(bottom) Fig. 789: 1986 Mar 23.60642 UT; exposure 0.5 minutes on Tri-X emulsion with no filter; r=1.05, Δ=0.72, β=114.4°; LSPN-440.
Photograph by P. Malloy/M. Spector, Tahiti, LSPN Island Network

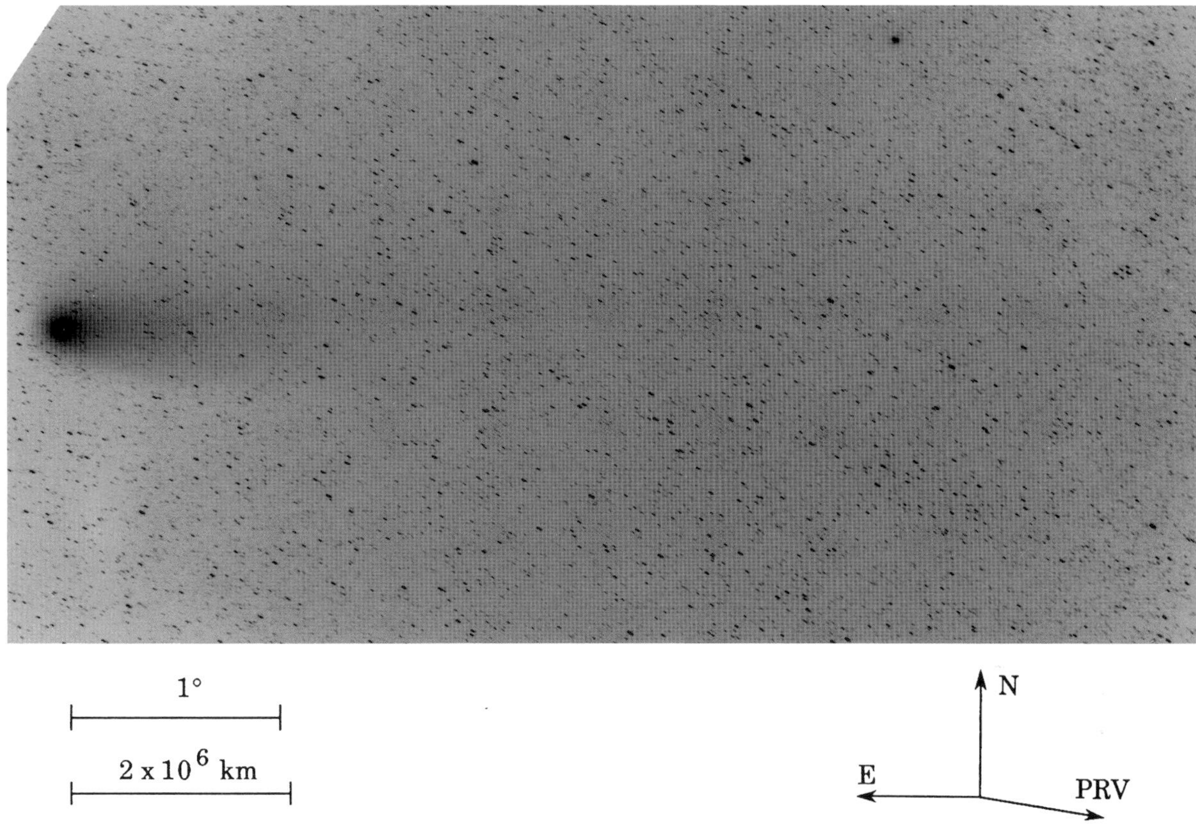

Fig. 790: 1986 Mar 26.48125 UT; exposure 16.0 minutes on Tech. Pan 2415 emulsion with Wratten 25 filter; r=1.09, Δ=0.65, β=115.9°; LSPN-1425.
Photograph by W. Liller, Easter Island, LSPN Island Network

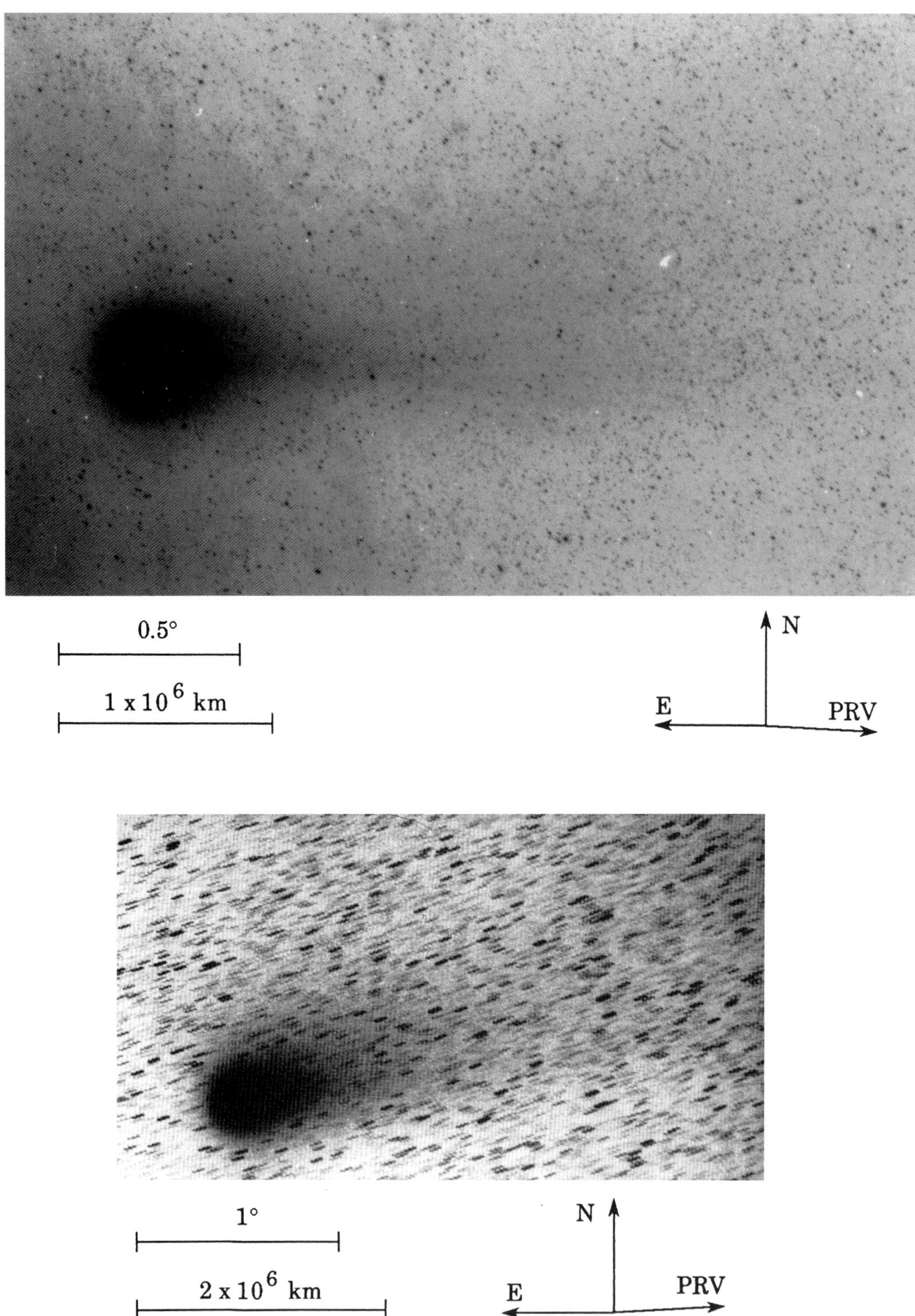

(top) Fig. 791: 1986 Mar 31.13258 UT; exposure 2.5 minutes on Tech. Pan 2415 emulsion with no filter; r=1.16, Δ=0.55, β=121.1°; LSPN-3490.
Photograph by C. Butler, University of Perugia/South African Astronomical Observatory.

(bottom) Fig. 792: 1986 Apr 2.40972 UT; exposure 10.0 minutes on Tri-X emulsion with no filter; r=1.20, Δ=0.50, β=125.1°; LSPN-490.
Photograph by P. Malloy/M. Spector, Tahiti, LSPN Island Network

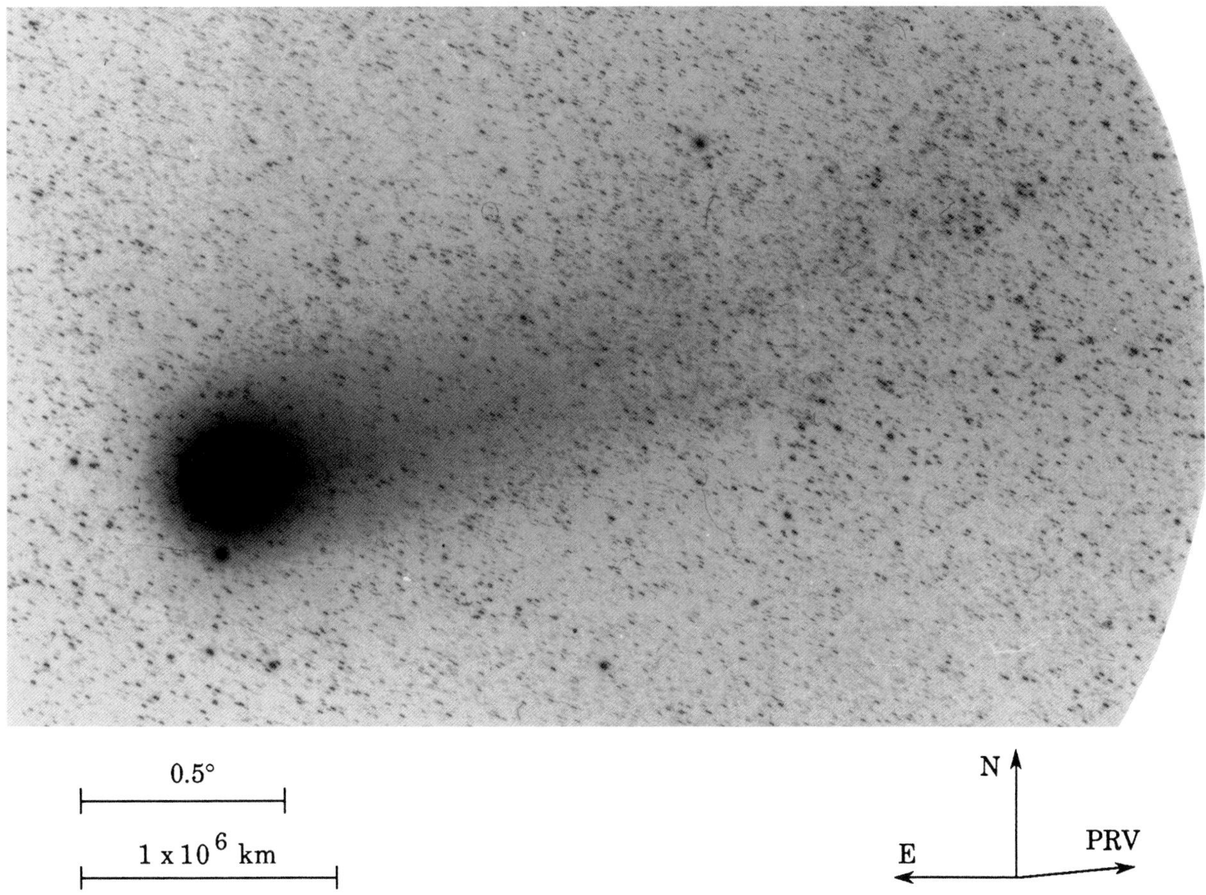

Fig. 793: 1986 Apr 3.07604 UT; exposure 5.0 minutes on Tech. Pan 2415 emulsion with no filter; r=1.21, Δ=0.49, β=126.4°; LSPN-3492.
Photograph by C. Butler, University of Perugia/South African Astronomical Observatory.

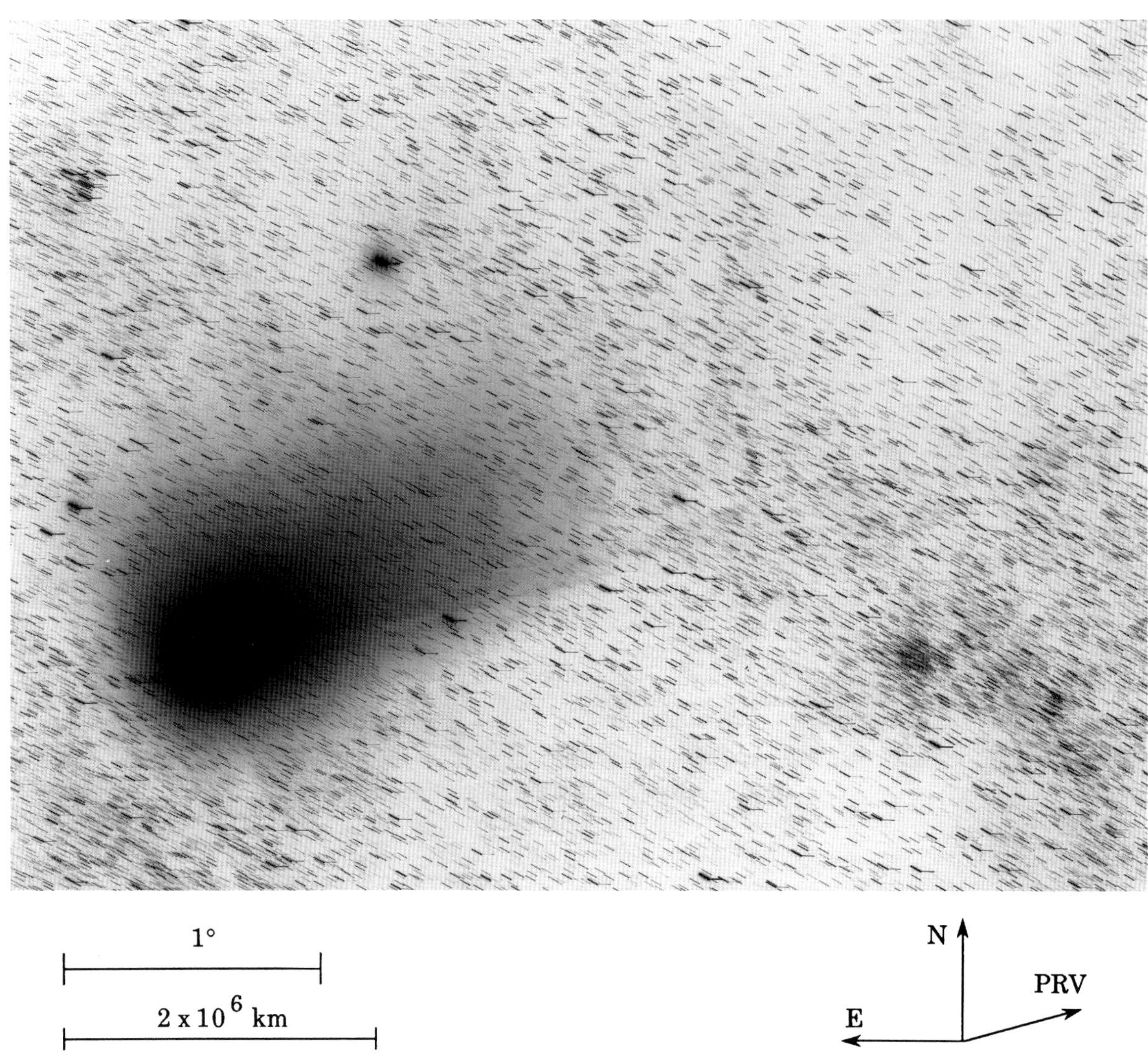

Fig. 794: 1986 Apr 5.37778 UT; exposure 20.0 minutes on IIa-O emulsion with no filter; r=1.25, Δ=0.46, β=131.9°; LSPN-1142.
Photograph by F. Miller, University of Michigan/CTIO.

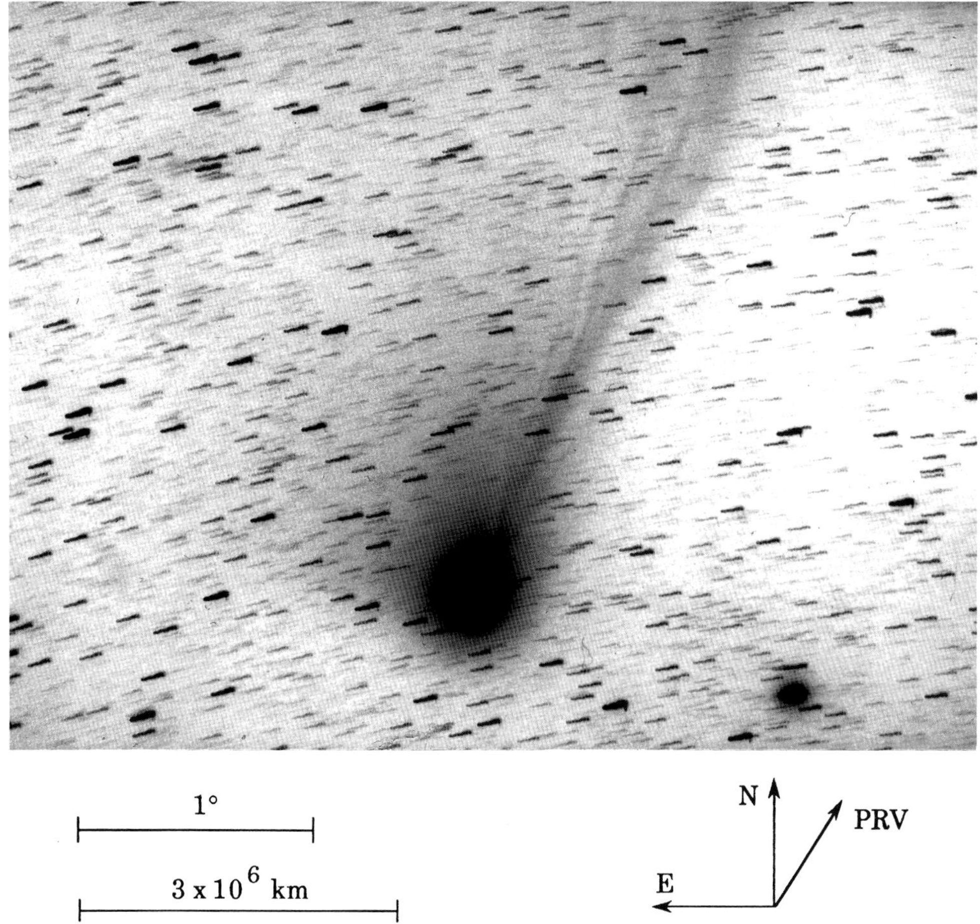

Fig. 795: 1986 Apr 11.33055 UT; exposure 30.0 minutes on Tech. Pan 2415 emulsion with Wratten 47B filter; r=1.34, Δ=0.42, β=149.1°; LSPN-559.
Photograph by P. Malloy/M. Spector, Tahiti, LSPN Island Network

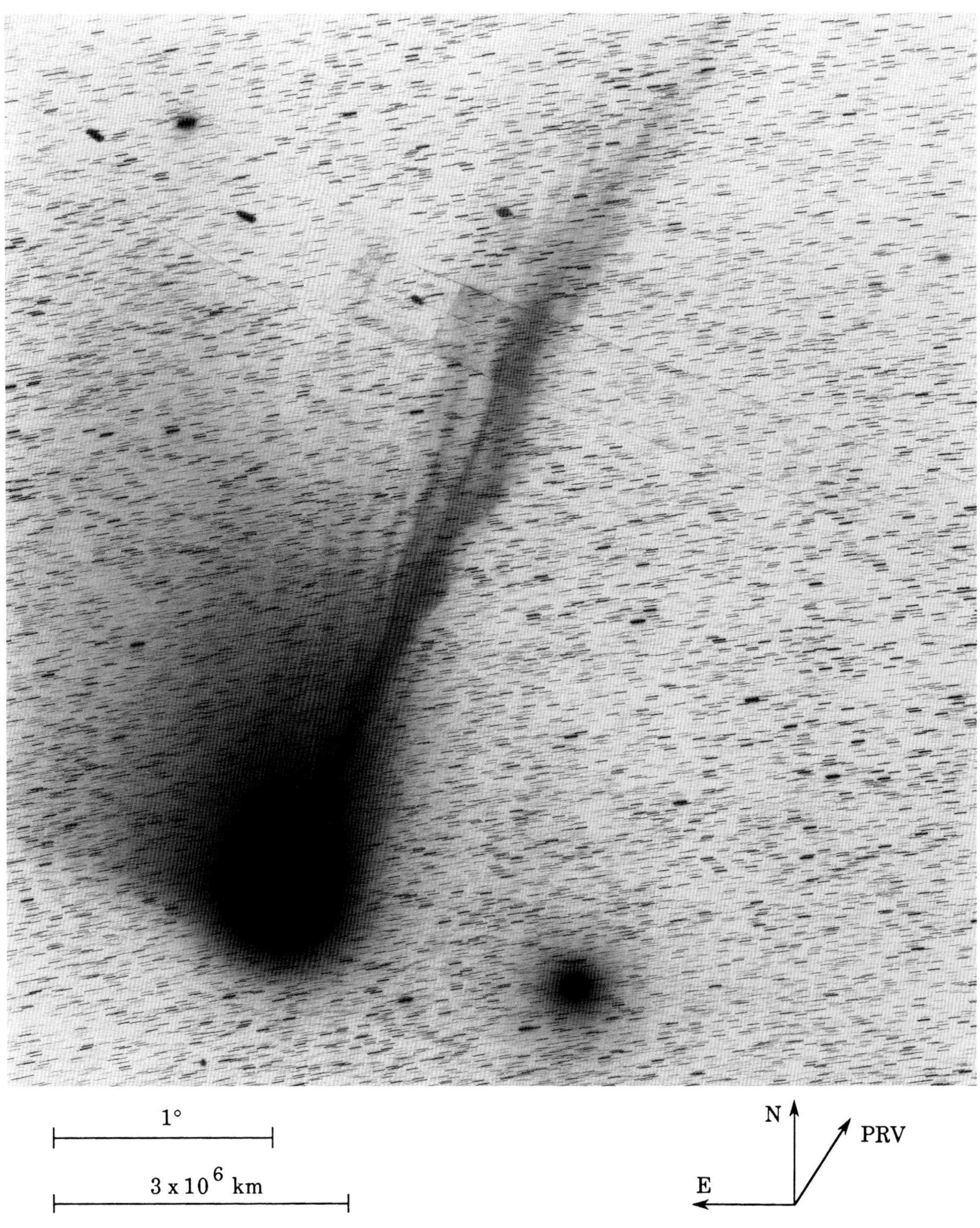

Fig. 796: 1986 Apr 11.34028 UT; exposure 20.0 minutes on 103a-O emulsion with no filter; r=1.34, Δ=0.42, β=149.1°; LSPN-1191.
Photograph by F. Miller, University of Michigan/CTIO.

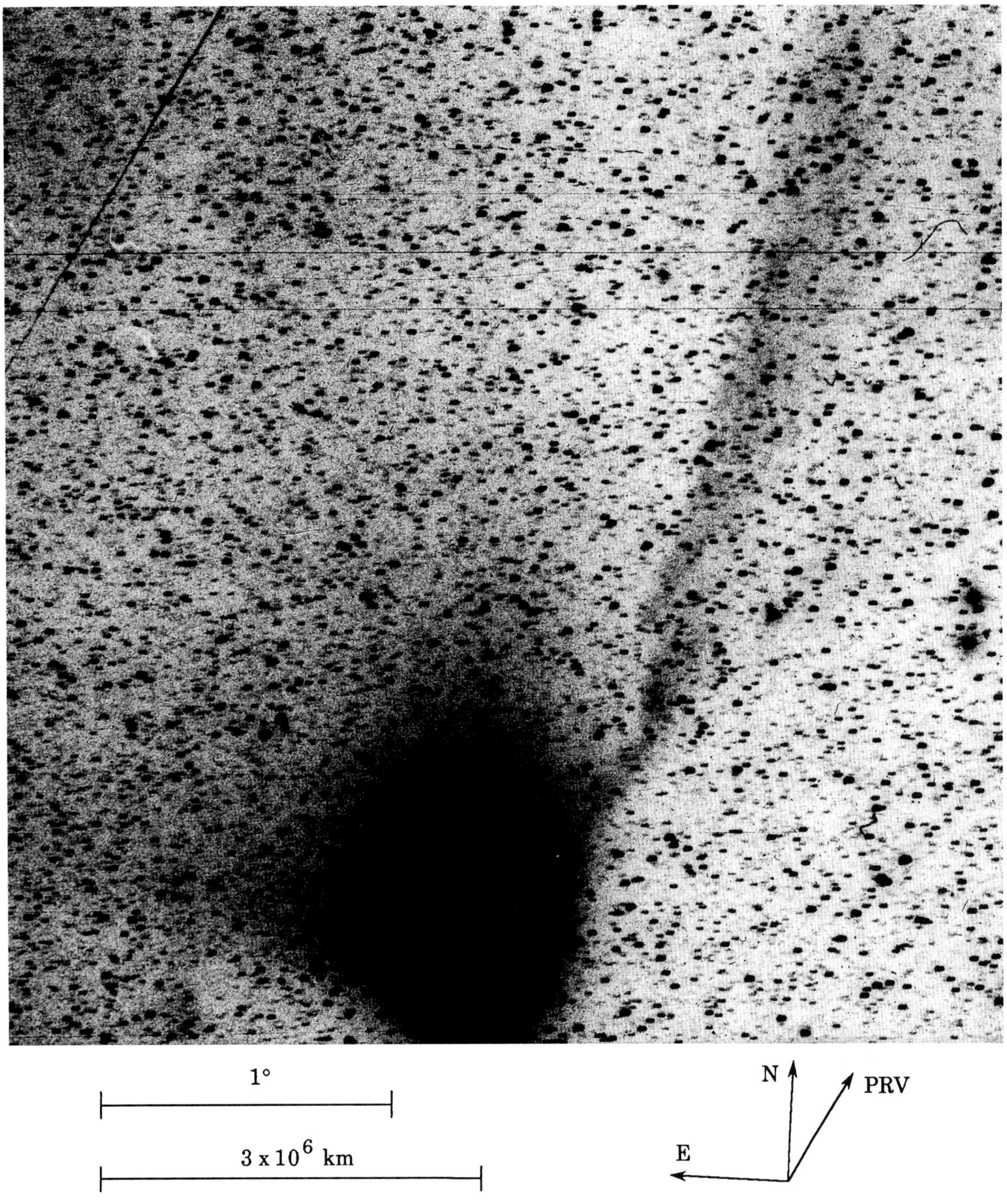

Fig. 797: 1986 Apr 11.87118 UT; exposure 9.0 minutes on Tech. Pan 2415 emulsion with no filter; r=1.35, Δ=0.42, β=150.6°; LSPN-608.
Photograph by P. Berge/G. Mahoux, Reunion Island, LSPN Island Network

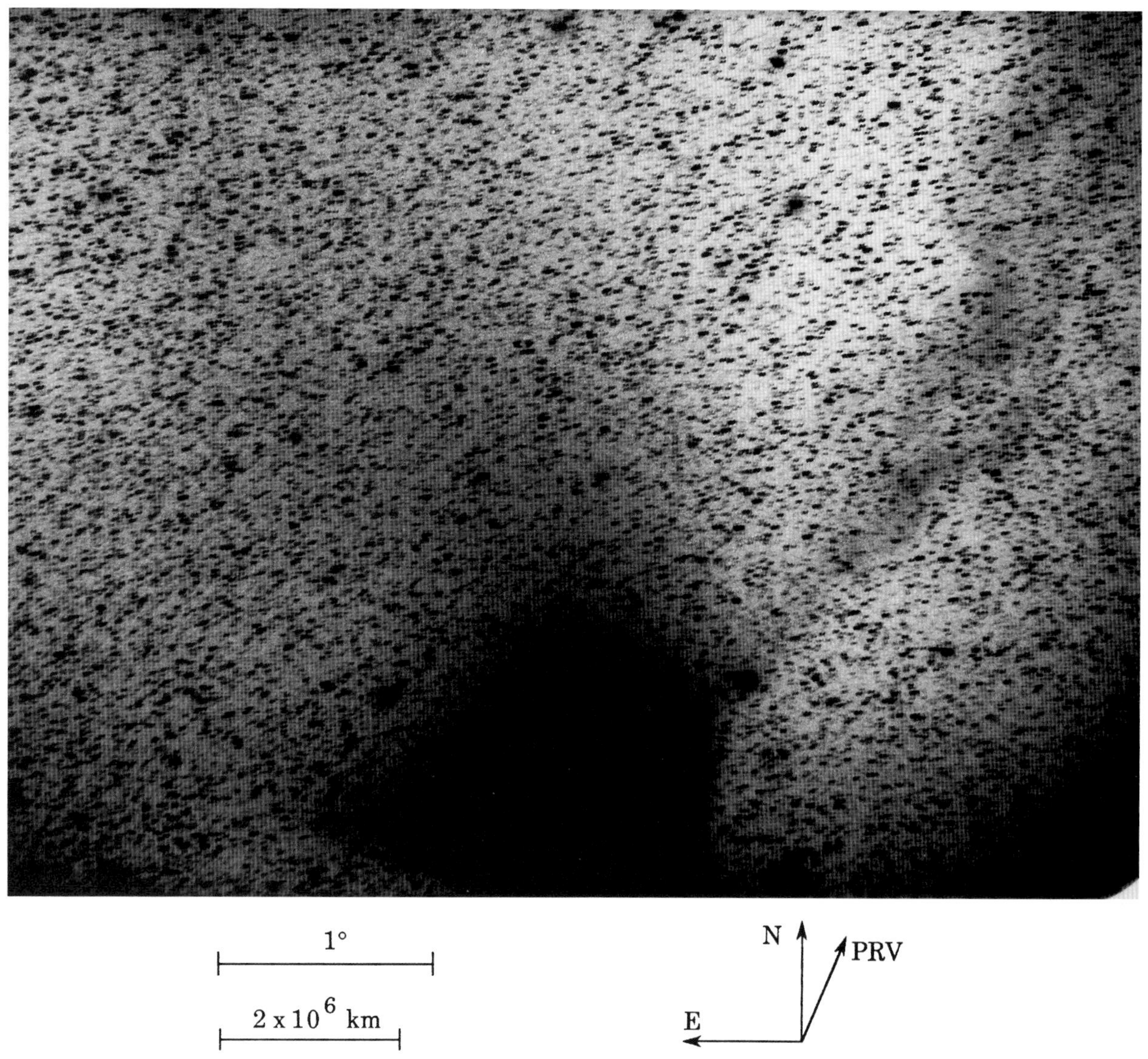

Fig. 798: 1986 Apr 12.34723 UT; exposure 8.0 minutes on Tech. Pan 2415 emulsion with no filter; r=1.35, Δ=0.42, β=151.8°; LSPN-570.
Photograph by P. Malloy/M. Spector, Tahiti, LSPN Island Network

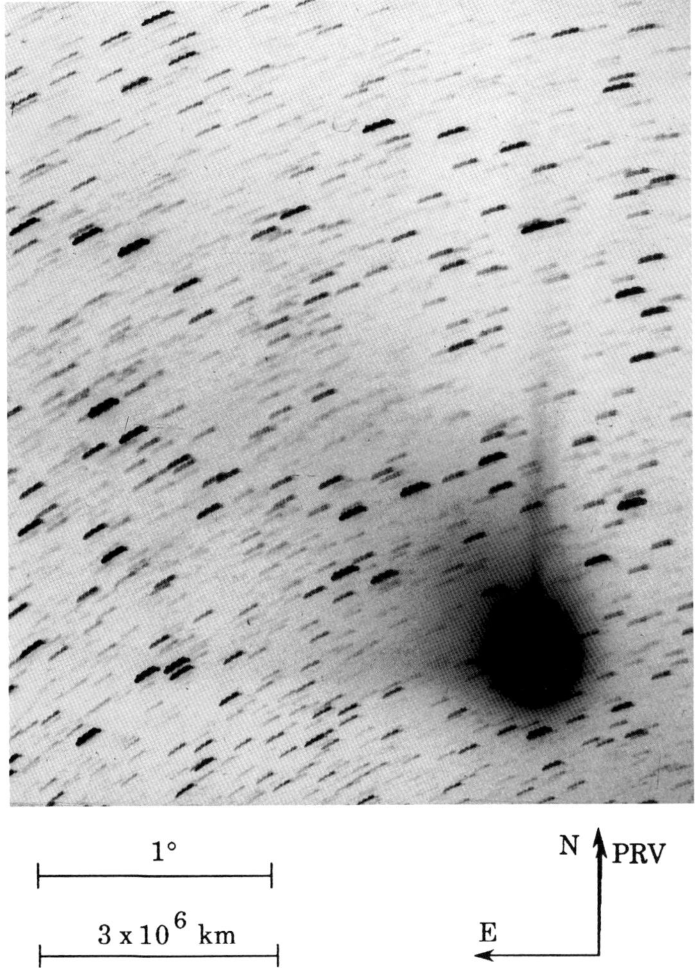

Fig. 799: 1986 Apr 14.51389 UT; exposure 30.0 minutes on Tech. Pan 2415 emulsion with no filter; r=1.39, Δ=0.43, β=156.4°; LSPN-587.
Photograph by P. Malloy/M. Spector, Tahiti, LSPN Island Network

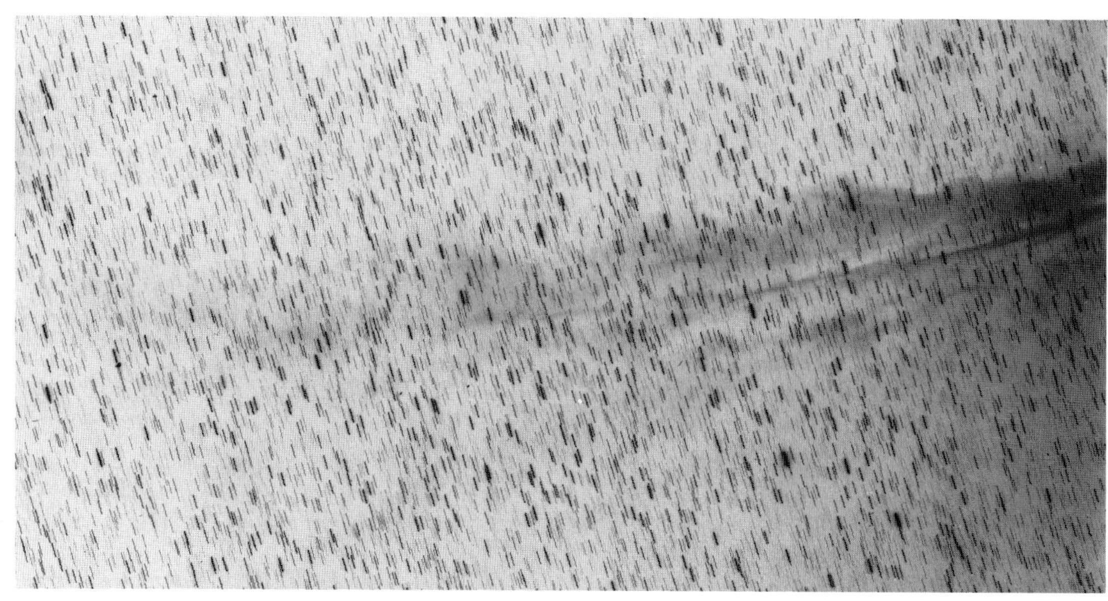

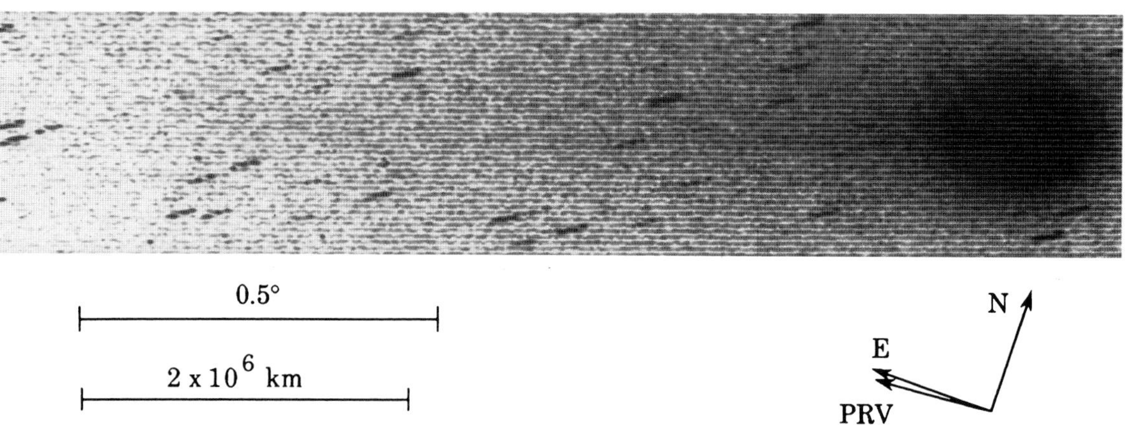

(top) Fig. 800: 1986 Apr 15.19792 UT; exposure 20.0 minutes on 103a-O emulsion with no filter; r=1.40, Δ=0.44, β=157.4°; LSPN-1225.
Photograph by F. Miller, University of Michigan/CTIO.

(bottom) Fig. 801: 1986 Apr 30.66667 UT; exposure 50.0 minutes on ORWO ZU-21 emulsion with no filter; r=1.63, Δ=0.79, β=151.3°; LSPN-3299.
Photograph by S. Gerasimenko, Gissar Observatory .

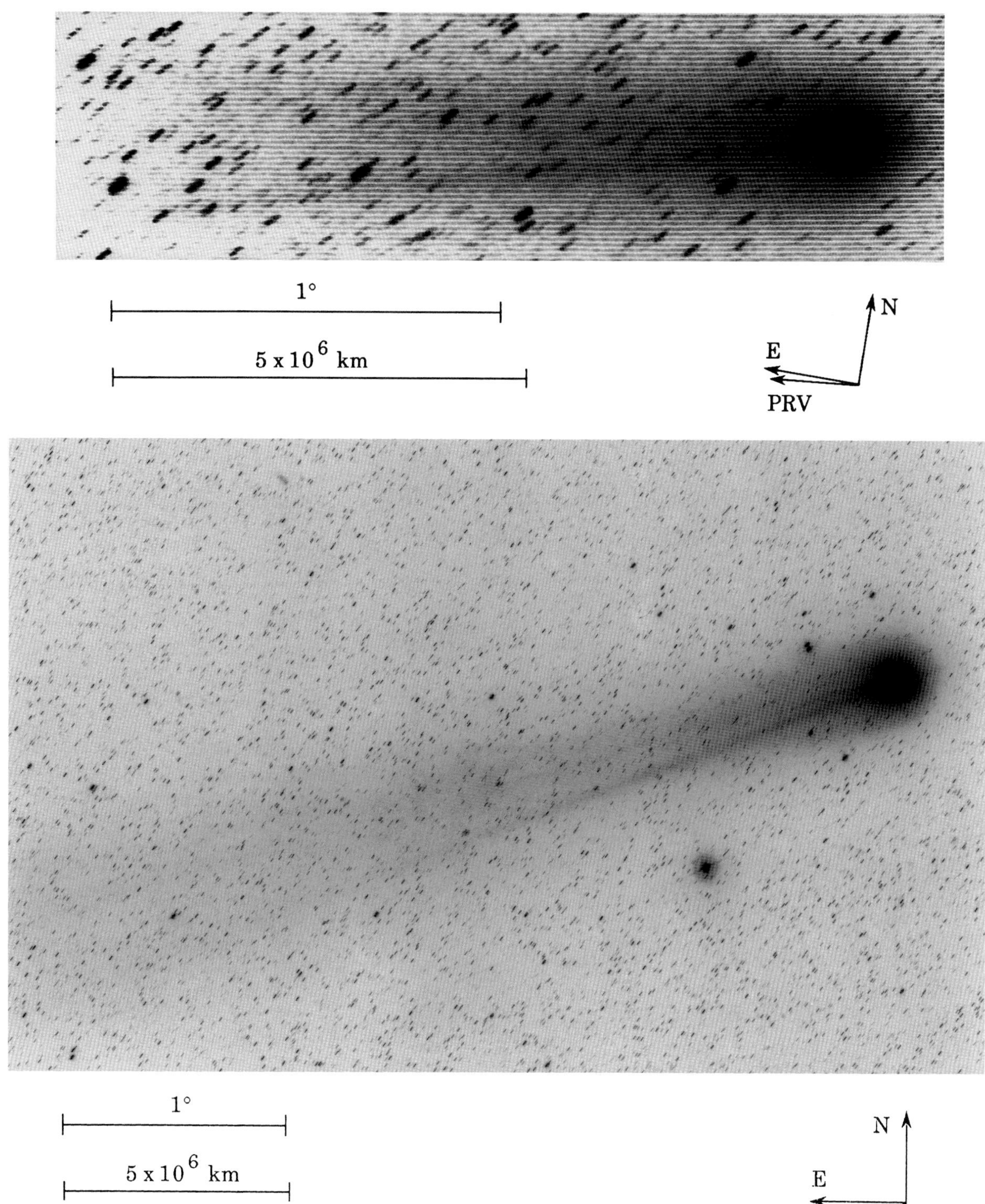

(top) Fig. 802: 1986 May 2.80191 UT; exposure 60.0 minutes on ORWO ZU-21 emulsion with no filter; r=1.66, Δ=0.86, β=150.4°; LSPN-3444.
Photograph by N. Chernykh, Crimean Astrophysical Observatory.

(bottom) Fig. 803: 1986 May 5.15322 UT; exposure 25.0 minutes on 103a-O emulsion with no filter; r=1.69, Δ=0.93, β=149.6°; LSPN-3038.
Photograph by K. Meech, University of Michigan/CTIO.

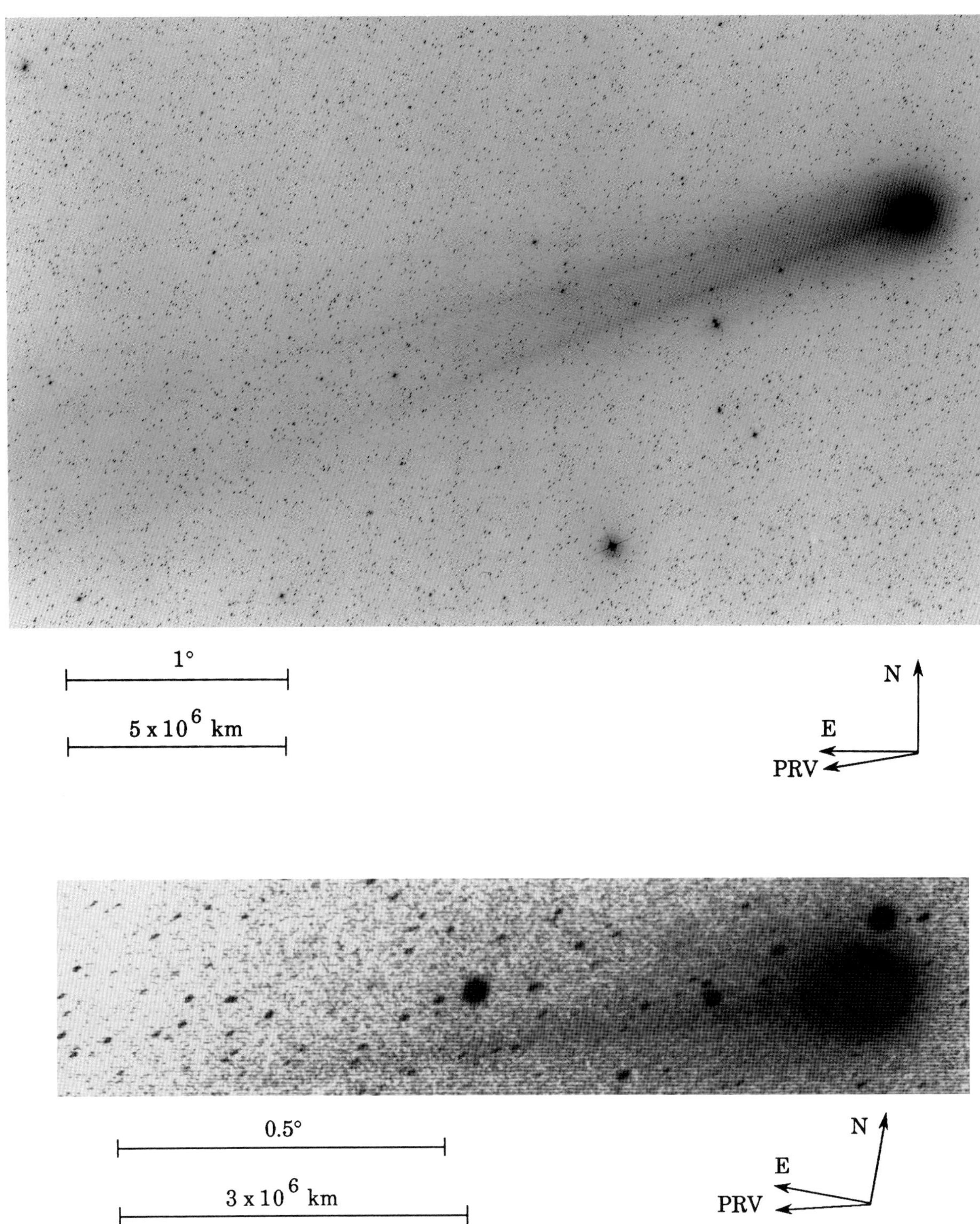

(top) Fig. 804: 1986 May 6.10198 UT; exposure 15.0 minutes on 103a-O emulsion with no filter; r=1.71, Δ=0.96, β=149.4°; LSPN-3044.
Photograph by K. Meech, University of Michigan/CTIO.

(bottom) Fig. 805: 1986 May 10.69793 UT; exposure 30.0 minutes on ORWO ZU-21 emulsion with no filter; r=1.77, Δ=1.10, β=148.7°; LSPN-3302.
Photograph by P. Pushnin, Gissar Observatory .

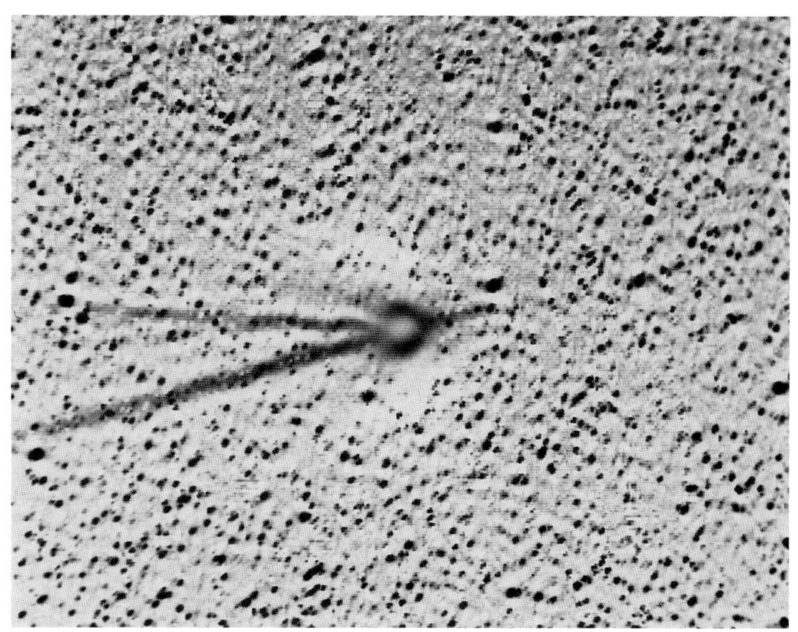

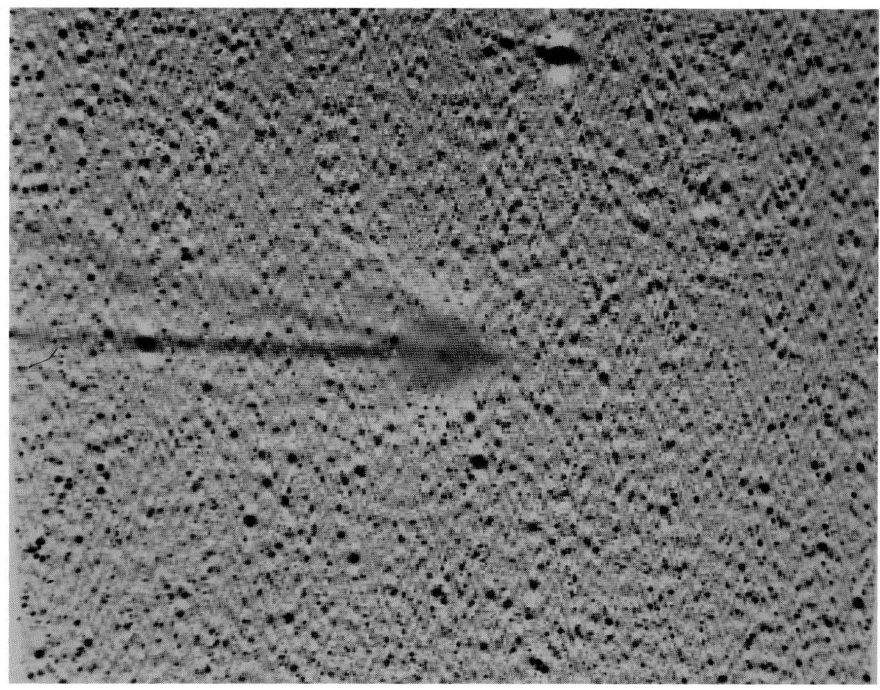

(top) Fig. 806: 1986 Apr 28.20382; 50 mm ~ 2 x 10^6 km; exposure 15 minutes on IIIa-J emulsion with GG385 filter. Image is processed; see Sekanina et al. (1987).
Photograph by E. Helin, Palomar Observatory.

(bottom) Fig. 807: 1986 Apr 30.17021; 50 mm ~ 2 x 10^6 km; exposure 5 minutes on IIa-D emulsion with no filter. Image is processed; see Sekanina et al. (1987).
Photograph by E. Helin, Palomar Observatory.

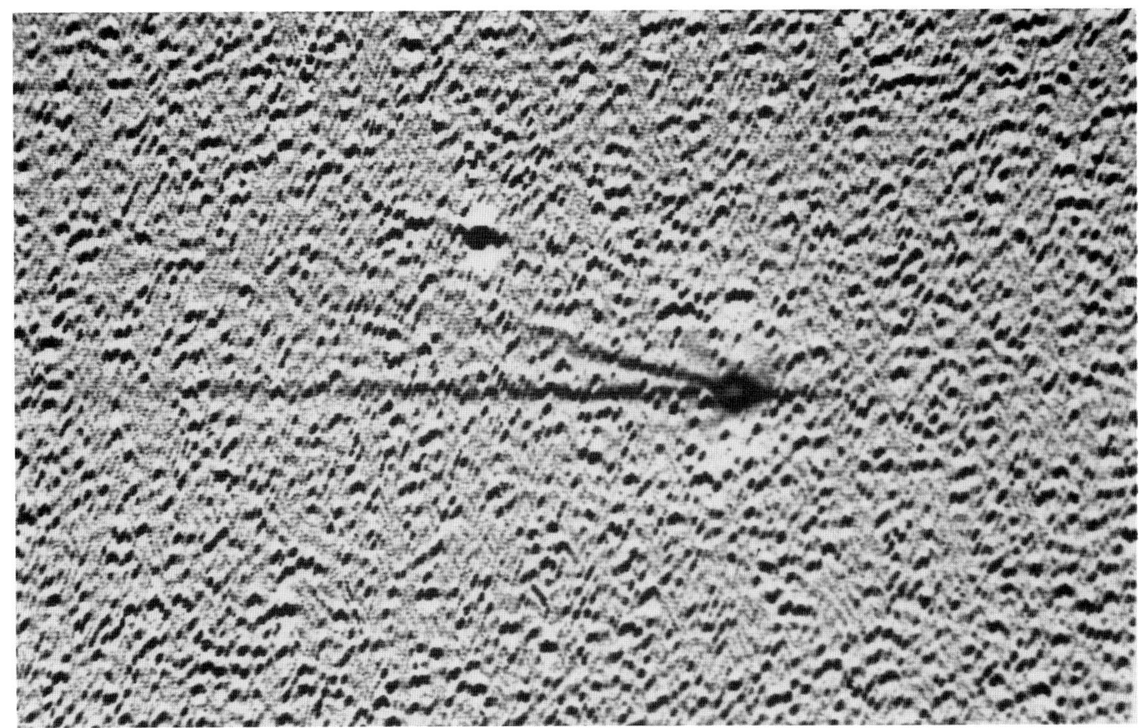

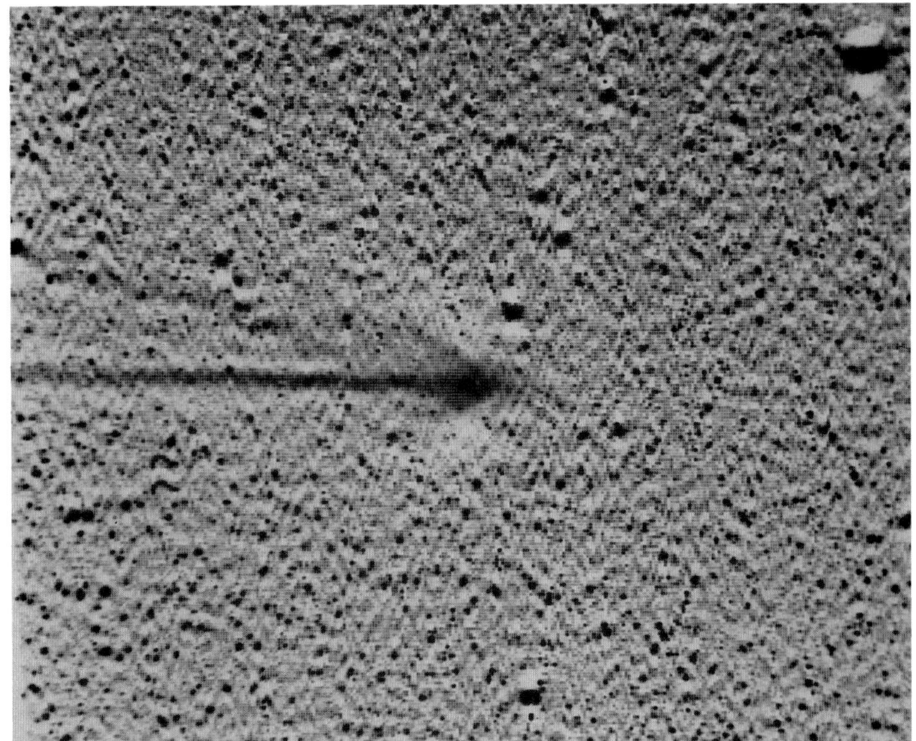

(top) Fig. 808: 1986 May 1.16500; 50 mm ~ 2 x 10^6 km; exposure 32 minutes on Tech. Pan 2415 emulsion with no filter. Image is processed; see Sekanina et al. (1987).
Photograph by G. Emerson, Barnard Observatory.

(bottom) Fig. 809: 1986 May 2.25590; 50 mm ~ 2 x 10^6 km; exposure 8 minutes on IIa-D emulsion with no filter. Image is processed; see Sekanina et al. (1987).
Photograph by E. Helin, Palomar Observatory.

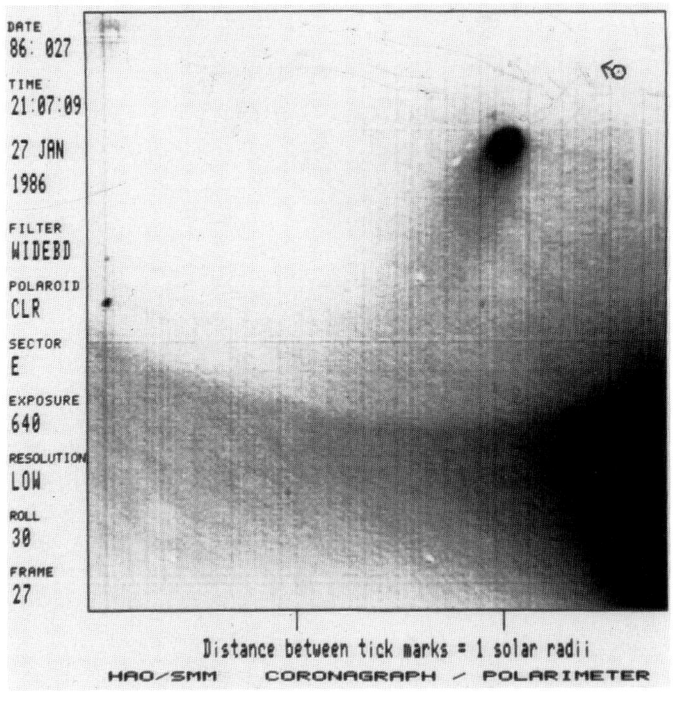

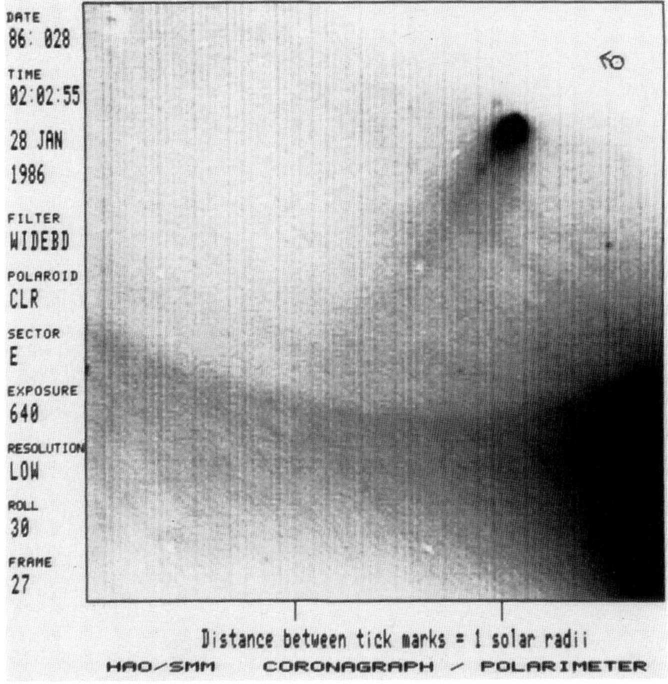

(top) Fig. 810: 1986 Jan 27.87997; arrow indicates solar north.
Photograph courtesy of High Altitude Observatory/Solar Maximum Mission/Coronagraph-Polarimeter.

(bottom) Fig. 811: 1986 Jan 28.08536; arrow indicates solar north.
Photograph courtesy of High Altitude Observatory/Solar Maximum Mission/Coronagraph-Polarimeter.

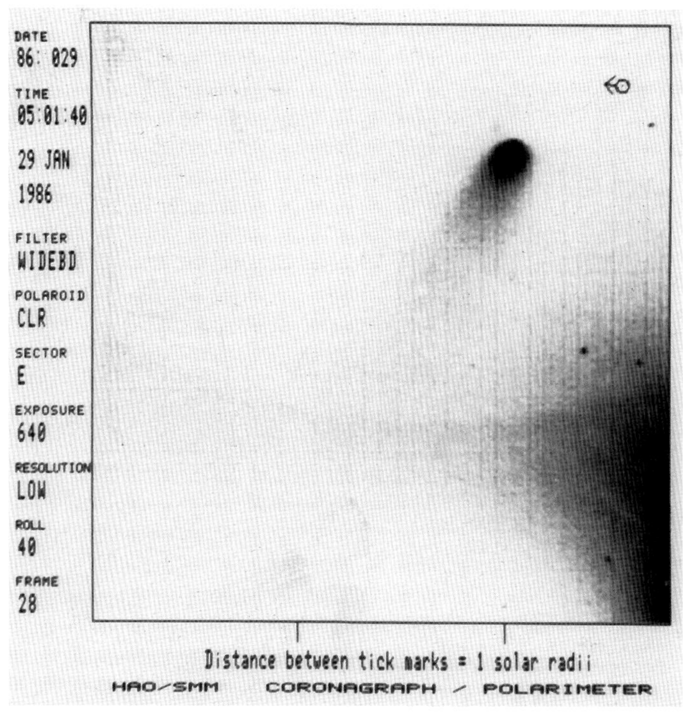

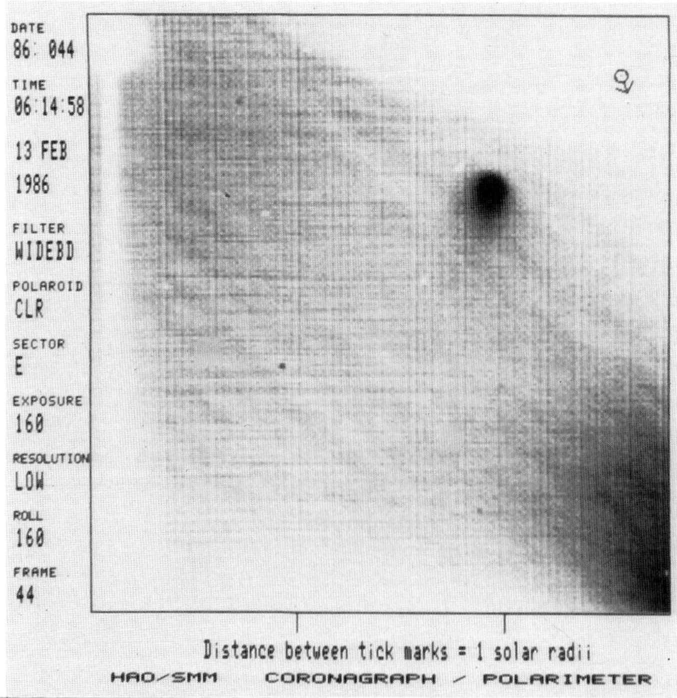

(top) Fig. 812: 1986 Jan 29.20949; arrow indicates solar north.
Photograph courtesy of High Altitude Observatory/Solar Maximum Mission/Coronagraph-Polarimeter.

(bottom) Fig. 813: 1986 Feb 13.26039; arrow indicates solar north.
Photograph courtesy of High Altitude Observatory/Solar Maximum Mission/Coronagraph-Polarimeter.

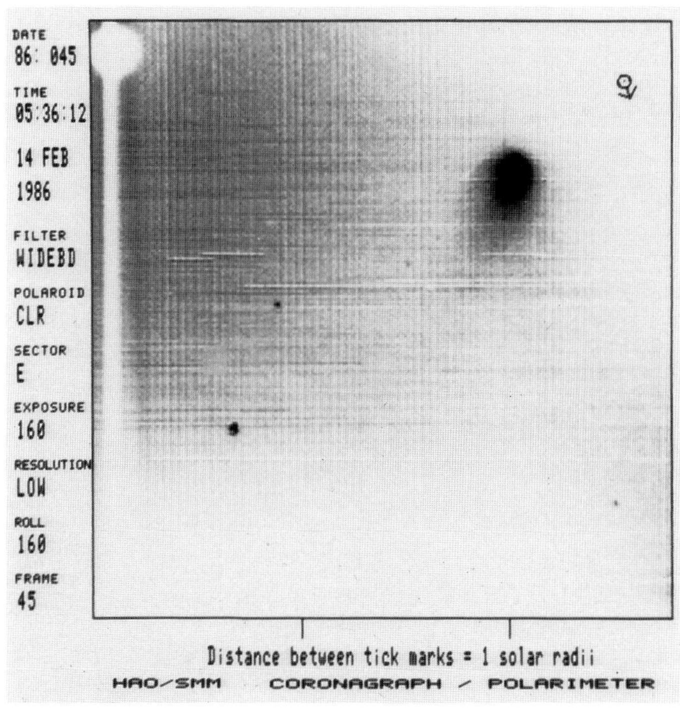

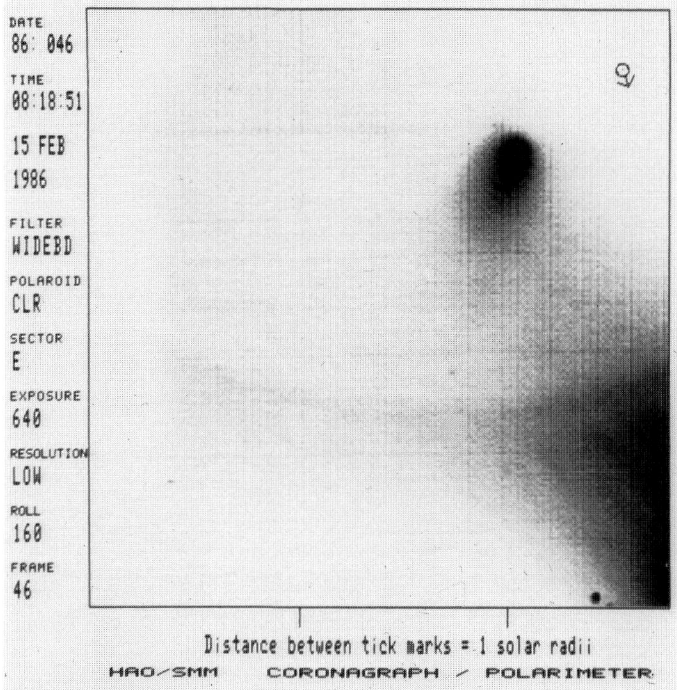

(top) Fig. 814: 1986 Feb 14.23347; arrow indicates solar north.
Photograph courtesy of High Altitude Observatory/Solar Maximum Mission/Coronagraph-Polarimeter.

(bottom) Fig. 815: 1986 Feb 15.34642; arrow indicates solar north.
Photograph courtesy of High Altitude Observatory/Solar Maximum Mission/Coronagraph-Polarimeter.

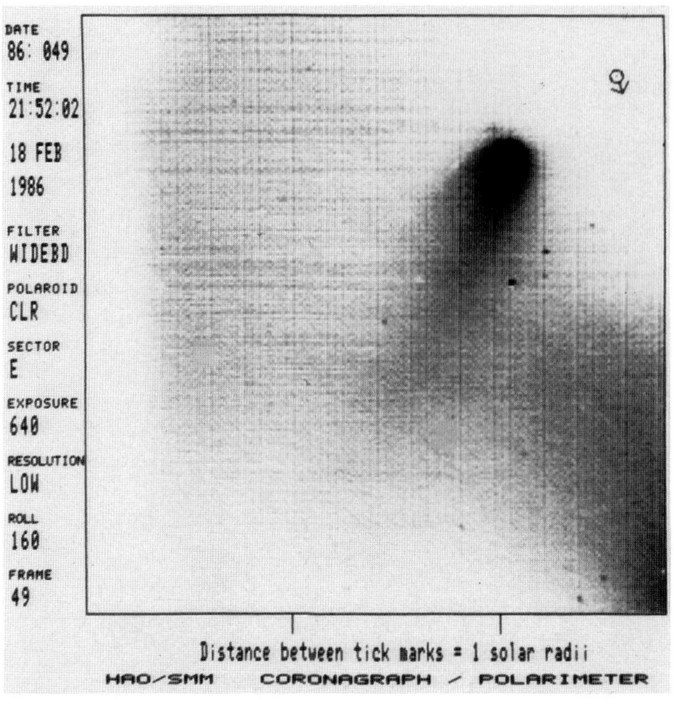

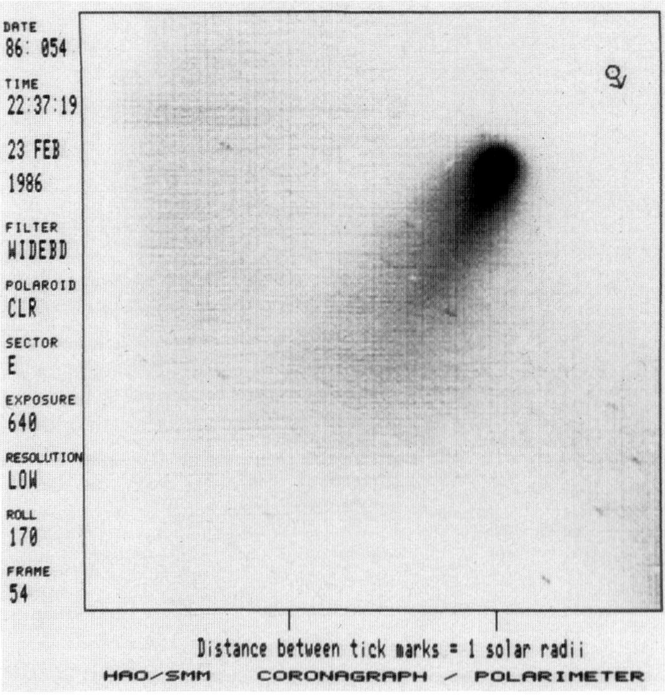

(top) Fig. 816: 1986 Feb 18.91113; arrow indicates solar north.
Photograph courtesy of High Altitude Observatory/Solar Maximum Mission/Coronagraph-Polarimeter.

(bottom) Fig. 817: 1986 Feb 23.94258; arrow indicates solar north.
Photograph courtesy of High Altitude Observatory/Solar Maximum Mission/Coronagraph-Polarimeter.

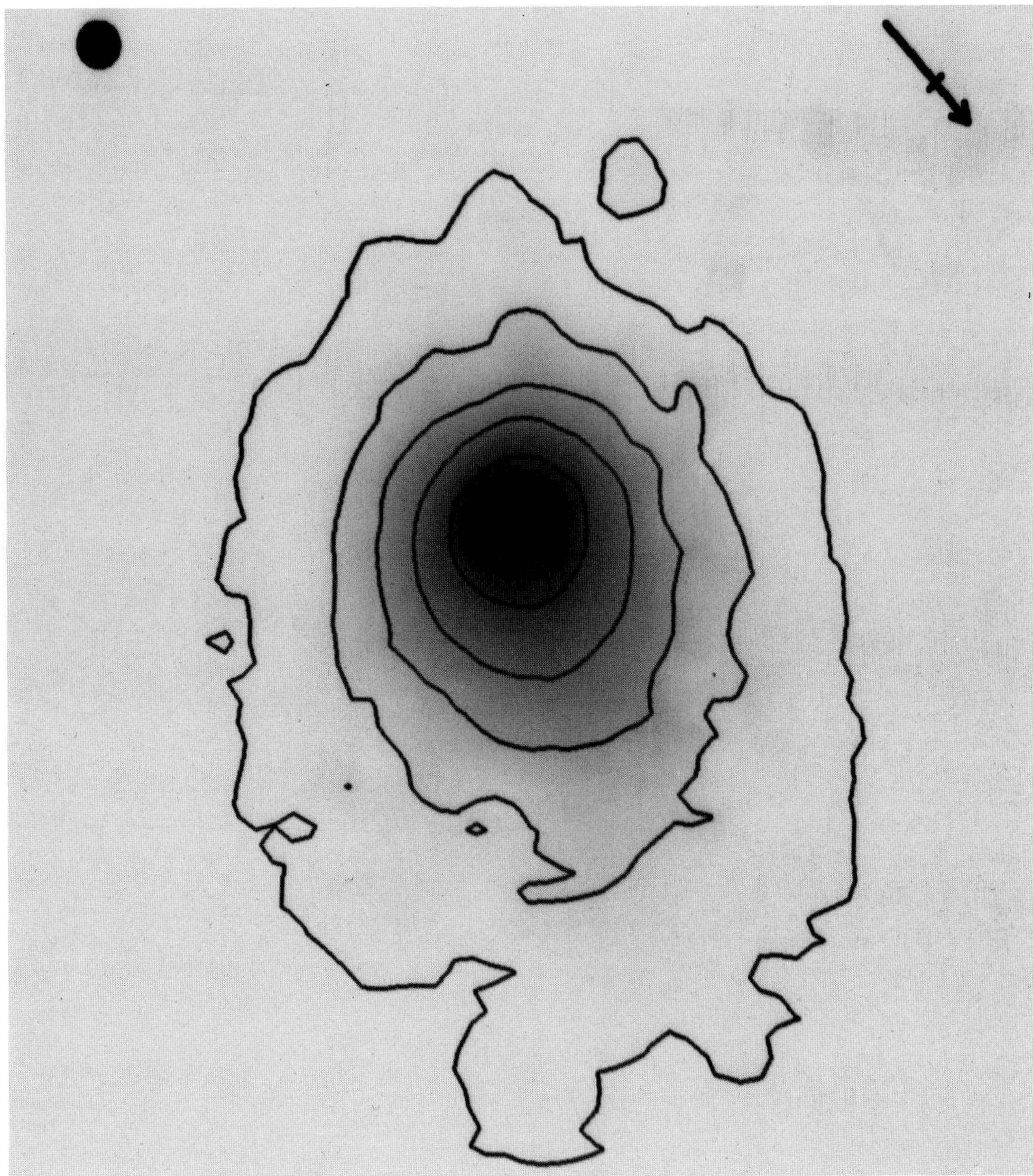

Fig. 818: Lyman-α image of Comet P/Halley obtained by the Ultraviolet Spectrometer experiment on the Pioneer Venus Orbiter spacecraft, during the period 2nd-6th February 1986. The comet was 0.60 A.U. (89 million km) from the sun. A background signal of about 0.4 kiloRayleighs due to solar Lyman-α scattered from interplanetary hydrogen has been subtracted. The isophotes are drawn at surface brightnesses of 0.2, 1, 2, 5, 10, and 20 kiloRayleighs (1 kilo-Rayleigh = $10^9/4\pi$ photons/cm^2/sec/ster). The arrow indicates ecliptic north. The filled circle to the upper left of the image represents the sun's disk to scale at the comet's distance.
Courtesy of A.I.F. Stewart, University of Colorado and M.R. Combi, AER. Inc.

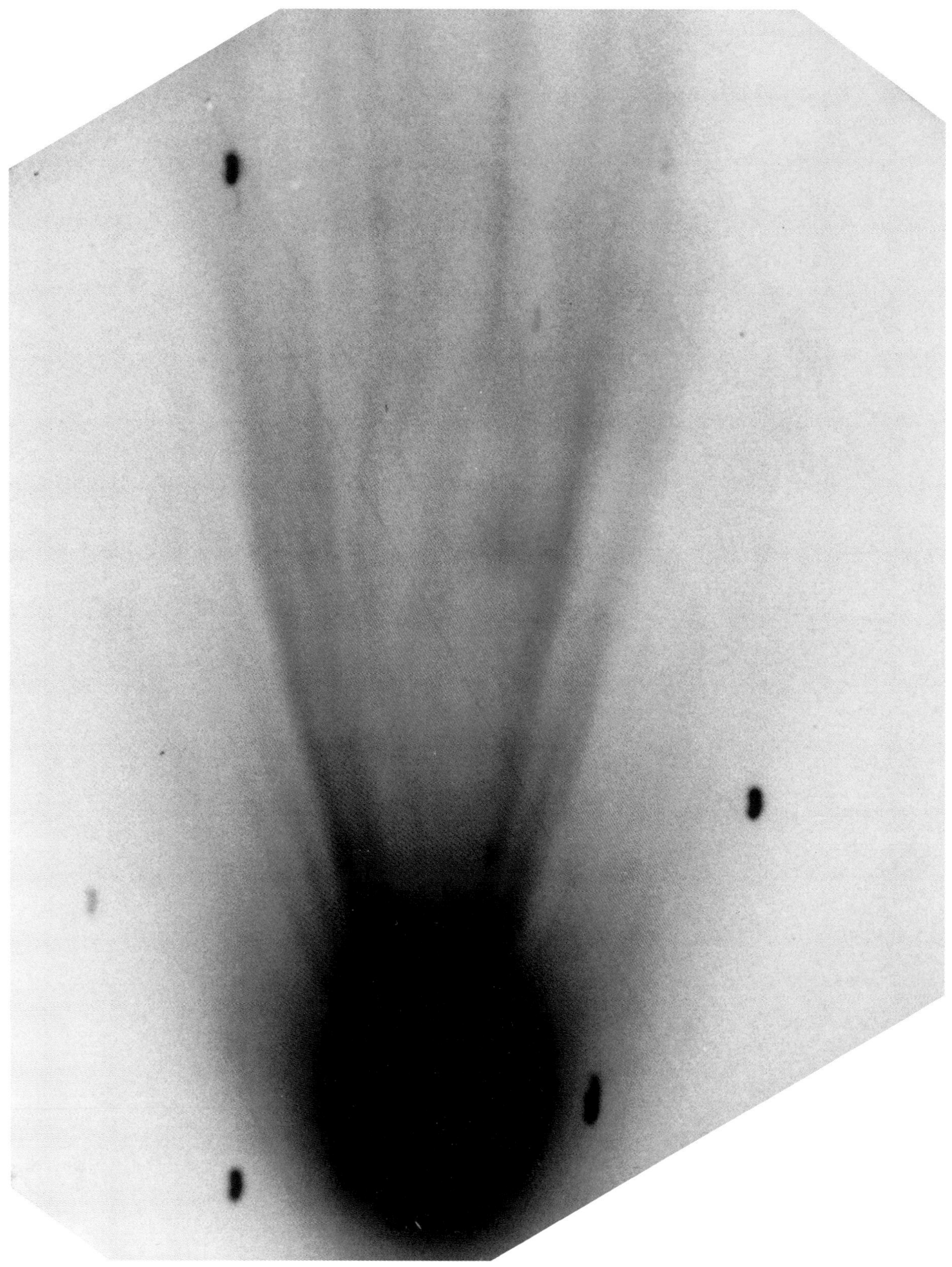

Fig. 819: 1986 Jan 8.09020 UT; 20mm ~ 50 arcsec; exposure 5.0 minutes on TI CCD with H_2O^+ filter; r=0.90, Δ=1.29, β=130.5°; NNSN-405502.
Photograph by U. Fink/A. Schultz, Catalina Observatory.

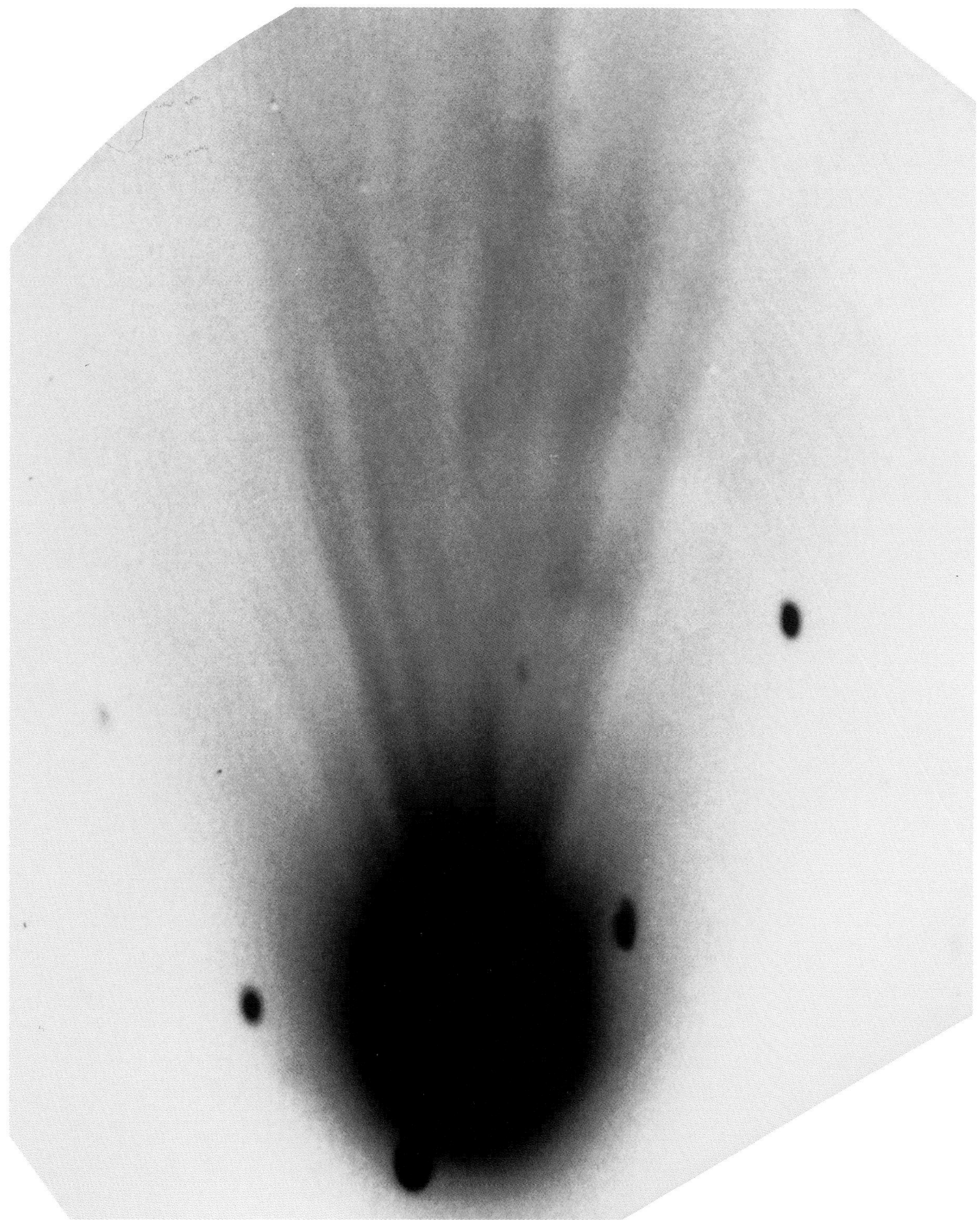

Fig 820: 1986 Jan 8.11204 UT; 20 mm ~ 50 arcsec; exposure 5.0 minutes on TI CCD with H_2O^+ filter; r=0.90, Δ=1.29, β=130.5°; NNSN=405505.
Photograph by U. Fink/A. Schultz, Catalina Observatory.

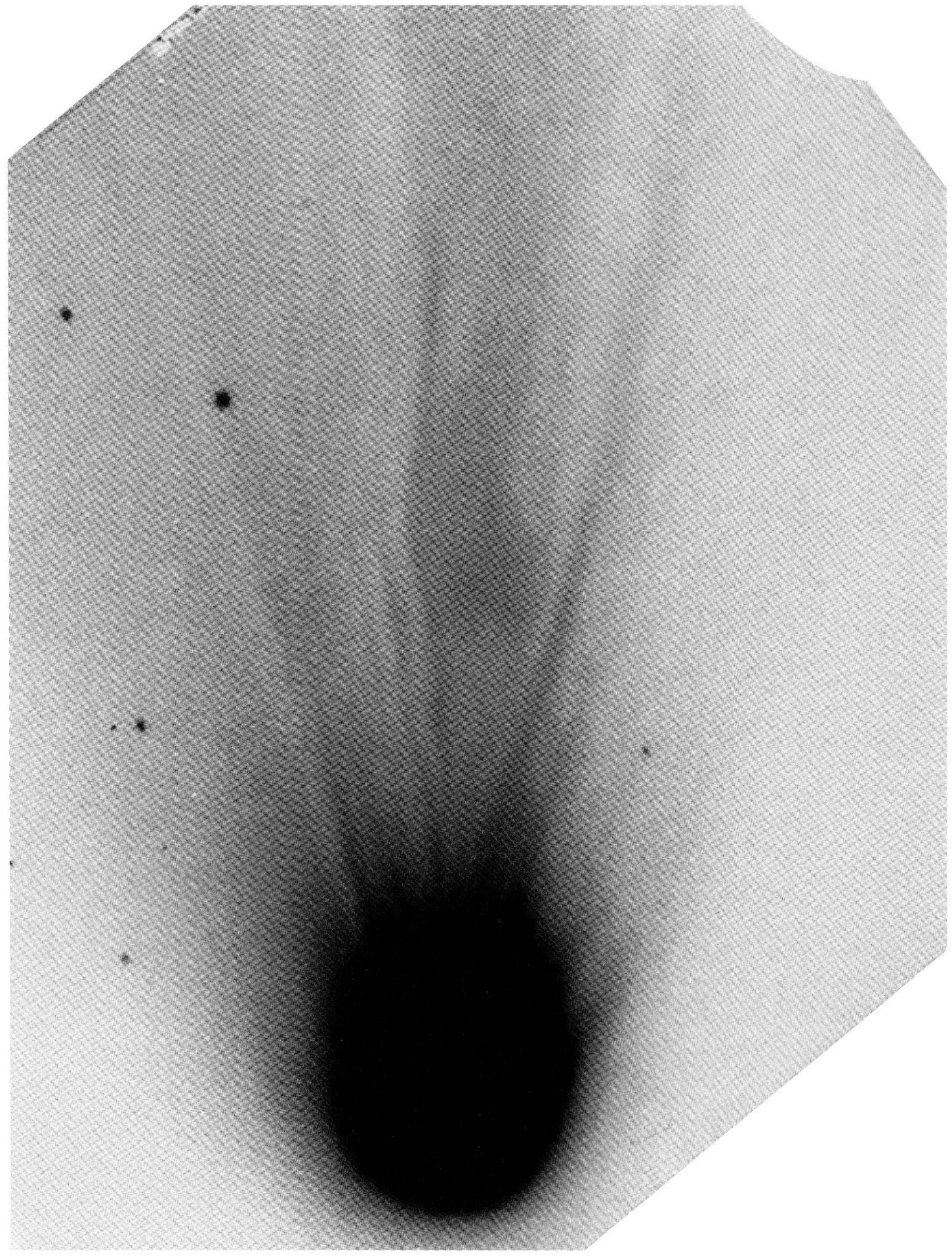

Fig 821: 1986 Jan 20.08551 UT; 20 mm ~ 50 arcsec; exposure 1.0 minute on TI CCD with H_2O^+ filter; r=0.73, Δ=1.47, β=143.4°; NNSN-405521.
Photograph by U. Fink / R. Fink, Catalina Observatory.

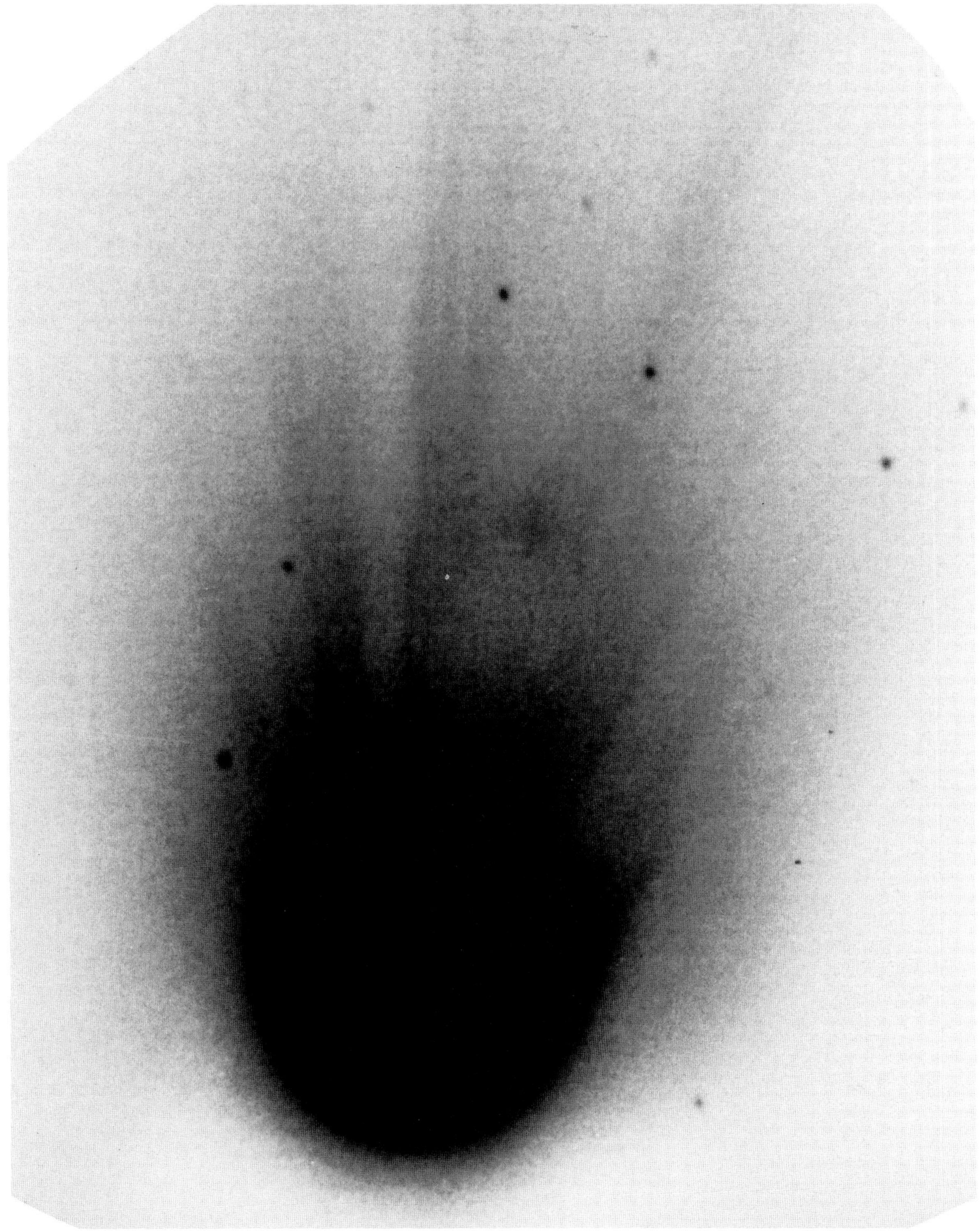

Fig 822: 1986 Mar 6.52520 UT; 20 mm ~ 50 arcsec; exposure 0.8 minutes on TI CCD with H_2O^+ filter; r=0.80, Δ=1.14, β=121.7°; NNSN-405525.
Photograph by G. Sill /M. Disanti, Catalina Observatory.

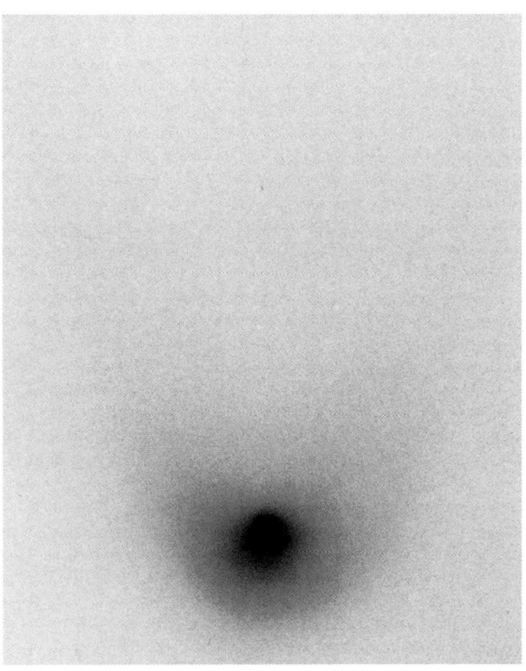

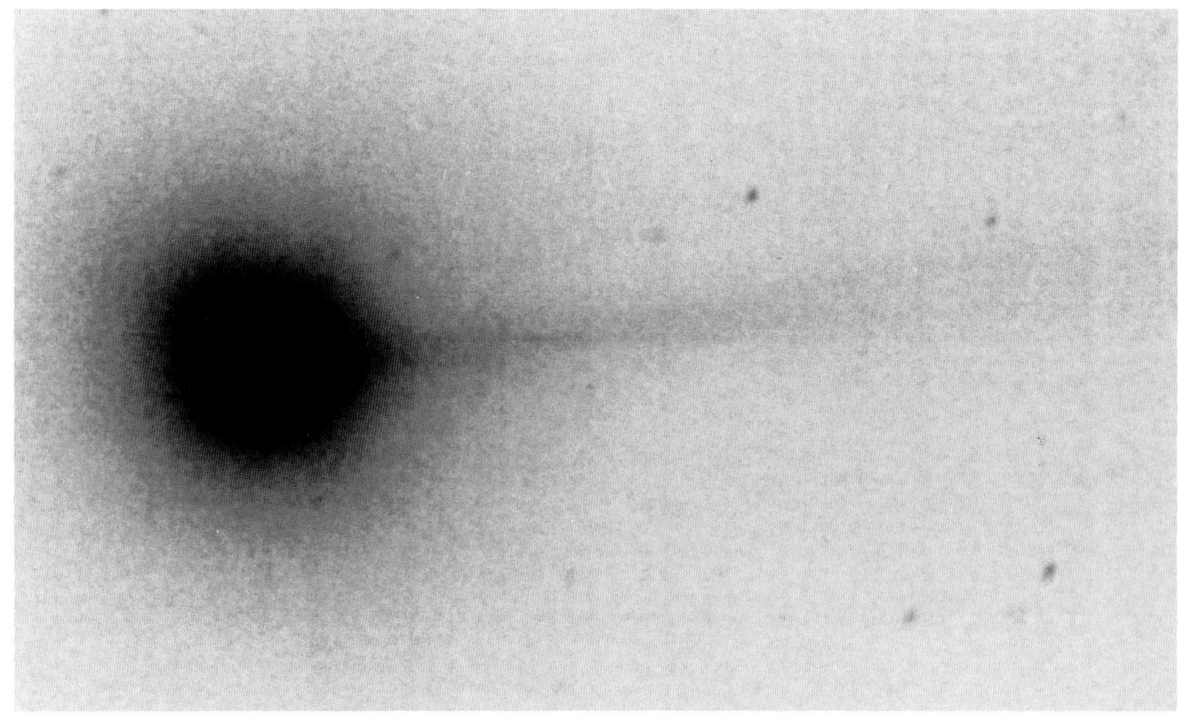

(top) Fig 823: 1986 Mar 6.52703 UT; 20 mm ~ 50 arcsec; exposure 0.6 minutes on TI CCD with continuum filter; r=0.80, Δ=1.14, β=121.7°; NNSN-405526.
Photograph by G. Sill/M. Disanti, Catalina Observatory.

(bottom) Fig 824*: 1986 May 10.18487 UT; 20 mm ~ 50 arcsec; exposure 3.0 minutes on TI CCD with H_2O^+ filter; r=1.77, Δ=1.09, β=148.7°; NNSN-405544.
Photograph by U. Fink/A. Schultz, Catalina Observatory.

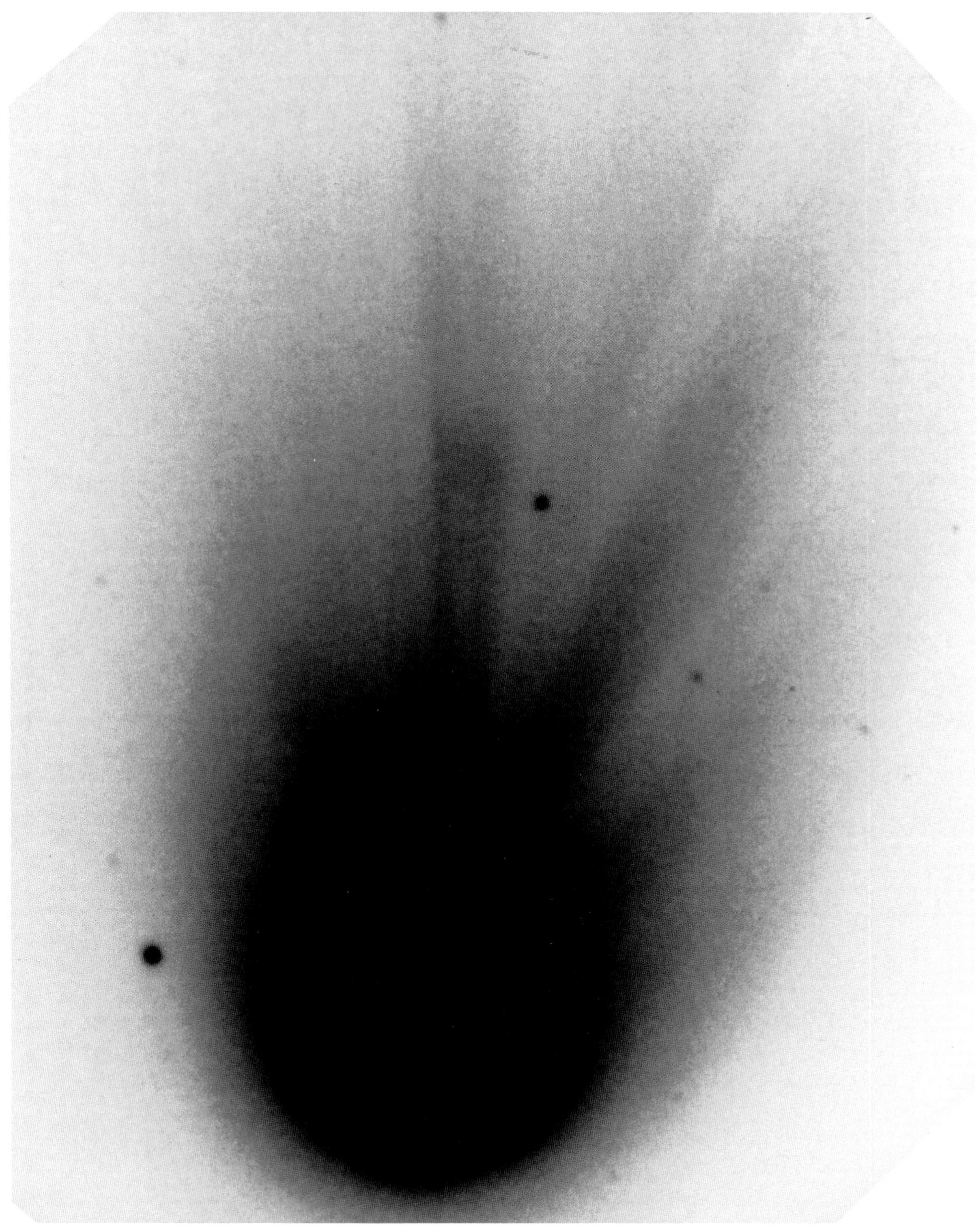

Fig 825: 1986 Mar 8.52902 UT; 20 mm ~ 50 arcsec; exposure 1.0 minute on TI CCD with H_2O^+ filter; r=0.82, Δ=1.10, β=119.7°; NNSN-405535.
Photograph by R. Marcialis et al., Catalina Observatory.

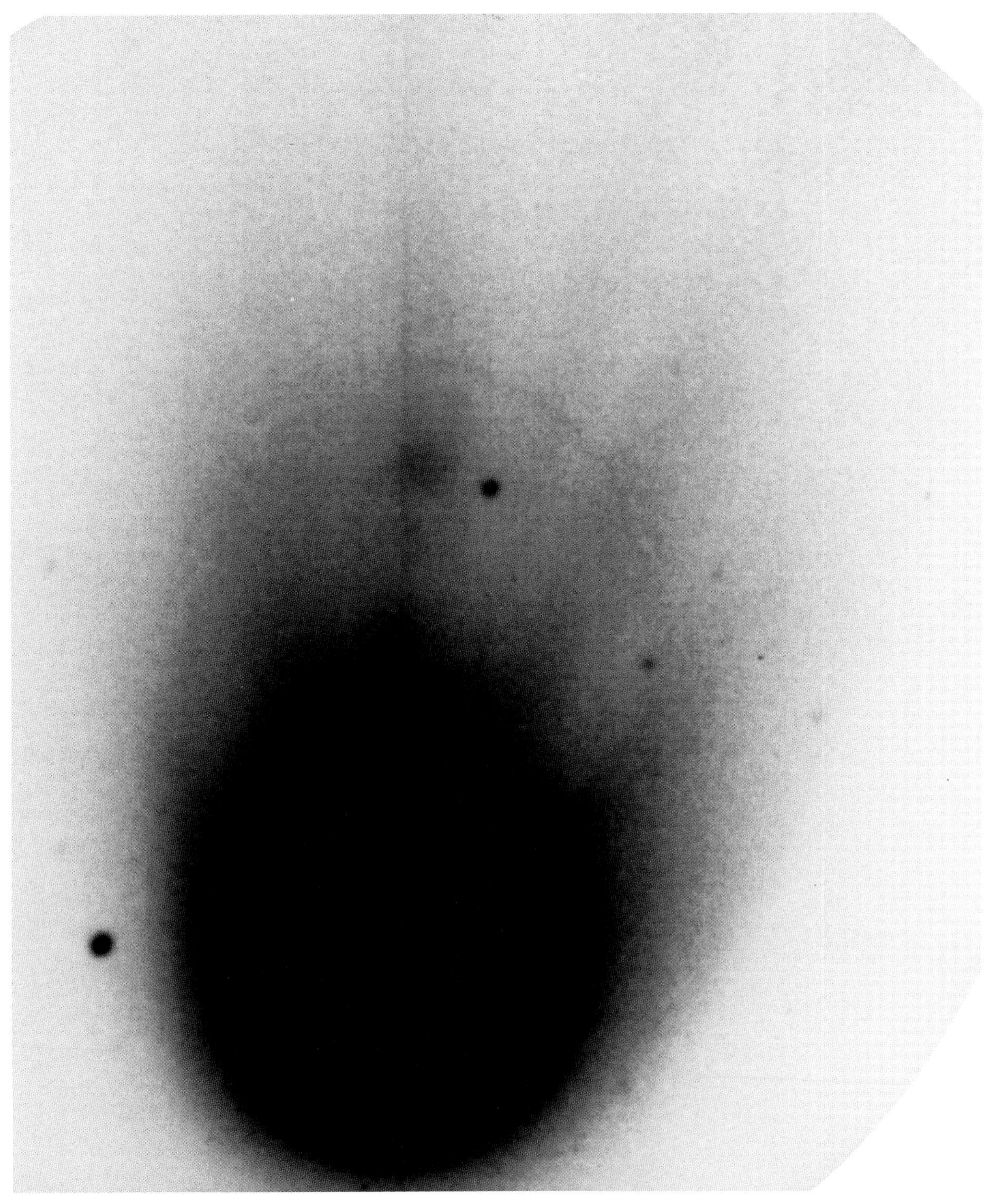

Fig 826: 1986 Mar 8.53328 UT; 20 mm ~ 50 arcsec; exposure 2.7 minutes on TI CCD with continuum filter; r=0.82, Δ=1.10, β=119.7°; NNSN-405537.
Photograph by R. Marcialis et al., Catalina Observatory.

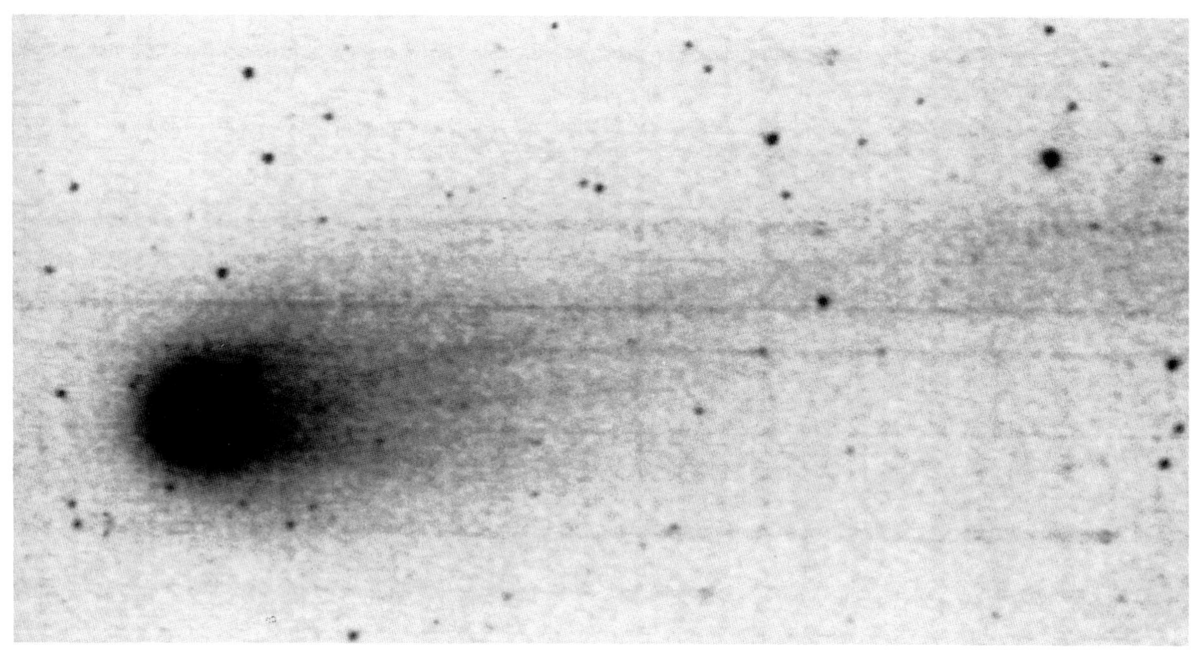

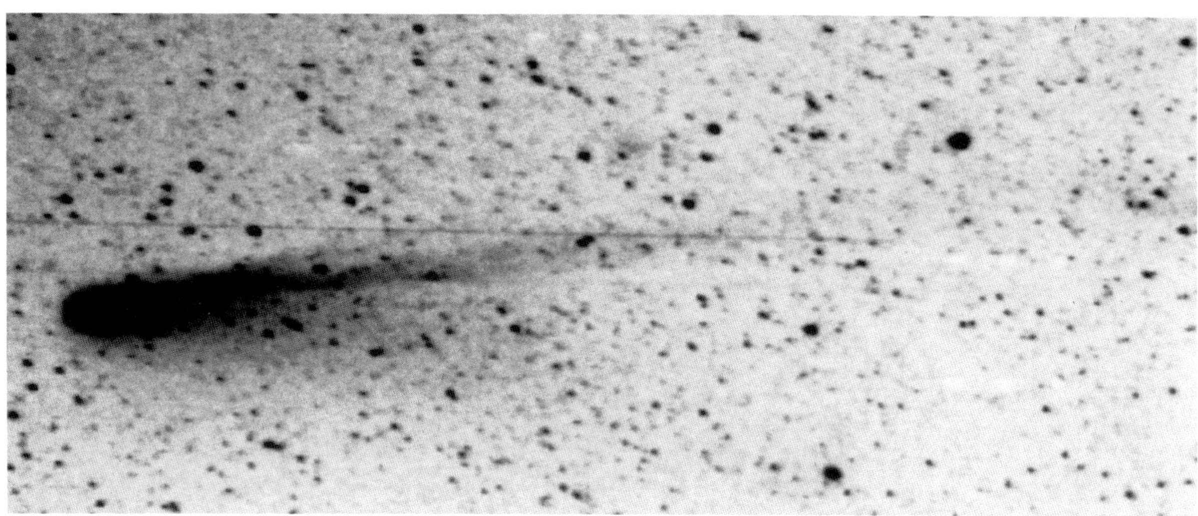

(top) Fig. 827: 1986 Mar 22.50000 UT; 40 mm ~ 10 arcmin; exposure 3.0 minutes on TI CCD with H_2O^+ filter; r=1.03, Δ=0.75, β=114.1°.
Photograph by U. Fink/A. Schultz, Catalina Observatory.

(bottom) Fig. 828: 1986 Mar 23.48470 UT; 40 mm ~ 10 arcmin; exposure 5.0 minutes on TI CCD with H_2O^+ filter; r=1.05, Δ=0.72, β=114.4°; LSPN-308.
Photograph by U. Fink/A. Schultz, Catalina Observatory.

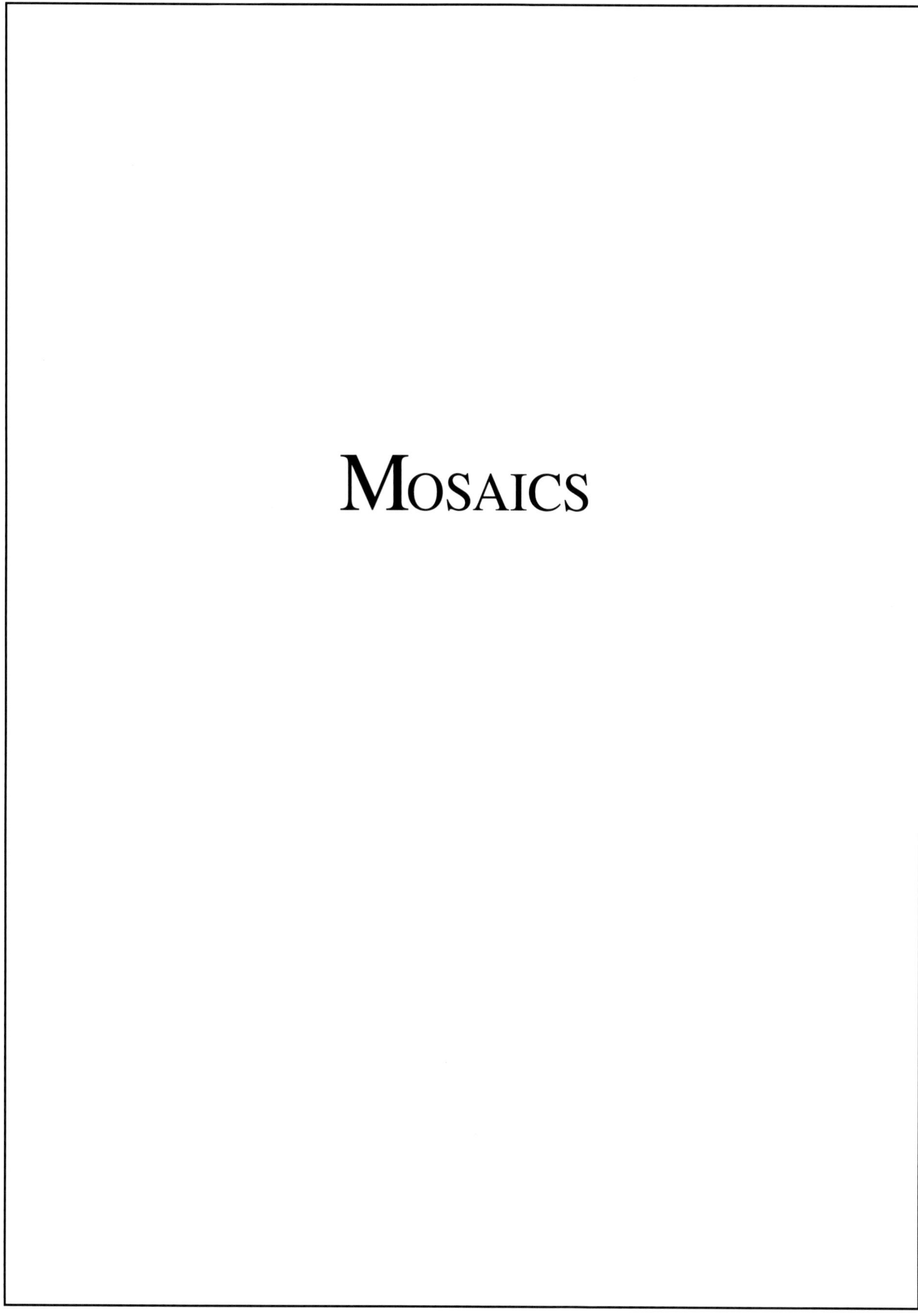

Mosaics

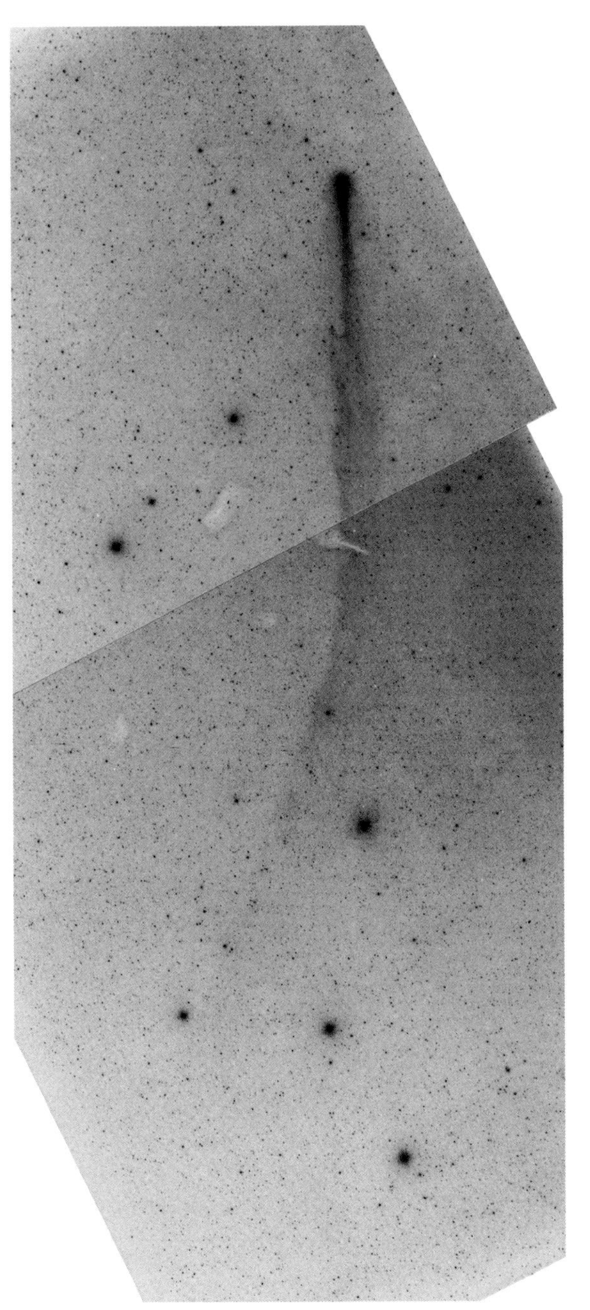

Fig. 829: 11 January 1986, LSPN 2096/2097

Fig. 830: 12 January 1986, LSPN 2100/2101

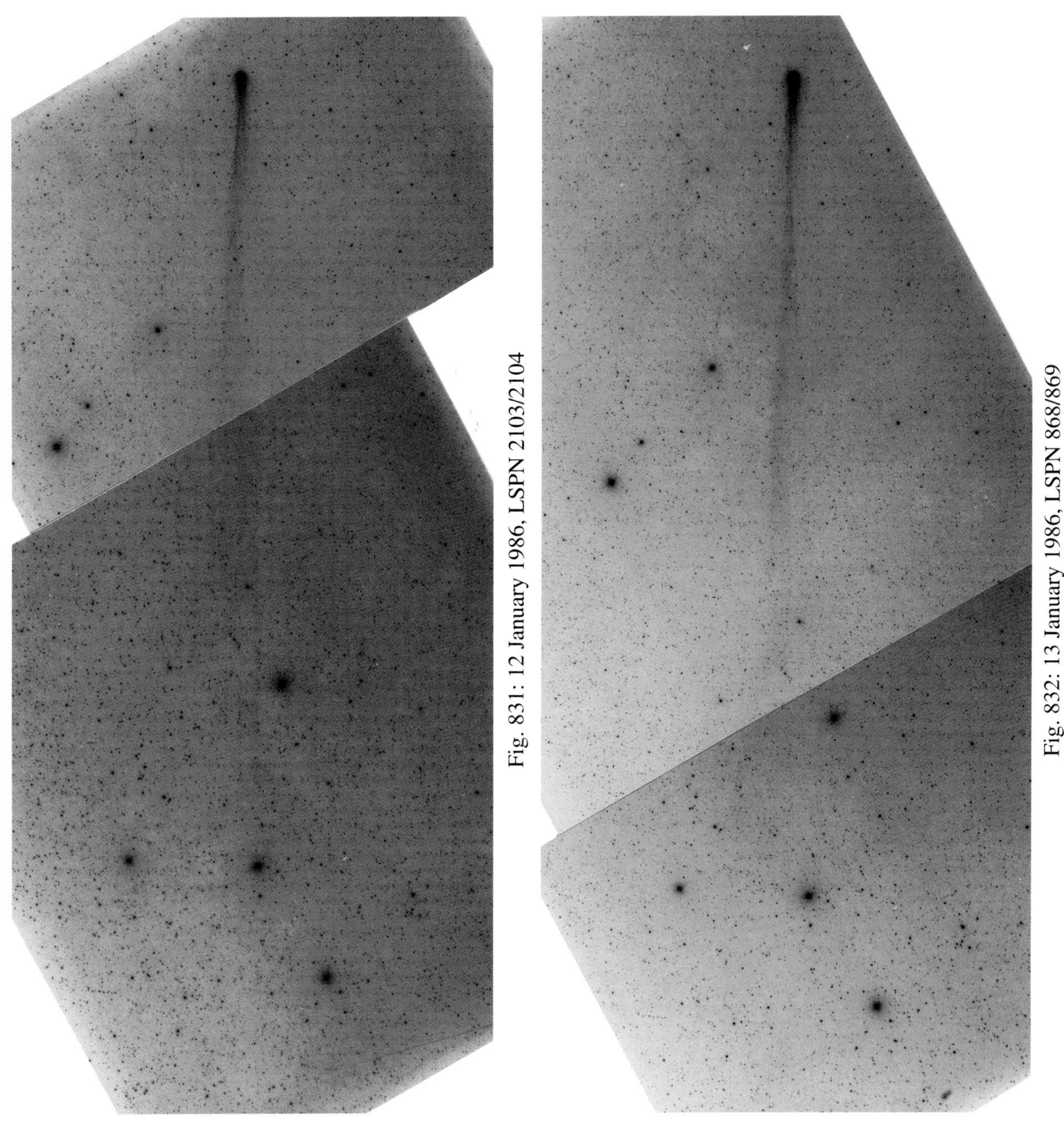

Fig. 831: 12 January 1986, LSPN 2103/2104

Fig. 832: 13 January 1986, LSPN 868/869

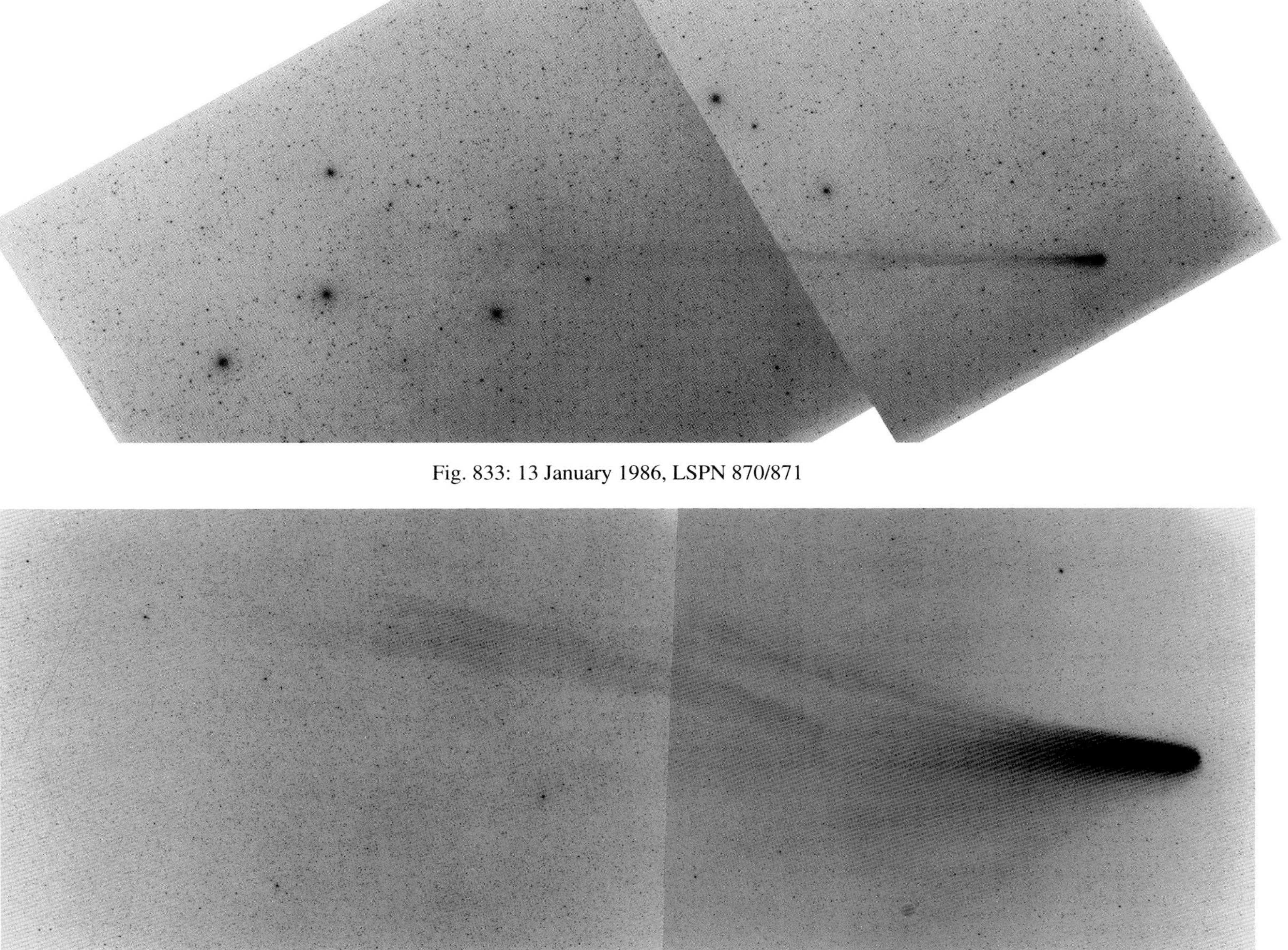

Fig. 833: 13 January 1986, LSPN 870/871

Fig. 834: 10 March 1986, LSPN 2383/2384

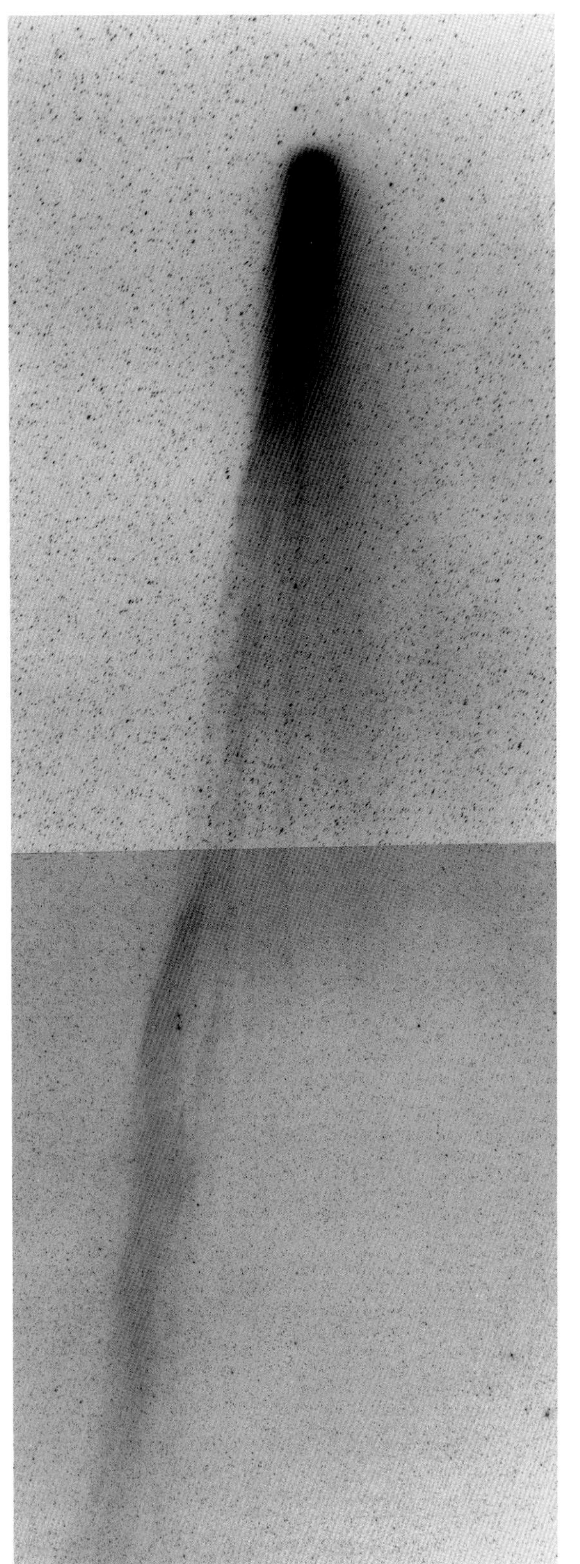

Fig. 835: 15 March 1986, LSPN 1291/1292

Fig. 836: 16 March 1986, LSPN 892/893

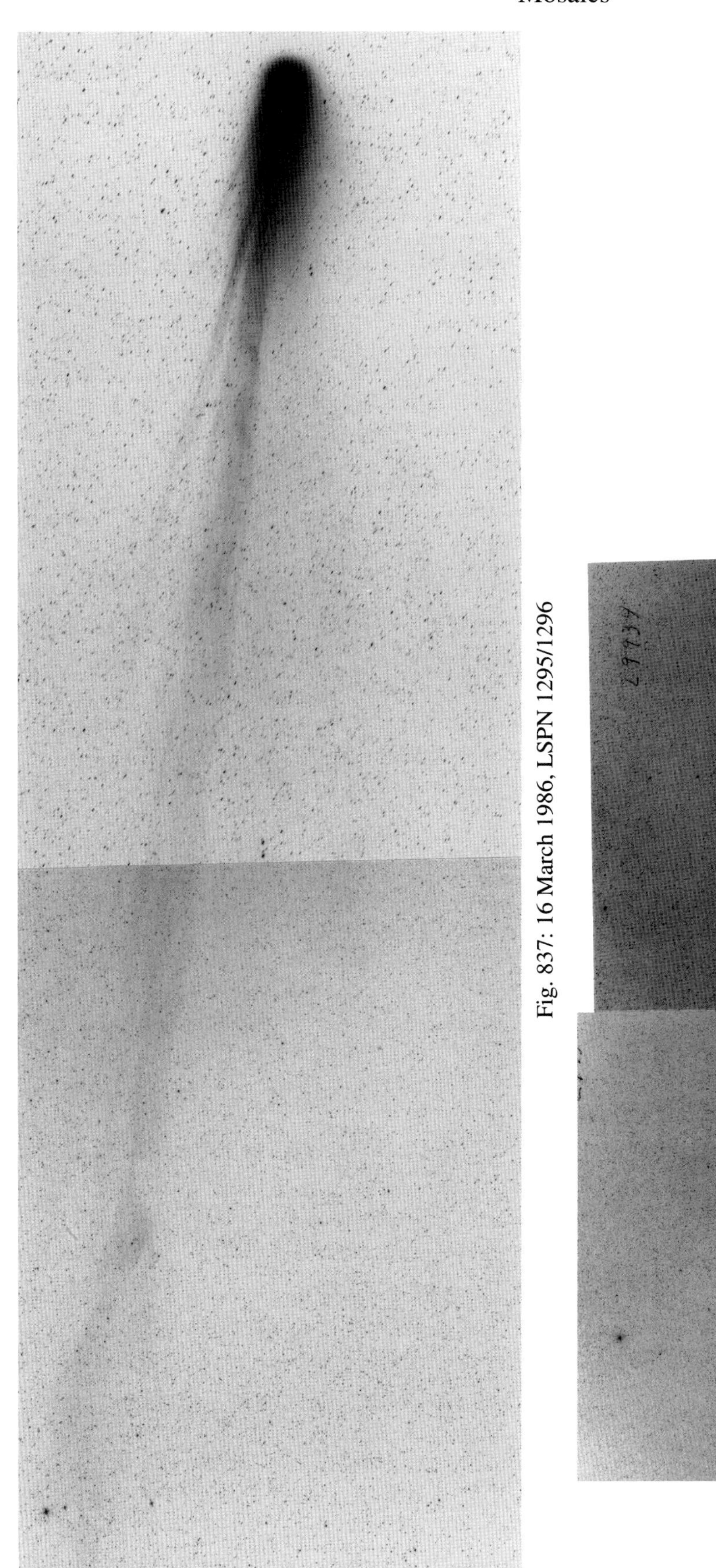

Fig. 837: 16 March 1986, LSPN 1295/1296

Fig. 838: 17 March 1986, LSPN 1300/1301/1302

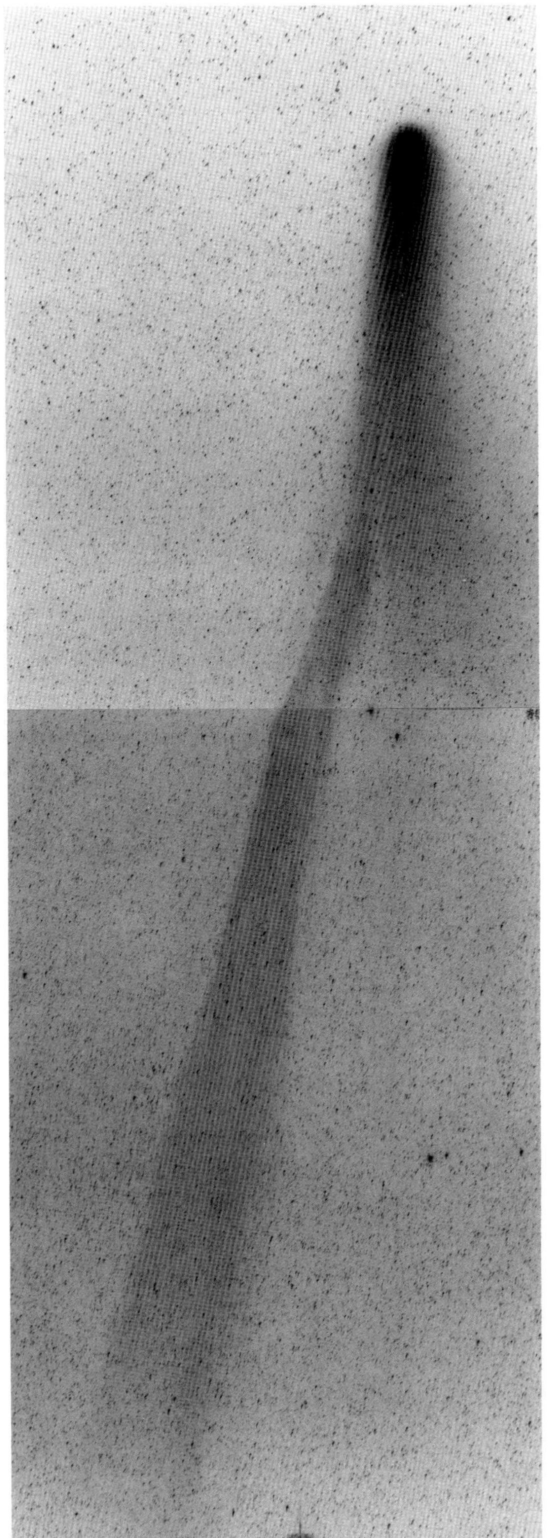

Fig. 839: 18 March 1986, LSPN 1304/1305

Fig. 840: 18 March 1986, LSPN 2387/2388/2389

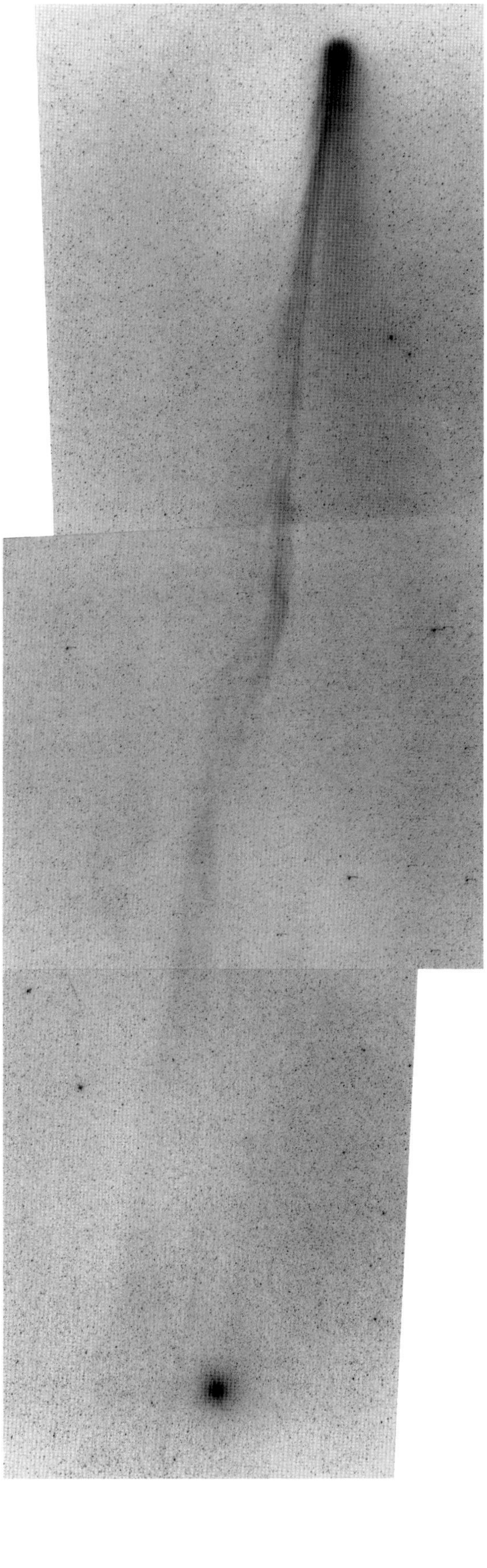

Fig. 841: 19 March 1986, LSPN 1310/1311/1312

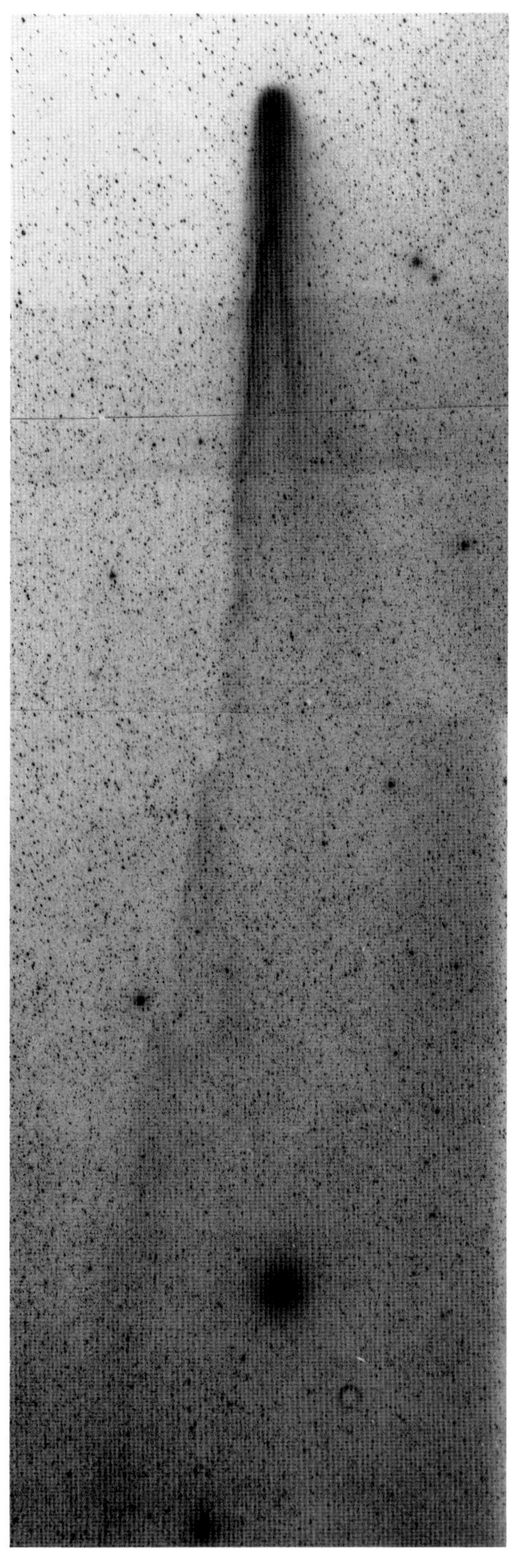

Fig. 842: 20 March 1986, LSPN 896/897

Fig. 843: 20 March 1986, LSPN 899/900

Fig. 844: 20 March 1986, LSPN 1103/1104/1105

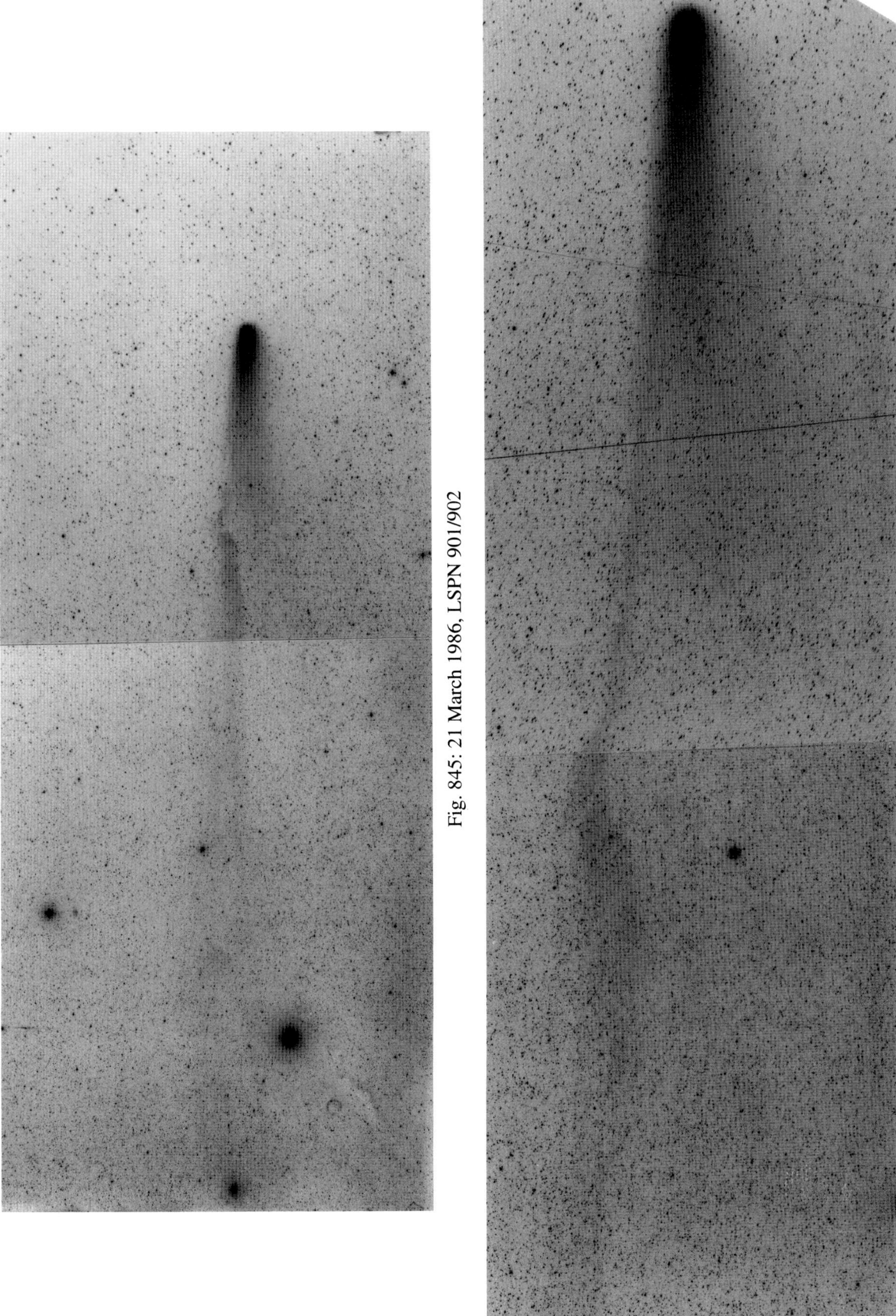

Fig. 845: 21 March 1986, LSPN 901/902

Fig. 846: 22 March 1986, LSPN 1415/1416

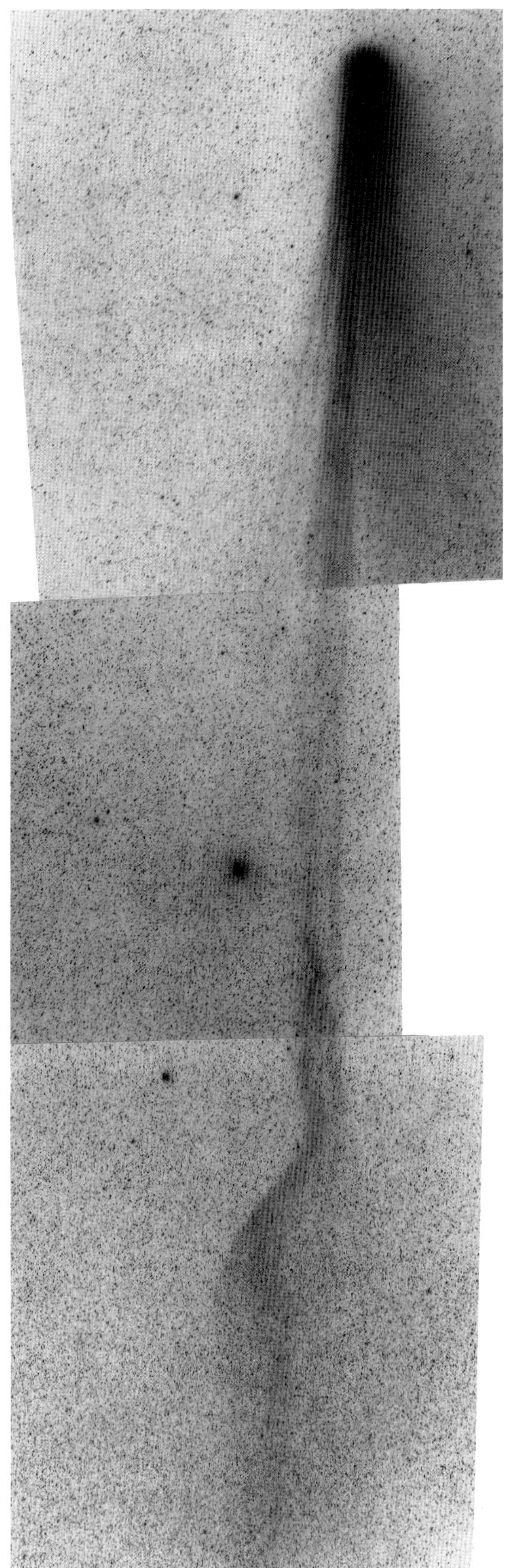

Fig. 847: 23 March 1986, LSPN 1122/1123/1124

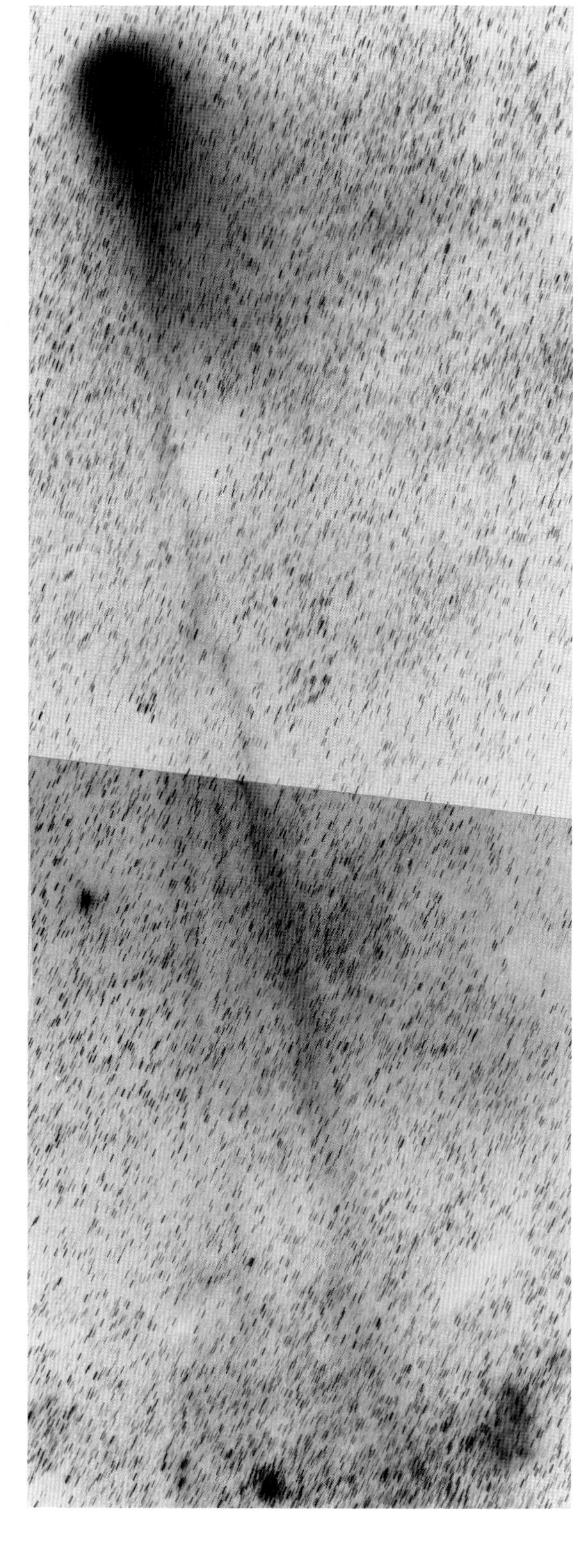

Fig. 848: 4 April 1986, LSPN 1130/1131

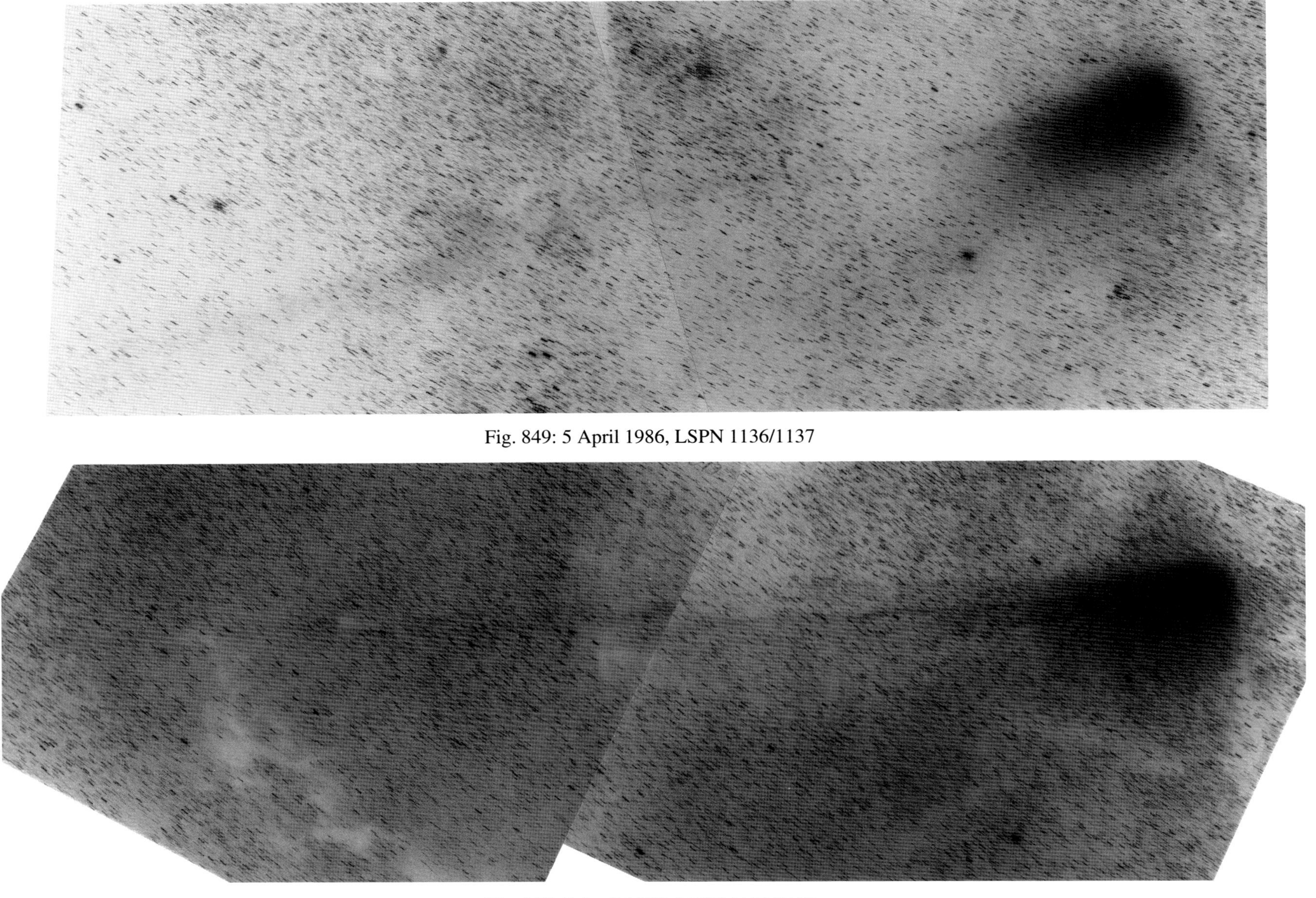

Fig. 849: 5 April 1986, LSPN 1136/1137

Fig. 850: 7 April 1986, LSPN 1151/1153

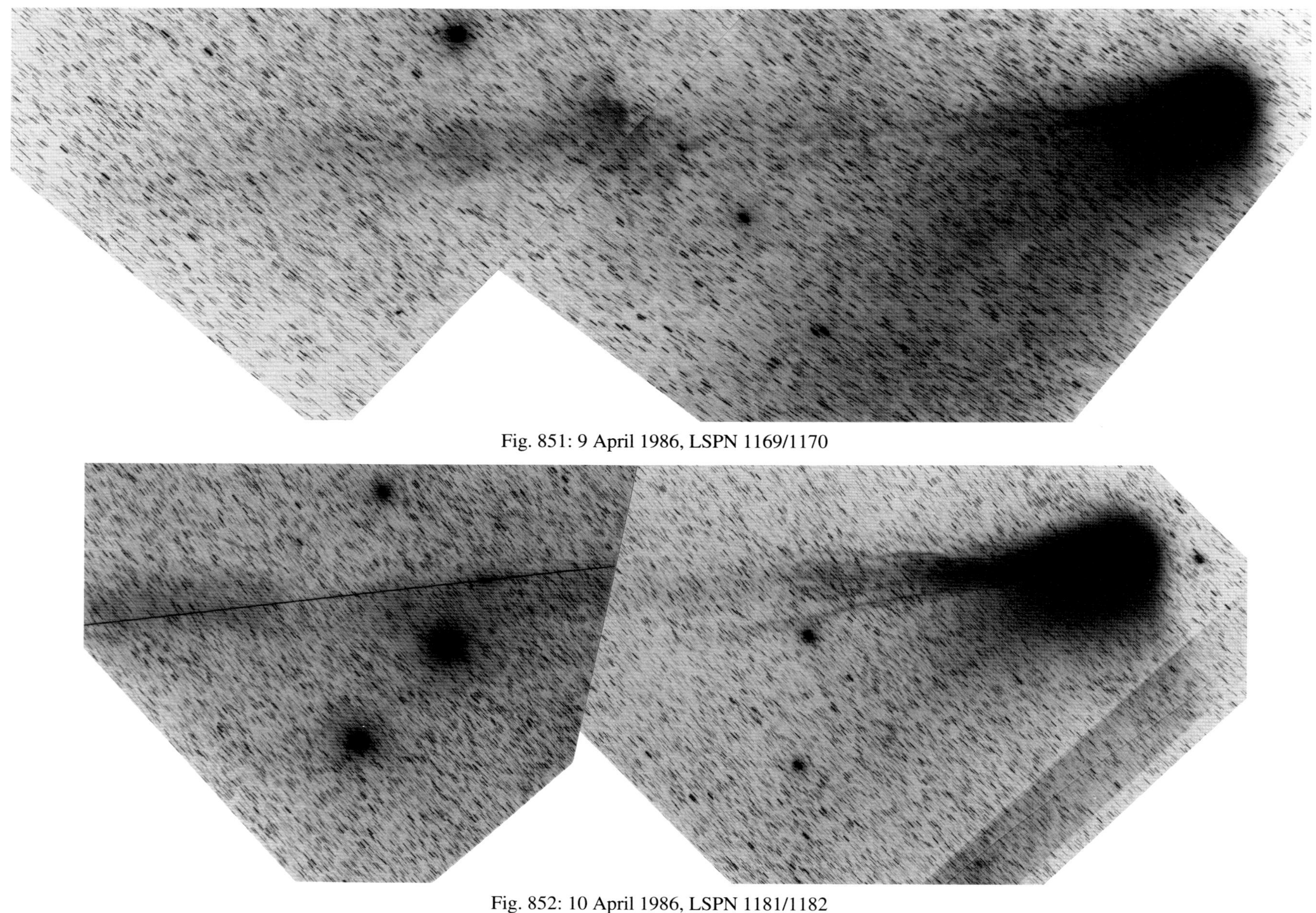

Fig. 851: 9 April 1986, LSPN 1169/1170

Fig. 852: 10 April 1986, LSPN 1181/1182

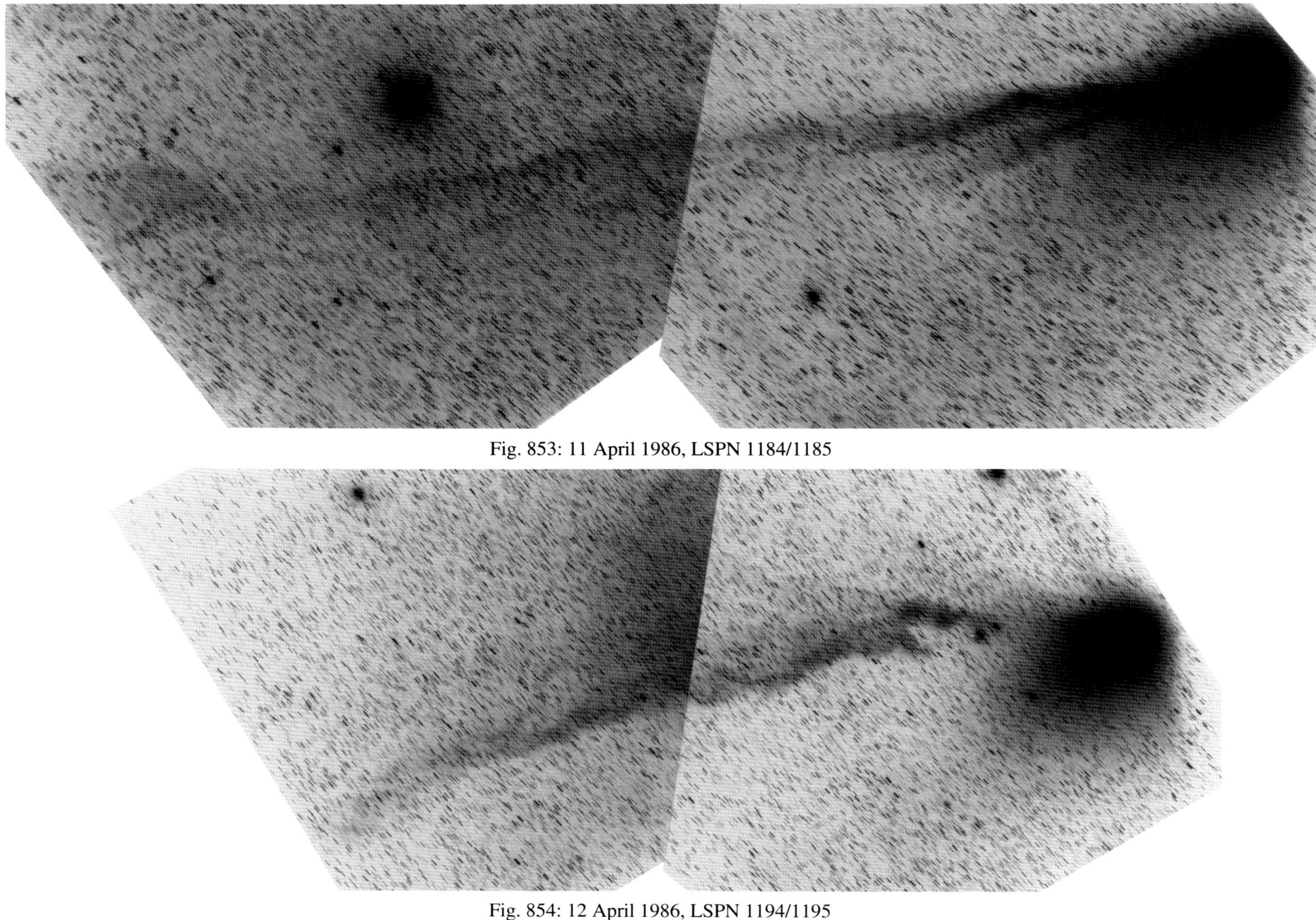

Fig. 853: 11 April 1986, LSPN 1184/1185

Fig. 854: 12 April 1986, LSPN 1194/1195

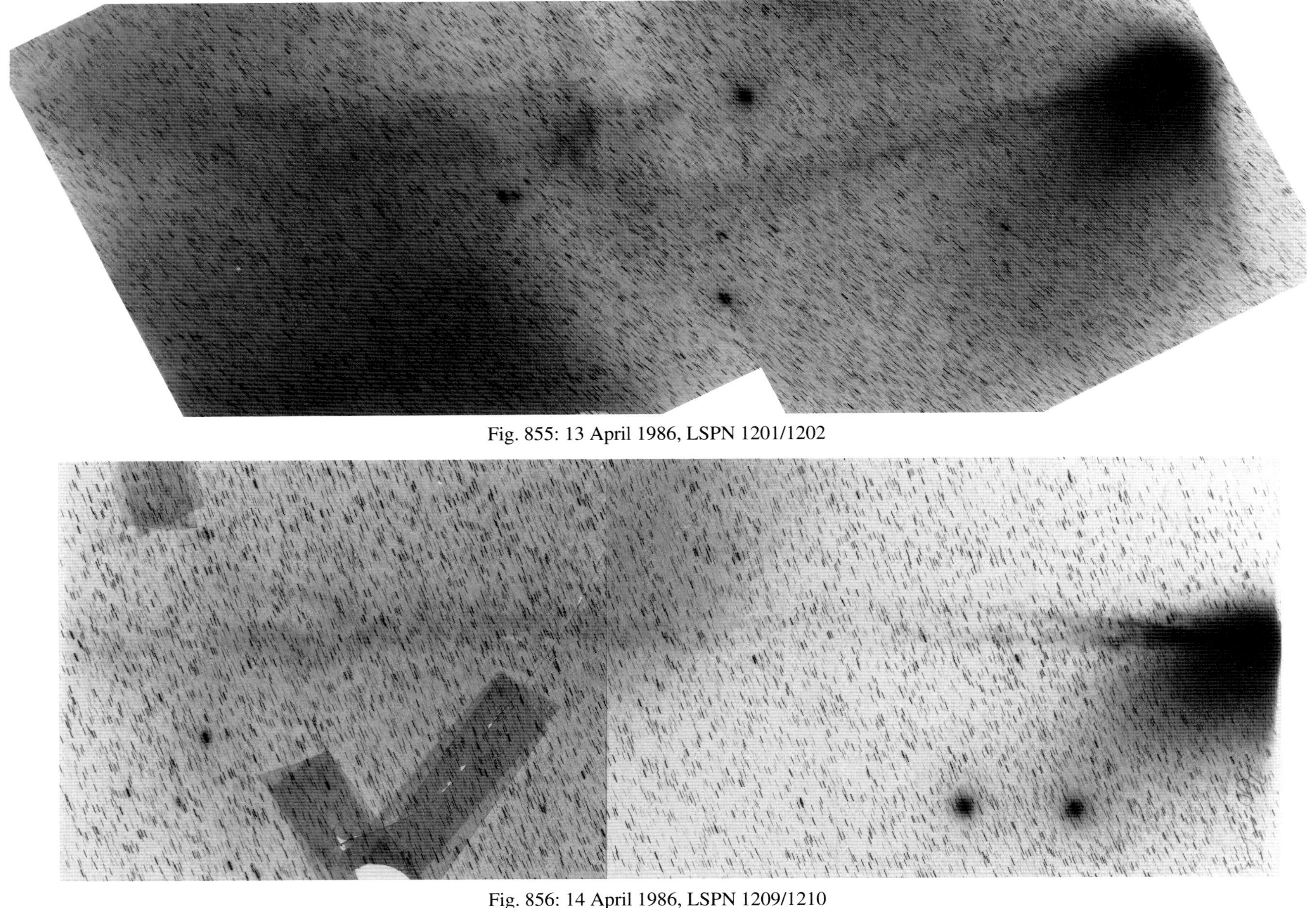

Fig. 855: 13 April 1986, LSPN 1201/1202

Fig. 856: 14 April 1986, LSPN 1209/1210

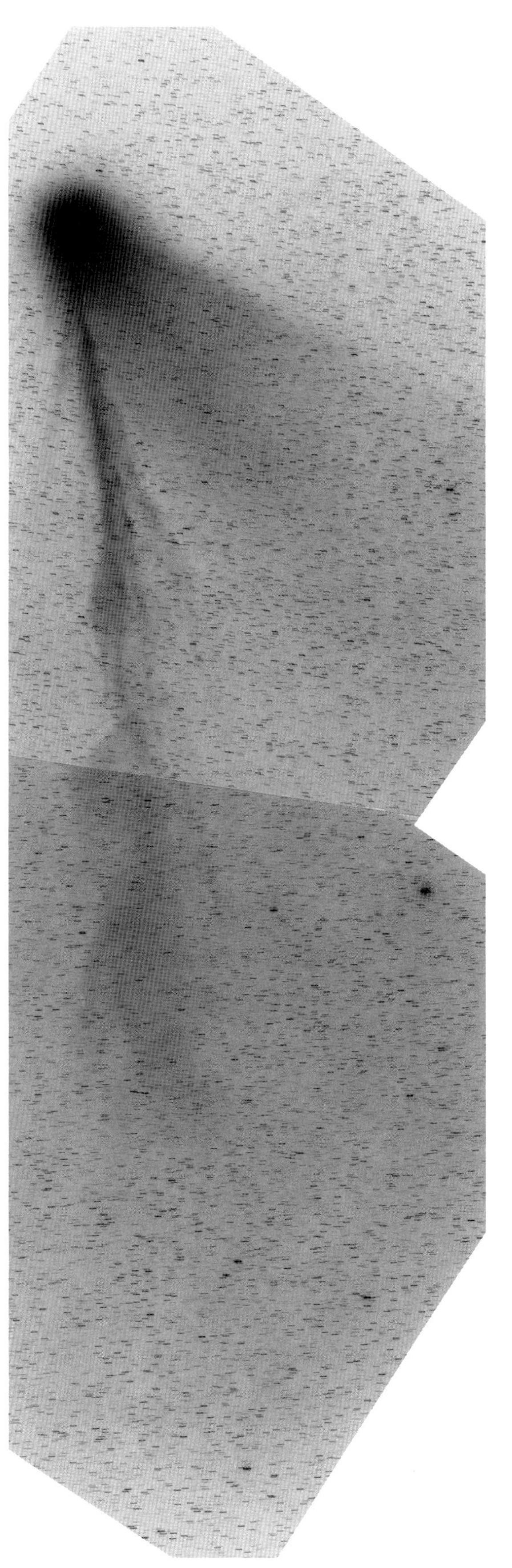

Fig. 857: 18 April 1986, LSPN 1256/1257

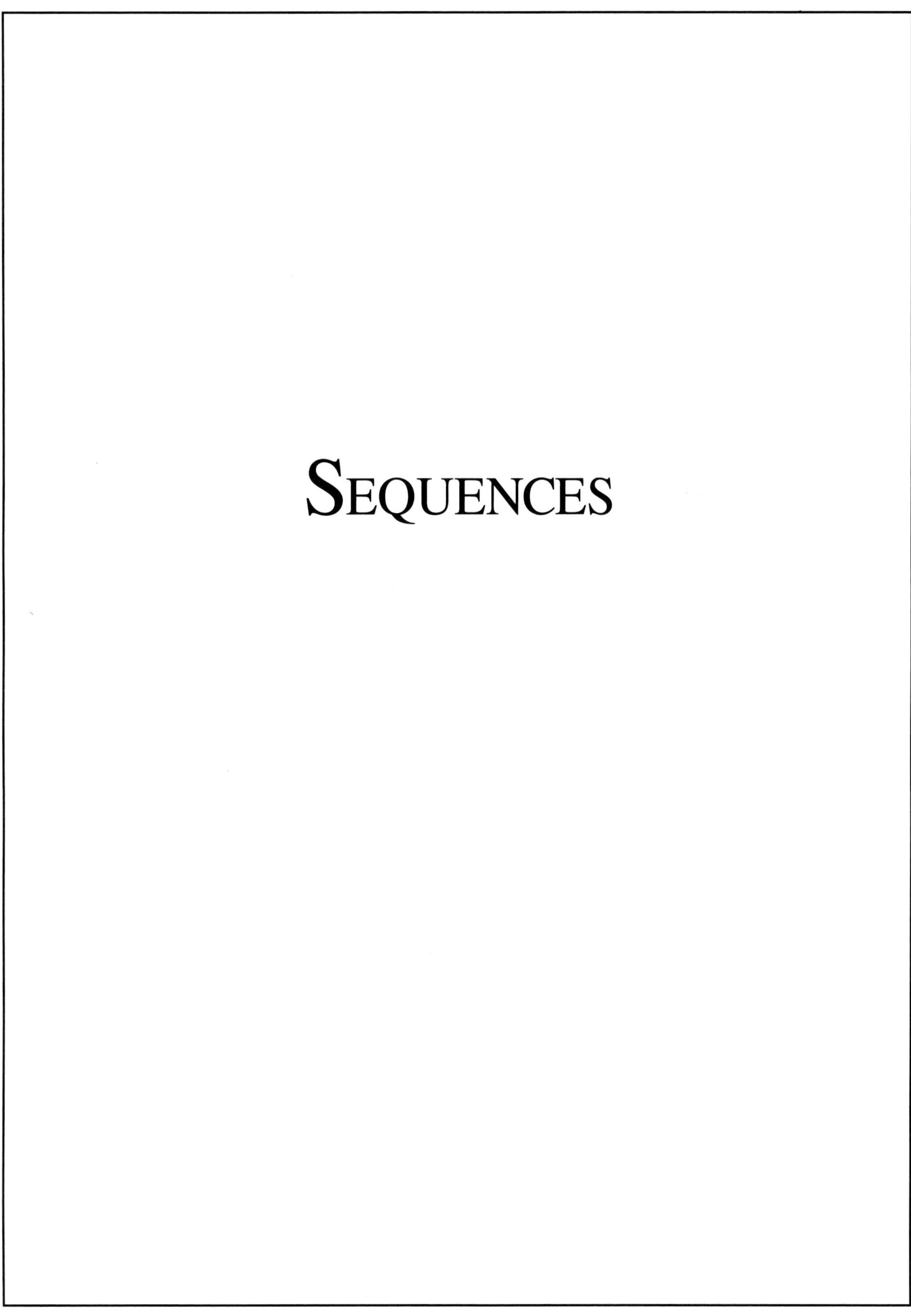

SEQUENCES

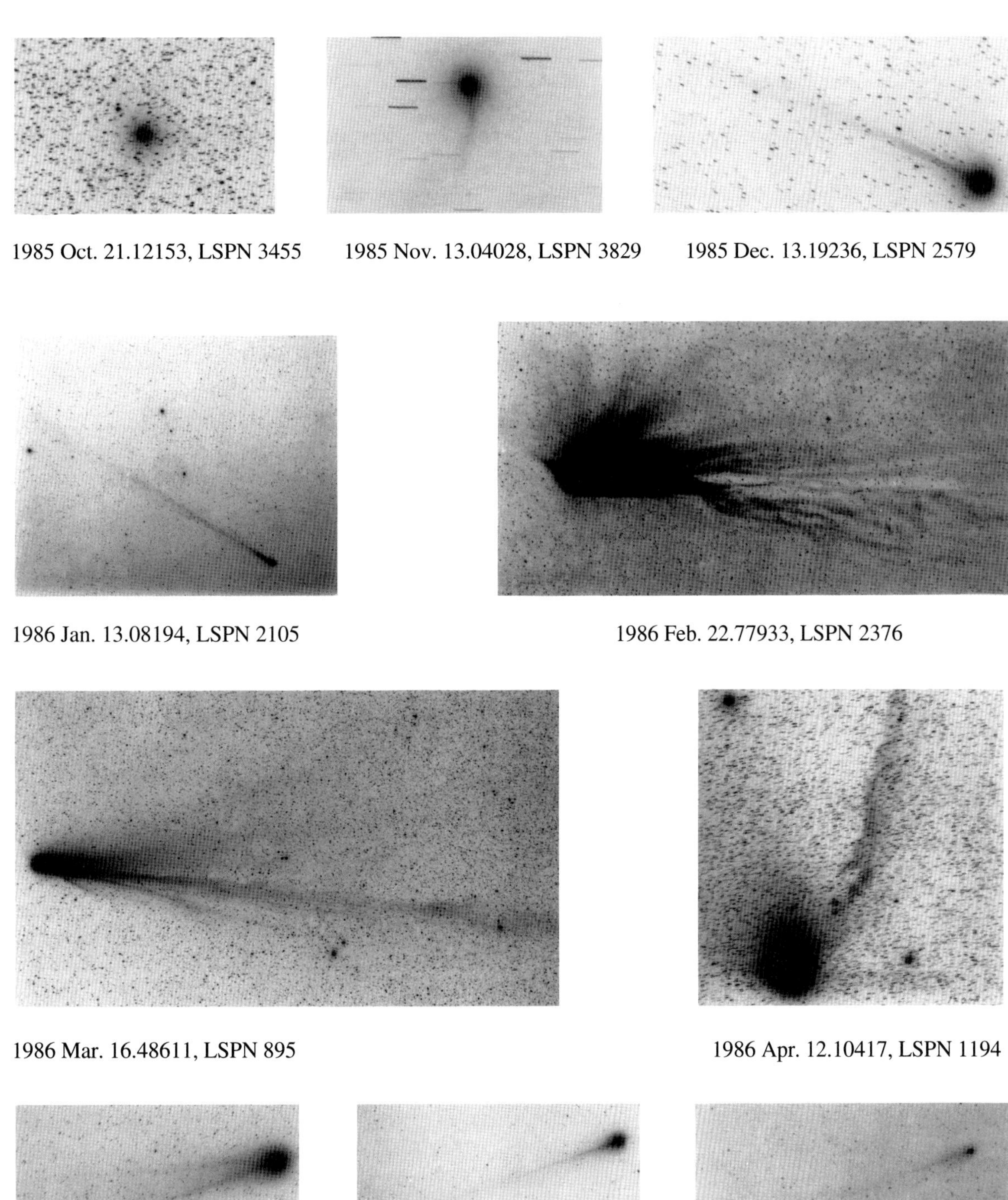

1985 Oct. 21.12153, LSPN 3455 1985 Nov. 13.04028, LSPN 3829 1985 Dec. 13.19236, LSPN 2579

1986 Jan. 13.08194, LSPN 2105 1986 Feb. 22.77933, LSPN 2376

1986 Mar. 16.48611, LSPN 895 1986 Apr. 12.10417, LSPN 1194

1986 May 2.00660, LSPN 2923 1986 June 1.06250, LSPN 759 1986 July 6.39137, LSPN 2453

Fig. 858: A monthly summary, October 1985-July 1986.

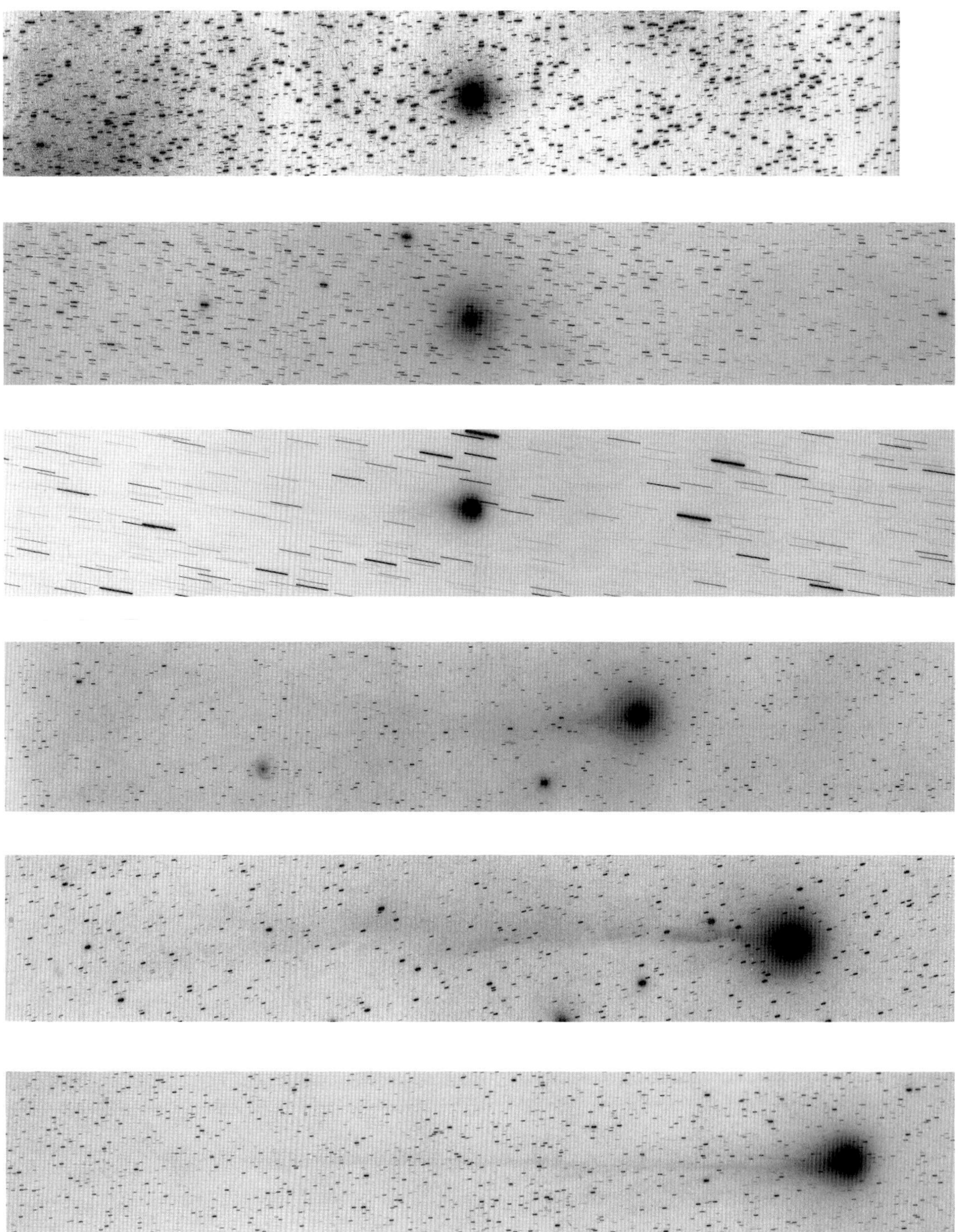

Fig. 859: weekly sequence, (1° ~ 33 mm) (top to bottom) 1985 Nov 8.95625, LSPN-2455, *V. Ivanova, et al., National Astronomical Observatory, Bulgaria*; 1985 Nov 16.33403, LSPN-2578, *R. Hill, Warner and Swasey Observatory;* 1985 Nov 21.24549, LSPN-153, *W. Liller, University of Michigan/CTIO;* 1985 Nov 29.03125; LSPN-165, *W. Liller, University of Michigan/CTIO*; 1985 Dec 7.17500, LSPN-1672, *E. Moore/L. Bair, Joint Observatory for Cometary Research;* 1985 Dec 16.93332, LSPN-3848, *L. Kohoutek, Calar Station, National Ast. Observatory.*

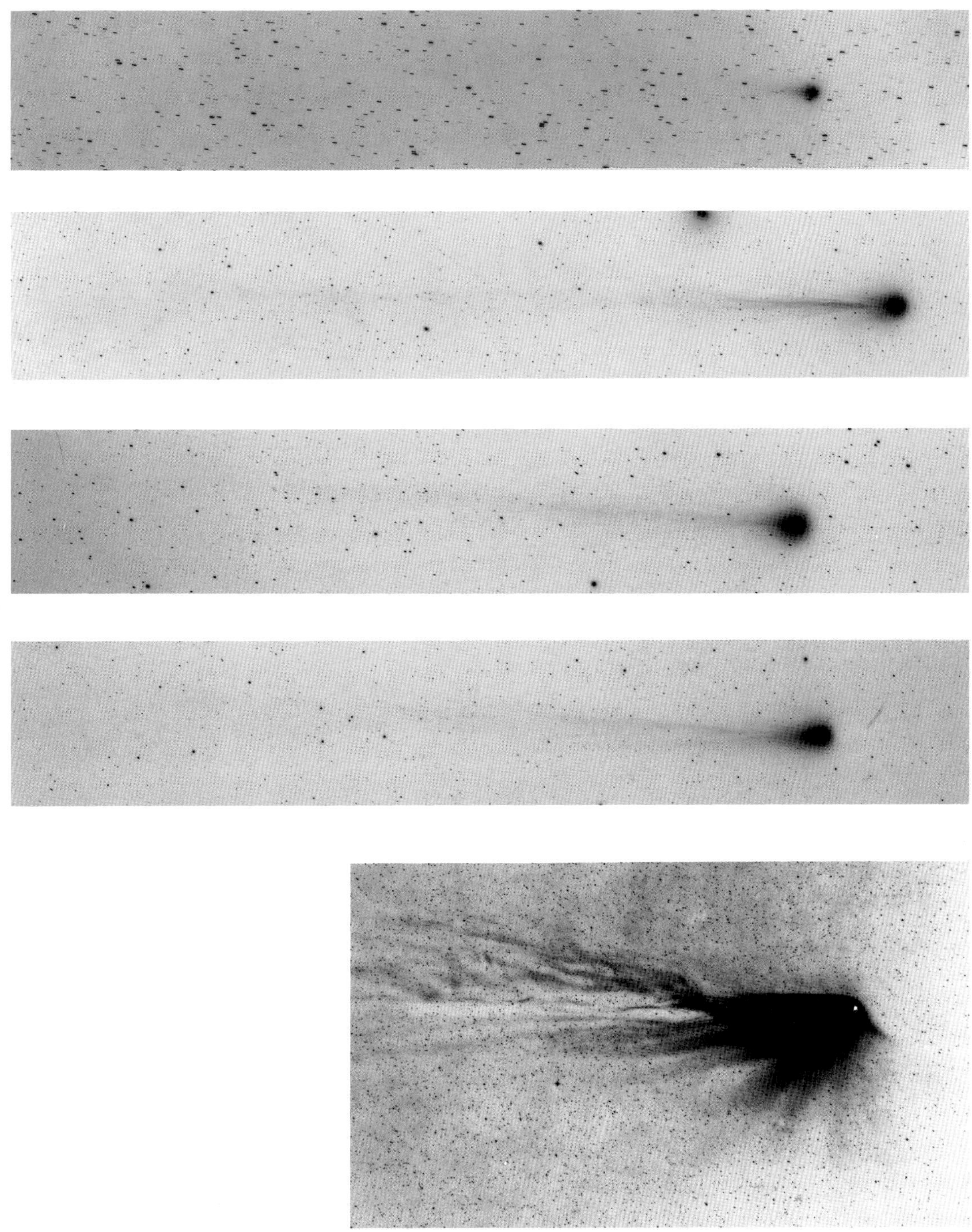

Fig. 860: weekly sequence, (1° ~ 33 mm) (top to bottom) 1985 Dec 21.07187, LSPN-239, *W. Liller, University of Michigan/CTIO*; 1985 Dec 31.38922, LSPN-1620, *H. Maehara, Kiso Observatory*; 1986 Jan 7.08715, LSPN-3570; 1986 Jan 12.07986, LSPN-3574, *H. Giclas, Lowell Observatory*; 1986 Feb 22.77933, LSPN-2376, *UK Schmidt Telescope Unit.*

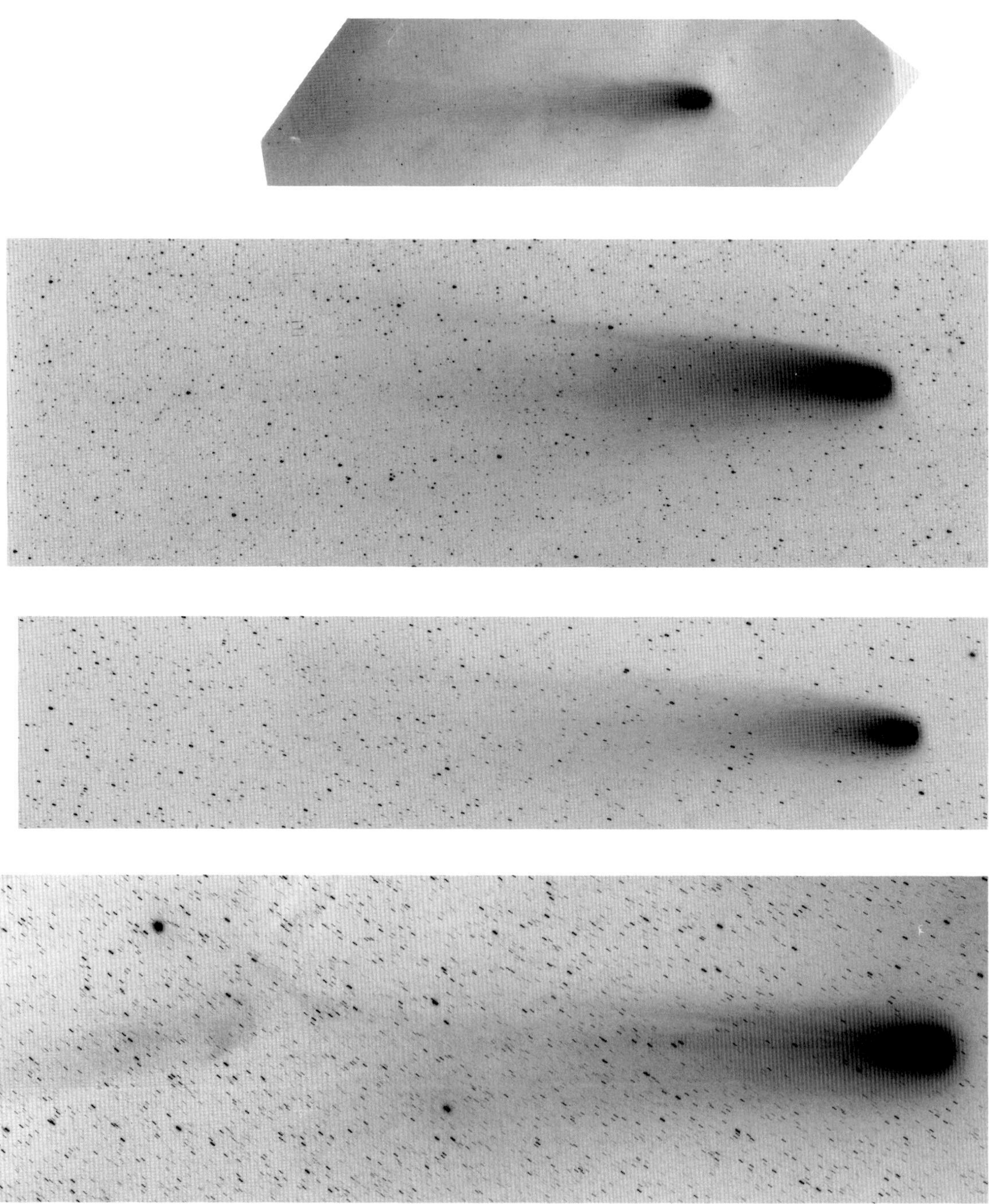

Fig. 861: weekly sequence, (1° ~ 33 mm) (top to bottom) 1986 Mar 1.85278, LSPN-3066, *M. Candy, Perth Observatory*; 1986 Mar 8.45313, LSPN-1377; 1986 Mar 13.45903, LSPN-1391, *W. Liller, Easter Island, LSPN Island Network*; 1986 Mar 21.50938, LSPN-3581, *H. Giclas, Lowell Observatory.*

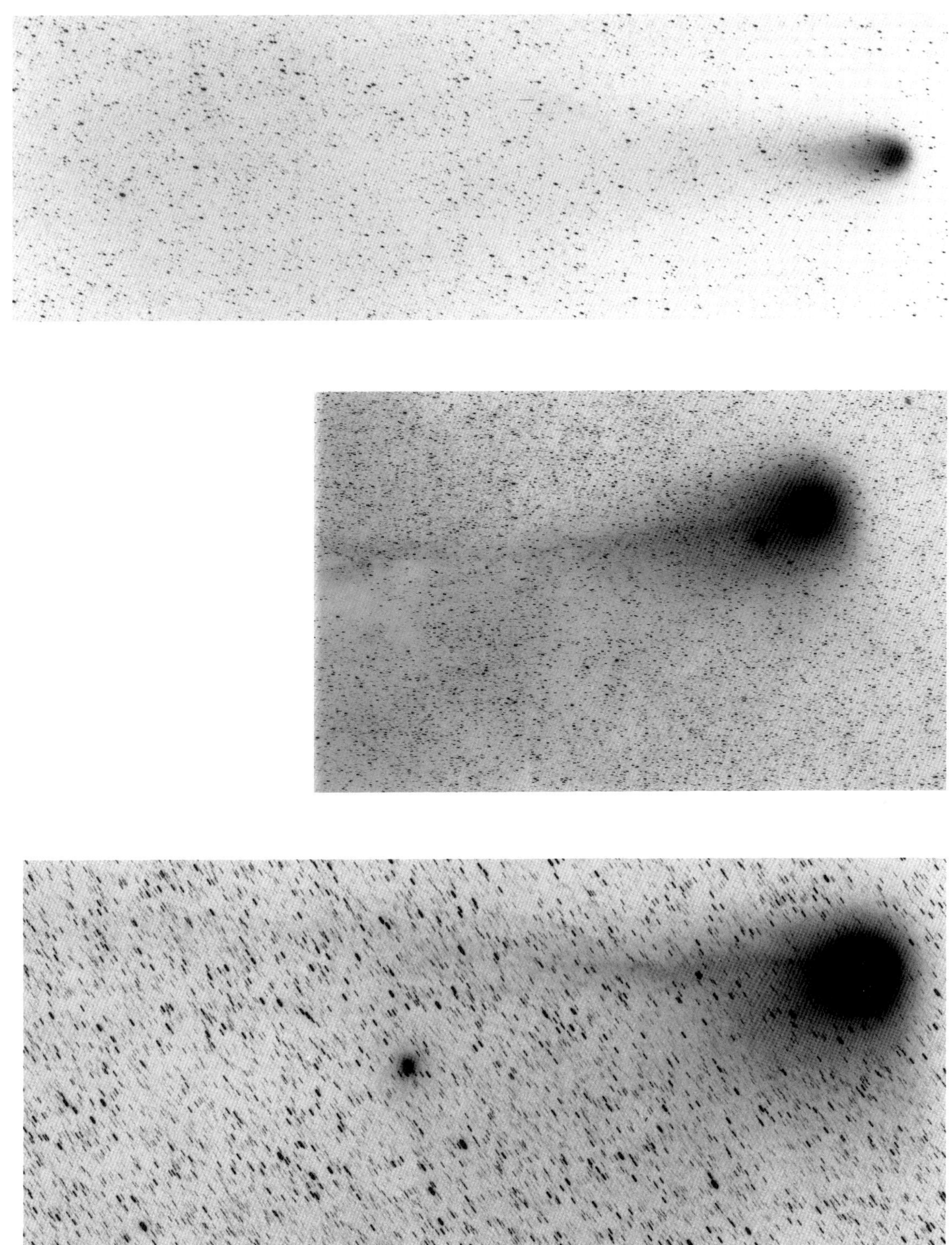

Fig 862: weekly sequence, (1° ~ 33 mm) (top to bottom) 1986 Mar 26.46771, LSPN-1424, *W. Liller, Easter Island, LSPN Island Network*; 1986 Apr 4.15660, LSPN-1129, *F. Miller, University of Michigan/CTIO*; 1986 Apr 10.39549, LSPN-912, *E. Moore, Joint Observatory for Cometary Research.*

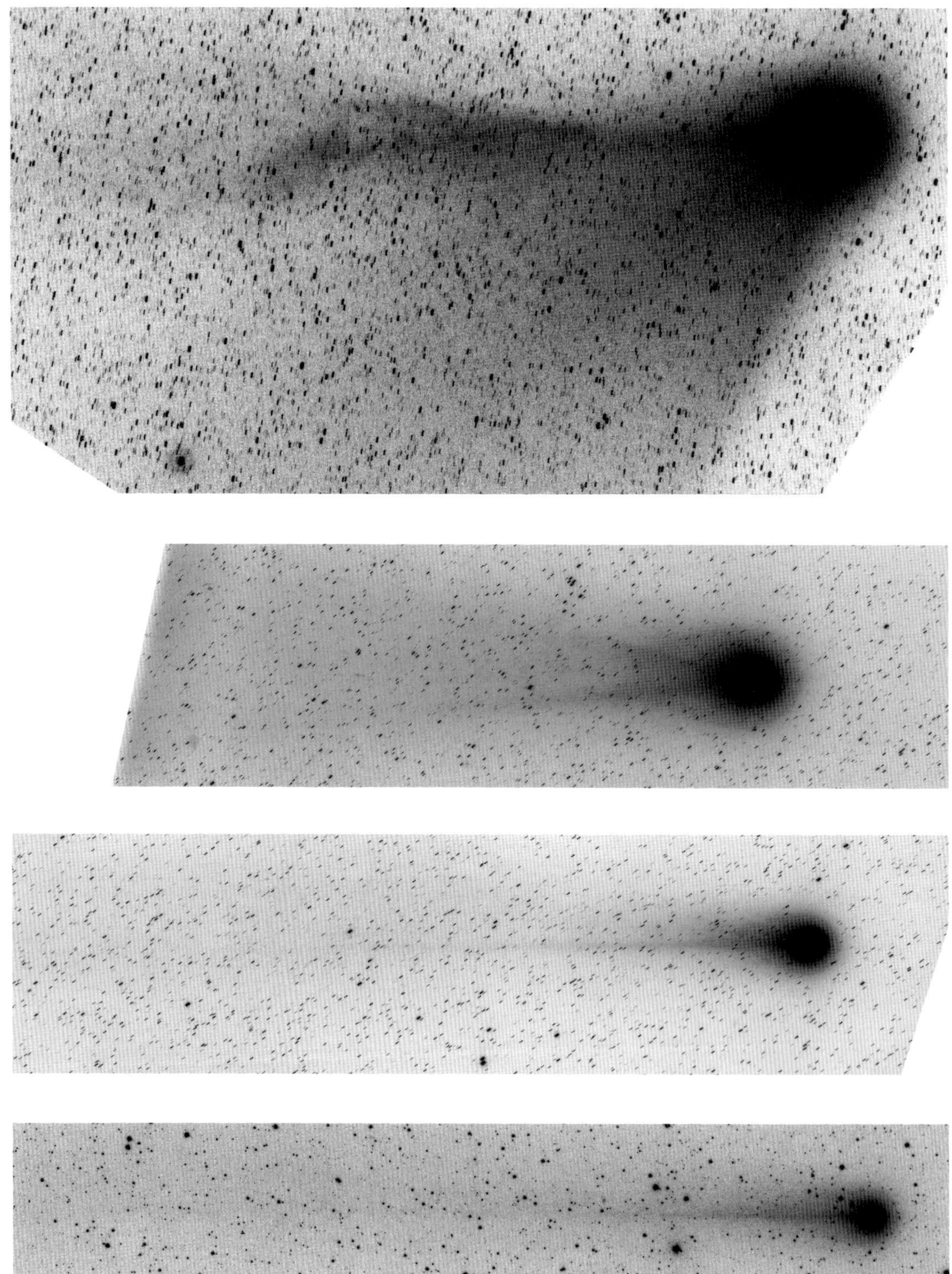

Fig. 863: weekly sequence, (1° ~ 33 mm) (top to bottom) 1986 Apr 16.41389, LSPN-1492, *W. Liller, Easter Island, LSPN Island Network*; 1986 Apr 28.02044, LSPN-2985; 1986 May 6.97870, LSPN-3052, *K. Meech, University of Michigan/CTIO*; 1986 May 13.15555, LSPN-952, *E. Moore/E. Marr, Joint Observatory for Cometary Research.*

Fig. 864: weekly sequence, (1° ~ 33 mm) (top to bottom) 1986 June 2.41043, LSPN-2450, *UK Schmidt Telescope Unit;* 1986 June 11.18368, LSPN-968, *E. Moore/E. Marr, Joint Observatory for Cometary Research*; 1986 June 25.37428, LSPN-2451; 1986 July 6.39137, LSPN-2453; *UK Schmidt Telescope Unit.*

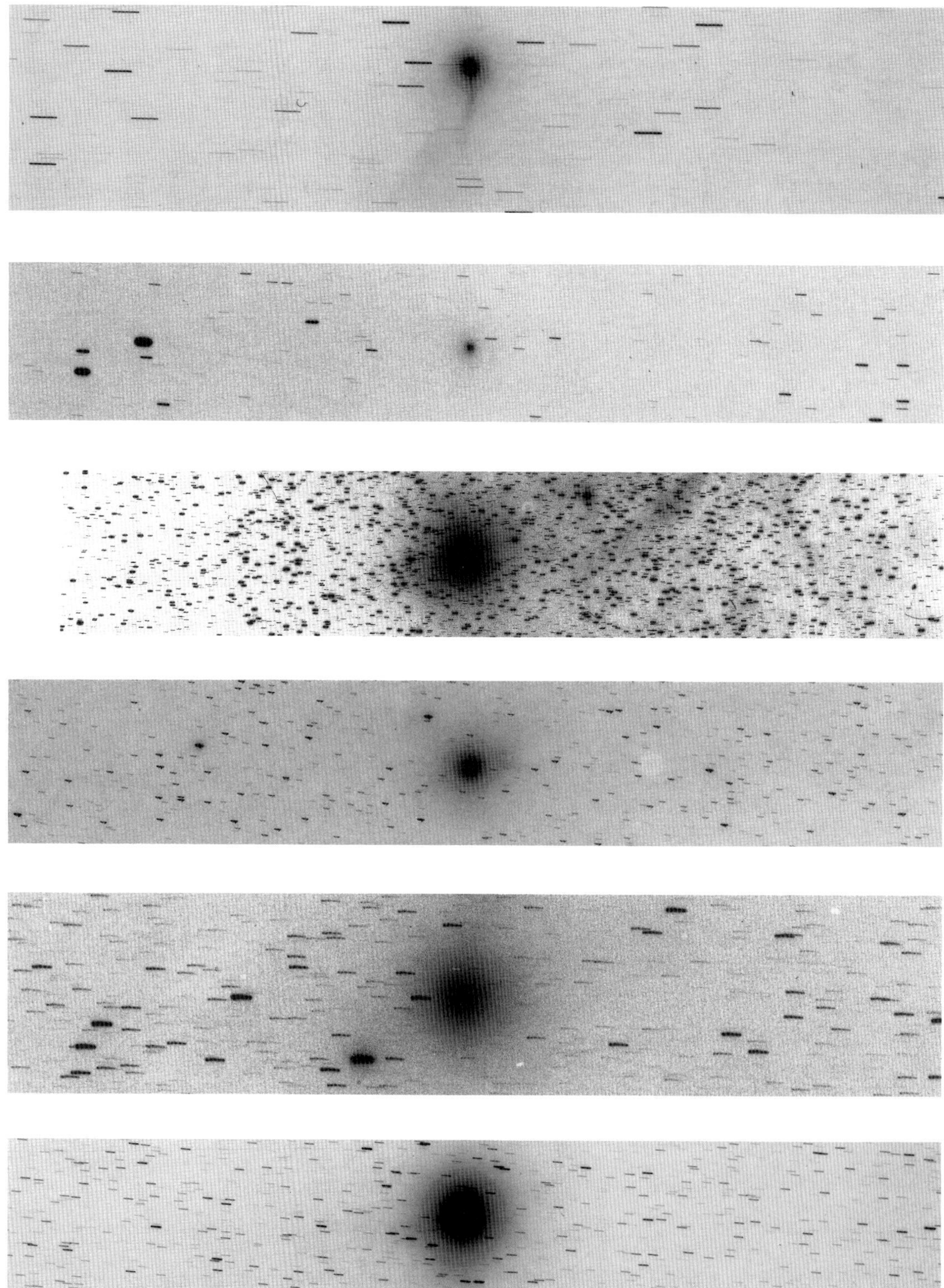

Fig. 865: turn-on sequence, (0.5° ~ 33 mm) (top to bottom) 1985 Nov 13.04028, LSPN-3829, *L. Kohoutek, Calar Station, National Ast. Observatory*; 1985 Nov 14.94410, LSPN-2840, *G. Malcolm/A. Van Den Heever, Boyden Observatory;* 1985 Nov 15.89375, LSPN-2459, *V. Ivanova, et al.*; 1985 Nov 16.06597, LSPN-722, *C. Torres/H. Wroblewski, Cerro el Roble Astronomical Station*; 1985 Nov 17.19201, LSPN-1688, *E. Moore, et al., Joint Observatory for Cometary Research;* 1985 Nov 18.09931, LSPN-3833, *L. Kohoutek, Calar Station, National Ast. Observatory.*

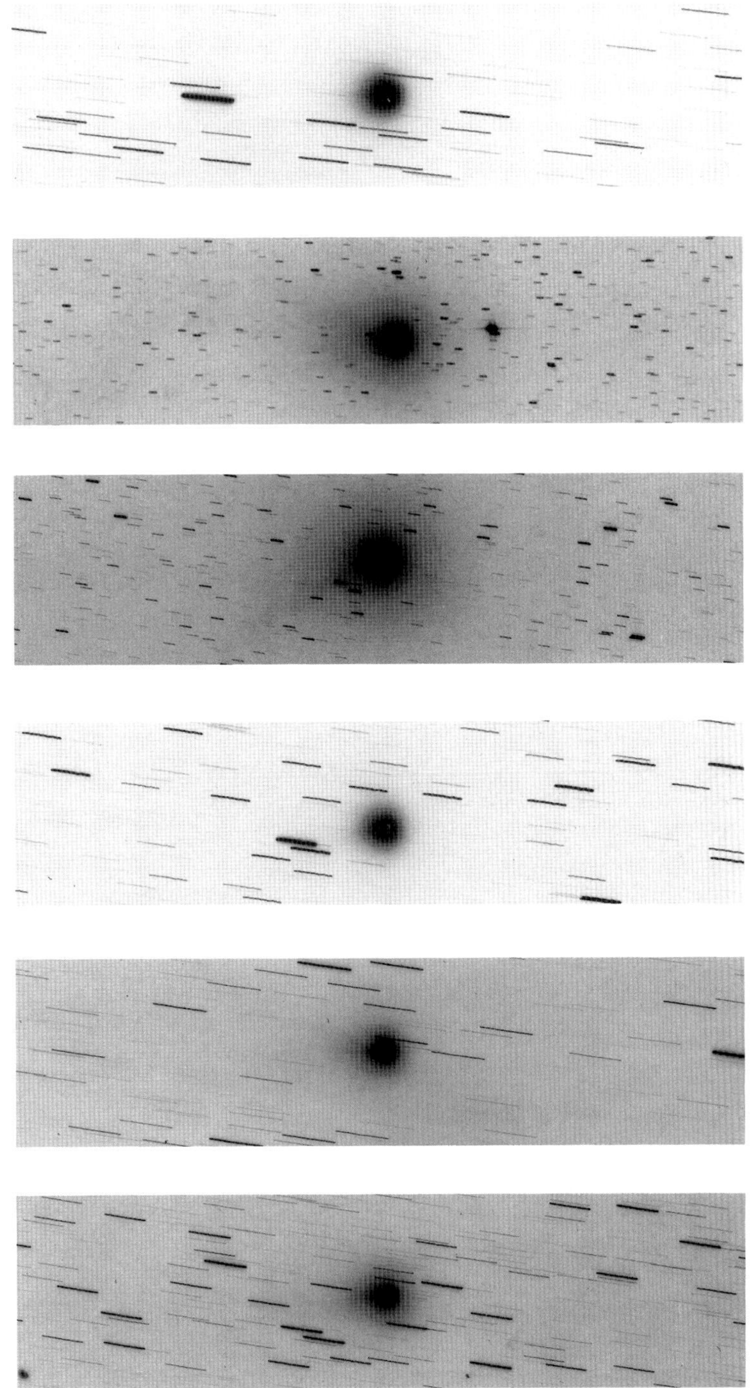

Fig. 866: turn-on sequence, (0.5° ~ 33 mm) (top to bottom) 1985 Nov 19.25694, LSPN-148, W. Liller, *University of Michigan/CTIO;* 1985 Nov 19.62470, LSPN-1600, *H. Maehara, Kiso Observatory;* 1985 Nov 20.26285, LSPN-150; 1985 Nov 20.30104, LSPN-151;1985 Nov 21.24549, LSPN-153; 1985 Nov 22.26805, LSPN-156, *W. Liller, University of Michigan/CTIO.*

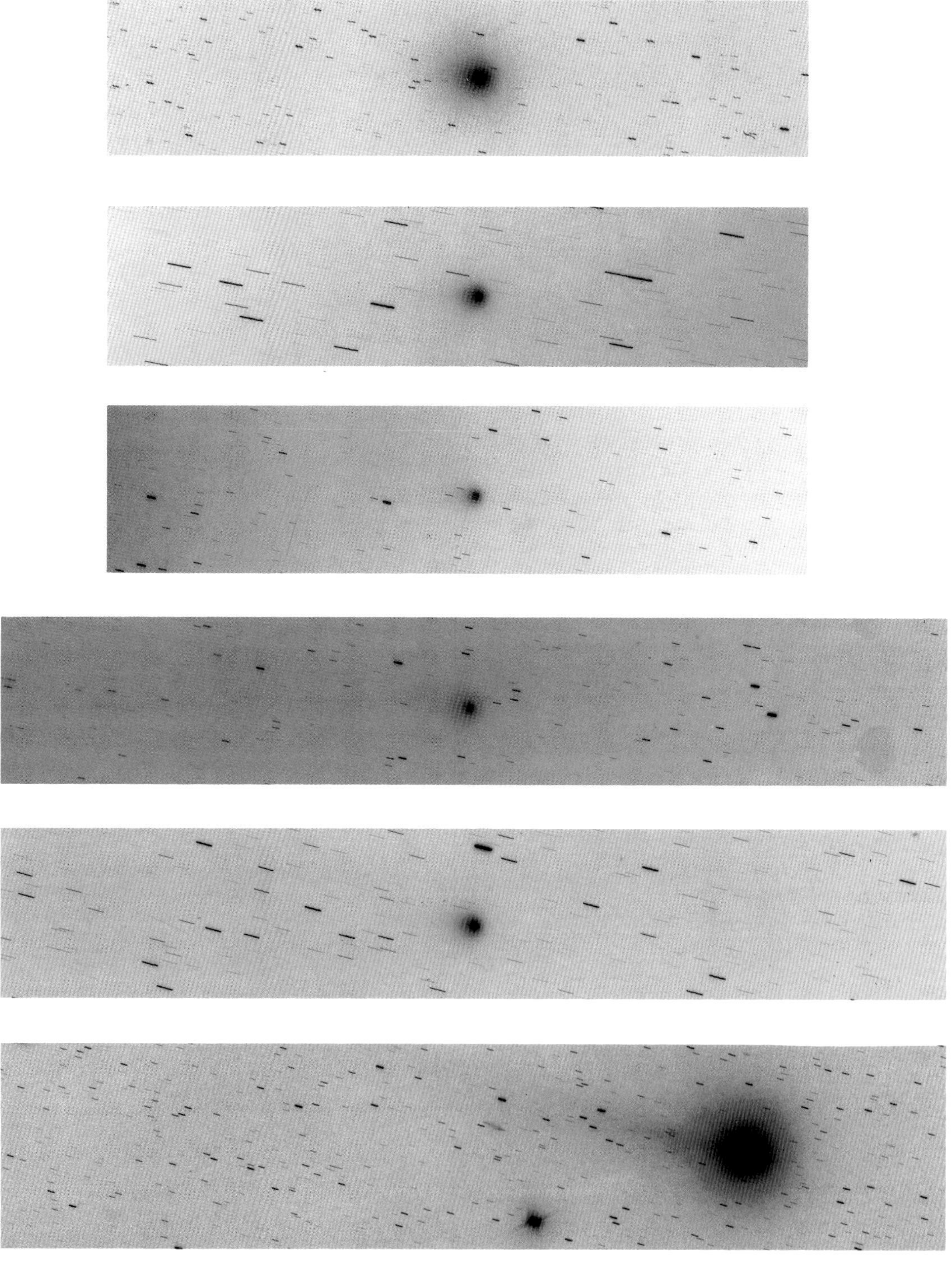

Fig. 867: turn-on sequence, (0.5° ~ 33 mm) (top to bottom) 1985 Nov 23.31319, LSPN-159; 1985 Nov 24.15000, LSPN-160; 1985 Nov 26.12292, LSPN-162; 1985 Nov 27.10486, LSPN-163; 1985 Nov 28.10347, LSPN-164; 1985 Nov 29.03125, LSPN-165.
All photos by W. Liller, University of Michigan/CTIO.

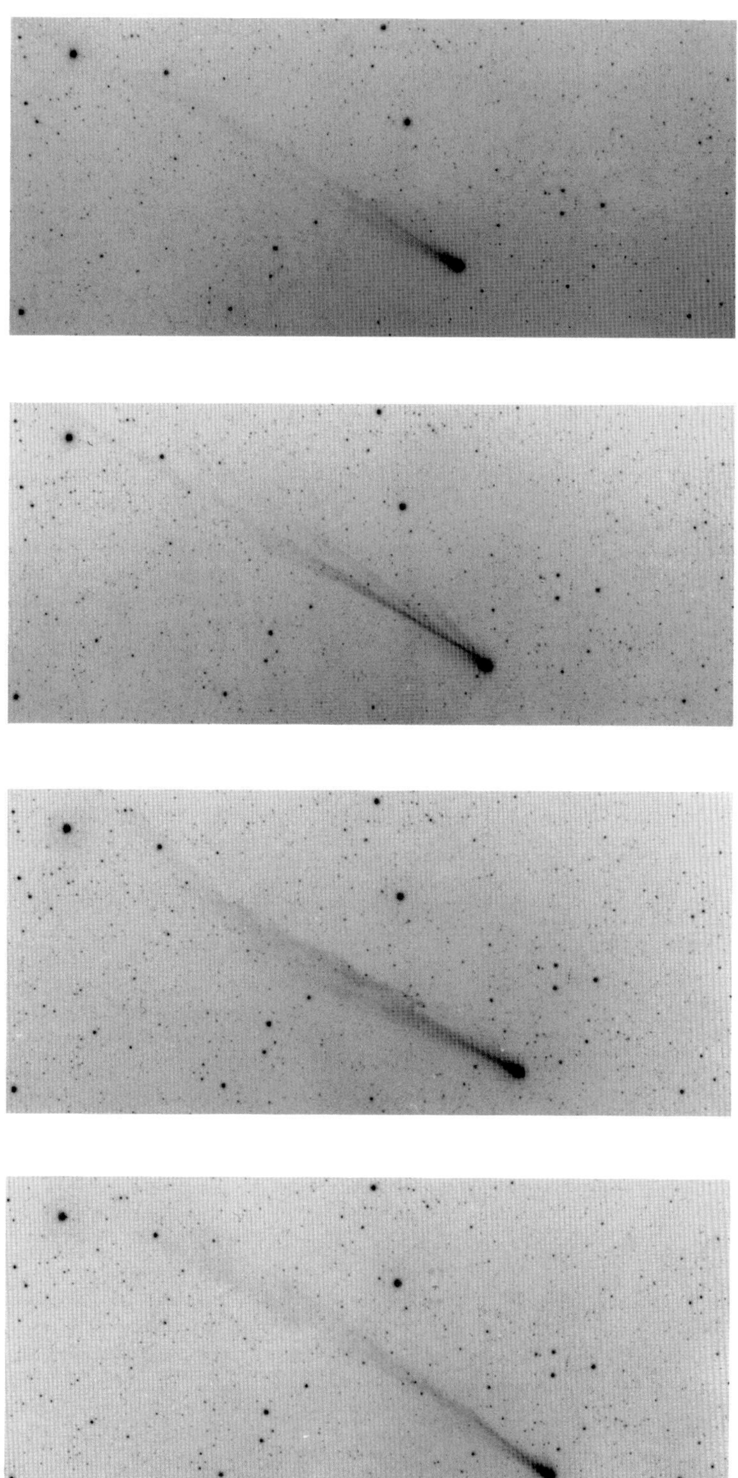

Fig. 868: daily sequence, Lowell Observatory, (1°~ 10 mm) (top to bottom) 1986 Jan 9.08194, LSPN-3585; 1986 Jan 10.09028, LSPN-3583; 1986 Jan 11.10069, LSPN-3584; 1986 Jan 12.09861, LSPN-3586.
All photographs by H. Giclas, Lowell Observatory.

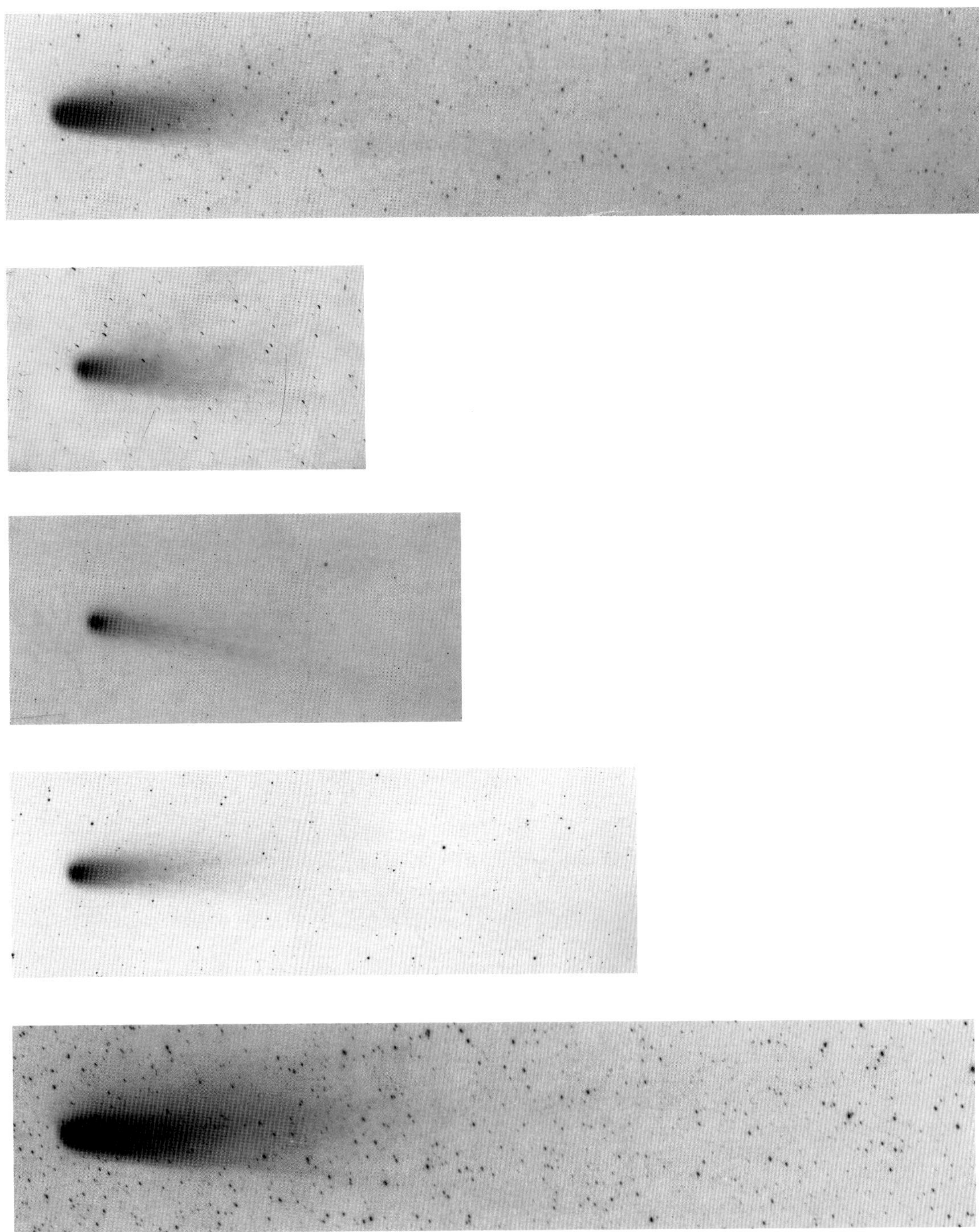

Fig. 869: Armada week, (1°~ 25 mm) (top to bottom) 1986 Mar 6.62266, LSPN-1903, *D. Cruikshank/ M. Buie, Mauna Kea Observatory*; 1986 Mar 6.97396, LSPN-3619, *K. Sivaraman, Indian Institute for Astrophysics, Kodaikanal Station*; 1986 Mar 7.39410, LSPN-1268, *F. Miller, University of Michigan/CTIO*; 1986 Mar 7.52500, LSPN-3577, *H. Giclas, Lowell Observatory*; 1986 Mar 8.63759, LSPN-1906, *D. Cruikshank, Mauna Kea Observatory.*

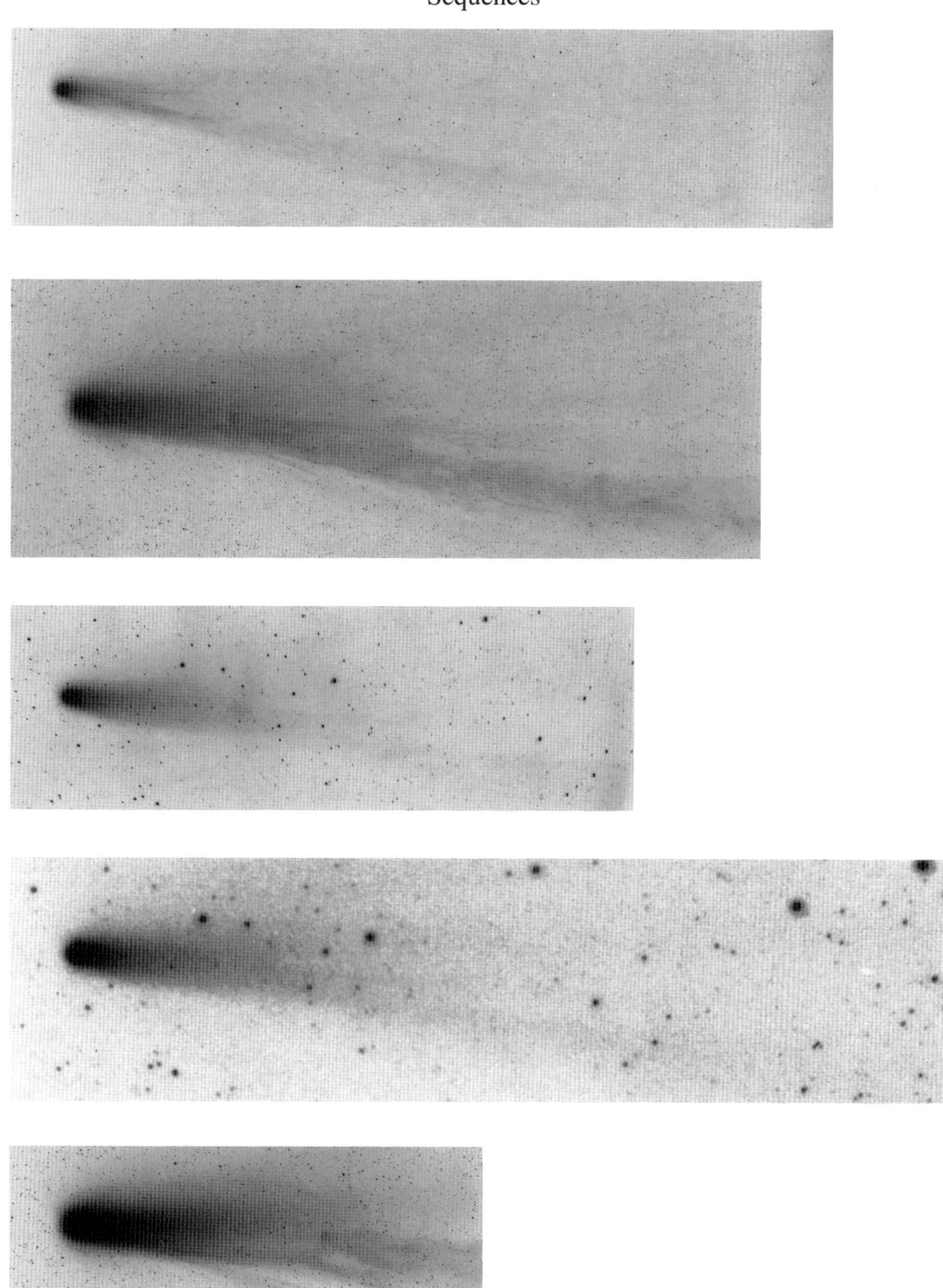

Fig. 870: Armada week, (1°~ 25 mm) (top to bottom) 1986 Mar 8.82484, LSPN-1628, *H. Maehara, Kiso Observatory;* 1986 Mar 9.37083, LSPN-1274, *F. Miller, University of Michigan/CTIO*; 1986 Mar 9.47570, LSPN-3414; 1986 Mar 9.47570, LSPN-3418, *G. Emerson, E.E. Barnard Observatory Station*; 1986 Mar 9.75708, LSPN-2116, *P. Magnusson, Uppsala Southern Station.*

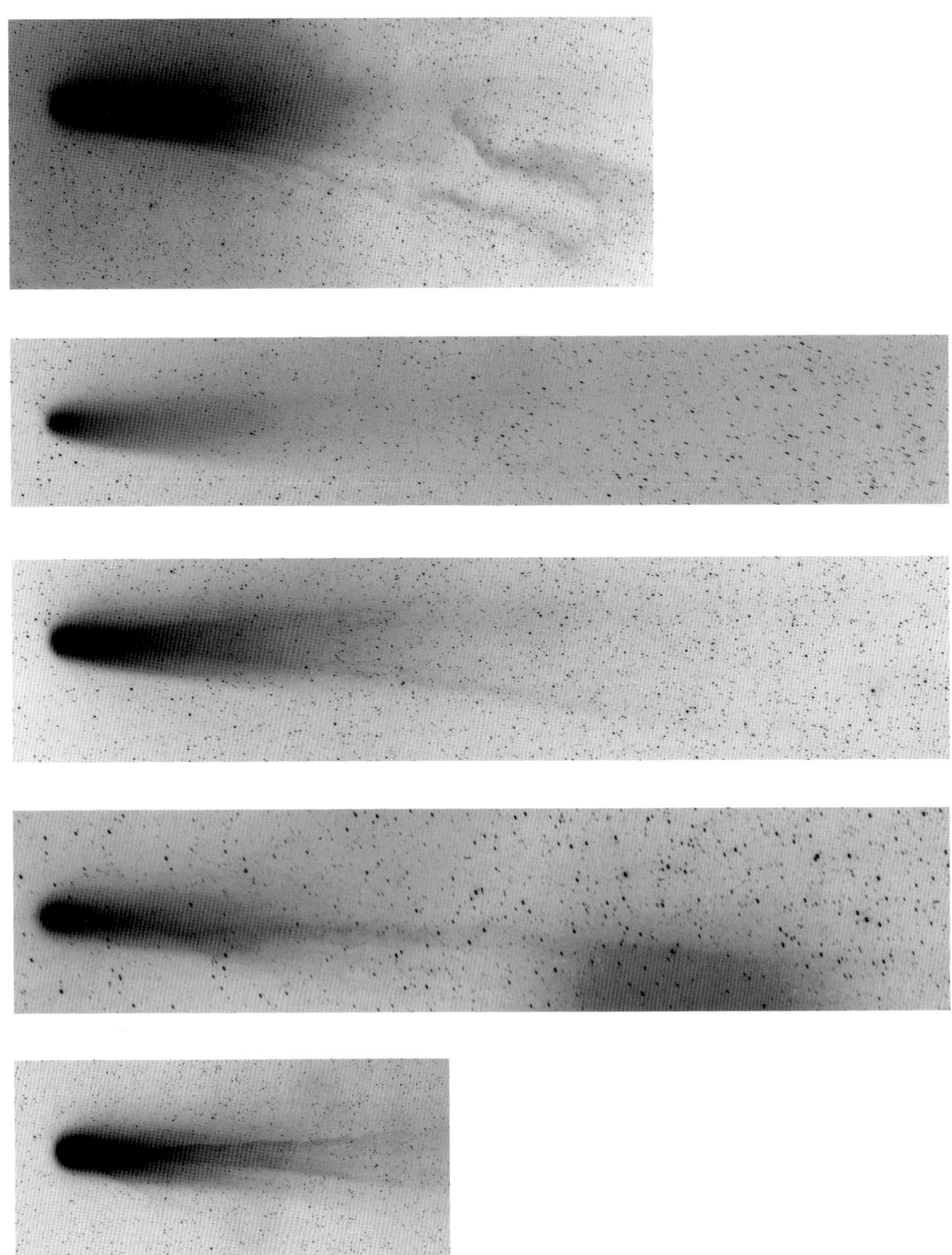

Fig. 871: Armada week, (1°~ 25 mm) (top to bottom) 1986 Mar 10.34896, LSPN-1275, *F. Miller, University of Michigan/CTIO*; 1986 Mar 10.42500, LSPN-1383; 1986 Mar 11.45660, LSPN-1386, *W. Liller, Easter Island, LSPN Island Network*; 1986 Mar 11.67118, LSPN-1909, *A. Storrs, Mauna Kea Observatory*; 1986 Mar 11.78590, LSPN-2117, *P. Magnusson, Uppsala Southern Station.*

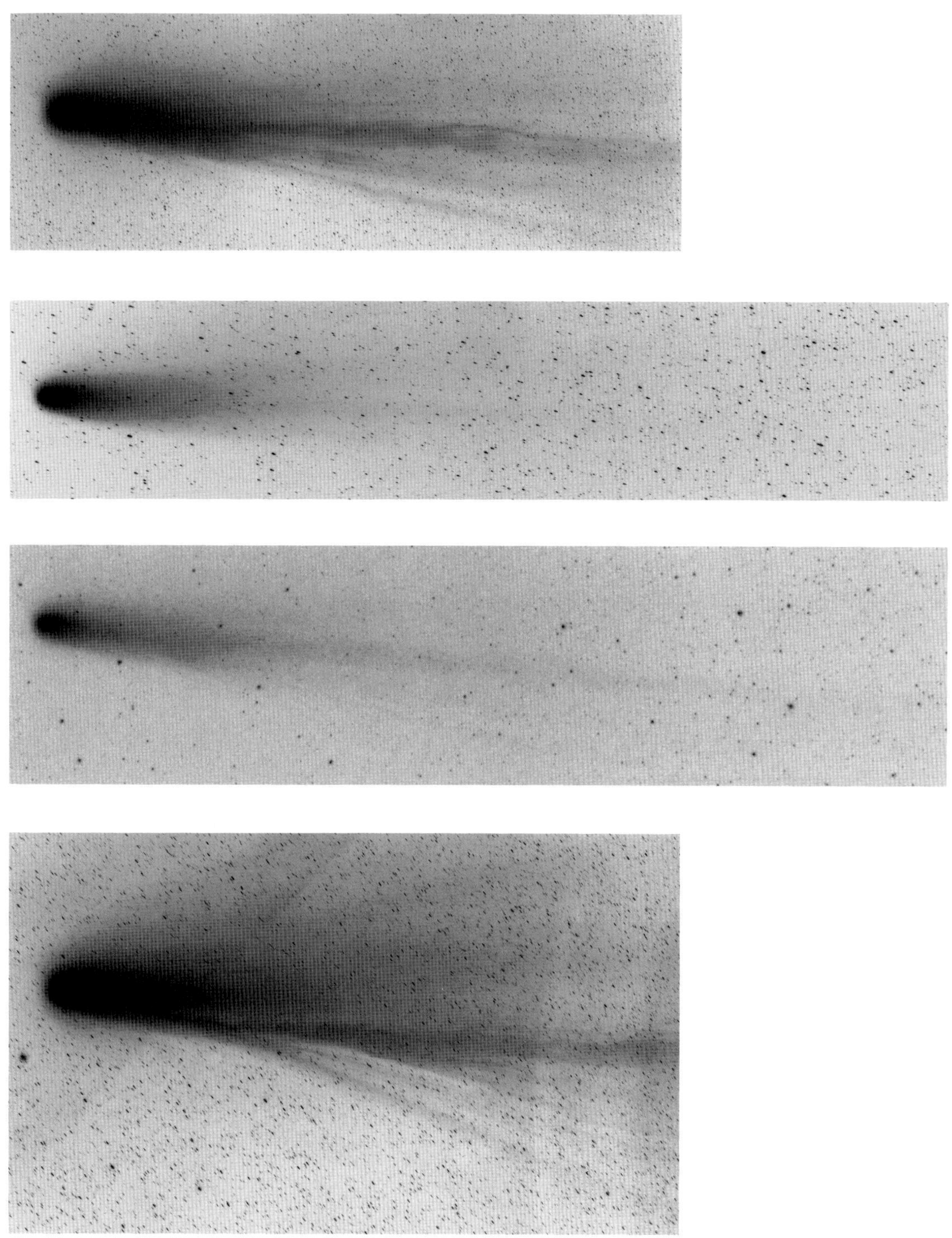

Fig. 872: Armada week, (1°~ 25 mm) (top to bottom) 1986 Mar 12.36319, LSPN-1281, *F. Miller, University of Michigan/CTIO;* 1986 Mar 12.45694, LSPN-1389, *W. Liller, Easter Island, LSPN Island Network*; 1986 Mar 12.51805, LSPN-3588, *H. Giclas, Lowell Observatory*; 1986 Mar 13.34792, LSPN-1285, *F. Miller, University of Michigan/CTIO.*

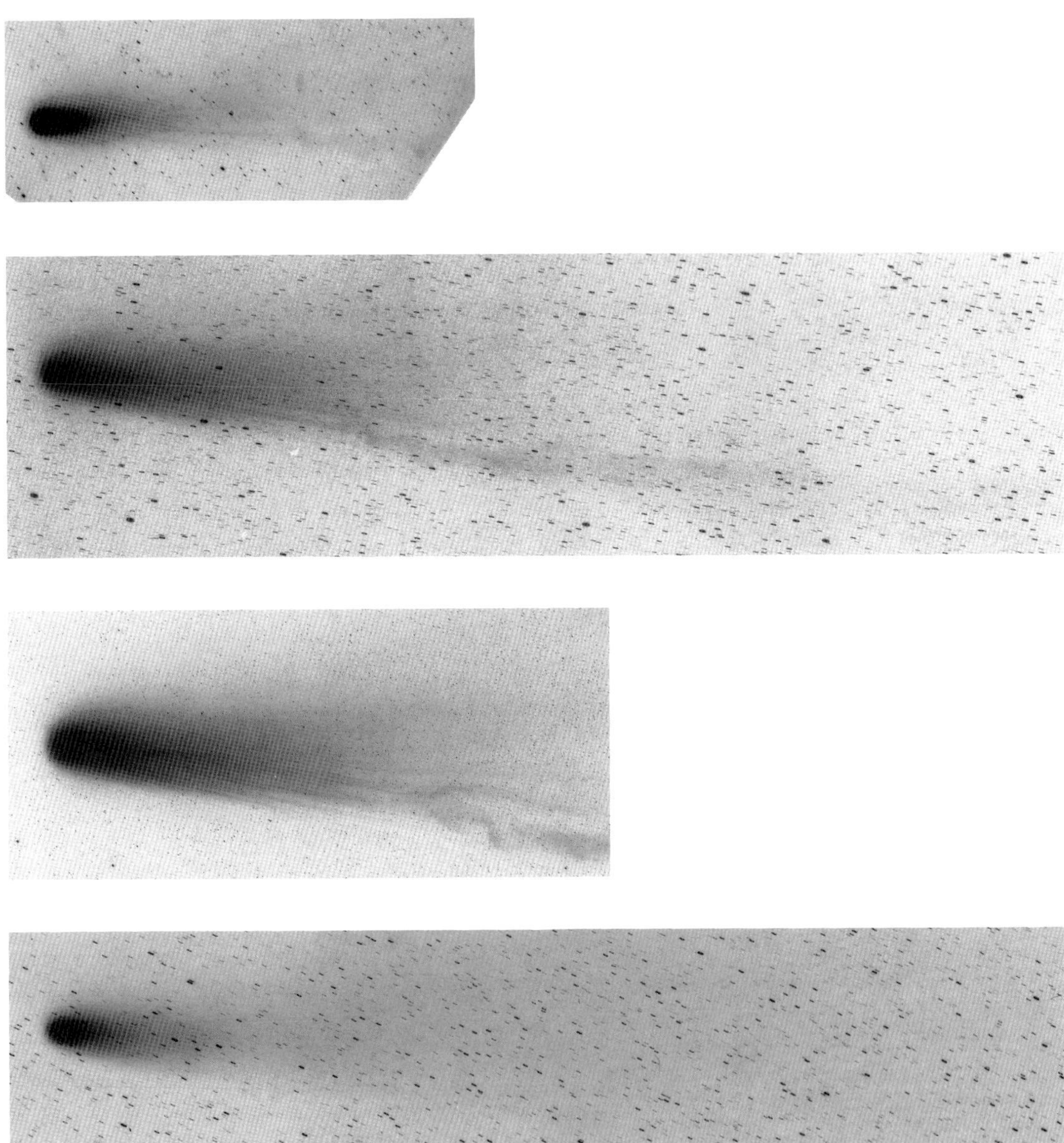

Fig. 873: Armada week, (1°~ 25mm) (top to bottom) 1986 Mar 13.87153, LSPN-3077, *P. Jekabsons, Perth Observatory*; 1986 Mar 14.10104, LSPN-2847, *G. Malcolm, Boyden Observatory*; 1986 Mar 14.35278, LSPN-727, *C. Torres/H. Wroblewski, Cerro el Roble Astronomical Station;* 1986 Mar 14.46128, LSPN-1393, *W. Liller, Easter Island, LSPN Island Network.*

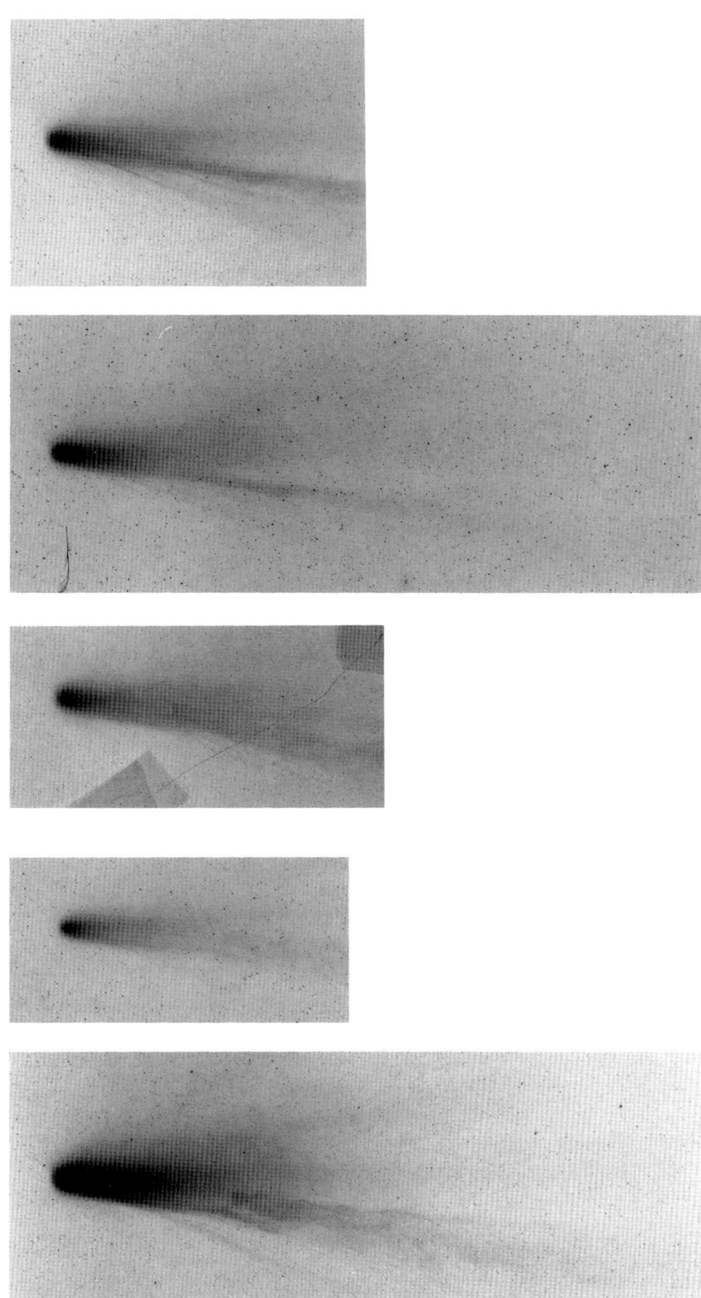

Fig. 874: a disconnection event, (1°~ 16 mm) (top to bottom) 1986 Mar 8.38507, LSPN-1271; *F. Miller University of Michigan/CTIO*; 1986 Mar 8.50567, LSPN-1380, *W. Liller, Easter Island, LSPN Island Network*; 1986 Mar 9.35660, LSPN-1273, *F. Miller, University of Michigan/CTIO*; 1986 Mar 9.70098, LSPN-2115, *P. Magnusson, Uppsala Southern Station;* 1986 Mar 9.77716, LSPN-2382, *UK Schmidt Telescope Unit.*

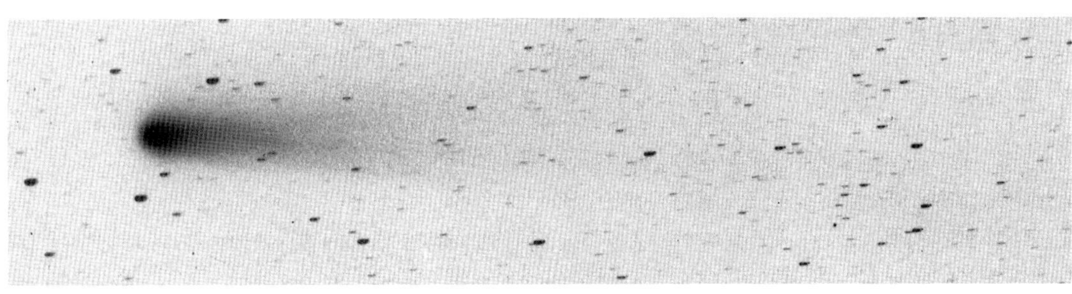

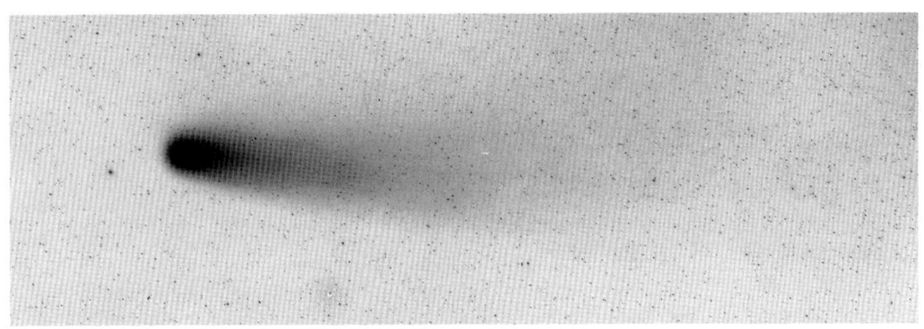

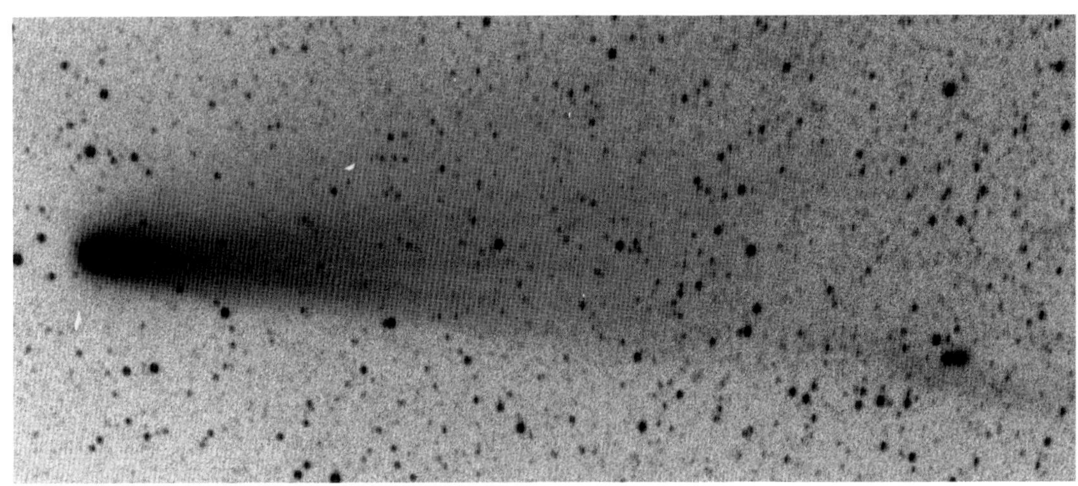

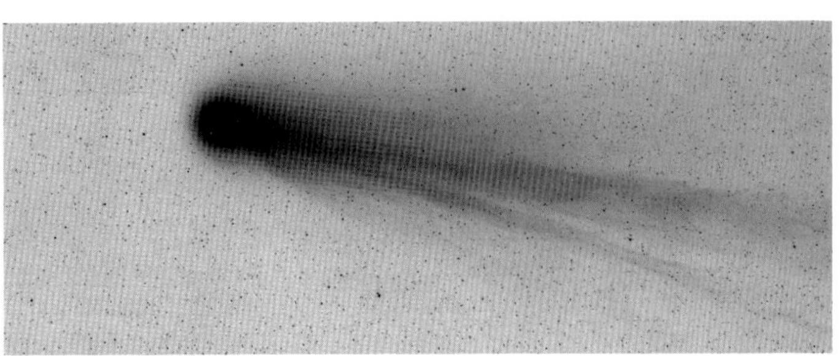

Fig. 875: mid-March 1986 activity, (1°~ 28 mm) (top to bottom) 1986 Mar 15.0000, LSPN-3650, *K. Sivaraman, Indian Institute for Astrophysics, Kodaikanal Station*;1986 Mar 15.34167; LSPN-728, *C. Torres/H. Wroblewski, Cerro el Roble Astronomical Station*;1986 Mar 15.46810, LSPN-3420, *G. Emerson, E.E. Barnard Observatory Station*; Mar 16.36215, LSPN-1297, *F. Miller, University of Michigan/CTIO.*

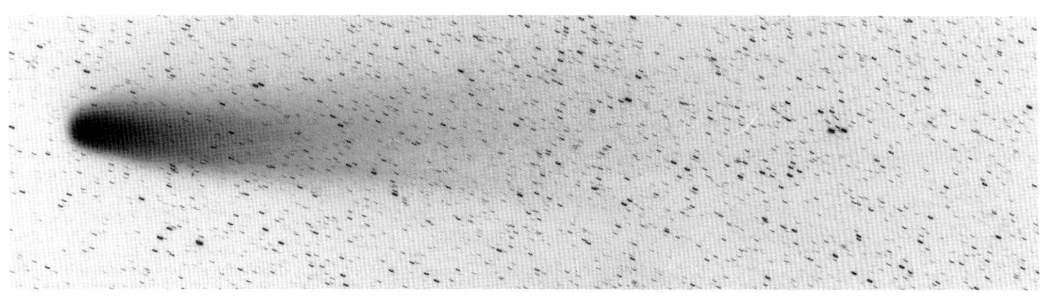

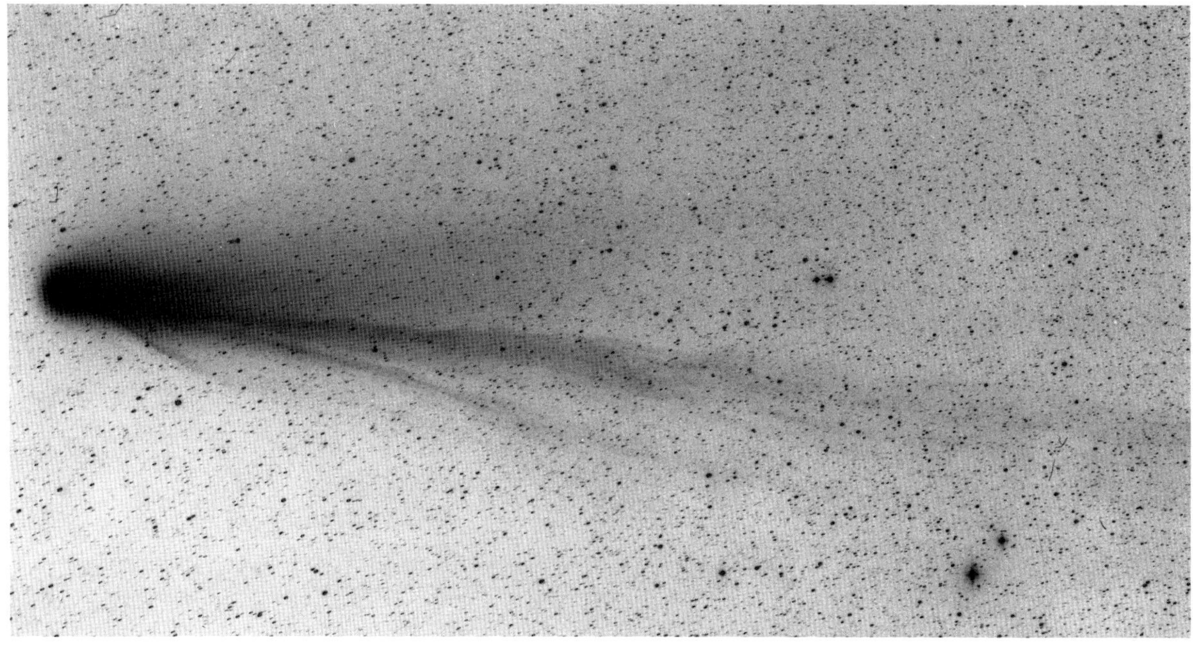

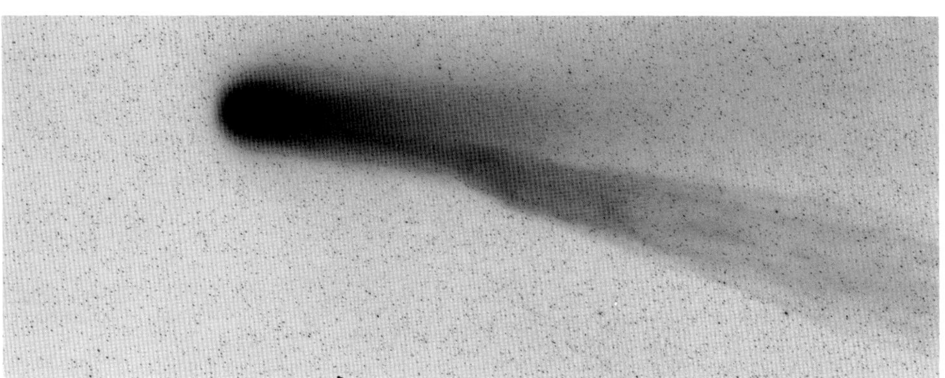

Fig. 876: mid-March 1986 activity, (1°~ 28 mm) (top to bottom) 1986 Mar 16.47326, LSPN-1398, *W. Liller, Easter Island, LSPN Island Network*; 1986 Mar 16.47708, LSPN-894, *E. Moore, Joint Observatory for Cometary Research*; 1986 Mar 17.35208, LSPN-731, *C. Torres/H. Wroblewski, Cerro el Roble Astronomical Station.*

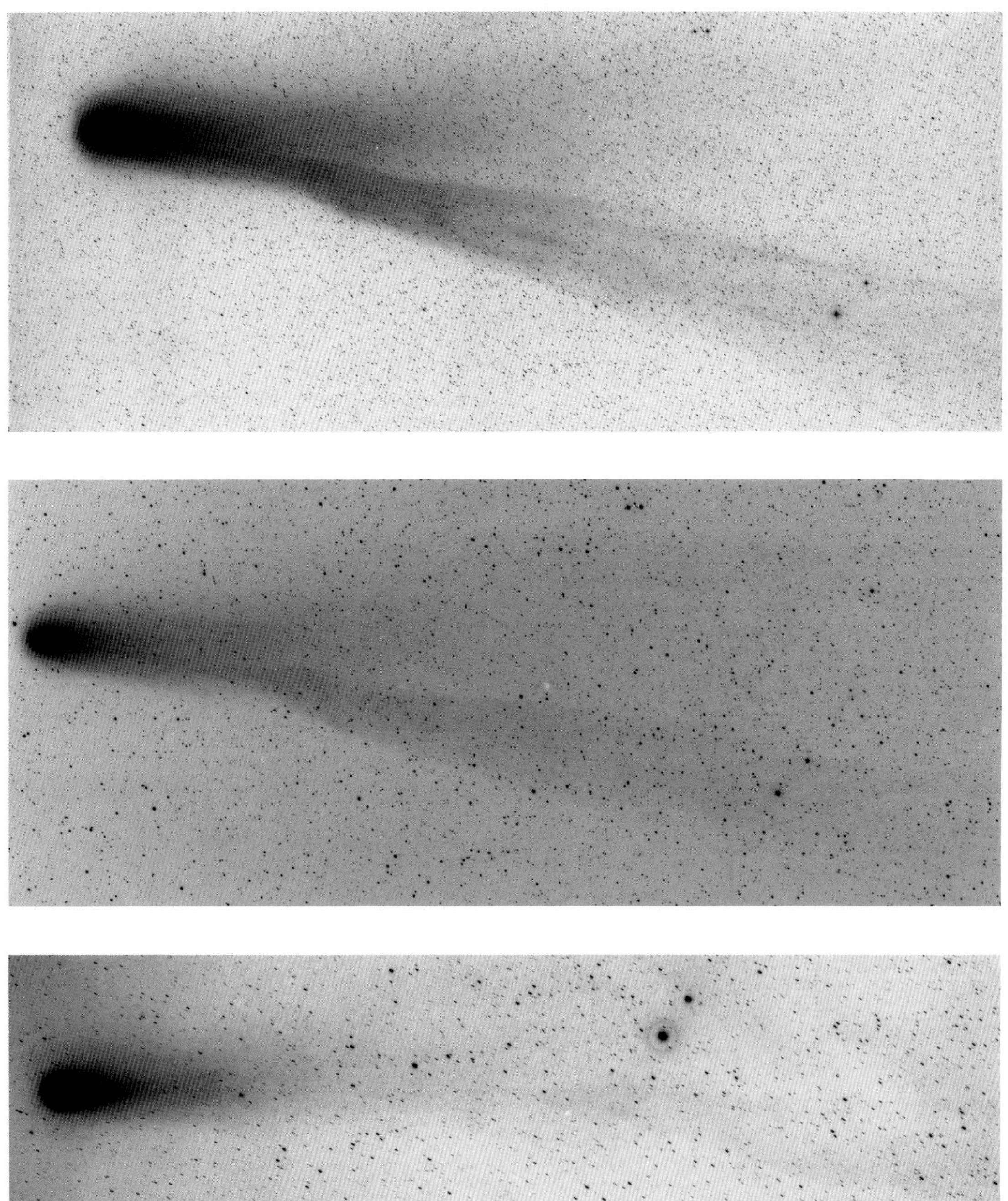

Fig. 877: mid-March 1986 activity, (1°~ 28 mm) (top to bottom) 1986 Mar 17.36910, LSPN-2899, *G. Pizarro, European Southern Observatory*; 1986 Mar 17.49896, LSPN-1402, *W. Liller, Easter Island, LSPN Island Network*; 1986 Mar 18.51424, LSPN-3579, *H. Giclas, Lowell Observatory.*

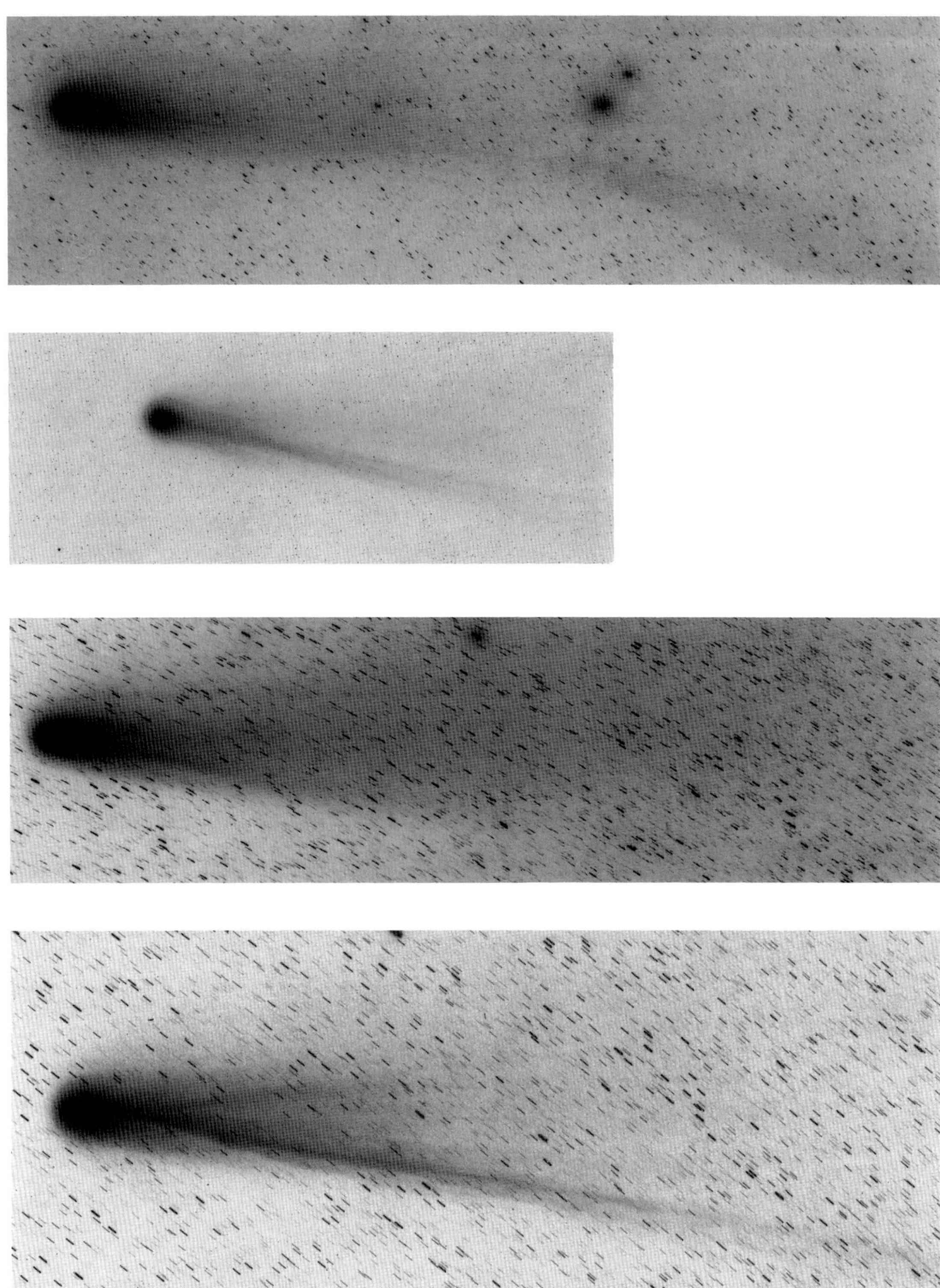

Fig. 878: mid-March 1986 activity, (1°~ 28 mm) (top to bottom) 1986 Mar 18.63634, LSPN-1916, *M. Buie, Mauna Kea Observatory*; 1986 Mar 19.28229, LSPN-1309, *F. Miller, University of Michigan/CTIO*; 1986 Mar 19.44306, LSPN-1405, *W. Liller, Easter Island, LSPN Island Network*; 1986 Mar 20.10417, LSPN-2849, *G. Malcolm/A. Jarrett, Boyden Observatory.*

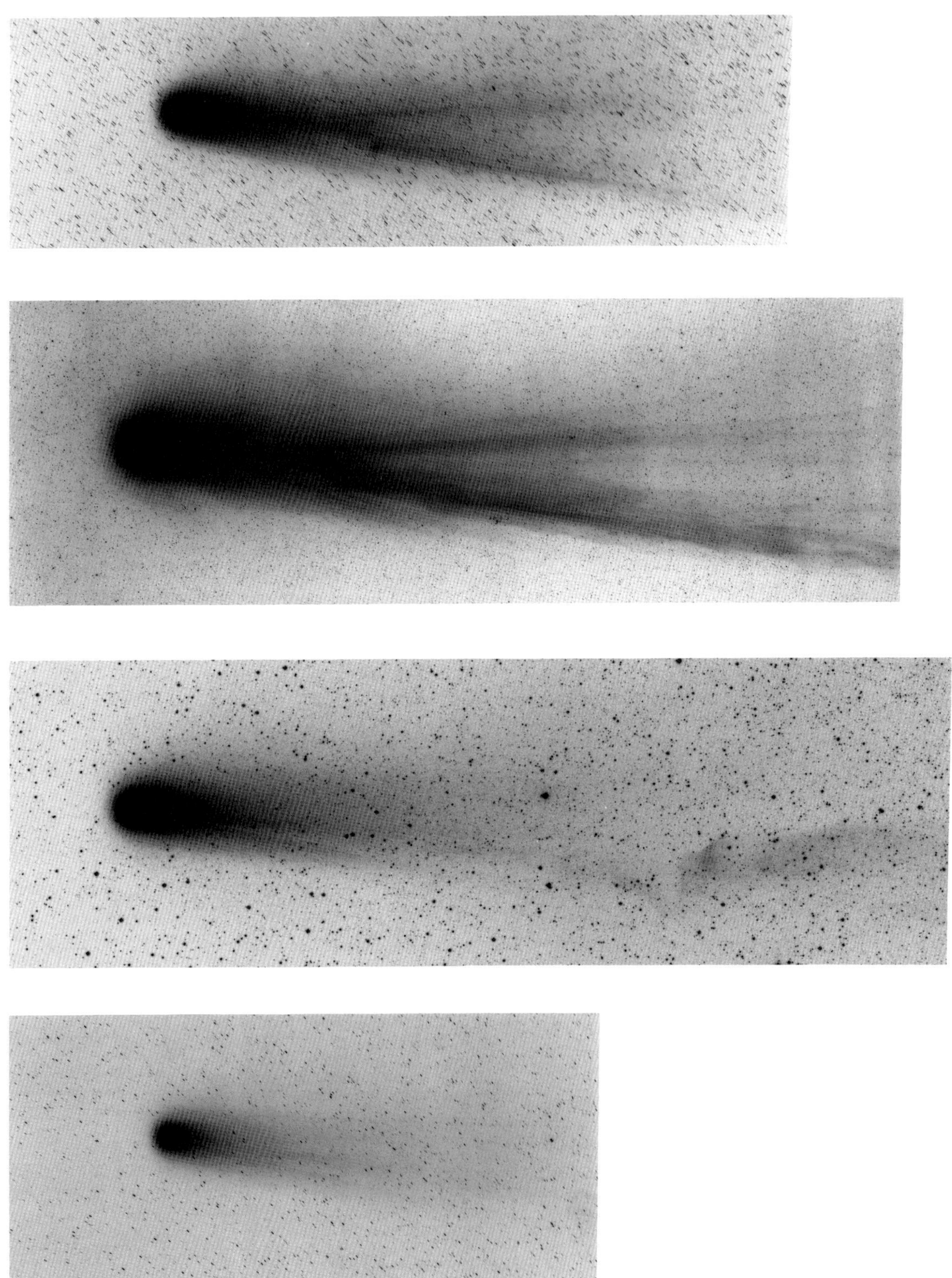

Fig. 879: mid-March 1986 activity, (1°~ 28 mm) (top to bottom) 1986 Mar 20.34028, LSPN-738, *C. Torres/H. Wroblewski, Cerro el Roble Astronomical Station;* 1986 Mar 20.39097, LSPN-1107, *F. Miller, University of Michigan/CTIO*; 1986 Mar 21.47326, LSPN-905, *E. Moore, Joint Observatory for Cometary Research*; 1986 Mar 21.47650, LSPN 2971, *K. Meech/D. Jewitt, Warner and Swasey Observatory.*

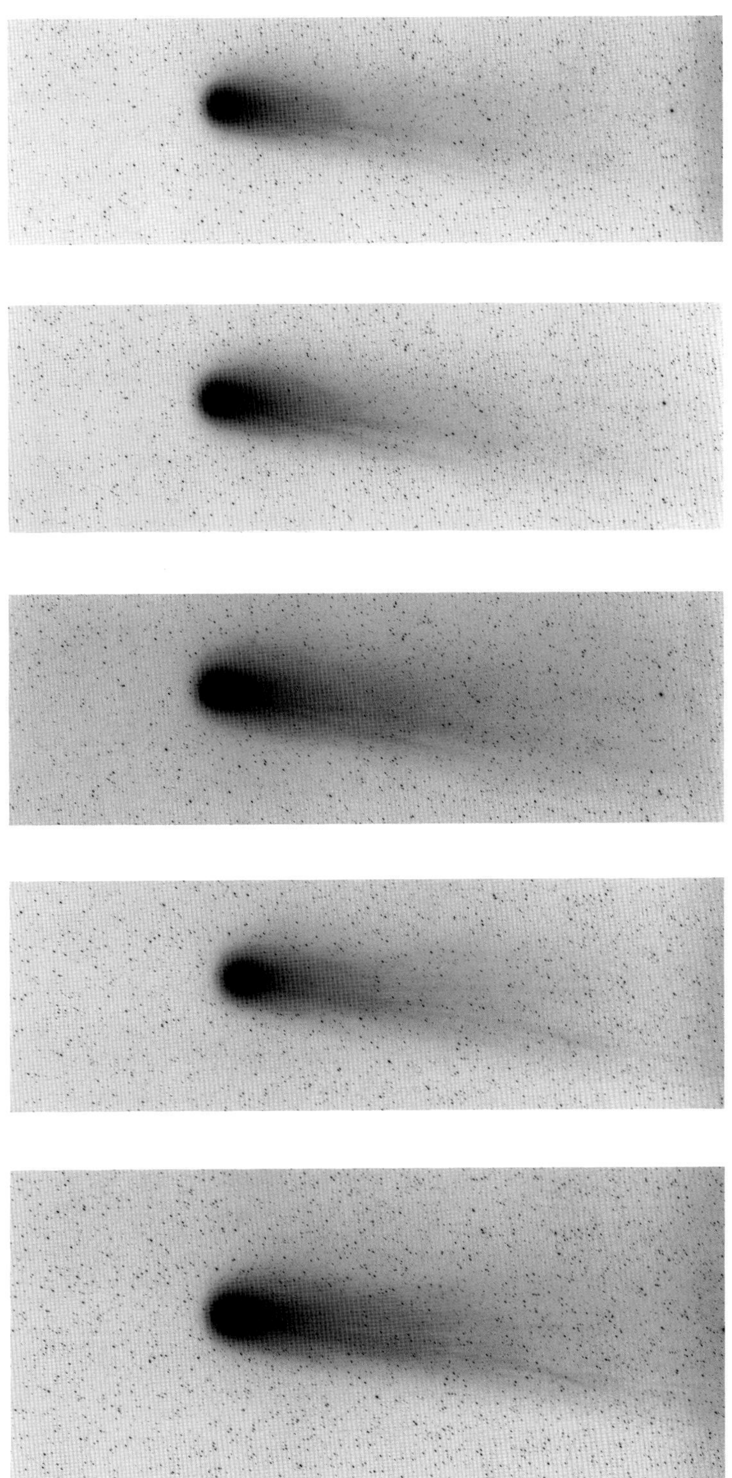

Fig. 880: mid-March 1986 activity, (1°~ 28 mm) (top to bottom) 1986 Mar 21.49080, LSPN-2972; 1986 Mar 21.50314, LSPN-2973; 1986 Mar 21.51476, LSPN-2974; 1986 Mar 22.47836, LSPN-2975; 1986 Mar 22.49044, LSPN-2976.
All photographs by K. Meech/D. Jewitt, Warner and Swasey Observatory.

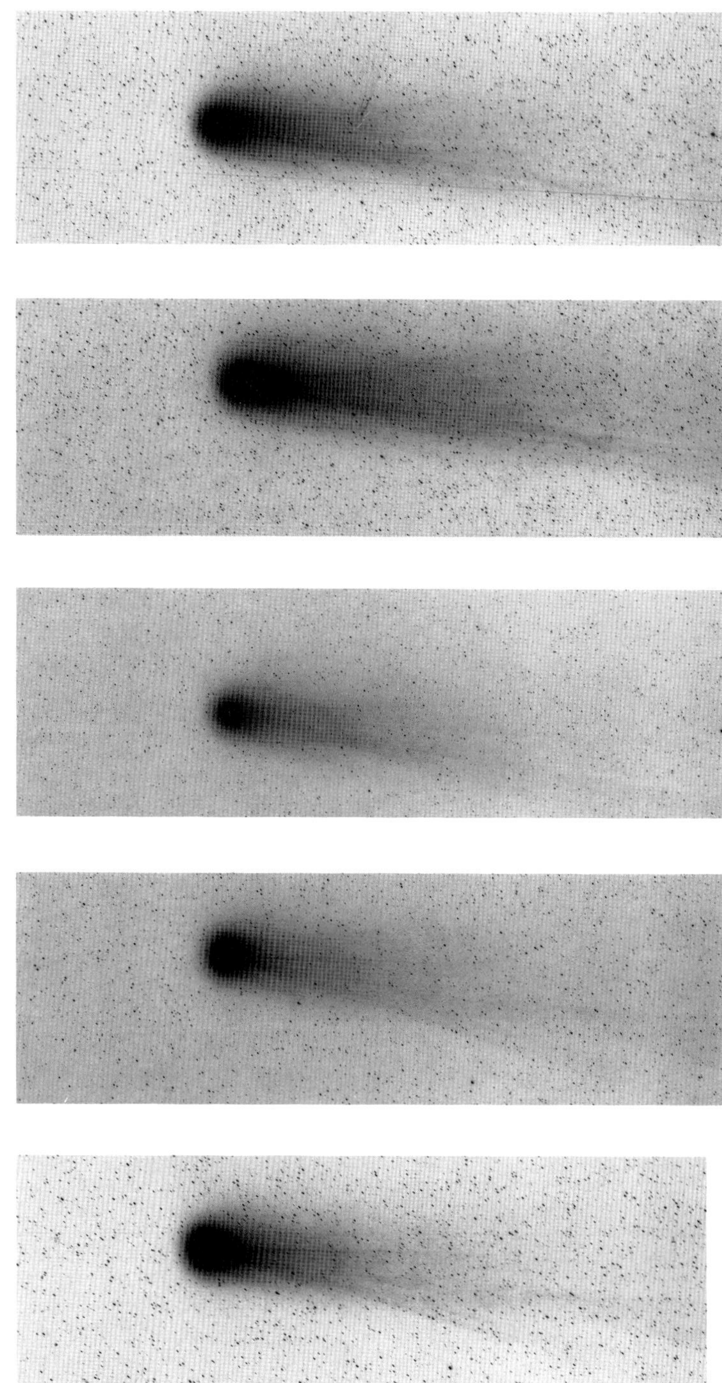

Fig. 881: mid-March 1986 activity, (1°~ 28 mm) (top to bottom) 1986 Mar 22.50170, LSPN-2977; 1986 Mar 22.51229, LSPN-2978; 1986 Mar 22.52051, LSPN-2979; 1986 Mar 23.49865, LSPN-2982; 1986 Mar 23.50794, LSPN-2983.
All photographs by K. Meech/D. Jewitt, Warner and Swasey Observatory.

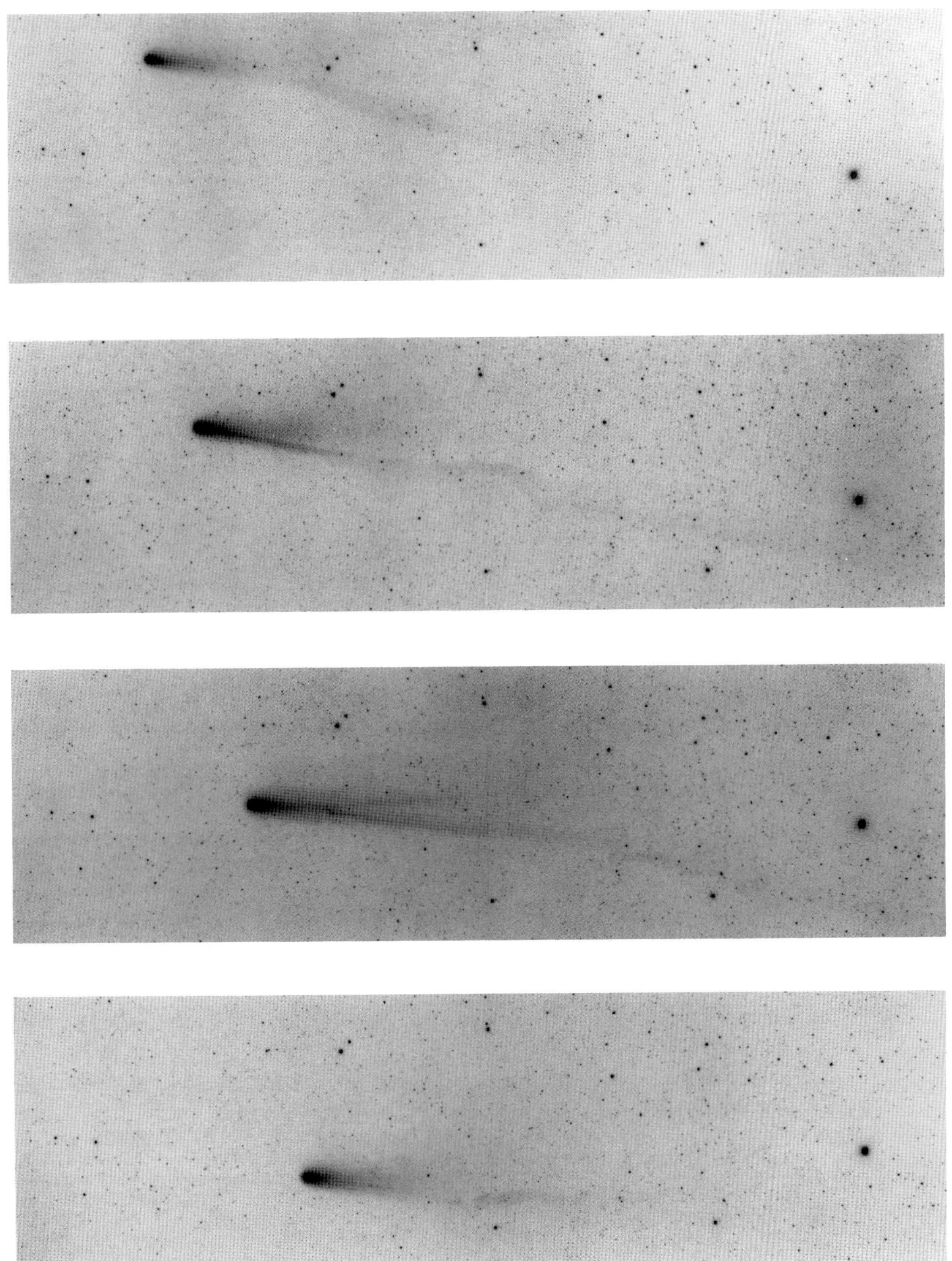

Fig. 882: daily sequence, Lowell Observatory, (1°~ 9 mm) (top to bottom) 1986 Mar 18.51319, LSPN-3589; 1986 Mar 19.50833, LSPN-3590; 1986 Mar 20.50972, LSPN-3592; 1986 Mar 21.50938, LSPN-3591.
All photographs by H. Giclas, Lowell Observatory.

Fig. 883: late-March 1986 sequence, (1° ~ 29 mm; 26 mm ~ 3 x 10^6 km) (top to bottom) 1986 Mar 25.50000, LSPN-1423; 1986 Mar 27.45014, LSPN-1426; 1986 Mar 27.47078, LSPN-1427; 1986 Mar 27.48819, LSPN-1428; 1986 Mar 28.42188, LSPN-1429; 1986 Mar 28.43403, LSPN-1430; 1986 Mar 30.44792, LSPN-1434.
All photographs by W. Liller, Easter Island, LSPN Island Network.

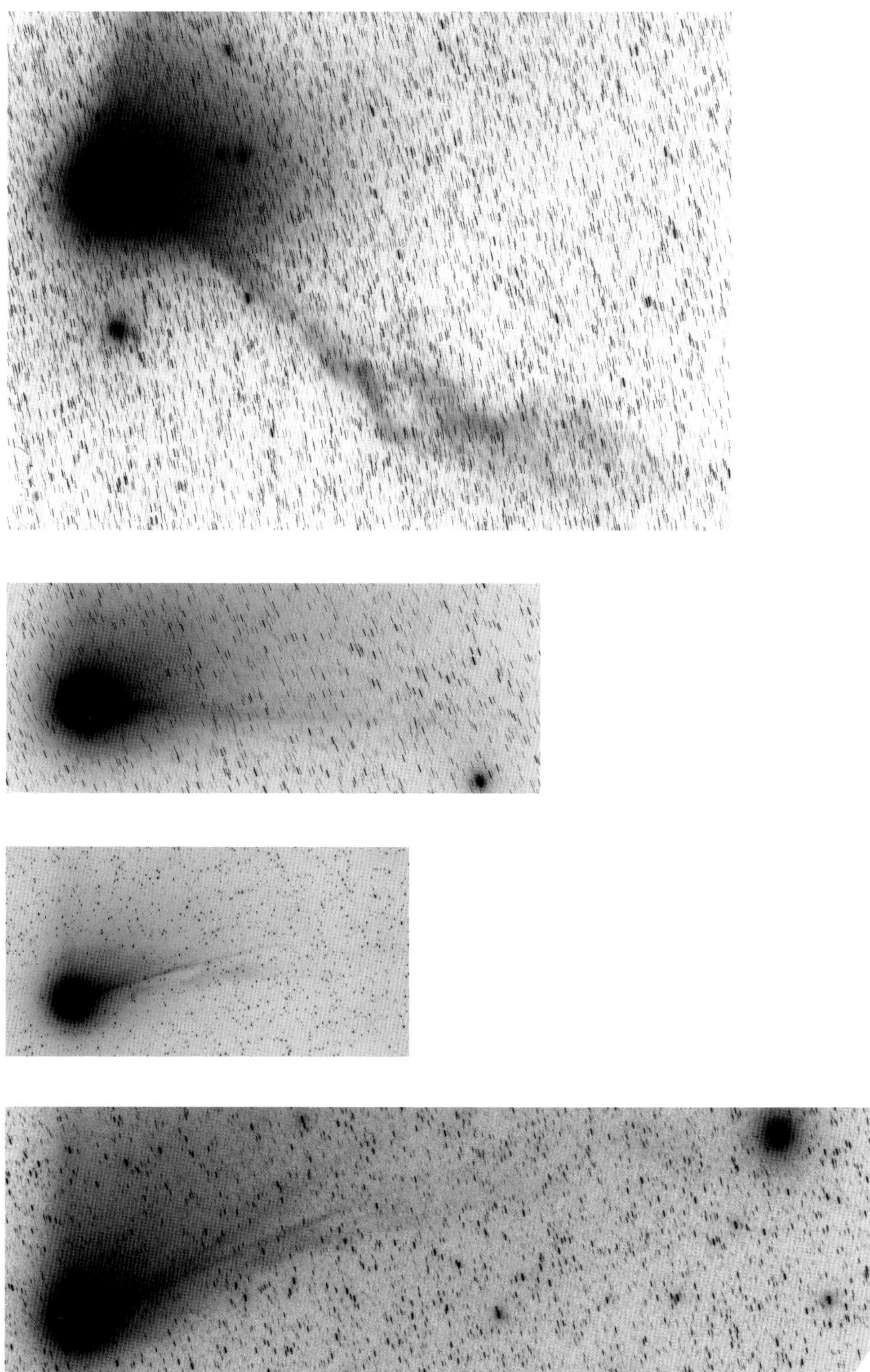

Fig. 884: rapid change in tail orientation; north is to the right, (22 mm ~ 2 x 10^6 km) (top to bottom) 1986 Apr 12.30903, LSPN-1198, *F. Miller, University of Michigan/CTIO*; 1986 Apr 13.66913, LSPN-2159, *P. Magnusson, Uppsala Southern Station*; 1986 Apr 14.99826, LSPN-1217; *F. Miller, University of Michigan/CTIO*; 1986 Apr 15.45313, LSPN-1487, *W. Liller, Easter Island, LSPN Island Network.*

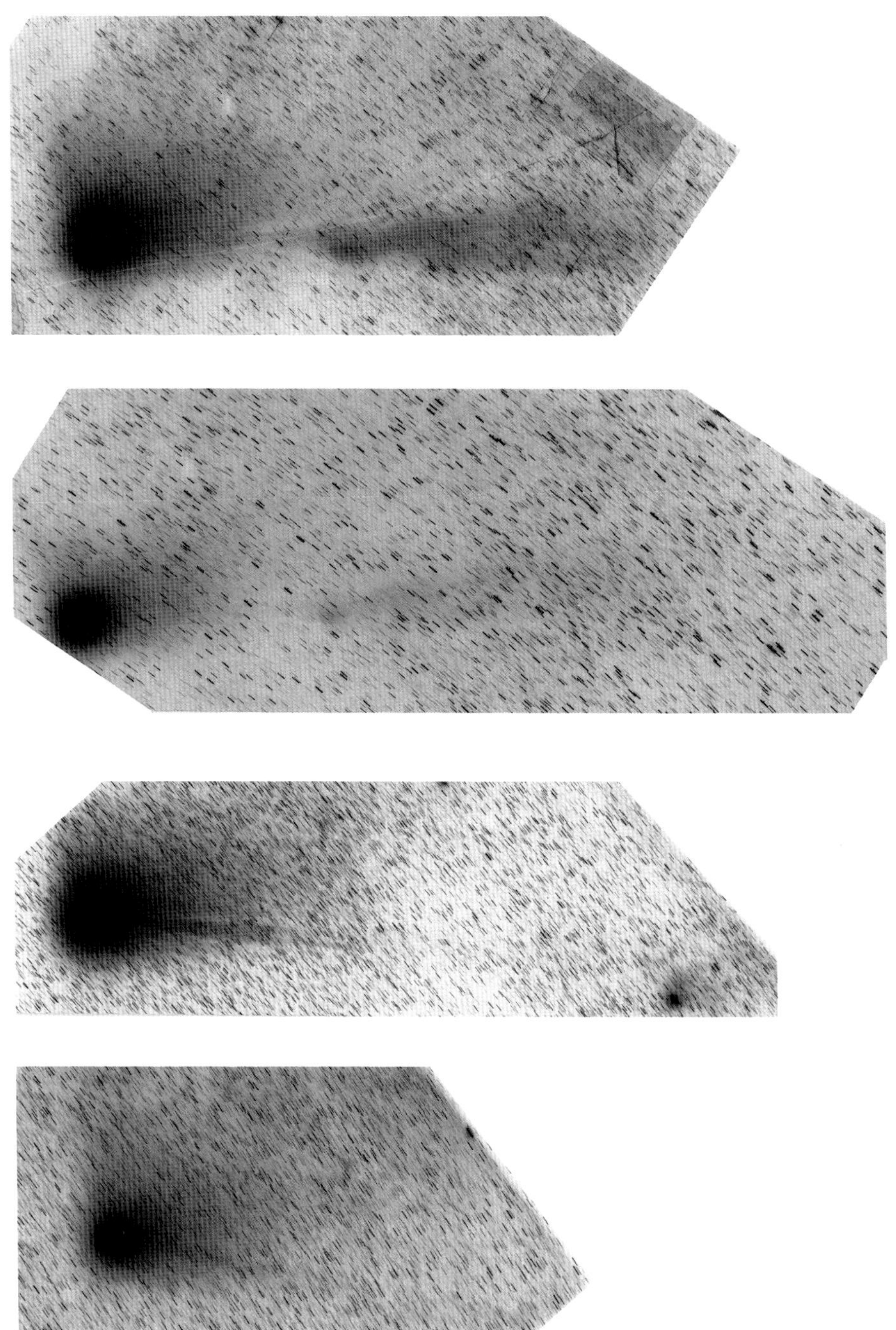

Fig. 885: mid-April 1986 activity, (1° ~ 18 mm) (top to bottom) 1986 Apr 8.36319, LSPN-1166, *F. Miller, University of Michigan/CTIO;* 1986 Apr 8.39132, LSPN-1457, *W. Liller, Easter Island, LSPN Island Network*; 1986 Apr 9.36250, LSPN-1176, *F. Miller, University of Michigan/CTIO*; 1986 Apr 9.38403, LSPN-2590, *R. Hill, Warner and Swasey Observatory.*

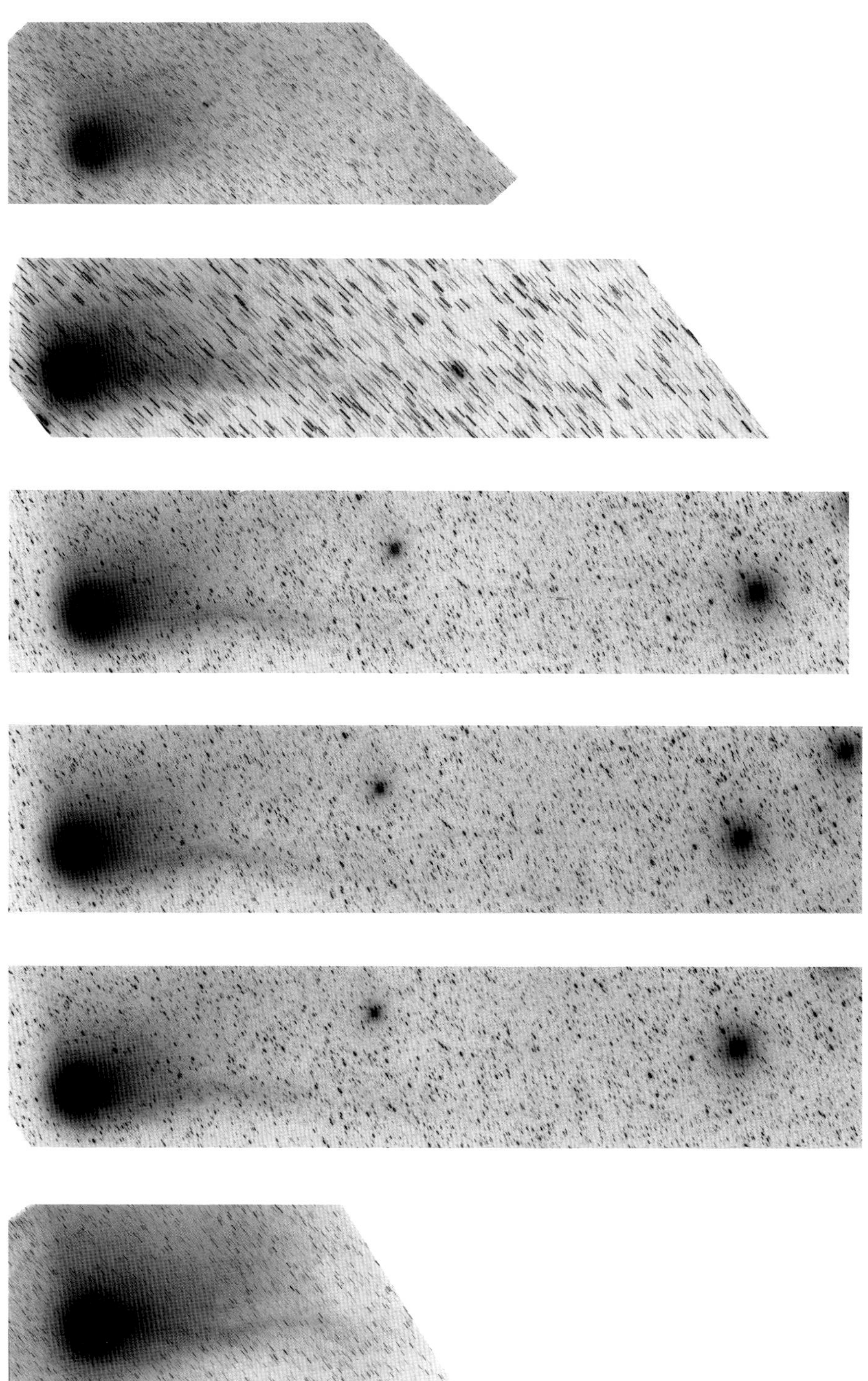

Fig. 886: mid-April 1986 activity, (1° ~ 18 mm) (top to bottom) 1986 Apr 9.74687, LSPN-2137, *P. Magnusson, Uppsala Southern Station*; 1986 Apr 10.22778, LSPN-1465, *W. Liller, Easter Island, LSPN Island Network*; 1986 Apr 10.40556, LSPN-913; 1986 Apr 10.41458, LSPN-914; 1986 Apr 10.42361, LSPN-915, *E. Moore, Joint Observatory for Cometary Research;* 1986 Apr 10.73422, LSPN-2143, *P. Magnusson, Uppsala Southern Station.*

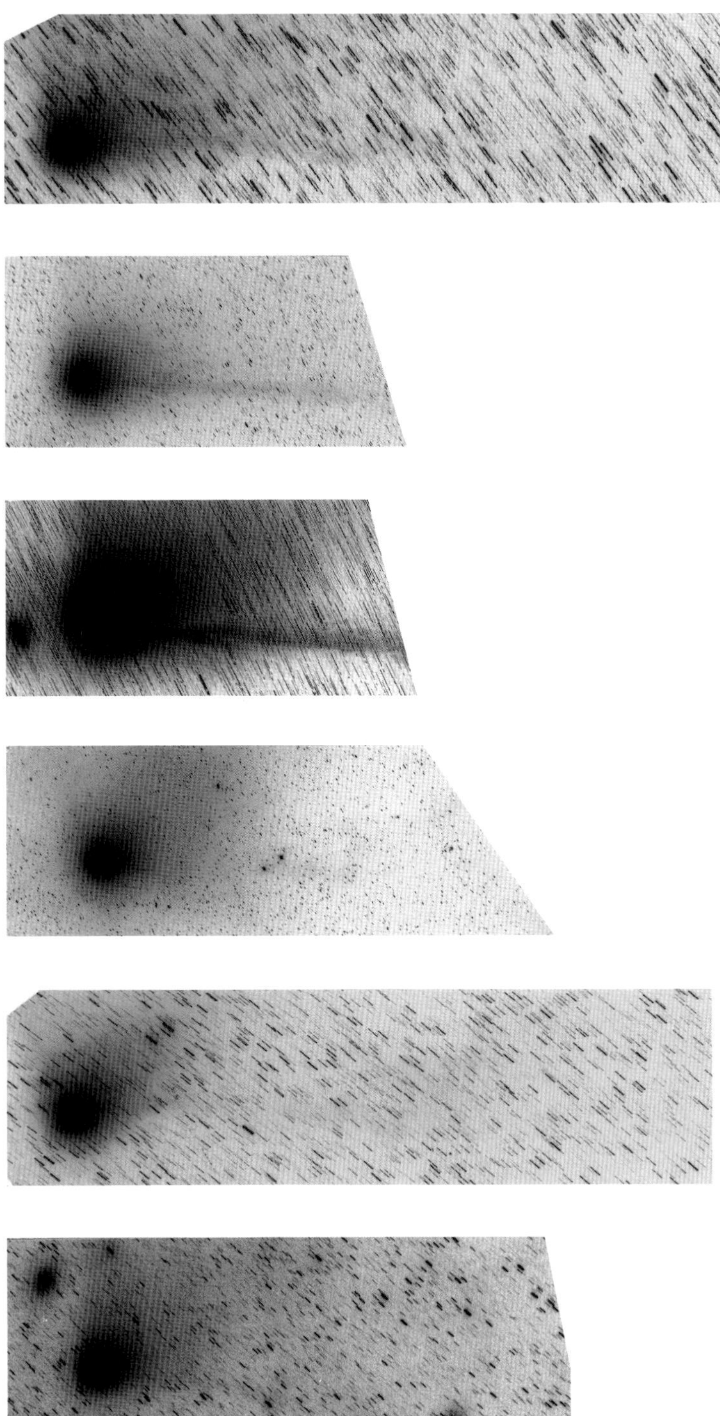

Fig. 887: mid-April 1986 activity, (1° ~ 18 mm) (top to bottom) 1986 Apr 11.28542, LSPN-1469, *W. Liller, Easter Island, LSPN Island Network;* 1986 Apr 11.53644, LSPN-2145; 1986 Apr 11.64441, LSPN-2147, *P. Magnusson, Uppsala Southern Station*; 1986 Apr 12.08715, LSPN-1193, *F. Miller, University of Michigan/CTIO;* 1986 Apr 12.27431, LSPN-1475, *W. Liller, Easter Island, LSPN Island Network*; 1986 Apr 12.79688, LSPN-3661, *K. Sivaraman, Indian Institute for Astrophysics, Kodaikanal Station.*

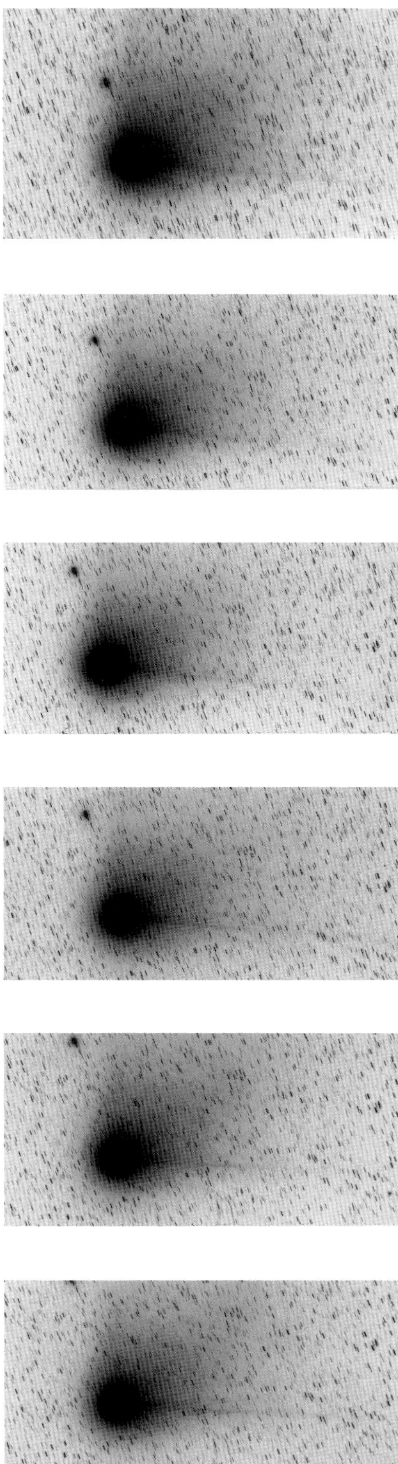

Fig. 888: mid-April 1986 activity, (1° ~ 18 mm) (top to bottom) 1986 Apr 13.52819, LSPN-2149; 1986 Apr 13.54447, LSPN-2150; 1986 Apr 13.55844, LSPN-2151; 1986 Apr 13.57216, LSPN-2152; 1986 Apr 13.59988, LSPN-2154; 1986 Apr 13.61373, LSPN-2155.
All photographs by P. Magnusson, Uppsala Southern Station.

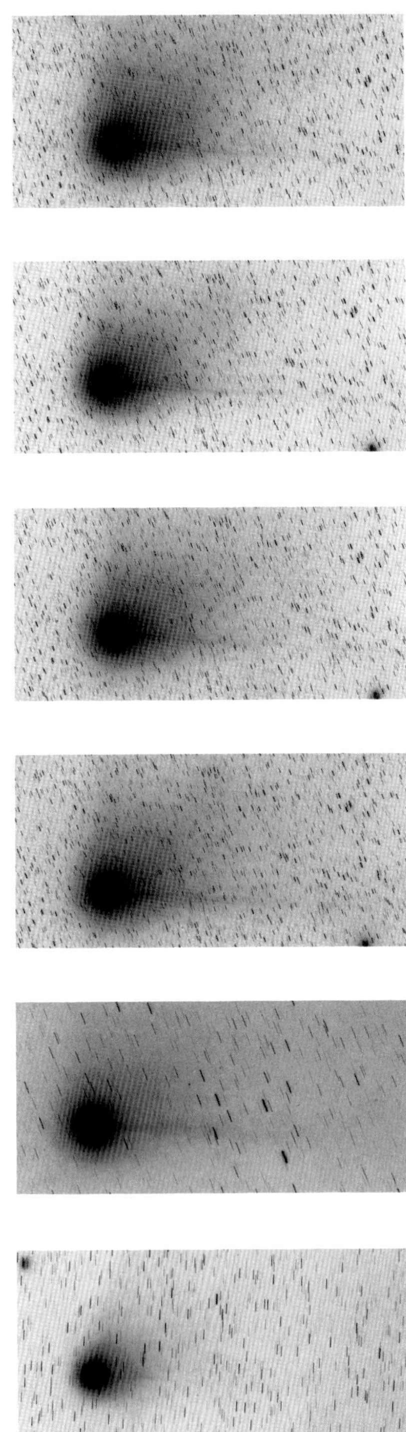

Fig. 889: mid-April 1986 activity, (1° ~ 18 mm) (top to bottom) 1986 Apr 13.62758, LSPN-2156; 1986 Apr 13.64143, LSPN-2157; 1986 Apr 13.65528, LSPN-2158; 1986 Apr 13.66913, LSPN-2159, *P. Magnusson, Uppsala Southern Station*; 1986 Apr 14.32951, LSPN-3582, *H. Giclas, Lowell Observatory;* 1986 Apr 15.06042, LSPN-1220, *F. Miller, University of Michigan/CTIO.*

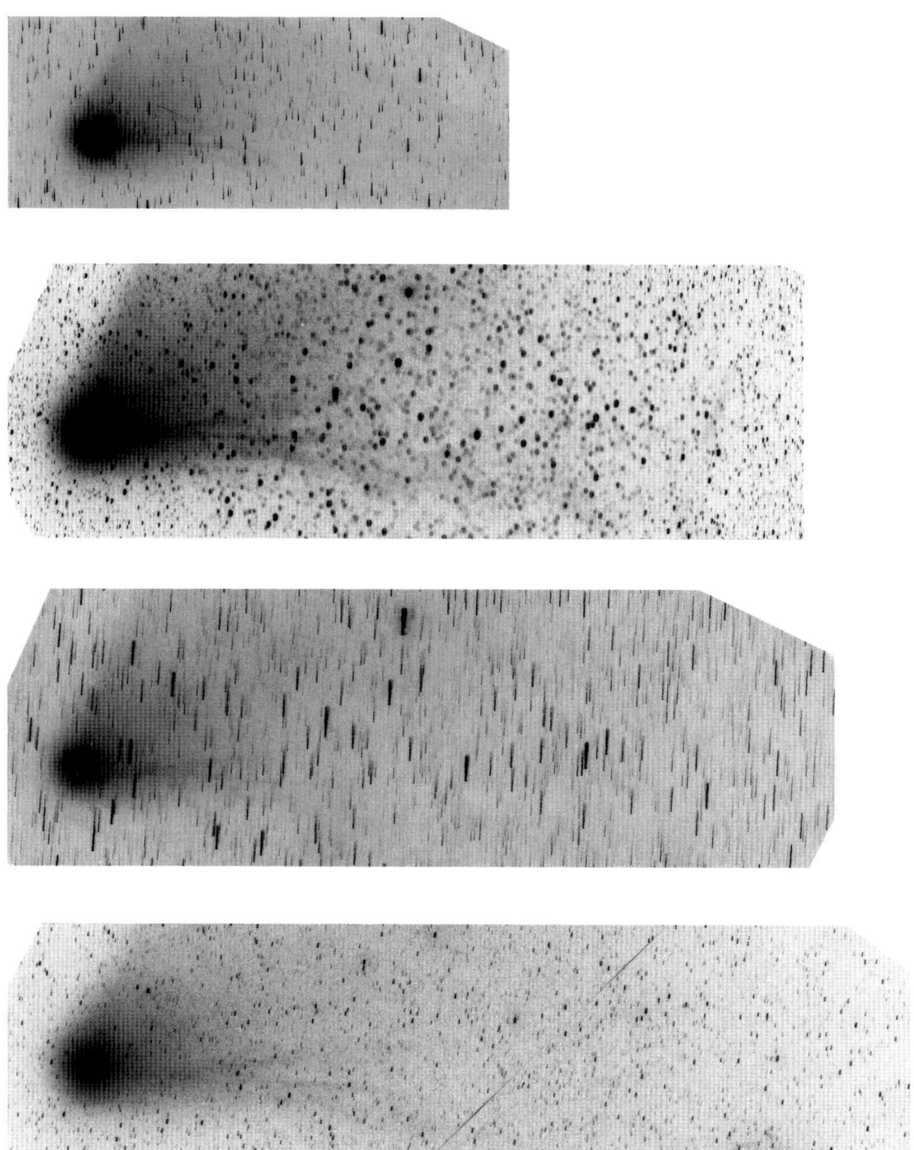

Fig. 890: mid-April 1986 activity, (1° ~ 18 mm) (top to bottom) 1986 Apr 15.24965, LSPN-1484; 1986 Apr 15.38056, LSPN-1485; 1986 Apr 15.40868, LSPN-1486; 1986 Apr 15.51285, LSPN-1489.
All photographs by W. Liller, Easter Island, LSPN Island Network.

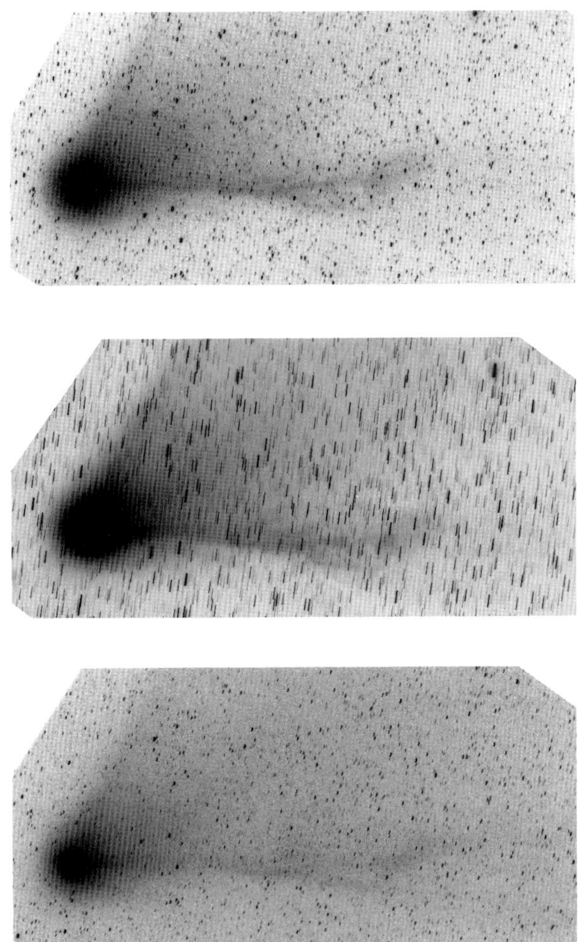

Fig. 891: mid-April 1986 activity, (1° ~ 18 mm) (top to bottom) 1986 Apr 16.37569, LSPN-1490; 1986 Apr 16.39757, LSPN-1491; 1986 Apr 16.48715, LSPN-1494.
All photographs by W. Liller, Easter Island, LSPN Island Network.

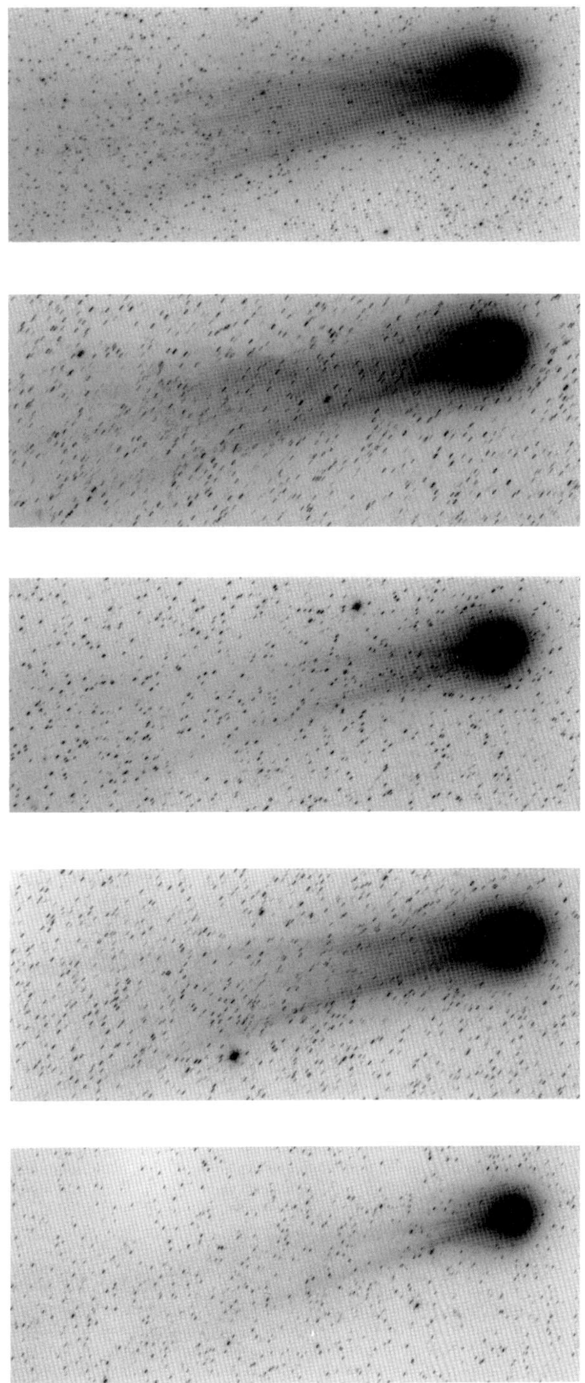

Fig. 892: daily sequence, University of Michigan/CTIO, (1° ~ 37 mm) (top to bottom) 1986 Apr 29.04439, LSPN-2992; 1986 Apr 30.06815, LSPN-3003; 1986 May 1.11451, LSPN-3014; 1986 May 2.01389, LSPN-3020; 1986 May 3.11019, LSPN-3028.
All photographs by K. Meech, University of Michigan/CTIO.

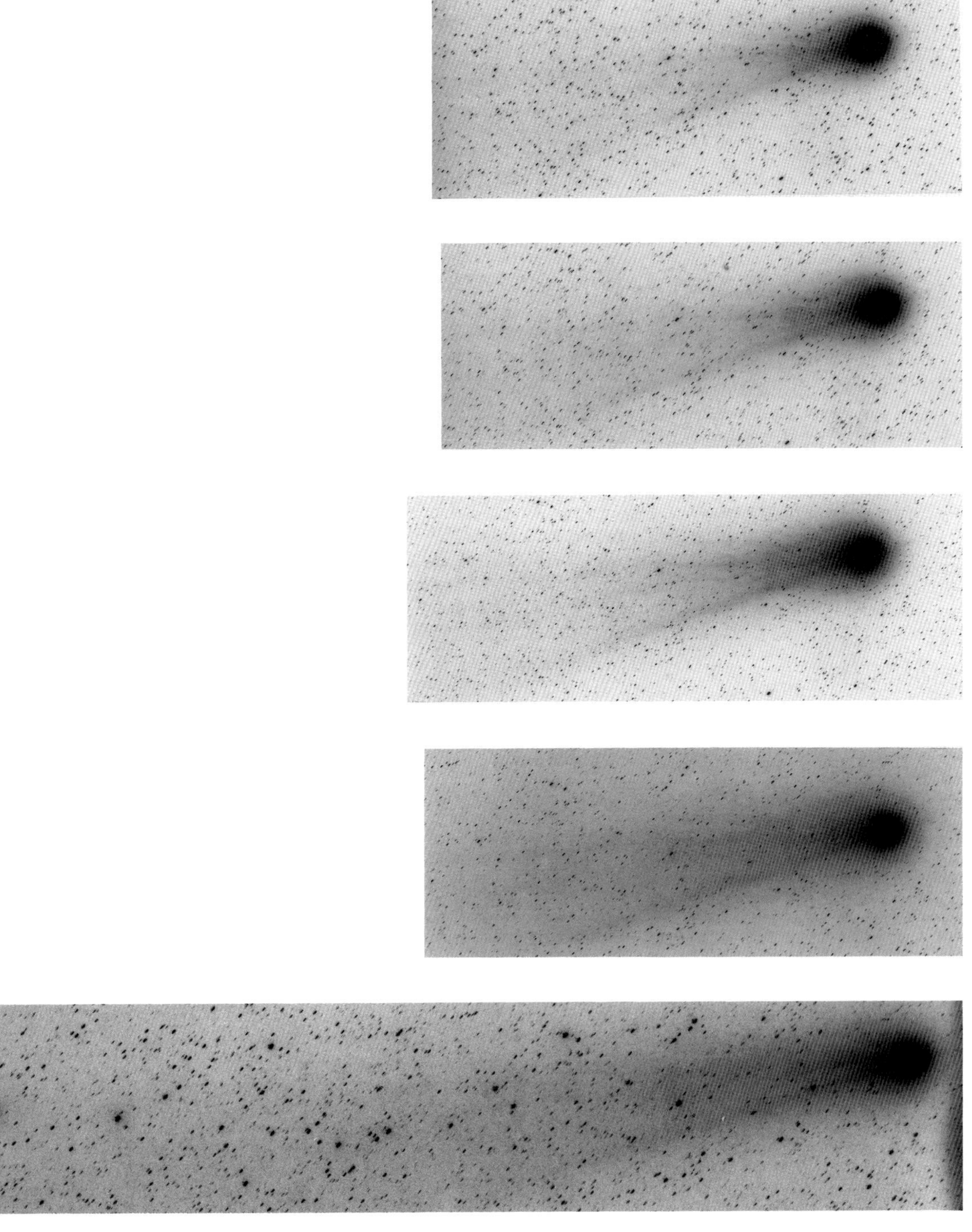

Fig. 893: turn-off sequence, (1° ~ 28 mm) (top to bottom) 1986 Apr 29.00945, LSPN-2990; 1986 Apr 29.02729, LSPN-2991; 1986 Apr 29.04439, LSPN-2992; 1986 Apr 29.09064, LSPN-2995, *K. Meech, University of Michigan/CTIO*; 1986 Apr 29.23958, LSPN-939, *E. Moore et al., Joint Observatory for Cometary Research.*

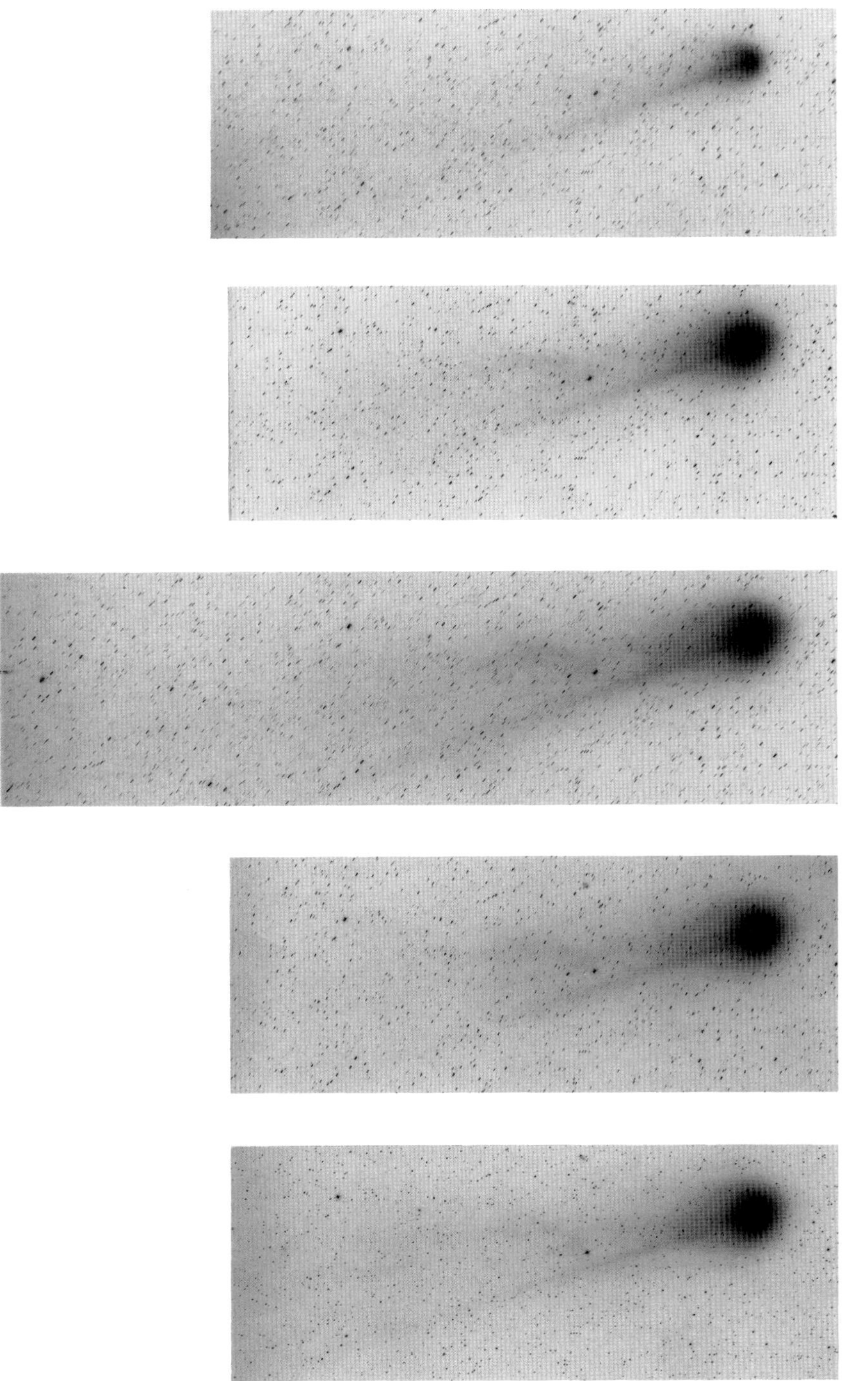

Fig. 894: turn-off sequence, (1° ~ 28 mm) (top to bottom) 1986 Apr 29.98264, LSPN-2918, *O. Pizarro, European Southern Observatory*; 1986 Apr 29.99919, LSPN-2999, *K. Meech, University of Michigan/CTIO*; 1986 Apr 30.01597, LSPN-2919, *O. Pizarro, European Southern Observatory*; 1986 Apr 30.03125, LSPN-3001; 1986 Apr 30.04618, LSPN-3002, *K. Meech, University of Michigan/CTIO.*

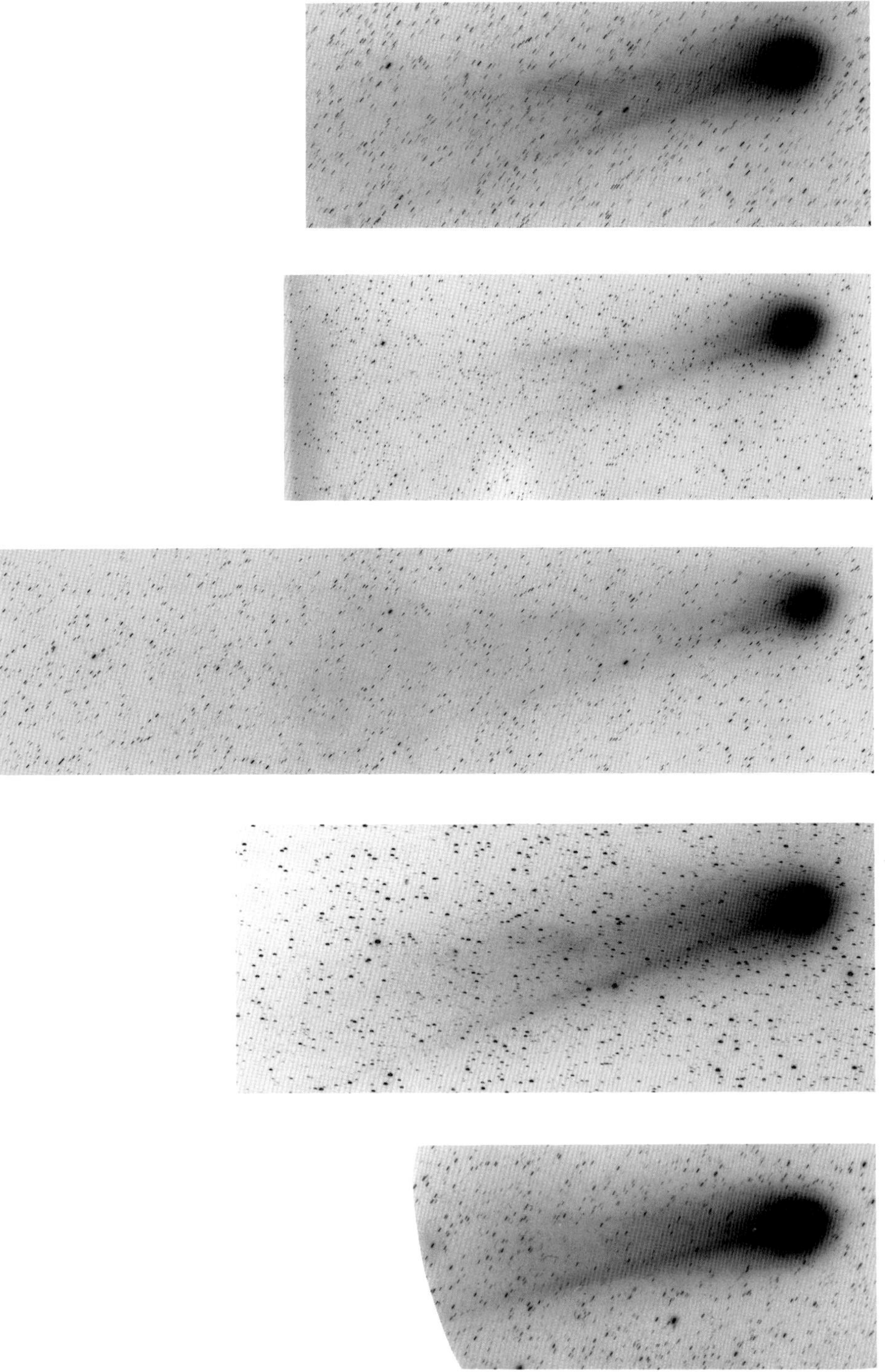

Fig. 895: turn-off sequence, (1° ~ 28 mm) (top to bottom) 1986 Apr 30.06815, LSPN-3003; 1986 Apr 30.12720, LSPN-3006; 1986 Apr 30.14757, LSPN-3007, *K. Meech, University of Michigan/CTIO*; 1986 Apr 30.22187, LSPN-1503, *W. Liller, Easter Island, LSPN Island Network*; 1986 Apr 30.73229, LSPN-3509, *D. O'Donoghue, University of Perugia.*

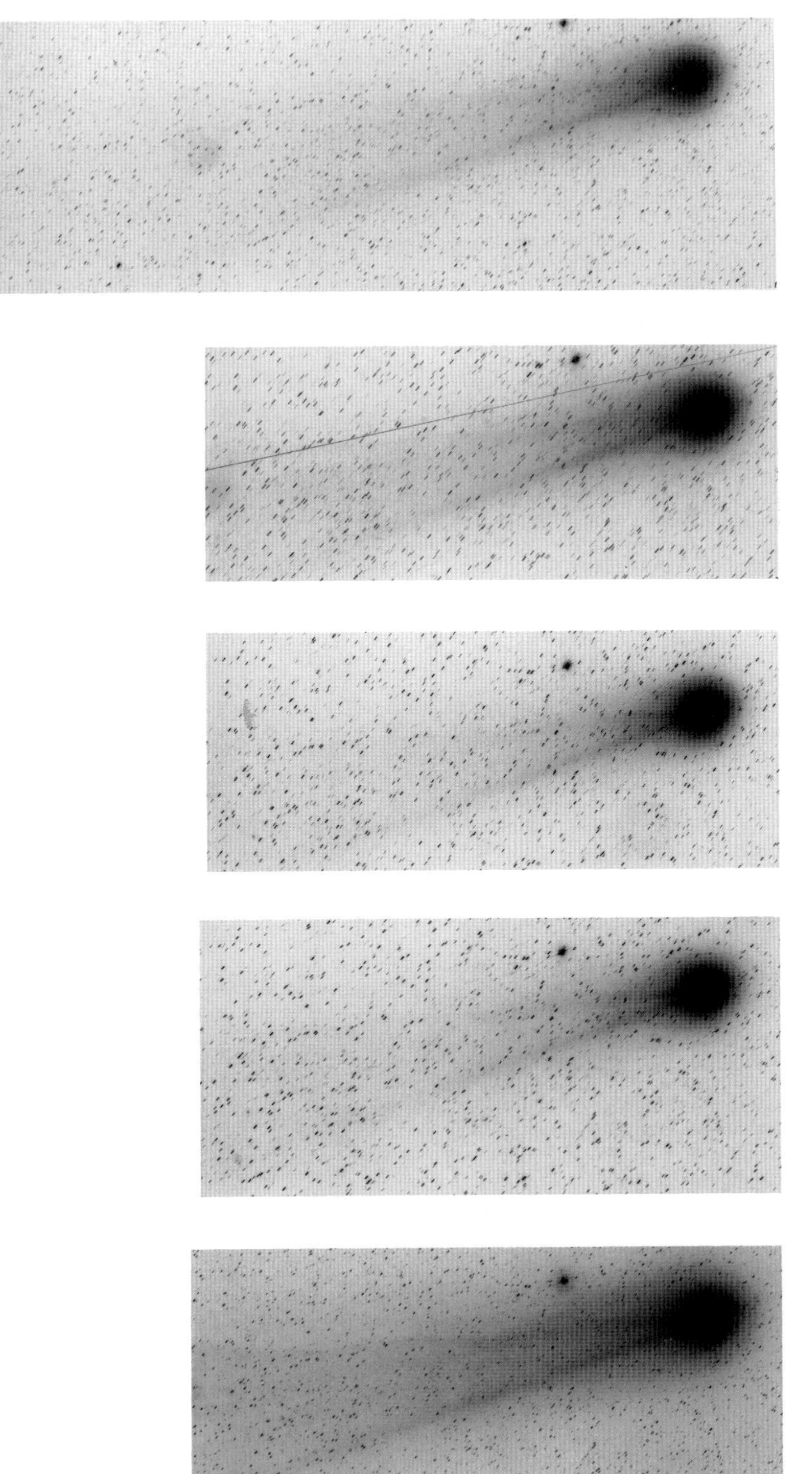

Fig. 896: turn-off sequence, (1° ~ 28 mm) (top to bottom) 1986 May 1.02014, LSPN-2921, *European Southern Observatory*; 1986 May 1.03576, LSPN-3011; 1986 May 1.07564, LSPN-3013; 1986 May 1.11451, LSPN-3014; 1986 May 1.13067, LSPN-3015, *K. Meech, University of Michigan/CTIO.*

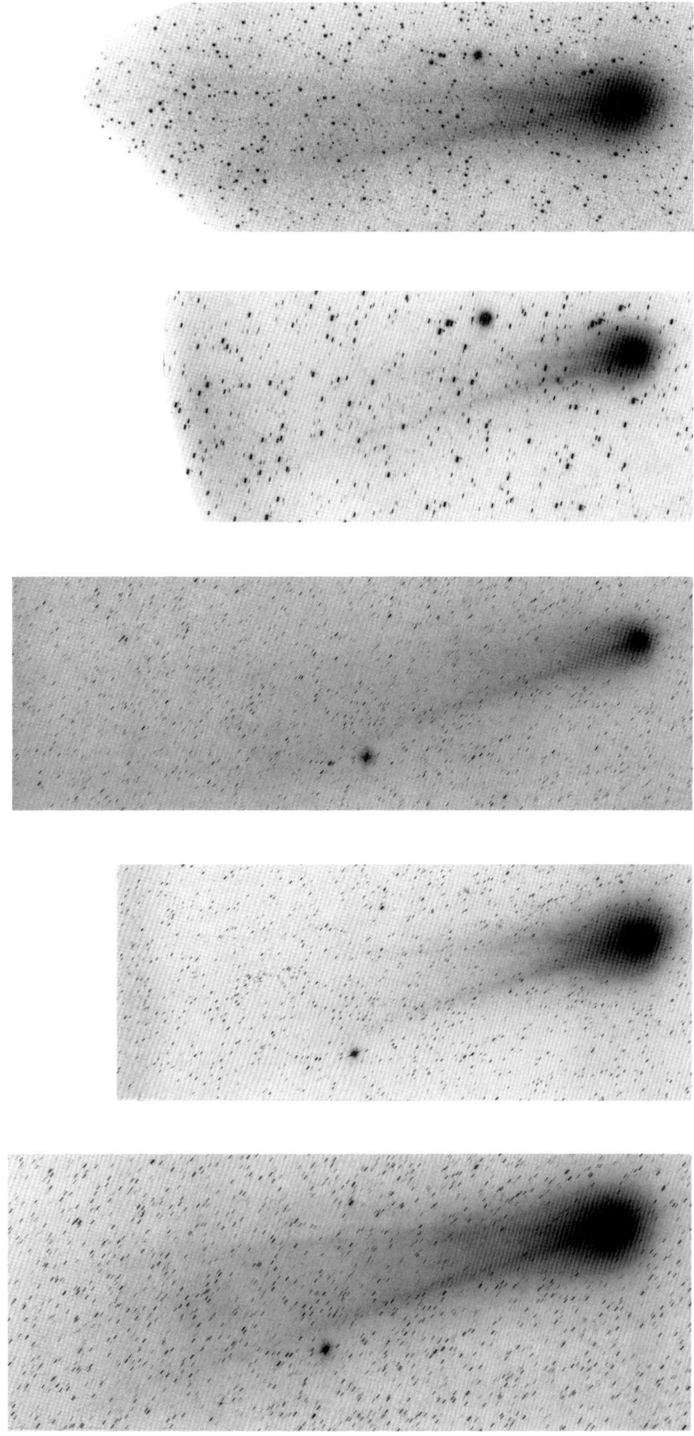

Fig. 897: turn-off sequence, (1° ~ 28 mm) (top to bottom) 1986 May 1.16458, LSPN-981; 1986 May 1.18472, LSPN-982, *G. Emerson, E.E. Barnard Observatory*; 1986 May 1.97396, LSPN-2922, *O. Pizarro, European Southern Observatory*; 1986 May 1.98351, LSPN-3019; 1986 May 2.01389, LSPN-3020, *K. Meech, University of Michigan/CTIO.*

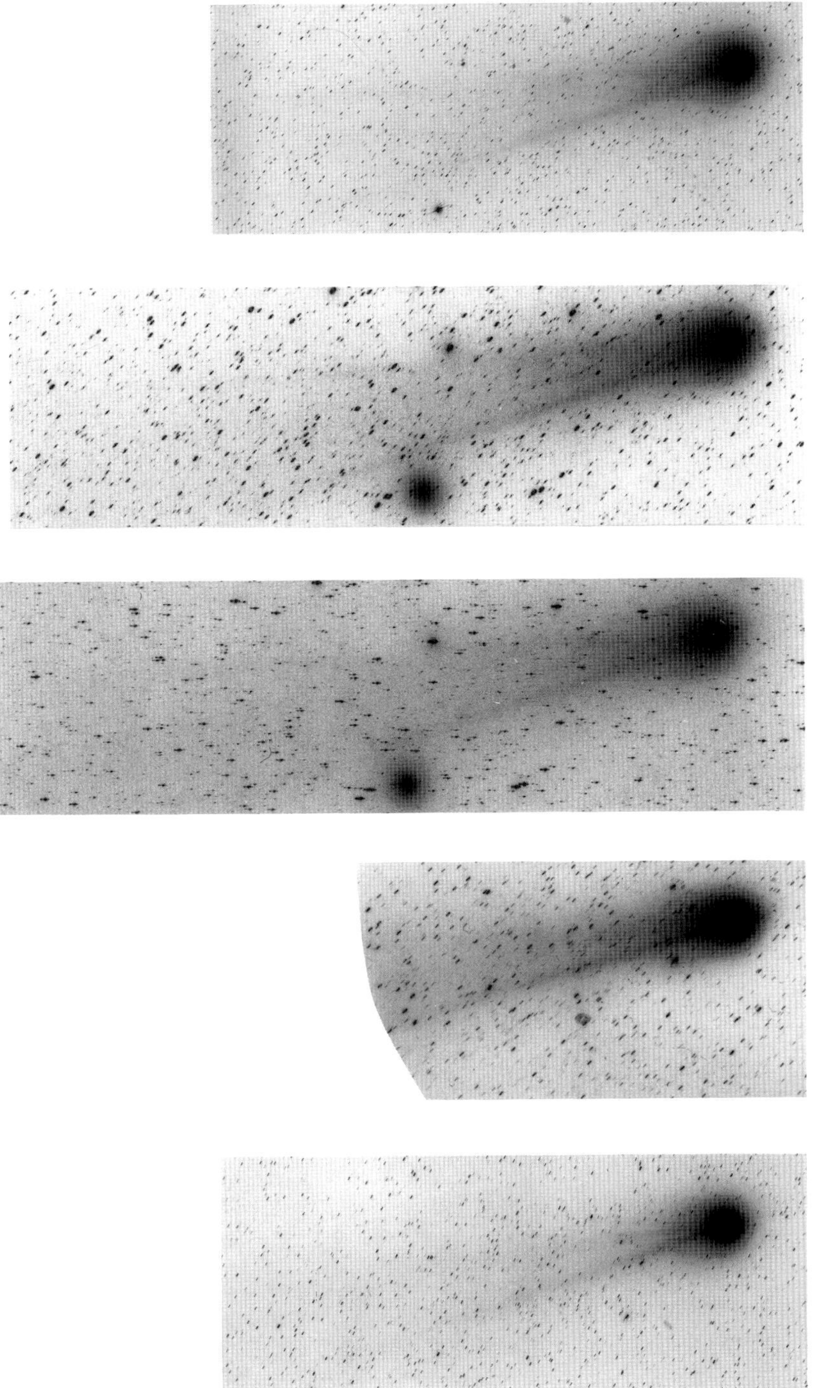

Fig. 898: turn-off sequence, (1° ~ 28 mm) (top to bottom) 1986 May 2.11944, LSPN-3022, *K. Meech, University of Michigan/CTIO*; 1986 May 2.17222, LSPN-941; 1986 May 2.18889, LSPN-942, *E. Moore, Joint Observatory for Cometary Research*; 1986 May 2.82118, LSPN-3510, *D. O'Donoghue, University of Perugia*; 1986 May 3.11018, LSPN-3028, *K.Meech, University of Michigan/CTIO.*

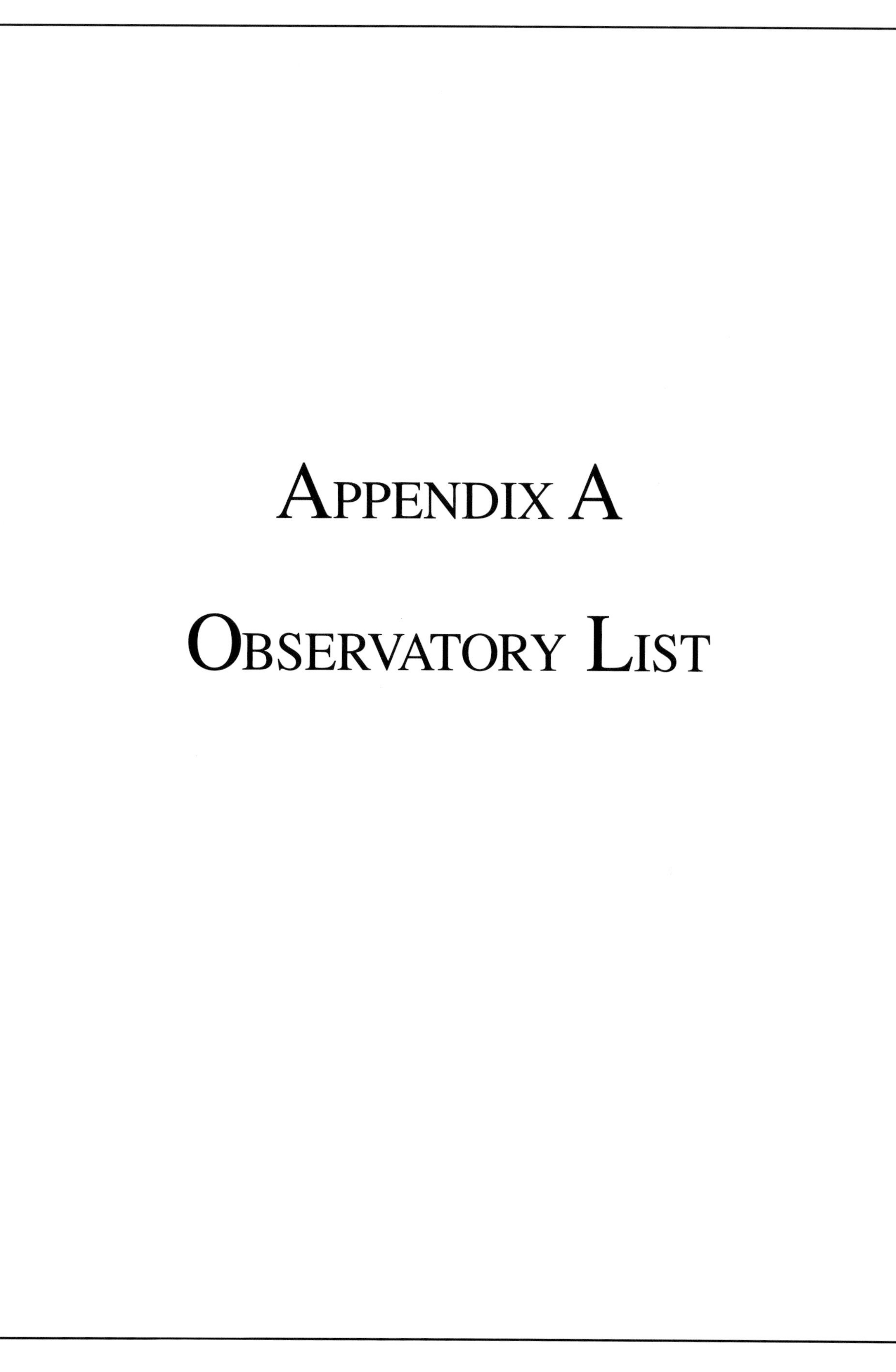

Appendix A

Observatory List

Appendix A

The observing sites for images used in this Atlas are listed in the following table; the equipment used is also listed, along with the longitude and latitude of each site. Figure A shows these sites on a world map.

The observing sites are assigned numbers and these numbers are used in the Ephemeris. The Kuiper Airborne Observatory (No. 83) is listed by the location of the airport used.

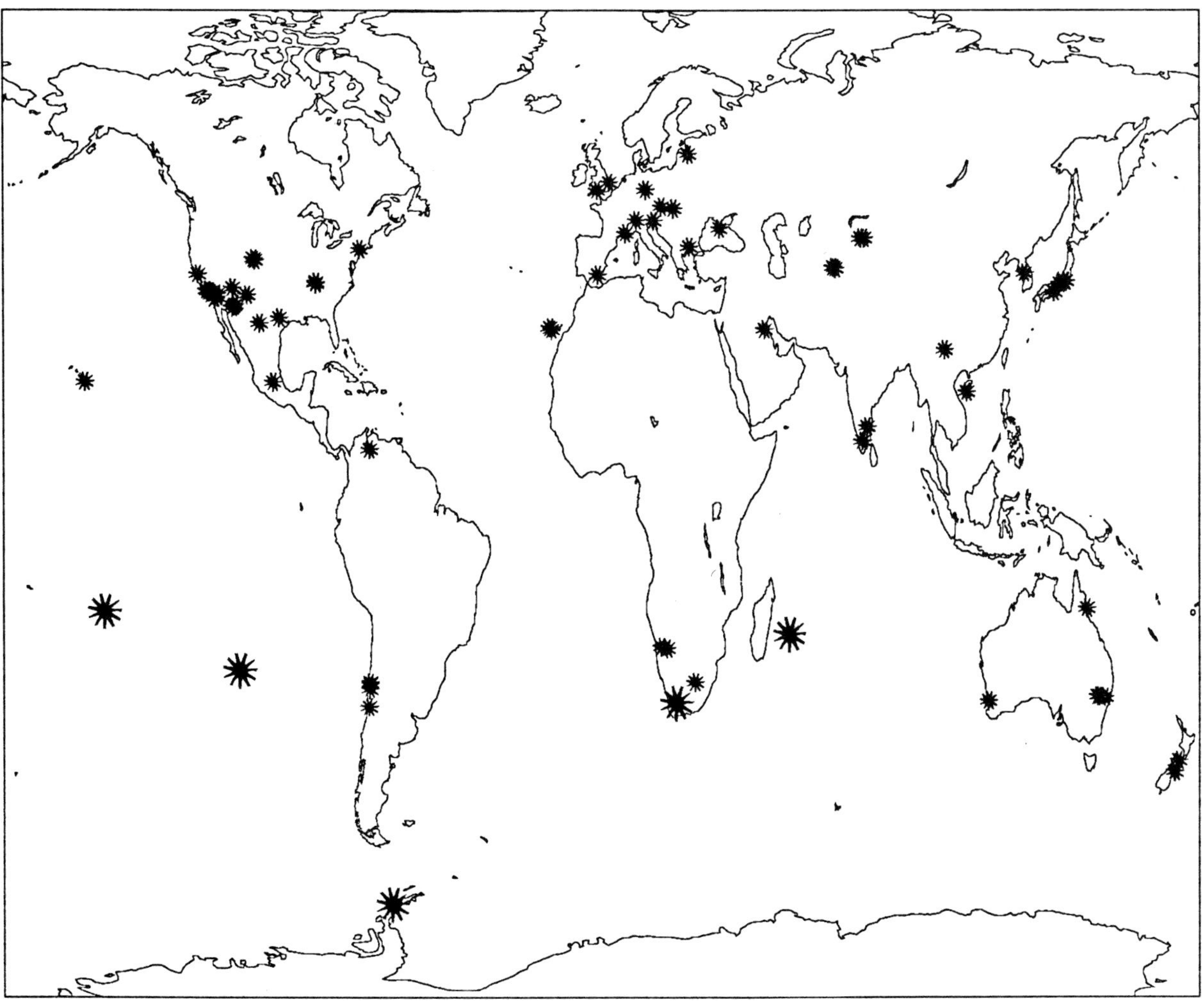

Fig. A. The observing sites for the images used in this Atlas. The sites for the LSPN Island Network are marked by the larger asterisms.

Observatory List

#	Observatory	Instrument	Country	Long.	Lat
1	Mauna Kea Observatory	Schmidt Camera	USA	204.5	19.8
2	LSPN Island Network, Papeete	Celestron Schmidt	Tahiti, France	210.4	-17.5
3	Lick Observatory	Crossley Reflector	USA	238.4	37.3
4	S. J. Edberg (USA) Site 4	f/2 Camera	USA	240.9	34.7
5	B. Yen (USA) Site 2	Celestron Schmidt	USA	240.9	34.8
6	S. J. Edberg (USA) Site 1	f/5.2 Camera	USA	241.2	34.7
7	H. Lazerson (USA)	f/4 Camera	USA	241.5	34.5
8	S. J. Edberg (USA) Site 2	f/5.2 Camera	USA	242.0	34.5
9	Palomar Observatory	18 in. Schmidt	USA	243.1	33.4
10	Palomar Observatory	5.1 meter	USA	243.1	33.4
11	Palomar Observatory	f/2 Schmidt Camera	USA	243.1	33.4
12	Palomar Observatory	48 in. Schmidt	USA	243.1	33.4
13	B. Yen (USA) Site 1	Celestron Schmidt	USA	244.0	33.7
14	S. J. Edberg (USA) Site 3	f/5.2 Camera	USA	244.0	34.1
15	Lowell Observatory	Lowell Astrograph	USA	248.3	35.2
16	Lowell Observatory	Cogshall Camera	USA	248.3	35.2
17	Warner and Swasey Observatory	Burrell Schmidt	USA	248.4	32.0
18	D. Levy (USA)	20 in. reflector	USA	249.0	32.2
19	Mt. Lemmon Observatory	6 inch reflector	USA	249.2	32.4
20	Catalina Observatory	Catalina Reflector	USA	249.3	32.4
21	LSPN Island Network, Easter Island	Celestron Schmidt	Chile	250.6	-27.2
22	Joint Observatory for Cometary Research	Schmidt Camera	USA	252.8	34.0
23	E. E. Barnard Observatory	Schmidt Camera	USA	254.6	39.9
24	R. B. Minton (USA)	f/8 Camera	USA	255.1	39.6
25	McDonald Observatory	Patrol Telescope	USA	256.0	30.7
26	E. E. Barnard Observatory Station	Astro Camera	USA	256.5	29.4
27	E. E. Barnard Observatory Station	Astrograph	USA	256.5	29.4
28	A. Levy (USA)	f/3.5 Camera	Mexico	260.5	19.8
29	A. J. Dyer Observatory	Schmidt Camera	USA	273.2	36.1
30	G. Polus (USA)	f/1.7 Camera	USA	273.2	35.8
31	R. Dilsizian (USA)	Celestron Schmidt	USA	286.2	41.4
32	Cerro el Roble Astronomical Sta.	Maksutov Astrograph	Chile	289.0	-33.0
33	University of Michigan/CTIO	Curtis Schmidt	Chile	289.2	-30.2
34	I. Ferrin (Venezuela)	f/2.8 Camera	Venezuela	289.2	8.9
35	European Southern Observatory	Minolta XD7 Camera	Chile	289.3	-29.3
36	European Southern Observatory	Schmidt Camera	Chile	289.3	-29.3
37	Mount Wilson/Las Campanas Observatories	Du Pont Telescope	Chile	289.3	-29.0
38	British Antarctic Survey	Celestron Schmidt	Antarctica	296.0	-65.0
39	M. Jaeger (Austria)	Schmidt Camera	Tenerife	343.0	28.7
40	Teide Observatory	300mm Zeiss lens	Canary Islands	343.5	28.3
41	H. B. Ridley (UK)	f/5 Camera	UK	357.3	50.9
42	Calar Alto Sta., German Nat. Ast. Obs.	Schmidt Camera	Spain	357.5	37.2
43	M. Mobberley (UK)	f/5 Camera	UK	0.8	52.2
44	Observatory of Haute-Provence	Schmidt Camera	France	5.7	43.9
45	V. Belli (Italy)	f/4 Camera	Italy	9.0	46.0
46	Karl Schwarzschild Observatory	Schmidt Camera	Germany	11.7	51.0
47	H. Mikuz (Yugoslavia)	f/4 Camera	Yugoslavia	14.1	45.9
48	R. Conrad (Austria)	f/4.7 Camera	Austria	16.3	48.2

Observatory List

#	Observatory	Instrument	Country	Long.	Lat
49	Volkssternwarte Frankfurt	f/6.0 Camera	Namibia	16.4	-23.3
50	B. Koch (Germany)	f/1.7 Camera	Namibia	18.0	-23.5
51	Konkoly Observatory, Piszkesteto Sta.	Schmidt Camera	Hungary	19.9	47.9
52	LSPN Island Network, Sutherland	Celestron Schmidt	South Africa	20.8	-32.4
53	University of Perugia/S. A. A. O.	Schmidt Camera	South Africa	20.8	-32.4
54	Riga Radio-Astrophysical Observatory	Schmidt Camera	Latvian S.S.R.	24.4	56.8
55	National Astronomical Obs.	Schmidt Camera	Bulgaria	24.7	41.7
56	Boyden Observatory	Ross-Fecker Patrol	South Africa	26.4	-29.0
57	Boyden Observatory	Metcalf Triplet	South Africa	26.4	-29.0
58	Crimean Astrophys. Obs.	Double Astrograph	U.S.S.R.	34.0	44.5
59	Wafra Observatory	Schmidt Camera	Kuwait	47.6	28.4
60	LSPN Island Network, Reunion Island	Celestron Schmidt	France	55.0	-21.3
61	Gissar Observatory	Zeiss Refractor	U.S.S.R.	68.6	38.5
62	Sanglok Observatory	1-M RCC	U.S.S.R.	69.3	38.3
63	Mountain Observatory	Maksutov Astrograph	U.S.S.R.	76.9	43.2
64	Assa Observatory	1-M Zeiss Refl.RCC	U.S.S.R.	77.0	43.2
65	Indian Inst. Astrophys. Kodaikanal Sta.	Long Focus Camera	India	77.5	10.2
66	Indian Inst. Astrophys. Kodaikanal Sta.	Tessar Lens	India	77.5	10.2
67	Indian Inst. Astrophys. Kavalur Sta.	Schmidt Camera	India	78.8	12.6
68	Yunnan Observatory	Tracking Camera	China	102.8	25.0
69	Shanghai Observatory Mobile Station	Astrograph	China	109.5	18.2
70	Perth Observatory	Astrograph	Australia	116.1	-32.0
71	Yonsei University Observatory	f/1.7 Camera	South Korea	127.0	37.5
72	Univ. Kyoto Observatory	Schmidt Camera	Japan	136.0	34.5
73	Kiso Observatory	Schmidt Camera	Japan	137.6	35.8
74	T. Niijima (Japan)	f/5.8 Camera	Japan	139.3	36.2
75	T. Kojima (Japan)	f/3.4 Camera	Japan	139.5	36.2
76	R. Royer (Australia)	f/6 Camera	Australia	145.8	-17.0
77	Uppsala Southern Station	Schmidt Camera	Australia	149.1	-31.3
78	A. Ward (Australia)	f/5.6 Camera	Australia	149.1	-31.3
79	Royal Observatory/UK Schmidt Telescope	UK Schmidt	Australia	149.1	-31.3
80	McNaught (UK)	f/1.8 Camera	Australia	149.7	-31.3
81	G. Garradd (Australia)	f/2.8 Camera	Australia	151.3	-31.4
82	G. Robertson (New Zealand)	f/2.8 Camera	New Zealand	172.3	-43.5
83	Kuiper Airborne Observatory	Camera Lens	New Zealand	172.7	-43.5
84	U.S. Naval Observatory Station	Twin Astrograph	New Zealand	173.8	-41.7

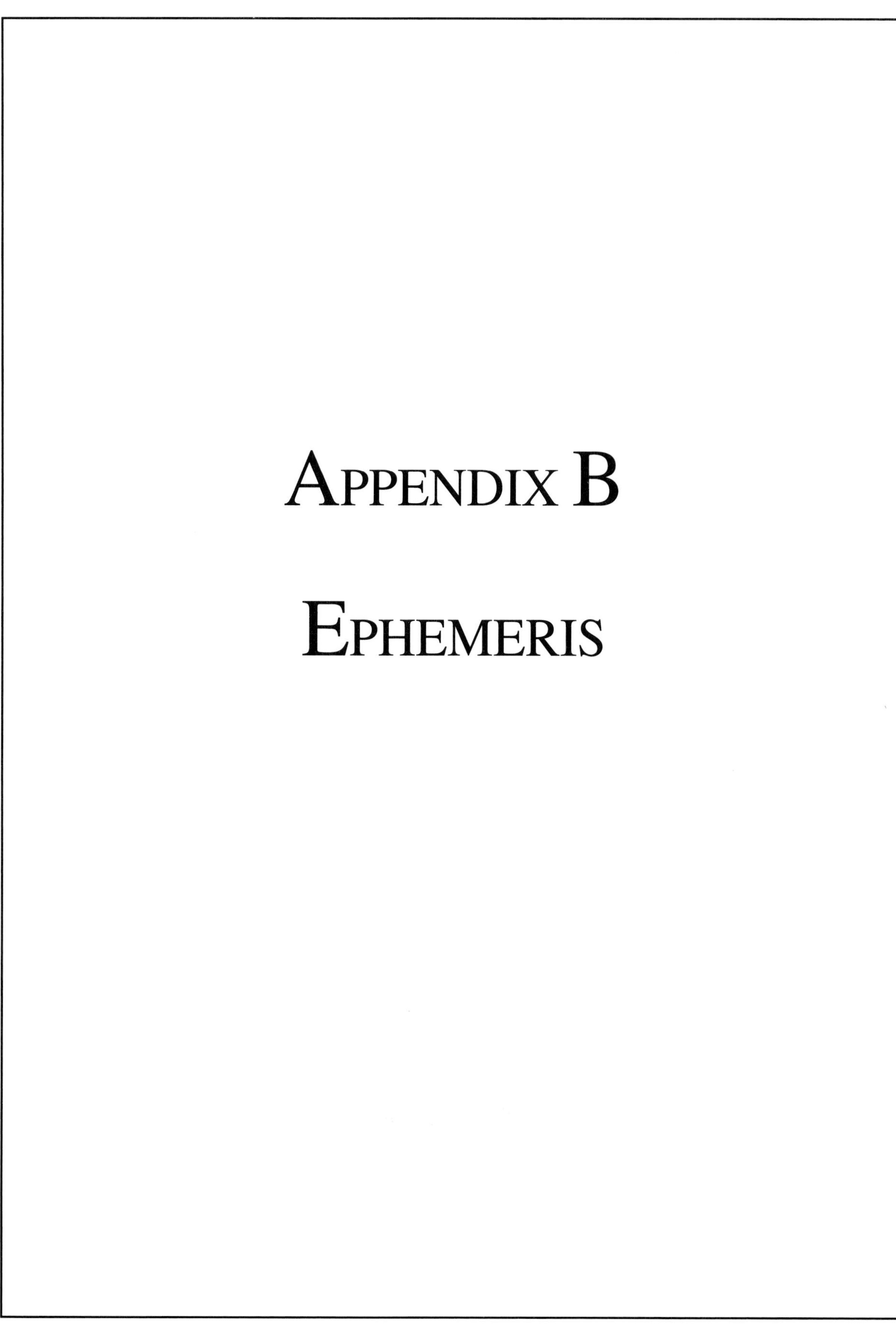

Appendix B

Ephemeris

Ephemeris

As an aid to interpretation of the images, we give the approximate wavelength response for emulsions (Table A) and emulsion/filter combinations (Table B).

TABLE A

Emulsion	Color	Greatest Sensitivity
098-04	Orange	650-680
103a-D	Yellow	580-630
103a-F	Orange	650-680
103a-G	Green	500-580
103a-O	Blue	→ 500
153-01	Panchromatic	
Fuji 400	Panchromatic	
IIIa-F	Orange	650-680
IIIa-J	Blue	450-530
IIa-D	Yellow	580-630
IIa-F	Orange	650-680
IIa-O	Blue	→ 500
Orwo ZU-21	Blue	→ 500
Tech. Pan 2415	Panchromatic	
Tri-X	Panchromatic	

TABLE B

Emulsion	Filter	Filter Wavelength Range	Combination
098-04	GG-495	500-700	Panchromatic
098-04	EF-610	610-700	yellow ↔ red
098-04	RG-630	635-700	orange-red
Tech. Pan 2415	Wratten 25	600-750	yellow ↔ red
Tech. Pan 2415	Wratten 2B	500-750	green ↔ red
103a-D	GG-385	385-500	blue-violet
103a-D	GG-495	495-650	blue ↔ yellow
103a-F	Y-50	500-650	yellow
153-01	RG-630	630→	orange-red
153-01	RG-610	610→	yellow ↔ red
IIa-D	GG-495	495-650	violet ↔ yellow
IIa-O	GG-495	495-500	violet-blue
IIa-O	GG-395	395-650	violet-blue
IIa-O	GG-385	385-500	violet-blue
IIIa-F	RG-630	630-700	orange
IIIa-F	AAO-643	643-700	orange
IIIa-F	Wratten 25	600-700	yellow-orange
IIIa-F	Wratten 2B	450-700	blue ↔ orange
IIIa-F	RG-610	610-700	yellow-orange
IIIa-J	UG-1	300-420	violet-blue
IIIa-J	GG-395	395-550	violet ↔ green
IIIa-J	GG-385	385-500	violet ↔ green
IIa-D	GG-495	495-650	violet ↔ yellow
IIa-O	GG-385	385-500	violet-blue
IIa-O	GG-13	385-500	violet-blue
IIa-O	GG-395	395-500	violet-blue
Tri-X	Wratten 47	400-500	violet-blue
Tri-X	Wratten 25	600-750	yellow ↔ red
Orwo ZU-21	BS-8	385-500	violet-blue

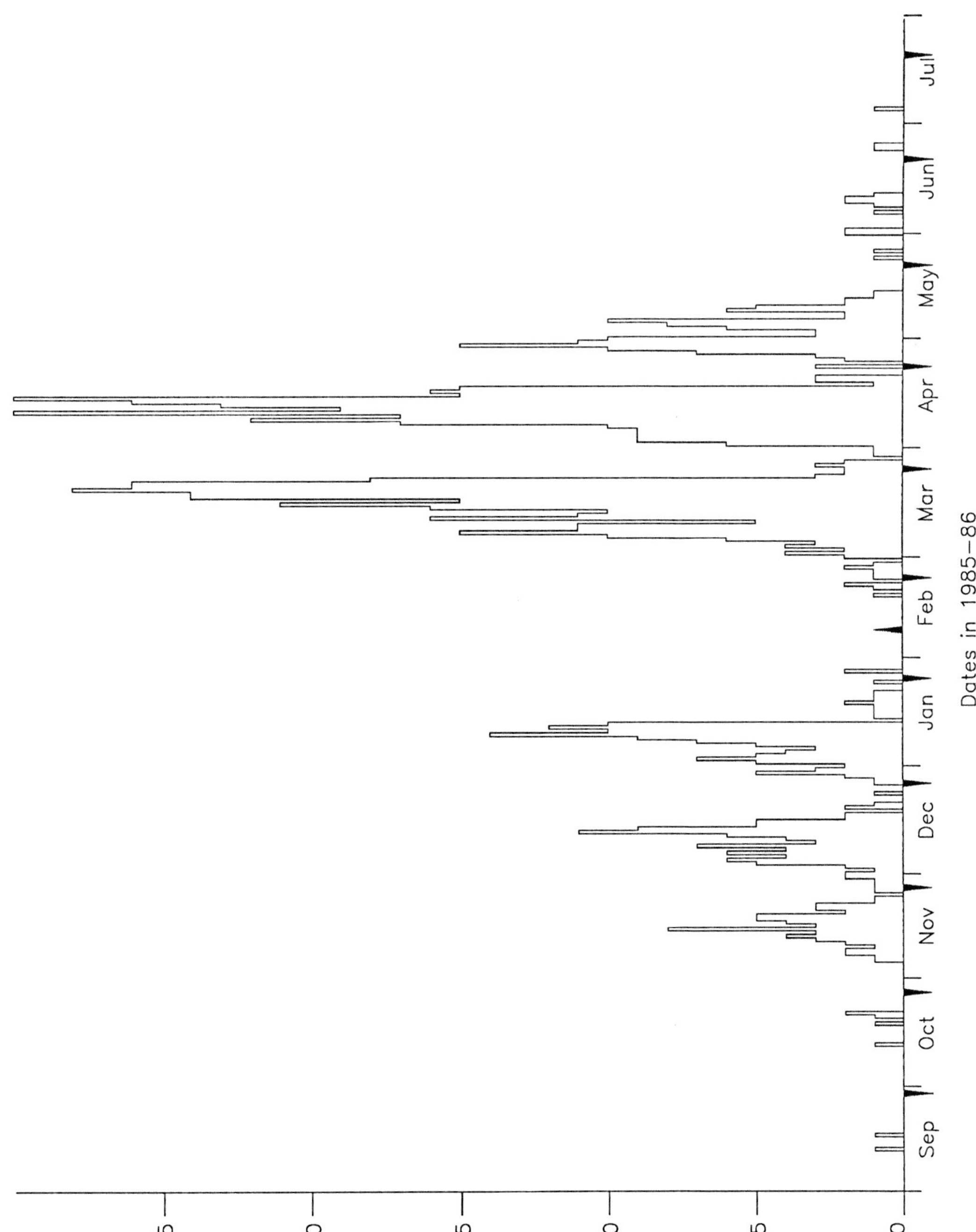

Figure B — Histogram showing the distribution of dates for all images in the Atlas except the recovery image of Oct. 16, 1982 and the outburst image on Feb. 12, 1991. The ordinate unit is images per day. The solid spike extending upward from the abscissa marks the date of perihelion passage and solid spikes extending downward from the abscissa mark the dates of full moon.

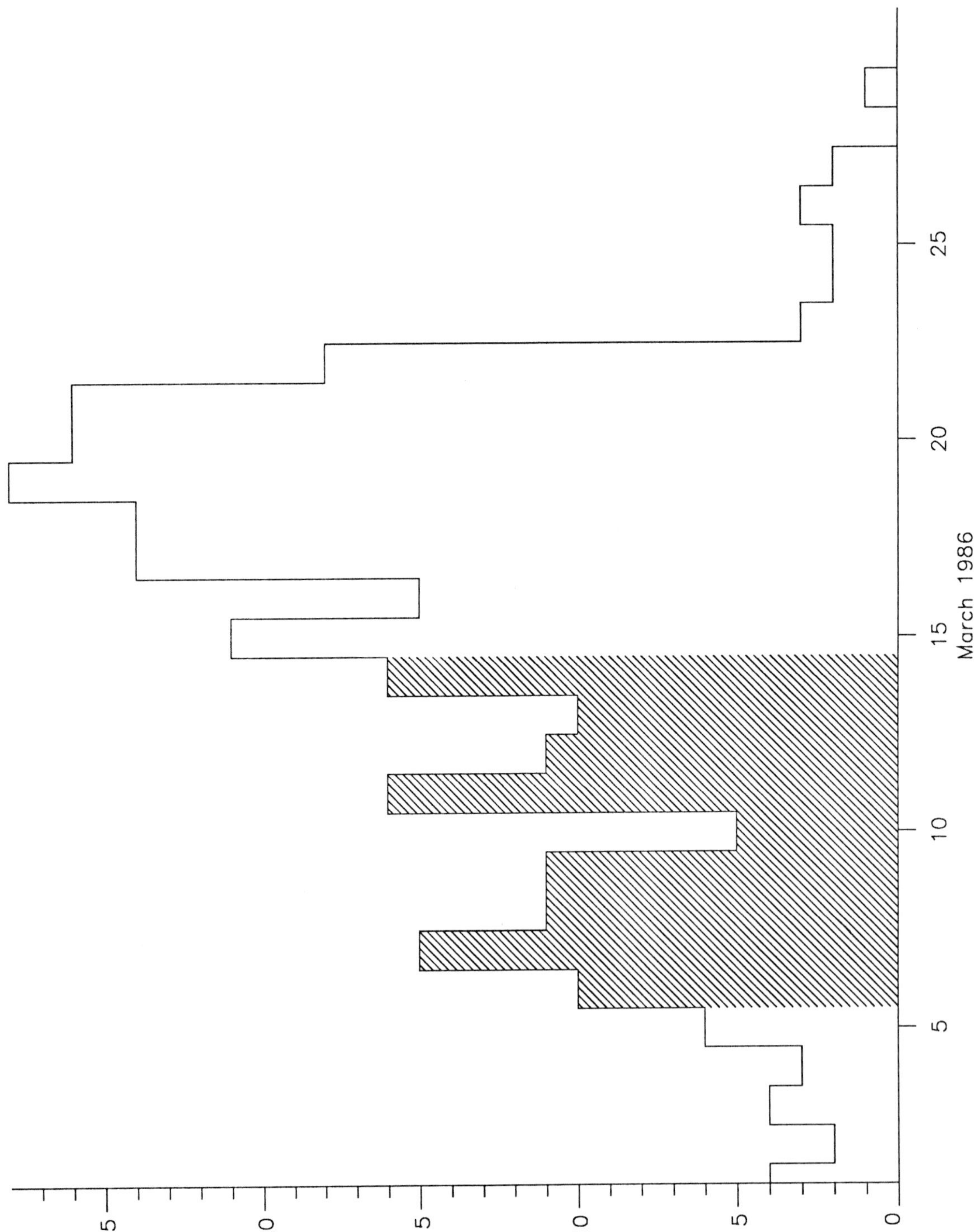

Figure C — Histogram showing the Atlas coverage for Armada Week (shaded area) defined as the interval March 6 through March 14, 1986. The ordinate unit is images per day.

Ephemeris

The orbital circumstances of the Earth and Halley's comet in 1985 and 1986 are shown in Figure D. The quantities r, Δ, and β are illustrated in Figure E. The quantities in the ephemeris by column, left-to-right are: the year, month, and decimal date; the reference number of the observation (LSPN, NNSN, AON); the observatory number (see Observatory List); the exposure in hours; the right ascension (α) and the declination (δ) in degrees, the geocentric distance of the comet in AU, Δ; the heliocentric distance of the comet in AU, r; the foreshortening angle, β, see Figure E; the position angle on the sky, measured from north through east, of the prolonged radius vector ϕ; and, the position angle on the sky, measured north through east, of the comet's negative velocity vector, φ.

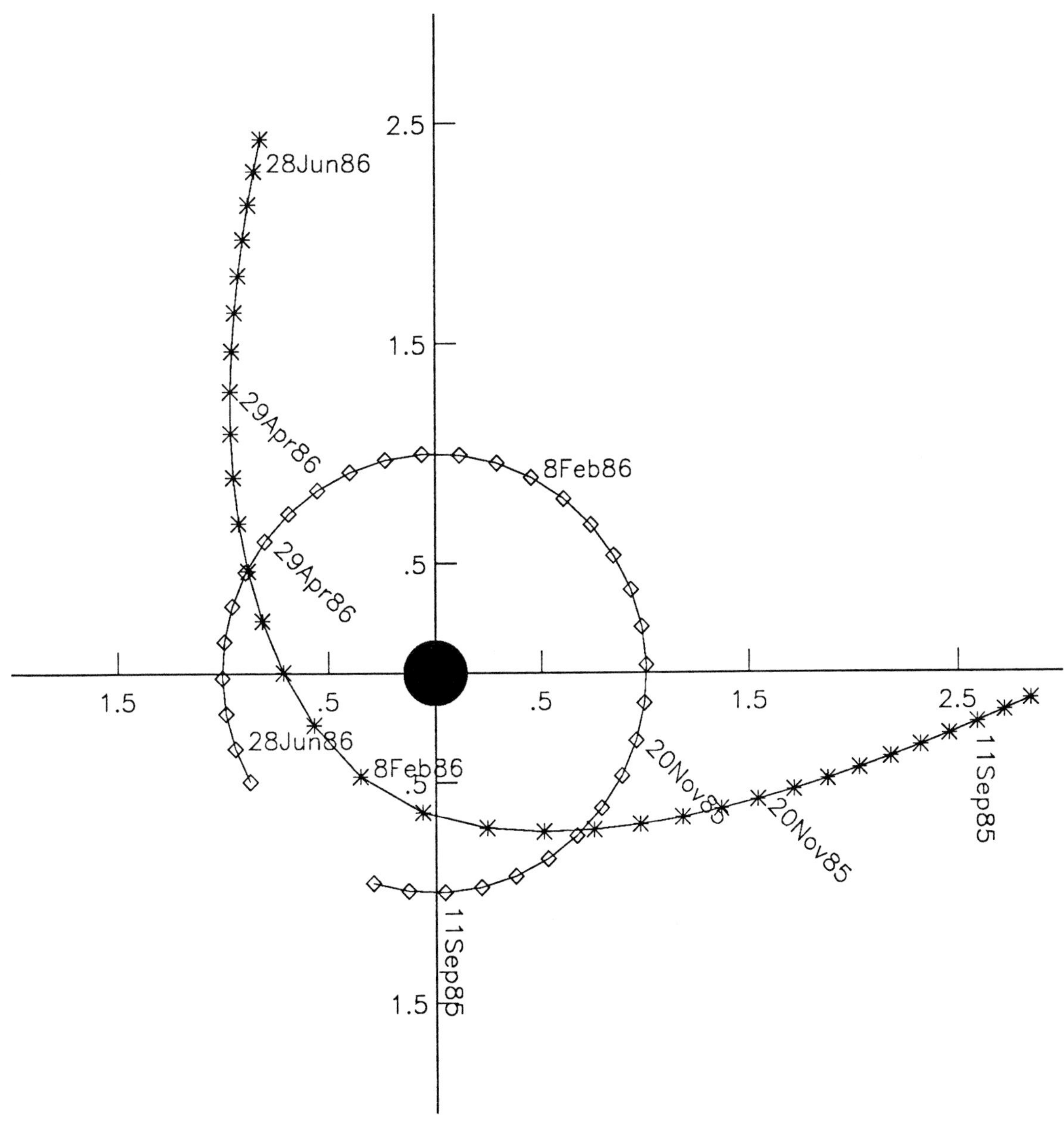

Figure D — The orbits of Earth and comet Halley as projected on the plane of the ecliptic. Tick marks are at 10-day intervals and the axes have units of AU.

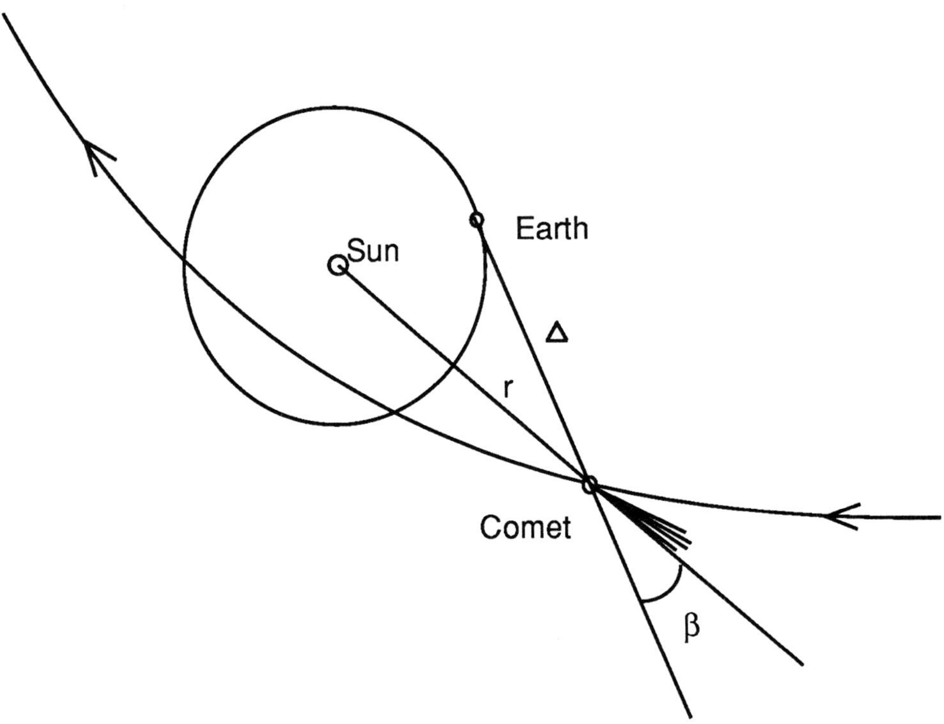

Figure E — Schematic illustrating the quantities r, Δ, and β.

Ephemeris

The following list gives ephemeris data for all images in the Atlas except the Pioneer Venus Orbiter image in the Lyman-α (Fig. 818) and the images obtained from the Solar Maximum Mission. The grand distribution of Atlas images is shown in Figure B. and the distribution for the time of the Halley Armada is show in Figure C.

Year	Mon	Day	Ref #	Obs #	Exp	α	δ	Δ	r	β	ϕ	φ
1982	10	16.49306		10	0.133	107.8	9.6	10.93	11.04	174.8	278.0	113.4
1985	9	13.42986	LSPN-43	22	0.167	93.1	19.6	2.60	2.57	157.6	270.3	144.2
1985	9	17.46736	LSPN-768	3	0.500	93.2	19.6	2.47	2.52	156.8	270.7	146.4
1985	10	13.50000	LSPN-771	3	0.500	90.7	20.5	1.64	2.18	154.3	271.3	147.7
1985	10	19.26040	LSPN-720	32	0.167	88.8	20.9	1.46	2.10	154.8	270.8	142.4
1985	10	21.12153	LSPN-3455	46	0.467	88.0	21.0	1.40	2.07	155.2	270.6	140.1
1985	10	22.02434	LSPN-3431	51	0.500	87.5	21.1	1.37	2.06	155.4	270.4	139.0
1985	10	22.04418	LSPN-3432	51	0.250	87.5	21.1	1.37	2.06	155.4	270.4	138.9
1985	11	6.29514	LSPN-777	3	0.500	74.6	22.1	0.93	1.85	163.3	264.8	114.4
1985	11	7.28125	LSPN-778	3	0.500	73.2	22.2	0.91	1.83	164.2	264.1	112.8
1985	11	8.45000		19	0.633	72.2	22.2	0.89	1.82	165.0	263.4	111.6
1985	11	8.95625	LSPN-2455	55	0.333	70.7	22.2	0.86	1.81	166.0	262.5	110.0
1985	11	9.04579	LSPN-1963	54	0.333	70.6	22.2	0.86	1.81	166.1	262.5	109.8
1985	11	9.14604	AON-850104	31	0.050	70.4	22.2	0.86	1.80	166.2	262.4	109.6
1985	11	10.57361	LSPN-1323	72	0.333	68.1	22.2	0.83	1.78	167.9	260.7	107.3
1985	11	11.01736	AON-850133	43	0.306	67.3	22.2	0.82	1.78	168.5	260.1	106.5
1985	11	11.91042	LSPN-2310	55	0.333	65.7	22.2	0.80	1.76	169.7	258.7	105.1
1985	11	12.06493	LSPN-3458	46	1.017	65.4	22.2	0.79	1.76	169.9	258.5	104.8
1985	11	12.91100	AON-850154	41	0.500	63.8	22.1	0.78	1.75	171.1	256.8	103.5
1985	11	12.98703	AON-850155	43	0.767	63.6	22.1	0.78	1.75	171.2	256.7	103.3
1985	11	13.00069	LSPN-3828	42	0.233	63.6	22.1	0.78	1.75	171.3	256.6	103.3
1985	11	13.04028	LSPN-3829	42	0.833	63.5	22.1	0.77	1.75	171.3	256.6	103.2
1985	11	13.06806	LSPN-2312	55	0.333	63.4	22.1	0.77	1.75	171.4	256.5	103.2
1985	11	13.66944	LSPN-1328	72	0.500	62.2	22.1	0.76	1.74	172.3	255.0	102.2
1985	11	14.08958	LSPN-2313	55	0.400	61.4	22.0	0.75	1.73	172.9	253.7	101.6
1985	11	14.22500	LSPN-779	3	0.500	61.1	22.0	0.75	1.73	173.1	253.3	101.3
1985	11	14.94410	LSPN-2840	57	0.350	59.5	21.9	0.74	1.72	174.3	250.4	100.2
1985	11	15.02708	LSPN-2314	55	0.333	59.4	21.9	0.74	1.72	174.4	250.0	100.1
1985	11	15.09306	LSPN-721	32	0.167	59.2	21.9	0.74	1.72	174.5	249.7	99.9
1985	11	15.28090	LSPN-2576	17	0.250	58.8	21.9	0.73	1.72	174.8	248.6	99.7
1985	11	15.89375	LSPN-2459	55	0.333	57.4	21.8	0.72	1.71	175.8	244.3	98.7
1985	11	15.93715	LSPN-2460	55	0.218	57.3	21.8	0.72	1.71	175.9	243.9	98.6
1985	11	15.94094	LSPN-1965	54	0.333	57.3	21.8	0.72	1.71	175.9	243.8	98.6
1985	11	15.95863	LSPN-1966	54	0.283	57.3	21.8	0.72	1.71	176.0	243.7	98.6
1985	11	15.96597	LSPN-2461	55	0.333	57.2	21.8	0.72	1.71	176.0	243.6	98.6
1985	11	16.06597	LSPN-722	32	0.167	57.0	21.8	0.72	1.70	176.1	242.6	98.4
1985	11	16.18958	LSPN-1686	22	0.500	56.7	21.7	0.72	1.70	176.3	241.3	98.2
1985	11	16.33403	LSPN-2578	17	0.333	56.4	21.7	0.71	1.70	176.6	239.6	98.0
1985	11	17.04097	LSPN-3460	46	0.833	54.7	21.6	0.70	1.69	177.7	225.8	96.9
1985	11	17.19201	LSPN-1688	22	0.500	54.4	21.5	0.70	1.69	177.9	220.9	96.7
1985	11	17.20174	LSPN-143	33	0.250	54.3	21.5	0.70	1.69	178.0	220.6	96.6
1985	11	17.91007	LSPN-2841	57	0.750	52.6	21.3	0.69	1.68	178.6	178.1	95.5

Ephemeris

Year	Mon	Day	Ref #	Obs #	Exp	α	δ	Δ	r	β	ϕ	φ
1985	11	18.09694	LSPN-3834	42	0.233	52.1	21.3	0.69	1.67	178.6	162.5	95.3
1985	11	18.09931	LSPN-3833	42	0.233	52.1	21.3	0.69	1.67	178.6	162.3	95.3
1985	11	18.22951	LSPN-146	33	0.250	51.8	21.3	0.68	1.67	178.6	152.0	95.1
1985	11	18.55208		71	0.167	51.0	21.2	0.68	1.67	178.2	131.4	94.6
1985	11	18.56597		71	0.250	50.9	21.2	0.68	1.67	178.2	130.7	94.5
1985	11	19.08403	LSPN-3838	42	0.233	49.6	21.0	0.67	1.66	177.4	112.1	93.8
1985	11	19.25694	LSPN-148	33	1.000	49.2	21.0	0.67	1.66	177.1	108.2	93.5
1985	11	19.29618	LSPN-149	33	0.250	49.1	20.9	0.67	1.66	177.0	107.5	93.4
1985	11	19.62470	LSPN-1600	73	0.167	48.2	20.8	0.67	1.65	176.4	102.2	93.0
1985	11	19.66042	LSPN-1331	72	0.167	48.1	20.8	0.67	1.65	176.3	101.7	92.9
1985	11	20.26285	LSPN-150	33	0.250	46.5	20.6	0.66	1.64	175.1	95.6	92.0
1985	11	20.30104	LSPN-151	33	0.750	46.4	20.6	0.66	1.64	175.0	95.3	92.0
1985	11	21.12569	LSPN-3179	44	0.667	44.1	20.2	0.65	1.63	173.3	90.3	90.8
1985	11	21.24549	LSPN-153	33	1.017	43.8	20.2	0.65	1.63	173.0	89.7	90.6
1985	11	21.28924	LSPN-154	33	0.250	43.7	20.2	0.65	1.63	172.9	89.5	90.5
1985	11	22.26805	LSPN-156	33	0.833	41.0	19.7	0.64	1.61	170.7	86.0	89.2
1985	11	22.28090	LSPN-783	3	0.333	40.9	19.7	0.64	1.61	170.7	86.0	89.1
1985	11	22.30729	LSPN-157	33	0.250	40.8	19.7	0.64	1.61	170.6	85.9	89.1
1985	11	23.31319	LSPN-159	33	0.167	38.0	19.2	0.63	1.60	168.3	83.3	87.7
1985	11	24.15000	LSPN-160	33	0.500	35.6	18.7	0.63	1.59	166.3	81.5	86.7
1985	11	26.12292	LSPN-162	33	0.167	29.9	17.4	0.62	1.56	161.6	78.3	84.3
1985	11	27.10486	LSPN-163	33	0.167	27.1	16.7	0.62	1.54	159.3	77.0	83.2
1985	11	28.10347	LSPN-164	33	0.333	24.2	15.9	0.62	1.53	156.9	75.8	82.1
1985	11	29.03125	LSPN-165	33	0.167	21.6	15.2	0.62	1.51	154.7	74.8	81.2
1985	11	30.06007	LSPN-168	33	0.250	18.8	14.4	0.63	1.50	152.3	73.8	80.3
1985	11	30.38681	LSPN-1333	72	0.167	18.0	14.1	0.63	1.49	151.6	73.5	80.0
1985	12	1.08194	LSPN-170	33	0.333	16.1	13.6	0.63	1.48	150.1	72.9	79.5
1985	12	1.26037	LSPN-1862	1	0.417	15.7	13.4	0.63	1.48	149.7	72.7	79.4
1985	12	2.72153	LSPN-2465	55	0.333	11.9	12.2	0.64	1.46	146.6	71.6	78.3
1985	12	3.07778	LSPN-175	33	0.333	11.1	11.9	0.65	1.45	145.8	71.4	78.1
1985	12	3.49847	LSPN-1606	73	0.333	10.1	11.6	0.65	1.45	145.0	71.1	77.9
1985	12	4.05417	LSPN-178	33	0.333	8.7	11.1	0.66	1.44	143.9	70.7	77.5
1985	12	4.45830	LSPN-2371	79	0.500	7.8	10.8	0.66	1.43	143.1	70.5	77.3
1985	12	4.71084	LSPN-2468	55	0.333	7.2	10.6	0.66	1.43	142.7	70.3	77.2
1985	12	4.75830	LSPN-2469	55	0.333	7.1	10.6	0.66	1.43	142.6	70.3	77.2
1985	12	4.79163	LSPN-2470	55	0.333	7.0	10.5	0.66	1.43	142.5	70.3	77.1
1985	12	5.06042	LSPN-181	33	0.333	6.4	10.3	0.67	1.42	142.0	70.1	77.0
1985	12	5.15139	LSPN-1675	22	0.500	6.2	10.2	0.67	1.42	141.8	70.1	77.0
1985	12	5.75830	LSPN-2472	55	0.367	4.9	9.8	0.67	1.41	140.8	69.8	76.7
1985	12	5.77236	LSPN-3435	58	0.750	4.9	9.7	0.68	1.41	140.7	69.7	76.7
1985	12	5.80483	LSPN-2473	55	0.333	4.8	9.7	0.68	1.41	140.7	69.7	76.7
1985	12	5.92149	LSPN-2476	55	0.333	4.5	9.6	0.68	1.41	140.5	69.7	76.6
1985	12	6.72897	LSPN-3436	58	1.000	2.8	9.0	0.69	1.40	139.1	69.3	76.3
1985	12	6.73052	LSPN-2477	55	0.333	2.8	9.0	0.69	1.40	139.1	69.3	76.3
1985	12	6.77705	LSPN-2478	55	0.333	2.7	9.0	0.69	1.40	139.0	69.3	76.2
1985	12	6.89163	LSPN-2481	55	0.333	2.5	8.9	0.69	1.39	138.8	69.2	76.2
1985	12	7.09028	LSPN-186	33	0.333	2.1	8.7	0.69	1.39	138.5	69.1	76.1
1985	12	7.13194	LSPN-187	33	0.333	2.0	8.7	0.69	1.39	138.5	69.1	76.1

Ephemeris

Year	Mon	Day	Ref #	Obs #	Exp	α	δ	Δ	r	β	ϕ	φ
1985	12	7.16667	LSPN-1673	22	0.167	1.9	8.7	0.69	1.39	138.4	69.1	76.1
1985	12	7.17361	LSPN-188	33	0.333	1.9	8.7	0.69	1.39	138.4	69.1	76.1
1985	12	7.17500	LSPN-1672	22	0.167	1.9	8.7	0.69	1.39	138.4	69.1	76.1
1985	12	7.18333	LSPN-122	23	0.217	1.9	8.7	0.69	1.39	138.4	69.1	76.1
1985	12	8.23021	LSPN-1671	22	0.250	359.8	7.9	0.71	1.37	136.8	68.6	75.7
1985	12	8.68607	LSPN-1970	54	0.500	359.0	7.6	0.71	1.37	136.1	68.4	75.6
1985	12	8.69087	LSPN-3437	58	1.000	359.0	7.6	0.71	1.37	136.1	68.4	75.6
1985	12	8.74579	LSPN-1971	54	0.333	358.9	7.5	0.72	1.37	136.0	68.4	75.6
1985	12	9.05556	LSPN-191	33	0.333	358.3	7.3	0.72	1.36	135.6	68.3	75.5
1985	12	9.13750	LSPN-193	33	0.333	358.2	7.2	0.72	1.36	135.5	68.3	75.5
1985	12	9.44444		79	0.300	357.6	7.0	0.73	1.35	135.0	68.2	75.4
1985	12	9.44845	LSPN-2372	79	0.417	357.6	7.0	0.73	1.35	135.0	68.2	75.4
1985	12	9.50940	LSPN-2374	79	0.417	357.5	7.0	0.73	1.35	135.0	68.1	75.3
1985	12	9.73537	LSPN-2319	55	0.333	357.1	6.8	0.73	1.35	134.7	68.1	75.3
1985	12	9.78190	LSPN-2320	55	0.333	357.0	6.8	0.73	1.35	134.6	68.1	75.3
1985	12	10.05625	LSPN-195	33	0.333	356.5	6.6	0.74	1.35	134.2	68.0	75.2
1985	12	10.39634	LSPN-1610	73	0.333	355.9	6.4	0.74	1.34	133.8	67.8	75.1
1985	12	10.68730	LSPN-3438	58	1.000	355.4	6.2	0.75	1.34	133.5	67.8	75.1
1985	12	11.07639	LSPN-201	33	0.333	354.8	5.9	0.75	1.33	133.0	67.6	75.0
1985	12	11.11736	LSPN-203	33	0.333	354.7	5.9	0.75	1.33	132.9	67.6	75.0
1985	12	11.13785	LSPN-1663	22	0.117	354.7	5.9	0.75	1.33	132.9	67.6	75.0
1985	12	11.21875	LSPN-1867	1	0.500	354.6	5.8	0.75	1.33	132.8	67.6	74.9
1985	12	12.07431	LSPN-723	32	0.167	353.2	5.3	0.77	1.31	131.9	67.3	74.8
1985	12	12.08056	LSPN-206	33	0.333	353.2	5.3	0.77	1.31	131.9	67.3	74.8
1985	12	12.17535	LSPN-1662	22	0.117	353.0	5.2	0.77	1.31	131.8	67.3	74.7
1985	12	12.18924	LSPN-1660	22	0.117	353.0	5.2	0.77	1.31	131.8	67.3	74.7
1985	12	12.47875	LSPN-1611	73	0.133	352.6	5.1	0.78	1.31	131.5	67.2	74.7
1985	12	12.50898	LSPN-1612	73	0.333	352.5	5.0	0.78	1.31	131.4	67.2	74.7
1985	12	13.05139	LSPN-210	33	0.333	351.7	4.7	0.79	1.30	130.9	67.1	74.6
1985	12	13.07014	LSPN-211	33	0.333	351.7	4.7	0.79	1.30	130.9	67.1	74.6
1985	12	13.13125	LSPN-214	33	0.333	351.6	4.7	0.79	1.30	130.8	67.1	74.6
1985	12	13.19236	LSPN-2579	17	0.333	351.5	4.6	0.79	1.30	130.7	67.1	74.6
1985	12	13.20972	LSPN-2580	17	0.333	351.5	4.6	0.79	1.30	130.7	67.0	74.6
1985	12	13.75139	LSPN-3186	44	0.167	350.7	4.3	0.80	1.29	130.2	66.9	74.5
1985	12	13.76528	LSPN-3187	44	0.333	350.7	4.3	0.80	1.29	130.2	66.9	74.5
1985	12	13.78351	AON-850459	39	0.075	350.6	4.3	0.80	1.29	130.2	66.9	74.5
1985	12	13.78403	LSPN-3839	42	0.233	350.6	4.3	0.80	1.29	130.2	66.9	74.5
1985	12	13.80000	LSPN-1769	52	0.167	350.6	4.3	0.80	1.29	130.2	66.9	74.5
1985	12	13.81146	LSPN-1770	52	0.250	350.6	4.3	0.80	1.29	130.2	66.9	74.5
1985	12	14.05069	LSPN-215	33	0.333	350.3	4.1	0.80	1.28	130.0	66.8	74.4
1985	12	14.06944	LSPN-216	33	0.333	350.2	4.1	0.80	1.28	129.9	66.8	74.4
1985	12	14.08819	LSPN-217	33	0.333	350.2	4.1	0.81	1.28	129.9	66.8	74.4
1985	12	14.09931	LSPN-2582	17	0.333	350.2	4.1	0.81	1.28	129.9	66.8	74.4
1985	12	14.10833	LSPN-218	33	0.333	350.2	4.1	0.81	1.28	129.9	66.8	74.4
1985	12	14.12986	LSPN-219	33	0.333	350.2	4.1	0.81	1.28	129.9	66.8	74.4
1985	12	14.13715	LSPN-1723	22	0.150	350.1	4.1	0.81	1.28	129.9	66.8	74.4
1985	12	14.81000	AON-850472	39	0.083	349.2	3.7	0.82	1.27	129.3	66.7	74.3
1985	12	14.91285	LSPN-3843	42	0.333	349.1	3.7	0.82	1.27	129.2	66.6	74.3

Ephemeris

Year	Mon	Day	Ref #	Obs #	Exp	α	δ	Δ	r	β	ϕ	φ
1985	12	15.05104	LSPN-220	33	0.250	348.9	3.6	0.82	1.27	129.1	66.6	74.3
1985	12	15.08715	LSPN-222	33	0.250	348.9	3.6	0.82	1.27	129.1	66.6	74.3
1985	12	15.67307	LSPN-1972	54	0.367	348.1	3.3	0.83	1.26	128.7	66.5	74.2
1985	12	15.69784	LSPN-2489	55	0.333	348.1	3.2	0.83	1.26	128.6	66.5	74.2
1985	12	15.89096	LSPN-3846	42	0.233	347.8	3.1	0.84	1.26	128.5	66.4	74.2
1985	12	16.05382	LSPN-225	33	0.250	347.6	3.1	0.84	1.25	128.4	66.4	74.2
1985	12	16.12257	LSPN-229	33	0.250	347.5	3.0	0.84	1.25	128.3	66.4	74.2
1985	12	16.16736	LSPN-1718	22	0.167	347.5	3.0	0.84	1.25	128.3	66.4	74.2
1985	12	16.80868	LSPN-1776	52	0.083	346.7	2.7	0.86	1.24	127.9	66.3	74.1
1985	12	16.93332	LSPN-3848	42	0.233	346.5	2.6	0.86	1.24	127.8	66.2	74.1
1985	12	17.08958	LSPN-231	33	0.667	346.3	2.5	0.86	1.24	127.7	66.2	74.1
1985	12	17.57957	LSPN-3397	64	0.050	345.8	2.3	0.87	1.23	127.4	66.1	74.0
1985	12	18.05556	LSPN-232	33	0.167	345.2	2.1	0.88	1.22	127.2	66.0	74.0
1985	12	18.08889	LSPN-233	33	0.667	345.2	2.1	0.88	1.22	127.2	66.0	74.0
1985	12	20.05868	LSPN-240	33	0.417	343.1	1.2	0.92	1.19	126.3	65.6	73.8
1985	12	20.07465	LSPN-237	33	0.417	343.1	1.2	0.92	1.19	126.3	65.6	73.8
1985	12	21.07187	LSPN-239	33	0.417	342.1	0.8	0.94	1.18	126.0	65.5	73.8
1985	12	24.05903	LSPN-243	33	0.267	339.4	-0.3	1.00	1.13	125.4	65.0	73.6
1985	12	27.68100	AON-850539	48	0.017	336.6	-1.4	1.07	1.07	125.5	64.3	73.4
1985	12	28.89210	AON-850541	31	0.038	335.7	-1.8	1.10	1.05	125.7	64.1	73.4
1985	12	29.05021	AON-850545	24	0.850	335.6	-1.8	1.10	1.05	125.7	64.1	73.4
1985	12	29.97400	AON-850552	31	0.033	335.0	-2.1	1.12	1.04	125.9	63.9	73.3
1985	12	30.08438	LSPN-126	23	0.050	335.0	-2.1	1.12	1.04	126.0	63.9	73.3
1985	12	30.09444	LSPN-844	22	0.100	334.9	-2.1	1.12	1.04	126.0	63.9	73.3
1985	12	30.09931	LSPN-845	22	0.100	334.9	-2.1	1.12	1.04	126.0	63.9	73.3
1985	12	30.58681	LSPN-3287	61	0.667	334.6	-2.2	1.13	1.03	126.1	63.8	73.3
1985	12	30.80208	LSPN-3195	44	0.500	334.5	-2.3	1.14	1.03	126.2	63.7	73.3
1985	12	31.38922	LSPN-1620	73	0.167	334.1	-2.4	1.15	1.02	126.4	63.6	73.2
1985	12	31.40644	LSPN-1621	73	0.167	334.1	-2.4	1.15	1.02	126.4	63.6	73.2
1985	12	31.53924	LSPN-3063	70	0.550	334.0	-2.4	1.15	1.01	126.4	63.6	73.2
1986	1	1.07153	LSPN-127	23	0.100	333.7	-2.6	1.16	1.01	126.6	63.5	73.2
1986	1	1.58681	LSPN-3269	61	0.833	333.4	-2.7	1.17	1.00	126.8	63.4	73.2
1986	1	2.11458	LSPN-850	22	0.167	333.1	-2.8	1.18	0.99	127.0	63.2	73.1
1986	1	2.58966	LSPN-3266	61	0.833	332.8	-2.9	1.19	0.98	127.2	63.1	73.1
1986	1	2.60694	LSPN-3615	65	0.867	332.8	-2.9	1.19	0.98	127.2	63.1	73.1
1986	1	2.71615	LSPN-3462	46	0.117	332.7	-3.0	1.19	0.98	127.3	63.1	73.1
1986	1	2.80764	LSPN-1779	52	0.167	332.7	-3.0	1.19	0.98	127.3	63.1	73.1
1986	1	3.05694	LSPN-128	23	0.100	332.5	-3.0	1.20	0.98	127.4	63.0	73.1
1986	1	3.23055	LSPN-1872	1	0.500	332.4	-3.1	1.20	0.97	127.5	63.0	73.1
1986	1	3.26493	LSPN-1873	1	0.500	332.4	-3.1	1.20	0.97	127.5	63.0	73.1
1986	1	3.62015	LSPN-3288	61	0.250	332.2	-3.2	1.21	0.97	127.7	62.9	73.0
1986	1	3.71528	LSPN-3463	46	0.167	332.2	-3.2	1.21	0.97	127.8	62.9	73.0
1986	1	3.71790	AON-850595	39	0.000	332.2	-3.2	1.21	0.97	127.8	62.9	73.0
1986	1	3.75069	LSPN-2491	55	0.500	332.2	-3.2	1.21	0.96	127.8	62.9	73.0
1986	1	4.09375	LSPN-855	22	0.167	332.0	-3.3	1.22	0.96	128.0	62.8	73.0
1986	1	4.11250	LSPN-2584	17	0.333	332.0	-3.3	1.22	0.96	128.0	62.8	73.0
1986	1	4.21215	LSPN-1875	1	0.250	331.9	-3.3	1.22	0.96	128.0	62.7	73.0
1986	1	4.23351	LSPN-1874	1	0.500	331.9	-3.3	1.22	0.96	128.0	62.7	73.0

Ephemeris

Year	Mon	Day	Ref #	Obs #	Exp	α	δ	Δ	r	β	ϕ	φ
1986	1	4.26112	LSPN-1876	1	0.500	331.9	-3.3	1.22	0.96	128.1	62.7	73.0
1986	1	5.09306	LSPN-863	22	0.167	331.4	-3.5	1.24	0.94	128.5	62.5	72.9
1986	1	5.09583	LSPN-129	23	0.167	331.4	-3.5	1.24	0.94	128.5	62.5	72.9
1986	1	5.23594	LSPN-1877	1	0.500	331.4	-3.5	1.24	0.94	128.6	62.5	72.9
1986	1	5.26146	LSPN-1878	1	0.500	331.3	-3.5	1.24	0.94	128.6	62.5	72.9
1986	1	6.00907	LSPN-137	29	0.167	330.9	-3.7	1.25	0.93	129.1	62.3	72.9
1986	1	6.25417	LSPN-1879	1	0.500	330.8	-3.7	1.26	0.93	129.3	62.2	72.9
1986	1	6.58683	LSPN-3290	61	0.667	330.6	-3.8	1.26	0.92	129.5	62.1	72.8
1986	1	7.08715	LSPN-3570	15	0.383	330.4	-3.9	1.27	0.91	129.8	62.0	72.8
1986	1	7.22899	LSPN-1881	1	0.500	330.3	-3.9	1.28	0.91	129.9	61.9	72.8
1986	1	7.38760	LSPN-1624	73	0.167	330.2	-4.0	1.28	0.91	130.0	61.9	72.8
1986	1	7.40412	LSPN-1625	73	0.333	330.2	-4.0	1.28	0.91	130.0	61.9	72.8
1986	1	7.55127	LSPN-3369	63	0.167	330.2	-4.0	1.28	0.91	130.1	61.8	72.8
1986	1	8.09020	NNSN-405502	20	0.083	329.9	-4.1	1.29	0.90	130.5	61.7	72.7
1986	1	8.10764	LSPN-3571	15	0.500	329.9	-4.1	1.29	0.90	130.5	61.7	72.7
1986	1	8.11204	NNSN-405505	20	0.083	329.9	-4.1	1.29	0.90	130.5	61.7	72.7
1986	1	8.23023	LSPN-1883	1	0.500	329.8	-4.1	1.29	0.90	130.6	61.6	72.7
1986	1	8.26235	LSPN-1882	1	0.500	329.8	-4.1	1.29	0.90	130.7	61.6	72.7
1986	1	8.57965	LSPN-3372	63	0.147	329.6	-4.2	1.30	0.89	130.9	61.5	72.7
1986	1	8.58333	LSPN-3292	61	0.333	329.6	-4.2	1.30	0.89	130.9	61.5	72.7
1986	1	9.04896	LSPN-130	23	0.117	329.4	-4.3	1.31	0.89	131.3	61.4	72.6
1986	1	9.06250	LSPN-131	23	0.117	329.4	-4.3	1.31	0.89	131.3	61.4	72.6
1986	1	9.08194	LSPN-3585	16	0.333	329.4	-4.3	1.31	0.89	131.3	61.4	72.6
1986	1	9.08194	LSPN-3572	15	0.333	329.4	-4.3	1.31	0.89	131.3	61.4	72.6
1986	1	9.10625	LSPN-2088	22	0.167	329.4	-4.3	1.31	0.88	131.3	61.4	72.6
1986	1	9.11458	LSPN-2089	22	0.167	329.4	-4.3	1.31	0.88	131.3	61.4	72.6
1986	1	9.22708	LSPN-1884	1	0.500	329.3	-4.3	1.31	0.88	131.4	61.3	72.6
1986	1	9.25208	LSPN-1885	1	0.500	329.3	-4.3	1.31	0.88	131.4	61.3	72.6
1986	1	9.66928	LSPN-3439	58	0.467	329.1	-4.4	1.32	0.88	131.8	61.2	72.6
1986	1	10.01919	LSPN-140	29	0.167	328.9	-4.5	1.32	0.87	132.1	61.0	72.6
1986	1	10.08056	LSPN-2091	22	0.167	328.9	-4.5	1.33	0.87	132.1	61.0	72.6
1986	1	10.08889	LSPN-2092	22	0.167	328.9	-4.5	1.33	0.87	132.1	61.0	72.6
1986	1	10.09028	LSPN-3569	15	0.500	328.9	-4.5	1.33	0.87	132.1	61.0	72.6
1986	1	10.09028	LSPN-3583	16	0.500	328.9	-4.5	1.33	0.87	132.1	61.0	72.6
1986	1	10.09722	LSPN-2093	22	0.167	328.9	-4.5	1.33	0.87	132.1	61.0	72.6
1986	1	10.10556	LSPN-2094	22	0.167	328.9	-4.5	1.33	0.87	132.1	61.0	72.6
1986	1	10.11389	LSPN-2095	22	0.167	328.9	-4.5	1.33	0.87	132.1	61.0	72.6
1986	1	10.12080	AON-850728	6	0.333	328.9	-4.5	1.33	0.87	132.1	61.0	72.6
1986	1	10.38750	AON-850739	75	0.167	328.8	-4.5	1.33	0.87	132.4	60.9	72.5
1986	1	10.39063	LSPN-1341	72	0.083	328.8	-4.5	1.33	0.87	132.4	60.9	72.5
1986	1	10.40104	LSPN-1342	72	0.083	328.8	-4.6	1.33	0.87	132.4	60.9	72.5
1986	1	10.58718	LSPN-3293	61	0.667	328.7	-4.6	1.33	0.86	132.6	60.8	72.5
1986	1	10.73900	AON-850749	39	0.000	328.6	-4.6	1.34	0.86	132.7	60.8	72.5
1986	1	11.01535	LSPN-141	29	0.200	328.5	-4.7	1.34	0.86	132.9	60.7	72.5
1986	1	11.07917	LSPN-2096	22	0.167	328.4	-4.7	1.34	0.86	133.0	60.7	72.5
1986	1	11.08889	LSPN-2097	22	0.167	328.4	-4.7	1.34	0.86	133.0	60.7	72.5
1986	1	11.09722	LSPN-2098	22	0.167	328.4	-4.7	1.34	0.86	133.0	60.7	72.5
1986	1	11.10069	LSPN-3573	15	0.500	328.4	-4.7	1.34	0.86	133.0	60.7	72.5

Ephemeris

Year	Mon	Day	Ref #	Obs #	Exp	α	δ	Δ	r	β	ϕ	φ
1986	1	11.10069	LSPN-3584	16	0.500	328.4	-4.7	1.34	0.86	133.0	60.7	72.5
1986	1	11.10556	LSPN-2099	22	0.167	328.4	-4.7	1.34	0.86	133.0	60.7	72.5
1986	1	11.24236	LSPN-1889	1	0.500	328.4	-4.7	1.34	0.85	133.1	60.6	72.5
1986	1	11.73229	LSPN-2492	55	0.167	328.1	-4.8	1.35	0.85	133.6	60.4	72.4
1986	1	11.75694	LSPN-3196	44	0.400	328.1	-4.8	1.35	0.85	133.6	60.4	72.4
1986	1	12.01422	LSPN-142	29	0.167	328.0	-4.9	1.36	0.84	133.9	60.3	72.4
1986	1	12.07292	LSPN-2100	22	0.167	328.0	-4.9	1.36	0.84	133.9	60.3	72.4
1986	1	12.07986	LSPN-3574	15	0.133	328.0	-4.9	1.36	0.84	133.9	60.3	72.4
1986	1	12.08472	LSPN-2101	22	0.167	328.0	-4.9	1.36	0.84	133.9	60.3	72.4
1986	1	12.09306	LSPN-2102	22	0.167	328.0	-4.9	1.36	0.84	133.9	60.3	72.4
1986	1	12.09861	LSPN-3575	15	0.467	328.0	-4.9	1.36	0.84	133.9	60.3	72.4
1986	1	12.09861	LSPN-3586	16	0.467	328.0	-4.9	1.36	0.84	133.9	60.3	72.4
1986	1	12.10139	LSPN-2103	22	0.167	328.0	-4.9	1.36	0.84	133.9	60.3	72.4
1986	1	12.11111	LSPN-2104	22	0.167	328.0	-4.9	1.36	0.84	133.9	60.3	72.4
1986	1	12.24236	LSPN-1892	1	0.500	327.9	-4.9	1.36	0.84	134.1	60.2	72.4
1986	1	12.76111	LSPN-3197	44	0.667	327.7	-5.0	1.37	0.83	134.6	60.0	72.3
1986	1	12.78438	LSPN-3198	44	0.250	327.7	-5.0	1.37	0.83	134.6	60.0	72.3
1986	1	13.07222	LSPN-868	22	0.167	327.5	-5.0	1.37	0.83	134.9	59.9	72.3
1986	1	13.08194	LSPN-2105	22	0.167	327.5	-5.0	1.37	0.83	134.9	59.9	72.3
1986	1	13.08866	LSPN-3576	15	0.250	327.5	-5.1	1.37	0.83	134.9	59.8	72.3
1986	1	13.08958	LSPN-869	22	0.167	327.5	-5.1	1.37	0.83	134.9	59.8	72.3
1986	1	13.09792	LSPN-870	22	0.167	327.5	-5.1	1.37	0.83	134.9	59.8	72.3
1986	1	13.10764	LSPN-871	22	0.167	327.5	-5.1	1.37	0.83	134.9	59.8	72.3
1986	1	13.24844	LSPN-1894	1	0.450	327.4	-5.1	1.38	0.83	135.1	59.8	72.3
1986	1	13.30937	LSPN-1893	1	0.500	327.4	-5.1	1.38	0.82	135.1	59.8	72.3
1986	1	13.75833	LSPN-3199	44	0.333	327.2	-5.2	1.38	0.82	135.6	59.6	72.2
1986	1	13.77326	LSPN-3200	44	0.250	327.2	-5.2	1.38	0.82	135.6	59.5	72.2
1986	1	15.75880	AON-850871	41	0.133	326.3	-5.5	1.41	0.79	137.8	58.6	72.0
1986	1	16.75000	AON-850844	45	0.067	325.9	-5.7	1.43	0.78	139.0	58.0	71.9
1986	1	17.72900	AON-850886	47	0.050	325.4	-5.9	1.44	0.76	140.2	57.4	71.8
1986	1	18.97104	AON-850891	7	0.050	324.9	-6.1	1.46	0.75	141.9	56.6	71.6
1986	1	19.05270	AON-850893	28	0.000	324.9	-6.1	1.46	0.75	142.0	56.5	71.6
1986	1	19.81807	LSPN-3349	62	0.167	324.5	-6.3	1.47	0.74	143.0	55.9	71.5
1986	1	20.08551	NNSN-405521	20	0.017	324.4	-6.3	1.47	0.73	143.4	55.7	71.5
1986	1	21.05347	AON-850909	24	0.167	324.0	-6.5	1.48	0.72	144.8	54.9	71.3
1986	1	22.37708	AON-850910	74	0.017	323.4	-6.7	1.50	0.71	146.8	53.6	71.2
1986	1	25.74480	AON-850911	74	0.003	321.9	-7.3	1.53	0.67	152.3	49.1	70.7
1986	1	28.08450		18	0.017	320.9	-7.8	1.54	0.65	156.5	44.4	70.3
1986	1	28.70800	AON-850912	39	0.003	320.6	-7.9	1.55	0.64	157.6	42.8	70.2
1986	2	19.41667	AON-850913	34	0.033	310.7	-13.3	1.44	0.63	145.8	267.5	65.4
1986	2	21.76700	AON-850914	81	0.050	309.7	-14.0	1.40	0.65	141.1	265.0	64.8
1986	2	22.12569	LSPN-1781	52	0.033	309.6	-14.1	1.40	0.65	140.4	264.6	64.7
1986	2	22.77933	LSPN-2376	79	0.033	309.3	-14.3	1.39	0.66	139.1	264.1	64.5
1986	2	24.55934	AON-850923	76	0.013	308.5	-14.8	1.36	0.67	135.9	262.7	64.1
1986	2	25.55934	AON-850925	76	0.017	308.1	-15.2	1.34	0.68	134.2	262.1	63.9
1986	2	26.13194	LSPN-1783	52	0.033	307.8	-15.3	1.33	0.69	133.2	261.7	63.8
1986	2	27.12188	LSPN-1786	52	0.083	307.4	-15.7	1.31	0.70	131.6	261.2	63.5
1986	2	27.78472	LSPN-2377	79	0.050	307.1	-15.9	1.29	0.71	130.6	260.8	63.4

Ephemeris

Year	Mon	Day	Ref #	Obs #	Exp	α	δ	Δ	r	β	ϕ	φ
1986	2	28.12569	LSPN-1789	52	0.033	307.0	-16.0	1.29	0.71	130.0	260.7	63.3
1986	3	1.12500	LSPN-1790	52	0.033	306.6	-16.4	1.26	0.73	128.6	260.2	63.1
1986	3	1.85278	LSPN-3066	70	0.167	306.2	-16.6	1.25	0.73	127.5	259.9	62.9
1986	3	2.00000	LSPN-1357	21	0.033	306.2	-16.7	1.25	0.74	127.3	259.9	62.9
1986	3	2.47847	LSPN-1354	21	0.100	306.0	-16.9	1.24	0.74	126.7	259.7	62.8
1986	3	2.49010	LSPN-1356	21	0.042	306.0	-16.9	1.24	0.74	126.6	259.7	62.8
1986	3	2.50148	LSPN-1358	21	0.004	306.0	-16.9	1.23	0.74	126.6	259.7	62.8
1986	3	3.47951	LSPN-1363	21	0.033	305.5	-17.2	1.21	0.75	125.3	259.3	62.6
1986	3	3.64514	LSPN-1898	1	0.333	305.4	-17.3	1.21	0.76	125.1	259.3	62.5
1986	3	4.10694	LSPN-1794	52	0.167	305.2	-17.5	1.20	0.76	124.5	259.1	62.4
1986	3	4.11354	LSPN-1795	52	0.083	305.2	-17.5	1.20	0.76	124.5	259.1	62.4
1986	3	4.69502	LSPN-380	84	0.167	305.0	-17.7	1.19	0.77	123.8	259.0	62.3
1986	3	4.77419	LSPN-2378	79	0.033	304.9	-17.7	1.18	0.77	123.7	258.9	62.3
1986	3	5.37674	LSPN-724	32	0.083	304.7	-18.0	1.17	0.78	123.0	258.8	62.2
1986	3	5.62677	LSPN-1900	1	0.333	304.5	-18.1	1.17	0.78	122.7	258.7	62.1
1986	3	5.64190	LSPN-1901	1	0.167	304.5	-18.1	1.16	0.78	122.7	258.7	62.1
1986	3	6.52529	NNSN-405525	20	0.013	304.1	-18.4	1.14	0.80	121.7	258.5	61.9
1986	3	6.52703	NNSN-405526	20	0.010	304.1	-18.4	1.14	0.80	121.7	258.5	61.9
1986	3	6.62266	LSPN-1903	1	0.333	304.1	-18.5	1.14	0.80	121.6	258.4	61.9
1986	3	6.63985	LSPN-1902	1	0.333	304.1	-18.5	1.14	0.80	121.6	258.4	61.9
1986	3	6.74623	LSPN-2381	79	0.117	304.0	-18.5	1.14	0.80	121.5	258.4	61.9
1986	3	6.97396	LSPN-3619	65	0.750	303.9	-18.6	1.13	0.80	121.3	258.4	61.9
1986	3	7.36319	LSPN-1266	33	0.167	303.7	-18.8	1.12	0.81	120.9	258.3	61.8
1986	3	7.38125	LSPN-1267	33	0.167	303.7	-18.8	1.12	0.81	120.8	258.3	61.8
1986	3	7.39410	LSPN-1268	33	0.050	303.7	-18.8	1.12	0.81	120.8	258.3	61.8
1986	3	7.45833		30	0.006	303.7	-18.8	1.12	0.81	120.8	258.3	61.8
1986	3	7.49062	LSPN-1375	21	0.050	303.7	-18.9	1.12	0.81	120.7	258.2	61.8
1986	3	7.50000	LSPN-1376	21	0.133	303.7	-18.9	1.12	0.81	120.7	258.2	61.8
1986	3	7.52500	LSPN-3577	15	0.233	303.6	-18.9	1.12	0.81	120.7	258.2	61.8
1986	3	7.62425	LSPN-1905	1	0.333	303.6	-18.9	1.12	0.81	120.6	258.2	61.7
1986	3	7.82486	LSPN-1627	73	0.083	303.5	-19.0	1.11	0.81	120.4	258.2	61.7
1986	3	7.97639	LSPN-3621	65	1.000	303.4	-19.1	1.11	0.82	120.3	258.2	61.7
1986	3	8.35486	LSPN-1269	33	0.167	303.2	-19.2	1.10	0.82	119.9	258.1	61.6
1986	3	8.37292	LSPN-1270	33	0.167	303.2	-19.2	1.10	0.82	119.9	258.1	61.6
1986	3	8.38507	LSPN-1271	33	0.050	303.2	-19.2	1.10	0.82	119.9	258.1	61.6
1986	3	8.45313	LSPN-1377	21	0.283	303.2	-19.3	1.10	0.82	119.8	258.1	61.6
1986	3	8.47326	LSPN-1725	21	0.067	303.2	-19.3	1.10	0.82	119.8	258.1	61.6
1986	3	8.48958	LSPN-1379	21	0.533	303.2	-19.3	1.10	0.82	119.8	258.1	61.6
1986	3	8.50567	LSPN-1380	21	0.050	303.2	-19.3	1.10	0.82	119.7	258.0	61.6
1986	3	8.52902	NNSN-405535	20	0.017	303.2	-19.3	1.10	0.82	119.7	258.0	61.6
1986	3	8.53328	NNSN-405537	20	0.045	303.1	-19.3	1.10	0.82	119.7	258.0	61.6
1986	3	8.61840	LSPN-1907	1	0.333	303.1	-19.3	1.09	0.82	119.6	258.0	61.6
1986	3	8.63759	LSPN-1906	1	0.363	303.1	-19.4	1.09	0.82	119.6	258.0	61.6
1986	3	8.68264	LSPN-382	84	0.167	303.1	-19.4	1.09	0.83	119.6	258.0	61.6
1986	3	8.82484	LSPN-1628	73	0.117	303.0	-19.4	1.09	0.83	119.4	258.0	61.5
1986	3	8.90532	LSPN-2729	69	0.417	303.0	-19.5	1.09	0.83	119.4	258.0	61.5
1986	3	8.92654	LSPN-2608	68	0.217	302.9	-19.5	1.09	0.83	119.4	258.0	61.5
1986	3	9.11944	LSPN-1800	52	0.200	302.9	-19.6	1.08	0.83	119.2	257.9	61.5

Year	Mon	Day	Ref #	Obs #	Exp	α	δ	Δ	r	β	ϕ	φ
1986	3	9.14236	LSPN-1802	52	0.167	302.8	-19.6	1.08	0.83	119.2	257.9	61.5
1986	3	9.34444	LSPN-1272	33	0.167	302.7	-19.7	1.08	0.83	119.0	257.9	61.4
1986	3	9.35660	LSPN-1273	33	0.050	302.7	-19.7	1.08	0.83	119.0	257.9	61.4
1986	3	9.37083	LSPN-1274	33	0.167	302.7	-19.7	1.08	0.84	119.0	257.9	61.4
1986	3	9.47570	LSPN-3414	27	0.250	302.7	-19.7	1.07	0.84	118.9	257.9	61.4
1986	3	9.47570	LSPN-3418	26	0.250	302.7	-19.7	1.07	0.84	118.9	257.9	61.4
1986	3	9.70098	LSPN-2115	77	0.200	302.6	-19.8	1.07	0.84	118.7	257.9	61.4
1986	3	9.75708	LSPN-2116	77	0.200	302.5	-19.9	1.07	0.84	118.6	257.8	61.4
1986	3	9.77716	LSPN-2382	79	0.133	302.5	-19.9	1.07	0.84	118.6	257.8	61.4
1986	3	9.90397	LSPN-2610	68	0.333	302.4	-19.9	1.06	0.84	118.5	257.8	61.4
1986	3	10.34896	LSPN-1275	33	0.250	302.2	-20.1	1.05	0.85	118.1	257.8	61.3
1986	3	10.37604	LSPN-1277	33	0.250	302.2	-20.2	1.05	0.85	118.1	257.8	61.3
1986	3	10.42500	LSPN-1383	21	0.333	302.2	-20.2	1.05	0.85	118.1	257.8	61.3
1986	3	10.43958	LSPN-1384	21	0.167	302.2	-20.2	1.05	0.85	118.1	257.7	61.3
1986	3	10.45532	LSPN-1726	21	0.067	302.2	-20.2	1.05	0.85	118.1	257.7	61.3
1986	3	10.47570	LSPN-3415	27	0.417	302.1	-20.2	1.05	0.85	118.0	257.7	61.3
1986	3	10.62014	LSPN-1908	1	0.333	302.1	-20.3	1.05	0.85	117.9	257.7	61.2
1986	3	10.73808	LSPN-2383	79	0.250	302.0	-20.3	1.04	0.85	117.8	257.7	61.2
1986	3	10.75781	LSPN-2384	79	0.167	302.0	-20.3	1.04	0.85	117.8	257.7	61.2
1986	3	10.77582	LSPN-2385	79	0.167	302.0	-20.4	1.04	0.86	117.8	257.7	61.2
1986	3	10.91308	LSPN-2612	68	0.667	301.9	-20.4	1.04	0.86	117.7	257.7	61.2
1986	3	11.45660	LSPN-1386	21	0.250	301.6	-20.7	1.03	0.87	117.3	257.6	61.1
1986	3	11.48229	LSPN-1727	21	0.100	301.6	-20.7	1.03	0.87	117.3	257.6	61.1
1986	3	11.67118	LSPN-1909	1	0.333	301.5	-20.8	1.02	0.87	117.1	257.6	61.1
1986	3	11.78590	LSPN-2117	77	0.083	301.4	-20.9	1.02	0.87	117.0	257.6	61.1
1986	3	11.82273	LSPN-1629	73	0.167	301.4	-20.9	1.02	0.87	117.0	257.6	61.1
1986	3	12.04510		59	0.083	301.3	-21.0	1.01	0.87	116.9	257.6	61.0
1986	3	12.21680		79	0.250	301.2	-21.1	1.01	0.88	116.8	257.6	61.0
1986	3	12.32917	LSPN-1280	33	0.333	301.1	-21.1	1.00	0.88	116.7	257.6	61.0
1986	3	12.36319	LSPN-1281	33	0.333	301.1	-21.2	1.00	0.88	116.7	257.6	61.0
1986	3	12.36771	LSPN-726	32	0.083	301.1	-21.2	1.00	0.88	116.7	257.6	61.0
1986	3	12.38750	LSPN-1282	33	0.333	301.1	-21.2	1.00	0.88	116.6	257.6	61.0
1986	3	12.43958	LSPN-1728	21	0.117	301.0	-21.2	1.00	0.88	116.6	257.5	61.0
1986	3	12.45694	LSPN-1389	21	0.500	301.0	-21.2	1.00	0.88	116.6	257.5	61.0
1986	3	12.49583	LSPN-1390	21	0.267	301.0	-21.2	1.00	0.88	116.6	257.5	61.0
1986	3	12.51805	LSPN-3588	16	0.367	301.0	-21.2	1.00	0.88	116.6	257.5	61.0
1986	3	12.51927	LSPN-3578	15	0.350	301.0	-21.2	1.00	0.88	116.6	257.5	61.0
1986	3	12.62708	LSPN-1910	1	0.500	300.9	-21.3	1.00	0.88	116.5	257.5	61.0
1986	3	12.71146		80	0.083	300.9	-21.3	0.99	0.88	116.4	257.5	60.9
1986	3	12.86667	LSPN-3075	70	0.600	300.8	-21.4	0.99	0.89	116.3	257.5	60.9
1986	3	12.95903	LSPN-3622	65	1.000	300.7	-21.5	0.99	0.89	116.3	257.5	60.9
1986	3	12.98194	LSPN-3663	67	0.133	300.7	-21.5	0.99	0.89	116.3	257.5	60.9
1986	3	13.09618	LSPN-2843	56	0.250	300.7	-21.5	0.99	0.89	116.2	257.5	60.9
1986	3	13.09618	LSPN-2844	57	0.250	300.7	-21.5	0.99	0.89	116.2	257.5	60.9
1986	3	13.11493	LSPN-2845	57	0.417	300.6	-21.5	0.98	0.89	116.2	257.5	60.9
1986	3	13.30972	LSPN-1283	33	0.333	300.5	-21.7	0.98	0.89	116.1	257.5	60.9
1986	3	13.34792	LSPN-1285	33	0.500	300.5	-21.7	0.98	0.89	116.0	257.5	60.9
1986	3	13.37569	LSPN-1286	33	0.333	300.5	-21.7	0.98	0.89	116.0	257.5	60.9

Ephemeris

Year	Mon	Day	Ref #	Obs #	Exp	α	δ	Δ	r	β	ϕ	φ
1986	3	13.45903	LSPN-1391	21	0.500	300.4	-21.7	0.98	0.89	116.0	257.5	60.9
1986	3	13.48438	LSPN-1392	21	0.217	300.4	-21.7	0.98	0.89	116.0	257.5	60.9
1986	3	13.49549	LSPN-1729	21	0.117	300.4	-21.8	0.98	0.89	115.9	257.5	60.9
1986	3	13.62014	LSPN-1911	1	0.500	300.3	-21.8	0.97	0.90	115.9	257.5	60.8
1986	3	13.87153	LSPN-3077	70	0.467	300.2	-22.0	0.97	0.90	115.7	257.5	60.8
1986	3	14.10104	LSPN-2847	56	0.750	300.0	-22.1	0.96	0.90	115.6	257.5	60.8
1986	3	14.10104	LSPN-2846	57	0.750	300.0	-22.1	0.96	0.90	115.6	257.5	60.8
1986	3	14.31667	LSPN-1287	33	0.333	299.9	-22.2	0.95	0.91	115.5	257.5	60.8
1986	3	14.35278	LSPN-727	32	0.167	299.9	-22.2	0.95	0.91	115.5	257.5	60.8
1986	3	14.35556	LSPN-1289	33	0.500	299.9	-22.2	0.95	0.91	115.5	257.5	60.8
1986	3	14.46128	LSPN-1393	21	0.708	299.8	-22.3	0.95	0.91	115.4	257.5	60.8
1986	3	14.48438	LSPN-1730	21	0.117	299.8	-22.3	0.95	0.91	115.4	257.5	60.8
1986	3	14.52670	AON-851064	8	0.083	299.8	-22.3	0.95	0.91	115.4	257.5	60.8
1986	3	14.95660	LSPN-3666	67	0.083	299.5	-22.6	0.94	0.92	115.2	257.5	60.7
1986	3	14.95660	LSPN-3665	67	0.083	299.5	-22.6	0.94	0.92	115.2	257.5	60.7
1986	3	15.00000	LSPN-3650	66	0.250	299.5	-22.6	0.94	0.92	115.2	257.5	60.7
1986	3	15.11649	LSPN-3479	53	0.500	299.4	-22.7	0.93	0.92	115.1	257.5	60.7
1986	3	15.32917	LSPN-1291	33	0.333	299.2	-22.8	0.93	0.92	115.0	257.5	60.7
1986	3	15.33924	LSPN-2891	36	0.250	299.2	-22.8	0.93	0.92	115.0	257.5	60.7
1986	3	15.34167	LSPN-728	32	0.167	299.2	-22.8	0.93	0.92	115.0	257.5	60.7
1986	3	15.35208	LSPN-1292	33	0.333	299.2	-22.8	0.93	0.92	115.0	257.5	60.7
1986	3	15.35903	LSPN-729	32	0.167	299.2	-22.8	0.93	0.92	115.0	257.5	60.7
1986	3	15.37066	LSPN-2892	36	0.250	299.2	-22.8	0.93	0.92	115.0	257.5	60.7
1986	3	15.38924	LSPN-2893	36	0.250	299.2	-22.8	0.93	0.92	115.0	257.5	60.7
1986	3	15.40521	LSPN-2564	37	0.083	299.2	-22.8	0.93	0.92	115.0	257.5	60.7
1986	3	15.46810	LSPN-3420	26	0.583	299.1	-22.9	0.93	0.92	115.0	257.5	60.7
1986	3	15.46810	LSPN-3416	27	0.583	299.1	-22.9	0.93	0.92	115.0	257.5	60.7
1986	3	15.47812	LSPN-1731	21	0.117	299.1	-22.9	0.93	0.92	114.9	257.5	60.7
1986	3	15.96562	LSPN-3667	67	0.333	298.8	-23.2	0.91	0.93	114.7	257.5	60.6
1986	3	15.97292	LSPN-3625	65	0.500	298.8	-23.2	0.91	0.93	114.7	257.5	60.6
1986	3	15.98125	LSPN-3668	67	0.333	298.8	-23.2	0.91	0.93	114.7	257.5	60.6
1986	3	16.05970		8	0.200	298.7	-23.2	0.91	0.93	114.7	257.5	60.6
1986	3	16.29236	LSPN-1294	33	0.500	298.6	-23.4	0.90	0.94	114.6	257.5	60.6
1986	3	16.32292	LSPN-1295	33	0.333	298.5	-23.4	0.90	0.94	114.6	257.5	60.6
1986	3	16.34409	LSPN-730	32	0.250	298.5	-23.4	0.90	0.94	114.6	257.5	60.6
1986	3	16.34583	LSPN-1296	33	0.333	298.5	-23.4	0.90	0.94	114.6	257.5	60.6
1986	3	16.36215	LSPN-1297	33	0.050	298.5	-23.4	0.90	0.94	114.6	257.5	60.6
1986	3	16.41597	LSPN-1397	21	0.267	298.5	-23.5	0.90	0.94	114.6	257.5	60.6
1986	3	16.43932	LSPN-1732	21	0.121	298.5	-23.5	0.90	0.94	114.6	257.5	60.6
1986	3	16.44900		25	0.167	298.5	-23.5	0.90	0.94	114.6	257.5	60.6
1986	3	16.45695	LSPN-892	22	0.167	298.4	-23.5	0.90	0.94	114.6	257.5	60.6
1986	3	16.46150	LSPN-3417	27	0.583	298.4	-23.5	0.90	0.94	114.6	257.5	60.6
1986	3	16.46150	LSPN-3421	26	0.583	298.4	-23.5	0.90	0.94	114.6	257.5	60.6
1986	3	16.46806	LSPN-893	22	0.167	298.4	-23.5	0.90	0.94	114.6	257.5	60.6
1986	3	16.47326	LSPN-1398	21	0.650	298.4	-23.5	0.90	0.94	114.6	257.5	60.6
1986	3	16.47708	LSPN-894	22	0.167	298.4	-23.5	0.90	0.94	114.6	257.5	60.6
1986	3	16.48611	LSPN-895	22	0.167	298.4	-23.5	0.90	0.94	114.6	257.5	60.6
1986	3	16.61296	LSPN-1912	1	0.500	298.3	-23.6	0.90	0.94	114.5	257.5	60.6

Ephemeris

Year	Mon	Day	Ref #	Obs #	Exp	α	δ	Δ	r	β	ϕ	φ
1986	3	16.63646	LSPN-1913	1	0.367	298.3	-23.6	0.90	0.94	114.5	257.5	60.6
1986	3	16.82516	LSPN-1631	73	0.167	298.2	-23.7	0.89	0.94	114.4	257.6	60.6
1986	3	16.96111	LSPN-3669	67	0.250	298.1	-23.8	0.89	0.95	114.4	257.6	60.6
1986	3	16.96667	LSPN-3626	65	1.000	298.1	-23.8	0.89	0.95	114.4	257.6	60.6
1986	3	17.28229	LSPN-1298	33	0.050	297.8	-24.0	0.88	0.95	114.3	257.6	60.6
1986	3	17.32153	LSPN-1300	33	0.333	297.8	-24.0	0.88	0.95	114.3	257.6	60.6
1986	3	17.34653	LSPN-1301	33	0.333	297.8	-24.1	0.88	0.95	114.3	257.6	60.6
1986	3	17.35208	LSPN-731	32	0.167	297.8	-24.1	0.88	0.95	114.3	257.6	60.6
1986	3	17.36910	LSPN-2899	36	0.250	297.8	-24.1	0.88	0.95	114.3	257.6	60.6
1986	3	17.37361	LSPN-1302	33	0.500	297.8	-24.1	0.88	0.95	114.3	257.6	60.6
1986	3	17.39410	LSPN-2566	37	0.083	297.7	-24.1	0.88	0.95	114.3	257.6	60.6
1986	3	17.40747	LSPN-2567	37	0.125	297.7	-24.1	0.88	0.95	114.3	257.6	60.6
1986	3	17.45417	LSPN-1401	21	0.133	297.7	-24.1	0.88	0.95	114.3	257.6	60.6
1986	3	17.48680	LSPN-1733	21	0.100	297.7	-24.2	0.87	0.95	114.2	257.6	60.6
1986	3	17.49896	LSPN-1402	21	0.117	297.7	-24.2	0.87	0.95	114.2	257.6	60.6
1986	3	17.60841	LSPN-1915	1	0.417	297.6	-24.2	0.87	0.96	114.2	257.6	60.6
1986	3	17.63251	LSPN-1914	1	0.500	297.6	-24.3	0.87	0.96	114.2	257.6	60.6
1986	3	17.77091	LSPN-2386	79	0.250	297.5	-24.3	0.87	0.96	114.2	257.7	60.6
1986	3	17.99132	LSPN-3628	65	0.750	297.3	-24.5	0.86	0.96	114.1	257.7	60.6
1986	3	18.29167	LSPN-1304	33	0.333	297.0	-24.7	0.85	0.97	114.1	257.7	60.6
1986	3	18.31458	LSPN-1305	33	0.333	297.0	-24.7	0.85	0.97	114.1	257.7	60.6
1986	3	18.32396	LSPN-732	32	0.250	297.0	-24.7	0.85	0.97	114.1	257.7	60.6
1986	3	18.32813	LSPN-2900	36	0.250	297.0	-24.7	0.85	0.97	114.1	257.7	60.6
1986	3	18.34375	LSPN-733	32	0.167	297.0	-24.7	0.85	0.97	114.1	257.7	60.6
1986	3	18.35000	LSPN-2901	36	0.250	297.0	-24.7	0.85	0.97	114.1	257.7	60.6
1986	3	18.36111	LSPN-1307	33	0.333	297.0	-24.7	0.85	0.97	114.1	257.7	60.6
1986	3	18.36181	LSPN-734	32	0.167	297.0	-24.7	0.85	0.97	114.1	257.7	60.6
1986	3	18.37049	LSPN-2902	36	0.250	297.0	-24.8	0.85	0.97	114.0	257.7	60.6
1986	3	18.40208	LSPN-2568	37	0.167	296.9	-24.8	0.85	0.97	114.0	257.7	60.6
1986	3	18.41042	LSPN-1404	21	0.267	296.9	-24.8	0.85	0.97	114.0	257.7	60.6
1986	3	18.48820	LSPN-1734	21	0.133	296.9	-24.8	0.85	0.97	114.0	257.8	60.6
1986	3	18.50139	LSPN-1403	21	0.233	296.9	-24.8	0.85	0.97	114.0	257.8	60.6
1986	3	18.51319	LSPN-3589	16	0.400	296.9	-24.9	0.85	0.97	114.0	257.8	60.6
1986	3	18.51424	LSPN-3579	15	0.383	296.9	-24.9	0.85	0.97	114.0	257.8	60.6
1986	3	18.52920	AON-851117	8	0.113	296.8	-24.9	0.85	0.97	114.0	257.8	60.6
1986	3	18.63634	LSPN-1916	1	0.417	296.8	-24.9	0.85	0.97	114.0	257.8	60.6
1986	3	18.68230	LSPN-2387	79	0.250	296.7	-25.0	0.84	0.97	114.0	257.8	60.6
1986	3	18.69245	LSPN-2118	77	0.083	296.7	-25.0	0.84	0.97	114.0	257.8	60.6
1986	3	18.70654	LSPN-2388	79	0.250	296.7	-25.0	0.84	0.97	114.0	257.8	60.6
1986	3	18.72281	LSPN-2120	77	0.333	296.7	-25.0	0.84	0.97	114.0	257.8	60.6
1986	3	18.72697	LSPN-2389	79	0.133	296.7	-25.0	0.84	0.97	114.0	257.8	60.6
1986	3	18.93750	LSPN-3652	66	1.000	296.5	-25.2	0.84	0.98	114.0	257.8	60.6
1986	3	18.96042	LSPN-3629	65	1.000	296.5	-25.2	0.84	0.98	114.0	257.8	60.6
1986	3	19.02778	LSPN-1803	52	0.100	296.4	-25.2	0.84	0.98	113.9	257.8	60.6
1986	3	19.03958		59	0.200	296.4	-25.2	0.83	0.98	113.9	257.8	60.6
1986	3	19.04271	LSPN-1804	52	0.417	296.4	-25.2	0.83	0.98	113.9	257.8	60.6
1986	3	19.05347	LSPN-602	60	0.133	296.4	-25.2	0.83	0.98	113.9	257.9	60.6
1986	3	19.06285	LSPN-1805	52	0.083	296.4	-25.2	0.83	0.98	113.9	257.9	60.6

Ephemeris

Year	Mon	Day	Ref #	Obs #	Exp	α	δ	Δ	r	β	ϕ	φ
1986	3	19.11875	LSPN-1806	52	0.500	296.3	-25.3	0.83	0.98	113.9	257.9	60.6
1986	3	19.28229	LSPN-1309	33	0.050	296.2	-25.4	0.83	0.98	113.9	257.9	60.6
1986	3	19.30000	LSPN-1310	33	0.333	296.2	-25.4	0.83	0.98	113.9	257.9	60.6
1986	3	19.31042	LSPN-735	32	0.167	296.2	-25.4	0.83	0.98	113.9	257.9	60.6
1986	3	19.32430	LSPN-1311	33	0.333	296.2	-25.4	0.83	0.98	113.9	257.9	60.6
1986	3	19.33125	LSPN-736	32	0.333	296.2	-25.4	0.83	0.98	113.9	257.9	60.6
1986	3	19.34722	LSPN-1312	33	0.333	296.1	-25.5	0.83	0.98	113.9	257.9	60.6
1986	3	19.35208	LSPN-737	32	0.167	296.1	-25.5	0.83	0.98	113.9	257.9	60.6
1986	3	19.37431	LSPN-1313	33	0.500	296.1	-25.5	0.83	0.98	113.9	257.9	60.6
1986	3	19.40000	LSPN-2569	37	0.167	296.1	-25.5	0.83	0.98	113.9	257.9	60.6
1986	3	19.44306	LSPN-1405	21	1.000	296.1	-25.5	0.82	0.98	113.9	257.9	60.6
1986	3	19.47292	LSPN-1406	21	0.250	296.0	-25.5	0.82	0.98	113.9	257.9	60.6
1986	3	19.48472	LSPN-1735	21	0.167	296.0	-25.6	0.82	0.98	113.9	257.9	60.6
1986	3	19.49757	LSPN-1407	21	0.267	296.0	-25.6	0.82	0.99	113.9	257.9	60.6
1986	3	19.50833	LSPN-3590	16	0.500	296.0	-25.6	0.82	0.99	113.9	257.9	60.6
1986	3	19.53670	AON-851146	8	0.083	296.0	-25.6	0.82	0.99	113.9	257.9	60.6
1986	3	19.63542	LSPN-1917	1	0.500	295.9	-25.7	0.82	0.99	113.9	258.0	60.6
1986	3	19.82604	LSPN-3080	70	0.750	295.7	-25.8	0.81	0.99	113.9	258.0	60.6
1986	3	19.96771	LSPN-3671	67	0.250	295.6	-25.9	0.81	0.99	113.9	258.0	60.7
1986	3	20.05069		59	0.133	295.5	-26.0	0.81	0.99	113.9	258.1	60.7
1986	3	20.10417	LSPN-2848	57	1.000	295.5	-26.0	0.81	0.99	113.9	258.1	60.7
1986	3	20.10417	LSPN-2849	56	1.000	295.5	-26.0	0.81	0.99	113.9	258.1	60.7
1986	3	20.29028	LSPN-1103	33	0.333	295.3	-26.2	0.80	1.00	113.9	258.1	60.7
1986	3	20.31250	LSPN-1104	33	0.333	295.3	-26.2	0.80	1.00	113.9	258.1	60.7
1986	3	20.32500	LSPN-2903	36	0.167	295.3	-26.2	0.80	1.00	113.9	258.1	60.7
1986	3	20.33542	LSPN-1105	33	0.333	295.2	-26.2	0.80	1.00	113.9	258.1	60.7
1986	3	20.34028	LSPN-738	32	0.500	295.2	-26.2	0.80	1.00	113.9	258.1	60.7
1986	3	20.34861	LSPN-2904	36	0.167	295.2	-26.2	0.80	1.00	113.9	258.1	60.7
1986	3	20.36458	LSPN-1106	33	0.500	295.2	-26.2	0.80	1.00	113.9	258.1	60.7
1986	3	20.36667	LSPN-2905	36	0.167	295.2	-26.2	0.80	1.00	113.9	258.1	60.7
1986	3	20.37986	LSPN-739	32	0.667	295.2	-26.2	0.80	1.00	113.9	258.1	60.7
1986	3	20.38403	LSPN-2570	37	0.300	295.2	-26.2	0.80	1.00	113.9	258.1	60.7
1986	3	20.39097	LSPN-1107	33	0.333	295.2	-26.2	0.80	1.00	113.9	258.2	60.7
1986	3	20.40139	LSPN-2571	37	0.167	295.2	-26.2	0.80	1.00	113.9	258.2	60.7
1986	3	20.41806	LSPN-1408	21	0.350	295.2	-26.3	0.80	1.00	113.9	258.2	60.7
1986	3	20.44583	LSPN-896	22	0.167	295.1	-26.3	0.80	1.00	113.9	258.2	60.7
1986	3	20.45903	LSPN-897	22	0.167	295.1	-26.3	0.80	1.00	113.9	258.2	60.7
1986	3	20.46806	LSPN-898	22	0.167	295.1	-26.3	0.80	1.00	113.9	258.2	60.7
1986	3	20.46807		30	0.006	295.1	-26.3	0.80	1.00	113.9	258.2	60.7
1986	3	20.47430		25	1.000	295.1	-26.3	0.80	1.00	113.9	258.2	60.7
1986	3	20.47708	LSPN-899	22	0.167	295.1	-26.3	0.80	1.00	113.9	258.2	60.7
1986	3	20.48500	AON-851187	14	0.333	295.1	-26.3	0.80	1.00	113.9	258.2	60.7
1986	3	20.48958	LSPN-900	22	0.167	295.1	-26.3	0.80	1.00	113.9	258.2	60.7
1986	3	20.50972	LSPN-3592	16	0.433	295.1	-26.3	0.80	1.00	113.9	258.2	60.7
1986	3	20.51007	LSPN-3580	15	0.400	295.1	-26.3	0.80	1.00	113.9	258.2	60.7
1986	3	20.52590	AON-851186	14	0.167	295.1	-26.3	0.80	1.00	113.9	258.2	60.7
1986	3	20.62570	LSPN-1918	1	0.500	295.0	-26.4	0.79	1.00	113.9	258.2	60.7
1986	3	21.04826	LSPN-2850	57	0.250	294.6	-26.8	0.78	1.01	113.9	258.3	60.8

Ephemeris

Year	Mon	Day	Ref #	Obs #	Exp	α	δ	Δ	r	β	ϕ	φ
1986	3	21.06771	LSPN-2853	57	0.250	294.5	-26.8	0.78	1.01	113.9	258.3	60.8
1986	3	21.06771	LSPN-1807	52	0.417	294.5	-26.8	0.78	1.01	113.9	258.3	60.8
1986	3	21.30694	LSPN-1110	33	0.333	294.3	-27.0	0.78	1.01	113.9	258.4	60.8
1986	3	21.32986	LSPN-2906	36	0.167	294.3	-27.0	0.78	1.01	113.9	258.4	60.8
1986	3	21.34167	LSPN-2907	36	0.167	294.3	-27.0	0.78	1.01	113.9	258.4	60.8
1986	3	21.36181	LSPN-1112	33	0.333	294.2	-27.0	0.78	1.01	113.9	258.4	60.8
1986	3	21.38264	LSPN-741	32	0.167	294.2	-27.0	0.78	1.01	113.9	258.4	60.8
1986	3	21.40000		35	1.167	294.2	-27.0	0.77	1.01	113.9	258.5	60.8
1986	3	21.41319	LSPN-1409	21	0.267	294.2	-27.1	0.77	1.01	113.9	258.5	60.8
1986	3	21.44201	LSPN-1410	21	0.250	294.2	-27.1	0.77	1.01	113.9	258.5	60.8
1986	3	21.45174	LSPN-901	22	0.083	294.1	-27.1	0.77	1.02	113.9	258.5	60.8
1986	3	21.45868	LSPN-902	22	0.083	294.1	-27.1	0.77	1.02	113.9	258.5	60.8
1986	3	21.47326	LSPN-905	22	0.083	294.1	-27.1	0.77	1.02	113.9	258.5	60.8
1986	3	21.47650	LSPN-2971	17	0.250	294.1	-27.1	0.77	1.02	113.9	258.5	60.8
1986	3	21.49080	LSPN-2972	17	0.167	294.1	-27.1	0.77	1.02	113.9	258.5	60.8
1986	3	21.49896	LSPN-1412	21	0.267	294.1	-27.1	0.77	1.02	113.9	258.5	60.8
1986	3	21.50000		25	0.000	294.1	-27.1	0.77	1.02	113.9	258.5	60.8
1986	3	21.50314	LSPN-2973	17	0.167	294.1	-27.1	0.77	1.02	113.9	258.5	60.8
1986	3	21.50937	LSPN-3581	15	0.417	294.1	-27.1	0.77	1.02	113.9	258.5	60.8
1986	3	21.50937	LSPN-3591	16	0.450	294.1	-27.1	0.77	1.02	113.9	258.5	60.8
1986	3	21.51476	LSPN-2974	17	0.167	294.1	-27.1	0.77	1.02	113.9	258.5	60.8
1986	3	21.52220	AON-851222	14	0.167	294.1	-27.1	0.77	1.02	113.9	258.5	60.8
1986	3	21.69315	LSPN-2125	77	0.500	293.9	-27.3	0.77	1.02	113.9	258.6	60.9
1986	3	21.76934	LSPN-2395	79	0.333	293.8	-27.3	0.77	1.02	113.9	258.6	60.9
1986	3	21.77083	LSPN-2128	77	0.417	293.8	-27.3	0.77	1.02	113.9	258.6	60.9
1986	3	22.06424	LSPN-2855	57	1.000	293.5	-27.6	0.76	1.02	114.0	258.7	60.9
1986	3	22.06424	LSPN-2854	56	1.000	293.5	-27.6	0.76	1.02	114.0	258.7	60.9
1986	3	22.10868	LSPN-2856	56	1.000	293.4	-27.6	0.76	1.03	114.0	258.7	60.9
1986	3	22.10868	LSPN-2857	57	1.000	293.4	-27.6	0.76	1.03	114.0	258.7	60.9
1986	3	22.32292	LSPN-2908	36	0.167	293.2	-27.8	0.75	1.03	114.0	258.8	61.0
1986	3	22.34236	LSPN-2909	36	0.167	293.2	-27.8	0.75	1.03	114.0	258.8	61.0
1986	3	22.34722	LSPN-1118	33	0.333	293.2	-27.8	0.75	1.03	114.0	258.8	61.0
1986	3	22.36042	LSPN-2910	36	0.167	293.2	-27.8	0.75	1.03	114.1	258.8	61.0
1986	3	22.44410	LSPN-1413	21	0.250	293.1	-27.9	0.75	1.03	114.1	258.8	61.0
1986	3	22.47500	LSPN-906	22	0.167	293.0	-27.9	0.75	1.03	114.1	258.8	61.0
1986	3	22.47836	LSPN-2975	17	0.167	293.0	-28.0	0.75	1.03	114.1	258.9	61.0
1986	3	22.48125	LSPN-1415	21	0.250	293.0	-28.0	0.75	1.03	114.1	258.9	61.0
1986	3	22.48542	LSPN-907	22	0.167	293.0	-28.0	0.75	1.03	114.1	258.9	61.0
1986	3	22.49044	LSPN-2976	17	0.167	293.0	-28.0	0.75	1.03	114.1	258.9	61.0
1986	3	22.49132	LSPN-1736	21	0.067	293.0	-28.0	0.75	1.03	114.1	258.9	61.0
1986	3	22.49306	LSPN-908	22	0.133	293.0	-28.0	0.75	1.03	114.1	258.9	61.0
1986	3	22.50000		20	0.050	293.0	-28.0	0.75	1.03	114.1	258.9	61.0
1986	3	22.50035	LSPN-1416	21	0.250	293.0	-28.0	0.75	1.03	114.1	258.9	61.0
1986	3	22.50170	LSPN-2977	17	0.167	293.0	-28.0	0.75	1.03	114.1	258.9	61.0
1986	3	22.51229	LSPN-2978	17	0.167	293.0	-28.0	0.75	1.03	114.1	258.9	61.0
1986	3	22.52051	LSPN-2979	17	0.083	293.0	-28.0	0.75	1.03	114.1	258.9	61.0
1986	3	22.72851	LSPN-2130	77	0.500	292.8	-28.2	0.74	1.03	114.1	259.0	61.1
1986	3	22.75760	LSPN-2121	77	0.500	292.7	-28.2	0.74	1.04	114.1	259.0	61.1

Year	Mon	Day	Ref #	Obs #	Exp	α	δ	Δ	r	β	ϕ	φ
1986	3	22.78080	LSPN-2122	77	0.417	292.7	-28.2	0.74	1.04	114.1	259.0	61.1
1986	3	22.90383	LSPN-2618	68	0.500	292.6	-28.3	0.74	1.04	114.2	259.0	61.1
1986	3	22.99028	LSPN-3674	67	0.567	292.5	-28.4	0.74	1.04	114.2	259.1	61.1
1986	3	23.00000	LSPN-3653	66	1.250	292.4	-28.4	0.74	1.04	114.2	259.1	61.1
1986	3	23.02431		59	0.083	292.4	-28.4	0.73	1.04	114.2	259.1	61.1
1986	3	23.32639	LSPN-1121	33	0.500	292.1	-28.7	0.73	1.04	114.3	259.2	61.2
1986	3	23.35347	LSPN-1122	33	0.333	292.0	-28.7	0.73	1.04	114.3	259.2	61.2
1986	3	23.35833	LSPN-2911	36	0.167	292.0	-28.7	0.73	1.04	114.3	259.2	61.2
1986	3	23.37500	LSPN-1123	33	0.333	292.0	-28.8	0.73	1.04	114.3	259.3	61.2
1986	3	23.37708	LSPN-2912	36	0.167	292.0	-28.8	0.73	1.04	114.3	259.3	61.2
1986	3	23.39653	LSPN-1124	33	0.333	292.0	-28.8	0.73	1.04	114.3	259.3	61.2
1986	3	23.47049	LSPN-1418	21	0.333	291.9	-28.8	0.72	1.05	114.3	259.3	61.2
1986	3	23.48470	LSPN-308	20	0.083	291.9	-28.9	0.72	1.05	114.4	259.3	61.2
1986	3	23.48785	LSPN-1419	21	0.333	291.9	-28.9	0.72	1.05	114.4	259.3	61.2
1986	3	23.49865	LSPN-2982	17	0.117	291.8	-28.9	0.72	1.05	114.4	259.3	61.2
1986	3	23.50625	LSPN-1737	21	0.067	291.8	-28.9	0.72	1.05	114.4	259.3	61.2
1986	3	23.50794	LSPN-2983	17	0.167	291.8	-28.9	0.72	1.05	114.4	259.3	61.2
1986	3	23.60642	LSPN-440	2	0.008	291.7	-29.0	0.72	1.05	114.4	259.4	61.3
1986	3	23.74575	LSPN-2131	77	0.500	291.5	-29.1	0.72	1.05	114.4	259.4	61.3
1986	3	23.77304	LSPN-2123	77	0.500	291.5	-29.1	0.72	1.05	114.5	259.5	61.3
1986	3	23.79123	LSPN-2398	79	0.250	291.5	-29.1	0.72	1.05	114.5	259.5	61.3
1986	3	24.50174	LSPN-1421	21	0.317	290.6	-29.8	0.70	1.06	114.8	259.9	61.6
1986	3	24.72306	LSPN-2399	79	0.333	290.3	-30.0	0.69	1.07	114.9	260.0	61.6
1986	3	24.78296	LSPN-2400	79	0.250	290.2	-30.1	0.69	1.07	114.9	260.1	61.7
1986	3	25.50000	LSPN-1423	21	0.333	289.1	-30.8	0.67	1.08	115.3	260.5	62.0
1986	3	25.77711	LSPN-2402	79	0.167	288.7	-31.1	0.67	1.08	115.4	260.8	62.1
1986	3	26.46771	LSPN-1424	21	0.150	287.6	-31.9	0.65	1.09	115.9	261.3	62.5
1986	3	26.48125	LSPN-1425	21	0.267	287.6	-31.9	0.65	1.09	115.9	261.3	62.5
1986	3	27.45014	LSPN-1426	21	0.140	286.0	-33.0	0.63	1.11	116.7	262.3	63.1
1986	3	27.47078	LSPN-1427	21	0.208	285.9	-33.0	0.63	1.11	116.7	262.3	63.1
1986	3	27.48820	LSPN-1428	21	0.300	285.9	-33.0	0.63	1.11	116.7	262.3	63.1
1986	3	28.42188	LSPN-1429	21	0.250	284.1	-34.1	0.60	1.12	117.6	263.4	63.8
1986	3	28.43403	LSPN-1430	21	0.167	284.1	-34.1	0.60	1.12	117.6	263.4	63.8
1986	3	30.44792	LSPN-1434	21	0.133	279.6	-36.7	0.56	1.15	120.1	266.3	65.9
1986	3	31.13258	LSPN-3490	53	0.042	277.8	-37.6	0.55	1.16	121.1	267.6	66.8
1986	4	1.51830	AON-851295	78	0.167	273.7	-39.5	0.52	1.19	123.4	270.7	69.1
1986	4	2.03090	LSPN-3491	53	0.050	272.0	-40.2	0.51	1.19	124.3	272.1	70.0
1986	4	2.17674	LSPN-1125	33	0.083	271.5	-40.4	0.51	1.20	124.6	272.5	70.3
1986	4	2.25590	LSPN-1440	21	0.250	271.2	-40.5	0.50	1.20	124.8	272.7	70.5
1986	4	2.40972	LSPN-490	2	0.167	270.7	-40.7	0.50	1.20	125.1	273.1	70.8
1986	4	2.51840	AON-851301	78	0.250	270.3	-40.9	0.50	1.20	125.3	273.4	71.0
1986	4	2.86944	LSPN-3676	67	0.333	269.0	-41.4	0.49	1.21	126.0	274.5	71.8
1986	4	3.07604	LSPN-3492	53	0.083	268.2	-41.7	0.49	1.21	126.4	275.2	72.3
1986	4	3.16146	LSPN-1127	33	0.083	267.9	-41.8	0.49	1.21	126.6	275.5	72.5
1986	4	3.20486	LSPN-1128	33	0.333	267.7	-41.8	0.49	1.21	126.7	275.6	72.6
1986	4	3.32917	LSPN-1444	21	0.133	267.2	-42.0	0.49	1.21	127.0	276.0	72.9
1986	4	3.52153	LSPN-1922	1	0.500	266.4	-42.3	0.48	1.22	127.4	276.7	73.4
1986	4	3.55065	LSPN-2132	77	0.500	266.3	-42.3	0.48	1.22	127.5	276.8	73.5

Ephemeris

Year	Mon	Day	Ref #	Obs #	Exp	α	δ	Δ	r	β	ϕ	φ
1986	4	3.58130	LSPN-2124	77	0.500	266.2	-42.4	0.48	1.22	127.6	276.9	73.5
1986	4	3.69853	LSPN-2409	79	0.333	265.7	-42.5	0.48	1.22	127.8	277.3	73.9
1986	4	3.72103	LSPN-2410	79	0.250	265.6	-42.5	0.48	1.22	127.9	277.4	73.9
1986	4	4.03646	LSPN-3493	53	0.250	264.3	-43.0	0.47	1.22	128.6	278.6	74.8
1986	4	4.15660	LSPN-1129	33	0.083	263.8	-43.1	0.47	1.23	128.9	279.1	75.1
1986	4	4.17500	LSPN-1130	33	0.333	263.7	-43.1	0.47	1.23	128.9	279.1	75.1
1986	4	4.19653	LSPN-1131	33	0.333	263.6	-43.2	0.47	1.23	129.0	279.2	75.2
1986	4	4.32361	LSPN-1740	21	0.083	263.0	-43.3	0.47	1.23	129.3	279.7	75.6
1986	4	4.37292	LSPN-1134	33	0.333	262.8	-43.4	0.47	1.23	129.4	279.9	75.7
1986	4	4.68412	LSPN-2134	77	0.500	261.4	-43.8	0.46	1.23	130.2	281.2	76.6
1986	4	4.68679	LSPN-2411	79	0.333	261.4	-43.8	0.46	1.23	130.2	281.2	76.7
1986	4	4.70929	LSPN-2412	79	0.250	261.3	-43.8	0.46	1.24	130.2	281.3	76.7
1986	4	5.00625	LSPN-2860	57	1.000	259.9	-44.2	0.46	1.24	131.0	282.6	77.7
1986	4	5.00625	LSPN-2859	56	1.000	259.9	-44.2	0.46	1.24	131.0	282.6	77.7
1986	4	5.17639	LSPN-1136	33	0.333	259.1	-44.4	0.46	1.24	131.4	283.4	78.2
1986	4	5.20000	LSPN-1137	33	0.333	259.0	-44.4	0.46	1.24	131.5	283.5	78.3
1986	4	5.21736	LSPN-2913	36	0.067	258.9	-44.5	0.46	1.24	131.5	283.6	78.3
1986	4	5.37778	LSPN-1142	33	0.333	258.1	-44.6	0.46	1.25	131.9	284.3	78.9
1986	4	5.43889	LSPN-1450	21	0.450	257.8	-44.7	0.45	1.25	132.1	284.6	79.1
1986	4	5.72007	LSPN-2414	79	0.333	256.3	-45.0	0.45	1.25	132.8	286.0	80.1
1986	4	5.99306		59	0.133	254.9	-45.3	0.45	1.26	133.6	287.4	81.0
1986	4	6.02083	LSPN-3495	53	0.167	254.8	-45.4	0.45	1.26	133.6	287.5	81.1
1986	4	6.15243	LSPN-1143	33	0.083	254.1	-45.5	0.45	1.26	134.0	288.2	81.6
1986	4	6.17014	LSPN-1144	33	0.333	254.0	-45.5	0.45	1.26	134.0	288.3	81.7
1986	4	6.24931	LSPN-1147	33	0.333	253.5	-45.6	0.44	1.26	134.3	288.7	82.0
1986	4	6.30208	LSPN-742	32	0.667	253.3	-45.7	0.44	1.26	134.4	289.0	82.2
1986	4	6.30417	LSPN-1148	33	0.333	253.2	-45.7	0.44	1.26	134.4	289.0	82.2
1986	4	6.32708	LSPN-1149	33	0.333	253.1	-45.7	0.44	1.26	134.5	289.2	82.3
1986	4	6.47188	LSPN-1742	21	0.083	252.3	-45.8	0.44	1.26	134.9	290.0	82.8
1986	4	6.89271	LSPN-3637	65	1.150	250.0	-46.2	0.44	1.27	136.1	292.4	84.5
1986	4	7.14618	LSPN-1150	33	0.083	248.5	-46.4	0.44	1.27	136.8	293.9	85.5
1986	4	7.16319	LSPN-1151	33	0.333	248.4	-46.4	0.44	1.27	136.8	294.0	85.6
1986	4	7.20556	LSPN-743	32	0.167	248.1	-46.5	0.43	1.27	137.0	294.2	85.8
1986	4	7.22014	LSPN-1153	33	0.333	248.1	-46.5	0.43	1.27	137.0	294.3	85.9
1986	4	7.24236	LSPN-1154	33	0.333	247.9	-46.5	0.43	1.27	137.1	294.5	85.9
1986	4	7.27222	LSPN-744	32	0.167	247.8	-46.5	0.43	1.27	137.1	294.7	86.1
1986	4	7.28472	LSPN-1155	33	0.333	247.7	-46.5	0.43	1.28	137.2	294.7	86.1
1986	4	7.30556	LSPN-1156	33	0.333	247.6	-46.6	0.43	1.28	137.2	294.9	86.2
1986	4	7.34653	LSPN-1157	33	0.333	247.3	-46.6	0.43	1.28	137.4	295.1	86.4
1986	4	7.35938	LSPN-1158	33	0.083	247.2	-46.6	0.43	1.28	137.4	295.2	86.4
1986	4	8.02292	LSPN-2863	57	0.717	243.2	-47.0	0.43	1.29	139.3	299.5	89.3
1986	4	8.06736	LSPN-2864	57	1.000	242.9	-47.0	0.43	1.29	139.5	299.8	89.5
1986	4	8.12951	LSPN-1159	33	0.083	242.5	-47.1	0.43	1.29	139.7	300.3	89.8
1986	4	8.15208	LSPN-1160	33	0.333	242.4	-47.1	0.43	1.29	139.7	300.4	89.9
1986	4	8.21701	LSPN-1743	21	0.083	242.0	-47.1	0.43	1.29	139.9	300.9	90.2
1986	4	8.23264	LSPN-1454	21	0.467	241.9	-47.1	0.43	1.29	140.0	301.0	90.3
1986	4	8.24722	LSPN-745	32	0.500	241.8	-47.1	0.43	1.29	140.0	301.1	90.4
1986	4	8.25451	LSPN-1163	33	0.083	241.8	-47.1	0.43	1.29	140.0	301.1	90.4

Ephemeris

Year	Mon	Day	Ref #	Obs #	Exp	α	δ	Δ	r	β	ϕ	φ
1986	4	8.27153	LSPN-746	32	0.167	241.7	-47.1	0.43	1.29	140.1	301.3	90.5
1986	4	8.27361	LSPN-1164	33	0.333	241.6	-47.1	0.43	1.29	140.1	301.3	90.5
1986	4	8.28681	LSPN-747	32	0.167	241.6	-47.1	0.43	1.29	140.1	301.4	90.5
1986	4	8.31319	LSPN-1165	33	0.333	241.4	-47.2	0.43	1.29	140.2	301.6	90.7
1986	4	8.36319	LSPN-1166	33	0.333	241.1	-47.2	0.43	1.29	140.4	301.9	90.9
1986	4	8.37743	LSPN-1167	33	0.083	241.0	-47.2	0.43	1.29	140.4	302.0	91.0
1986	4	8.39132	LSPN-1457	21	0.350	240.9	-47.2	0.43	1.29	140.4	302.1	91.0
1986	4	8.56555	LSPN-1995	83	0.089	239.8	-47.3	0.42	1.29	141.0	303.4	91.8
1986	4	8.72921	LSPN-2416	79	0.500	238.7	-47.3	0.42	1.30	141.5	304.6	92.6
1986	4	9.12118	LSPN-1168	33	0.083	236.2	-47.4	0.42	1.30	142.6	307.5	94.5
1986	4	9.13958	LSPN-1169	33	0.333	236.1	-47.4	0.42	1.30	142.7	307.6	94.6
1986	4	9.16111	LSPN-748	32	0.500	235.9	-47.4	0.42	1.30	142.8	307.8	94.7
1986	4	9.16389	LSPN-1170	33	0.333	235.9	-47.4	0.42	1.30	142.8	307.8	94.7
1986	4	9.23542	LSPN-1459	21	0.417	235.5	-47.4	0.42	1.31	143.0	308.4	95.0
1986	4	9.26319	LSPN-749	32	0.167	235.3	-47.4	0.42	1.31	143.1	308.6	95.1
1986	4	9.28403	LSPN-1173	33	0.333	235.1	-47.4	0.42	1.31	143.1	308.8	95.2
1986	4	9.32361	LSPN-1174	33	0.333	234.9	-47.4	0.42	1.31	143.2	309.1	95.4
1986	4	9.36250	LSPN-1176	33	0.333	234.6	-47.4	0.42	1.31	143.4	309.4	95.6
1986	4	9.37604	LSPN-1177	33	0.083	234.5	-47.4	0.42	1.31	143.4	309.5	95.7
1986	4	9.38403	LSPN-2590	17	0.333	234.5	-47.4	0.42	1.31	143.4	309.6	95.7
1986	4	9.40104	LSPN-2591	17	0.250	234.4	-47.4	0.42	1.31	143.5	309.7	95.8
1986	4	9.42153	LSPN-1463	21	0.467	234.2	-47.4	0.42	1.31	143.5	309.8	95.9
1986	4	9.59696	LSPN-2417	79	0.333	233.1	-47.4	0.42	1.31	144.1	311.2	96.8
1986	4	9.68306	LSPN-2135	77	1.000	232.5	-47.4	0.42	1.31	144.3	311.9	97.2
1986	4	9.72091	LSPN-2136	77	0.500	232.3	-47.4	0.42	1.31	144.4	312.3	97.4
1986	4	9.72508	LSPN-2420	79	0.333	232.2	-47.4	0.42	1.31	144.4	312.3	97.4
1986	4	9.74687	LSPN-2137	77	0.250	232.1	-47.4	0.42	1.31	144.5	312.5	97.5
1986	4	9.75049	LSPN-2421	79	0.157	232.1	-47.4	0.42	1.31	144.5	312.5	97.5
1986	4	9.83854	LSPN-3642	65	1.250	231.5	-47.4	0.42	1.31	144.8	313.2	97.9
1986	4	9.87674	LSPN-3657	66	1.250	231.2	-47.4	0.42	1.31	144.9	313.5	98.1
1986	4	9.95139	LSPN-3688	67	1.000	230.7	-47.4	0.42	1.32	145.1	314.2	98.5
1986	4	10.02570	LSPN-1953	38	0.333	230.3	-47.4	0.42	1.32	145.3	314.8	98.9
1986	4	10.10521	LSPN-1178	33	0.083	229.7	-47.4	0.42	1.32	145.6	315.4	99.3
1986	4	10.12917	LSPN-1179	33	0.333	229.6	-47.4	0.42	1.32	145.6	315.6	99.4
1986	4	10.22778	LSPN-1465	21	0.500	228.9	-47.3	0.42	1.32	145.9	316.5	99.9
1986	4	10.23125	LSPN-750	32	0.167	228.9	-47.3	0.42	1.32	145.9	316.5	99.9
1986	4	10.31597	LSPN-751	32	0.167	228.3	-47.3	0.42	1.32	146.2	317.2	100.3
1986	4	10.33194	LSPN-752	32	0.167	228.2	-47.3	0.42	1.32	146.2	317.4	100.4
1986	4	10.33299	LSPN-1180	33	0.083	228.2	-47.3	0.42	1.32	146.2	317.4	100.4
1986	4	10.34792	LSPN-1181	33	0.333	228.1	-47.3	0.42	1.32	146.3	317.5	100.4
1986	4	10.36736	LSPN-1182	33	0.333	228.0	-47.3	0.42	1.32	146.3	317.7	100.5
1986	4	10.39549	LSPN-912	22	0.183	227.8	-47.3	0.42	1.32	146.4	317.9	100.7
1986	4	10.40556	LSPN-913	22	0.167	227.7	-47.3	0.42	1.32	146.5	318.0	100.7
1986	4	10.41458	LSPN-914	22	0.167	227.7	-47.3	0.42	1.32	146.5	318.1	100.8
1986	4	10.42361	LSPN-915	22	0.167	227.6	-47.3	0.42	1.32	146.5	318.2	100.8
1986	4	10.64134	LSPN-2422	79	0.667	226.2	-47.2	0.42	1.33	147.1	320.1	101.9
1986	4	10.66289	LSPN-2140	77	0.250	226.0	-47.2	0.42	1.33	147.2	320.3	102.0
1986	4	10.73422	LSPN-2143	77	0.250	225.6	-47.1	0.42	1.33	147.4	320.9	102.3

Year	Mon	Day	Ref #	Obs #	Exp	α	δ	Δ	r	β	ϕ	φ
1986	4	11.00764	LSPN-2865	57	1.000	223.8	-47.0	0.42	1.33	148.2	323.4	103.7
1986	4	11.02465	LSPN-3497	53	0.250	223.6	-47.0	0.42	1.33	148.2	323.5	103.8
1986	4	11.08715	LSPN-1183	33	0.083	223.2	-47.0	0.42	1.33	148.4	324.1	104.1
1986	4	11.10625	LSPN-1184	33	0.333	223.1	-46.9	0.42	1.33	148.5	324.3	104.2
1986	4	11.12986	LSPN-1185	33	0.333	223.0	-46.9	0.42	1.33	148.5	324.5	104.3
1986	4	11.19201	LSPN-1187	33	0.083	222.5	-46.9	0.42	1.34	148.7	325.0	104.6
1986	4	11.19410	LSPN-1961	38	0.333	222.5	-46.9	0.42	1.34	148.7	325.1	104.6
1986	4	11.21181	LSPN-1467	21	0.233	222.4	-46.9	0.42	1.34	148.8	325.2	104.7
1986	4	11.22812	LSPN-753	32	0.250	222.3	-46.9	0.42	1.34	148.8	325.4	104.8
1986	4	11.22917	LSPN-1746	21	0.083	222.3	-46.9	0.42	1.34	148.8	325.4	104.8
1986	4	11.23056	LSPN-1189	33	0.333	222.3	-46.9	0.42	1.34	148.8	325.4	104.8
1986	4	11.24479	LSPN-754	32	0.250	222.2	-46.8	0.42	1.34	148.9	325.5	104.8
1986	4	11.25972	LSPN-1468	21	0.167	222.1	-46.8	0.42	1.34	148.9	325.7	104.9
1986	4	11.28542	LSPN-1469	21	0.700	221.9	-46.8	0.42	1.34	149.0	325.9	105.0
1986	4	11.30208	LSPN-1190	33	0.333	221.8	-46.8	0.42	1.34	149.0	326.1	105.1
1986	4	11.33055	LSPN-559	2	0.500	221.6	-46.8	0.42	1.34	149.1	326.3	105.3
1986	4	11.34028	LSPN-1191	33	0.333	221.6	-46.8	0.42	1.34	149.1	326.4	105.3
1986	4	11.35035	LSPN-1470	21	0.250	221.5	-46.8	0.42	1.34	149.2	326.5	105.3
1986	4	11.35312	LSPN-1192	33	0.050	221.5	-46.8	0.42	1.34	149.2	326.5	105.4
1986	4	11.41944	LSPN-1471	21	0.233	221.1	-46.7	0.42	1.34	149.3	327.2	105.7
1986	4	11.48646	LSPN-1473	21	0.500	220.6	-46.7	0.42	1.34	149.5	327.8	106.0
1986	4	11.53644	LSPN-2145	77	0.150	220.3	-46.6	0.42	1.34	149.7	328.3	106.2
1986	4	11.57833	LSPN-2146	77	0.100	220.0	-46.6	0.42	1.34	149.8	328.6	106.4
1986	4	11.64441	LSPN-2147	77	0.136	219.6	-46.5	0.42	1.34	150.0	329.3	106.8
1986	4	11.82190		49	0.333	218.5	-46.4	0.42	1.34	150.4	331.0	107.6
1986	4	11.84549	LSPN-3659	66	1.250	218.3	-46.4	0.42	1.35	150.5	331.2	107.7
1986	4	11.86111	AON-851697	49	0.333	218.2	-46.4	0.42	1.35	150.5	331.4	107.8
1986	4	11.87118	LSPN-608	60	0.150	218.1	-46.3	0.42	1.35	150.6	331.4	107.8
1986	4	11.87396	LSPN-3498	53	0.283	218.1	-46.3	0.42	1.35	150.6	331.5	107.9
1986	4	11.95556	LSPN-2867	57	1.000	217.6	-46.3	0.42	1.35	150.8	332.3	108.2
1986	4	12.00521	LSPN-2868	57	1.083	217.3	-46.2	0.42	1.35	150.9	332.8	108.5
1986	4	12.01042		40	0.050	217.3	-46.2	0.42	1.35	150.9	332.8	108.5
1986	4	12.03299	AON-851705	39	0.083	217.1	-46.2	0.42	1.35	151.0	333.0	108.6
1986	4	12.04583	AON-851707	39	0.100	217.0	-46.2	0.42	1.35	151.0	333.1	108.7
1986	4	12.05035	LSPN-2869	57	0.750	217.0	-46.2	0.42	1.35	151.0	333.2	108.7
1986	4	12.08715	LSPN-1193	33	0.083	216.8	-46.1	0.42	1.35	151.1	333.6	108.9
1986	4	12.09722	AON-851708	39	0.100	216.7	-46.1	0.42	1.35	151.2	333.7	108.9
1986	4	12.10417	LSPN-1194	33	0.333	216.7	-46.1	0.42	1.35	151.2	333.7	108.9
1986	4	12.12569	LSPN-1195	33	0.333	216.5	-46.1	0.42	1.35	151.2	333.9	109.0
1986	4	12.23160	LSPN-1474	21	0.117	215.9	-46.0	0.42	1.35	151.5	335.0	109.5
1986	4	12.25208	LSPN-1197	33	0.333	215.7	-46.0	0.42	1.35	151.6	335.2	109.6
1986	4	12.25313	LSPN-1747	21	0.083	215.7	-46.0	0.42	1.35	151.6	335.2	109.6
1986	4	12.27431	LSPN-1475	21	0.467	215.6	-46.0	0.42	1.35	151.6	335.4	109.7
1986	4	12.30903	LSPN-1198	33	0.333	215.4	-45.9	0.42	1.35	151.7	335.7	109.9
1986	4	12.32951	LSPN-1199	33	0.050	215.2	-45.9	0.42	1.35	151.7	335.9	110.0
1986	4	12.34723	LSPN-570	2	0.133	215.1	-45.9	0.42	1.35	151.8	336.1	110.1
1986	4	12.79688	LSPN-3661	66	1.250	212.4	-45.4	0.42	1.36	152.9	340.7	112.1
1986	4	12.89236	LSPN-3692	67	1.500	211.8	-45.2	0.42	1.36	153.1	341.6	112.5

Ephemeris

Year	Mon	Day	Ref #	Obs #	Exp	α	δ	Δ	r	β	ϕ	φ
1986	4	12.98715	LSPN-2870	57	0.833	211.2	-45.1	0.42	1.36	153.3	342.6	112.9
1986	4	13.02708	LSPN-2871	57	0.833	211.0	-45.1	0.42	1.36	153.4	343.0	113.1
1986	4	13.04931	LSPN-1200	33	0.067	210.8	-45.0	0.42	1.36	153.5	343.3	113.2
1986	4	13.06319	LSPN-1201	33	0.333	210.8	-45.0	0.42	1.36	153.5	343.4	113.3
1986	4	13.09375	LSPN-1202	33	0.333	210.6	-45.0	0.42	1.36	153.6	343.7	113.4
1986	4	13.18542	LSPN-1204	33	0.333	210.0	-44.9	0.42	1.37	153.8	344.7	113.8
1986	4	13.34549	LSPN-755	32	0.250	209.1	-44.6	0.43	1.37	154.1	346.4	114.5
1986	4	13.44514	LSPN-1479	21	0.167	208.5	-44.5	0.43	1.37	154.3	347.4	114.9
1986	4	13.52819	LSPN-2149	77	0.250	208.0	-44.4	0.43	1.37	154.5	348.3	115.3
1986	4	13.54447	LSPN-2150	77	0.200	207.9	-44.4	0.43	1.37	154.6	348.5	115.3
1986	4	13.55844	LSPN-2151	77	0.206	207.9	-44.3	0.43	1.37	154.6	348.6	115.4
1986	4	13.57216	LSPN-2152	77	0.200	207.8	-44.3	0.43	1.37	154.6	348.8	115.4
1986	4	13.58683	LSPN-2153	77	0.239	207.7	-44.3	0.43	1.37	154.6	348.9	115.5
1986	4	13.59988	LSPN-2154	77	0.200	207.6	-44.3	0.43	1.37	154.7	349.0	115.6
1986	4	13.61373	LSPN-2155	77	0.200	207.5	-44.3	0.43	1.37	154.7	349.2	115.6
1986	4	13.62758	LSPN-2156	77	0.200	207.5	-44.2	0.43	1.37	154.7	349.3	115.7
1986	4	13.64144	LSPN-2157	77	0.200	207.4	-44.2	0.43	1.37	154.8	349.5	115.7
1986	4	13.65528	LSPN-2158	77	0.200	207.3	-44.2	0.43	1.37	154.8	349.6	115.8
1986	4	13.66913	LSPN-2159	77	0.200	207.2	-44.2	0.43	1.37	154.8	349.8	115.9
1986	4	13.68297	LSPN-2160	77	0.200	207.1	-44.2	0.43	1.37	154.8	349.9	115.9
1986	4	13.74618	LSPN-3649	65	1.250	206.8	-44.1	0.43	1.37	155.0	350.6	116.2
1986	4	13.76736	LSPN-3693	67	1.500	206.7	-44.0	0.43	1.37	155.0	350.8	116.3
1986	4	13.82639	LSPN-3694	67	1.000	206.3	-44.0	0.43	1.38	155.1	351.5	116.5
1986	4	13.84132	LSPN-3500	53	0.250	206.3	-43.9	0.43	1.38	155.1	351.6	116.6
1986	4	14.05868	LSPN-1208	33	0.083	205.0	-43.6	0.43	1.38	155.6	354.0	117.4
1986	4	14.06250	LSPN-1209	33	0.033	205.0	-43.6	0.43	1.38	155.6	354.0	117.4
1986	4	14.09722	LSPN-1210	33	0.333	204.8	-43.5	0.43	1.38	155.6	354.4	117.6
1986	4	14.18333	LSPN-1212	33	0.333	204.4	-43.4	0.43	1.38	155.8	355.3	117.9
1986	4	14.19410	LSPN-756	32	0.250	204.3	-43.4	0.43	1.38	155.8	355.4	118.0
1986	4	14.25972	LSPN-1215	33	0.333	203.9	-43.3	0.43	1.38	155.9	356.2	118.2
1986	4	14.27465	LSPN-1216	33	0.083	203.9	-43.3	0.43	1.38	156.0	356.3	118.3
1986	4	14.32951	LSPN-3582	15	0.350	203.6	-43.2	0.43	1.38	156.0	356.9	118.5
1986	4	14.46285	LSPN-1482	21	0.117	202.9	-43.0	0.43	1.39	156.3	358.4	119.0
1986	4	14.51389	LSPN-587	2	0.500	202.6	-42.9	0.43	1.39	156.4	358.9	119.2
1986	4	14.51528	LSPN-1748	21	0.083	202.6	-42.9	0.43	1.39	156.4	359.0	119.2
1986	4	14.64285	LSPN-2162	77	0.133	201.9	-42.7	0.44	1.39	156.6	0.4	119.7
1986	4	14.66155	LSPN-2163	77	0.500	201.8	-42.6	0.44	1.39	156.6	0.6	119.7
1986	4	14.67090	LSPN-2426	79	0.083	201.8	-42.6	0.44	1.39	156.6	0.7	119.8
1986	4	14.72876	LSPN-2428	79	0.667	201.5	-42.5	0.44	1.39	156.7	1.3	120.0
1986	4	14.76685	LSPN-2429	79	0.667	201.3	-42.4	0.44	1.39	156.8	1.7	120.1
1986	4	14.79479	LSPN-3695	67	1.050	201.1	-42.4	0.44	1.39	156.8	2.0	120.2
1986	4	14.85069	LSPN-3696	67	1.500	200.8	-42.3	0.44	1.39	156.9	2.7	120.4
1986	4	14.90278		59	0.167	200.6	-42.2	0.44	1.39	157.0	3.2	120.6
1986	4	14.90417	LSPN-3501	53	0.500	200.6	-42.2	0.44	1.39	157.0	3.3	120.6
1986	4	14.92360	AON-851830	49	0.467	200.5	-42.2	0.44	1.39	157.0	3.5	120.7
1986	4	14.94410	LSPN-2873	57	1.000	200.4	-42.1	0.44	1.39	157.0	3.7	120.8
1986	4	14.94410	LSPN-2872	56	1.000	200.4	-42.1	0.44	1.39	157.0	3.7	120.8
1986	4	14.97500	AON-851833	50	0.167	200.2	-42.1	0.44	1.39	157.1	4.0	120.9

Ephemeris

Year	Mon	Day	Ref #	Obs #	Exp	α	δ	Δ	r	β	ϕ	φ
1986	4	14.98368	LSPN-2874	57	0.750	200.2	-42.1	0.44	1.39	157.1	4.1	120.9
1986	4	14.99826	LSPN-1217	33	0.083	200.1	-42.1	0.44	1.39	157.1	4.3	121.0
1986	4	15.03715	AON-851839	39	0.417	199.9	-42.0	0.44	1.39	157.2	4.7	121.1
1986	4	15.05382	AON-851841	39	0.117	199.8	-42.0	0.44	1.39	157.2	4.9	121.2
1986	4	15.06042	LSPN-1220	33	0.500	199.8	-41.9	0.44	1.39	157.2	5.0	121.2
1986	4	15.19792	LSPN-1225	33	0.333	199.1	-41.7	0.44	1.40	157.4	6.5	121.7
1986	4	15.22326	LSPN-1749	21	0.083	199.0	-41.7	0.44	1.40	157.4	6.8	121.8
1986	4	15.24965	LSPN-1484	21	0.417	198.8	-41.6	0.44	1.40	157.4	7.1	121.9
1986	4	15.29375	LSPN-1227	33	0.333	198.6	-41.5	0.44	1.40	157.5	7.6	122.0
1986	4	15.30660	LSPN-1228	33	0.083	198.6	-41.5	0.44	1.40	157.5	7.7	122.1
1986	4	15.38056	LSPN-1485	21	0.133	198.2	-41.4	0.44	1.40	157.6	8.5	122.3
1986	4	15.40868	LSPN-1486	21	0.917	198.1	-41.3	0.44	1.40	157.6	8.8	122.4
1986	4	15.44861	LSPN-424	84	1.000	197.9	-41.3	0.44	1.40	157.7	9.3	122.5
1986	4	15.45313	LSPN-1487	21	0.133	197.9	-41.3	0.44	1.40	157.7	9.3	122.6
1986	4	15.47463	LSPN-2430	79	0.500	197.8	-41.2	0.44	1.40	157.7	9.6	122.6
1986	4	15.50198	LSPN-2431	79	0.483	197.6	-41.2	0.44	1.40	157.7	9.9	122.7
1986	4	15.51285	LSPN-1489	21	0.117	197.6	-41.1	0.45	1.40	157.8	10.0	122.8
1986	4	15.55012	LSPN-2432	79	0.500	197.4	-41.1	0.45	1.40	157.8	10.4	122.9
1986	4	15.57851	LSPN-2433	79	0.500	197.3	-41.0	0.45	1.40	157.8	10.7	123.0
1986	4	15.59514	LSPN-3093	70	0.333	197.2	-41.0	0.45	1.40	157.9	10.9	123.0
1986	4	15.60829	LSPN-2434	79	0.500	197.1	-41.0	0.45	1.40	157.9	11.1	123.1
1986	4	15.63737	LSPN-2435	79	0.500	197.0	-40.9	0.45	1.40	157.9	11.4	123.2
1986	4	15.67061	LSPN-2436	79	0.500	196.8	-40.9	0.45	1.40	157.9	11.8	123.3
1986	4	15.69900	LSPN-2437	79	0.500	196.7	-40.8	0.45	1.40	158.0	12.1	123.4
1986	4	15.72743	LSPN-3698	67	1.417	196.6	-40.8	0.45	1.40	158.0	12.4	123.5
1986	4	15.72880	LSPN-2438	79	0.500	196.6	-40.8	0.45	1.40	158.0	12.4	123.5
1986	4	15.75788	LSPN-2439	79	0.500	196.4	-40.7	0.45	1.40	158.0	12.7	123.6
1986	4	15.78315	LSPN-2440	79	0.350	196.3	-40.7	0.45	1.41	158.1	13.0	123.6
1986	4	15.78819	LSPN-3699	67	1.000	196.3	-40.6	0.45	1.41	158.1	13.1	123.7
1986	4	15.83403	LSPN-1828	52	0.167	196.1	-40.6	0.45	1.41	158.1	13.6	123.8
1986	4	15.89410	LSPN-3700	67	1.417	195.8	-40.5	0.45	1.41	158.2	14.2	124.0
1986	4	15.97010	AON-851884	39	0.667	195.5	-40.3	0.45	1.41	158.2	15.1	124.2
1986	4	16.00104	LSPN-1229	33	0.083	195.3	-40.3	0.45	1.41	158.3	15.4	124.3
1986	4	16.13958	LSPN-1233	33	0.333	194.7	-40.0	0.45	1.41	158.4	16.9	124.8
1986	4	16.29965	LSPN-1236	33	0.083	194.0	-39.7	0.46	1.41	158.5	18.7	125.3
1986	4	16.33264	LSPN-1237	33	0.333	193.8	-39.6	0.46	1.41	158.5	19.1	125.4
1986	4	16.35833	LSPN-918	22	0.167	193.7	-39.6	0.46	1.41	158.6	19.3	125.4
1986	4	16.37569	LSPN-1490	21	0.100	193.7	-39.6	0.46	1.41	158.6	19.5	125.5
1986	4	16.39757	LSPN-1491	21	0.500	193.6	-39.5	0.46	1.41	158.6	19.8	125.6
1986	4	16.41389	LSPN-1492	21	0.133	193.5	-39.5	0.46	1.41	158.6	19.9	125.6
1986	4	16.48715	LSPN-1494	21	0.117	193.2	-39.4	0.46	1.42	158.6	20.7	125.8
1986	4	16.56771	LSPN-426	84	0.950	192.8	-39.2	0.46	1.42	158.7	21.6	126.1
1986	4	16.74826	LSPN-3701	67	1.250	192.1	-38.9	0.46	1.42	158.8	23.6	126.6
1986	4	16.80208	LSPN-3502	53	0.167	191.9	-38.8	0.46	1.42	158.8	24.1	126.7
1986	4	16.81944	LSPN-3702	67	1.000	191.8	-38.7	0.46	1.42	158.8	24.3	126.8
1986	4	16.85590	LSPN-2875	56	1.000	191.6	-38.7	0.46	1.42	158.9	24.7	126.9
1986	4	16.99653	LSPN-3503	53	0.500	191.1	-38.4	0.46	1.42	158.9	26.2	127.3
1986	4	17.05764	LSPN-1240	33	0.167	190.8	-38.3	0.47	1.42	159.0	26.9	127.4

Ephemeris

Year	Mon	Day	Ref #	Obs #	Exp	α	δ	Δ	r	β	ϕ	φ
1986	4	17.06007	AON-851892	39	0.083	190.8	-38.3	0.47	1.42	159.0	26.9	127.4
1986	4	17.10833	LSPN-1241	33	0.067	190.6	-38.2	0.47	1.43	159.0	27.4	127.6
1986	4	17.11875	LSPN-1242	33	0.167	190.6	-38.2	0.47	1.43	159.0	27.5	127.6
1986	4	17.16910	LSPN-1243	33	0.083	190.4	-38.1	0.47	1.43	159.0	28.0	127.7
1986	4	17.18403	LSPN-1244	33	0.333	190.3	-38.0	0.47	1.43	159.0	28.2	127.8
1986	4	17.28889	LSPN-1495	21	0.467	189.9	-37.8	0.47	1.43	159.0	29.3	128.1
1986	4	17.29688	LSPN-1248	33	0.083	189.9	-37.8	0.47	1.43	159.0	29.4	128.1
1986	4	17.30451	LSPN-1751	21	0.100	189.9	-37.8	0.47	1.43	159.0	29.4	128.1
1986	4	17.31250	LSPN-1925	1	0.500	189.8	-37.8	0.47	1.43	159.0	29.5	128.1
1986	4	17.31597	LSPN-1249	33	0.333	189.8	-37.8	0.47	1.43	159.0	29.6	128.1
1986	4	17.34097	LSPN-1496	21	0.133	189.7	-37.7	0.47	1.43	159.1	29.8	128.2
1986	4	17.40347	LSPN-1926	1	0.500	189.5	-37.6	0.47	1.43	159.1	30.5	128.4
1986	4	17.45278	LSPN-1927	1	0.500	189.3	-37.5	0.47	1.43	159.1	31.0	128.5
1986	4	17.65617	LSPN-2441	79	0.667	188.5	-37.1	0.48	1.43	159.1	33.0	129.0
1986	4	17.96007	LSPN-3504	53	0.583	187.4	-36.6	0.48	1.44	159.2	36.1	129.7
1986	4	18.08611	LSPN-1252	33	0.167	187.0	-36.3	0.48	1.44	159.2	37.3	130.0
1986	4	18.15625	LSPN-1253	33	0.067	186.7	-36.2	0.48	1.44	159.2	38.0	130.2
1986	4	18.16667	LSPN-1254	33	0.167	186.7	-36.2	0.48	1.44	159.2	38.1	130.2
1986	4	18.24792	LSPN-1256	33	0.333	186.4	-36.0	0.49	1.44	159.1	38.9	130.4
1986	4	18.28472	LSPN-1257	33	0.333	186.3	-36.0	0.49	1.44	159.1	39.2	130.5
1986	4	18.31389	LSPN-1258	33	0.500	186.2	-35.9	0.49	1.44	159.1	39.5	130.6
1986	4	18.35278	LSPN-1500	21	0.500	186.0	-35.8	0.49	1.44	159.1	39.9	130.7
1986	4	18.35625	LSPN-1933	1	0.500	186.0	-35.8	0.49	1.44	159.1	39.9	130.7
1986	4	18.35764	LSPN-1259	33	0.333	186.0	-35.8	0.49	1.44	159.1	39.9	130.7
1986	4	18.37917	LSPN-927	22	0.167	186.0	-35.8	0.49	1.44	159.1	40.1	130.7
1986	4	18.38073	LSPN-1752	21	0.075	185.9	-35.8	0.49	1.44	159.1	40.1	130.7
1986	4	18.39549	LSPN-1501	21	0.117	185.9	-35.7	0.49	1.44	159.1	40.3	130.8
1986	4	18.43472	LSPN-1929	1	0.500	185.8	-35.7	0.49	1.45	159.1	40.6	130.8
1986	4	18.44444	LSPN-1502	21	0.100	185.7	-35.7	0.49	1.45	159.1	40.7	130.9
1986	4	18.48565	LSPN-1931	1	0.500	185.6	-35.6	0.49	1.45	159.1	41.1	131.0
1986	4	19.00995	LSPN-3505	53	0.344	183.9	-34.6	0.50	1.45	159.0	45.9	132.1
1986	4	20.04167	LSPN-3506	53	0.667	180.8	-32.7	0.52	1.47	158.6	54.3	134.1
1986	4	20.70891	LSPN-2445	79	0.667	179.1	-31.5	0.53	1.48	158.2	59.2	135.2
1986	4	20.75116	LSPN-2446	79	0.667	179.0	-31.4	0.53	1.48	158.1	59.5	135.3
1986	4	21.08125	LSPN-3507	53	0.500	178.1	-30.9	0.54	1.49	157.9	61.7	135.8
1986	4	21.74980	AON-851908	76	0.100	176.6	-29.8	0.56	1.50	157.4	65.8	136.8
1986	4	21.74980	AON-851910	76	0.100	176.6	-29.8	0.56	1.50	157.4	65.8	136.8
1986	4	24.53640	AON-851915	76	0.250	171.2	-25.5	0.62	1.54	155.2	79.0	139.9
1986	4	24.54340	LSPN-1345	72	0.250	171.2	-25.5	0.62	1.54	155.2	79.0	139.9
1986	4	24.55555	AON-851921	82	0.000	171.2	-25.5	0.62	1.54	155.2	79.0	140.0
1986	4	26.68056	LSPN-3296	61	0.667	168.1	-22.7	0.68	1.57	153.6	85.8	141.6
1986	4	26.78514	LSPN-3440	58	0.667	168.0	-22.6	0.68	1.57	153.6	86.1	141.7
1986	4	27.22410	AON-851925	13	0.100	167.4	-22.1	0.69	1.58	153.3	87.2	141.9
1986	4	27.74896	LSPN-3508	53	0.417	166.8	-21.5	0.71	1.59	152.9	88.5	142.2
1986	4	27.99306	LSPN-2914	36	0.333	166.5	-21.2	0.72	1.59	152.8	89.0	142.3
1986	4	28.01736	LSPN-2915	36	0.333	166.5	-21.2	0.72	1.59	152.7	89.1	142.4
1986	4	28.02044	LSPN-2985	33	0.250	166.5	-21.2	0.72	1.59	152.7	89.1	142.4
1986	4	28.02292	LSPN-2917	36	0.367	166.5	-21.2	0.72	1.59	152.7	89.1	142.4

Ephemeris

Year	Mon	Day	Ref #	Obs #	Exp	α	δ	Δ	r	β	ϕ	φ
1986	4	28.20382		12	0.250	166.3	-21.0	0.72	1.59	152.6	89.5	142.4
1986	4	28.80566	LSPN-3442	58	1.000	165.7	-20.4	0.74	1.60	152.3	90.7	142.7
1986	4	28.99340	LSPN-2989	33	0.267	165.5	-20.2	0.74	1.60	152.2	91.1	142.8
1986	4	28.99514	LSPN-2916	36	0.367	165.5	-20.2	0.74	1.60	152.2	91.1	142.8
1986	4	29.00944	LSPN-2990	33	0.250	165.5	-20.2	0.75	1.60	152.2	91.2	142.8
1986	4	29.02729	LSPN-2991	33	0.333	165.4	-20.2	0.75	1.60	152.1	91.2	142.8
1986	4	29.04439	LSPN-2992	33	0.250	165.4	-20.1	0.75	1.60	152.1	91.2	142.8
1986	4	29.05932	LSPN-2993	33	0.250	165.4	-20.1	0.75	1.60	152.1	91.3	142.8
1986	4	29.07535	LSPN-2994	33	0.250	165.4	-20.1	0.75	1.60	152.1	91.3	142.8
1986	4	29.09064	LSPN-2995	33	0.250	165.4	-20.1	0.75	1.61	152.1	91.3	142.8
1986	4	29.21458	LSPN-936	22	0.167	165.3	-20.0	0.75	1.61	152.0	91.6	142.8
1986	4	29.23958	LSPN-939	22	0.167	165.2	-19.9	0.75	1.61	152.0	91.6	142.9
1986	4	29.98264	LSPN-2918	36	0.333	164.5	-19.2	0.77	1.62	151.6	92.9	143.1
1986	4	29.99919	LSPN-2999	33	0.333	164.5	-19.2	0.77	1.62	151.6	93.0	143.1
1986	4	30.01580	LSPN-3000	33	0.250	164.5	-19.2	0.77	1.62	151.6	93.0	143.1
1986	4	30.01597	LSPN-2919	36	0.333	164.5	-19.2	0.77	1.62	151.6	93.0	143.1
1986	4	30.03125	LSPN-3001	33	0.250	164.5	-19.2	0.77	1.62	151.6	93.0	143.1
1986	4	30.04618	LSPN-3002	33	0.250	164.5	-19.2	0.78	1.62	151.6	93.1	143.1
1986	4	30.06815	LSPN-3003	33	0.500	164.5	-19.1	0.78	1.62	151.6	93.1	143.1
1986	4	30.09225	LSPN-3004	33	0.250	164.4	-19.1	0.78	1.62	151.6	93.1	143.1
1986	4	30.11123	LSPN-3005	33	0.250	164.4	-19.1	0.78	1.62	151.6	93.2	143.1
1986	4	30.12720	LSPN-3006	33	0.250	164.4	-19.1	0.78	1.62	151.6	93.2	143.1
1986	4	30.14757	LSPN-3007	33	0.250	164.4	-19.1	0.78	1.62	151.5	93.2	143.1
1986	4	30.17021		9	0.083	164.4	-19.0	0.78	1.62	151.5	93.3	143.1
1986	4	30.22188	LSPN-1503	21	0.250	164.3	-19.0	0.78	1.62	151.5	93.4	143.1
1986	4	30.23264	LSPN-1753	21	0.133	164.3	-19.0	0.78	1.62	151.5	93.4	143.1
1986	4	30.66667	LSPN-3299	61	0.833	163.9	-18.6	0.79	1.63	151.3	94.1	143.2
1986	4	30.73229	LSPN-3509	53	0.417	163.9	-18.5	0.80	1.63	151.3	94.2	143.2
1986	4	30.97569	LSPN-2920	36	0.333	163.7	-18.3	0.80	1.63	151.1	94.6	143.2
1986	5	1.01128	LSPN-3010	33	0.333	163.6	-18.3	0.80	1.63	151.1	94.6	143.3
1986	5	1.02014	LSPN-2921	36	0.333	163.6	-18.3	0.80	1.63	151.1	94.6	143.3
1986	5	1.03576	LSPN-3011	33	0.500	163.6	-18.3	0.80	1.63	151.1	94.7	143.3
1986	5	1.07564	LSPN-3013	33	0.333	163.6	-18.2	0.81	1.63	151.1	94.7	143.3
1986	5	1.11451	LSPN-3014	33	0.333	163.6	-18.2	0.81	1.64	151.1	94.8	143.3
1986	5	1.13067	LSPN-3015	33	0.250	163.6	-18.2	0.81	1.64	151.1	94.8	143.3
1986	5	1.16458	LSPN-981	23	0.533	163.5	-18.1	0.81	1.64	151.1	94.9	143.3
1986	5	1.16500		23	0.533	163.5	-18.1	0.81	1.64	151.1	94.9	143.3
1986	5	1.18472	LSPN-982	23	0.167	163.5	-18.1	0.81	1.64	151.0	94.9	143.3
1986	5	1.97396	LSPN-2922	36	0.417	162.9	-17.5	0.83	1.65	150.7	96.0	143.3
1986	5	1.98351	LSPN-3019	33	0.333	162.9	-17.5	0.83	1.65	150.7	96.0	143.3
1986	5	2.00660	LSPN-2923	36	0.417	162.9	-17.4	0.83	1.65	150.7	96.1	143.3
1986	5	2.01389	LSPN-3020	33	0.500	162.9	-17.4	0.83	1.65	150.7	96.1	143.3
1986	5	2.11944	LSPN-3022	33	0.333	162.8	-17.3	0.84	1.65	150.6	96.2	143.3
1986	5	2.17222	LSPN-941	22	0.500	162.8	-17.3	0.84	1.65	150.6	96.3	143.3
1986	5	2.18889	LSPN-942	22	0.167	162.7	-17.3	0.84	1.65	150.6	96.3	143.3
1986	5	2.20703	LSPN-3023	33	0.250	162.7	-17.3	0.84	1.65	150.6	96.3	143.3
1986	5	2.22766	LSPN-3024	33	0.333	162.7	-17.3	0.84	1.65	150.6	96.4	143.3
1986	5	2.25590		11	0.133	162.7	-17.2	0.84	1.65	150.6	96.4	143.3

Ephemeris

Year	Mon	Day	Ref #	Obs #	Exp	α	δ	Δ	r	β	ϕ	φ
1986	5	2.80191	LSPN-3444	58	1.000	162.3	-16.8	0.86	1.66	150.4	97.1	143.2
1986	5	2.82118	LSPN-3510	53	0.417	162.3	-16.8	0.86	1.66	150.4	97.1	143.2
1986	5	3.11018	LSPN-3028	33	0.333	162.1	-16.6	0.87	1.66	150.3	97.5	143.2
1986	5	3.21667	LSPN-2592	17	0.333	162.0	-16.5	0.87	1.67	150.2	97.6	143.2
1986	5	3.23472	LSPN-2593	17	0.333	162.0	-16.5	0.87	1.67	150.2	97.6	143.2
1986	5	4.19444	LSPN-983	23	0.500	161.4	-15.8	0.90	1.68	149.9	98.7	142.9
1986	5	4.24514	LSPN-2595	17	0.333	161.4	-15.7	0.90	1.68	149.9	98.8	142.9
1986	5	4.96910	LSPN-2924	36	0.417	161.0	-15.2	0.92	1.69	149.7	99.5	142.6
1986	5	5.01042	LSPN-3034	33	0.500	160.9	-15.2	0.92	1.69	149.7	99.6	142.6
1986	5	5.03438	LSPN-3035	33	0.250	160.9	-15.2	0.93	1.69	149.7	99.6	142.5
1986	5	5.11111	LSPN-757	32	0.500	160.9	-15.2	0.93	1.69	149.6	99.7	142.5
1986	5	5.15322	LSPN-3038	33	0.417	160.9	-15.1	0.93	1.69	149.6	99.7	142.5
1986	5	5.21146	LSPN-3040	33	0.250	160.8	-15.1	0.93	1.70	149.6	99.8	142.5
1986	5	5.98229	LSPN-2926	36	0.417	160.4	-14.6	0.96	1.71	149.4	100.5	142.0
1986	5	6.00689	LSPN-3043	33	0.250	160.4	-14.6	0.96	1.71	149.4	100.5	142.0
1986	5	6.00938	LSPN-2927	36	0.417	160.4	-14.6	0.96	1.71	149.4	100.5	142.0
1986	5	6.10198	LSPN-3044	33	0.250	160.4	-14.5	0.96	1.71	149.4	100.6	141.9
1986	5	6.13056	LSPN-3045	33	0.500	160.4	-14.5	0.96	1.71	149.4	100.6	141.9
1986	5	6.15660	LSPN-3046	33	0.250	160.3	-14.5	0.96	1.71	149.4	100.6	141.9
1986	5	6.17569	LSPN-944	22	0.500	160.3	-14.5	0.96	1.71	149.4	100.7	141.9
1986	5	6.92465	LSPN-3511	53	0.683	160.0	-14.0	0.98	1.72	149.2	101.3	141.3
1986	5	6.97870	LSPN-3052	33	0.500	159.9	-14.0	0.99	1.72	149.2	101.3	141.2
1986	5	7.00132	LSPN-3053	33	0.250	159.9	-14.0	0.99	1.72	149.2	101.4	141.2
1986	5	7.00660	LSPN-2929	36	0.417	159.9	-14.0	0.99	1.72	149.2	101.4	141.2
1986	5	7.02292	LSPN-3054	33	0.500	159.9	-14.0	0.99	1.72	149.2	101.4	141.2
1986	5	7.05306	LSPN-3055	33	0.500	159.9	-13.9	0.99	1.72	149.2	101.4	141.1
1986	5	7.10625	LSPN-3057	33	0.250	159.9	-13.9	0.99	1.72	149.2	101.5	141.1
1986	5	7.13264	LSPN-3058	33	0.500	159.9	-13.9	0.99	1.72	149.2	101.5	141.1
1986	5	7.16944	LSPN-3059	33	0.500	159.9	-13.9	0.99	1.72	149.2	101.5	141.0
1986	5	7.20747	LSPN-3060	33	0.500	159.8	-13.9	0.99	1.72	149.2	101.5	141.0
1986	5	7.51701	LSPN-1349	72	0.250	159.7	-13.7	1.00	1.73	149.1	101.8	140.7
1986	5	7.98368	LSPN-2930	36	0.417	159.5	-13.4	1.02	1.74	149.0	102.1	140.2
1986	5	8.01632	LSPN-2931	36	0.417	159.5	-13.4	1.02	1.74	149.0	102.2	140.2
1986	5	8.11250	LSPN-660	29	0.167	159.5	-13.4	1.02	1.74	149.0	102.2	140.1
1986	5	9.42227	LSPN-2447	79	1.000	158.9	-12.7	1.06	1.76	148.8	103.2	138.4
1986	5	9.53598	LSPN-1641	73	0.250	158.9	-12.6	1.07	1.76	148.8	103.3	138.2
1986	5	10.18472	LSPN-984	23	0.133	158.7	-12.3	1.09	1.77	148.7	103.7	137.2
1986	5	10.18487	NNSN-405544	20	0.050	158.7	-12.3	1.09	1.77	148.7	103.7	137.2
1986	5	10.21944	LSPN-946	22	0.500	158.7	-12.3	1.09	1.77	148.7	103.7	137.1
1986	5	10.22920	AON-852104	4	0.167	158.7	-12.3	1.09	1.77	148.7	103.7	137.1
1986	5	10.48160	LSPN-1642	73	0.250	158.6	-12.2	1.10	1.77	148.7	103.9	136.7
1986	5	10.69793	LSPN-3302	61	0.500	158.5	-12.1	1.10	1.77	148.7	104.0	136.3
1986	5	11.04896	LSPN-2932	36	0.417	158.4	-11.9	1.12	1.78	148.7	104.2	135.6
1986	5	11.18194	LSPN-947	22	0.167	158.4	-11.8	1.12	1.78	148.7	104.3	135.3
1986	5	11.40465	LSPN-2165	77	0.450	158.3	-11.7	1.13	1.79	148.7	104.4	134.9
1986	5	11.41163	LSPN-2448	79	1.250	158.3	-11.7	1.13	1.79	148.7	104.4	134.9
1986	5	11.86875	LSPN-3512	53	0.900	158.1	-11.5	1.14	1.79	148.6	104.7	133.9
1986	5	12.84310	AON-852130	39	0.000	157.9	-11.1	1.17	1.81	148.6	105.2	131.7

Ephemeris

Year	Mon	Day	Ref #	Obs #	Exp	α	δ	Δ	r	β	ϕ	φ
1986	5	12.89375	LSPN-3513	53	0.667	157.9	-11.1	1.18	1.81	148.6	105.3	131.5
1986	5	13.15556	LSPN-952	22	0.167	157.8	-11.0	1.18	1.81	148.6	105.4	130.9
1986	5	13.15833	LSPN-953	22	0.500	157.8	-11.0	1.18	1.81	148.6	105.4	130.9
1986	5	14.17292	LSPN-955	22	0.167	157.6	-10.6	1.22	1.82	148.6	105.9	128.1
1986	5	15.15104	LSPN-957	22	0.117	157.3	-10.3	1.25	1.84	148.6	106.4	125.2
1986	5	25.83130	AON-852143	39	0.000	156.1	-7.5	1.60	1.99	149.6	110.3	90.1
1986	5	27.37936	LSPN-2449	79	1.000	156.0	-7.2	1.65	2.01	149.9	110.8	86.8
1986	6	1.06250	LSPN-759	32	0.500	156.0	-6.5	1.80	2.08	150.8	112.0	80.5
1986	6	1.21493	LSPN-964	22	0.183	156.0	-6.5	1.81	2.08	150.8	112.1	80.4
1986	6	2.03125	LSPN-760	32	0.500	156.0	-6.4	1.84	2.09	151.0	112.3	79.7
1986	6	2.41043	LSPN-2450	79	1.250	156.0	-6.4	1.85	2.09	151.0	112.4	79.5
1986	6	7.61220	AON-852153	6	0.000	156.2	-5.8	2.02	2.17	152.2	113.6	77.5
1986	6	9.27050	AON-852164	5	0.083	156.3	-5.7	2.07	2.19	152.6	114.0	77.4
1986	6	10.16076	LSPN-971	22	0.050	156.3	-5.6	2.10	2.20	152.8	114.2	77.4
1986	6	10.17500	LSPN-965	22	0.500	156.3	-5.6	2.10	2.20	152.8	114.2	77.4
1986	6	11.16944	LSPN-967	22	0.333	156.4	-5.6	2.13	2.21	153.1	114.4	77.5
1986	6	11.18368	LSPN-968	22	0.250	156.4	-5.6	2.13	2.21	153.1	114.4	77.5
1986	6	12.16528	LSPN-969	22	0.167	156.4	-5.5	2.16	2.23	153.3	114.6	77.5
1986	6	25.37428	LSPN-2451	79	1.000	157.6	-5.0	2.57	2.40	156.8	117.5	80.9
1986	6	26.38197	LSPN-2452	79	1.500	157.7	-5.0	2.60	2.42	157.0	117.7	81.2
1986	7	6.39137	LSPN-2453	79	1.500	159.0	-5.0	2.90	2.54	159.8	120.0	84.4
1991	2	12.29373		36	7.049	140.5	-4.6	13.37	14.31	178.7	12.1	114.8

BIBLIOGRAPHY

Bibliography

1. Brandt, J.C., Friedman, L.D., Newburn, R.L., Yeomans, D.K. (1981), *The International Halley Watch - Report of the Science Working Group*, Pasadena: NASA/JPL, Pub. TM 82181(400-88).

2. Sekanina, Z., Fry, L. (1991), *The Comet Halley Archive — Summary Volume*, Pasadena: NASA/JPL 400-450, 8/91.

3. Newburn, R., Rahe, J. (1991), "The Organizational History of the International Halley Watch," in *The Comet Halley Archive — Summary Volume*, Pasadena, NASA/JPL 400-450, 8/91, pp. 1-14.

4. Donn, B., Rahe, J., Brandt, J.C. (1986), *Atlas of Comet Halley 1910II*, NASA SP-488.

5. Niedner, M.B., Jr. (1991), "Large-Scale Phenomena Network," in *The Comet Halley Archive — Summary Volume,* Pasadena, NASA/JPL 400-450, 8/91, pp. 107-146.

6. Usher, P.D. (1990), "Photometric Theory for Wide-Angle Phenomena," *Icarus* **86**, pp. 93-99.

7. Brandt, J.C., Rahe, J., Niedner, M.B., Jr., (1982), "Large Scale Phenomena," *The International Halley Watch Newsletter*, Issue No. 1, pp. 13-15.

8. Brandt, J.C., Rahe, J., Niedner, M.B., Jr., (1982), "Large Scale Phenomena," *The International Halley Watch Newsletter*, Issue No. 2, pp. 6-7.

9. Brandt, J.C., Niedner, M.B., Jr., Rahe, J., (1983), "Large Scale Phenomena," *The International Halley Watch Newsletter,* Issue No. 3, pp. 24-25.

10. Brandt, J.C., Niedner, M.B., Jr., Rahe, J., (1984), "Large Scale Phenomena," *The International Halley Watch Newsletter*, Issue No. 4, pp. 18-19.

11. Brandt, J.C., Niedner, M.B., Jr., Rahe, J., (1984), "Large Scale Phenomena," *The International Halley Watch Newsletter*, Issue No. 5, pp. 43-45.

12. Brandt, J.C., Niedner, M.B., Jr., Rahe, J., (1985), "Large Scale Phenomena," *The International Halley Watch Newsletter*, Issue No. 6, pp. 32-33.

13. Brandt, J.C., Niedner, M.B., Jr., and Rahe, J. (1986), "Large Scale Phenomena". *The International Halley Watch Newsletter*, Issue No. 9, pp. 37-38.

14. Niedner, M.B., Jr. (1984), "Plasma and Dust Tail Orientations for Halley's Comet in 1985-1986," *The International Halley Watch Newsletter*, Issue No. 4, pp. 5-7.

15. Niedner, M.B., Jr., Brandt, J.C., Rahe, J., (1987), "Large-Scale Phenomena," *The International Halley Watch Newsletter*, Issue No. 10, pp. 15-16.

16. Niedner, M.B., Jr., Liller, W. (1987), "The IHW Island Network," *Sky & Telescope* **73**, pp. 258-263.

17. Yeomans, D.K. (1983), *The Comet Halley Handbook (1983)*, 2nd Ed, Pasadena, NASA/JPL 400-91, Rev. 1, 6/83.

18. Frost, K.J., Woodgate, B.E., House, L.L. (1984), "Solar Maximum Mission Observations of Halley's Comet," *The International Halley Watch Newsletter*, Issue No. 5, pp. 19-23.

19. Brandt, J.C., Niedner, M.B., Jr., Rahe, J., (1987) "The Halley's Comet Archive — Last Call for Photographs Showing Large-Scale Phenomena," *Bulletin of the American Astronomical Society* **19**, No. 4, p. 1136.

20. Brandt, J.C. (1988), "Halley Photos Wanted," *Sky & Telescope* **75**, No. 1, p. 5.

21. Brandt, J.C. (1990), "Periodic Comet Halley (1986 III)," *IAU Circular* No. 4961.

22. Jewitt, D., Danielson, G.E. (1984), "Charge-coupled Device Photometry of Comet P/Halley," *Icarus* **60**, pp. 435-444.

23. Niedner, M.B., Jr. (1986), "First Impressions of Plasma-Tail Activity in Halley's Comet," *The International Halley Watch Newsletter*, Issue No. 9, pp. 2-8.

24. Brandt, J.C., Niedner, M.B., Jr. (1985), "Tail Phenomena," *Advances in Space Research* **5**, No. 12, pp. 247-253.

25. Brandt., J.C., Niedner, M.B., Jr. (1987), "Plasma structures in comets P/Halley and Giacobini-Zinner," *Astronomy and Astrophysics* **187**, pp. 281-286.

26. Brandt, J.C. (1990), "The large-scale plasma structure of Halley's comet, 1985-86," in *Comet Halley: Investigations, Results, Interpretations* **1**, John Mason, ed. Ellis Horwood Ltd., pp. 33-55.

27. Ershkovich, A.I. (1980), "Kelvin-Helmholtz instability in type-1 comet tails and associated phenomena," *Space Science Reviews* **25**, pp. 3-34.

28. West, R.M., Pedersen, H., Monderen, P., Vio, R., Grosbol, P. (1986), "Post-perihelion imaging of comet Halley at ESO," *Nature* **321**, pp. 363-365.

29. Niedner, M.B., Jr. Schwingenschuh, K. (1987), "Plasma-tail activity at the time of the VEGA encounters," *Astronomy and Astrophysics* **187**, pp. 103-108.

30. Cremonese, G., Fulle, M. (1989), "Photometrical Analysis of the Neck-Line Structure of Comet Halley," *Icarus* **80**, No. 2, pp. 267-279.

31. Sekanina, Z., Larson, S.M., Emerson, G., Helin, E.F., Schmidt., R.F. (1987), "The Sunward Spike of Halley's Comet," *Astronony and Astrophysics* **197**, pp. 645-649.

32. West, R.M., Hainaut, O., Smette, A. (1991), "Post-perihelion observations of P/Halley, *Astronomy and Astrophysics* **246**, pp. L77-L80.